ENDOCANNABINOIDS

The Brain and Body's Marijuana and Beyond

ENDOCANNABINOIDS

The Brain and Body's Marijuana and Beyond

Edited by
Emmanuel S. Onaivi
Takayuki Sugiura
Vincenzo Di Marzo

Taylor & Francis
Taylor & Francis Group

Boca Raton London New York

A CRC title, part of the Taylor & Francis imprint, a member of the Taylor & Francis Group, the academic division of T&F Informa plc.

Published in 2006 by
CRC Press
Taylor & Francis Group
6000 Broken Sound Parkway NW, Suite 300
Boca Raton, FL 33487-2742

Printed in the United States of America on acid-free paper
10 9 8 7 6 5 4 3 2 1

International Standard Book Number-10: 0-415-30008-8 (Hardcover)
International Standard Book Number-13: 978-0-415-30008-7 (Hardcover)
Library of Congress Card Number 2005048536

Library of Congress Cataloging-in-Publication Data

Endocannabinoids : the brain and body's marijuana and beyond / edited by Emmanuel S. Onaivi, Takayuki Sugiura, Vincenzo di Marzo.
p. cm.
Includes bibliographical references and index.
ISBN 0-415-30008-8 (alk. paper)
1. Cannabinoids—Physiological effect. 2. Cannabinoids—Receptors. I. Onaivi, Emmanuel S. II. Sugiura, Takayuki. III. Di Marzo, Vincenzo.

QP801.C27E53 2005
615'.7827—dc22 2005048536

Preface

Marijuana remains one of the most widely used and abused drugs in the world. Over the past decade there have been major advances in understanding the mechanisms whereby marijuana interacts with the brain, producing psychoactive and potentially therapeutic effects. Recent findings indicate the existence of an endocannabinoid physiological control system (EPCS) whose role is just beginning to become understood.

This book presents these remarkable advances in the discovery and existence of naturally occurring marijuana-like substances in the human body. The discovery of specific genes coding for cannabinoid receptors (CBrs) that are activated by smoking marijuana and the finding of endogenous cannabinoids that also activate CBrs have transformed marijuana-cannabinoid research into mainstream science with significant implications in human health and disease. The cannabinoid receptors, endocannabinoid ligands, and enzymes for the biosynthesis and degradation of these substances represent this endocannabinoid physiological control system (EPCS). An overwhelming body of scientific evidence now indicates the existence of this elaborate and previously unknown but ubiquitous EPCS whose fundamental role in human development, health, and disease is unfolding. Although cannabinoid effects on the brain have received most attention, the endogenous cannabinoid system has been found to be associated with the immune, endocrine, and reproductive systems as well. Because of the ubiquity of the EPCS, it is not surprising that this endocannabinoid signaling system has been suggested to play several roles in animals and humans, including the regulation of cell development and growth, nervous functions, modulation of immune function, reproduction and feeding behavior, and in a number of biological functions from movement, memory, learning, appetite stimulation, and pain, to emotions, blood pressure, intraocular pressure, etc. Furthermore, the presence of endogenous cannabinoids in some of the simplest multicellular organisms in the animal kingdom as well as in mammalian cells at the earliest stages of development may suggest a fundamental role in human biology. The aims and scope of this book are therefore broad in nature and address the biology of this recently discovered endocannabinoid signaling system. This system appears to exert a powerful modulatory action on retrograde signaling associated with cannabinoid inhibition of synaptic transmission. The promiscuous action and distribution of CBrs in most biological systems provide the EPCS extensive signaling capabilities for cross talk within and possibly between receptor families, which may explain the numerous behavioral effects associated with smoking marijuana. The advances in marijuana-cannabinoid research have already shed light on issues concerning marijuana dependence in vulnerable individuals. For example, there is evidence that a missense in human fatty acid amide hydrolase, which inactivates endocannabinoids (anandamide) and related lipids, is associated with alcoholism and other forms of drug dependence.

This remarkable progress in understanding the biological actions of marijuana and endocannabinoids is fueling genomic research and raising a number of critical issues on the molecular mechanisms of cannabinoid-induced behavioral and biochemical alterations. While these advances have enhanced our understanding of the molecular mechanisms associated with the behavioral effects of marijuana use, the molecular identity of other cannabinoid receptor subtypes and transporters (if any), along with the growing number of endocannabinoids, will allow specific therapeutic targeting of different components of the EPCS in health and disease.

This book on the biology of endocannabinoids should be especially helpful to the clinician who requires an understanding of the actions of marijuana in order to properly evaluate patients using this drug. It is on the cutting edge of research, covering ground that few books have touched upon before. It will be particularly important as new medications are developed from knowledge of the endocannabinoid system. Already one such medication, rimonabant, a cannabinoid 1 (CB_1) receptor antagonist, has been reported to both reduce appetite for food and cravings for cigarettes. This medication is progressing through the FDA approval process and should be followed by other medications based on novel mechanisms explored in the pages of this volume.

Charles P. O'Brien, M.D., Ph.D.

Acknowledgments

Many thanks to Barbara Ellen Norwitz, publisher, for inspiration, and to the staff at Taylor and Francis for their patience and support throughout the process of this book project. I am indebted to my colleagues in the biology department, the dean of the College of Science, Dr. Sandra DeYoung, the Center for Research, and the provost office of William Paterson University for their continuous support and release time. We are also enormously thankful to Mary Pfeiffer, Chief, Library Services, NIDA–NIH, for editorial assistance and patience.

Dr. Emmanuel S. Onaivi

About the Editors

Emmanuel S. Onaivi is professor of biology at William Paterson University in New Jersey and a guest scientist at the National Institute on Drug Abuse at the National Institutes of Health in Baltimore, Maryland. He worked as a research professor at Vanderbilt University in Tennessee prior to joining the faculty at William Paterson University in New Jersey. He received a M.Sc. degree in pharmacology and a Ph.D. degree in neuropharmacology, both from the University of Bradford, England. He is also a pharmacist, with a B. Pharm. (Hons.) degree from the University of Benin, and he completed the management science program at the Imperial College of Science and Technology in London. Although his doctoral research was on drug abuse, he became intrigued by cannabinoid research during his first postdoctoral internship in the laboratory of Dr. Billy Martin at the Medical College of Virginia–Virginia Commonwealth University in Richmond. There he worked on cannabinoids and anxiety in animal models and completed work on the validation of the two-compartment black box for analysis of anxiolytic and anxiogenic compounds. He became an instructor at Louisiana State University Pennington Biomedical Research Center and later was a visiting research scientist at Stanford University. He then was appointed to the faculty of Meharry Medical College, where his laboratory was the first to clone and sequence the mouse CB_1 cannabinoid receptor gene. At Vanderbilt University, his team reported racial and gender differences in the expression of CB_1 cannabinoid receptors. His current research is on cannabinoid genomics and proteomics on behavior. He is also interested in changes in anxiety and depression as a basis for addiction. During the past 15 years he has published several papers and contributed book chapters, edited *The Biology of Marijuana from Gene to Behavior*, and supervised doctoral and postdoctoral fellows. His student research was selected to be presented on Capitol Hill in Washington, D.C.

Takayuki Sugiura is professor of biochemistry and molecular pharmacology, Department of Molecular Health Science, Faculty of Pharmaceutical Sciences, Teikyo University, Japan. Dr. Sugiura obtained his Ph.D. degree from the University of Tokyo, Faculty of Pharmaceutical Sciences; his thesis was on the "Biochemistry of Ether Lipids in Macrophages." He joined the Department of Molecular Health Science, Faculty of Pharmaceutical Sciences, Teikyo University, as a research associate in 1978. After a brief research fellowship position at the University of Texas Health Science Center in San Antonio with Professor Donald J. Hanahan, who discovered phospholipase D in 1940s and the structure of PAF in 1979, Dr. Sugiura conducted studies on the biochemistry and pharmacology of lysophosphatidic acid. He returned to the Department of Molecular Health Science, Faculty of Pharmaceutical Sciences, Teikyo University, where he later became professor of biochemistry and pharmacology. Throughout most of his career he has worked on the biochemistry and pharmacology of bioactive lipids and jointly discovered the endocannabinoid, 2-arachidonyl glycerol (2-AG). Other major findings of his studies include (1) that alkyl ether-linked choline glycerophospholipid (a PAF precursor) is abundantly present in leukocytes; (2) that arachidonic acid is rapidly transferred between phospholipids in living cells; (3) that CoA-independent transacylation and CoA-dependent transacylation are different enzyme reactions; (4) development of the first selective antagonist of lysophosphatidic acid receptor; (5) identification of 2-AG as an endogenous cannabinoid receptor ligand; and (6) final elucidation of a biosynthetic pathway for anandamide (N-acyl PE pathway). His current studies are directed mainly towards 2-AG. He is the author of many publications, book chapters, and abstracts.

Vincenzo Di Marzo was born in Naples, Italy, in 1960. He is a First Researcher and group leader at the Institute of Biomolecular Chemistry of the National Research Council (ICB-CNR) in Pozzuoli, Naples, Italy, and an adjunct associate professor in the Department of Pharmacology and Toxicology at the Medical College of Virginia — Virginia Commonwealth University. He was awarded a master's degree in chemistry from the University of Naples in 1983, and a Ph.D. in biochemistry and molecular pharmacology from the Imperial College in London in 1988. He completed post-doctoral studies in lipid biochemistry and natural product chemistry at ICB-CNR in 1990. Dr. Di Marzo is coauthor of more than 200 articles published in peer-reviewed journals and is senior author of several reviews on lipids in cell signaling and on endocannabinoids. In addition, he is on the editorial board of *Prostaglandins, Leukotrienes and Essential Fatty Acids*, the *Journal of Cannabis Therapeutics*, *Critical Reviews in Neurobiology*, and the *British Journal of Pharmacology* in the U.K. His research interests in the last decade are the biosynthesis, metabolism, and pharmacological actions of the endocannabinoids, and the development of new potential therapeutic drugs from the cannabinoid and vanilloid signaling systems. A member of the International Cannabinoid Research Society (ICRS) since 1997, Dr. Di Marzo founded in 1995 the Endocannabinoid Research Group (ERG, www.e-r-g.it), a multidisciplinary research group that puts together different specialties in the Naples area with the common effort of understanding the physiological and pathological role of endogenous cannabimimetic substances in both central and peripheral systems. He was elected to serve as president of the ICRS from June 2004 to June 2005.

Contributors

Mary E. Abood
Forbes Norris MDA/ALS Research Center
California Pacific Medical Center Research Institute
San Francisco, California

Babatunde E. Akinshola
Department of Pharmacology
Howard University College of Medicine
Washington, D.C.

Evgeny V. Berdyshev
University of Chicago
Chicago, Illinois

Lani J. Burkman
Department of Gynecology and Obstetrics
School of Medicine and Biomedical Sciences
University at Buffalo
State University of New York
Buffalo, New York

Sumner Burstein
Department of Biochemistry and Molecular Biology
University of Massachusetts Medical School
Worcester, Massachusetts

Maribel Cebeira
Departamento de Bioquimica y Biologia Molecular
Facultad de Medicina
Universidad Complutense
Madrid, Spain

Frédérique Chaperon
Novartis Pharma, AG
Basel, Switzerland

Sanika S. Chirwa
Department of Physiology
Meharry Medical College
Nashville, Tennessee

Nissar A. Darmani
Department of Basic Medical Sciences
College of Osteopathic Medicine of the Pacific
Western University of Health Sciences
Pomona, California

Rosario de Miguel
Departamento de Bioquimica y Biologia Molecular
Facultad de Medicina
Universidad Complutense
Madrid, Spain

Luciano De Petrocellis
Endocannabinoid Research Group
Istituto di Cibernetica
Consiglio Nazionale delle Ricerche
Pozzuoli (Napoli), Italy

Vincenzo Di Marzo
Endocannabinoid Research Group
Istituto di Chimica Biomolecolare
Consiglio Nazionale delle Ricerche
Pozzuoli (Napoli), Italy

Javier Fernández-Ruiz
Departamento de Bioquimica y Biologia Molecular
Facultad de Medicina
Universidad Complutense
Madrid, Spain

Ester Fride
Department of Behavioral Sciences and Molecular Biology
College of Judea and Samaria
Ariel, Israel

Michelle Glass
Faculty of Medicine and Human Services
University of Auckland
Auckland, New Zealand

Jianping Gong
Molecular Neurobiology Branch
NIDA–NIH
Baltimore, Maryland

Harald S. Hansen
Department of Pharmacology
Danish University of Pharmaceutical Sciences
Copenhagen, Denmark

Henrik H. Hansen
Department of Pharmacology
Danish University of Pharmaceutical Sciences
Copenhagen, Denmark
and
Department of Pediatric Neurology
The Charite–Virchow Clinics
Humboldt University
Berlin, Germany

Mariluz Hernandez
Departamento de Psicobiologia
Facultad de Psicologia
Universidad Complutense
Madrid, Spain

Alexander F. Hoffman
Cellular Neurobiology Branch
Section on Cellular Neurophysiology
NIDA–NIH
Baltimore, Maryland

Shinobu Ikeda
Faculty of Pharmaceutical Sciences
Teikyo University
Sagamiko, Kanagawa, Japan

Hiroki Ishiguro
Department of Medical Genetics
University of Tsukuba
Tsukuba, Ibaraki, Japan

Yoshio Ishima
Ishima Institute for Neurosciences
Kunitachi, Tokyo, Japan

Claire M. Leonard
Department of Biology
William Paterson University
Wayne, New Jersey

Alessia Ligresti
Endocannabinoid Research Group
Istituto di Chimica Biomolecolare
Consiglio Nazionale delle Ricerche
Pozzuoli (Napoli), Italy

Zhicheng Lin
Molecular Neurobiology Branch
NIDA–NIH
Baltimore, Maryland

Carl R. Lupica
Cellular Neurophysiology Branch
NIDA–NIH
Baltimore, Maryland

Mauro Maccarrone
Department of Biomedical Sciences
University of Teramo
Teramo, Italy

Sean D. McAllister
Forbes Norris MDA/ALS Research Center
California Pacific Medical Center
San Francisco, California

John M. McPartland
GW Pharmaceuticals, Ltd
Salisbury, Wiltshire, U.K.
and
Faculty of Health and Environmental Sciences
UNITEC
Auckland, New Zealand

Raphael Mechoulam
Department of Medicinal Chemistry and Natural Products
Hebrew University
Ein Kerem, Jerusalem, Israel

Marisela Morales
Cellular Neurophysiology Branch
NIDA–NIH
Baltimore, Maryland

Laura L. Murphy
Department of Physiology
Southern Illinois University School of Medicine
Carbondale, Illinois

Charles P. O'Brien
University of Pennsylvania
Philadelphia VAMC/MIRECC
Philadelphia, Pennsylvania

Saori Oka
Faculty of Pharmaceutical Sciences
Teikyo University
Sagamiko, Kanagawa, Japan

Emmanuel S. Onaivi
Department of Biology
William Paterson University
Wayne, New Jersey
Molecular Neurobiology Branch
NIDA–NIH
Baltimore, Maryland

Gitte Petersen
Department of Pharmacology
Danish University of Pharmaceutical Sciences
Copenhagen, Denmark

José A. Ramos
Departamento de Bioquimica y Biologia Molecular
Facultad de Medicina
Universidad Complutense
Madrid, Spain

Michael D. Randall
School of Biomedical Sciences
University of Nottingham Medical School
Queen's Medical Centre
Nottingham, U.K.

Patricia H. Reggio
Department of Chemistry and Biochemistry
University of North Carolina at Greensboro
Greensboro, North Carolina

Ruth A. Ross
Department of Biomedical Sciences
Institute of Medical Sciences
University of Aberdeen
Foresterhill, Aberdeen, Scotland

Ethan Russo
GW Pharmaceuticals
Missoula, Montana

Herbert Schuel
Division of Anatomy and Cell Biology
Department of Pathology and Anatomical Sciences
School of Medicine and Biomedical Sciences
University at Buffalo
State University of New York
Buffalo, New York

Takayuki Sugiura
Faculty of Pharmaceutical Sciences
Teikyo University
Sagamiko, Kanagawa, Japan

Marie-Hélène Thiébot
Faculty of Medicine
Pitié-Salpêtrière
Paris, France

George R. Uhl
Molecular Neurobiology Branch
NIDA–NIH
Baltimore, Maryland

Keizo Waku
Faculty of Pharmaceutical Sciences
Teikyo University
Sagamiko, Kanagawa, Japan

Ping Wu Zhang
Molecular Neurobiology Branch
NIDA–NIH
Baltimore, Maryland

Table of Contents

Part I

Historical Aspects and Chemistry

1 Look Back in Ananda—40 Years of Research on Cannabinoids

Raphael Mechoulam

In the autumn of 1960 I returned to Israel after a postdoctoral stay at the Rockefeller Institute in New York, where I had worked on the structure of triterpenes. I accepted a junior position in the chemistry department of the Weizmann Institute in Rehovot. In order to advance and earn tenure, I was expected to publish papers that would make an impact in a scientific field of my choice. I was interested in the chemistry of natural products, but how was I to choose an exciting topic that had not been well trod before? An established route was to pick a plant from a family known to contain alkaloids or exotic terpenes and isolate a few unknown constituents. This seemed a promising way to publish numerous papers, but does a young scientist want to spend his years just adding to a catalog of plant constituents? A more exciting field seemed to be the investigation of plants used as drugs over millennia by the many tribes and nations in the Middle East.

The Assyrians, who ruled parts of Mesopotamia for centuries, left us a legacy of clay tablets with hundreds of prescriptions, nearly all of which were based on local plants. And the Egyptian papyri are a treasure trove of many yet-uninvestigated plants used in numerous medical conditions — some of which, unfortunately, have no parallel in modern diagnosis. But the most valuable ancient text is undoubtedly that of Pedanius Dioscorides, a Greek physician in the Roman army during the first century A.D., describing about 600 medicinal plants of the Middle East. For over 18 centuries the *Dioscorides Herbal* was one of the most influential drug books (for an overview of these ancient sources see Mechoulam, 1986). Are there medicinal plants in these sources which, from the viewpoint of modern chemistry and pharmacology, have not been well investigated? The answer was yes then and is still yes today. One of plants I decided to look into was *Cannabis sativa*.

A species of what is commonly known as hemp, it was well known in the ancient world. It was used for fiber, food, and medicine and in social and religious rites. The Assyrians had several names for it. *Gan zi gun nu*, translated as "the drug which takes away the mind," was one of them. *Azallu*, another name, was a drug for a yet unidentified disease translated as "hand of the ghost." It was also a drug for "the poison of all limbs" (presumably arthritis). The Egyptians used it as an oral medication for mothers and children, possibly during childbirth, and Dioscorides recommended it for ear pain and for inflammation.

By the end of the 1950s there were already hundreds of papers on cannabis, many of them in German, French, and Russian, and many in obscure, defunct journals of the 19th century (Eddy, 1965). My mid-European background and schooling, with its compulsory teaching of languages, was of great help. After reading most of the published papers and quite a few unpublished Ph.D. theses, I realized that the active constituent had never been isolated in pure form and had therefore not been investigated. Fractions, which caused various effects, had been isolated throughout the 19th century. In the 1930s several outstanding research groups had reported important advances. The structure of the inactive constituent cannabinol was elucidated, and later a second inactive compound, named cannabidiol (CBD), was isolated, but its structure was not fully determined. The active components were assumed to be unidentified "tetrahydrocannabinols." (For a review on the early work on cannabis chemistry, see Todd, 1946.) From the point of view of a then-modern, natural products chemist, the field needed a reinvestigation.

In 1962 I started research on cannabis with the expectation that it would be a minor project to be completed within 6 months. Forty years hence, I am still working in the field, having published more than 250 articles on it. My first goal was to reisolate and elucidate the structure of CBD, which had previously been isolated by Todd and by Adams. Mostly by use of the then-innovative NMR method, Yuval Shvo, a colleague, and I reported in 1963 the structure of CBD (Mechoulam and Shvo, 1963), and in parallel, Yechiel Gaoni and I started work on the rest of the constituents. Fractions from cannabis were tested on monkeys and ultimately we were able, by column chromatography, to identify a pure compound that caused ataxia in dogs and a sleepy state in monkeys (Gaoni and Mechoulam, 1964). These assays were done by the late Habib Edery, who had a monkey colony for his research (Edery et al., 1971). The structure of the active compound turned out to be closely related to that of CBD except that it was a tricyclic compound, while CBD was a bicyclic one. We named this compound Δ^1-tetrahydrocannabinol (Δ^1-THC), later to be changed to Δ^9-THC, following a different nomenclature. It is to be regretted that by the use of the Δ^9 nomenclature, THC and CBD have different numberings. Numerous other plant cannabinoids were isolated, and their structures were elucidated. So cannabigerol, cannabichromene, cannabielsoic acid, and others came into being (Mechoulam, 1970). The absolute configuration of THC was determined, being (R) at both asymmetric positions (Mechoulam and Gaoni, 1967). A few other groups had now entered the field, and a German group reported the presence of Δ^9-THC acid (Korte et al., 1965). Then, we isolated a second Δ^9-THC acid (now called THC acid B) (Mechoulam et al., 1969). Both acids probably serve as THC precursors in the plant as they easily decarboxylate to THC. THC acid B is crystalline, and several years later we sent it to a Norwegian group that was able to take an X-ray picture of it — the first crystallographic study in this area (Rosenqvist and Ottersen, 1975). It confirmed the almost-planar structure of THC.

In the mid-1960s I was invited to establish a laboratory of natural products at the Hebrew University in Jerusalem, where I continued the cannabis project. I was still quite excited by our work, but apparently nobody else was. We could not get grants — be they local, Israeli, or foreign. We had assumed that the National Institutes of Health (NIH) would be interested in cannabis, but my friends there let me know that NIH had never awarded a grant on cannabis and did not plan to do so in the future. "Marijuana smoking is a South American problem of little relevance to us," I was told by colleagues at NIH. In the mid-1960s, however, the winds changed. Apparently up till then, marijuana use was limited to jazz players and other characters outside the main stream of American life. When, however, marijuana suddenly became the drug of choice in American schools and one of the symbols of the flower generation, NIH became interested. I gave Dr. Dan Ephron, the Head of Pharmacology, National Institute of Mental Health, about 10 grams of THC, which was then used for many of the initial research projects in the U.S. This THC was prepared from natural CBD by a simple cyclization, a method we still employ (Gaoni and Mechoulam, 1971). However, NIH preferred to have purely synthetic THC, and they used the method reported by Petrzilka in Switzerland (Petrzilka and Sikemeier, 1967), which was a variation of a method described by us (Mechoulam et al., 1967). I believe that most of the THC used today as a medicinal agent is still made by this method, although THC cyclized from crystalline CBD is stereochemically purer.

Although THC is relatively easy to synthesize, it seems that it has never been used illegally, or at least there have not been any reports on such use. The reason presumably is that cannabis smokers enjoy the smell of a cannabis "joint," rather than the sterile taste of an oily synthetic product. Besides, the effect of smoking appears almost immediately, while orally consumed THC produces its effects after 2 to 3 h. I am also not aware of any illegal use of intravenously administered THC or its derivatives. These compounds are not readily soluble in water solutions, and an IV injection could be dangerous in the hands of an amateur.

The next step in the cannabis saga was the investigation of the metabolism of its constituents. In the late 1960s and early 1970s the time was apparently ripe for such research, as within several months 4 groups, including ours, announced the initial metabolic oxidation of THC to 11-hydroxy-THC. This compound, both in the Δ^8 and in the Δ^9 series, turned out to be active in all the tests in which THC had been investigated previously. It is generally assumed that in the body, both THC and 11-hydroxy-THC contribute to the "high." A huge number of metabolites were identified later (Agurell et al., 1986). Most of them are inactive, but some parallel THC activity. The final stage of cannabinoid metabolism turned out to be the conversion into various cannabinoid acids, conjugated to glucuronic acid. These final metabolites appear in urine, and one of them is widely used in assays for the determination of marijuana use as it is excreted over a period of many weeks. In this field I collaborated with a group of young Swedes in Uppsala. My close friend Stig Agurell, who headed the group, told me that he chose cannabis as a research project as "obviously cannabinoids are of no therapeutic interest" and would not interfere with his consultantship at Astra. Indeed, later when he became the director of research of Astra, the company never undertook any cannabis work!

My collaboration with the very bright and industrious Swedes not only taught me how to investigate metabolic pathways but also brought me many times to Uppsala to serve as foreign "opponent" to the Ph.D. theses of the students — a great opportunity to be wined and dined over several days and to visit the lovely small island of the Agurells.

During the late 1970s and early 1980s most of our cannabis work was devoted to investigations on the structure-activity requirements (SAR) for cannabinoid activity and initial work on the therapeutic opportunities opened by the cannabinoids (Mechoulam and Carlini, 1978). Together with friends in South America, we found that some cannabinoids — including the nonpsychotropic CBD — were potent antiepileptics. We prepared nearly 1 kg of CBD, which was used in a clinical trial in Brazil (Cunha et al., 1980). The results were positive, but unfortunately this work has not been further pursued, and CBD is not in clinical use.

The SAR work led to some, rather unexpected developments. We knew that hydroxylation at the 11-position and the exchange of the pentyl side chain with a dimethylheptyl moiety enhance activity. We combined both SAR leads and prepared 11-hydroxy-dimethylheptyl –Δ^8-THC (HU210) (Mechoulam et al., 1988). This new compound turned out to be extremely active in all cannabinoid tests, *in vitro* and *in vivo*. It is widely used in cannabinoid research. Unfortunately, the U.S. Drug Enforcement Administration (DEA) has put it on the prohibited list, and it is not sold anymore in the U.S.; all work on it is now done in Europe and Canada. Then, we prepared the (+) enantiomer, code-named HU211, which was completely inactive in all cannabinoid tests (Mechoulam et al., 1988). Jeff Feigenbaum, a bright young pharmacologist who worked in my lab at that time, observed that HU211 seemed to act as a glutamate (NMDA) antagonist. We asked two pharmacologists in Tel Aviv, M. Sokolovski and Y. Kloog, who were working on glutamates, to look at this phenomenon, and, indeed, they confirmed the initial observations (Feigenbaum et al., 1989). Later it was found that HU211 is a potent antioxidant and inhibitor of TNF-α (Shohami et al., 1997). We assumed that it can be a novel neuroprotectant, and a new, small American–Israeli company, Pharmos, took over its development. Over the years, we were gratified to see that HU211 passed all the *in vitro*, *in vivo*, and toxicological tests which Pharmos could think of. After brain injury in mice, it helped to recover the function of the blood brain barrier, it reduced neurological defects, and showed no side effects. In brain ischemia in gerbils, it induced higher survival of hippocampal cells, and in ischemia in rats, it reduced the infarct volume (for a review on the pharmacology of HU211, see Shohami and Mechoulam, 2000). HU211 also very significantly reduces the formation of brain lesions caused by the nerve gas Soman (Filbert et al., 1999).

In human patients, HU211, administered up to 6 h after severe head injury, significantly reduced the elevation of intracranial pressure and showed a consistent trend toward better outcome in the severe patient subgroup (Knoller et al., 2002). HU211 (now named Dexanabinol) is in phase 3 clinical trials against brain trauma in Europe and Israel. We expect that the clinical trials will be completed in 2004.

The synthesis and pharmacological evaluation of HU211 was also of considerable significance outside the therapeutic area. As mentioned earlier, a huge amount of cannabinoid research had been done in many laboratories for nearly two decades since the identification of Δ^9-THC. We learned much from their pharmacology, biochemistry, and physiological effects; however, their mode of action remained an enigma.

The reasons for this baffling situation were both technical and conceptual. On the technical side, it was assumed that THC is active in both enantiomeric forms (though with a different level of potency), and this was incompatible with action on a receptor, which will usually bind one stereoisomer only.

But all work on the stereospecificity of cannabinoid action had been done with (+) THC synthesized according to a procedure published by our group based on commercial α-pinene. And we knew that commercial pinene is not stereochemically pure, and therefore would lead to stereochemically impure products. Hence, the lack of stereospecificity could be due to the presence of varying amounts of the active (–) stereoisomer in the presumed pure (+) isomer. So we repeated the synthesis with stereochemically pure (+) α-pinene and tested the (+) THC produced. It had no (–) THC-like activity, as expected. Then we evaluated the activity in the 11-hydroxy-dimethylheptyl series. As mentioned earlier (–) HU210, is very potent — at least 100 times more than THC. Its enantiomer, (+) HU211, turned out to be many thousands of times less active than HU210 in a wide series of tests done in collaboration with my friends Billy Martin, Toby Jarbe, and Allyn Howlett. The stereochemical hurdle was thus overcome. (For a review on the cannabinoid stereochemistry and biological activity, see Mechoulam et al., 1992).

The conceptual problem related to THC activity had been raised by the late William Paton of Oxford, who had pointed out that the cannabinoids belong to the group of biologically active lipophiles and that their effects should be compared with the chronic effects of anesthetics and solvents.

Hence, following this line of thought, it was possible to explain the action of THC without postulating the existence of a specific cannabinoid receptor. However, it slowly emerged that cannabinoid action had both specific and nonspecific aspects, and it seemed possible that, while the nonspecific actions could be explained by Paton's proposals, the specific ones could be due to action on receptors or enzymes.

In 1988, Allyn Howlett, with her then-graduate student Bill Devane, brought out the first evidence that a cannabinoid receptor exists in the brain (Devane et al., 1988). We assumed that such a receptor is not formed for the sake of a plant that has compounds that bind to it, but for endogenous brain ligands. Bill Devane had by this time completed his Ph.D. thesis and applied for a post doctorate position in my lab. He wanted to learn some synthetic chemistry. I had other plans for him. We first synthesized a novel, highly active radioactive probe (so Bill got synthetic experience) (Devane, Breuer, et al., 1992) and then proceeded towards the identification of a cannabinoid ligand. Later, a visiting fellow from Czechoslovakia, Lumir Hanus, joined my group. Over more than a year, Bill and Lumir tried to solve the isolation problems associated with the cannabinoid ligand. Because of the lipophilic nature of the plant cannabinoids, we assumed that the brain ligands are lipids. Or maybe we just wanted them to be lipids, as the lab had experience with such compounds, but we didn't know how to deal with peptides.

The isolation problems were tremendous. As soon as the fraction that bound to the cannabinoid receptor was purified, it started losing its activity. We know now that this was due to the lack of stability of anandamide. Ultimately, we had a very small amount of what looked like pure material (on TLC at least) and tried to get a NMR spectrum.

This was not simple to do on the 300-MHz machine that was available to us at that time, but we let the spectrum be run over a weekend. We ended up with a curve that contained more impurities than actual material.

However, two peaks were quite clear: one of them was obviously due to 8 double bond protons, and these were coupled to 10 allylic protons. The spectrum looked very much like that of arachidonic acid. For us, this was the breakthrough. We knew we were dealing with a fatty acid derivative.

Then, with the help of a colleague, Asher Mandelbaum, at the Technion in Haifa, we got a high-resolution mass spectrum, which indicated that the molecule contains a nitrogen, certainly not a common feature in fatty acids. However, the structure was now close at hand. Some more mass spectra and a better nuclear magnetic resonance (NMR) led to a final formulation of the ligand as arachidonoyl ethanol amide (Devane, Hanus, et al., 1992).

We were also interested in some tests, which were closer to physiological reality than just binding to the receptor. But with the miniscule amounts of material we had from the brain, we obviously could not do *in vivo* work. However Roger Pertwee in Scotland had reported experiments with cannabinoids on inhibition of the electrically evoked twitch response of mouse vas deferens, which required a very small amount of material.

We sent him some of the (impure) material, and within a few days he happily informed us that this material paralleled THC in activity. However, when we sent him pure anandamide for the first time, it was inactive! It turned out that pure anandamide had oxidized on its trip from Jerusalem to Aberdeen; the impure material obviously contained an antioxidant!

We decided to name the new brain ligand *anandamide*. Bill Devane was learning Sanskrit at the time and suggested *ananda* ("supreme joy" in this ancient tongue), and this portion of the name certainly fits the "amide" moiety of the structure. I looked for a suitable Hebrew equivalent, but nothing came to my mind. There are lots of synonyms for "sorrow" in Hebrew, but considerably fewer for "joy."

Sean Munro in Cambridge had identified a cannabinoid receptor in spleen, which was absent in brain (Munro et al., 1993). I asked a new Ph.D. student, Shimon Ben-Shabat, to try to find the peripheral ligand that activates this receptor. In a few months he had an active mixture which, however, bound to both receptors. This mixture contained no fatty acid amides, but three fatty acid glycerol esters, only one of which — the arachidonoyl one, obviously — was found to bind to the receptors. Its binding potency was much lower than that of anandamide, and we were uncertain of its role as a natural ligand.

By "we" I mean our team of 15 authors: my group, that of Zvi Vogel in Rehovot, who did some of the binding; that of Norb Kaminski who worked on cAMP; Roger Pertwee who looked at the contractions of vas deferens; and that of Billy Martin who did the animal pharmacology (Mechoulam et al., 1995).

Today, we know that the low-binding potency was due to the unsuitable *in vitro* conditions we used. Now, we have values which parallel those of anandamide, and we also know that the inactive fatty acid glycerol esters, which accompany 2-arachidonyl glycerol (2-AG), strongly enhance its activity, and that this "entourage" effect may be a general one for endogenous cannabinoid ligands (Ben Shabat et al., 1998).

Over the last decade, work on the cannabinoids has expanded tremendously. The endogenous cannabinoids have been found to affect many physiological systems, as will be presented in the following chapters. Anandamide and 2-AG have become central players in pharmacology and physiology. Many pharmaceutical companies are developing cannabinoid drugs in the fields of obesity and neuroprotection. Our original paper on anandamide has been cited more than 1500 times.

Our group has continued work on the isolation of additional endocannabinoids (Hanus et al., 2001) and has branched out, in collaboration with many friends and colleagues, into various pharmacological areas: neuroprotection (with Esty Shohami and our joint Ph.D. student David Panikashvili; Panikashvili et al., 2001); feeding (with Elliot Berry and Yosepha Avraham; Hanus et al., 2003); suckling (with Ester Fride; Fride et al., 2001); immunology (with Ruth Gallily and M. Feldmann; Malfait et al., 2000); prevention of emesis by nonpsychotropic cannabinoids (with Linda Parker; Parker et al., 2003); and even bone remodeling (with Itai Bab; patent submitted).

May I end by saying thanks to my students over three decades, to my devoted collaborator in the laboratory, Dr Aviva Breuer, to my long time secretary Rachel Heifetz, and to my numerous colleagues in many countries. Feeling at home and having friends in so many places around the globe is a great dividend in doing research.

REFERENCES

Agurell, S., Halldin, M., Lindgren, J.E., Ohlsson, A., Widman, M., Gillespie, H., and Hollister, L. (1986). Pharmacokinetics and metabolism of delta-1-tetrahydrocannabinol and other cannabinoids with emphasis on man. *Pharmacol. Rev.*, 38: 21–43.

Ben-Shabat, S., Fride, E., Sheskin, T., Tamiri, T., Rhee, M.H., Vogel, Z., Bisogno, T., De-Petrocellis, L., Di Marzo, V., and Mechoulam, R. (1998). An entourage effect: inactive endogenous fatty acid glycerol esters enhance 2-arachidonyl-glycerol cannabinoid activity. *Eur. J. Pharmacol.,* 353: 23–31.

Cunha, J.M., Carlini, E.A., Pereira, A.E., Ramos, O.L., Pimentel, C., Gagliardi, R., Sanvito, W.L., Lander, N., and Mechoulam, R. (1980). Chronic administration of cannabidiol to healthy volunteers and epileptic patients. *Pharmacology,* 21: 175–185.

Devane, W.A., Dysarz, F.A., 3rd, Johnson, M.R., Melvin, L.S., and Howlett, A.C. (1988). Determination and characterization of a cannabinoid receptor in rat brain. *Mol. Pharmacol.,* 34: 605–13.

Devane, W.A., Breuer, A., Sheskin, T., Jarbe, T.U.C., Eisen, M.S., and Mechoulam, R. (1992). A novel probe for the cannabinoid receptor. *J. Med. Chem.,* 35: 2065–2069.

Devane, W.A., Hanus, L., Breuer, A., Pertwee, R.G., Stevenson, L.A., Griffin, G., Gibson, D., Mandelbaum, A., Etinger, A., and Mechoulam, R. (1992). Isolation and structure of a brain constituent that binds to the cannabinoid receptor. *Science,* 18: 1946–1949.

Eddy, N.B. (1965). The question of cannabis. Cannabis bibliography. U.N. Economic and Social Council publication E/CN. 7/479.

Edery, H., Grunfeld, Y., Ben-Zvi, Z., and Mechoulam, R. (1971). Structural requirements for cannabinoid activity. *Ann. N.Y. Acad. Sci.,* 191: 40–53.

Feigenbaum, J.J., Bergmann, F., Richmond, S.A., Mechoulam, R., Nadler, V., Kloog, Y., and Sokolovsky, M. (1989). A non-psychotropic cannabinoid acts as a functional N-methyl-D-asparate (NMDA) receptor blocker. *Proc. Natl. Acad. Sci.,* 86: 9584–9587.

Filbert, M.G., Forster, J.S., Smith, C.D., and Ballough, G.P. (1999). Neuroprotective effects of HU-211 on brain damage resulting from soman-induced seizures. *Ann. N.Y. Acad. Sci.,* 890: 505–514.

Fride, E., Ginzburg, Y., Breuer, A., Bisogno, T., Di Marzo, V., and Mechoulam, R. (2001). Critical role of the endogenous cannabinoid system in mouse pup suckling and growth. *Eur. J. Pharmacol.,* 419: 207–214.

Gaoni, Y. and Mechoulam, R. (1971). The isolation and structure of Δ^1-THC and other neutral cannabinoids from hashish. *J. Am. Chem. Soc.,* 93: 217–224.

Gaoni, Y. and Mechoulam, R. (1964). Isolation, structure and partial synthesis of an active constituent of hashish. *J. Am. Chem. Soc.,* 86: 1946.

Hanus, L., Abu-Lafi, S., Fride, E., Breuer, A., Vogel, Z., Shalev, D.E., Kustanovich, I., and Mechoulam, R. (2001). 2-Arachidonyl glyceryl ether, an endogenous agonist of the cannabinoid CB_1 receptor. *Proc. Natl. Acad. Sci. USA.,* 98: 3662–3665.

Hanus, L., Avraham, Y., Ben-Shushan, D., Zolotarev, O., Berry, E.M., and Mechoulam, R. (2003). Short term fasting and prolonged semistarvation have opposite effect on 2-AG levels in mouse brain. *Brain Res.,* 983: 144–151.

Knoller, N., Levi, L., Shoshan, I., Reichenthal, E., Razon, N., Rappaport, Z.H., and Biegon, A. (2002). Dexanabinol (HU-211) in the treatment of severe closed head injury: a randomized, placebo-controlled, phase II clinical trial. *Crit. Care Med.,* 30: 548–554.

Korte, F., Haag, M., and Claussen U. (1965). Tetrahydrocannabinolcarboxylic acid, a component of hashish. *Angew. Chem. Int.* 4: 872.

Malfait, A.M., Gallily, R., Sumariwalla, P.F., Malik, A.S., Andreakos, E., Mechoulam, R., and Feldmann, M. (2000). The nonpsychoactive cannabis constituent cannabidol is an oral anti-arthritic therapeutic in murine collagen-induced arthritis. *Proc. Natl. Acad. Sci USA.,* 97: 9561–9566.

Mechoulam, R. (1970). Marihuana chemistry. *Science,* 168: 1159–1166.

Mechoulam, R. (1986). The Pharmacohistory of Cannabis sativa. in *Cannabinoids as Therapeutic Agents.* R. Mechoulam, Ed., CRC Press, Boca Raton, FL., pp. 1–19.

Mechoulam, R. and Carlini, E.A. (1978). Toward drugs derived from cannabis. *Naturwissenchaften,* 65: 174–179.

Mechoulam, R. and Shvo, Y. (1963). The structure of cannabidiol. *Tetrahedron,* 19: 2073–2078.

Mechoulam, R., Ben-Shabat, S., Hanus, L., Ligumsky, M., Kaminski, N.E., Schatz, A.R., Gopher, A., Almog, S., Martin, B.R., Compton, D.R. et al. (1995). Identification of an endogenous 2-monoglyceride, present in canine gut, that binds to cannabinoid receptors. *Biochem. Pharmacol.,* 50: 83–90.

Mechoulam, R., Ben-Zvi, Z., Yagnitinsky, B., and Shani, A. (1969). A new tetrahydrocannabinolic acid. *Tet. Lett.,* 2339–2341.

Mechoulam, R., Braun, P., and Gaoni, Y. (1967). A stereospecific synthesis of (-)-Δ^1- and (-)-Δ^6-tetrahydrocannabinol. *J. Am. Chem. Soc.,* 89: 4552–4554.

Mechoulam, R., Devane, W.A., and Glaser, R. (1992). Cannabinoid geometry and biological activity in *Marijuana/Cannabinoids: Neurobiology and Neurophysiology.* L. Murphy and A. Bartke, Eds., CRC Press, Boca Raton, FL., pp. 1–33.

Mechoulam, R., Feigenbaum, J.J., Lander, N., Segal, M., Jarbe, T.U.C., Hiltunen, A.J., and Consroe, P. (1988). Enantiomeric cannabinoids: stereospecificity of psychotropic activity. *Experientia,* 44: 762–764.

Mechoulam, R. and Gaoni, Y. (1967). The absolute configuration of Δ^1-tetrahydrocannabinol, the major active constituent of hashish. *Tet. Lett.,* 1109–1111.

Munro, S., Thomas, K.L., and Abu-Shaar, M. (1993). Molecular characterization of a peripheral receptor for cannabinoids. *Nature,* 365: 61–65.

Panikashvili, D., Sineonidou, C., Ben-Shabat, S., Hanus, L., Breuer, A., Mechoulam, R., and Shohami, E. (2001). An endogenous cannabinoid (2-AG) is neuroprotective after brain injury, *Nature,* 413: 527–532.

Parker, L.A., Mechoulam, R., Schlievert, C., Abbott, L., Fudge, M.L., and Burton, P. (2003). Effects of cannabinoids on lithium-induced conditioned rejection reactions in a rat model of nausea. *Psychopharmacology,* 166: 156–162.

Petrzilka, T. and Sikemeir, C. (1967). Umwandlung von (-)-delta-6, 1-3,4-trans-Tetrahydrocannabinol in (-)-delta-1,2,-3,4-trans-Tetrahydrocannabinol. *Helv. Chim. Acta.,* 50: 2111–2113.

Rosenqvist, E. and Ottersen, T. (1975). The crystal and molecular structure of delta-9-tetrahydrocannabinolic acid b. *Acta Chem. Scand. B.,* 29: 379–384.

Shohami, E. and Mechoulam, R. (2000). Dexanabinol (HU-211): A nonpsychotropic cannabinoid with neuroprotective properties. *Drug Dev. Res.,* 50: 211–215.

Shohami, E., Gallily, R., Mechoulam, R., Bass, R., and Ben-Hur, T. (1997). Cytokine production in the brain following closed head injury: dexanabinol (HU-211) is a novel TNF-alpha inhibitor and an effective neuroprotectant. *J. Neuroimmunol.* 72: 169–77.

Todd, A.R. (1946). Hashish. *Experientia,* 2: 55–60.

2 The Relationship between Endocannabinoid Conformation and Endocannabinoid Interaction at the Cannabinoid Receptors

Patricia H. Reggio

CONTENTS

INTRODUCTION

The cannabinoid CB_1 and CB_2 receptors are G-protein-coupled receptors (GPCRs) that belong to the rhodopsin GPCR family (Class A). The CB_1 receptor transduces signals in response to the endocannabinoids anandamide (*N*-arachidonoylethanolamine, AEA, **1**), sn-2-arachidonoylglycerol (2-AG, **2**), and noladin ether (**3**). The CB_1 receptor also tranduces signals in response to the CNS-active constituents of *Cannabis sativa*, such as the classical cannabinoid (–)-trans-delta-9-tetrahydrocannabinol ((–)-Δ^9-THC, **4**) and to two other structural classes of ligands, the nonclassical cannabinoids typified by (1R,3R,4R)-3-[2-hydroxy-4-(1,1-dimethylheptyl)phenyl]-4-(3-hydroxypropyl) cyclohexanol-1-(CP55940, **5**) (Devane et al., 1988; Melvin et al., 1995), the aminoalkylindoles (AAIs) typified by (R)-[2,3-dihydro-5-methyl-3-[(4-morpholinyl)methyl]pyrrolo[1,2,3-de]-1,4-benzoxazin-6-yl] (1-naphthalenyl)methanone (WIN55212-2, **6**) (Compton et al., 1992; D'Ambra et al., 1992; Ward et al., 1991).

The first CB_1 antagonist, *N*-(piperidin-1-yl)-5-(4-chlorophenyl)-1-(2,4-dichlorophenyl)-4-methyl-1H-pyrazole-3-carboxamide [SR141716A (**7**; Chart 2)] was developed by M. Rinaldi-Carmona and co-workers at Sanofi Recherche (Rinaldi-Carmona et al., 1994). SR141716A displays nanomolar CB_1 affinity ($K_i = 1.98 \pm 0.13$ n*M*), but very low affinity for CB_2. *In vitro*, SR141716A antagonizes the inhibitory effects of cannabinoid agonists on both mouse vas deferens (MVD) contractions and adenylyl cyclase activity in rat brain membranes. SR141716A also antagonizes the pharmacological and behavioral effects produced by CB_1 agonists after intraperitoneal (IP) or oral administration (Rinaldi-Carmona et al., 1994). Several other CB_1 antagonists have been reported, AM630 (Hosohata, Quock, Hosohata, Burkey et al., 1997; Hosohata, Quock, Hosohata, Makriyannis et al., 1997; Pertwee et al., 1995), LY320135 (Felder et al., 1998), O-1184 (Ross et al., 1998), CP272871 (Meschler et al., 2000), and a class of benzocycloheptapyrazoles (Stoit et al., 2002).

SR141716A (**7**) has been shown to act as a competitive antagonist and inverse agonist in host cells transfected with exogenous CB_1 receptor, as well as in biological preparations endogenously expressing CB_1 (Bouaboula et al., 1997; Meschler et al., 2000; Pan et al., 1998). SR144528 (**8**)

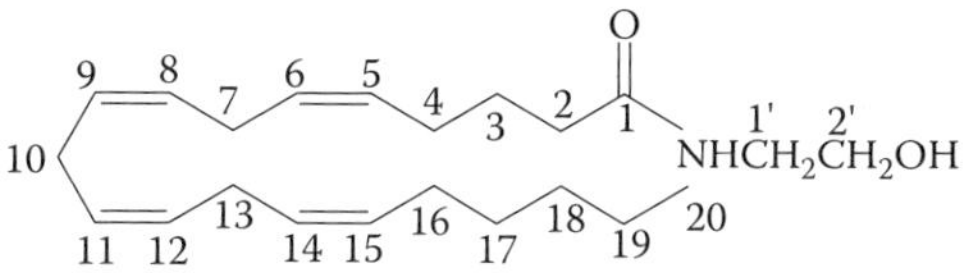

Anandamide
1

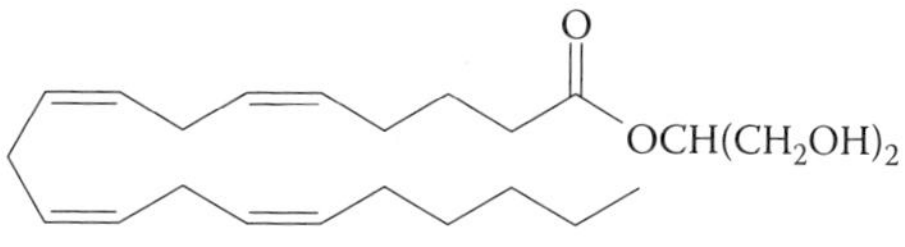

2-AG
2

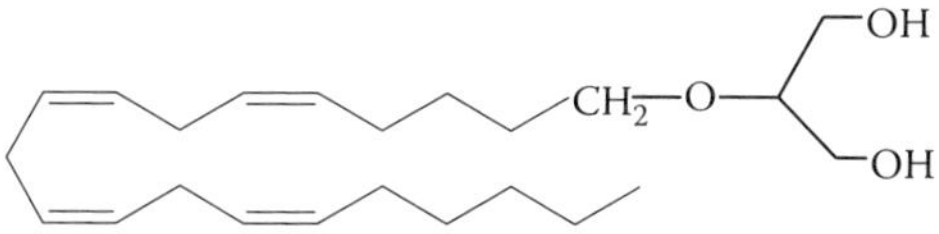

Noladin Ether
3

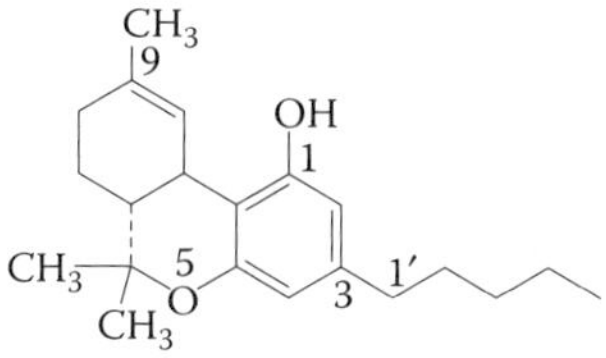

delta-9-THC
4

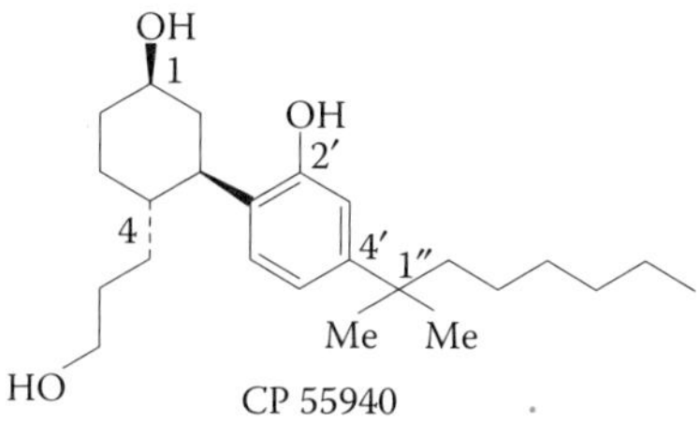

CP 55940
5

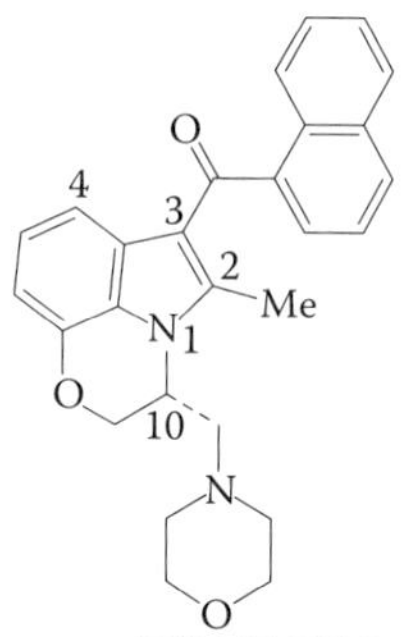

WIN 55,212-2
6

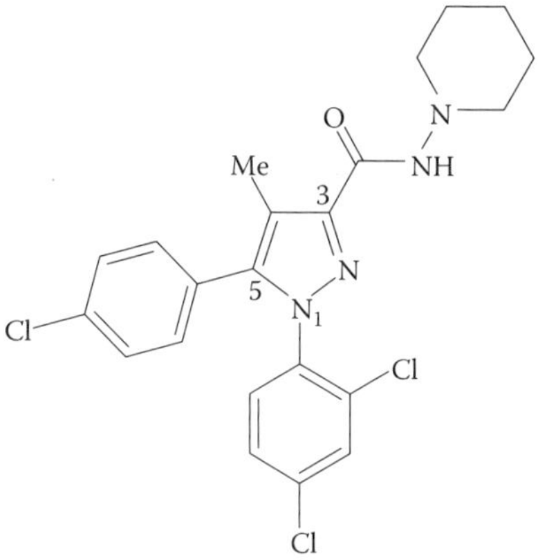

SR 141716A
(CB1)
7

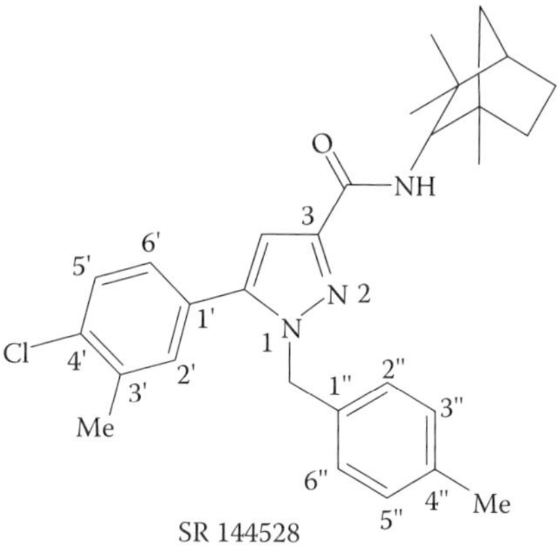

SR 144528
(CB2)
8

was the first identified competitive antagonist/inverse agonist of the CB_2 receptor (Bouaboula et al., 1999; Rinaldi-Carmona et al., 1998). SR144528 displays subnanomolar affinity for both the rat spleen and cloned human CB_2 receptors (K_i = 0.60 ± 0.13 n*M*). SR144528 displays a 700-fold lower affinity for both the rat brain and cloned human CB_1 receptors. JTE-907 has also been identified as an inverse agonist at CB_2 (Iwamura et al., 2001). CB_2-receptor-transfected Chinese hamster cells exhibit high constitutive activity as well. This activity can be blocked by the CB_2-selective ligand, SR144528, working as an inverse agonist (Bouaboula et al., 1999).

Although many Class A GPCRs have endogenous ligands that are hydrophilic cations (e.g., the serotonin and dopamine receptors), the cannabinoid receptors have neutral, highly lipophilic ligands derived from the fatty acid, arachidonic acid (AA). These ligands include AEA (**1**), 2-AG (**2**), and noladin ether (**3**) (Reggio, 2002). Much is known about the synthesis, storage, release, and transport of the cationic neurotransmitters. However, information about the endocannabinoid system is only now emerging in the literature. The major focus of this chapter is on endocannabinoid conformation and interaction with the cannabinoid (CB) receptors. The chapter begins more generally, however, by discussing the entire endocannabinoid system, of which the CB receptors are a part. To understand endocannabinoid action, one needs an appreciation of the environments for which these ligands have been designed and the conformational changes these ligands must undergo in order to act on the CB receptors.

THE ENDOCANNABINOID SYSTEM

The CB_1 receptor is one of the most abundant neuromodulatory receptors in the brain and is expressed at high levels in the hippocampus, cortex, cerebellum, and basal ganglia (Herkenham et al., 1991; Matsuda et al., 1993; Tsou et al., 1998; Wilson and Nicoll, 2001; Wilson and Nicoll, 2002). Recently, the endogenous cannabinoids, AEA (**1**) and 2-AG (**2**) have been proposed to be retrograde signaling molecules that mediate the suppression of GABA release from presynaptic terminals following depolarization of a hippocampal CA1 pyramidal neuron. This process is termed *depolarization-induced suppression of inhibition* (DSI) (Wilson and Nicoll, 2001). Pharmacological and kinetic evidence suggests that CB_1 activation inhibits presynaptic Ca^{2+} channels through direct G protein inhibition. The involvement of the cannabinoid system in DSI has been supported by experiments which show that DSI is absent in mice that lack CB_1 (i.e., CB_1 knockout mice) (Wilson et al., 2001). Paired recordings show that endocannabinoids selectively inhibit a subclass of synapses distinguished by their fast kinetics and large unitary conductance. Furthermore, cannabinoid-sensitive inputs are unusual among central nervous system (CNS) synapses in that they use *N*-, but not P/Q-type Ca^{2+}, channels for neurotransmitter release. These results indicate that endocannabinoids are highly selective, rapid modulators of hippocampal inhibition (Wilson et al., 2001).

The life cycle of endogenous cannabinoids is quite different from that of the endogenous ligands of the GPCR Class A cationic neurotransmitters such as the dopamine or serotonin receptors. In the brain, cationic neurotransmitter endogenous ligands (e.g., dopamine or 5-HT) are confined spatially by membrane-delineated compartments for controlled storage, release, and uptake. Once released from vesicle stores in the presynaptic cell, these hydrophilic ligands are thought to approach the receptor (which is located on the postsynaptic cell) from the extracellular milieu of the synaptic cleft and after receptor interaction are thought to be transported by a specific transporter protein from the extracellular milieu back into the presynaptic cell to be re-stored in synaptic vesicles (see Figure 2.1).

The molecular mechanisms for regulating lipid-based signaling events such as cannabinoid receptor signaling are not yet well understood. Because lipids and their derivatives can readily partition into and diffuse throughout cellular membranes, lipid messengers such as AEA (**1**) are not easily contained by such physical boundaries as those of the membrane vesicles that store cationic neurotransmitters. Instead, AEA (**1**) appears to be rapidly synthesized from lipid by neurons in response to depolarization and consequent Ca^{2+} influx (Di Marzo et al., 1998; Piomelli et al., 1998) (see Figure 2.2A). Endocannabinoid synthesis can also be triggered by activation of group I metabotropic glutamate receptors (mGluRs). It was previously known that activation of group I mGluRs can suppress neurotransmitter release by acting at a presynaptic locus, even though group I mGluRs are localized almost exclusively to postsynaptic structures (Lujan et al., 1997). This paradox has been resolved with the findings of two studies — one focusing on hippocampal GABAergic

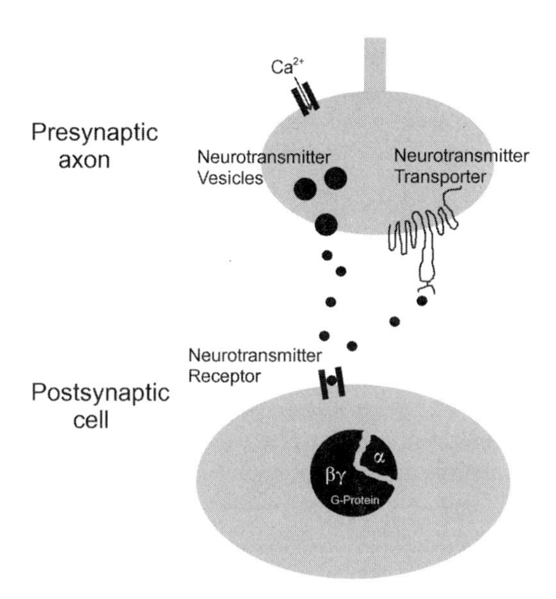

FIGURE 2.1 The controlled storage, release, and uptake for the signaling of G-protein-coupled cationic neurotransmitter endogenous ligands (e.g., the dopamine or 5-HT receptors). In the brain, cationic neurotransmitter endogenous ligands (e.g., dopamine or serotonin) are confined spatially by membrane delineated compartments for controlled storage, release, and uptake. Once released from vesicle stores in the presynaptic cell, these hydrophilic ligands are thought to approach the receptor (which is located on the postsynaptic cell) from the extracellular milieu of the synaptic cleft and after receptor interaction are thought to be transported by a specific transporter protein from the extracellular milieu back into the presynaptic cell to be re-stored in synaptic vesicles.

synapses (Varma et al., 2001) and the other on cerebellar climbing fiber synapses (Maejima et al., 2001) — in which a CB_1 antagonist was shown to block the effects of group I mGluR activation. Furthermore, inhibition of hippocampal GABA release by group I mGluRs is absent in CB_1 knockout (–/–) mice (Varma et al., 2001). Ca^{2+} does not seem to be required for endocannabinoid synthesis mediated by this pathway because a postsynaptic Ca^{2+} chelator does not block the effects of a group I mGluR agonist (Maejima et al., 2001). Thus, mGluRs and depolarization appear to be two independent pathways to endocannabinoid synthesis, (see Figure 2.2A) which, together, can additively increase the magnitude of DSI (Varma et al., 2001).

Because the biosynthetic enzymes that produce endocannabinoids are still being characterized, it is not yet possible to localize endocannabinoid production to a particular subcellular compartment. It has been suggested, however, that endocannabinoids might be released from neuronal somata and dendrites (Egertova et al., 1998; Levenes et al., 1998). On the basis of biochemical data, it has been suggested that endocannabinoids might diffuse widely through brain tissue and affect brain regions remote from their site of release (Bisogno et al., 1999). Electrophysiological recordings in hippocampal slice, however, have indicated that the distance that endocannabinoids can diffuse is only 20 μm (Wilson and Nicoll, 2001). In cerebellar slices, mGluR-dependent endocannabinoid synthesis was not observed to have any heterosynaptic effects (Maejima et al., 2001). Together, these results indicate that endocannabinoids are quite local signals.

In retrograde signaling, it is still unclear how newly synthesized endocannabinoids are induced to leave the postsynaptic plasma membrane to interact with CB_1, which is located presynaptically. Endocannabinoids may be secreted by simple diffusion; alternatively, passive (energy-independent) carrier proteins may be required to extrude endocannabinoids (Di Marzo et al., 1994; Piomelli et al., 1998). Endocannabinoids are thought to bind within the binding-site crevice formed by the seven

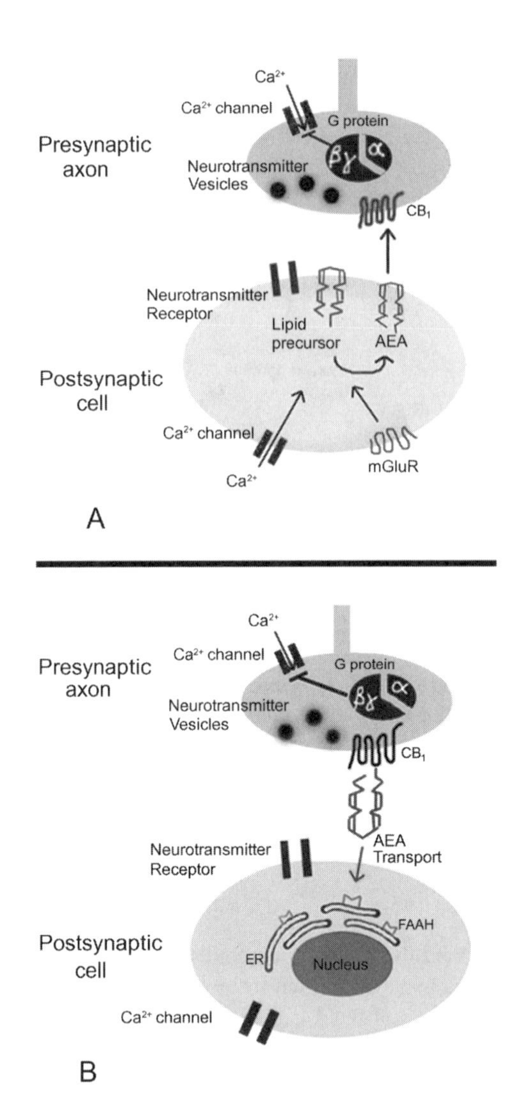

FIGURE 2.2 The life cycle of anandamide (AEA). (A) AEA (**1**) is rapidly synthesized from lipid by neurons in response to depolarization and consequent Ca^{2+} influx or via activation of group I metabotropic glutamate receptors (mGluRs). Newly synthesized AEA is possibly induced to leave the postsynaptic plasma membrane either by simple diffusion or by passive (energy-independent) carrier proteins in order to interact with the CB1 receptor, which is located presynaptically. AEA is thought to bind within the binding site crevice formed by the seven transmembrane helices of CB1. The subsequent activation of CB1 by AEA results in the inhibition of Ca^{2+} channels in the presynaptic cell. Because the release of cationic neurotransmitters from vesicles (in the presynaptic cell) is Ca^{2+}-dependent, activation of CB1 results in the suppression of neurotransmitter release by decreasing the local release probability of synaptic vesicles. (B) AEA uptake into the postsynaptic cell after CB1 interaction has been proposed to be mediated by a specific transport protein or to be via a simple diffusion process. After being transported into the cell, AEA is subsequently broken down into arachidonic acid and ethanolamine by a membrane-bound enzyme called fatty-acid amide hydrolase (FAAH) that has been shown by immunohistochemistry to be localized to the endoplasmic reticulum.

transmembrane helices of CB_1 (Barnett-Norris, Hurst, Lynch et al., 2002). The subsequent activation of CB_1 by endocannabinoids results in the inhibition of Ca^{2+} channels in the presynaptic cell. Because the release of cationic neurotransmitters from vesicles (in the presynaptic cell) is Ca^{2+}-dependent (Hoffman and Lupica, 2000), activation of CB_1 results in the suppression of neurotransmitter release by decreasing the local release probability of synaptic vesicles (Hoffman and Lupica, 2000; Kreitzer and Regehr, 2001; Levenes et al., 1998; Ohno-Shosaku et al., 2001; Takahashi and Linden, 2000; Wilson and Nicoll, 2001) (see Figure 2.2A).

What happens to endocannabinoids immediately after exiting CB_1 is still a matter for debate. Its ability to readily partition means that termination of the signaling capacity of AEA (**1**) may not rely on cellular uptake and compartmentalization. Instead, termination may rely primarily on chemical transformation, possibly within the bilayer itself. After receptor interaction, it has been proposed that endocannabinoids are released into the extracellular space to be transported by a specific transport protein on both neurons and glia that mediate endocannabinoid uptake (Di Marzo et al., 1994; Piomelli et al., 1998). An antagonist of this transporter, AM404, potentiates the effect of exogenous AEA on cultured neurons (Beltramo et al., 1997). After being transported into the cell, AEA is subsequently broken down into AA and ethanolamine by a membrane-bound enzyme called fatty acid amide hydrolase (FAAH) (Di Marzo et al., 1998; Piomelli et al., 1998) that has been shown by immunohistochemistry to be localized to the endoplasmic reticulum (Arreaza and Deutsch, 1999; Giang and Cravatt, 1997) (see Figure 2.2B). Mice deficient in FAAH show significantly increased levels of brain AEA, implying that FAAH helps regulate endogenous cannabinoid tone (Cravatt et al., 2001). An alternative postreceptor interaction scenario proposed by Deutsch and co-workers (Glaser et al., 2003) is that AEA uptake after receptor interaction is a process of simple diffusion. In this case, the uptake process would be driven by the metabolism of AEA by FAAH and other downstream events, rather than by a specific membrane-associated AEA carrier. These investigators have found that transport inhibitors such as AM404 are actually inhibitors of FAAH *in vitro*. Therefore, the likely mechanism by which the transport inhibitors raise AEA levels to exert pharmacological effects is by inhibiting FAAH, rather than by inhibiting a transporter protein. Cellular uptake may also be facilitated by caveolae/lipid raft-related endocytosis (McFarland et al., 2004). Termination of AEA signaling appears then to rely on chemical transformation, possibly within the bilayer itself. The crystal structure of FAAH is consistent with this idea, as FAAH is an integral membrane protein and AEA's direct access to the enzyme's active site appears to be via the bilayer (Bracey et al., 2002).

To assess more widely the functions of FAAH in the brain and the potential impact of FAAH activity on the spatiotemporal dynamics of endocannabinoid signaling in different regions of the brain, Egertova and co-workers (2003) employed immunocytochemistry to compare the distribution of FAAH and CB_1 throughout the mouse brain, using FAAH(–/–) knockout mice as negative controls to validate the specificity of FAAH-immunoreactivity observed in wild-type animals. In many regions of the brain, a complementary pattern of FAAH and CB_1 expression was observed, with FAAH-immunoreactive neuronal somata and dendrites surrounded by CB_1-immunoreactive fibers. In these regions of the brain, FAAH may regulate postsynaptic formation of AEA, thereby influencing the spatiotemporal dynamics of retrograde endocannabinoid signaling, as was discussed earlier and illustrated in Figure 2.2. However, in some regions of the brain, such as the globus pallidus and substantia nigra pars reticulata, CB_1 receptors were found to be abundant but with little or no associated FAAH expression. In these brain regions, the spatial impact and duration of endocannabinoid signaling may be less restricted than in regions enriched with FAAH. A more complex situation arises in several regions of the brain where both FAAH and CB_1 are expressed, but in a noncomplementary pattern, with FAAH located in neurons and oligodendrocytes that are proximal but not postsynaptic to CB_1-expressing axon fibers. Here, FAAH may nevertheless influence endocannabinoid signaling, though more remotely. Finally, there are regions of the brain where FAAH-immunoreactive neurons or oligodendrocytes occur in the absence of CB_1-immunoreactive fibers. Here, FAAH may be involved in regulation of signaling mediated by other endocannabinoid receptors or by receptors for other fatty acid amide signaling molecules. Consistent with this, Cravatt and coworkers have reported that although FAAH converts AEA to AA and ethanolamine, it can also accept other fatty acid amides as substrates, including oleamide (Cravatt et al., 1996).

Although there are many yet unresolved issues concerning the life cycle of AEA and other endocannabinoids, it is clear that the endocannabinoid story is intimately associated with the lipid milieu, for AEA is synthesized from lipid and ultimately must enter lipid in order to be hydrolyzed by FAAH. Between these initial and final steps, AEA must interact with one (CB_1) or more (AEA transporter) proteins and possibly diffuse in an aqueous environment. It can be expected, then, that the endocannabinoids may need to be highly flexible ligands which can change conformation to adapt to very different environments. In the next section, the inherent structural flexibility built into endocannabinoid structures (and the structure of their parent fatty acid) is discussed. The conformations adopted by these ligands alone or in complexes with various proteins as revealed by X-ray crystal studies are discussed. Also discussed are the conformations that are most likely adopted by these ligands in nonpolar or polar solvents, and in a lipid bilayer, as revealed by computational and biophysical studies.

ENDOCANNABINOID CONFORMATION AS A FUNCTION OF ENVIRONMENT

Endocannabinoid/Parent Fatty Acid, Low-Energy Conformations

Fatty Acid Nomenclature

There are two accepted shorthand notation systems for long-chain unsaturated fatty acids in use in the fatty acid literature. In one system, the acyl moiety 20:4, $\Delta^{5,8,11,14}$ indicates the presence of a 20-carbon atom chain; four cis homoallylic double bonds, with the double bonds at C5–C6, C8–C9, C11–C12, and C14–C15 (see drawing of AEA (**1**), for numbering system). In a second system, the acyl moiety (20:4, n-6) indicates the presence of a 20-carbon atom chain; four cis homoallylic double bonds with the last sp^2 hybridized carbon being the sixth carbon (i.e., C15) from the last carbon of the acyl chain (C20).

Arachidonic Acid and Related Fatty Acids: Conformational Analysis

Conformational analysis of highly flexible ligands such as the endogenous cannabinoids is a challenging task due to the large number of conformations available to the ligand. The AA acyl chain of AEA and of 2-AG contains three different types of C–C bonds:C_{sp2}–C_{sp2}, C_{sp2}–C_{sp3}, and C_{sp3}–C_{sp3}, of which the last two types permit bond rotation and thus control the conformational changes possible for the molecule.

C_{sp2}–C_{sp3} Bond Rotation

The C5 to C15 portion of the AA acyl chain of AEA and of 2-AG contains four cis homoallylic double bonds (i.e., cis double bonds separated by methylene carbons). One important feature of this chain is the great torsional mobility about the two torsion angles involving each methylene carbon between adjacent pairs of double bonds (vinyl groups) in the acyl chain (for example, the C8–C9–C10–C11 and C9–C10–C11–C12 torsion angles, for which rotation would occur about the C9–C10 and C10–C11 bonds. See drawing of AEA (**1**) for the numbering system). This involves rotation about C_{sp2}–C_{sp3} bonds. Rabinovich and Ripatti (Rabinovich and Ripatti, 1991) reported that polyunsaturated acyl chains in which double bonds are separated by one methylene group (as in AA) are characterized by the highest equilibrium flexibility compared with other unsaturated acyl chains. Rich (Rich, 1993) reports that a broad domain of low-energy conformational freedom exists for these C–C bonds. Feller and coworkers have reported that the rotational barriers for isomerization around methylenes connecting the vinyl groups in DHA (22:6, $\Delta^{4,7,10,13,16,19}$) are extremely low (<1 kcal/mol) (Feller et al., 2002). Additionally, the torsional minima for the rotatable bonds in DHA are very broad. Results of the biased sampling phase from conformational memories (CM) calculations of AA are consistent with all of these results (Barnett-Norris et al., 1998), as they reveal a relatively broad distribution of populated torsional space about the classic skew angles of −119°(s′) and 119°(s) for the

C8–C9–C10–C11 torsion angle in AEA, for example. Thus, C_{sp2}–C_{sp3} bond rotation introduces great flexibility into an acyl chain, and regions of acyl chains that contain two torsion angles involving a methylene carbon between adjacent pairs of double bonds (vinyl groups) are regions that can introduce curvature in the overall acyl chain shape (Barnett-Norris, Hurst, Lynch et al., 2002). The fact that the AEA acyl chain contains four cis double bonds separated by methylene carbons (like its parent AA) suggests that AEA is engineered to easily adapt and change shape. This may facilitate AEA interaction with targets of importance.

C_{sp3}–C_{sp3} Bond Rotation

The C2 to C4 and C16 to C20 portion of the acyl chain of AEA and 2-AG, as well as their parent fatty acid AA, is saturated (i.e., contains C_{sp3}–C_{sp3} bonds). Feller and co-workers have reported that in contrast to the low rotational energy barrier for rotation about the C_{sp2}–C_{sp3} in docosahexaenoic acid (DHA 22:6, $\Delta^{4,7,10,13,16,19}$) (<1 kcal/mol), the rotational barrier for rotation about the C_{sp3}–C_{sp3} bonds in the saturated portion of the DHA acyl chain is about 3.5 kcal/mol (Feller et al., 2002). Rich reported similar results for AA. These results indicated that for saturated acyl chains, freedom of motion is restricted to trans (±180°) or gauche (±60°) configurations (Rich, 1993). Results of the biased sampling phase from CM calculations of AA are consistent with these results (Barnett-Norris et al., 1998), as they reveal a relatively narrow distribution of populated torsional space at 180° (trans), 60°, and –60° (gauche), with 180° being the highest populated torsion angle. These results indicate that saturated fatty acids or saturated portions of fatty acid chains are not very flexible and favor extended (all trans) conformations.

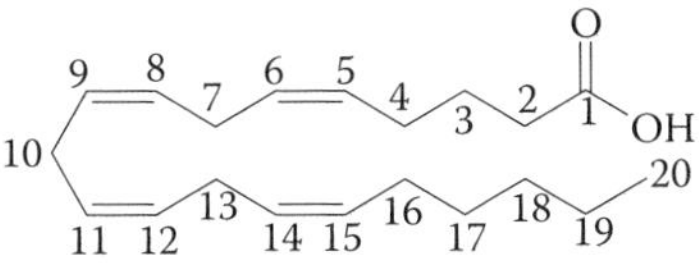

Arachidonic acid
9

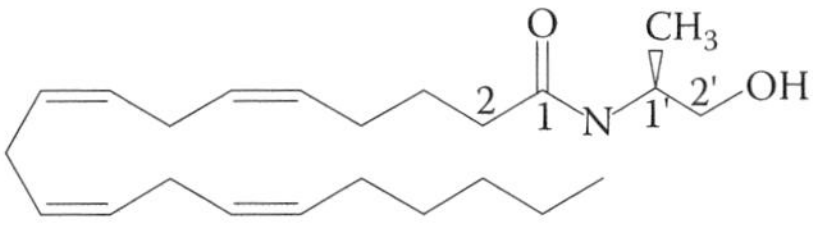

(R)-(+)-methanandamide
13

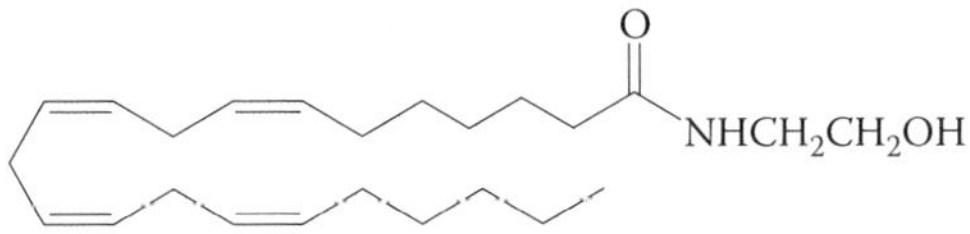

N-docosatetraenoyl-ethanolamine
10

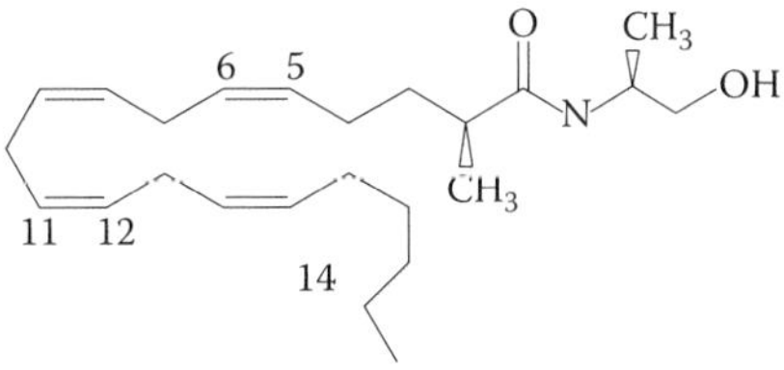

14

N-homo-γ-linolenoyl-ethanolamine
11

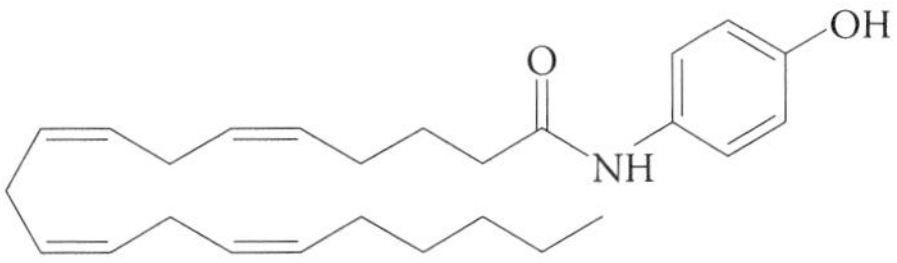

AM404
15

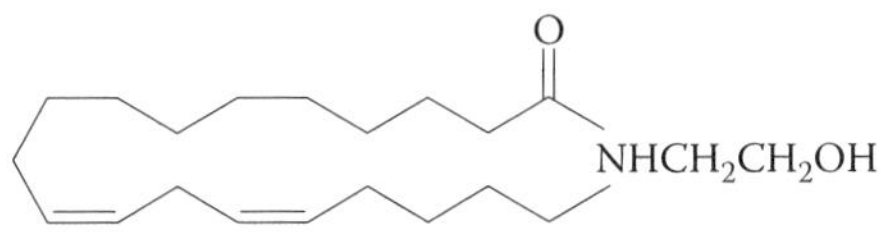

N-11,14-eicosadienoyl-ethanolamine
12

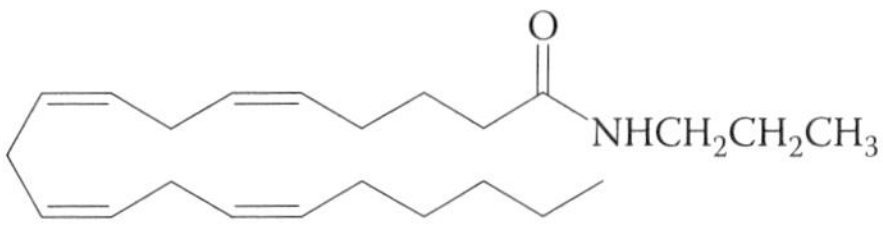

16

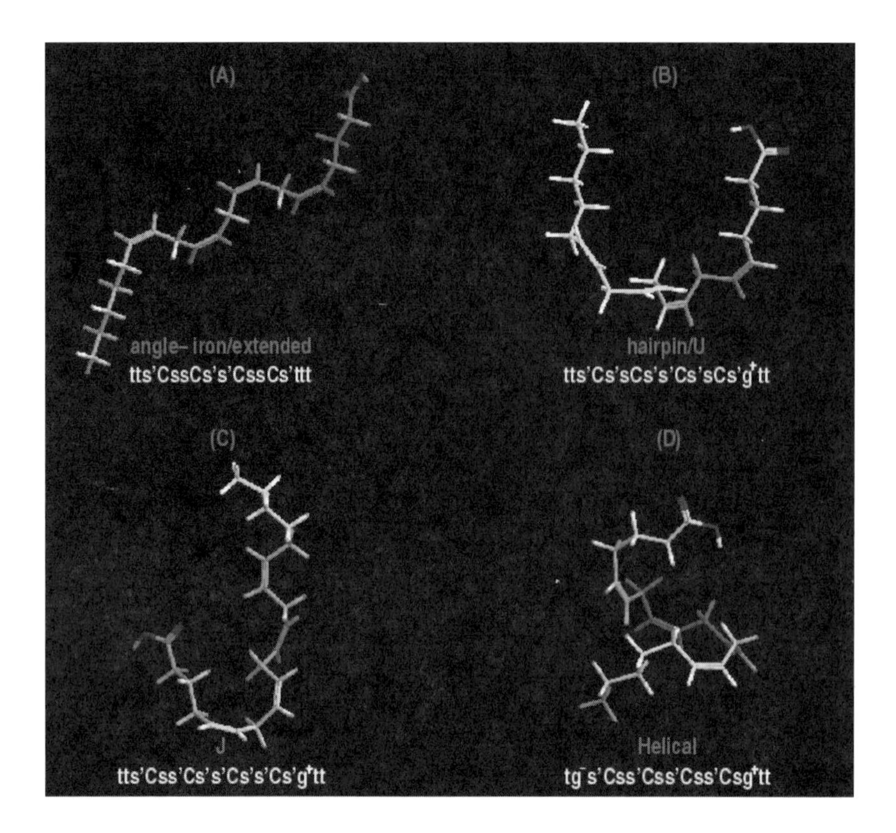

FIGURE 2.3 The general shapes/conformers commonly identified for arachidonic acid in structural and computational studies: (A) the angle-iron or extended conformation; (B) the hairpin or U-shaped conformation; (C) the J-shaped conformation; and (D) the helical conformation. Beneath the label for each conformation is a summary of torsion angle values beginning with ω_1 = C1–C2–C3–C4 and ending with ω_{17} = C17–C18–C19–C20 torsion angle of AA (see structure **9** for numbering system). The letter codes correspond to the following ideal torsion angles: cis, C ($\omega = 0°$); gauche, g$^-$ ($\omega = 60°$) and g$^+$ ($\omega = -60°$); skew, s ($\omega = 120°$) and s′ ($\omega = -120°$); and trans, t ($\omega = 180°$). (Barnett-Norris J, Guarnieri F, Hurst DP, and Reggio PH [1998] Exploration of biologically relevant conformations of anandamide, 2-arachidonylglycerol, and their analogues using conformational memories. *J Med Chem* 41: 4861–72. With permission.)

What shapes are the acyl chain in AEA and its analogs likely to adopt? Discussed in the following subsections are both computational and experimental results that assess what conformations can be adopted by AEA and related molecules as a function of the environment. Figure 2.3 illustrates the general shapes reported in the literature to be adopted by AA (**9**) and its derivatives.

Calculations *In Vacuo*

The two most commonly used methods in the literature for studying flexible ligands such as AA and AEA have been the Monte Carlo and molecular dynamics (MD) techniques. In their Monte Carlo study, Rabinovich and Ripatti (Rabinovich and Ripatti, 1991) found that polyunsaturated fatty acids whose double bonds are separated by one methylene carbon assume an extended (angle-iron) conformation when all molecules are efficiently packed below the phase-transition temperature. MD or molecular dynamics/simulated annealing (MD/SA) studies (*in vacuo*) of AA (**9**) and of other polyunsaturated fatty acids related to AA have been published by several groups

(Bonzom et al., 1997; Corey et al., 1983; Corey et al., 1979; Leach and Prout, 1990; Rabinovich and Ripatti, 1991; Rich, 1993; Wilson et al., 1988). These computational studies of AA have primarily found looped or back-folded conformations to be low-energy conformers of AA. Rich conducted a quenched MD study of AA *in vacuo* (Rich, 1993). The two lowest enthalpy conformers found for AA were J-shaped conformers in which the carboxylic acid group is in close proximity to the C14–C15 π bond. This same J-shape was reported by Corey and co-workers (1983) as one type of low-energy minimum identified in their conformational analysis of AA. Corey suggested that such a J-shaped conformation in solution would be energetically favorable and would be consistent with the chemistry of peroxyarachidonic acid for which an internal epoxidation leads to 14, 15-epoxyarachidonic acid (Corey et al., 1979).

Using the A* algorithm followed by minimization in the BatchMin module of MacroModel, Leach and Prout (Leach and Prout, 1990) found right- and left-handed looped or "helical-like" conformations of AA to be its lowest energy conformations. A simulated annealing study of AA identified a looped conformation (Wilson et al., 1988), while a MD study *in vacuo* found low-energy structures of AA to be hairpin, helical, and crown shaped (i.e., U-shaped conformers in which the ends of the chain come close to each other) (Bonzom et al., 1997). Lopez and co-workers also employed MD simulations in their study of AA, finding two low-energy conformations, one of which was U-shaped and was proposed as the bioactive conformation at its cyclooxygenase binding site (Lopez M. et al., 1993). Bonechi and co-workers performed an *in vacuo* Monte Carlo conformational analysis of AEA. Results indicated that the dominant conformation of AEA was a hybrid of which the extended and U-shaped forms are the limiting structures (Bonechi et al., 2001).

Calculations in Nonpolar ($CHCl_3$) vs. Polar (Aqueous) Environments

A new conformational analysis method, CM (Guarnieri and Weinstein, 1996; Guarnieri and Wilson, 1995), was employed by Barnett-Norris and co-workers (1998) to study the conformations of AA in solvent. The CM technique employs multiple Monte Carlo simulated annealing (MC/SA) random walks using the MM3 force field and the generalized Born/surface area (GB/SA) continuum solvation model for chloroform or water as implemented in the MacroModel molecular modeling package (Mohamadi et al., 1990). CM has been shown to achieve complete sampling of the conformational space of highly flexible molecules, to converge in a very practical number of steps, and to be capable of overcoming energy barriers efficiently (Guarnieri and Weinstein, 1996). The program, XCluster in MacroModel, is used to group the conformers generated by CM into families of like conformation called *clusters*. Conformers are grouped according to their increasing RMS deviation from the first structure output at 310 K. Because XCluster rearranges the conformers so that the RMS deviation between the nearest neighbors is minimized, any large jump in RMS deviation is indicative of a large conformational change and hence identifies a new conformational family or cluster (Cls). Conformational families that are identified by CM to be highly populated are those that have low free energies, whereas conformations that outlie these major groupings are those whose free energies are not sufficiently low enough to be visited frequently during the simulation.

CM calculations reported by Barnett-Norris and co-workers (1998) identified both extended conformers and folded or U-shaped conformers to be major conformational families of AA. The more compact (U-shaped) structure was found to predominate in water, whereas the extended shape was found to predominate in $CHCl_3$ (and *in vacuo*). The existence of J-shaped conformers as suggested in the literature (Corey et al., 1983; Corey et al., 1979; Rich, 1993) was confirmed by CM. In this case, the J-shaped family was found to comprise a smaller but still significant conformational family of AA. These results suggest that the CM method resulted in a much broader sampling of the conformational space of AA than had been performed previously using MD or MD/SA techniques (Applegate and Glomset, 1986; Bonzom et al., 1997; Corey et al., 1983; Leach and Prout, 1990; Lopez M. et al., 1993; Rich, 1993; Wilson et al., 1988). As discussed above, these previous studies of AA found primarily only compact structures (Us, Js, and helicals) as energy minima. Although no other theoretical studies of AA in solvent have been reported, the trends in the CM results reported by Barnett-Norris et al. (1998) for AA in $CHCl_3$ and in H_2O are consistent with the idea that in H_2O, AA would minimize the exposure of its hydrophobic portions by forming a more compact (U)-shape.

Barnett-Norris and co-workers reported CM results for a series of fatty acid ethanolamides, including AEA [22:4, $\Delta^{7,10,13,16}$ (**10**); 20:4, $\Delta^{5,8,11,14}$ (AEA; 1); 20:3, $\Delta^{8,11,14}$ (**11**); 20:2, $\Delta^{11,14}$ (**12**)] (Barnett-Norris, Hurst, Lynch et al., 2002) in $CHCl_3$. CM studies indicated that each analog could form both U-/J-shaped and extended families of conformers. For the analogs with three or four homoallylic double bonds, the U- or J-shaped family was higher in population (see Figure 2.4). When the number of homoallylic double bonds decreased to two, a shift in the population was seen, with the extended conformer family now higher in population for the 20:2, $\Delta^{11,14}$ (**12**) analog. In addition, radius of curvature calculations for the U- or J-shaped family of each analog indicated that the radius increased as the number of double bonds decreased. The average radii of curvature (with their 95% confidence intervals) were found to be 4.0 Å (3.6–4.5) for 22:4, $\Delta^{7,10,13,16}$; 4.0 Å (3.7–4.2) for 20:4, $\Delta^{5,8,11,14}$; 4.4 Å (4.1–4.7) for 20:3, $\Delta^{8,11,14}$; and 5.8 Å (5.3–6.2) for 20:2, $\Delta^{11,14}$. These calculations suggested that for fatty acid acyl chains that contain homoallylic double bonds, chains with decreasing amounts of unsaturation tend to show a decreasing tendency to form folded structures but still tend to curve in the acyl chain regions in which unsaturation is present.

Calculations in Protein Environments

A recent modeling study of the binding of AA in the cyclooxygenase active site of mouse prostaglandin endoperoxide synthase-2 (COX-2) predicts that AA will orient in a kinked or L-shaped conformation. In this conformation, the carboxylate moiety binds to Arg^{120}, while the ω-end is positioned above Ser^{530} in a region termed the *top channel* (Rowlinson et al., 1999). In their theoretical conformational analysis of the complex of AA with α-tocopherol, Shamovsky and Yarovskaya (Shamovsky and Yarovskaya, 1990) found that AA assumes an extended conformation. A computer docking study of (R)-*N*-(1-methyl-2-hydroxyethyl)-2-(R)-methyl-arachidonamide (**14**) in a computer model of the transmembrane helix (TMH) bundle of CB_1 (activated state, R*) suggested that this molecule adopts a tightly curved U-shaped conformation at CB_1, and suggested that the TMH 2-3-6-7 region is the endocannabinoid binding region at CB_1 (Barnett-Norris, Hurst, Lynch et al., 2002). A subsequent docking study of AEA in a computer model of the CB_1 activated state suggested that AEA adopts a tightly curved conformation as well, to bind in the TMH 2-3-6-7 region of CB_1. The orientation of AEA was slightly different from that of (R)-*N*-(1-methyl-2-hydroxyethyl)-2-(R)-methyl-arachidonamide (14) mentioned earlier, due to the differences in the head groups of these two ligands (McAllister et al., 2003).

Experimental X-Ray Crystal Structures: Arachidonic Acid and Arachidonic Acid Derivatives

The X-ray crystal literature is now replete with numerous experimental examples of the natural plasticity of the AA acyl chain. In its X-ray crystal structure, AA has been found to exist in an extended (angle-iron) conformation in which double bonds 1 and 3 and double bonds 2 and 4 are coplanar, while the planes of adjacent double bonds are perpendicular to one another (see Figure 2.5A) (Ernst et al., 1979; Rich, 1993). An X-ray crystal study of arachidonate ion complexed with adipocyte-lipid-binding protein revealed that AA binds within the β barrel cavity of this protein. Here the carboxylate group of AA is engaged in a strong electrostatic interaction with Arg^{126}, Tyr^{128}, and Arg^{106} through an intervening water molecule, while the arachidonate acyl chain assumes a hairpin (i.e., U-shaped) conformation for which the first double bond of AA is distorted out of the plane formed by the other three double bonds (see Figure 2.5B) (LaLonde et al., 1994). In the 3 Å resolution structure of prostaglandin H synthase-1 (Figure 2.5C), AA

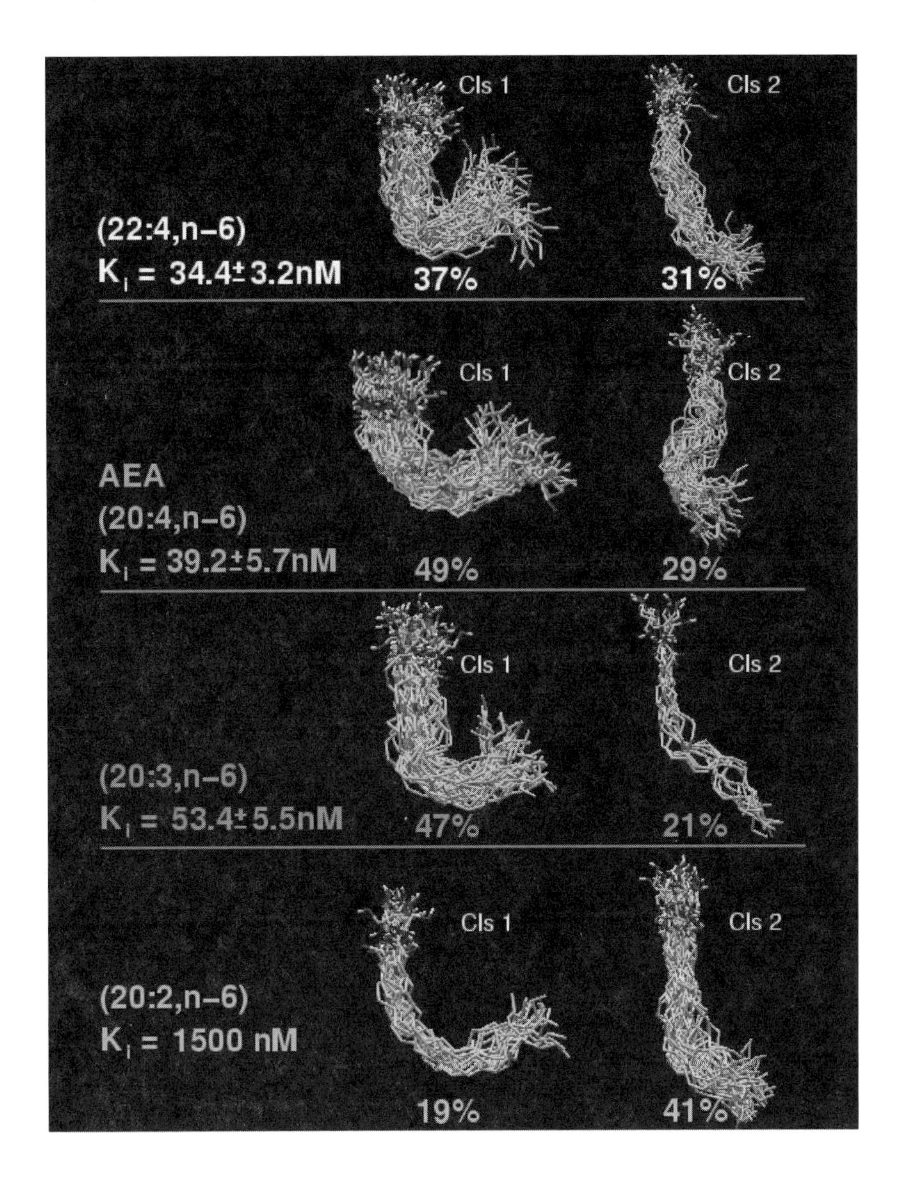

FIGURE 2.4 Conformational memories (CM) results for a series of fatty acid ethanolamides [22:4, $\Delta^{7,10,13,16}$ (**10**); 20:4, $\Delta^{5,8,11,14}$ (AEA, **1**); 20:3, $\Delta^{8,11,14}$ (**11**); and 20:2, $\Delta^{11,14}$ (**12**)] in $CHCl_3$. (From Barnett-Norris J, Hurst DP, Lynch DL, Guarnieri F, Makriyannis A, and Reggio PH [2002] Conformational memories and the endocannabinoid binding site at the cannabinoid CB1 receptor. *J Med Chem* 45: 3649–59. With permission.) CM studies indicated that each analog could form both U- or J-shaped and extended families of conformers. For the analogs with three or four homoallylic double bonds (**1**, **10**, **11**), the U- or J-shaped family was higher in population. When the number of homoallylic double bonds decreased to two, a shift in the population was seen, with the extended conformer family now being higher in population for the 20:2, $\Delta^{11,14}$ analog (**12**).

adopts an extended L-shaped conformation that positions the 13proS hydrogen of AA for abstraction by Tyr^{385}, the likely radical donor at this enzyme's active site (Malkowski et al., 2000). In the recently elucidated X-ray crystal structure of FAAH, the AA derivative, methoxy arachidonyl phosphonate (MAP) is covalently bound to Ser 241 (Bracey et al., 2002) and the conformation of the arachidonyl chain can best be described as helical (see Figure 2.5D).

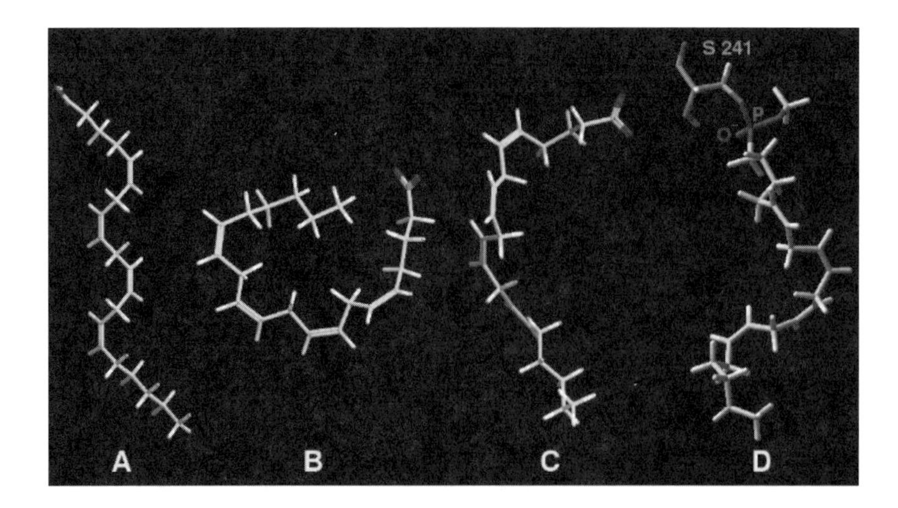

FIGURE 2.5 X-ray crystal structure results for arachidonic acid (AA) and its derivatives alone or complexed with various proteins. (A) In its x-ray crystal structure, AA (**9**) has been found to exist in an extended (angle-iron) conformation in which double bonds 1 and 3 and double bonds 2 and 4 are coplanar, whereas the planes of adjacent double bonds are perpendicular to one another. (From Ernst J, Sheldrick WS, and Fuhrop JH [1979] The structures of the essential unsaturated fatty acids. Crystal structure of linoleic acid and evidence for the crystal structures of alpha-linoleic and arachidonic acid. *Z. Naturforsch* 34b: 706–711; Rich MR [1993] Conformational analysis of arachidonic and related fatty acids using molecular dynamics simulations. *Biochim Biophys Acta* 1178: 87–96.) (B) An X-ray crystal study of arachidonate ion (AA) complexed with adipocyte-lipid-binding protein revealed that the arachidonate acyl chain assumes a hairpin (i.e., U-shaped) conformation for which the first double bond of AA is distorted out of the plane formed by the other three double bonds. (From LaLonde JM, Levenson MA, Roe JJ, Bernlohr DA, and Banaszak LJ [1994] Adipocyte lipid-binding protein complexed with arachidonic acid. Titration calorimetry and X-ray crystallographic studies. *J Biol Chem* 269: 25339–47.) (C) In the 3 Å resolution structure of prostaglandin H synthase-1, AA adopts an extended L-shaped conformation. (From Malkowski MG, Ginell SL, Smith WL, and Garavito RM [2000] The productive conformation of arachidonic acid bound to prostaglandin synthase. *Science* 289: 1933–7.) (D) In the recent X-ray crystal structure of FAAH, the AA derivative, methoxy arachidonyl phosphonate (MAP) is covalently bound to Ser241 (From Bracey MH, Hanson MA, Masuda KR, Stevens RC, and Cravatt BF [2002] Structural adaptations in a membrane enzyme that terminates endocannabinoid signaling. *Science* 298: 1793–6.) and the conformation of the arachidonyl chain can best be described as helical.

Experimental NMR Solution Studies of AEA and AEA Analogs

Bonechi and co-workers have assessed the solution conformation of AEA in DMSO-d_6 at 298 K using ^{1}H and ^{13}C NMR combined with Monte Carlo calculations (Bonechi et al., 2001). These investigators found that the acyl chain of AEA was predominantly in an extended conformation in solution, whereas the head group of AEA was involved in an intramolecular hydrogen bond between the hydroxyl group hydrogen and the oxygen of the carboxamide group. The study of the dynamic properties of AEA in solution showed clearly that the molecule is in fast motion conditions and that only the C18, C19, and C20 carbons (see drawing of AEA (**1**) for numbering system) show values of the dynamic parameter τ_C, indicating the presence of a free tumbling motion in the terminal part of the acyl chain.

The CM calculations reported by Barnett-Norris and co-workers (1998) discussed earlier, identified both extended conformers and folded or U-shaped conformers to be major conformational families of AA. The more compact (U-shaped) structure was found to predominate in water, whereas the extended shape was found to predominate in $CHCl_3$ (and *in vacuo*). These results were taken to be consistent with the idea that in H_2O, AA would minimize the exposure of its hydrophobic portions by forming a more compact (U-shape) conformation. Van der Stelt and co-workers (2002) recently studied the solution conformation of a conjugated/hydroxylated analog of AEA, 15(S)-hydroxy-eicosa-5Z,8Z,11Z,13E-tetraenoic acid in D_2O and in $CDCl_3$. Analysis of ^1H NMR spectra revealed that the acyl chain of 15(S)-hydroxy-eicosa-5Z,8Z,11Z,13E-tetraenoic acid did not assume notably different conformations in the polar (water) vs. nonpolar (chloroform) environments. This result is likely because the location of the conjugation (C11 = C12 – C13 = C14; which introduces rigidity into the acyl chain) is in the region of the acyl chain that would need to curve to permit the adoption of a more compact (U-shape) conformation in water.

Calculations in a Lipid Environment

In order to explore the behavior of AEA in a lipid environment, we recently undertook MD simulations of AEA in a dioleoylphosphatidylcholine (DOPC) bilayer (see Figure 2.6) (Reggio et al., 2003; Lynch and Reggio, 2005). A fully hydrated DOPC bilayer (28 waters/lipid) was built in order to generate a reasonable representation for the lipid bilayer. The calculations were run using an NPAT ensemble and the CHARMM27 set of all atom parameters designed for lipid simulations (Feller et al., 2002). Four AEAs were placed in the equilibrated bilayer, and the system was heated to 310 K and equilibrated. Two separate 5 nsec NPAT trajectories were simulated.

These simulations revealed that AEA assumes an extended conformation in the bilayer, with its C16–C20 alkyl tail portion (located near the bilayer center) capable of considerable motion. During the simulation, AEA assumed many of the conformations discussed above for AA, including

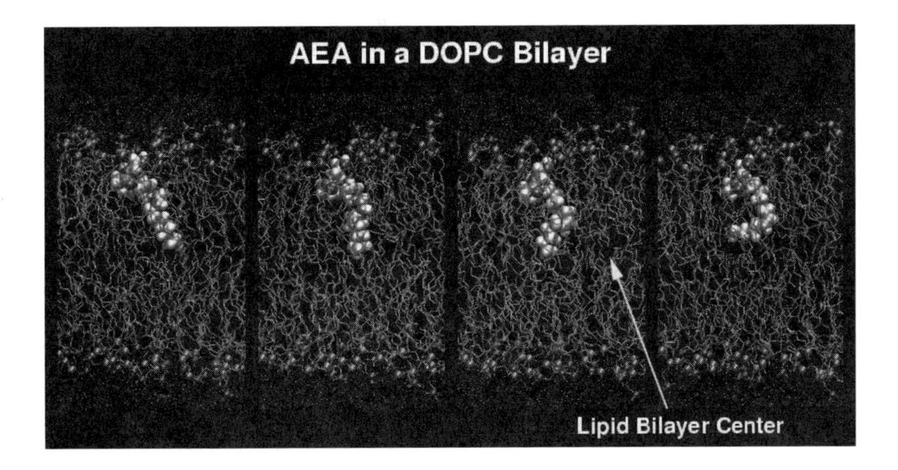

FIGURE 2.6 Conformations adopted by anandamide (AEA) in a dioleoylphoshatidylcholine (DOPC) bilayer during a molecular dynamics (MD) simulation. (From Reggio P, Hurst DP, Ballesteros JA, and Lynch DL [2003] The CB1 V6.43/I6.46 groove is optimally placed in the lipid bilayer to recognize endocannabinoids in lipid, in *2003 Symposium on the Cannabinoids,* p. 8, International Cannabinoid Research Society, Cornwall, Ontario, Canada.) This simulation employed a fully hydrated DOPC bilayer (28 waters/lipid) and the CHARMM27 set of all atom parameters designed for lipid simulations. (From Feller SE, Gawrisch K, and MacKerell AD, Jr. [2002] Polyunsaturated fatty acids in lipid bilayers: intrinsic and environmental contributions to their unique physical properties. *J Am Chem Soc* 124: 318–26.) AEA was placed in the equilibrated bilayer, and the system was heated to 310 K and equilibrated for 200 psec using an NVT ensemble. The simulation then was switched to NPT ensemble conditions (P = 1 atm, T = 310 K) and run for 2.5 nsec. This simulation revealed that the ethanolamide head group is located at the same depth as the DOPC glycerol group and overlaps substantially with the choline group of DOPC. AEA assumes an extended conformation in the bilayer, with its C16–C20 alkyl tail portion (located near the bilayer center) capable of considerable motion. During the simulation, AEA assumed many of the conformations discussed in this chapter for AA, including extended and J-shaped conformations.

extended and J-shaped conformations (see Figure 2.6). The ethanolamide head group was located at the same depth as the glycerol group and overlapped substantially with the choline group of DOPC. The radial distribution function of various lipid functional groups with the C2 hydrogens of AEA revealed that the species closest to the C2 hydrogens of AEA were water, choline, and glycerol hydrogens. These results were not dependent on the starting position of AEA in the simulation, as simulations in which the head group was pushed higher into the polar head group region or lower into the nonpolar portion of the bilayer resulted in the repositioning of the head group to the glycerol region (as described earlier) later in these simulations.

Biophysical Studies of Cannabinoids in a Lipid Environment

Similar to many other membrane-active lipophilic molecules, cannabinoids preferentially partition into the membrane, where they assume a thermodynamically favorable orientation and location (Makriyannis, 1993). Makriyannis and co-workers were the first to study the interaction of cannabinoids with the membrane. These investigators proposed that the membrane may serve to orient cannabinoid ligands for entry into the cannabinoid receptor. Using solid-state [^{2}H]-NMR spectroscopy, these investigators have found that the amphipathic Δ^9-THC (**4**) uses its phenolic hydroxyl to anchor itself at the interface of the amphipathic membrane bilayer, while it orients the long axis of its tricyclic system perpendicular to the bilayer; small-angle x-ray results confirmed this interpretation (Makriyannis et al., 1989). Cannabinoids with two hydroxyl groups orient in a manner that allows both hydroxyls to be near the interface, facing the polar surface of the membrane (Yang et al., 1991). However, if the phenolic hydroxyl of Δ^9-THC is replaced with a methoxy group, the cannabinoid orients with its natural axis parallel to the lipid acyl chains (Yang et al., 1992).

As described in the introduction to this chapter, AEA also is a highly lipophilic ligand that is synthesized from lipid in the lipid bilayer (Di Marzo et al., 1998; Piomelli et al., 1998) and is hydrolyzed after entering FAAH from lipid (Bracey et al., 2002). Gawrisch and co-workers have recently reported NMR studies of anandamide in a model membrane composed of 1-stearoyl-2-oleoyl-sn-glycero-3-phosphocholine (SOPC) that are very consistent with the calculations illustrated in Figure 2.6. NOESY cross-relaxation rates indicated that AEA aligns itself parallel to lipids with its ethanolamine group in the lipid–water interface and its acyl chain in the hydrophobic region of the bilayer (Eldho and Gawrisch, 2003). By exploiting the bilayer lattice as an internal three-dimensional reference grid, the Makriyannis lab determined the conformation and location of anandamide in a dipalmitoylphosphatidylcholine (DPPC) multilamellar model membrane bilayer system by measuring selected inter- and intra-molecular distances between strategically introduced isotopic labels using rotational echo double resonance (REDOR) NMR method (Tian et al., 2005). A molecular model was proposed to represent the structural features of the anandamide/lipid system and was subsequently used in calculating the multi-spin dephasing curves. Results demonstrated that anandamide adopts an extended conformation within the membrane with its headgroup at the level of the phospholipid polar group and its terminal methyl group near the bilayer center. These results are very consistent with results from AEA/DOPC simulations (Lynch and Reggio, 2005). Parallel static ^{2}H-NMR experiments further confirmed these findings and provided evidence that anandamide experiences dynamic properties similar to those of the membrane phospholipids and produces no perturbation to the bilayer (Tian et al., 2005).

SUMMARY

Computational and experimental studies of AEA and its parent fatty acid AA in a variety of environments indicate that these molecules are highly flexible and able to adopt both extended and folded shapes. Environment plays a strong role in the shape that AEA or AA takes, for, despite the fact that the shape of the binding cavity/crevice of the proteins with which AEA or AA complexes are quite different, these molecules can easily alter their conformation to adapt to each individual site. This ability to alter conformation clearly has been built into the acyl chain of AEA or AA via the arrangement of the cis homoallylic double bonds in the acyl chain. Furthermore, this ability to alter conformation would appear to be key, because the life cycle of AEA illustrated in Figure 2.2A and Figure 2.2B clearly would require multiple conformational changes from the time of AEA's initial synthesis in the bilayer to its final hydrolysis by FAAH.

ENDOGENOUS CANNABINOID STRUCTURE–ACTIVITY RELATIONSHIPS

Cannabinoid structure-activity requirements (SAR) literature has primarily focused on CB_1-receptor affinities of fatty acid ethanolamides, such as AEA and its analogs. Recently, however, Sugiura has begun to develop an SAR for 2-AG and its analogs based on a functional assay of Ca^{2+} mobilization (Sugiura et al., 1999; Sugiura and Waku, 2000). Because of their difference in basis, each of these SARs will be presented separately here.

FATTY ACID ETHANOLAMIDES

CB_1 SAR for Fatty Acid Ethanolamides

It is important to note that because CB_1 binding assays of AEA analogs can be performed using either a filtration- or centrifugation-based assay on either tissue homogenates or transfected cells, there is variation in reported K_i values for endocannabinoids between laboratories. Earlier, this variation was even more dramatic between assays that used or did not use an amidase inhibitor such as PMSF. This was particularly a concern for filtration-based assays (Sheskin et al., 1997). For these reasons, in the sections which follow, an attempt has been made to provide the K_i value for AEA obtained in a given laboratory as a benchmark against which to compare the relative affinities of other analogs.

Endocannabinoid SAR at the CB receptors can be divided into studies of head group- and acyl chain requirements for high-affinity binding.

Head Group SAR: Fatty Acid Ethanolamides

Arachidonamide and simple alkyl esters of AA do not show significant CB_1 affinity (Pinto et al., 1994). Cyclization of the head group into an oxazoline ring diminishes affinity (Lin et al., 1998). Methylation at the C-1 position in the AEA head group results in an 1-*R*-methyl isomer (AM356, *R*-methanandamide, 13), which has a four-fold higher CB_1 affinity than AEA, whereas the 1-*S*-methyl isomer has a two-fold lower CB_1 affinity than AEA. AM356 also was found to be resistant to enzymatic breakdown ($K_i = 20 \pm 1.6$ n*M* with PMSF; $K_i = 28 \pm 3$ n*M* without PMSF) (Abadji et al., 1994). Methylation at the 2′-position also produced some stereoselectivity, as the S(+) isomer was found to have two- to fivefold higher CB_1 affinity than the R(−) isomer (Abadji et al., 1994; Berglund et al., 1998). Goutopoulos and co-workers (Goutopoulos et al., 2001) reported that introduction of a single methyl at the 2-position (R or S) on the acyl chain of AEA leads to compounds that show moderately improved affinities for CB_1 relative to AEA, but with limited enantioselectivity [AEA, $K_i = 78 \pm 1.6$ n*M*; (R)-*N*-(2-hydroxyethyl)-2-methyl-arachidonamide $K_i = 54.1 \pm 5.2$ n*M*; (S)-*N*-(2-hydroxyethyl)-2-methyl-arachidonamide $K_i = 35.3 \pm 4.3$ n*M*] and only modestly improved metabolic stability. The CB_1 affinity of the 2,2-dimethyl analog of AEA ($K_i = 72.2 \pm 6.3$ n*M*) is unchanged from that of AEA. This compound did show somewhat enhanced metabolic stability, however. Introduction of larger alkyl groups at the 2 position has a detrimental effect on CB_1 affinity (Adams, Ryan, Singer, Razdan et al., 1995). A high degree of enantio- and diastereoselectivity was observed for the 2, 1′-dimethyl analogs. (R)-*N*-(1-methyl-2-hydroxyethyl)-2-(R)-methyl-arachidonamide (**14**) (Goutopoulos et al., 2001) exhibits the highest CB_1 affinities in this series with a $K_i = 7.42 \pm 0.86$ n*M*, a tenfold improvement on AEA ($K_i = 78 \pm 1.6$ n*M*).

Enlargement of the ethanolamine head group by insertion of methylene groups revealed that the *N*-propanol analog has slightly higher CB_1 affinity than AEA, whereas higher homologs have reduced CB_1 affinity (Pinto et al., 1994; Sheskin et al., 1997). Alkyl branching of the alcoholic head group leads to lower-affinity analogs (Sheskin et al., 1997), as does the incorporation of hydroxyphenyl groups such as in AM404 (**15**) (Lang et al., 1999).

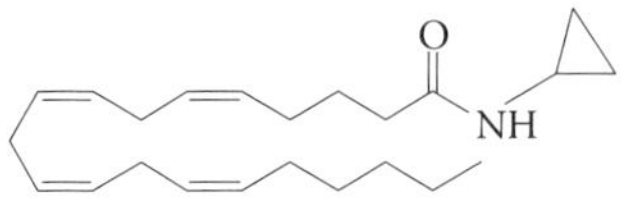

ACPA
17

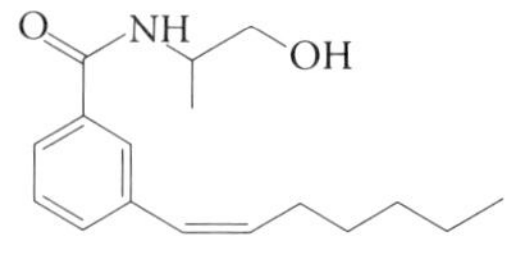

20

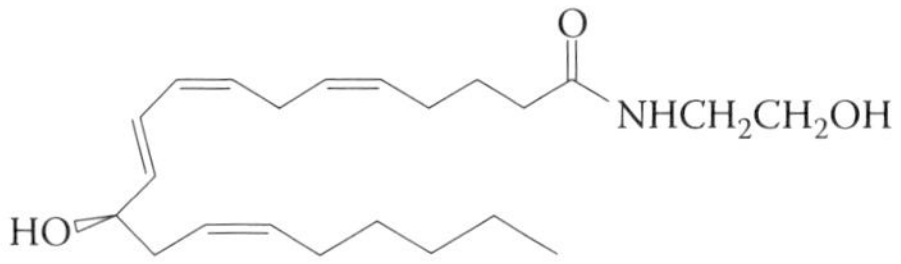

12S-HAEA
18

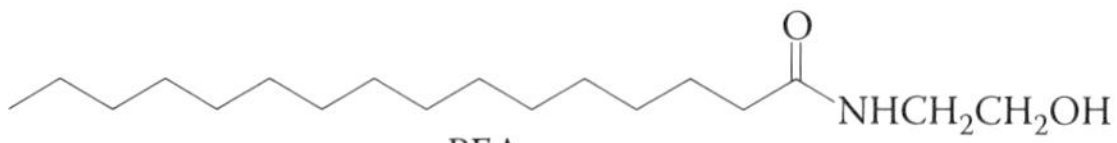

PEA
21

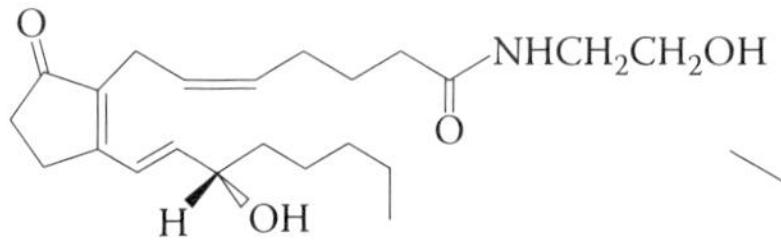

PGB2-EA
19

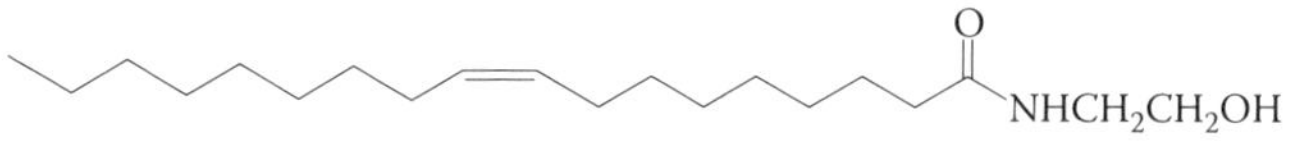

OEA
22

N-(propyl) arachidonylamide (**16**), in which the hydroxyl group is replaced with a methyl group, possesses higher CB_1 affinity ($K_i = 7.3$ n*M*) than AEA itself ($K_i = 22$ n*M*) (Pinto et al., 1994; Sheskin et al., 1997). However, when this same substitution was performed in *R*-methanandamide (**13**, $K_i = 17.9$ n*M*), a lower-affinity analog resulted ($K_i = 73.6$ n*M*) (Lin et al., 1998). Substitution of an *N*-cyclopropyl group for the ethanolamine head group of AEA leads to a very high-affinity CB_1 compound, arachidonylcyclopropylamide, ACPA (**17**; $K_i = 2.2 \pm 0.4$ n*M*), whereas methylene insertion between the amide nitrogen and the cyclopropyl ring leads to reduced affinity ($K_i > 100$ n*M*) (Hillard et al., 1999; Jarrahian et al., 2000). Substitution of an *N*-allylarachidonamide ($K_i = 9.9$ n*M*) and *N*-propargylarachidonamide ($K_i = 10.8$ n*M*) shows approximately six-fold higher affinity for CB_1 than AEA ($K_i = 61.0$ n*M*) (Lin et al., 1998). Substitution of an *N*-phenyl group for the ethanolamine head group of AEA results in a compound that retains CB_1 affinity ($K_i = 109$ n*M*) similar to AEA ($K_i = 143 \pm 37$ n*M*), whereas substitution of an *N*-adamantyl group results in an extreme loss of affinity ($K_i > 1000$ n*M*) (Jarrahian et al., 2000). These results suggest that there may exist a hydrophobic subsite for the AEA head group, such that the hydroxyl of AEA may not be necessary for receptor interaction (Lin et al., 1998). These results also suggest that this hydrophobic subsite is of limited size, as large hydrophobic groups such as adamantyl are not well tolerated.

Replacement of the hydroxyl group of AEA with a halogen such as F or Cl increases CB_1 affinity as well (Adams, Ryan, Singer, Thomas et al., 1995; Hillard et al., 1999; Lin et al., 1998). Substitution of the 2-hydroxyethyl group of AEA with a phenolic group or pyridine group decreases affinity for CB_1(Edgemond et al., 1995; Jarrahian et al., 2000; Khanolkar et al., 1996); however, the insertion of a methylene group between the amide nitrogen and a 4-phenol group results in an analog with modest CB_1 affinity ($K_i = 217 \pm 3$ n*M* vs. AEA $K_i = 78 \pm 2$ n*M*) (Khanolkar et al., 1996). Substitution with a 3,4-dihydroxy-ethyl-phenyl results in an analog with greater CB_1 affinity than AEA (Bisogno et al., 2000).

Amido Group Modifications

For high-affinity binding to the CB_1 receptor to occur and for agonist binding to activate G proteins, the carbonyl group of the AEA amide head group cannot be replaced with a methylene group (Berglund et al., 1998). Retro-anandamide, in which the position of the carbonyl and the NH of the amide group in AEA are reversed, retains receptor affinity and exhibits stability with regard to hydrolysis by FAAH (K_i = 115 n*M* with PMSF and 134 n*M* without PMSF)(Lin et al., 1998). Arachidonylethers, carbamates, and norarachidonylcarbamates have poor CB_1 affinity (Ng et al., 1999). However, norarachidonyl ureas generally show good binding affinities to the CB_1 receptor (K_i = 55 – 746 n*M*; AEA K_i = 89 ± 10 n*M*). Norarachidonyl 2-fluoroethyl urea (**10**; K_i = 55 ± 8 n*M*) has higher CB_1 affinity than AEA (K_i = 89 ± 10 n*M*)(Ng et al., 1999). Some of the weaker-affinity analogs in this series produce potent pharmacological activity. These analogs show hydrolytic stability toward amidase enzymes as well (Ng et al., 1999).

Head Group SAR Summary

Taken together, all of these results suggest that the carboxamide oxygen in the AEA head group is essential, whereas the hydroxyl is not essential for receptor interaction. The cannabinoid receptor can accommodate both hydrophobic and hydrophilic head groups, possibly in two different subsites. Although the hydrophilic site accommodates slightly larger head groups, the sizes of the cavities in which the head group binds still appear to be relatively small as only conservative variations on the head group permit the retention of high-affinity binding.

Acyl Chain SAR: Fatty Acid Ethanolamides

Effect of Unsaturation

Sheskin et al. (1997) reported that *N*-docosatetraenoyl-ethanolamine 22:4, $\Delta^{7,10,13,16}$ (**10**, K_i = 34.4 ± 3.2 n*M*) and *N*-homo-γ-linolenoyl-ethanolamine 20:3, $\Delta^{8,11,14}$ (**11**, K_i = 53.4 ± 5.5 n*M*) have CB_1 affinities comparable to AEA 20:4, $\Delta^{5,8,11,14}$ (1, K_i = 39.2 ± 5.7 n*M*). As unsaturation is decreased to two cis double bonds in *N*-11,14 eicosadienoyl ethanolamine 20:2, $\Delta^{11,14}$ analog, CB_1 affinity is greatly diminished (**12**, K_i = 1500 n*M*). A saturated fluoro-derivative of AEA (20:0), O-586 was found not to bind to CB_1 receptors (K_i > 10,000 n*M* with or without PMSF). This compound also does not produce cannabimimetic discriminative stimulus effects (Wiley et al., 1998).

Figure 2.4 illustrates CM results for an n-6 series of ethanolamides (**1**, **10–12**) in $CHCl_3$ with varying degrees of unsaturation in their acyl chains (Barnett-Norris, Hurst, Lynch et al., 2002). As discussed earlier, CM identified two major clusters for **1**, **10–12**, a curved cluster in which the ends of the molecule are brought closer together in a U or J shape (illustrated in the left column, Cls 1, Figure 2.4) and an extended-shape cluster in which the ends of the molecule are far from each other (illustrated in the right column, Cls 2, Figure 2.4). The percentage of the structures in each family is also indicated. Although it is clear from the CM results illustrated in Figure 2.4 that each analog can form both U- or J-shaped (Cls 1) and extended families (Cls 2) of conformers, these families differ in two important ways. First, for the 20:2, n-6 ethanolamide (also called 20:2 $\Delta^{11,14}$ ethanolamide, **12**) the higher populated cluster is the extended conformer cluster (Cls 2), whereas for the other analogs in the series, the U- or J-shaped cluster (Cls 1) has the higher population. Second, although the 20:2, n-6 ethanolamide (**12**) has a curved cluster, this cluster is not as tightly curved as are those of the other analogs studied. This is evident in the radii of curvature (with 95% confidence intervals) measured for the Cls 1 conformers in this series: 5.8 Å (5.3–6.2) for 20:2, n-6 (**12**); 4.4 Å (4.1–4.7) for 20:3,n-6 (**11**); 4.0 Å (3.7–4.2) for 20:4,n-6 (**1**); and 4.0 Å (3.6–4.5) for 22:4,n-6 (**10**) (Barnett-Norris, Hurst, Lynch et al., 2002). These results illustrate that as the number of homoallylic double bonds in the acyl chain decreases, the distance between the head group and acyl tail increases, favoring extended conformers when only two homoallylic double bonds remain. Furthermore, as illustrated in Figure 2.4, CB_1 affinity decreases (Sheskin et al., 1997) as the number of homoallylic double bonds decreases, suggesting that higher CB_1 affinity is associated with endocannabinoids that can form tightly curved structures.

Effect of Terminal Alkyl Tail Length

The degree of unsaturation is not the only acyl chain requirement for recognition at the CB_1 receptor. Sheskin and co-workers (1997) found that in an n-3 series, highly unsaturated analogs had poor to no CB_1 affinity as revealed by the following series of n-3 ethanolamide derivatives: 20:5, $\Delta^{5,8,11,14,17}$ (K_i = 162.3 ± 13.6 n*M*) and 22:6, $\Delta^{4,7,10,13,16,19}$ (K_i = 324.1 ± 9.2 n*M*); 18:4, $\Delta^{6,9,12,15}$ (K_i > 1000 n*M*), 20:3, $\Delta^{11,14,17}$ (K_i > 10,000 n*M*); 18:3, $\Delta^{9,12,15}$ (no activity). The fact that this n-3 series shows low to no CB_1 affinity reveals that flexibility cannot be the only acyl chain requirement for CB_1 recognition, as all of these n-3 analogs will have great flexibility due to their levels of unsaturation. However, these analogs may lack an essential part of the cannabinoid pharmacophore: a saturated alkyl tail of sufficient length.

An analogy has been drawn by several groups in the literature between the C16–C20 saturated portion of AEA and the C-3 pentyl side chain of the classical cannabinoid, Δ^9-THC (**4**) (Ryan et al., 1997; Seltzman et al., 1997; Thomas et al., 1996; Tong et al., 1998). In classical cannabinoid SAR, the C-3 pentyl side chain (see Δ^9-THC, **4**) is considered to have the minimum length needed to produce cannabinoid activity; whereas a 1, 1-dimethylheptyl side chain appears to be optimum (Razdan, 1986). Replacement of the pentyl side chain of a classical or nonclassical cannabinoid with a 1, 1-dimethylheptyl side chain has been reported to result in as much as a 75-fold enhancement in CB_1 affinity (Compton et al., 1993). Although the replacement of the C16–C20 saturated portion of AEA with a 1′, 1′-dimethylheptyl chain does not result in as great an enhancement as in the classical cannabinoids, it does produce a 13-fold enhancement in CB_1 affinity (Ryan et al., 1997; Seltzman et al., 1997). This enhancement has been taken as support for the hypothesis that the C16–C20 saturated region of AEA corresponds to the C-3 pentyl chain of Δ^9-THC for interaction at the CB_1 receptor. Therefore, from this correspondence, one might conclude that the drop in affinity which accompanies the shortening of the saturated tail of AEA in the n-3 series may be related to SAR requirements for classical cannabinoid binding at CB_1.

Sheskin and co-workers (1997) also found that the 18:3, $\Delta^{6,9,12}$ (n-6) ethanolamide has low CB_1 affinity (K_i = 4600 ± 300 n*M*). This result suggests that a C18 acyl chain is too short to be recognized by CB_1, because this analog has both a long enough saturated tail (five carbons) and three cis homoallylic double bonds separated by methylene carbons.

Effects of Branching in Acyl Chain

Mono- (K_i = 53 ± 11 n*M*) or di- (K_i = 47 ± 2 n*M*) methylation at C2 on the acyl chain of AEA produces analogs with slightly higher CB_1 affinities than AEA (K_i = 89 ± 10 n*M*); however, larger or branched alkyl substituents at C2 lead to low CB_1 affinities (Adams, Ryan, Singer, Thomas, et al., 1995).

Effects of Conjugation in Acyl Chain

Whereas most acyl chain SAR studies have focused on unsaturated cis polyenes, AEA analogs with altered double bond patterns have also been reported. All of these analogs possess regions in which some double bond conjugation is present. Conjugated triene anandamide (CTA) has been reported to have six-fold lower CB_1 affinity (K_i = 607 ± 44 n*M*) than AEA (K_i = 97 ± 7 n*M*) (Wise et al., 1996). Edgemond and co-workers have described two oxygenated derivatives of AEA that are products of lipoxygenase activity on AEA, 12(S)-hydroxyeicosatetraenoyl-ethanolamide (12(S)-HAEA) and

12(R)-hydroxyeicosatetraenoyl-ethanolamide (12(R)-HAEA). Although in these investigators' hands, AEA has a K_i of 107 ± 54 n*M* at CB_1, the S stereoisomer (**18**, K_i = 207 ± 52 n*M*) was found to have one-half the CB_1 affinity of AEA, and the R stereoisomer (K_i = 416 ± 14 n*M*) was found to have one-quarter the CB_1 affinity of AEA (Edgemond et al., 1998).

Van der Stelt and co-workers (2002) probed the importance of acyl chain flexibility in interaction of endocannabinoids with cannabinoid receptors, membrane transporter protein, and FAAH by the introduction of an 1-hydroxy-2Z,4E-pentadiene system at various positions in the AEA structure using different lipoxygenases. Although the displacement constants were modest, 15(S)-hydroxy-eicosa-5Z,8Z,11Z,-13E-tetraenoyl-*N*-(2-hydroxyethyl)amine (K_i = 600 ± 120 n*M* CB_1; K_i > 1000 n*M*, CB_2) was found to bind selectively to the CB_1 receptor, whereas its 13(S)-hydroxy-octadeca-9Z,11E-dienoyl- *N*-(2-hydroxyethyl)amine (K_i > 1000 n*M* CB_1; K_i = 600 ± 120 n*M*, CB_2) could selectively bind to the CB_2 receptor. 11(S)-Hydroxy-eicosa- 5Z,8Z,12E,14Z-tetraenoyl-*N*-(2-hydroxy-ethyl)amine did not bind to either receptor, whereas 12(S)-hydroxy-eicosa-5Z,8Z,10E,14Z-tet-raenoyl-*N*-(2-hydroxyethyl)amine (**18**, K_i = 150 ± 30 n*M* CB_1; K_i = 500 ± 60 n*M*, CB_2) did bind to both CB receptors with an affinity similar to that of AEA (K_i = 90 ± 20 n*M*, CB_1; K_i = 360 ± 50 n*M*, CB_2). This latter result is consistent with earlier work reported by Edgemond and co-workers (1998). All oxygenated AEA derivatives were good inhibitors of FAAH (low micromolar K_i), but were ineffective on the AEA transporter.

Pinto et al. (1994) investigated a series of arachidonyl amides and esters in addition to a series of "rigid hairpin" conformations typified by *N*-(2-hydroxyethyl)-prostaglandin amides to determine the structural requirements for binding to the CB_1 receptor. Two-dimensional drawings of AEA and PGB_2-EA (**19**) make the shapes of these two compounds look similar, and, therefore, it is possible that PGB_2-EA (**19**) may be a rigid (hairpin- or U-shaped) analog of AEA. However, all of the rigid prostaglandin analogs synthesized by Pinto et al. (1994) failed to alter [^{3}H]CP55940 binding to CB_1 even in concentrations as great as 100 μ*M*. α-alkylamide derivatives of PGE_2 and $PGF_{2\alpha}$ showed enhanced binding affinity relative to the prostaglandin ethanolamides tested previously (Pinto et al., 1994); however, none of the affinities of the α-alkylamide derivatives were comparable with that of AEA. In addition, although the PGE_2 amides were able to activate G proteins, including G_s, these signal transduction events were likely not mediated by the CB receptors (Berglund et al., 1999).

Barnett-Norris and co-workers (1998) reported CM results for PGB_2-EA (**19**) which showed an attenuated ability to adopt either an extended or a U-shaped conformation like AEA and 2-AG. Instead, CM results showed a single family/cluster for PGB_2-EA (**19**) in $CHCl_3$ that was L-shaped. These results suggest that the conjugation of the acyl chain with the ring double bond introduces a "stiffness" into this part of the molecule. The authors propose that the low CB_1 affinity of PGB_2-EA may be caused by its lack of conformational flexibility. Alternatively, the steric bulk introduced by the five-membered ring in PGB_2-EA may produce a steric clash at CB_1 and prevent ligand binding.

Berglund and co-workers designed a series of conformationally restricted analogs that contained ortho- and meta-disubstituted benzene derivatives with alkylamide and alkyl substituents to test whether an endocannabinoid hairpin conformation was recognized at the CB_1 receptor. The majority of these analogs displayed CB_1 affinity in low to high micromolar concentrations in rat membranes (Berglund et al., 2000). Several of the analogs inhibited basal binding of GTPγS in rat brain membranes at micromolar concentrations, but none of the analogs affected either basal or stimulated adenylate cyclase activity in N18TG2 membranes. Two meta-substituted derivatives, however, had CB_1 K_is < 100 n*M*. One of these derivatives (**20**) showed ≥ 2-fold [^{35}S]GTPγS binding compared to basal and > 20% inhibition of forskolin stimulated adenylyl cyclase in N18TG2.

Acyl Chain SAR Summary

In summary, these results suggest that high-affinity receptor binding requires high flexibility of the acyl chain, as a minimum of three homoallylic double bonds are necessary for binding to CB_1.

The endocannabinoid acyl chain may need to adopt a folded conformation to interact with CB_1, or the endocannabinoid may need to adopt several different conformations in order to reach, be recognized by, and trigger the receptor. Only those analogs with high degrees of equilibrium flexibility may be able to achieve these conformations easily.

Acyl chain SAR results also indicate that a minimum of five carbons must be present in the saturated, terminal region of the acyl chain, and that a minimum of 20 carbons in the acyl chain are required for binding to the cannabinoid CB_1 receptor. It is difficult to reconcile these requirements based solely on conformational mobility; yet any pharmacophore for AEA interaction with CB_1 should be able to account for these two requirements as well. Hypotheses concerning the origin of these two SAR requirements are discussed in the section titled "Pharmacophore for Endocannabinoid Interaction with Cannabinoid Receptors" later in this chapter.

CB_2 SAR for Fatty Acid Ethanolamides

AEA exhibits higher affinity for the cannabinoid CB_1 receptor (K_i CB_1 = 89 ± 10 n*M*) than for the CB_2 receptor (K_i CB_2 = 371 ± 102 n*M*) (Showalter et al., 1996). AEA analogs tend to be CB_1 selective with CB_2 affinities comparable to or lower than the CB_2 affinity of AEA (Goutopoulos et al., 2001; Hillard et al., 1999; Lang et al., 1999; Lin et al., 1998). Berglund and co-workers found that α-alkylamide derivatives of PGE_2 and $PGF_{2\alpha}$ showed a higher binding affinity for the CB_2 receptor than for the CB_1 receptor; however, the affinity was reduced relative to AEA's affinity for CB_2 (Berglund et al., 1999). Van der Stelt reported that 13(S)-hydroxy-octadeca-9Z,11E-dienoyl-*N*-(2-hydroxyethyl)amine (K_i > 1000 n*M* CB_1; K_i = 600 ± 120 n*M*, CB_2) could selectively bind to the CB_2 receptor (van der Stelt et al., 2002).

Facci and co-workers reported that palmitoylethanolamide (16:0, PEA, **21**), a C16 saturated derivative, behaves as an endogenous agonist of the CB_2 receptor on rat mast cells (RBL-2H3 cells). The IC_{50} for competitive inhibition of the AAI cannabinoid agonist [^{3}H]WIN55212-2 binding to RBL-2H3 cell membranes by PEA was reported to be 1.0 ± 0.6 n*M*, while that of AEA was reported to be 33 ± 2.9 n*M* (Facci et al., 1995). However, PEA was found to displace only 10% of nonclassical cannabinoid agonist [^{3}H]CP55940 binding from human cloned CB_2 receptors at concentrations up to 10 μ*M* (Showalter et al., 1996). PEA differs from AEA not only in its lack of unsaturation, but also in its shorter acyl chain (C16 vs. C20 for AEA). Reggio and Traore have reported that CM calculations suggest that PEA adopts predominantly only one conformation, an extended conformation (83 out of 100 conformers in $CHCl_3$) (Reggio and Traore, 2000). Thus, PEA is not a very flexible molecule, in marked contrast to AEA.

Lambert and co-workers synthesized a series of 10 *N*-palmitoylethanolamine (PEA, **21**) homologs and analogs, varying by the elongation of the fatty acid chain from caproyl to stearoyl and by the nature of the amide substituent, respectively; they evaluated the affinity of these compounds for cannabinoid receptors in the rat spleen, in RBL-2H3 cells and in CHO-CB_1 and CHO-CB_2 receptor transfected cells. No displacement of [^{3}H]CP55940 or [^{3}H]WIN55212-2 by PEA derivatives was observed in rat spleen slices. In RBL-2H3 cells, no binding of [^{3}H]CP55940 or [^{3}H]WIN55212-2 could be observed, and, conversely, no inhibitory activity of PEA derivatives and analogs was measurable. These investigators concluded that it seemed unlikely that PEA is an endogenous agonist of the CB_2 receptor (Lambert et al., 1999).

Most recently, Jonsson and co-workers (2001) studied homologs and analogs of PEA (**21**). Oleoylethanolamide at 100 μ*M* showed 64.6 ± 1.5% inhibition of [^{3}H] WIN55212-2 binding to human CB_2 receptors expressed on CHO cells. Other compounds had lesser effects at CB_2. PEA, palmitoylisopropylamide, and R-palmitoyl-(2-methyl)ethanolamide had modest effects upon [^{3}H] CP55940 binding to human CB_1 receptors expressed on CHO cells.

PEA has been reported to be orally active in reducing edema formation and inflammatory hyperalgesia by downmodulating mast cell activation (Mazzari et al., 1996). PEA has also been shown to act in synergy with AEA. Recently, Di Marzo and co-workers showed that PEA potently enhances the antiproliferative effect of AEA on human breast cancer cells (HBCCs), in part by inhibiting the expression of the FAAH enzyme (Di Marzo et al., 2001).

Oleamide

Oleamide (18:1, Δ^9 cis; **22**) isolated from the cerebrospinal fluid (CSF) of sleep-deprived cats has been shown to induce physiological sleep in rats (Cravatt et al., 1995). Cheer and co-workers (1999) reported that oleamide occupied CB_1 receptors on rat brain membranes labeled with the nonclassical cannabinoid agonist [^{3}H]CP55940 with an IC_{50} of 10 μM. However, oleamide has been reported previously by several other groups as not binding to recombinant CB_1 or CB_2 receptors overexpressed in host cells (Boring et al., 1996; Mechoulam et al., 1997; Sheskin et al., 1997). Oleamide, at a 50 μM concentration, potentiates AEA binding to CB_1 receptors by an order of magnitude (Mechoulam et al., 1997). This effect was not observed in the presence of the FAAH inhibitor PMSF (200 μM), which alone also potentiates AEA binding, thus suggesting that the facilitatory action of oleamide is due to inhibition of AEA hydrolysis in membrane preparations. This is not surprising as oleamide and AEA have both been shown to be substrates for the FAAH enzyme (Cravatt et al., 1996).

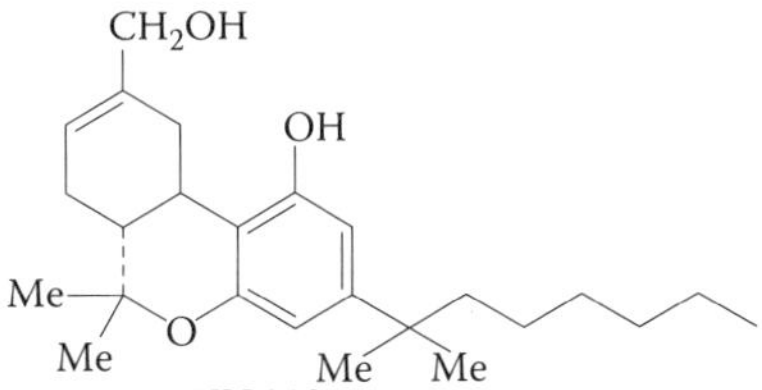

HU 210
23

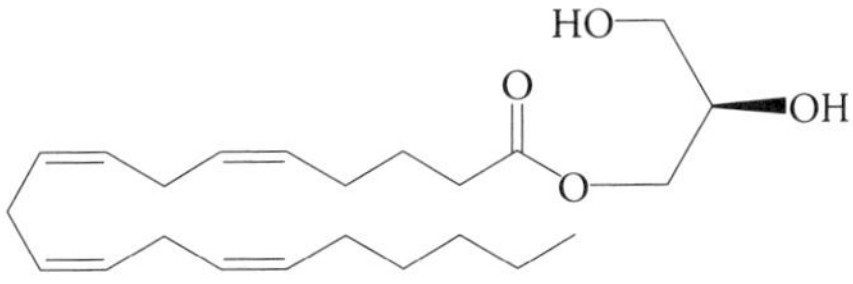

3-AG
25

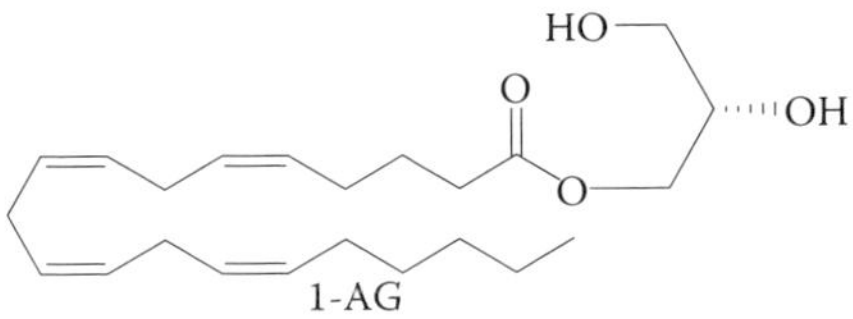

1-AG
24

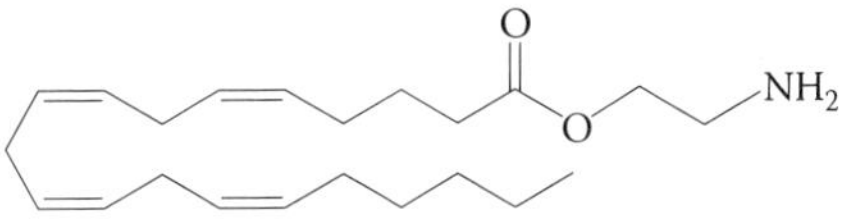

Virodhamine
26

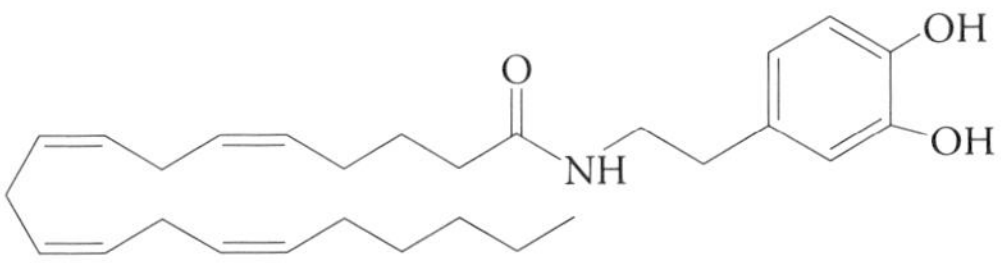

NADA
27

SN-2-Arachidonylglycerol and Derivatives

SN-2-arachidonylglycerol (2-AG; **2**) has been isolated from intestinal tissue and shown to be a second endogenous CB ligand (CB_1 $K_i = 472 \pm 55$ nM; CB_2 $K_i = 1400 \pm 172$ nM)(Mechoulam et al., 1995; Sugiura et al., 1995). 2-AG has been found to be present in the brain at concentrations 170 times greater than that of AEA (Stella et al., 1997). 2-AG's interactions with the cannabinoid receptors result in the activation of G proteins (Berglund et al., 1998; Gonsiorek et al., 2000), particularly of the $G_{i/o}$ family, and in the inhibition of adenylyl cyclase (Gonsiorek et al., 2000). 2-AG induces a rapid, transient elevation of intracellular free Ca^{2+} in neuroblastoma × glioma hybrid NG108-15 cells which express the CB_1 receptor (Sugiura et al., 1996) and in N18TG2 cells (Sugiura et al., 1997). This action of 2-AG is blocked by pretreatment with pertussis toxin, indicating that $G_{i/o}$ is involved in the response. This action of 2-AG is also blocked by treatment with the CB_1 antagonist SR141716A, suggesting that this effect may be CB_1 receptor mediated. The elevation of intracellular Ca^{2+} is also produced by the cannabinoid agonists, WIN55212-2 (**6**), HU210 (**23**), CP55940 (**5**), AEA (**1**), and R-methanandamide (**13**); however, these agonists produce much lower maximal responses than do 2-AG and its congeners. The relative efficacy profile differs from that described for [^{35}S]GTPγS binding and inhibition of adenylyl cyclase, suggesting that the signal transduction pathway for this intracellular Ca^{2+} regulation must differ, perhaps in the type of G protein that tranduces the response (Sugiura et al., 1997; Sugiura et al., 1996; Sugiura et al., 1999). A similar effect in HL60 cells which express the CB2 receptor has also been demonstrated (Sugiura et al., 2000). 2-AG has also been found to induce the activation of p42/44 MAP kinase in HL60 cells (Sugiura and Waku, 2000).

Similar to AEA, 2-AG is thought to be eliminated through a two-step process consisting of carrier-mediated transport into cells and subsequent enzymatic hydrolysis (for a review, see Sugiura and Waku, 2002). 2-AG has been shown to compete with AEA for uptake by the AEA transporter (Jarrahian et al., 2000; Piomelli et al., 1999). FAAH can function as a general hydrolytic enzyme not only for AEA and other fatty ethanolamides, but also for fatty esters such as 2-AG. This unusually broad substrate selectivity, which was demonstrated in experiments with membrane and purified enzyme (Gonsiorek et al., 2000; Hillard, 2000; Savinainen et al., 2001), has led researchers to suggest that FAAH may terminate the biological actions of both AEA and 2-AG. In fact, the FAAH enzyme has been reported to hydrolyze 2-AG four times faster than it hydrolyzes AEA (Goparaju et al., 1998).

Two findings, however, argue against the idea that 2-AG is hydrolyzed by FAAH. First, a 2-AG hydrolase distinct from FAAH has been partially purified from porcine brain (Bisogno et al., 1997). Second, in intact astrocytoma cells, inhibition of FAAH activity prevents the hydrolysis of AEA but not that of 2-AG (Stella et al., 1997). Thus, although 2-AG can be hydrolyzed by FAAH *in vitro*, different enzymes may be responsible for its degradation *in vivo*. A possible candidate for this role is monoglyceride lipase (MGL), a serine hydrolase that cleaves 2- and 1-monoglycerides into fatty acid and glycerol (Sugiura et al., 1998). To test this hypothesis, Piomelli and co-workers (Dinh et al., 2002) cloned rat brain MGL, determined its anatomical distribution, and used adenovirus-mediated gene transfer to investigate its role in neuronal 2-AG inactivation. Northern blot and *in situ* hybridization analyses revealed that MGL mRNA is heterogeneously expressed in the rat brain, with the highest levels in those regions where CB_1 cannabinoid receptors are also present (i.e., hippocampus, cortex, anterior thalamus, and cerebellum). Immunohistochemical studies in the hippocampus showed that MGL distribution has laminar specificity, suggesting a presynaptic localization of the enzyme. Adenovirus-mediated transfer of MGL cDNA into rat cortical neurons increased MGL expression and attenuated *N*-methyl-D-aspartate/carbachol-induced 2-AG accumulation in these cells. No such effect was observed on the accumulation of AEA. The results suggest that hydrolysis by means of MGL is a primary mechanism for 2-AG inactivation in intact neurons.

2-AG rapidly isomerizes both *in vitro* and *in vivo*. The rate of this process is increased by high temperature and by acidic or basic pH. Two stereoisomers can be formed, i.e., 1-arachidonoyl-sn-glycerol (1-AG, **24**) and 3-arachidonoyl-sn-glycerol (3-AG, **25**). In endocannabinoid analysis, usually 10–40% of a racemic mixture of 1(3)-AG is found. To date, most interaction studies of 2-AG with proteins of the endocannabinoid system have not taken into account the isomerization of 2-AG into 1(3)-AG during the incubation period. Van der Stelt and co-workers (2002) have reported that 2-AG inhibited potently the binding of [^{3}H]CP55940 to CB_1 and CB_2 receptors with a K_i of 100 nM (with PMSF). 2-AG also inhibited [^{3}H]AEA transport with a K_i of 3 μM and was twofold more potent than AEA in inhibiting FAAH. However, neither 1-AG nor 3-AG bound to either CB receptor or interfered with AEA transport. Thus, 2-AG is inactivated by isomerization. This uncontrolled isomerization may account to some extent for the large differences in K_i values reported in the literature for 2-AG binding to CB receptors and for its inhibition of AEA transport.

CB_1 SAR for 2-AG

Sugiura and co-workers have reported that 2-AG (**2**) and other cannabinoid ligands such as AEA (**1**) and Δ^9-THC (**4**) induce rapid transient increases in Ca^{2+} in NG108-15 cells through a cannabinoid CB_1-receptor-dependent mechanism (Sugiura et al., 1997; Sugiura et al., 1996; Sugiura et al., 1999). 2-AG was the most potent compound for inducing these transient increases, its activity being detectable even at concentrations as low as 0.3 n*M*. The maximal response induced by 2-AG exceeded the responses induced by other CB_1 agonists. Activities of the CB_1 agonists HU210 (**23**) and CP55940 (**5**) were also detectable from levels as low as 0.3 n*M*, whereas, the maximal responses induced by these compounds were low compared with 2-AG. AEA was also found to act as a partial agonist in this system. AA (**9**) failed to elicit a response, whereas noladin ether (**3**) possessed appreciable activity, although its activity was apparently lower than that of 2-AG (Sugiura et al., 1999).

Sugiura and co-workers (1999) have begun to generate an SAR based on this transient increase in Ca^{2+}. Glycerol has been found to be the most suitable head group, and the 2-isomer is preferable over the 1(3)-isomer. AA is the most preferred fatty acid moiety, although the activity of eicosatrienoic acid (n-9)-containing species was almost comparable to that of the AA containing species. Because the activities of 2-eicosatrienoyl(20:3, $\Delta^{8,11,14}$, n-6) glycerol, 2-eicosatrienoyl(20:3, $\Delta^{11,14,17}$, n-3) glycerol, and 2-docosatetraenoyl(22:4, $\Delta^{7,10,13,16}$, n-6) glycerol are lower than those of 2-eicosatrienoyl(20:3, $\Delta^{5,8,11}$, n-9) and 2-eicosapentaenoyl(20:5, $\Delta^{5,8,11,14,17}$, n-3) glycerol, it appears that the presence of a double bond at the Δ^5 position, rather than further toward the end of the acyl chain, is crucially important, probably for folding or curvature to occur nearer the head group.

CB_2 SAR for 2-AG

Sugiura and co-workers (2000) have also found that 2-AG can induce rapid transient increases in Ca^{2+} in HL60 cells. It was evident in this case that the response was mediated by the CB_2 receptor, but not the CB_1 receptor, because the CB_2 antagonist SR144528 (**8**), but not the CB_1 antagonist SR141716A (**7**), blocked the response.

2-AG was found to have more activity than its 1(3)-isomer. The ester and ether analogs, including nolandin ether (**3**) showed appreciable activity, albeit less than that of 2-AG. AEA was found to be a weak partial agonist toward the CB_2 receptor. The cannabinoid agonist Δ^9-THC (**4**) exhibited weak agonistic activity for the CB_2 receptor in the transient Ca^{2+} assay. 2-glycerols with fatty acid chains of 20 to 22 carbons showed the strongest CB_2 agonist activity. For C20 fatty acids, activity was best for the 20:3, $\Delta^{5,8,11}$ and 20:4, $\Delta^{5,8,11,14}$ (2-AG, **2**), suggesting that like the 2-AG SAR generated for CB_1, a double bond at Δ^5 was important (Sugiura et al., 2000).

Different mammalian lipoxygenases are capable of using AEA and 2-AG as substrates in *in vitro* systems (Edgemond et al., 1998; Hampson et al., 1995; Moody et al., 2001; Ueda et al., 1995). Van der Stelt and co-workers restricted acyl chain flexibility by the introduction of an 1-hydroxy-2Z,4E-pentadiene system in 2-AG (K_i = 100 ± 20 n*M* CB_1; K_i = 100 ± 20 n*M*) at various positions using different lipoxygenases as biocatalysts. This brought about selectivity and attenuated the binding potency of 2-AG. Neither 1- nor 3-arachidonoyl-sn-glycerol bound to CB_1 or CB_2. However, introduction of the 15-hydroxy-11Z,13E diene system in 1-AG to produce 15-HO-1-AG increased its CB_2 binding affinity (K_i > 1000 n*M*, CB_1; K_i = 550 ± 80 n*M*, CB_2) as compared to its parent compound 1-AG (K_i > 1000 n*M*, CB_1; K_i > 1000 n*M*, CB_2), but this was not the case for 3-AG. Both 15-HO-1-AG and 15-HO-3-AG could inhibit AEA transport at low micromolar concentrations, but were worse inhibitors of [^{3}H]AEA hydrolysis than the parent compounds.

Production of an "Entourage Effect" by Related Glycerol Derivatives

Ben-Shabat and co-workers (1998) have reported that in the spleen, as in the brain and gut, 2-AG is accompanied by several 2-acyl-glycerol esters, two major ones being 2-linoleoyl-glycerol (2-Lino-Gl) and 2-palmitoyl-glycerol (2-Palm-Gl). These two esters do not bind to the cannabinoid receptors, nor do they inhibit adenylyl cyclase via either CB_1 or CB_2; however, they significantly potentiate the apparent binding of 2-AG and its apparent capacity to inhibit adenylyl cyclase. Together these esters have also been shown to significantly potentiate 2-AG inhibition of motor behavior, immobility on a ring, analgesia on a hot plate, and hypothermia caused by 2-AG in mice. 2-Lino-Gl, but not 2-Palm-Gl, significantly inhibits the inactivation of 2-AG by neuronal and basophilic cells. These data indicate that the biological activity of 2-AG can be increased by related endogenous 2-acylglycerols, which show no significant activity when used alone in any of the tests. This effect (entourage effect) may represent a novel route for molecular regulation of endogenous cannabinoid activity.

Noladin Ether

An ether-type endocannabinoid, 2-arachidonyl glyceryl ether (noladin ether, **3**) has been isolated from porcine brain (Hanus et al., 2001) and identified to be another endogenous cannabinoid ligand. 2-arachidonyl glyceryl ether was found to bind to the CB_1 receptor ($K_i = 21.2 \pm 0.5$ n*M*) and cause sedation, hypothermia, intestinal immobility, and mild antinociception in mice — effects typically produced by cannabinoid agonists. Noladin ether was found to bind weakly to CB_2 ($K_i > 3$ μ*M*). This compound had been synthesized previously by the Mechoulam lab in a program to develop compounds less prone to enzymatic hydrolysis (Mechoulam et al., 1998). Independently, Sugiura and co-workers (1999) prepared this compound and examined its effects on Ca^{2+} levels in cells (Suhara et al., 2000). Noladin ether was found by these investigators to exhibit appreciable agonistic activity, although its activity was significantly lower than that of 2-AG. Fezza and co-workers (2002) have also reported that a small amount of 2-AG ether (**3**) is present in rat brain. However, Sugiura and co-workers (Oka et al., 2003) have reported that ether bonds are located exclusively at the 1-position of the glycerol backbone in mammalian tissues, and that they could not find 2-AG ether in the brains of several mammalian species. Thus, although the affinity of 2-AG ether (**3**) for the CB receptors is not in dispute, the existence of noladin ether as an endogenous cannabinoid in mammalian tissues is still a subject of debate.

Virodhamine—Endogenous Cannabinoid with Antagonist Properties at CB_1 and Agonist Properties at CB_2

Felder and co-workers have reported the identification and pharmacological characterization of a novel endocannabinoid, virodhamine, with antagonist properties at the CB_1 cannabinoid receptor (Porter et al., 2002). Virodhamine is AA and ethanolamine joined by an ester linkage (**26**). Concentrations of virodhamine measured by liquid chromatography-atmospheric pressure chemical ionization-tandem mass spectrometry in rat brain and human hippocampus were similar to that of AEA. In peripheral tissues that express the CB_2 cannabinoid receptor, virodhamine concentrations were two- to nine-fold higher than that of AEA. In contrast to previously described endocannabinoids, virodhamine is a partial agonist with *in vivo* antagonist activity at the CB_1 receptor. However, at the CB_2 receptor, virodhamine acted as a full agonist. Transport of [^{14}C] AEA by RBL-2H3 cells was inhibited by virodhamine. Virodhamine produced hypothermia in mice and acted as an antagonist in the presence of AEA both *in vivo* and *in vitro*.

NADA

N-arachidonoyl-dopamine (NADA, **27**) is an endogenous "capsaicin-like" substance in mammalian nervous tissues. NADA activates cannabinoid CB_1 receptors, but not dopamine D1 and D2 receptors (Bezuglov et al., 2001; Bisogno et al., 2000). NADA occurs in nervous tissues, with the highest concentrations being found in the striatum, hippocampus, and cerebellum and the lowest concentrations in the dorsal root ganglion (Chu et al., 2003). Proposed mechanisms of NADA biosynthesis include the condensation of AA with tyrosine and the subsequent conversion of *N*-arachidonoyl-tyrosine to NADA by tyrosine hydroxylase and L-aromatic amino acid decarboxylase.

ENDOCANNABINOID INTERACTION WITH THE CANNABINOID CB_1 RECEPTOR

THE CANNABINOID RECEPTORS

Both the CB_1 and the CB_2 receptors belong to the Class A rhodopsin-like family of GPCRs. The cloning and expression of a complementary DNA from a rat cerebral cortex cDNA library that encodes the cannabinoid receptor subtype, CB_1, was first reported by L.A. Matsuda and co-workers (1990). Subsequently, the primary amino acid sequences of an amino terminus variant CB_1 receptor (Shire et al., 1995), as well as the CB_1 sequence in mouse and in human were reported (Abood et al., 1997; Gerard et al., 1991). Figure 2.7 provides a helix net representation of the human CB_1 sequence. In addition to being found in the CNS, mRNA for CB_1 has also been identified in testis (Gerard et al., 1991). The CB_1 receptor has been shown to have a high level of ligand-independent activation (i.e., constitutive activity) in transfected cell lines, as well in cells that naturally express the CB_1 receptor (Bouaboula et al., 1997; Pan et al., 1998; Mato et al., 2002; Meschler et al., 2000). Kearn and co-workers (1999) have estimated that in a population of wild-type (WT) CB_1 receptors, 70% exist in the inactive state (R) and 30% exist in the activated state (R*).

The second cannabinoid receptor subtype, CB_2, was derived from a human promyelocytic leukemia cell HL60 cDNA library (Munro et al., 1993). The human CB_2 receptor exhibits 68% identity to the human CB_1 receptor within the transmembrane regions and 44% identity throughout the whole protein. The CB_2 receptor in both rat (Griffin et al., 2000) and mouse (Shire et al., 1996) has been cloned as well. Figure 2.8 presents a helix net representation of the human CB_2 sequence. Unlike the CB_1 receptor, which is highly conserved across human, rat, and mouse, the CB_2 receptor is much more divergent. Sequence analysis of the coding region of the rat CB_2 genomic clone indicates 93% amino acid identity between rat and mouse and 81% amino acid identity between rat and human. CB_2-receptor-transfected Chinese hamster ovary cells exhibit high constitutive activity (Bouaboula et al., 1999). Evidence for the presence of other cannabinoid receptors is mounting in the literature (Breivogel et al., 2001; Di Marzo et al., 2000; Jarai et al., 1999; Wagner et al., 1999). A recent meeting report suggests that the orphan GPCR, GPR55 may be a new cannabinoid receptor sub-type (Sjogren et al., 2005).

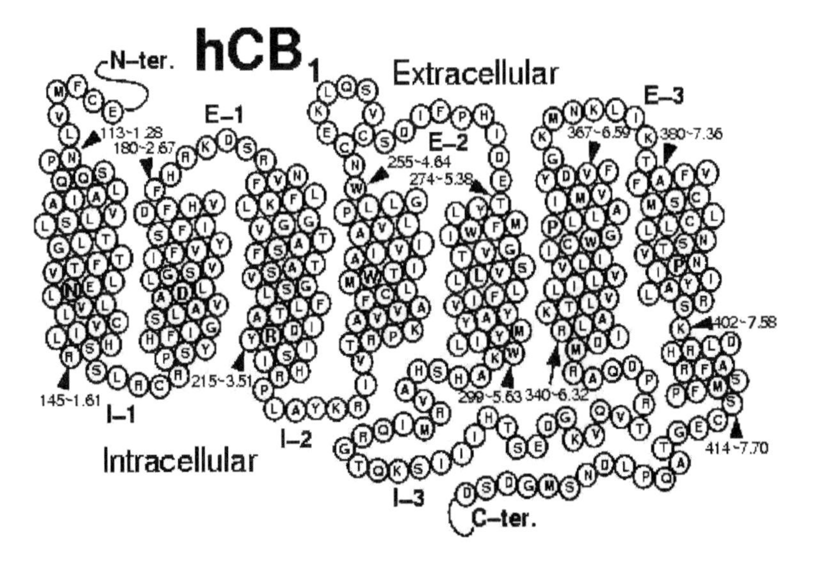

FIGURE 2.7 A helix net representation of the human CB_1 receptor sequence.

Tong Model

Tong and co-workers (Tong et al., 1998) used constrained conformational searching and CoMFA to generate an endocannabinoid pharmacophore model using superposition rules that differed from those proposed by Thomas and co-workers (1996). In the Tong model, 9-nor-9β-OH-HHC (**28**) was used as the template to which AEA and its analogs were fitted. The conformation identified for AEA was a helical conformation in which (1) the oxygen of the carboxamide group overlaid the C-1 phenolic hydroxyl group of 9-nor-9β-OH-HHC, (2) the head group hydroxyl overlaid the C9 hydroxyl of 9-nor-9β-OH-HHC, (3) the alkyl tail of AEA overlaid the C-3 alkyl side chain of 9-nor-9β-OH-HHC, and (4) the polyolefin loop overlaid the tricyclic ring structure of 9-nor-9β-OH-HHC. Tong et al. (1998) supported their use of a helical-shaped AEA by citing a recently described X-ray crystallographic structure that shows that AA adopts a helical conformation when it is a substrate for cyclooxygenase (Stegeman et al., 1998). The authors cite the close matching of common pharmacophoric elements of AEA and 9-nor-9β-OH-HHC as persuasive evidence of the biological relevance of this helical conformer. A 3D-QSAR model derived using CoMFA for a training set of 29 classical and nonclassical cannabinoids rationalized the binding affinity in terms of the steric and electrostatic properties.

The Tong model of structural correspondences between classical cannabinoids and AEA has been employed by several groups in attempts to understand the differences between high- and low-affinity AEA analogs. Howlett and co-workers (Berglund et al., 2000) used the same set of structural correspondences as Tong and co-workers to design a series of monocyclic and bicyclic alkyl amides. The bend in the U-shaped conformation of AEA was approximated with incorporation of a phenyl or naphthyl ring, and the importance of a flat ring was tested by incorporation of a cyclohexyl ring. Aspects of the Tong pharmacophore that were found to be important (i.e., the alkyl tail and carbonyl of the amide) were included or excluded in the series. Highest affinity was associated with phenyl analogs, and among these analogs, meta-substitution on the phenyl ring yielded the highest-affinity compounds, presumably because it places the amide group and the alkyl side chain at the best distance. Using the pharmacophoric elements proposed to be important in the Tong model, the investigators calculated the distances between the pharmacophoric elements (carbocyclic ring-hydroxyl, phenolic hydroxyl, and alkyl side chain) of a series of high-affinity, moderate-affinity, and low-affinity analogs relative to these distances in the high-affinity nonclassical cannabinoid CP55244 (**29**). However, the authors found it difficult to establish a clear relationship between relative binding affinities and their corresponding pharmacophoric distances due to the high flexibility of the compounds.

Van der Stelt and co-workers (2002) evaluated a series of AEA and 2-AG lipoxygenase products for their CB_1 and CB_2 affinities, as well as their ability to inhibit AEA hydrolysis at the FAAH enzyme and to inhibit AEA transport. Several of these compounds have previously been reported by Hillard and co-workers (Edgemond et al., 1998). Conformational analysis was performed using NMR solution studies, as well as MD calculations. Conformational analysis results for the hydoxylated AEAs were probed for consistency of placement of the key pharmacophoric elements identified by Tong and co-workers (Tong et al., 1998) vs. a CP55940 (**5**) template. However, the overlapping regions of CP55940 and the hydroxylated-AEA series did not reveal great differences between analogs with high or low CB_1 affinities. These investigators concluded that a tight-folded conformation of the acyl chain (as required by the backbone constraints of the pharmacophore) is not the only requirement for CB_1 binding.

An Alternative Approach to Endocannabinoid Pharmacophore Construction

Whereas the pharmacophores discussed above have focused on key areas of possible correspondence between endocannabinoids such as AEA and a rigid classical or nonclassical cannabinoid, none of these pharmacophores have addressed the more subtle aspects of endocannabinoid SAR for interaction with the CB_1 receptor. These aspects include requirements for the number of carbon atoms in the acyl chain and for the number of cis homoallylic double bonds in this chain. An alternate approach to the construction of an endogenous cannabinoid pharmacophore for binding at the cannabinoid receptor has been taken by Reggio and co-workers. In this approach, an independent pharmacophore for the endocannabinoids has been developed based upon the endocannabinoid SAR literature reviewed. This alternate strategy carries with it the tacit assumption that the binding sites of cannabinoid ligands from different ligand structural classes can be overlapping but are likely not identical. This assumption accounts for the ability of cannabinoid ligands to displace one another but puts the focus more squarely on the endocannabinoid SAR literature itself. Progress on the development of this pharmacophore is detailed in the following subsection.

Endocannabinoid Interaction with CB_1: Reggio Model

Reggio and co-workers have proposed a hypothesis about endocannabinoid interaction with CB_1 based upon their CM results (Barnett-Norris et al., 1998; Barnett-Norris, Hurst, Lynch et al., 2002; Reggio et al., 2003). This hypothesis requires AEA to assume more than one conformation during its approach toward and interaction with CB_1. In the discussion of receptor residues that follows, the amino acid numbering scheme proposed by Ballesteros and Weinstein (Ballesteros and Weinstein, 1995) has been used. In this numbering system, the most highly conserved residue in each TMH is assigned a locant of 0.50. This number is preceded by the TMH number and followed in parentheses by the sequence number. All other residues in a TMH are numbered relative to this residue. In this numbering system, for example, the most highly conserved residue in TMH 2 of the human CB_1 receptor is D2.50 (163). The residue that immediately precedes it is A2.49 (162). Figure 2.7 serves as a reference for this numbering system.

AEA Approaches CB_1 from Lipid in an Extended Conformation and Interacts with the Lipid Face of CB_1

The high lipophilicity of endocannabinoids and the short distances that endocannabinoids have been shown to diffuse (Wilson and Nicoll, 2001), suggest that AEA, once synthesized, may partition readily into the presynaptic membrane (see Figure 2.2A). It is possible, then, that AEA reaches CB_1 through diffusion in the lipid bilayer and that the first contact that it has with CB_1 is with its lipid face. Reggio and co-workers have hypothesized that AEA approaches the lipid face of CB_1 in an extended conformation, being so oriented by the lipid. This hypothesis is consistent with the hypothesis proposed by the Makriyannis group that the lipid bilayer serves to orient classical cannabinoids for their initial interaction with the receptor (Makriyannis, 1993; Makriyannis et al., 1989; Yang et al., 1991; Yang et al., 1992).

AEA/DOPC simulations illustrated in Figure 2.6 and recent NMR data from the Gawrisch lab (Eldho and Gawrisch, 2003; Lynch and Reggio, 2005) and the Makriyannis lab (Tian et al., 2005) point to AEA's ability to exist in a lipid bilayer in an extended conformation. In this conformation, the acyl tail of AEA is near the bilayer center (see Figure 2.6). If AEA approaches CB_1 in lipid, then the initial endocannabinoid or CB_1 interaction site may be an "exosite" (i.e., a site on the outside of the receptor, its lipid face). The Reggio lab has identified a groove on the lipid face of CB_1 TMH6 into which the acyl tail of AEA may fit (Barnett-Norris, Hurst, Buehner et al., 2002). This groove is formed by the beta-branching residues, Val6.43(351) and Ile6.46(354), located near the bilayer center. At least five saturated carbons at the end of the acyl chain are needed for optimal interaction with the Val6.43(351) and Ile6.46(354) groove. It is hypothesized that this groove serves as a filter for the acyl chain variants of AEA, as only those with a sufficient number of carbons (20–22) and degree of saturation (3 to 4 cis homoallylic double bonds at C5–C6, C8–C9, C11–C12, and C14–C15 or C8–C9, C11–C12, and C14-C15 or their equivalents in a 22 carbon fatty acid chain) are at the correct depth in the membrane and have the correct angle to interact with this groove (Reggio et al., 2003; and unpublished studies). Thus, the lipid face of CB_1 at TMH6 may serve as an initial CB_1 interaction site for endocannabinoids, much like the entry binding site characterized recently for cis-retinal and rhodopsin which is also on the lipid face of rhodopsin (albeit at a different TMH helix) (Schadel et al., 2003). This hypothesis concerning the initial site of AEA/CB_1 interaction, then, would account for the endocannabinoid SAR requirements for acyl chain length, the need for unsaturation in the last five carbons of the acyl chain, as well as the requirements for the number of cis homoallylic double bonds in the acyl chain.

Interaction with a TMH6 Lipid Face Groove May Result in Entry of AEA Into the Binding Site Crevice of CB_1

Docosohexaenoic acid (DHA; 22:6, $\Delta^{4,7,10,13,16,19}$, n-3) possesses six homoallylic double bonds in its acyl chain, making it a molecule with a flexibility similar to that of AEA. In their computational study of the interaction of a 1-stearoyl-2-docosahexaenoyl-sn-glycero-3-phosphatidylcholine (SDPC) lipid bilayer with rhodopsin (Rho), Feller and co-workers reported that DHA is able to adopt the rugged surface of the alpha-helical chains on the lipid face of Rho and to penetrate between them at little or no intramolecular energy cost (Feller et al., 2003). CM results for helix 6 in CB_1 suggest that alkyl chain interaction with the Val6.43(351)/Ile6.46(354) groove may induce a conformational change in TMH6 that would move the intracellular end of TMH6 away from TMH3 (Barnett-Norris, Hurst, Buehner et al., 2002). Such a change has been proposed to be the beginning of the R (inactive state) to R* (active state) transition for the β_2-adrenergic receptor (Ballesteros et al., 2001; Jensen et al., 2001) and for rhodopsin (Farrens et al., 1996), both G-protein-coupled receptors. This process has also been shown to involve a counterclockwise rotation of TMH6 (Ghanouni et al., 2001; Javitch et al., 1997) in the β_2-adrenergic receptor. Reggio and co-workers have hypothesized that given the natural plasticity built into the AEA acyl chain (see section titled "Experimental X-Ray Crystal Structures: Arachidonic Acid and Arachidonic Acid Derivatives"), AEA and its congeners that are recognized by the TMH6 groove residues are able to conform to the shape of the lipid face of TMH6 and be rotated into the binding-site crevice of CB_1 when CB_1 activates.

AEA May Assume a Folded/U-Shape Once It is Within the CB_1 Binding Site Crevice

Once in the protein interior, AEA is hypothesized to fold into a U-shape. Figure 2.9 illustrates the results of docking studies for the endogenous agonist AEA in a model of the CB_1 activated state (CB_1 R*) (McAllister et al., 2003). Docking studies suggest that in its binding site in the CB_1 TMH 2-3-6-7 region, AEA is engaged in a hydrogen bonding interaction with K3.28. In this interaction, K3.28 forms a hydrogen bond with the amide oxygen of AEA (N to amide O distance = 2.64 Å, N-H—O angle = 158°). At the same time, the head group hydroxyl of AEA is engaged in an intramolecular hydrogen bond with the amide oxygen (O to O distance = 2.65 Å, O—H-O angle = 130°). The formation of such an intramolecular hydrogen bond in AEA helps the hydroxyl exist in a hydrophobic region and still satisfy the requirement for a hydrogen bond.

The residues that line the AEA binding pocket are largely hydrophobic, including F2.57, F3.25, L3.29, V3.32, F6.60, F7.35, A7.36, Y6.57, S7.39 (hydrogen bonded back to its own backbone carbonyl oxygen), and L7.43. The interaction with F3.25 is a C-H•••π interaction with the C5–C6 double bond in AEA. In the R* model, F2.57 has an interaction with the amide oxygen. Here, an aromatic ring hydrogen interacts with one of the lone pairs of electrons of the amide oxygen (C to O distance = 3.69 Å, C-H—O angle = 168°, ring center to O distance = 5.08 Å). A similar interaction is evident in the X-ray crystal structure of bovine rhodopsin (Palczewski et al., 2000) in which F5.38 points its positive edge into the backbone carbonyl oxygen of P4.60 (C to O distance = 3.31 Å, C-H—O = 147°, ring center to O distance = 4.62 Å).

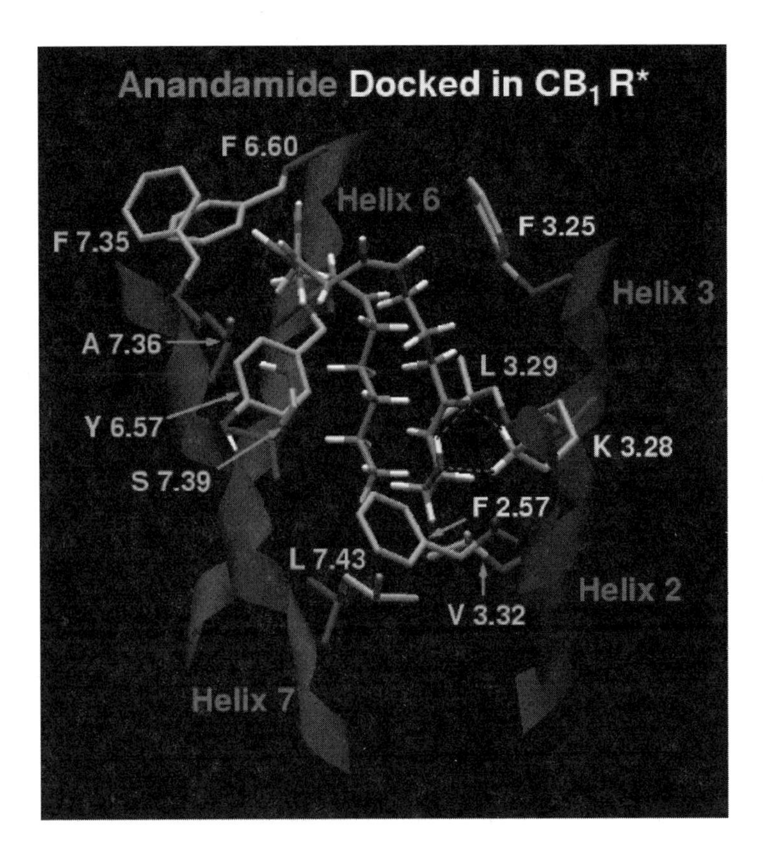

FIGURE 2.9 The anandamide binding site in the TMH2-3-6-7 region of the activated state of CB1 as identified by modeling studies. (From McAllister SA, Gulrukh Rizvi G, Anavi-Goffer S, Hurst DP, Barnett-Norris J, Lynch DL, Reggio PH, and Abood ME [2003] An aromatic microdomain at the cannabinoid CB1 receptor constitutes an agonist/inverse agonist binding region. *J Med Chem*, 46: 5139–5152.) K3.28 forms a hydrogen bond with the amide oxygen of anandamide. At the same time, the head-group hydroxyl of anandamide is engaged in an intramolecular hydrogen bond with its amide oxygen. The anandamide binding pocket is lined with residues that are largely hydrophobic, including L3.29, V3.32, F6.60, F7.35, A7.36, Y6.57, S7.39 (hydrogen bonded back to its own backbone carbonyl oxygen), and L7.43. F3.25 has a C-H•••π interaction with the C5–C6 double bond of anandamide; F2.57 has an interaction with the amide oxygen of anandamide.

As has been discussed earlier in this chapter, the arrangement of homoallylic double bonds in the acyl chain of AEA makes it a very flexible molecule. Although there are no references to crystal structures of AEA bound to CB_1 available in the literature, there is an X-ray crystal structure of anandamide's parent acid, AA (20:4, n-6) complexed with adipocyte-lipid-binding protein (LaLonde et al., 1994). In this structure, AA clearly adopts a curved/U-shaped conformation (see Figure 2.5B) that is consistent with the conformations identified for AEA (20:4, n-6) in CM calculations (Barnett-Norris et al., 1998; Barnett-Norris, Hurst, Lynch et al., 2002) and is consistent with the conformation of AEA in CB_1 illustrated in Figure 2.9. The importance of K3.28 as a direct interaction site for AEA is supported by the work of Song and Bonner (Song and Bonner, 1996), who reported that AEA was unable to compete for [^{3}H]WIN55212-2 binding in a human CB_1 K3.28(192)A mutant and that the potency of AEA in inhibiting cAMP accumulation was reduced by >100-fold in this mutant. This loss of affinity could occur with the loss of a strong hydrogen bonding interaction.

The binding site interactions illustrated for AEA in Figure 2.9 agree with the results first reported by Pinto and co-workers (1994), which showed that the hydroxyl group in the head-group region of AEA could be replaced by a methyl group without a loss in CB_1 affinity. This result suggests that the hydroxyl group is not essential for AEA binding and also that this hydroxyl may exist in a hydrophobic region of CB_1. This result has been echoed in later endocannabinoid structure–activity relationship studies that showed, for example, that a cyclopropyl head group results in a very high-CB_1-affinity ligand (see compound **17**) (Hillard et al., 1999). In the binding site, identified for AEA in our model, the head-group hydroxyl is located in a hydrophobic pocket and satisfies its hydrogen bonding potential by forming an intramolecular hydrogen bond with the amide oxygen. This result is consistent with recent NMR solution studies of AEA reported by Bonechi and co-workers (2001), who found that this intramolecular hydrogen bond in AEA persists in solution.

C-H•••π interactions are moderate but nevertheless important interactions that contribute to protein stability (Brandl et al., 2001). Modeling studies illustrated in Figure 2.9 suggest that F3.25(190) has a C-H•••π interaction with the C5–C6 double bond in the AEA acyl chain. This interaction is consistent with a recent mutation study which showed that a F3.25(190). A mutation resulted in a 6-fold loss in affinity for AEA [CB_1 WT $K_i = 0.3$ μ*M* (0.1–0.6); CB_1 F3.25A $K_i = 1.8$ μ*M* (0.6–5.6)], suggesting that F3.25 is part of the AEA binding site. The F3.25 C-H•••π interaction with the C5–C6 bond of AEA is consistent with the recently described crystal structure of FAAH in which the arachidonyl inhibitor, MAP is bound (Bracey et al., 2002). Here, several aromatic amino acids (F194, F244, Y335, F381, F432, and W531) line the substrate binding pocket surrounding the arachidonyl chain and F194, F381, and F432 are engaged in C-H•••π interactions with the first through third double bonds, respectively, of the MAP arachidonyl acyl chain. The C-H•••π interaction between F3.25 and AEA is also consistent with the recently described crystal structure of AA bound to prostaglandin synthase (Malkowski et al., 2000). This crystal structure also shows C-H•••π interactions between AA acyl chain double bonds and aromatic residues.

Taken together, this proposed mechanism for endocannabinoid interaction with CB_1 accounts for both acyl chain and head group SAR requirements discussed earlier in this chapter. This hypothesis also provides a reason why the design of rigidified AEA analogs has not resulted in ligands with high affinity or efficacy at CB_1, for these ligands are not capable of achieving both extended and compact conformations. It must be emphasized, however, that although the Reggio model does show consistency with the experimental literature cited here, validation of the model requires further experimental testing. Such testing is presently under way.

ACKNOWLEDGMENTS

The author wishes to thank Dow Hurst and Beverly Brookshire for their technical assistance in the preparation of this chapter. This work was supported by the National Institute on Drug Abuse (Grants DA03934 and DA00489).

REFERENCES

Abadji V, Lin S, Taha G, Griffin G, Stevenson LA, Pertwee RG, and Makriyannis A (1994) (R)-methanandamide: a chiral novel anandamide possessing higher potency and metabolic stability. *J Med Chem* 37: 1889–93.

Abood ME, Ditto KE, Noel MA, Showalter VM, and Tao Q (1997) Isolation and expression of a mouse CB1 cannabinoid receptor gene. Comparison of binding properties with those of native CB1 receptors in mouse brain and N18TG2 neuroblastoma cells. *Biochem Pharmacol* 53: 207–14.

Adams IB, Ryan W, Singer M, Razdan RK, Compton DR, and Martin BR (1995) Pharmacological and behavioral evaluation of alkylated anandamide analogs. *Life Sci* 56: 2041–8.

Adams IB, Ryan W, Singer M, Thomas BF, Compton DR, Razdan RK, and Martin BR (1995) Evaluation of cannabinoid receptor binding and in vivo activities for anandamide analogs. *J Pharmacol Exp Ther* 273: 1172–81.

Applegate KR and Glomset JA (1986) Computer-based modeling of the conformation and packing properties of docosahexaenoic acid. *J Lipid Res* 27: 658–680.

Arreaza G and Deutsch DG (1999) Deletion of a proline-rich region and a transmembrane domain in fatty acid amide hydrolase. *FEBS Lett* 454: 57–60.

Ballesteros JA, Jensen AD, Liapakis G, Rasmussen SG, Shi L, Gether U, and Javitch JA (2001) Activation of the beta 2-adrenergic receptor involves disruption of an ionic lock between the cytoplasmic ends of transmembrane segments 3 and 6. *J Biol Chem* 276: 29171–7.

Ballesteros JA and Weinstein H (1995) Integrated methods for the construction of three dimensional models and computational probing of structure function relations in G protein-coupled receptors, in *Methods in Neuroscience* (Sealfon PMCaSC ed), pp. 366–428, chap. 19.

Barnett-Norris J, Guarnieri F, Hurst DP, and Reggio PH (1998) Exploration of biologically relevant conformations of anandamide, 2-arachidonylglycerol, and their analogues using conformational memories. *J Med Chem* 41: 4861–72.

Barnett-Norris J, Hurst DP, Buehner K, Ballesteros JA, Guarnieri F, and Reggio PH (2002) Agonist alkyl tail interaction with cannabinoid CB1 receptor V6.43/I6.46 groove induces a Helix 6 active conformation. *Int J Quantum Chem* 88: 76–86.

Barnett-Norris J, Hurst DP, Lynch DL, Guarnieri F, Makriyannis A, and Reggio PH (2002) Conformational memories and the endocannabinoid binding site at the cannabinoid CB1 receptor. *J Med Chem* 45: 3649–59.

Beltramo M, Stella N, Calignano A, Lin SY, Makriyannis A, and Piomelli D (1997) Functional role of high-affinity anandamide transport, as revealed by selective inhibition. *Science* 277: 1094–7.

Ben-Shabat S, Fride E, Sheskin T, Tamiri T, Rhee MH, Vogel Z, Bisogno T, De Petrocellis L, Di Marzo V, and Mechoulam R (1998) An entourage effect: inactive endogenous fatty acid glycerol esters enhance 2-arachidonoyl-glycerol cannabinoid activity. *Eur J Pharmacol* 353: 23–31.

Berglund BA, Boring DL, and Howlett AC (1999) Investigation of structural analogs of prostaglandin amides for binding to and activation of CB1 and CB2 cannabinoid receptors in rat brain and human tonsils. *Adv Exp Med Biol* 469: 527–33.

Berglund BA, Boring DL, Wilken GH, Makriyannis A, Howlett AC, and Lin S (1998) Structural requirements for arachidonylethanolamide interaction with CB1 and CB2 cannabinoid receptors: pharmacology of the carbonyl and ethanolamide groups. *Prostaglandins Leukot Essent Fatty Acids* 59: 111–8.

Berglund BA, Fleming PR, Rice KC, Shim JY, Welsh WJ, and Howlett AC (2000) Development of a novel class of monocyclic and bicyclic alkyl amides that exhibit CB1 and CB2 cannabinoid receptor affinity and receptor activation. *Drug Des Discov* 16: 281–94.

Bezuglov V, Bobrov M, Gretskaya N, Gonchar A, Zinchenko G, Melck D, Bisogno T, Di Marzo V, Kuklev D, Rossi JC, Vidal JP, and Durand T (2001) Synthesis and biological evaluation of novel amides of polyunsaturated fatty acids with dopamine. *Bioorg Med Chem Lett* 11: 447–9.

Bisogno T, Berrendero F, Ambrosino G, Cebeira M, Ramos JA, Fernandez-Ruiz JJ, and Di Marzo V (1999) Brain regional distribution of endocannabinoids: implications for their biosynthesis and biological function. *Biochem Biophys Res Commn* 256: 377–80.

Bisogno T, Melck D, Bobrov M, Gretskaya NM, Bezuglov VV, De Petrocellis L, and Di Marzo V (2000) N-acyl-dopamines: novel synthetic CB(1) cannabinoid-receptor ligands and inhibitors of anandamide inactivation with cannabimimetic activity in vitro and in vivo. *Biochem J* 351 (Pt. 3): 817–24.

Bisogno T, Sepe N, Melck D, Maurelli S, De Petrocellis L, and Di Marzo V (1997) Biosynthesis, release and degradation of the novel endogenous cannabimimetic metabolite 2-arachidonoylglycerol in mouse neuroblastoma cells. *Biochem J* 322 (Pt. 2): 671–7.

Bonechi C, Brizzi A, Brizzi V, Francoli M, Donati A, and Rossi C (2001) Conformational analysis of N-arachidonylethanolamide (anandamide) using nuclear magnetic R\resonance and theoretical calculations. *Magn Reson Chem* 39: 432–437.

Bonzom PM, Zloh M, Reid RE, and Gibbons WA (1997) Lipid mimetics: the design and properties of conformationally-restricted arachidonic acid lipidic and peptidic analogues. *Biochem Soc Trans* 25: 26S.

Boring DL, Berglund BA, and Howlett AC (1996) Cerebrodiene, arachidonyl-ethanolamide, and hybrid structures: potential for interaction with brain cannabinoid receptors. *Prostaglandins Leukot Essent Fatty Acids* 55: 207–10.

Bouaboula M, Desnoyer N, Carayon P, Combes T, and Casellas P (1999) Gi protein modulation induced by a selective inverse agonist for the peripheral cannabinoid receptor CB2: implication for intracellular signalization cross-regulation. *Mol Pharmacol* 55: 473–80.

Bouaboula M, Perrachon S, Milligan L, Canat X, Rinaldi-Carmona M, Portier M, Barth F, Calandra B, Pecceu F, Lupker J, Maffrand JP, Le Fur G, and Casellas P (1997) A selective inverse agonist for central cannabinoid receptor inhibits mitogen-activated protein kinase activation stimulated by insulin or insulin-like growth factor 1. Evidence for a new model of receptor/ligand interactions. *J Biol Chem* 272: 22330–9.

Bouaboula M, Poinot-Chazel C, Bourrie B, Canat X, Calandra B, Rinaldi-Carmona M, Le Fur G, and Casellas P (1995) Activation of mitogen-activated protein kinases by stimulation of the central cannabinoid receptor CB1. *Biochem J* 312 (Pt 2): 637–41.

Bouaboula M, Poinot-Chazel C, Marchand J, Canat X, Bourrie B, Rinaldi-Carmona M, Calandra B, Le Fur G, and Casellas P (1996) Signaling pathway associated with stimulation of CB2 peripheral cannabinoid receptor. Involvement of both mitogen-activated protein kinase and induction of Krox-24 expression. *Eur J Biochem* 237: 704–11.

Bracey MH, Hanson MA, Masuda KR, Stevens RC, and Cravatt BF (2002) Structural adaptations in a membrane enzyme that terminates endocannabinoid signaling. *Science* 298: 1793–6.

Brandl M, Weiss MS, Jabs A, Suhnel J, and Hilgenfeld R (2001) C-H ... pi-interactions in proteins. *J Mol Biol* 307: 357–77.

Breivogel CS, Griffin G, Di Marzo V, and Martin BR (2001) Evidence for a new G protein-coupled cannabinoid receptor in mouse brain. *Mol Pharmacol* 60: 155–63.

Caulfield MP, and Brown DA (1992) Cannabinoid receptor agonists inhibit Ca current in NG108-15 neuroblastoma cells via a pertussis toxin-sensitive mechanism. *Br J Pharmacol* 106: 231–2.

Cheer JF, Cadogan AK, Marsden CA, Fone KC, and Kendall DA (1999) Modification of 5-HT2 receptor mediated behaviour in the rat by oleamide and the role of cannabinoid receptors. *Neuropharmacology* 38: 533–41.

Chu CJ, Huang SM, De Petrocellis L, Bisogno T, Ewing SA, Miller JD, Zipkin RE, Daddario N, Appendino G, Di Marzo V, and Walker JM (2003) N-oleoyldopamine, a novel endogenous capsaicin-like lipid that produces hyperalgesia. *J Biol Chem* 278: 13633–9.

Compton DR, Gold LH, Ward SJ, Balster RL, and Martin BR (1992) Aminoalkylindole analogs: cannabimimetic activity of a class of compounds structurally distinct from delta 9-tetrahydrocannabinol. *J Pharmacol Exp Ther* 263: 1118–26.

Compton DR, Rice KC, De Costa BR, Razdan RK, Melvin LS, Johnson MR, and Martin BR (1993) Cannabinoid structure-activity relationships: correlation of receptor binding and in vivo activities. *J Pharmacol Exp Ther* 265: 218–26.

Corey EJ, Iguchi S, Albright J, and De B (1983) Studies on the conformational mobility of arachidonic acid. Facile macrolactonization of 20-hydroxyarachidonic acid. *Tet Lett* 24: 37–40.

Corey EJ, Niwa H, and Falck JR (1979) Selective Epoxidation of Eicosa-cis-5, 8, 11, 14-tetraenoic (Arachidonic) Acid and Eicosa-cis-8, 11,14-trienoic Acid. *J Am Chem Soc* 101: 1586–1587.

Cravatt BF, Demarest K, Patricelli MP, Bracey MH, Giang DK, Martin BR, and Lichtman AH (2001) Supersensitivity to anandamide and enhanced endogenous cannabinoid signaling in mice lacking fatty acid amide hydrolase. *Proc Natl Acad Sci U S A* 98: 9371–6.

Cravatt BF, Giang DK, Mayfield SP, Boger DL, Lerner RA, and Gilula NB (1996) Molecular characterization of an enzyme that degrades neuromodulatory fatty-acid amides. *Nature* 384: 83–7.

Cravatt BF, Prospero-Garcia O, Siuzdak G, Gilula NB, Henriksen SJ, Boger DL, and Lerner RA (1995) Chemical characterization of a family of brain lipids that induce sleep. *Science* 268: 1506–9.

D'Ambra TE, Estep KG, Bell MR, Eissenstat MA, Josef KA, Ward SJ, Haycock DA, Baizman ER, Casiano FM, Beglin NC, and et al. (1992) Conformationally restrained analogues of pravadoline: nanomolar potent, enantioselective, (aminoalkyl)indole agonists of the cannabinoid receptor. *J Med Chem* 35: 124–35.

Derkinderen P, Toutant M, Burgaya F, Le Bert M, Siciliano JC, de Franciscis V, Gelman M, and Girault JA (1996) Regulation of a neuronal form of focal adhesion kinase by anandamide. *Science* 273: 1719–22.

Devane WA, Dysarz FA, 3rd, Johnson MR, Melvin LS, and Howlett AC (1988) Determination and characterization of a cannabinoid receptor in rat brain. *Mol Pharmacol* 34: 605–13.

Di Marzo V, Breivogel CS, Tao Q, Bridgen DT, Razdan RK, Zimmer AM, Zimmer A, and Martin BR (2000) Levels, metabolism, and pharmacological activity of anandamide in CB(1) cannabinoid receptor knockout mice evidence for non-CB(1), non-CB(2) receptor-mediated actions of anandamide in mouse brain. *J Neurochem* 75: 2434–44.

Di Marzo V, Fontana A, Cadas H, Schinelli S, Cimino G, Schwartz JC, and Piomelli D (1994) Formation and inactivation of endogenous cannabinoid anandamide in central neurons. *Nature* 372: 686–91.

Di Marzo V, Melck D, Bisogno T, and De Petrocellis L (1998) Endocannabinoids: endogenous cannabinoid receptor ligands with neuromodulatory action. *Trends Neurosci* 21: 521–8.
Di Marzo V, Melck D, Orlando P, Bisogno T, Zagoory O, Bifulco M, Vogel Z, and De Petrocellis L (2001) Palmitoylethanolamide inhibits the expression of fatty acid amide hydrolase and enhances the anti-proliferative effect of anandamide in human breast cancer cells. *Biochem J* 358: 249–55.
Dinh TP, Carpenter D, Leslie FM, Freund TF, Katona I, Sensi SL, Kathuria S, and Piomelli D (2002) Brain monoglyceride lipase participating in endocannabinoid inactivation. *Proc Natl Acad Sci U S A* 99: 10819–24.
Edgemond WS, Campbell WB, and Hillard CJ (1995) The binding of novel phenolic derivatives of anandamide to brain cannabinoid receptors. *Prostaglandins Leukot Essent Fatty Acids* 52: 83–6.
Edgemond WS, Hillard CJ, Falck JR, Kearn CS, and Campbell WB (1998) Human platelets and polymorphonuclear leukocytes synthesize oxygenated derivatives of arachidonylethanolamide (anandamide): their affinities for cannabinoid receptors and pathways of inactivation. *Mol Pharmacol* 54: 180–8.
Egertova M, Cravatt BF, and Elphick MR (2003) Comparative analysis of fatty acid amide hydrolase and cb(1) cannabinoid receptor expression in the mouse brain: evidence of a widespread role for fatty acid amide hydrolase in regulation of endocannabinoid signaling. *Neuroscience* 119: 481–96.
Egertova M, Giang DK, Cravatt BF, and Elphick MR (1998) A new perspective on cannabinoid signalling: complementary localization of fatty acid amide hydrolase and the CB1 receptor in rat brain. *Proc R Soc Lond B Biol Sci* 265: 2081–5.
Eldho NV and Gawrisch K (2003) Location, conformation and dynamics of anandamide in biomembranes. *Biophys J* 84: 49a.
Ernst J, Sheldrick WS, and Fuhrop JH (1979) The structures of the essential unsaturated fatty acids. Crystal structure of linoleic acid and evidence for the crystal structures of alpha-linoleic and arachidonic acid. *Z Naturforsch* 34b: 706–711.
Facci L, Toso RD, Romanello S, Buriani A, Skaper SD, and Leon A (1995) Mast cells express a peripheral cannabinoid receptor with differential sensitivity to anandamide and palmitoylethanolamide. *Proc Natl Acad Sci U S A* 92: 3376–3380.
Farrens DL, Altenbach C, Yang K, Hubbell WL, and Khorana HG (1996) Requirement of rigid-body motion of transmembrane helices for light-activation of rhodopsin. *Science* 274: 768–770.
Felder CC, Joyce KE, Briley EM, Glass M, Mackie KP, Fahey KJ, Cullinan GJ, Hunden DC, Johnson DW, Chaney MO, Koppel GA, and Brownstein M (1998) LY320135, a novel cannabinoid CB1 receptor antagonist, unmasks coupling of the CB1 receptor to stimulation of cAMP accumulation. *J Pharmacol Exp Ther* 284: 291–7.
Felder CC, Joyce KE, Briley EM, Mansouri J, Mackie K, Blond O, Lai Y, Ma AL, and Mitchell RL (1995) Comparison of the pharmacology and signal transduction of the human cannabinoid CB1 and CB2 receptors. *Mol Pharmacol* 48: 443–50.
Feller SE, Gawrisch K, and MacKerell AD, Jr. (2002) Polyunsaturated fatty acids in lipid bilayers: intrinsic and environmental contributions to their unique physical properties. *J Am Chem Soc* 124: 318–26.
Feller SE, Gawrisch K, and Woolf TB (2003) Rhodopsin exhibits a preference for solvation by polyunsaturated docosohexaenoic acid. *J Am Chem Soc* 125: 4434–5.
Fezza F, Bisogno T, Minassi A, Appendino G, Mechoulam R, and Di Marzo V (2002) Noladin ether, a putative novel endocannabinoid: inactivation mechanisms and a sensitive method for its quantification in rat tissues. FEBS *Lett* 513: 294–8.
Fimiani C, Mattocks D, Cavani F, Salzet M, Deutsch DG, Pryor S, Bilfinger TV, and Stefano GB (1999) Morphine and anandamide stimulate intracellular calcium transients in human arterial endothelial cells: coupling to nitric oxide release. *Cell Signal* 11: 189–93.
Galiegue S, Mary S, Marchand J, Dussossoy D, Carriere D, Carayon P, Bouaboula M, Shire D, Le Fur G, and Casellas P (1995) Expression of central and peripheral cannabinoid receptors in human immune tissues and leukocyte subpopulations. *Eur J Biochem* 232: 54–61.
Gebremedhin D, Lange AR, Campbell WB, Hillard CJ, and Harder DR (1999) Cannabinoid CB1 receptor of cat cerebral arterial muscle functions to inhibit L-type Ca2+ channel current. *Am J Physiol* 276: H2085–93.
Gerard CM, Mollereau C, Vassart G, and Parmentier M (1991) Molecular cloning of a human cannabinoid receptor which is also expressed in testis. *Biochem J* 279 (Pt. 1): 129–34.
Ghanouni P, Steenhuis JJ, Farrens DL, and Kobilka BK (2001) Agonist-induced conformational changes in the G-protein-coupling domain of the beta 2 adrenergic receptor. *Proc Natl Acad Sci U S A* 98: 5997–6002.

Giang DK and Cravatt BF (1997) Molecular characterization of human and mouse fatty acid amide hydrolases. *Proc Natl Acad Sci U S A* 94: 2238–42.

Glaser ST, Abumrad NA, Fatade F, Kaczocha M, Studholme KM, and Deutsch DG (2003) Evidence against the presence of an anandamide transporter. *Proc Natl Acad Sci U S A* 100: 4269–74.

Glass M and Felder CC (1997) Concurrent stimulation of cannabinoid CB1 and dopamine D2 receptors augments cAMP accumulation in striatal neurons: evidence for a Gs linkage to the CB1 receptor. *J Neurosci* 17: 5327–33.

Gonsiorek W, Lunn C, Fan X, Narula S, Lundell D, and Hipkin RW (2000) Endocannabinoid 2-arachidonyl glycerol is a full agonist through human type 2 cannabinoid receptor: antagonism by anandamide. *Mol Pharmacol* 57: 1045–50.

Goparaju SK, Ueda N, Yanaguchi H, and Yamamoto S (1998) Anandamide amidohydrolase reacting with 2-arachidonylglycerol, another cannabinoid receptor ligand. *FEBS Lett* 422: 69–73.

Goutopoulos A, Fan P, Khanolkar AD, Xie XQ, Lin S, and Makriyannis A (2001) Stereochemical selectivity of methanandamides for the CB1 and CB2 cannabinoid receptors and their metabolic stability. *Bioorg Med Chem* 9: 1673–84.

Griffin G, Tao Q, and Abood ME (2000) Cloning and pharmacological characterization of the rat CB(2) cannabinoid receptor. *J Pharmacol Exp Ther* 292: 886–94.

Guarnieri F and Weinstein H (1996) Conformational memories and the exploration of biologically relevant peptide conformations: an illustration for the gonadotropin-releasing hormone. *J Am Chem Soc* 118: 5580–5589.

Guarnieri F and Wilson SR (1995) Conformational memories and a simulated annealing program that learns: application to LTB4. *J Comp Chem* 16: 648–653.

Hampson AJ, Hill WA, Zan-Phillips M, Makriyannis A, Leung E, Eglen RM, and Bornheim LM (1995) Anandamide hydroxylation by brain lipoxygenase: metabolite structures and potencies at the cannabinoid receptor. *Biochim Biophys Acta* 1259: 173–9.

Hanus L, Abu-Lafi S, Fride E, Breuer A, Vogel Z, Shalev DE, Kustanovich I, and Mechoulam R (2001) 2-arachidonyl glyceryl ether, an endogenous agonist of the cannabinoid CB1 receptor. *Proc Natl Acad Sci U S A* 98: 3662–5.

Henry DJ and Chavkin C (1995) Activation of inwardly rectifying potassium channels (GIRK1) by co-expressed rat brain cannabinoid receptors in *Xenopus* oocytes. *Neurosci Lett* 186: 91–4.

Herkenham M, Lynn AB, Johnson MR, Melvin LS, de Costa BR, and Rice KC (1991) Characterization and localization of cannabinoid receptors in rat brain: a quantitative in vitro autoradiographic study. *J Neurosci* 11: 563–83.

Hillard CJ (2000) Biochemistry and pharmacology of the endocannabinoids arachidonylethanolamide and 2-arachidonylglycerol. *Prostaglandins Other Lipid Mediat* 61: 3–18.

Hillard CJ, Manna S, Greenberg MJ, DiCamelli R, Ross RA, Stevenson LA, Murphy V, Pertwee RG, and Campbell WB (1999) Synthesis and characterization of potent and selective agonists of the neuronal cannabinoid receptor (CB1). *J Pharmacol Exp Ther* 289: 1427–33.

Ho BY, Uezono Y, Takada S, Takase I, and Izumi F (1999) Coupling of the expressed cannabinoid CB1 and CB2 receptors to phospholipase C and G protein-coupled inwardly rectifying K+ channels. *Receptors Channels* 6: 363–74.

Hoffman AF and Lupica CR (2000) Mechanisms of cannabinoid inhibition of GABA(A) synaptic transmission in the hippocampus. *J Neurosci* 20: 2470–9.

Hosohata K, Quock RM, Hosohata Y, Burkey TH, Makriyannis A, Consroe P, Roeske WR, and Yamamura HI (1997) AM630 is a competitive cannabinoid receptor antagonist in the guinea pig brain. *Life Sci* 61: PL115–8.

Hosohata Y, Quock RM, Hosohata K, Makriyannis A, Consroe P, Roeske WR, and Yamamura HI (1997) AM630 antagonism of cannabinoid-stimulated [35S]GTP gamma S binding in the mouse brain. *Eur J Pharmacol* 321: R1–3.

Iwamura H, Suzuki H, Ueda Y, Kaya T, and Inaba T (2001) In vitro and in vivo pharmacological characterization of JTE-907, a novel selective ligand for cannabinoid CB2 receptor. *J Pharmacol Exp Ther* 296: 420–5.

Jarai Z, Wagner JA, Varga K, Lake KD, Compton DR, Martin BR, Zimmer AM, Bonner TI, Buckley NE, Mezey E, Razdan RK, Zimmer A, and Kunos G (1999) Cannabinoid-induced mesenteric vasodilation through an endothelial site distinct from CB1 or CB2 receptors. *Proc Natl Acad Sci U S A* 96: 14136–41.

Jarrahian A, Manna S, Edgemond WS, Campbell WB, and Hillard CJ (2000) Structure-activity relationships among N-arachidonylethanolamine (Anandamide) head group analogues for the anandamide transporter. *J Neurochem* 74: 2597–606.
Javitch JA, Fu D, Liapakis G, and Chen J (1997) Constitutive activation of the beta2 adrenergic receptor alters the orientation of its sixth membrane-spanning segment. *J Biol Chem* 272: 18546–9.
Jensen AD, Guarnieri F, Rasmussen SG, Asmar F, Ballesteros JA, and Gether U (2001) Agonist-induced conformational changes at the cytoplasmic side of transmembrane segment 6 in the beta 2 adrenergic receptor mapped by site-selective fluorescent labeling. *J Biol Chem* 276: 9279–90.
Jonsson KO, Vandevoorde S, Lambert DM, Tiger G, and Fowler CJ (2001) Effects of homologues and analogues of palmitoylethanolamide upon the inactivation of the endocannabinoid anandamide. *Br J Pharmacol* 133: 1263–75.
Kearn CS, Greenberg MJ, DiCamelli R, Kurzawa K, and Hillard CJ (1999) Relationships between ligand affinities for the cerebellar cannabinoid receptor CB1 and the induction of GDP/GTP exchange. *J Neurochem* 72: 2379–87.
Khanolkar AD, Abadji V, Lin S, Hill WA, Taha G, Abouzid K, Meng Z, Fan P, and Makriyannis A (1996) Head group analogs of arachidonylethanolamide, the endogenous cannabinoid ligand. *J Med Chem* 39: 4515–9.
Kreitzer AC and Regehr WG (2001) Retrograde inhibition of presynaptic calcium influx by endogenous cannabinoids at excitatory synapses onto Purkinje cells. *Neuron* 29: 717–27.
LaLonde JM, Levenson MA, Roe JJ, Bernlohr DA, and Banaszak LJ (1994) Adipocyte lipid-binding protein complexed with arachidonic acid. Titration calorimetry and X-ray crystallographic studies. *J Biol Chem* 269: 25339–47.
Lambert DM, DiPaolo FG, Sonveaux P, Kanyonyo M, Govaerts SJ, Hermans E, Bueb J, Delzenne NM, and Tschirhart EJ (1999) Analogues and homologues of N-palmitoylethanolamide, a putative endogenous CB(2) cannabinoid, as potential ligands for the cannabinoid receptors. *Biochim Biophys Acta* 1440: 266–74.
Lang W, Qin C, Lin S, Khanolkar AD, Goutopoulos A, Fan P, Abouzid K, Meng Z, Biegel D, and Makriyannis A (1999) Substrate specificity and stereoselectivity of rat brain microsomal anandamide amidohydrolase. *J Med Chem* 42: 896–902.
Leach AR and Prout K (1990) Automated conformational analysis: directed conformational search using the A* algorithm. *J Comp Chem* 11: 1193–1205.
Levenes C, Daniel H, Soubrie P, and Crepel F (1998) Cannabinoids decrease excitatory synaptic transmission and impair long-term depression in rat cerebellar Purkinje cells. *J Physiol* 510 (Pt. 3): 867–79.
Lin S, Khanolkar AD, Fan P, Goutopoulos A, Qin C, Papahadjis D, and Makriyannis A (1998) Novel analogues of arachidonylethanolamide (anandamide): affinities for the CB1 and CB2 cannabinoid receptors and metabolic stability. *J Med Chem* 41: 5353–61.
Lopez M, Lozano JJ, Ruiz J, and RP (1993) Conformational search and transitional states study of arachidonic acid using molecular dynamics and AM1 semi-empirical calculations, in *9th European Symposium on Structure-Activity Relationships: QSAR and Molecular Modeling*, pp. 429–430, Leiden, Netherlands.
Lujan R, Roberts JD, Shigemoto R, Ohishi H, and Somogyi P (1997) Differential plasma membrane distribution of metabotropic glutamate receptors mGluR1 alpha, mGluR2 and mGluR5, relative to neurotransmitter release sites. *J Chem Neuroanat* 13: 219–41.
Lynch DL and Reggio PH (2005) Molecular dynamics simulations of the endocannabinoid N-arachidonoylethanolamine (anandamide) in a phospholipid bilayer: probing structure and dynamics. *J Med Chem* **48**: 4824–4833.
Maccarrone M, Bari M, Lorenzon T, Bisogno T, Di Marzo V, and Finazzi-Agro A (2000) Anandamide uptake by human endothelial cells and its regulation by nitric oxide. *J Biol Chem* 275: 13484–92.
Mackie K and Hille B (1992) Cannabinoids inhibit N-type calcium channels in neuroblastoma-glioma cells. *Proc Natl Acad Sci U S A* 89: 3825–9.
Mackie K, Lai Y, Westenbroek R, and Mitchell R (1995) Cannabinoids activate an inwardly rectifying potassium conductance and inhibit Q-type calcium currents in AtT20 cells transfected with rat brain cannabinoid receptor. *J Neurosci* 15: 6552–61.
Maejima T, Hashimoto K, Yoshida T, Aiba A, and Kano M (2001) Presynaptic inhibition caused by retrograde signal from metabotropic glutamate to cannabinoid receptors. *Neuron* 31: 463–75.
Makriyannis A (1993) Probes for the cannabinoid sites of action. *NIDA Res Monogr* 134: 253–67.

Makriyannis A, Banijamali A, Jarrell HC, and Yang DP (1989) The orientation of (-)-delta 9-tetrahydrocannabinol in DPPC bilayers as determined from solid-state 2H-NMR. *Biochim Biophys Acta* 986: 141–5.

Malkowski MG, Ginell SL, Smith WL, and Garavito RM (2000) The productive conformation of arachidonic acid bound to prostaglandin synthase. *Science* 289: 1933–7.

Maneuf YP and Brotchie JM (1997) Paradoxical action of the cannabinoid WIN 55,212-2 in stimulated and basal cyclic AMP accumulation in rat globus pallidus slices. *Br J Pharmacol* 120: 1397–8.

Mato S, Pazos A, and Valdizan EM (2002) Cannabinoid receptor antagonism and inverse agonism in response to SR141716A on cAMP production in human and rat brain. *Eur J Pharmacol* 443: 43–6.

Matsuda LA, Bonner TI, and Lolait SJ (1993) Localization of cannabinoid receptor mRNA in rat brain. *J Comp Neurol* 327: 535–50.

Matsuda LA, Lolait SJ, Brownstein MJ, Young AC, and Bonner TI (1990) Structure of a cannabinoid receptor and functional expression of the cloned cDNA. *Nature* 346: 561–4.

Mazzari S, Canella R, Petrelli L, Marcolongo G, and Leon A (1996) N-(2-hydroxyethyl)hexadecanamide is orally active in reducing edema formation and inflammatory hyperalgesia by down-modulating mast cell activation. *Eur J Pharmacol* 300: 227–36.

McAllister SA, Gulrukh Rizvi G, Anavi-Goffer S, Hurst DP, Barnett-Norris J, Lynch DL, Reggio PH, and Abood ME (2003) An aromatic microdomain at the cannabinoid CB1 receptor constitutes an agonist/inverse agonist binding region. *J Med Chem* 46: 5139–5152.

McFarland MJ, Porter AC, Rakhshan FR, Rawat DS, Gibbs RA, and Barkar EL (2004) A role for caveolae/lipid rafts in the uptake and recycling of the endogeneous cannabinoid anandamide. *J Biol Chem* 279: 41991–41997.

Mechoulam R, Ben-Shabat S, Hanus L, Ligumsky M, Kaminski NE, Schatz AR, Gopher A, Almog S, Martin BR, Compton DR, and et al. (1995) Identification of an endogenous 2-monoglyceride, present in canine gut, that binds to cannabinoid receptors. *Biochem Pharmacol* 50: 83–90.

Mechoulam R, Fride E, Ben-Shabat S, Meiri U, and Horowitz M (1998) Carbachol, an acetylcholine receptor agonist, enhances production in rat aorta of 2-arachidonoyl glycerol, a hypotensive endocannabinoid. *Eur J Pharmacol* 362: R1–3.

Mechoulam R, Fride E, Hanus L, Sheskin T, Bisogno T, Di Marzo V, Bayewitch M, and Vogel Z (1997) Anandamide may mediate sleep induction. *Nature* 389: 25–6.

Melvin LS, Milne GM, Johnson MR, Wilken GH, and Howlett AC (1995) Structure-activity relationships defining the ACD-tricyclic cannabinoids: cannabinoid receptor binding and analgesic activity. *Drug Des Discov* 13: 155–66.

Meschler JP, Kraichely DM, Wilken GH, and Howlett AC (2000) Inverse agonist properties of N-(piperidin-1-yl)-5-(4-chlorophenyl)-1-(2, 4-dichlorophenyl)-4-methyl-1H-pyrazole-3-carboxamide HCl (SR141716A) and 1-(2-chlorophenyl)-4-cyano-5-(4-methoxyphenyl)-1H-pyrazole-3-carboxylic acid phenylamide (CP-272871) for the CB(1) cannabinoid receptor. *Biochem Pharmacol* 60: 1315–23.

Mohamadi F, Richards NGJ, Guida WC, Lizkamp R, Lipton M, Caulfield C, Chang G, Hendickson T, and Still WC (1990) Macromodel — An Integrated Software System for Modeling Organic and Bioorganic Molecules Using Molecular Mechanics. *J Comp Chem* 11: 440–467.

Mombouli JV, Schaeffer G, Holzmann S, Kostner GM, and Graier WF (1999) Anandamide-induced mobilization of cytosolic Ca2+ in endothelial cells. *Br J Pharmacol* 126: 1593–600.

Moody JS, Kozak KR, Ji C, and Marnett LJ (2001) Selective oxygenation of the endocannabinoid 2-arachidonylglycerol by leukocyte-type 12-lipoxygenase. *Biochemistry* 40: 861–6.

Munro S, Thomas KL, and Abu-Shaar M (1993) Molecular characterization of a peripheral receptor for cannabinoids. *Nature* 365: 61–5.

Ng EW, Aung MM, Abood ME, Martin BR, and Razdan RK (1999) Unique analogues of anandamide: arachidonyl ethers and carbamates and norarachidonyl carbamates and ureas. *J Med Chem* 42: 1975–81.

Ohno-Shosaku T, Maejima T, and Kano M (2001) Endogenous cannabinoids mediate retrograde signals from depolarized postsynaptic neurons to presynaptic terminals. *Neuron* 29: 729–38.

Oka S, Tsuchie A, Tokumura A, Muramatsu M, Suhara Y, Takayama H, Waku K, and Sugiura T (2003) Ether-linked analogue of 2-arachidonoylglycerol (noladin ether) was not detected in the brains of various mammalian species. *J Neurochem* 85: 1374–81.

Palczewski K, Kumasaka T, Hori T, Behnke CA, Motoshima H, Fox BA, Le Trong I, Teller DC, Okada T, Stenkamp RE, Yamamoto M, and Miyano M (2000) Crystal structure of rhodopsin: a G protein-coupled receptor. *Science* 289: 739–45.

Pan X, Ikeda SR, and Lewis DL (1996) Rat brain cannabinoid receptor modulates N-type Ca2+ channels in a neuronal expression system. *Mol Pharmacol* 49: 707–14.

Pan X, Ikeda SR, and Lewis DL (1998) SR 141716A acts as an inverse agonist to increase neuronal voltage-dependent Ca2+ currents by reversal of tonic CB1 cannabinoid receptor activity. *Mol Pharmacol* 54: 1064–72.

Pertwee R, Griffin G, Fernando S, Li X, Hill A, and Makriyannis A (1995) AM630, a competitive cannabinoid receptor antagonist. *Life Sci* 56: 1949–55.

Pinto JC, Potie F, Rice KC, Boring D, Johnson MR, Evans DM, Wilken GH, Cantrell CH, and Howlett AC (1994) Cannabinoid receptor binding and agonist activity of amides and esters of arachidonic acid. *Mol Pharmacol* 46: 516–522.

Piomelli D, Beltramo M, Giuffrida A, and Stella N (1998) Endogenous cannabinoid signaling. *Neurobiol Dis* 5: 462–73.

Piomelli D, Beltramo M, Glasnapp S, Lin SY, Goutopoulos A, Xie XQ, and Makriyannis A (1999) Structural determinants for recognition and translocation by the anandamide transporter. *Proc Natl Acad Sci U S A* 96: 5802–7.

Porter AC, Sauer JM, Knierman MD, Becker GW, Berna MJ, Bao J, Nomikos GG, Carter P, Bymaster FP, Leese AB, and Felder CC (2002) Characterization of a novel endocannabinoid, virodhamine, with antagonist activity at the CB1 receptor. *J Pharmacol Exp Ther* 301: 1020–4.

Rabinovich AL and Ripatti PO (1991) On the conformational, physical properties and function of polyunsaturated acylchains. *Biochim Biophys Acta* 1085: 53–56.

Razdan RK (1986) Structure-activity relationships in cannabinoids. *Pharmacol Rev* 38: 75–149.

Reggio P, Hurst DP, Ballesteros JA, and Lynch DL (2003) The CB1 V6.43/I6.46 groove is optimally placed in the lipid bilayer to recognize endocannabinoids in lipid, in *2003 Symposium on the Cannabinoids*, p. 8, International Cannabinoid Research Society, Cornwall, Ontario, Canada.

Reggio PH (2002) Endocannabinoid structure-activity relationships for interaction at the cannabinoid receptors. *Prostaglandins Leukot Essent Fatty Acids* 66: 143–60.

Reggio PH and Traore H (2000) Conformational requirements for endocannabinoid interaction with the cannabinoid receptors, the anandamide transporter and fatty acid amidohydrolase. *Chem Phys Lipids* 108: 15–35.

Rhee MH, Bayewitch M, Avidor-Reiss T, Levy R, and Vogel Z (1998) Cannabinoid receptor activation differentially regulates the various adenylyl cyclase isozymes. *J Neurochem* 71: 1525–34.

Rich MR (1993) Conformational analysis of arachidonic and related fatty acids using molecular dynamics simulations. *Biochim Biophys Acta* 1178: 87–96.

Rinaldi-Carmona M, Barth F, Heaulme M, Shire D, Calandra B, Congy C, Martinez S, Maruani J, Neliat G, Caput D, et al. (1994) SR141716A, a potent and selective antagonist of the brain cannabinoid receptor. *FEBS Lett* 350: 240–4.

Rinaldi-Carmona M, Barth F, Millan J, Derocq JM, Casellas P, Congy C, Oustric D, Sarran M, Bouaboula M, Calandra B, Portier M, Shire D, Breliere JC, and Le Fur GL (1998) SR 144528, the first potent and selective antagonist of the CB2 cannabinoid receptor. *J Pharmacol Exp Ther* 284: 644–50.

Ross RA, Brockie HC, Fernando SR, Saha B, Razdan RK, and Pertwee RG (1998) Comparison of cannabinoid binding sites in guinea-pig forebrain and small intestine. *Br J Pharmacol* 125: 1345–51.

Rowlinson SW, Crews BC, Lanzo CA, and Marnett LJ (1999) The binding of arachidonic acid in the cyclooxygenase active site of mouse prostaglandin endoperoxidase synthase -2 (COX-2). *J Biol Chem* 274: 23305–23310.

Ryan WJ, Banner WK, Wiley JL, Martin BR, and Razdan RK (1997) Potent anandamide analogs: the effect of changing the length and branching of the end pentyl chain. *J Med Chem* 40: 3617–25.

Savinainen JR, Jarvinen T, Laine K, and Laitinen JT (2001) Despite substantial degradation, 2-arachidonoylglycerol is a potent full efficacy agonist mediating CB(1) receptor-dependent G-protein activation in rat cerebellar membranes. *Br J Pharmacol* 134: 664–72.

Schadel SA, Heck M, Maretzki D, Filipek S, Teller DC, Palczewski K, and Hofmann KP (2003) Ligand channeling within a G-protein-coupled receptor. The entry and exit of retinals in native opsin. *J Biol Chem* 278: 24896–903.

Seltzman HH, Fleming DN, Thomas BF, Gilliam AF, McCallion DS, Pertwee RG, Compton DR, and Martin BR (1997) Synthesis and pharmacological comparison of dimethylheptyl and pentyl analogs of anandamide. *J Med Chem* 40: 3626–34.

Shamovsky IL and Yarovskaya IY (1990) A theoretical conformational analysis of the complex of tocopherol and arachidonic acid molecules. *Biol Mem* 4: 844–852.

Sheskin T, Hanus L, Slager J, Vogel Z, and Mechoulam R (1997) Structural requirements for binding of anandamide-type compounds to the brain cannabinoid receptor. *J Med Chem* 40: 659–67.

Shire D, Calandra B, Rinaldi-Carmona M, Oustric D, Pessegue B, Bonnin-Cabanne O, Le Fur G, Caput D, and Ferrara P (1996) Molecular cloning, expression and function of the murine CB2 peripheral cannabinoid receptor. *Biochim Biophys Acta* 1307: 132–6.

Shire D, Carillon C, Kaghad M, Calandra B, Rinaldi-Carmona M, Le Fur G, Caput D, and Ferrara P (1995) An amino-terminal variant of the central cannabinoid receptor resulting from alternative splicing. *J Biol Chem* 270: 3726–31.

Showalter VM, Compton DR, Martin BR, and Abood ME (1996) Evaluation of binding in a transfected cell line expressing a peripheral cannabinoid receptor (CB2): identification of cannabinoid receptor subtype selective ligands. *J Pharmacol Exp Ther* 278: 989–99.

Sjogren S, Ryberg E, Lindblom A, Larrson N, Astrand A, Hjorth S, Andersson A-K, Groblewski T, and Greasley P (2005) A New Receptor for Cannabinoid Ligands, in *2005 Symposium on the Cannabinoids* pp 106, International Cannabinoid Research Society, Clearwater, FL.

Skaper SD, Buriani A, Dal Toso R, Petrelli L, Romanello S, Facci L, and Leon A (1996) The ALIAmide palmitoylethanolamide and cannabinoids, but not anandamide, are protective in a delayed post-glutamate paradigm of excitotoxic death in cerebellar granule neurons. *Proc Natl Acad Sci USA* 93: 3984–9.

Song ZH and Bonner TI (1996) A lysine residue of the cannabinoid receptor is critical for receptor recognition by several agonists but not WIN55212-2. *Mol Pharmacol* 49: 891–6.

Stefano GB, Liu Y, and Goligorsky MS (1996) Cannabinoid receptors are coupled to nitric oxide release in invertebrate immunocytes, microglia, and human monocytes. *J Biol Chem* 271: 19238–42.

Stefano GB, Salzet M, Magazine HI, and Bilfinger TV (1998) Antagonism of LPS and IFN-gamma induction of iNOS in human saphenous vein endothelium by morphine and anandamide by nitric oxide inhibition of adenylate cyclase. *J Cardiovasc Pharmacol* 31: 813–20.

Stegeman R, Pawlitz J, Stevens A, Gierse J, Stallings W, and Kurumbail R (1998) American Crystallographic Association, Abstract.

Stella N, Schweitzer P, and Piomelli D (1997) A second endogenous cannabinoid that modulates long-term potentiation. *Nature* 388: 773–8.

Stoit AR, Lange JHM, den Hartog AP, Ronken E, Tipker K, van Stuivenberg HH, Dijksman JAR, Wals HC, and Kruse CG (2002) Design, synthesis and biological activity of rigid cannabinoid CB1 receptor antagonists. *Chem Pharm Bull* 50: 1109–1113.

Sugiura T, Kodaka T, Kondo S, Nakane S, Kondo H, Waku K, Ishima Y, Watanabe K, and Yamamoto I (1997) Is the cannabinoid CB1 receptor a 2-arachidonoylglycerol receptor? Structural requirements for triggering a Ca2+ transient in NG108-15 cells. *J Biochem* (Tokyo) 122: 890–5.

Sugiura T, Kodaka T, Kondo S, Tonegawa T, Nakane S, Kishimoto S, Yamashita A, and Waku K (1996) 2-Arachidonoylglycerol, a putative endogenous cannabinoid receptor ligand, induces rapid, transient elevation of intracellular free Ca2+ in neuroblastoma x glioma hybrid NG108-15 cells. *Biochem Biophys Res Commn* 229: 58–64.

Sugiura T, Kodaka T, Nakane S, Kishimoto S, Kondo S, and Waku K (1998) Detection of an endogenous cannabimimetic molecule, 2-arachidonoylglycerol, and cannabinoid CB1 receptor mRNA in human vascular cells: is 2-arachidonoylglycerol a possible vasomodulator? *Biochem Biophys Res Commn* 243: 838–43.

Sugiura T, Kodaka T, Nakane S, Miyashita T, Kondo S, Suhara Y, Takayama H, Waku K, Seki C, Baba N, and Ishima Y (1999) Evidence that the cannabinoid CB1 receptor is a 2-arachidonoylglycerol receptor. Structure-activity relationship of 2-arachidonoylglycerol, ether-linked analogues, and related compounds. *J Biol Chem* 274: 2794–801.

Sugiura T, Kondo S, Kishimoto S, Miyashita T, Nakane S, Kodaka T, Suhara Y, Takayama H, and Waku K (2000) Evidence that 2-arachidonoylglycerol but not N-palmitoylethanolamine or anandamide is the physiological ligand for the cannabinoid CB2 receptor. Comparison of the agonistic activities of various cannabinoid receptor ligands in HL-60 cells. *J Biol Chem* 275: 605–12.

Sugiura T, Kondo S, Sukagawa A, Nakane S, Shinoda A, Itoh K, Yamashita A, and Waku K (1995) 2-Arachidonoylglycerol: a possible endogenous cannabinoid receptor ligand in brain. *Biochem Biophys Res Commn* 215: 89–97.

Sugiura T and Waku K (2000) 2-Arachidonoylglycerol and the cannabinoid receptors. *Chem Phys Lipids* 108: 89–106.

Sugiura T and Waku K (2002) Cannabinoid receptors and their endogenous ligands. *J Biochem* (Tokyo) 132: 7–12.

Suhara Y, Takayama H, Nakane S, Miyashita T, Waku K, and Sugiura T (2000) Synthesis and biological activities of 2-arachidonoylglycerol, an endogenous cannabinoid receptor ligand, and its metabolically stable ether-linked analogues. *Chem Pharm Bull* (Tokyo) 48: 903–7.

Takahashi KA and Linden DJ (2000) Cannabinoid receptor modulation of synapses received by cerebellar Purkinje cells. *J Neurophysiol* 83: 1167–80.

Thomas BF, Adams IB, Mascarella SW, Martin BR, and Razdan RK (1996) Structure-activity analysis of anandamide analogs: relationship to a cannabinoid pharmacophore. *J Med Chem* 39: 471–9.

Tian X, Guo J, Yao F, Yang DP, and Makriyannis A (2005) The conformation, location, and dynamic properties of the endocannabinoid ligand anandamide in a membrane bilayer. *J Biol Chem.* In press.

Tong W, Collantes ER, Welsh WJ, Berglund BA, and Howlett AC (1998) Derivation of a pharmacophore model for anandamide using constrained conformational searching and comparative molecular field analysis. *J Med Chem* 41: 4207–15.

Tsou K, Brown S, Sanudo-Pena MC, Mackie K, and Walker JM (1998) Immunohistochemical distribution of cannabinoid CB1 receptors in the rat central nervous system. *Neuroscience* 83: 393–411.

Ueda N, Yamamoto K, Yamamoto S, Tokunaga T, Shirakawa E, Shinkai H, Ogawa M, Sato T, Kudo I, Inoue K, et al. (1995) Lipoxygenase-catalyzed oxygenation of arachidonylethanolamide, a cannabinoid receptor agonist. *Biochim Biophys Acta* 1254: 127–34.

van der Stelt M, van Kuik JA, Bari M, van Zadelhoff G, Leeflang BR, Veldink GA, Finazzi-Agro A, Vliegenthart JF, and Maccarrone M (2002) Oxygenated metabolites of anandamide and 2-arachidonoylglycerol: conformational analysis and interaction with cannabinoid receptors, membrane transporter, and fatty acid amide hydrolase. *J Med Chem* 45: 3709–20.

Varma N, Carlson GC, Ledent C, and Alger BE (2001) Metabotropic glutamate receptors drive the endocannabinoid system in hippocampus. *J Neurosci* 21: RC188.

Wagner JA, Varga K, Jarai Z, and Kunos G (1999) Mesenteric vasodilation mediated by endothelial anandamide receptors. *Hypertension* 33: 429–34.

Ward SJ, Baizman E, Bell M, Childers S, D'Ambra T, Eissenstat M, Estep K, Haycock D, Howlett A, Luttinger D, and et al. (1991) Aminoalkylindoles (AAIs): a new route to the cannabinoid receptor? *NIDA Res Monogr* 105: 425–6.

Wiley JL, Ryan WJ, Razdan RK, and Martin BR (1998) Evaluation of cannabimimetic effects of structural analogs of anandamide in rats. *Eur J Pharmacol* 355: 113–8.

Wilson RI, Kunos G, and Nicoll RA (2001) Presynaptic specificity of endocannabinoid signaling in the hippocampus. *Neuron* 31: 453–62.

Wilson RI and Nicoll RA (2001) Endogenous cannabinoids mediate retrograde signalling at hippocampal synapses. *Nature* 410: 588–92.

Wilson RI and Nicoll RA (2002) Endocannabinoid signaling in the brain. *Science* 296: 678–82.

Wilson SR, Cui W, Moskowitz JW, and Schmidt KE (1988) Conformational analysis of flexible molecules: location of the global minimum conformation by the simulated annealing method. *Tet Lett* 29: 4373–4376.

Wise ML, Soderstrom K, Murray TF, and Gerwick WH (1996) Synthesis and cannabinoid receptor binding activity of conjugated triene anandamide, a novel eicosanoid. *Experientia* 52: 88–92.

Yang DP, Banijamali A, Charalambous A, Marciniak G, and Makriyannis A (1991) Solid state 2H-NMR as a method for determining the orientation of cannabinoid analogs in membranes. *Pharmacol Biochem Behav* 40: 553–7.

Yang DP, Mavromoustakos T, Beshah K, and Makriyannis A (1992) Amphipathic interactions of cannabinoids with membranes. A comparison between delta 8-THC and its O-methyl analog using differential scanning calorimetry, x-ray diffraction and solid state 2H-NMR. *Biochim Biophys Acta* 1103: 25–36.

Part II

Cannabinoid Receptor Genetics and Signal Transduction

3 Endocannabinoid Receptor Genetics and Marijuana Use

Emmanuel S. Onaivi, Hiroki Ishiguro, Ping Wu Zhang, Zhicheng Lin, Babatunde E. Akinshola, Claire M. Leonard, Sanika S. Chirwa, Jianping Gong, and George R. Uhl

CONTENTS

INTRODUCTION

Certain parts of this chapter represent an update of our continuous quest for an understanding of the biological basis of marijuana use. A more comprehensive review of cannabinoid receptor genetics was previously presented as "Endocannabinoids and Cannabinoid Receptor Genetics" in Onaivi et al., 2002. Here, we provide updates and new data on the endocannabinoid genetic basis of marijuana use. Endocannabinoids are endogenous marijuana-like substances that are synthesized in animals, including humans, and induce pharmacological responses that are similar to those produced by cannabinoids, e.g., THC, which exists in the marijuana plant (*Cannabis sativa*). Usually receptors are named after the endogenous molecules that activate them, such as dopamine for dopamine receptors or serotonin for serotonin receptors, etc. Endocannabinoid receptors could therefore be the nomenclature used for receptors activated by endocannabinoids, which have been identified in a variety of invertebrates and vertebrate species. Therefore, endocannabinoid and cannabinoid receptors are used interchangeably in this manuscript. Marijuana has remained one of the most widely used and abused drugs in the world. The recreational and medicinal properties of cannabis-derived preparations have been known for centuries. Cannabinoids are the constituents of the marijuana plant of which the principal psychoactive ingredient is Δ^9-tertrahydrocannabinol (Δ^9-THC). Although research on the molecular and neurobiological bases of the physiological and neurobehavioral effects of marijuana use was slowed by the lack of specific tools and technology for many decades, much progress has been achieved in marijuana-cannabinoid and endocannabinoid research. A central feature of this progress has been the elucidation of the cDNAs and genes that encode G-protein-coupled receptor (GPCR) CBRs. This has facilitated discoveries of endogenous ligands (endocannabinoids), which in turn has led to the use of endocannabinoids to help to define other potential GPCR and even ligand-gated channel cannabinoid receptors. The current knowledge about CBRs has been associated with mediating most of the psychoactive effects of marijuana, other neurobehavioral alterations, and the bulk of the cellular, biochemical, and physiological effects of cannabinoids (Childers and Breivogel, 1998). The signaling pathways associated with cannabinoids and endocannabinoids is complex with conflicting examples. Briefly, both CBRs inhibit adenylate cyclase and N- and Q- type calcium channel activity, and stimulate potassium channel conductance. In addition to cAMP-dependent MAP kinase activation, cannabinoid and endocannabinoid-induced signaling events are further complicated by the variability in potency to link the receptors with G_i- and G_o- proteins (Berdyshev, 2000).

Two cannabinoid receptor subtypes have been cloned. However, numerous studies have recently indicated the existence of many novel cannabinoid receptor subtypes CBn or CNRn (where n = 1 – 12; with as many as 15 subtypes of cannabinoid receptors without molecular but functional identity). This complexity is currently due to our lack of understanding of the regulation of the cannabinoid receptor genes and the molecular identity of other subtypes and splice variants. The known and cloned cannabinoid receptors are designated as CNR1 and CNR2 or CB_1R and CB_2R. They belong to the large super family of receptors that couple to guanine–nucleotide-binding proteins and that thread through cell membranes seven times (heptahelical receptors). The CB_1R is predominantly expressed in the brain and spinal cord, and thus is often referred to as the brain CBr. A spliced variant of human CB_1, cDNA, CB_1A, was isolated and characterized (Rinaldi-Carmona et al., 1996; Shire et al., 1995). It appears that the CB_1A resulted from the excision of an intron at the 5′-extremity of the coding region of the human receptor mRNA (Shire et al., 1995). These subtypes of CBRs were stably expressed in Chinese hamster ovary cell lines, and amino-truncated and modified CB_1 isoform CB_1A was shown to exhibit all the pharmacological properties of CB_1R to a slightly attenuated extent (Rinaldi-Carmona et al., 1996). A study of the distribution of the human CB_1A mRNA by reverse transcriptase PCR showed the presence of minor quantities of this isozyme together with CB_1R throughout the brain and in all the peripheral tissues examined (Shire et al., 1996). Although the presence of CB_1A has not been detected in rodents, evidence for the existence of unknown cannabinoid receptor subtypes in mouse brain has been described (Onaivi et al., 1996; Pertwee, 1999; Hajos et al., 2001; Breivogel et al., 2001; Wiley and Martin, 2002; Kofalvi et al., 2003) and in this chapter. The CB_2R is, at times, referred to as the peripheral CNR because of its largely peripheral expression in immune cells. Results of new studies indicate that CB_2Rs or CB_2-like receptors are selectively expressed in the normal brain and when required, particularly during pathological events (Benito et al., 2003) and in mast cells (Samson et al., 2003). CB_1 and CB_2 receptor gene products are expressed in relative abundance in specific tissues and cell types (Matsuda et al., 1993; Herkenham et al., 1991; and Bouaboula et al., 1993). The CB_1R is expressed at relatively high levels in brain regions such as hippocampus and cerebellum, and expressed at low levels in peripheral tissues including spleen, testis, and leucocytes. Das et al. (1995) demonstrated that the CB_1 mRNA but not CB_2 mRNA are expressed in the mouse uterus, where endocannabinoids can also be synthesized. Although there is extensive distribution and localization of the CB_1 receptors in the brain and peripheral tissues and in the immune system, the predominant cannabinoid receptors in the immune tissues are CB_2Rs. The abundant expression of CB_2Rs in peripheral tissues including white blood cells (Munro et al., 1993; Facci et al., 1995), B cells, and in natural killer (NK) cells may be related to the established alteration of the function of these cells by cannabinoids (Matias et al., 2002). cDNA sequences encoding the rat (Matsuda et al., 1990), human (Gerard et al., 1991; Munro et al., 1993), mouse (Chakrabarti et al., 1995; Abood et al., 1997), bovine (Wessner, Genebank submission, 1997), feline (Gebremedhin et al., Genebank submission, 1999), puffer fish (Yamaguchi et al., 1996a), leech (Stefano et al., 1997), and newt (Soderstrom et al., Genebank submission, 2000) CB_1- or CB_2-like receptors have been reported. The CB_1R is highly conserved across species, whereas the CB_2R shows more cross-species variation. Human CB_1 and CB_2 receptors share 44% overall amino acid identity. Although this might suggest significant overall evolutionary divergence, the receptors' amino acid identities range from 35% to as high as 82% in different CB_1 transmembrane regions (Shire et al., 1999).

The identification of endogenous ligands for CBRs has focused on modified eicosanoid-like fatty acids. Devane et al. (1992) named the first endocannabinoid to be discovered "anandamide" after the Sanskrit word for "bliss." The second principal endocannabinoid ligand 2-arachidonyl-glycerol (2-AG) was identified by Mechoulam et al., 1995, and Sugiura et al., 1995. The third endocannabinoid in the ether series is 2-arachidonyl glyceryl ether (noladin ether), isolated from porcine brain by Hanus et al., 2001. An endocannabinoid virodhamine, with antagonist activity at the CB_1 receptor, has also been identified and characterized (Porter et al., 2002). Virodhamine was shown to be present in rat and human brain and in peripheral tissues that express CB_2Rs where it acted as a full agonist (Porter et al., 2002). Potent synthetic cannabinoid agonists and antagonists have also been developed (Shire et al., 1999). Examples of natural and synthetic cannabinoids and endocannabinoids are presented in Table 3.1.

This chapter discusses the current state of description of the genes encoding CBRs, from their serendipitous identification to the existence of an EPCS. This previously unknown but ubiquitous EPCS consists of the membrane cannabinoid receptors, their ligands, endocannabinoids that are known to act as retrograde messengers, and the associated proteins for their biosynthesis, e.g., phospholipase D, and for their inactivation, e.g., fatty acid amide hydrolase (FAAH) and monoacylglycerols.

This chapter examines the endocannabinoid genetic basis of marijuana use; we also define some of the limitations of current knowledge. The molecular biology and pharmacology of the CBRs is still less well understood than that of many GPCRs. Only scanty information describes how these cannabinoid receptor genes are regulated. A moderate store of data describes some of the complex signal transduction pathways engaged by cannabinoid receptor activation (Onaivi et al., 2002). However, many topics, including the ways in which the abundant CB_1 receptors could alter activities mediated through other coexpressed GPCRs by sequestering G proteins and other means, are still in their infancy. As we discuss these GPCR–CBrs, we also need to be aware of the possible ligand-gated ion channels influenced by cannabinoids and interactions at the vanilloid receptors. We need to bear in mind data suggesting that cannabinoids can exert receptor-independent effects on biological (Hillard et al., 1985, and Makriyannis et al., 1989) and enzyme systems such as protein kinase C (PKC). Despite these caveats, it is impressive to consider the large amount of data about cannabinoid receptors amassed over the last decade, which has enhanced our knowledge of the physiological factors that trigger and contribute to the cycle of marijuana use.

Cannabinoid research and the use of cannabis products continue to attract significant attention. The current dramatic advances in molecular biology and technology that increased scientific knowledge in cannabinoid research will certainly contribute to a better policy on the medical use of marijuana. For example, preliminary studies with CBR antagonists have contributed to resolving

TABLE 3.1

	Marijuana Cannabinoids		Endocannabinoids	
	Natural (Plant)	Synthetic	Natural	Synthetic
Agonist	Δ^9-THC	Dronabinol	Anandamide (AEA)	Methanandamide
	Δ^8-THC	Nabilone	2-Arachidonyl-Glycerol (2-AG).	Linoleyl-ethanolamide
	Cannabinol	1′1dimethylhepyl-Δ^8-THC-oic-acid (CT3)	Noladine ether	2-methylarachidonoyl-2′-fluoroethylamide.
	Cannabidiol	WIN55212-2	*N*-archidonyl-dopamine(NADA)	ACEA
	Cannabigerol	CP55940	2-sciadonoylglycerol (2-SG)	ACPA
		HU210		O1812
				L759633
				L759656
				JWH133
				JWH015
				HU308
				Indomethacin-morpholinylamide
Antagonist		SR144528	Virodhamine	
		Rimonabant		
		LY320135		
		SR147778		
		AM630		
		AM251		
Entourage compounds	Palmitoylethanolamide; oleoylethanolamide			

the long-standing debate about addiction to marijuana. Specifically, the controversial question of physical dependence on psychoactive cannabinoids has now been resolved, using the antagonist rimonabant, also known as SR141716A, to precipitate withdrawal reactions in rats injected with increasing doses of Δ^9-THC. Aceto et al. (1995) reported a precipitated withdrawal syndrome that was absent in control animals, providing evidence that Δ^9-THC could produce physical dependence. In general, it had been claimed that the psycho activity and euphoria induced by cannabinoids limit their use in the clinic for numerous therapeutic applications for which they are currently being evaluated. With the availability of these genes, gene products, and other CBR research tools, it is speculated that the properties of these genes and regulation will be intensely studied so as to reveal how the psycho activity can be dissociated from the therapeutic properties of marijuana and cannabinoids, or to demonstrate that certain therapeutic actions of marijuana and cannabinoids cannot be separated from their psycho activity.

BIOLOGY OF ENDOCANNABINOIDS

The earlier hypothesis concerning the mechanism of action of Δ^9-THC, the major psychoactive component in marijuana, by general membrane perturbation may not be completely abandoned with the discovery of specific cannabinoid receptors on cell membranes that are activated by endogenous ligands. This is because these endogenous ligands and the cannabinoid receptors they activate may be derived from lipid precursors that are associated with lipid raft domains in the plasma membrane. If some components of the EPCS are localized in lipid raft microdomains, this may be important for the complex nature and multiple signaling mechanisms associated with the regulation of cannabinoid receptors following the activation of these receptors or after smoking marijuana and release of endocannabinoids. The characterization of lipid raft microdomains in the action of cannabinoids *in vivo* is poorly understood, but the identification of mouse transient-receptor homologs, including VR1 and lipid rafts from spermatogenic cells and sperm has been reported (Trevino et al., 2001). *In vitro* and in some cells, lipid rafts contain, besides sphingolipids and cholesterol, all classes of glycerophospholipids (Rouquette-Jazdanian et al., 2002). Thus, lipid rafts that contain glycerophospholipids from which some components of the EPCS are derived may be involved in the signaling pathways associated with mediating the effects of endocannabinoids, cannabinoids, or marijuana use. Many proteins are also partitioned to lipid rafts that can affect their function (Brown, 2000), and the possible involvement of lipid rafts in the apoptosis induced by anandamide in a variety of cells has been demonstrated (Sarker and Maruyama, 2003), thus supporting a role for specialized membrane microdomains in anandamide signaling. A previous study had shown that smoking hashish alters various blood cell and human platelet phospholipids (Kalofoutis et al., 1980). It also showed that phosphatidylethanolamine was increased after hashish intake, and we now know that these may be endogenous marijuana-like substances called endocannabinoids, such as anandamide and 2-AG. Therefore, membrane lipid rafts may be involved in mediating some of the biological actions of endocannabinoids after smoking marijuana. Anandamide is an endogenous ligand at both VR1 and cannabinoid receptors, and a component of the EPCS that may be associated with lipid rafts. In addition to the noncannabinoid-receptor-mediated effects, the unraveling of novel signaling pathways linked to the activation of different components of the EPCS along with the cross talk of endocannabinoids with a number of ion channels may warrant a fresh look at cannabinoid-induced membrane perturbation. Since the first plant-derived cannabinoid, Δ^9-THC, was structurally defined and synthesized over 30 yr ago, analogs have been synthesized exogenously which include synthetic cannabinoids CP55940, HU210, and WIN55212 (an aminoalkylindole), all of which bind to receptors (CB_1, CB_2...CB_n), inducing cannabimimetic activity (see Table 3.1). There was considerable speculation as to whether or not endogenous ligands existed for the cloned cannabinoid receptors. Endogenous ligands for mammalian cannabinoid receptors and their entourage ligands have been discovered and characterized. These endocannabinoids include anandamide, 2-AG, noladin ether, and virodhamine (Table 3.1). The occurrence of a novel cannabimimetic molecule 2-sciadonoylglycerol (2-SG) in the plant seeds of umbrella pine (*Sciadopitys verticillata*) has also been reported (Nakane et al., 2000). 2-SG was found to have effects on the CB_1R similar to, but with lower activity than, 2-AG, demonstrating the occurrence of these interesting molecules, not only in plants and animals but also in disparate organisms such as ticks. This widespread occurrence of endocannabinoids and related fatty acid amides and their receptors appears to be highly conserved in nature, indicating a fundamental role in biological systems. For example, the salivary glands of ticks, which are ectoparasitic and obligate blood-feeding arthropods, can make endocannabinoids and their congeners with analgesic and anti-inflammatory activity, which possibly participate in the inhibition of the host defense reactions (Fezza et al., 2003). Apparently, the EPCS plays a critical role in the survival and mechanisms of cell death. Previously, the existence of anandamide analogs in chocolate had been demonstrated (di Tomaso et al., 1996). It is thought that chocolate and cocoa contain N-acylethanolamines, which are chemically and pharmacologically related to anandamide. These lipids could mimic cannabinoid ligands either directly by activating CBRs or indirectly by increasing anandamide levels (Bruinsma and Taren, 1999). These observations demonstrate that endocannabinoid analogs exist in plants and animals and further illustrate the evolutionary conservation of the cannabinoid system in nature. In this section, we will briefly review the properties and functions of these endocannabinoids. Thus, the EPCS represented by CBrs, endocannabinoids, and enzymes for the biosynthesis and degradation of these ligands is conserved throughout evolution. Endocannabinoids are present in peripheral as well as in brain tissues and have recently been demonstrated to be present in breast milk. In addition, the recent demonstration of the expression of functional CB_1R in the preimplantation embryo and synthesis of anandamide in the pregnant uterus of mice suggested that cannabinoid ligand-receptor signaling is operative in the regulation of preimplantation embryo development and implantation (Paria and Dey, 2000). 2-AG has been characterized as a unique molecular species of the monoacylglycerol isolated from rat brain and canine gut as an endogenous CBR ligand (Sugiura and Waku, 2000). 2-AG also exhibits a variety of cannabimimetic activities *in vitro* and *in vivo*, and clearly further studies are necessary to determine the relative importance of 2-AG and anandamide in the human body and brain. This is because the levels of anandamide (800 times lower than the levels of 2-AG) found by some investigators in several mammalian tissues, and its production mainly in the postmortem period in the brain, have led to questions about the physiological significance of anandamide, especially in the brain, despite its high-affinity binding to CBRs (Sugiura and Waku, 2000). These research findings undoubtedly have advanced cannabis research and have allowed us to hypothesize that the EPCS consists of a previously unrecognized but elaborate network of endocannabinoid neuromodulators complete with their accompanying biosynthetic, uptake, and degradation pathways just like the monoaminergic or opioidergic systems.

Because of its rapid progress and the transformation of marijuana research into mainstream science, a number of reviews of the effects of endocannabinoids in various *in vivo* and *in vitro* systems have appeared (Kozak et al., 2004; Walter and Stella, 2004; Park et al., 2003; Castellano et al., 2004; Piomelli, 2003). Briefly, a role for endocannabinoids has been suggested in brain development through the activation of second-messenger-coupled cannabinoid receptors (Fernandez-Ruiz et al., 2000). The evidence for a role of endocannabinoids in neural development was derived from studies in which the presence of CBRs and endocannabinoids during brain development and in neuronal or fetal glial cell cultures was shown to produce cellular responses. These responses, along with the neurotoxicological changes produced in pups from pregnant rats treated with phytocannabinoids, indicate the existence of a endocannabinoid system early in the development of the CNS (Fernandez-Ruiz et al., 2000). The stimuli for the production and release and the general physiological roles and significance of endocannabinoids are incompletely understood. Their interaction with CBRs produces a wide range of effects. The administration of endocannabinoids to experimental animals produces several of the pharmacological and behavioral actions associated with cannabinoids

(Onaivi et al., 1996; Martin et al., 1999; Salzet et al., 2000). Nevertheless, there appears to be cannabinoid-dependent and cannabinoid-independent actions of the endocannabinoids. Evidence for non-CB_1, non-CB_2, CNR-mediated actions of anandamide in CB_1 knockout mouse brain was demonstrated by Di Marzo et al. (2000a). They reported that anandamide levels in the hippocampus and the striatum were lower in the CB_1 knockout homozygotes (CB_1–/–) than in the wild-type (CB_1+/+), whereas there was no change in the 2-AG levels in both the homozygotes and wild-type controls. In addition, the effects of anandamide, unlike Δ^9-THC, were not decreased in CB_1–/– mice. Furthermore, anandamide, but not Δ^9-THC, stimulated GTPγS binding in brain membranes from CB_1–/– mice, and this stimulation was insensitive to CB_1 and CB_2 antagonist. It was, therefore, suggested that non-CB_1, non-CB_2, G-protein-coupled receptors might mediate some of the behavioral actions of anandamide in mice. Thus, exo- and endogenous cannabinoids exert pleiotropic actions in the human brain and body.

Other supporting evidence that some of these effects involve non-CB_1 and non-CB_2 actions of anandamide is its activation of the capsaicin, the VR1 vanilloid receptor. It is also conceivable that there are other CBRs (e.g., CB_3, CB_4, CB_5...CB_n), which are yet to be identified and characterized, as discussed in this chapter. In the absence of such multiplicity of cannabinoid receptor subtypes, perhaps posttranscriptional modification of cannabinoid receptor gene expression yields splice variants that are involved in the myriad behavioral, physiological, and biochemical effects of smoking marijuana or follow the administration of cannabinoids. Interestingly, there are some pharmacological differences between the plant-derived THC and endocannabinoids that could be due to the pharmacodynamic or pharmacokinetics profiles. Some of the similarities between the classical cannabinoids and the endocannabinoids, which are structurally dissimilar, have been reviewed (Salzet et al., 2000). These include neuromodulatory effects through which they influence motor behavior, memory and learning, and sensory, autonomic, and neuroendocrine responses. A pathological consequence of marijuana use is the modulation of immune responses, and endocannabinoids are physiological immune regulators (Samson et al., 2003). Endocannabinoids also induce hypotension and bradycardia, inhibit cell growth, affect energy metabolism, and modulate immune and inflammatory responses (Onaivi et al., 1996; Martin et al., 1999; Salzet et al., 2000; Sugiura and Waku, 2000).

The pathways for the synthesis and degradation of endocannabinoids have also been studied intensively. The endocannabinoids are unique from other neurotransmitters in that they are not known to be stored in vesicles but, rather, are synthesized on demand. In the nervous and immune systems the endocannabinoids, 2-AG, and anandamide, appear to be derived from the hydrolysis of phospholipid precursors from membrane phosphoglycerides. Anandamide can be generated by hydrolysis of N-arachidonoyl phoshatidylethanolamine (N-arachidonoyl PE) catalyzed by phospholipase D (PLD). An N-acyl transferase activity (NAT) mediates the synthesis of new N-arachidonoyl PE by detaching an arachidonate moiety from the sn-1 position of other phospholipids — such as phosphatidylcholine (PC) — and by transferring it to the primary amino group of PE (Giuffrida and Piomelli, 2000). The NAT/PLD action is known to give rise to a family of saturated and monounsaturated acylethanolamides, such as palmitoylethanolamide and oleylethanolamide (Schmid et al., 1996). The anandamide formed is released into extracellular space where it can activate the CBR. As described below, functional studies show that anandamide is inactivated by a carrier-mediated transport (AT), which can be inhibited by the transport inhibitor AM404. In the cells, anandamide is hydrolyzed into arachidonic acid and ethanolamine by a membrane-bound anandamide amidohydrolase (AAH).

Although anandamide has been extensively studied as a ligand for CBRs, Sugiura and Waku (2000) and co-workers believe that 2-AG, but not anandamide, may be the natural ligand for both the CB_1 and the CB_2 CBRs. Based on structure activity relationships and the ability of endocannabinoids to induce Ca^{2+} transients in cells expressing CBRs, Sugiura and Waku (2000) surmised that 2-AG appears to be the intrinsic optimum ligand among the compounds tested at the CB_1 and CB_2 CBRs. The generation of 2-AG from different types of cells under a number of conditions had been known for several years before it was recently established as endocannabinoid ligand

(Sugiura and Waku 2000). Therefore studies of the formation and inactivation of 2-AG indicate that the biosynthetic pathways for 2-AG appear to differ, depending on the types of tissues and cells and the types of stimuli (Sugiura and Waku, 2000; Giuffrida and Piomelli, 2000). 2-AG may be degraded like other monoacylglycerols or like anandamide, perhaps under different conditions. Unlike the classical neurotransmitters, 2-AG and anandamide may be produced upon demand by receptor-stimulated cleavage of membrane lipid precursors and released from cells after their production, using similar but distinct receptor-dependent pathways, (Piomelli et al., 2000). It can, therefore be suggested that the nonsynaptic release of endocannabinoids and their short lifespans may be one determinant of their functional roles *in vivo*.

Nuclear lipid signaling is an established, widespread mechanism that operates in multiple cellular processes in response to a variety of stimuli (Gilmore and Mitchell, 2001). Knowledge of the role of endocannabinoids and their precursors in nuclear lipid signaling is scanty. A better understanding of their involvement will shed light on the significance of the EPCS. The possible involvement of lipid rafts in the action of endocannabinoids, along with the localization of ion channels to lipid raft domains, implies a role for lipid rafts in the biological and physiological effects of endolipids and endocannabinoids that is poorly understood. If receptors and ion channels activated by endocannabinoids, marijuana use, and other cannabinoids are in these lipid rafts, then Simons and Toomre (2000) suggest that localization and protection of cell signaling will allow these rafts to effectively act as platforms for receptors activated by lipid-signaling molecules like endocannabinoids, endovanilloids, and other endolipids. Although it is attractive to speculate on how cannabinoid receptors will behave if partitioned into lipid rafts, anandamide-induced apoptosis may be independent of cannabinoid or vanilloid receptors, but has been shown to involve lipid rafts (Sarker and Maruyama, 2003). Others have suggested a caveolae-related endocytic mechanism for the cellular uptake of anandamide because the depletion of cholesterol that disrupts caveolae and lipid rafts reduces the uptake of anandamide (McFarland and Barker, 2004). It has also been suggested that methanandamide induces cyclooxygenase-2 expression in human neuroglioma cells via a pathway linked to lipid raft microdomains (Hinz et al., 2004). The physiological significance is that this may help to explain the diverse biological processes that have been associated with the effects of marijuana use and possible therapeutic benefits. There is also increasing awareness that lipid-signaling molecules and endogenous lipid modulators (endolipids) may be involved in a variety of processes associated with substance abuse and dependency. These endogenous "lipid-based" molecules, such as endocannabinoids, endovanilloids, lysophospholipids, and the acyldopamine and arachidonoylglycine family of lipids, act on ligand-gated channels or GPCRs to regulate a number of peripheral and CNS functions. Furthermore, these endolipid mediators are involved in a variety of functions such as neuronal migration, neurogenesis, and neuroinflammatory effects. We have initiated studies to determine the nature of the interaction between cannabinoids and these endolipid molecules in drug and alcohol addiction.

Endocannabinoids have now been localized in the developing rat brain from early gestational (fetal) stages, suggesting a specific role for these ligands as part of the EPCS in brain development. It was quantitated in peripheral tissues such as spleen and skin, in cells such as peritoneal macrophages and lymphocytes, and in very small amounts in such body fluids as serum and cerebrospinal fluid, suggesting that it is metabolized in tissues where it is synthesized and the mechanism of action is not hormonal in nature. Anandamide levels in the brain may be equivalent to those of other neurotransmitters such as dopamine and serotonin, whereas less than that of GABA and glutamate. As anandamide has been shown to compete for binding to CB_1 in the brain in RIA assays, and as it was found in human and rat spleen, which express high levels of CB_2R, it is considered an agonist at both CB_1 and CB_2 CBrs. There are, however, some reports of vast affinity differences that govern the binding of anandamide to CB_1 and CB_2, with a much greater Ki for ligand binding to CB_1. With binding to the CB_1R, anandamide has been shown to inhibit adenylate cyclase *in vitro* in cell lines where the receptor occurs naturally and in transfected cell lines. In addition, binding of this ligand is capable of inhibiting

N-type calcium channels in cell lines, although to a much lesser extent than synthetic cannabinoids, suggesting that anandamide displays the properties of a partial cannabinoid CB_1 receptor agonist. Reports of the binding of anandamide to CB_2R are controversial, and inhibition of adenylate cyclase in CB_2R expression in transfected CHO cells has not been observed.

A review of the plausible functions of anandamide will include functional aspects that range from involvement in the immune responses of the body to acting as a potent neuromodulator. It has a role as a developmental regulator during the formation of the brain, as an analgesic, and in neurobehavioral modification. The question immediately arising after the discovery of anandamide was to what extent this fatty acid amide shares common pharmacological properties with the plant-derived Δ^9-THC. In summary, the many reports addressing this question concluded that anandamide shares many of the biological activities exhibited by THC, and the biological effects of anandamide include decreasing spontaneous motor activity, immobility, and production of hypothermia and analgesia. There are pharmacokinetic differences between the two substances. For example, when examining its effects on motor activity, anandamide has a faster onset and a shorter duration of action, most likely due to the rapid reuptake into neurons and astrocytes, and subsequent enzymatic breakdown. Anandamide was shown to be up to 20-fold less potent in producing hypomobility or decreased mobility than Δ^9-THC. It was demonstrated that THC potentiates analgesic effects produced by anandamide but not *vice versa* (Welch and Eads, 1999). It, therefore, has been suggested that anandamide and THC bind to the same receptor but activate it in distinctively different ways. A future direction for endocannabinoid research will likely include the elucidation of anandamide transduction mechanisms that may involve activation of Gs and Gi proteins, similar to that reported for the opioid receptors. The analgesic properties of cannabinoids have commanded considerable attention, because this class of analgesics has been shown to interact with other analgesics like opioids. In acute pain models in rodents, anandamide produces analgesia after systemic or local (paw) administration. This data proved that anandamide attenuates pain by interacting with CB_1-like receptors located outside the CNS as the CB_1 and not CB_2 CBR antagonist inhibited the analgesic effects. In addition, it was also shown that after coadministration of both anandamide and palmitylethanolamide, the antinociceptive potency of each compound was increased 100-fold. Palmitylethanolamide is an analgesic agent and a CB_1 agonist, which was found to be released together with anandamide from its putative precursor in the neurons of the PNS. Together, these data indicate that the simultaneous activation of peripheral CB_1 and CB_2 CBRs results in a synergistic inhibition of peripheral pain transmission, and it has been shown by mass spectrometry that the skin contains 5 to 10 times more anandamide and palmitylethanolamide than in the brain, high enough to cause activation of local cannabinoid receptors.

A number of biological activities by 2-AG have been reported, in immune function, in cell proliferation, embryo development, long-term potentiation (LTP) in the hippocampus, neuroprotection and neuromodulation, cardiovascular function, and inflammatory responses (for a review, see Sugiura and Waku, 2000), most of which have been demonstrated for anandamide. For example, anandamide and 2-AG are known to protect cerebral cortical neurons subjected to 8 h hypoxia and glucose deprivation, through a mechanism independent of CBrs. In an animal model of Parkinson's disease, the enhanced levels and presence of endocannabinoids in the basal ganglia were associated with movement disorders (Di Marzo et al., 2000b). This is not surprising because there is substantial evidence supporting a role for the cannabinoid system as a modulator of dopaminergic activity in the basal ganglia (for a review, see Giuffrida and Piomelli, 2000). Thus, available data point to a key role of the endogenous cannabinoid system in the regulation of psychomotor activity and suggest that this system may offer a therapeutic target in pathologies involving a dysregulation of dopamine neurotransmission (Giuffrida and Piomelli, 2000). Thus, in the last decade, rapid progress has been achieved with increased understanding of the EPCS. The endocannabinoids that serve as the endogenous marijuana-like substances have been shown to participate in a broad array of physiological and pathological processes. Whereas attention has been directed at defining the mechanism responsible for endocannabinoid synthesis, much research is still needed to unravel the molecular identity of the transport of these molecules. The exact role of endocannabinoids and other endolipids in vulnerability to marijuana use and dependence remains to be determined.

Biology of Cannabinoid Signaling

One area in cannabinoid research that has accumulated significant information and is associated with a tangled web of signaling is that of the signal transduction pathways which are involved with the actions of endocannabinoids, marijuana, and other cannabinoids. This may be related to the "promiscuous" actions of marijuana and cannabinoids on most biological systems in the human body and brain. This is not surprising as there is increasing evidence that the EPCS is abundantly distributed in the human body and brain. Although the biology of cannabinoid signaling is not our major focus in this chapter, attempts are made to provide an overview of the major signal transducers activated by endocannabinoids and cannabinoids that may be relevant to marijuana use. For a detailed review, see Pertwee (1997), Howlett and Mukhopdhyay (2000), Elphick and Egertova (2001), and Freund et al. (2003). Briefly, both CB_1 and CB_2 CBRs are coupled with Gi/o protein, negatively to adenylate cyclase and positively to mitogen-activated PK. CB_1R coupling to the G protein signal transduction pathways in presynaptic nerve terminals transduces the cannabinoid stimulation of mitogen-activated protein kinase (MAP kinase) and the inhibition of adenylate cyclase, thereby attenuating the production of cyclic AMP. CB_1Rs are also coupled to ion channels through Gi/o proteins, positively to A-type and inwardly rectifying potassium channels, and negatively to N-type and P/Q-type calcium channels and to D-type potassium channels (Pertwee, 1997; Howlett and Mukhopadhyay, 2000). The coupling to A-type and D-type potassium channels is thought to be through adenylate cyclase (Mu et al., 1999). These are also stimulated via the inhibition of adenylate cyclase by cannabinoids. Due to the decrease in cAMP accumulation, cAMP-dependent PK (PKA) is inhibited by CB_1R activation. In the absence of cannabinoids, PKA phosphorylates the potassium channel protein, thereby exerting decreased outward potassium current. In the presence of cannabinoids, however, the phosphorylation of the channel by PKA is reduced, which leads to enhanced outward potassium current. In addition, cannabinoids can close sodium channels, but whether or not this effect is receptor mediated has yet to be established. Other cellular responses triggered with CB_1 receptor activation include other PK pathways such as the p38 MAKP, JUN N-terminal kinase, (JNK), focal adhesion kinase (FAK), and phosphoinositide-3 kinase (Akt) signaling, which might mediate effects on apoptosis (Molina-Holgado et al., 2002). CB_1R stimulation has also been linked to activation of the extracellular, signal-regulated kinase cascade through ceramide signaling, as discussed below. Based on these findings, it has been suggested that endocannabinoids play a role in regulation of neurotransmitter releases, also discussed below.

The CB_2R is coupled, in addition, to a G protein and thereby negatively coupled to adenylate cyclase. Inwardly rectifying potassium channels can also serve as a signaling mechanism for the CB_2R, at least in *Xenopus* oocytes that have been transfected with such channels together with the CB_2R (Ho et al., 1999; McAllister et al., 1999). There is also evidence from experiments with rat hippocampal CA1 pyramidal neurons that CB_1Rs are negatively coupled to M-type potassium channels (Schweitzer, 2000). CB_1Rs may also mobilize arachidonic acid and close 5-HT3 receptor ion channels (Pertwee, 1997), and under certain conditions couple to Gs proteins to activate adenylate cyclase (Calandra, et al., 1999) and/or to reduce outward potassium K current, possibly through arachidonic acid-mediated stimulation of PKC (Hampson et al., 2000). The questions of whether CB_1R coupling to Gs proteins has physiological importance and of whether such coupling increases after Gi/o protein sequestration by colocalized noncannabinoid Gi/o protein-coupled receptors have yet to be resolved. CB_1Rs have also been reported to be positively coupled to phospholipase C through G proteins in COS7 cells cotransfected with CB_1Rs and $G\alpha$ subunits (Ho et al., 1999) and negatively coupled to voltage-gated L-type calcium channels in cat cerebral arterial smooth muscle cells (Gebremedhin et al., 1999). CB_1Rs on cultured cerebellar granule neurons can operate through phospholipase C-sensitive mechanism to enhance NMDA-elicited calcium release from inositol 1,4,5-triphosphate-gated intracellular stores (Netzeband et al., 1999).

Besides the well-established cannabinoid signal transduction mechanisms, Guzman et al., (2001) have shown that the CB_1R triggers the activation of other G-protein-coupled signaling systems such as PKB/Akt and C-Jun N-terminal kinase (Gomez Del Pulgar et al., 2000). In addition, they have found that CB_1R activation leads to the generation of ceramide (Guzman et al., 2001). This ubiquitous lipid second messenger is known to play an important role in the control of cell fate in the CNS. The studies showed that cannabinoid-dependent ceramide generation occurs via a G-protein-independent process and involves two different metabolic pathways: sphingomyelin hydrolysis and ceramide synthesis *de novo*. Ceramide in turn mediates cannabinoid-induced apoptosis, as evidenced by *in vitro* and *in vivo* studies. Thus, the CB_1R of astrocytes was shown to be coupled to sphingomyelin hydrolysis through the adaptor protein (FAN), a factor associated with neutral sphingomyelinase activation (Sanchez et al., 2001).

Cannabinol (CBN), a ligand with selective affinity for the CB_2R-mediated inhibition of CREB, NF-kB, and IL-2, can inhibit several signaling pathways in activated T cells (Herring et al., 2001). Another important pathway involved with the action of endocannabinoids and cannabinoids is the nitric oxide pathway. The differential role of the nitric oxide pathway on Δ^9-THC-induced effects has been investigated in nitric oxide synthase (nNOS) knockout mice and wild-type control (Azad et al., 2001), in invertebrate immunocytes, microglia, and human monocytes (Stefano et al., 1996). These studies demonstrate that CBRs are also coupled to nitric oxide release in some but not all of the effects of cannabinoids.

Retrograde synaptic signaling, which has long been recognized as a fundamental feature of neural systems, has now been established as a mechanism of synaptic regulation in the brain, and endocannabinoids have emerged as one of the classes of retrograde messengers involved in the regulation of synaptic transmission (see Alger, 2002, for a review). Thus, endocannabinoids mediate the rapid retrograde suppression of both excitatory and inhibitory synapses. Functionally, endocannabinoids are released from depolarized postsynaptic neurons in a calcium-dependent manner and act retrogradely onto presynaptic CBRs to suppress neurotransmitter release (Kano et al., 2002). While a physiological role of retrograde signaling by endocannabinoids may be to provide a mechanism by which neurons can rapidly regulate the strength of their inputs (Kreitzer and Regehr, 2002), the mechanisms of the regulation, as well as the relationship between retrograde signal release and physiological activity remain to be determined (Alger, 2002). The molecular logic of endocannabinoid signaling was reviewed by Piomelli (2003), and their ability to modulate synaptic efficacy in the brain occurs following the activation of cannabinoid receptors on axon terminals to regulate ion channel activity and neurotransmitter release (Piomelli, 2003). It is conceivable that smoking marijuana may induce similar physiological changes that may be exploited for a number of therapeutic possibilities.

Neurobehavioral and *In Vitro* Actions of Cannabinoids

The gateway hypothesis of marijuana, which proposes that marijuana use leads to the use of harder drugs such as heroin and cocaine, may indeed have a biological basis. However, solid scientific data is required to link marijuana use with the use of other addictive substances. If it is established that initial marijuana use activates the reward system with the involvement of the EPCS for subsequent transition to substance use, then this EPCS may be part of the solution in the treatment of the compulsive use of addictive substances. This is because different levels of the EPCS, e.g., at the genetic, receptor, uptake, or enzymatic hydrolysis levels, could become therapeutic targets for treating addictive behaviors. This may also provide a basis to obtain deeper insight into an endocannabinoid hypothesis of reward and reinforcement. Accumulating evidence supporting a central role of an endocannabinoid system in the rewarding effects of abused substances further weakens the traditional dopamine hypothesis. Perhaps the EPCS might be a better target in the treatment of addictive behaviors than manipulating different components of the dopamine system, which has not yielded any successful therapeutic target for substance abuse treatment after over 40 yr of research. For example, WIN55212-2 has been shown to decrease the reinforcing actions of cocaine through CB_1R stimulation (Vlachou et al., 2003), and overeating and alcohol and sucrose consumption is known to be decreased in CB_1R-deleted mice (Poncelet et al., 2003), implicating a major role for the EPCS in the effects of drug and alcohol consumption. It appears that inducing the release of dopamine in the brain is not a sufficient requirement mediating addiction, because mutant mice without dopamine receptors and transporters continue to self-administer cocaine, whereas mutant mice without the mGluR5 are completely unresponsive to cocaine even though their dopaminergic systems remain intact (Chiamulera et al., 2001). Other studies show that *in vivo* blockade of the cannabinoid CB_1 receptor or ablation of the CB_1R gene alters the rewarding effects of abused substances (Onaivi et al., 2002) and that smoking marijuana interferes with short-term memory. CB_1R also plays some, but not an exclusive, role in the regulation of appetitive behavior as cannabinoid receptor agonists stimulate food consumption in animals and humans (see Black, 2004). The brain levels of endocannabinoids increase with greater demand for food in rodents and rimonabant; the CB_1R antagonists decrease food intake in rodents and animals and have shown promise in decreasing appetite and weight gain in humans (Black, 2004). This new knowledge and the advances in substance abuse research and identification of the EPCS may help drive this field forward.

We and others have attempted to define the behavioral effects of endocannabinoids and cannabinoids in the mouse model, and are intrigued by the larger ambition of laying the groundwork for future evaluation of the behavioral, molecular, and genetic basis for the interaction between cannabinoids with other abused substances. As reviewed by Martin et al. (1999), the immediate question that arose following the discovery of endocannabinoids was whether or not they share similar pharmacological actions with the prototypical plant-derived cannabinoid, Δ^9-THC. There is now overwhelming evidence that the endocannabinoids interact with the CBRs and share some of the biological properties of other cannabinoids with significant differential effects. These significant differential effects involve other non-CBR systems and unknown CBRs as described below. Despite the decades of extensive investigations and recent developments in cannabinoid research, the identification of specific mechanisms for the actions of cannabinoids has been slow to emerge. We, therefore, do not attempt to provide a comprehensive account of the numerous *in vivo* and *in vitro* effects of cannabinoids but just a few examples from our studies and those of others. The discovery of endocannabinoids such as anandamide and 2-AG, and the widespread localization of CBRs in the brain and peripheral tissues, suggest that EPCS represents a previously unrecognized, ubiquitous network in the nervous system, whose biology and function is unfolding. We have tested the hypothesis that some of the actions of anandamide are independent of a CBR mechanism (Akinshola et al., 1999b). In the first series of experiments, the effects of anandamide or methanadamide on behavior and CB_1 CBR gene expression in three mouse strains were determined. This was accomplished by the use of cannabinoid agonist and antagonist interaction in *in vitro* and *in vivo* test systems. The effects of acute administration of anandamide to C57BL/6, DBA/2, and ICR mice were evaluated in motor function and emotionality tests. The C57BL/6 and ICR mouse strains were more sensitive than the DBA/2 strain to the depression of locomotor and stereotyped behavior caused by anandamide. Although anandamide produced catalepsy in all three strains, anandamide induced ataxia in the minus-maze test only in the C57BL/6 animals and at the lowest dose used. In the plus-maze test, anandamide produced a mild aversive response, which became an intense aversion to the open arms of the plus-maze following repeated daily treatment. Northern analysis data using the CB_1 cDNA as a probe indicated that there was a more abundant expression of the CB_1 gene in the whole brain of the ICR mouse than in the brains of the C57BL/6 and DBA/2 strains, with or without pretreatment with anandamide. Because the anandamide-induced neurobehavioral changes did not correspond to the CB_1 CNR gene expression in the mouse strains, it is unlikely that the CB_1R mediates all the cannabimimetic effects of anandamide in the brain.

In vitro, we used *Xenopus laevis* oocytes and two-voltage clamp technique in combination with differential display polymerase chain reaction to determine whether the differential display of genes following treatment with anandamide may be linked to AMPA glutamate receptor. The differential expression of genes *in vivo* after the subacute administration of anandamide could not be directly linked with AMPA glutamate receptor. In the *in vivo* studies using rimonabant, the CB_1R antagonist-induced anxiolysis that was dependent on the mouse strain used in the anxiety model blocked the anxiogenic effects of anandamide or methanandamide, whereas rimonabant had no effect on the anandamide inhibition of kainate-activated currents *in vitro*.

We tested another hypothesis that there might exist in the central nervous system a multiplicity of CBRs. The basis for the hypothesis had been the pleiotropic effects produced after smoking marijuana or the administration of cannabinoids to humans and animals. We therefore studied the neurobehavioral specificity of CB_1R gene expression and whether Δ^9-THC-induced neurobehavioral changes are attributable to genetic differences (Onaivi et al., 1996). We also examined whether specific brain regions in the mouse model mediated some of these neurobehavioral changes. We found that the differential sensitivity following the administration of Δ^9-THC to three mouse strains, C57BL/6, DBA/2, and ICR mice, indicated that some of the neurobehavioral changes might be attributable to genetic differences. The objective of the study was to determine the extent to which the CB_1R is involved in the behavioral changes following Δ^9-THC administration. This objective was addressed by experiments using the following strategies: DNA-PCR and reverse PCR; systemic administration of Δ^9-THC; and intracerebral microinjection of Δ^9-THC. The site specificity of the action of Δ^9-THC in the brain was determined using stereotaxic surgical approaches. The intracerebral microinjection of Δ^9-THC into the nucleus accumbens was found to induce catalepsy, while injection of Δ^9-THC into the central nucleus of amygdala resulted in the production of an anxiogenic-like response. The reverse PCR data showed two additional distinct CB_1 mRNAs in the C57BL/6 mouse which also differed in pain sensitivity and rectal temperature changes following the administration of Δ^9-THC (Onaivi et al., 1996). We, therefore, suggested that the diverse neurobehavioral alterations induced by Δ^9-THC may not be mediated by CB_1Rs in the brain and that the CB_1R gene may not be uniform in the mouse strains used. The potential of antisense oligonucleotides as research tools and therapeutic agents has been the subject of close scrutiny and attention, particularly the application of gene therapy in the clinic. Thus, a number of problems have been identified with their use as research tools. Our serendipitous use of CB_2 antisense oligonucleotide indicated that CB_2 might be present in the brain to influence behavior. The ICV administration of the CB_2 antisense induced a significant antiaversive response in the elevated plus-maze test of anxiety, a response similar to that following the administration of the CB_1R antagonist rimonabant. Knowing, that this might be a nonspecific effect, it is interesting and provocative that CB_2R or CB_2-like CBRs might be in the brain. Although a number of laboratories have not been able to detect CB_2 expression in the brain, a demonstration of CB_2 expression in the rat microglial cells (Kearn and Hilliard, 1997) in cerebral granule cells (Skaper et al., 1996), mast cells (Facci et al., 1995), and recently in neurite plaques (Benito et al., 2003) have been reported. We utilized a CB_2 antisense, 5′-TGTCTCCCGGCATCCCTC-3′; CB_2 sense oligonucleotide was 5′-GAGGGATGCCGGGAGACA-3′ by stereotaxic intracerebral injection. This led to cannabinoid-induced anxiolysis (unpublished observation, Onaivi). While this observation requires confirmation, it reinforces the hypothesis of the presence of multiple CBRs in the brain as discussed below. The use of the antagonists that are selective for the CB_1 and CB_2 CBRs in these behavioral tests will also contribute to further understanding of the role of these CBR subtypes in the behavioral effects of cannabinoids. As discussed elsewhere, our recent studies show that CB_2 cannabinoid receptors, which are expressed in naïve mouse brain, are enhanced when the animals are subjected to chronic mild stress. If this is confirmed, this will add to our understanding of mechanisms associated with depression and drug addiction.

Other behavioral effects of cannabinoid agents in animal models have been reviewed by Chaperon and Thiebot (1999), and a review of the behavioral effects of endocannabinoids is presented in this book by Thiebot et al. Briefly, cannabinomimetics produce complex behavioral and pharmacological effects that probably involve numerous neuronal substrates. Interactions with acetylcholine, dopamine, serotonin, adrenergic, opiate, glutamatergic, and GABAergic systems have been demonstrated in several brain structures. In animals, cannabinoid agonists such as, WIN55212-2, and CP55940 produce a characteristic combination of four prototypic profiles, sometimes referred to as response to the tetrad tests, including catalepsy, analgesia, hypoactivity, and hypothermia. The selective CB_1R antagonist, rimonabant, providing evidence for the involvement of CB_1R-related mechanisms, reverses these effects. The fact that some investigators have reported the failure of rimonabant to antagonize the tetrad effects of anandamide has recently been explained. This is because the effects of the metabolically stable 2-methyl-2′-fluoroethylanandamide were blocked by rimonabant in the tetrad tests. This suggested that metabolism is at least partly responsible for anandamide refractoriness to the antagonist rimonabant (Martin et al., 1999).

Accumulating evidence indicates that endocannabinoids have cannabinoid and non-CBR mediated effects in these classical cannabinomimetic actions. CBR-related processes seem also involved in cognition, memory, anxiety, control of appetite, emesis, inflammatory, and immune responses. The cannabinoid agonist may induce biphasic effects; for example, hyperactivity at low doses and severe motor deficits at larger doses have been documented.

The conditioned place preference (CPP) paradigm has been used extensively to study brain mechanisms of reward and reinforcement. Marijuana (cannabinoid) interactions with the brain substrates for reward and reinforcement have been reviewed by Gardner (2002) and also by Tzschentke, 1998. Although the paradigm has been criticized because of some inherent methodological problems, it is clear that the place preference conditioning has become a valuable, firmly established, and widely used tool in addiction research (Tzschentke, 1998). The rewarding properties of cannabinoids and Δ^9-THC are difficult to demonstrate in rodents using standard place preference procedures, (Valjent and Maldonado, 2000; Tzschentke, 1998). Furthermore, only a few studies have examined the effects of marijuana and hashish, and inconsistent results have been reported. Sanudo-Pena et al. (1997) found no CPP at a low dose of THC (1.5 mg/kg) and a conditioned place aversion (CPA) at a high dose (15 mg/kg), whereas the CBR antagonist rimonabant induced a CPP at a low and high dose (0.5 and 5 mg/kg). In contrast, Mallet and Beninger (1998) found a significant CPA for the same low doses of THC (1.0 and 1.5 mg/kg), and neither CPP nor CPA was found for anandamide. Lepore et al. (1995), however, reported THC induced CPP for 2 and 4 mg/kg but not for 1 mg/kg, when animals received one conditioning session per day. However, when conditioning took place only every other day (to allow for a 24-h washout period for THC), the dose of 1 mg/kg THC was sufficient to produce CPP, whereas the higher doses (2 and 4 mg/kg) produced CPA. The synthetic cannabinoid CP55940 was reported to produce CPA, (McGregor et al., 1996). The synthetic CBR agonist WIN55212-2 produced a robust CPA, while the CB_1 antagonist SR141716A produced neither CPP nor CPA (Chaperon et al., 1998). The emerging consensus appears to be that cannabinoid antagonism produces CPP while cannabinoid agonism induces place aversion, (Sanudo-Pena et al., 1997). Taken together, the effects of cannabinoids in the CPP paradigm suggests that the effect of endocannabinoids, cannabinoids, and perhaps marijuana may be complex, and conclusions about their rewarding and aversive actions deserve further intensive study. It is now clear that marijuana is addictive in susceptible individuals, regardless of the inconclusive cannabinoid rewarding profile in the CPP/CPA paradigm.

Interaction between the Cannabinoid and Vanilloid Systems

The mammalian vanilloid subtype 1 capsaicin receptor (VR1r) has been cloned and shown to be activated by plant-derived agonists such as capsaicin (the pungent ingredient in hot chili pepper) and resiniferatoxin (Caterina et al., 1997). Anandamide has been shown to activate the VR1r, resulting in physiological responses that are capsazepine-sensitive and CBR-insensitive (Zygmunt et al., 1999 and Szolcsanyi, 2000). It has also been demonstrated that anandamide acts as full agonist at the human VR1 vanilloid receptor (Smart et al., 2000). The interaction between synthetic vanilloids and the endogenous cannabinoid system has been studied because of the structural similarity between some vanilloid agonists, e.g., olvanil, and endocannabinoid, e.g., anandamide (Di Marzo et al., 1998a). While impaired nociception and pain sensation and elimination of capsaicin sensitivity in mice lacking capsaicin (vanilloid) VR1 receptor has been demonstrated (Caterina et al., 2000), it can be hypothesized that these mice might be sensitive to the antinociceptive effects of CBR activation. The cannabinoid system is known to functionally interact with many other neurotransmitter systems in the brain, and the interactions with the dopaminergic and opioid systems are believed to be of primary importance for the expression of the rewarding effects of cannabinoids and development of cannabinoid physical dependence (Tanda and Goldberg, 2003). However, there are other numerous reports of cannabinoid interactions with serotonergic, adrenergic, GABAergic, glutaminergic, and cholinergic systems (Ameri, 1999; Doherty and Dingledine, 2001; Mechoulam and Parker, 2003; Onaivi et al., 2002). CB_1 and VR1 receptors are coexpressed on a subpopulation of primary sensory neurons whose activation by capsaicin induces anandamide production and release (Ahluwalia et al., 2000, 2003). Both CB_1 and VR1 receptors are responsive to anandamide, and it has been demonstrated that anandamide activates VR1 in hippocampal slices (Al-Hayani, 2001). It appears that possible therapeutic benefits may be exploited from the interactions of the EPCS with other neurochemical systems; for example, antinociception is an important area where the interaction between the cannabinoid and opioid systems may be exploited for possible therapeutic benefits. Furthermore, cannabinoid agonists have been shown to attenuate capsaicin-induced responses in animals (Richardson et al., 1998) and in human skin (Rukweid et al., 2003). The interaction between vanilloid VR1 and glutamate receptors in the central modulation of nociception was sensitive to the antagonists of group 1 mGluRs, NMDA, and capsaicin but not to rimonabant, the CB_1R antagonist (Palazzo et al., 2002).

Evidence also points to a possible involvement of the CB_1R in ethanol's effects, and a critical role for the CB_1Rs in alcohol dependence and stress-stimulated ethanol drinking has been demonstrated, indicating that ethanol withdrawal symptoms were completely absent in CB_1R deficient mice (Racz et al., 2003). In addition, the transsynaptic link between metabotropic glutamate and cannabinoid receptors may have profound implications, both for control of synaptic transmission and for novel therapeutic strategies (Doherty and Dingledine, 2001). The presynaptic localization of CB_1Rs and mediation of retrograde signaling in neuronal tissues involved in the inhibition of classical neurotransmitter release (Ohno-Shosaku et al., 2002; Wilson and Nicoll, 2001) may be one of the hallmarks of the promiscuous and pervasive action of this previously unknown but elaborate EPCS with other neurochemical systems, including the vanilloid system.

Whereas the interaction between EPCS and vanilloid systems is not well established and studied, we are excited by the idea of investigating whether the interaction between endocannabinoid and endovanilloid systems induced by the natural ligands could be the basis for why some people like hot chili peppers and others do not. Possible interactions between the cannabinoid and vanilloid signaling systems have been suggested because of the chemical similarity between some synthetic agonists of vanilloid receptors, such as olvanil and the endocannabinoid, anandamide (Di Marzo et al., 1998a). Anandamide was the first endocannabinoid to be discovered (Devane et al., 1992) and the most widely studied, yet this compound acts as a partial agonist at CB_1Rs and led some to question how an endogenous natural ligand could act as a partial agonist of its own receptor (Sugiura et al., 1995). They proposed that 2-AG may be the natural ligand for the CNRs as it is a full agonist at CB_1 and CB_2 receptors, and concluded that the physiological significance of anandamide may be as an endogenous ligand of a receptor other than CNRs or as a modulator of ion channels (Sugiura et al., 1995). Indeed, to date the only well characterized, noncannabinoid site of action for anandamide is at the VR1. Therefore, it is tempting to speculate that an interaction or a cross talk exists between the G-protein-coupled cannabinoid receptor and the vanilloid systems as depicted in the hypothetical sketch in Figure 3.1. The possibility for the existence of non-CB_1, non-CB_2, and non-VR1 in which some VR1 agonists may induce cannabinoid

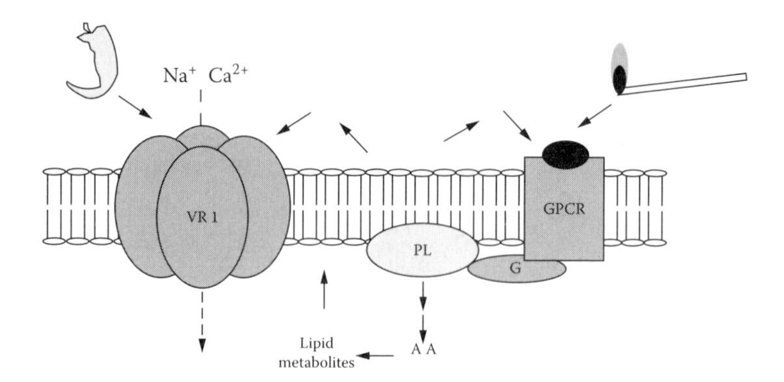

FIGURE 3.1 Cross talk between endocannabinoid and endovanilloid system. A hypothetical illustration of the cross talk between the VR1 ion channel receptor and G-protein-coupled CB_1R at the cellular and molecular level. Endocannabinoids released as byproducts of phospholipid metabolism are able to activate both receptors.

effects is also attractive. Whether or not the interaction between the vanilloid and cannabinoid systems is involved in the effects of abused substances remains to be firmly established, as does exploiting such interaction for the increased understanding marijuana use and also the pharmacotherapy of substance dependence.

Involvement of Endocannabinoids in Alcohol and Drug Abuse

Accumulating electrophysiological and neuroanatomical analyses indicate that some of the CB_1 CNR receptor is targeted to the presynaptic terminals of neurons where it acts to inhibit release of "classical" neurotransmitters as reviewed by Elphick and Egertova, (2001). It is therefore tempting to speculate that the endocannabinoid system may be a major player in the reward pathway, particularly as it is one of the most abundant neurochemical systems in the CNS. Other studies report that THC, the major psychoactive constituent in marijuana, inhibits the synaptosomal uptake of dopamine, serotonin, norepinephrine, acetylcholine, and GABA. This, therefore, warrants preclinical evaluations to determine the role(s) of the endocannabinoid system in drug and alcohol dependence and other neuropsychiatric conditions.

The acute effects of smoking marijuana have been associated with short-term disruption of memory and learning (Hollister, 1986) that may involve the actions of cannabinoids and endocannabinoids in LTP and synaptic plasticity in the brain. Both LTP and synaptic plasticity (Collingridge and Singer, 1990) are electrophysiological measurements of learning and memory associated with glutamate receptors. Glutamate is the major excitatory neurotransmitter in the CNS with NMDA and non-NMDA receptors of which AMPA receptors belong to the family of ligand-gated glutamate receptors (Akinshola et al., 1999a; Akinshola, 2001). Therefore, the involvement of the EPCS in glutamatergic transmission and consequences in alcohol and drug abuse and in other CNS function is being intensively investigated (Akinshola et al., 1999a; Rodriguez et al., 2001; Hoffman and Lupica, 2001). Thus, CB_1R activation was found to inhibit synaptic release of glutamate in rat dorsolateral striatum (Gerdeman and Lovinger, 2001). Furthermore, the interaction between alcohol and the endocannabinoid system has been suggested from recent investigations (Freedland et al., 2001) and also that cannabinoids modulate the motivation for beer via both cannabinoid and opioid receptors (Gallate et al., 1999). For example, it has also been demonstrated that the pharmacological actions of ethanol may involve alterations in the endocannabinoid-signaling system (Basavarajappa et al., 2000); administration of rimonabant decreased sucrose and ethanol intake in rats (Arnone et al., 1997), decreased alcohol consumption in alcohol-preferring rats (Colombo et al., 1998), and reduced breakpoints for beer and "near-beer" in animals responding under a progressive ratio schedule of reinforcement (Gallate and Mcgregor, 1999); and decreased operant ethanol self-administration in rats exposed to ethanol vapor (Rodriguez de Fonseca et al., 1999). Based on these results it appears that the CB_1R blockade may have potential utility for the treatment of alcoholism (Rodriguez de Fonseca et al., 1999). Previously, a blockade of CB_1 receptors by rimonabant was shown to dose-dependently suppress the stimulation of locomotor activity produced by d-amphetamine, cocaine, and morphine in habituated gerbils (Poncelet et al., 1999). Rimonabant suppressed the enhanced locomotor activity induced by each stimulant drug in habituated gerbils, but not in naïve, nonhabituated animals. The authors suggested that inhibition of CB_1Rs might result in dissociating putative motivational and cognitive effects from motor effects of psychostimulants (Poncelet et al., 1999). We have extended the antagonistic activity of rimonabant against the disruption of cognition or reward-enhancing properties of morphine, amphetamine, and cocaine (Poncelet et al., 1999) to ethanol and diazepam, as discussed below.

We have tested the hypothesis that the endocannabinoid control system is an integral component of the reward circuit using *in vivo* and *in vitro* test systems. We determined the influence of cannabinoid receptor antagonism on withdrawal anxiogenesis from chronic alcohol, cocaine, and diazepam treatment *in vivo*, using the plus-maze test. For the *in vitro* studies, we investigated the effects of anandamide or ethanol on the current generated in kainite-activated Xenopus oocytes expressing wild-type and several mutant AMPA glutamate receptor subunits. The CB_1R antagonist rimonabant blocked the behavioral aversions to the open arms of the plus-maze, which was precipitated from withdrawal from cocaine, diazepam, and ethanol. These data, therefore, provide additional evidence to support the existence and involvement of a cannabinoid physiological control system associated with the rewarding effects of alcohol and drug abuse. This rewarding effect was measured by withdrawal aversions in a mouse model of withdrawal anxiogenesis. Thus, the cannabinoid CB_1 antagonist, rimonabant, in the plus-maze test of anxiety, blocked withdrawal aversions from addictive substances. Similar to the withdrawal aversions in the plus-maze test, the CPP paradigm has been used to investigate the rewarding effects of cannabinoids. Rimonabant has been shown to counteract the CPP supported by classical reinforcers including food, cocaine, and morphine (Chaperon et al., 1998). This is in agreement with data that demonstrates the antagonistic activity of rimonabant against disruption of cognition or the reward-enhancing properties of morphine, amphetamine, and cocaine (Poncelet et al., 1999) that we have extended to ethanol and diazepam.

There is also an overwhelming accumulating body of evidence to support interaction of the cannabinoid and opioid systems (Valverde et al., 2000). For example, mutant mice with a cannabinoid CB_1 receptor deletion show a significant decrease in morphine self-administration and a slight but significant reduction in the withdrawal syndrome to chronic morphine (Ledent et al., 1999). Furthermore, stimulation of CBRs by THC increases endogenous opioids (Valverde et al., 2001). It has also been reported that naloxone precipitates abstinence signs in animals treated chronically with THC, and administration of the CB_1R antagonist, rimonabant, triggered withdrawal signs in chronically morphine-treated animals (Navarro et al., 1998). Anatomically, the ultrastructural localization of the CB_1R in μ-opioid receptor patches of the rat caudate putamen nucleus has been demonstrated by Rodriguez et al. (2001). The presence of CB_1R and μ-opioid receptors in some of the same spiny neurons suggested that the dual modulation of the output from these mainly GABAergic projection neurons may, in part, account for some of the common physiological effects produced by cannabinoids and opioids (Rodriguez et al., 2001). Interestingly, therefore, a number of behavioral, biochemical and anatomical localization data demonstrate the existence of a direct link between endogenous opioid and cannabinoid systems (Valverde et al., 2001). In our studies, using *in vivo* and *in vitro* techniques, we report and extend the link between this EPCS and opioids to the rewarding effects of alcohol, cocaine, and diazepam, all known to have abuse and dependence liabilities. *In vitro* electrophysiological recordings have demonstrated that endocannabinoid, anandamide, and ethanol share a similar pattern in the inhibition of kainate-activated currents in Xenopus oocytes expressing the AMPA glutamate receptor, although anandamide was 1,000-fold more potent at inhibiting AMPA receptor function than ethanol. This is in agreement with reports that ethanol inhibits the function of both NMDA and non-NMDA glutamate receptors (Lovinger, 1993; Akinshola, 2001). Furthermore, we have previously shown that anandamide inhibition of kainite-activated homomeric and heteromeric glutamate receptor subunits, which were specific and voltage-independent, may underlie the involvement of endocannabinoids in the modulation of fast synaptic transmission in the CNS. Therefore, the persisting consequences of compulsive, uncontrollable drug and alcohol use may be associated with memory formation during long-term ingestion of drugs and/or alcohol (Heyne et al., 2000). So, if the memory of drug use, effects, and dependency are associated with alcohol and drug addiction, then it remains to be determined if short-term memory disruptions from cannabis use that are involved in glutamatergic transmission can be exploited in the treatment of drug and alcohol addiction.

The blockade of the behavioral aversions by cannabinoid antagonists following chronic administration with alcohol, cocaine, and diazepam was interesting because cannabinoid-induced alterations in brain disposition of drugs of abuse was demonstrated by Reid and Bornheim (2001) to correlate with behavioral alterations in mice. Because marijuana is often consumed with other drugs and alcohol, Reid and Bornheim (2001) determined the influence of pretreatment with cannabinoids, e.g., THC and CBD on the brain levels of cocaine, PCP, morphine, methadone, and MDMA. They reported that THC or CBD pretreatment of mice increases brain levels (two- to fourfold) of subsequently administered cocaine and PCP, which correlated with behavioral tests. Our own previous data indicated that THC induced in both rats and mice increased aversion to the open arms of the plus-maze which was similar to that produced by anxiogenic agents (Onaivi et al., 1990). In that study it was demonstrated that pretreatment with the bi-directional inverse agonist carboline-3-carboxylate or diazepam and flumazenil, high-affinity BDZ receptor antagonists that lack intrinsic activity (at doses that did not modify mouse behavior), all blocked the aversions provoked by THC. Although, the mechanism by which cannabinoids alter mouse aversions in the plus-maze remains to be established, we provide evidence that withdrawal aversions from addictive substances was blocked by CBR antagonism, suggesting that the EPCS may be involved with drug and alcohol use. Additional evidence was shown by protein blotting, which demonstrated a down- and upregulation of CB_1R after chronic administration with methanandamide, a cannabinoid agonist, and rimonabant, a cannabinoid inverse agonist/antagonist, respectively. Curiously, chronic ethanol exposure downregulated the cannabinoid CB_1Rs in mouse brain synaptic plasma membrane (Basavarajappa and Hungund, 1999), which is similar to the data obtained in our study demonstrating that administration of methanadamide reduced CB_1R protein expression in mouse brain. It is also interesting that chronic ethanol exposure to cells *in vitro* which stimulate the release of endocannabinoids has been suggested as a mechanism of neuronal adaptation which may serve as a compensatory mechanism to counteract the continuous presence of ethanol (Basavarajappa et al., 2000; Basavarajappa and Hungund, 1999).

The existence and involvement of a cannabinoid physiological control system in the *in vivo* model is presented as additional evidence that manipulating the endocannabinoid system could be exploited in reducing the behavioral consequences of withdrawal from alcohol and drug dependency, and maintaining codependences in the many humans who coabuse cannabis along with cocaine, ethanol, or benzodiazepines. In fact, the existence of an endogenous cannabinoid tone involved in the regulation of emotional responses, motor activity, pain, memory, and improved social recognition in adult rats was demonstrated by the action of CB_1R antagonist rimonabant. This suggests that rimonabant, which inhibits the endogenous cannabinoid system and induces in rats a behavioral pattern opposite to those of CB_1R agonists (Costa and Colleoni, 1991), has additional inverse agonist properties. Furthermore, Yamamoto and Takada (2000) suggested endocannabinoid involvement in signal transduction system-mediating information processing of nociceptive and rewarding stimuli. Thus, the EPCS may be a directly important natural regulatory mechanism for reward in the brain and also contribute to reduction in aversive consequences of use of several classes of abused substances. We cannot, however, exclude the possibility that there may be other as yet unknown and uncharacterized CBRs (Breivogel et al., 2001; Di Marzo et al., 2000a) that are different from CB_1, CB_1A, and CB_2 CBRs that may constitute part of a well-sought-after common pathway mediating reinforcement of abused substances that may be involved in manifestation of the effects of marijuana, alcohol, and drug abuse. Thus, we provide further additional data to support the existence of an EPCS that is intricately involved with the reward circuits associated with marijuana and other abused substances.

CANNABINOID RECEPTOR GENE EXPRESSION

The expression of CB_1R in the CNS has been extensively studied (Onaivi et al., 1996, 2002). While significant progress has been achieved in many aspects of the biology of marijuana and cannabinoids, our knowledge of cannabinoid genomics and proteomics is increasing, although the nature of the regulation of known CBR genes is poor, despite the mapping of the human genome. The molecular complexity of the cannabinoid physiological control system is, therefore, associated with numerous signal transduction pathways culminating in CBR and non-CBR mediated effects. The CB_2R gene has been detected particularly in the immune system, and the expression of its transcripts has been found in spleen, tonsils, thymus, mast cells, and blood cells. CB_1 and CB_2 CBRs can be coexpressed in some of the same cells in which cannabimimetic effects can be mediated by their combination. The relative abundance of the endocannabinoids and the relatively large numbers of expressed cannabinoid receptors may allow these systems to influence many biochemical systems. It, thus, may not be surprising that intimate links between cannabinoid systems and dopaminergic, glutamatergic, serotonergic, opioidergic, and other important neurotransmitters can be readily identified.

The expression of the CB_1R genes has been detected in the brains from many species, including human, monkey, pig, dog, cat, cattle, guinea pig, rat, mouse, frog, zebra finch, puffer fish, and leech, but not in insects. CB_1R gene expression can be detected in regions that influence a number of key functions, including mood, motor coordination, autonomic function, memory, sensation, and cognition. Expression is more abundant in hippocampus, cerebral cortex, some olfactory regions, caudate, putamen, nucleus accumbens, and the horizontal limb of the diagonal band. In concert with the localization and distribution pattern of the CB_1R gene, the radioligand-binding sites to CB_1R protein reported in the rat brain mirrors the expression of mRNA products of the CB_1R gene. The highest densities of the expression of the CB_1Rs are consistent with the marked effects that cannabinoids exert on motor function tests, such as spontaneous locomotor activity and catalepsy in rodents (Chakrabarti et al., 1998; Onaivi et al., 1996). There is, however, relatively low abundance of CB_1Rs in the human cerebellum in comparison with rodents. The low abundance of CB_1Rs in the human cerebellum is consistent with the more subtle defects noted in human gross motor functioning following marijuana use.

Interaction between these receptors and alterations in mental and neurological disorders have been reviewed (Croxford, 2003). While the specific effects of CBR gene expression in mental and neurological function are incompletely understood, Gilles de la Tourette syndrome (GTS), obsessive compulsive disorder (OCD), Parkinson's disease, Alzheimer's disease, drug and alcohol addiction, and other neuropsychiatric or neurological disturbances are candidates to be influenced by possible variants in the CBRs gene (Gadzicki et al., 1999). Altered CB_1 expression has been reported and clinical trials begun in the use of cannabinoids in a number of mental disorders as well as in brain injury. Apparently, one emerging important physiological role of the endocannabinoids is neuroprotection. The processes leading to neuronal damage, including ischemia, hypoxia, and glutamate toxicity, have been shown to be attenuated by cannabinoids (Grotenhermen, 2004).

The expression of the CB_1 and to a lesser extent CB_2R genes have been studied at different stages in development using brain tissues and preimplantation embryo and in the aging brain. CB_1R gene expression can be detected in tissue from newborn infant (Mailleux et al., 1992). The ontogeny of rat CBR expression allows the receptor to be detected at postnatal day 2 and at even earlier time points in rat embryonic brains (McLaughlin and Abood, 1993; de Fonseca et al., 1993).

Preimplantation embryos from mouse express CB_1R and CB_2R genes (Belue et al., 1995). Whereas the CB_2R gene has been detected from single-cell embryo through the blastocyst stages, the CB_1R gene is expressed in the four-cell embryo. There appears to be a general decline in the expression of CB_1R genes with age in the human and rodent brain (Westlake et al., 1994), although conflicting data indicating decline are reported in aged rats.

Both CB_1 and CB_2 CBR genes are differentially expressed, but in the immune cells, spleen, and bone marrow, the CB_2R gene is more abundant (Galiegue et al., 1995). In many of these cells in peripheral tissues and some cell lines, the CB_1 message is detected only after PCR amplification. CB_1 receptor-specific PCR products can be obtained from human polymorphonuclear (PMN), monocytes, T4 and T8 cells, B cells, NK cells, T leukemia cells, lymphoma cells, B lymphoblasts and lymphocytes, immortalized monocytes, mouse NL-like cells, and T cells. By contrast, abundant levels of CB_2 message has been reported in human lung, uterus, pancreas, tonsils, thymus, peripheral blood mononuclear cells, NK cells, B cells, macrophages, PMN cells, mast cells, basophilic leukemia cells (RBL-2H3 cells), and T4 and T8 cells (Galiegue et al., 1995; Daaka et al., 1996). The expression of both CB_1 and CB_2 CBR genes in the human placenta was demonstrated by Kenney et al. (1999) and found to play a role in the regulation of serotonin transporter activity. As the human placenta is a direct target for cannabinoids, use of marijuana during pregnancy could affect the placental clearance of serotonin. Cannabinoids negatively regulate AP-1 activity through inhibition of c-fos and c-jun proteins (Faubert and Kaminski, 2000). They inhibit interleukin-2 gene transcription (Yea et al., 2000).

The emergence of novel research tools has accelerated cannabis research in the last decade, more so than at any time in the thousands of years of marijuana use in human history. Although it is not yet known why the marijuana (cannabinoid) system is so abundant in the nervous system, the analysis of the receptor proteins and genes encoding these CBRs may shed some light on the mode of action of cannabinoids and the biological role of these genes in the nervous system. In our studies, we have analyzed both CB_1 and CB_2 CBR genes in normal humans who do not use marijuana. The expression of the CBR proteins in different human population according to gender and ethnic background in Asians, blacks, and whites were compared. The finding that the expression of CBRs in humans varies according to gender and ethnic differences among whites, blacks, and Asian populations should be confirmed in a larger sample size. The implication and physiological relevance of this finding are only speculative and premature if unconfirmed. However, this is not surprising as numerous studies have linked genetic determinants and differences to the neurobehavioral responses of abused drugs in man and animals, (Le et al., 1994; Harada et al., 1996). For example, genetic differences in alcohol and compulsive drug-taking behavior have been demonstrated in animals and man (Le et al., 1994; Harada et al., 1996). Genetic variation in some receptor and enzyme systems, e.g., cholecystokinin and serotonin 1A receptors and liver enzymes, alcohol, and aldehyde dehydrogenase, may be associated with alcohol dependence due to the modified function in physiological and behavioral responses (Thomasson et al., 1991; Harada et al., 1996). Thus, the implication and relevance of the differential expression of CBRs in humans according to ethnic background remains to be determined. If it turns out that these levels are relevant to psychoactivity, toxicity, and perhaps therapeutic efficacy, then determination of the expression of cannabinoids in human blood may be used to predict the outcome of their actions. Endocannabinoids may also play important roles in the regulation and activation of CBRs and genes *in vivo*. With the identification of genes associated with human diseases and development of approaches to understand transcriptional regulation, modulation of gene activity for treating disease may be applied to the development of cannabinoid therapeutics. Additionally, this may be useful in identifying individuals who may be vulnerable to marijuana and other substance use and dependence.

Large-scale gene expression changes during long-term exposure to Δ^9-THC in rats have been analyzed. Zhuang et al. (1998) used cDNA microarrays to assess changes in expression levels of very large numbers of genes. They randomly selected and arrayed at high density 24,456 rat brain cDNA clones to investigate differential gene expression profiles following acute (24 h), short-term (7 d) and chronic (21 d) treatment with Δ^9-THC. They found a total of 64 different genes altered by Δ^9-THC; of these 43 were known, 10 had transcripts to homologous ESTs, and 11 transcripts had no homology to known sequences in the Genebank database. In addition, they found that a slightly higher percentage of altered genes were downregulated (58%) than upregulated (42%), whereas some genes showed both up and downregulation at different times during chronic Δ^9-THC treatment. The study indicated that utilizing large-scale screening demonstrated that different sets of genes were altered at different times during chronic exposure to Δ^9-THC. The complete identity of these altered genes when known may shine more light on the mechanism of action of cannabinoids. The same group investigated the effects of long-term exposure to Δ^9-THC on expression of CB_1R mRNA in different rat brain regions (Zhuang et al., 1998). They found that in the striatum the levels of CB_1 transcripts were significantly reduced from days 2 to 14 and returned to control levels by day 21. Thus, multiple neurotransmitters and neuromodulators are influenced by cannabinoids which may help explain the myriad effects associated with smoking marijuana and consequently turning on or off those genes that are sensitive to the actions of cannabinoids.

Molecular Characteristics of Cannabinoid Receptors (CNRs)

The marijuana CBR gene was elusive to clone, but evidence for the existence of the receptor had been demonstrated since the 1980s (Howlett et al., 1988; Devane et al., 1988). It has now been shown and recognized that cannabinoids have specific receptors with endogenous ligands showing inhibition of adenylate cyclase. The CB_1 receptors also modulate the activities of calcium and potassium channels. Although a number of approaches are now available for the cloning of genes encoding different receptors, the most common method previously used, which involved the purification to homogeneity of the gene protein product, did not work for the cannabinoid receptors.

Despite the wealth of information and major advances that have transformed cannabinoid research into mainstream science, little information is available at the molecular level about CBR gene structure, regulation, and polymorphisms. Therefore, much research remains to be conducted at the molecular level about the 5′ untranslated regions, particularly the cannabinoid-promoter structure and regulation, and the 3′ untranslated regions which, apart from containing several polyadenylation signals, may also play important regulatory roles (Shire et al., 1999). In order to begin characterizing the genomic structure of the CBRs, we have cloned, sequenced (Chakrabarti et al., 1995), constructed a 3D model of (Onaivi et al., 1998), and localized the mouse CB_1R gene to chromosome 4 (Stubbs et al., 1996). The currently available information on the genomic structure of CB_1 and CB_2 CBR gene is sketchy, and the regulation of these genes is poorly understood. The emerging putative structure for the CB_1 gene is depicted in Figure 3.2. As discussed below, the CB_1R gene structure is polymorphic with implications, not only for substance abuse but also for other neuropsychiatric disorders. Furthermore, the rat and human CB_1 cDNA sequences are very similar (Matsuda et al., 1990; Gerard et al., 1991). Unlike the CB_1R, which is highly conserved across the human, rat, and mouse species, the CB_2R is much more divergent (Griffin et al., 1999). This divergence in mouse, rat, and human CB_2R leading to differences in functional assays may be related to species specificity.

The existence of a subtype of CB_1R gene, originally designated as CB_1A (now designated CB_1b and described by Shire et al. (1995), has not been detected in any species *in vivo*. Therefore, while it is unlikely and doubtful that CB_1b exists in the form described by Shire et al. (1995), this does not mean that other CBR subtypes may not exist. The primary structure of the CB_1 and CB_2 CBRs are similar to those of other G-protein-coupled receptors with the characteristic features of typical seven hydrophobic domains with some highly conserved amino acid residues. A detailed comparison of the molecular properties of the human, rat, and mouse CB_1, and, where applicable, CB_2Rs, had been previously reviewed (Onaivi et al., 1996; Matsuda, 1997). These receptors mediate their

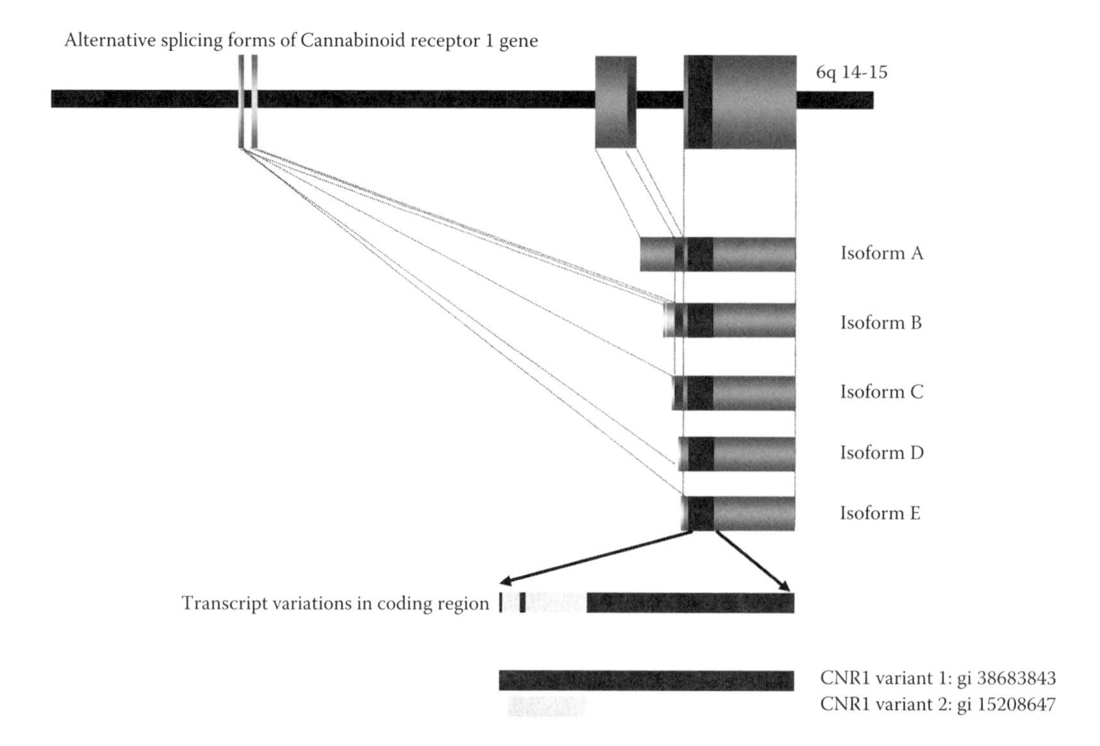

FIGURE 3.2 Structure of the CB_1 CNR gene. The emerging genetic structure of the CB_1 CNR gene mapped on two finished genomic sequences: AL121835 and AL136096. Complete mRNA sequence and exact transcript size is now known with continued characterization. The known sequences consist of four exons, one of which includes the whole coding region of the gene. A number of ESTs are identified and, surprisingly, some of them are between the exons of the published mRNA. A number of polymorphisms have been identified in the CB_1 CNR gene and discussed.

intracellular actions by a pathway that involves activation of one or more guanine, nucleotide-binding, regulatory proteins, which respond to cannabinoids, including the endocannabinoids. The conservatism of the CB_1R sequence contrasts with the variability seen with CB_2R. The composition and amino acid sequence alignments of CB_1 and CB_2 CBRs show considerable structural homology and distribution in CNS between species, with substantial amino acid conservation but with significant difference from the CB_2R; the presence of CB_2R in CNS is itself controversial. As with other GPCRs, the primary structures of CBR are characterized by the seven hydrophobic stretches of 20 to 25 amino acids estimated to form the transmembrane helices, connected by alternating extracellular and intracellular loops. In comparing the composition of the N-terminal 28 amino acids between human CB_1 and CB_2 CBRs and also between human, rat, and mouse, it has been reported by Onaivi et al. (1996) that: (1) the human and rat N-terminal 28 amino acids in the CB_1Rs are similar in the total number of nonpolar, polar, acidic, and basic amino acids; (2) the mouse N-terminal 28 amino acids differ from the rat and human CB_1Rs in number and composition of the total nonpolar and polar amino acids; (3) there are significant differences in the total nonpolar, polar, acidic, and basic amino acid composition of the N- terminal 28 amino acids between human CB_1 and CB_2 CBRs, and (4) the molecular weights of human, rat, and mouse CB_1Rs are similar. Therefore, the amino acid composition of the mammalian CB_1Rs shows strong conservatism in contrast to molecular weights and the amino acid composition of CB_2Rs.

Three-dimensional models of the helix bundle arrangement of human, rat, and mouse CB_1 and CB_2 receptors have been constructed and compared (Bramblett et al., 1995; Onaivi et al., 1996). The transmembrane helix bundle arrangement obtained for the CB_1Rs is consistent with that obtained for other GPCRs. Potential sites for N-glycosylation, and the action of PKC, cAMP-dependent PK, and Ca-calmodulin-dependent PK II in the derived amino acid sequence of the CNR proteins have been identified (Onaivi et al., 1996). Most but not all GPCRs are glycoproteins, and consensus sites for N-glycosylation are mainly concentrated at the N-terminus of the protein. There are three potential N-glycosylation sites highly conserved in human, rat, and mouse. The rodent CB_1R protein has an additional potential N-glycosylation site at the C-terminal segment that is absent in the human CB_1R protein. One potential N-glycosylation site is present in human and rat CB_1R protein, but that site is missing in mouse CB_1R. Whether all of these potential N-glycosylation sites are naturally glycosylated in CB_1R proteins or whether these N-glycosylation are essential for CBR function, and whether additional N-glycosylation in the CB_1R of different mammalian species imparts differential activity of this protein are yet to be determined. However, mutation of N-glycosylation sites in similar GPCRs, e.g., -adrenergic receptors and muscarinic receptors, abolishes glycosylation but has essentially no effect on receptor expression and function (Dohlman et al., 1991). The human CBR subtypes CB_1 and CB_2, with some similarities and differences in their receptor function, appear to differ in the number and distribution of their potential N-glycosylation sites. Due to the modification of the N-terminal region, CB_2R has only one potential N-glycosylation site, whereas CB_1R has five. There is no potential N-glycosylation site at the C-terminal segment of CB_2R. The biological significance (if any) of these differences is yet to be determined. The C-terminal regions and the third intracellular loop of GPCRs are known to be rich in serine and threonine residues. In the case of rhodopsin, -adrenergic, and some muscarinic receptors, some of these residues are targets of cAMP-dependent PK and other PKs (Strada et al., 1994). These phosphorylations are often agonist-dependent and result in desensitization and coupling of the receptor from the G protein. There are four clusters of potential cAMP-dependent PK and Ca-calmodulin-dependent-PK sites in CB_1R that are conserved in human, rat, and mouse proteins. There is a single potential PKC site that is also conserved in CB_1R, whereas CB_2R has no such site. The N-terminal potential cAMP clusters present in CB_1R appear to be conserved, and the CB_2R has two such potential sites. None of the CBRs have any potential PK site at the C-terminal regions. The biological significance of these potential protein phosphorylation sites in these receptor molecules is yet to be determined. In addition many members of GPCRs are known to contain conserved cysteine residues that appear to stabilize the tertiary structure of the receptor because of their involvement in an intramolecular disulfide bridge. In most receptors these cysteines occur in the extracellular domains that lie between hydrophobic domains two and three, and hydrophobic domains four and five (second and third extracellular domains, on the assumption that N-terminal domain is also extracellular). In CB_1 and CB_2 CBRs, no cysteines are found within the second extracellular domain, but the third extracellular domain contains two or more cysteines. One other deviation from most other GPCRs is that CB_1 and CB_2 CBRs lack a highly conserved proline residue in the fifth hydrophobic domain (Matsuda, 1997). The structural features of these proteins, which are critical for ligand binding and functional properties, have been evaluated in *in vivo* and *in vitro* models (Akinshola et al., 1999b). Heterologously, CBRs bind tritriated synthetic ligands such as WIN55212-2 in a saturable and competitive manner. The binding activities of CB_1 and CB_2 CBRs have been determined after transfection into CHO cells, COS7 cells, and mouse AtT20 cells (Felder et al., 1995; Slipetz et al., 1995). In other expression systems, the CB_1Rs have been examined in insect Sfi cells, (Pettit et al., 1994), Xenopus oocytes (Henry and Chavkin, 1995), mouse L cells, human embryonic kidney 293 cells (Song and Bonner, 1996), and dissociated rat superior cervical ganglion neurons (Pan et al., 1996). Furthermore, the performance of mutated CBRs with that of wild-type receptors have been compared in a number of functional assays and reviewed by Matsuda, 1997. In those studies mutant CBRs containing point mutation (a single amino acid substitution) or that have been modified by replacement of a series of amino acids from one receptor with that of another (chimeric receptor) have been expressed and studied (Matsuda, 1997). In more recent studies, findings on the structural, and molecular complexities of CB_1R continue to shed light on the multiplicity of cannabinoid-induced behavioral changes with linkage to the molecular basis of marijuana use and dependence, the understanding of which is elusive but unfolding. In the absence of CB_1R crystallized structure, it has been postulated that there are different conformations of CB_1R that bind different types of cannabinoid ligands under physiological conditions (Salo et al., 2004).

Cannabinoid Receptor Gene Knockout Mice

Because CB_2Rs were initially cloned and thought to be predominantly in immune cells (B cells, T cells, and macrophages), they were referred to as peripheral cannabinoid receptors. Although a number of laboratories have not been able to detect the presence of this so-called peripheral CB_2 cannabinoid receptor in the brain, there has been demonstration of CB_2 expression in the rat microglial cells, in cerebral granule cells, and in mast cells which are all part of the CNS. New reports also indicate that CB_2 receptors are expressed by a limited population of microglial cells in normal healthy and in neuritic plaque-associated glia in the Alzheimer's disease brain. We have investigated whether the so-called peripheral CB_2 cannabinoid receptors are present in naïve and diseased mouse brain. Our preliminary data using a real-time PCR system (TaqMan, ABI) with a CB_2 cannabinoid-receptor, gene-specific probe and primers indicated abundant expression of CB_2 mRNA in the striatum, cortex, and hippocampus in naïve and stressed mice.

There are two major experimental approaches that have been used to elucidate the biological roles of cannabinoids. One is the traditional pharmacological manipulation, including the use of the highly selective CBR antagonist to determine the involvement of cannabinoids in any biological processes or systems. The second approach is targeted gene disruption and manipulation of CBR genes. Targeting a specific gene, also referred to as homologous recombination, enables the study of the physiological consequences of invalidating the function of a specific gene. These approaches are being used to study the physiological role of cannabinoids. An important question is whether the CB_1R mediates all of the numerous central effects of cannabinoids. Thus, synthesis of selective and specific CBR receptor antagonists and the development of genetically modified strains of mice, in which the expression of the CB_1 and the CB_2 CBRs have been eliminated, have revealed significant information on the physiological role of the cannabinoid system. The CB_1 and CB_2 CBR knockout mice have been generated. Buckley et al., (2000) reported the development of CB_2R mutant mice. The CB_2R knockout mouse was generated in order to study the effects of cannabinoids on immune cells and immunomodulation. The CB_2R gene was invalidated by using homologous recombination in embryonic stem (ES) cells. The 3′ region of the CB_2 coding exon was replaced with phosphoglycerate kinase (PGK)-neomycin sequences through homologous recombination in the ES cells. This mutation eliminated part of intracellular CB_2R for the homologous recombination in 129 ES cells and was injected into C57BL/6 blastocysts and placed in foster mothers. The mice deficient in CB_2R gene were generally healthy, fertile, and cared for their offspring and *in situ* hybridization histochemistry demonstrated the absence of the CB_2 mRNA in the knockout mice (Buckley et al., 2000). Binding studies on intact spleens and splenic membranes using the highly specific $[^3H]$CP55940 showed significant binding in spleens derived from wild-type but was absent from the CB_2 CNR mutant mice (Buckley et al., 2000). Fluorescence-activated cell-sorting (FACS) analysis showed no differences in immune cell populations between cannabinoid CB_2R knockout and wild-type mice. In addition, the role of the CB_2R on Δ^9-THC inhibition of macrophage costimulatory activity was also determined. Buckley et al. (2000) reported that Δ^9-THC inhibits helper T cell activation through macrophages derived from wild-type, but not from the CB_2 mutant mice, indicating that cannabinoids inhibit macrophage costimulatory activity and T cell activation via the CB_2R. While these studies continue, mice deficient for the CB_2R gene demonstrated that the CB_2R is involved in cannabinoid-induced immunomodulation.

CB_1R gene knockout mice have been independently generated by two groups. These two groups produced mutant mice with disrupted CB_1R gene by standard homologous recombination techniques similar to that used in the production of the CB_2 mutant mice. The first report by Ledent et al.

(1999) showed that the spontaneous locomotor activity of these mutant mice was increased, and that they did not respond to cannabinoid drugs, suggesting that the CB_1 CNR was responsible for mediating the analgesic, reinforcement, hypothermic, hypolocomotive, and hypotensive effects of cannabinoids. In their CB_1 mutant mice, Ledent et al. (1999) also showed that the acute effects of opiates were unaffected, but that the reinforcing properties of morphine and the severity of the withdrawal syndrome were strongly reduced. The second report was by Zimmer et al. (1999) and Steiner et al. (1999), who showed that the CB_1 mutant mice appeared healthy and fertile, but had significantly high mortality rates and showed reduced spontaneous locomotor activity, increased immobility, and hypoanalgesia when compared to the wild-type litter mates. In these CB_1 mutant mice, Δ^9-THC-induced catalepsy, hypomobility, and hypothermia were absent, but it was reported that Δ^9-THC induced analgesia in the tail-flick test and other behavioral (licking of the abdomen) and physiological (diarrhea) responses to THC were still present. Thus, there were behavioral similarities and differences in the results of the second group by Zimmer et al. (1999) and Steiner et al. (1999) with those of Ledent et al. (1999). It appears that the differences in responses by the mutant mice from the two groups might be related to methodological differences due to different laboratory techniques used. The groups, however, differed in their findings on the baseline motility of the CB_1 mutants. Ledent et al. (1999) found that the CB_1 mutant mice exhibited higher levels of spontaneous locomotion, even when placed in fear-inducing novel environments (like in elevated plus-maze and open field). In contrast, Zimmer et al. (1999) found that the CB_1 mutant mice displayed reduced activity in the open-field test and an increased tendency to be cataleptic. In the basal ganglia, a brain structure with high levels of CB_1R important for sensorimotor and motivational aspects of behavior, it was shown by Steiner et al. (1999) that these mutant mice displayed significantly increased levels of the substances P, dynorphin, and enkephalin, and of GAD67 gene expression that may account for the alterations in spontaneous activity observed in the CB_1 mutant mice. These data, however, remain at variance with those of an apparently similar strain of mice tested by Ledent et al. (1999). Overall, however, these findings provide many valuable insights into cannabinoid mechanisms despite some differences among reports on CB_1R gene knockout mice. There is, therefore, a general agreement that CB_1R plays a key role in mediating most but not all CNS effects of cannabinoids.

The biological consequences of inactivating CB_1 and CB_2 CBR genes continue to be studied intensively. The availability of the cannabinoid knockout mice provides an excellent opportunity to study the biological roles of these genes. In a hippocampal model for synaptic changes that are believed to underlie memory at the cellular level, Bohme et al. (1999) examined the physiological properties of the Schaffer collateral–CA1 synapses in mutant mice lacking the CB_1R gene and found that these mice exhibit a half-larger LTP than wild-type controls, with other properties of these synapses, such as paired-pulse facilitation, remaining unchanged. They concluded that disrupting the CB_1R-mediated neurotransmission at the genome level produces mutant mice with an enhanced capacity to strengthen synaptic connections in a brain region crucial for memory formation (Bohme et al., 2000). Reibaud et al. (2000) have also used the CB_1R knockout mice in a two-trial object recognition test to assess the role of CBRs in memory. They showed that the CB_1R knockout mice were able to retain memory for at least 48 h after the first trial, whereas the wild-type controls lost their capacity to retain memory after 24 h. These data, along with previous findings of other investigators, suggest that the endogenous cannabinoid systems play a crucial role in the process of memory storage and retrieval. This finding is supported by previous data indicating enhanced LTP in mice lacking CB_1R gene (Bohme et al., 2000). These rapid advances in cannabinoid research have continued to add to our knowledge about the biology of marijuana (cannabinoids) in the vertebrate and invertebrate systems. Mice lacking CBR genes have also enabled scientists to investigate the interaction of cannabinoids with other neurochemical networks. The interaction between the cannabinoid and opioid systems was examined by Valverde et al. (2000) who demonstrated that the absence of CB_1R did not modify the antinociceptive effects induced by mu, delta, and kappa opioid agonists, but these mice exhibited a reduction in stress-induced analgesia. These results, therefore, indicate that the CB_1Rs are not involved in the antinociceptive responses to exogenous opioids, but that a physiological interaction between the opioid and cannabinoid systems is necessary to allow the development of opioid-mediated responses to stress. In a different study, Mascia et al. (1999) showed that morphine did not modify dopamine release in the nucleus accumbens of CB_1R knockout mice under conditions where it dose-dependently stimulates the release of dopamine in the corresponding wild-type mice, indicating that the CB_1Rs regulate mesolimbic dopaminergic transmission in brain areas known to be involved in the reinforcing effects of morphine (Mascia et al., 1999).

Other studies have started to look at the CB_1R knockout mice as an animal model of schizophrenia (Fritzsche, 2000) and perhaps as models in other neuropsychiatric conditions because of the ubiquity and diverse functions and numerous signal transducers involved in the actions of marijuana and other cannabinoids. The link and genetic homologies between CB_1R and D2 dopamine receptor (Fritzsche, 2000) and the presynaptic regulation of neurotransmitter release supports a closer investigation of the EPCS in addictions and other neuropyschiatric disorders. These genetic manipulations provide additional evidence for the involvement of specific genes encoding the known and other as yet unknown cannabinoid receptors with the physiological triggers for the rewarding basis of the continuous circle of marijuana use and dependence.

Other Cannabinoid Receptor Transgenic Models

Rapid advances in designing genetically engineered laboratory animals are producing not only research models but also models that are more effective research tools. As a result, the need for precision genetic characterization and definition of laboratory animals is of primary concern. However, the use of transgenic mouse models for the over-expression and other forms of modification of CBR genes (except for the CBR gene inactivation described above) to study the regulation and site-specific mechanisms of the action of cannabinoids is currently being explored. For example, cannabinoid transgenes, in which genomic regulatory sequences of interest are coupled to a reporter gene, can be used to probe further the mechanism of regulation of the CBR genes. But the current lack of information about the 5′ and 3′ untranslated regions and other regulatory elements of CBR genes makes it difficult to make necessary CBR gene construct modifications for generating such rodent CBR transgenic models. Obviously, the use of CBR transgenic animals will provide new *in vivo* systems for studying genetic regulation, development, and normal function and dysfunction associated with EPCS. Our current investigation and those of others indicate that a number of lines of evidence make the gene that encodes this G-protein-coupled cannabinoid receptors a strong candidate to harbor variants that might contribute to individual differences in human addiction vulnerability.

Fatty Acid Amide Hydrolase Knockout Mice

The currently known endocannabinoids, derived from membrane phospholipids that contain arachidonate (Mecholam et al., 1998) are metabolized by FAAH (Deutsch and Chin, 1993), which is a membrane-associated serine hydrolase enriched in brain and liver. There is an overlap of the distribution of FAAH and its activity in the rat brain with the expression of CB_1, which has led to the suggestion that FAAH is probably the major enzyme in the brain responsible for inactivation of fatty acid amides (Elphick and Egertova, 2001). Thus, in the human brain the distribution of CB_1R and the FAAH enzyme frequently overlap in many structures. As the identity of endocannabinoid transporters is still unknown, the generation of FAAH-mutant mice has shed some light into the activity of FAAH. Using homologous recombination, the role that FAAH plays in controlling fatty acid amide levels and activity *in vivo* was evaluated in mice that possess a targeted disruption of the FAAH gene (Cravatt et al., 2001). It was demonstrated that mice lacking FAAH were supersensitive to anandamide with an enhanced endogenous cannabinoid signaling (Cravatt et al., 2001). Thus mice lacking FAAH were severely impaired in their ability to degrade anandamide and when treated with this compound, the mutant animals exhibited an array of intense CB_1R-dependent behavioral responses. In addition, these workers reported that FAAH-mutant mice possess 15-fold augmented endogenous brain levels of anandamide and display reduced pain sensation, which was reversed by the CB_1 CB_1R antagonist rimonabant. The data from Cravatt et al. (2001) support an important role for FAAH in the metabolism of the amide series and perhaps also for other endocannabinoids. The mechanisms involved in the inactivation of endocannabinoids *in vivo* are not completely understood. However, functional studies indicate that the biological actions of endocannabinoids are probably terminated by a two-step inactivation process consisting of a carrier-mediated uptake and intracellular hydrolysis by FAAH (Piomelli et al., 1999; Di Marzo et al., 1999; Hillard, 2000). FAAH has been purified, cloned, sequenced from mouse, rat, and human, and thus fairly well characterized. It is a single-copy gene with 579 amino acids and a highly conserved primary structure, homologous in mouse, rat, and human species (Giang and Cravatt, 1997). This membrane-associated enzyme is 63 kDa and possesses the ability to hydrolyze a range of fatty acid amides including anandamide, 2-AG, and oleamide. The distribution of FAAH and CB_1R in rat brain is similar, with FAAH often occurring in neuronal somata that are postsynaptic CB_1R-expressing axons and therefore consistent with a potential role in the regulation of endocannabinoids (Egertova et al., 2000). Whereas FAAH has gene sequence homologous to FAAH enzymes in other species, it is the first mammalian member of this enzyme family. So far, this enzyme has been identified also in rat and mouse with more than >90% sequence homology to humans, and this homology indicates a general role for the fatty acid amides in mammalian neurobiology. In addition to PMSF, numerous compounds have been identified that block FAAH reversibly and irreversibly. Among these is ibuprofen, which is an active inhibitor, but not other nonsteroidal anti-inflammatory agents (NSAID) such as naproxen. The ability of neuronal tissue to synthesize and rapidly metabolize anandamide with the aid of a specific transport carrier mechanism suggests either a role for anandamide as an new member of fatty acid-derived neuromodulators or that it could act as a specific neurotransmitter. Enhanced NAE biosynthesis and turnover have been demonstrated in peritoneal macrophages from mice treated with a calcium ionophore (Kuwae et al., 1999). It was suggested that arachidonic acid mobilization induced by ionophore treatment of macrophages could result in the selective generation of anandamide. The biosynthesis and inactivation of endocannabinoids and other cannabimimetic fatty acid derivatives have been extensively reviewed by Di Marzo et al. (1998b). An acid amidase hydrolyzing anandamide and other N-acylethanolamines distinct from FAAH that can hydrolyze *N*-acylethanolamines have been reported (Ueda et al., 2000). There is evidence that the synthesis of 2-AG and anandamide can be independently regulated, even though the plasma membranes contain precursor molecules for both anandamide and 2-AG (Piomelli, 2003). Whereas the existence of cannabinoid transporters continues to be controversial (see text following), the differential intracellular hydrolysis of 2-AG and anandamide has been reported (Piomelli, 2003). Apparently, anandamide and 2-AG can be hydrolyzed by distinct serine hydrolases with anandamide predominantly hydrolyzed by FAAH, as discussed, and 2-AG by monoglyceride lipase (MGL), to yield inactive breakdown products (Piomelli, 2003). It appears that there is partial overlap in the distribution of FAAH and MGL in the CNS, with FAAH predominantly localized in postsynaptic structures and MGL mostly associated with nerve endings (Piomelli, 2003). Further research will undoubtedly continue to unravel the biochemical pathways associated with deactivation of endocannabinoids and the role of these processes in marijuana-smoking dependence.

Genes Encoding Endocannabinoid Transporters

Although there is evidence from functional studies for the existence of some form of cannabinoid transporter(s), their identity, sequence information, and biological characteristics at the molecular level are unknown. There is functional evidence that the transport of endocannabinoids such as anandamide and 2-AG across a biological membrane is accomplished via a protein carrier (Piomelli et al., 1999;

Di Marzo et al., 1999; Hillard, 2000). Further evidence has been shown for this carrier-mediated, transmembrane transport of anandamide in human neuroblastoma and lymphoma cells (Maccarone et al., 1998), in mouse macrophages and RBL-2H3 cells (Bisogno et al., 1998), and in neurons (Di Marzo et al., 1994; Hilliard et al., 1997). This transport process fulfills several criteria of a carrier-mediated process, including saturability, temperature dependence, high affinity, substrate selectivity, facilitated diffusion, and Na^+-independence (Piomelli et al., 1999; Di Marzo et al., 1999; Hillard, 2000). Some of these features make this process fundamentally different from other known transport carriers such as catecholamine and amino acid transporters. Using a relatively potent uptake inhibitor N-(4hydroxyphenyl) arachidonylamide, AM404, these investigators have demonstrated that a high-affinity transport system present in neurons and astrocytes has a role in anandamide uptake and subsequent inactivation by FAAH (Piomelli et al., 1999; Di Marzo et al., 1999; Hillard, 2000). AM404 has been reported to inhibit anandamide uptake by rat-cultured cortical neurons and astrocytes and to potentiate anandamide both *in vitro* and *in vivo* (Piomelli et al., 1999; Di Marzo et al, 2000c). When administered to rats by itself, AM404 increases plasma levels of anandamide and shares the ability of this endocannabinoid to decrease locomotor activity, depress plasma levels of prolactin, and alter tyrosine hydroxylase activity in different brain regions. The inhibitory effect of AM404 on locomotor activity has been found to be susceptible to antagonism by rimonabant. However, AM404 does not elicit typical cannabinoid responses of catalepsy in the Pertwee ring test, nor does it produce signs of analgesia in the hot plate test. While there is ample scientific evidence to support the concept that anandamide transport across membranes is protein-mediated, definitive evidence awaits its molecular characterization. Other studies have suggested that passive diffusion alone is sufficient for anandamide uptake, with intracellular hydrolysis by FAAH maintaining the concentration gradient (Glaser et al., 2003). But recent work using FAAH knockout mice show that anandamide uptake is the same in FAAH-mutant and wild-type mice and is blocked by anandamide transport inhibitors, suggesting that anandamide transport is independent of FAAH (Fegley et al., 2004). Although this work and the report of the blockade of endocannabinoid long-term depression (LTD) induction by anandamide uptake inhibitors (Ronesi et al., 2004) functionally support the existence of an anandamide transporter, the molecular identity of the anandamide transporter is still elusive. Furthermore, the differential uptake of the different endocannabinoids (for example, anandamide and 2-AG in different cell types) may indicate the possibility of different cannabinoid transporters for the different endocannabinoids (Di Marzo et al., 1999; Hillard, 2000). Further research will show whether the additional endocannabinoids being discovered, such as noladin ether, will be metabolized and taken up by similar metabolic and uptake inhibitors. This will not be unprecedented because the monoamines have different transporters for dopamine, serotonin, and norepinephrine.

Polymorphic Structure of Cannabinoid Receptor Genes

New information on the CBR gene and its allelic variants in humans and rodents can add to our understanding of vulnerabilities to addictions and other neuropsychiatric disorders and the genetic basis of marijuana use and addiction in vulnerable individuals. Little information is, however, available at the molecular level about CBR gene structure, regulation, and polymorphisms. Different human CBR gene polymorphisms have been reported. A silent mutation of a substitution from G to A at nucleotide position 1359 in codon 453 (Thr), which turned out to be a common polymorphism in the German population, was reported (Gadzicki et al., 1999). In this study, allelic frequencies of 1359(G/A) in genomic DNA samples from German GTS patients and controls were determined by screening the coding exon of the CB_1R gene using PCR single-stranded conformation polymorphism (PCR-SSCP) analysis (Gadzicki et al., 1999). This was accomplished by the use of a PCR-based assay by artificial creation of a MSP1 restriction site in amplified wild-type DNA (G-allele), which is destroyed by A-allele (Gadzicki et al., 1999). They found no significant differences of the allelic distributions between GTS patients and controls within the coding region of the CB_1R gene.

In our studies, the frequencies of this polymorphism are significantly different between the Caucasian, African-American, and Japanese population (Ishiguro et al., unpublished observation). There is a HindIII restriction fragment length polymorphism (RFLP) located in an intron approximately 14 kb in 5′ region of the initiation codon of the CB_1R gene. Caenazzo et al. (1991) genotyped 96 unrelated Caucasians, using hybridization of human DNA digested with HindIII, and identified two allele with bands at 5.5 (A1) and 3.3 kb (A2). The frequencies of these alleles were 0.23 and 0.77, respectively. Another polymorphism is a triplet repeat marker for the CB_1R gene. This is a simple sequence-repeat polymorphism (SSRP) consisting of nine alleles containing (AAT) 12-20- repeat sequences that was identified by Dawson, (1995). This polymorphism has been used in linkage and association studies of the CB_1R gene with mental illness and drug abuse in a different population. This CB_1R gene triplet repeat marker was used to test for a linkage with schizophrenia using 23 multiplex schizophrenia pedigrees (Dawson, 1996), and associations with heroin abuse in a Chinese population (Li et al., 2000) and intravenous (IV) drug use in Caucasians (Comings, 1997). There was no linkage and association of the marker with schizophrenia, indicating that the CB_1R gene is not a polymorphism of major etiological effect for schizophrenia, but a CB_1R gene might be a susceptibility locus in certain individuals with schizophrenia, particularly those whose symptoms are apparently precipitated or exacerbated by cannabis use (Dawson, 1996). Comings et al. (1997) hypothesized that genetic variants of the CB_1R gene might be associated with susceptibility to alcohol or drug dependence and analyzed the triplet repeat marker in the CB_1R gene. They found a significant association of the CB_1R gene with a number of different types of drug dependence (cocaine, amphe-tamine, cannabis) and IV drug use but no significant association with variables related to alcohol abuse and dependence in nonHispanic Caucasians. In addition, this group also reported that a significant association of the triplet repeat marker in the CB_1R gene alleles with the P300-event-related potential that has been implicated in substance abuse (Johnson et al., 1997). Li et al. (2000) attempted to replicate the finding of Comings et al. (1997) in a sample of Chinese heroin addicts and did not find any evidence that CB_1R gene AAT repeat polymorphism confers susceptibility to heroin abuse. CB_1R gene is located in human chromosome 6q14-q15, and it is interesting that previous reports showed evidence for suggestive linkage to schizophrenia with chromosome 6q markers (Martinez et al., 1999). Suggestive evidence also exists for a schizophrenia susceptibility locus on chromosome 6q (Cao et al., 1997). Although there was no linkage and association of the CB_1R gene triplet marker with schizophrenia, it remains to be determined if linkage and association to schizophrenia might exist with other unknown polymorphisms that might exist in the CB_1R gene structure, which is currently poorly characterized. Three other variants have been reported in the CB_1R gene of an epilepsy patient (Kathmann et al., 2000). This was obtained from PCR assay with cDNA from hippocampal tissue taken from patients undergoing neurosurgery for intractable epilepsy. They detected four mutations in the coding region of the CB_1R gene, with the first three mutations yielding amino acid substitutions.

We have initiated a series of studies to analyze CB_1R gene structure, regulation, and expression in the mouse and human models to determine genotypic and haplotypic associations of CB_1R gene with addictions and other neuropsychiatric disturbances. Genotypes at markers near the mouse Chr 4, 13.9 cM CB_1/CNR locus in 9 mouse strains reveal apparent haplotypes that extend from at least D4Mit213 to D4Mit90. These haplotypes can be correlated with strain differences in cannabinoid effects.

The human Chr 6, 91.8~96.1 cM CB_1R gene locus encodes at least four exons which account for 24 to 28 kb of sequence Figure 3.3. Examination of CB_1R gene sequence variations in distinct populations has revealed a G/A single nucleotide polymorphism (SNP) in CB_1 5′ flanking sequences. The initial values for linkage disequilibrium between these markers and genotypic frequencies of the markers in drug abusing and control populations were calculated. The A-allele of the SNP polymorphism was present in fewer African-Americans and Asians than in Caucasians. However, in the Caucasian and African-American samples used, no association between drug abuse and the 1359(G/A) polymorphism could be found (Ishiguro et al., unpublished observation). Furthermore, 1359(G/A)

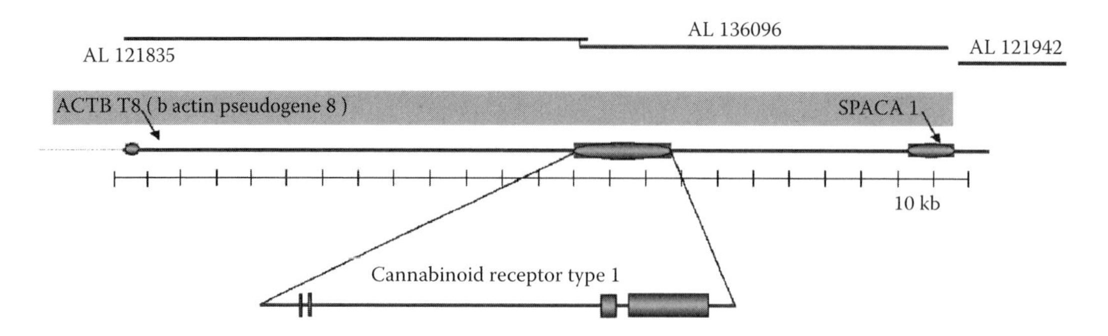

FIGURE 3.3 Part of human chromosome 6 genomic sequences (complete or draft) of a part of chromosome 6p. The completed sequence covers more than 200kb around the CB_1/CNR gene. The computer program GENSCAN (http://CCR-081.mit.edu/GENCAN.html) estimated four fairly possible gene sequences (including CB_1/CNR) in this region. These estimated sequences showed similarity to known protein sequences, revealed by computer program BLAST (http://www.ncbi.nlm.nih.gov/BLAST/).

polymorphism has been determined in healthy control male Japanese subjects (aged 20 to 30) for whom personality traits were measured with a temperament and character inventory (TCI). Although the statistical power is weak because of fewer frequency of allelic distribution, no association between TCI scores and this polymorphism could be found (Ishiguro et al., unpublished observation). While these studies continue, these findings add to the characterization of these CB_1R genes in species in which they can be tested for impact on substance abuse and other neuropsychiatric disorders. The amino acid sequence alignments and phylogenetic tree of known CBRs indicate some similarities and significant divergence between CB_1 and CB_2 CBRs. Polymorphisms at the CB_1R gene may be associated with the diverse actions of marijuana use. They may be a major factor which, when triggered by environmental, age, and metabolic factors, could lead to the continuous but mild circle of marijuana dependence and use.

Chromosomal Mapping of the CBR Genes

With *in situ* hybridization using a biotinylated cosmid probe, the CB_1R gene in humans was localized at 6q14-q15, thus confirming the linkage analysis and defining a precise alignment of the genetic and cytogenetic maps (Hoehe et al., 1991). These investigators found that the location of the human CB_1R gene is very near the gene encoding the alpha subunit of chorionic gonadotropin (CGA). It has been determined that the mouse CB_1 and CB_2 CBR genes are located in proximal chromosome 4 (Stubbs et al., 1996). This location is within a region in which other homologs of human 6q genes are located. In order to localize the mouse CB_1 and CB_2 CBR genes in the mouse genome, Stubbs et al. (1996) traced the inheritance of species-specific variants of the gene in 160 progeny of an interspecific backcross. Therefore, using the interspecific and four DNA probes we mapped the mouse CB_1 and CB_2 CBR genes to chromosome 4 with map positions calculated for Mos, Cntfr, Pax5, and Cd72, in excellent agreement with previously published results that clearly established linkage between CB_1R gene and other genes known to be located on mouse chromosome 4. The CB_1R gene, GABRR1, GABRR2, and Cga are linked together both in the mouse and on human chromosome 6q. The genes encoding the peripheral CB_2R and α-L-fucosidase have been shown to be located near a common virus integration site, Ev11 (Valk et al., 1997). They showed that Ev11 is located at the distal end of mouse chromosome 4 in a region that is synthenic with human 1p36, in agreement with our report (Onaivi et al., 2002), that the mouse CB_2R gene is also located at the distal end of mouse chromosome 4. The results of the chromosomal location of the human CB_1R gene (Hoehe et al., 1991) and the mouse (Stubbs et al., 1996; Onaivi et al., 2002) CBR genes add a new marker to this region of the mouse-human homology, and confirm the close linkage of

CBR genes in both species. The location of the rat CB_1R gene in the rat genome has been determined and localized at 5q13-q24 and fits the rodent-human homology as the CB_1R genes are highly conserved in the mammalian species. The physical and genetic localization of the bovine CB_1R gene has also been mapped to chromosome 9q22 using fluorescence *in situ* hybridization (FISH) and R-banding to identify the chromosome (Pfister-Genskow et al., 1997). The genetic mapping of the CB_1R gene on bovine chromosome 9q22 by *in situ* localization and linkage mapping of a dinucleotide repeat, D9S32, also adds to coverage of the bovine genome map and contributes to the mammalian comparative gene map (Pfister-Genskow et al., 1997). The location of the human CB_1 genomic sequences (using complete or draft) of a part of chromosome 6p is shown in Figure 3.3. The completed sequences cover more than 200kb around the CB_1R gene. The program GENESCAN was used to estimate four fairly possible gene sequences including CB_1R in this region. These estimated sequences showed similarity to known protein sequences, revealed by BLAST. As the neurobiological effects of marijuana and other cannabinoids suggest the involvement of the CBR genes in mental and neurological disturbances, the mapping of the genes will undoubtedly enhance our understanding of the linkage and possible cannabinoid genetic abnormalities. In the case of the cattle, research into the role of CBRs in mediating responses to natural and production-induced stressors could lead to improvement of production inefficiencies that exists in meat and milk animal systems (Pfister-Genskow et al., 1997). The chromosomal location and genomic structure of human and mouse FAAH genes have been mapped to chromosomes 1p34-p35 and 4, respectively (Wan et al., 1998). The localization of FAAH and CB_2R genes are in the same chromosomal regions in the mouse and human, and this again adds a new marker to mouse–human homology, confirming the close linkage of FAAH and CBR genes in both species.

Other Cannabinoid Receptors (CBRs): Functional Evidence for the Existence of Other CBRs

There is accumulating physiological evidence for the existence of other distinct functional cannabinoid receptors different from the currently known CB_1 and CB_2 CBRs that are putatively referred to as CB_3, CB_4, CB_5, CB_6, CB_7, CB_8, CB_9, CB_{10}, CB_{11}, CB_{12}, CB_{13} … CB_n. Whereas the molecular identities of other unknown cannabinoid receptors have been elusive, recent characterization of the human CB_1R gene indicates many structural features and splice variants that may be involved in the actions of smoked marijuana. The emerging functional evidences for the existence of putative CB_3, CB_4, CB_5, CB_6, CB_7, CB_8, CB_9, CB_{10}, CB_{11}, CB_{12}, CB_{13} … CB_n cannabinoid receptors are randomly classified below until their molecular identity is known:

1. CB_1: Fairly well characterized in the brain and peripheral organs and tissues. Many effects of marijuana and cannabinoids have been attributed to activation of this receptor. New information emerging on the complex structure and splice variations that exist.
2. CB_2: Also fairly well characterized and previously believed to be present in immune cells and in the periphery and not in the brain. New research shows expression in brain cells under varying conditions.
3. CB_3: This is derived mainly from the CB_1 mutant mice in which some cannabinoid effects are activated by WIN55212-2 or anandamide but not THC, CP55940, or HU210. This putative CB3r, which is a non-CB_1, non-CB_2, nonvanilloid receptor and is insensitive to rimonabant, may be specifically localized in some brain region to mediate specific physiological function.
4. CB_4: Endothelial cannabinoid receptor. A subtype that has been identified in arteries and human umbilical vein endothelial cells (HUVEC) and is present in vascular endothelium. It can be activated by anandamide, N-arachidonyl dopamine, virohdamine, and abnormal cannabidiol but not by the 2-AG and classical CB_1/CB_2 cannabinoid agonist to induce endothelium-dependent vasodilation that is present in CB_1 and CB_2 wild-type and mutant

mice (Begg et al., 2003; Mo et al., 2004). A similar observation was reported in the rat mesenteric artery RFERENCE.

5. CB_5: A subtype that is present in the brain that is activated by capsaicin, WIN55212-2, and CP55940. It is VR1-like and sensitive to capsazepine and rimonabant.
6. CB_6: A CB_2-like receptor that can be activated by palmitoylethanolamide but not anandamide is known to relieve inflammation. It is a subtype that may be nonvanilloid and rimonabant insensitive but sensitive to SR144528, a CB_2R antagonist.
7. CB_7: An EDG-encoded lysophospholipid/imidazoline receptor on sympathetic nerves (Yamaguchi et al., 1996b; Molderings et al., 2002) and may be present in the brain. It is CB_1-like G-protein-coupled and can be activated by CP55940, HU210, WIN55212-2 and sensitive to rimonabant.
8. CB_8: A subtype that is present on nerve terminals in the guinea pig ileum that are activated by anandamide. It appears to be non-CB_1, and non-CB_2 (Mang et al., 2001) but rimonabant sensitive.
9. CB_9: A non-CB_1, non-CB_2, and rimonabant insensitive allosteric sites on 5-HT3 receptors that are more sensitive to THC than to WIN55212-2, anandamide, or CP55940 at inhibiting 5-HT-induced current triggered by activation of 5-HT3 receptor (Barann et al., 2002).
10. CB_{10}: A cannabinoid subtype receptor in mouse vas deferens at which AM630 is 20 times more potent as an antagonist of THC than of anandamide.
11. CB_{11}: An allosteric non-CB_1, non-CB_2 sites on muscarinic M1 and M4 receptors with affinity for anandamide and methanandamide but not for WIN55212-2 or rimonabant.
12. CB_{12}: A non-CB_1, non-CB_2, cannabidiol-sensitive subtype discovered in Pertwee (2004) laboratory and present in isolated mouse vas deferens.
13. CB_{13}: An allosteric non-CB_1, non-CB_2 site on AMPA GLU_{A3} receptors with which anandamide but not WIN55212-2 can interact to inhibit kainate-activated currents (Akinshola et al., 1999a).

The cloning and characterization of these subtypes of cannabinoid receptors will add validity to the functional evidence for the existence of multiple cannabinoid receptor subtypes. In comparison to the monoaminergic system, the application of modern techniques to cannabinoid research is new. For example molecular cloning has revealed the presence of serotonin (5-hydroxytryptamine; 5-HT) receptor subtypes, which can be subdivided in seven subfamilies (Gerhardt and Van Heerikhuizen, 1997) and 15 serotonin (5-HT) receptor subtypes and growing. Similar to 5-HT receptors that include ligand-gated ion channels, the VR1 receptor, it has been suggested, should be reclassified as a cannabinoid receptor subtype as the endocannabinoid, anandamide, is a full agonist at VR1r, but acts as a partial agonist at CB_1R. New knowledge on cannabinoid posttranscriptional and posttranslational modifications, such as alternate splicing and perhaps RNA editing, may indicate that the formation of multiple proteins could unravel specific mechanisms associated with numerous behavioral and physiological effects of marijuana use. The cloning and sequencing of CB_1 gene from 62 species has also been reported by Murphy et al. (2001) and awaits full characterization. As predicted here, the identification and characterization of these putative CBR isozymes and different elements of the endocannabinoid system may reveal novel targets for medication development. However, the limitless signaling capabilities and the endless complexity of the cannabinoid system requires these continuous intensive investigation.

IS MARIJUANA USE ADDICTING?

The question of physical and psychological addiction to marijuana is no longer a debate as the significant progress in marijuana–cannabinoid research has come about through the application of modern techniques using new molecular probes that were previously unavailable to resolve the issue of marijuana addiction. In addition, animal models allowing the evaluation of genetic and neural correlates of marijuana-cannabinoid dependence and abuse potential have now been developed. Although the question of marijuana-cannabinoid dependence was once controversial, the discovery that rimonabant, the CB_1 cannabinoid receptor antagonist, precipitated withdrawal in rodent models is indicative of cannabinoid dependence (Cook et al., 1998). Similar questions are also no longer posed about cigarette and opiate dependence. As this review focuses on the molecular and genetic basis of the biological effects of marijuana and the cannabinoid system, we cannot dismiss the larger question and current debate about the consequences of medicinal and recreational use of marijuana. Thus, is there evidence for the existence of a diagnosable cannabis specific withdrawal syndrome in human users? This issue has been addressed by the review of the published literature on cannabis withdrawal symptoms in human users (Smith, 2002). Long-term marijuana use can lead to addiction for some people; that is, they use the drug compulsively even though it often interferes with family, school, work, and recreational activities. Along with the craving, withdrawal symptoms can make it hard for long-term marijuana smokers to stop using the drug. People trying to quit report irritability, difficulty sleeping, and anxiety. They also display increased aggression on psychological tests, peaking approximately 1 week after they last used the drug. Tolerance, which varies with effects of marijuana and cannabinoids, has been attributable to pharmacodynamic changes that may be associated with cannabinoid receptor downregulation and receptor desensitization (Grotenhermen, 2004). Although withdrawal symptoms from chronic dosing with high doses of THC in humans are usually mild, the risk for physical and psychic dependency is low compared to opiates, tobacco, alcohol, and benzodiazepines (Grotenhermen, 2004). As the smoke clears from such debates, it will become obvious that the elaborate network of the EPCS plays critical roles in the control of cell survival with effects on most biochemical processes in the human body, far beyond the psychology of dependence and withdrawal syndrome from marijuana use.

The Genetics of Marijuana Dependence

Marijuana or cannabis use has dependence liability because of the physical and psychological effects that lead to the classical relapse cycle of abused substances. The genetics of marijuana dependence may play a more important role in marijuana use than environmental factors when its use is initiated, because if marijuana use was not initiated, the susceptibility to continued use might not persist (Hollister et al., 1986). Emerging evidence comes from accumulating knowledge on cannabinoid genetics, animal modeling of key features of marijuana-cannabinoid dependence that can be evaluated in electrophysiological, pharmacological, and immunohistochemical studies. For example, a behavioral cannabinoid withdrawal syndrome and gene transcription along with hormonal alterations have been described in the mouse model following cessation of treatment from CP55940, a cannabinoid agonist (Oliva et al., 2003). Although the etiology of marijuana dependence is a complex interaction of psychological, environmental, and biological factors associated with genetic vulnerability, the study of the contribution of genetic and environmental factors in cannabis dependence is further complicated by polysubstance use. Comorbidity and genetic loading studies indicate that adults using marijuana on a daily basis have a high frequency of coabuse of other substances such as nicotine and alcohol. Therefore, an individual who becomes dependent on marijuana will exhibit tolerance, withdrawal reactions, craving, and relapse, preventing cessation of use that characterizes the addiction cycle. Comorbidity with other psychiatric disturbances such as depression, panic attacks, anxiety, antisocial personality disorders (Comings, 1996) have also been reported. The use of family, twin, and adoptee studies have demonstrated the likelihood of a genetic contribution to marijuana dependence. Many studies have, therefore, examined the genetic and environmental contributions to the risk of cannabis dependence. The notion of the combined use of multiple substances with dependence potential sharing a common genetic influence has been an enduring issue in addiction research (Crabbe, 2002). However, each abused substance appears to be linked to independent genetic influences that have made it difficult to identify addiction specific genes or even if addictive personality genes exist. The questions about marijuana dependence are what genes are involved in the continuous use, withdrawal syndrome, and relapse to cessation of use?

Our current knowledge shows that specific cannabinoid receptors, which are encoded by cannabinoid receptor genes, are activated by smoking marijuana or by the administration of cannabinoids. The myriad behavioral, physiological, and modulated programs of gene expression following activation or inhibition of the cannabinoid receptors are subjects of intense investigation. Although there is increasing evidence that genetic factors are involved in marijuana dependence through studies of familial transmission of marijuana use, abuse, and dependence (for example, Hopfer et al., 2003), such studies do not provide evidence of either the number, location, or identity of the genes involved (Crabbe, 2002). Therefore, risk for cannabinoid dependence is likely to be the result of a number of genes, each contributing a small fraction of the overall risk, with allelic variants contributing to different aspects of marijuana dependence. One commonly used approach to identify genetic influences on addictions is the candidate gene approach, which directly tests the effects of genetic variants of a potentially contributing gene in an association study (Kwon and Goate, 2000). The age of onset of marijuana dependence and smoking may also be influenced by genetic risk factors, because it may be unlikely for a 70-year-old man or woman who never initiated smoking marijuana to become dependent on marijuana. As with tobacco and alcohol dependence, application of oligogenic linkage analysis and determination of endophenotypes or trait markers (Tyndale, 2003) may improve the ability to identify specific genetic variants contributing to marijuana dependence. Perhaps, as with genetic association studies of alcoholism, similar problems may be encountered with the candidate gene approach in marijuana dependence because of the complexity of addictions (Buckland, 2001).

The CB_1 cannabinoid receptor gene, which is a target of marijuana smoke, is among the addiction "hot spots" in the human chromosomes. The central role of the EPCS in the effects of other abused substances, as demonstrated by CB_1 receptor mutant mice and from association studies, make CB_1 receptor locus a strong candidate with variants that might contribute to individual differences in drug abuse vulnerability (Zhang et al., unpublished). The retrograde action of endocannabinoids on the inhibition of classical neurotransmitter release (e.g., on glutamate, GABA, serotonin, dopamine neurotransmitters) may be differentially modified by genetic variants of the CB_1 receptors, which are the main molecular targets for the endocannabinoids and marijuana-cannabinoids. Until recently, many features of the CB_1 gene's structure, expression, regulatory regions, polymorphisms, haplotypes, and association with polysubstance abuse were poorly defined. Some studies have examined CB_1 genomic variation and association studies have used known markers and increasingly CB_1-associated single nucleotide polymorphisms (SNPS) have appeared (NCBI://www.ncbi.nlm.nih.gov). For example, two such transcripts variants derived from the 3′ end, encoding a longer isoform and short isoform, missing a segment near the 5′ end results in a frameshift with a different N-terminus as compared to the long isoform was reported. Our recent studies have now demonstrated the existence of novel exons, splice variants, and candidate-promoter region sequences that confer reporter gene expression in cells that express CB_1 receptors (Ishiguro, unpublished). The results from the association studies show common human CB_1 polymorphisms which reveal patterns of linkage disequilibrium in Caucasian and African-American individuals, whereas a 5′ CB_1/CNR1 "TAG" haplotype displays significant allelic frequency differences between substance abusers and controls in Caucasian, African-American, and Japanese samples. This new work of improved definition of the CB_1/CNR1 locus and its variants, therefore, adds to the interesting features of regulating the neural circuits important for the reinforcing effects of most abused substances with a critical role in addiction vulnerability (Zhang et al., unpublished). The involvement of CB_1R variants in other phenotypes may have significant roles in cannabinoid pleiotropy as the effects of activation of the CB_1R have been implicated in a number of behavioral functions. The emerging structure of the genomic CB_1R gene is complex with multiple exonic sequences, splice variations, polymorphisms, and regulatory regions that are the central features of the EPCS. Additional and compelling evidence indicated that the discovery of a polymorphism in human FAAH, which is over-represented in patients with problem drug use, might suggest that changes in FAAH function could influence addictive behavior (Cravatt and Lichtman, 2003). Certainly, understanding the molecular structure and regulation of genes involved with the EPCSs might be a first step in characterizing the genetic basis of smoking marijuana. Unfortunately, our knowledge of how genes affect complex traits is poorly understood.

Human Cannabinoid Receptor Genes

Most of the physiological effects of smoking marijuana are probably mediated via cannabinoid receptors and actions at other noncannabinoid receptors and nonreceptor mechanisms. Thus, the genetic basis of marijuana use and perhaps dependence may be associated at least in part with their effects on these G-protein-coupled cannabinoid receptors. The presently known cannabinoid receptors and their gene transcripts can now be analyzed in human blood samples (Onaivi et al., 1999), fetal brain (Wang et al., 2003; Mato et al., 2003), placenta (Park et al., 2003), uterus during pregnancy (Dennedy et al., 2004), Alzheimer's disease brains (Benito et al., 2003), and prefrontal cortex of depressed suicide victims (Hungund et al., 2004). The human CB_1 cDNA was isolated by Gerard et al. (1991) from a human brain stem cDNA library using a 600-bp DNA probe and polymerase chain reaction. The deduced amino acid sequences of the rat and human receptors showed that they encode protein residues of 473 and 472 amino acids, respectively, with 97.3% homology. These proteins share the seven hydrophobic transmembrane domains and residues common among the family of G protein receptors (Matsuda et al., 1990; Gerard et al., 1991; Shire et al., 1995). The human and rat CB_1 receptors also share pharmacological characteristics including the inhibition of adenylate cyclase activity via Gi/o in a stereoselective and pertussis-sensitive manner following activation by cannabinoids (Devane et al., 1988; Matsuda et al., 1990; Gerard et al; 1991). The CB_1Rs alter potassium channel conductance (Hampson et al., 1995) and decrease calcium channel conductance (Mackie and Hille, 1992). Expression of the CB_1 receptors is particularly dense in presynaptic terminals, and recent studies suggest that endocannabinoids act as retrograde messengers at many synapses in the CNS (Kreitzer and Regehr, 2002). Although the transcripts for both CB_1 and CB_2 receptors were reported to be present in the human placenta, the identification of CB_1Rs and FAAH in the human placenta may be targets for the effects of marijuana use during pregnancy (Park et al., 2003). Most of the studies examining the expression and role of CBRs in the reproductive system have used mice in which both CB_1 and CB_2 receptor gene transcripts were identified in mouse preimplantation embryos (Das et al., 1995). The presence of cannabinoid receptors in the human uterus during pregnancy has also been reported, and *in vitro* cannabinoids were reported to induce relaxation on human pregnant myometrium through the CB_1 receptor (Dennedy et al., 2004). Thus, from mouse model to humans, the CBRs is a placental and uterine site of action of endocannabinoids and marijuana use during pregnancy and parturition. The identification of CBRs in the uterus and placenta, and the high densities of CB_1Rs, functionally coupled to G protein in prenatal developmental stages throughout the human brain, suggest the involvement of the cannabinoid system in neural development (Mato et al., 2003). It has also been suggested that the high expression of CB_1 mRNA in the human fetal limbic structures may render such brain structures more vulnerable to prenatal cannabis exposure (Wang et al., 2003). In adult diseased brains, an upregulation of CB_1 receptors and agonist-stimulated [^{35}S]GTPγS binding in the prefrontal cortex of diseased suicide victims was reported in comparison to matched controls (Hungund et al., 2004). This and other observations of the involvement of the EPCS in a variety of CNS disorders render different component of the EPCS possible targets in the treatment of a variety of CNS mental and neurological disturbances. A trinucleotide repeat (AAT) polymorphism of the CB_1R gene has been associated with appetite disorders involving the bingeing and purging type of anorexia nervosa, which is a severe and disabling psychiatric disorder characterized by profound weight loss and body image disturbance (Siegfried et al., 2004).

The assembly of the human CB_1 receptors and the impact of its long N-terminal tail has been examined by Andersson et al., (2003). They found that the long N-tail of CB_1Rs that lack a typical signal sequence, and cannot be efficiently translocated across endoplasmic reticulum membrane, might not be relevant to ligand binding or activation of the receptor (Andersson et al., 2003). With the share abundance of CBRs, the functional significance of the existence of a CB_1R with long N-terminal and a shorter N-terminal tail in CB_1A receptor is not immediately clear, but may be related to rapid formation of different tonically active states allowing for sequestration of G proteins, making them unavailable to couple to other receptors (Vasquez and Lewis, 1999). One of the major advantages for the physiological actions of these cannabinoids receptors may be associated with the quick response needed for the sequestration of G proteins and the retrograde signaling of endocannabinoids on presynaptic CB_1 receptors to inhibit neurotransmitter release.

The second cannabinoid CB_2R clone was isolated from myeloid cells by PCR and degenerate primers using cDNA template from human promyelocytic leukemic line HL60 (Munro et al., 1993). This clone was shown to be related to the rat CB_1 and was therefore used to screen the HL60 library. The primary structure of the CB_1 was essential in the identification of this subtype of CBR gene. Most importantly, there was the similar hydrophobic domains 1, 2, 5, 6, and 7, in which 50% or greater of the amino acids were identical between CB_1 and CB_2Rs. The extracellular domain also contained sequence motifs that were common to both clones (Matsuda, 1997). The protein encoded by the CB_2 shows 44% identity with the human CB_1R. A number of functional and expression studies have been performed with the CB_2 gene, and the results indicate that the CB_2 is the predominant CBR in the immune system, where it is expressed in B and T cells (Gurwitz and Kloog, 1998; Schatz et al., 1997; Munro et al., 1993). Previous studies have shown that CB_2Rs are expressed in granule and Purkinje cells of the mouse cerebellum (Skaper et al., 1996) and in rat microglial cells (Carlisle et al., 2002), where CB_2Rs have been known to be upregulated when microglial cells are activated (Carlisle et al., 2002). CB_2Rs have also been shown to be involved with microglial migration (Walter et al., 2003). Thus, the report of the presence of CB_2Rs and FAAH in neuritic plaque-associated glia in Alzheimer's disease brains (Benito et al., 2003) is in agreement with data from CB_2 mutant mice indicating that CB_2Rs are involved in cannabinoid-induced immunomodulation and inflammatory response. The hypothesis that CB_2 or other as yet unidentified subtype of cannabinoid receptors or components of the EPCS are expressed in the brains of addicted individuals remains to be determined. The relevance, if any, to problem and compulsive marijuana use is unknown. However, a specific enzyme activity by a phosphodiesterase of the phospholipase D-type responsible for generating endocannabinoids such as anandamide has been identified, cloned from mouse, rat, and humans and characterized with the expression of the mRNA and proteins widely distributed in murine brain, kidney, and testis (Okamoto et al., 2004). The direct contribution of this enzyme and its influence in the effects of abused substances and marijuana dependence have not been determined, but the levels of N-acylethanolamines are known to increase in a variety of animal models of tissue degeneration (Okamoto et al., 2004). Furthermore, cells and tissues involved in neuroinflammation express functional cannabinoid receptors with an upregulation of components of the endocannabinoid-signaling system in the inflammatory response (Walter and Stella, 2004). As the cannabinoid signaling machinery is involved in immune function, it is yet unknown if neuroinflammatory processes contribute to continued marijuana use and dependence, once initiated. Other as yet unknown components of the cannabinoid system have been shown to be involved in neuroinflammatory processes that are sensitive to the nonpsychotropic cannabinoid receptors whose molecular identity is currently unknown. This additional evidence about cannabinoid regulation of neuroinflammation and immunomodulation has been provided by 2-AG stimulation of microglial cell migration that is blocked by nonpsychotropic cannabinoid receptor antagonists, cannabinoid and abnormal cannabidiol (Walter et al, 2003). It appears, therefore, that the vital role played by the cannabinoids system in the neuroinflammation and immunomodulation may be exploited in the development of immunopharmacotherapy against drugs of abuse by targeting some component of the EPCS.

The study of cannabinoid dependence in animal models has been reviewed to clarify the consequences of chronic exposure to cannabinoid agonists and their abuse liability (Maldonado, 2002). Based on animal studies, the potential ability of the endocannabinoids to induce physical dependence will be unprecedented as most endogenous ligands that are substrates for the rewarding properties of abused substances are labile. In humans, the genetic liability following fetal exposure to marijuana during pregnancy might be a predisposing vulnerability factor to use of marijuana in adulthood, although those genetic markers have not been identified. However, earlier studies have reported on the capability of marijuana use to induce genotoxity (see Li and Lin, 1998, for a review).

NEUROBIOLOGY OF ENDOCANNABINOID MODULATION OF OTHER RECEPTOR SYSTEMS

Endocannabinoid ligands (Di Marzo et al., 1998b; Mechoulam et al., 1998; Hanus et al., 2001) for CBRs modulate other CNS neurotransmitter systems such as GABA, DA, 5-HT, NE, Ach, opiates, and the glutamate receptors. The ability of these CBRs to interact with multiple systems provides limitless signaling capabilities of cross talk within and possibly between receptor families. It has been suggested that anandamide and, to a lesser extent, 2-arachidonyl glycerol, modulate other CNS neurotransmitter systems in drug addiction or reinforcing properties and memory impairment through presynaptic CBR receptor inputs on neurons releasing different neurotransmitters (for a review, see Onaivi et al., 2002; Schlicker and Kathmann, 2001). Evidence of presynaptic localization of CBRs on native neuron in the CNS has been confirmed through retrograde signaling by endocannabinoid action at presynaptic CB_1Rs to suppress neurotransmitter release in neurons (Kreitzer and Regehr, 2002; Kano et al., 2002; Trettel and Levine 2003.). The CBR is a member of the slow-acting G-protein-coupled receptors, known to modulate the fast-acting ligand-gated ion channel receptors, and it is just becoming evident that this control is important in diseased states from drug abuse. The existence of the EPCS and an endocannabinoid hypothesis of substance are discussed below.

This section provides information on endocannabinoid modulation of AMPA glutamate receptors with implication in learning and memory processes in marijuana habituation. Glutamate is the major excitatory neurotransmitter in the CNS, and the majority of fast glutamatergic synaptic transmissions are mediated by AMPA receptors (Jonas and Sakmann, 1992). NMDA, AMPA, and KA are the 3-receptor subtypes that constitute the excitatory glutamate neurotransmitter system in the CNS. Earlier reports of CBR agonist effects on glutamate receptors were focused on inhibition of NMDA neurotransmission in brain neuronal cultures such as the cerebellum (Hampson et al., 1998), basal ganglia (Glass et al., 1997), hippocampus (Terranova et al., 1995), and forebrain (Nadler et al., 1993). Anandamide is reported to have dual effects on the NMDA receptor by a reduction in NMDA mediated intracellular calcium flux in rat brain slices and potentiation of NMDA intracellular calcium flux in rat brain (cortical, cerebellar, hippocampal slices) in the presence of CB_1 antagonists (Hampson et al., 1998).

A direct effect of anandamide in the augmentation of NMDA-stimulated currents was recorded in Xenopus oocytes expressing the recombinant receptor in the same study by Hampson et al., 1998. There is a paucity of data on the physiological effects of the action of endocannabinoids on glutamatergic neurotransmission beyond their inhibition and potentiation of glutamate-induced LTP and LTD in hippocampal (Rouach and Nicoll, 2003), cerebellar (Brown et al., 2003), and striatal neurons (Ronesi et al., 2004) in relation to learning and memory formation (Chevaleyre and Castillo, 2003) and drug abuse (Derkinderen et al., 2003). A significant role however, is being ascribed to metabotropic glutamate receptors (mGluR) in the presynaptic inhibition of glutamate receptors by endocannabinoids (Maejima et al., 2001; Varma et al., 2001; Ohno-Shosaku et al., 2002; Rouach and Nicoll, 2003). It has now been proposed with ample evidence in the hippocampus that in retrograde signaling in the brain, there is a cooperative production of endocannabinoids by metabotropic glutamate (mGlur) receptor activation, leading to the inhibition of inhibitory postsynaptic currents (IPSCs) in neurons (Varma et al., 2001; Ohno-Shosaku et al., 2002).

Most of the work on endocannabinoid interaction with glutamatergic system is done in brain slices and to a lesser extent in acutely dissociated neurons and neuronal cultures. There are very few reports in the literature on the effect of endocannabinoids on recombinant glutamate receptors despite the advantages of a well-defined *in vitro* system free of unknown regulatory factors that may be present in brain slices and neuronal cultures. There is currently no literature cited on the

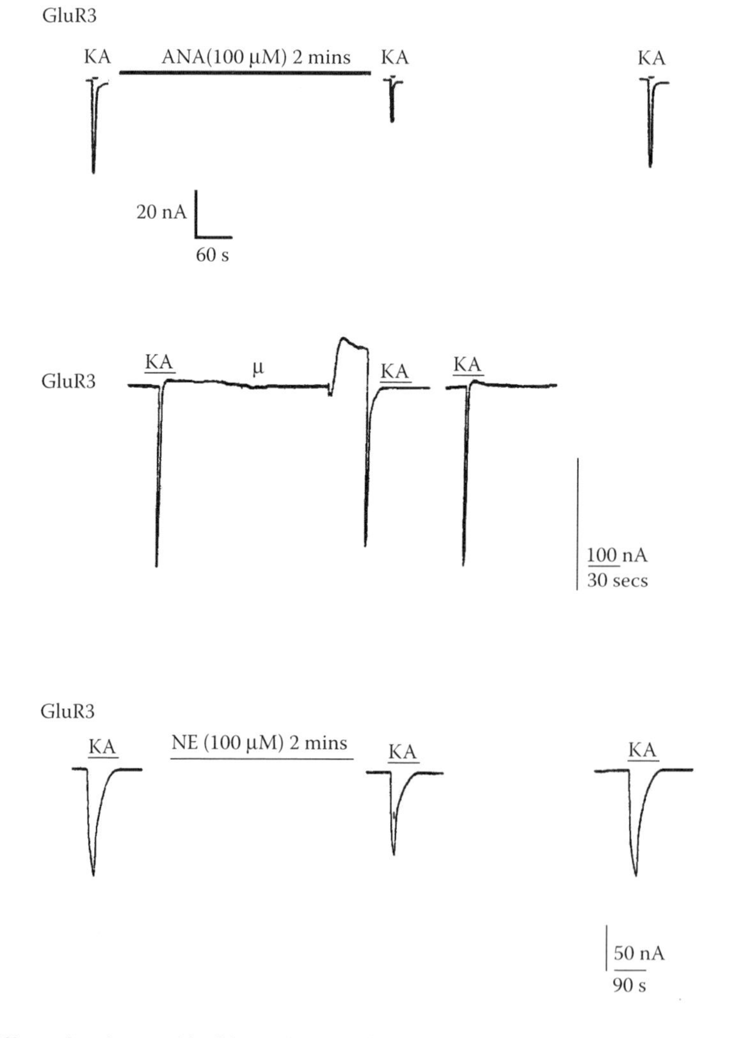

FIGURE 3.4 Effect of endocannabinoids on AMPA glutamate receptor. In Xenopus oocytes expressing the recombinant rat GluR3 subunit of AMPA receptor, anandamide, 2-arachidonyl glycerol, and noladin ether inhibited kainate-activated currents. The order of potency of the endocannabinoids in inhibiting receptor currents in oocytes was anandamide > 2-arachidonyl glycerol > noladin ether.

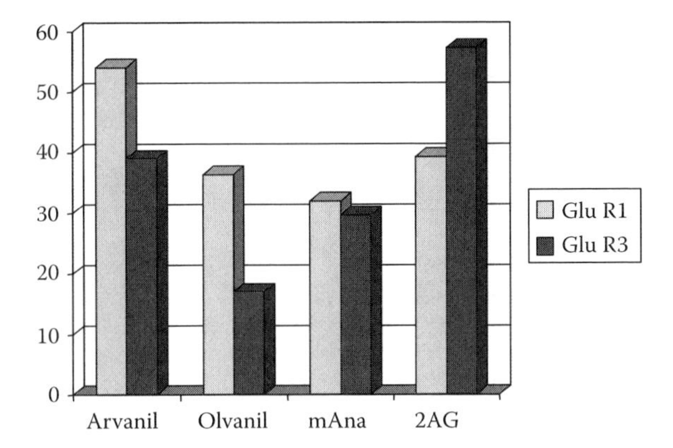

FIGURE 3.5 Percent inhibition of kainate activated current. The similar inhibitory effects of 2-AG and the synthetic endocannabinoid methanandamide are shown in oocytes expressing homomeric GluR1 and GluR3 AMPA receptor. The effects of the endocannabinoids were compared to arvanil and olvanil (vanilloid agonist) effects in the AMPA receptors and expressed as percentage inhibition of receptor currents in oocytes.

effect of endocannabinoids on kainate receptors. Few reports of endocannabinoid inhibition of recombinant AMPA receptor currents *in vitro* has been reported in the Xenopus oocyte expression system (Akinshola et al., 1999b; Onaivi et al., 2002). All the homomeric and heteromeric AMPA receptor subunits expressed in oocytes were reported to be functionally inhibited by anandamide (Akinshola et al., 1999b). This result allows for a definitive conclusion that endocannabinoids profoundly modulate AMPA receptor function *in vivo* and *in vitro* (Akinshola et al., 1999).

A follow-up question is whether the other endocannabinoids, — 2-AG and noladin ether — have similar effects to those that anandamide has on glutamate receptors *in vivo* and *in vitro*. Neuronal data suggest that 2-AG produces effects to those similar to those of anandamide in the modulation of LTD and retrograde signaling. In *Xenopus* oocytes expressing the recombinant rat GluR3 subunit of AMPA receptor, anandamide, 2-AG and noladin ether inhibited kainate-activated currents as seen in Figure 3.4. The order of potency of the endocannabinoids in inhibiting receptor currents in oocytes was anandamide > 2-AG > noladin ether. In the figure, 2-AG appears to temporarily reverse the GluR3 inward current in oocytes. The endocannabinoid effects were not blocked by rimonabant and therefore were independent of a CB_1 effect. A similar inhibitory effect of 2-AG and the synthetic endocannabinoid methanandamide is shown in oocytes expressing homomeric GluR1 and GluR3 AMPA receptor subunits in Figure 3.5. The effects of the endocannabinoids were compared to arvanil and olvanil (vanilloid agonist) effects in the AMPA receptors and expressed as percentage inhibition of receptor currents in oocytes. Arvanil and 2-AG had similar potencies, just as olvanil and methanandamide — all at concentrations of 100 μM — exhibited similar potencies in inhibiting GluR1 and GluR3 receptor currents in oocytes. Whereas capsazepine, the vanilloid antagonist, also inhibited AMPA receptor currents in oocytes, SR141716A, the CB_1 receptor antagonist, has no effect on AMPA receptor currents or endocannabinoid inhibition of receptor currents *in vitro* (Akinshola et al., 2002). It can be concluded from these *in vitro* reports that endocannabinoids can directly modulate glutamatergic neurotransmission *in vitro* via CB_1R independent mechanism.

The amino acid sequence alignments (Figure 3.6) and construction of the phylogenetic tree of known CBRs (Figure 3.7A) and tree depicting closely related lyosophosphatidic and vanilloid receptors (Figure 3.7B) indicate some similarities and significant divergence between CB_1 and closely related CB_2, LPA, and VR1 receptors. Thus, there is indication of a natural variation in gene expression, which is extensive in human and other organisms.

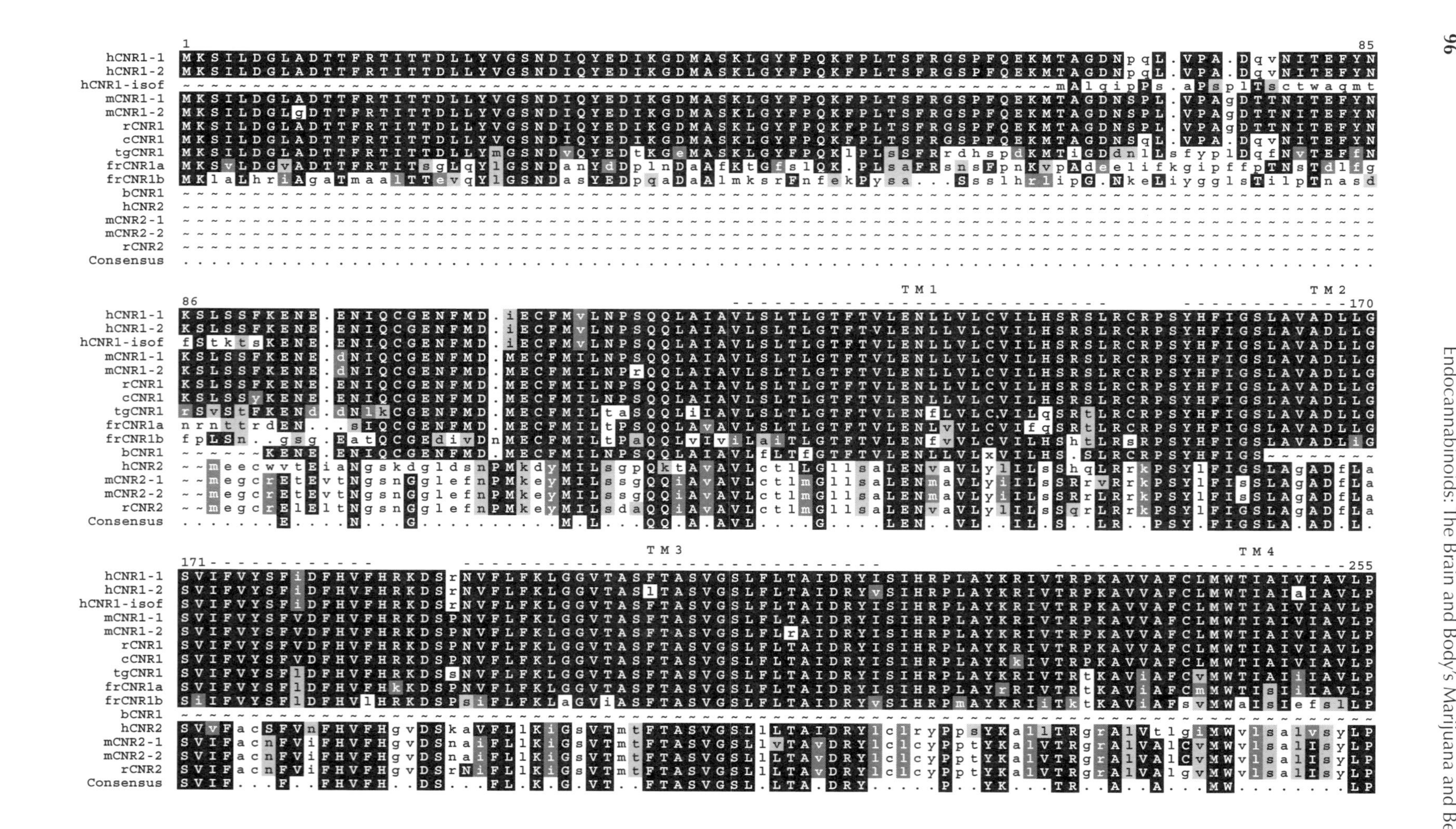
hCNR1-1
hCNR1-2
hCNR1-isof
mCNR1-1
mCNR1-2
rCNR1
cCNR1
tgCNR1
frCNR1a
frCNR1b
bCNR1
hCNR2
mCNR2-1
mCNR2-2
rCNR2
Consensus
TM 1
TM 2
TM 3
TM 4

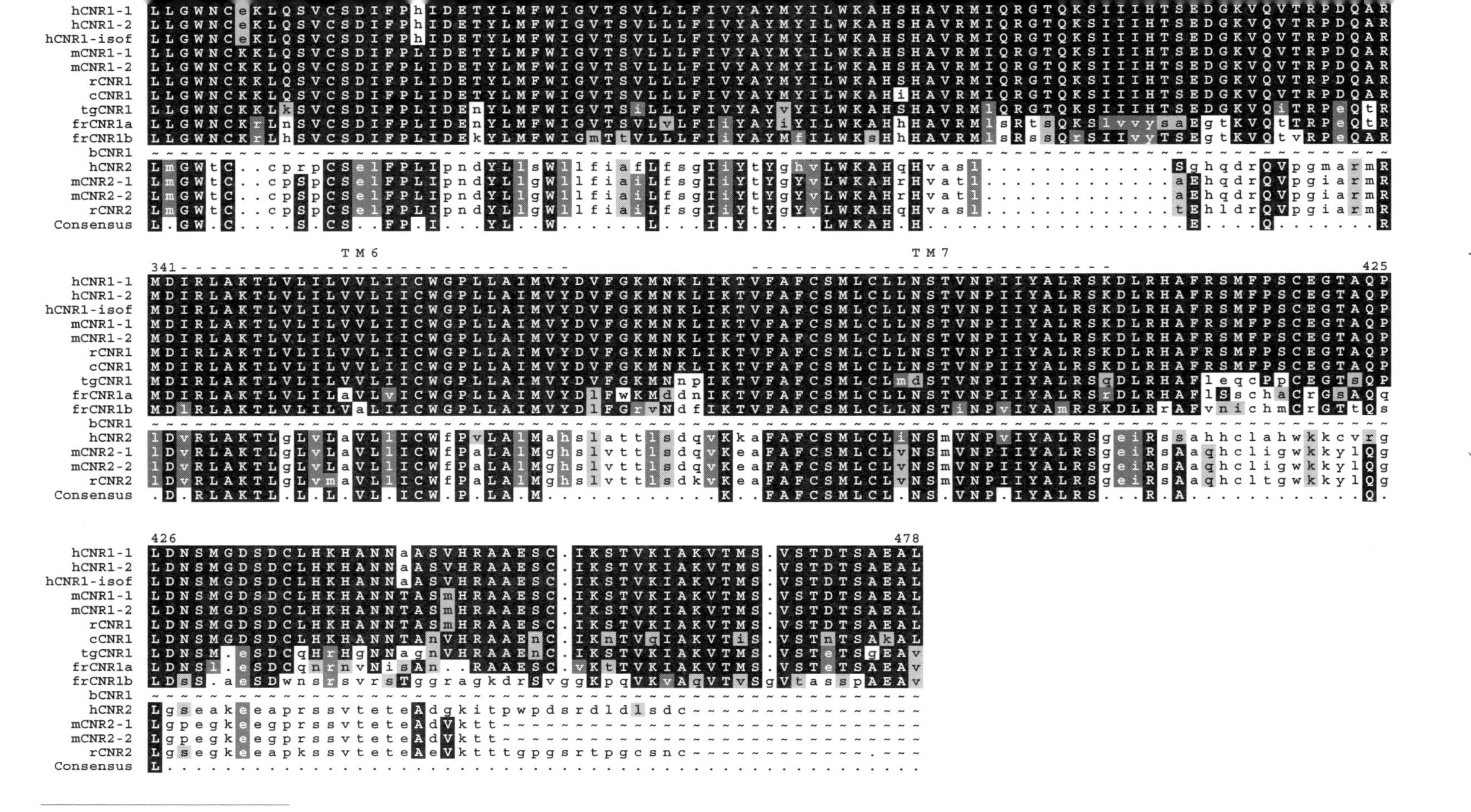

FIGURE 3.6 Amino acid sequence alignments of CB_1 and CB_2 CNRs. hCNR1, mCNR1, and mCNR2 have different entries in the Genbank. The plurality used is 4 and the consensus sequence is shown at the bottom of the alignments. Identical amino acid residues are in black, closely related ones in dark gray, less-closely related ones in light gray, and unrelated in white. The seven putative transmembrane domainsTM are indicated at the top of the alignments. The Genbank accession numbers are U73304 (hCNR1, human), AF107262 (hCNR1-2, human), X81121, U22948 (mCNR1-1, mouse), U17985 (mCNR1-2, mouse), X55812 (rCNR1, rat), U94342 (cCNR1, cat), AF181894 (tgCNR1, *Taricha granulosa*), X94401 (frCNR1a, type a from *F. rubripes*), X94402 (frCNR1b, type b), U77348 (bCNR1, partial bovine sequence), X74328 (hCNR2), X93168 (mCNR2-1), X86405 (mCNR2-2), and AF176350 (rCNR2).

MARIJUANA USE IN CELL SURVIVAL, NEUROGENESIS, AND APOPTOSIS

Marijuana use and the effects of exogenous synthetic cannabinoid administration, including the activation of endocannabinoids and multiple signaling capabilities, have been shown to control cell survival and death by inhibiting or stimulating cell growth. Thus, cannabinoids induce growth arrest or apoptosis in a number of transformed neural and nonneural cells in culture (Guzman et al., 2002). Cannabinoids of all classes also have the ability to protect neurons from a variety of insults that are believed to underlie delayed neuronal death after traumatic brain injury, including excitotoxicity, calcium influx, free radical formation, neurodegeneration, and neuroinflammation (Biegon, 2004). Specifically, an important physiological role of cannabinoids is neuroprotection, and in animal studies, inhibition of calcium influx into cells, antioxidant properties that reduce damage caused by oxygen radicals, and modulation of vascular tone has been shown (Grotenhermen, 2004). It has been proposed that an endogenous cannabinoid signaling may represent a key component of cell survival programs mobilized in the injured brain (Jin et al., 2004). *In vitro*, it was shown that the neuroprotective effect of cannabidiol on β-amyloid-induced toxicity in PC12 cells might be associated with caspase-3 signaling pathway (Esposito et al., 2004). There is currently no effective or beneficial pharmacotherapy for traumatic brain injury but the pathways and experimental models supporting a neuroprotective role for the various classes of cannabinoids has been identified as a potential therapeutic target in traumatic brain injury (Biegon, 2004). On the other hand, cannabinoid administration induces regression of malignant gliomas in rodents by a mechanism that may involve sustained ceramide generation and extracellular signal-regulated kinase activation (Guzman et al., 2002). In addition to cannabinoid-receptor-independent action of antitumor effects, abundant evidence now exist that phytocannabinoids and endocannabinoids inhibit the proliferation of some tumors and cancer cells by mechanisms involving inhibition of angiogenesis and the activity of K-ras oncogene product. It has been shown that the levels of endocannabinoids and of their known receptors can change in some tumors and cancer cells compared to the corresponding healthy tissues and cells, suggesting that endocannabinoids might tonically control cancer cell growth. One hypothesis to explain the dual effect of cannabinoid ligands as neurotoxic or neuroprotective agents is dependent on the dose and duration of the treatment. It was speculated that while acute administration will protect, chronic administration would induce neuronal death (Sarne and Keren, 2004).

With cannabinoids involved in cell death and cell survival, it is not clear how chronic marijuana use decreases adult neurogenesis in susceptible brain regions in those exposed to marijuana smoke. But cannabinoids and marijuana use have dose dependently been associated with apoptosis and neuroprotection, and marijuana use has been shown to regulate acute hippocampal memory function. A case for the influence of chronic use of abused substances, including smoking marijuana on hippocampal neurogenesis, has been proposed. Most drugs of abuse, including marijuana, are known to induce a decrease in "adult neurogenesis" in the hippocampus that is associated with learning and memory. The consequences of marijuana use in the regulation of hippocampal neurogenesis are unknown and whether inhibition of adult neurogenesis may contribute to reported effects of marijuana use on cognition remains to be determined.

Epigenetic regulation may play a role in the effects of cannabinoids as marijuana smoking may alter DNA methylation patterns. Preliminary reports indicate that marijuana smoking may lead to hypermethylation of a DNA repair gene (Tashkin et al., 2004) that may be associated with bronchial and buccal carcinogenesis. Cannabinoid epigenesis may explain individual differences and susceptibility to marijuana use. It is therefore of significant interest to determine whether the pathways associated with genetic basis of marijuana use may be linked with the effects of marijuana use and cell survival/death and adult neurogenesis. Interestingly, low doses of cannabinoids may enhance cell proliferation, whereas high doses usually induce growth apoptosis (Guzman et al., 2002). Other experimental evidence indicates that cannabinoids may protect normal neurons from toxic insults, such as glutamatergic overstimulation, ischemia, and oxidative damage (Guzman et al., 2002). The

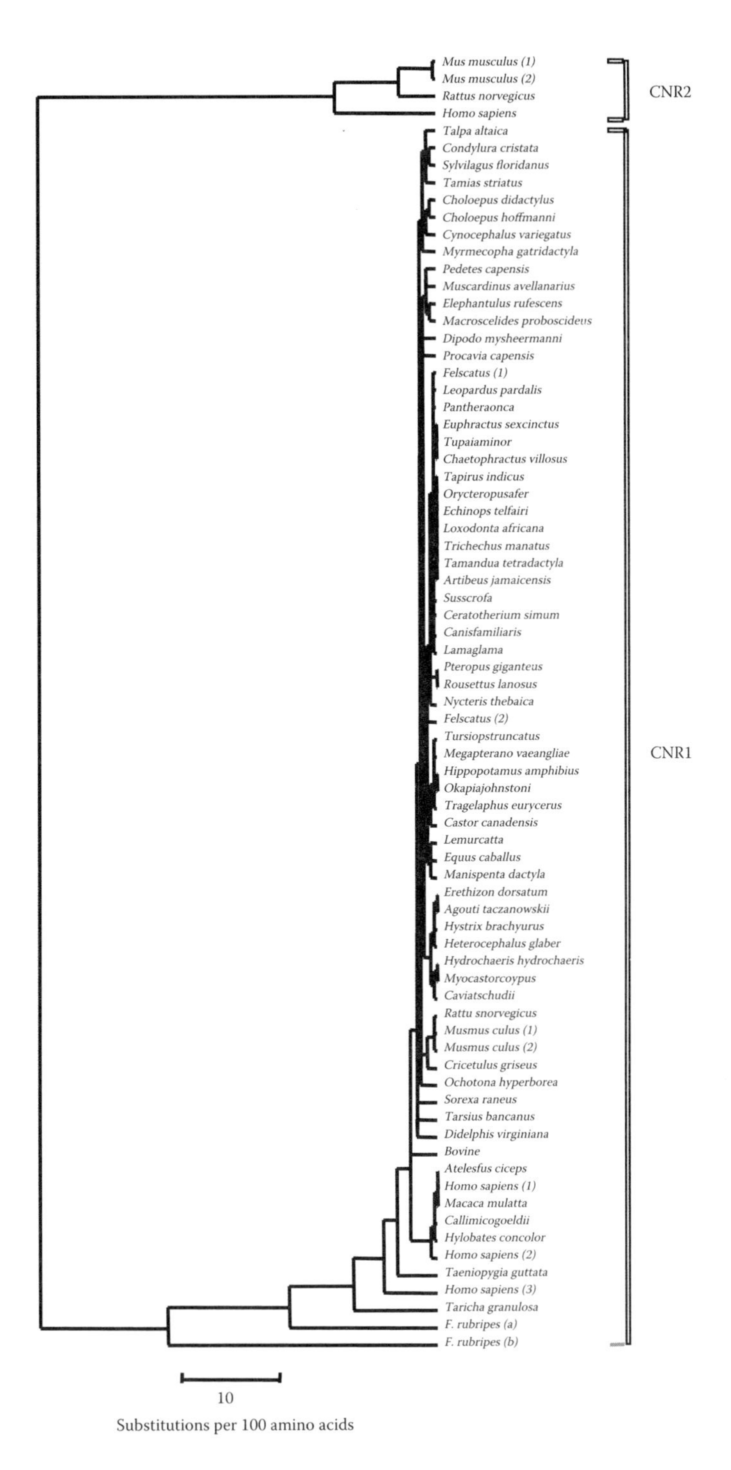

FIGURE 3.7A Phylogenetic tree of CNRs. GCG programs Distances and GrowTree (UPGMA) generated the graphical representation of CNRs phylogenetic tree. In the GeneBank for the same species, 1 and 2 are different entries; 3 is an isoform of the human CB_1 CNR protein; a, b: type A and type B. Murphy et al. (2001) utilized the CB_1 CNR gene in a phylogenetic analysis of mammals, and cloned and sequenced the CB_1 CNR gene from 62 species, sampled across all extant orders of placental mammals, as well as from a marsupial. The 62 CB_1 CNR gene from these species was included in the current CNR phylogenetic tree.

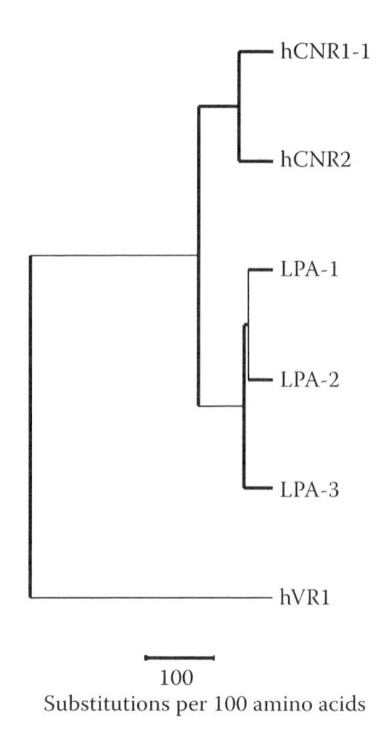

FIGURE 3.7B Phylogenetic relationship among human CNRs, lysophosphatidic, and vanilloid (VR1) receptors. Protein entry numbers used are Q92633 (LPA-1), Q9HBW0 (LPA-2), Q9UBY5 (LPA-3), and JC7621 (VRI). The homology between hCNR1 and hLPA-1 is about 40%.

mechanisms associated with cannabinoid protection of cells from oxidative cell death have also been linked to a receptor-independent mechanism (Chen and Buck, 2000). Taken together with previous studies, it appears that cannabinoid-receptor-dependent and -independent effects are involved in cell death and survival. When misregulated, apoptosis can contribute to various diseases including cancer and autoimmune and neurodegenerative diseases, all of which can be associated with activation of different components of the complex cannabinoid signal transduction pathway. Conversely, cannabinoids' role in cell survival and their potential as neuroprotectants in human traumatic brain injury may involve the activation of multiple protective mechanisms as demonstrated by dexanabinol (Biegon, 2004). The high level of expression of cannabinoid receptors and the display of natural constitutive activity in the mammalian brain may be associated with the intracellular translocation of the receptors (Leterrier et al., 2004). The relevance and physiological significance of the constitutive recycling of the cannabinoid receptors remains an open question. However, it is unknown if the effect of constitutive trafficking of cannabinoid receptors may be associated with lipid rafts during the long-term use of marijuana. Such a link with lipid rafts may be related to neuroplasticity, and such neuronal restructuring through lipid rafts may be involved in the mild withdrawal effects associated with marijuana use.

The contribution of environmental factors associated with cannabinoid genetics and epigenetics and neuroprotection and neurogenesis in the pleiotropic actions of cannabinoids and especially in the smoking of marijuana remains to be determined. Certainly, the discovery of the existence of the EPCS has enhanced our knowledge of the genetic basis of marijuana use.

AN ENDOCANNABINOID HYPOTHESIS OF SUBSTANCE ABUSE

There is accumulating evidence indicating a central role of an EPCS in the regulation of the rewarding effects of abused substances. Thus, a role for the EPCS in addiction has been proposed, based on the assumption of a role of the EPCS in brain reward mechanisms. The ubiquitous effects of cannabinoids and therefore marijuana use on almost any biochemical system support this hypothesis. Specifically, it appears that manipulation of some components of the EPCS might be effective at attenuating drug-seeking behavior or other signs of dependence and withdrawal to opiates, psychostimulants, or alcohol. With our current knowledge it has been difficult to determine all the relevant changes associated with or caused by addiction, but the discovery of EPCS, which is activated by marijuana use, certainly plays a crucial role in the manifestation of the effects of other abused substances. It has been suggested that the EPCS might be a component of the brain reward circuitry and thus play a role not only in cannabinoid tolerance and dependence, but also in dependence and withdrawal to other drugs of abuse (Gonzalez et al., 2003). An important and exciting discovery with further implication in the effects of other abused substances is retrograde signaling mediated by endocannabinoids in the CNS. As discussed above, endocannabinoids act as retrograde messengers at many synapses in the CNS to influence both synaptic plasticity and classical neurotransmitter release that are modulated by abused substances; this may have an impact in understanding and treating addictions. One of the enzymes that degrade endocannabinoids such as anandamide and similar fatty acid amide signaling ligands of the CBRs is FAAH, and a mutation in the FAAH gene, which is an important component of the EPCS, has been associated with problem drug and alcohol use (Sipe et al., 2002).

It is tantalizing in view of the above and our current knowledge to postulate that the EPCS may be a major integrator of the central network associated with the modifications of the effects of abused substances and perhaps the long-sought-after common substrate for addictions. Exogenous cannabinoids, and perhaps smoking marijuana, modulate several neurotransmitters and their receptors that are the targets of abused drugs. Furthermore, in drug-abuse-induced changes in the brain, the core features of addiction pathology include the cell-signaling and synaptic mechanisms associated with learning; memory and reward processes modulated by abused substances are, in part, regulated by EPCS. However, the nature of the interactions between the EPCS and pathways modulated by abused substances has continued to be a focus of intensive investigation. The major roles of some components of the EPCS have been described for the major classes of abused substances: thus a critical role for the cannabinoid CB_1Rs in alcohol dependence and stress-stimulated ethanol drinking was revealed in CB_1 mutant mice that were devoid of stress-induced ethanol consumption and withdrawal symptoms (Racz et al., 2003). Other studies have shown that endocannabinoid levels are increased in mouse brains chronically exposed to alcohol (Hungund and Basavarajappa, 2000) and a reduction occurred in CB_1Rs and transcripts in alcohol-preferring fawn hooded rats in comparison to alcohol-nonpreferring Wistar rats (Ortiz et al., 2004). These data and the modification of alcohol preference and self-administration by cannabinoid ligands demonstrate the involvement of the EPCS in the effects of alcohol and provide a basis for the targeting of some components of the EPCS in the treatment of alcohol dependence.

Further evidence not only indicates the central role of the EPCS in the reward circuitry but also in the dependency and withdrawal from other abused substances. In the case of opioids and cannabinoids, a number of studies have demonstrated a link and interaction between these two naturally occurring plant constituents in human biological systems, particularly in the reward and pain pathways in the CNS. Opioids and cannabinoids also have endogenous ligands that mimic the effects of the natural drugs. Thus, the cross talk between cannabinoid and opioids in animal models of opiate dependence and colocalization of CB_1R and mu-opioid receptor mRNAs in the brain support the existence of a cross interaction between opioid and cannabinoid systems in the addiction process (Navarro et al., 1998). Interestingly, both the opioid and cannabinoid systems produce endogenous ligands, and treatment with many abused substances including psychostimulants, alcohol, and THC induces the release of β-endorphin, an endogenous opioid in the reward centers (Solinas et al., 2004). In one study there were no changes in the single endocannabinoid, anandamide, that was measured in the brain regions analyzed in morphine-dependent rats, although changes in the binding of the CB_1 receptors were reported in the cerebral cortex and brain stem of these morphine-dependent rats (Gonzalez et al., 2003). Rimonabant reduced the rewarding effects of morphine in mice in the CPP paradigm (Mas-Nieto et al., 2001), which supports the hypothesis of a potential role of cannabinoid receptors in the neurobiological changes that culminate in opioid (Navarro et al., 1998) and substance dependence. It has also been observed that maternal exposure to THC increases morphine place preference in adult male offspring, suggesting the influence of the endocannabinoid system in the rewarding effects of morphine (Manzanedo et al., 2004).

Dopaminergic neurotransmission has been implicated in the mechanisms associated with psychostimulant dependence and many substances of abuse, including marijuana (Cheer et al., 2004). The presence of CB_1 receptors in dopaminergic neurons strongly suggests that cannabinoids play a modulatory role in dopaminergic neuronal pathways, and the release of dopamine prompted the speculation of cross talk between endocannabinoids and dopamine-dependent reward mechanisms (Wenger and Furst, 2004). Cannabinoid agonist was reported to provoke relapse to cocaine seeking after prolonged withdrawal periods whereas the antagonist rimonabant attenuates relapse induced by re-exposure to cocaine, implicating an endocannabinoid mechanism in relapse to cocaine seeking (De Vries, 2001). The cannabinoid mechanism in relapse to cocaine taken together with alteration in endocannabinoid levels in animal models of cocaine and nicotine dependence suggest an important role of the EPCS in the neuronal processes underlying relapse to cocaine seeking (De Vries, 2001).

We have continued studies on the interaction between the EPCS and substances with dependence potential using mouse strains in the plus-maze and CPP tests. The *in vivo* pharmacological interaction was evaluated by the coadministration of cannabinoid and abused substances. Acutely, the cannabinoid agonist WIN55212-2 induced mouse aversions in the plus-maze test similar to those induced by capsaicin, which were long lasting and persistent after 2 d of withdrawal from chronic treatment with capsaicin. While 30-min pretreatment with WIN55212-2 enhanced the aversions induced by capsaicin, the vanilloid receptor antagonist capsazepine blocked the aversions induced by capsaicin and WIN55212-2, respectively. Data obtained from the alcohol studies also implicate an interaction between the cannabinoid and vanilloid systems. Taken together with the evidence reviewed above, our data support the possibility of cross talk between cannabinoids and abused substances. The data also indicate that cannabinoid CB_1R antagonism reduced behavioral aversions following withdrawal from abused substances. It is therefore suggested that the EPCS may be playing a central role in reward pathways in the nervous system. Several preclinical studies provide evidence for the interaction between the EPCS and dependence on abused substances and provides a rationale for the trial of cannabinoid ligands in the treatment of drug and alcohol dependence.

IMPLICATION FOR THE MEDICAL USE OF MARIJUANA

The use of cannabis for both recreational and medicinal purposes dates back for thousands of years. In recent times, there has been increased attention that marijuana should be legalized for medicinal use in AIDS, cancer, obesity, multiple sclerosis, and other medical conditions where patients might benefit from the pharmacological effects of cannabis. Synthetic cannabinoids such as dronabinol, marinol, and nabilone already have an established use as antiemetics in nausea and vomiting associated with cancer chemotherapy. The reported beneficial effects in cancer and AIDS patients might be reflected in improved weight gain, owing to the well-documented antiemetic and appetite-stimulating effects of cannabinoids. This might be a major advantage for cancer patients undergoing rigorous chemotherapy or for advanced AIDS patients. Interestingly, although cannabis is widely used as a recreational drug in humans, only a few studies have revealed an appetite stimulant potential of cannabinoids in animals. However, evidence for the aversive effects of Δ^9-THC, WIN55212-2, and CP55940 on appetite is more readily obtained in a variety of tests. The selective blockade of CB_1R rimonabant impaired the perception of the appetitive value of positive reinforcers (cocaine, morphine, and food) and reduced the motivation for sucrose, beer, and alcohol consumption, indicating that positive incentive and motivational processes could be under the permissive control of CNR-related mechanisms. With the implication of the EPCS in appetite regulation and critical roles in reward pathways, it appears that blockade of the CB_1 receptors could suppress the powerful cravings that drive the cycle of addiction. Taken together, the appearance of the endocannabinoid system in development, its presence in breast milk, and its activation by drug and alcohol use, including smoking marijuana, may provide a target to control cravings for food and chemical dependency. In recent clinical trials, rimonabant, an antagonist at the CB_1 cannabinoid receptor, was reported to be effective in suppressing the powerful cravings that drive people to overeat and smoke. The abundance of EPCS in the human body and brain in comparison to other G-protein-coupled-receptors supports the critical role of this system in most physiological and biochemical functions that affect behavioral changes after smoking marijuana. Therefore, several cannabinoid-mediated effects are of interest for therapeutic applications. There is now excitement and interest in the potential use of cannabinoid ligands and compounds derived from cannabinoids such as dexanabinol for the treatment of brain injury. The use of marijuana can be addicting in some individuals and, interestingly, the pharmacological antagonism of the CB_1 cannabinoid receptors has shown promise in blocking the craving for food and addictive substances in preliminary studies. However, whether or not the antagonism of the cannabinoid receptors will be effective in treating individuals with problematic and uncontrollable marijuana-smoking addiction and who wish to quit remains to be demonstrated. Furthermore, endocannabinoids and other endogenous fatty acid ethanolamides with their roles in sleep and inflammation are emerging as new, important biological signaling molecules and targets for therapeutic intervention (Boger et al., 1998). Other targets for potential development of therapeutic agents are cannabinoid uptake inhibitors, if they indeed exist. When the physiological roles of CNRs and endocannabinoids are established, it is likely that other therapeutic targets may be uncovered. Based on what we currently know about the anatomical distribution of CNRs and endocannabinoids, and from marijuana users themselves, disorders associated with memory and motor coordination may benefit from novel cannabinoids. Similarly, CNRs and endocannabinoids in the periphery can be targeted in immune disorders and blood pressure regulation (Mechoulam, 1999; Varga et al., 1998). This is to be expected as many novel drugs are based on chemical modifications of transmitters. Therefore, it seems reasonable to expect that, as with other receptor-transmitter systems, excess or lack of CNR or endogenous ligands may be the cause for disorders in the CNS or those associated with the immune system (Mechoulam, 1999) and other peripheral organs.

CONCLUDING REMARKS AND FUTURE DIRECTIONS

The recent significant progress in marijuana-cannabinoid research has led to a new understanding about the biological effects and the unique therapeutic possibilities of targeting the EPCS, which is activated by THC and marijuana use. In just a few years since the discovery of endocannabinoids that serve as cannabinoids *in vivo*, these lipid-signaling molecules and endolipids have been shown to participate in a broad array of physiological and pathological processes. However, the impact of "lipidomics" and these new lipid-signaling pathways in modulating behavioral neurobiology and aspects of drug addiction is of current interest. It is now clear that marijuana use is addicting in vulnerable populations. Just like most medications that are not benign, benzodiazepines, opiates, and marijuana have side effects that accompany their therapeutic use. Unlike opiates, the intensity of the craving and withdrawal reactions appear to be mild and vary from individual to individual. Thus, we can look forward to a bright future of discoveries concerning the role of endocannabinoids in brain function, immune function, reproductive function, and emotional behavior, as well as the discovery of potential medicines to improve the health of many who suffer from various disorders. One notable future direction is in brain injury and stroke for which there are currently no major drugs, and one cannabinoid based drug, HU211 (dexanabinol), has successfully completed phase II trials. The success of rimonabant in clinical trials for obesity and cessation from smoking is heightening interest in the therapeutic potential of manipulating the endogenous marijuana system. Marijuana (cannabinoid) research appears to offer a solid scientific background as a significant contribution in understanding human biology and in the development of cannabinoid based therapeutics. The only medical uses for which there has been rigorous scientific evidence are in the treatment of sickness associated with cancer chemotherapy and to counter the loss of appetite and wasting syndrome in AIDS (Iversen, 2000). There is, however, scientific evidence for the potential therapeutic use of cannabinoids for the treatment of psychomotor disorders, including spasticity, multiple sclerosis, epilepsy, disk prolapse, and attention deficit hyperactivity disorder (ADHD), and its pain relief properties are no longer a matter of speculation. It is only a matter of time before new medications for the pharmacological manipulation of the link between endocannabinoid and other neurotransmitter systems will be developed as targets for the numerous indications of marijuana and endocannabinoid ligands. In the absence of clinical data on the abuse liability of endocannabinoids, the preclinical data indicate that as with marijuana, the endocannabinoids may have minimal abuse liability in humans. The safety profile of THC, the active ingredient of cannabis, is good. It has low toxicity in both the short and long term. As concluded by Iverson (2000), marijuana is not a completely benign substance. It is a powerful drug with a variety of effects. However, except for the harm associated with smoking, the adverse effects of marijuana use are within the range of the side effects tolerated for other medications. Unraveling the genetics of the EPCS that is a target for smoking marijuana is dependent on further understanding of the complex CB_1, CB_2, and CB_n gene structure and the associated regulatory elements that are still poorly understood.

ACKNOWLEDGMENTS

G. R. Uhl's laboratory is supported by NIDA/NIH intramural program. B. E. Akinshola acknowledges the support of grant AA13415 from NIAAA. E.S. Onaivi wishes to especially thank the Center for Research at William Paterson University for research and support of students working with him; he also acknowledges the continued Guest Scientist support by NIDA/NIH.

REFERENCES

Abood, M.E., Ditto, K.E., Noel, M.A., Showalter, V.M., Tao, Q. (1997) Isolation and expression of a CB1 cannabinoid receptor gene. Comparison of binding properties with those of native CB1 receptors in mouse brain and N18TG2 neuroblastoma cells. *Biochem. Pharmacol.* 24, 207–214.

Aceto, M.D., Martin, B.R., Scates, S.M., Loew, J. (1995) Cannabinoid precipitated withdrawal: Induction by the antagonist SR141716A. *FASEB J.* 9, 3, 2320.

Acousta-Urquidi, J., Chase, R. (1975) The effects of delta-9-tetrahydrocannabinol on action potentials in the mollusk *Aplysia. Can. J. Physiol. Pharmacol.* 53, 793–798.

Ahluwalia, J., Urban, L., Capogna, M., Bevan, S., Nagy, I. (2000) Cannabinoid 1 receptors are expressed in nociceptive primary sensory neurons. *Neuroscience* 100, 685–688.

Ahluwalia. J., Yaqoob, M., Urban, L., Bevan, S., Nagy, I. (2003) Activation of capsaicin-sensitive primary sensory neurones induces anandamide production and release. *J. Neurochem.* 84, 585–591.

Akinshola B.E., Chakrabarti A., Onaivi E.S. (1999a) In vitro and in vivo action of cannabinoids. *Neurochem. Res.* 24, 1233–1240.

Akinshola B.E., Taylor R.E., Ogunseitan A.B., Onaivi E.S. (1999b). Anandamide inhibition of recombinant AMPA receptor subunits in Xenopus oocytes is increased by forskolin and 8-bromo-cyclic AMP. *Naunyn-Schmiedeberg's Arch. Pharmacol.* 360, 242–248.

Akinshola B.E., Fryar E.B., Taylor R.E., Onaivi E.S. (2002) Cannabinoid and vanilloid agonist sensitivities of AMPA glutamate receptor subunits. *Drug Alcohol. Depend.* 66, S5; 12.

Akinshola, B.E. (2001) Straight-chain alcohols exhibit a cutoff in potency for the inhibition of recombinant glutamate receptor subunits. *Br. J. Pharmacol.* 133, 651–658.

Alger, B.E. (2002) Retrograde signaling in the regulation of synaptic transmission: focus on endocannabinoids. *Prog. Neurobiol.* 68, 247–286.

Al-Hayani, A., Wease, K.N., Ross, R.A., Pertwee, R.G., Davies, S.N. (2001) The endogenous cannabinoid anandamide activates vanilloid receptors in the rat hippocampal slice. *Neuropharmacology.* 418, 1000–1005.

Ameri, A. (1999) The effects of cannabinoids on the brain. *Prog. Neurobiol.* 58, 315–348.

Andersson, H., D'Antona, A.A., Kendall, D.A., Heijne, G.V., Chin, C.-N. (2003) Membrane assembly of the cannabinoid receptor 1: impact of a long N-terminal tail. *Mol. Pharmacol.* 64, 570–577.

Arnone, M., Maruani, J., Chaperon, F., Thiebot, M.H., Poncelet, M., Soubrie, P., Le Fur, G. (1997) Selective inhibition of sucrose and ethanol intake by SR141716, an antagonist of central cannabinoid (CB1) receptors. *Psychopharmacology* 132, 104–106.

Azad, S.C., Marsicano, G., Eberlein, I., Putzke, J., Zieglgansberger, W., Spanagel, R., Lutz, B. (2001) Differential role of the nitric oxide pathway that is involved in some, but not all of the actions of Δ^9-THC-induced central nervous effects in the mouse. *Eur. J. Neurosci.* 13, 561–568.

Barann, M., Molderings, G., Bruss, M., Bonisch, H., Urban, B.W., Gothert, M. (2002) Direct inhibition by cannabinoids of human 5-HT3A receptors: probable involvement of an allosteric modulatory site. *Br. J. Pharmacol.* 137, 589–596.

Basavarajappa, B.S., Hungund, B.L. (1999) Down-regulation of cannabinoid receptor agonist-stimulated [35S]GTP gamma S binding in synaptic plasma membrane from chronic ethanol exposed mouse *Brain Res.* 815, 89–97.

Basavarajappa, B.S., Saito, M., Cooper, T.B., Hungund, B.L. (2000) Stimulation of cannabinoid receptor agonist 2-arachidonylglycerol by chronic ethanol and its modulation by specific neuromodulators in cerebellar granule neurons. *Biochim. Biophys. Acta* 1535, 78–86.

Begg, M., Mo, F-M., Offertaler, L., Batkai, S., Pacher, P., Razdan, R.K., Lovinger, D.M., Kunos, G. (2003) G protein-coupled endothelial receptor for atypical cannabinoid ligands modulates a Ca2+-dependent K+ current. *J. Biol. Chem.* 278, 46188–46194.

Biegon, A. (2004) Cannabinoids as neuroprotective agents in traumatic brain injury. *Curr. Pharm. Design* 10, 2177–2183.

Belue, R.C., Howlett, A.C., Westlake, T.M., Hutchings, D.E. (1995) The ontogeny of cannabinoid receptors in the brain of postnatal and aging rats. *Neurotoxicol. Teratol.* 17, 25–30.

Benito, C., Numez, E., Tolon, R.M., Carrier, E.J., Rabano, A., Hillard, C.J., Romero, J. (2003). Cannabinoid CB2 receptors and fatty acid amide hydrolase are selectively overexpressed in neuritic plaque-associated glia in Alzheimer's disease brains. *J. Neurosci.* 23, 11136–11141.

Berdyshev, E.V. (2000) Cannabinoid receptors and the regulation of immune response. *Chem. Phys. Lipids.* 108, 169–190.

Bisogno, T., Katayama, K., Melck, D., Ueda, N., De Petrocellis, L., Yamamoto, S., Di Marzo, V. (1998) Biosynthesis and degradation of bioactive fatty acid amides in human breast cancer and rat pheochromocytoma cells — implication for cell proliferation and differentiation. *Eur. J. Biochem.* 254, 634–642.

Bisogno, T., Ventriglia, M., Milone, A., Mosca, M., Cimino, G., Di Mazo, V. (1997) Occurrence and metabolism of anandamide and related acyl-ethanolamides in ovaries of the sea urchin *Paracentrotus lividus*. *Biochim. Biophys. Acta* 1345, 338–348.

Black, C.M. (2004) Cannabinoid receptor antagonists and obesity. *Curr. Opin. Invest. Drugs* 5, 389–394.

Boger, D.L., Henriksen S.J., Cravatt, B.F. (1998) Oleamide: An endogenous sleep-inducing lipid and prototypical member of a new class of biological signaling molecules. *Curr. Pharm. Design* 4, 303–314.

Bohme, G.A., Laville, M., Ledent, C., Parmentier, M., Imperato, A. (2000) Enhanced long-term potentiation in mice lacking cannabinoid CB1 receptors. *Neuroscience* 95, 5–7.

Brady, R.O., Carbone, E. (1973) Comparison of the effects of 9-tetrahydrocannabinol, 11-hydroxy-9-tetrahydrocannabinol, and ethanol on the electrophysiological activity of the giant axon of the squid. *Neuropharmacology* 12, 601–605.

Bouaboula, M., Rinaldi, M., Carayon, P., Carrilon, C., Delpech, B., Shire, D., Le Fur, G., Casellas, P. (1993) Cannabinoid receptor expression in human leucocytes. *Eur. J. Biochem.* 214, 173–180.

Bramblett, R.D., Panu, A.M., Ballesteros, J.A., Reggio, P.H. (1995) Construction of a 3D model of the cannabinoids CB1 receptor: Determination of the helix ends and helix orientation. *Life Sci.* 56, 1971–1982.

Breivogel, C.S., Di Marzo, V., Zimmer, A.M., Zimmer, A., Martin, B.R. (2000) Characterization of an unknown cannabinoid receptor in CB1 knockout mouse brain membranes. 2000 Symposium on the Cannabinoids, Burlington, Vermont, ICRS, p. 6.

Breivogel, C.S., Griffin, G., Di Marzo, V., Martin, B.R. (2001) Evidence for a new G protein-coupled cannabinoid receptor in mouse brain. *Mol. Pharmacol.* 60, 155–163.

Brown, S.P., Brenowitz, S.D., Gegehr, W.G. (2003). Brief presynaptic bursts evoke synapse-specific retrograde inhibition mediated by endogenous cannabinoids. *Nat. Neurosci.* 10, 1048–1057.

Brown, D. (2002) Structure and function of membrane rafts. *Int. J. Med. Microbiol.*, 291, 433–437.

Bruinsma, K., Taren, D.L. (1999) Chocolate: Food or drug? *J. Am. Diet Assoc.* 99, 1249–1256.

Buckland P.R. (2001) Genetic association studies of alcoholism — problems with the candidate gene approach. *Alcohol Alcohol.* 36, 99–103.

Buckley, N.E., Hansson, S., Harta, G., Mezey, E. (1998) Expression of the CB1 and CB2 receptor messenger RNAs during embryonic development in the rat. *Neuroscience* 82, 1131–1149.

Buckley, N.E., McCoy, K.L., Mezey, E., Bonner, T., Zimmer, A., Felder, C.C., Glass, M., Zimmer, A. (2000) Immunomodulation by cannabinoids is absent in mice deficient for the cannabinoid CB2 receptor. *Eur. J. Pharmacol.* 396, 141–149.

Buckley, N.E., Mezey, E., Bonner, T., Zimmer, A., Felder, C.C., Glass, M., Zimmer, A. (1997) Development of a CB2 knockout mouse. 1997 Symposium on the Cannabinoids, Burlington, Vermont, International Cannabinoid Research Society, 57.

Caenazzo, L., Hoehe, M.R., Hsieh, W.-T., Berrettini, W.H., Bonner, T.I., Gershon, E.S. (1991) HindIII identifies a two allele DNA polymorphism of the human cannabinoid receptor gene (CNR). *Nucleic Acids Res.* 19, 4798.

Calandra, B., Portier, M., Kerneis, A., Delpech, M., Carillon, C., Le Fur, G., Ferrara, P., Shire, D. (1999) Dual intracellular signaling pathways mediated by the human cannabinoid CB1 receptor. *Eur. J. Pharmacol.* 374, 445–455.

Calignano, A., La Rana, G., Giuffrida, A., Piomelli, D. (1998) Control of pain initiation by endogenous cannabinoids. *Nature* 394, 277–281.

Cao, Q., Martinez, M., Zhang, J., Sanders, A.R., Badner, J.A., Cravchik, A., Markey, C.J., Beshah, E., Guroff, J.J., Maxwell, M.E., Kazuba, D.M., Whiten, R., Goldin, L.R., Gershon, E.S., Gejman, P.V. (1997) Suggestive evidence for a schizophrenia susceptibility locus on chromosome 6q and a confirmation in an independent series of pedigrees. *Genomics* 43, 1–8.

Carlisle, S.J., Marciano-Cabral, F., Staab, A., Ludwick, C., Cabral, G.A. (2002) Differential expression of the CB2 cannabinoid receptor by rodent macrophages and macrophage — like cells in relation to cell activation. *Int. Immunopharmacol.* 2, 69–82.

Castellano, C., Rossi-Arnaud, C., Cestari, V., Costanzi, M. (2004) Cannabinoids and memory: animal studies. *Curr. Drug Target CNS Neurol. Disord.* 2, 389–402.

Caterina, M.J., Leffler, A., Malmberg, A.B., Martin, W.J., Trafton, J., Petersen-Zeitz, K.R., Koltzenburg, M., Basbaum, A.I., Julius, D. (2000) Impaired nociception and pain sensation in mice lacking the capsaicin receptor. *Science* 288, 306–313.

Caterina, M.J., Schumacher, M.A., Tominaga, M., Rosen, T.A., Levine, J.D., Julius, D. (1997) The capsaicin receptor: a heat-activated ion channel in the pain pathway. *Nature* 389, 816–824.

Chakrabarti, A., Ekuta, J.E., Onaivi, E.S. (1998) Neurobehavioral effects of anandamide and cannabinoid receptor gene expression in mice. *Brain Res. Bull.* 45, 67–74.

Chakrabarti, A., Onaivi, E.S., Chaudhuri, G. (1995) Cloning and sequencing of a cDNA encoding the mouse brain-type cannabinoid receptor protein. *DNA Sequence* 5, 385–388.

Chaperon, F., Thiebot, M.-H. (1999) Behavioral effects of cannabinoid agents in animals. *Crit. Rev. Neurobiol.* 13, 243–281.

Chaperon, F., Soubrie, P., Puech, A.J., Thiebot, M.H. (1998) Involvement of central cannabinoid (CB1) in the establishment of place conditioning in rats. *Psychopharmacology* 135, 324–332.

Cheer, J.F., Wassum, K.M., Heien, M.L., Phillips, P.E., Wightman, R.M. (2004) Cannabinoids enhance subsecond dopamine release in the nucleus accumbens of awake rats. *J. Neurosci.* 24, 4393–4400.

Chen, Y., Buck, J. (2000) Cannabinoids protect cells from oxidative cell death: A receptor-independent mechanism. *J. Phanrmacol. Exp. Ther.* 293, 807–812.

Chevaleyre V., Castillo P.E. (2003) Heterosynaptic LTD of hippocampal GABAergic synapses: a novel role of endocannabinoids in regulating excitability. *Neuron* 38, 461–472.

Chiamulera, C., Epping-Jordan, M.P., Zocchi, A., Marcon, C., Cottiny, C., Tacconi, S., Corsi, M., Orzi, F., Conquet, F. (2001) Reinforcing and locomotor stimulant effects of cocaine are absent in mGluR5 null mutant mice. *Nat. Neurosci.* 4, 873–874.

Childers, S.R., Breivogel, C.S. (1998) Cannabis and endogenous cannabinoid systems. *Drug Alcohol. Depend.* 51, 173–187.

Collingridge, G.L., Singer, W. (1990) Excitatory amino acid receptors and synaptic plasticity. *Trends Pharmacol. Sci.* 7, 290–296.

Colombo, G., Agabio, R., Fa, M., Guano, L., Lobina, C., Loche, A., Reali, R., Gessa, G.L. (1998) Reduction of voluntary ethanol intake in ethanol-preferring sP rats by the cannabinoid antagonist SR141716. *Alcohol Alcohol.* 33, 126–130.

Comings, D.E. (1996) Genetic factors in drug abuse and dependence. *NIDA Res. Monogr.* 159, 16–48.

Comings, D.E., Muhleman, D., Gade, R., Johnson, P., Verde, R., Saucier, G., MacMurray, J. (1997) Cannabinoid receptor gene (CNR1): association with IV drug use. *Mol. Psychiatry* 2, 161–168.

Cook, S.A., Lowe, J.A., Martin, B.R. (1998) CB1 receptor antagonist precipitates withdrawal in mice exposed to Delta9-tetrahydrocannabinol. *J. Pharmacol. Exp. Ther.* 285, 1150–1156.

Costa, B., Colleoni, M. (1999) SR141716A induces in rats a behavioral pattern opposite to that of CB1 receptor agonist. *Zhongguo Yao Li Xue Bao* 20, 1103–1108.

Crabbe, J.C. (2002) Genetic contributions to addictions. *Annu. Rev. Psychol.* 53, 435–462.

Cravatt, B.F., Lichtman, A.R. (2001) Fatty acid amide hydrolase: an emerging therapeutic target in the endocannabinoid system. *Curr. Opin Chem. Biol.* 7, 469–475.

Cravatt, B.F., Demarest, K., Patricelli, M.P., Bracey, M.H., Giang, D.K., Martin, B.R., Lichtman, A. *Proc. Natl. Acad. Sci. U S A* 98, 9371–9376.

Croxford, J.L. (2003) Therapeutic potential of cannabinoids in CNS disease. *CNS Drugs.* 17, 179–202.

Daaka, Y., Friedman, H., Klein T.W. (1996) Cannabinoid receptor proteins are increased in Jurkat human-T cell line after mitogen activation. *J. Pharmacol. Exp. Ther.* 276, 776–783.

Das, S.K., Paria, B.C., Chakraborty, I., Dey, S.K. (1995) Cannabinoid ligand-receptor signaling in the mouse uterus. *Proc. Natl. Acad. Sci. U S A* 92, 4332–4336.

Dawson, E. (1995) Identification of a polymorphic triplet marker for the brain cannabinoid receptor gene: use in linkage and association studies of schizophrenia. *Psychiatr. Genet.* 5, S50.

de Fonseca, F.R., Ramos, J.A., Bonnin, A., Fernandez-Ruiz, J.J. (1993) Presence of cannabinoid binding sites in the brain from early postnatal ages. *Neuroreport* 4, 135–138.

De Petrocellis L., Melck, D., Bisogno, T., Milone, A., Di Marzo, V. (1999) Finding of the endocannabinoid signaling system in Hydra, a very primitive organism: possible role in the feeding response. *Neuroscience* 92, 377–387.

Dennedy, M.C., Friel, A.M., Houlihan, D.D., Broderick, V.M., Smith, T., Morrison, J.J. (2004) Cannabinoids and human uterus during pregnancy. *Am. J. Obstet. Gynecol.* 190, 2–9.

Derkinderen P., Valjent E., Toutant M., Corvol J.C., Enslen H., Ledent C., Trzaskos J., Caboche J., Girault J.H. (2003) Regulation of extracellular signal-regulated kinase by cannabinoids in hippocampus. *J. Neurosci.* 23, 2371–2382.

Deutsch, D.G., Chin, S. (1993) Enzymatic synthesis and degradation of anandamide, a cannabinoid receptor antagonist. *Biochem. Pharmacol.* 46, 791–796.

Devane, W.A., Dysarz, F.A., Johnson, M.R., Melvin, L.S., Howlett, A.C. (1988) Determination and characterization of cannabinoid receptor in rat brain. *Mol. Pharmacol.* 34, 605–613.

Devane, W.A., Hanus, L., Breuer, A, Pertwee, R.G., Stevenson, L.A., Griffin, G., Gibson, D., Mandelbaum, A., Etinger, A., Mechoulam, R. (1992) Isolation and structure of a brain constituent that binds to the cannabinoid receptor. *Science* 258, 1946–1949.

De Vries, T.J., Shaham, Y., Homberg, J.R., Crombag, H., Schuurman, K., Dieben, J., Vanderschuren, L.J., Schoffelmeer, A.N. (2001) A cannabinoid mechanism in relapse to cocaine seeking. *Nat. Med.* 7, 1151–1154.

Di Marzo, V., Bisogno, T., Melck, D., Ross, R., Brockie, H., Stevenson, L., Pertwee, R. G. and De Petrocellis, L. (1998a) Interactions between synthetic vanilloids and the endogenous cannabinoid system. *FEBS Lett.* 436, 449–454.

Di Marzo V., Melck D., Bisogno T., De Petrocellis L. (1998b). Endocannabinoids: endogenous cannabinoid receptor ligands with neuromodulatory action. *TINS* 21, 521–528.

Di Marzo, V., Breivogel, C.S., Tao, Q., Bridgen, D.T., Razdan, R.K., Zimmer, A.M., Zimmer, A., Martin, B.R. (2000a) Levels, metabolism, and pharmacological activity of anandamide in CB1 cannabinoid receptor knockout mice: Evidence for non-CB1, non-CB2 receptor-mediated actions of anandamide in mouse brain. *J. Neurochem.* 75, 2434–2444.

Di Marzo, V., Hill, M.P., Bisogno, T., Crossman, A.R., Brotie, J.M. (2000b) Enhanced levels of endogenous cannabinoids in the globus pallidus are associated with a reduction in movement in an animal model of Parkinson's disease. *FASEB J.* 14, 1432–1438.

Di Marzo, V., Bisogno, T., De Petrocellis, L. (2000c) Endocannabinoids: New targets for drug development. *Curr. Pharm. Design.* 6, 1361–1380.

Di Marzo, V., Bisogno, T., De Petrocellis, L., Melck, D., Martin B.R. (1999) Cannabimimetic fatty acid derivatives: The anandamide family and other endocannabinoids. *Curr. Med. Chem.* 6, 721–744.

Di Marzo, V., Fontana, A., Cadas, H., Schinelli, S., Cimino, G., Schwartz, J.C., Piomelli, D. (1994) Formation and inactivation of endogenous cannabinoid anandamide in central neurons. *Nature* 372, 686–691.

di Tomaso, E., Beltramo, M., Piomelli, D. (1996) Brain cannabinoids in chocolate. *Nature* 382, 677–678.

Doherty, J., Dingledine, R. (2001) Functional interactions between cannabinoid and metabotropic glutamate receptors in the central nervous system. *Curr. Opin. Pharmacol.* 3, 46–53.

Dohlman, H.G., Thorner, J., Caron, M.G., Lefkowitz, R.J. (1991) Model systems for the study of seven-transmembrane-segment receptors. *Rev. Biochem.* 60, 653–688.

Egertova, M., Cravatt, B.F., Elphick, M.R. (2000) Fatty acid amide hydrolase expression in rat choroids plexus: possible role in regulation of the sleep-inducing action of oleamide. *Neurosci. Lett.* 282, 13–16.

Elphick, M.R. (1998) An invertebrate G-protein coupled receptor is a chimeric cannabinoid/melanocortin receptor. *Brain Res.* 780, 170–173.

Elphick, M.R., Egertova, M. (2001) The neurobiology and evolution of cannabinoid signaling. *Phil. Trans. R. Soc. London B* 356, 381–408.

Esposito, G., Iuvone, T., Esposito, R., Santamaria, R., Izzo, A.A. (2004) Neuroprotective effect of cannabidiol on β-amyloid-induced toxicity in PC12 cell. ICRS Abstracts, 46.

Facci, L., Dal Toso, R., Romanello, S., Buriani, A., Skaper, S.D., Leon, A. (1995) Mast cells express a peripheral cannabinoid receptor with differential sensitivity to anandamide and palmitoylethanolamide. *Proc. Natl. Acad. Sci. U S A* 92, 3376–3380.

Faubert, B.L., Kaminski, N.E. (2000) AP-1 activity is negatively regulated by cannabinol through inhibition of its protein components, c-fos and c-jun. *J. Leukoc. Biol.* 67, 259–266.

Fegley, D., Kathuria, S., Mercier, R., Li, C., Goutopoulos, A., Makriyannis, A., Piomelli, D. (2004). Anandamide transport is independent of fatty-acid amide hydrolase activity and is blocked by the hydrolysis-resistant inhibitor AM1172. *Proc. Natl. Acad. Sci. USA.* 101, 8756–8761.

Felder, C.C., Veluz, J.S., Williams, H.L., Briley, E.M., Matsuda, L.A. (1992) Cannabinoid agonists stimulate both receptor- and non-receptor-mediated signal transduction pathways in cells transfected with and expressing cannabinoid receptor clones. *Mol. Pharmacol.* 42, 838–845.

Fernandez-Ruiz, J., Berrendero, F., Hernandez, M.L., Ramos, J.A. (2000) The endogenous cannabinoid system and brain development. *Trends Neurosci.* 23 14–20.

Fezza, F., Dillwith, J.W., Bisogno, T., Tucker, J.S., Di Marzo, V., Sauer, J.R. (2003) Endocannabinoids and related fatty acid amides, and their regulation, in salivary glands of the lone star tick. *Biochim. Biophys. Acta* 1633, 61–67.

Fowler, C.J., Nilson, O., Andersson, M., Disney, G., Jacobsson, S.O.P., Tiger, G. (2001) Pharmacological properties of cannabinoid receptors in the avian brain: Similarity of rat and chicken $cannabinoid_1$ receptor recognition sites and expression of $cannabinoid_2$ receptor-like immunoreactivity in the embryonic chick brain. *Pharmacol. Toxicol.* 88, 213–222.

Freedland, C.S., Sharpe, A.L., Samson, A.H., Porrino, L.J. (2001). Effects of SR141716A on ethanol and sucrose self-administration. *Alcohol. Clin. Exp. Res.* 25, 277–282.

Freund, T.F., Katona, I., Piomelli, D. (2003) Role of endogenous cannabinoids in synaptic signaling. *Physiol. Rev.* 83, 1017–1066.

Frischknecht, H.R., Waser P.G. (1980) Actions of hallucinogens on ants (*Formica pratensis*) — III. Social behavior under the influence of LSD and tetrahydrocannabinol. *Gen. Pharmacol.* 11, 97–106.

Fritzsche, M. (2000) Are cannabinoid receptor knockout mice animal models of schizophrenia. *Medical Hypothesis* 56, 638–643.

Gadzicki, D., Muller-Vahl, K., Stuhrmann, M. (1999) A frequent polymorphism in the coding exon of the human cannabinoid receptor (CNR1) gene. *Mol. Cell. Probes* 13, 321–323.

Galiegue, S., Mary, S., Marchand, J., Dussossoy, D., Carriere, D., Carayon, P., Bouaboula, M., Shire, D., Le Fur, G., Casellas, P. (1995) Expression of central and peripheral cannabinoid receptors in human immune tissues and leukocyte subpopulations. *Eur. J. Biochem.* 232, 54–61.

Gallate, J.E. McGregor, I.S. (1999) The motivation for beer in rats: effects of ritanserin, naloxone, and SR141716. *Psychopharmacology* 142, 302–308.

Gallate, J.E., Saharov, T., Mallet, P.E., McGregor, I.S. (1999) Increased motivation for beer in rats following administration of a cannabinoid CB1 receptor antagonist. *Eur. J. Pharmacol.* 370, 233–240.

Gardner, E.L. (2002) Addictive potential of cannabinoids: the underlying neurobiology. *Chem. Phys. Lipids* 121, 267–290.

Gebremedhin, D., Lange, A.R., Campbell, W.B., Hillard, C.J., Harder, D.R. (1999) Cannabinoid CB1 receptor of cat cerebral arterial muscle functions to inhibit L-type Ca2+ channel current. *Am. J. Physiol.* 276, H2085–2093.

Gerard, C.M., Mollereau, C., Vassart, G., Parmentier, M. (1991) Molecular cloning of a human cannabinoid receptor which is also expressed in the testis. *Biochem. J.* 279, 129–134.

Gerdeman, G., Lovinger, D.M. (2001) CB1 cannabinoid receptor inhibits synaptic release of glutamate in rat dorsolateral striatum. *J. Neurophysiol.* 85, 468–471.

Gerhardt, C.C., van Heerikhuizen, H. (1997) Functional characteristics of heterologously expressed 5-HT receptors. *Eur. J. Pharmacol.* 334, 1–23.

Giang, D.K., Cravatt, B.F. (1997) Molecular characterization of human and mouse fatty acid amide hydrolase. *Proc. Natl. Acad. Sci. USA* 94, 2238–2242.

Gilmore, R.S., Mitchell, M.D. (2001) Nuclear lipid signaling: Novel role of eicosanoids. *Exp. Biol. Med.* 226, 1–4.

Giuffrida, A., Piomelli, D. (2000) The endocannabinoid system: a physiological perspective on its role in psychomotor control. *Chem. Phys. Lipids* 108, 151–158.

Glaser, S.T., Abumrad, N.A., Fatade, F., Kaczocha, M., Studholme, K.M., Deutsch, D.G. (2003) Evidence against the presence of an anandamide transporter. *Proc. Natl. Acad. Sci. USA.* 100, 4269–4274.

Glass, M., Brotchie, J.M., Maneuf, Y.P. (1997) Modulation of neurotransmission by cannabinoids in the basal ganglia. *Eur. J. Neurosci.* 9, 199–203.

Gomez Del Pulgar, T., Velasco, G., Guzman, M. (2000) The CB1 cannabinoid receptor is coupled to the activation of protein kinase B/Akt. *Biochem. J.* 347, 369–373.

Gonzalez, S., Schmid, P.C., Fernandez-Ruiz, J., Krebsbach, R., Schmid, H.H.O., Ramos, J.A. (2003) Region-dependent changes in endocannabinoid transmission in the brain of morphine-dependent rats. *Addiction Biol.* 8, 159–166.

Griffin, G., Fernando, S.R., Ross, R.A., McKay, N.G., Ashford, M,L., Shire, D., Huffman, J.W., Yu, S., Lainton, J.A., Pertwee, R.G. (1997) Evidence for the presence of CB2-like cannabinoid receptors on peripheral nerve terminals. *Eur. J. Pharmacol.* 339, 53–61.

Griffin, G., Tao, Q., Abood, M.E. (1999) Cloning and pharmacological characterization of the rat CB2 cannabinoid receptor. *J. Pharmacol. Exp. Ther.* 292, 886–894.

Grotenhermen, F. (2004) Pharmacology of cannabinoids. *Neuroendocrinol. Lett.* 25, 14–23.

Gurwitz, D., Kloog, Y. (1998) Do endogenous cannabinoids contribute to HIV-mediated immune failure. *Mol. Med. Today* 4, 196–200.

Guzman, M., Galve-Roperh, I., Sanchez, C. (2001) Ceramide: a new second messenger of cannabinoid action. *Trends Pharmacol. Sci.* 22, 19–22.

Guzman, M., Sanchez, C., Galve-Roperth, I. (2002) Cannabinoids and cell fate. *Pharmacol. Ther.* 95, 175–184.

Hajos, N., Ledent, C., Freund, T.F. (2001) Novel cannabinoid sensitive receptor mediates inhibition of glutamatergic synaptic transmission in the hippocampus. *Neuroscience* 106, 1–4.

Hampson, A.J., Bornheim, L.M., Scanziani, M., Yost, C.S., Gray, A.T., Hansen, B.M., Leonoudakis, D.J., Bickler, P.E. (1998) Dual effects of anandamide on NMDA receptor-mediated responses and neurotransmission. *J. Neurochem.* 70, 671–676.

Hampson, R.E., Mu, J., Deadwyler, S.A. (2000) Cannabinoid and kappa opioid receptors reduce potassium K current via activation of G(s) proteins in cultured hippocampal neurons. *J. Neurophysiol.* 84, 2356–2364.

Hampson, R.E., Evans, G.J., Mu, J., Zhuang, S.Y., King, V.C., Childers, S.R., Deadwyler, S.A. (1995) Role of cyclic AMP dependent protein kinase in cannabinoid receptor modulation of potassium "A-current" in cultured rat hippocampal neurons. *Life Sci.* 56, 2081–2088.

Hanus L., Abu-Lafi S., Fride E., Breuer A., Vogel Z., Shalev D.E., Kustanovich I., Mechoulam R. (2001) 2-archidonyl glyceryl ether, an endogenous agonist of the cannabinoid CB1 receptor. *Proc. Natl. Acad. Sci. U S A* 98, 3662–3665.

Harada, S., Okubo, T., Tsutsumi, M., Takase, S., Muramatsu, T. (1996) Investigation of genetic risk factors associated with alcoholism. *Alcohol.Clin. Exp. Res.* 20, 293A–296A.

Henry, D.J., Chavkin, C. (1995) Activation of inwardly rectifying potassium channels (GIRK1) by co-expressed rat brain cannabinoid receptors in Xenopus oocytes. *Neurosci. Lett.* 186, 91–94.

Herkenham, M., Lynn, A.B., Johnson, M.R., Melvin, L.S., de Costa, B.R., Rice, K.C. (1991) Characterization and localization of cannabinoid receptors in rat brain: A quantitative in vitro autoradiography study. *J. Neurosci.* 11, 563–583.

Herring, A.C., Faubert Kaplan, B.L., Kaminski, N.E. (2001) Modulation of CREB and NF-kB signal transduction by cannabinol in activated thymocytes. *Cell. Signal.* 13, 241–250.

Heyne, A., May, T., Goll, P., Wolffgramm, J. (2000) Persisting consequences of drug intake: towards a memory of addiction. *J. Neural. Transm.* 107, 613–638.

Hillard C.J. (2000) Biochemistry and pharmacology of the endocannabinoids arachidonylethanolamide and 2-arachidonylglycerol. *Prostaglandins Other Lipid Mediat.* 61, 3–18.

Hillard, C.J., Auchampach, J.A. (1994) In vitro inactivation of brain protein kinase C by the cannabinoids. *Biochim. Biophys. Acta* 1220, 163–170.

Hillard, C.J., Edgemond, W.S., Jarrhian, A., Campbell, W.B. (1997) Accumulation of N-arachidonoyethanolamine (anandamide) into cerebral granule cells occurs via facilitated diffusion. *J. Neurochem.* 69, 631–638.

Hillard, C.J., Harris, R.A., Bloom, A.S. (1985) Effects of the cannabinoids on physical properties of brain membranes and phospholipid vesicles: Fluorescence studies. *J. Pharmacol. Exp. Ther.* 232, 579–588.

Hinz, B., Ramer, R., Eichele, K., Weinzierl, U., Brune, K. (2004) R(+)-Methandamide-induced cyclooxygenase-2 expression in H4 human neuroglioma cells: possible involvement of membrane lipid rafts. *Biochim. Biophys. Res. Commn.* 324, 621–626.

Ho, B.Y., Uezono, Y., Takada, S., Takase, I., Izumi, F. (1999) Coupling of the expressed cannabinoid CB1 and CB2 receptors to phospholipase C and G protein-coupled inwardly rectifying K+ channels. *Receptors Channels* 6, 363–374.

Hoehe, M.R., Caenazzo, L., Martinez, M.M., Hsieh, W-T., Modi, W.S., Gershon, E.S., Bonner T.I. (1991) Genetic and physical mapping of the human cannabinoid receptor gene to chromosome 6q14-q15. *New Biol.* 3, 880–885.

Hoffman, A.F., Lupica, C.R. (2001) Direct actions of cannabinoids on synaptic transmission in the nucleus accumbens: A comparison with opioids. *J. Neurophysiol.* 85, 72–83.

Hollister, L.E. (1986) Health aspects of cannabis. *Pharmacol. Rev.* 38, 1–20.

Hopfer, C.J., Stallings, M.C., Hewith, J.K., Crowley, T.J. (2003) Family transmission of marijuana use, abuse, and dependence. *J. Am. Acad. Child Adolesc. Psychiatry* 42, 834–841.

Howlett, A.C., Mukhopadhyay, S. (2000) Cellular signal transduction by anandamide and 2-arachidonoylglycerol. *Chem. Phys. Lipids.* 108, 53–70.

Howlett, A.C., Mukhopadhyay, S., Wilken, G.H., Nechameyer, W.S. (2000) A cannabinoid receptor in Drosophila is pharmacologically unique. *Soc. Neurosci. Abstr.* 26, 2165.

Howlett, A.C., Johnson, M.R., Melvin, L.S., Milne, G.M. (1988) Nonclassical cannabinoid analgetics inhibit adenylate cyclase: development of a cannabinoid receptor model. *Mol. Pharmacol.* 33, 297–302.

Hungund, B.L., Basavarajappa, B.S. (2000) Are anandamide and cannabinoid receptors involved in ethanol tolerance? A review of the evidence. *Alcohol Alcohol.* 35, 126–133.

Iversen, L.L. (1993) Medical uses of marijuana? *Nature* 365, 12–13.

Iversen, L.L. (2000) *The Science of Marijuana.* Oxford University Press, NY.

Jacobson, S.O.P., Rongard, E., Stridh, M., Tiger, G., Fowler, C.J. (2000) Serum-dependent effects of tamoxifen and cannabinoids upon C6 glioma cell viability. *Biochem. Pharmacol.* 60, 1807–1813.

Jin, K., Xie, L., Kim, S.H., Parmentier-Batteur, S., Sun, Y., Mao, X.O., Childs, J., Greenberg, D.A. (2004) Defective adult neurogenesis in CB1 cannabinoid receptor knockout mice. *Mol. Pharmacol.* 66, 204–208.

Johnson, J.P., Muhleman, D., MacMurray, J., Gade, R., Verde, R., Ask, M., Kelley, J., Comings, D.E. (1997) Association between the cannabinoid receptor gene (CNR1) and the P300 event-related potential. *Mol. Psychiatry* 2, 169–171.

Jonas P., Sakmann B. (1992). Glutamate receptor channels in isolated patches from CA1 and CA3 pyramidal cells of rat hippocampal slices. *J. Physiol.* 455, 143–171.

Kalofoutis, A., Lekakis, J., Koutselinis A. (1980) Effects of d9-THC on human platelet phospholipids. *Pharmacol. Biochem. Behav.* 12, 697–699.

Kano, M., Ohno-Shosaku, T., Maejima, T. (2002). Retrograde signaling at central synapses via endogenous cannabinoids. *Mol. Psychiatry* 7, 234–235.

Kathmann, M., Haug, K., Heils, A., Nothen, M.M., Schlicker, E. (2000) Exchange of three amino acids in the cannabinoid CB1 receptor (CNR1) of an epilepsy patient. 2000 Symposium on the cannabinoids, Burlington, Vermont, ICRS, p. 18.

Kearn, C.S., Hilliard, C.J. (1997) Rat microglial cell express the peripheral-type cannabinoid receptor (CB2) which is negatively coupled to adenylyl cyclase. 1997 Symposium on the Cannabinoids, Burlington, Vermont, International Cannabinoid Research Society, 57.

Kenney, S.P., Kekuda, R., Prasad P.D., Leibach, L.D., Devoe, L.D., Ganapathy, V. (1999) Cannabinoid receptors and their role in the regulation of the serotonin transporter in human placenta. *Am. J. Obstet. Gynecol.* 181, 491–497.

Kittler J.J., Clayton, C., Zhuang S-Y., Trower, M.M., Wallace, D., Hampson, R., Grigorenko, E.V., Deadwyler, S. (1999) Large scale gene expression changes during long-term exposure to Δ^9-THC in rats, 1999 Symposium on the Cannabinoids, Burlington, Vermont, ICRS, p. 79.

Kofalvi, A., Vizi, E.S., Ledent, C., Sperlagh, B. (2003) Cannabinoids inhibit the release of [3H] glutamate from rodent hippocampal synaptosomes via a novel CB1 receptor-independent action. *Eur. J. Neurosci.* 18, 1973–1978.

Kozak, K.R., Prusakiewicz, J.J., Marnett, L.J. (2004) Oxidative metabolism of endocannabinoids by COX-2. *Curr. Pharm. Des.* 10, 659–67.

Kreitzer, A.C., Regehr (2002) Retrograde signaling by endocannabinoids. *Curr. Opin. Neurobiol.* 12, 324–330.

Kunos, G., Jarai, Z., Batkai, S., Goparaju, S.K., Ishac, E.J., Liu, J., Wang, L., Wagner, J.A. (2000) Endocannabinoids as cardiovascular modulators. *Chem. Phys. Lipids* 108, 159–168.

Kuwae, T., Shiota, Y., Schmid, P.C., Krebsbach, R., Schmid, H.H. (1999) Biosynthesis and turnover of anandamide and other N-acylethanolamines in peritoneal macrophages. *FEBS Lett.* 459, 123–127.

Kwon, J.M., Goate, A.M. (2000) The candidate gene approach. *Alcohol. Res. Health* 24, 164–168.

Lambowitz, A.M., Belfort, M. (1993) Introns as mobile genetic elements. *Ann. Rev. Biochem.* 62, 587–622.

Le, A.D., Ko, J., Chow, S., Quan, B. (1994) Alcohol consumption by C57BL/6 Balb/c and DBA/2 mice in a limited access paradigm. *Pharmacol. Biochem. Behav.* 47, 375–378.

Ledent, C., Valverde, O., Cossu, G., Petitet, F., Aubert, J-F., Beslot, F., Bohme, G. A., Imperato, A., Pedrazzinni, T., Roques, B. P., Vassart, G., Fratta, W., Parmentier, M. (1999) Unresponsiveness to cannabinoids and reduced addictive effects of opiates in CB1 receptor knockout mice. *Science* 283, 401–404.

Lenicque, P.M., Paris, M.R., Poulot, M. (1972) Effects of some components of cannabis sativa on the regenerating planarian worm *Dugesia tigrina*. *Experientia* 28, 1399–1400.

Lepore, M., Vorel, S.R., Lowinson, J., Gardner, E.L. (1995) Conditioned place preference induced by delta 9-tetrahydrocannabinol: comparison with cocaine, morphine, and food reward. *Life Sci.* 56, 2073–2080.

Leterrier, C., Bonnard, D., Carrel, D., Rossier, J., Lenkei, Z. (2004) Constitutive endocytic cycle of the CB1 cannabinoid receptor. *J. Biol. Chem.* 279, 36013–36021.

Li, J-H., Lin, L-F. (1998) Genetic toxicology of abused drugs: a brief review. *Mutagenesis* 13, 557–565.

Li, T., Liu, X., Zhu, Z-H., Zhao, J., Hu, X., Ball, D.M., Sham, P.C., Collier, D.A. (2000) No association between (AAT)n repeats in the cannabinoid receptor gene (CNR1) and heroin abuse in a Chinese population. *Mol. Psychiatry* 5, 128–130.

Lovinger, D.M. (1993) High ethanol sensitivity of recombinant AMPA-type glutamate receptors expressed in mammalian cells. *Neurosci. Lett.* 159, 83–87.

Lu, Q., Straiker, A., Lu, Q., Maguire, G. (2000) Expression of CB2 cannabinoid receptor mRNA in adult rat retina. *Vis. Neurosci.* 17, 91–95.

Maccarrone, M., Van der Stelt, M., Rossi, A., Veldink, G.A., Vliegenthart, Agro, A.F. (1998) Anandamide hydrolysis by human cells in culture and brain. *J. Biol. Chem.* 273, 32332–2339.

Mackie, K., Hille, B. (1992) Cannabinoids inhibit N-type calcium channels in neuroblastoma-glioma cells. *Proc. Natl. Acad. Sci.* 89, 3825–3829.

Mackie, K., Lai, Y., Westenbroek, R., Mitchell, R. (1995) Cannabinoids activate an inwardly rectifying potassium conductance and inhibit Q-type calcium currents in AtT20 cells transfected with rat brain cannabinoid receptor. *J. Neurosci.* 10, 6552–6561.

Maejima T., Hashimoto K., Yoshida T., Aiba A., Kano M., 2001. Presynaptic inhibition caused by retrograde signal from metabotropic glutamate to cannabinoid receptors. *Neuron* 31, 463–475.

Mailleux, P., Parmentier, M., Vanderhaeghen, J. (1992) Distribution of cannabinoid receptor messenger RNA in the human brain: an in situ hybridization histochemistry with oligonucleotides. *Neurosci. Lett.* 143, 200–203.

Makriyannis, A., Banijamali, A., Jarrell, H.C., Yang, D.–P. (1989) The orientation of (-)-delta-9-tetrahydrocannabinol in DPPC bilayers as determined by solid state 2H-NMR. *Biochim. Biophys. Acta* 986, 141–145.

Maldonado, R. (2002) Study of cannabinoid dependence in animals. *Pharmacol. Ther.* 95, 153–164.

Mallet, P.E., Beninger, R.J. (1998) Delta-9-tetrahydrocannabinol, but not the endogenous cannabinoid receptor ligand anandamide, produces conditioned place avoidance. *Life Sci.* 62, 2431–2439.

Mang, C.F., Erbelding, D., Kilbinger, H. (2001) Differential effects of anandamide on acetylcholine release in the guinea-pig ileum mediated via vanilloid and non-CB1 cannabinoid receptors. *Br. J. Pharmacol.* 134, 161–167.

Manzanedo, C., Aguilar, M.A., Rodriguez-Arias, M., Navarro, M., Minarro, J. (2004) Cannabinoid agonist-induced sensitization to morphine place preference in mice. *Neuroreport* 15, 1373–1377.

Martin R.S., Luong, L.A., Welsh, N.J., Eglen, R.M., Martin, G.R., Maclennan, J. (2000) Effects of cannabinoid receptor agonists on neuronally-evoked contractions of urinary bladder tissues isolated from rat, mouse, pig, dog, monkey and human. *Br. J. Pharmacol.* 219, 1707–1775.

Martin, B.R., Mechoulam, R., Razdan, R.K. (1999) Discovery and characterization of endogenous cannabinoids. *Life Sci.* 65, 573–595.

Martin, B.R. (1986) Cellular effects of cannabinoids. *Pharmacol. Rev.* 38, 45–74.

Martinez, M., Goldin, L.R., Cao, Q., Zhang, J., Sanders, A.R., Nancarrow, D.J., Taylor, J.M., Levinson, D.F., Kirby, A., Crowe, R.R., Andreasen, N.C., Black, D.W., Silverman, J.M., Lennon, D.P., Nertney, D.A., Brown, D.M., Mowry, B.J., Gershon, E.S., Gejman, P.V. (1999) Follow-up study on a susceptibility locus for schizophrenia on chromosome 6q. *Am. J. Med. Genet.* 88, 337–343.

Mascia, M.S., Obinu, M.C., Ledent, C., Parmentier, M., Bohme, G.A., Imperato, A., Fratta, W. (1999) Lack of morphine-induced dopamine release in the nucleus accumbens of cannabinoid CB1 receptor knockout mice. *Eur. J. Pharmacol.* 383, R1-R2.

Mas-Nieto, M., Pommier, B., Tzavara, E.T., Caneparo, A., Da Nascimento, S.L., Fur, G., Roques, B.P., Noble, F. (2001) Reduction of opioid dependence by the CB1 antagonist SR141716A in mice: evaluation of the interest in pharmacotherapy of opioid addiction. *Br. J. Pharmacol.* 132, 1809–1816.

Matias, I., Pochard, P., Orlando, P., Salzet, M., Pestel, J., Di Marzo, V (2002) Presence and regulation of the endocannabinoid system in human dendritic cells. *Eur. J. Biochem.* 269, 3771–3778.

Mato, S., Olmo, E.D., Pazos, A. (2003) Ontogenetic development of cannabinoid receptor expression and signal transduction functionality in the human brain. *Eur. J. Neurosci.* 17, 1747–1754.

Matsuda, L. A. (1997) Molecular aspects of cannabinoid receptors. *Crit. Rev. Neurobiol.* 11, 143–166.

Matsuda, L.A., Bonner, T.I., Lolait, S.J. (1993) Localization of cannabinoid mRNA in rat brain. *J. Comp. Neurol.* 327, 535–550.

Matsuda, L.A., Lolait, T.I., Brownstein, M.J., Young, A.C., Bonner, T.I. (1990) Structure of a cannabinoid receptor and functional expression of the cloned cDNA. *Nature* 346, 561–564.

McAllister, S.D., Griffin, G., Satin, L.S., Abood, M.E. (1999) Cannabinoid receptors can activate and inhibit G protein-coupled inwardly rectifying potassium channels in a *Xenopus* oocyte expression system. *J. Pharmacol. Exp. Ther.* 291, 618–626.

McClean, D.K., Zimmerman, A.M. (1976). Action of delta 9-tetrahydrocannabinol on cell division and macromolecular synthesis in division-synchronized protozoa. *Pharmacology* 14, 307–321.

McFarland, M.J., Barker, E.L. (2004) Anandamide transport. *Pharmacol. Ther.* 104, 117–135.

McGregor, I.S., Issakidis, C.N., Prior, G. (1996) Aversive effects of the synthetic cannabinoid receptor agonist CP 55, 940 in rats. *Pharmacol. Biochem. Behav.* 53, 657–664.

McLaughlin, C.R., Abood, M.E. (1993) Developmental expression of cannabinoid receptor mRNA. *Dev. Brain Res.* 76, 75–81.

McPartland, J., Di Marzo, V., De Petrocellis, L., Mercer, A., Glass, M. (2001) Cannabinoid receptors are absent in insects. *J. Comp. Neurol.* 436, 423–429.

Mecholam, R., Fride, E., Di Marzo, V. (1998) Endocannabinoids. *Eur. J. Pharm.* 359, 1–18.

Mechoulam, R. (1999) Recent advances in cannabinoid research. *Forsch Komplementarmed* 6, 16–20.

Mechoulam, R., Ben-Shabat S., Hanus, L., Ligumsky, M., Kaminski, N.E., Schatz, N.E., Gopher, A., Almog, S., Martin, B.R., Compton, D.R., Pertwee, R.G., Griffin, G., Bayewitch, M., Barge, J., Vogel, Z. (1995) Identification of an endogenous 2-monoglyceride, present in canine gut, that binds to cannabinoid receptors. *Biochem. Pharmacol.* 50, 83–90.

Mo, F.M., Offertaler, L., Kunos, G. (2004) Atypical cannabinoid stimulates endothelial cell migration via a Gi/Go-coupled receptor distinct from CB1, CB2 or EDG-1. *Eur. J. Pharmacol.* 489, 21–27.

Molderings, G.J., Bonisch, H., Hammermann, R., Gothert, M., Bruss, M. (2002) Noradrenaline release-inhibiting receptors on PC12 cells devoid of α_2- and CB1 receptors: similarities to presynaptic imidazoline and edg receptors. *Neurochem. Int.* 40, 157–167.

Molina-Holgado, E., Vela, J.M., Arevalo-Martin, A., Almazan G., Molinda-Holgado, F., Borell, J., Guaza, C. (2002) Cannabinoids promote oligodendrocyte progenitor survival: Involvement of cannabinoid receptors and phosphatidylinositol-3 kinase/Akt signaling. *J. Neurosci.* 22, 9742–9753.

Mu, J., Zhuang, S.Y., Kirby, M.T., Hampson, R.E., Deadwyler, S.A. (1999) Cannabinoid receptors differentially modulate potassium A and D currents in hippocampal neurons in culture. *J. Pharmacol. Exp. Ther.* 291, 893–902.

Munro, S., Thomas, K.L., Abu-Shaar, M. (1993) Molecular characterization of a peripheral cannabinoid receptor. *Nature* 365, 61–65.

Murphy, W.J., Elzirik, E., Johnson, W.E., Zhang, Y.P., Ryder, O.A., O'Brien, S.J. (2001) Molecular phylogenetics and origins of placental mammals. *Nature* 409, 614–618.

Nadler, V., Mechoulam, R., Sokolovsky, M. (1993) Blockade of 45Ca2+ influx through the N methyl-D-aspartate receptor ion channel by the non-psychoactive cannabinoid HU-211. *Brain Res.* 622, 79–85.

Nakane, S., Tanaka, T., Satouchi, K., Kobayashi, Y., Waku, K., Sugiura, T. (2000) Occurrence of a novel cannabimimetic molecule 2-sciadonoylglycerol (2-eicosa-5,11,14-trienoylglycerol) in the umbrella pine Sciadopitys verticillata seeds. *Biol. Pharm. Bull.* 23, 758–761.

Navarro, M., Chowen, J., Carrera, M. R. A., del Arco, I., Villanua, M. A., Martin, Y., Roberts, A. J., Koob, G. B., Rodriguez de Fonseca, F (1998) CB1 cannabinoid receptor antagonist-induced opiate withdrawal in morphine dependent rats. *Neuroreport* 9, 3397–3402.

Netzeband, J.G., Conroy, S.M., Parsons, K.L., Gruol, D.L. (1999) Cannabinoids enhance NMDA-elicited Ca2+ signals in cerebellar granule neurons in culture. *J. Neurosci.* 19, 8765–8777.

O'Dowd, B.F. (1993) Structures of dopamine receptors. *J. Neurochem.* 60, 804–816.

Ohno-Shosaku T., Shosaku J., Tsubokawa H., Kano M. (2002) Cooperative endocannabinoid production by neuronal depolarization and group 1 metabotropic glutamate receptor activation. *Eur. J. Neurosci.* 15, 953–961.

Okamoto, Y., Morishita, J., Tsuboi, K., Tonai, T., Ueda, N. (2004) Molecular characterization of a phospholipase D generating anandamide and its congeners. *J. Biol. Chem.* 279, 5298–5305.

Oliva, J.M., Ortiz, S., Palomo, T., Manzanares, J. (2003) Behavioral and gene transcription alterations induced by spontaneous cannabinoid withdrawal in mice. *J. Neurochem.* 85, 94–104.

Onaivi, E.S., Leonard, C.M., Ishiguro, H., Zhang, P.W., Lin, Z., Akinshola, B.E., Uhl, G.R. (2002). Endocannabinoids and cannabinoid receptor genetics. *Prog. Neurobiol.* 66, 307–344.

Onaivi, E.S., Chakrabarti, A., Chaudhuri, G. (1996) Cannabinoid receptor genes. *Prog. Neurobiol.* 48, 275–305.

Onaivi, E.S., Chakrabarti, A., Gwebu, E.T., Chaudhuri, G. (1995) Neurobehavioral effects of delta-9-THC and cannabinoid receptor gene expression in mice. *Behav. Brain Res.* 72, 115–125.

Onaivi, E.S., Green, M.R., Martin, B.R. (1990) Pharmacological characterization of cannabinoids in the elevated plus maze. *J. Pharmacol. Exp. Ther.* 253, 1002–1009.

Onaivi, E.S., Mora, Z., Perchuk, A., Brandoni, C., Leonard, C.M., Uhl, G.R., Akinshola, B.E. (2004) *FASEB J.* 395.5.

Onaivi, E.S., Stubbs, L., Chakrabarti, A., Chittenden, L., Hurst, D.P., Akinshola, B.E., Shire, D., Reggio, P.H. (1998) Murine cannabinoid receptor genetics. *FASEB J.* 12(4), A194.

Ortiz, S., Olivia, J.M., Perez-Rial, S., Palomo, T., Manzanares, J. (2004) Differences in basal cannabinoid CB1 receptor function in selective brain areas and vulnerability to voluntary alcohol consumption in Fawn Hooded and Wistar rats. *Alcohol Alcohol.* 39, 297–302.

P (2001) d-9-tetrahydrocannabinol releases and facilitates the effects of endogenous enkephalins: reduction in morphin withdrawal syndrome without change in rewarding effect. *Eur. J. Neurosci.* 13, 1816–1824.

Pan, X., Ikeda, S.R., Lewis, D.L. (1996) Rat brain cannabinoid receptor modulates N-type Ca2+ channels in a neuronal expression system. *Mol. Pharmacol.* 49, 707–714.

Paria, B.C. Dey, S.K. (2000) Ligand-receptor signaling with endocannabinoids in preimplantation embryo development and implantation. *Chem. Phys. Lipids* 108, 211–220.

Park, B., Gibbons, H.M., Mitchell, M.D., Glass, M. (2003) Identification of the CB1 cannabinoid receptor and fatty acid amide hydrolase (FAAH) in the human placenta. *Placenta* 24, 990–995.

Pertwee, R.G. (1997) Pharmacology of cannabinoid CB1 and CB2 receptors. *Pharmacol. Ther.* 74, 129–180.

Pertwee, R.G. (1999) Evidence for the presence of CB1 cannabinoid receptors on peripheral neurons and for the existence of neuronal non-CB1 cannabinoid receptors. *Life Sci.* 65, 597–605.

Pertwee, R.G., Gibson, T.M., Stevenson, L.A., Ross, R.A., Banner, W.K., Saha, B., Razdan, R.K., Martin, B.R. (2000) O-1057, a potent water-soluble cannabinoid receptor agonist with antinociceptive properties. *Br. J. Pharmacol.* 129, 1577–1584.

Pettit, D.A., Showalter, V.M., Abood, M.E., Cabral, G.A. (1994) Expression of a cannabinoid receptor in baculovirus-infected insect cells. *Biochem. Pharmacol.* 48, 1231–1243.

Pfister-Genskow, M., Weesner, G.D., Hayes, H., Eggen, A., Bishop M.D. (1997) Physical and genetic localization of the bovine cannabinoid receptor (CNR1) gene to bovine chromosome 9. *Mamm. Genome.* 8, 301–302.

Piomelli, D. (2003) The molecular logic of endocannabinoid signaling. *Nat. Rev. Neurosci.* 4, 873–884.

Piomelli, D., Giuffrida, A., Calignano, A., Rodriguez de Fonseca, F. (2000) The endocannabinoid system as a target for therapeutic drugs. *TIPS* 21, 218–224.

Piomelli, D., Beltramo, M., Glasnapp, S., Lin, S.Y., Goutopoulos, A., Xie, X-Q., Makriyannis, A. (1999) Structural determinants for recognition and translocation by the anandamide transporter. *Proc. Natl. Acad. Sci. USA* 96, 5802–5807.

Poncelet, M., Barnouin, M-C., Breliere, J.-C., Le Fur, G., Soubrie, P (2003) Blockade of cannabinoid (CB1) receptors by SR141716 selectively antagonizes drug-induced reinstatement of exploratory behavior in gerbils. *Psychopharmacology* 144, 144–150.

Porter, A.C., Sauer, J-M., Knierman, M.D., Becker, G.D., Berna, M.J., Bao, J., Nomikos, G.G., Carter, P., Bymaster, F.P., Leese, A.B., Felder, C.C. (2002) Characterization of a novel endocannabinoid, virodhamine, with antagonist activity at the CB1 receptor. *JPETH* 301, 1020–1024.

Pringle, H.L., Bradley S.G., Harris L.S. (1979) Susceptibility of *Naegleria fowleri* to delta-9-tetrahydrocannabinol. *Antimicrob. Agents Chemother.* 16, 674–679.

Racz, I., Bilkei-Gorzo, A., Toth, Z.E., Michel, K., Palkovits, M., Zimmer, A. (2003) A critical role for the cannabinoid CB1 receptors in alcohol dependence and stress-stimulated ethanol drinking. *J. Neurosci.* 23, 2453–2458.

Reibaud, M., Obinu, M.C., Ledent, C., Parmentier, M., Bohme, G.A., Imperato, A. (1999) Enhancement of memory in cannabinoid CB1 receptor knockout mice. *Eur. J. Pharmacol.* 379, R1–R2.

Reid, M.J., Bornheim, L.M. (2001) Cannabinoid-induced alterations in brain disposition of drugs of abuse. *Biochem. Pharmacol.* 61, 1357–1367.

Richardson, J.D., Kilo, S., Hargreaves, K.M. (1998) Cannabinoids reduce hyperalgesia and inflammation via interaction with peripheral CB1 receptors. *Pain.* 75, 111–119.

Rinaldi-Carmona, M., Calandra, B., Shire, D., Bouaboula, M., Oustric, D., Barth, F., Casellas, P., Ferrara, P., Le Fur, G. (1996) Characterization of two cloned CB1 cannabinoid receptor isoforms. *J. Pharmacol. Exp. Ther.* 278, 871–878.

Rodriguez de Fonseca, F., Roberts, A.J., Bilbao, A., Koob, G.F., Navarro, M. (1999) Cannabinoid receptor antagonist SR141716A decreases operant ethanol self-administration in rats exposed to ethanol-vapor chambers. *Acta Pharmacol. Sin.* 20, 1109–1114.

Rodriguez, J.J., Mackie, K., Pickel, V.M. (2001) Ultrastructural localization of the CB1 cannabinoid receptor in u-opioid receptor patches of the rat caudate putamen nucleus. *J. Neurosci.* 21, 823–833.

Ronesi, J., Gerdeman, G.L., Lovinger, D.M., (2004) Disruption of endocannabinoid release and striatal long-term depression by postsynaptic blockade of endocannabinoid membrane transport. *J. Neurosci.* 24, 1673–1679.

Rouach, N., Nicoll, R.A. (2003) Endocannabinoid contributes to short-term but not long-term mGluR-induced depression in the hippocampus. *Eur. J. Neurosci.* 4, 1017–1020.

Rouquette-Jazdanian, A., Pelassy, C., Breittmayer, J-P., Cousin J-C., Aussel, C. (2002) Metabolic labeling of membrane microdomains/rafts in Jurcat cells indicates the presence of glycerophospholipids implicated in signal transduction by the CD3 T-cell receptor. *Biochem. J.* 363, 645–655.

Rukwied, R., Watkinson, A., McGlone, F., Dvorak, M.(2003) Cannabinoid agonists attenuate capsaicin-induced responses in human skin. *Pain.* 102, 283–288.

Salo, O.M.H., Lahtela-Kakkonen, Gynther, J., Jarvinen, T., Poso, A. (2004) Development of a 3D model for the human cannabinoid receptor. *J. Med. Chem.* 47, 3048–3057.

Salzet, M., Breton, C., Bisogno, T., Di Marzo, V. (2000) Comparative biology of the endocannabinoid system. Possible role in the immune response. *Eur. J. Biochem.* 267, 4917–4927.

Samson, M.T., Small-Howard, A., Shimoda, L.M., Koblan-Huberson, M., Stokes, A.J., Turner, H. (2003) Differential roles of CB1 and CB2 cannabinoid receptors in mast cells. *J. Immunol.* 170, 4953–4962.

Sanchez, C., Rueda, D., Segui, B., Galve-Roperh, I., Levade, T., Guzman, M. (2001) The CB1 cannabinoid receptor of astrocytes is coupled to sphingomyelin hydrolysis through the adaptor protein fan. *Mol. Pharmacol.* 59, 955–959.

Sanudo-Pena, M.C., Tsou, K., Delay, E.R., Hohman, A.G., Force, M., Walker, J.M. (1997) Endogenous cannabinoids as an aversive or counter-rewarding system in the rat. *Neurosci. Lett.* 223, 125–128.

Sarker, K.P., Maruyama, I. (2003) Anandamide induces cell death independently of cannabinoid receptors or vanilloid receptor 1: possible involvement of lipid rafts. *Cell. Mol. Life Sci.* 60, 1200–1208.

Sarne, Y., Keren, O. (2004) Are cannabinoid drugs neurotoxic or neuroprotective? *Med. Hypotheses.* 63, 187–192.

Schatz, A.R., Lee, M., Condie, R.B., Pulaski, J.T., Kamiski, N.E. (1997) Cannabinoid receptors CB1 and CB2: A characterization of expression and adenylate cyclase modulation within the immune system. *Toxicol. Appl. Pharmacol.* 142, 278–287.

Schlicker E., Kathmann M. (2001) Modulation of transmitter release via presynaptic cannabinoid receptors. *TIPS* 22, 565–572.

Schmid, H.H.O., Schmid, P.C., Natarajan, V. (1996) The N-acylation-phosphodiesterase pathway and cell signaling. *Chem. Phys. Lipids* 80, 133–142.

Schuel, H., Goldstein, E., Mechoulam, R., Zimmerman, A.M., Zimmerman, S. (1994) Anandamide (arachidonylethanolamide), a brain cannabinoid receptor agonist, reduces sperm fertilizing capacity in sea urchins by inhibiting the acrosome reaction. *Proc. Natl. Acad. Sci. U S A* 91, 7678–7682.

Schuel, H., Schuel, R., Zimmerman, A.M., Zimmerman, S. (1987) Cannabinoids reduce fertility of sea urchin sperm. *Biochem. Cell Biol.* 65, 130–136.

Schweitzer, P. (2000) Cannabinoids decrease the K(+) M-current in hippocampal CA1 neurons. *J. Neurosci.* 20, 51–58.

Sepe, N., De Petrocellis, L., Montanaro, F., Cimino, G., Di Marzo, V. (1998) Bioactive long chain N-acylethanolamines in five species of edible bivalve mollusks. Possible implications for mollusk physiology and sea food industry. *Biochem. Biophys. Acta* 1389, 101–111.

Shen, M., Piser, T.M., Seybold, V.S., Thayer, S.A. (1996) Cannabinoid receptor agonists inhibit glutamatergic synaptic transmission in rat hippocampal cultures. *J. Neurosci.* 16, 4322–4334.

Shire, D., Calandra, B., Bouaboula, M., Barth, F., Rinaldi-Carmona, M., Casellas, P., Ferrara, P. (1999) Cannabinoid receptor interactions with the antagonists SR 141716A and SR 144528. *Life Sci.* 65, 627–635.

Shire, D., Calendra, B., Rinaldi-Carmona, M., Oustric, D., Pessegue, B., Bonnin-Cabanne, O., Le Fur, G., Caput, D., Ferrara, P. (1996) Molecular cloning, expression and function of the murine CB2 peripheral cannabinoid receptor. *Biochim. Biophys. Acta* 1307, 132–136.

Shire, D., Carillon, C., Kaghad, M., Rinaldi-Carmona, M., Le Fur, G., Caput, D., Ferrara, J., (1995) An amino terminal variant of the central cannabinoid receptor resulting from alternative splicing. *J. Biol. Chem.* 270, 3726–3731.

Siegfried, Z., Kanyas, K., Latzer, Y., Karni, O., Bloch, M., Lerer, B., Berry, E.M. (2004) Association study of cannabinoid receptor gene (CNR1) alleles and anorexia nervosa: differences between restricting and binging/purging subtypes.

Simons, K., Toomre, D. (2000) Lipid rafts and signal transduction. *Nat. Rev. Mol. Cell. Biol.* 1, 31–39.

Sipe, J.C., Chiang, K., Gerber, A.L., Beutler, E., Cravatt, B.F. (2002) A missense mutation in human fatty acid amide hydrolase associated with problem drug use. *Proc. Natl. Acad. Sci. USA.* 99, 8394–8399.

Skaper, S.D., Buriani, A., Dal Toso, R., Petrelli, L., Romanello, S., Facci, L and Leon, A. (1996) The ALIAmide palmitoylethanolamide and cannabinoids, but not anandamide, are protective in a delayed postglutamate paradigm of excitotoxic death in cerebellar granule neurons. *Proc. Natl. Acad. Sci. USA* 93, 3984–3989.

Slipetz, D.M., O'Neill, G.P., Favreau, L., Dufresne, C., Gallant, M., Gareau, Y., Guay, D., Labelle, M., Metters, K.M. (1995) Activation of the human peripheral cannabinoid receptor results in inhibition of adenylyl cyclase. *Mol. Pharmacol.* 48, 352–361.

Smart, D., Gunthorpe, M.J., Jerman, J.C., Nasir, S., Gray, J., Muir, A.I., Chambers, J.K., Randall, A.D., Davis, J.B. (2000) The endogenous lipid anandamide is a full agonist at the human vanilloid receptor (hVR1). *Br. J. Pharmacol.* 129, 227–230.

Smith, N.T. (2002) A review of the published literature into cannabis withdrawal symptoms in human users. *Addiction* 97, 621–632.

Soderstrom, K., Johnson, F. (2000) CB1 cannabinoid receptor expression in brain regions associated with zebra finch song control. *Brain Res.* 857, 151–157.

Soderstrom, K., Johnson, F. (2001) Zebra Finch CB1 cannabinoid receptor: pharmacology and in vivo and in vitro effects of activation. *J. Pharmacol. Exp. Ther.* 297, 189–197.

Soderstrom, K., Leid, M., Moore, F.L. Murray, T.F. (2000) Behavioral, pharmacological, and molecular characterization of an amphibian cannabinoid receptor. *J. Neurochem.* 75, 413–423.

Solinas, M., Zangen, A., Thiriet, N., Goldberg, S.R. (2004) β-endorphin elevations in the ventral tegmental area regulate the discriminative effects of d-9-tetrahydrocannabinoid. *Eur. J. Neurosci.* 19, 3183–3192.

Soma, M.R, Baetta, R., Bergamaschi, S., De Renzis, M.R., Davegna, C., Battaini, F., Fumagalli, R., Govoni, S. (1994) PKC activity in rat C6 glioma cells: changes associated with cell cycle and simvastatin treatment. *Biochem. Biophys. Res. Commn.* 200, 1143–1149.

Song, Z.H., Bonner, T.I. (1996) A lysine residue of the cannabinoid receptor is critical for receptor recognition by several agonists but not WIN55212-2. *Mol. Pharmacol.* 49, 891–896.

Staiker, A., Stella, N., Piomelli, D, Mackie, K, Karten, H.J., Maguire, G. (1999) Cannabinoid CB1 receptors and ligands in vertebrate retina: localization and function of an endogenous signaling system. *Proc. Natl. Acad. Sci. U S A* 96, 14565–14570.

Stefano, G.B., Liu, Y.U., Goligorsky, M.S. (1996) Cannabinoid receptors are coupled to nitric oxide release in invertebrate immunocytes, microglia, and human monocytes. *J. Biol. Chem.* 271, 19238–19242.

Stefano, G.B., Salzet, B., Salzet, M. (1997) Identification and characterization of the leech CNS cannabinoid receptor: coupling to the nitric oxide release. *Brain Res.* 753, 219–224.

Steiner, H., Bonner, T.I., Zimmer, A.M., Kitai, S.T., Zimmer, A. (1999) Altered gene expression in striatal projection neurons in CB1 cannabinoid receptor knockout mice. *Proc. Natl. Acad. Sci. USA.* 96, 5786–5790.

Stelt, V.D., Di Marzo, V. (2003) The endocannabinoid system in the basal ganglia and in the mesolimbic reward system: implications for neurological and psychiatric disorders. *Eur. J. Pharmacol.* 480, 133–150.

Strada, C.D., Fong, T.M., Tota, M.R., Underwood, D., Dixon, R.A.F. (1994) Structure and function of G-coupled receptors. *Ann. Rev. Biochem.* 63, 101–132.

Striem, S., Bar-Joseph, A., Berkovitch, Y and Biegon, A. (1997) Interaction of dexanabinol (HU-211), a novel NMDA receptor antagonist, with the dopaminergic system. *Eur. J. Pharmacol.* 338, 205–213.

Stubbs, L., Chittenden, L., Chakrabarti, A, Onaivi, E.S. (1996) The mouse cannabinoid receptor gene is located in proximal chromosome 4. *Mamm. Genome.* 7, 165–166.

Sugiura, T., Kondo, S., Sukagawa, A., Nakane, A., Shinoda, A., Itoh, K., Yamashita, A., Waku, K. (1995) 2-Arachidonoylglycerol: a possible endogenous cannabinoid ligand in brain. *Biochem. Biophys. Res. Commn.* 215, 89–97.

Sugiura, T and Waku, K (2000) 2-Arachidonoylglycerol and cannabinoid receptors. *Chem. Phys. Lipids* 108, 89–106.

Szolcsanyi, J. (2000) Are cannabinoids endogenous ligands for the VR1 capsaicin receptor. *TIPS* 21, 41–42.

Tanda, G., Goldberg, S.R. (2003) Cannabinoids: reward, dependence, and underlying neurochemical mechanisms — a review of recent preclinical data. *Psychopharmacology (Berl):* 169, 115–134.

Tashkin, D.P., Cao, W., Morganstern, H., Greenland, S., Roth, M.D., Simmons, M., Zhang, Z-F (2004) Marijuana smoking and hypermethylation of the promoter region of the O6 (6)-methylguanine DNA methyltransferase gene (MGMT) DNA repair gene MGMT. ICRS Abstracts 73.

Terranova, J.P., Michaud, J.C., Le Fur, G., Soubrie, P. (1995) Inhibition of long-term potentiation in rat hippocampal slices by anandamide and WIN55212-2: reversal by SR141716 A, a selective antagonist of CB1 cannabinoid receptors. *Naunyn. Schmiedebergs Arch. Pharmacol.* 352, 576–579.

Thomasson, H.R., Edenberg, H.J., Crabb, D.W., Mai, X.L., Jerome, R.E., Li, T.K., Wang, S.P., Lin, Y.T., Lu, R.B., Yin, S.J. (1991) Alcohol and alcohol dehydrogenase genotypes and alcoholism in Chinese men. *Am. J. Hum. Genet.* 48, 677–681.

Trettel J., Levine E.S. (2003) Endocannabinoids mediate rapid retrograde signaling at interneuron → pyramidal neuron synapses of the neocortex. *J. Neurophysiol.* 89, 2334–2338.

Trevino, C.L., Serrano, C.J., Beltran, C., Felix, R., Darszon, A. (2001) Identification of mouse trp homologs and lipid rafts from spermatogenic cells and sperm. *FEBS Lett.* 509, 119–125.

Tyndale, R.F. (2003) Genetics of alcohol and tobacco use in humans. *Ann. Med.* 35, 94–121.

Tzschentke, T.M. (1998) Measuring reward with the conditioned place preference paradigm: a comprehensive review of drug effects, recent progress and new issues. *Prog. Neurobiol.* 56, 613–672.

Ueda, N., Yamanaka, K., Terasawa, Y., Yamamoto, S. (2000) An acid amidase hydrolyzing anandamide and other N-acylethanolamines. 2000 Symposium on the cannabinoids, Burlington, Vermont, ICRS, p. 13.

Valjent, E., Maldonado (2000) A behavioral model to reveal place preference to Δ^9-tetrahydrocannabinol in mice. *Psychopharmacology* 147, 436–438.

Valk, P.J.M., Hol, S., Vankin, Y., Ihle, J.N., Askew, D., Jenkins, N.A., Gilbert, D.J., Copeland, N.G., DE Both, N.J., Lowenberg, B., Delwel, R. (1997) The genes encoding the peripheral cannabinoid receptor and a-L-fucosidase are located near a newly identified common virus integration site, Evi11. *J. Virol.* 71, 6796–6804.

Valverde, O., Ledent, C., Beslot, F., Parmentier, M and Roques, B. P (2000) Reduction of stress-induced analgesia but not of exogenous opioid effects in mice lacking CB1 receptors. *Eur. J. Neurosci.* 12, 533–539.

Valverde, O., Noble, F., Beslot, F., Dauge, V., Fournie-Zaluski, M.C., Roques, B.P. (2001) Delta9-tetrahydrocannabinol releases and facilitates the effects of endogenous enkephalins: reduction in morphine withdrawal syndrome without change in rewarding effect. *Eur. J. Neurosci.* 13, 1816–1824.

Varga, K., Wagner, J.A., Bridgen, D.T and Kunos, G. (1998) Platelet- and macrophage-derived endogenous cannabinoids are involved in endotoxin-induced hypotension. *FASEB J.* 12, 1035–1044.

Varma, N., Carlson, G.C., Ledent, C., Alger, B.E. (2001). Metabotropic glutamate receptors drive the endocannabinoid system in hippocampus. *J. Neurosci.* 21, RC188.

Vasquez, C., Lewis, D.L. (1999) The CB1 cannabinoid receptor can sequester G-proteins, making them unavailable to couple to other receptors. *J. Neurosci.* 19, 9271–9280.

Vaughan, C.W., McGregor, I.S., Christie, M.J. (1999) Cannabinoid receptor activation inhibits GABAergic neurotransmission in rostral ventromedial medulla neurons in vitro. *Br. J. Pharmacol.* 127, 935–940.

Vlachou, S., Nomikos, G.G., Panagis, G. (2003) WIN 55, 212-2 decreases the reinforcing actions of cocaine through CB cannabinoid receptor stimulation. *Behav. Brain Res.* 141, 215–222.

Walter, L., Stella, N. (2004) Cannabinoids and neuroinflammation. *Br. J. Pharmacol.* Advance online publication.

Walter, L., Franklin, A., Witting, A., Wade, C., Xie, Y., Kunos, G., Mackie, K., Stella, N. (2003) Nonpsychotic cannabinoid receptors regulate microglial cell migration. *J. Neurosci.* 23, 1398–1405.

Wan, M., Cravatt, B.F., Ring H.Z., Zhang, X., Franke, U. (1998) Conserved chromosomal location and genomic structure of human and mouse fatty-acid amide hydrolase genes and evaluation of clasper as a candidate neurological mutation. *Genomics* 54, 408–414.

Wang, X., Dow-Edwards, D., Keller, E., Hurd, Y.L. (2003) Preferential limbic expression of the cannabinoid receptor mRNA in the human fetal brain. *Neuroscience* 118, 681–694.

Welch, S.P., Eads, M. (1999) Synergistic interactions of endogenous opioids and cannabinoid systems. *Brain Res.* 848, 183–190.

Wenger, T., Furst, S. (2004) Role of endogenous cannabinoids in cerebral reward mechanisms. *Neuropsychopharmacology* 6, 26–29.

Westlake, T.M., Howlett, A.C., Bonner, T.L., Matsuda, L.A., Herkenham, M. (1994) Cannabinoid receptor binding and messenger RNA expression in human brain: an in vitro receptor autoradiography and in situ hybridization histochemistry study of normal aged and Alzheimer's brains. *Neuroscience* 63, 637–652.

Wiley, J.L., Martin, B.R. (2002) Cannabinoid pharmacology: implications for additional cannabinoid receptor subtypes. *Chem. Phys. Lipids* 121, 57–63.

Wilson, R.I., Nicoll, R.A. (2001) Endogenous cannabinoids mediate retrograde signalling at hippocampal synapses. *Nature* 410, 588–592.

Wilson, R., Ainscough, R., Anderson, K., Baynes, C., Berks, M., Bonfield, J. et al. (1994) 2.2 Mb of contiguous nucleotide sequence from chromosome III of *C. elegans*. *Nature* 368, 32–38.

Yamaguchi, F., Tokuda, M., Hatase, O., Brenner, S. (1996a) Molecular cloning of the novel human G protein-coupled receptor (GPCR) gene mapped on chromosome 9. *Biochem. Biophys. Res. Commn.* 227, 608–614.

Yamaguchi, F., Macrae A.D., Brenner, S. (1996b) Molecular cloning of two cannabinoid type1-like receptor genes from the puffer fish *Fugu rubripes*. *Genomics* 35, 603–605.

Yamamoto, T., Takada, K. (2000) Role of cannabinoid receptor in the brain as it relates to drug reward. *Jpn. J. Pharmacol.* 84, 229–236.

Yazulla, S., Studholme, K.M., McIntosh, H.H., Deutsch, D.G. (1999) Immunocytochemical localization of cannabinoid CB1 receptor and fatty acid amide hydrolase in rat retina. *J. Comp. Neurol.* 415, 80–90.

Yazulla, S., Studholme, McIntosh, H.H., Fan, S.-H (2000) Cannabinoid receptors on goldfish retinal bipolar cells: Electron-microscope immunocytochemistry and whole-cell recordings. *Vis. Neurosci.* 17, 391–401.

Yea, S.S., Yang, K.H., Kaminski, N.E. (2000) Role of nuclear factor of activated T-cells and activator protein-1 in the inhibition of interleukin-2 gene transcription by cannabinol in EL4 T-cells. *J. Pharmacol. Exp. Ther.* 292, 597–605.

Zhuang, S-Y., Kittler, J., Grigorenko, E.V., Kirby, M.T., Sim, L.J., Hampson, R.E., Childers, S.R and Deadwyler S.A. (1998) Effects of long-term exposure to Δ^9-THC on expression of cannabinoid (CB1) mRNA in different rat brain regions. *Mol. Brain Res.* 62, 141–149.

Zimmer, A., Zimmer, A.M., Hohmann, A.G., Herkenham, M., Bonner, T.I. (1999) Increased mortality, hypoactivity, and hypoalgesia in cannabinoid CB1 receptor knockout mice. *Proc. Natl. Acad. Sci. U S A* 96, 5780–5785.

Zygmunt, P.M., Petersson, J., Andersson, D.A., Chuang, H.-H., Sorgard M., Di Marzo, V., Julius, D., Hogestatt, E.D. (1999) Vanilloid receptors on sensory nerves mediate the vasodilator action of anandamide. *Nature* 400, 452–457.

4 Endocannabinoids and Intracellular Signaling

Sean D. McAllister and Mary E. Abood

CONTENTS

INTRODUCTION

Marijuana is one of the most widely used drugs throughout the world. The principle psychoactive ingredient and prototypic cannabinoid compound in marijuana is Δ^9-tetrahydrocannabinol (Δ^9-THC). In recent years, a family of endogenous cannabinoids, termed *endocannabinoids*, has been elucidated. Pharmacological effects of cannabinoids include euphoria, analgesia, antiemesis, antispasmotic and cardiovascular effects, and immunomodulation. That the pharmacological effects of cannabinoids are mediated through cell surface receptors can be demonstrated with specific cannabinoid agonists and antagonists (reviewed in Howlett, 1995; Pertwee, 1997).

Two cannabinoid receptors have been isolated: CB_1, which is predominantly expressed in the central nervous system (CNS), and CB_2, which is largely restricted to cells of immune origin. Isolation of a cDNA clone for the rat CB_1 receptor (Matsuda et al., 1990) led to the subsequent identification of human CB_1 receptor cDNAs (Gerard et al., 1991) and human CB_2 receptor cDNAs (Munro et al., 1993). The cloned cannabinoid receptors are members of the G-protein-coupled receptor family. As implied by the name, these receptors transduce the signal produced by the receptor-binding ligand into function via GTP-binding or G proteins. The human CB_1 and CB_2

receptors share only 44% amino acid identity overall, which rises to 68% in the transmembrane domains (Munro et al., 1993). However, most ligands do not discriminate between the receptor subtypes (Munro et al., 1993, Felder et al., 1995; Showalter et al., 1996). The International Union of Pharmacology report on the classification of cannabinoid receptors provides a thorough review of this field (Howlett, 1995). There is an excellent correlation between binding affinities at the cloned CB_1 receptor and radioligand binding in brain homogenates (Felder et al., 1992). Devane et al. (1988) reported that a selected series of analogs exhibited an excellent correlation between antinociceptive potency and affinity for the site in brain homogenates. Compton et al. (1993) extended this correlation to include 60 cannabinoids and the following behavioral measures: hypoactivity, antinociception, hypothermia, and catalepsy. A high degree of correlation was found between the Ki values and *in vivo* potency in the mouse. Thus, this receptor appears to be sufficient to mediate many of the known pharmacological effects of cannabinoids. There is also evidence for additional receptor subtypes.

The CB_1 receptor gene has been inactivated (by in-frame deletion of most of the coding region), through homologous recombination, in two laboratories (Ledent et al., 1999; Zimmer et al., 1999). Significantly, whereas most pharmacological effects of cannabinoids are abolished, anandamide effects remain, indicating the existence of additional receptor subtypes (Zimmer et al., 1999; Jarai et al., 1999; Di Marzo et al., 2000; Monory et al., 2002). CB_2 knockout mice have also been generated (Buckley et al., 2000); these mice have been useful in determining the role of the CB_2 receptor in the immune system.

Endogenous ligands for the cannabinoid receptors have been identified (Devane et al., 1988). Devane et al. (1992) isolated arachidonic acid ethanolamide (AEA), called anandamide, from porcine brain and demonstrated that it competed for binding to the cannabinoid receptor and inhibited electrically stimulated contractions of the mouse vas deferens in the same manner as Δ^9-THC. Additional fatty acid ethanolamides have been isolated, as well as a 2-arachidonyl glycerol (2-AG) with cannabimimetic properties, suggesting the existence of a family of endogenous cannabinoids, which may interact with additional cannabinoid receptor subtypes (Mechoulam et al., 1994; Mechoulam et al., 1995; Sugiura et al., 1995). Indeed, an ether-type endocannabinoid has been isolated, 2-arachidonyl glyceryl ether or noladin ether (Hanus et al., 2001). Additionally, virodhamine, arachidonic acid and ethanolamine joined by an ester linkage, has been isolated (Porter et al., 2002). N-arachidonylglycine, a conjugate of arachidonic acid and glycine, present in bovine and rat brain, as well as other tissues, suppresses tonic inflammatory pain (Huang et al., 2002). N-arachidonyl-dopamine (NADA), although primarily a vanilloid receptor agonist, has some activity at the CB_1 receptors as well (Huang et al., 2002). Palmitoylethanolamide (PEA) has been suggested as a possible endogenous ligand at the CB_2 receptor (Facci et al., 1995). However, subsequent studies showed no affinity for PEA at the CB_2 receptor (Showalter et al., 1996; Lambert et al., 1999; Griffin et al., 2000). Instead, PEA seems to increase the potency of AEA, in part, by inhibiting fatty acid amide hydrolase (FAAH), the enzyme responsible for breakdown of AEA (Di Marzo et al., 2001). Finally, in addition to actions at cannabinoid receptors, AEA, 2-AG, virodhamine, noladin ether, and NADA also act at the VR1 vanilloid receptor, a ligand-gated ion channel (recently reviewed by Di Marzo et al., 2002). This review will focus on endocannabinoid signaling via cannabinoid receptors.

ENDOCANNABINOID SIGNALING THROUGH CB_1

Regulation of Intracellular cAMP Levels

Modulation of cAMP production was the first second messenger system convincingly demonstrated to be mediated by cannabinoids via the CB_1 receptor. Both exo- and endocannabinoids inhibit adenylyl cyclase via G proteins. Numerous studies have shown the involvement of cannabinoids in the modulation of cAMP levels in cells of various types in culture, as well as in homogenates of brain regions.

In the neuroblastoma (N18TG2) or neuroblastoma X glioma (NG108-15) cell lines, cannabinoid-induced decreases in cAMP formation were shown not to be due to interaction with prostanoid, opioid, muscarinic, or adrenergic systems (reviewed by Howlett, 1995). Studies in NG108-15 cells showed that cannabinoid-induced inhibition of cyclic-AMP formation is rapid and reversible, occurs at low concentrations of the cannabinoids, and follows a structure–activity relationship, and its stereoselectivity is similar to that observed for behavioral measures (with the exception of the anticonvulsant effects of the cannabinoids). Furthermore, the effect on cAMP was not blocked by antagonists of other classical neurotransmitters (binding is not displaced by classical neurotransmitters) and is mediated via coupling to the Gi protein, because pertussis toxin attenuates activity of the commonly used synthetic cannabinoid agonist, CP55940. These studies, along with supporting work evaluating the stereoselectivity and the Hill coefficient for binding of the synthetic bicyclic cannabinoid, CP55940 (Devane et al., 1988), established the CB_1 cannabinoid receptor linkage through a G_i protein to the modulation of cyclic AMP.

The potency of numerous cannabinoids to inhibit cAMP formation in the neuroblastoma cells was found to correlate to the antinociceptive effects of the drugs *in vivo* (Howlett et al., 1988). The cannabinoid-induced antinociception was proposed to be due to the modulation of adenylyl cyclase. The potency of various cannabinoids to displace CP55940 binding and to inhibit adenylyl cyclase has been shown to be similar in rank order to the production of not only antinociception but also hypothermia, spontaneous activity, and catalepsy by the cannabinoids (Compton et al., 1993; Little et al., 1988). The aminoalkylindole analogs—WIN55212-2 is the prototype—have also been shown to inhibit adenylyl cyclase activity in rat brain membranes and compete for cannabinoid-binding sites (Pacheco et al., 1991). These compounds are interesting in that they were initially developed as nonsteroidal anti-inflammatory agents, and some analogs have demonstrable *in vitro* cannabinoid antagonist effects.

Soon after its discovery, AEA was shown to inhibit cAMP accumulation in CB_1-transfected CHO cells and N18TG2 cells (Felder et al., 1993; Vogel et al., 1993). Similarly, in the initial discovery of 2-AG, Mechoulam and coauthors reported its ability to inhibit adenylyl cyclase activity in CB_1- and CB_2-transfected COS cells (Mechoulam et al., 1995). Both AEA and 2-AG are full (or nearly full) agonists in the inhibition of adenylyl cyclase via the CB_1 receptor (Stella et al., 1997).

Early studies of forskolin-stimulated cAMP in mouse brain synaptosomes showed biphasic responses to Δ^9-THC and Δ^8-THC (Little and Martin, 1991). Some cannabinoids did not alter cAMP levels. These early studies suggesting biphasic effects of cannabinoids have been confirmed, both in brain and in CB_1-receptor-transfected cells (Glass and Felder, 1997; Bonhaus et al., 1998). Furthermore, a mutational study points to selective G-protein activation as a possible mechanism for the biphasic effects of cannabinoids, because a point mutation produced a constitutively active CB_1 receptor that coupled to G_s instead of G_i (Abadji et al., 1999). AEA is a partial agonist in activation of this stimulatory pathway (Bonhaus et al., 1998). 2-arachidonylglycerol has not been examined for stimulatory effects.

The determinants of which cAMP signaling pathway is activated by CB_1 may include agonist-specific G-protein interactions, as well as the isoform of adenylyl cyclase present in the cell type examined. There is mounting evidence that CB_1 and CB_2 receptors interact with different sets of G proteins (Mackie et al., 1995; Glass and Northup, 1999; McAllister et al., 1999). In addition, it has recently been shown in studies from several laboratories that different ligands promote interactions with different G proteins (Selley et al., 1996; Bonhaus et al., 1998; Griffin et al., 1998; Glass and Northup, 1999; Kearn et al., 1999; Tao et al., 1999; Mukhopadhyay et al., 2000). Furthermore, one ligand can activate several G proteins (Glass and Northup, 1999; Mukhopadhyay et al., 2000; Prather et al., 2000). In addition, coexpression of adenylyl cyclase isoforms 1,3,5,6, or 8 with CB_1 or CB_2 receptors resulted in inhibition of cAMP, whereas isoforms 2,4 or 7 resulted in increases in cAMP formation (Rhee et al., 1998).

Regulation of Ion Channels

The recent discovery that endocannabinoids serve as retrograde messengers in multiple brain regions has allowed for the integration of past data concerning endocannabinoid influence on ion currents into a specific unit in a complete functional neurotransmitter circuit (Ohno-Shosaku et al., 2001; Kreitzer and Regehr, 2001; Wilson and Nicoll, 2001). Many important initial studies laid the foundation for these pathways, as detailed in the following paragraphs.

Inhibition of Voltage-Dependent Calcium Channels

The first experiments utilizing an endocannabinoid to affect ion channel activity demonstrated the inhibition of N- and Q-type voltage-dependent calcium channels (VDCC) in cell lines transfected with the cDNA for CB_1 receptors or containing the native protein (Mackie et al., 1993; Mackie et al., 1995). The inhibition produced by AEA was blocked by pertussis toxin and not because of the activation of the cAMP pathway (Mackie et al., 1993). The effect observed was likely the result of a direct interaction of G subunits with calcium channels (Ikeda, 1996).

Numerous studies using synthetic cannabinoids have shown that activation of CB_1 receptors on axonal terminals leads to inhibition of calcium channels and a subsequent decrease in the release of neurotransmitter (reviewed in Pertwee, 1997). As CB_1 receptors have been localized to terminals containing acetylcholine, GABA, glutamate, or norepinephrine, the physiological outcome produced by receptor activation clearly depends upon which cell populations are being activated. The presynaptic inhibition of calcium channels on axon terminals that synapse on pyramidal neurons of the hippocampus correlates with the ability of endocannabinoids to inhibit long-term potentiation (Terranova et al., 1995; Stella et al., 1997; Misner and Sullivan, 1999). Cannabinoids may alter learning and memory through this mechanism. This point has been reiterated in recent investigations focusing on retrograde signaling in the hippocampus and cerebellum and is discussed in the following paragraphs.

The distribution of presynaptic cannabinoid receptors, the postsynaptic localization of both precursors to endocannabinoid synthesis, and the enzyme involved in their metabolism suggested that endocannabinoids might act as retrograde signaling molecules; these molecules would be released from postsynaptic cells to inhibit presynaptic transmitter release via CB_1 receptor activation (Di Marzo et al., 1994; Stella et al., 1997; Egertova et al., 1998; Elphick and Egertova, 2001). An intact system demonstrating the interplay between all these components was discovered by groups that were not focused on endocannabinoids but on the phenomenon termed depolarization-induced suppression of inhibition or excitation (DSI or DSE).

Retrograde signaling by endocannabinoids results in DSI and DSE and was first demonstrated in the hippocampus and cerebellum (Ohno-Shosaku et al., 2001; Kreitzer and Regehr, 2001; Wilson and Nicoll, 2001). The investigations showed that activation of postsynaptic neurons resulted in the release of endocannabinoids from these neurons. The endogenous ligands then acted as retrograde signaling molecules to inhibit presynaptic calcium influx in axonal terminals and subsequently reduced the release of neurotransmitter.

Using CB_1 knockout animals, a study in hippocampal slices definitively showed that activation of the CB_1 receptor was responsible for the downstream inhibition of presynaptic VDCC in cells in which DSI was produced (Wilson et al., 2001). The investigators also showed that a specific subtype of GABAergic interneuron is targeted by endocannabinoids. In this case, activation of CB_1 receptors led to the specific inhibition of N- but not P/Q- type calcium channels, and this resulted in decreased release of neurotransmitter.

In contrast, in GABAergic interneurons which synapse with Purkinje cells in the cerebellum, an interaction between CB_1 receptors and VDCC does not lead to decreased transmitter release (Kreitzer et al., 2002). The authors demonstrated that activation of a K^+ current by CB_1 receptors in GABAergic interneurons led to the inhibition of firing in these cells. Therefore, in presynaptic terminals, the interaction between CB_1 receptors and ion channels in the endocannabinoid retrograde signaling pathway can change substantially depending on the cell type involved.

Another calcium channel subtype inhibited by endocannabinoids via CB_1 receptors includes the native L-type Ca^{2+} channels in cat cerebral vascular smooth muscle cells (VSMC) (Gebremedhin et al., 1999). The inhibition of the L-type calcium current by AEA in VSMC may relate to the ability of AEA to relax preconstricted cerebral vessels in similar tissue.

Synthetic cannabinoid agonists were also shown to produce inhibition of L-type calcium channels of bipolar cells in retinal slices from larval tiger salamander (Straiker et al., 1999). The presence of 2-AG, PEA, and oleylethanolamide, but not AEA, was reported in the tissue used in this study but the direct application of these ligands was not tested in regard to L-type calcium channel activity. Interestingly, a follow-up study by Straiker and Sullivan found that cannabinoids enhanced L-type calcium channel activity in rods but inhibited channel activity in cones (Straiker and Sullivan, 2003). In both rods and cones, a potassium current was also inhibited (Straiker and Sullivan, 2003). The enhancement of L-type calcium currents in rods was due to the modulation of the cAMP pathway by CB_1, however, the signaling pathways leading to the other effects were not examined. Future studies focusing on the endocannabinoid signaling pathway in this tissue will help explain the functional relevance of differential ionic effects produced by CB_1 receptors in the eye.

Modulation of Potassium Currents

As discussed earlier, endocannabinoids can enhance K^+ currents in cerebellar slices leading to decreased neurotransmitter release in the retrograde signaling pathway (Kreitzer et al., 2002). The first indication that AEA could modulate potassium channels was shown in AtT20 cells transfected with the CB_1 receptor (Mackie et al., 1995). In this investigation, application of AEA caused an enhancement of a G-protein-coupled inwardly rectifying potassium current (K_{ir} /GIRK). Cannabinoid ligands have also been shown to enhance A-type potassium currents in hippocampal cells, an effect that was indirectly mediated by inhibition of cAMP accumulation (Deadwyler et al., 1995). In the cerebellar slices in which endocannabinoid retrograde signaling was investigated, it was determined that CB_1 receptors most likely modulated a K_{ir}-type potassium current vs. K_A; the exact type of current affected still needs to be elucidated (Kreitzer et al., 2002). In the nucleus accumbens, CB_1 receptors located on glutametergic afferents can also potentially modulate K_A and K_{ir} currents leading to decreased neurotransmitter release, but endocannabinoids have not been studied in this pathway (Robbe et al., 2001).

AEA can enhance potassium currents in *Xenopus* oocytes transiently expressing CB_1 receptors and GIRK channels (Henry and Chavkin, 1995; McAllister et al., 1999). The efficient coupling between the receptors and channels has allowed for a convenient system to study structure–activity relationships of the CB_1 receptor (Jin et al., 1999; McAllister et al., 2002).

In the brain, almost all investigations point to a presynaptic locus for CB_1 receptor modulation of ionic channels (Elphick and Egertova, 2001). However, one study carried out in hippocampal slices suggests that a postsynaptic interaction between CB_1 receptors and specific potassium channels can occur. In CA1 hippocampal neurons, methanandamide acting through CB_1 receptors decreased postsynaptic K^+ M-current (I_M) (Schweitzer, 2000). The author did not suggest a direct interaction of CB_1 receptor-activated G proteins with the channels but rather hypothesized that CB_1-mediated stimulation of intracellular calcium stores, leading to increase in intracellular calcium, could be one of the mechanisms behind the inhibition of I_M. Regardless, this study suggests an interesting potential addition to the endocannabinoid retrograde signaling pathway. If a postsynaptic CA1 pyramidal cell is activated, releasing endocannabinoids, it may potentially have two ways to modulate its activity: the indirect decreased release of neurotransmitter from presynaptic terminals via CB_1 and a direct change in excitability through activation of postsynaptic CB_1 receptors. If I_M

is inhibited in a cell, it makes it harder for the cell to repolarize after an action potential (Marrion, 1997). In general, CB_1 receptors have a presynaptic localization in the hippocampus and cerebellum (Elphick and Egertova, 2001; Tsou et al., 1998). Future studies will be needed to determine if an intact *in vivo* system with these characteristics exists.

Modulation of Intracellular Calcium

Early studies of endocannabinoid influence over intracellular calcium levels reported non-receptor-mediated effects in brain tissue and cells transfected with CB_1 receptors (Felder et al., 1992; Felder et al., 1995; Mombouli et al., 1999). However, more recent studies in cell lines and tissue with native CB_1 receptors have demonstrated CB_1-receptor-mediated activation of intracellular calcium stores. The key determinant between the production of receptor and nonreceptor effects appears to be the cellular background used.

In rat brain slices and in cerebeller granule neurons in culture, endocannabinoids were shown to modulate calcium flux through NMDA channels (Hampson et al,1998; Netzeband et al., 1999). In rat brain slices, AEA inhibited calcium influx brought about by addition of NMDA (Hampson et al., 1998). In this model, addition of NMDA alone caused an increase in intracellular calcium. The proposed mechanism responsible for the endocannabinoid effect was an opposing inhibition of calcium entry into the cells brought about by CB_1-receptor inhibition of voltage-dependent P/Q-type calcium channels.

In primary cerebellar cultures and acutely isolated cerebellar granule neurons, methanandamide has been demonstrated to enhance NMDA-evoked calcium release (Netzeband et al., 1999). This effect was due to activation of the CB_1 receptor. In a study using inhibitors of VDCC, a variety of compounds known to modulate intracellular calcium release, and PKC and PKA inhibitors, it was determined that the enhancement of NMDA-evoked calcium release was due to CB_1-receptor activation of the phospholipase C pathway; this led to the downstream release of calcium from intracellular stores. In this study, it was also noted that blockade of the phospholipase C pathway unmasked a CB_1-mediated inhibition of the NMDA-evoked calcium release (Netzeband et al., 1999). This is consistent with the study reported by Hampson et al. (1998) in rat brain slices, referred to earlier.

To relate the findings reported by both groups, Netzeband et al. (1999) hypothesized the existence of an activity-dependent pathway that is differentially modulated by endocannabinoids. They proposed that, under conditions of strong cellular stimulation, an inhibitory effect on voltage-gated calcium channels by endocannabinoids would be apparent, whereas under periods of lower stimulation, the PLC-mediated pathway would dominate.

After the discovery of the endocannabinoid 2-AG (Sugiura et al., 1995; Mechoulam et al., 1995), Sugiura and colleagues demonstrated a rapid and transient calcium increase that was produced by 2-AG but not other structural analogs such as 2-palmityol-glycerol, 2-oleoyl-glycerol, and 2-linoleoyl-glycerol (Sugiura et al., 1995; Mechoulam et al., 1995). AEA was a partial agonist in this system, as were the potent synthetic analogs HU210 and CP55940. The cell line used was a neuroblastoma–glioma hybrid (NG108-15) that contained endogenous CB_1 receptors. The endocannabinoid-mediated increase in transient Ca^{2+} release was both CB_1 and G_i/G_o mediated and could be produced by the synthetic agonist WIN55212-2 but not the inactive isomer WIN55212-3.

The investigators suggested CB_1 receptors could activate PLC through $G\beta\gamma$ subunits as these subunits have previously been demonstrated to stimulate $PLC\beta$ in NG108-15 cells (Jin et al., 1994). It would be interesting to determine if 2-AG produces similar calcium transients in neuronal populations in the CNS where CB_1 receptors exist. Perhaps, future studies will be able to relate this effect in NG108-15 cells to endogenous signaling in native neuronal tissue.

Recently, the CB_1 receptor was shown to couple the activated FGF receptor to an axonal growth response via increased calcium influx into cerebellar granule neurons (Williams et al., 2003). CB_1 receptor antagonists inhibited the neurite outgrowth response stimulated by N-cadherin and FGF2.

Synthetic AEA and 2-AG analogs, arachidonyl-2-chloroethylamide and 2-arachidonylglycerol ether, were shown to enhance neurite outgrowth. Furthermore, activation of the FGF receptor resulted in activation of phospholipase C with the subsequent hydrolysis of DAG (Williams et al., 2003). The initial hydrolysis of DAG at the *sn*-1 position will generate 2-AG (Stella et al., 1997), the proposed endogenous neuromodulator of neurite outgrowth (Williams et al., 2003). In this case, activation of the CB_1 receptor required calcium influx into the neurons; the agonist-induced neurite outgrowth response was inhibited by N- and L-type calcium channel antagonists (Williams et al., 2003).

Nitric Oxide

Anandamide and 2-arachidonylglycerol can directly stimulate nitric oxide release via CB_1 (Stefano et al., 1996; Stefano et al., 2000). This direct stimulation of constitutive nitric oxide synthase has been shown in immunocytes, microglia, and monocytes, as well as in endothelial cells from blood vessels (Stefano et al., 2000). As constitutive nitric oxide synthase isoforms are Ca^{2+}-calmodulin dependent, an intermediate step in this pathway may be increased intracellular calcium, as described earlier.

CB_1 receptors also appear to mediate inhibition of inducible nitric oxide synthase. Studies in microglia showed a dose-dependent inhibition of lipopolysaccharide-induced and/or γ-interferon-induced nitric oxide release that was antagonized by the CB_1 receptor antagonist SR141716A (Waksman et al., 1999). Recently, AEA has been demonstrated to inhibit γ-interferon- and HIV-1 Tat protein–induced release of nitric oxide in rat C6 glioma cells (Esposito et al., 2002). AEA also inhibited lipopolysaccharide-stimulated release of nitric oxide from glia in primary culture; however, in this case, the CB_2 receptor also appears to be involved (Molina-Holgado et al., 2002). The pathway activated in glia involved the subsequent release of interleukin-1, indicating a role of AEA in the modulation of brain injury (Esposito et al., 2002; Molina-Holgado et al., 2003).

Regulation of Intracellular Kinases

Early studies using synthetic cannabinoid agonists showed a CB_1-receptor-dependent stimulation of the two MAP kinases (MAPK), p42 and p44 kDa (Bouaboula, Poinot-Chazel, et al., 1995). This effect on MAPK was independent of the cAMP pathway and led to downstream changes in the immediate-early gene *Krox*-24 (Bouaboula, Bourrie, et al., 1995). Futhermore, phosphatidylinositol 3-kinase (PI3K) was an intermediate in the pathway leading to CB_1 receptor upregulation of MAPK (Bouaboula et al., 1997). More recent studies have expanded this pathway in the presence of endocannabinoids. AEA activation of p42 and p44 kDa MAPK, now more commonly termed *extracellular signal-regulated kinase* (ERK) 1 and 2 or ERK, led to the induction of chemotaxis and chemokinesis in HEK cells transfected with CB_1 receptors (Song and Zhong, 2000).

AEA can reduce progenitor cell differentiation by inhibiting the Rap1/B-Raf-ERK pathway through CB_1 receptors in developing neurons (Rueda et al., 2002). The inhibition observed *in vitro* was also correlated with *in vivo* activity of the compounds in the hippocampus. Interestingly, although intraperitoneal administration of methanandamide did not decrease the total number of dividing cells in the subgranular zone of the dentate gyrus of the adult rat, it did inhibit the ability of neuronal progenitors to reach a mature phenotype in this region (Rueda et al., 2002).

Endocannabinoids, working through the CB_1 receptor, also activate ERK in hippocampal slices through a pathway that involves $G\alpha$-cAMP but not $G\beta\gamma$-PI3K (Derkinderen, Ledent, et al., 2001; Derkinderen et al., 2003). The same group also showed that AEA and 2-AG increase tyrosine phosphorylation of focal adhesion kinase (FAK) in rat hippocampal slices (Derkinderen, Toutant, et al., 2001). This effect was mediated downstream of CB_1 receptor inhibition of cAMP (Derkinderen et al., 1996). The tyrosine kinase Fyn, which associated with FAK during endocannabinoid treatments, appeared to be critical to the effects produced, because in Fyn knockout mice, stimulation of FAK by 2-AG was lost (Derkinderen, Toutant, et al., 2001). Another intriguing finding by Derkinderen

(2003) (Derkinderen et al., 2003) was that in the Fyn knockout mice, 2-AG stimulation of ERK was lost. This suggests that Fyn is a common link between ERK and FAK activation in hippocampal slices. Pathways utilizing ERK and FAK have been associated with changes in long-term potentiation (LTP) in the hippocampus (Grant et al., 1995; Kojima et al., 1997). These proteins may be an important part of endocannabinoid signaling, associated with effects produced downstream of acute actions such as DSI.

Antitumor Activity

Endocannabinoids have been shown to reduce nerve growth factor–induced and basal cell proliferation of human breast cancer cells (De Petrocellis et al., 1998; Melck et al., 1999; Melck et al., 2000). Early studies using EFM-19 and MCF-7 cells confirmed a CB_1-receptor-dependent effect that resulted in the downregulation of both prolactin receptors (PRLr) and a downstream breast-cancer-susceptibility gene, *brca1* (De Petrocellis et al., 1998). Prolactin has been shown to act as major proliferative hormone in breast cancer cell lines (Fuh and Wells, 1995). The ability of AEA to inhibit PRLr and *brac1* was correlated with the observed decrease in cells in the S phase of their mitotic cycle (De Petrocellis et al., 1998). Further investigations showed the involvement of a CB_1-receptor-dependent pathway where cAMP is inhibited and Raf-1 and MAPK activity is increased. These changes preceded the downregulation of PRLr as well as *trk* NGF receptors (Melck et al., 1999; Melck et al., 2000). The endocannabinoid 2-methyl-arachidonyl-2′-fluoro-ethylamide (Met-F-AEA), a stable analog of the endocannabinoid, has also been shown to inhibit epithelial tumors through a CB_1-receptor-dependent pathway (Bifulco et al., 2001). This study suggested that Met-F-AEA inhibits *ras* oncogene-dependent tumor growth *in vivo* through CB_1 cannabinoid receptors.

AEA has been demonstrated to activate protein kinase B (PKB)/Akt in both CB_1-transfected CHO cells and in a human astrocytomas cell line U373 MG; the latter cell line naturally expresses CB_1 (Gomez del Pulgar et al., 2000). A later study by Gomez del Pulgar et al., (2002) also linked cannabinoid modulation of PKB to the inhibitory effects of the drugs on rat C6 glioma cell growth. PKBs play an important signaling role in differentiating cells, and the upregulation of these proteins has been linked to cancer (Coffer et al., 1998).

ENDOCANNABINOID SIGNALING VIA CB_2 RECEPTORS

Regulation of Intracellular cAMP Levels

As with CB_1, modulation of cAMP production was the first second messenger system convincingly demonstrated to be mediated by cannabinoids via the CB_2 receptor in transfected cell lines (Bayewitch et al., 1995, Felder et al., 1995, Slipetz et al., 1995). Inhibition of cAMP production has been shown in cells that natively express the CB_2 receptor, including thymocytes and splenocytes (Herring et al., 1998). AEA is a partial agonist for forskolin-stimulated cAMP production at the CB_2 receptor, whereas 2-AG is a full agonist, inhibiting adenylyl cyclase activity via CB_2 receptors (Gonsiorek et al., 2000).

Modulation of Intracellular Calcium

As was observed with CHO cells transfected with CB_1, when these cells were transfected with CB_2 receptors, cannabinoids did not produce receptor-mediated changes in intracellular calcium (Gonsiorek et al., 2000). However, in response to 2-AG, Sugiura et al. (2000) demonstrated rapid transient increases in intracellular free Ca^{2+} concentrations in HL60 cells that naturally express the CB_2 receptor. This was similar to what was reported on the NG108-15 cells at the CB_1 receptor (Sugiura et al., 1996). It would be of interest to see if similar changes in intracellular free Ca^{2+} occur in the presence of 2-AG in nonimmortalized cell lines or native tissues that contain CB_2 receptors.

Regulation of Intracellular Kinases

CB_2-mediated activation of MAPK (ERK) has been demonstrated in CB_2-transfected CHO cells as well as in HL60 cells, which possess endogenous CB_2 receptors, and from which CB_2 was first isolated (Bouaboula et al., 1996; Kobayashi et al., 2001). These studies established a pertussis-toxin-sensitive, cAMP-independent pathway for CB_2-receptor activation (Bouaboula et al., 1996). In a mouse microglial cell line, BV-2 and 2-AG but not AEA stimulated MAPK through CB_2 receptors, and this led to increased cell migration. Pathological stimulation of the microglia cells specifically caused the release of 2-AG, suggesting the presence of a feed-back circuit (Walter et al., 2003).

In mast cells, CB_2-receptor agonists and nonselective CB_1/CB_2 agonists activate the ERK AKT and a selected subset of AKT targeted by Samson et al. (2003). However, activation of AKT or PI3K by endocannabinoids via the CB_2 receptor has not yet been demonstrated.

ENDOCANNABINOID SIGNALING VIA CB_1 AND CB_2 RECEPTORS

Antitumor Activity

The antiproliferative effects of AEA and 2-AG have been demonstrated, using rat C6 glioma cell lines *in vitro*. Interestingly, this effect was reversed only by a combination of CB_1 and CB_2 antagonists together (Jacobsson et al., 2001). This was first observed in a study carried out by Galve-Roperh et al. (2000), in which a synthetic CB_1/CB_2 agonist WIN55212-2 was used. The reduction in cell viability produced by this compound could only be antagonized by the use of both CB_1 and CB_2 antagonists. Targeting the CB_2 receptor population alone, however, could produce similar antiproliferative effects *in vivo*. The lack of euphoric effects produced by a CB_2 agonist make it a better candidate as an antitumor agent, but it would be interesting to see how a more pure CB_1 agonist compares to a CB_2 agonist in this model. The CB_1/CB_2-dependent pathways responsible for the effects of endocannabinoids on glioma cell lines are still being investigated. This endeavor is also complicated by the fact that endocannabinoids interact with cannabinoid and vanilloid receptors in rat C6 glioma cells (Jacobsson et al., 2001). Activation of each receptor population leads to antiproliferative effects in this cell line.

The pathway responsible for the antiproliferative and apoptotic effects of THC and WIN55212 in the rat C6 glioma cell line has been examined in numerous studies (Sanchez et al., 1998; Galve-Roperh et al., 2000; Gomez Del Pulgar et al., 2002). It involves a cannabinoid-receptor-dependent increase in ceramide synthesis and a long-term upregulation of ERK. This pathway was not related to the initial modulation of cAMP by cannabinoid receptors (Galve-Roperh et al., 2000). In the study carried out by Jacobsson et al. (2001), an inhibitor of ceramide synthesis did reduce the antiproliferative effects of AEA and 2-AG, suggesting this pathway could be involved. However, more work needs to be done to determine if this is indeed the case or if the pathways involved are similar to those described for endocannabinoids in breast cancer cell lines (i.e., cAMP-dependent pathways).

SIGNALING PATHWAYS IN OTHER CANNABINOID RECEPTOR SUBTYPES

There is mounting evidence for additional cannabinoid receptors. The CB_1-receptor gene has been inactivated (by in-frame deletion of most of the coding region) through homologous recombination, in two laboratories (Ledent et al., 1999; Zimmer et al., 1999). Significantly, although the CB_1 receptor knockout mice lost responsiveness to most cannabinoids, Δ^9-THC still produced antinociception in the tail-flick test of analgesia (Zimmer et al., 1999). Further characterization of this non-CB_1 Δ^9-THC response suggests the presence of a novel cannabinoid receptor–ion channel in the pain pathway (Zygmunt et al., 2002).

AEA produces the full range of behavioral effects (antinociception, catalepsy, and impaired locomotor activity) in CB_1 receptor knockout mice (Di Marzo et al., 2000). Futhermore, AEA-stimulated GTPγS activity can be elicited in brain membranes from these mice (Breivogel et al., 2001). These effects were not sensitive to inhibition by SR141716A. This same phenomenon has also been demonstrated in a second strain of CB_1 receptor knockout mice (Monory et al., 2002).

In another set of studies, cannabinoids including AEA were found to elicit cardiovascular effects via peripherally located CB_1 receptors (Ishac et al., 1996; Jarai et al., 1999; Wagner et al., 1999). Abnormal cannabidiol (abn-cbd), a neurobehaviorally inactive cannabinoid, which does not bind to CB_1 receptors, caused hypotension and mesenteric vasodilation in wild-type mice and in mice lacking either CB_1 receptors only or both CB_1 and CB_2 receptors (Jarai et al., 1999). In contrast to the studies described earlier, these cardiovascular and endothelial effects were SR141716A sensitive. A stable analog of AEA, methanandamide, also produced SR141716A-sensitive hypotension in CB_1/CB_2 knockout mice. These effects were not due to activation of vanilloid receptors, which also interact with AEA (Zygmunt et al., 1999). This subtype is referred to as the *abnormal cannabidiol receptor* (Walter et al., 2003). A selective antagonist, O-1918, has recently been developed; it inhibits the vasorelaxant effects of abn-cbd and AEA (Offertaler et al., 2003).

Signal transduction pathways for the abn-cbd receptor have been studied in human umbilical endothelial cells (HUVEC) (Offertaler et al., 2003). Abn-cbd induces phosphorylation of ERK and PKB/Akt via a PI3 kinase-dependent pertussis–toxin-sensitive pathway; these effects were blocked by O-1918 (Offertaler et al., 2003). The abn-cbd receptor subtype also appears to be present in microglia (Walter et al., 2003). AEA and 2-AG triggered migration in BV-2 cells, a microglial cell line; their effects were blocked with O-1918. 2-AG also induced phosphorylation of ERK1/2 in BV-2 cells (Walter et al., 2003). These data suggest a common signaling pathway for the abn-cbd receptor in endothelial cells and microglia.

SUMMARY

CB_1 and CB_2 receptors transduce endocannabinoid binding to a diverse set of second messenger systems. The cellular environment is a key determinant to the pathway activated; factors in this environment include the types of G proteins involved as well as subsequent effector enzymes. CB_1 receptors are capable of activating more signal transduction pathways than the CB_2 receptors. The features underlying the promiscuity of the CB_1 receptor have begun to be elucidated.

The multiple signaling pathways known to date link the actions of endocannabinoids to a multitude of pharmacological effects, including learning and memory alteration, antinociception, neuroprotection, cardiovascular effects, and antitumor activity. With the discovery of additional endocannabinoids and receptor subtypes will come the elucidation of further physiological effects of the endocannabinoid signaling pathways.

REFERENCES

Abadji, V., Lucas-Lenard, J., Chin, C., and Kendall, D. (1999) *J Neurochem,* **72,** 2032–2038.

Bayewitch, M., Avidor-Reiss, T., Levy, R., Barg, J., Mechoulam, R., and Vogel, Z. (1995) *FEBS Lett,* **375,** 143–147.

Bifulco, M., Laezza, C., Portella, G., Vitale, M., Orlando, P., De Petrocellis, L., and Di Marzo, V. (2001) *FASEB J,* **15,** 2745–2747.

Bonhaus, D., Chang, L., Kwan, J., and Martin, G. (1998) *J Pharmacol Exp Ther,* **287,** 884–888.

Bouaboula, M., Bourrie, B., Rinaldi-Carmona, M., Shire, D., Le Fur, G., and Casellas, P. (1995) *J Biol Chem,* **270,** 13973–13980.

Bouaboula, M., Perrachon, S., Milligan, L., Canat, X., Rinaldi-Carmona, M., Portier, M., Barth, F., Calandra, B., Pecceu, F., Lupker, J., Maffrand, J.P., Le Fur, G., and Casellas, P. (1997) *J Biol Chem,* **272,** 22330–22339.

Bouaboula, M., Poinot-Chazel, C., and Casellas, P. (1995) *J. Biol. Chem,* **312,** 637–641.

Bouaboula, M., Poinot-Chazel, C., and Casellas, P. (1996) *Eur J Pharmacol,* **237,** 704–711.
Breivogel, C.S., Griffin, G., Di Marzo, V., and Martin, B.R. (2001) *Mol Pharmacol,* **60,** 155–163.
Buckley, N.E., McCoy, K.L., Mezey, E., Bonner, T., Zimmer, A., Felder, C.C., and Glass, M. (2000) *Eur J Pharmacol,* **396,** 141–149.
Coffer, P.J., Jin, J., and Woodgett, J.R. (1998) *Biochem J,* **335,** 1–13.
Compton, D.R., Rice, K.C., De Costa, B.R., Razdan, R.K., Melvin, L.S., Johnson, M.R., and Martin, B.R. (1993) *J Pharmacol Exp Ther,* **265,** 218–226.
De Petrocellis, L., Melck, D., Palmisano, A., Bisogno, T., Laezza, C., Bifulco, M., and Di Marzo, V. (1998) *Proc Natl Acad Sci U S A,* **95,** 8375–8380.
Deadwyler, S.A., Hampson, R.E., Mu, J., Whyte, A., and Childers, S. (1995) *J Pharmacol Exp Ther,* **273,** 734–743.
Derkinderen, P., Ledent, C., Parmentier, M., and Girault, J.A. (2001) *J Neurochem,* **77,** 957–960.
Derkinderen, P., Toutant, M., Burgaya, F., Le Bert, M., Siciliano, J.C., de Franciscis, V., Gelman, M., and Girault, J.A. (1996) *Science,* **273,** 1719–1722.
Derkinderen, P., Toutant, M., Kadare, G., Ledent, C., Parmentier, M., and Girault, J.A. (2001) *J Biol Chem,* **276,** 38289–96.
Derkinderen, P., Valjent, E., Toutant, M., Corvol, J.C., Enslen, H., Ledent, C., Trzaskos, J., Caboche, J., and Girault, J.A. (2003) *J Neurosci,* **23,** 2371–82.
Devane, W.A., Dysarz, I.F.A., Johnson, M.R., Melvin, L.S., and Howlett, A.C. (1988) *Mol Pharmacol,* **34,** 605–613.
Di Marzo, V., Breivogel, C.S., Tao, Q., Bridgen, D.T., Razdan, R.K., Zimmer, A.M., Zimmer, A., and Martin, B.R. (2000) *J Neurochem,* **75,** 2434–44.
Di Marzo, V., De Petrocellis, L., Fezza, F., Ligresti, A., and Bisogno, T. (2002) *Prostaglandins Leukot Essent Fatty Acids,* **66,** 377–391.
Di Marzo, V., Fontana, A., Cadas, H., Schinelli, S., Cimino, G., Schwartz, J.C., and Piomelli, D. (1994) *Nature,* **372,** 686–691.
Di Marzo, V., Melck, D., Orlando, P., Bisogno, T., Zagoory, O., Bifulco, M., Vogel, Z., and De Petrocellis, L. (2001) *Biochem J,* **358,** 249–255.
Egertova, M., Giang, D.K., Cravatt, B.F., and Elphick, M.R. (1998) *Proc R Soc Lond B Biol Sci,* **265,** 2081–2085.
Elphick, M.R. and Egertova, M. (2001) *Philos Trans R Soc Lond B Biol Sci,* **356,** 381–408.
Esposito, G., Ligresti, A., Izzo, A.A., Bisogno, T., Ruvo, M., Di Rosa, M., Di Marzo, V., and Iuvone, T. (2002) *J Biol Chem,* **277,** 50348–50354.
Facci, L., Dal Toso, R., Romanello, S., Buriani, A., Skaper, S.D., and Leon, A. (1995) *Proc Natl Acad Sci U S A,* **92,** 3376–3380.
Felder, C.C., Briley, E.M., Axelrod, J., Simpson, J.T., Mackie, K., and Devane, W.A. (1993a) *Proc Natl Acad Sci U S A,* **90,** 7656–7660.
Felder, C.C., Joyce, K.E., Briley, E.M., Mansouri, J., Mackie, K., Blond, O., Lai, Y., Ma, A.L., and Mitchell, R.L. (1995) *Mol Pharmacol,* **48,** 443–450.
Felder, C.C., Veluz, J.S., Williams, H.L., Briley, E.M., and Matsuda, L.A. (1992) *Mol Pharmacol,* **42,** 838–845.
Fuh, G. and Wells, J.A. (1995) *J Biol Chem,* **270,** 13133–13137.
Galve-Roperh, I., Sanchez, C., Cortes, M.L., del Pulgar, T.G., Izquierdo, M., and Guzman, M. (2000) *Nat Med,* **6,** 313–319.
Gebremedhin, D., Lange, A.R., Campbell, W.B., Hillard, C.J., and Harder, D.R. (1999) *Am J Physiol,* **276,** H2085–93.
Gerard, C.M., Mollereau, C., Vassart, G., and Parmentier, M. (1991) *Biochem J,* **279,** 129–134.
Glass, M. and Felder, C.C. (1997) *J Neurosci,* **17,** 5327–5333.
Glass, M. and Northup, J. (1999) *Mol Pharmacol,* **56,** 1362–1369.
Gomez Del Pulgar, T., De Ceballos, M.L., Guzman, M., and Velasco, G. (2002) *J Biol Chem,* **277,** 36527–36533.
Gomez del Pulgar, T., Velasco, G., and Guzman, M. (2000) *Biochem J,* **347,** 369–373.
Gonsiorek, W., Lunn, C., Fan, X., Narula, S., Lundell, D., and Hipkin, R.W. (2000) *Mol Pharmacol,* **57,** 1045–1050.
Grant, S.G., Karl, K.A., Kiebler, M.A., and Kandel, E.R. (1995) *Genes Dev,* **9,** 1909–1921.
Griffin, G., Tao, Q., and Abood, M. (2000) *J Pharmacol Exp Ther,* **292.**
Griffin, G.R., Atkinson, P.J., Showalter, V.M., Martin, B.R., and Abood, M.E. (1998) *J Pharmacol Exp Ther,* **285,** 553–560.
Hampson, A.J., Bornheim, L.M., Scanziani, M., Yost, C.S., Gray, A.T., Hansen, B.M., Leonoudakis, D.J., and Bickler, P.E. (1998) *J Neurochem,* **70,** 671–676.
Hanus, L., Abu-Lafi, S., Fride, E., Breuer, A., Vogel, Z., Shalev, D.E., Kustanovich, I., and Mechoulam, R. (2001) *Proc Natl Acad Sci U S A,* **98,** 3662–3665.

Henry, D.J. and Chavkin, C. (1995) *Neurosci Lett,* 91–94.

Herring, A.C., Koh, W.S., and Kaminski, N.E. (1998) *Biochem Pharmacol,* **55,** 1013–1023.

Howlett, A.C. (1995) *Ann Rev Pharmacol Toxicol,* **35,** 607–634.

Howlett, A.C., Johnson, M. R., Melvin, L. S., and Milne, G. M. (1988) *Mol Pharmacol,* **33,** 297–302.

Huang, S.M., Bisogno, T., Trevisani, M., Al-Hayani, A., De Petrocellis, L., Fezza, F., Tognetto, M., Petros, T.J., Krey, J.F., Chu, C.J., Miller, J.D., Davies, S.N., Geppetti, P., Walker, J.M., and Di Marzo, V. (2002) *Proc Natl Acad Sci U S A,* **99,** 8400–8405.

Ikeda, S. R. (1996) *Nature,* **380,** 255–8.

Ishac, E.J.N., Jiang, L., Lake, K.D., Varga, K., Abood, M.E., and Kunos, G. (1996) *Br J Pharmacol,* **118,** 2023–2028.

Jacobsson, S.O., Wallin, T., and Fowler, C.J. (2001) *J Pharmacol Exp Ther,* **299,** 951–959.

Jarai, Z., Wagner, J., Varga, K., Lake, K., Compton, D., Martin, B., Zimmer, A., Bonner, T., Buckley, N., Mezey, E., Razdan, R., Zimmer, A., and Kunos, G. (1999) *Proc Natl Acad Sci U S A,* **96,** 14136–14141.

Jin, W., Brown, S., Roche, J. P., Hsieh, C., Celver, J. P., Kovoor, A., Chavkin, C., and Mackie, K. (1999) *J Neurosci,* **19,** 3773–3780.

Jin, W., Lee, N.M., Loh, H.H., and Thayer, S.A. (1994) *J Neurosci,* **14,** 1920–1929.

Kearn, C., Greenberg, M., DiCamelli, R., Kurzawa, K., and Hillard, C. (1999) *J Neurochem,* **72,** 2379–2387.

Kobayashi, Y., Arai, S., Waku, K., and Sugiura, T. (2001) *J Biochem (Tokyo),* **129,** 665–669.

Kojima, N., Wang, J., Mansuy, I.M., Grant, S.G., Mayford, M., and Kandel, E.R. (1997) *Proc Natl Acad Sci U S A,* **94,** 4761–4765.

Kreitzer, A.C., Carter, A.G., and Regehr, W.G. (2002) *Neuron,* **34,** 787–796.

Kreitzer, A.C. and Regehr, W.G. (2001) *Neuron,* **29,** 717–727.

Lambert, D., DiPaolo, F., Sonveaux, P., Kanyonyo, M., Govaerts, S., Hermans, E., Bueb, J., Delzenne, N., and Tschirhart, E. (1999) *Biochim Biophys Acta,* **1440,** 266–274.

Ledent, C., Valverde, O., Cossu, G., Petitet, F., Aubert, J., Beslot, F., Bohme, G., Imperato, A., Pedrazzini, T., Roques, B., Vassart, G., Fratta, W., and Parmentier, M. (1999) *Science,* **283,** 401–404.

Little, P.J., Compton, D.R., Johnson, M.R., Melvin, L.S., and Martin, B.R. (1988) *J Pharmacol Exp Ther,* **247,** 1046–1051.

Little, P.J. and Martin, B.R. (1991) *Life Sci,* **48,** 1133–1141.

Mackie, K., Devane, W.A., and Hille, B. (1993) *Mol Pharmacol,* **44,** 498–503.

Mackie, K., Lai, Y., Westenbroek, R., and Mitchell, R. (1995) *J Neurosci,* **15,** 6552–6561.

Marrion, N.V. (1997) *Annu Rev Physiol,* **59,** 483–504.

Matsuda, L.A., Lolait, S.J., Brownstein, M.J., Young, A.C., and Bonner, T.I. (1990) *Nature,* **346,** 561–564.

McAllister, S., Griffin, G., Satin, L., and Abood, M. (1999) *J Pharmacol Exp Ther,* **291,** 618–626.

McAllister, S.D., Tao, Q., Barnett-Norris, J., Buehner, K., Hurst, D.P., Guarnieri, F., Reggio, P.H., Nowell Harmon, K.W., Cabral, G.A., and Abood, M.E. (2002) *Biochem Pharmacol,* **63,** 2121–2136.

Mechoulam, R., Ben-Shabat, S., Hanus, L., Ligumsky, M., Kaminski, N.E., Schatz, A.R., Gopher, A., Almog, S., Martin, B.R., Compton, D.R., Pertwee, R.G., Griffin, G., Bayewitch, M., Barg, J., and Vogel, Z. (1995) *Biochem Pharmacol,* **50,** 83–90.

Mechoulam, R., Hanus, L., Ben-Shabat, S., Fride, E., and Weidenfeld, J. (1994) *Neuropsychopharmacology* **10,** 145S–145S.

Melck, D., De Petrocellis, L., Orlando, P., Bisogno, T., Laezza, C., Bifulco, M., and Di Marzo, V. (2000) *Endocrinology,* **141,** 118–126.

Melck, D., Rueda, D., Galve-Roperh, I., De Petrocellis, L., Guzman, M., and Di Marzo, V. (1999) *FEBS Lett,* **463,** 235–240.

Misner, D.L. and Sullivan, J.M. (1999) *J Neurosci,* **19,** 6795–6805.

Molina-Holgado, F., Molina-Holgado, E., Guaza, C., and Rothwell, N.J. (2002) *J Neurosci Res,* **67,** 829–836.

Molina-Holgado, F., Pinteaux, E., Moore, J.D., Molina-Holgado, E., Guaza, C., Gibson, R.M., and Rothwell, N.J. (2003) *J Neurosci,* **23,** 6470–6474.

Mombouli, J. V., Schaeffer, G., Holzmann, S., Kostner, G.M., and Graier, W.F. (1999) *Br J Pharmacol,* **126,** 1593–1600.

Monory, K., Tzavara, E.T., Lexime, J., Ledent, C., Parmentier, M., Borsodi, A., and Hanoune, J. (2002) *Biochem Biophys Res Commn,* **292,** 231–235.

Mukhopadhyay, S., McIntosh, H., Houston, D., and Howlett, A. (2000) *Mol Pharmacol,* **57,** 162–170.

Munro, S., Thomas, K.L., and Abu-Shaar, M. (1993) *Nature,* **365,** 61–65.

Netzeband, J.G., Conroy, S.M., Parsons, K.L., and Gruol, D.L. (1999) *J Neurosci,* **19,** 8765–8777.

Offertaler, L., Mo, F.M., Batkai, S., Liu, J., Begg, M., Razdan, R.K., Martin, B.R., Bukoski, R.D., and Kunos, G. (2003) *Mol Pharmacol,* **63,** 699–705.

Ohno-Shosaku, T., Maejima, T., and Kano, M. (2001) *Neuron,* **29,** 729–738.

Pacheco, M., Childers, S.R., Arnold, R., Casiano, F., and Ward, S.J. (1991) *J Pharmacol Exp Ther,* **257,** 170–183.

Pertwee, R.G. (1997) *Pharmacol Ther,* **74,** 129–180.

Porter, A.C., Sauer, J.M., Knierman, M.D., Becker, G.W., Berna, M.J., Bao, J., Nomikos, G.G., Carter, P., Bymaster, F.P., Leese, A.B., and Felder, C.C. (2002) *J Pharmacol Exp Ther,* **301,** 1020–1024.

Prather, P.L., Martin, N.A., Breivogel, C.S., and Childers, S.R. (2000) *Mol Pharmacol,* **57,** 1000–1010.

Rhee, M.H., Bayewitch, M., Avidor-Reiss, T., Levy, R., and Vogel, Z. (1998) *J Neurochem,* **71,** 1525–1534.

Robbe, D., Gerard, A., Florence, D., Joel, B., and Manzoni, O.J. (2001) *J Neurosci,* **21,** 109–116.

Rueda, D., Navarro, B., Martinez-Serrano, A., Guzman, M., and Galve-Roperh, I. (2002) *J Biol Chem,* **277,** 46645–50.

Samson, M.T., Small-Howard, A., Shimoda, L.M., Koblan-Huberson, M., Stokes, A.J., and Turner, H. (2003) *J Immunol,* **170,** 4953–4962.

Sanchez, C., Galve-Roperh, I., Canova, C., Brachet, P., and Guzman, M. (1998) *FEBS Lett,* **436,** 6–10.

Schweitzer, P. (2000) *J Neurosci,* **20,** 51–58.

Selley, D.E., Stark, S., Sim, L.J., and Childers, S.R. (1996) *Life Sci,* **59,** 659–668.

Showalter, V.M., Compton, D.R., Martin, B.R., and Abood, M.E. (1996) *J Pharmacol Exp Ther,* **278,** 989–999.

Slipetz, D.M., O'Neill, G.P., Favreau, L., Dufresne, C., Gallant, M., Gareau, Y., Guay, D., Labelle, M., and Metters, K.M. (1995) *Mol Pharmacol,* **48,** 352–361.

Song, Z.H. and Zhong, M. (2000) *J Pharmacol Exp Ther,* **294,** 204–209.

Stefano, G.B., Bilfinger, T.V., Rialas, C.M., and Deutsch, D.G. (2000) *Pharmacol Res,* **42,** 317–322.

Stefano, G.B., Liu, Y., and Goligorsky, M.S. (1996) *J Biol Chem,* **271,** 19238–19242.

Stella, N., Schweitzer, P., and Piomelli, D. (1997) *Nature,* **388,** 773–778.

Straiker, A., Stella, N., Piomelli, D., Mackie, K., Karten, H. J., and Maguire, G. (1999) *Proc Natl Acad Sci U S A,* **96,** 14565–14570.

Straiker, A. and Sullivan, J.M. (2003) *J Neurophysiol,* **89,** 2647–2654.

Sugiura, T., Kodaka, T., Kondo, S., Tonegawa, T., Nakane, S., Kishimoto, S., Yamashita, A., and Waku, K. (1996) *Biochem Biophys Res Commn,* **229,** 58–64.

Sugiura, T., Kondo, S., Kishimoto, S., Miyashita, T., Nakane, S., Kodaka, T., Suhara, Y., Takayama, H., and Waku, K. (2000) *J Biol Chem,* **275,** 605–612.

Sugiura, T., Kondo, S., Sukagawa, A., Nakane, S., Shinoda, A., Itoh, K., Yamashita, A., and Waku, K. (1995) *Biochem Biophys Res Commn,* **215,** 89–97.

Tao, Q., McAllister, S., Andreassi, J., Nowell, K., Cabral, G., Hurst, D., Bachtel, K., Ekman, M., Reggio, P., and Abood, M. (1999) *Mol Pharmacol,* **55,** 605–613.

Terranova, J.P., Michaud, J.C., Le Fur, G., and Soubrie, P. (1995) *Naunyn Schmiedebergs Arch Pharmacol,* **352,** 576–579.

Tsou, K., Brown, S., Sanudo-Pena, M. C., Mackie, K., and Walker, J.M. (1998) *Neuroscience,* **83,** 393–411.

Vogel, Z., Barg, J., Levy, R., Saya, D., Heldman, E., and Mechoulam, R. (1993) *J Neurochem,* **61,** 352–355.

Wagner, J., Varga, K., Jarai, Z., and Kunos, G. (1999) *Hypertension,* **33,** 429–434.

Waksman, Y., Olson, J.M., Carlisle, S.J., and Cabral, G.A. (1999) *J Pharmacol Exp Ther,* **288,** 1357–1366.

Walter, L., Franklin, A., Witting, A., Wade, C., Xie, Y., Kunos, G., Mackie, K., and Stella, N. (2003) *J Neurosci,* **23,** 1398–1405.

Williams, E.J., Walsh, F.S., and Doherty, P. (2003) *J Cell Biol,* **160,** 481–486.

Wilson, R.I., Kunos, G., and Nicoll, R.A. (2001) *Neuron,* **31,** 453–462.

Wilson, R.I. and Nicoll, R.A. (2001) *Nature,* **410,** 588–592.

Zimmer, A., Zimmer, A., Hohmann, A., Herkenham, M., and Bonner, T. (1999) *Proc Natl Acad Sci U S A,* 5780–5785.

Zygmunt, P., Petersson, J., Andersson, D., Chuang, H., Sorgard, M., DiMarzo, V., Julius, D., and Hogestatt, E. (1999) *Nature,* **400,** 452–457.

Zygmunt, P.M., Andersson, D.A., and Hogestatt, E.D. (2002) *J Neurosci,* **22,** 4720–4727.

5 Endocannabinoids as Retrograde Messengers in Synaptic Transmission

Saori Oka, Yoshio Ishima, Keizo Waku, and Takayuki Sugiura

CONTENTS

INTRODUCTION

Marijuana has been used as a traditional medicine and a pleasure-inducing drug for thousands of years. The major pharmacologically active constituent of marijuana is Δ^9-tetrahydrocannabinol (Δ^9-THC). The administration of Δ^9-THC to experimental animals and humans elicits a variety of biological responses in various tissues and organs, particularly in the central nervous system. For example, Δ^9-THC induces reduced spontaneous motor activity, immobility, analgesia, impairment of short-term memory, and hypothermia in experimental animals and altered perception, euphoria, and hallucination in humans (Dewey, 1986). The mechanisms of these actions of Δ^9-THC remained elusive until recently.

The presence of a specific binding site for cannabinoids had long been postulated by a number of investigators, although binding experiments were not satisfactory due to the highly lipophilic properties of cannabinoids. Finally, Howlett and co-workers (Devane et al., 1988) succeeded in providing evidence that a specific binding site for cannabinoids is present in rat brain synaptosomes using [^{3}H]CP55940, a radiolabeled synthetic cannabinoid, as a ligand. This strongly suggested that Δ^9-THC exerts its diverse biological activities, if not all, by acting on this specific binding site. Soon after, Matsuda et al. (1990) reported the cloning of a cDNA encoding a seven-transmembrane, G protein-coupled receptor, which is now known as the cannabinoid CB_1 receptor. The CB_1 receptor contains 472 (human) or 473 (rat) amino acid residues and is expressed abundantly in the central nervous system and in various peripheral tissues. It is noteworthy that the whole-brain cannabinoid receptor density is similar to whole-brain densities of receptors for glutamate and gamma-aminobutyric acid (GABA) (Herkenham, 1995). Among the various brain regions, the CB_1 receptor is highly abundant in the substantia nigra, globus pallidus, molecular layer of the cerebellum, hippocampus, and cerebral cortex (Herkenham et al., 1990). Based on these observations as well as numerous other experimental results, the CB_1 receptor is assumed to be involved in the regulation of motor activity, memory, and cognition in mammals (For reviews, Matsuda and Bonner, 1995; Herkenham, 1995; Pertwee, 1997; Di Marzo et al., 1998; Giuffrida and Piomelli, 2000; Pertwee and Ross, 2002; Fride, 2002; Fernandez-Ruiz et al., 2002; Lichtman et al., 2002).

Another cDNA that encodes a cannabinoid receptor was cloned by Munro et al. (1993). This cannabinoid receptor (the CB_2 receptor) is also a seven-transmembrane, G protein-coupled receptor and consists of 360 amino acids. The CB_2 receptor is mainly expressed in the immune system such as the spleen and tonsil and lymph nodes. The CB_2 receptor is abundantly expressed in several types of leukocytes, such as B cells, macrophages/monocytes, and natural killer cells, and is assumed to participate in the regulation (stimulation or inhibition) of inflammatory reactions and immune responses. (For reviews, see Berdyshev, 2000; Di Marzo, Melck, et al., 2000; De Petrocellis et al., 2000; Parolaro et al., 2002; Cabral, 2002; and Roth et al., 2002.)

To date, two types of endogenous cannabinoid receptor ligands (endocannabinoids) have been identified. Both compounds are arachidonic-acid-containing molecules. The first endocannabinoid to be found was N-arachidonoylethanolamine (anandamide). This compound was isolated by Devane et al. (1992) from the pig brain as an endogenous ligand of the cannabinoid receptors. The second endocannabinoid to be found was 2-AG. We isolated 2-AG from the rat brain (Sugiura et al., 1995), and Mechoulam et al. (1995) isolated it from the canine gut as an endogenous cannabinoid receptor ligand. Later, Stella et al. (1997) also confirmed that 2-AG is present in the rat brain. Numerous studies have been carried out on these endocannabinoids, and evidence is gradually accumulating that the endocannabinoid system plays physiologically essential roles in diverse biological systems including the central nervous system. (For reviews, see Felder and Glass, 1998; Di Marzo, 1998; Di Marzo et al., 1998; Piomelli et al., 1998, 2000; Mechoulam et al., 1998; Hillard, 2000; Giuffrida and Piomelli 2000; Sugiura and Waku, 2000, 2002; Di Marzo et al., 2002; Sugiura et al., 2002; Pertwee and Ross, 2002; Fride, 2002; Fernandez-Ruiz et al., 2002; Lichtman et al., 2002; and Schmid et al., 2002.)

In this chapter, we focus on the physiological significance of endocannabinoids in mammalian brains, especially their possible roles as messenger molecules in the modulation of synaptic transmission.

ENDOCANNABINOIDS IN THE NERVOUS SYSTEM

Anandamide binds to the CB_1 receptor, abundantly expressed in the brain, with a high affinity (Devane et al., 1992). A number of investigators have demonstrated that anandamide exhibits a variety of cannabimimetic activities in the nervous system, such as reduced spontaneous motor activities, immobility, hypothermia, analgesia, impairment of memory, stimulation of appetite when administered to experimental animals, inhibition of adenylyl cyclase, inhibition of voltage-gated Ca^{2+} channels, activation of an inwardly rectifying K^+ current, reduction of gap junction permeability, inhibition of neurotransmitter release, and the inhibition of long-term potentiation (LTP) in hippocampal slices *in vitro* (Pertwee, 1997; Hillard and Campbell, 1997; Felder and Glass, 1998; Di Marzo, 1998; Mechoulam et al., 1998; Piomelli et al., 1998, 2000; Hillard, 2000; Di Marzo et al., 2002; Schmid et al., 2002, for reviews).

Despite these pharmacological activities, the physiological significance of anandamide in the nervous system of living animals is not yet fully elucidated. Concerning the roles of anandamide as the endogenous natural ligand for the cannabinoid receptors in the nervous system, the following issues still remain enigmas:

1. Fresh brain usually contains only small amounts of anandamide (in the order of pmol/g tissue) (see Chapter 6), although the possibility that anandamide is localized in certain areas in the brain cannot be excluded.
2. There appears to be no selective or efficient synthetic pathway for anandamide in the brain as mentioned in the following text. Anandamide can be formed from free arachidonic acid and ethanolamine through the reverse reaction of an anandamide amidohydrolase/

fatty acid amide hydrolase (FAAH) (for reviews, Hillard and Campbell, 1997; Di Marzo, 1998; Piomelli et al., 1998; Di Marzo et al., 1999; Sugiura et al., 2002; Di Marzo et al., 2002; Schmid et al., 2002) (see Chapter 6). However, this pathway is probably not physiologically relevant, because high concentrations of substrates are required for the synthetic reaction. Anandamide can also be formed from pre-existing *N*-arachidonoyl phosphatidylethanolamine (PE) through the action of a phosphodiesterase. (For reviews, see Hillard and Campbell, 1997; Di Marzo, 1998; Piomelli et al., 1998; Di Marzo et al., 1999, 2002; Sugiura et al., 2002; Schmid, 2000; Schmid et al., 2002.) (See also Chapter 6.) However, this pathway also does not appear to be able to generate a large amount of anandamide, because the level of *N*-arachidonyl PE in the brain is fairly low (Sugiura, et al., 1996; Cadas et al., 1997; for reviews, Schmid et al., 1990; Hansen et al., 2000). Di Marzo et al. (1994) previously demonstrated that anandamide was produced in rat cortical neurons when stimulated with ionomycin or with several membrane-depolarizing agents such as kainate, high K^+ and 4-aminopyridine; nevertheless, the levels of anandamide generated were found to be very low.

3. Anandamide acts as a partial agonist toward the CB_1 receptor as well as the CB_2 receptor (Mackie et al., 1993; Fride et al., 1995; Burkey et al., 1997; Sugiura, 1996; Sugiura et al., 1997, 1999, 2000; Breivogel et al., 1998; Kearn et al., 1999; McAllister et al., 1999; Hillard, 2000; Gonsiorek et al., 2000). It is unusual that an endogenous natural ligand is a partial agonist of its own receptor. Various kinds of neurotransmitters, such as glutamate and acetylcholine, are known to act as full agonists at their receptors. It is thus rather questionable that anandamide acts as an effective ligand for the cannabinoid receptors (CB_1 and CB_2) with profound physiological significance *in vivo*, yet there remains the possibility of the presence of specific receptors for anandamide distinct from the CB_1 receptor and CB_2 receptor, such as the CB_3 receptor. In fact, Breivogel et al. (2001) demonstrated the presence of a binding site for anandamide, WIN55212-2 and SR141716A in $CB_1^{-/-}$ mouse brain. In any case, the physiological significance of anandamide may exist mainly in its being an endogenous ligand of a receptor other than the CB_1 receptor and CB_2 receptor or as an endogenous modulator of certain ion channels.

In contrast to anandamide, 2-AG fulfills the requirements for a natural ligand of the cannabinoid receptors (CB_1 and CB_2) as follows:

1. 2-AG possesses pharmacological activities similar to those of anandamide and Δ^9-THC (Di Marzo, 1998; Mechoulam et al., 1998; Piomelli et al., 1998, 2000; Hillard, 2000; Sugiura and Waku, 2000, 2002; Sugiura et al., 2002; Di Marzo et al., 2002; Schmid et al., 2002, for reviews).
2. 2-AG acts as a full agonist toward both the CB_1 receptor and CB_2 receptor (Sugiura et al., 1996, 1997, 1999, 2000; Gonsiorek et al., 2000; Hillard, 2000; Savinainen et al., 2001), which is different from the case of anandamide and Δ^9-THC (Mackie et al., 1993; Fride et al., 1995; Burkey et al., 1997; Sugiura et al., 1996, 1997, 1999, 2000; Breivogel et al., 1998; Kearn et al., 1999; McAllister et al., 1999; Hillard, 2000; Gonsiorek et al., 2000).
3. 2-AG can be rapidly formed from inositol phospholipids (through combined actions of phospholipase C and diacylglycerol lipase or combined actions of phospholipase A_1 and phospholipase C) or from other arachidonic-acid-containing phospholipids in a variety of tissues and cells upon stimulation (Sugiura et al., 1995; Stella et al., 1997; for reviews, Di Marzo, 1998; Piomelli et al., 1998, 2000; Di Marzo et al., 1999, 2002;

Hillard, 2000; Sugiura and Waku, 2000, 2002; Sugiura et al., 2002; Schmid et al., 2002) (see Chapter 6).

4. A significant portion of the 2-AG generated was rapidly released into the extracellular fluid.
5. Tissue levels of 2-AG were 10 to 100 times higher than those of anandamide in the same tissue (see Chapter 6). Taken together, it is logical to assume that 2-AG rather than anandamide is the true natural ligand of the cannabinoid receptors (CB_1 and CB_2) (Sugiura et al., 1997, 1999, 2000; for review, Sugiura and Waku 2000).

Noticeably, as mentioned in the preceding text, several investigators have demonstrated that 2-AG was rapidly produced in various nervous tissues and cells when challenged with a variety of stimuli. An important aspect is that Ca^{2+} plays an essential role in the generation of 2-AG (Bisogno et al., 1997; Stella et al., 1997; Kondo et al., 1998). Bisogno et al. (1997) demonstrated that N18TG2 cells generate 2-AG when stimulated with ionomycin and that a large portion of 2-AG was released into the extracellular medium. Di Marzo, Hill, et al. (2000) also reported that the injection of reserpine induced an increase in the level of 2-AG in the globus pallidus in rats. Stella et al. (1997) reported the generation of 2-AG in electrically stimulated rat hippocampal slices and ionomycin-stimulated neurons. They also detected the generation of 2-AG in NMDA-stimulated rat cortical neurons (Stella and Piomelli, 2001). On the other hand, we found that the administration of picrotoxinin, a central nervous system stimulant, induces a rapid generation of 2-AG in the rat brain *in vivo* (Sugiura et al., 2000a). We also found that a substantial amount of 2-AG was generated and released from rat brain synaptosomes upon depolarization (Oka, S and Sugiura, T., unpublished results). The rapid generation of 2-AG in stimulated nervous tissues and cells is in good contrast to the generation of anandamide in the same tissues and cells. Such a characteristic of 2-AG would be favorable in acting as a fast signaling molecule in the nervous system.

Importantly, the cannabinoid CB_1 receptor is mainly located presynaptically, and various cannabimimetic molecules have been shown to inhibit the release of neurotransmitters from the presynapses (Pertwee, 1997; Di Marzo et al., 1998; Schlicker and Kathmann, 2001; Pertwee and Ross, 2002; Lutz, 2002, for reviews). In fact, the addition of SR141716A, a CB_1 receptor antagonist, to rat brain synaptosomes diminished the release of glutamate from the synaptosomes upon depolarization (Oka, S. and Sugiura. T., unpublished results). These observations strongly suggested that the endogenous ligands of the cannabinoid receptors play important regulatory roles in the neurotransmission in synapses.

We have focused on 2-AG since the mid-1990s and have proposed that 2-AG, generated through increased inositol phospholipid metabolism upon synaptic transmission, plays an essential role in attenuating subsequent neurotransmitter release from the presynaptic terminals by acting on the CB_1 receptor (Sugiura et al., 1998, 1999, 2000a; for reviews, Sugiura and Waku, 2000, 2002). In this scenario, 2-AG was rapidly formed from inositol phospholipids in neurons (postsynaptic neurons and possibly presynaptic terminals as well) during accelerated synaptic transmission. The 2-AG generated was rapidly released into the synaptic cleft, because 2-AG is a membrane-permeable molecule. The released 2-AG then binds to the cannabinoid CB_1 receptor and stimulates Gi/Go to inhibit voltage-gated Ca^{2+} channels and reduce the intracellular free Ca^{2+} concentration, thereby diminishing subsequent neurotransmitter release (Sugiura et al., 1997, 1998, 1999, 2002; Sugiura and Waku, 2000, 2002) (Figure 5.1). We have already provided evidence that 2-AG suppresses the depolarization-induced rapid elevation of the intracellular free Ca^{2+} concentration using differentiated NG108-15 cells (Sugiura et al.,1997). Such a negative feedback mechanism should be effective in calming stimulated neurons after excitation (Sugiura et al., 1998, 1999, 2000a,b, 2002; Sugiura and Waku, 2000, 2002). Presumably, 2-AG plays such an important role in cooperation with other inhibitory neurotransmitters and neuromodulators. Several investigators have also reported

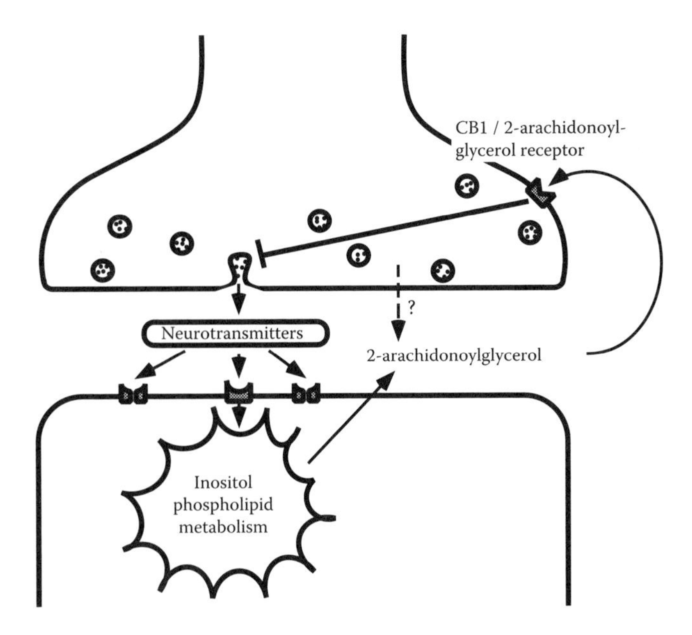

FIGURE 5.1 Proposed model for the possible role of 2-arachidonoylglycerol (2-AG) in the synapse. (From Sugiura, T., Kondo, S., Kodaka, T., Nakane, S., Yamashita, A., Kishimoto, S., Waku, K., and Ishima, Y. [1998] Biosynthesis and actions of anandamide and 2-AG: two endogenous cannabimimetic molecules, in *Essential Fatty Acids and Eicosanoids*, Eds., R.A. Riemesma, R. Armstrong, W. Kelly, and R. Wilson, Champaign: AOCS Press, pp. 380–384. With permission.)

neuroregulatory roles for 2-AG. Stella et al. (1997) reported that 2-AG inhibits the induction of LTP in rat hippocampal slices. The effect of 2-AG was blocked by SR141716A, a CB_1 receptor antagonist, indicating that the effect of 2-AG was mediated through the CB_1 receptor. Ameri and Simmet (2000) also reported that 2-AG reduces neuronal excitability in rat hippocampal slices in a CB_1 receptor-dependent manner. Thus, evidence is gradually accumulating which indicates that 2-AG plays an important role in the attenuation of neurotransmitter release in the brain. In any case, rapid shutoff mechanisms would be of great physiological significance for neurons, because sustained activation of neurons is known to cause cell exhaustion and may lead to cell death. In this context, it is worth mentioning that Sinor et al. (2000) and Panikashvili et al. (2001) have recently reported that 2-AG plays possible neuroprotective roles in the brain.

Concerning the physiological significance of 2-AG in mammalian tissues, there is an important issue to be noted; that is, 2-AG links increased inositol phospholipid turnover in stimulated tissues and cells with the function of the CB_1 receptor (Sugiura et al., 1995). This is distinct from the case for anandamide, which is known to be formed primarily from *N*-arachidonoyl PE through the action of a phosphodiesterase. In contrast, 2-AG can be easily formed through enhanced inositol phospholipid metabolism. Noticeably, the CB_1 receptor is particularly abundant in the cerebellum, striatum, globus pallidus, and hippocampus (Herkenham et al., 1990; Herkenham, 1995), where inositol phospholipids and their metabolites such as inositol 1,4,5-trisphosphate (IP_3) are assumed to play important roles in the signal transduction mechanism (Sugiura et al., 1998). Herkenham (1995) previously pointed out that the brain regions where the CB_1 receptor is highly expressed are those where signal transduction actively takes place. All these observations support the hypothesis that 2-AG, rather than anandamide, is an effective signaling molecule in the modulation of synaptic transmission in the brain.

DEPOLARIZATION-INDUCED SUPPRESSION OF INHIBITION

Independent from the studies on endocannabinoids (anandamide and 2-AG) mentioned above, a noticeable phenomenon concerning the regulation of synaptic transmission was found in the early 1990s through electrophysiological studies, that is, the suppression of synaptic transmission upon the depolarization of postsynaptic neurons. This phenomenon was first reported by Pitler and Alger (1992) and Marty and colleagues (Llano et al., 1991; Vincent et al., 1992) and referred to as DSI. Pitler and Alger (1992) demonstrated that the spontaneous GABA-mediated inhibitory postsynaptic potentials (IPSPs) were transiently suppressed following a train of action potentials in rat hippocampal CA1 pyramidal cells. Noticeably, the addition of dihydropyridine methyl-1,4 dihydro-2,6-dimethyl-3-nitro-4-(2-trifluoromethyl-phenyl-pyridine-5-carboxylate) (BAY K8644), a Ca^{2+} channel agonist, enhanced the suppression of IPSPs, whereas buffering changes in the postsynaptic responses to intracellular free Ca^{2+} concentrations ($[Ca^{2+}]_i$) with EGTA or 1,2-bis(2-aminophenoxy) ethane-N,N,N′,N′-tetraacetic acid) (BAPTA) prevented it. Importantly, iontophoretically applied GABA did not change significantly, suggesting that the inhibition produced by pyramidal cell depolarization is mainly presynaptic. They concluded that localized postsynaptic $[Ca^{2+}]_i$ potently modulate synaptic $GABA_A$ inputs and that this modulation may be an important regulatory mechanism in mammalian brains.

Similar results were reported by Marty and colleagues (Llano et al., 1991; Vincent et al., 1992) for rat cerebellar Purkinje cells. They demonstrated that spontaneous synaptic currents in rat cerebellar Purkinje cells were reduced by depolarizing voltage pulses, with a half recovery time of about 20 sec. The inhibition was largely explained by a decrease in the frequency of synaptic events, suggesting that the primary location of the effect was presynaptic. They assumed that a Ca^{2+} rise in the postsynapse triggered a retrograde inhibition of presynaptic terminals by releasing a diffusible Ca^{2+}-dependent messenger (Llano et al., 1991; Vincent et al., 1992). The involvement of a retrograde messenger molecule in the inhibition has also been postulated by Pitler and Alger (1994; for review, see Alger and Pitler, 1995). They observed a short time lag (1 sec) before the initiation of DSI, which seems to be consistent with the involvement of the second messenger system.

The mechanisms underlying DSI were studied further by several investigators. Pitler and Alger (1994) examined the effect of pertussis toxin (PTX)-treatment on DSI. They found that PTX-treatment abolished DSI, suggesting that Gi/Go is involved in DSI. They demonstrated that perturbation of the G protein function in the pyramidal cells by omitting guanosine triphosphate (GTP) from the recording pipette or by including GTPS or GDPS did not affect DSI, which clearly indicates that the relevant G protein is presynaptic. Ohno-Shosaku et al. (1998) investigated DSI using cultured rat hippocampal neurons. They demonstrated that DSI can be observed with a dissociated cell culture system and confirmed that the presence of extracellular Ca^{2+} is essential for the induction of DSI. They also demonstrated the presence of a lag before peak suppression, no involvement of axonal firing, and the presynaptic site of expression. Noticeably, both inhibitory and excitatory neurons can induce depolarization-induced suppression of synaptic transmission, suggesting that this is a rather general phenomenon. They also investigated the mechanism underlying DSI and found that a process sensitive to phorbol esters is involved in DSI. On the other hand, presynaptic opioid receptors did not participate in DSI, because the application of naloxane did not affect DSI.

Several investigators have proposed that the DSI is mediated by glutamate or a glutamate-like substance (Glitsch et al., 1996; Morishita et al., 1998). In this scenario, glutamate or a glutamate-like substance acts on metabotropic glutamate receptors (mGluRs) expressed on the inhibitory presynaptic terminals to reduce GABA release (Glitsch et al., 1996; Morishita et al., 1998). However, it is apparent that glutamate is not the major mediator of DSI, because the DSI in the cerebellum was not affected by a mGluR antagonist, α-methyl-4-carboxyphenylglycine (MCPG) (Glitsch et al., 1996), and DSI in the hippocampus was blocked only in part by MCPG even at a high concentration (Morishita et al., 1998). Thus, the exact molecular mechanisms of DSI remained unsolved until recently.

ENDOCANNABINOIDS AS RETROGRADE SYNAPTIC MESSENGERS

In 2001, there was a breakthrough. Three groups independently and concurrently provided evidence that endocannabinoids act as retrograde messengers in the synaptic transmission in the brain (Ohno-Shosaku et al., 2001; Wilson and Nicoll, 2001; Kreitzer and Regehr, 2001a; for reviews, Montgomery and Madison, 2001; Christie and Vaughan, 2001; Wilson and Nicoll, 2002; Kreitzer and Regehr, 2002). Ohno-Shosaku et al. (2001) examined inhibitory postsynaptic currents (IPSCs) in neuron pairs in rat hippocampal cultures (Figure 5.2). They demonstrated that the application of WIN55212-2, a cannabinoid receptor agonist, induced a remarkable suppression of IPSCs in 16 of 26 pairs. The presynaptic locus for the action of WIN55212-2 was confirmed by the following criteria: (1) In a pair where IPSCs were greatly suppressed by WIN55212-2, inhibitory autaptic currents (IACs) were also depressed; (2) WIN55212-2 induced a clear increase in the paired-pulse ratio of IPSCs; and (3) WIN55212-2 did not alter the postsynaptic sensitivity to iontophoretically applied GABA. They demonstrated that the depolarization of the postsynaptic neuron induced the elevation of $[Ca^{2+}]_i$ as well as DSI. Importantly, in all neuron pairs where DSI was observed, WIN55212-2 suppressed IPSCs. Moreover, DSI was totally eliminated by CB_1 receptor antagonists, AM281 and SR141716A. These results strongly suggested that DSI is mediated through endocannabinoids. In support of this notion, DSI was not affected by various types of mGluR antagonists such as MCPG, (RS)-1-aminoindan-1,5-dicarboxylic acid (AIDA), 2-methyl-6-(phenylethynyl)pyridine (MPEP), and (RS)-α-cyclopropyl-4-phosphonophenylglycine (CPPG). A $GABA_B$ receptor antagonist (2S)-3-[[(1S)-1-(3,4-dichlorophenyl) ethyl]amino-2-hydroxypropyl](phenylmethyl)phosphinic acid (CGP55845A) also failed to affect DSI.

Wilson and Nicoll (2001) reported similar results using rat hippocampal slices. They demonstrated that DSI in a CA1 pyramidal neuron was eliminated by incubating slices with a CB_1 receptor-specific antagonist AM251 or SR141716A. On the other hand, the effect of DSI was mimicked by WIN55212-2. WIN55212-2 did not affect the DSI-resistant component of the evoked IPSC. These results indicated that DSI is mediated by endocannabinoids. DSI does not appear to involve the vesicular release of endocannabinoids, because both botulinum toxin B and E, which prevent vesicular fusion, exerted no effect on DSI. They then examined the effect of 2-AG. They found that 2-AG had only a small effect compared with WIN55212-2. They assumed that this may be attributed to rapid removal from the extracellular space by an endogenous transporter. In fact, AM404, a structural analog of anandamide known as an inhibitor of the endocannabinoid transporter, depressed the baseline evoked IPSC amplitude without affecting the DSI-resistant component of the evoked IPSC. Regarding such an effect of AM404, however, it seems necessary to examine whether AM404 actually blocked the action of an endocannabinoid transporter, thereby exhibiting such an activity, because Glaser et al. (2003) recently reported that the uptake of endocannabinoids is not mediated by a specific transporter or carrier protein. Wilson and Nicoll (2001) also examined the effect of augmented cytoplasmic $[Ca^{2+}]_i$ in the postsynapse on spontaneous IPSCs. They demonstrated that the effect of uncaging Ca^{2+} from a photolabile chelator is indistinguishable from the effect of a depolarizing step, indicating that the elevation of $[Ca^{2+}]_i$ in the postsynapse is sufficient to trigger DSI. They also examined the effects of mGluR antagonists, (2S)-2-amino-2-[(1S,2S)-2-carboxycycloprop-1-yl]-3-(xanth-9-yl)propanoic acid (LY341495), and MCPG on DSI. Neither LY34195 nor MCPG was found to affect DSI, this being consistent with the notion that endocannabinoids but not excitatory amino acids are responsible for the induction of DSI. These results are in general agreement with the results of Ohno-Shosaku et al. (2001) on cultured hippocampal neurons mentioned before.

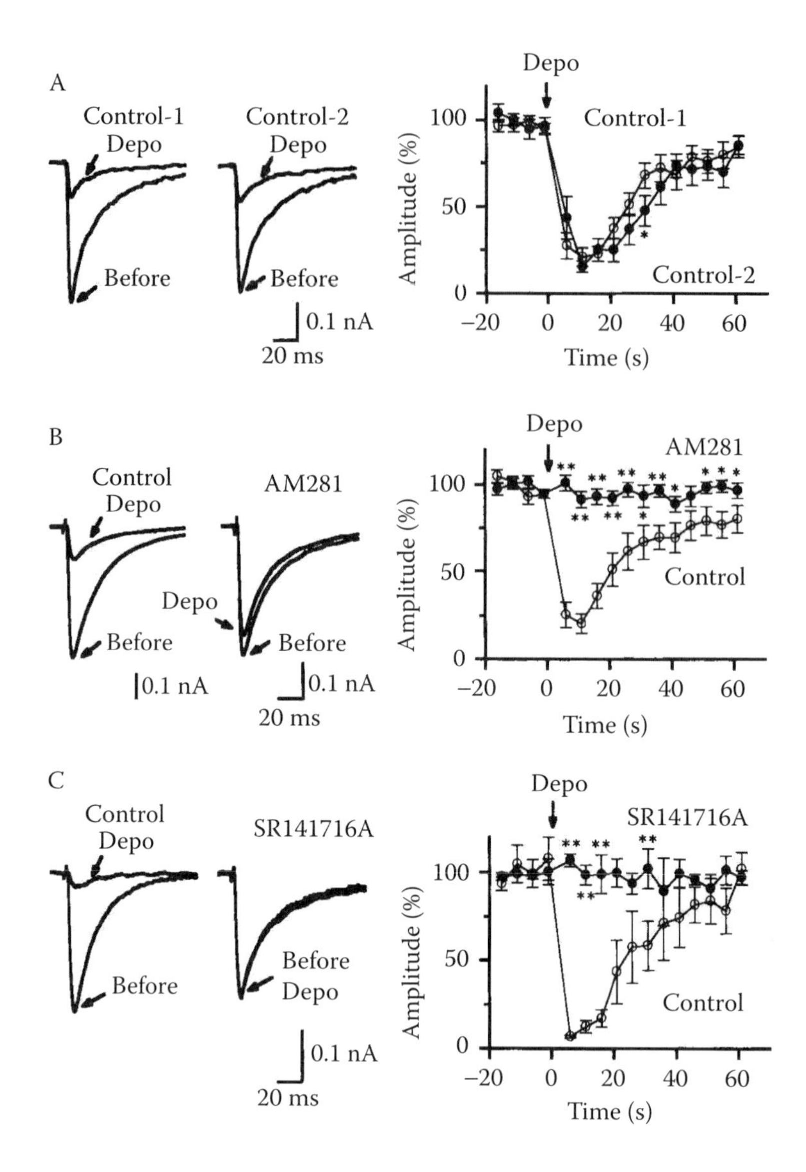

FIGURE 5.2 Blockade of depolarization-induced suppression of IPSCs by cannabinoid antagonists. (A): Examples of IPSCs (left) and the summary (right) of the results showing that the depolarization-induced suppression can be elicited repeatedly without any run-down of its magnitude. Traces acquired before and 6 sec after the first (control-1) or the second (control-2) depolarization in the normal external solution are shown. Averaged time courses of the changes in IPSC amplitudes induced by the first (open circles) and the second (closed circles) depolarization (n = 10). (B) and (C): Examples of IPSCs (left) and the summary (right) of the results showing the blockade of the depolarization-induced suppression by 0.3 μ*M* AM281 ([B]; n = 11) and 0.3 μ*M* SR141716A ([C]; n = 3). IPSC traces and averaged time courses of the depolarization-induced changes in IPSC amplitudes are shown in the similar manners to (A). The asterisks attached to data points represent statistically significant differences from the control (asterisk, $p < 0.05$; double asterisks, $p < 0.01$; paired t test). (From Ohno-Shosaku, T., Maejima, T., and Kano, M. [2001] Endogenous cannabinoids mediate retrograde signals from depolarized postsynaptic neurons to presynaptic terminals, *Neuron* 29: 729–738. With permission.)

Possible roles of endocannabinoids as retrograde messengers in the cerebellum were reported by Kreitzer and Regehr (2001a). They examined the effects of postsynaptic depolarization of Purkinje cells on excitatory synaptic transmission between: (1) the parallel fiber (cerebellar granule cells) and Purkinje cells and (2) the climbing fiber (originating in the inferior olive) and Purkinje cells. They found that the depolarization of Purkinje cells induced a reduction of parallel fiber excitatory postsynaptic currents (EPSCs). Similar inhibition was observed with climbing fiber EPSCs. They referred to this phenomenon as DSE. The paired-pulse ratio was increased at both the parallel fiber synapse and the climbing fiber synapse, suggesting that the postsynaptic depolarization induced the reduced release of excitatory amino acids from both the parallel fiber and the climbing fiber. They next examined whether a rise in postsynaptic Ca^{2+} is required for DSE. They found that the inclusion of a Ca^{2+} chelator BAPTA in the postsynaptic recording pipette completely blocked DSE at both the parallel fiber synapse and the climbing fiber synapse. This clearly indicates that the elevation of postsynaptic Ca^{2+} is essential for the induction of DSE at both synapses. They also found that a reduction of Ca^{2+} influx in climbing fiber takes place after the postsynaptic depolarization. The reduction of Ca^{2+} influx, followed the same time course as the inhibition of the EPSC. Noticeably, the inclusion of BAPTA in Purkinje cells resulted in the failure of the inhibition of the presynaptic Ca^{2+} influx. These results strongly suggested that the increase in postsynaptic Ca^{2+} induces the decrease in presynaptic Ca^{2+} influx, thereby reducing the release of excitatory amino acids, and that retrograde messengers must be involved in such a regulatory mechanism. Excitatory amino acids, GABA, and adenosine appeared not to be involved in DSE, because a group II mGluR agonist, L-carboxycyclopropylglycine (L-CCG-1), a group II mGluR antagonist, LY34195, a $GABA_B$ receptor antagonist, CGP55845A, and an adenosine A_1 receptor antagonist, 1,3-dipropyl-8-cyclopentyladenosine (DPCPX), did not affect DSE. On the other hand, the application of AM251, a CB_1 receptor antagonist, greatly reduced DSE. In contrast, the application of WIN55212-2, a CB_1 receptor agonist, inhibited EPSC at both the parallel fiber synapse and the climbing fiber synapse. These results strongly suggested that endocannabinoids act as retrograde messengers in these synapses and play crucial roles in the induction of DSE.

Details of endocannabinoid-mediated retrograde suppression of synaptic transmission (DSI and DSE) were further investigated by a number of investigators. Kreitzer and Regehr (2001b) examined DSI in the rat cerebellum and reported that it is mediated by endocannabinoids as in the case of DSI in the hippocampus and of DSE in the cerebellum. Endocannabinoid-mediated DSI in the cerebellum was also reported by Diana et al. (2002). Kreitzer et al. (2002) recently demonstrated that endocannabinoids released from cerebellar Purkinje cell dendrites suppress the spontaneous firing of nearby interneurons.

Wilson et al. (2001) investigated DSI in the mouse hippocampus. They showed that DSI is absent in the hippocampus of CB_1 receptor-deficient mice (CB_1 receptor$^{-/-}$ mice), providing further concrete evidence that the CB_1 receptor is crucially involved in DSI. The absence of DSI in the hippocampus of CB_1 receptory$^{-/-}$ mice was also reported by Varma et al. (2001). DSI was impaired in the cerebellum of CB_1 receptor$^{-/-}$ mice as well (Yoshida et al., 2002). Wilson et al. (2001) reported that CB_1 receptor activation inhibits presynaptic Ca^{2+} channels possibly through direct action of the $G\beta\gamma$ protein on voltage-dependent Ca^{2+} channels (VDCC) and that endocannabinoids exhibit a striking specificity in targeting a distinct class of interneurons, which are distinguished by their profile and exclusive use of the N-type Ca^{2+} channels for the neurotransmitter release. This is consistent with the finding that DSI is initiated mainly by N-type Ca^{2+} channels (Lenz et al., 1998), although Lenz et al. (1998) assumed at the time that postsynaptic rather than presynaptic N-type Ca^{2+} channel activation is necessary for DSI. Later, the same group reported that the block of presynaptic N-type Ca^{2+} channels is probably a major mechanism of DSI, yet they also described that the block of N-type Ca^{2+} channels does not fully explain the DSI in the hippocampus (Varma et al., 2002).

Ohno-Shosaku, Tsubokawa, et al. (2002) demonstrated that not only DSI but also DSE can be induced in the rat and mouse hippocampus. DSE in the hippocampus was mediated by the presynaptic CB_1 receptor and endocannabinoids released from depolarized postsynaptic neurons. They compared DSE and DSI in slices from the same animals and found that DSE was much less prominent than DSI. For the induction of DSE, the necessary duration of depolarization was longer than for DSI, and the magnitude of DSE was much smaller than that of DSI. They showed that excitatory synapses were homogeneous and had moderate sensitivities to WIN55212-2, whereas inhibitory synapses were dichotomized into two distinct populations, one with a high sensitivity to WIN55212-2 and the other with no sensitivity. They concluded that the presynaptic cannabinoid sensitivity is a major factor that determines the extent of DSI and DSE.

Noticeably, the depolarization of the postsynaptic neurons is not necessarily required for the reduction of the neurotransmitter release from the presynaptic terminals. Maejima et al. (2001) reported that activation of the mGluR subtype 1 (mGluR1) expressed in cerebellar Purkinje cells with dihydroxyphenylglycol (DHPG), a group I mGluR agonist, reduced neurotransmitter release from climbing fibers similar to the case of DSE. This phenomenon required the activation of G proteins but not Ca^{2+} elevation in the postsynaptic Purkinje cells. Thus, it is apparent that not only depolarization but also the activation of G proteins in Purkinje cells can induce the backward regulation of presynaptic neurotransmitter release. Ohno-Shosaku, Shosaku, et al. (2002) also investigated this phenomenon using rat hippocampal neurons. They demonstrated that activation of group I mGluRs by DHPG suppressed IPSCs in about half of the neuron pairs. WIN55212-2 also suppressed IPSCs in all DHPG-sensitive pairs but not in most of the DHPG-insensitive pairs. Both DSI and DHPG- or WIN55212-2-induced suppression of IPSCs were abolished by treatment with AM281 or SR141716A, indicating that these responses were mediated through the CB_1 receptor.

Interestingly, Varma et al. (2001) found that the bath application of DHPG or a group I and II mGluR agonist aminocyclopentane-1S, 3R-dicarboxylic acid (ACPD) induced marked acceleration of DSI. On the other hand, a mGluRs antagonist LY34195 reduced DSI. Ohno-Shosaku, Shosaku, et al. (2002) also reported that DSI was significantly enhanced in the presence of DHPG. This enhancement was much more prominent than expected from the simple summation of depolarization-induced and group I mGluR-induced endocannabinoid release; it is probable that postsynaptic depolarization and the activation of group I mGluRs worked in concert to generate sufficient amounts of endocannabinoids in postsynaptic neurons.

Importantly, such cooperation in the generation of endocannabinoids is not restricted to the case of depolarization and activation of mGluRs. Recently, Kim et al. (2002) reported that activation of muscarinic acetylcholine receptors also enhances DSI in the rat hippocampus. They demonstrated that the application of carbacol enhanced DSI, and this effect of carbacol was blocked by atropine. The addition of eserine, a cholinesterase inhibitor, also enhanced DSI, implying that muscarinergic acetylcholine receptor–mediated endocannabinoid release is a physiologically relevant event. Ohno-Shosaku et al. (2003) demonstrated that postsynaptic M1 and M3 receptors are responsible for such an enhancement in the hippocampus. These observations, together with previous findings, suggested that the retrograde modulation by endocannabinoids is an important and widespread mechanism in the brain by which the activity of postsynaptic neurons can influence the functions of various types of inhibitory and excitatory presynaptic neurons, although the neuroregulatory roles of endocannabinoids have already been postulated by a number of investigators as mentioned previously (for reviews, see Pertwee, 1997; Di Marzo, 1998; Di Marzo et al., 1998; Mechoulam et al., 1998; Piomelli et al., 1998, 2000; Sugiura et al., 1998; Sugiura and Waku, 2000; Schlicker and Kathmann, 2001; Davies et al., 2002; Fride, 2002; Lichtman et al., 2002; Pertwee and Ross, 2002; Schweitzer, 2002).

What then are the physiological implications of endocannabinoid-mediated DSI and DSE *in vivo*? This important issue has not yet been fully clarified. There are several possibilities concerning the physiological meanings of DSI and DSE. In the hippocampus and neocortex, the CB_1 receptor is mainly expressed on interneurons which form GABAergic synapses with fast kinetics (Wilson and Nicoll, 2001). Such interneurons are assumed to be responsible for regulating gamma oscillations which are synchronized over long distances in the brain and have been proposed to be involved in binding simultaneous perceptions. Endocannabinoids may be closely involved in the process of cognition by attenuating GABA release in these brain regions through DSI. In support of this, Hajos et al. (2000) have reported that WIN55212-2 interferes with kainate-induced gamma oscillations in hippocampal slices. It is also plausible that DSE is one of the intrinsically fast inhibitory mechanisms of certain types of excitatory neurons. As mentioned previously, we have proposed that 2-AG, generated during increased synaptic transmission, plays an important role in attenuating subsequent neurotransmitter release by acting on the CB_1 receptor expressed in the presynaptic terminals (Sugiura et al., 1998, 2000; Sugiura and Waku, 2000). Such a negative feedback mechanism would greatly contribute to the homeostasis of neurons and synaptic transmission.

Another possibility is that DSI and DSE are involved in the modification of synaptic plasticity such as LTP and long-term depression (LTD). For example, suppression of inhibitory neurotransmitter release by endocannabinoids (DSI) leads to hyperactivation of postsynaptic neurons, which appears to be favorable for stimulation or promotion of LTP. In fact, the modulation of LTP (potentiation or suppression) by endocannabinoids has already been demonstrated by several investigators (Terranova et al., 1995; Collin et al., 1995; Stella et al., 1997; Carlson et al., 2002). In addition to LTP, endocannabinoids may also participate in LTD. Robbe et al. (2002) recently provided evidence that endocannabinoids, derived from the postsynapse, induce LTD in the nucleus accumbens. Several investigators have also demonstrated the involvement of endocannabinoids in LTD in the amygdala (Marsicano et al., 2002), the striatum (Gerdeman et al., 2002), and the hippocampus (Chevaleyre and Castillo, 2003). Thus, it seems conceivable that DSI and DSE are actually involved in several physiological phenomena *in vivo*, yet details still remain to be determined. A thorough elucidation of the physiological significance of DSI and DSE as well as other endocannabinoid-dependent neural cell functions awaits future investigations.

FUTURE DIRECTIONS AND CONCLUDING REMARKS

As mentioned above, there have been two independent research flows concerning endocannabinoid-induced neuromodulation. One involves biochemical and pharmacological studies on endocannabinoids and cannabinoid receptors, the other electrophysiological studies on DSI and DSE using brain slices and dissociated neurons. Now, these two flows have merged to provide a promising novel research field.

Currently, a number of investigators are involved in studies on the physiological roles of endocannabinoids in the nervous system. There is a growing body of experimental data concerning the neuroregulatory roles of endocannabinoids. However, several important issues remain rather ambiguous. For example, which endocannabinoid is responsible for DSI and DSE, anandamide or 2-AG or both? In relation to this, we have previously provided clear evidence that 2-AG rather than anandamide is the true natural ligand for the cannabinoid receptors (Sugiura, et al., 1996, 1997, 1999, 2000). Detailed studies are thus required to determine whether 2-AG actually acts as a principal effector molecule in DSI and DSE.

The precise mechanisms underlying the biosynthesis of endocannabinoids in neurons are also yet to be fully elucidated. It has already been demonstrated that multiple pathways can be involved in the biosynthesis of either anandamide or 2-AG (see Chapter 6). However, the properties of individual enzyme activities involved in the biosynthesis of these endocannabinoids are not yet clarified at the molecular level. Moreover, the role allotments as well as subcellular distribution of these enzyme activities remain uncertain. Further studies are necessary to clarify these important issues. Studies on the mechanisms of inactivation and uptake of endocannabinoids are also essential, because endocannabinoid levels in the synaptic cleft must be strictly regulated. As for the inactivation, the genes encoding FAAH and monoacylglycerol lipase, both of which are involved in the catabolism of endocannabinoids, have already been cloned (Cravatt et al., 1996; Dinh et al., 2002). Details regarding the degradation of endocannabinoids will be clarified in the near future.

Information concerning the cellular origin of endocannabinoids is also important. It is apparent that endocannabinoids are released from the postsynaptic neurons in DSI and DSE in the hippocampus and cerebellum. However, the possibility that endocannabinoids are also released from the presynaptic terminals in some cases cannot be ruled out. In addition, it is possible that significant amounts of endocannabinoids are released from proximal glial cells as well. These possibilities should carefully be examined.

Finally, it is necessary to mention the possibility raised by several investigators that a non-CB_1, non-CB_2 cannabinoid receptor plays some role in the brain. Several lines of evidence suggested that there is another type of cannabinoid receptor, different from the CB_1 and CB_2 receptors, in the brain, although its level is thought to be very low if at all present. Hajos et al. (2001) recently demonstrated that a putative novel type of cannabinoid receptor is involved in the regulation of glutamatergic neurotransmission in the hippocampus. Further studies are needed to confirm whether such a novel type of cannabinoid receptor actually takes part in neuromodulation.

Cannabinoid receptors are abundantly expressed in various brain regions, particularly those involved in advanced functions such as the control of movement, memory, and cognition, and have been assumed to play important neuroregulatory roles in DSI and DSE. Their endogenous ligand 2-AG is also a rather common molecule and transmits the signal of augmented synaptic transmission involving increased inositol phospholipid turnover to the presynaptic sensor, that is, the cannabinoid receptor. It seems very likely that the endocannabinoid system is one of the general and ubiquitous neuroregulatory mechanisms in mammalian brains, although details of their physiological significance are not yet fully elucidated. Thus, further intensive studies on the endocannabinoid system are essential to obtain insight into various advanced brain functions in mammals.

Note Added in Proof

Recently, Bisogno et al. (2003) cloned a cDNA for diacylglycerol lipase. (Bisogno, T., Howell. F., Williams, G., Minassi, A., Cascio, M.G., Ligresti, A., Matias, I., Schiano-Moriello, A., Paul, P.,Williams, E.J., Gangadharan, U., Hobbs, C., Di Marzo, V. and Doherty, P. (2003) *Journal of Cell Biology,* 163, 463–468). Evidence is gradually accumulating which suggests that 2-AG is the intrinsic messenger molecule of DSI or DSE. (Galante, M. and Diana, M. A., (2004) *Journal of Neuroscience* 24, 4865–4874; Melis, M., Perra, S., Muntoni, A. L., Pillolla, G., Lutz, B., Marsicano, G., Di Marzo, V., Gessa, G. L. and Perra, S., Muntoni, A. L., Pillolla, G., Lutz, B., Marsicano, G., Di Marzo, V., Gessa, G. L., and Pistis, M. (2004) *Journal of Neuroscience* 24, 10707–10715; Hashimotodani, Y., Ohno-Shosaku, T., Tsubokawa, H., Ogata, H., Emoto, K., Maejima, T., Araishi, K., Shin, H-S. and Kano, M. (2005) *Neuron,* 45, 257–268; Maejima, T., Oka, S., Hashimotodani, Y., Ohno-Shosaku, T., Aiba, A., Wu, D., Sugiura T., and Kanoh, M. (2005) *Journal of Neuroscience,* in press.)

REFERENCES

Alger, B.E. and Pitler, T.A. (1995) Retrograde signaling at GABAA-receptor synapses in the mammalian CNS, *Trends in Neurosciences* 18: 333–340.

Ameri, A. and Simmet, T. (2000) Effects of 2-arachidonylglycerol, an endogenous cannabinoid, on neuronal activity in rat hippocampal slices, *Naunyn Schmiedebergs Archives of Pharmacology* 361: 265–272.

Berdyshev, E.V. (2000) Cannabinoid receptors and the regulation of immune response, *Chemistry and Physics of Lipids* 108: 169–190.

Bisogno, T., Sepe, N., Melck, D., Maurelli, S., De Petrocellis, L., and Di Marzo, V. (1997) Biosynthesis, release and degradation of the novel endogenous cannabimimetic metabolite 2-arachidonoylglycerol in mouse neuroblastoma cells, *Biochemical Journal* 322: 671–677.

Breivogel, C.S., Selley, D.E., and Childers, S.R. (1998) Cannabinoid receptor agonist efficacy for stimulating [^{35}S]GTPγS binding to rat cerebellar membranes correlates with agonist-induced decreases in GDP affinity, *Journal of Biological Chemistry* 273: 16865–16873.

Breivogel, C.S., Griffin, G., Di Marzo, V., and Martin, B.R. (2001) Evidence for a new G protein coupled cannabinoid receptor in mouse brain, *Molecular Pharmacology* 60: 155–163.

Burkey, T.H., Quock, R.M., Consroe, P., Ehlert, F.J., Hosohata, Y., Roeske, W.R., and Yamamura, H.I. (1997) Relative efficacies of cannabinoid CB_1 receptor agonists in the mouse brain, *European Journal of Pharmacology* 336: 295–298.

Cabral, G.A. (2002) Marijuana and cannabinoid effects in immunity and AIDS, in *Biology of Marijuana,* E.S. Onaivi, Ed., London: Taylor and Francis, pp. 282–307.

Cadas, H., Di Tomaso, E., and Piomelli, D. (1997) Occurrence and biosynthesis of endogenous cannabinoid precursor, N-arachidonoyl phosphatidylethanolamine, in rat brain, *Journal of Neuroscience* 17: 1226–1242.

Carlson, G., Wang, Y., and Alger, B.E. (2002) Endocannabinoids facilitate the induction of LTP in the hippocampus, *Nature Neuroscience* 5: 723–724.

Chevaleyre, V. and Castillo, P.E. (2003) Heterosynaptic LTD of hippocampal GABAergic synapses: a novel role of endocannabinoids in regulating excitability, *Neuron* 38: 461–472.

Christie, M.J. and Vaughan, C.W. (2001) Cannabinoids act backwards, *Nature* 410: 527–530.

Collin, C., Devane, W.A., Dahl, D., Lee, C.J., Axelrod, J., and Alkon, D.L. (1995) Long-term synaptic transformation of hippocampal CA1 gamma-aminobutyric acid synapses and the effect of anandamide, *Proceedings of the National Academy of Sciences of the USA* 92: 10167–10171.

Cravatt, B.F., Giang, D.K., Mayfield, S.P., Boger, D.L., Lerner, R.A., and Gilula, N.B. (1996) Molecular characterization of an enzyme that degrades neuromodulatory fatty-acid amides, *Nature* 384: 83–87.

Davies, S.N., Pertwee, R.G., and Riedel, G. (2002) Functions of cannabinoid receptors in the hippocampus, *Neuropharmacology* 42: 993–1007.

De Petrocellis, L., Melck, D., Bisogno, T., and Di Marzo, V. (2000) Endocannabinoids and fatty acid amides in cancer, inflammation and related disorders, *Chemistry and Physics of Lipids* 108: 191–209.

Devane, W.A., Dysarz, F.A., III, Johnson, M.R., Melvin, L.S., and Howlett, A.C. (1988) Determination and characterization of a cannabinoid receptor in rat brain, *Molecular Pharmacology* 34: 605–613.

Devane, W.A., Hanus, L., Breuer, A., Pertwee, R.G., Stevenson, L.A., Griffin, G., Gibson, D., Mandelbaum, A., Etinger, A., and Mechoulam, R. (1992) Isolation and structure of a brain constituent that binds to the cannabinoid receptor, *Science* 258: 1946–1949.

Dewey, W.L. (1986) Cannabinoid pharmacology, *Pharmacological Review* 38: 151–178.

Diana, M.A., Levenes, C., Mackie, K., and Marty, A. (2002) Short-term retrograde inhibition of GABAergic synaptic currents in rat Purkinje cells is mediated by endogenous cannabinoids, *Journal of Neuroscience* 22: 200–208.

Di Marzo, V., Fontana, A., Cadas, H., Schinelli, S., Cimino, G., Schwartz, J.C., and Piomelli, D. (1994) Formation and inactivation of endogenous cannabinoid anandamide in central neurons, *Nature* 372: 686–691.

Di Marzo, V. (1998) "Endocannabinoids" and other fatty acid derivatives with cannabimimetic properties: biochemistry and possible physiopathological relevance, *Biochimica et Biophysica Acta* 1392: 153–175.

Di Marzo, V., Melck, D., Bisogno, T., and De Petrocellis, L. (1998) Endocannabinoids: endogenous cannabinoid receptor ligands with neuromodulatory action, *Trends in Neurosciences* 21: 521–528. Erratum *Trends in Neurosciences* (1999) 22: 80.

Di Marzo, V., De Petrocellis, L., Bisogno, T., and Melck, D. (1999) Metabolism of anandamide and 2-arachidonoylglycerol: an historical overview and some recent developments, *Lipids* 34: S319–325.

Di Marzo, V., Melck, D., De Petrocellis, L., and Bisogno, T. (2000) Cannabimimetic fatty acid derivatives in cancer and inflammation, *Prostaglandins and Other Lipid Mediators* 61: 43–61.

Di Marzo, V., Hill, M.P., Bisogno, T., Crossman, A.R., and Brotchie, J.M. (2000) Enhanced levels of endogenous cannabinoids in the globus pallidus are associated with a reduction in movement in an animal model of Parkinson's disease, *FASEB Journal* 14: 1432–1438.

Di Marzo, V., De Petrocellis, L., Bisogno, T., Berger, A., and Mechoulam, R. (2002) Biology of endocannabinoids, in *Biology of Marijuana*, E.S. Onaivi, Ed., London: Taylor and Francis, pp. 125–173.

Dinh, T.P., Carpenter, D., Leslie, F.M., Freund, T.F., Katona, I., Sensi, S.L., Kathuria, S., and Piomelli, D. (2002) Brain monoglyceride lipase participating in endocannabinoid inactivation, *Proceedings of the National Academy of Sciences of the USA* 99: 10819–10824. Erratum *Proceedings of the National Academy of Sciences of the USA* (2002) 99: 13961.

Felder, C.C. and Glass, M. (1998) Cannabinoid receptors and their endogenous agonists, *Annual Review of Pharmacology and Toxicology* 38: 179–200.

Fernandez-Ruiz, J., Lastres-Becker, I., Cabranes, A., Gonzalez, S., and Ramos, J.A. (2002) Endocannabinoids and basal ganglia functionality, *Prostaglandins, Leukotrienes and Essential Fatty Acids* 66: 257–267.

Fride, E., Barg, J., Levy, R., Saya, D., Heldman, E., Mechoulam, R., and Vogel, Z. (1995) Low doses of anandamides inhibit pharmacological effects of Δ^9-tetrahydrocannabinol, *Journal of Pharmacology and Experimental Therapeutics* 272: 699–707.

Fride, E. (2002) Endocannabinoids in the central nervous system — an overview, *Prostaglandins, Leukotrienes and Essential Fatty Acids* 66:221–233.

Gerdeman, G.L., Ronesi, J., and Lovinger, D.M. (2002) Postsynaptic endocannabinoid release is critical to long-term depression in the striatum, *Nature Neuroscience* 5: 446–451.

Giuffrida, A. and Piomelli, D. (2000) The endocannabinoid system: a physiological perspective on its role in psychomotor control, *Chemistry and Physics of Lipids* 108: 151–158.

Glaser, S.T., Abumrad, N.A., Fatade, F., Kaczocha, M., Studholme, K.M., and Deutsch, D.G. (2003) Evidence against the presence of an anandamide transporter, *Proceedings of the National Academy of Sciences of the USA*. 100: 4269–4274.

Glitsch, M., Llano, I., and Marty, A. (1996) Glutamate as a candidate retrograde messenger at interneurone-Purkinje cell synapses of rat cerebellum, *Journal of Physiology* 497: 531–537.

Gonsiorek, W., Lunn, C., Fan, X., Narula, S., Lundell, D., and Hipkin, R.W. (2000) Endocannabinoid 2-arachidonyl glycerol is a full agonist through human type 2 cannabinoid receptor: antagonism by anandamide, *Molecular Pharmacology* 57: 1045–1050.

Hajos, N., Katona, I., Naiem, S.S., MacKie, K., Ledent, C., Mody, I., Freund, T.F. (2000) Cannabinoids inhibit hippocampal GABAergic transmission and network oscillations, *European Journal of Neuroscience* 12: 3239–3249.

Hajos, N., Ledent, C., and Freund, T.F. (2001) Novel cannabinoid-sensitive receptor mediates inhibition of glutamatergic synaptic transmission in the hippocampus, *Neuroscience* 106: 1–4.

Hansen, H.S., Moesgaard, B., Hansen, H.H., and Petersen, G. (2000) N-Acylethanolamines and precursor phospholipids — relation to cell injury, *Chemistry and Physics of Lipids* 108: 135–150.

Herkenham, M., Lynn, A.B., Little, M.D., Johnson, M.R., Melvin, L.S., de Costa, B.R., Rice, K.C. (1990) Cannabinoid receptor localization in brain, *Proceedings of the National Academy of Sciences of the USA* 87:1932–1936.

Herkenham, M. (1995) Localization of cannabinoid receptors in brain and periphery, in *Cannabinoid Receptors*, R.G. Pertwee, Ed., London: Academic Press, pp. 45–166.

Hillard, C.J. and Campbell, W.B. (1997) Biochemistry and pharmacology of arachidonylethanolamide, a putative endogenous cannabinoid, *Journal of Lipid Research* 38: 2383–2398.

Hillard, C.J. (2000) Biochemistry and pharmacology of the endocannabinoids arachidonylethanolamide and 2-arachidonylglycerol, *Prostaglandins and Other Lipid Mediators* 61: 3–18.

Kearn, C.S., Greenberg, M.J., DiCamelli, R., Kurzawa, K., and Hillard, C.J. (1999) Relationships between ligand affinities for the cerebellar cannabinoid receptor CB1 and the induction of GDP/GTP exchange, *Journal of Neurochemistry* 72: 2379–2387.

Kim, J., Isokawa, M., Ledent, C., and Alger, B.E. (2002) Activation of muscarinic acetylcholine receptors enhances the release of endogenous cannabinoids in the hippocampus, *Journal of Neuroscience* 22: 10182–10191.

Kondo, S., Kondo, H., Nakane, S., Kodaka, T., Tokumura, A., Waku, K., and Sugiura, T. (1998) 2-Arachidonoylglycerol, an endogenous cannabinoid receptor agonist: identification as one of the major species of monoacylglycerols in various rat tissues, and evidence for its generation through Ca^{2+}-dependent and -independent mechanisms, *FEBS Letters* 429: 152–156.

Kreitzer, A.C. and Regehr, W.G. (2001a) Retrograde inhibition of presynaptic calcium influx by endogenous cannabinoids at excitatory synapses onto Purkinje cells, *Neuron* 29: 717–727.

Kreitzer, A.C. and Regehr, W.G. (2001b) Cerebeller depolarization-induced suppression of inhibition is mediated by endogenous cannabinoids, *Journal of Neuroscience* 21: RC174.

Kreitzer, A.C. and Regehr, W.G. (2002) Retrograde signaling by endocannabinoids, *Current Opinion in Neurobiology* 12: 324–330.

Kreitzer, A.C., Carter, A.G., and Regehr, W.G. (2002) Inhibition of interneuron firing extends the spread of endocannabinoid signaling in the cerebellum, *Neuron* 34:787–796.

Lenz, R.A., Wagner, J.J., and Alger, B.E. (1998) N- and L-type calcium channel involvement in depolarization-induced suppression of inhibition in rat hippocampal CA1 cells, *Journal of Physiology* 512: 61–73.

Lichtman, A.H., Varvel, S.A., and Martin, B.R. (2002) Endocannabinoids in cognition and dependence, *Prostaglandins, Leukotrienes and Essential Fatty Acids* 66: 269–285.

Llano, I., Leresche, N., and Marty, A. (1991) Calcium entry increases the sensitivity of cerebellar Purkinje cells to applied GABA and decreases inhibitory synaptic currents, *Neuron* 6: 565–574.

Lutz, B. (2002) Molecular biology of cannabinoid receptors, *Prostaglandins, Leukotrienes and Essential Fatty Acids* 66: 123–142.

Mackie, K., Devane, W.A., and Hille, B. (1993) Anandamide, an endogenous cannabinoid, inhibits calcium currents as a partial agonist in N18 neuroblastoma cells, *Molecular Pharmacology* 44: 498–503.

Maejima, T., Hashimoto, K., Yoshida, T., Aiba, A., and Kano, M. (2001) Presynaptic inhibition caused by retrograde signal from metabotropic glutamate to cannabinoid receptors, *Neuron* 31: 463–475.

Marsicano, G., Wotjak, C.T., Azad, S.C., Bisogno, T., Rammes, G., Cascio, M.G., Hermann, H., Tang, J., Hofmann, C., Zieglgansberger, W., Di Marzo, V., and Lutz, B. (2002) The endogenous cannabinoid system controls extinction of aversive memories, *Nature* 418: 530–534.

Matsuda, L.A., Lolait, S.J., Brownstein, M.J., Young, A.C., and Bonner, T.I. (1990) Structure of a cannabinoid receptor and functional expression of the cloned cDNA, *Nature* 346: 561–564.

Matsuda, L.A. and Bonner, T.I. (1995) Molecular biology of the cannabinoid receptor,in *Cannabinoid Receptors*, Ed. R.G. Pertwee, London: Academic Press, pp. 117–143.

McAllister, S.D., Griffin, G., Satin, L.S., and Abood, M.E. (1999) Cannabinoid receptors can activate and inhibit G protein-coupled inwardly rectifying potassium channels in a *Xenopus* oocyte expression system, *Journal of Pharmacology and Experimental Therapeutics* 291: 618–662.

Mechoulam, R., Ben-Shabat, S., Hanus, L., Ligumsky, M., Kaminski, N.E., Schatz, A.R., Gopher, A., Almog, S., Martin, B.R., Compton, D.R., Pertwee, R.G., Griffin, G., Bayewitch, M., Barg, J., and Vogel, Z. (1995) Identification of an endogenous 2-monoglyceride, present in canine gut, that binds to cannabinoid receptors, *Biochemical Pharmacology* 50: 83–90.

Mechoulam, R., Fride, E., and Di Marzo, V. (1998) Endocannabinoids, *European Journal of Pharmacology* 359: 1–18.

Montgomery, J.M. and Madison, D.V. (2001) The grass roots of synapse suppression, *Neuron* 29: 567–570.

Morishita, W., Kirov, S.A., and Alger, B.E. (1998) Evidence for metabotropic glutamate receptor activation in the induction of depolarization-induced suppression of inhibition in hippocampal CA1, *Journal of Neuroscience* 18: 4870–4882.

Munro, S., Thomas, K.L., and Abu-Shaar, M. (1993) Molecular characterization of a peripheral receptor for cannabinoids, *Nature* 365: 61–65.

Ohno-Shosaku, T., Sawada, S., and Yamamoto, C. (1998) Properties of depolarization-induced suppression of inhibitory transmission in cultured rat hippocampal neurons, *Pflugers Archiv European Journal of Physiology* 435: 273–279.

Ohno-Shosaku, T., Maejima, T., and Kano, M. (2001) Endogenous cannabinoids mediate retrograde signals from depolarized postsynaptic neurons to presynaptic terminals, *Neuron* 29: 729–738.

Ohno-Shosaku, T., Shosaku, J., Tsubokawa, H., and Kano, M. (2002) Cooperative endocannabinoid production by neuronal depolarization and group I metabotropic glutamate receptor activation, *European Journal of Neuroscience* 15: 953–961.

Ohno-Shosaku, T., Tsubokawa, H., Mizushima, I., Yoneda, N., Zimmer, A., and Kano, M. (2002) Presynaptic cannabinoid sensitivity is a major determinant of depolarization-induced retrograde suppression at hippocampal synapses, *Journal of Neuroscience* 22: 3864–3872.

Ohno-Shosaku, T., Matsui, M., Fukudome, Y., Shosaku, J., Tsubokawa, H., Taketo, M.M., Manabe, T., and Kano, M. (2003) Postsynaptic M1 and M3 receptors are responsible for the muscarinic enhancement of retrograde endocannabinoid signalling in the hippocampus, *European Journal of Neuroscience* 18: 109–116.

Panikashvili, D., Simeonidou, C., Ben-Shabat, S., Hanus, L., Breuer, A., Mechoulam, R., and Shohami, E. (2001) An endogenous cannabinoid (2-AG) is neuroprotective after brain injury, *Nature* 413: 527–531.

Parolaro, D., Massi, P., Rubino, T., and Monti, E. (2002) Endocannabinoids in the immune system and cancer, *Prostaglandins, Leukotrienes and Essential Fatty Acids* 66: 319–332.

Pertwee, R.G. (1997) Pharmacology of cannabinoid CB1 and CB2 receptors, *Pharmacology and Therapeutics* 74: 129–180.

Pertwee, R.G. and Ross, R.A. (2002) Cannabinoid receptors and their ligands, *Prostaglandins, Leukotrienes and Essential Fatty Acids* 66: 101–121.

Piomelli, D., Beltramo, M., Giuffrida, A., and Stella, N. (1998) Endogenous cannabinoid signaling, *Neurobiology of Disease* 5: 462–473.

Piomelli, D., Giuffrida, A., Calignano, A., and Rodriguez de Fonseca, F. (2000) The endocannabinoid system as a target for therapeutic drugs, *Trends in Pharmacological Sciences* 21: 218–224.

Pitler, T.A. and Alger, B.E. (1992) Postsynaptic spike firing reduces synaptic GABAA responses in hippocampal pyramidal cells, *Journal of Neuroscience* 12: 4122–4132.

Pitler, T.A. and Alger, B.E. (1994) Depolarization-induced suppression of GABAergic inhibition in rat hippocampal pyramidal cells: G protein involvement in a presynaptic mechanism, *Neuron* 13: 1447–1455.

Robbe, D., Kopf, M., Remaury, A., Bockaert, J., and Manzoni, O.J. (2002) Endogenous cannabinoids mediate long-term synaptic depression in the nucleus accumbens, *Proceedings of the National Academy of Sciences of the USA* 99: 8384–8388.

Roth, M.D., Baldwin, G.C., and Tashkin, D.P. (2002) Effects of delta-9-tetrahydrocannabinol on human immune function and host defense, *Chemistry and Physics of Lipids* 121: 229–239.

Savinainen, J.R., Jarvinen, T., Laine, K., and Laitinen, J.T. (2001) Despite substantial degradation, 2-arachidonoylglycerol is a potent full efficacy agonist mediating CB(1) receptor-dependent G-protein activation in rat cerebellar membranes, *British Journal of Pharmacology* 134: 664–672.

Schlicker, E. and Kathmann, M. (2001) Modulation of transmitter release via presynaptic cannabinoid receptors, *Trends in Pharmacological Sciences* 22: 565–572.

Schmid, H.H.O., Schmid, P.C., and Natarajan, V. (1990) N-acylated glycerophospholipids and their derivatives, *Progress in Lipid Research* 29: 1–43.

Schmid, H.H.O. (2000) Pathways and mechanisms of N-acylethanolamine biosynthesis: can anandamide be generated selectively? *Chemistry and Physics of Lipids* 108: 71–87.

Schmid, H.H.O., Schmid, P.C., and Berdyshev, E.V. (2002) Cell signaling by endocannabinoids and their congeners: questions of selectivity and other challenges, *Chemistry and Physics of Lipids* 121: 111–134.

Schweitzer, P. (2002) Electrophysiological actions of marijuana, in *Biology of Marijuana*, Ed. E.S. Onaivi, London: Taylor and Francis, pp. 523–541

Sinor, A.D., Irvin, S.M., and Greenberg, D.A. (2000) Endocannabinoids protect cerebral cortical neurons from in vitro ischemia in rats, *Neuroscience Letters* 278: 157–160.

Stella, N., Schweitzer, P., and Piomelli, D. (1997) A second endogenous cannabinoid that modulates long-term potentiation, *Nature* 388: 773–778.

Stella, N. and Piomelli, D. (2001) Receptor-dependent formation of endogenous cannabinoids in cortical neurons, *European Journal of Pharmacology* 425: 189–196.

Sugiura, T., Kondo, S., Sukagawa, A. Nakane, S., Shinoda, A., Itoh, K., Yamashita, A., and Waku, K. (1995) 2-Arachidonoylglycerol: a possible endogenous cannabinoid receptor ligand in brain, *Biochemical and Biophysical Research Communications* 215: 89–97.

Sugiura, T., Kodaka, T., Kondo, S., Tonegawa, T., Nakane, S., Kishimoto, S., Yamashita, A., and Waku, K. (1996) 2-Arachidonoylglycerol, a putative endogenous cannabinoid receptor ligand, induces rapid, transient elevation of intracellular free Ca^{2+} in neuroblastoma x glioma hybrid NG108-15 cells, *Biochemical and Biophysical Research Communications* 229: 58–64.

Sugiura, T., Kondo, S., Sukagawa, A., Tonegawa, T., Nakane, S., Yamashita, A., Ishima, Y., and Waku, K. (1996) Transacylase-mediated and phosphodiesterase-mediated synthesis of N-arachidonoylethanolamine, an endogenous cannabinoid-receptor ligand, in rat brain microsomes. Comparison with synthesis from free arachidonic acid and ethanolamine, *European Journal of Biochemistry* 240: 53–62.

Sugiura, T., Kodaka, T., Kondo, S., Tonegawa, T., Nakane, S., Kishimoto, S., Yamashita, A., and Waku, K. (1997) Inhibition by 2-arachidonoylglycerol, a novel type of possible neuromodulator, of the depolarization-induced increase in intracellular free calcium in neuroblastoma x glioma hybrid NG108-15 cells, *Biochemical and Biophysical Research Communications* 233: 207–210.

Sugiura, T., Kodaka, T., Kondo, S., Nakane, S., Kondo, H., Waku, K., Ishima, Y., Watanabe, K., and Yamamoto, I. (1997) Is the cannabinoid CB1 receptor a 2-arachidonoylglycerol receptor? Structural requirements for triggering a Ca^{2+} transient in NG108-15 cells, *Journal of Biochemistry* 122: 890–895.

Sugiura, T., Kondo, S., Kodaka, T., Nakane, S., Yamashita, A., Kishimoto, S., Waku, K., and Ishima, Y. (1998) Biosynthesis and actions of anandamide and 2-arachidonoylglycerol: two endogenous cannabimimetic molecules, in *Essential Fatty Acids and Eicosanoids*, Eds., R.A. Riemesma, R. Armstrong, W. Kelly, and R. Wilson, Champaign: AOCS Press, pp. 380–384.

Sugiura, T., Kodaka, T., Nakane, S., Miyashita, T., Kondo, S., Suhara, Y., Takayama, H., Waku, K., Seki, C., Baba, N., and Ishima, Y. (1999) Evidence that the cannabinoid CB1 receptor is a 2-arachidonoylglycerol receptor. Structure-activity relationship of 2-arachidonoylglycerol, ether-linked analogues, and related compounds, *Journal of Biological Chemistry* 274: 2794–2801.

Sugiura, T. and Waku, K. (2000) 2-Arachidonoylglycerol and the cannabinoid receptors, *Chemistry and Physics of Lipids* 108: 89–106.

Sugiura, T., Yoshinaga, N., Kondo, S., Waku, K., and Ishima, Y. (2000a) Generation of 2-arachidonoylglycerol, an endogenous cannabinoid receptor ligand, in picrotoxinin-administered rat brain, *Biochemical and Biophysical Research Communications* 271: 654–658.

Sugiura, T., Kondo, S., Kishimoto, S., Miyashita, T., Nakane, S., Kodaka, T., Suhara, Y., Takayama, H., and Waku, K. (2000b) Evidence that 2-arachidonoylglycerol but not N-palmitoylethanolamine or anandamide is the physiological ligand for the cannabinoid CB2 receptor. Comparison of the agonistic activities of various cannabinoid receptor ligands in HL-60 cells, *Journal of Biological Chemistry* 275: 605–612.

Sugiura, T. and Waku, K. (2002) Cannabinoid receptors and their endogenous ligands, *Journal of Biochemistry* 132: 7–12.

Sugiura, T., Kobayashi, Y., Oka, S., and Waku, K. (2002) Biosynthesis and degradation of anandamide and 2-arachidonoylglycerol and their possible physiological significance, *Prostaglandins, Leukotrienes and Essential Fatty Acids* 66: 173–192.

Terranova, J.P., Michaud, J.C., Le Fur, G., and Soubrie, P. (1995) Inhibition of long-term potentiation in rat hippocampal slices by anandamide and WIN55212-2: reversal by SR141716 A, a selective antagonist of CB1 cannabinoid receptors, *Naunyn Schmiedebergs Archives of Pharmacology* 352: 576–579.

Varma, N., Carlson, G.C., Ledent, C., and Alger, B.E. (2001) Metabotropic glutamate receptors drive the endocannabinoid system in hippocampus, *Journal of Neuroscience* 21: RC188.

Varma, N., Brager, D., Morishita, W., Lenz, R.A., London, B., and Alger, B. (2002) Presynaptic factors in the regulation of DSI expression in hippocampus, *Neuropharmacology* 43: 550–562.

Vincent, P., Armstrong, C.M., and Marty, A. (1992) Inhibitory synaptic currents in rat cerebellar Purkinje cells: modulation by postsynaptic depolarization, *Journal of Physiology* 456: 453–471.

Wilson, R.I. and Nicoll, R.A. (2001) Endogenous cannabinoids mediate retrograde signalling at hippocampal synapses, *Nature* 410: 588–592.

Wilson, R.I., Kunos, G., and Nicoll, R.A. (2001) Presynaptic specificity of endocannabinoid signaling in the hippocampus, *Neuron* 31: 453–462.

Wilson, R.I. and Nicoll, R.A. (2002) Endocannabinoid signaling in the brain, *Science* 296: 678–682.

Yoshida, T., Hashimoto, K., Zimmer, A., Maejima, T., Araishi, K., and Kano, M. (2002) The cannabinoid CB1 receptor mediates retrograde signals for depolarization-induced suppression of inhibition in cerebellar Purkinje cells, *Journal of Neuroscience* 22: 1690–1697.

6 Non-CB_1, Non-CB_2 Receptors for Endocannabinoids

Vincenzo Di Marzo and Luciano De Petrocellis

CONTENTS

INTRODUCTION

Endocannabinoids are by definition (Di Marzo, 1998; Di Marzo and Fontana, 1995) endogenous compounds binding to and functionally activating the two receptor subtypes identified so far for *Cannabis*' psychoactive principle Δ^9-tetrahydrocannabinol, the CB_1 and CB_2 receptors. These two receptors are, therefore, responsible for many of the pharmacological effects of the four endocannabinoids discovered to date, i.e., anandamide (arachidonoylethanolamide, AEA) (Devane et al., 1992), 2-arachidonoyl glycerol (2-AG) (Mechoulam et al., 1995; Sugiura et al., 1995), 2-arachidonyl glyceryl ether (2-AGE) (Hanus et al., 2001), and virodhamine (Porter et al., 2002). However, the four endocannabinoids appear to activate to a varying extent the two cannabinoid receptors. AEA and 2-AGE are more selective for CB_1 receptors, and 2-AG is equally potent and efficacious at both binding sites, whereas virodhamine activates preferentially the CB_2 receptor. Several qualitative and quantitative differences between the pharmacological profile of endocannabinoids, and of AEA in particular, and that of natural and synthetic cannabinoid receptor ligands, such as THC, HU210, CP55940, and WIN55212-2, have been observed (Khanolkar et al., 2000; Mechoulam and Fride, 1995; Pertwee, 1999).

Whereas at least part of these differences can be ascribed to AEA and 2-AG metabolic instability *in vivo* and *in vitro*, or to the fact that AEA is a partial cannabinoid receptor agonist in several functional assays, recent data have suggested the existence for this compound of other molecular targets, including non-CB_1, G-protein-coupled receptors (GPCRs) non-CB_2 and various ion channels. These data, as well as the possibility that endocannabinoid oxidative metabolism can generate ligands for new or already known receptors, will be critically discussed in this article and seem to suggest that endocannabinoid biochemistry and pharmacology might be much more complex than anticipated after the discovery of AEA.

NON-CB_1, NON-CB_2 RECEPTORS FOR AEA

Evidence in Favor of the Existence of Non-CB_1, Non-CB_2 Receptors for AEA

Pharmacological data suggesting that AEA might act *in vivo* on non-CB_1, non-CB_2 receptors have been reviewed previously (Di Marzo et al., 1999; Di Marzo, Bisogno, and De Petrocellis, 2001; Pertwee and Ross, 2002). Possibly, the first of this evidence was the observation that three typical cannabimimetic effects of AEA, the inhibition of spontaneous activity in an "open field," the induction of immobility on a "ring," and the antinociception in the "hot plate" test, can still be observed in transgenic mice lacking the CB_1 receptor, and are not blocked by the CB_1 receptor antagonist SR141716A (Adams et al., 1998; Di Marzo et al., 2000). Conversely, with THC or synthetic cannabinoids, these effects are efficaciously counteracted by SR141716A in wild-type mice and cannot be observed in CB_1 receptor "knockout" mice (Ledent et al., 1999; Zimmer et al., 1999). In view of the lack of any upregulation mechanisms for CB_2 receptors in the brain following inactivation of CB_1 receptors, these findings might suggest the presence of at least one additional receptor type in mouse brain mediating some of AEA neurobehavioral effects. However, no molecular evidence for the presence of these receptors has been reported as yet, and the issue is further complicated by the observations that when AEA enzymatic hydrolysis by fatty acid amide hydrolase (FAAH) is blocked, either by enzyme inhibitors (Wiley et al., 2000) or by disruption of the FAAH-encoding gene (Cravatt et al., 2001), the susceptibility to SR141716A of AEA effects on locomotion, nociception, and body temperature is restored. Thus, AEA hydrolysis to arachidonic acid (AA) *in vivo* may prevent AEA direct activation of CB_1 receptors and, at the same time, lead to the formation of compounds capable of inducing some cannabimimetic actions via non-CB_1 receptors. Although this possibility is indeed supported by the observation that AA, AEA, and 2-AG metabolites may induce some THC-like effect, if metabolism *in vivo* is so rapid that it can immediately prevent AEA from activating CB_1 receptors, then it is not clear why other cannabis-like effects of systemic AEA in mice are efficaciously blocked by SR141716A (Calignano et al., 1997; Calignano et al., 2001; Jarai et al., 2000; Souilhac et al., 1995). It is, therefore, also possible that this endocannabinoid induces some cannabimimetic effects *in vivo* by acting via multiple alternative mechanisms, including CB_1 receptors as well as non-CB_1 targets for both AEA and its FAAH-derived metabolites. When one of these mechanisms is blocked, the others might become more important.

Known GPCRs that Interact with AEA

Recent data indicate that AEA can interact with known GPCRs for other neuronal mediators or with novel GPCRs. AEA was found to exhibit inhibitory activity on 5-hydroxy-tryptamine receptors (Kimura et al., 1998), and both muscarinic and nicotinic acetylcholine receptors (Lagalwar et al., 1999; Oz et al., 2003). Except for the inhibitory effect on nACh alpha7 nicotinic receptors (IC_{50} = 218 n*M*), however, AEA appears to modulate the binding and functional activity of 5-HT and acetylcholine only at concentrations ($\geq$ 5 μM) that are not likely to occur *in vivo*. Regarding muscarinic receptors, AEA was found to inhibit the binding of high affinity ligands for both M_1 and, particularly, M_4 receptors for acetylcholine at 1-10 μM concentrations and in a noncompetitive manner (Christopoulos and Wilson, 2001). However, these effects do not appear to be due to metabolites formed following AEA hydrolysis because the more metabolically stable AEA analog (*R*)-methanandamide is usually found to be equipotent to, or even slightly more potent than, AEA, and inhibitors of AEA hydrolysis do not reduce the extent of these actions (Christopoulos and Wilson, 2001; Lagalwar et al., 1999). The effects are also quite selective as they cannot be mimicked by other cannabinoid receptor ligands such as WIN55212-2. It remains to be established if any of the *in vivo* pharmacological actions of AEA that are not reduced by CB_1 receptor antagonists are due to these interactions with acetylcholine and 5-HT receptors.

Novel and As Yet Uncharacterized Non-CB_1, Non-CB_2 Receptors in the Brain

At least two types of receptors for AEA, distinct from CB_1 and CB_2, have been proposed to exist in rodents, based on pharmacological and biochemical evidence. The first possibly novel GPCR for AEA was suggested to exist in mouse astrocytes, where the compound was shown to inhibit isoproterenol-induced cAMP accumulation, a typical cannabinoid receptor-induced response, in a manner insensitive to both CB_1 and CB_2 receptor antagonists (Sagan et al., 1999). The effect was blocked by pertussis toxin and was also observed with WIN55212-2. Later, a possibly similar GPCR was also described to occur in rat brain membranes prepared from CB_1 receptor knockout mice (Di Marzo et al., 2000), where AEA stimulated GTPγS-binding with an EC_{50} ~3 μM and in a manner insensitive to CB_1 and CB_2 receptor antagonists. When this effect of AEA was assessed also in brain membranes from wild-type mice it could only be reduced by a CB_1 antagonist down to a level identical to that observed in knockout mice in the absence of antagonists. As reported in a subsequent study carried out again by using brain membranes from CB_1 gene-disrupted mice, also WIN55212-2, but not THC, CP55940, or HU210, appears to exert the same effect in a more efficacious and potent manner than AEA, again in a way insensitive to CB_1 and CB_2 receptor blockade (Breivogel et al., 2001). The distribution of these AEA- and WIN55212-2-stimulated GPCRs in mouse brain was shown to be somehow different from that of CB_1 receptors, with the highest levels in the cortex, hippocampus, and brainstem, and no activity being detected in two regions with a high concentration of CB_1 receptors, the basal ganglia and the cerebellum. Binding sites for tritiated WIN55212-2 were also found in membranes from CB_1 receptor knockout mice and correlated to a great extent with WIN55212-2-stimulated GTPγS-binding. These findings have been confirmed, with some differences regarding the distribution of the novel receptor, by a more recent investigation (Monory et al., 2001) using a different strain of CB_1 knockout mice. It remains now to be assessed whether these GPCRs are responsible at least in part for AEA and WIN55212-2 effects in some neurobehavioral assays, and if they are the same as those described in mouse astrocytes (Sagan et al., 1999). More importantly, these GPCRs might be the same receptors mediating WIN55212-2 inhibition of glutamatergic postsynaptic currents in the hippocampus of CB_1 knockout mice (Hajos et al., 2001). In this latter case, the same non-CB_1-mediated effect on glutamate release was reported also for natural or synthetic compounds containing a vanillyl moiety, and capable of binding to vanilloid VR1 receptors (see following text) such as capsaicin (Hajos and Freund 2002). This important finding might explain why many of VR1 agonists induce cannabimimetic effects *in vivo* via non-CB_1, non-VR1-mediated mechanisms (Di Marzo, Bisogno, De Petrocellis, Brandi, et al., 2001; Di Marzo, Lastres-Becker, et al., 2001; Di Marzo et al., 2002). However, unlike the GPCR found in the brain of CB_1 knockout mice (Breivogel et al., 2001), the receptor responsible for the non-CB_1-mediated effect on glutamate release was also activated by the cannabinoid CP55940.

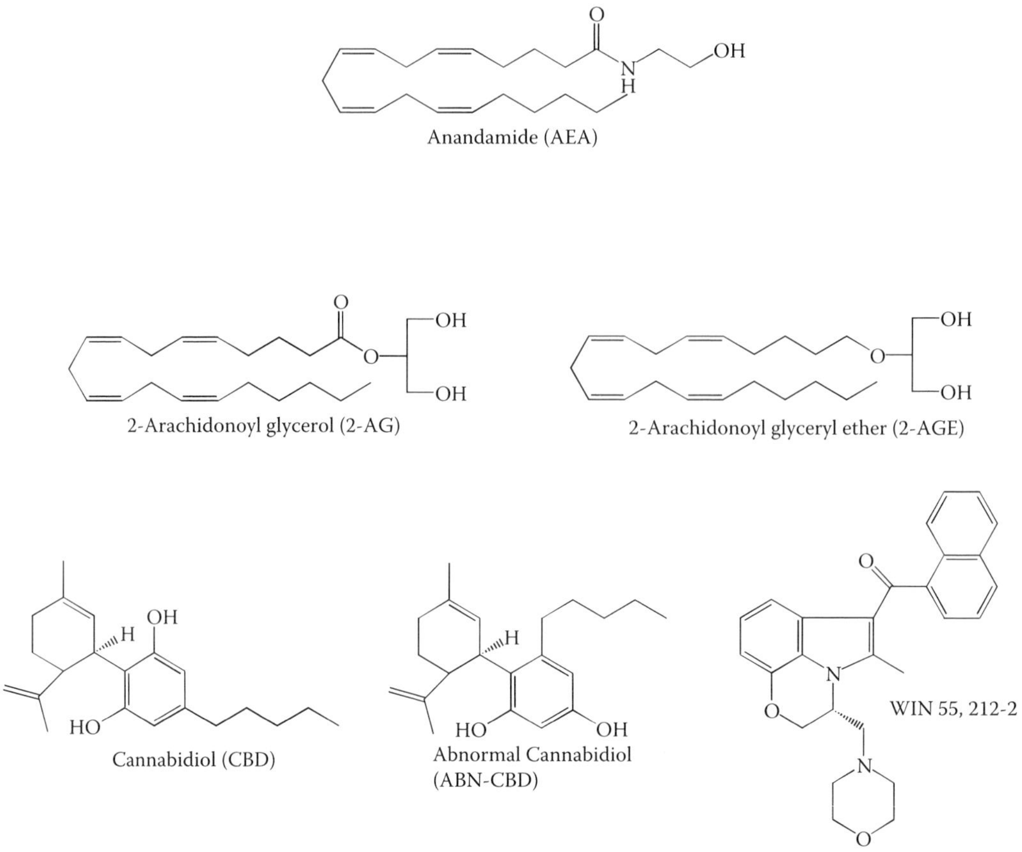

Anandamide (AEA)

2-Arachidonoyl glycerol (2-AG)

2-Arachidonoyl glyceryl ether (2-AGE)

Cannabidiol (CBD)

Abnormal Cannabidiol (ABN-CBD)

WIN 55, 212-2

FIGURE 6.1 Chemical structures of endocannabinoids and of possible templates for the development of ligands of putative novel cannabinoid receptors.

Novel and As Yet Uncharacterized Non-CB$_1$, Non-CB$_2$ Receptors in the Endothelium

Strong pharmacological and biochemical evidence exists also for non-CB$_1$, non-CB$_2$ AEA receptors in vascular endothelium (Járai et al., 1999; Wagner et al., 1999). Járai and co-workers showed that AEA can induce an endothelium-dependent relaxation of rat and mouse mesenteric arteries via novel endothelial sites of action activated by the nonpsychotropic cannabinoid "abnormal cannabidiol," as well as by some of its analogs, and antagonized by the natural cannabinoid, cannabidiol (Járai et al., 1999) (Figure 6.1). Interestingly, the endothelium-dependent action of both AEA and abnormal cannabidiol on the mesenteric artery was partly antagonized by a rather "selective" concentration (0.5 μ*M*) of the CB$_1$ antagonist SR141716A. More recent observations showed that the novel endothelial sites of action of AEA are: (1) coupled to G-proteins, as they are blocked by pertussis toxin pretreatment of the rat mesenteric artery, (2) coupled to BK(Ca) calcium channels and to the PI3 kinase/Akt signaling pathway, and (3) selectively blocked by the synthetic cannabidiol analog O-1918 (Offertaler et al., 2003; Ho and Hiley, 2003a). They do not seem to be involved in the hypotensive effect of systemic AEA as this effect is absent in CB$_1$ receptor knockout (as well as in CB$_1$/CB$_2$ receptor double-knockout) mice (Járai et al., 1999), nor in the psychotropic effects of AEA, as abn-cbd exhibits no such effects. Indirect evidence has been recently published for the occurrence of the novel endothelial GPCR in microglial cells, where it might be involved in mediating part of the stimulatory effects of 2-AG on the migration of these cells (Walter et al., 2003). However, the effect of this endocannabinoid on mesenteric artery dilation had been previously described not to be due to interaction with the novel receptors (Járai et al., 2000).

DIRECT INTERACTION WITH ION CHANNELS

Ca^{2+} CHANNELS AND IONOTROPIC RECEPTORS

AEA and 2-AG, by acting at CB_1 receptors and $G_{i/o}$ proteins, inhibit N and P/Q type voltage-operated Ca^{2+} channels (VOCCs) in central neurons, and L-type VOCCs in cerebral artery smooth muscle cells (Gebremedhin et al., 1999). AEA, unlike other *N*-acylethanolamines, also inhibits L-type channels directly and non-competitively in rat cardiac and brain membranes and in rabbit skeletal membranes, whereas the direct effect of 2-AG on these channels has never been tested. AEA appears to bind to 1,4-dihydropyridine, 1,5-benzodiazepine, and phenylalkylamine binding sites (Jarrahian and Hillard, 1997; Johnson et al., 1993; Shimasue et al., 1996), and, despite the high concentrations needed (IC_{50} between 4 and 29 μM), AA possibly produced from AEA hydrolysis does not appear to mediate this effect. Direct inhibition of unspecified smooth muscle VOCCs seems to underlie in part also the endothelium-independent vasodilatory action of AEA in the rat mesenteric artery and in rabbit T-tubule membranes (Ho and Hiley, 2003b; Oz et al., 2000). Further studies should be carried out in order to understand whether these direct actions mediate any of the known biological actions of AEA on excitable cells.

AEA can modulate the intracellular Ca^{2+} concentration also by inhibiting astrocyte gap junctions (Venance et al., 1995). This effect also can be exerted more or less efficaciously by other *N*-acylethanolamines and fatty primary amides (Boger et al., 1999) in a way attenuated by pretreatment of cells with pertussis toxin. In neuronal cells that do not express CB_1 receptors, or when these receptors are blocked by pertussis toxin or SR141716A, AEA can also directly activate glutamate ionotropic receptors, the NMDA-coupled Ca^{2+} channels, an effect that opposes the CB_1-mediated counteraction of glutamate effects via inhibition of VOCCs (Hampson et al., 1998). On the other hand, AEA inhibits the AMPA receptors for glutamate, although at concentrations far from physiological (Akinshola et al., 1999). Also in this case, however, too little work has been done to assess the physiological importance of these effects, and the contribution of gap junctions and NMDA or AMPA receptors in the complex effects of AEA on Ca^{2+} homeostasis is not known.

Finally, both cloned [alpha(1G), alpha(1H) and alpha(1I) subunits] and native T-type Ca^{2+} channels are blocked by AEA at submicromolar concentrations, independently from the activation of CB_1/CB_2 receptors, G-proteins, phospholipases, and protein kinase (PK) pathways. Block of AEA transport into cells via the AEA membrane transporter (AMT) prevented T-current inhibition, suggesting that AEA acts at an intracellular site (Chemin et al., 2001).

INHIBITION OF K^+ CHANNELS

Endocannabinoids are capable of activating, via CB_1 receptors and $G_{i/o}$ proteins, G-protein-coupled inwardly rectifying K^+ channels (GIRK) and to inhibit A-type K^+ channels (via inhibition of cAMP formation). However, it has now become known that AEA can interact directly with other K^+ channels, thereby leading to their inhibition. Thus, AEA, together with THC and other polyunsaturated *N*-acylethanolamines, inhibits the Shaker-related voltage-gated channels, an effect exerted at low μM concentrations (Poling et al., 1996). Recently, it was found that TASK-1 channels overexpressed in CHO cells, unlike two other pore-domain K^+ channels, are directly blocked by submicromolar concentrations of AEA and methanandamide independently of G-proteins (Maingret et al., 2001). TASK-1 is an acid- and anesthetic-sensitive background K^+ channels, which sets the resting membrane potential of both cerebellar granule neurons and somatic motoneurons. This effect of AEA was also observed in cerebellar granule neurons, where it leads to membrane depolarization (Maingret et al., 2001).

AEA, methanandamide, and WIN55212-2 also inhibit the delayed-rectifier K^+ current in rat arterial myocytes (van den Bossche and Vanheel, 2000), AEA being the most potent of the three compounds with half-maximal effect (corresponding to 25% inhibition of the current) at 0.6 μ*M*. This effect might be exerted at an extracellular site as it was not produced by intracellular applied AEA. In this case, AA also exerted the same effect, even though at slightly lower concentrations. However, it is unlikely that AEA effects were due to hydrolysis to AA because (1) AA but not AEA, also elevated Ca^{2+}-activated K^+ currents in the same cells and (2) methanandamide, although being more metabolically stable that AEA, was still active as an inhibitor. The authors concluded that AEA, methanandamide, and WIN55212-2 inhibit the delayed-rectifier K^+ current in a CB_1 receptor-independent way and at an extracellular site on or near the K^+ channel (van den Bossche and Vanheel, 2000). These findings indicate a certain capability of AEA and some of its analogs to exert a depolarizing effect on excitable cells by inhibiting K^+ channels. The issue of whether these effects result in some biological actions *in vivo*, however, has not been yet investigated.

ENDOCANNABINOIDS AND VANILLOID VR1 RECEPTORS

AEA and Vanilloid VR1 Receptors

A six-trans-membrane domain, nonselective cation channel, the vanilloid receptor of type 1 (VR1 or TRPV1) (Caterina et al., 1997) is the first and only member of the family of the TRP channels discovered so far to be activated by capsaicin, the pungent component of "hot" chili peppers (Holzer, 1991), as well as by another plant toxin, resiniferatoxin (RTX) (Szallasi and Blumberg, 1999) (Figure 6.2). VR1 acts as a ligand-, proton-, and heat-activated molecular integrator of nociceptive stimuli (Tominaga et al., 1998) and, as such, is mostly expressed in sensory C (and, to a lesser extent, A-δ) fibers (Guo, 1999). Although originally identified as a plasma membrane protein, VR1 was recently shown to be also present on the endoplasmic reticulum, where it might participate in intracellular calcium mobilization (Liu et al., 2003). VR1 receptor knockout mice (Caterina et al., 2000; Davis, 2000) exhibited lower sensitivity to inflammatory and thermal pain, thus demonstrating the involvement of this protein in the transduction of inflammatory and thermal hyperalgesia. However, activation of VR1 by capsaicin and, particularly, RTX is almost immediately followed by desensitization. For this reason, some synthetic long-chain capsaicin analogs do not exhibit pungent activity and have been proposed as oral anti-inflammatory and analgesic compounds (Dray, 1992). VR1 is also expressed in several brain areas of primates and rodents, including the hippocampus, striatum, hypothalamus, substantia nigra compacta, and locus coeruleus (Cortright et al., 2001; Hayes et al., 2000; Mezey et al., 2000). This finding suggested the existence of endogenous ligands for vanilloid receptors (Kwak et al., 1998), as the function of VR1 as a nociceptor in the brain is rather unlikely, and instead a possible role in ligand-activated neurotransmitter modulation could be proposed (Szallasi and Di Marzo, 2000). Indeed, recent investigations have shown that VR1 activation in the brain leads to glutamate release in the locus coeruleus and substantia nigra (Marinelli et al., 2002, 2003), as well as in the periaqueductal gray (Palazzo et al., 2002; McGaraughty et al., 2003).

The chemical similarity between AEA and capsaicin, both of these two compounds being fatty acid amides, suggested that they might have a molecular target in common. The observation that the C18:1 capsaicin analog, olvanil (Figure 6.2) (Dray, 1992), inhibits AEA cellular uptake via the anandamide membrane transporter (AMT) (Di Marzo et al., 1998; Beltramo and Piomelli, 1999) strengthened this suggestion. Based on this finding, an AEA/capsaicin structural "hybrid" named arvanil (Melck et al., 1999) (Figure 6.2), was synthesized and found to act as a partial agonist at CB_1 receptors and a full agonist at VR1 receptors, and to be one of the most potent AMT inhibitors developed to date. These data suggested the existence of a partial overlap of the ligand recognition properties of VR1 and CB_1 receptors and of VR1 and the AMT. Indeed, subsequent studies showed that AEA could produce a typical capsaicin-induced and calcitonin gene-related peptide (CGRP)-mediated response, i.e., the

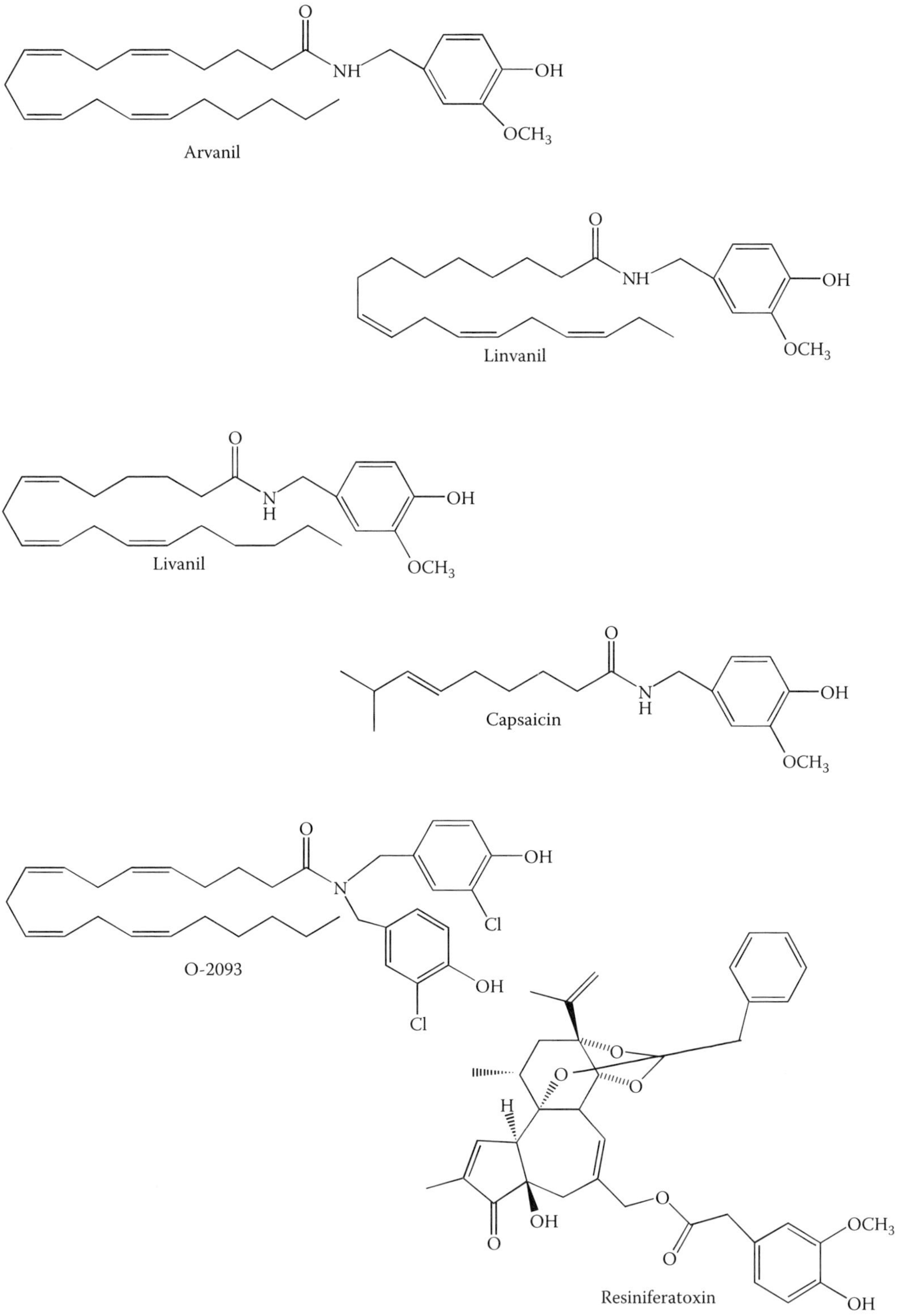

FIGURE 6.2 Chemical structures of (1) some VR1 receptors agonists, (2) CB_1/VR1 "hybrid" agonists, and (3) proposed ligands of non-CB_1, non-VR1 receptors in the brain.

relaxation of rodent small arteries *in vitro*, via a mechanism that was CB_1- and endothelium-independent and blocked by previous treatment of arteries with capsaicin and by the VR1 antagonist, capsazepine (Zygmunt et al., 1999). Accordingly, AEA activated the recombinant rat VR1 receptor overexpressed in either human embryonic kidney (HEK-293) cells or *Xenopus* oocytes, although less potently than with native vanilloid receptors in the rat mesenteric artery (Zygmunt et al., 1999). The subsequent discovery that a similar action was exerted more efficaciously and potently also at recombinant human VR1 receptors (Smart et al., 2000) triggered a series of studies aimed at investigating whether other pharmacological activities of this compound might be mediated by vanilloid receptors. There are now tens of reports indicating that AEA might work as an activator of native vanilloid receptors and possibly as an "endovanilloid," under both physiological and, particularly, pathological conditions, *in vitro* and *in vivo*. Neonatal AEA treatment can even mimic the prolonged mitochondrial damage of VR1-containing trigeminal neurons caused by capsaicin (Szoke et al., 2002). In peripheral sensory neurons, AEA induces the release of either tachykinins or CGRP or both, and causes diverse effects, including:

- Relaxation of small arteries (Zygmunt et al., 1999), particularly of the mesenteric artery, where most studies have been performed (Vanheel and Van de Voorde, 2001) but also of rabbit aortic rings (Ho and Hiley, 2003b; Harris et al., 2002; Mukhopadhyay et al., 2002; Ralevic et al., 2002 for review); these effects, or most likely the VR1-mediated stimulation of vagal fibers (see the following text) might be responsible for the transient, initial phase of the hypotensive effects of AEA *in vivo* in anesthetized normal and hypertensive rats (Malinowska et al, 2001; Li et al., 2003). By contrast, White et al., (2001) found that AEA relaxes the serotonin-precontracted rat coronary artery via a non-VR1, non-CB_1 mechanism, even though the same authors confirmed that AEA relaxes the rat mesenteric artery via VR1 receptors and at low concentrations (EC_{50} = 200 n*M*).
- Stimulation of pulmonary vagal C-fibers in rats and guinea pigs (Lin and Li, 2002; Kagaya et al., 2002; Undem and Kollarik, 2002), or the stimulation of neurons of the nucleus solitary tract (Geraghty and Mazzone, 2002). The effects are likely to cause the cardiovascular and respiratory reflexes consisting, for example, of a fall in blood pressure and an increase in ventilation, observed after AEA injection intra-arterially in the hind limb of anesthetized rats (Smith and McQueen, 2001).
- Contraction of guinea pig bronchi, an effect that independent studies from many different laboratories (Tucker et al., 2001; De Petrocellis, Harrison, et al., 2001; Craib et al., 2001) have ascribed to activation of vanilloid receptors and which leads to cough (Jia et al., 2002).
- Inhibition of the contraction induced by endogenously released acetylcholine, or stimulation of neuropeptide release, in rat tracheae (Nemeth et al., 2003; Nieri et al., 2003). Interestingly, the latter can be observed only at higher doses, whereas an inhibitory effect due to CB_1 receptors is observed at lower doses (Nemeth et al., 2003).
- Relaxation of electrically-stimulated mouse vas deferens (Ross et al., 2001), an effect that is, in part, due also to a CB_1-receptor-mediated mechanism; the involvement of sensory neuropeptides in this case was not investigated.
- Constriction of guinea pig ileum, an action likely to be mediated by an increase of acetylcholine release from myenteric neurones (Mang et al., 2001).

AEA stimulates the release of the sensory neuropeptides CGRP and substance P, also from rat dorsal root ganglia (DRG) (Tognetto et al., 2001). Whereas in the case of perivascular sensory neurons, AEA is quite potent in its action (Zygmunt et al., 1999), in the case of DRG neurons, however, the effect is seemingly observed only at high concentrations, and lower doses of AEA appear to be sufficient to exert a CB_1-mediated inhibitory action on neuropeptide release (Tognetto et al., 2001). Very probably, for this reason a stronger stimulatory action, comparable in both potency and efficacy to the CB_1-mediated inhibitory one, is instead observed with AEA when these experiments are performed in the presence of the CB_1 antagonist SR141716A (Ahluwalia, Yaqoob, et al., 2003). These findings show good correlation with electrophysiological and *in vivo* studies where AEA activates nociceptive afferents innervating rat knee joints in both normal and arthritic animals, at high nanomolar doses (up to 800 nmol) (Gauldie et al., 2001), whereas it inhibits spinal neuronal responses in noninflamed and carrageenan-inflamed rats, at lower doses (up to 144 nmol) (Harris et al., 2000). Thus, while stimulation of CB_1 receptors by synthetic agonists with no action on VR1 can lead to inhibition of capsaicin nociceptive and inflammatory effects (Richardson et al., 1998; Rukwied et al., 2003), exogenous AEA can induce either pro- or antinociceptive effects, depending on its concentration and, possibly, on the presence or absence of cannabinoid CB_1 receptors on sensory neurons. Also, if one looks at the currents induced by AEA via VR1 in sensory and brain neurons, the potency for this effect is usually 1 to 2 orders of magnitude lower than that of capsaicin (Roberts et al., 2002; Marinelli et al., 2003; Jennings et al., 2003). However, this is not always true, because some DRG neuron preparations appear to respond to low concentrations of AEA only with VR1-mediated hyperactivity (Greffrath et al., 2002). Also the VR1-mediated effects of AEA on cytosolic calcium can be exerted at concentrations much lower than 1 μM, depending on the experimental conditions (see the following text). For example, a recent study (Hermann et al., 2003) showed that if CB_1 and VR1 receptors are overexpressed in HEK cells (1) pre-stimulation with a CB_1 agonist leads to a significant twofold potentiation of capsaicin effect on intracellular calcium, and (2) AEA is more potent in these cells as a VR1 agonist than in cells only expressing VR1 receptors. This system may have a physiological correspondent in DRG neurons, where CB_1 and VR1 receptors are colocalized to some extent, not only when the neurons are in culture (Ahluwalia et al., 2000; Farquhar-Smith et al., 2000) but also in normal tissue (Bridges et al., 2003).

AEA, VR1, and Cell Proliferation and Apoptosis

By activating native vanilloid receptors, AEA may also induce tumor cell apoptosis (Maccarrone et al., 2000) or cause inhibition of tumor cell proliferation (Fowler et al., 2001). In the former case the effect was completely blocked by capsazepine, thus suggesting that AEA was exclusively acting via vanilloid receptors. Accordingly, AEA binding distinctly displaced by capsaicin but not by SR141716A was detected in membranes from human neuroblastoma cells, whose apoptosis was induced by AEA. Interestingly, when tumor cells also express cannabinoid receptors, as in the case of C6 glioma cells, the apoptotic effect of AEA was smaller and was increased by SR141716A, thus pointing to a possible protective action of cannabinoid receptors against apoptosis (Maccarrone et al., 2000). This possibility would be in contrast with previous investigations carried out with the same cell line (Sanchez et al., 2001). In a more recent investigation with C6 cells, again somewhat different results were obtained (Fowler et al., 2001). In this case, the antiproliferative effect of AEA was blocked by a combination of SR141716A, the CB_2 receptor antagonist SR144528 and capsazepine, thus indicating a cannabinoid-receptor- and vanilloid-receptor-mediated effect. Finally, in human breast cancer cells, the cytostatic effect of AEA was uniquely mediated by CB_1 receptors, even though the CB_1/VR1 "hybrid" agonist arvanil exerted a more potent effect that was antagonized by both SR141716A and capsazepine (Melck et al., 1999). Further investigations, particularly aimed at finding molecular evidence for the actual presence of VR1 receptors in C6 cells, will be required to understand the respective role of CB_1 and VR1 receptors in glioma cell proliferation vs. apoptosis. We have found, by using the reverse transcriptase-polymerase chain reaction methodology, that C6 cells do indeed express an mRNA transcript corresponding to the sequence of rat VR1 mRNA (L. De Petrocellis and V. Di Marzo, unpublished data). Interestingly, very recently AEA was suggested to induce endothelial cell apoptosis via a VR1-mediated action, and evidence for the presence of VR1 in these cells was again provided by using the RT-PCR technique (Yamaji et al., 2003).

In Vivo Effects of AEA and Capsaicin: Differences and Similarities

Apart from the above mentioned cardiovascular and respiratory effects of AEA *in vivo*, the effect of intraperitoneal AEA, capsaicin, and piperine on gastrointestinal transit was investigated in mice.

Although in this case capsaicin did not induce an efficacious response per se, it did desensitize the animals to the inhibitory action of piperine but not AEA. On the other hand, also capsazepine did not appear to efficaciously block the effect of piperine, whereas SR141716A fully antagonized AEA inhibitory effect. The authors suggested that the effect of piperine involves capsaicin-sensitive neurones but not vanilloid receptors, whereas the effect of AEA involves cannabinoid CB_1 but not vanilloid receptors (Izzo et al., 2001). However, more recently it was shown that AEA is produced in the gut during toxin A-induced intoxication and participates in small intestine inflammation by activating VR1 receptors (McVey et al., 2002). The VR1 antagonist capsazepine inhibited both toxin A- and AEA-induced inflammation. Intriguingly, these observations were found to apply also to 2-AG, despite the fact that this compound is a very weak agonist at CB_1. This suggests either the existence for 2-AG of non-VR1 receptors sensitive to capsazepine, or else that for this compound, as with AEA and *N*-oleoyl-ethanolamine (see the following text), regulatory factors occurring during inflammation significantly enhance its potency at VR1, thus possibly unmasking a VR1 agonist activity that cannot be observed under physiological conditions.

Intraperitoneal AEA, capsaicin and livanil (Figure 6.2) were found to inhibit locomotion and induce immobility in rats. However, while the effect of capsaicin was effectively counteracted by capsazepine, thus pointing to a role of "central" vanilloid receptors in the control of locomotion, the hypolocomotor effects of either AEA or livanil were not antagonized by either capsazepine or SR141716A, thus suggesting the existence of a non-CB_1, non-VR1 site of action for AEA and long-chain capsaicin analogs (see the preceding text) (Di Marzo, Lastres-Becker, et al., 2001). It must be pointed out, however, that a more recent investigation instead implicated VR1 in AEA-induced rat hypolocomotion and inhibition of dopamine release in rat striatum (de Lago et al., 2003). Regarding antinociception, capsaicin and AEA, administered to mice intrathecally, induce antinociception with a different time-course, the action of AEA being maximal immediately after administration, whereas that of capsaicin peaked 24 h after treatment (Di Marzo et al., 2000). In agreement with a different mechanism of action for the two substances in this case, capsazepine was found to antagonize only the effect of capsaicin. Thus, if one also takes into account the hypothermic effect of capsaicin, VR1 agonists appear to share with AEA and CB_1 agonists all four cannabimimetic responses in the mouse "tetrad" (Di Marzo, Lastres-Becker, et al., 2001; Di Marzo et al., 2002), i.e., induction of hypolocomotion, catalepsy, hypothermia, and antinociception. Indeed, very potent VR1 agonists devoid of CB_1 receptor activity are extremely potent in these four tests (Di Marzo et al., 2002). It would be worthwhile determining whether AEA can still induce a positive response in such tests when administered to VR1 receptor knockouts (Caterina et al., 2000; Davis, 2000). These transgenic animals will certainly prove very useful to help providing a conclusive answer to the controversial issue of AEA as an "endovanilloid."

Regulation of AEA Activity at VR1 Receptors

The possibility that AEA could be a physiological agonist of native vanilloid receptors in mammals was initially opposed due to the low potency observed with this effect as compared to the CB_1-mediated pharmacological actions of AEA in heterologous expression systems (Zygmunt et al., 1999; Smart et al., 2000). Yet, there are now numerous studies where AEA actions on native vanilloid receptor are observed at concentrations (< 1 μ*M*) similar to those required to observe overt CB_1-mediated effects. Thus, AEA induces relaxation of rat mesenteric and hepatic arteries and of electrically-stimulated mouse vas deferens, and contracts the guinea pig ileum, with ED_{50} ranging between 10 and 500 n*M* (Zygmunt et al., 1999; Ross et al., 2001; Mang et al., 2001). On the other hand, in the case of the vanilloid-receptor-mediated contraction of guinea pig bronchi, or the release of SP and CGRP from DRG slices, high μ*M* concentrations are necessary to observe a response. Interestingly, the very poor effect of AEA on guinea pig bronchi is not simply due to low affinity for VR1 in this species, because if one uses the guinea pig urinary bladder instead, a much higher potency and efficacy can be again observed for AEA (Harrison et al., 2003). Even when using the same type of assay, i.e., the enhancement of intracellular calcium concentrations, and recombinant VR1 from the same animal species, but different experimental procedures, the reported potencies for AEA range between 0.4 and 5 μM. There are several possible explanations for the varying potency of AEA at vanilloid receptors. First, a heterogeneous population of vanilloid receptors may exist in sensory neurons, as it was hypothesized in the past (Holzer, 1991; Szallasi and Blumberg, 1999). Second, regulatory factors affecting vanilloid receptor sensitivity to ligands have been known for several years (Holzer, 1991; Szallasi and Blumberg, 1999) and might also render AEA more efficacious or potent at VR1. Finally, varying experimental conditions might affect AEA potency and efficacy at recombinant VR1. The only molecular support that has been found to date for the first of these possibilities consists of the finding that TRPV3, a VR1-like receptor which is not directly gated by vanilloids or AEA, regulates the activity of VR1 by forming different types of heterodimers with this receptor, thus resulting in a possible molecular and pharmacological heterogeneity of VR1 (Smith et al., 2002). The other two possibilities explaining the heterogeneity of VR1 responses to AEA have been, instead, recently supported by numerous experimental data. In particular:

- AEA activity at recombinant VR1 receptors is dramatically enhanced by inhibition of FAAH (Ross et al., 2001; De Petrocellis, Bisogno, et al., 2001) and by activation of the AMT by nitric oxide (NO) donors (De Petrocellis, Bisogno, et al., 2001). Furthermore, selective AMT inhibitors block AEA activity at VR1 (De Petrocellis, Bisogno, et al., 2001). These findings were obtained both in isolated cells and tissue preparations expressing VR1 (De Petrocellis, Bisogno, et al., 2001; Andersson et al., 2002). Together with previous data suggesting that the capsaicin-binding domain on its receptor is intracellular (Jung et al., 1999), they suggest that AEA activates this receptor by acting from the cytosolic side of the cell and that the AMT plays a permissive role in the interaction of AEA with VR1. This suggestion was confirmed using both a molecular approach (Jordt and Julius, 2002) and, more recently, by demonstrating that by allowing direct access of AEA into DRG neurons, this compound becomes ten times more efficacious on cation currents in these neurons than when it is applied extracellularly (Evans et al., 2003). More importantly, they indicate that if AEA is produced intracellularly by cells expressing VR1 in local concentrations that might be significantly higher than those measured in whole tissues, it may be able to directly activate this receptor. Finally, the permissive role of the AMT in AEA effects on VR1: (1) might open the possibility that overstimulation of NO release, such as that occurring during inflammation, creates the conditions for extracellular AEA to act on vanilloid receptors, (2) holds true also for some long-chain analogs of capsaicin, but not for capsaicin itself, and (3) might explain why some tissue preparations are more representative than others of the actions of the endocannabinoid on vanilloid receptors (Andersson et al., 2002).
- AEA activity at recombinant VR1 receptors is significantly enhanced when protein kinase C (PKC) or protein kinase A (PKA) are stimulated (Premkumar and Ahern, 2000; De Petrocellis, Harrison, et al., 2001; Vellani et al., 2001). Activation of these two enzymes by phorbol esters and cAMP, respectively, also greatly enhances those vanilloid-receptor-mediated actions of AEA that are otherwise observed only with high micromolar concentrations of the compound, i.e., guinea pig bronchoconstriction and SP release from DRG slices (De Petrocellis, Harrison, et al., 2001). Interestingly, PKC, by promoting their internalization, also downregulates CB_1 receptors (Garcia et al., 1998). On the other hand, activation of PKA enhances the biosynthesis of AEA phospholipid precursor, the *N*-arachidonoyl-phosphatidyl-ethanolamine (Cadas et al., 1996; Di Marzo et al., 1994). These findings indicate, therefore, that conditions and stimuli leading to PKC and PKA activation,

such as those occurring during neurodegeneration and inflammation, might, together with NO release (see the preceding text), "redirect" AEA from CB_1 to VR1 receptors

- Palmitoylethanolamide, an AEA congener coreleased with AEA by cells and proposed to act as an endogenous enhancer of AEA activity (the "entourage" effect; Lambert and Di Marzo, 1999), significantly potentiates AEA activity at recombinant human VR1 (De Petrocellis, Davis, et al., 2001) when coadministered at a molar ratio with AEA similar to that found to occur in mammalian tissues and cells. Other saturated or monounsaturated AEA congeners, and particularly lauroylethanolamide, exert a similar action (Smart et al., 2002), and this effect has been recently the subject of a detailed structure–activity relationship study (Vandevoorde et al., 2003). Palmitoylethanolamide was found to enhance also the activity on human VR1 of low concentrations of capsaicin and, particularly, RTX (De Petrocellis, Davis, et al., 2001), and to potentiate the antiproliferative effects of these two compounds on human breast cancer cells (De Petrocellis et al., 2002). It is possible that palmitoylethanolamide acts as an allosteric activator of VR1 because it produces a leftward shift in the dose-dependent displacement by AEA of [^{3}H]RTX binding to membranes containing human VR1 (De Petrocellis, Davis, et al., 2001).
- AEA bronchoconstricting activity in guinea pigs was found to be inhibited by lipoxygenase blockers, thus suggesting that these enzymes act on intracellular AEA to catalyze the formation of AEA products capable of activating vanilloid receptors, and that co-expression of lipoxygenases is necessary to observe stronger VR1-mediated effects with exogenous AEA (Craib et al., 2001). It has also been suggested that, in the same preparation, AEA might activate phospholipase A_2, AA release, and the formation of hydroperoxy-derivatives of AA (Kagaya et al., 2002), which are known to activate VR1 receptors (Hwang et al., 2000). The possibility that AEA lipoxygenase products mediate AEA activity at VR1 has been addressed by testing three different hydroperoxy-ethanolamides on Ca^{2+} influx into intact HEK cells overexpressing the human VR1 receptor (De Petrocellis, Bisogno, et al., 2001). None of the compounds was a potent or efficacious VR1 agonist under these conditions, although it is possible that exogenous hydroperoxy-ethanolamides are not recognized by the AMT and hence cannot easily interact with VR1 in intact cells. It remains therefore to be investigated if indeed lipoxygenase plays a role in regulating AEA activity at VR1 receptors. Interestingly, a recent study demonstrated that AEA, following its hydrolysis to AA and the formation of 5,6-epoxyeicosatrienoic acid, activates TRPV4 channels, which are related to VR1 receptors but are not gated by capsaicin or AEA directly (Watanabe et al., 2003).
- Acidic pH (5.5) also enhances AEA potency and efficacy at both recombinant and native vanilloid receptors (Olah, Karai, et al., 2001), whereas an increase in the assay temperature from 22 to between 37 and 50°C enhances AEA efficacy but reduces its potency at recombinant VR1 (Sprague et al., 2001). Therefore, it can be speculated that AEA might act as an endovanilloid during pathological states, such as inflammation and fever, which enhance temperature and increase the pH.
- Bovine serum albumin (BSA), used by most researchers in assay buffers to prevent AEA from sticking to the plastic ware, dramatically reduces AEA capability to activate recombinant VR1 in intact cells (De Petrocellis, Davis, et al., 2001). This phenomenon is possibly due to prevention of AEA cellular uptake, which is necessary to the compound to interact with the intracellular site on vanilloid receptors (see preceding text), and it explains why AEA potency at recombinant human VR1 is five- to tenfold higher in the absence (De Petrocellis et al., 2000; De Petrocellis, Bisogno, et al., 2001) than in the presence (Zygmunt et al., 1999; Smart et al., 2000; Ross et al., 2001; Olah, Karai, et al., 2001) of BSA.

These findings indicate that there are at least four separate ways of regulating AEA activity at VR1 receptors: (1) by regulating the activity of the AMT, (2) by PK-catalyzed phosphorylation of

VR1 and its subsequent sensitization (or inhibition of desensitization; see paper by Rathee et al., 2002), (3) by low pH, and (4) by coproduction of AEA with its congeners. Separately or together, these events may make of AEA a physiological agonist of vanilloid receptors. It is possible that, for example during inflammation or cell damage, when PKC is activated, pH is decreased and palmitoylethanolamide is biosynthesized by cells (Bisogno et al., 1997; Hansen et al., 1995), AEA becomes more active at vanilloid receptors than at CB_1 or CB_2 receptors, thus possibly producing different effects on pain perception, inflammation, and cell survival, depending on whether it only activates or also immediately desensitizes VR1. In addition, it has been shown that the levels of AEA can be enhanced following stimulation of VR1, possibly as a consequence of the intracellular calcium rise induced by this receptor (Di Marzo, Lastres-Becker, et al., 2001; Ahluwalia, Urban, et al., 2003). Hence, AEA produced during VR1-mediated inflammatory hyperalgesia may feed-back on nociception either negatively or positively by activating CB_1 or VR1 receptors, respectively.

Other Endocannabinoids and AEA Congeners, and VR1 Receptors

The other proposed and best-studied endocannabinoid, 2-AG, unlike AEA, is not an efficacious activator of vanilloid receptors (Zygmunt et al., 1999). Other *N*-acylethanolamines also appear to be far less effective than AEA (Smart et al., 2002), although a recent study underscored how the potency of these AEA congeners, and in particular of *N*-oleoyl-ethanolamine (Ahern, 2003) may very much depend on the occurrence of regulatory factors, such as co-stimulation of PKC (see previous text). At any rate, the different behavior of 2-AG and AEA may result in different roles of these two compounds during retrograde modulation of inhibitory or excitatory neurotransmission, a function recently proposed for endocannabinoids (Maejima et al., 2001). Two studies have, in fact, demonstrated how the two major endocannabinoids may exert different, if not opposing, roles in short-term and long-term synaptic plasticity. In the case of paired-pulse depression in hippocampal slices, 2-AG produces the inhibitory action expected from activation of CB_1 receptors, whereas AEA induces a dose-dependent increase due to activation of VR1 receptors (Davies and Al-Hayani, 2001). This latter effect might be due to stimulation of GABAergic interneurons in the hippocampus, because VR1 gating, as opposed to CB_1 activation, has been found to lead to enhanced release of neurotransmitters (generally glutamate) in the brain (Palazzo et al., 2002; Marinelli et al., 2002, 2003; McGaraughty et al., 2003). In the other study, 2-AG, but not AEA, was suggested to intervene as a retrograde messenger in heterosynaptic long-term depression, again in the hippocampus (Chevaleyre and Castillo, 2003). Also in the case of central neurons, the possible expression of CB_1 and VR1 receptors in the same or neighboring cells may potentially result in a regulation of synaptic signaling even finer than we can appreciate at the present date.

Recently, a synthetic AEA analog, NADA, previously shown to activate CB_1 receptors as efficaciously as AEA (Bisogno et al., 2000), was identified in bovine and rat nervous tissues, with the highest levels found in the striatum and hippocampus, and relatively low levels in sensory neurons. Most important, NADA was shown to potently activate both human and rat recombinant VR1 (Huang et al., 2002). By acting on native rat VR1 receptors, NADA was also more potent than AEA at inducing CGRP and substance P release from spinal cord slices, and at causing intracellular calcium rises in dorsal root ganglion neurons, and was almost as active as capsaicin at enhancing paired-pulse inhibition in the rat hippocampus. A possible NADA metabolic product, *O*-methyl-NADA, was much less active as a VR1 agonist (Huang et al., 2002) and did not affect paired-pulse inhibition (Al-Hayani et al., 2003), thus suggesting that methylation by catechol *O*-methyl transferase might be a way to terminate the effect of NADA at VR1. More recently (Toth et al., 2003), the effect of NADA on rat VR1 has been reexamined by using a different methodology and was found to be less strong than in the previous study. Furthermore, while NADA appears to induce a potent VR1-dependent dilation of the rat mesenteric artery (personal communication by E. Hogestatt, and O'Sullivan et al., 2003) and a significant VR1-mediated constriction of guinea pig bronchi (Harrison et al., 2003), this compound, unlike AEA and capsaicin, is a rather weak

VR1 agonist compared to AEA and, particularly, capsaicin, in assays where the tone of the urinary bladder from either the rat or guinea pig is measured (Harrison et al., 2003). These recent findings indicate that the potency of NADA might depend upon both the experimental procedure and the type of tissue preparation used to assess VR1 agonist activity. It is possible that also in the case of this compound, the presence of an active membrane transporter in the several cell types that separate the tissue preparations used from the corresponding sensory neurons, or the concomitant activation of regulatory factors, might play a crucial role in determining its potency in complex *in vitro* systems and *in vivo* (see preceding text).

Finally, the effect on VR1 of the other two proposed endocannabinoids, virodhamine and noladin ether, were assessed in two different preliminary studies (Duncan et al., 2003; Gough et al., 2003). Both compounds were found to be inactive.

NEW RECEPTORS FOR AEA AND 2-AG OXIDATION PRODUCTS

Both AEA and 2-AG are substrates for enzymes of the AA cascade, in particular for 12- and 15-lipoxygenases, for cytochrome p450 oxygenases, and for cycloxygenase-2 (COX-2). However, the biological meaning of these reactions, that only in a few cases have been found to occur in intact cell systems, is not known (Kozak and Marnett, 2002, for review). Lipoxygenase products (Figure 6.3) are usually still capable to bind to cannabinoid receptors to some extent or to inhibit FAAH (van der Stelt et al., 2002), whereas prostanoid derivatives of both AEA and 2-AG, obtained from the transformation of the respective endoperoxide products originated from COX-2 action on the two compounds, are weakly active or entirely inactive on all cannabinoid and prostanoid receptor types known to date. There is, however, some indirect evidence suggesting that some of these oxidation products of the two endocannabinoids might bind to novel, and yet to be characterized, receptors.

AEA Oxidation Products

The AEA oxidation products produced from the initial COX-2-derived products have been named prostamides (after prostaglandin-ethanolamides) (Figure 6.3) (Woodward et al., 2000). Prostamides are 2 to 3 orders of magnitude less active than the corresponding prostaglandins (PGs) in both binding and functional assays of the several prostanoid receptors known to date (Ross et al., 2002; Woodward et al., 2001; 2003). For example, prostamide E_2 is 100 to 1000-fold less potent than PGE_2 in binding assays carried out with EP_1, EP_2, EP_3, and EP_4 receptor-containing membranes. However, although this compound is, as expected, about 100-fold less potent than the free acid in functional assays of EP_3 and EP_4 receptors, it is also surprisingly only 15 times less potent than PGE_2 in a functional assay of EP_2 receptors, the relaxation of histamine-induced contractions of guinea pig trachea (Ross et al., 2002). Prostamide F_2 and its 17-phenyl-derivative, bimatoprost (AGN192024, Lumigan®) (Figure 6.3) exhibit very low affinity for, and potency at, the FP receptor, although they both exert a powerful contraction of the isolated feline iris sphincter and potently reduce the intraocular pressure in ocular normotensive dogs (Woodward et al., 2000; 2001). Bimatoprost and prostamide F_2 were suggested to act on yet uncharacterized novel receptors because they exert also a strong contraction of feline lung parenchyma at low nanomolar concentrations, although they have an altogether different pharmacological profile from PGF_2 in all the other available assays for FP-receptor-mediated activity, and exert no activity in a wide range of binding assays for other known receptors (Woodward et al., 2001; 2003).

Regarding AEA metabolites obtained from lipoxygenases, their potential capability of stimulating VR1 receptors, hypothesized so far only on the basis of pharmacological evidence, has been mentioned in the preceding text. Interestingly, a preliminary report has suggested that the leukotriene B_4 derivative of AEA (Figure 6.3) might also activate VR1 with the same potency as LTB_4 (McHugh et al., 2003).

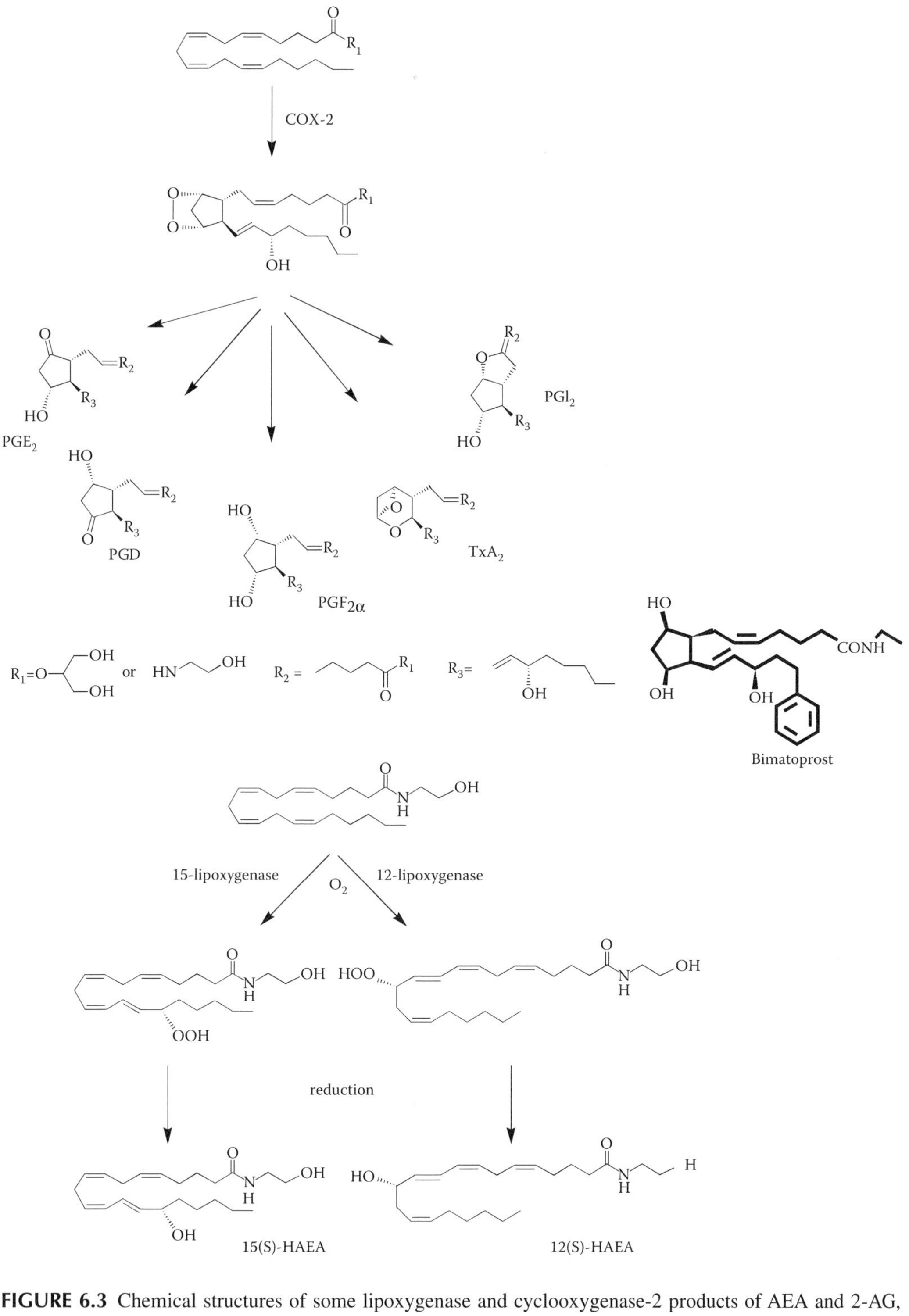

FIGURE 6.3 Chemical structures of some lipoxygenase and cyclooxygenase-2 products of AEA and 2-AG, of prostamides and their synthetic analog, bimatoprost, and of prostaglandin glyceryl esters.

2-AG Oxidation Products

Recent evidence suggested that the 15-lipoxygenase derivative of 2-AG (Figure 6.3), but not 2-AG itself, can interact with the peroxisome proliferator-activated receptor (PPAR)-α (Kozak et al., 2002). Although the potential implications of this important finding are numerous, also in view of the high intracellular concentration of 2-AG, its physiological relevance remains to be established because the dose of 2-AG necessary to achieve maximal stimulation of (PPAR)-α was around 10 μM. More recently, 2-AG itself was suggested to activate PPAR-γ in T-cells, although at even higher concentrations (approaching 50 μM) (Rockwell and Kaminski, 2003). In this case, the involvement of lipoxygenase products of 2-AG was not investigated.

Finally, at the 2003 meeting of the ARVO, Rouzer and collaborators reported that prostaglandin E_2 glyceryl ester (Figure 6.3), which can be obtained from the action of COX-2 and prostaglandin E_2 synthase on 2-AG, is capable of activating yet to be characterized, as well as possibly novel, GPCRs. It will be interesting to find out if the receptors hypothesized for prostamide F_2 and this compound are the same.

CONCLUDING REMARKS

Although at the start the suggestion that AEA could behave as one ligand for more receptors appeared as a confusing pharmacological mishap or even as a bizarre mistake of Nature, the evidence for the existence of several AEA receptors is now gaining more and more strength as well as physiopathological relevance. Although it is true that, in most cases, the actual role and contribution of these proteins in determining the effects of endogenous or exogenous AEA has not been yet investigated, some of these molecular targets, such as the L-type Ca^{2+} channels and the various K^+ channels, keep on emerging from time to time. Others, such as the two or more non-CB_1, non-CB_2 GPCRs, have not been characterized yet. Clearly, these receptors, if they exist, are likely to be much more structurally different from CB_1 than CB_2 itself, because otherwise they would have been already identified by means of the very sophisticated cloning-by-homology techniques available nowadays. The possible role of vanilloid receptors in AEA pharmacology is now being investigated very actively, and this effort seems to confirm, at least for the time being, the original hypothesis that VR1 should be referred to as an "ionotropic" receptor for AEA, as opposed to "metabotropic" receptors for this compound, which include the cannabinoid CB_1 receptors and the other uncharacterized GPCRs described here. This hypothesis is supported further by the finding that other fatty acid amides, i.e., NADA, *N*-oleoyl-dopamine, and *N*-oleoyl-ethanolamine (Huang et al., 2002; Chu et al., 2003; Ahern, 2003), are also capable of activating either CB_1 or VR1 or both receptors to a varying extent. Clearly, this classification would not exclude the possibility that other endogenous compounds may act more specifically only at one of the two classes of receptors. In fact, 2-AG appears be a selective agonist of cannabinoid receptors, although this compound or its lipoxygenase derivative might act at PPARs, whereas some oxidation products of AA seem to be selective agonists of VR1 or of other VR1-related ion channels. Also, the two most abundant plant cannabinoids, THC, a selective agonist for cannabinoid receptors, and cannabidiol, a weak although selective agonist for VR1 (Bisogno et al., 2001), might support this dichotomy. As for metabotropic and ionotropic receptors of other transmitters, e.g., glutamate, GABA, and purines, distinct functions might be associated with the two major classes of anandamide receptors. Metabotropic receptors would mediate the inhibitory effects of locally released AEA and NADA, such as retrograde inhibition of depolarization-induced neurotransmission in the CNS (Maejima et al., 2001), whereas ionotropic receptors would underlie either excitatory inputs or, if immediately desensitized by AEA, inhibitory effects on subsequent excitatory stimuli.

In conclusion, it can be now sustained that the discovery of AEA in 1992 did not represent merely the finding of a ligand for the until-then-orphan cannabinoid receptors. However, a further effort will be necessary to identify those molecular targets for this compound that still await molecular characterization.

REFERENCES

Adams, I.B., Compton, D.R., and Martin, B.R. (1998) Assessment of anandamide interaction with the cannabinoid brain receptor: SR 141716A antagonism studies in mice and autoradiographic analysis of receptor binding in rat brain, *J. Pharmacol. Exp. Ther.,* 284: 1209–17.

Ahern, G.P. (2003) Activation of TRPV1 by the satiety factor oleoylethanolamide. *J. Biol. Chem.,* 25: 5109–5116.

Ahluwalia, J., Urban, L., Capogna, M., Bevan, S., and Nagy, I. (2000) Cannabinoid 1 receptors are expressed in nociceptive primary sensory neurons, *Neuroscience,* 100: 685–88.

Ahluwalia, J., Urban, L., Bevan, S., and Nagy, I. (2003) Anandamide regulates neuropeptide release from capsaicin-sensitive primary sensory neurons by activating both the cannabinoid 1 receptor and the vanilloid receptor 1 in vitro, *Eur. J. Neurosci.,* 17: 2611–8.

Ahluwalia, J., Yaqoob, M., Urban, L., Bevan, S., and Nagy, I. (2003) Activation of capsaicin-sensitive primary sensory neurones induces anandamide production and release, *J. Neurochem.,* 84: 585–91.

Akinshola, B.E., Taylor, R.E., Ogunseitan, A.B., and Onaivi, E.S. (1999) Anandamide inhibition of recombinant AMPA receptor subunits in Xenopus oocytes is increased by forskolin and 8-bromo-cyclic AMP. *Naunyn. Schmiedebergs Arch. Pharmacol.,* 360: 242–8.

Al-Hayani, A., Di Marzo, V., and Davies S. (2003) Homo-N-arachidonoyl-dopamine mimics vanilloid receptor agonists in enhancing paired pulse depression of hippocampal population spikes, *Int. Cannabinoid Res. Soc. Symposium,* p. 83.

Andersson, D.A., Adner, M., Hogestatt, E.D., and Zygmunt, P.M. (2002) Mechanisms underlying tissue selectivity of anandamide and other vanilloid receptor agonists, *Mol. Pharmacol.,* 62: 705–13.

Beltramo, M. and Piomelli, D. (1999) Anandamide transport inhibition by the vanilloid agonist olvanil, *Eur. J. Pharmacol.,* 364: 75–8.

Bisogno, T., Hanus, L., De Petrocellis, L., Tchilibon, S., Ponde, D.E., Brandi, I., Schiano Moriello A., Davis, J.B., Mechoulam, R., and Di Marzo, V. (2001) Molecular targets for cannabidiol and its synthetic analogues: effect on vanilloid VR1 receptors and on the cellular uptake and enzymatic hydrolysis of anandamide, *Br. J. Pharmacol.,* 134: 845–52.

Bisogno, T., Maurelli, S., Melck, D., De Petrocellis, L., and Di Marzo, V. (1997) Biosynthesis, uptake, and degradation of anandamide and palmitoylethanolamide in leukocytes, *J. Biol. Chem.,* 272: 3315–23.

Bisogno, T., Melck, D., Bobrov, M.Yu., Gretskaya, N.M., Bezuglov, V.V., De Petrocellis, L., and Di Marzo, V. (2000) N-acyl-dopamines: novel synthetic CB(1) cannabinoid-receptor ligands and inhibitors of anandamide inactivation with cannabimimetic activity in vitro and in vivo, *Biochem. J.,* 351: 817–24.

Boger, D.L., Sato, H., Lerner, A.E., Guan, X., and Gilula, N.B. (1999) Arachidonic acid amide inhibitors of gap junction cell-cell communication, *Bioorg. Med. Chem. Lett.,* 9: 1151–4.

Breivogel, C.S., Griffin, G., Di Marzo, V., and Martin, B.R. (2001) Evidence for a new G-protein-coupled cannabinoid receptor in mouse brain, *Mol. Pharmacol.,* 60: 155–63.

Bridges, D., Rice, A.S., Egertova, M., Elphick, M.R., Winter, J., and Michael, G.J. (2003) Localisation of cannabinoid receptor 1 in rat dorsal root ganglion using in situ hybridisation and immunohistochemistry. *Neuroscience,* 119:803–12.

Cadas, H., Gaillet, S., Beltramo, M., Venance, L., and Piomelli, D. (1996) Biosynthesis of an endogenous cannabinoid precursor in neurons and its control by calcium and cAMP, *J. Neurosci.,* 16: 3934–42.

Calignano, A., La Rana, G., Makriyannis, A., Lin, S.Y., Beltramo, M., and Piomelli, D. (1997) Inhibition of intestinal motility by anandamide, an endogenous cannabinoid, *Eur. J. Pharmacol.,* 340: R7–8.

Calignano, A., La Rana, G., and Piomelli, D. (2001) Antinociceptive activity of the endogenous fatty acid amide, palmitylethanolamide, *Eur. J. Pharmacol.,* 419: 191–8.

Caterina, M.J., Leffler, A., Malmberg, A.B., Martin, W.J., Trafton, J., Petersen-Zeitz, K.R., Koltzenburg, M., Basbaum, AI., and Julius, D. (2000) Impaired nociception and pain sensation in mice lacking the capsaicin receptor, *Science,* 288: 306–13.

Caterina, M.J., Schumacher, M.A., Tominaga, M., Rosen, T.A., Levine, J.D., and Julius, D. (1997) The capsaicin receptor: a heat-activated ion channel in the pain pathway, *Nature,* 389: 816–24.

Chemin, J., Monteil, A., Perez-Reyes, E., Nargeot, J., and Lory, P. (2001) Direct inhibition of T-type calcium channels by the endogenous cannabinoid anandamide. *EMBO J.*, 20: 7033–40.

Chevaleyre, V. and Castillo, P.E. (2003) Heterosynaptic LTD of hippocampal GABAergic synapses: a novel role of endocannabinoids in regulating excitability, *Neuron*, 38: 461–72.

Christopoulos, A. and Wilson, K. (2001) Interaction of anandamide with the M(1) and M(4) muscarinic acetylcholine receptors, *Brain Res.*, 915: 70–8.

Chu, C.J., Huang, S.M., De Petrocellis, L., Bisogno, T., Ewing, S.A., Miller, J.D., Zipkin, R.E., Daddario, N., Appendino, G., Di Marzo, V., and Walker, J.M. (2003) N-oleoyldopamine, a novel endogenous capsaicin-like lipid that produces hyperalgesia, *J. Biol. Chem.*, 278: 13633–9.

Cortright, D.N., Crandall, M., Sanchez, J.F., Zou, T., Krause, J.E., and White, G. (2001) The tissue distribution and functional characterization of human VR1, *Biochem. Biophys. Res. Commn.*, 281: 1183–9.

Craib, S.J., Ellington, H.C., Pertwee, R.G., and Ross, R.A. (2001) A possible role of lipoxygenase in the activation of vanilloid receptors by anandamide in the guinea-pig bronchus, *Br. J. Pharmacol.*, 134: 30–7.

Cravatt, B.F., Demarest, K., Patricelli, M.P., Bracey, M.H., Giang, D.K., Martin, B.R., and Lichtman, A.H. (2001) Supersensitivity to anandamide and enhanced endogenous cannabinoid signaling in mice lacking fatty acid amide hydrolase, *Proc. Natl. Acad. Sci. U.S.A.*, 98: 9371–6.

Davies, S.N. and Al-Hayani, A. (2001) Actions of putative endogenous cannabinoids on synaptic transmission in the rat hippocampal slice, *Int. Cannabinoid Res. Soc. Symposium*, p. 27.

Davis, J.B. (2000) Vanilloid receptor-1 is essential for inflammatory thermal hyperalgesia, *Nature*, 405: 183–8.

de Lago, E., de Miguel, R., Hernàndez, M., Cebeira, M., Ramos, J.A., and Fernàndez-Ruiz, J. (2003) Anandamide effects on motor behavior and nigrostriatal dopaminergic activity are mediated by the action of vanilloid VR1 receptors, *Int. Cannabinoid Res. Soc. Symposium*, p. 92.

De Petrocellis, L., Bisogno, T., Davis, J.B., Pertwee, R.G., and Di Marzo, V. (2000) Overlap between the ligand recognition properties of the anandamide transporter and the VR1 vanilloid receptor: inhibitors of anandamide uptake with negligible capsaicin-like activity, *FEBS Lett.*, 483: 52–6.

De Petrocellis, L., Bisogno, T., Maccarone, M., Davis, J.D., Finazzi-Agrò, A., and Di Marzo, V. (2001) The activity of anandamide at vanilloid VR1 receptors requires facilitated transport across the cell membrane and is limited by intracellular metabolism, *J. Biol. Chem.*, 276: 12856–63.

De Petrocellis, L., Bisogno, T., Ligresti, A., Bifulco, M., Melck, D., and Di Marzo, V. (2002) Effect on cancer cell proliferation of palmitoylethanolamide, a fatty acid amide interacting with both the cannabinoid and vanilloid signalling systems, *Fundam. Clin. Pharmacol.*, 16: 297–302.

De Petrocellis, L., Davis, J.B., and Di Marzo, V. (2001) Palmitoylethanolamide enhances anandamide stimulation of human vanilloid VR1 receptors, *FEBS Lett.*, 506: 253–6.

De Petrocellis, L., Harrison, S., Bisogno, T., Tognetto, M., Brandi, I., Smith, G.D., Creminon, C., Davis, J.B., Geppetti, P., and Di Marzo, V. (2001) The vanilloid receptor (VR1)-mediated effects of anandamide are potently enhanced by the cAMP-dependent protein kinase, *J. Neurochem.*, 77: 1660–3.

Devane, W.A., Hanus, L., Breuer, A., Pertwee, R.G., Stevenson, L.A., Griffin, G., Gibson, D., Mandelbaum, A., Etinger, A., and Mechoulam R. (1992) Isolation and structure of a brain constituent that binds to the cannabinoid receptor, *Science*, 258: 1946–9.

Di Marzo, V. (1998) 'Endocannabinoids' and other fatty acid derivatives with cannabimimetic properties: biochemistry and possible physiopathological relevance, *Biochim. Biophys. Acta*, 1392: 153–75.

Di Marzo, V. and Fontana, A. (1995) Anandamide, an endogenous cannabinomimetic eicosanoid: killing two birds with one stone, *Prostaglandins Leukot. Essent. Fatty Acids*, 53: 1–11.

Di Marzo, V., Bisogno, T., and De Petrocellis, L. (2001) Endocannabinoids Part I: Molecular basis of endocannabinoid formation, action and inactivation and development of selective inhibitors, *Emerging Therapeutic Targets*, 5: 241–65.

Di Marzo, V., Bisogno, T., De Petrocellis, L., Brandi, I., Jefferson, R.G., Winckler, R.L., Davis, J.B., Dasse, O., Mahadevan, A., Razdan, R.K., and Martin, B.R. (2001) Highly selective CB(1) cannabinoid receptor ligands and novel CB(1)/VR(1) vanilloid receptor "hybrid" ligands, *Biochem. Biophys. Res. Commn.*, 281: 444–51.

Di Marzo, V., Bisogno, T., De Petrocellis, L., Melck, D., and Martin, B.R. (1999) Cannabimimetic fatty acid derivatives: the anandamide family and other endocannabinoids, *Curr. Med. Chem.*, 6: 721–44.

Di Marzo, V., Bisogno, T., Melck, D., Ross, R., Brockie, H., Stevenson, L., Pertwee, R., and De Petrocellis, L. (1998) Interactions between synthetic vanilloids and the endogenous cannabinoid system, *FEBS Lett.*, 436: 449–54.

Di Marzo, V., Breivogel, C.S., Tao, Q., Bridgen, D.T., Razdan, R.K., Zimmer, A.M., Zimmer, A., and Martin, B.R. (2000) Levels, metabolism and pharmacological activity of anandamide in CB(1) cannabinoid receptor knockout mice: evidence for non-CB(1), non-CB(2) receptor-mediated actions of anandamide in mouse brain, *J. Neurochem.,* 75: 2434–44.

Di Marzo, V., Fontana, A., Cadas, H., Schinelli, S., Cimino, G., Schwartz, J.C., and Piomelli, D. (1994) Formation and inactivation of endogenous cannabinoid anandamide in central neurons, *Nature,* 372: 686–91.

Di Marzo, V., Griffin, G., De Petrocellis, L., Brandi, I., Bisogno, T., Williams, W., Grier, M.C., Kulasegram, S., Mahadevan, A., Razdan, R.K., and Martin, B.R. (2002) A structure/activity relationship study on arvanil, an endocannabinoid and vanilloid hybrid, *J. Pharmacol. Exp. Ther.,* 300: 984–91.

Di Marzo, V., Lastres-Becker, I., Bisogno, T., De Petrocellis, L., Milone, A., Davis, J.B., and Fernandez-Ruiz, J.J. (2001) Hypolocomotor effects in rats of capsaicin and two long chain capsaicin homologues, *Eur. J. Pharmacol.,* 420: 123–31.

Dray, A. (1992) Neuropharmacological mechanisms of capsaicin and related substances, *Biochem Pharmacol.,* 44: 611–5.

Duncan M., Millns P., Kendall D., and Ralevic V. (2003) Noladin ether attenuates sensory neurotransmission in the rat isolated mesenteric arterial bed via a novel Gi/o linked cannabinoid receptor, *Int. Cannabinoid Res. Soc. Symposium,* p. 93.

Evans, R., Scott, R., and Ross, R. (2003) Anandamide is more potent as a TRPV1 agonist when applied to the intracellular compared with the extracellular environment of rat cultured DRG neurones, *Int. Cannabinoid Res. Soc. Symposium,* p. 18.

Farquhar-Smith. W.P., Egertova, M., Bradbury, E.J., McMahon, S.B., Rice, A.S., and Elphick, M.R. (2000) Cannabinoid CB(1) receptor expression in rat spinal cord, *Mol. Cell. Neurosci.,* 15: 510– 21.

Fowler, C.J., Wallin, T., and Jacobsson, S.O.P. (2001) Comparison of the antiproliferative effects of anandamide, 2-arachidonoylglycerol and olvanil on C6 cells, *Int. Cannabinoid Res. Soc. Symposium,* p. 42.

Garcia, D.E., Brown, S., Hille, B., and Mackie, K. (1998) Protein kinase C disrupts cannabinoid actions by phosphorylation of the CB1 cannabinoid receptor, *J. Neurosci.,* 18: 2834–41.

Gauldie, S.D., McQueen, D.S., Pertwee, R., and Chessell, I.P. (2001) Anandamide activates peripheral nociceptors in normal and arthritic rat knee joints, *Br. J. Pharmacol.,* 132: 617–21.

Gebremedhin, D., Lange, A.R., Campbell, W.B., Hillard, C.J., and Harder D.R. (1999) Cannabinoid CB_1 receptor of cat cerebral arterial muscle functions to inhibit L-type Ca^{2+} channel current, *Am. J. Physiol.,* 276: H2085–93.

Geraghty, D.P. and Mazzone, S.B. (2002) Respiratory actions of vanilloid receptor agonists in the nucleus of the solitary tract: comparison of resiniferatoxin with non-pungent agents and anandamide, *Br. J. Pharmacol.,* 137: 919–27.

Gough, W., Bensinger, J.W., Li, J., Baker, A., and Porter, A.C. (2003). Pharmacological characterization of five endogenous cannabinoids. *Int. Cannabinoid Res. Soc. Symposium,* p. 80.

Greffrath, W., Fischbach, T., Nawrath, H., and Treede R.D. (2002) Anandamide predominantly excites nociceptide dorsal root ganglion neurons of rats. World Congress on Pain, 381–P15, IASP Press, USA.

Guo, A., (1999) Immunocytochemical localization of the vanilloid receptor 1 (VR1): relationship to neuropeptides, the P2X3, purinoceptor and IB4 binding sites, *Eur. J. Neurosci.,* 11: 946–58.

Hajos, N. and Freund, T.F. (2002) Pharmacological separation of cannabinoid sensitive receptors on hippocampal excitatory and inhibitory fibers. *Neuropharmacology,* 43: 503–10.

Hajos, N., Ledent, C., and Freund, T.F. (2001) Novel cannabinoid-sensitive receptor mediates inhibition of glutamatergic synaptic transmission in the hippocampus, *Neuroscience,* 106: 1–4.

Hampson, A.J., Bornheim, L.M., Scanziani, M., Yost, C.S., Gray, A.T., Hansen, B.M., Leonoudakis, D.J., and Bickler, P.E. (1998) Dual effects of anandamide on NMDA receptor-mediated responses and neurotransmission, *J. Neurochem.,* 70: 671–6.

Hansen, H.S., Lauritzen, L., Strand, A.M., Moesgaard, B., and Frandsen, A. (1995) Glutamate stimulates the formation of N-acylphosphatidylethanolamine and N-acylethanolamine in cortical neurons in culture, *Biochim. Biophys. Acta,* 1258: 303–8.

Hanus, L., Abu-Lafi, S., Fride, E., Breuer, A., Vogel, Z., Shalev, D.E., Kustanovich, I., and Mechoulam, R. (2001) 2-arachidonyl glyceryl ether, an endogenous agonist of the cannabinoid CB1 receptor, *Proc. Natl. Acad. Sci. U.S.A.,* 98: 3662–5.

Harris, D., McCulloch, A.I., Kendall, D.A., and Randall, M.D. (2002) Characterization of vasorelaxant responses to anandamide in the rat mesenteric arterial bed, *J. Physiol.,* 539: 893–902.

Harris, J., Drew, L.J., and Chapman, V. (2000) Spinal anandamide inhibits nociceptive transmission via cannabinoid receptor activation in vivo, *Neuroreport,* 11: 2817–9.

Harrison, S., De Petrocellis, L., Trevisani, M., Benvenuti, F., Bifulco, M., Geppetti, P., and Di Marzo, V. (2003) Capsaicin-like effects of N-arachidonoyl-dopamine in the isolated guinea pig bronchi and urinary bladder, *Eur. J. Pharmacology,* 475: 107–114.

Hayes, P., Meadows, H.J., Gunthorpe, M.J., Harries, M.H., Duckworth, D.M., Cairns, W., Harrison, D.C., Clarke, C.E., Ellington, K., Prinjha, R.K., Barton, A.J, Medhurst, A.D., Smith, G.D., Topp, S., Murdock, P., Sanger, G.J., Terrett, J., Jenkins, O., Benham, C.D., Randall, A.D., Gloger, I.S., and Davis, J.B. (2000) Cloning and functional expression of a human orthologue of rat vanilloid receptor-1, *Pain,* 88: 205–5.

Hermann, H., De Petrocellis, L., Bisogno, T., Schiano Moriello, A., Lutz, B., and Di Marzo, V. (2003) Dual effect of cannabinoid CB1 receptor stimulation on a vanilloid VR1 receptor-mediated response, *Cell. Mol. Life Sci.,* 60: 607–16.

Ho, W.S. and Hiley, C.R. (2003a) Vasodilator actions of abnormal-cannabidiol in rat isolated small mesenteric artery. *Br. J. Pharmacol.,* 138: 1320–32.

Ho, W.S. and Hiley, C.R. (2003b) Endothelium-independent relaxation to cannabinoids in rat-isolated mesenteric artery and role of Ca^{2+} influx, *Br. J. Pharmacol.,* 139: 585–97.

Ho W.-S.V. and Hiley C.R. (2003c) Vasorelaxation to the novel endocannabinoid virodhamine, *Int. Cannabinoid Res. Soc. Symposium,* p. 89.

Holzer, P. (1991) Capsaicin: cellular targets, mechanisms of action, and selectivity for thin sensory neurons, *Pharmacol. Rev.,* 43: 143–201.

Huang, S.M., Bisogno, T., Trevisani, M., Al-Hayani, A., De Petrocellis, L., Fezza, F., Tognetto, M., Petros, T.J., Krey, J.F., Chu, C.J., Miller, J.D., Davies, S.N., Geppetti, P., Walker, J.M., and Di Marzo, V. (2002) An endogenous capsaicin-like substance with high potency at recombinant and native vanilloid VR1 receptors, *Proc. Natl. Acad. Sci U.S.A.,* 99: 8400–5.

Hwang, S.W., Cho, H., Kwak, J., Lee, S.Y., Kang, C.J., Jung, J., Cho, S., Min, K.H., Suh, Y.G., Kim, D., and Oh, U. (2000) Direct activation of capsaicin receptors by products of lipoxygenases: endogenous capsaicin-like substances, *Proc. Natl. Acad. Sci. U.S.A.,* 97: 6155–60.

Izzo, A.A., Capasso, R., Pinto, L., Di Carlo, G., Mascolo, N., and Capasso, F. (2001) Effect of vanilloid drugs on gastrointestinal transit in mice, *Br. J. Pharmacol.,* 132: 1411–6.

Jarai, Z., Wagner, J.A., Varga, K., Lake, K.D., Compton, D.R., Martin, B.R., Zimmer, A.M., Bonner, T.I., Buckley, N.E., Mezey, E., Razdan, R.K., Zimmer, A., and Kunos, G. (1999) Cannabinoid induced mesenteric vasodilatation through an endothelial site distinct from CB_1 or CB_2 receptors, *Proc. Natl. Acad. Sci. U.S.A.,* 96: 14136–41.

Jarai, Z., Wagner, J.A., Goparaju, S.K., Wang, L., Razdan, R.K., Sugiura, T., Zimmer, A.M., Bonner, T.I., Zimmer, A., and Kunos, G. (2000) Cardiovascular effects of 2-arachidonoyl glycerol in anesthetized mice, *Hypertension,* 35: 679–84.

Jarrahian, A. and Hillard, C.J. (1997) Arachidonylethanolamide (anandamide) binds with low affinity to dihydropyridine binding sites in brain membranes, *Prostaglandins Leukot. Essent. Fatty Acids,* 57: 551–4.

Jennings, E.A., Vaughan, C.W., Roberts, L.A., and Christie, M.J. (2003) The actions of anandamide on rat superficial medullary dorsal horn neurons in vitro, *J. Physiol.,* 548: 121–9.

Jia, Y., McLeod, R.L., Wang, X., Parra, L.E., Egan, R.W., and Hey, J.A. (2002) Anandamide induces cough in conscious guinea-pigs through VR1 receptors, *Br. J. Pharmacol.,* 137: 831–6.

Johnson, D.E., Heald, S.L., Dally, R.D., and Janis, R.A. (1993) Isolation, identification and synthesis of an endogenous arachidonic amide that inhibits calcium channel antagonist 1,4-dihydropyridine binding, *Prostaglandins Leukot. Essent. Fatty Acids,* 48: 429–37.

Jordt, S.E. and Julius, D. (2002) Molecular basis for species-specific sensitivity to "hot" chili peppers, *Cell,* 108: 421–30.

Jung, J., Hwang, S.W., Kwak, J., Lee, S.Y., Kang, C.J., Kim, W.B., Kim, D., and Oh, U. (1999) Capsaicin binds to the intracellular domain of the capsaicin-activated ion channel, *J. Neurosci.,* 19: 529–38.

Kagaya, M., Lamb, J., Robbins, J., Page, C.P., and Spina, D. (2002) Characterization of the anandamide induced depolarization of guinea-pig isolated vagus nerve, *Br. J. Pharmacol.,* 137: 39–48.

Khanolkar A.D., Palmer S.L., and Makriyannis A. (2000) Molecular probes for the cannabinoid receptors, *Chem. Phys. Lipids,* 108: 37–52.

Kimura, T., Ohta, T., Watanabe, K., Yoshimura, H., and Yamamoto, I. (1998) Anandamide, an endogenous cannabinoid receptor ligand, also interacts with 5-hydroxytryptamine (5-HT) receptor, *Biol. Pharm. Bull.,* 21: 224–6.

Kozak, K.R., Gupta, R.A., Moody, J.S., Ji, C., Boeglin, W.E., DuBois, R.N., Brash, A.R., and Marnett, L.J. (2002) 15-Lipoxygenase metabolism of 2-arachidonylglycerol. Generation of a peroxisome proliferator-activated receptor alpha agonist, *J. Biol. Chem.,* 277: 23278–86.

Kozak KR. and Marnett LJ. (2002) Oxidative metabolism of endocannabinoids, *Prostaglandins Leukot. Essent. Fatty Acids,* 66: 211–20.

Kwak, J.Y., Jung, J.Y., Hwang, S.W., Lee, W.T., and Oh, U. (1998) A capsaicin-receptor antagonist, capsazepine, reduces inflammation-induced hyperalgesic responses in the rat: evidence for an endogenous capsaicin-like substance, *Neuroscience,* 86: 619–26.

Lagalwar, S., Bordayo, E.Z., Hoffmann, K.L., Fawcett, J.R., and Frey, W.H. 2nd. (1999) Anandamides inhibit binding to the muscarinic acetylcholine receptor, *J. Mol. Neurosci.,* 13: 55–61.

Lambert, D.M. and Di Marzo, V. (1999) The palmitoylethanolamide and oleamide enigmas: are these two fatty acid amides cannabimimetic?, *Curr. Med. Chem.,* 6: 757–73.

Ledent, C., Valverde, O., Cossu, G., Petitet, F., Aubert, J.F., Beslot, F., Bohme, G.A., Imperato, A., Pedrazzini, T., Roques, B.P., Vassart, G., Fratta, W., and Parmentier, M. (1999) Unresponsiveness to cannabinoids and reduced addictive effects of opiates in CB1 receptor knockout mice, *Science,* 283: 401–4.

Li, J., Kaminski N.E., and Wang, D.H. (2003) Anandamide-induced depressor effect in spontaneously hypertensive rats: role of the vanilloid receptor, *Hypertension,* 41: 757–62.

Lin, Y.S. and Lee, L.Y. (2002) Stimulation of pulmonary vagal C-fibres by anandamide in anaesthetized rats: role of vanilloid type 1 receptors, *J. Physiol.,* 539: 947–55.

Liu, M., Liu, M.C., Magoulas, C., Priestley, J.V., and Willmott, N.J. (2003) Versatile regulation of cytosolic Ca^{2+} by vanilloid receptor I in rat dorsal root ganglion neurons, *J. Biol. Chem.,* 278: 5462–72.

Maccarone, M., Lorenzon, T., Bari, M., Melino, G., and Finazzi-Agrò, A. (2000) Anandamide induces apoptosis in human cells via vanilloid receptors. Evidence for a protective role of cannabinoid receptors, *J. Biol. Chem.,* 275: 31938–45.

Maejima, T., Hashimoto, K., Yoshida, T., Aiba, A., and Kano, M. (2001) Presynaptic inhibition caused by retrograde signal from metabotropic glutamate to cannabinoid receptors, *Neuron,* 31: 463–75.

Maingret, F., Patel, A.J., Lazdunski, M., and Honore, E. (2001) The endocannabinoid anandamide is a direct and selective blocker of the background K(+) channel TASK-1, *EMBO J.,* 20: 47–54.

Malinowska, B., Kwolek, G., and Gothert, M. (2001) Anandamide and methanandamide induce both vanilloid VR1- and cannabinoid CB1 receptor-mediated changes in heart rate and blood pressure in anaesthetized rats, *Naunyn Schmiedebergs Arch. Pharmacol.,* 364: 562–9.

Mang, C.F., Erbelding, D., and Kilbinger, H. (2001) Differential effects of anandamide on acetylcholine release in the guinea-pig ileum mediated via vanilloid and non-CB1 cannabinoid receptors, *Br. J. Pharmacol.,* 134: 161–7.

Marinelli, S., Di Marzo, V., Berretta, N., Matias, I., Maccarrone, M., Bernardi, G., and Mercuri, N.B. (2003) Presynaptic facilitation of glutamatergic synapses to dopaminergic neurons of the rat substantia nigra by endogenous stimulation of vanilloid receptors, *J. Neurosci.,* 23: 3136–44.

Marinelli, S., Vaughan, C.W., Christie, M.J., and Connor, M. (2002) Capsaicin activation of glutamatergic synaptic transmission in the rat locus coeruleus in vitro, *J. Physiol.,* 543: 531–40.

McGaraughty, S., Chu, K.L., Bitner, R.S., Martino, B., El Kouhen, R., Han, P., Nikkel, A.L., Burgard, E.C., Faltynek, C.R., and Jarvis, M.F. (2003). Capsaicin infused into the periaqueductal gray affects rat tail flick responses to noxious heat and alters neuronal firing in the rostral ventromedial medulla. *J. Neurophysiol.,* published ahead of print 10.1152/jn.00433.2003.

McHugh, D., McMaster, S., and Ross, R. (2003) Pharmacological characterisation of LTB_4 ethanolamide: interaction with leukotriene (BLT) and vanilloid (TRPV1) receptors, *Int. Cannabinoid Res. Soc. Symposium,* p. 121.

McVey, D.C., Schmid, P.C., Schmid, H.H., and Vigna, S.R. (2003) Endocannabinoids induce ileitis in rats via the capsaicin receptor (VR1). *J. Pharmacol. Exp. Ther.,* 304: 713–722.

Mechoulam, R., Ben-Shabat, S., Hanus, L., Ligumsky, M., Kaminski, N.E., Schatz, A.R., Gopher, A., Almog, S., Martin, B.R., Compton, D.R., Pertwee, R.G., Griffin, G., Bayewitch, M., Barg, J., and Vogel Z. (1995) Identification of an endogenous 2-monoglyceride, present in canine gut, that binds to cannabinoid receptors, *Biochem. Pharmacol.,* 50: 83–90.

Mechoulam, R. and Fride, E. (1995) in *Cannabinoid Receptors,* Pertwee R., Ed., Academic Press, London, pp. 233–58.

Melck, D., Bisogno, T., De Petrocellis, L. Chuang, H., Bifulco, M., Julius, D., and Di Marzo, V. (1999) Unsaturated long-chain N-acyl-vanillyl-amides (N-AVAMs): vanilloid receptor ligands that inhibit anandamide-facilitated transport and bind to CB_1 cannabinoid receptor, *Biochem. Biophys. Res. Commn.*, 262: 275–84.

Mezey, E., Toth, Z.E., Cortright, D.N., Arzubi, M.K., Krause, J.E., Elde, R., Guo, A., Blumberg, P.M., and Szallasi, A. (2000) Distribution of mRNA for vanilloid receptor subtype 1 (VR1), and VR1-like immunoreactivity, in the central nervous system of the rat and human, *Proc. Natl. Acad. Sci .U.S.A.*, 97: 3655–60.

Monory, K., Tzavara, E.T., Ledent, C., Parmentier, M., and Hanoune, J. (2001) Nonconventional cannabinoid stimulated [^{35}S]GTPγS binding in the CB1 knockout mice, *Int. Cannabinoid Res. Soc. Symposium*, p. 63.

Mukhopadhyay, S., Chapnick, B.M., and Howlett, A.C. (2002) Anandamide-induced vasorelaxation in rabbit aortic rings has two components: G protein dependent and independent, *Am. J. Physiol. Heart Circ. Physiol.*, 282: H2046–54.

Nemeth, J., Helyes, Z., Than, M., Jakab, B., Pinter, E., and Szolcsanyi, J. (2003) Concentration-dependent dual effect of anandamide on sensory neuropeptide release from isolated rat tracheae, *Neurosci. Lett.*, 336: 89–92.

Nieri, P., Martinotti, E., Testai, L., Adinolfi, B., Calderone, V., and Breschi, M.C. (2003) R(+)-methanandamide inhibits tracheal response to endogenously released acetylcholine via capsazepine-sensitive receptors, *Eur. J. Pharmacol.*, 459: 75–81.

Offertaler, L., Mo, F.M., Batkai, S., Liu, J., Begg, M., Razdan, R.K., Martin, B.R., Bukoski, R.D., and Kunos, G. (2003) Selective ligands and cellular effectors of a G protein-coupled endothelial cannabinoid receptor, *Mol. Pharmacol.*, 63: 699–705.

Olah, Z., Karai, L., and Iadarola, M.J. (2001) Anandamide activates vanilloid receptor 1 (VR1) at acidic pH in dorsal root ganglia neurons and cells ectopically expressing VR1, *J. Biol. Chem.*, 276: 31163–70.

Olah, Z., Szabo, T., Karai, L., Hough, C., Fields, R.D., Caudle, R.M., Blumberg, P.M., and Iadarola, M.J. (2001) Ligand-induced dynamic membrane changes and cell deletion conferred by vanilloid receptor 1, *J. Biol. Chem.*, 276: 11021–30.

O'Sullivan S., Kendall D., and Randall M. (2003) Vasorelaxant properties of the novel endocannabinoid N-arachidonoyl-dopamine (NADA), *Int. Cannabinoid Res. Soc. Symposium*, p. 13.

Oz, M., Ravindran, A., Diaz-Ruiz, O., Zhang, L., and Morales, M. (2003) The endogenous cannabinoid anandamide inhibits {alpha}7–nicotinic acetylcholine receptor-mediated responses in Xenopus oocytes, *J. Pharmacol. Exp. Ther.*, epub ahead of print.

Oz, M., Tchugunova, Y.B., and Dunn, S.M. (2000) Endogenous cannabinoid anandamide directly inhibits voltage-dependent Ca^{2+} fluxes in rabbit T-tubule membranes, *Eur. J. Pharmacol.*, 404: 13–20.

Palazzo, E., de Novellis, V., Marabese, I., Cuomo, D., Rossi, F., Berrino, L., Rossi, F., and Maione, S. (2002) Interaction between vanilloid and glutamate receptors in the central modulation of nociception, *Eur. J. Pharmacol.*, 439: 69–75.

Pertwee, R.G. (1999) Pharmacology of cannabinoid receptor ligands, *Curr. Med. Chem.*, 6: 635–64.

Pertwee, R.G. and Ross, R.A. (2002) Cannabinoid receptors and their ligands, *Prostaglandins Leukot. Essent. Fatty Acids*, 66: 101–21.

Poling, J.S., Rogawski, M.A., Salem, N. JR., and Vicini, S. (1996) Anandamide, an endogenous cannabinoid, inhibits Shaker-related voltage-gated K^+ channels. *Neuropharmacology*, 35: 983–91.

Porter, A.C., Sauer, J.M., Knierman, M.D., Becker, G.W., Berna, M.J., Bao, J., Nomikos, G.G., Carter, P., Bymaster, F.P., Leese, A.B., and Felder C.C. (2002) Characterization of a novel endocannabinoid, virodhamine, with antagonist activity at the CB1 receptor, *J. Pharmacol. Exp. Ther.*, 301:1020–4.

Premkumar, L.S. and Ahern, G.P. (2000) Induction of vanilloid receptor channel activity by protein kinase C, *Nature*, 408: 985–90.

Ralevic, V., Kendall D.A., Randall, M.D., and Smart, D. (2002) Cannabinoid modulation of sensory neurotransmission via cannabinoid and vanilloid receptors: roles in regulation of cardiovascular function, *Life Sci.*, 71: 2577–94.

Rathee, P.K., Distler, C., Obreja, O., Neuhuber, W., Wang, G.K., Wang, S.Y., Nau, C., and Kress, M. (2002) PKA/AKAP/VR-1 module: A common link of Gs-mediated signaling to thermal hyperalgesia, *J. Neurosci.*, 22: 4740–5.

Richardson, J.D., Aanonsen, L., and Hargreaves, K.M. (1998) Antihyperalgesic effects of spinal cannabinoids, *Eur. J. Pharmacol.*, 345: 145–53.

Roberts, L.A., Christie, M.J., and Connor, M. (2002) Anandamide is a partial agonist at native vanilloid receptors in acutely isolated mouse trigeminal sensory neurons, *Br. J. Pharmacol.*, 137: 421–8.

Rockwell, C. and Kaminski, N. (2003) Inhibition of interleukin-2 production by 2-arachidonoylglycerol is partially mediated by peroxisome proliferator activated receptor gamma (PPAγ), *Int. Cannabinoid Res. Soc. Symposium,* pp. 32.

Ross, R.A., Craib, S.J., Stevenson, L.A., Pertwee, R.G., Henderson, A., Toole, J., and Ellington, H.C. (2002) Pharmacological characterization of the anandamide cyclooxygenase metabolite: prostaglandin E2 ethanolamide, *J. Pharmacol. Exp. Ther.,* 301: 900–7.

Ross, R.A., Gibson, T.M., Brockie, H.C., Leslie, M., Pashmi, G., Craib, S.J., Di Marzo, V., and Pertwee, R.G. (2001) Structure-activity relationship for the endogenous cannabinoid, anandamide, and certain of its analogues at vanilloid receptors in transfected cells and vas deferens, *Br. J. Pharmacol.,* 132: 631–40.

Rukwied, R., Watkinson, A., McGlone, F., and Dvorak, M. (2003) Cannabinoid agonists attenuate capsaicin-induced responses in human skin, *Pain,* 102: 283–8.

Sagan, S., Venance, L., Torrens, Y., Cordier, J., Glowinski, J., and Giaume, C. (1999) Anandamide and WIN 55212-2 inhibit cyclic AMP formation through G-protein-coupled receptors distinct from CB1 cannabinoid receptors in cultured astrocytes, *Eur. J. Neurosci.,* 11: 691–9.

Sanchez, C., de Ceballos, M.L., del Pulgar, T.G., Rueda, D., Corbacho, C., Velasco, G., Galve-Roperh, I., Huffman, J.W., Ramon y Cajal, S., and Guzman, M. (2001) Inhibition of glioma growth in vivo by selective activation of the CB(2) cannabinoid receptor, *Cancer Res.,* 61: 5784–9.

Shimasue, K., Urushidani, T., Hagiwara, M., and Nagao, T. (1996) Effects of anandamide and arachidonic acid on specific binding of (+) -PN200-110, diltiazem and (-) -desmethoxyverapamil to L-type Ca^{2+} channel, *Eur. J. Pharmacol.,* 296: 347–50.

Smart, D., Gunthorpe, M.J., Jerman, J.C., Nasir, S., Gray, J., Muir, A.I., Chambers, J.K., Randall, A.D., and Davis, J.B. (2000) The endogenous lipid anandamide is a full agonist at the human vanilloid receptor (hVR1), *Br. J Pharmacol.,* 129: 227–30.

Smart, D., Jonsson, K.O., Vandevoorde, S., Lambert, D.M., and Fowler, C.J. (2002) 'Entourage' effects of N-acyl ethanolamines at human vanilloid receptors. Comparison of effects upon anandamide-induced vanilloid receptor activation and upon anandamide metabolism, *Br. J. Pharmacol.,* 136: 452–8.

Smith, G.D., Gunthorpe, M.J., Kelsell, R.E., Hayes, P.D., Reilly, P., Facer, P., Wright, J.E., Jerman, J.C., Walhin, J.P., Ooi, L., Egerton, J., Charles, K.J., Smart, D., Randall, A.D., Anand, P., and Davis, J.B. (2002) TRPV3 is a temperature-sensitive vanilloid receptor-like protein, *Nature,* 418: 186–90.

Smith, P.J. and McQueen, D.S. (2001) Anandamide induces cardiovascular and respiratory reflexes via vasosensory nerves in the anaesthetized rat, *Br. J. Pharmacol.,* 134: 655–63.

Souilhac, J., Poncelet, M., Rinaldi-Carmona, M., Le Fur, G., and Soubrie, P. (1995) Intrastriatal injection of cannabinoid receptor agonists induced turning behavior in mice, *Pharmacol. Biochem. Behav.,* 51: 3–7.

Sprague, J., Harrison, C., Rowbotham, D.J., Smart, D., and Lambert, D.G. (2001) Temperature-dependent activation of recombinant rat vanilloid VR1 receptors expressed in HEK293 cells by capsaicin and anandamide, *Eur. J. Pharmacol.,* 423: 121–5.

Sugiura, T., Kondo, S., Sukagawa, A., Nakane, S., Shinoda, A., Itoh, K., Yamashita, A., and Waku K. (1995) 2-Arachidonoylglycerol: a possible endogenous cannabinoid receptor ligand in brain, *Biochem. Biophys. Commn.,* 215: 89–97.

Szallasi, A. and Blumberg, P.M., (1999) Vanilloid (Capsaicin) receptors and mechanisms. *Pharmacol. Rev.* 51: 159–212.

Szallasi, A. and Di Marzo, V. (2000) New perspectives on enigmatic vanilloid receptors, *Trends Neurosci.,* 23: 491–97.

Szoke, E., Czeh, G., Szolcsanyi, J., and Seress, L. (2002) Neonatal anandamide treatment results in prolonged mitochondrial damage in the vanilloid receptor type 1-immunoreactive B-type neurons of the rat trigeminal ganglion, *Neuroscience,* 115: 805–14.

Tognetto, M., Amadesi, S., Harrison, S., Creminon, C., Trevisani, M., Carreras, M., Matera, M., Geppetti, P., and Bianchi, A. (2001) Anandamide excites central terminals of dorsal root ganglion neurons via vanilloid receptor-1 activation, *J. Neurosci.,* 21: 1104–9.

Tominaga, M., Caterina, M.J., Malmberg, A.B., Rosen, T.A., Gilbert, H., Skinner, K., Raumann, B.E., Basbaum, A.I., and Julius, D. (1998) The cloned capsaicin receptor integrates multiple pain-producing stimuli, *Neuron,* 21: 531–43.

Toth, A., Kedei, N., Wang Y., and Blumberg P.M. (2003) Arachidonyl dopamine as a ligand for the vanilloid receptor VR1 of the rat, *Life Sci.,* 73: 487–98.

Tucker, R.C., Kagaya, M., Page, C.P., and Spina, D. (2001) The endogenous cannabinoid agonist, anandamide stimulates sensory nerves in guinea-pig airways, *Br. J. Pharmacol.,* 132: 1127–35.

Undem, B.J. and Kollarik, M. (2002) Characterization of the vanilloid receptor 1 antagonist iodo-resiniferatoxin on the afferent and efferent function of vagal sensory C-fibers, *J. Pharmacol. Exp. Ther.,* 303: 716–22.

Van den Bossche, I. and Vanheel, B. (2000) Influence of cannabinoids on the delayed rectifier in freshly dissociated smooth muscle cells of the rat aorta, *Br. J. Pharmacol.,* 131: 85–93.

van der Stelt, M., van Kuik, J.A., Bari, M., van Zadelhoff, G., Leeflang, B.R., Veldink, G.A., Finazzi-Agro, A., Vliegenthart, J.F., and Maccarrone, M. (2002) Oxygenated metabolites of anandamide and 2-arachidonoylglycerol: conformational analysis and interaction with cannabinoid receptors, membrane transporter, and fatty acid amide hydrolase, *J. Med. Chem.,* 45: 3709–20.

Vandevoorde, S., Lambert, D.M., Smart, D., Jonsson, K.O., and Fowler, C.J. (2003) N-Morpholino- and N-diethyl-analogues of palmitoylethanolamide increase the sensitivity of transfected human vanilloid receptors to activation by anandamide without affecting fatty acid amidohydrolase activity, *Bioorg. Med. Chem.,* 11: 817–25.

Vanheel, B. and Van De Voorde, J. (2001) Regional differences in anandamide- and, methanandamide-induced membrane potential changes in rat mesenteric arteries, *J. Pharmacol. Exp. Ther.,* 296: 322–8.

Vellani, V., Mapplebeck, S., Moribondo, A., Davis, J.B., and McNaughton, P.A. (2001) Protein kinase C activation potentiates gating of the vanilloid receptor VR1 by capsaicin, protons, heat and anandamide, *J. Physiol.,* 534: 813–25.

Venance, L., Piomelli, D., Glowinski, J., and Giaume, C., (1995) Inhibition by anandamide of gap junctions and intercellular calcium signalling in striatal astrocytes, *Nature,* 372: 590–4.

Wagner, J.A., Varga, K., Jarai, Z., and Kunos, G. (1999) Mesenteric vasodilation mediated by endothelial anandamide receptors, *Hypertension,* 33: 429–34.

Walter, L., Franklin, A., Witting, A., Wade, C., Xie, Y., Kunos, G., Mackie, K., and Stella, N. (2003) Nonpsychotropic cannabinoid receptors regulate microglial cell migration, *J. Neurosci.,* 23: 1398–405.

Watanabe, H., Vriens, J., Prenen, J., Droogmans, G., Voets, T., and Nilius B. (2003) Anandamide and arachidonic acid use epoxyeicosatrienoic acids to activate TRPV4 channels. *Nature,* 424: 434–8.

White, R., Ho, W.S., Bottrill, F.E., Ford, W.R., and Hiley, C.R. (2001) Mechanisms of anandamide-induced vasorelaxation in rat isolated coronary arteries, *Br. J. Pharmacol.,* 134: 921–9.

Wiley, J.L., Dewey, M.A., Jefferson, R.G., Winckler, R.L., Bridgen, D.T., Willoughby, K.A., and Martin, B.R. (2000) Influence of phenylmethylsulfonyl fluoride on anandamide brain levels and pharmacological effects, *Life Sci.,* 67: 1573–83.

Woodward, D.F., Krauss, A.H., Chen, J., Gil, D.W., Kedzie, K.M., Protzman, C.E., Shi, L., Chen, R., Krauss, H.A., Bogardus, A., Dinh, H.T., Wheeler, L.A., Andrews, S.W., Burk, R.M., Gac, T., Roof, M.B., Garst, M.E., Kaplan, L.J., Sachs, G., Pierce, K.L., Regan, J.W., Ross, R.A., and Chan, M.F. (2000) Replacement of the carboxylic acid group of prostaglandin f(2alpha) with a hydroxyl or methoxy substituent provides biologically unique compounds, *Br. J. Pharmacol.,* 130: 1933–43.

Woodward, D.F., Krauss, A.H., Chen, J., Lai, R.K., Spada, C.S., Burk, R.M., Andrews, S.W., Shi, L., Liang, Y., Kedzie, K.M., Chen, R., Gil, D.W., Kharlamb, A., Archeampong, A., Ling, J., Madhu, C., Ni, J., Rix, P., Usansky, J., Usansky, H., Weber, A., Welty, D., Yang, W., Tang-Liu, D.D., Garst, M.E., Brar, B., Wheeler, L.A., and Kaplan, L.J. (2001) The pharmacology of bimatoprost (Lumigan), *Surv. Ophthalmol.,* 45 (Suppl. 4): S337–45.

Woodward, D.F., Krauss, A.H., Chen, J., Liang, Y., Li, C., Protzman, C.E., Bogardus, A., Chen, R., Kedzie, K.M., Krauss, H.A., Gil, D.W., Kharlamb, A., Wheeler, L.A., Babusis, D., Welty, D., Tang-Liu, D.D., Cherukury, M., Andrews, S.W., Burk, R.M., and Garst, M.E. (2003) Pharmacological characterization of a novel antiglaucoma agent, Bimatoprost (AGN 192024), *J. Pharmacol. Exp. Ther.,* 305: 772–85.

Yamaji, K., Pada Sarker, K., Kawahara, K., Iino, S., Yamakuchi, M., Abeyama, K., Hashiguchi, T., and Maruyama, I. (2003) Anandamide induces apoptosis in human endothelial cells: its regulation system and clinical implications, *Thromb. Haemost.,* 89: 875–84.

Zimmer, A., Zimmer, A.M., Hohmann, A.G., Herkenham, M., and Bonner, T.I. (1999) Increased mortality, hypoactivity, and hypoalgesia in cannabinoid CB1 receptor knockout mice, *Proc. Natl. Acad. Sci. U.S.A.,* 96: 5780–5.

Zygmunt, P.M., Petersson, J., Andersson, D.A., Chuang, H., Sorgard, M., Di Marzo, V., Julius, D., and Hogestatt, E.D. (1999) Vanilloid receptors on sensory nerves mediate the vasodilator action of anandamide, *Nature,* 400: 452–7.

Part III

Biochemistry of the Endocannabinoid System

7 Occurrence, Biosynthesis, and Metabolism of Endocannabinoids

Takayuki Sugiura, Saori Oka, Shinobu Ikeda, and Keizo Waku

CONTENTS

INTRODUCTION

Δ^9-THC (Figure 7.1) is a major psychoactive constituent of marijuana and is known to exhibit a variety of pharmacological activities. The mechanisms of action of Δ^9-THC had long remained obscure until the late 1980s. A breakthrough was then made by Devane et al. (1988). They provided clear evidence for the occurrence of specific binding sites for cannabinoids in rat brain synaptosomes using a radiolabeled synthetic cannabinoid, [^{3}H]CP55940. Thereafter, Matsuda et al. (1990) cloned a cDNA encoding a cannabinoid receptor (CB_1 receptor) from a rat brain cDNA library. The CB_1 receptor is present in various mammalian tissues, especially in the nervous tissues, and is assumed to be involved in the attenuation of neurotransmission (for reviews, see Pertwee, 1997; Di Marzo, Melck, et al., 1998; Schlicker and Kathmann, 2001; Pertwee and Ross, 2002; Lutz, 2002). The CB_1 receptor is a seven-transmembrane, G-protein-coupled receptor, which contains 472 (human) or 473 (rat) amino acids. Later, Munro et al. (1993) cloned a cDNA encoding another type of cannabinoid receptor (CB_2 receptor) from an HL60 cell cDNA library. The CB_2 receptor is also a seven-transmembrane,

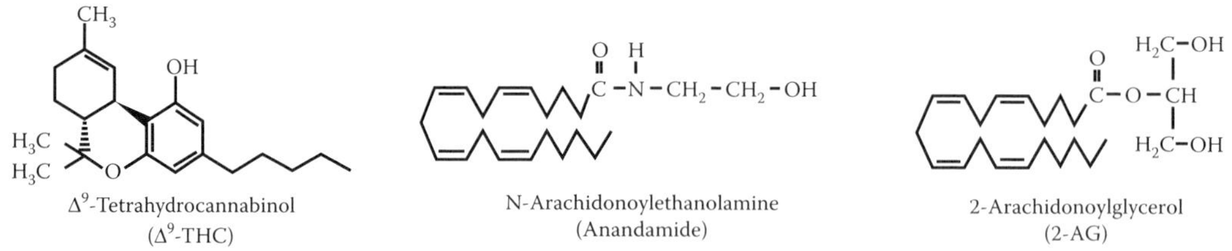

Δ^9-Tetrahydrocannabinol (Δ^9-THC)

N-Arachidonoylethanolamine (Anandamide)

2-Arachidonoylglycerol (2-AG)

FIGURE 7.1 Chemical structures of Δ^9-THC, anandamide, and 2-AG.

G-protein-coupled receptor, which consists of 360 amino acids. The CB_1 and CB_2 receptors share 44% overall identity (68% identity for the transmembrane domains). The CB_2 receptor is mainly present in the immune system (e.g., spleen and tonsil) and is supposed to be involved in the regulation of immune responses and/or inflammatory reactions (for reviews, see Pertwee, 1997; Berdyshev, 2000; Di Marzo, Melck, et al., 2000; De Petrocellis et al., 2000; Parolaro et al., 2002; Cabral, 2002; Roth et al., 2002), although details of the physiological functions of the CB_2 receptor still remain to be determined. In any case, it has generally been assumed that various cannabinoids including Δ^9-THC exert their biological activities mainly by acting on these cannabinoid receptors (CB_1 and CB_2).

The presence of specific binding sites for cannabinoids stimulated the search for endogenous ligands. Evans et al. (1992) demonstrated that an endogenous substance that binds to the cannabinoid receptor was released from the A23187-stimulated rat brain synaptosomes, although its chemical structure was not determined in their pioneering study. Finally, Devane et al. (1992) isolated a unique molecular species of *N*-acylethanolamine, i.e., *N*-arachidonoylethanolamine (Figure 7.1) from pig brain and reported that this compound is an endogenous cannabinoid receptor ligand. They demonstrated that this compound possesses strong binding activity toward the cannabinoid receptors and named it *anandamide*. It has already been demonstrated by numerous investigators that anandamide induces a variety of cannabimimetic activities *in vitro* and *in vivo* (for reviews, see Pertwee, 1997; Hillard and Campbell, 1997; Felder and Glass, 1998; Di Marzo, 1998b; Mechoulam, Fride, and Di Marzo, 1998; Piomelli et al., 1998; Hillard, 2000; Piomelli et al., 2000; Di Marzo, De Petrocellis, Bisogno, et al., 2002; Di Marzo, De Petrocellis, Fezza, et al., 2002; Schmid, H.H.O., et al., 2002), although anandamide acted as a partial agonist at the cannabinoid receptors in many cases. Several investigators recently provided evidence indicating that anandamide also interacts with binding sites other than the cannabinoid receptors, such as the vanilloid receptors (e.g., Zygmunt et al., 1999; Smart et al., 2000; De Petrocellis, Bisogno, et al., 2001; De Petrocellis, Harrison, et al., 2001; Ross et al., 2001; Tognetto et al., 2001; Olah et al., 2001; for reviews, see Di Marzo, De Petrocellis, Bisogno, et al., 2002; Di Marzo, De Petrocellis, Fezza, et al., 2002). Thus, it is possible that some of the actions of anandamide are mediated through receptors or binding sites other than the cannabinoid receptors.

On the other hand, we isolated 2-AG (Figure 7.1) from rat brain, and found that it acts as an endogenous cannabinoid receptor ligand (Sugiura et al., 1995). Mechoulam et al. (1995) also isolated 2-AG from the canine gut as an endogenous cannabinoid receptor ligand. It has been shown that 2-AG binds to the cannabinoid receptors (CB_1 and CB_2) and exhibits several cannabimimetic activities *in vitro* and *in vivo* (for reviews, see Di Marzo, 1998a, 1998b; Piomelli et al., 1998; Mechoulam, Fride, and Di Marzo, 1998; Hillard, 2000; Piomelli et al., 2000; Sugiura and Waku, 2000, 2002; Sugiura et al., 2002). Notably, 2-AG was found to act as a full agonist at the cannabinoid receptors (Sugiura, Kodaka, et al., 1996; Sugiura, Kodaka, Kondo, Nakane, et al., 1997; Sugiura, Kodaka, et al., 1999; Sugiura, Kondo, et al., 2000; Gonsiorek et al., 2000; Hillard, 2000). 2-AG is the most efficacious naturally occurring agonist so far examined. Recently, the physiological significance of 2-AG has received increasing attention, and evidence is gradually accumulating which indicates that 2-AG is a physiologically essential molecule (see Sugiura and Waku, 2000, 2002; Sugiura et al., 2002, for reviews).

In this review, we focused on the tissue levels, biosynthesis, and metabolism, as well as the possible physiological significance of these endocannabinoids in mammalian tissues.

ANANDAMIDE

IDENTIFICATION OF ANANDAMIDE AS AN ENDOGENOUS CANNABINOID RECEPTOR LIGAND

Devane et al. (1992) isolated anandamide from pig brain lipid extracts by column chromatography, TLC, and HPLC. The structure of anandamide was confirmed by GC/MS and ^{1}H-NMR. They demonstrated that anandamide binds to the brain cannabinoid receptor with a high affinity (K_i = 52 nM) and induces inhibition of the mouse twitch response. They also showed that anandamide elicits reduced spontaneous motor activities, immobility, hypothermia, and analgesia when administered to mice (Fride and Mechoulam, 1993). These observations were later confirmed by a number of investigators (e.g., Smith et al., 1994; Crawley et al., 1993). To date, it has been shown that anandamide exhibits diverse biological activities *in vitro* and *in vivo*, such as the inhibition of adenylyl cyclase, the inhibition of voltage-gated Ca^{2+} channels, the activation of an inwardly rectifying K^+ current, the inhibition of neurotransmitter release, the stimulation of [^{35}S]GTPγS binding to G proteins, the activation of various kinases such as the p42/44 mitogen-activated protein kinase (MAPK), the p38 MAPK, the c-Jun N-terminal kinase, the neural form of focal adhesion kinase and protein kinase B, release of arachidonic acid, Ca^{2+} transients, reduction of gap junction permeability, inhibition of long-term potentiation, stimulation of constitutive nitric oxide synthase, inhibition of inducible nitric oxide synthase, apoptosis, inhibition of sperm acrosomal reaction, inhibition of the growth of human breast and prostate cancer cells, the acceleration of the growth of various mouse hematopoietic cell lines, stimulation of appetite, inhibition of memory, modulation of pain, vasodilation, hypotension, hemodynamic changes, and neuroprotection (for reviews, see Pertwee, 1997; Felder and Glass, 1998; Di Marzo, 1998b; Mechoulam, Fride, and Di Marzo, 1998; Piomelli et al., 1998; Hillard, 2000; Piomelli et al., 2000; Di Marzo, De Petrocellis, Bisogno, et al., 2002; Schmid, H.H.O., et al., 2002), although it is apparent that some of these actions of anandamide are not mediated via the cannabinoid receptors.

In any case, the discovery of anandamide was a breakthrough in the field of cannabinoid research. It was surprising that anandamide exhibited pharmacological activities similar to those of cannabinoids, despite the fact that their chemical structures are quite different. The presence of arachidonic acid as the constituent of the fatty acyl moiety is crucially important. *N*-acylethanolamines containing saturated monoenoic and dienoic fatty acids such as palmitic acid, stearic acid, and linoleic acid did not show any cannabimimetic activities. On the other hand, various *N*-acylethanolamines containing trienoic, tetraenoic, pentaenoic, and hexanoic fatty acids were shown to exhibit cannabimimetic activities comparable to or weaker than those of anandamide. Several investigators have provided evidence indicating that *N*-acylethanolamines containing C20 and C22 polyunsaturated fatty acids other than arachidonic acid, such as eicosa-8,11,14-trienoic acid (Hanus et al., 1993), eicosa-5,8,11,14,17-pentaenoic acid (Berger et al., 2001), docosa-7,10,13,16-tetraenoic acid (Hanus et al., 1993; Bisogno, Delton-Vandenbroucke, et al., 1999; Berger et al., 2001), docosa-4,7,10,13,16-pentaenoic acid (Sugiura, Kondo, Sukagawa, Tonegawa, Nakane, Yamashita, and Waku, 1996; Kondo, Sugiura, et al., 1998), docosa-7,10,13,16,19-pentaenoic acid (Berger et al., 2001), and docosa-4,7,10,13,16,19-hexanoic acid (Sugiura, Kondo, Sukagawa, Tonegawa, Nakane, Yamashita, Ishima, and Waku, 1996; Bisogno, Delton-Vandenbroucke, et al., 1999; Berger et al., 2001), are present in small amounts in several mammalian tissues.

Tissue Levels of Anandamide

Schmid et al. (1995) reported the results of a GC/MS analysis of *N*-acylethanolamines obtained from the brain. They demonstrated the occurrence of anandamide in the brains of several animal species. The level of anandamide in sheep brain was negligible, whereas low levels were found in pig and cow brains (17 and 10 pmol/g tissue, respectively). These values were, however, significantly lower than the levels of anandamide in pig brain (370 pmol/g tissue) reported by Devane et al. (1992). There is a possibility that a considerable amount of anandamide was generated during the postmortem period in the latter case. Schmid et al. (1995) demonstrated that the levels of anandamide in the isolated pig brain were dramatically augmented when kept at ambient temperature. Kempe et al. (1996) and Felder et al. (1996) also demonstrated the generation of anandamide in the brain during the postmortem period. We detected a small amount of anandamide in fresh rat brain (4.3 pmol/g tissue), which accounts for only 0.7% of the total *N*-acylethanolamine (Sugiura, Kondo, Sukagawa, Tonegawa, Nakane, Yamashita, Ishima, and Waku, 1996). The predominant species of *N*-acylethanolamine detected in fresh rat brain were the *N*-palmitoyl (50.6%), *N*-stearoyl (19.4%), *N-cis*-vaccenoyl (12.6%), and *N*-oleoyl (12.2%) species.

The levels of anandamide in the brains of various mammalian species have also been reported by a number of investigators. The levels of anandamide in the rat brain were 3.4 to 15 pmol/g tissue and those in the mouse brain were 10 to 15 pmol/g tissue (Table 7.1). Anandamide was detected in various regions of the human and rat brains, such as the cortex, limbic forebrain, hippocampus, thalamus, hypothalamus, diencephalon, striatum, mesencephalon, cerebellum, brainstem, medulla, and pituitary (Table 7.1). Anandamide was also detected in the male and female mice spinal cord (27.6 pmol/g tissue and 20.2 pmol/g tissue, respectively) (Di Marzo, Breivogel, et al., 2000). Di Marzo, Hill, et al. (2000) reported that the levels of anandamide in the rat substantia nigra and globus pallidus were several times higher than those in other brain regions. On the other hand, Gonzalez et al. (1999) reported that anandamide was detected in the rat anterior pituitary and hypothalamus, although the levels were below the lower quantifiable amount.

Notably, the level of anandamide varied depending on the physiological and pathophysiological conditions of the brains. Berger et al. (2001) demonstrated that the levels of anandamide in several brain regions were elevated in piglets fed an arachidonic-acid-containing diet. There are several reports describing the changes in the levels of anandamide in the hypothalamus. The level of anandamide in the hypothalamus peaks immediately before the onset of puberty in female rats (Wenger et al., 2002). Di Marzo et al. (2001) found that the administration of leptin reduced the levels of anandamide in the hypothalamus. Interestingly, Matias et al. (2003) recently reported that maternal undernutrition resulted in a decrease in the hypothalamic level of anandamide. On the other hand, Giuffrida et al. (1999) reported that a D_2-like dopamine receptor activation induces the release of anandamide in the dorsal striatum of freely moving rats. They also showed that the levels of anandamide in the cerebrospinal fluid of schizophrenia patients are higher than those of normal subjects (Leweke et al., 1999). As for neurodegenerative diseases, Di Marzo, Hill, et al. (2000) investigated the levels of anandamide in various brain regions in a rat model of Parkinson's disease induced by the injection of reserpine. They observed increases in the amounts of anandamide in the substantia nigra and striatum, although these changes were not statistically significant. Recently, Maccarrone, Gubellini, et al. (2003) and Gubellini et al. (2002) reported that the striatal level of anandamide was increased in a rat model of Parkinson's disease induced by a unilateral nigral lesion with 6-hydroxydopamine. On the other hand, Lastres-Becker et al. (2001) detected a decreased amount of anandamide in the striatum and an increased amount of anandamide in the mesencephalon in a rat model of Huntington's disease induced by the intrastriatal injection of 3-nitropropionic acid. Baker et al. (2001) reported increased levels of anandamide in the brain and spinal cord in spastic mice with chronic relapsing experimental allergic encephalomyelitis, a model of multiple sclerosis.

Interestingly, Di Marzo, Breivogel, et al. (2000) demonstrated that the levels of anandamide in the hippocampus and striatum in CB_1-receptor-knockout mice were lower than those in the

TABLE 7.1
Tissue Levels of Anandamide

Pig brain	17 pmol/g tissue	Schmid et al. (1995)
Cow brain	10 pmol/g tissue	Schmid et al. (1995)
Rat brain	4.3 pmol/g tissue	Sugiura et al. (1996)
Rat brain	11 pmol/g tissue	Cadas et al. (1997)
Rat brain	15 pmol/g tissue	Berrendero et al. (1999)
Rat brain	3.4 pmol/g tissue	Arai et al. (2000)
Rat brain	90000 pmol/g tissue	Maccarrone, Attina, et al. (2001)
Mouse brain	10–15 pmol/g tissue	Schmid, Paria, et al. (1997)
Rat cortex	13.6 pmol/g tissue	Bisogno, Berrendero, et al. (1999)
Rat cortex	11.0–16.7 pmol/g tissue	Gonzalez et al. (2002)
Rat cortex	32 pmol/g tissue	Hansen, Ikonomidou, et al. (2001)
Rat limbic forebrain	28.1 pmol/g tissue	Bisogno, Berrendero, et al. (1999)
Rat limbic forebrain	32.6–34.1 pmol/g tissue	Gonzalez et al. (2002)
Human hippocampus	148 pmol/g tissue	Felder et al. (1996)
Rat hippocampus	29 pmol/g tissue	Felder et al. (1996)
Rat hippocampus	32.9–76.2 pmol/g tissue	Koga et al. (1997)
Rat hippocampus	45.8 pmol/g tissue	Bisogno, Berrendero, et al. (1999)
Rat hippocampus	78.9–87.6 pmol/g tissue	Gonzalez et al. (2002)
Rat thalamus	20 pmol/g tissue	Felder et al. (1996)
Rat hypothalamus	9 pmol/g tissue	Di Marzo et al. (2001)
Rat hypothalamus	0.7 pmol/mg of lipid extract	Matias et al. (2003)
Rat diencephalon	10.2 pmol/g tissue	Bisogno, Berrendero, et al. (1999)
Rat diencephalon	14.3–15.3 pmol/g tissue	Gonzalez et al. (2002)
Rat striatum	51.5 pmol/g tissue	Bisogno, Berrendero, et al. (1999)
Rat striatum	62.1–85.6 pmol/g tissue	Gonzalez et al. (2002)
Rat striatum	250 pmol/mg protein	Gubellini et al. (2002)
Rat mesencephalon	30.2 pmol/g tissue	Bisogno, Berrendero, et al. (1999)
Rat midbrain	33.8–49.5 pmol/g tissue	Gonzalez et al. (2002)
Human cerebellum	25 pmol/g tissue	Felder et al. (1996)
Rat cerebellum	14.8 pmol/g tissue	Bisogno, Berrendero, et al. (1999)
Rat cerebellum	10.5–73.1 pmol/g tissue	Koga et al. (1997)
Rat cerebellum	12.9–15.3 pmol/g tissue	Gonzalez et al. (2002)
Rat brainstem	87 pmol/g tissue	Bisogno, Berrendero, et al. (1999)
Rat brainstem	20.2–24.0 pmol/g tissue	Gonzalez et al. (2002)
Rat medulla	44.9 pmol/g tissue	Bisogno, Berrendero, et al. (1999)
Human pituitary	14.6–28.1 pmol/g tissue	Pagotto et al. (2001)
Human eye (neurosensory retina)	4.48 pmol/g tissue	Stamer et al. (2001)
(ciliary process)	49.42 pmol/g tissue	Stamer et al. (2001)
(trabecular meshwork)	3.08 pmol/g tissue	Stamer et al. (2001)
Mouse spinal cord (male)	27.6 pmol/g tissue	Di Marzo, Breivogel, et al. (2000)
Bovine retina	64 pmol/g tissue	Bisogno, Delton-Vandenbroucke, et al. (1999)
Human heart	10 pmol/g tissue	Felder et al. (1996)
Rat liver	19.7–77.1 pmol/g tissue	Koga et al. (1997)
Human spleen	15 pmol/g tissue	Felder et al. (1996)
Rat spleen	6 pmol/g tissue	Felder et al. (1996)
Rat spleen	0.34 pmol/ μmol lipid P	Schmid et al. (2000)
Rat kidney	63.7–164 pmol/g tissue	Koga et al. (1997)
Rat kidney	8 pmol/g tissue	Deutsch et al. (1997)
Rat kidney	0.32–0.35 pmol/μmol lipid P	Schmid et al. (2000)
Rat thymus	40.6–137 pmol/g tissue	Koga et al. (1997)

(*continued*)

TABLE 7.1
Tissue Levels of Anandamide (Continued)

Rat testis	6 pmol/g tissue	Sugiura, Kondo, et al. (1996)
Rat testis	2.9–43.5 pmol/g tissue	Koga et al. (1997)
Rat testis	5 pmol/g tissue	Kondo, Sugiura, et al. (1998)
Rat testis	0.25–0.31 pmol/μmol lipid P	Schmid et al. (2000)
Mouse uterus (periimplantation, day 1)	289 pmol/μmol lipid P	Schmid, Paria, et al. (1997)
Mouse small intestine (normal)	25 pmol/g tissue	Mascolo et al. (2002)
(ileus)	56.3 pmol/g tissue	Mascolo et al. (2002)
Rat skin	23 pmol/g tissue	Felder et al. (1996)
Rat skin	49 pmol/g tissue	Calignano et al. (1998)
Rat skin	0.69 pmol/mg of lipid extract	Beaulieu et al. (2000)
Rat blood	0.69 pmol/ml	Burstein et al. (2002)
Rat plasma (collected by decapitation)	144 pmol/ml	Giuffrida, Rodriguez de Fonseca, et al. (2000)
(collected by cardiac puncture)	3.1 pmol/ml	Giuffrida, Rodriguez de Fonseca, et al. (2000)
Human sera (normal)	4 pmol/ml	Wang et al. (2001)
(endotoxin shock)	18 pmol/ml	Wang et al. (2001)
Human cerebrospinal fluid	0.30 pmol/ml	Leweke et al. (1999)
Human milk	0.15 pmol/ml	Fride et al. (2001)
Human seminal plasma	12.1 pmol/ml	Schuel et al. (2002)
Human mid-cycle oviductal fluid	10.7 pmol/ml	Schuel et al. (2002)
Human follicular fluid	2.9 pmol/ml	Schuel et al. (2002)
Human prostate tumor	18.3–77.4 pmol/g tissue	Schmid, Wold, Krebsbach, et al. (2002)
Benign prostate	9.65–23.4 pmol/g tissue	Schmid, Wold, Krebsbach, et al. (2002)
Human bladder tumor	7.40 pmol/g tissue	Schmid, Wold, Krebsbach, et al. (2002)
Benign bladder	4.82 pmol/g tissue	Schmid, Wold, Krebsbach, et al. (2002)
Human left anterior thigh tumor	9.19 pmol/g tissue	Schmid, Wold, Krebsbach, et al. (2002)
Benign thigh muscle	1.92 pmol/g tissue	Schmid, Wold, Krebsbach, et al. (2002)
Human endometrial tumor	11.86 pmol/g tissue	Schmid, Wold, Krebsbach, et al. (2002)
Benign cervix	3.86 pmol/g tissue	Schmid, Wold, Krebsbach, et al. (2002)
Human stomach tumor	8.41 pmol/g tissue	Schmid, Wold, Krebsbach, et al. (2002)
Benign stomach	5.71 pmol/g tissue	Schmid, Wold, Krebsbach, et al. (2002)
Human terminal ileum tumor	5.64 pmol/g tissue	Schmid, Wold, Krebsbach, et al. (2002)
Benign ileum	5.24 pmol/g tissue	Schmid, Wold, Krebsbach, et al. (2002)

wild-type control mice, although Maccarrone, Attina, Bari, et al. (2001) did not find any difference in the amounts of anandamide in the hippocampus and cortex between control and CB_1-receptor-knockout mice. Di Marzo, Berrendero, et al. (2000) reported that the levels of anandamide in the limbic forebrain of Δ^9-THC-tolerant rats were higher than those of normal rats, and the levels of anandamide and 2-AG in the striatum of the Δ^9-THC-tolerant rats were lower than those of normal animals. Recently, Gonzalez et al. (2002) also reported changes in the levels of anandamide (increase or decrease) in various brain regions of the rat following chronic exposure to nicotine, ethanol, and cocaine.

Anandamide was detected in various peripheral tissues as well, such as the bovine retina, the human heart, human spleen, rat heart, rat liver, rat spleen, rat kidney, rat thymus, rat testis, mouse uterus, rat skin, and rat paw skin (Table 7.1). Schmid, Kuwae, et al. (1997) reported that mouse peritoneal macrophages contain a small amount of anandamide, and Matias et al. (2002) demonstrated that human dendritic cells contain both anandamide and 2-AG. Several human reproductive fluids such as seminal plasma, mid-cycle oviductal fluid, follicular fluid, and amniotic fluid were shown to contain measurable amounts of anandamide (Schuel et al., 2002). A small amount of anandamide was also found in human milk (Fride et al., 2001). In addition, anandamide was detected in rat plasma collected by decapitation or by cardiac puncture and in the sera from normal donors and patients with endotoxin shock (Wang et al., 2001). Giuffrida, Rodriguez de Fonseca, Nava, et al. (2000) reported that the administration of AM404, an anandamide transport inhibitor, to rats induced the elevation of the plasma anandamide level. On the other hand, Yang et al. (1999) described that the levels of anandamide in the rat brain, spleen, testis, liver, lung, and heart were below the detection limit (<0.1 pmol/mg protein). Anandamide was also detected in several tumor cells. Bisogno et al. (1998) detected appreciable amounts of anandamide in human breast cancer EFM-19 cells and rat pheochromocytoma PC-12 cells. Schmid, P.C., et al. (2002) reported that there is variation in the contents of *N*-acylethanolamines including anandamide in various human tumors; the levels of *N*-acylethanolamines were 3.88 to 254.5 pmol/μmol lipid P, with anandamide representing 1.5 to 48% of total *N*-acylethanolamines. Previously, Maccarrone, Attina, Cartoni, et al. (2001) demonstrated that high amounts of anandamide are present in the tumoral human brain (meningioma), human neuroblastoma CHP100 cells, and lymphoma U937 cells. However, Schmid, H.H.O. et al., (2002) recently pointed out that these values might be overestimated, because the membranes were obtained by centrifugation following the addition of an internal standard to the homogenates.

Very recently, the occurrence of several novel analogs of anandamide was reported; these are virodhamine and *N*-arachidonoyldopamine. Porter et al. (2002) demonstrated that an ester-linked isomer of anandamide is present in the rat brain and human hippocampus. They named this compound virodhamine. Virodhamine was shown to act as an antagonist or a partial agonist toward the CB_1 receptor and as a full agonist toward the CB_2 receptor. Microdialysates of the rat striatum contained 60 fmol/ml of virodhamine, whereas the level of anandamide was below the detection limit (<8.7 fmol/ml). *N*-arachidonoyldopamine was isolated by Huang et al. (2002) from the bovine brain (on the order of pmol/g tissue). *N*-arachidonoyldopamine binds to both the cannabinoid receptors and the vanilloid receptor and induces analgesia upon systemic administration and hyperalgesia when intradermally injected (for a review, see Huang et al., 2002; Walker et al., 2002). It is not clear at present, however, whether these anandamide analogs, virodhamine and *N*-arachidonoyldopamine, actually play specific physiological roles as signaling molecules in living animals. Further investigation is required to answer this question. On the other hand, Rodriguez de Fonseca et al. (2001) recently demonstrated that *N*-oleoylethanolamine, which is known to be present in the brain as mentioned before, induces anorexia in rats. Yet, additional studies appear to be necessary to clarify the mechanism of action and to examine whether this is also the case in various other animal species.

Biosynthesis of Anandamide

From the late 1970s to the early 1980s, Schmid and co-workers (for reviews, see Schmid et al., 1990, 1996, Schmid, H.H.O., et al., 2002; Schmid, 2000; Schmid and Berdyshev, 2002) found that large amounts of *N*-acylethanolamines were produced in degenerating tissues such as infarcted hearts and ischemic brains. In these studies, they reported the generation of saturated monoenoic and dienoic species of *N*-acylethanolamines in several mammalian tissues. However, they did not describe the generation of tetraenoic species, i.e., anandamide. This is because these studies were conducted before the discovery of anandamide. The generation of anandamide in stimulated cells was first reported by Di Marzo et al. (1994). They demonstrated that rat brain neurons generated anandamide when stimulated with ionomycin or with several membrane-depolarizing agents such as kainate, high K^+, and 4-aminopyridine. Di Marzo and co-workers also demonstrated the generation of anandamide in ionomycin-treated J774 macrophages (Di Marzo, De Petrocellis, Sepe, et al., 1996; Bisogno, Maurelli, et al., 1997), ionomycin-treated RBL-2H3 cells (Bisogno, Maurelli, et al., 1997), and phospholipase-D-treated N18TG2 neuroblastoma cells (Di Marzo, De Petrocellis, Sepe, et al., 1996). On the other hand, Hansen et al. (1995) investigated the generation of *N*-acylethanolamine in glutamate- or A23187-stimulated mouse cortical neurons in culture. They reported that the generation of *N*-acylethanolamine occurs in stimulated cells prelabeled with [^{3}H]ethanolamine. However, they failed to detect the generation of anandamide when the cells were prelabeled with [^{3}H]arachidonic acid. Later, they succeeded in detecting the formation of anandamide in NaN_3-treated neurons (Hansen et al., 1997; Hansen, H.H., et al., 2000). The generation of anandamide has also been detected in Δ^9-THC-stimulated N18TG2 cells (Burstein and Hunter, 1995), in ionomycin-stimulated rat macrophages (Wagner et al., 1997), in LPS-, platelet-activating factor-, and Δ^9-THC-stimulated RAW264.7 mouse macrophages (Pestonjamasp and Burstein, 1998), in *N*-arachidonoylglycine-stimulated RAW264.7 cells (Burstein et al., 2002), in mouse peritoneal macrophages in culture supplemented with ethanolamine (Kuwae et al., 1999), in rat testis following the injection of cadmium chloride (Kondo, Sugiura, et al., 1998), in ratridine-, 4-aminopyridine-, and A23187-stimulated SK-N-SH neuroblastoma cells (Basavarajappa and Hungund, 1999), in the periaqueductal gray region of the rat brain following electrical stimulation and the subcutaneous injection of formalin (Walker et al., 1999), in rat brain injected intracerebrally with *N*-methyl-D-aspartate (NMDA) (Hansen, Ikonomidou, et al., 2001; Hansen, Schmid, et al., 2001), in rat cortical neurons following simultaneous activation of the NMDA receptor and the acetylcholine receptor (Stella and Piomelli, 2001), in capsaicin-stimulated rat sensory neurons (Ahluwalia et al., 2003), in the medial prefrontal cortex of mice reexposed to the tone 24 h after conditioning (Marsicano et al., 2002), and in *Clostridium difficile* toxin-A-treated rat ileum (McVey et al., 2003). Walter et al. (2002) also demonstrated that the level of anandamide was augmented in endothelin-1-stimulated mouse astrocytes. On the other hand, Berdyshev et al. (2001) reported that the treatment of human platelets or P388D1 macrophages with platelet-activating factor did not affect the cellular levels of anandamide. Beaulieu et al. (2000) also reported that there was no significant difference between the levels of anandamide in the control rat paw skin and inflamed paw skin.

There are two independent pathways for the biosynthesis of anandamide (for reviews, see Hillard and Campbell, 1997; Di Marzo, 1998b; Piomelli et al., 1998; Di Marzo, De Petrocellis, et al., 1999; Sugiura et al., 2002; Di Marzo, De Petrocellis, Bisogno, et al., 2002; Schmid, H.H.O., et al., 2002). The first pathway is the direct *N*-acylation of ethanolamine (the condensation pathway). The second pathway is the synthesis through the combined actions of a transacylase and a phosphodiesterase (Schmid pathway) (Figure 7.2). In the 1960s, Udenfriend and co-workers (Colodzin et al., 1963; Bachur and Udenfriend, 1966) demonstrated that *N*-acylethanolamine can be enzymatically formed from free fatty acids and ethanolamine, although they did not mention the case of arachidonic acid. The enzymatic formation of anandamide (*N*-arachidonoylethanolamine) from free arachidonic acid and ethanolamine was first reported by Deutsch and Chin (1993). Several investigators have also demonstrated that anandamide can be enzymatically formed from free arachidonic acid and ethanolamine (Devane and Axelrod, 1994; Kruszka and Gross, 1994; Ueda, Kurahashi, et al., 1995). We also confirmed that various fatty acids, including arachidonic acid, act as substrates in the formation of *N*-acylethanolamines through this pathway in rat brain microsomes (Sugiura, Kondo, Sukagawa, Tonegawa, Nakane, Yamashita, Ishima, and Waku, 1996). Notably, the fatty acid specificity of this enzyme reaction is not high. The decapitated rat brain contains a substantial amount of unesterified arachidonic acid in addition to other fatty acids (Sugiura, Kondo, Sukagawa, Tonegawa, Nakane, Yamashita, Ishima, and Waku, 1996). However, the fatty acid profile of the *N*-acyl moiety of *N*-acylethanolamine is considerably different from that of the free fatty acids (Figure 7.3). Furthermore, high amounts of substrates, especially ethanolamine, are required to form anandamide through this pathway (Sugiura, Kondo, Sukagawa, Tonegawa, Nakane, Yamashita, Ishima, and Waku, 1996; Devane and Axelrod, 1994; Kruszka and Gross, 1994; Ueda, Kurahashi, et al., 1995; Katayama et al., 1997; Kurahashi et al., 1997; Katayama et al., 1999; Schmid et al., 1998). It is now believed that the formation of anandamide through this condensation pathway is catalyzed by a FAAH operating in reverse (Ueda, Kurahashi, et al., 1995; Kurahashi et al., 1997). Therefore, it seems likely that the formation of anandamide through this pathway is not physiologically relevant, although there remains the possibility that a significant amount of anandamide can be formed via this pathway if high concentrations of arachidonic acid and ethanolamine are colocalized at some sites within the

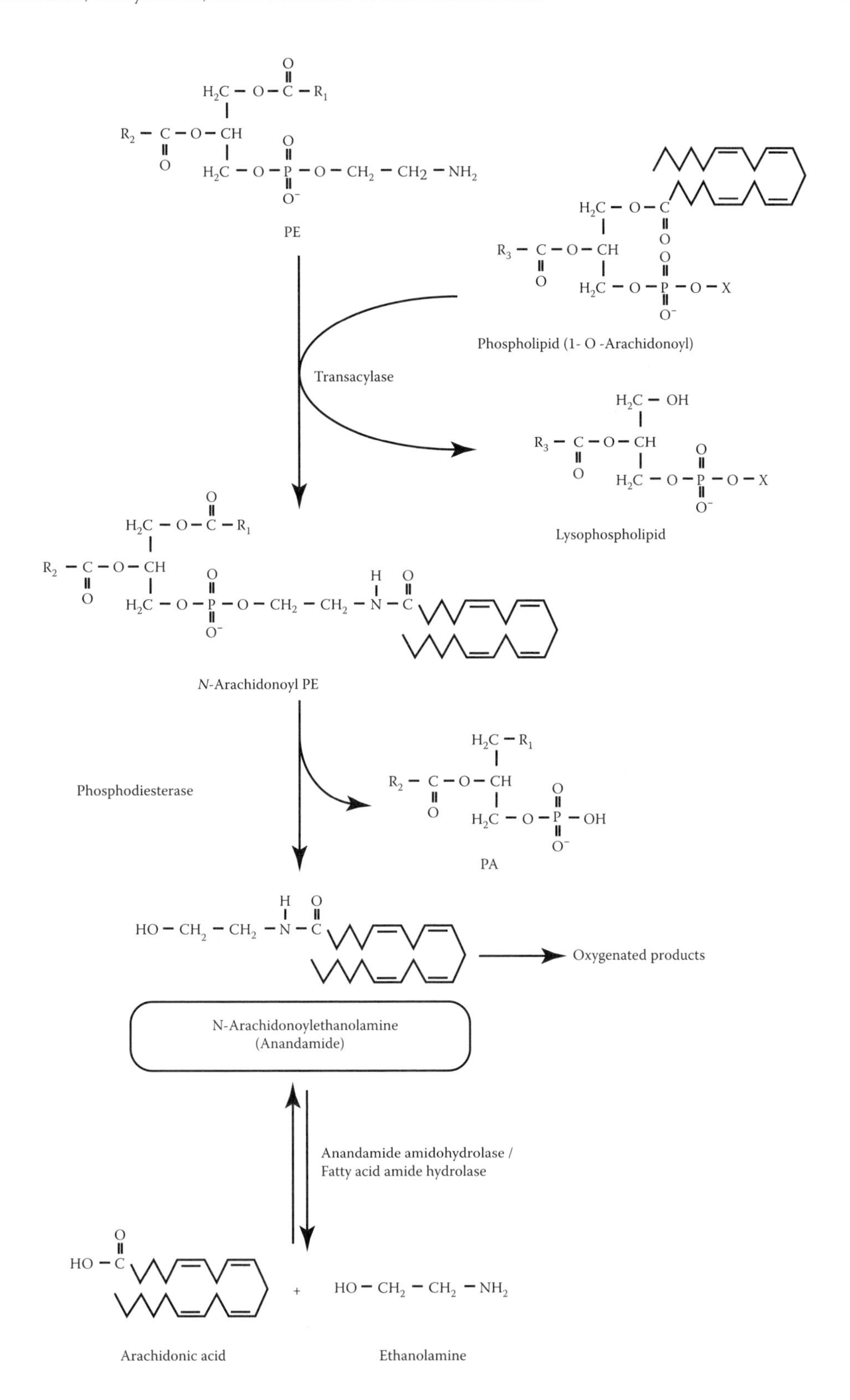

FIGURE 7.2 Pathways for the biosynthesis and degradation of anandamide.

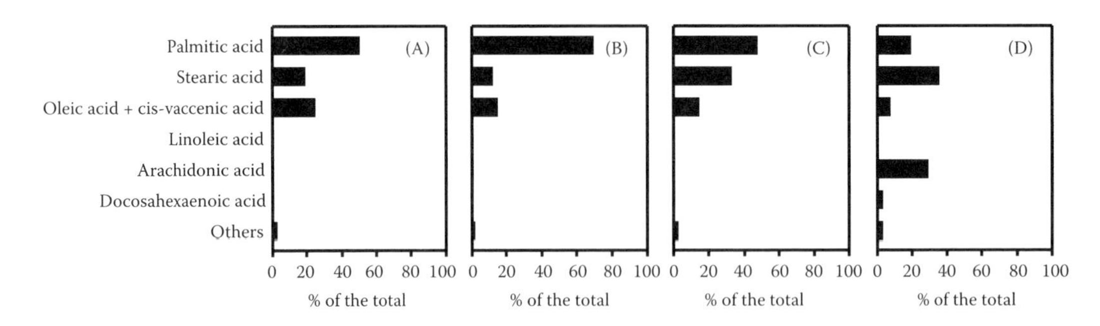

FIGURE 7.3 Comparison of the *N*-acyl moieties of *N*-acylethanolamines (A), *N*-acyl moieties of *N*-acyl PEs (B), fatty acids esterified at the 1-position of phosphatidylcholine (C) and free fatty acids (D) in rat brain. Values are the means of four to six determinations. (From Sugiura, T., Kondo, S., Sukagawa, A., Tonegawa, T., Nakane, S., Yamashita, A., Ishima, Y., and Waku, K. (1996) Transacylase-mediated and phosphodiesterase-mediated synthesis of *N*-arachidonoylethanolamine, an endogenous cannabinoid receptor ligand, in rat brain microsomes. Comparison with synthesis from free arachidonic acid and ethanolamine, *European Journal of Biochemistry* **240**: 53–62. With permission.)

cell. In any case, it is reasonable to consider that a large portion, if not all, of *N*-acylethanolamine is produced through pathways other than the condensation pathway mentioned earlier.

The second pathway for the biosynthesis of anandamide is the formation from preexisting *N*-arachidonoyl PE through the action of a phosphodiesterase (Figure 7.2). This enzyme reaction has long been assumed to be the major synthetic route for various *N*-acylethanolamines such as *N*-palmitoyl- and *N*-stearoyl-ethanolamine in mammalian tissues (for reviews, see Schmid et al., 1990, 1996, Schmid, H.H.O., et al., 2002; Schmid, 2000; Schmid and Berdyshev, 2002). Schmid and co-workers (Epps et al., 1979, 1980; Natarajan et al., 1982, 1983, 1986; Berdyshev et al., 2000) have demonstrated the accumulation of various species of *N*-acyl PE in addition to *N*-acylethanolamines in several degenerating or injured tissues and cells. They found a phosphodiesterase (phospholipase-D-type) activity catalyzing the formation of *N*-acylethanolamine from the corresponding *N*-acyl PE (Schmid et al., 1983). The properties of this enzyme reaction were further investigated by the same group. The addition of Triton X-100 stimulated the enzyme activity, whereas sodium dodecyl sulfate and alkyltrimethylammonium bromide were inhibitory (Schmid et al., 1983). Ca^{2+} above 1 m*M* was inhibitory, whereas Ca^{2+} up to 0.5 m*M* slightly stimulated the enzyme activity. In the absence of detergents, *N*-acyl lysoPE and glycerophospho (*N*-acyl)ethanolamine acted as better substrates than *N*-acyl PE. However, they did not examine whether the arachidonic-acid-containing species, i.e., anandamide, could be formed via this pathway, because their studies were carried out before the discovery of anandamide.

Soon after the discovery of anandamide, Di Marzo et al. (1994) provided evidence that rat brain neurons contain *N*-arachidonoyl PE and that it can be hydrolyzed by a phosphodiesterase to yield anandamide. This was further confirmed using N18TG2 cells and J774 cells (Di Marzo, De Petrocellis, Sepe, et al., 1996). We confirmed that the rat brain and testis contain substantial amounts of *N*-arachidonoyl PE (Sugiura, Kondo, Sukagawa, Tonegawa, Nakane, Yamashita, and Waku, 1996; Sugiura, Kondo, Sukagawa, Tonegawa, Nakane, Yamashita, Ishima, and Waku, 1996). We also confirmed that the rat brain and testis contain a phosphodiesterase activity to form anandamide from *N*-arachidonoyl PE (Sugiura, Kondo, Sukagawa, Tonegawa, Nakane, Yamashita, and Waku, 1996; Sugiura, Kondo, Sukagawa, Tonegawa, Nakane, Yamashita, Ishima, and Waku, 1996). Importantly, the fatty acid composition of the *N*-acyl moiety of *N*-acyl PE resembles that of *N*-acylethanolamine present in the same tissue (Figure 7.3), reinforcing the hypothesis that a large portion of *N*-acylethanolamine present in the tissues is derived from the corresponding *N*-acyl PE through the action of a phosphodiesterase. Enzymatic hydrolysis of *N*-acyl PE was also studied by several investigators (Sasaki and Chang, 1997; Petersen and Hansen, 1999; Moesgaard et al., 2000, 2002). Petersen and Hansen (1999) reported that the *N*-acyl-PE-hydrolyzing phosphodiesterase lacks the ability to transphosphatidylate. Recently, Ueda, Liu, and Yamanaka (2001) partially purified *N*-acyl-PE-hydrolyzing phosphodiesterase from the rat heart. They reported that the enzyme activity of the partially purified enzyme dramatically increased up to 30-fold by millimolar order of Ca^{2+} and other divalent cations. Interestingly, *N*-acyl-PE-hydrolyzing phosphodiesterase was also markedly stimulated by polyamines (Liu et al., 2002). Moesgaard et al. (2000) reported that the *N*-acyl-PE-hydrolyzing phosphodiesterase activity substantially increased during the early development of the rat brain. They also demonstrated that there are substantial species differences in the activity of *N*-acyl-PE-hydrolyzing phosphodiesterase in heart tissues (Moesgaard et al., 2002).

The enzymatic mechanism underlying the formation of *N*-acyl PE was first investigated by Schmid and co-workers in the early 1980s. Natarajan et al. (1982, 1983, 1986) demonstrated using the dog heart, dog brain, and rat brain that the fatty acids esterified at the 1-position of the glycerophospholipids are transferred to the amino group of PE through the action of a transacylase to form *N*-acyl PE. Notably, Ca^{2+} is required for this transacylase activity, suggesting that the entry of Ca^{2+} into the cells may trigger the formation of *N*-acyl PE. It is also worth mentioning that the earlier-mentioned transacylase activity in the rat brain was very high at birth but declined shortly thereafter (Natarajan et al., 1986; Moesgaard et al., 2000). Moesgaard et al. (2002) also reported that there are marked species differences in the enzyme activity in heart tissues. In any case, the transacylation reaction has been assumed to be responsible for the formation of various species of *N*-acyl PE that accumulate in several degenerating tissues. However, until the mid-1990s, it remained to be determined whether this enzyme reaction was involved in the formation of the arachidonic-acid-containing species of *N*-acyl PE, i.e., *N*-arachidonoyl PE.

In 1996, we provided evidence that microsomal fractions obtained from the rat brain and testis contain a Ca^{2+}-dependent transacylation activity that catalyzes the formation of *N*-arachidonoyl PE from PE and arachidonic acid esterified at the 1-position of phospholipids (Figure 7.2) (Sugiura, Kondo, Sukagawa, Tonegawa, Nakane, Yamashita, and Waku, 1996; Sugiura, Kondo, Sukagawa, Tonegawa, Nakane, Yamashita, Ishima, and Waku, 1996). We showed that various types of fatty acids esterified at the 1-position were transferred to PE to form *N*-acyl PE via this pathway. On the contrary, fatty acids esterified at the 2-position were not transferred. Di Marzo, De Petrocellis, Sugiura, et al. (1996) confirmed that the N18TG2 cell homogenate contains an enzyme activity catalyzing the formation of *N*-arachidonoyl PE from PE and arachidonic acid esterified at the 1-position of phospholipids. Cadas et al. (1997) also found this enzyme activity in the rat brain particulate fraction. They reported the enhanced formation of *N*-arachidonoyl PE in ionomycin-stimulated neurons (Cadas et al., 1996) and potentiation of the Ca^{2+}-dependent *N*-acyl PE synthesis by agents that augment the level of cyclic AMP (Cadas et al., 1996).

An important issue regarding this pathway is that arachidonic acid esterified at the 1-position of the glycerophospholipids is utilized for the formation of *N*-arachidonoyl PE. Therefore, significant amounts of arachidonic acid must be present at the 1-position of glycerophospholipids for the formation of significant amounts of *N*-arachidonoyl PE. It is well known, however, that arachidonic acid is usually esterified at the 2-position of glycerophospholipids in mammalian tissues. We examined whether arachidonic acid is esterified at the 1-position of glycerophospholipids in the rat brain and testis. We confirmed that small amounts of arachidonic acid are present at the 1-position of phosphatidylcholine in the rat brain and testis (0.3% and 3.4% of the total, respectively) (Sugiura, Kondo, Sukagawa, Tonegawa, Nakane, Yamashita, and Waku, 1996; Sugiura, Kondo, Sukagawa, Tonegawa, Nakane, Yamashita, Ishima, and Waku, 1996). Cadas et al. (1997) also confirmed that 0.5% of the fatty acyl moiety of the 1-position of glycerophospholipids in the rat brain is accounted for by arachidonic acid. Thus, the rat brain and testis actually contain an acyl donor, albeit at low levels, for the transacylase-mediated synthesis of *N*-arachidonoyl PE, a stored precursor form of anandamide. It is apparent from these results that the rat brain and testis are able to generate anandamide through the earlier-mentioned transacylase-phosphodiesterase pathway

(Schmid pathway). Presumably, this is the main pathway for the synthesis of anandamide in various mammalian tissues. However, it is not possible to generate a large amount of anandamide via this pathway, because the levels of arachidonic acid esterified at the 1-position are usually very low. This is consistent with the observation that the tissue levels of anandamide are generally low except in a few cases as already mentioned.

Degradation of Anandamide

Anandamide has been shown to be metabolized by a variety of tissues and cells to yield free arachidonic acid and ethanolamine (e.g., Di Marzo et al., 1994; Bisogno, Maurelli, et al., 1997; Maccarrone et al., 1998). The enzyme activity catalyzing the hydrolysis of *N*-acylethanolamine was first characterized by Schmid et al. (1985). They called this enzyme *N*-acylethanolamine amidohydrolase. This enzyme activity was shown to be associated with the membranes of microsomes and mitochondria in the rat liver. *N*-acylethanolamine amidohydrolase did not require the presence of divalent cations for the activity and was inhibited by several SH reagents and detergents. Amides of hydroxypropanolamine, propanolamine, or higher homologs were hydrolyzed at slower rates. This enzyme catalyzed the hydrolysis of various species of the *N*-acylethanolamines. However, the hydrolysis of arachidonic-acid-containing species was not examined in these pioneering studies. The hydrolysis of anandamide (*N*-arachidonoylethanolamine) was first described by Deutsch and Chin (1993). They detected the enzyme activity catalyzing the hydrolysis of anandamide in several cell lines such as the N18TG2 mouse neuroblastoma cells, C6 rat glioma cells, H358 human lung carcinoma cells, and in several rat tissues. The enzyme activity was inhibited by phenylmethylsulfonyl fluoride (PMSF), a serine esterase inhibitor. They called this enzyme activity anandamide amidase. Anandamide amidase or anandamide amidohydrolase was partially purified by Ueda, Kurahashi, et al. (1995). The activity of the partially purified enzyme was inhibited by arachidonyl trifluoromethyl ketone (ATFMK), *p*-chloromercuribenzoic acid, diisopropylfluorophosphate (DFP), and PMSF. The properties of this enzyme have also been studied by a number of investigators (e.g., Maurelli et al., 1995; Desarnaud et al., 1995; Hillard et al., 1995; Omeir et al., 1995; Giang and Cravatt, 1997; Maccarrone et al., 1998; Watanabe et al., 1998; for reviews, see Ueda et al., 2000; Ueda and Yamamoto, 2000; Deutsch et al., 2002; Cravatt and Lichtman, 2002). The optimal pH was 8.5 to 10.0, and the K_m value for anandamide was 0.8 to 180 μM. Anandamide was the most effective substrate among the saturated and unsaturated *N*-acylethanolamines of the C16-C21 fatty chain. Introduction of a methyl group into the ethanolamine moiety of anandamide yields a substrate that is hydrolyzed more slowly than anandamide. This compound (methanandamide) has been used as a valuable experimental tool by many investigators.

Shortly after the identification of anandamide amidase/anandamide amidohydrolase, Cravatt et al. (1996) cloned an enzyme catalyzing the hydrolysis of oleamide, a sleep-inducing lipid, and demonstrated that this enzyme activity is also able to hydrolyze anandamide and other *N*-acylethanolamines. They called this enzyme the FAAH. It is now widely accepted that *N*-acylethanolamine hydrolase, anandamide amidase/anandamide amidohydrolase, and FAAH are the same enzyme. FAAH consists of 579 amino acids, and the molecular weight is approximately 63,300. FAAH contains an amidase signature sequence (amino acids 215 to 257) and is distinct from other serine proteases. It has been shown that Ser-241 is the catalytic nucleophile. The enzyme activity was blocked by several fatty acid derivatives such as ATFMK, arachidonoyldiazomethyl ketone (ADMK), palmitylsulfonyl fluoride (AM374), and methyl arachidonoyl fluorophosphonate (MAFP). Among them, MAFP was the most potent inhibitor; the IC_{50} value of MAFP was as low as 2.5 nM. FAAH is widely distributed in various mammalian tissues such as the liver, small intestine, brain, testis, and kidney (Deutsch and Chin, 1993; Desarnaud et al., 1995; Cravatt et al., 1996; Katayama et al., 1997). On the other hand, the activity was hardly detectable in the heart and muscle of rats. In contrast, FAAH activity was detected in the skeletal muscle of humans. Immunohistochemical analysis showed that FAAH is present in various regions of the brain. The highest expression was observed in large principal neurons such as the pyramidal cells in the cerebral cortex and hippocampus, Purkinje cells in the cerebellar cortex, and mitral cells in the olfactory bulb. FAAH was also detected in the eyes of several animal species (Matsuda et al., 1997; Bisogno, Delton-Vandenbroucke, et al., 1999) and mouse uterus (Paria et al., 1996). The expression of FAAH in the mouse uterus was shown to be regulated during the course of pregnancy (Maccarrone et al., 2000). It was demonstrated that FAAH is upregulated by progesterone (Maccarrone, Valensise, et al., 2001) and leptin (Maccarrone, Di Rienzo, et al., 2003) in human lymphocytes.

FAAH is able to catalyze not only the hydrolysis but also the synthesis of *N*-acylethanolamine from free fatty acids and ethanolamine as mentioned before, although high amounts of the substrates are required for the synthetic reaction. Notably, Deutsch et al. (2001) recently demonstrated that the breakdown of anandamide by FAAH is a driving force for the uptake of anandamide by cells.

Unequivocally, knockout mice are useful tools to investigate the physiological significance of functional proteins such as enzymes and receptors. Recently, Cravatt et al. (2001) created a mouse lacking FAAH (FAAH$^{-/-}$). Notably, the tissues of the FAAH$^{-/-}$ mice contained dramatically reduced levels of hydrolytic activity for *N*-acylethanolamines including anandamide and other fatty acid amides. This strongly suggested that FAAH plays a principal role in the degradation of various *N*-acylethanolamines and other fatty acid amides such as oleamide. In supporting this conclusion, they found that the levels of anandamide and other *N*-acylethanolamines were elevated over tenfold in FAAH$^{-/-}$ mice relative to the FAAH$^{+/+}$ mice. They also investigated the effects of the administration of anandamide to FAAH$^{-/-}$ mice and compared them with those in the FAAH$^{+/+}$ mice. They found that the FAAH$^{-/-}$ mice and FAAH$^{+/+}$ mice displayed remarkably different responses to anandamide. Anandamide failed to produce significant behavioral responses in the FAAH$^{+/+}$ mice, whereas it produced robust dose-dependent behavioral responses in the FAAH$^{-/-}$ mice. These observations highlighted the physiological significance of FAAH in the metabolism of anandamide and other *N*-acylethanolamines *in vivo*. Interestingly, Sipe et al. (2002) recently reported that genetic mutations in FAAH may constitute important risk factors for drug abuse.

In addition to FAAH, there is another type of amidohydrolase. Ueda et al. (1999) reported the presence of such an amidohydrolase in human megakaryoblastic leukemia cells (CMK). The optimal pH of this new enzyme was 5.0. The enzyme activity was markedly reduced at alkaline pH, the preferred pH range for FAAH. This enzyme is resistant to several inhibitors of FAAH, such as PMSF and MAFP. They referred to this new enzyme as *N*-palmitoylethanolamine hydrolase. *N*-palmitoylethanolamine hydrolase is present in various rat organs; the rank order of the specific activity was lung > spleen > small intestine > thymus > cecum. High enzyme activity was also detected in macrophages. Recently, Ueda, Yamanaka, and Yamamoto (2001) purified the enzyme protein to apparent homogeneity (31 kDa) from rat lung. *N*-palmitoylethanolamine hydrolase is assumed to be involved in the catabolism of *N*-acylethanolamines, including anandamide to some extent *in vivo*, because mice lacking FAAH (FAAH$^{-/-}$) contain very weak but measurable *N*-acylethanolamine-hydrolyzing activity. Details of the exact physiological roles of the *N*-palmitoylethanolamine hydrolase remain to be clarified in the future.

Oxygenation of Anandamide

In addition to being hydrolyzed by amidohydrolase such as FAAH, anandamide has been shown to be metabolized by various types of oxygenases. Ueda, Yamamoto, et al. (1995) demonstrated that anandamide acts as a substrate for porcine leukocyte 12-lipoxygenase and rabbit reticulocyte, and soybean 15-lipoxygenases to yield the 12- and 15-hydroperoxy derivatives of anandamide, respectively. Anandamide was also metabolized by human platelet 12-lipoxygenase albeit at a much lower rate, whereas porcine leukocyte 5-lipoxygenase was almost inactive. Several investigators also reported the enzymatic hydroxylation of anandamide (Hampson et al., 1995; Edgemond et al., 1998). Anandamide has also been shown to serve as a substrate for cyclooxygenase-2 (COX-2). Yu et al. (1997) reported that anandamide was metabolized by COX-2 (not by COX-1) to the prostaglandin H_2 (PGH_2) derivative of anandamide, which was further converted to the PGE_2 derivative of anandamide. Anandamide was also metabolized to the PGI_2 or thromboxane A_2 (TXA_2) derivative of anandamide when COX-2 was added simultaneously with prostacyclin synthase or TX synthase (Kozak, Crews, et al., 2002). Bornheim et al. (1993, 1995) also reported that anandamide can be metabolized by murine hepatic cytochrome P450s to yield the monooxygenated derivatives of anandamide. However, anandamide was found to be a poor substrate for these oxygenases, except in a few cases, in comparison with free arachidonic acid.

The physiological or pathophysiological significance of the enzymatic oxygenation of anandamide is not yet clarified (Ross et al., 2002). The binding affinities of oxygenated products of anandamide to the cannabinoid receptors are generally markedly lower than that of anandamide; it seems unlikely that the oxygenated products of anandamide have roles in acting as endogenous cannabinoid receptor ligands *in vivo*. The physiological significance of the oxygenation reaction of anandamide may exist in the inactivation of anandamide, because the oxygenated products of anandamide are less potent compared with anandamide, as already mentioned. Another possibility is that the oxygenated derivatives of anandamide may interact with some unknown receptors, thereby eliciting pharmacological responses, although little information is so far available regarding this possibility. Thus, further studies are necessary to answer the question of whether these oxygenated derivatives of anandamide actually possess intrinsic physiological roles in living animals (for reviews, see Burstein et al., 2000; Kozak and Marnett, 2002).

Physiological Significance of Anandamide

Numerous studies have been carried out on anandamide since 1992. It has been established that anandamide exhibits a variety of cannabimimetic activities *in vitro* and *in vivo*. Cravatt et al. (2001) recently reported that the level of anandamide in the brain of $FAAH^{-/-}$ mice was markedly higher than that of $FAAH^{+/+}$ mice. They found that $FAAH^{-/-}$ mice exhibited reduced pain sensitivity, which was restored by treatment of the mice with SR141716A, a CB_1 receptor antagonist, although SR141716A failed to affect the pain sensitivity in the $FAAH^{+/+}$ mice. These results suggested that endogenous anandamide modulates pain perception in the $FAAH^{-/-}$ mice. Nevertheless, there are several enigmatic issues as to the physiological significance of anandamide. The levels of anandamide in tissues are very low (for reviews, see Hansen H.S., et al., 2000; Sugiura et al., 2002) except for a few cases such as the mouse uterus (Schmid, Paria, et al., 1997). No selective and efficient synthetic pathway for anandamide has hitherto been found (for reviews, see Schmid, 2000; Schmid, H.H.O., et al., 2002; Schmid and Berdyshev, 2002; Sugiura et al., 2002). In addition, anandamide, as well as Δ^9-THC, acted as partial agonists in many cases (Mackie et al., 1993; Fride et al., 1995; Burkey et al., 1997; Sugiura, Kodaka, et al.,1996; Sugiura, Kodaka, Kondo, Nakane, et al., 1997; Sugiura, Kodaka, et al., 1999; Sugiura, Kondo, et al., 2000; Breivogel et al., 1998; Kearn et al., 1999; McAllister et al., 1999; Hillard, 2000; Gonsiorek et al., 2000). It is curious that an endogenous natural ligand acts as a partial agonist at its own receptor. Further studies are thus required to answer the question of whether anandamide actually acts as an endogenous cannabinoid receptor ligand (CB_1 and CB_2) with profound physiological significance.

The possible roles of anandamide as an endogenous ligand of the receptors other than cannabinoid receptors (CB_1 and CB_2) or as a modulator of ion channels should also be intensively examined. Several investigators have demonstrated that anandamide is a potent agonist of the vanilloid receptor (e.g., Zygmunt et al., 1999; Smart et al., 2000; De Petrocellis, Bisogno, et al., 2001; De Petrocellis, Harrison, et al., 2001; Ross et al., 2001; Tognetto et al., 2001; Olah et al., 2001; for reviews, see Di Marzo, De Petrocellis, Bisogno, et al., 2002; Di Marzo, De Petrocellis, Fezza, et al., 2002). Anandamide is also known to regulate several other ion channels such as the K^+ channels and Ca^{2+} channels (Johnson et al., 1993; Shimasue et al., 1996; Jarrahian and Hillard, 1997; Oz et al., 2000; Maingret et al., 2001). The modulation of pain by anandamide and its structural analogs may be partly attributed to the modulation of the functions of these ion channels, including the vanilloid receptor. In addition to these possibilities, anandamide may also act through unidentified binding sites such as the putative non-CB_1, non-CB_2 anandamide receptors (e.g., Sagan et al., 1999; Jarai et al., 1999; Wagner et al., 1999; Di Marzo, Breivogel, et al., 2000; Breivogel et al., 2001; Hajos et al., 2001; for review, see Di Marzo, De Petrocellis, Fezza, et al., 2002), yet details still remain elusive.

More than 10 years have passed since the discovery of anandamide. However, its exact physiological significance has not yet been fully understood; anandamide still exists as a rather mysterious molecule. A thorough elucidation of the physiological significance of anandamide in mammalian tissues has to await future investigation.

2-ARACHIDONOYLGLYCEROL

Identification of 2-AG as an Endogenous Cannabinoid Receptor Ligand

2-AG, an arachidonic-acid-containing monoacylglycerol, is an intriguing molecule from a variety of viewpoints. About a decade ago, we found that *N*-acylethanolamine phosphate and lysophosphatidic acid (LPA) interact with a common receptor site (LPA receptor) on human platelets and exhibit similar biological activities (Sugiura, Tokumura, et al., 1994). This prompted us to postulate that an analog of anandamide containing a glycerol backbone, i.e., 2-AG, possesses cannabimimetic activities. We synthesized 2-AG and examined its binding activity toward the cannabinoid receptor in rat brain synaptosomes. We found that 2-AG possesses binding activity toward the cannabinoid receptor (K_i = 15 μ*M*), although its activity was significantly lower than that of anandamide. We also found that AG is actually present in rat brain in the order of nmol/g tissue. We pointed out that arachidonic-acid-containing monoacylglycerol may function as an endogenous cannabinoid receptor ligand at some sites in the brain (Sugiura, Itoh, et al., 1994). The reason for the relatively high apparent K_i value of 2-AG in that study may be due, at least in part, to possible hydrolysis of 2-AG during incubation. We then examined the effect of the addition of DFP, an esterase inhibitor. The K_i value of 2-AG estimated in the presence of DFP was 2.4 μ*M* and that of anandamide was 99 n*M*, indicating that the binding activity of 2-AG was 24 times less potent than that of anandamide. We also confirmed that the rat brain contains 3.25 nmol/g tissue of AG, a level about 800 times higher than that of anandamide in the same tissue. We published these results in 1995 (Sugiura et al., 1995). Independently and concurrently, Mechoulam et al. (1995) also reported that 2-AG is an endogenous cannabinoid receptor ligand. They isolated 2-AG from the canine gut and demonstrated that 2-AG possesses various biological activities such as binding activity toward cannabinoid receptors expressed on COS-7 cells transfected with cannabinoid receptor genes, the inhibition of adenylyl cyclase in mouse spleen cells, twitch response in the mouse vas deferens, and induction of hypothermia, reduced spontaneous activity, analgesia, and immobility in mice (Mechoulam et al., 1995). In general, the biological activities of 2-AG resemble those of anandamide (for reviews, see Di Marzo, 1998a, 1998b; Mechoulam, Fride, and Di Marzo, 1998; Piomelli et al., 1998; Hillard, 2000; Piomelli et al., 2000; Sugiura and Waku, 2000, 2002; Sugiura et al., 2002; Di Marzo, De Petrocellis, Bisogno, et al., 2002; Schmid, H.H.O., et al., 2002). Contrary to our expectation, however, 2-AG did not receive much attention at that time. This is probably due to the fact that the interests of many investigators were mostly directed to the first endocannabinoid, anandamide.

Tissue Levels of 2-AG

We have previously reported that 3.25 nmol/g tissue of AG (Sugiura et al., 1995) or 3.36 nmol/g tissue of 2-AG (Kondo, S., Kondo, H., et al., 1998) is present in the rat brain. Indeed, 2-AG was one of the major molecular species of monoacylglycerol (31.2% of the total) (Kondo, S., Kondo, H., et al., 1998). Stella et al. (1997) also detected 4.0 nmol/g tissue of AG (2.4 nmol/g tissue of 2-AG and 1.6 nmol/g tissue of 1(3)-AG) in the rat brain. Bisogno, Berrendero, et al. (1999) reported that 2-AG is distributed among various brain regions such as the cortex, limbic forebrain, hippocampus, hypothalamus, diencephalon, striatum, mesencephalon, cerebellum, brainstem, medulla, and pituitary. 2-AG was also detected in the rat anterior pituitary (Gonzalez et al., 1999), human pituitary (Pagotto et al., 2001), and rat hypothalamus (Gonzalez et al., 1999), although Giuffrida et al. (1999) did not detect 2-AG in the microdialysates from the dorsal striatum of freely moving rats. The levels of 2-AG in the whole brain and various brain regions are summarized in Table 7.2.

The levels of 2-AG in the brain changed following several treatments of experimental animals. Di Marzo, Hill, et al. (2000) demonstrated that the levels of 2-AG in the globus pallidus were markedly augmented in rats treated with reserpine, which induces immobility and Parkinson's-disease-like symptoms by causing a depletion of cathecholamines including dopamine from the striatum. On the other hand, Lastres-Becker et al. (2001) reported that the level of 2-AG in the striatum was reduced and that in the mesencephalon was augmented in a rat model of Huntington's disease. Baker et al. (2001) demonstrated that the levels of 2-AG in the brain and spinal cord were elevated in spastic mice with chronic relapsing experimental allergic encephalomyelitis as in the case of anandamide. Furthermore, Gonzalez et al. (2002) recently reported that the chronic treatments of rats with nicotine, ethanol, and cocaine induced a number of changes in the amounts of 2-AG (increase or decrease) in various brain regions. Notably, the levels of 2-AG as well as anandamide in the rat hypothalamus were markedly reduced in mice injected with leptin compared with the control (Di Marzo et al., 2001), suggesting that these endocannabinoids are involved in the regulation of appetite. Kirkham et al. (2002) demonstrated that fasting increased the level of 2-AG in the limbic forebrain and to a lesser extent in the hypothalamus, although hypothalamic 2-AG declined as the animals ate. In contrast to the case of anandamide, the levels of 2-AG in various brain regions of CB_1-knockout mice are not very different from those in the wild-type control mice (Di Marzo, Breivogel, et al., 2000). This was also confirmed by other investigators (Maccarrone, Attina, Bari, et al., 2001).

It should be kept in mind that there may be possible changes in the amounts of endocannabinoids in the brains depending on the age of the animals. In fact, Berrendero et al. (1999) reported that the level of 2-AG as well as that of anandamide in the rat brain varied during the course of development. Further studies are thus required to examine in detail possible age-dependent changes in the levels of endocannabinoids in various brain regions.

2-AG was also detected in the peripheral nervous system. Huang et al. (1999) reported that 2-AG is present in the rat sciatic nerve, lumbar spinal cord, and lumbar dorsal root ganglion, and Di Marzo, Breivogel, et al. (2000) demonstrated the occurrence of 2-AG in the mouse spinal cord. 2-AG has also been shown to occur in the rat retina and bovine retina (Straiker et al., 1999; Bisogno, Delton-Vandenbroucke, et al., 1999). Several tumors of nervous system origin have also been shown to contain 2-AG; Maccarrone, Attina, Cartoni, et al. (2001) reported that 2-AG was detected in the tumoral human brain (meningioma) and human neuroblastoma CHP100 cells.

We have also estimated the levels of 2-AG in several rat tissues other than the brain and found that 2-AG is present in the liver, spleen, lung, kidney, and plasma (Kondo, S., Kondo, H., et al., 1998) (Table 7.2). Tissue levels of 2-AG in rats have also been reported by Schmid et al. (2000). They detected 2-AG in the heart, liver, spleen, kidney, and testis (Table 7.2). 2-AG was also detected in human milk (Di Marzo, Sepe, et al., 1998), rat paw skin (Beaulieu et al., 2000), and in the sera from normal donors and patients with endotoxin shock (Wang et al., 2001) (Table 7.2).

Substantial amounts of 1(3)-AG, which is also known to activate the cannabinoid receptors, always coexist with 2-AG. The levels of 1(3)-AG were 1.39 nmol/g tissue (brain), 0.55 nmol/g tissue (liver), 0.63 nmol/g tissue (spleen), 0.59 nmol/g tissue (lung), 0.48 nmol/g tissue (kidney), and 0.004 nmol/ml (plasma) (Kondo, S., Kondo, H., et al., 1998). Presumably, 1(3)-AG was mainly formed from 2-AG through nonenzymatic acyl migration; the 1(3)-acyl isomer is more stable than the 2-acyl isomer in aqueous solution.

TABLE 7.2
Tissue Levels of 2-AG

Rat brain	3.25 nmol/g tissue[a]	Sugiura et al. (1995)
Rat brain	4.0 nmol/g tissue[a]	Stella et al. (1997)
Rat brain	3.36 nmol/g tissue	Kondo, S., Kondo, H., et al. (1998)
Rat brain (frozen immediately after decapitation)	0.34 nmol/g tissue	Sugiura et al. (2001)
Rat brain (killed in liquid nitrogen)	0.23 nmol/g tissue	Sugiura, Yoshinaga, et al. (2000)
Rat brain	65 nmol/g tissue	Maccarrone, Attina, et al. (2001)
Mouse brain	5–7 nmol/g tissue	Baker et al. (2001)
Rat cortex	4.3 nmol/g tissue	Bisogno, Berrendero, et al. (1999)
Rat cortex	2.1 nmol/g tissue	Hansen, Schmid, et al. (2001)
Rat cortex	8.9–10.2 nmol/g tissue	Gonzalez et al. (2002)
Rat limbic forebrain	10.0 nmol/g tissue	Bisogno, Berrendero, et al. (1999)
Rat limbic forebrain	3.3–3.6 nmol/g tissue	Gonzalez et al. (2002)
Rat hippocampus	12.6 nmol/g tissue	Bisogno, Berrendero, et al. (1999)
Rat hippocampus	4.6–5.7 nmol/g tissue	Gonzalez et al. (2002)
Rat hypothalamus	7.4 nmol/g tissue	Di Marzo et al. (2001)
Rat hypothalamus	0.60 nmol/mg of lipid extract	Matias et al. (2003)
Rat hypothalamus	3.1 nmol/g tissue	Gonzalez et al. (1999)
Rat diencephalon	2.0 nmol/g tissue	Bisogno, Berrendero, et al. (1999)
Rat diencephalon	3.9–5.8 nmol/g tissue	Gonzalez et al. (2002)
Rat striatum	10.7 nmol/g tissue	Bisogno, Berrendero, et al. (1999)
Rat striatum	4.3–4.9 nmol/g tissue	Gonzalez et al. (2002)
Rat striatum	1.25 nmol/mg protein	Gubellini et al. (2002)
Rat mesencephalon	4.0 nmol/g tissue	Bisogno, Berrendero, et al. (1999)
Rat mid brain	7.7–8.7 nmol/g tissue	Gonzalez et al. (2002)
Rat cerebellum	3.5 nmol/g tissue	Bisogno, Berrendero, et al. (1999)
Rat cerebellum	6.8–18.2 nmol/g tissue	Gonzalez et al. (2002)
Rat brainstem	14.0 nmol/g tissue	Bisogno, Berrendero, et al. (1999)
Rat brainstem	7.2–14.1 nmol/g tissue	Gonzalez et al. (2002)
Rat medulla	10.5 nmol/g tissue	Bisogno, Berrendero, et al. (1999)
Human pituitary	0.32–1.32 nmol/g tissue	Pagotto et al. (2001)
Rat anterior pituitary	3.7 nmol/g tissue	Gonzalez et al. (1999)
Rat sciatic nerve	0.052 nmol/g tissue	Huang et al. (1999)
Rat lumber spinal cord	0.432 nmol/g tissue	Huang et al. (1999)
Rat lumber dorsal root ganglion	0.370 nmol/g tissue	Huang et al. (1999)
Bovine retina	1.63 nmol/g tissue	Bisogno, Delton-Vandenbroucke, et al. (1999)
Rat retina	2.97 nmol/g tissue	Straiker et al. (1999)
Rat heart	10.94 pmol/μmol lipid P	Schmid et al. (2000)
Rat liver	1.15 nmol/g tissue	Kondo, S., Kondo, H., et al. (1998)
Rat liver	5.05 pmol/μmol lipid P	Schmid et al. (2000)
Rat spleen	1.17 nmol/g tissue	Kondo, S., Kondo, H., et al. (1998)
Rat spleen	29.03 pmol/μmol lipid P	Schmid et al. (2000)
Rat lung	0.78 nmol/g tissue	Kondo, S., Kondo, H., et al. (1998)
Rat kidney	0.98 nmol/g tissue	Kondo, S., Kondo, H., et al. (1998)
Rat kidney	6.19–13.04 pmol/μmol lipid P	Schmid et al. (2000)
Rat testis	3.05–3.21 pmol/μmol lipid P	Schmid et al. (2000)
Mouse small intestine (normal)	31.1 nmol/g tissue	Mascolo et al. (2002)
(ileus)	37.8 nmol/g tissue	Mascolo et al. (2002)

(*continued*)

TABLE 7.2
Tissue Levels of 2-AG (Continued)

Rat skin	51.1 pmol/mg of lipid extract	Beaulieu et al. (2000)
Rat plasma	0.012 nmol/ml	Kondo, S., Kondo, H., et al. (1998)
Human milk	0.83 nmol/ml	Fride et al. (2001)

[a] 2 AG + 1(3) AG

Concerning the quantitative analysis of 2-AG, there is an important issue to be addressed as described in the following text. Previously, we found that the amount of 2-AG in the brains obtained from rats sacrificed by immersion in liquid nitrogen was 0.23 nmol/g tissue (Sugiura, Yoshinaga, et al., 2000), this value being about one fifteenth of the amount of 2-AG in the brains obtained from rats by decapitation without freezing (Kondo, S., Kondo, H., et al., 1998). This strongly suggests that a substantial amount of 2-AG was rapidly produced in the brain during the postmortem period. We confirmed that the rapid generation of 2-AG takes place in the rat brain immediately after decapitation (Sugiura et al., 2001). To estimate the exact tissue level of 2-AG under physiological conditions, it is therefore essential to minimize the postmortem changes in the levels of 2-AG. Even in the cases of frozen samples, however, there is a possibility of the generation of a large amount of 2-AG when frozen samples thaw in organic solvents. We found that a large amount of 2-AG was produced in frozen brains following the addition of chloroform:methanol unless lipid extraction was carried out very quickly (Sugiura et al., 2001). Thus, special care has to be exercised to minimize the possible artificial formation of 2-AG in the analysis of 2-AG in mammalian tissues, especially the brain.

Finally, it is necessary to mention the hydrolyzing-enzyme-resistant analog of 2-AG. Previously, we (Sugiura, Kodaka, et al., 1999) and Mechoulam, Fride, Ben-Shabat, et al., 1998) developed an ether-linked analog of 2-AG (2-AG ether or HU310). This compound is a useful tool in exploring the possible biological activities of 2-AG, especially *in vivo*, because this compound is quite stable against hydrolyzing enzymes. Recently, Hanus et al. (2001) reported that 2-AG ether is present in the pig brain (0.6 nmol/g tissue). They renamed it *noladin ether*. Fezza et al. (2002) also reported that a small amount of 2-AG ether is present in the rat brain (25.4 pmol/g tissue). However, we did not detect 2-AG ether in the brains of various mammalian species such as the rat, mouse, hamster, and pig (Oka et al., 2003). As for the ether-linked glycerolipids, it is well known that the ether bond is exclusively located at the 1-position of the glycerol backbone in mammalian tissues (for reviews, see Horrocks, 1970; Horrocks and Sharma, 1982; Sugiura and Waku, 1987). Accordingly, it is questionable that 2-AG ether is present in appreciable amounts in mammalian brains and acts as an endogenous ligand for the cannabinoid receptors.

Biosynthesis of 2-AG

About two decades ago, Prescott and Majerus (1983) reported the generation of AG in thrombin-stimulated platelets. The generation of AG in platelet-derived growth-factor-stimulated Swiss 3T3 cells (Hasegawa-Sasaki, 1985) and in bradykinin-stimulated rat dorsal ganglion neurons (Gammon et al., 1989) has also been reported. However, at that time, it was not known that 2-AG has an essential role acting as an endogenous ligand for the cannabinoid receptors. The generation of 2-AG as an endogenous cannabinoid receptor ligand was first described in ionomycin-stimulated N18TG2 cells (Bisogno, Sepe, et al., 1997), electrically-stimulated rat hippocampal slices, and ionomycin-stimulated neurons (Stella et al., 1997). We also investigated the generation of 2-AG and found that the rapid generation of 2-AG occurs in the rat brain homogenate during incubation in the presence of Ca^{2+} (Kondo, S., Kondo, H., et al., 1998), in thrombin- or A23187-stimulated human umbilical vein endothelial cells (Sugiura, Kodaka, Nakane, et al., 1998) and in the picrotoxinin-stimulated rat

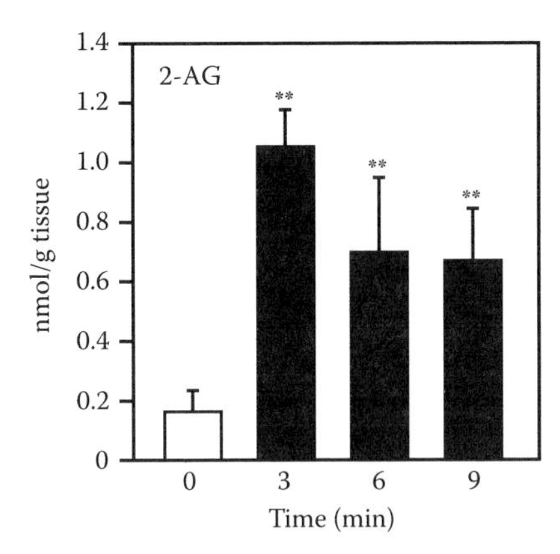

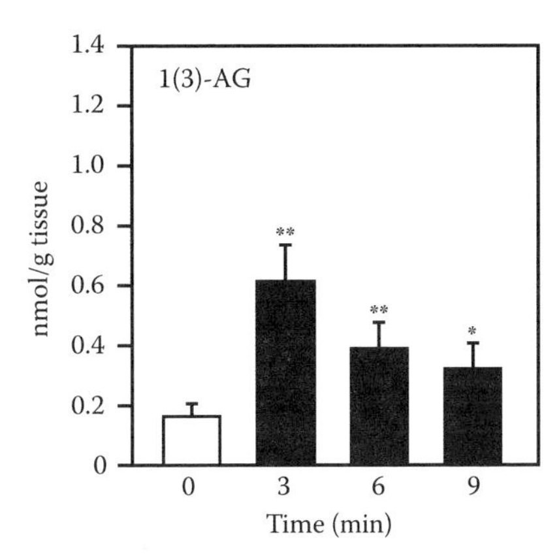

FIGURE 7.4 Rapid generation of 2-AG in rat brain following the injection of picrotoxinin. Rats were injected intraperitoneally with picrotoxinin (10 mg/kg). After the indicated time periods, rats were killed by immersion in liquid nitrogen. The data are the means ± SD (n = 7). $^{**}P < 0.001$, $^{*}P < 0.01$ (Student's *t*-test). (From Sugiura, T., Yoshinaga, N., Kondo, S., Waku, K., and Ishima, Y. (2000) Generation of 2-arachidonoylglycerol, an endogenous cannabinoid receptor ligand, in picrotoxinin-administered rat brain, *Biochemical and Biophysical Research Communications* **271**: 654–658. With permission.)

brain (Sugiura, Yoshinaga, et al., 2000) (Figure 7.4). The generation of 2-AG was also observed in the carbachol-treated rat aorta (Mechoulam, Fride, Ben-Shabat, et al., 1998), in the ethanol-treated cerebellar granule neurons in culture (Basavarajappa et al., 2000), in the mouse brain following traumatic brain injury (Panikashvili et al., 2001), in the NMDA-stimulated rat cortical neurons (Stella and Piomelli, 2001), in the medial prefontal cortex of mice reexposed to the tone 24 h after conditioning (Marsicano et al., 2002), and in the *Clostridium difficile* toxin-A-treated rat ileum (McVey et al., 2003). Several types of blood cells or inflammatory cells also produce 2-AG upon stimulation; 2-AG has been shown to be produced in LPS-stimulated rat platelets (Varga et al., 1998), in LPS-stimulated rat macrophages, in LPS- or ionomycin-stimulated J774 macrophage-like cells (Di Marzo, Bisogno, et al., 1999), in platelet-activating factor-stimulated human platelets (Berdyshev et al., 2001), and in platelet-activating factor-stimulated P388D1 macrophages (Berdyshev et al., 2001). On the other hand, Beaulieu et al. (2000) reported that the levels of 2-AG in rat paw skin did not markedly differ between the control and formalin-induced inflammation groups. It is noteworthy that substantial amounts of 2-AG were released from the cells into the extracellular fluid upon stimulation (Hasegawa-Sasaki, 1985; Bisogno, Sepe, et al., 1997; Sugiura, Kodaka, Nakane, et al., 1998).

As for the biosynthetic pathways for 2-AG, we have pointed out that 2-AG can be formed from arachidonic-acid-enriched membrane phospholipids such as inositol phospholipids through the combined actions of phospholipase C and diacylglycerol lipase or through the combined actions of phospholipase A_1 and phospholipase C (Figure 7.5) (Sugiura et al., 1995). We also suggested that 2-AG can also be formed from arachidonic-acid-containing LPA through the action of a phosphatase (Figure 7.5) (Sugiura et al., 1995). The first pathway, involving the rapid hydrolysis of inositol phospholipids by phospholipase C and subsequent hydrolysis of the resultant diacylglycerol by diacylglycerol lipase, was described by Prescott and Majerus (1983) as a degradation pathway for arachidonic-acid-containing diacylglycerols in platelets. Stella et al. (1997) demonstrated that these enzyme activities are involved in the ionomycin-induced generation of 2-AG in cultured neurons using metabolic inhibitors. We confirmed that this pathway is important for the Ca^{2+}-induced generation of 2-AG in the rat brain homogenate (T. Sugiura, unpublished results). The properties of phospholipase C (for reviews, see Rhee and Bae, 1997; Rebecchi and Pentyala, 2000) and diacylglycerol lipase (Bell et al., 1979; Farooqui et al., 1990, 1993; Moriyama et al., 1999) have already been studied by a number of investigators. Evidence is accumulating that this pathway can operate in

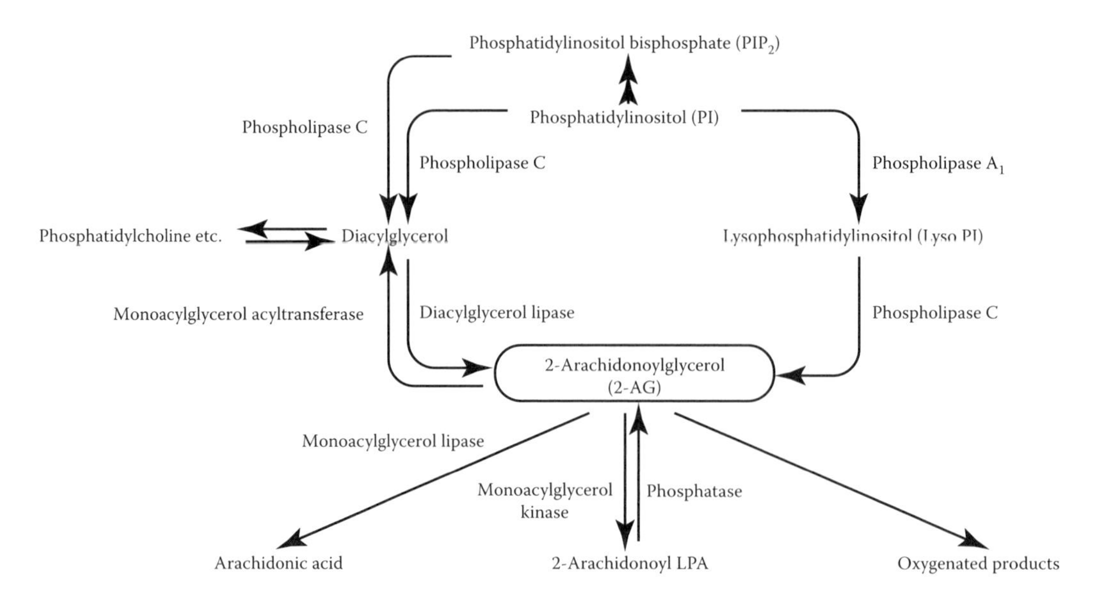

FIGURE 7.5 Pathways for the biosynthesis and degradation of 2-AG.

various stimulated tissues and cells (for reviews, see Di Marzo, 1998a, 1998b; Piomelli et al., 1998; Di Marzo, De Petrocellis, et al., 1999; Hillard, 2000; Piomelli et al., 2000; Sugiura and Waku, 2000, 2002; Sugiura et al., 2002; Di Marzo, De Petrocellis, Bisogno, et al., 2002; Schmid, H.H.O. et al., 2002).

The enzyme activities involved in the second pathway, i.e., hydrolysis of phosphatidylinositol (PI) by phospholipase A_1 and hydrolysis of the resultant lysoPI by a specific phospholipase C, were investigated by Okuyama and co-workers several years ago (Ueda et al., 1993; Tsutsumi et al., 1994, 1995), although they did not know at the time that these enzyme activities might be implicated in the generation of an endogenous cannabinoid receptor ligand. Notably, lysoPI-specific phospholipase C is distinct from various types of phospholipase C that act on other inositol phospholipids, and is localized in the synaptosomes (Tsutsumi et al., 1994, 1995). It is possible, therefore, that this unique enzyme is involved in the metabolism of lysoPI and the generation of lysoPI-derived lipid mediators such as 2-AG in the synapses.

Alternative pathways for the generation of 2-AG are the conversion of 2-arachidonoyl LPA to 2-AG and the conversion of 2-arachidonoyl PA to 2-AG (Figure 7.5). We have detected a substantial amount of arachidonic-acid-containing LPA in the rat brain (0.84 nmol/g tissue) (Sugiura, Nakane, et al., 1999). We confirmed that 63% of the arachidonic acid is esterified at the 2-position (Nakane et al., 2002). We have also detected a phosphatase activity that hydrolyzes 2-arachidonoyl LPA to yield 2-AG in a rat brain homogenate (Nakane et al., 2002). There is a possibility, therefore, that 2-arachidonoyl LPA acts as a substrate for the synthesis of 2-AG under certain conditions in the brain. On the other hand, Bisogno, Melck, et al. (1999) demonstrated that 2-AG is formed from PA in ionomycin-stimulated N18TG2 neuroblastoma cells employing several metabolic inhibitors. In this case, 2-arachidonoyl PA was first converted to 1-acyl-2-arachidonoylglycerol and then to 2-AG. They described that inositol phospholipid breakdown is not involved in the generation of 2-AG. A part of 2-AG may also be derived from other arachidonic-acid-containing phospholipids such as 1-acyl-2-arachidonoyl-*sn*-glycero-3-phosphocholine (Di Marzo, De Petrocellis, Sugiura, et al., 1996). The biosynthetic pathways for 2-AG appear to differ, depending on the types of tissues and cells and the types of stimuli. Further detailed studies are still required for a full understanding of the mechanism and regulation of the biosynthesis of 2-AG in mammalian tissues.

DEGRADATION OF 2-AG

It has been shown that 2-AG is rapidly metabolized by a variety of cells to yield arachidonic acid and glycerol (e.g., Jarai et al., 2000; Bisogno, Sepe, et al., 1997; Di Marzo, Bisogno, et al., 1998, Di Marzo, Bisogno, et al., 1999; Beltramo and Piomelli, 2000; Maccarrone, Bari, et al., 2001). The most ubiquitous mechanism of the degradation of 2-AG is that by a monoacylglycerol lipase. Konrad et al. (1994) demonstrated that 2-AG can be hydrolyzed in a porcine islets homogenate. Goparaju et al. (1999) also demonstrated that 2-AG was hydrolyzed by a monoacylglycerol-lipase-like activity present in the porcine brain cytosol and particulate fractions. An enzyme activity catalyzing the hydrolysis of 2-AG was also found and partially characterized in macrophages (Di Marzo, Bisogno, et al., 1999). The properties of the monoacylglycerol lipase activity in the brain were also investigated by Horrocks and co-workers (Farooqui et al., 1990, 1993). Recently, Piomelli and co-workers (Dinh, Freund, et al., 2002; Dinh, Carpenter, et al., 2002) cloned a monoacylglycerol lipase from a rat brain cDNA library. This monoacylglycerol lipase contained 303 amino acids, and the molecular weight was calculated to be 33,367. Northern blot and *in situ* hybridization analyses showed that monoacylglycerol lipase mRNA is expressed in various regions of the rat brain, the highest levels being expressed in regions in which the CB_1 receptor is abundant, such as the hippocampus, cortex, anterior thalamus, and cerebellum. Importantly, the monoacylglycerol lipase is presynaptically expressed. They demonstrated that the increased expression of monoacylglycerol lipase in cortical neurons attenuated the level of 2-AG in these cells upon stimulation. These results strongly suggested a primary role of this enzyme in the inactivation of 2-AG in neurons.

In addition to being hydrolyzed by monoacylglycerol lipase, Di Marzo, Bisogno, et al. (1998) and Goparaju et al. (1998) provided evidence that 2-AG is metabolized by FAAH as well. 2-AG may be degraded by FAAH in addition to monoacylglycerol lipase under some circumstances.

2-AG can also be metabolized by several anabolic enzymes. For example, 2-AG can be metabolized to 2-arachidonoyl LPA through the action of a kinases (Figure 7.5). The enzyme activity involved in the formation of 2-acyl LPA from the corresponding 2-monoacylglycerol has already been studied by several investigators (e.g., Kanoh et al., 1986; Shim et al., 1989). This pathway is probably important in recycling 2-AG to form glycerophospholipids such as PI. Simpson et al. (1991) presented evidence that [^{3}H]glycerol-labeled 2-AG, but not free [^{3}H]glycerol, was gradually incorporated into PI when added to Swiss 3T3 cells. They demonstrated that these cells contain a monoacylglycerol kinase activity and the resultant 2-arachidonoyl LPA can be metabolized to 1-stearoyl-2-arachidonoyl PA. 1-Stearoyl-2-arachidonoyl PA then enters the "PI cycle" or the *de novo* synthesis of PC and PE. Another possible metabolic pathway for 2-AG is the enzymatic acylation of 2-AG, which is well known to take place in the intestine. Di Marzo, Bisogno, et al. (1998) and Di Marzo, Bisogno, et al. (1999) demonstrated that [^{3}H]arachidonic-acid-containing 2-AG was gradually converted to phospholipids prior to its hydrolysis to arachidonic acid in N18TG2 cells, in RBL-2H3 cells, and in murine macrophages. In this case, it seems likely that a part of 2-AG was first metabolized into the lipid intermediates mentioned in the preceding text (2-arachidonoyl LPA or diacylglycerol) and then converted to phospholipids.

OXYGENATION OF 2-AG

An additional pathway for the metabolism of 2-AG is the enzymatic oxygenation of the molecule. Recently, Marnett and co-workers (Kozak et al., 2000, Kozak, Prusakiewicz, et al., 2001) demonstrated that 2-AG can be oxygenated by COX-2 to yield PGH_2 glycerol ester (PGH_2-G), and to a lesser extent, hydroxyeicosatetraenoic acid glycerol esters (HETE-G). The resultant PGH_2-G isomerizes to provide PGE_2-G and PGD_2-G. In contrast to COX-2, COX-1 failed to metabolize 2-AG. 2-AG was also shown to be metabolized to PGH_2-G and then to PGD_2-G when added to RAW264.7 cells. In addition, the conversion of 2-AG to PGI_2-G or TXA_2-G was found to occur when COX-2 was added together with prostacyclin synthase or TX synthase (Kozak,

Crews, et al., 2002). Notably, human COX-2 and murine COX-2 metabolize 2-AG as efficiently as arachidonic acid. Among the various arachidonoyl esters, 2-AG is the most preferred substrate for COX-2. These results suggest that 2-AG acts as a natural COX-2 substrate in mammalian tissues, although concrete evidence has not yet been obtained as to whether such an enzymatic reaction actually takes place *in vivo*.

2-AG has also been shown to be metabolized by lipoxygenases *in vitro*. The incubation of 2-AG with soybean 15-lipoxygenase produced 15-HETE-G (Kozak, Gupta, et al., 2002). The leukocyte-type 12-lipoxygenase metabolized 2-AG 40% as efficiently as arachidonic acid to yield 12(S)-hydroperoxyeicosatetraenoic acid glycerol esters (12-HpETE-G) (Moody et al., 2001). On the other hand, platelet-type 12-lipoxygenase did not metabolize 2-AG effectively as in the case with anandamide (Moody et al., 2001).

It is not known whether the oxygenated derivatives of 2-AG have specific physiological or pathophysiological functions distinct from those of 2-AG and eicosanoids such as PGs. Further studies are necessary to answer this question. One possibility is that these oxygenated derivatives of 2-AG may act as a kind of prodrug in tissues. These oxygenated derivatives of 2-AG are metabolically more stable compared with PGs derived from free arachidonic acid and may have long half-lives (Kozak, Crews, et al., 2001). These oxygenated derivatives of 2-AG are transferred from the site of generation to remote target tissues in which hydrolases are expressed and then release bioactive eicosanoids such as PGs (See Kozak and Marnett, 2002, for review). The elucidation of whether this hypothesis is the case awaits further investigation.

Physiological Significance of 2-AG

As mentioned earlier, 2-AG possesses several notable features as a signaling molecule: (1) 2-AG exhibits potent cannabimimetic activities *in vitro* and *in vivo*; (2) importantly, unlike anandamide, 2-AG acts as a full agonist at both the CB_1 receptor and the CB_2 receptor; (3) 2-AG is rapidly and selectively produced in a variety of cells upon stimulation and is easily released into the extracellular milieu; and (4) different from anandamide, 2-AG does not bind to receptors other than the cannabinoid receptors, such as the vanilloid receptor. Based on these observations, we have previously proposed that 2-AG is the true natural ligand of the cannabinoid receptors (Sugiura, Kodaka, Kondo, Nakane, et al., 1997; Sugiura, Kodaka, et al., 1999; Sugiura, Kondo, et al., 2000; for review, see Sugiura and Waku 2000).

It seems unlikely that 2-AG induces various psychedelic reactions in normal living animals. The CB_1 receptor is mainly present in the presynapses and is assumed to be involved in the attenuation of neurotransmission. We have previously proposed that the physiological role of 2-AG, the natural ligand of the CB_1 receptor, in the synapse is as follows: 2-AG generated through an increased phospholipid metabolism, especially inositol phospholipid breakdown, in neurons (presynapses and/or postsynapses) during accelerated synaptic transmission plays an important role in calming the excitation of neuronal cells by acting at the CB_1 receptor, thereby diminishing any subsequent neurotransmitter release (Sugiura, Kondo, et al., 1998; Sugiura and Waku, 2000). We have already demonstrated that 2-AG inhibited the depolarization-induced rapid elevation of the intracellular free Ca^{2+} concentration in NG108-15 cells (Sugiura, Kodaka, Kondo, Tonegawa, et al., 1997). Such a negative-feedback-regulation mechanism should be effective in calming stimulated neurons after excitation. In this context, it is noteworthy that several investigators have recently provided evidence indicating that endogenous cannabinoid receptor ligands derived from the postsynapse as retrograde messenger molecules play important roles in the attenuation of neurotransmission (Wilson and Nicoll, 2001; Kreitzer and Regehr, 2001; Ohno-Shosaku et al., 2001). Stella et al. (1997) also previously reported that 2-AG suppresses long-term potentiation in rat hippocampal slices, and Ameri and Simmet (2000) demonstrated that 2-AG reduces neuronal excitability in rat hippocampal slices in a cannabinoid CB_1-receptor-dependent manner. In addition, Sinor et al. (2000) and Panikashvili et al. (2001)

reported that 2-AG plays an important protective role in the brain. Thus, the elucidation of the exact physiological functions of 2-AG in the synapses is essential to better understand the details of the regulatory mechanisms of neurotransmission in the mammalian nervous system.

In addition to the roles in the central nervous system, 2-AG is also suggested to play important physiological roles in other biological systems such as the immune system (for reviews, see Sugiura and Waku, 2000, 2002; Sugiura et al., 2002; Sugiura et al., 2003). Kaminski and co-workers (Lee et al., 1995; Ouyang et al., 1998) have conducted pioneering studies concerning the effects of 2-AG on the functions of murine lymphocytes. Gallily et al. (2000) reported that 2-AG suppressed the production of TNF- in LPS-stimulated mouse macrophages *in vitro* and in LPS-administered mice *in vivo*. However, the precise physiological roles of 2-AG as a CB_2 receptor agonist in acute and chronic inflammation and/or immune responses have not yet been fully elucidated. Recently, we found that 2-AG induces the activation of MAP kinases (Kobayashi et al., 2001) and the acceleration of the generation of chemokines such as IL-8 and MCP-1 in HL60 cells (Kishimoto et al., 2004). We also found that 2-AG induces the migration of HL60 cells differentiated into macrophage-like cells (Kishimoto et al., 2003). 2-AG-induced migration has also been observed with myeloid leukemia cells and mouse splenocytes (Jorda et al., 2002) as well as with mouse microglia cells (Walter et al., 2003). Notably, Iwamura et al. (2001) demonstrated that JTE-907, a CB_2 receptor antagonist/inverse agonist, suppresses inflammatory reactions *in vivo*. Taking these results together, it is tempting to assume that 2-AG stimulates inflammatory reactions through acting at the CB_2 receptor, at least under some circumstances. Whether 2-AG actually plays important stimulative roles during the course of inflammation and immune responses *in vivo* should be clarified in the near future.

There is also growing evidence showing that endogenous cannabinoid receptor ligands play some essential roles in the cardiovascular system (for reviews, see Kunos et al., 2000; Hogestatt and Zygmunt, 2002; Randall et al., 2002; Kunos et al., 2002). Human umbilical vein endothelial cells generate 2-AG when stimulated with thrombin, and the CB_1 receptor mRNA is present in human aorta smooth muscle cells (Sugiura, Kodaka, Nakane, et al., 1998). The administration of 2-AG or 2-AG ether elicited hypotension in experimental animals (Mechoulam, Fride, Ben-Shabat, et al., 1998; Jarai et al., 2000). 2-AG also induced the relaxation of blood vessels *in vitro* (Kagota et al., 2001) and affected the norepinephrine release from rat heart sympathetic nerves (Kurihara et al., 2001). 2-AG has also been shown to counteract with endothelin-1-induced cerebral microvascular endothelial responses (Chen et al., 2000). Moreover, Stefano et al. (2000) reported that 2-AG induced the generation of nitric oxide from human vascular tissues. Despite these previous data, however, details of the physiological significance of 2-AG in the cardiovascular system still remain to be determined. Further studies are necessary for a full understanding of the physiological roles of 2-AG in the cardiovascular system as well as in other biological systems, such as the reproductive system and the endocrine system, under various physiological and pathophysiological conditions.

CONCLUDING REMARKS

Cannabinoid receptors are widely distributed among various mammalian tissues and cells. It is becoming evident that cannabinoid receptors play physiologically essential roles in diverse biological systems such as the nervous system and the immune system. Studies on endogenous cannabinoid receptor ligands, i.e., endocannabinoids, are therefore essential. Numerous studies have already been carried out on endocannabinoids. Nevertheless, the physiological significance as well as the detailed molecular mechanisms of the biosynthesis and degradation of endocannabinoids are not yet fully understood. For example, several lines of evidence have suggested that 2-AG rather than anandamide is the true natural ligand for the cannabinoid receptors. However, this still remains a matter of argument. 2-AG can be produced via several different metabolic pathways; however, the relative importance of these pathways has not yet been fully elucidated. Mouse uterus contains a large amount of anandamide; yet, the mechanism underlying the accumulation of anandamide in this tissue is not known. FAAH and monoacylglycerol lipase have recently been cloned, and the properties of these enzymes have been clarified at the molecular level. However, the molecular properties of other enzymes are still far from being fully understood. Thus, further intensive studies are indispensable to clarify the detailed molecular mechanisms of the biosynthesis and degradation as well as metabolic regulation of endocannabinoids. Such studies should provide insights into the physiological significance of the endocannabinoid system in mammals.

Note Added in Proof

Recently, Bisogno et al. (2003) cloned a cDNA for diacylglycerol lipase. (Bisogno, T., Howell, F., Williams, G., Minassi, A., Cascio, M.G., Ligresti, A., Matias, I., Schiano-Moriello, A., Paul, P.,Williams, E.J., Gangadharan, U., Hobbs, C., Di Marzo, V. and Doherty, P. (2003) *Journal of Cell Biology,* 163, 463–468). Evidence is gradually accumulating which suggests that 2-AG is the intrinsic messenger molecule of DSI or DSE. (Galante, M. and Diana, M. A., (2004) *Journal of Neuroscience* 24, 4865–4874; Melis, M., Perra, S., Muntoni, A. L., Pillolla, G., Lutz, B., Marsicano, G., Di Marzo, V., Gessa, G. L. and Perra, S., Muntoni, A. L., Pillolla, G., Lutz, B., Marsicano, G., Di Marzo, V., Gessa, G. L., and Pistis, M. (2004) *Journal of Neuroscience* 24, 10707–10715; Hashimotodani, Y., Ohno-Shosaku, T., Tsubokawa, H., Ogata, H., Emoto, K., Maejima, T., Araishi, K., Shin, H-S. and Kano, M. (2005) *Neuron*, 45, 257–268; Maejima, T., Oka, S., Hashimotodani, Y., Ohno-Shosaku, T., Aiba, A., Wu, D., Sugiura T., and Kanoh, M. (2005) *Journal of Neuroscience,* in press.)

REFERENCES

Ahluwalia, J., Yaqoob, M., Urban, L., Bevan, S., and Nagy, I. (2003) Activation of capsaicin-sensitive primary sensory neurons induces anandamide production and release, *Journal of Neurochemistry* **84**: 585–591.

Ameri, A. and Simmet, T. (2000) Effects of 2-arachidonylglycerol, an endogenous cannabinoid, on neuronal activity in rat hippocampal slices, *Naunyn Schmiedebergs Archives of Pharmacology* **361**: 265–272.

Arai,Y., Fukushima, T., Shirao, M., Yang, X., and Imai, K. (2000) Sensitive determination of anandamide in rat brain utilizing a coupled-column HPLC with fluorimetric detection, *Biomedical Chromatography* **14**: 118–124.

Bachur, N.R. and Udenfriend, S. (1966) Microsomal synthesis of fatty acid amides, *Journal of Biological Chemistry* **241**: 1308–1313.

Baker, D., Pryce, G., Croxford, J.L., Brown, P., Pertwee, R.G., Makriyannis, A., Khanolkar, A., Layward, L., Fezza, F., Bisogno, T., and Di Marzo, V. (2001) Endocannabinoids control spasticity in a multiple sclerosis model, *FASEB Journal* **15**: 300–302.

Basavarajappa, B.S. and Hungund, B.L. (1999) Chronic ethanol increases the cannabinoid receptor agonist anandamide and its precursor *N*-arachidonoylphosphatidylethanolamine in SK-N-SH cells, *Journal of Neurochemistry* **72**: 522–528.

Basavarajappa, B.S., Saito, M., Cooper, T.B., and Hungund, B.L. (2000) Stimulation of cannabinoid receptor agonist 2-arachidonylglycerol by chronic ethanol and its modulation by specific neuromodulators in cerebellar granule neurons, *Biochimica et Biophysica Acta* **1535**: 78–86.

Beaulieu, P., Bisogno, T., Punwar, S., Farquhar-Smith, W.P., Ambrosino, G., Di Marzo, V., and Rice, A.S. (2000) Role of the endogenous cannabinoid system in the formalin test of persistent pain in the rat, *European Journal of Pharmacology* **396**: 85–92.

Bell, L., Kennerly, D.A., Stanford, N., and Majerus, P.W. (1979) Diglyceride lipase: a pathway for arachidonate release from human platelets, *Proceedings of the National Academy of Sciences of the USA* **76**: 3238–3241.

Beltramo, M. and Piomelli, D. (2000) Carrier-mediated transport and enzymatic hydrolysis of the endogenous cannabinoid 2-arachidonylglycerol, *Neuroreport* **11**: 1231–1235.

Berdyshev, E.V. (2000) Cannabinoid receptors and the regulation of immune response, *Chemistry and Physics of Lipids* **108**: 169–190.

Berdyshev, E.V., Schmid, P.C., Dong, Z., and Schmid, H.H.O. (2000) Stress-induced generation of *N*-acylethanolamines in mouse epidermal JB6 P^+ cells, *Biochemical Journal* **346**: 369–374.

Berdyshev, E.V., Schmid, P.C., Krebsbach, R.J., and Schmid, H.H.O. (2001) Activation of PAF receptors results in enhanced synthesis of 2-arachidonoylglycerol (2-AG) in immune cells, *FASEB Journal*. **15**: 2171–2178.

Berger, A., Crozier, G., Bisogno, T., Cavaliere, P., Innis, S., and Di Marzo, V. (2001) Anandamide and diet: inclusion of dietary arachidonate and docosahexaenoate leads to increased brain levels of the corresponding *N*-acylethanolamines in piglets, *Proceedings of the National Academy of Sciences of the USA* **98**: 6402–6406.

Berrendero, F., Sepe, N., Ramos, J.A., Di Marzo, V., and Fernandez-Ruiz, J.J. (1999) Analysis of cannabinoid receptor binding and mRNA expression and endogenous cannabinoid contents in the developing rat brain during late gestation and early postnatal period, *Synapse* **33**: 181–191.

Bisogno, T., Maurelli, S., Melck, D., De Petrocellis, L., and Di Marzo, V. (1997) Biosynthesis, uptake, and degradation of anandamide and palmitoylethanolamide in leukocytes, *Journal of Biological Chemistry* **272**: 3315–3323.

Bisogno, T., Sepe, N., Melck, D., Maurelli, S., De Petrocellis, L., and Di Marzo, V. (1997) Biosynthesis, release and degradation of the novel endogenous cannabimimetic metabolite 2-arachidonoylglycerol in mouse neuroblastoma cells, *Biochemical Journal* **322**: 671–677.

Bisogno, T., Katayama, K., Melck, D., Ueda, N., De Petrocellis, L., Yamamoto, S., and Di Marzo, V. (1998) Biosynthesis and degradation of bioactive fatty acid amides in human breast cancer and rat pheochromocytoma cells. Implications for cell proliferation and differentiation, *European Journal of Biochemistry* **254**: 634–642.

Bisogno, T., Melck, D., De Petrocellis, L., and Di Marzo, V. (1999) Phosphatidic acid as the biosynthetic precursor of the endocannabinoid 2-arachidonoylglycerol in intact mouse neuroblastoma cells stimulated with ionomycin. *Journal of Neurochemistry* **72**: 2113–2119.

Bisogno, T., Delton-Vandenbroucke, I., Milone, A., Lagarde, M., and Di Marzo, V. (1999) Biosynthesis and inactivation of *N*-arachidonoylethanolamine (anandamide) and *N*-docosahexaenoylethanolamine in bovine retina, *Archives of Biochemistry and Biophysics* **370**: 300–307.

Bisogno, T., Berrendero, F., Ambrosino, G., Cebeira, M., Ramos, J.A., Fernandez-Ruiz, J.J., and Di Marzo, V. (1999) Brain regional distribution of endocannabinoids: implications for their biosynthesis and biological function, *Biochemical and Biophysical Research Communications* **256**: 377–380.

Bornheim, L.M., Kim, K.Y., Chen, B., and Correia, M.A. (1993) The effect of cannabidiol on mouse hepatic microsomal cytochrome P450-dependent anandamide metabolism, *Biochemical and Biophysical Research Communications* **197**: 740–746.

Bornheim, L.M., Kim, K.Y., Chen, B., and Correia, M.A. (1995) Microsomal cytochrome P450-mediated liver and brain anandamide metabolism, *Biochemical Pharmacology* **50**: 677–686.

Breivogel, C.S., Selley, D.E., and Childers, S.R. (1998) Cannabinoid receptor agonist efficacy for stimulating [^{35}S]GTPγS binding to rat cerebellar membranes correlates with agonist-induced decreases in GDP affinity, *Journal of Biological Chemistry* **273**: 16865–16873.

Breivogel, C.S., Griffin, G., Di Marzo, V., and Martin, B.R. (2001) Evidence for a new G protein coupled cannabinoid receptor in mouse brain, *Molecular Pharmacology* **60**: 155–163.

Burkey, T.H., Quock, R.M., Consroe, P., Ehlert, F.J., Hosohata, Y., Roeske, W.R., and Yamamura, H.I. (1997) Relative efficacies of cannabinoid CB_1 receptor agonists in the mouse brain, *European Journal of Pharmacology* **336**: 295–298.

Burstein, S.H. and Hunter, S.A. (1995) Stimulation of anandamide biosynthesis in *N*-18TG2 neuroblastoma cells by delta 9-tetrahydrocannabinol (THC), *Biochemical Pharmacology* **49**: 855–858.

Burstein, S.H., Rossetti, R.G., Yagen, B., and Zurier, R.B. (2000) Oxidative metabolism of anandamide, *Prostaglandins and Other Lipid Mediators* **61**: 29–41.

Burstein, S.H., Huang, S.M., Petros, T.J., Rossetti, R.G., Walker, J.M., and Zurier, R.B. (2002) Regulation of anandamide tissue levels by *N*-arachidonylglycine, *Biochemical Pharmacology* **64**: 1147–1150.

Cabral, G.A. (2002) Marijuana and cannabinoid effects in immunity and AIDS, in *Biology of Marijuana*, Ed. E.S. Onaivi, London: Taylor and Francis, pp. 282–307.

Cadas, H., Gaillet, S., Beltramo, M., Venance, L., and Piomelli, D. (1996) Biosynthesis of an endogenous cannabinoid precursor in neurons and its control by calcium and cAMP, *Journal of Neuroscience* **16**: 3934–3942.

Cadas, H., Di Tomaso, E., and Piomelli, D. (1997) Occurrence and biosynthesis of endogenous cannabinoid precursor, *N*-arachidonoyl phosphatidylethanolamine, in rat brain, *Journal of Neuroscience* **17**: 1226–1242.

Calignano, A., La Rana, G., Giuffrida, A., and Piomelli, D. (1998) Control of pain initiation by endogenous cannabinoids, *Nature* **394**: 277–281.

Chen, Y., McCarron, R.M., Ohara, Y., Bembry, J., Azzam, N., Lenz, F.A., Shohami, E., Mechoulam, R., and Spatz, M. (2000) Human brain capillary endothelium: 2-arachidonoglycerol (endocannabinoid) interacts with endothelin-1, *Circulation Research* **87**: 323–327.

Colodzin, M., Bachur, N.R., Weissbach, H., and Udenfriend, S. (1963) Enzymatic formation of fatty acid amides of ethanolamine by rat liver microsomes, *Biochemical and Biophysical Research Communications* **10**: 165–171.

Cravatt, B.F., Giang, D.K., Mayfield, S.P., Boger, D.L., Lerner, R.A., and Gilula, N.B. (1996) Molecular characterization of an enzyme that degrades neuromodulatory fatty-acid amides, *Nature* **384**: 83–87.

Cravatt, B.F., Demarest, K., Patricelli, M.P., Bracey, M.H., Giang, D.K., Martin, B.R., and Lichtman, A.H. (2001) Supersensitivity to anandamide and enhanced endogenous cannabinoid signaling in mice lacking fatty acid amide hydrolase, *Proceedings of the National Academy of Sciences of the USA* **98**: 9371–9376.

Cravatt, B.F. and Lichtman, A.H. (2002) The enzymatic inactivation of the fatty acid amide class of signaling lipids, *Chemistry and Physics of Lipids* **121**: 135–148.

Crawley, J.N., Corwin, R.L., Robinson, J.K., Felder, C.C., Devane, W.A., and Axelrod, J. (1993) Anandamide, an endogenous ligand of the cannabinoid receptor, induces hypomotility and hypothermia in vivo in rodents, *Pharmacology Biochemistry and Behavior* **46**: 967–972.

De Petrocellis, L., Melck, D., Bisogno, T., and Di Marzo, V. (2000) Endocannabinoids and fatty acid amides in cancer, inflammation and related disorders, *Chemistry and Physics of Lipids* **108**: 191–209.

De Petrocellis, L., Bisogno, T., Maccarrone, M., Davis, J.B., Finazzi-Agro, A., and Di Marzo, V. (2001) The activity of anandamide at vanilloid VR1 receptors requires facilitated transport across the cell membrane and is limited by intracellular metabolism, *Journal of Biological Chemistry* **276**: 12856–12863.

De Petrocellis, L., Harrison, S., Bisogno, T., Tognetto, M., Brandi, I., Smith, G.D., Creminon, C., Davis, J.B., Geppetti, P., and Di Marzo, V. (2001) The vanilloid receptor (VR1)-mediated effects of anandamide are potently enhanced by the cAMP-dependent protein kinase, *Journal of Neurochemistry* **77**: 1660–1663.

Desarnaud, F., Cadas, H., and Piomelli, D. (1995) Anandamide amidohydrolase activity in rat brain microsomes. Identification and partial characterization, *Journal of Biological Chemistry* **270**: 6030–6035.

Deutsch, D.G. and Chin, S.A. (1993) Enzymatic synthesis and degradation of anandamide, a cannabinoid receptor agonist, *Biochemical Pharmacology* **46**: 791–796.

Deutsch, D.G., Goligorsky, M.S., Schmid, P.C., Krebsbach, R.J., Schmid, H.H.O., Das, S.K., Dey, S.K., Arreaza, G., Thorup, C., Stefano, G., and Moore, L.C. (1997) Production and physiological actions of anandamide in the vasculature of the rat kidney, *Journal of Clinical Investigation* **100**: 1538–1546.

Deutsch, D.G., Glaser, S.T., Howell, J.M., Kunz, J.S., Puffenbarger, R.A., Hillard, C.J., and Abumrad, N. (2001) The cellular uptake of anandamide is coupled to its breakdown by fatty-acid amide hydrolase, *Journal of Biological Chemistry* **276**: 6967–6973.

Deutsch, D.G., Ueda, N., and Yamamoto, S. (2002) The fatty acid amide hydrolase (FAAH), *Prostaglandins, Leukotrienes and Essential Fatty Acids* **66**: 201–210. Erratum *Prostaglandins, Leukotrienes and Essential Fatty Acids* (2003) **68**: 69.

Devane, W.A., Dysarz, F.A., III, Johnson, M.R., Melvin, L.S., and Howlett, A.C. (1988) Determination and characterization of a cannabinoid receptor in rat brain, *Molecular Pharmacology* **34**: 605–613.

Devane, W.A., Hanus, L., Breuer, A., Pertwee, R.G., Stevenson, L.A., Griffin, G., Gibson, D., Mandelbaum, A., Etinger, A., and Mechoulam, R. (1992) Isolation and structure of a brain constituent that binds to the cannabinoid receptor, *Science* **258**: 1946–1949.

Devane, W.A. and Axelrod, J. (1994) Enzymatic synthesis of anandamide, an endogenous ligand for the cannabinoid receptor, by brain membranes, *Proceedings of the National Academy of Sciences of the USA* **91**: 6698–6701.

Di Marzo, V., Fontana, A., Cadas, H., Schinelli, S., Cimino, G., Schwartz, J.C., and Piomelli, D. (1994) Formation and inactivation of endogenous cannabinoid anandamide in central neurons, *Nature* **372**: 686–691.

Di Marzo, V., De Petrocellis, L., Sepe, N., and Buono, A. (1996) Biosynthesis of anandamide and related acylethanolamides in mouse J774 macrophages and N18 neuroblastoma cells, *Biochemical Journal* **316**: 977–984.

Di Marzo, V., De Petrocellis, L., Sugiura, T., and Waku, K. (1996) Potential biosynthetic connections between the two cannabimimetic eicosanoids, anandamide and 2-arachidonoyl-glycerol, in mouse neuroblastoma cells, *Biochemical and Biophysical Research Communications* **227**: 281–288.

Di Marzo, V. (1998a) 2-Arachidonoyl-glycerol as an endocannabinoid: limelight for a formerly neglected metabolite, *Biochemistry (Moscow)* **63**: 13–21.

Di Marzo, V. (1998b) 'Endocannabinoids' and other fatty acid derivatives with cannabimimetic properties: biochemistry and possible physiopathological relevance, *Biochimica et Biophysica Acta* **1392**: 153–175.

Di Marzo, V., Melck, D., Bisogno, T., and De Petrocellis, L. (1998) Endocannabinoids: endogenous cannabinoid receptor ligands with neuromodulatory action, *Trends in Neurosciences* **21**: 521–528. Erratum *Trends in Neurosciences* (1999) **22**: 80.

Di Marzo, V., Bisogno, T., Sugiura, T., Melck, D., and De Petrocellis, L. (1998) The novel endogenous cannabinoid 2-arachidonoylglycerol is inactivated by neuronal- and basophil-like cells: connections with anandamide, *Biochemical Journal* **331**: 15–19.

Di Marzo, V., Sepe, N., De Petrocellis, L., Berger, A., Crozier, G., Fride, E., and Mechoulam, R. (1998) Trick or treat from food endocannabinoids? *Nature* **396**: 636–637.

Di Marzo, V., De Petrocellis, L., Bisogno, T., and Melck, D. (1999) Metabolism of anandamide and 2-arachidonoylglycerol: an historical overview and some recent developments, *Lipids* **34**: S319–325.

Di Marzo, V., Bisogno, T., De Petrocellis, L., Melck, D., Orlando, P., Wagner, J.A., and Kunos, G. (1999) Biosynthesis and inactivation of the endocannabinoid 2-arachidonoylglycerol in circulating and tumoral macrophages, *European Journal of Biochemistry* **264**: 258–267.

Di Marzo, V., Melck, D., De Petrocellis, L., and Bisogno, T. (2000) Cannabimimetic fatty acid derivatives in cancer and inflammation, *Prostaglandins and Other Lipid Mediators* **61**: 43–61.

Di Marzo, V., Hill, M.P., Bisogno, T., Crossman, A.R., and Brotchie, J.M. (2000) Enhanced levels of endogenous cannabinoids in the globus pallidus are associated with a reduction in movement in an animal model of Parkinson's disease, *FASEB Journal* **14**: 1432–1438.

Di Marzo, V., Berrendero, F., Bisogno, T., Gonzalez, S., Cavaliere, P., Romero, J., Cebeira, M., Ramos, J.A., and Fernandez-Ruiz, J.J. (2000) Enhancement of anandamide formation in the limbic forebrain and reduction of endocannabinoid contents in the striatum of Δ^9-tetrahydrocannabinol-tolerant rats, *Journal of Neurochemistry* **74**: 1627–1635.

Di Marzo, V., Breivogel, C.S., Tao, Q., Bridgen, D.T., Razdan, R.K., Zimmer, A.M., Zimmer, A., and Martin, B.R. (2000) Levels, metabolism, and pharmacological activity of anandamide in CB_1 cannabinoid receptor knockout mice: evidence for non-CB_1, non-CB_2 receptor-mediated actions of anandamide in mouse brain, *Journal of Neurochemistry* **75**: 2434–2444.

Di Marzo, V., Goparaju, S.K., Wang, L. Liu, J., Batkai, S., Jarai, Z., Fezza, F., Miura, G.I., Palmiter, R.D., Sugiura, T., and Kunos, G. (2001) Leptin-regulated endocannabinoids are involved in maintaining food intake, *Nature* **410**: 822–825.

Di Marzo, V., De Petrocellis, L., Bisogno, T., Berger, A., and Mechoulam, R. (2002) Biology of endocannabinoids, in *Biology of Marijuana*, Ed. E.S. Onaivi, London: Taylor and Francis, pp. 125–173.

Di Marzo, V., De Petrocellis, L., Fezza, F., Ligresti, A., and Bisogno, T. (2002) Anandamide receptors, *Prostaglandins, Leukotrienes and Essential Fatty Acids* **66**: 377–391.

Dinh, T.P., Freund, T.F., and Piomelli, D. (2002) A role for monoglyceride lipase in 2-arachidonoylglycerol inactivation, *Chemistry and Physics of Lipids* **121**: 149–158.

Dinh, T.P., Carpenter, D., Leslie, F.M., Freund, T.F., Katona, I., Sensi, S.L., Kathuria, S., and Piomelli, D. (2002) Brain monoglyceride lipase participating in endocannabinoid inactivation, *Proceedings of the National Academy of Sciences of the USA* **99**: 10819–10824. Erratum *Proceedings of the National Academy of Sciences of the USA* (2002) **99**: 13961.

Edgemond, W.S., Hillard, C.J., Falck, J.R., Kearn, C.S., and Campbell, W.B. (1998) Human platelets and polymorphonuclear leukocytes synthesize oxygenated derivatives of arachidonylethanolamide (anandamide): their affinities for cannabinoid receptors and pathways of inactivation, *Molecular Pharmacology* **54**: 180–188.

Epps, D.E., Schmid, P.C., Natarajan, V., and Schmid, H.H.O. (1979) *N*-Acylethanolamine accumulation in infarcted myocardium, *Biochemical and Biophysical Research Communications* **90**: 628–633.

Epps, D.E., Natarajan, V., Schmid, P.C., and Schmid, H.H.O. (1980) Accumulation of *N*-acylethanolamine glycerophospholipids in infarcted myocardium, *Biochimica et Biophysica Acta* **618**: 420–430.

Evans, D.M., Johnson, M.R., and Howlett, A.C. (1992) Ca^{2+}-dependent release from rat brain of cannabinoid receptor binding activity, *Journal of Neurochemistry* **58**: 780–782.

Farooqui, A.A., Anderson, D.K., Flynn, C., Bradel, E., Means, E.D., and Horrocks, L.A. (1990) Stimulation of mono- and diacylglycerol lipase activities by bradykinin in neural cultures, *Biochemical and Biophysical Research Communications* **166**: 1001–1009.

Farooqui, A.A., Anderson, D.K., and Horrocks, L.A. (1993) Effect of glutamate and its analogs on diacylglycerol and monoacylglycerol lipase activities of neuron-enriched cultures, *Brain Research* **604**: 180–184.

Felder, C.C., Nielsen, A., Briley, E.M., Palkovits, M., Priller, J., Axelrod, J., Nguyen, D.N., Richardson, J.M., Riggin, R.M., Koppel, G.A., Paul, S.M., and Becker, G.W. (1996) Isolation and measurement of the endogenous cannabinoid receptor agonist, anandamide, in brain and peripheral tissues of human and rat, *FEBS Letters* **393**: 231–235.

Felder, C.C. and Glass, M. (1998) Cannabinoid receptors and their endogenous agonists, *Annual Review of Pharmacology and Toxicology* **38**: 179–200.

Fezza, F., Bisogno, T., Minassi, A., Appendino, G., Mechoulam, R., and Di Marzo, V. (2002) Noladin ether, a putative novel endocannabinoid: inactivation mechanisms and a sensitive method for its quantification in rat tissues, *FEBS Letters* **513**: 294–298.

Fride, E. and Mechoulam, R. (1993) Pharmacological activity of the cannabinoid receptor agonist, anandamide, a brain constituent, *European Journal of Pharmacology* **231**: 313–314.

Fride, E., Barg, J., Levy, R., Saya, D., Heldman, E., Mechoulam, R., and Vogel, Z. (1995) Low doses of anandamides inhibit pharmacological effects of Δ^9-tetrahydrocannabinol, *Journal of Pharmacology and Experimental Therapeutics* **272**: 699–707.

Fride, E., Ginzburg, Y., Breuer, A., Bisogno, T., Di Marzo, V., and Mechoulam, R. (2001) Critical role of the endogenous cannabinoid system in mouse pup suckling and growth, *European Journal of Pharmacology* **419**: 207–214.

Gallily, R., Breuer, A., and Mechoulam, R. (2000) 2-Arachidonylglycerol, an endogenous cannabinoid, inhibits tumor necrosis factor-α production in murine macrophages, and in mice, *European Journal of Pharmacology* **406**: R5–7.

Gammon, C.M., Allen, A.C., and Morell, P. (1989) Bradykinin stimulates phosphoinositide hydrolysis and mobilization of arachidonic acid in dorsal root ganglion neurons, *Journal of Neurochemistry* **53**: 95–101.

Giang, D.K. and Cravatt, B.F. (1997) Molecular characterization of human and mouse fatty acid amide hydrolases, *Proceedings of the National Academy of Sciences of the USA* **94**: 2238–2242.

Giuffrida, A., Parsons, L.H., Kerr, T.M., Rodriguez de Fonseca, F., Navarro, M., and Piomelli, D. (1999) Dopamine activation of endogenous cannabinoid signaling in dorsal striatum, *Nature Neuroscience* **2**: 358–363.

Giuffrida, A., Rodriguez de Fonseca, F., and Piomelli, D. (2000) Quantification of bioactive acylethanolamides in rat plasma by electrospray mass spectrometry, *Analytical Biochemistry* **280**: 87–93.

Giuffrida, A., Rodriguez de Fonseca, F., Nava, F., Loubet-Lescoulie, P., and Piomelli, D. (2000) Elevated circulating levels of anandamide after administration of the transport inhibitor, AM404, *European Journal of Pharmacology* **408**: 161–168.

Gonsiorek, W., Lunn, C., Fan, X., Narula, S., Lundell, D., and Hipkin, R.W. (2000) Endocannabinoid 2-arachidonyl glycerol is a full agonist through human type 2 cannabinoid receptor: antagonism by anandamide, *Molecular Pharmacology* **57**: 1045–1050.

Gonzalez, S., Manzanares, J., Berrendero, F., Wenger, T., Corchero, J., Bisogno, T., Romero, J., Fuentes, J.A., Di Marzo, V., Ramos, J.A., and Fernandez-Ruiz, J. (1999) Identification of endocannabinoids and cannabinoid CB_1 receptor mRNA in the pituitary gland, *Neuroendocrinology* **70**: 137–145.

Gonzalez, S., Cascio, M.G., Fernandez-Ruiz, J., Fezza, F., Di Marzo, V., and Ramos, J.A. (2002) Changes in endocannabinoid contents in the brain of rats chronically exposed to nicotine, ethanol or cocaine, *Brain Research* **954**: 73–81.

Goparaju, S.K., Ueda, N., Yamaguchi, H., and Yamamoto, S. (1998) Anandamide amidohydrolase reacting with 2-arachidonoylglycerol, another cannabinoid receptor ligand, *FEBS Letters* **422**: 69–73.

Goparaju, S.K., Ueda, N., Taniguchi, K., and Yamamoto, S. (1999) Enzymes of porcine brain hydrolyzing 2-arachidonoylglycerol, an endogenous ligand of cannabinoid receptors, *Biochemical Pharmacology* **57**: 417–423.

Gubellini, P., Picconi, B., Bari, M., Battista, N., Calabresi, P., Centonze, D., Bernardi, G., Finazzi-Agro, A., and Maccarrone, M. (2002) Experimental parkinsonism alters endocannabinoid degradation: implications for striatal glutamatergic transmission, *Journal of Neuroscience* **22**: 6900–6907.

Hajos, N., Ledent, C., and Freund, T.F. (2001) Novel cannabinoid-sensitive receptor mediates inhibition of glutamatergic synaptic transmission in the hippocampus, *Neuroscience* **106**: 1–4.

Hampson, A.J., Hill, W.A., Zan-Phillips, M., Makriyannis, A., Leung, E., Eglen, R.M., and Bornheim, LM. (1995) Anandamide hydroxylation by brain lipoxygenase:metabolite structures and potencies at the cannabinoid receptor, *Biochimica et Biophysica Acta* **1259**: 173–179.

Hansen, H.S., Lauritzen, L., Strand, A.M., Moesgaard, B., and Frandsen, A. (1995) Glutamate stimulates the formation of *N*-acylphosphatidylethanolamine and *N*-acylethanolamine in cortical neurons in culture, *Biochimica et Biophysica Acta* **1258**: 303–308.

Hansen, H.S., Lauritzen, L., Strand, A.M., Vinggaard, A.M., Frandsen, A., and Schousboe, A. (1997) Characterization of glutamate-induced formation of *N*-acylphosphatidylethanolamine and *N*-acylethanolamine in cultured neocortical neurons, *Journal of Neurochemistry* **69**: 753–761.

Hansen, H.S., Moesgaard, B., Hansen, H.H., and Petersen, G. (2000) *N*-Acylethanolamines and precursor phospholipids — relation to cell injury, *Chemistry and Physics of Lipids* **108**: 135–150

Hansen, H.H., Hansen, S.H., Schousboe, A., and Hansen, H.S. (2000) Determination of the phospholipid precursor of anandamide and other *N*-acylethanolamine phospholipids before and after sodium azide-induced toxicity in cultured neocortical neurons, *Journal of Neurochemistry* **75**: 861–871.

Hansen, H.H., Ikonomidou, C., Bittigau, P., Hansen, S.H., and Hansen, H.S. (2001) Accumulation of the anandamide precursor and other *N*-acylethanolamine phospholipids in infant rat models of in vivo necrotic and apoptotic neuronal death, *Journal of Neurochemistry* **76**: 39–46.

Hansen, H.H., Schmid, P.C., Bittigau, P., Lastres-Becker, I., Berrendero, F., Manzanares, J., Ikonomidou, C., Schmid, H.H.O., Fernandez-Ruiz, J.J., and Hansen, H.S. (2001) Anandamide, but not 2-arachidonoylglycerol, accumulates during in vivo neurodegeneration, *Journal of Neurochemistry* **78**: 1415–1427.

Hanus, L., Gopher, A., Almog, S., and Mechoulam, R. (1993) Two new unsaturated fatty acid ethanolamides in brain that bind to the cannabinoid receptor, *Journal of Medicinal Chemistry* **36**: 3032–3034.

Hanus, L., Abu-Lafi, S., Fride, E., Breuer, A., Vogel, Z., Shalev, D.E., Kustanovich, I., and Mechoulam, R. (2001) 2-Arachidonyl glyceryl ether, an endogenous agonist of the cannabinoid CB1 receptor, *Proceedings of the National Academy of Sciences of the USA* **98**: 3662–3665.

Hasegawa-Sasaki, H. (1985) Early changes in inositol lipids and their metabolites induced by platelet-derived growth factor in quiescent Swiss mouse 3T3 cells, *Biochemical Journal* **232**: 99–109.

Hillard, C.J., Wilkison, D.M., Edgemond, W.S., and Campbell, W.B. (1995) Characterization of the kinetics and distribution of *N*-arachidonylethanolamine (anandamide) hydrolysis by rat brain, *Biochimica et Biophysica Acta* **1257**: 249–256.

Hillard, C.J. and Campbell, W.B. (1997) Biochemistry and pharmacology of arachidonylethanolamide, a putative endogenous cannabinoid, *Journal of Lipid Research* **38**: 2383–2398.

Hillard, C.J. (2000) Biochemistry and pharmacology of the endocannabinoids arachidonylethanolamide and 2-arachidonylglycerol, *Prostaglandins and Other Lipid Mediators* **61**: 3–18.

Hogestatt, E.D. and Zygmunt, P.M. (2002) Cardiovascular pharmacology of anandamide, *Prostaglandins, Leukotrienes and Essential Fatty Acids* **66**: 343–351.

Horrocks, L.A. (1970) Content, composition, and metabolism of mammalian and avian lipids that contain ether groups, in *Ether Lipids*, F. Snyder, Ed., New York: Academic Press, pp. 177–272.

Horrocks, L.A. and Sharma, M. (1982) Plasmalogens and *O*-alkyl glycerophospholipids, in *Phospholipids*, J.N. Hawthorne and G.B. Ansell, Eds., Amsterdam: Elsevier, pp. 51–93.

Huang, S.M., Strangman, N.M., and Walker, J.M. (1999) Liquid chromatographic-mass spectrometric measurement of the endogenous cannabinoid 2-arachidonylglycerol in the spinal cord and peripheral nervous system, *Zhongguo Yao Li Xue Bao* **20**: 1098–1102.

Huang, S.M., Bisogno, T., Trevisani, M., Al-Hayani, A., De Petrocellis, L., Fezza, F., Tognetto, M., Petros, T.J., Krey, J.F., Chu, C.J., Miller, J.D., Davies, S.N., Geppetti, P., Walker, J.M., and Di Marzo, V. (2002) An endogenous capsaicin-like substance with high potency at recombinant and native vanilloid VR1 receptors, *Proceedings of the National Academy of Sciences of the USA* **99**: 8400–8405.

Iwamura, H., Suzuki, H., Ueda, Y., Kaya, T., and Inaba, T. (2001) In vitro and in vivo pharmacological characterization of JTE-907, a novel selective ligand for cannabinoid CB2 receptor, *Journal of Pharmacology and Experimental Therapeutics* **296**: 420–425.

Jarai, Z., Wagner, J.A., Varga, K., Lake, K.D., Compton, D.R., Martin, B.R., Zimmer, A.M., Bonner, T.I., Buckley, N.E., Mezey, E., Razdan, R.K., Zimmer, A., and Kunos, G. (1999) Cannabinoid-induced mesenteric vasodilation through an endothelial site distinct from CB1 or CB2 receptors, *Proceedings of the National Academy of Sciences of the USA* **96**: 14136–14141.

Jarai, Z., Wagner, J.A., Goparaju, S.K., Wang, L., Razdan, R.K., Sugiura, T., Zimmer, A.M., Bonner, T.I., Zimmer, A., and Kunos, G. (2000) Cardiovascular effects of 2-arachidonoyl glycerol in anesthetized mice, *Hypertension* **35**: 679–684.

Jarrahian, A. and Hillard, C.J. (1997) Arachidonylethanolamide (anandamide) binds with low affinity to dihydropyridine binding sites in brain membranes, *Prostaglandins, Leukotrienes and Essential Fatty Acids* **57**: 551–554.

Johnson, D.E., Heald, S.L., Dally, R.D., and Janis, R.A. (1993) Isolation, identification and synthesis of an endogenous arachidonic amide that inhibits calcium channel antagonist 1,4-dihydropyridine binding, *Prostaglandins, Leukotrienes and Essential Fatty Acids* **48**: 429–437.

Jorda, M.A., Verbakel, S.E., Valk, P.J., Vankan-Berkhoudt, Y.V., Maccarrone, M., Finazzi-Agro, A., Lowenberg, B., and Delwel, R. (2002) Hematopoietic cells expressing the peripheral cannabinoid receptor migrate in response to the endocannabinoid 2-arachidonoylglycerol, *Blood* **99**: 2786–2793.

Kagota, S., Yamaguchi, Y., Nakamura, K., Sugiura, T., Waku, K., and Kunitomo, M. (2001) 2-Arachidonoylglycerol, a candidate of endothelium-derived hyperpolarizing factor, *European Journal of Pharmacology* **415**: 233–238.

Kanoh, H., Iwata, T., Ono, T., and Suzuki, T. (1986) Immunological characterization of *sn*-1,2-diacylglycerol and *sn*-2-monoacylglycerol kinase from pig brain, *Journal of Biological Chemistry* **261**: 5597–5602.

Katayama, K., Ueda, N., Kurahashi, Y., Suzuki, H., Yamamoto, S., and Kato, I. (1997) Distribution of anandamide amidohydrolase in rat tissues with special reference to small intestine, *Biochimica et Biophysica Acta* **1347**: 212–218.

Katayama, K., Ueda, N., Katoh, I., and Yamamoto, S. (1999) Equilibrium in the hydrolysis and synthesis of cannabimimetic anandamide demonstrated by a purified enzyme, *Biochimica et Biophysica Acta* **1440**: 205–214.

Kearn, C.S., Greenberg, M.J., DiCamelli, R., Kurzawa, K., and Hillard, C.J. (1999) Relationships between ligand affinities for the cerebellar cannabinoid receptor CB1 and the induction of GDP/GTP exchange, *Journal of Neurochemistry* **72**: 2379–2387.

Kempe, K., Hsu, F.F., Bohrer, A., and Turk, J. (1996) Isotope dilution mass spectrometric measurements indicate that arachidonylethanolamide, the proposed endogenous ligand of the cannabinoid receptor, accumulates in rat brain tissue post mortem but is contained at low levels in or is absent from fresh tissue, *Journal of Biological Chemistry* **271**: 17287–17295.

Kirkham, T.C., Williams, C.M., Fezza, F., and Di Marzo, V. (2002) Endocannabinoid levels in rat limbic forebrain and hypothalamus in relation to fasting, feeding and satiation: stimulation of eating by 2-arachidonoyl glycerol, *The British Journal of Pharmacology* **136**: 550–557.

Kishimoto, S., Gokoh, M., Oka, S., Muramatsu, M., Kajiwara, T., Waku, K., and Sugiura, T. (2003) 2-Arachidonoylglycerol induces the migration of HL-60 cells differentiated into macrophage-like cells and human peripheral blood monocytes through the cannabinoid CB2 receptor-dependent mechanism, *Journal of Biological Chemistry* **278**: 24469–24475.

Kishimoto, S., Kobayashi, Y., Oka, S., Gokoh, M., Waku, K., and Sugiura, T. (2004) 2-Arachidonoylglycerol, an endogenous cannabinoid receptor ligand, induces accelerated production of chemokines in HL-60 cells through the cannabinoid CB2 receptor-dependent machanism, *Journal of Biochemistry* **135**: 517–524.

Kobayashi, Y., Arai, S., Waku, K., and Sugiura, T. (2001) Activation by 2-arachidonoylglycerol, an endogenous cannabinoid receptor ligand, of p42/44 mitogen-activated protein kinase in HL-60 cells, *Journal of Biochemistry* **129**: 665–669.

Koga, D., Santa, T., Fukushima, T., Homma, H., and Imai, K. (1997) Liquid chromatographic-atmospheric pressure chemical ionization mass spectrometric determination of anandamide and its analogs in rat brain and peripheral tissues, *Journal of Chromatography B, Biomedical Sciences and Applications* **690**: 7–13.

Kondo, S., Sugiura, T., Kodaka, T., Kudo, N., Waku, K., and Tokumura, A. (1998) Accumulation of various *N*-acylethanolamines including *N*-arachidonoylethanolamine (anandamide) in cadmium chloride-administered rat testis, *Archives of Biochemistry and Biophysics* **354**: 303–310.

Kondo, S., Kondo, H., Nakane, S., Kodaka, T., Tokumura, A., Waku, K., and Sugiura, T. (1998) 2-Arachidonoylglycerol, an endogenous cannabinoid receptor agonist: identification as one of the major species of monoacylglycerols in various rat tissues, and evidence for its generation through Ca^{2+}-dependent and -independent mechanisms, *FEBS Letters* **429**: 152–156.

Konrad, R.J., Major, C.D., and Wolf, B.A. (1994) Diacylglycerol hydrolysis to arachidonic acid is necessary for insulin secretion from isolated pancreatic islets: sequential actions of diacylglycerol and monoacylglycerol lipases, *Biochemistry* **33**: 13284–13294.

Kozak, K.R., Rowlinson, S.W., and Marnett, L.J. (2000) Oxygenation of the endocannabinoid, 2-arachidonylglycerol, to glyceryl prostaglandins by cyclooxygenase-2, *Journal of Biological Chemistry* **275**: 33744–33749.

Kozak, K.R., Prusakiewicz, J.J., Rowlinson, S.W., Schneider, C., and Marnett, L.J. (2001) Amino acid determinants in cyclooxygenase-2 oxygenation of the endocannabinoid 2-arachidonylglycerol, *Journal of Biological Chemistry* **276**: 30072–30077.

Kozak, K.R., Crews, B.C., Ray, J.L., Tai, H.H., Morrow, J.D., and Marnett, L.J. (2001) Metabolism of prostaglandin glycerol esters and prostaglandin ethanolamides in vitro and in vivo, *Journal of Biological Chemistry* **276**: 36993–36998.

Kozak, K.R. and Marnett, L.J. (2002) Oxidative metabolism of endocannabinoids, *Prostaglandins, Leukotrienes and Essential Fatty Acids* **66**: 211–220.

Kozak, K.R., Gupta, R.A., Moody, J.S., Ji, C., Boeglin, W.E., DuBois, R.N., Brash, A.R., and Marnett, L.J. (2002) 15-Lipoxygenase metabolism of 2-arachidonylglycerol. Generation of a peroxisome proliferator-activated receptor alpha agonist, *Journal of Biological Chemistry* **277**: 23278–23286.

Kozak, K.R., Crews, B.C., Morrow, J.D., Wang, L.H., Ma, Y.H., Weinander, R., Jakobsson, P.J., and Marnett, L.J. (2002) Metabolism of the endocannabinoids, 2-arachidonylglycerol and anandamide, into prostaglandin, thromboxane, and prostacyclin glycerol esters and ethanolamides, *Journal of Biological Chemistry* **277**: 44877–44885.

Kreitzer, A.C. and Regehr, W.G. (2001) Retrograde inhibition of presynaptic calcium influx by endogenous cannabinoids at excitatory synapses onto Purkinje cells, *Neuron* **29**: 717–727.

Kruszka, K.K. and Gross, R.W. (1994) The ATP- and CoA-independent synthesis of arachidonoylethanolamide. A novel mechanism underlying the synthesis of the endogenous ligand of the cannabinoid receptor, *Journal of Biological Chemistry* **269**: 14345–14348.

Kunos, G., Jarai, Z., Batkai, S., Goparaju, S.K., Ishac, E.J., Liu, J., Wang, L., and Wagner, J.A. (2000) Endocannabinoids as cardiovascular modulators, *Chemistry and Physics of Lipids* **108**: 159–168.

Kunos, G., Batkai, S., Offertaler, L., Mo, F., Liu, J., Karcher, J., and Harvey-White, J. (2002) The quest for a vascular endothelial cannabinoid receptor, *Chemistry and Physics of Lipids* **121**: 45–56.

Kurahashi, Y., Ueda, N., Suzuki, H., Suzuki, M., and Yamamoto S. (1997) Reversible hydrolysis and synthesis of anandamide demonstrated by recombinant rat fatty-acid amide hydrolase, *Biochemical and Biophysical Research Communications* **237**: 512–515.

Kurihara, J., Nishigaki, M., Suzuki, S., Okubo, Y., Takata, Y., Nakane, S., Sugiura, T., Waku, K., and Kato, H. (2001) 2-Arachidonoylglycerol and anandamide oppositely modulate norepinephrine release from the rat heart sympathetic nerves, *Japanese Journal of Pharmacology* **87**: 93–96.

Kuwae, T., Shiota, Y., Schmid, P.C., Krebsbach, R., and Schmid, H.H.O. (1999) Biosynthesis and turnover of anandamide and other *N*-acylethanolamines in peritoneal macrophages, *FEBS Letters* **459**: 123–127.

Lastres-Becker, I., Fezza, F., Cebeira, M., Bisogno, T., Ramos, J.A., Milone, A., Fernandez-Ruiz, J., and Di Marzo, V. (2001) Changes in endocannabinoid transmission in the basal ganglia in a rat model of Huntington's disease, *Neuroreport* **12**: 2125–2129.

Lee, M., Yang, K.H., and Kaminski, N.E. (1995) Effects of putative cannabinoid receptor ligands, anandamide and 2-arachidonyl-glycerol, on immune function in B6C3F1 mouse splenocytes, *Journal of Pharmacology and Experimental Therapeutics* **275**: 529–536.

Leweke, F.M., Giuffrida, A., Wurster, U., Emrich, H.M., and Piomelli, D. (1999) Elevated endogenous cannabinoids in schizophrenia, *Neuroreport* **10**: 1665–1669.

Liu, Q., Tonai, T., and Ueda, N. (2002) Activation of *N*-acylethanolamine-releasing phospholipase D by polyamines, *Chemistry and Physics of Lipids* **115**: 77–84.

Lutz, B. (2002) Molecular biology of cannabinoid receptors, *Prostaglandins, Leukotrienes and Essential Fatty Acids* **66**: 123–142.

Maccarrone, M., Van der Stelt, M., Rossi, A., Veldink, G.A., Vliegenthart J.F., and Agro, A.F. (1998) Anandamide hydrolysis by human cells in culture and brain, *Journal of Biological Chemistry* **273**: 32332–32339.

Maccarrone, M., De Felici, M., Bari, M., Klinger, F., Siracusa, G., and Finazzi-Agro, A. (2000) Down-regulation of anandamide hydrolase in mouse uterus by sex hormones, *European Journal of Biochemistry* **267**: 2991–2997.

Maccarrone, M., Attina, M., Cartoni, A., Bari, M., and Finazzi-Agro, A. (2001) Gas chromatography-mass spectrometry analysis of endogenous cannabinoids in healthy and tumoral human brain and human cells in culture, *Journal of Neurochemistry* **76**: 594–601.

Maccarrone, M., Bari, M., Menichelli, A., Giuliani, E., Del Principe, D., and Finazzi-Agro, A. (2001) Human platelets bind and degrade 2-arachidonoylglycerol, which activates these cells through a cannabinoid receptor, *European Journal of Biochemistry* **268**: 819–825.

Maccarrone, M., Valensise, H., Bari, M., Lazzarin, N., Romanini, C., and Finazzi-Agro, A. (2001) Progesterone up-regulates anandamide hydrolase in human lymphocytes: role of cytokines and implications for fertility, *Journal of Immunology* **166**: 7183–7189.

Maccarrone, M., Attina, M., Bari, M., Cartoni, A., Ledent, C., and Finazzi-Agro, A. (2001) Anandamide degradation and *N*-acylethanolamines level in wild-type and CB1 cannabinoid receptor knockout mice of different ages, *Journal of Neurochemistry* **78**: 339–348.

Maccarrone, M., Di Rienzo, M., Finazzi-Agro, A., and Rossi, A. (2003) Leptin activates the anandamide hydrolase promoter in human T lymphocytes through STAT3, *Journal of Biological Chemistry* **278**: 13318–13324.

Maccarrone, M., Gubellini, P., Bari, M., Picconi, B., Battista, N., Centonze, D., Bernardi, G., Finazzi-Agro, A., and Calabresi, P. (2003) Levodopa treatment reverses endocannabinoid system abnormalities in experimental parkinsonism, *Journal of Neurochemistry* **85**: 1018–1025.

Mackie, K., Devane, W.A., and Hille, B. (1993) Anandamide, an endogenous cannabinoid, inhibits calcium currents as a partial agonist in N18 neuroblastoma cells, *Molecular Pharmacology* **44**: 498–503.

Maingret, F., Patel, A.J., Lazdunski, M., and Honore, E. (2001) The endocannabinoid anandamide is a direct and selective blocker of the background K^+ channel TASK-1, *EMBO Journal* **20**: 47–54.

Marsicano, G., Wotjak, C.T., Azad, S.C., Bisogno, T., Rammes, G., and Cascio, M.G., Hermann, H., Tang, J., Hofmann, C., Zieglgansberger, W., Di Marzo, V., and Lutz, B. (2002) The endogenous cannabinoid system controls extinction of aversive memories, *Nature* **418**: 530–534.

Mascolo, N., Izzo, A.A., Ligresti, A., Costagliola, A., Pinto, L., Cascio, M.G., Maffia, P., Cecio, A., Capasso, F., and Di Marzo, V. (2002) The endocannabinoid system and the molecular basis of paralytic ileus in mice, *FASEB Journal* **16**: 1973–1975.

Matias, I., Pochard, P., Orlando, P., Salzet, M., Pestel, J., and Di Marzo, V. (2002) Presence and regulation of the endocannabinoid system in human dendritic cells, *European Journal of Biochemistry* **269**: 3771–3778.

Matias, I., Leonhardt, M., Lesage, J., De Petrocellis, L., Dupouy, J.P., Vieau, D., and Di Marzo, V. (2003) Effect of maternal under-nutrition on pup body weight and hypothalamic endocannabinoid levels, *Cellular and Molecular Life Sciences* **60**: 382–389.

Matsuda, L.A., Lolait, S.J., Brownstein, M.J., Young, A.C., and Bonner, T.I. (1990) Structure of a cannabinoid receptor and functional expression of the cloned cDNA, *Nature* **346**: 561–564.

Matsuda, S., Kanemitsu, N., Nakamura, A., Mimura, Y., Ueda, N., Kurahashi, Y., and Yamamoto, S. (1997) Metabolism of anandamide, an endogenous cannabinoid receptor ligand, in porcine ocular tissues, *Experimental Eye Research* **64**: 707–711.

Maurelli, S., Bisogno, T., De Petrocellis, L., Di Luccia, A., Marino, G., and Di Marzo, V. (1995) Two novel classes of neuroactive fatty acid amides are substrates for mouse neuroblastoma 'anandamide amidohydrolase', *FEBS Letters* **377**: 82–86.

McAllister, S.D., Griffin, G., Satin, L.S., and Abood, M.E. (1999) Cannabinoid receptors can activate and inhibit G protein-coupled inwardly rectifying potassium channels in a xenopus oocyte expression system, *Journal of Pharmacology and Experimental Therapeutics* **291**: 618–62.

McVey, D.C., Schmid, P.C., Schmid, H.H.O., and Vigna, S.R. (2003) Endocannabinoids induce ileitis in rats via the capsaicin receptor (VR1), *Journal of Pharmacology and Experimental Therapeutics* **304**: 713–722.

Mechoulam, R., Ben-Shabat, S., Hanus, L., Ligumsky, M., Kaminski, N.E., Schatz, A.R., Gopher, A., Almog S., Martin, B.R., Compton, D.R., Pertwee, R.G., Griffin, G., Bayewitch, M., Barg, J., and Vogel, Z. (1995) Identification of an endogenous 2-monoglyceride, present in canine gut, that binds to cannabinoid receptors, *Biochemical Pharmacology* **50**: 83–90.

Mechoulam, R., Fride, E., and Di Marzo, V. (1998) Endocannabinoids, *European Journal of Pharmacology* **359**: 1–18.

Mechoulam, R., Fride, E., Ben-Shabat, S., Meiri, U., and Horowitz, M. (1998) Carbachol, an acetylcholine receptor agonist, enhances production in rat aorta of 2-arachidonoyl glycerol, a hypotensive endocannabinoid, *European Journal of Pharmacology* **362**: R1–3.

Moesgaard, B., Petersen, G., Jaroszewski, J.W., and Hansen, H.S. (2000) Age dependent accumulation of *N*-acyl-ethanolamine phospholipids in ischemic rat brain. A ^{31}P NMR and enzyme activity study, *Journal of Lipid Research* **41**: 985–990.

Moesgaard, B., Petersen, G., Mortensen, S.A., and Hansen, H.S. (2002) Substantial species differences in relation to formation and degradation of *N*-acyl-ethanolamine phospholipids in heart tissue: an enzyme activity study, *Comparative Biochemistry and Physiology B, Biochemistry and Molecular Biology* **131**: 475–482.

Moody, J.S., Kozak, K.R., Ji, C., and Marnett, L.J. (2001) Selective oxygenation of the endocannabinoid 2-arachidonylglycerol by leukocyte-type 12-lipoxygenase, *Biochemistry* **40**: 861–866.

Moriyama, T., Urade, R., and Kito, M. (1999) Purification and characterization of diacylglycerol lipase from human platelets, *Journal of Biochemistry* **125**: 1077–1085.

Munro, S., Thomas, K.L., and Abu-Shaar, M. (1993) Molecular characterization of a peripheral receptor for cannabinoids, *Nature* **365**: 61–65.

Nakane, S., Oka, S., Arai, S., Waku, K., Ishima, Y., Tokumura, A., and Sugiura, T. (2002) 2-Arachidonoyl-*sn*-glycero-3-phosphate, an arachidonic acid-containing lysophosphatidic acid: occurrence and rapid enzymatic conversion to 2-arachidonoyl-*sn*-glycerol, a cannabinoid receptor ligand, in rat brain, *Archives of Biochemistry and Biophysics* **402**: 51–58.

Natarajan, V., Reddy, P.V., Schmid, P.C., and Schmid, H.H.O. (1982) *N*-Acylation of ethanolamine phospholipids in canine myocardium, *Biochimica et Biophysica Acta* 1982; **712**: 342–355.

Natarajan, V., Schmid, P.C., Reddy, P.V., Zuzarte-Augustin, M.L., and Schmid, H.H.O. (1983) Biosynthesis of *N*-acylethanolamine phospholipids by dog brain preparations, *Journal of Neurochemistry* **41**: 1303–1312.

Natarajan, V., Schmid, P.C., and Schmid, H.H.O. (1986) *N*-acylethanolamine phospholipid metabolism in normal and ischemic rat brain, *Biochimica et Biophysica Acta* **878**: 32–41.

Ohno-Shosaku, T., Maejima, T., and Kano, M. (2001) Endogenous cannabinoids mediate retrograde signals from depolarized postsynaptic neurons to presynaptic terminals, *Neuron* **29**: 729–738.

Oka, S., Tsuchie, A., Tokumura, A., Muramatsu, M., Suhara, Y., Takayama, H., Waku, K., and Sugiura, T. (2003) Ether-linked analogue of 2-arachidonoylglycerol (noladin ether) was not detected in the brains of various mammalian species, *Journal of Neurochemistry* **85**: 1374–1381.

Olah, Z., Karai, L., and Iadarola, M.J. (2001) Anandamide activates vanilloid receptor 1 (VR1) at acidic pH in dorsal root ganglia neurons and cells ectopically expressing VR1, *Journal of Biological Chemistry* **276**: 31163–31170.

Omeir, R.L., Chin, S., Hong, Y., Ahern, D.G., and Deutsch, D.G. (1995) Arachidonoyl ethanolamide-[1,2-^{14}C] as a substrate for anandamide amidase, *Life Sciences* **56**: 1999–2005.

Ouyang, Y., Hwang, S.G., Han, S.H., and Kaminski, N.E. (1998) Suppression of interleukin-2 by the putative endogenous cannabinoid 2-arachidonyl-glycerol is mediated through down-regulation of the nuclear factor of activated T cells, *Molecular Pharmacology* **53**: 676–683.

Oz, M., Tchugunova, Y.B., and Dunn, S.M. (2000) Endogenous cannabinoid anandamide directly inhibits voltage-dependent Ca^{2+} fluxes in rabbit T-tubule membranes, *European Journal of Pharmacology* **404**: 13–20.

Pagotto, U., Marsicano, G., Fezza, F., Theodoropoulou, M., Grubler, Y., Stalla, J., Arzberger, T., Milone, A., Losa, M., Di Marzo, V., Lutz, B., and Stalla, G.K. (2001) Normal human pituitary gland and pituitary adenomas express cannabinoid receptor type 1 and synthesize endogenous cannabinoids: first evidence for a direct role of cannabinoids on hormone modulation at the human pituitary level, *Journal of Clinical Endocrinology and Metabolism* **86**: 2687–2696.

Panikashvili, D., Simeonidou, C., Ben-Shabat, S., Hanus, L., Breuer, A., Mechoulam, R., and Shohami, E. (2001) An endogenous cannabinoid (2-AG) is neuroprotective after brain injury, *Nature* **413**: 527–531.

Paria, B.C., Deutsch, D.D., and Dey, S.K. (1996) The uterus is a potential site for anandamide synthesis and hydrolysis: differential profiles of anandamide synthase and hydrolase activities in the mouse uterus during the periimplantation period, *Molecular Reproduction and Development* **45**: 183–192.

Parolaro, D., Massi, P., Rubino, T., and Monti, E. (2002) Endocannabinoids in the immune system and cancer, *Prostaglandins, Leukotrienes and Essential Fatty Acids* **66**: 319–332.

Pertwee, R.G. (1997) Pharmacology of cannabinoid CB1 and CB2 receptors, *Pharmacology and Therapeutics* **74**: 129–180.

Pertwee, R.G., and Ross, R.A. (2002) Cannabinoid receptors and their ligands, *Prostaglandins, Leukotrienes and Essential Fatty Acids* **66**: 101–121.

Pestonjamasp, V.K. and Burstein, S.H. (1998) Anandamide synthesis is induced by arachidonate mobilizing agonists in cells of the immune system, *Biochimica et Biophysica Acta* **1394**: 249–260.

Petersen, G. and Hansen, H.S. (1999) *N*-acylphosphatidylethanolamine-hydrolysing phospholipase D lacks the ability to transphosphatidylate, *FEBS Letters* **455**: 41–44.

Piomelli, D., Beltramo, M., Giuffrida, A., and Stella, N. (1998) Endogenous cannabinoid signaling, *Neurobiology of Disease* **5**: 462–473.

Piomelli, D., Giuffrida, A., Calignano, A., and Rodriguez de Fonseca, F. (2000) The endocannabinoid system as a target for therapeutic drugs, *Trends in Pharmacological Sciences* **21**: 218–224.

Porter, A.C., Sauer, J.M., Knierman, M.D., Becker, G.W., Berna, M.J., Bao, J., Nomikos, G.G., Carter, P., Bymaster, F.P., Leese, A.B., and Felder, C.C. (2002) Characterization of a novel endocannabinoid, virodhamine, with antagonist activity at the CB1 receptor, *Journal of Pharmacology and Experimental Therapeutics* **301**: 1020–1024.

Prescott, S.M. and Majerus, P.W. (1983) Characterization of 1,2-diacylglycerol hydrolysis in human platelets. Demonstration of an arachidonoyl-monoacylglycerol intermediate, *Journal of Biological Chemistry* **258**: 764–769.

Randall, M.D., Harris, D., and Kendall, D.A. (2002) The vascular pharmacology of endocannabinoids, in *Biology of Marijuana*, E.S. Ed., Onaivi, London: Taylor and Francis, pp. 542–553.

Rebecchi, M.J. and Pentyala, S.N. (2000) Structure, function, and control of phosphoinositide-specific phospholipase C, *Physiological Reviews* **80**: 1291–1335.

Rhee, S.G. and Bae, Y.S. (1997) Regulation of phosphoinositide-specific phospholipase C isozymes, *Journal of Biological Chemistry* **272**: 15045–15048.

Rodriguez de Fonseca, F., Navarro, M., Gomez, R., Escuredo, L., Nava, F., Fu, J., Murillo-Rodriguez, E., Giuffrida, A., LoVerme, J., Gaetani, S., Kathuria, S., Gall, C., and Piomelli, D. (2001) An anorexic lipid mediator regulated by feeding, *Nature* **414**: 209–212.

Ross, R.A., Gibson, T.M., Brockie, H.C., Leslie, M., Pashmi, G., Craib, S.J., Di Marzo, V., and Pertwee, R.G. (2001) Structure-activity relationship for the endogenous cannabinoid, anandamide, and certain of its analogues at vanilloid receptors in transfected cells and vas deferens, *British Journal of Pharmacology* **132**: 631–640.

Ross, R.A., Craib, S.J., Stevenson, L.A., Pertwee, R.G., Henderson, A., Toole, J., and Ellington, H.C. (2002) Pharmacological characterization of the anandamide cyclooxygenase metabolite: prostaglandin E_2 ethanolamide, *Journal of Pharmacology and Experimental Therapeutics* **301**: 900–907.

Roth, M.D., Baldwin, G.C., and Tashkin, D.P. (2002) Effects of delta-9-tetrahydrocannabinol on human immune function and host defense, *Chemistry and Physics of Lipids* **121**: 229–239.

Sagan, S., Venance, L., Torrens, Y., Cordier, J., Glowinski, J., and Giaume, C. (1999) Anandamide and WIN 55212-2 inhibit cyclic AMP formation through G-protein-coupled receptors distinct from CB1 cannabinoid receptors in cultured astrocytes, *European Journal of Neuroscience* **11**: 691–699.

Sasaki, T. and Chang, M.C. (1997) *N*-arachidonylethanolamine (anandamide) formation from *N*-arachidonylphosphatidylethanolamine in rat brain membranes, *Life Sciences* **61**: 1803–1810.

Schlicker, E. and Kathmann, M. (2001) Modulation of transmitter release via presynaptic cannabinoid receptors, *Trends in Pharmacological Sciences* **22**: 565–572.

Schmid, H.H.O., Schmid, P.C., and Natarajan, V. (1990) *N*-acylated glycerophospholipids and their derivatives, *Progress in Lipid Research* **29**: 1–43.

Schmid, H.H.O., Schmid, P.C., and Natarajan, V. (1996) The *N*-acylation-phosphodiesterase pathway and cell signalling, *Chemistry and Physics of Lipids* **80**: 133–142.

Schmid, H.H.O. (2000) Pathways and mechanisms of *N*-acylethanolamine biosynthesis: can anandamide be generated selectively? *Chemistry and Physics of Lipids* **108**: 71–87.

Schmid, H.H.O., Schmid, P.C., and Berdyshev, E.V. (2002) Cell signaling by endocannabinoids and their congeners: questions of selectivity and other challenges, *Chemistry and Physics of Lipids* **121**: 111–134.

Schmid, H.H.O. and Berdyshev, E.V. (2002) Cannabinoid receptor-inactive *N*-acylethanolamines and other fatty acid amides: metabolism and function, *Prostaglandins, Leukotrienes and Essential Fatty Acids* **66**: 363–376.

Schmid, P.C., Reddy, P.V., Natarajan, V., and Schmid, H.H.O. (1983) Metabolism of *N*-acylethanolamine phospholipids by a mammalian phosphodiesterase of the phospholipase D type, *Journal of Biological Chemistry* **258**: 9302–9306.

Schmid, P.C., Zuzarte-Augustin, M.L., and Schmid, H.H.O. (1985) Properties of rat liver *N*-acylethanolamine amidohydrolase, *Journal of Biological Chemistry* **260**: 14145–14149.

Schmid, P.C., Krebsbach, R.J., Perry, S.R., Dettmer, T.M., Maasson, J.L., and Schmid H.H.O. (1995) Occurrence and postmortem generation of anandamide and other long-chain *N*-acylethanolamines in mammalian brain, *FEBS Letters* **375**: 117–120. Erratum *FEBS Letters* (1996) **385**: 124–130

Schmid, P.C., Kuwae, T., Krebsbach, R.J., and Schmid, H.H.O. (1997) Anandamide and other *N*-acylethanolamines in mouse peritoneal macrophages, *Chemistry and Physics of Lipids* **87**: 103–110.

Schmid, P.C., Paria, B.C., Krebsbach, R.J., Schmid, H.H.O., and Dey, S.K. (1997) Changes in anandamide levels in mouse uterus are associated with uterine receptivity for embryo implantation, *Proceedings of the National Academy of Sciences of the USA* **94**: 4188–4192.

Schmid, P.C., Schwindenhammer, D., Krebsbach, R.J., and Schmid, H.H.O. (1998) Alternative pathways of anandamide biosynthesis in rat testes, *Chemistry and Physics of Lipids* **92**: 27–35.

Schmid, P.C., Schwartz, K.D., Smith, C.N., Krebsbach, R.J., Berdyshev, E.V., and Schmid, H.H.O. (2000) A sensitive endocannabinoid assay. The simultaneous analysis of *N*-acylethanolamines and 2-monoacylglycerols, *Chemistry and Physics of Lipids* **104**: 185–191.

Schmid, P.C., Wold, L.E., Krebsbach, R.J., Berdyshev, E.V., and Schmid, H.H.O. (2002) Anandamide and other *N*-acylethanolamines in human tumors, *Lipids* **37**: 907–912.

Schuel, H., Burkman, L.J., Lippes, J., Crickard, K., Forester, E., Piomelli, D., and Giuffrida, A. (2002) *N*-Acylethanolamines in human reproductive fluids, *Chemistry and Physics of Lipids* **121**: 211–227.

Shim, Y.H., Lin, C.H., and Strickland, K.P. (1989) The purification and properties of monoacylglycerol kinase from bovine brain, *Biochemistry and Cell Biology* **67**: 233–241.

Shimasue, K., Urushidani, T., Hagiwara, M., and Nagao, T. (1996) Effects of anandamide and arachidonic acid on specific binding of (+)-PN200-110, diltiazem and (-)-desmethoxyverapamil to L-type Ca^{2+} channel, *European Journal of Pharmacology* **296**: 347–350.

Simpson, C.M., Itabe, H., Reynolds, C.N., King, W.C., and Glomset, J.A. (1991) Swiss 3T3 cells preferentially incorporate *sn*-2-arachidonoyl monoacylglycerol into *sn*-1-stearoyl-2-arachidonoyl phosphatidylinositol, *Journal of Biological Chemistry* **266**: 15902–15909.

Sinor, A.D., Irvin, S.M., and Greenberg, D.A. (2000) Endocannabinoids protect cerebral cortical neurons from in vitro ischemia in rats, *Neuroscience Letters* **278**: 157–160.

Sipe, J.C., Chiang, K., Gerber, A.L., Beutler, E., and Cravatt, B.F. (2002) A missense mutation in human fatty acid amide hydrolase associated with problem drug use, *Proceedings of the National Academy of Sciences of the USA* **99**: 8394–8399.

Smart, D., Gunthorpe, M.J., Jerman, J.C., Nasir, S., Gray, J., Muir, A.I., Chambers, J.K., Randall, A.D., and Davis, J.B. (2000) The endogenous lipid anandamide is a full agonist at the human vanilloid receptor (hVR1), *British Journal of Pharmacology* **129**: 227–230.

Smith, P.B., Compton, D.R., Welch, S.P., Razdan, R.K., Mechoulam, R., and Martin, B.R. (1994) The pharmacological activity of anandamide, a putative endogenous cannabinoid, in mice, *Journal of Pharmacology and Experimental Therapeutics* **270**: 219–227.

Stamer, W.D., Golightly, S.F., Hosohata, Y., Ryan, E.P., Porter, A.C., Varga, E., Noecker, R.J., Felder, C.C., and Yamamura, H.I. (2001) Cannabinoid CB(1) receptor expression, activation and detection of endogenous ligand in trabecular meshwork and ciliary process tissues, *European Journal of Pharmacology* **431**: 277–286.

Stefano, G.B., Bilfinger, T.V., Rialas, C.M., and Deutsch, D.G. (2000) 2-Arachidonyl-glycerol stimulates nitric oxide release from human immune and vascular tissues and invertebrate immunocytes by cannabinoid receptor 1, *Pharmacological Research* **42**: 317–322.

Stella, N., Schweitzer, P., and Piomelli, D. (1997) A second endogenous cannabinoid that modulates long-term potentiation, *Nature* **388**: 773–778.

Stella, N. and Piomelli, D. (2001) Receptor-dependent formation of endogenous cannabinoids in cortical neurons, *European Journal of Pharmacology* **425**: 189–196.

Straiker, A., Stella, N., Piomelli, D., Mackie, K., Karten, H.J., and Maguire, G. (1999) Cannabinoid CB1 receptors and ligands in vertebrate retina: localization and function of an endogenous signaling system, *Proceedings of the National Academy of Sciences of the USA* **96**: 14565–14570.

Sugiura, T. and Waku, K. (1987) Composition of alkyl ether-linked phospholipids in mammalian tissues, in *Platelet-Activating Factor and Related Lipid Mediators*, F. Snyder, Ed., New York: Plenum Press, pp. 55–85.

Sugiura, T., Itoh, K., Waku, K., and Hanahan, D.J. (1994) Biological activities of *N*-acylethanolamine and *N*-acylethanolamine phosphate analogues (in Japanese), *Proceedings of Japanese Conference on the Biochemistry of Lipids* **36**: 71–74.

Sugiura, T., Tokumura, A., Gregory, L, Nouchi, T., Weintraub, S.T., and Hanahan, D.J. (1994) Biochemical characterization of the interaction of lipid phosphoric acids with human platelets: comparison with platelet activating factor, *Archives of Biochemistry and Biophysics* **311**: 358–368.

Sugiura, T., Kondo, S., Sukagawa, A. Nakane, S., Shinoda, A., Itoh, K., Yamashita, A., and Waku, K. (1995) 2-Arachidonoylglycerol: a possible endogenous cannabinoid receptor ligand in brain, *Biochemical and Biophysical Research Communications* **215**: 89–97.

Sugiura, T., Kondo, S., Sukagawa, A., Tonegawa, T., Nakane, S., Yamashita, A., and Waku, K. (1996) Enzymatic synthesis of anandamide, an endogenous cannabinoid receptor ligand, through *N*-acylphosphatidyle thanolamine pathway in testis: involvement of Ca^{2+}-dependent transacylase and phosphodiesterase activities, *Biochemical and Biophysical Research Communications* **218**: 113–117.

Sugiura, T., Kodaka, T., Kondo, S., Tonegawa, T., Nakane, S., Kishimoto, S., Yamashita, A., and Waku, K. (1996) 2-Arachidonoylglycerol, a putative endogenous cannabinoid receptor ligand, induces rapid, transient elevation of intracellular free Ca^{2+} in neuroblastoma x glioma hybrid NG108-15 cells, *Biochemical and Biophysical Research Communications* **229**: 58–64.

Sugiura, T., Kondo, S., Sukagawa, A., Tonegawa, T., Nakane, S., Yamashita, A., Ishima, Y., and Waku, K. (1996) Transacylase-mediated and phosphodiesterase-mediated synthesis of *N*-arachidonoylethanolamine, an endogenous cannabinoid-receptor ligand, in rat brain microsomes. Comparison with synthesis from free arachidonic acid and ethanolamine, *European Journal of Biochemistry* **240**: 53–62.

Sugiura, T., Kodaka, T., Kondo, S., Tonegawa, T., Nakane, S., Kishimoto, S., Yamashita, A., and Waku, K. (1997) Inhibition by 2-arachidonoylglycerol, a novel type of possible neuromodulator, of the depolarization-induced increase in intracellular free calcium in neuroblastoma x glioma hybrid NG108-15 cells, *Biochemical and Biophysical Research Communications* **233**: 207–210.

Sugiura, T., Kodaka, T., Kondo, S., Nakane, S., Kondo, H., Waku, K., Ishima, Y., Watanabe, K., and Yamamoto, I. (1997) Is the cannabinoid CB1 receptor a 2-arachidonoylglycerol receptor? Structural requirements for triggering a Ca^{2+} transient in NG108-15 cells, *Journal of Biochemistry* **122**: 890–895.

Sugiura, T., Kodaka, T., Nakane, S., Kishimoto, S., Kondo, S., and Waku, K. (1998) Detection of an endogenous cannabimimetic molecule, 2-arachidonoylglycerol, and cannabinoid CB1 receptor mRNA in human vascular cells: is 2-arachidonoylglycerol a possible vasomodulator? *Biochemical and Biophysical Research Communications* **243**: 838–843.

Sugiura, T., Kondo, S., Kodaka, T., Nakane, S., Yamashita, A., Kishimoto, S., Waku, K., and Ishima, Y. (1998) Biosynthesis and actions of anandamide and 2-arachidonoylglycerol: two endogenous cannabimimetic molecules, in *Essential Fatty Acids and Eicosanoids*, R.A. Riemesma, R. Armstrong, W. Kelly, and R. Wilson, Eds., Champaign: AOCS Press, pp. 380–384.

Sugiura T., Nakane S., Kishimoto S., Waku, K., Yoshioka, Y., Tokumura, A., and Hanahan, D.J. (1999) Occurrence of lysophosphatidic acid and its alkyl ether-linked analog in rat brain and comparison of their biological activities toward cultured neural cells, *Biochimica et Biophysica Acta* **1440**: 194–204.

Sugiura, T., Kodaka, T., Nakane, S., Miyashita, T., Kondo, S., Suhara, Y., Takayama, H., Waku, K., Seki, C., Baba, N., and Ishima, Y. (1999) Evidence that the cannabinoid CB1 receptor is a 2-arachidonoylglycerol receptor. Structure-activity relationship of 2-arachidonoylglycerol, ether-linked analogues, and related compounds, *Journal of Biological Chemistry* **274**: 2794–2801.

Sugiura, T. and Waku, K. (2000) 2-Arachidonoylglycerol and the cannabinoid receptors, *Chemistry and Physics of Lipids* **108**: 89–106.

Sugiura, T., Yoshinaga, N., Kondo, S., Waku, K., and Ishima, Y. (2000) Generation of 2-arachidonoylglycerol, an endogenous cannabinoid receptor ligand, in picrotoxinin-administered rat brain, *Biochemical and Biophysical Research Communications* **271**: 654–658.

Sugiura, T., Kondo, S., Kishimoto, S., Miyashita, T., Nakane, S., Kodaka, T., Suhara, Y., Takayama, H., and Waku, K. (2000) Evidence that 2-arachidonoylglycerol but not *N*-palmitoylethanolamine or anandamide is the physiological ligand for the cannabinoid CB2 receptor. Comparison of the agonistic activities of various cannabinoid receptor ligands in HL-60 cells, *Journal of Biological Chemistry* **275**: 605–612.

Sugiura, T., Yoshinaga, N., and Waku, K. (2001) Rapid generation of 2-arachidonoylglycerol, an endogenous cannabinoid receptor ligand, in rat brain after decapitation, *Neuroscience Letters* **297**: 175–178.

Sugiura, T. and Waku, K. (2002) Cannabinoid receptors and their endogenous ligands, *Journal of Biochemistry* **132**: 7–12.

Sugiura, T., Kobayashi, Y., Oka, S., and Waku, K. (2002) Biosynthesis and degradation of anandamide and 2-arachidonoylglycerol and their possible physiological significance, *Prostaglandins, Leukotrienes and Essential Fatty Acids* **66**: 173–192.

Sugiura, T., Kishimoto, S., Oka, S., Gokoh, M., and Waku, K. (2004) Metabolism and physiological significance of anandamide and 2-arachidonoylglycerol, endogenous cannabinoid receptor ligands, in *Arachidonate Remodeling and Inflammation*, A.N. Fonteh and R.L. Wykle, Eds., Basel: Birkhauser Verlag.

Tognetto, M., Amadesi, S., Harrison, S., Creminon, C., Trevisani, M., Carreras, M., Matera, M., Geppetti, P., and Bianchi, A. (2001) Anandamide excites central terminals of dorsal root ganglion neurons via vanilloid receptor-1 activation, *Journal of Neuroscience* **21**: 1104–1109.

Tsutsumi, T., Kobayashi, T., Ueda, H., Yamauchi, E., Watanabe, S., and Okuyama, H. (1994) Lysophosphoinositide-specific phospholipase C in rat brain synaptic plasma membranes, *Neurochemical Research* **19**: 399–406.

Tsutsumi, T., Kobayashi, T., Miyashita, M., Watanabe, S., Homma, Y., and Okuyama, H. (1995) A lysophosphoinositide-specific phospholipase C distinct from other phospholipase C families in rat brain, *Archives of Biochemistry and Biophysics* **317**: 331–336.

Ueda, H., Kobayashi, T., Kishimoto, M., Tsutsumi, T., and Okuyama, H. (1993) A possible pathway of phosphoinositide metabolism through EDTA-insensitive phospholipase A_1 followed by lysophosphoinositide-specific phospholipase C in rat brain, *Journal of Neurochemistry* **61**: 1874–1881.

Ueda, N., Kurahashi, Y., Yamamoto, S., and Tokunaga, T. (1995) Partial purification and characterization of the porcine brain enzyme hydrolyzing and synthesizing anandamide, *Journal of Biological Chemistry* **270**: 23823–23827.

Ueda, N., Yamamoto, K., Yamamoto, S., Tokunaga, T., Shirakawa, E., Shinkai, H., Ogawa, M., Sato, T., Kudo, I., Inoue, K., Takizawa, H., Nagano, T., Hirobe, M., Matsuki, N., and Saito, H. (1995) Lipoxygenase-catalyzed oxygenation of arachidonylethanolamide, a cannabinoid receptor agonist, *Biochimica et Biophysica Acta* **1254**: 127–134.

Ueda, N., Yamanaka, K., Terasawa, Y., and Yamamoto, S. (1999) An acid amidase hydrolyzing anandamide as an endogenous ligand for cannabinoid receptors, *FEBS Letters* **454**: 267–270.

Ueda, N. and Yamamoto, S. (2000) Anandamide amidohydrolase (fatty acid amide hydrolase), *Prostaglandins and Other Lipid Mediators* **61**: 19–28.

Ueda, N., Puffenbarger, R.A., Yamamoto, S., and Deutsch, D.G. (2000) The fatty acid amide hydrolase (FAAH), *Chemistry and Physics of Lipids* **108**:107–121.

Ueda, N., Liu, Q., and Yamanaka, K. (2001) Marked activation of the *N*-acylphosphatidylethanolamine-hydrolyzing phosphodiesterase by divalent cations, *Biochimica et Biophysica Acta* **1532**: 121–127.

Ueda, N., Yamanaka, K., and Yamamoto, S. (2001) Purification and characterization of an acid amidase selective for N-palmitoylethanolamine, a putative endogenous anti-inflammatory substance, *Journal of Biological Chemistry* **276**: 35552–35557.

Varga, K., Wagner, J.A., Bridgen, D.T., and Kunos, G. (1998) Platelet- and macrophage-derived endogenous cannabinoids are involved in endotoxin-induced hypotension, *FASEB Journal* **12**: 1035–1044.

Wagner, J.A., Varga, K., Ellis, E.F., Rzigalinski, B.A., Martin, B.R., and Kunos, G. (1997) Activation of peripheral CB_1 cannabinoid receptors in haemorrhagic shock, *Nature* **390**: 518–521.

Wagner, J.A., Varga, K., Jarai, Z., and Kunos, G. (1999) Mesenteric vasodilation mediated by endothelial anandamide receptors, *Hypertension* **33**: 429–434.

Walker, J.M., Huang, S.M., Strangman, N.M., Tsou, K., and Sanudo-Pena, M.C. (1999) Pain modulation by release of the endogenous cannabinoid anandamide, *Proceedings of the National of Academy of Sciences of the USA* **96**: 12198–12203.

Walker, J.M., Krey, J.F., Chu, C.J., and Huang, S.M. (2002) Endocannabinoids and related fatty acid derivatives in pain modulation, *Chemistry and Physics of Lipids* **121**: 159–172.

Walter, L., Franklin, A., Witting, A., Moller, T., and Stella, N. (2002) Astrocytes in culture produce anandamide and other acylethanolamides, *Journal of Biological Chemistry* **277**: 20869–20876.

Walter, L., Franklin, A., Witting, A., Wade, C., Xie, Y., Kunos, G., Mackie, K., and Stella, N. (2003) Nonpsychotropic cannabinoid receptors regulate microglial cell migration, *Journal of Neuroscience* **23**: 1398–1405.

Wang, Y., Liu, Y., Ito, Y., Hashiguchi, T., Kitajima, I., Yamakuchi, M., Shimizu, H., Matsuo, S., Imaizumi, H., and Maruyama, I. (2001) Simultaneous measurement of anandamide and 2-arachidonoylglycerol by polymyxin b-selective adsorption and subsequent high-performance liquid chromatography analysis: increase in endogenous cannabinoids in the sera of patients with endotoxic shock, *Analytical Biochemistry* **294**: 73–82.

Watanabe, K., Ogi, H., Nakamura, S., Kayano, Y., Matsunaga, T., Yoshimura, H., and Yamamoto, I. (1998) Distribution and characterization of anandamide amidohydrolase in mouse brain and liver, *Life Sciences* **62**: 1223–1229.

Wenger, T., Gerendai, I., Fezza, F., Gonzalez, S., Bisogno, T., Fernandez-Ruiz, J., and Di Marzo, V. (2002) The hypothalamic levels of the endocannabinoid, anandamide, peak immediately before the onset of puberty in female rats, *Life Sciences* **70**: 1407–1414. Erratum *Life Sciences* **71**: 1349–1350.

Wilson, R.I. and Nicoll, R.A. (2001) Endogenous cannabinoids mediate retrograde signalling at hippocampal synapses, *Nature* **410**: 588–592.

Yang, H.Y., Karoum, F., Felder, C., Badger, H., Wang, T.C., and Markey, S.P. (1999) GC/MS analysis of anandamide and quantification of *N*-arachidonoylphosphatidylethanolamides in various brain regions, spinal cord, testis, and spleen of the rat, *Journal of Neurochemistry* **72**: 1959–1968.

Yu, M., Ives, D., and Ramesha, C.S. (1997) Synthesis of prostaglandin E_2 ethanolamide from anandamide by cyclooxygenase-2, *Journal of Biological Chemistry* **272**: 21181–21186.

Zygmunt, P.M., Petersson, J., Andersson, D.A., Chuang, H., Sorgard, M., Di Marzo, V., Julius, D., and Hogestatt, E.D. (1999) Vanilloid receptors on sensory nerves mediate the vasodilator action of anandamide, *Nature* **400**: 452–457.

8 Endocannabinoids and Eicosanoids: All in the Family

Sumner Burstein

CONTENTS

INTRODUCTION

The connections between endocannabinoids and eicosanoids occur in several areas. First, there is the obvious common structural feature based on arachidonic acid (eicosatetraeneoic acid) as well as other long-chain polyunsaturated acids. In the case of the eicosanoids, all of the members are either cyclization or oxidation products of arachidonic acid whereas in the endocannabinoids the fatty acid structure is preserved (Figure 8.1).

Other similarities are the biosynthetic and metabolic pathways in which each is involved. Several of the biosynthetic routes that lead to the eicosanoids also cause bioconversions of certain endocannabinoids into eicosanoid-like substances. This might suggest that endocannabinoids can serve as alternative substrates for eicosanoid synthesis. If such were the case, there would be important implications for physiological and pharmacological regulation by the endocannabinoids.

Areas where the similarities are more blurred occur in the biological effects of each family. Both display an impressive scope of actions, so it is inevitable that similarities can be perceived. Whether these similarities involve related mechanisms is something that remains to be proved in each instance.

Numerous reviews have been written on the endocannabinoids, such as recent examples by Di Marzo, Bisogno, et al. (1999), Sugiura et al. (2002), and Sugiura and Waku (2002), in which detailed presentations covering all aspects of these substances can be found. Comparisons between the eicosanoids and cannabinoids have also been previously published; however, these dealt primarily with exogenous cannabinoids (Burstein et al., 1995; Burstein, 2002).

BIOSYNTHESIS OF PGs AND LEUKOTRIENES

Although they were discovered in 1930, these lipid mediators are still actively studied, and many questions concerning their mode of action remain unanswered. Numerous reviews have been published on this subject, and a recent example can be found that gives a detailed description of

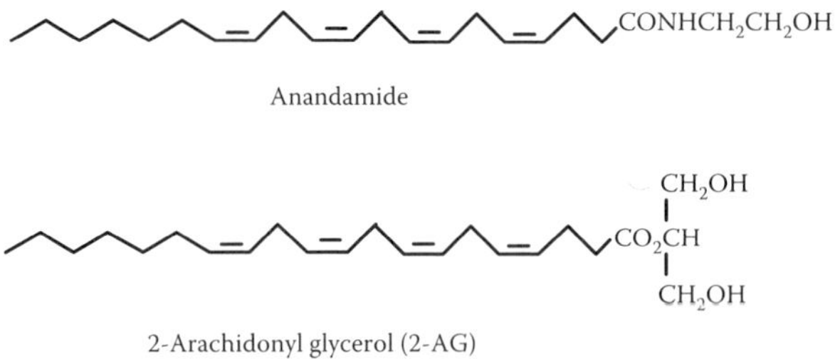

$CONHCH_2CH_2OH$

Anandamide

CH_2OH
CO_2CH
CH_2OH

2-Arachidonyl glycerol (2-AG)

FIGURE 8.1 The principal endogenous cannabinoids. Anandamide and 2-arachidonyl glycerol (2-AG) are the most well-studied endocannabinoids; however, conjugates with other fatty acids such as palmitic and eicosapenteneoic acid are well known. Novel structures such as virodhamine and noladin ether have also been reported. Arachidonic acid derivatives with other molecules such as dopamine (NADA) and glycine (NAGly) are also known.

the current status of our understanding of this complex story (Funk, 2001). A brief summary of the field will be given here for the purpose of comparison with the endocannabinoids.

The stereospecific introduction of oxygen into the 5-position of arachidonic acid is catalyzed by the nonheme iron dioxygenase 5-lipoxygenase. This initiates a cascade of transformations that result in the synthesis of an array of potent mediators of cellular activities, some of which are illustrated in Figure 8.2. Simple reduction of the hydroperoxy group leads to 5-hydroxyeicosatrienoic acid (HETE), which is a potent chemotactic agent as well as a promoter of islet cell insulin release. Epoxide formation leads to leukotriene (LT)A_4 which is a central precursor for a number of metabolites such as LTB_4 and LTC_4, which in turn lead to about two dozen known products (not shown).

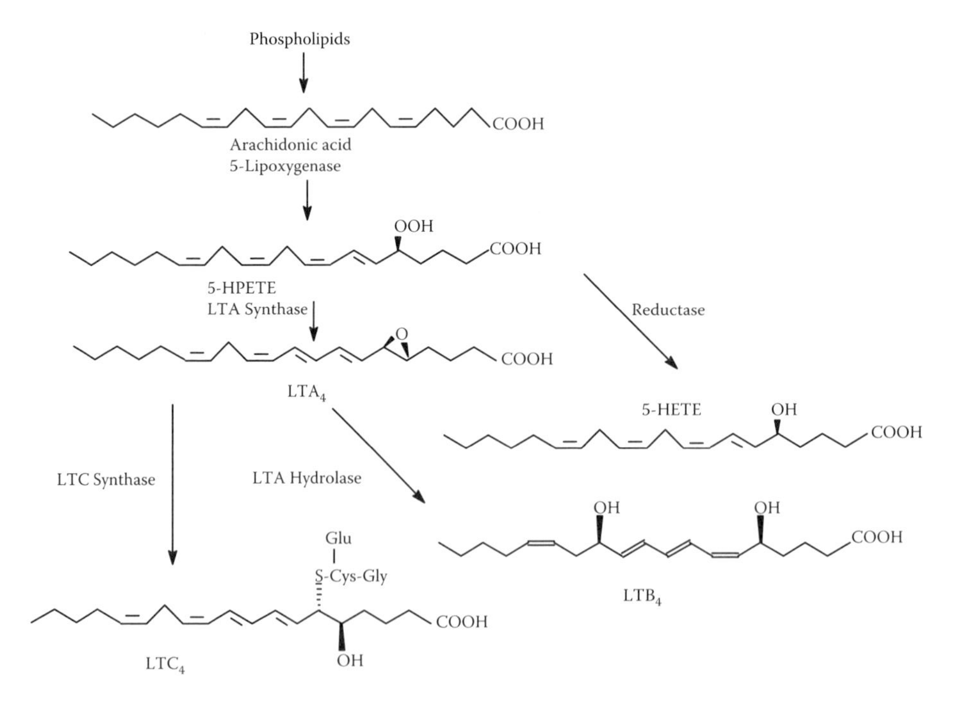

FIGURE 8.2 5-Lipoxygenase pathways. The key intermediate in this family of metabolites is 5-HPETE, which can serve as a precursor for three subsequent cascades of products of which only a few examples are shown here. LT = leukotriene.

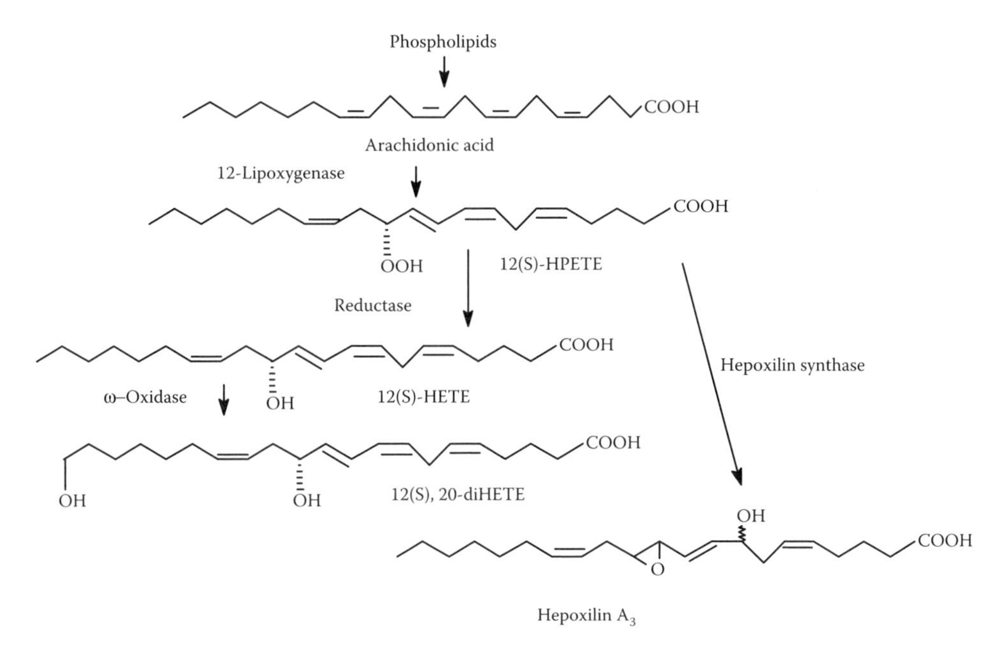

FIGURE 8.3 12-Lipoxygenase pathways. The key intermediate in this family of metabolites is 12(S)-HPETE, which can serve as a precursor for two subsequent cascades of products of which only a few examples are shown here.

In addition to chemotaxis, LTB_4 modulates adhesion, proliferation, and differentiation of cells and induces vascular permeability, hyperalgesia, NK cell activity, and bronchoconstriction through a variety of receptors. LTC_4 induces hypotension and stimulates angiogenesis, LH, LHRH, and prolactin release and opens muscarinic potassium channels. Its structure is somewhat unique being a chimer of a tripeptide and a long-chain fatty acid derivative.

Arachidonic acid is also metabolized at the 12-position as shown in Figure 8.3 to give rise to a group of mono- and poly-oxygenated products. The initial metabolite 12(S)-hydroperoxyeicosatrienoic acid (HPETE), mediates a number of biological responses and is produced in platelets and leucocytes. In aplysia sensory neurons, it mediates the FMRF amide inhibitory synaptic response, and in rat brain cortex it inhibits protein kinase II. It is rapidly reduced to 12(S)-HETE, which is a potent promoter of tumor cell adhesion. An alternate pathway mediated by hepoxilin synthase gives rise to hepoxilin A_3, which exhibits a variety of actions, including modulation of synaptic transmission in rat hippocampus.

A third lipoxygenase acts at the 15-position of arachidonic acid to produce 15-HPETE as shown in Figure 8.4. This product can be further metabolized via two pathways, a reductase leading to 15-HETE and a 12-lipoxygenase that gives rise to 14,15-LTA_4. The latter can be further transformed to the chimeric tripeptide lipid 14,15-LTC_4. These molecules are mainly involved in various aspects of leukocyte biology; however, less has been reported about their actions than is published about the 5- and 12-lipoxygenase metabolites.

Unlike the classical neurotransmitters, PGs are not stored prior to use but are rapidly synthesized upon demand from phospholipid precursors. Activation and translocation of cytosolic phospholipase A_2 causes the release of free arachidonic acid and the initiation of PG synthesis. A variety of endogenous as well as environmental stimuli have been shown to initiate this process, which takes place on the endoplasmic reticulum and Golgi apparatus of most cells.

The initial mediator is an enzyme complex known as cyclooxygenase (COX), which promotes the reaction of two oxygen molecules with free arachidonic acid to yield a transient hydroperoxy

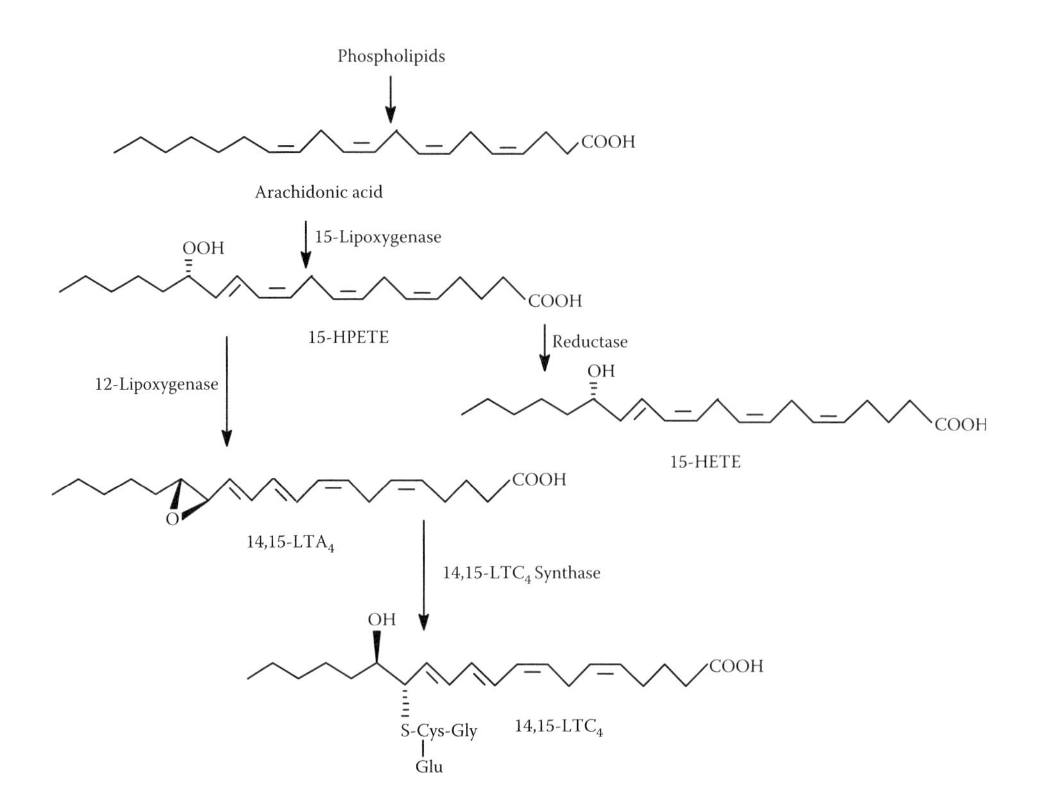

FIGURE 8.4 15-Lipoxygenase pathways. The key intermediate in this family of metabolites is 15-HPETE, which can serve as a precursor for two subsequent cascades of products, only a few examples of which are shown here. LT = leukotriene.

intermediate designated as PGG_2 (Figure 8.5). This is rapidly converted to a central precursor for all of the PGs called PGH_2. Three distinct COXs have been identified and cloned to date, COX-1, COX-2, and COX-3. In most cell types, COX-1 is responsible for constitutive PG synthesis, whereas COX-2 levels are rapidly induced in response to a variety of agents and play a major role in the inflammatory response. Aspirin-like drugs can inhibit either or both of these enzymes with varying degrees of selectivity. COX-3 is expressed primarily in brain and appears to be a major target for *p*-acetamidophenol, which is ineffective against either COX-1 or COX-2.

The conversion of PGH_2 into the primary PGs is mediated by a group of cell-specific synthases as shown in Figure 8.5. For example, thromboxane synthase is found in platelets where it promotes aggregation via TxA_2. An antagonist prostacyclin is synthesized in endothelial cells where it causes vasodilation and prevents excessive platelet aggregation. PGD synthetases are found in brain and mast cells. PGE_2 is made by most cells and acts on one or more of a family of receptors (EP_{1-4}) found on neighboring cells such as spinal neurons where it mediates a pain response.

Arachidonic acid is also a substrate for the cytochrome family of oxidative enzymes as might be expected from its polyunsaturated character. A number of cytochrome metabolites have been found, some of which are shown below (Figure 8.6). Interestingly, anandamide appears to exhibit a similar course of metabolism (*vide infra*).

BIOSYNTHESIS OF THE ENDOCANNABINOIDS

The details of the biosynthetic pathways leading to the endocannabinoids are well described elsewhere in this volume as well as in several excellent reviews such as the recent one by Sugiura et al. (2002) so that only a brief description will be given here for convenience. In the case of

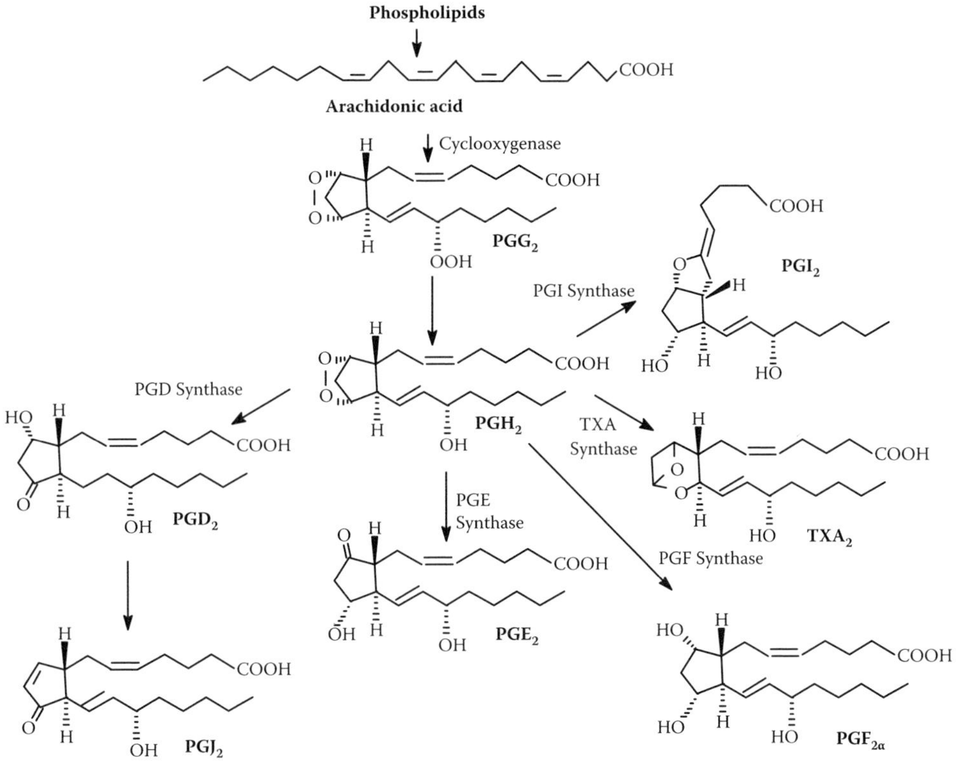

FIGURE 8.5 Cyclooxygenase pathways. The key intermediate in this pathway is PGG_2, which can be produced by the action of any one of three different cyclooxygenases, COX-1, COX-2 or COX-3. The relative importance of each will depend on tissue site and physiological state. PGG_2 is reduced to PGH_2, which serves as the precursor for each of the prostaglandins.

anandamide, two routes have been studied, and these are outlined in Figure 8.7. Route A has a slight resemblance to the eicosanoid pathways in that a phospholipid hydrolytic step is involved. However, in the case of anandamide biosynthesis, a phosphodiesterase is involved rather than a phospholipase such as PLA_2. The phospholipid substrate is believed to be generated from an arachidonyl containing phospholipid by a trans-acylation reaction that requires calcium.

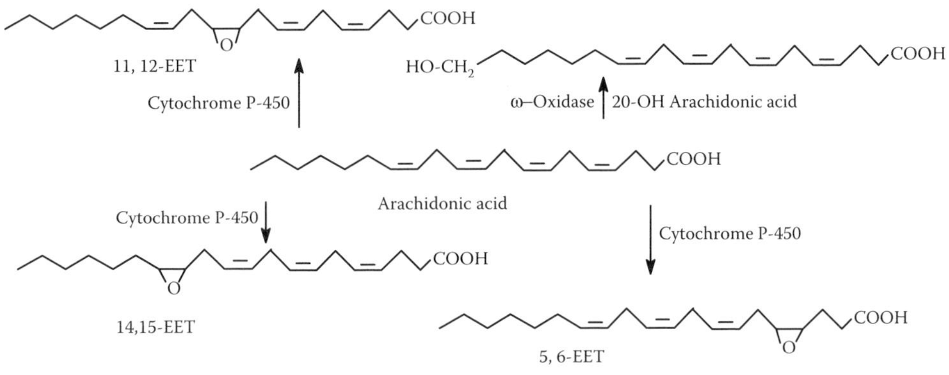

FIGURE 8.6 Metabolism of arachidonic acid by cytochrome P-450.

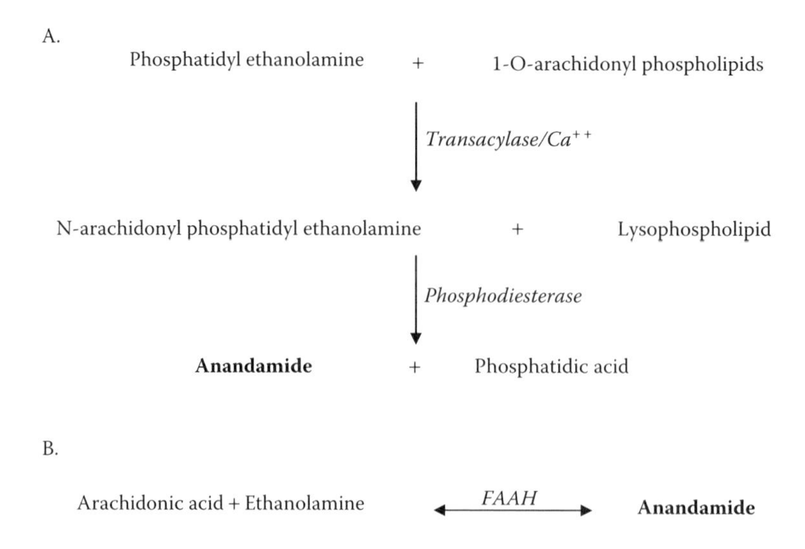

FIGURE 8.7 Biosynthesis of anandamide.

Pathway B is simply the reverse of the inactivation reaction for anandamide and is also mediated by fatty acid amide hydrolase (FAAH). Arguments have been raised against the physiological importance of this route because of the high concentrations of precursors required when studied *in vitro*. The events responsible for the initiation of anandamide production *in vivo* are not very well understood at this time, and much work remains to be done in this area.

The elucidation of the biosynthesis of 2-arachidonyl glycerol (AG) predates the discovery of its role as an endocannabinoid. In most tissues its abundance is greater than that of anandamide; however, it seems to have a lower affinity for the cannabinoid receptors. Thus, the relative importance of anandamide and 2-AG in various processes has been the subject of much research and discussion. The major precursor for 2-AG is phosphatidylinositol that contains an arachidonyl group at the 2-position. Two routes have thus far been described, and these are shown in Figure 8.8. In one, the action of a phospholipase C gives rise to a 2-arachidonyl diacylglyceride that itself is an important signaling molecule. Further action by a diacylglyceride lipase results in the generation of 2-AG. The second pathway is initiated by the action of phospholipase A_1 resulting in the production of a lysophosphatidylinositol as an intermediate. This is then converted by phospholipase C into 2-AG. Hydrolysis of 2-AG leads to free arachidonic acid as in the case of anandamide; however, re-esterification mediated by a monoacylglycerol transferase would result in diacylglyceride (DAG) formation. As mentioned above, DAG is an important signaling molecule that can act as a second messenger, suggesting that it is one possible mechanism for 2-AG action.

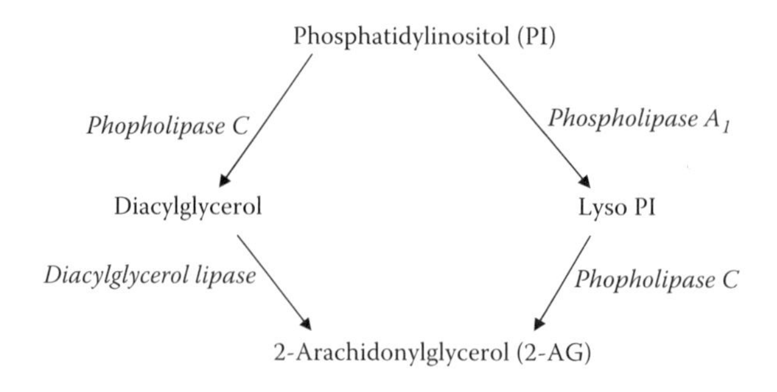

FIGURE 8.8 Biosynthesis of 2-arachidonyl glycerol.

A comparison of the biosyntheses of the eicosanoids and the endocannabinoids shows some general similarities; however, there are no reports that would suggest cross talk or interdependence among any of the processes. Both classes of molecules originate from phospholipid storage pools that are enriched in arachidonic acid. In the case of the eicosanoids, bioactive substances generally arise from the oxidative transformations of free arachidonic acid, leading to an extensive family of potent but short-lived substances. The endocannabinoids, on the other hand, in their active form are either amidated or esterified derivatives of arachidonic acid. This picture may change somewhat with the emergence of new data on the oxidative transformations of the endocannabinoids; however, relatively little has been reported on these processes or on the possible activities of the products.

METABOLISM OF THE ENDOCANNABINOIDS

There are currently five reported routes for the metabolism of anandamide, and these are outlined in Figure 8.9 (Di Marzo, De Petrocellis, et al., 1999). The principal route for the metabolism of anandamide is mediated by an amidase called FAAH that releases free arachidonic acid and ethanolamine (Ueda et al., 1997; Ueda, Katayama, et al., 1999; Ueda, Yamanaka, et al., 1999; Ueda et al., 2000; Ueda and Yamamoto, 2000; Ueda, 2002). The free acid could either be re-incorporated into phospholipids or transformed into eicosanoids with potent biological actions of their own. At this time there is little evidence to support the latter suggestion; however, should this turn out to be true, it would have interesting implications for the mechanism of anandamide action. This hydrolytic transformation of anandamide is considered to be an important physiological mechanism for regulating *in vivo* levels of anandamide, and FAAH has become a popular target for therapeutic intervention where it is desired to maintain or elevate anandamide tissue concentrations.

Other metabolic routes involve oxidative transformations that bear close resemblance to the processes that lead to the eicosanoids (Kozak et al., 2000; Kozak, Crews, et al., 2001; Kozak, Prusakiewicz, et al., 2001; Kozak, Crews, et al., 2002; Kozak and Marnett, 2002). For example, both 12- and 15-lipoxygenase products of anandamide are known (Edgemond et al., 1998; Moody et al., 2001; Kozak, Gupta, et al., 2002), as well as a group of P-450-mediated hydroxy anandamide metabolites (Bornheim et al., 1993; Bornheim et al., 1995). Free arachidonic acid is required for COX-1-mediated conversion to PGs, whereas the requirements for COX-2-mediated conversion

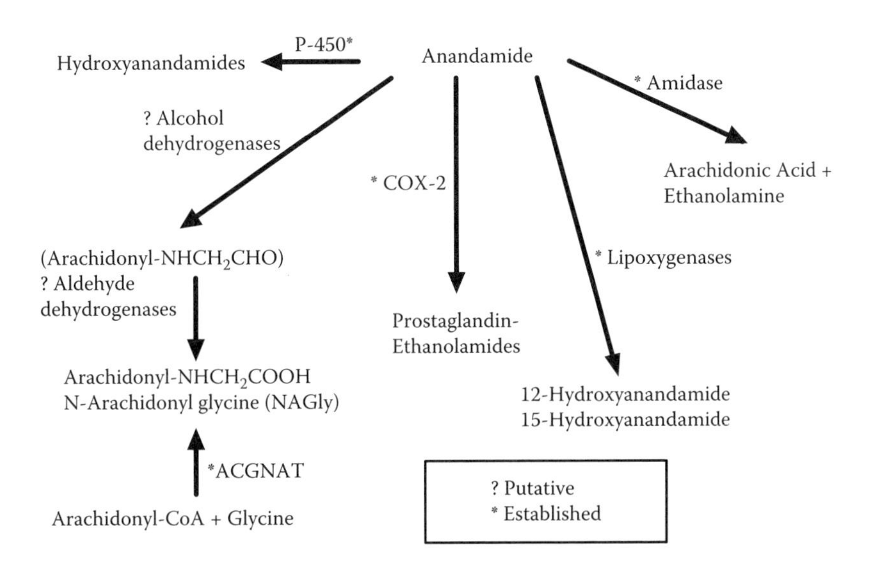

FIGURE 8.9 Metabolism of anandamide. The bioconversions shown are either taken from published reports (established), or are proposed in this review (putative). ACGNAT; acyl-CoA glycine *N*-acyltransferase.

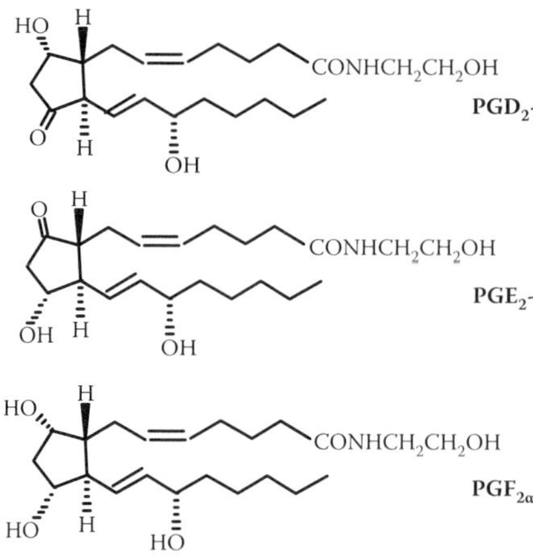

FIGURE 8.10 Structures of the prostaglandin ethanolamides. These novel molecules are products of the action of COX-2 on anandamide. Hydrolysis of the amide bond would give rise to a classical prostaglandin that would then result in a number of actions attributed to these compounds.

are less rigid (Kozak et al., 2003). A relevant example was recently reported in which it was shown that anandamide is a good substrate for COX-2 but not for COX-1 (Yu et al., 1997; So et al., 1998). The reaction was observed using either purified enzymes or intact cells. The major product produced by these models was the ethanolamide of PGE_2, a novel PG with no known biological function (Figure 8.10). The rate of COX-2 oxygenation of anandamide was slower in the presence of added arachidonic acid (So et al., 1998).

PGE_2 ethanolamide did not compete with CP55940 for binding to the CB_1 receptor at a concentration of 100 μM (Pinto et al., 1994). The nonphysiological ethanolamides of the PGA and PGB also showed no binding activity. It might be instructive to look at their binding activities with CB_2 or the appropriate PG receptors. An explanation for the lack of binding to CB_1 based on conformational analysis has been reported (Barnett-Norris et al., 1998).

Ross et al. (2002) reported that in dorsal root ganglion neurons in culture, 3 μM PGE_2 ethanolamide, the COX-2-mediated product of arachidonic acid (Figure 8.10) caused an increase in intracellular calcium concentration in a fraction of small-diameter capsaicin-sensitive neurons. This suggested that this compound is pharmacologically active; however, its physiological relevance remains to be established.

Several *N*-arachidonyl amino acid conjugates have recently been reported to be endogenous constituents of rat tissues (Huang, 2001). The best characterized is *N*-arachidonyl glycine (NAGly) which could conceivably arise via oxidation of the hydroxyl group in anandamide by a two-step process (Figure 8.9) (Burstein et al., 2000). An alternate route in which glycine is directly coupled to arachidonic acid is also possible, and experimental evidence has been obtained to support this pathway (Huang, 2001). Either route may function *in vivo* at different sites and under different conditions.

INTERRELATIONSHIPS BETWEEN ENDOCANNABINOIDS AND EICOSANOIDS

In several different models, treatment of cultured cells with anandamide as well as other cannabinoids results in increased arachidonic acid release and eicosanoid biosynthesis. Wartmann et al. (1995) observed such an effect and gave evidence that MAP kinase contributes to this response. Exposure of WI-38 fibroblasts to anandamide causes increased MAP kinase activity and increases the phosphorylation of cytoplasmic phospholipase A2 (cPLA2) resulting in its activation and the release of free arachidonic acid. Anandamide also induces arachidonate release in J774 mouse macrophages (Di Marzo et al., 1997).

Confluent monolayer cultures of astrocytes, labeled with [^{3}H]arachidonic acid, were incubated with anandamide in the presence or absence of thimerosal, a fatty acid acyl-CoA transferase inhibitor and phenylmethylsulfonyl fluoride (PMSF), an FAAH inhibitor (Shivachar et al., 1996). Anandamide induced a time- and concentration-dependent release of arachidonic acid in the presence but not in the absence of thimerosal. Anandamide-stimulated arachidonic acid release was pertussis-toxin-sensitive, indicating a receptor and G-protein involvement. The selective receptor antagonist, SR141716A, blocked the arachidonic acid release, suggesting a cannabinoid-receptor-mediated pathway. Also, an assay of FAAH activity indicated that degradation of anandamide into arachidonic acid and ethanolamine was negligible in this model. Their results suggest that anandamide stimulates cannabinoid-receptor-mediated release of arachidonic acid.

Inhibition of FAAH by anandamide in CHP100 cells was accompanied by arachidonic acid release, which was not observed upon treatment of cells with other fatty acid amides (Maccarrone et al., 2000). Cell treatment with anandamide or 2-AG also increased the activity of COX and 5-lipoxygenase, and the hydroperoxides generated by lipoxygenase were shown to inhibit FAAH, with inhibition constants in the low-micromolar range. Inhibitors of 5-lipoxygenase but not COX significantly antagonized the inhibition of FAAH by anandamide or 2-AG.

Anandamide and the vanilloid agonist capsaicin stimulated arachidonic acid release in rat pheochromocytoma PC12 cells (Someya et al., 2002). Both substances increased arachidonic acid release in a concentration-dependent manner from labeled cells even in the absence of extracellular $CaCl_2$. The effect of anandamide was not mimicked by cannabinoid agonists and was not inhibited by a cannabinoid antagonist. The effects of anandamide and capsaicin were inhibited by phospholipase A_2 inhibitors but not by an antagonist for vanilloid VR-1 receptor. Addition of anandamide or capsaicin synergistically enhanced arachidonic acid release by mastoparan in the absence of $CaCl_2$. Anandamide stimulated PGF_{2a} formation. These data suggest that anandamide and capsaicin can stimulate arachidonic acid metabolism in a cannabinoid or vanilloid VR-1 receptor-independent manner in PC12 cells.

Using H4 human neuroglioma cells, the effect of R(+)-methanandamide, a metabolically stable analog of the anandamide, on the expression of COX-2 was studied (Ramer et al., 2001). Incubation of cells with R(+)-methanandamide resulted in a concentration-dependent increase in COX-2 mRNA, COX-2 protein, and COX-2-dependent PGE_2 synthesis. Moreover, treatment of cells with R(+)-methanandamide in the presence of interleukin-1 led to an increased induction of COX-2 expression. Stimulation of both COX-2 mRNA expression and subsequent PGE_2 synthesis by R(+)-methanandamide were not affected by the selective CB_1 receptor antagonist AM251 or the G(i/o) protein inactivator pertussis toxin. Enhancement of COX-2 expression by R(+)-methanandamide was paralleled by time-dependent phosphorylations of p38 mitogen-activated protein kinase (MAPK) and p42/44 MAPK. In agreement with the activation of both kinases, R(+)-methanandamide-induced COX-2 mRNA expression and PGE_2 formation were reduced in the presence of specific inhibitors of p38 MAPK (SB203580) and p42/44 MAPK activation (PD98059). These findings demonstrate that R(+)-methanandamide induces COX-2 expression in human glioma cells via a cannabinoid-receptor-independent mechanism involving activation of the MAPK pathway. The authors concluded that induction of COX-2 expression may represent a novel mechanism by which endocannabinoids mediate PG-dependent effects in the central nervous system.

Anandamide initiated platelet aggregation at concentrations similar to those of arachidonic acid (Braud et al., 2000). The aggregating effect of anandamide was inhibited by aspirin but not by SR141716A, a cannabinoid receptor antagonist. Studies with radiolabeled material showed that exogenous anandamide is cleaved by platelets into arachidonic acid through a PMSF-sensitive pathway. In agreement with this observation, PMSF was shown to abolish the aggregating effect of anandamide. Thus, it might be concluded that anandamide can serve as a source of arachidonic acid for TxA_2 production.

Pratt et al. (1998) suggested that the vasodilatory effect of anandamide results from its hydrolysis to arachidonic acid followed by biotransformation to vasodilatory eicosanoids. In support of this hypothesis, they reported that bovine coronary arteries incubated with anandamide hydrolyzed 15% of added substrate; approximately 9% of the radiolabeled product was free arachidonic acid, and 6% co-migrated with the PGs and epoxyeicosatrienoic acids. A similar result was obtained in cultured bovine coronary endothelial cells. Inhibition of FAAH activity with diazomethylarachidonyl ketone prevented both the metabolism of anandamide and the vasodilatory action. Whole vessel and cultured endothelial cells prelabeled with arachidonic acid synthesized PGs and EETs, but not anandamide, in response to the calcium ionophore A-23187. Furthermore, SR141716A attenuated A-23187-stimulated release of arachidonic acid, suggesting that it may have actions other than inhibition of the CB_1 receptor. Similarly, Grainger and Boachie-Ansah (2001) reported data suggesting that the relaxant effects of anandamide in sheep coronary arteries are mediated in part via the endothelium and result from the cellular uptake and conversion of anandamide to a vasodilatory eicosanoid.

ACKNOWLEDGMENTS

This publication was made possible by support from NIDA. Its contents are solely the responsibility of the author and do not necessarily represent the official views of the National Institute on Drug Abuse. The assistance of Annette Stratton in preparing this chapter is gratefully acknowledged.

REFERENCES

Barnett-Norris J, Guarnieri F, Hurst DP, and Reggio PH (1998) Exploration of biologically relevant conformations of anandamide, 2-arachidonylglycerol, and their analogues using conformational memories. *J Med Chem* 41: 4861–4872.

Bornheim LM, Kim KY, Chen B, and Correia MA (1993) The effect of cannabidiol on mouse hepatic microsomal cytochrome P450-dependent anandamide metabolism. *Biochem Biophys Res Commn* 197: 740–746.

Bornheim LM, Kim KY, Chen B, and Correia MA (1995) Microsomal cytochrome P450-mediated liver and brain anandamide metabolism. *Biochem Pharmacol* 50: 677–686.

Braud S, Bon C, Touqui L, and Mounier C (2000) Activation of rabbit blood platelets by anandamide through its cleavage into arachidonic acid. *FEBS Lett* 471: 12–16.

Burstein S (2002) Cannabis and prostaglandins: an overview, in *Biology of Marijuana* (Onaivi ES, Ed.), Taylor and Francis, London. pp. 390–406.

Burstein SH, Rossetti RG, Yagen B, and Zurier RB (2000) Oxidative metabolism of anandamide. *Prostaglandins Other Lipid Mediat* 61: 29–41.

Burstein SH, Young JK, and Wright GE (1995) Relationships between eicosanoids and cannabinoids. Are eicosanoids cannabimimetic agents? *Biochem Pharmacol* 50: 1735–1742.

Di Marzo V, Bisogno T, De Petrocellis L, Melck D, and Martin BR (1999) Cannabimimetic fatty acid derivatives: the anandamide family and other endocannabinoids. *Curr Med Chem* 6: 721–744.

Di Marzo V, De Petrocellis L, Bisogno T, and Maurelli S (1997) The endogenous cannabimimetic eicosanoid, anandamide, induces arachidonate release in J774 mouse macrophages. *Adv Exp Med Biol* 407: 341–346.

Di Marzo V, De Petrocellis L, Bisogno T, and Melck D (1999) Metabolism of anandamide and 2-arachidonoylglycerol: an historical overview and some recent developments. *Lipids* 34 Suppl.: S319–325.

Edgemond WS, Hillard CJ, Falck JR, Kearn CS, and Campbell WB (1998) Human platelets and polymorphonuclear leukocytes synthesize oxygenated derivatives of arachidonylethanolamide (anandamide): their affinities for cannabinoid receptors and pathways of inactivation. *Mol Pharmacol* 54: 180–188.

Funk CD (2001) Prostaglandins and leukotrienes: advances in eicosanoid biology. *Science* 294: 1871–1875.

Grainger J and Boachie-Ansah G (2001) Anandamide-induced relaxation of sheep coronary arteries: the role of the vascular endothelium, arachidonic acid metabolites and potassium channels. *Br J Pharmacol* 134: 1003–1012.

Huang SM, Bisogno T, Petros TJ, Chang SY, Zavitsanos PA, Zipkin RE, Sivakumar R, Coop A, Maeda DY, De Petrocellis L, Burstein S, Di Marzo V, and Walker JM (2001) Identification of a new class of molecules, the arachidonyl amino acids, and characterization of one member that inhibits pain. *J Biol Chem* 276: 42639–42644.

Kozak KR, Crews BC, Morrow JD, Wang LH, Ma YH, Weinander R, Jakobsson PJ, and Marnett LJ (2002) Metabolism of the endocannabinoids, 2-arachidonylglycerol and anandamide, into prostaglandin, thromboxane, and prostacyclin glycerol esters and ethanolamides. *J Biol Chem* 277: 44877–44885.

Kozak KR, Crews BC, Ray JL, Tai HH, Morrow JD, and Marnett LJ (2001) Metabolism of prostaglandin glycerol esters and prostaglandin ethanolamides in vitro and in vivo. *J Biol Chem* 276: 36993–36998.

Kozak KR, Gupta RA, Moody JS, Ji C, Boeglin WE, DuBois RN, Brash AR, and Marnett LJ (2002) 15-Lipoxygenase metabolism of 2-arachidonylglycerol. Generation of a peroxisome proliferator-activated receptor alpha agonist. *J Biol Chem* 277: 23278–23286.

Kozak KR and Marnett LJ (2002) Oxidative metabolism of endocannabinoids. *Prostaglandins Leukot Essent Fatty Acids* 66: 211–220.

Kozak KR, Prusakiewicz JJ, Rowlinson SW, Prudhomme DR, and Marnett LJ (2003) Amino acid determinants in cyclooxygenase-2 oxygenation of the endocannabinoid anandamide. *Biochemistry* 42:9041–9049.

Kozak KR, Prusakiewicz JJ, Rowlinson SW, Schneider C, and Marnett LJ (2001) Amino acid determinants in cyclooxygenase-2 oxygenation of the endocannabinoid 2-arachidonylglycerol. *J Biol Chem* 276: 30072–30077.

Kozak KR, Rowlinson SW, and Marnett LJ (2000) Oxygenation of the endocannabinoid, 2-arachidonylglycerol, to glyceryl prostaglandins by cyclooxygenase-2. *J Biol Chem* 275: 33744–33749.

Maccarrone M, Salvati S, Bari M, and Finazzi A (2000) Anandamide and 2-arachidonoylglycerol inhibit fatty acid amide hydrolase by activating the lipoxygenase pathway of the arachidonate cascade. *Biochem Biophys Res Commn* 278: 576–583.

Moody JS, Kozak KR, Ji C, and Marnett LJ (2001) Selective oxygenation of the endocannabinoid 2-arachidonylglycerol by leukocyte-type 12-lipoxygenase. *Biochemistry* 40: 861–866.

Pinto JC, Potie F, Rice KC, Boring D, Johnson MR, Evans DM, Wilken GH, Cantrell CH, and Howlett AC (1994) Cannabinoid receptor binding and agonist activity of amides and esters of arachidonic acid. *Mol Pharmacol* 46: 516–522.

Pratt PF, Hillard CJ, Edgemond WS, and Campbell WB (1998) N-arachidonylethanolamide relaxation of bovine coronary artery is not mediated by CB1 cannabinoid receptor. *Am J Physiol* 274: H375–381.

Ramer R, Brune K, Pahl A, and Hinz B (2001) R(+)-methanandamide induces cyclooxygenase-2 expression in human neuroglioma cells via a non-cannabinoid receptor-mediated mechanism. *Biochem Biophys Res Commn* 286:1144–1152.

Ross RA, Craib SJ, Stevenson LA, Pertwee RG, Henderson A, Toole J, and Ellington HC (2002) Pharmacological characterization of the anandamide cyclooxygenase metabolite: prostaglandin E2 ethanolamide. *J Pharmacol Exp Ther* 301: 900–907.

Shivachar AC, Martin BR, and Ellis EF (1996) Anandamide- and delta9-tetrahydrocannabinol-evoked arachidonic acid mobilization and blockade by SR141716A [N-(Piperidin-1-yl)-5-(4-chlorophenyl)-1-(2,4-dichlorophenyl)-4-methyl-1H-pyrazole-3-carboximide hydrochloride]. *Biochem Pharmacol* 51: 669–676.

So OY, Scarafia LE, Mak AY, Callan OH, and Swinney DC (1998) The dynamics of prostaglandin H synthases. Studies with prostaglandin h synthase 2 Y355F unmask mechanisms of time-dependent inhibition and allosteric activation. *J Biol Chem* 273: 5801–5807.

Someya A, Horie S, and Murayama T (2002) Arachidonic acid release and prostaglandin F(2alpha) formation induced by anandamide and capsaicin in PC12 cells. *Eur J Pharmacol* 450: 131–139.

Sugiura T, Kobayashi Y, Oka S and Waku K (2002) Biosynthesis and degradation of anandamide and 2-arachidonoylglycerol and their possible physiological significance. *Prostaglandins Leukot Essent Fatty Acids* 66: 173–192.

Sugiura T and Waku K (2002) Cannabinoid receptors and their endogenous ligands. *J Biochem (Tokyo)* 132: 7–12.

Ueda N (2002) Endocannabinoid hydrolases. *Prostaglandins Other Lipid Mediat* 68–69: 521–534.

Ueda N, Katayama K, Kurahashi Y, Suzuki M, Suzuki H, Yamamoto S, Katoh I, Di Marzo V, and De Petrocellis L (1999) Enzymological and molecular biological studies on anandamide amidohydrolase. *Adv Exp Med Biol* 469: 513–518.

Ueda N, Kurahashi Y, Yamamoto K, and Yamamoto S (1997) Anandamide amidohydrolase from porcine brain. Partial purification and characterization. *Adv Exp Med Biol* 407: 323–328.

Ueda N, Puffenbarger RA, Yamamoto S, and Deutsch DG (2000) The fatty acid amide hydrolase (FAAH). *Chem Phys Lipids* 108: 107–121.

Ueda N and Yamamoto S (2000) Anandamide amidohydrolase (fatty acid amide hydrolase). *Prostaglandins Other Lipid Mediat* 61: 19–28.

Ueda N, Yamanaka K, Terasawa Y, and Yamamoto S (1999) An acid amidase hydrolyzing anandamide as an endogenous ligand for cannabinoid receptors. *FEBS Lett* 454: 267–270.

Wartmann M, Campbell D, Subramanian A, Burstein SH, and Davis RJ (1995) The MAP kinase signal transduction pathway is activated by the endogenous cannabinoid anandamide. *FEBS Lett* 359: 133–136.

Yu M, Ives D, and Ramesha, CS (1997) Synthesis of prostaglandin E2 ethanolamide from anandamide by cyclooxygenase-2. *J Biol Chem* 272: 21181–21186.

Part IV

Endocannabinoids in CNS Physiology

9 Man-Made Marijuana: Endocannabinoid Modulation of Synaptic Transmission and Implications for the Regulation of Synaptic Plasticity

Alexander F. Hoffman and Carl R. Lupica

CONTENTS

INTRODUCTION

This chapter will review the recent literature describing the physiological actions of endogenous cannabinoids at the major excitatory (glutamatergic) and inhibitory (GABAergic) neurotransmitter synapses in the central nervous system. Following a brief summary of the known effects of exogenously administered cannabinoid ligands, we will examine recent data identifying roles for endogenously produced cannabinoids in regulating short-term and long-term synaptic plasticity in several brain areas, including the hippocampus and nucleus accumbens.

SYNAPTIC EFFECTS OF EXOGENOUS CANNABINOID LIGANDS

The development of pharmacological agents that alter central cannabinoid receptor function has been indispensable in the elucidation of the physiological actions of cannabinoids. By combining these compounds with long-established intracellular and extracellular electrophysiological recording techniques, a number of laboratories have characterized the fundamental actions of cannabinoids on synaptic transmission in several different brain areas. Although the nonselective cannabinoid receptor agonist WIN55212-2 has been the most commonly employed ligand in our laboratory and others, similar results have been obtained with other compounds. Because the CB_1 receptor antagonist SR141716A (Rinaldi-Carmona et al., 1994) has typically been shown to block or reverse the actions of all putative CB_1 agonists, it is clear that the majority of the effects on synaptic physiology can be ascribed to activation of this receptor. An overview of some of the major findings is provided in Table 9.1.

CANNABINOID REGULATION OF SYNAPTIC GABA RELEASE

By far, the most consistently observed effect of cannabinoid agonist application in the CNS has been the presynaptic inhibition of the release of the primary inhibitory neurotransmitter γ-aminobutyric acid (GABA). In this respect, cannabinoid receptors are similar to a broader class of G-protein-coupled receptors that also inhibit neurotransmitter release, including those for dopamine (Nicola and Malenka, 1997; Miyazaki and Lacey, 1998), opiates (Cohen et al., 1992; Jiang and North, 1992; Lupica, 1995; Hoffman and Lupica, 2001), and $GABA_B$ auto receptors (Uchimura and North, 1991; Doze et al., 1995). Where the mechanisms have been investigated in detail (e.g., through the use of selective blockers of ion channels), cannabinoid-mediated inhibition of GABA release has been attributed, with few exceptions (Vaughan et al., 2000), to a presynaptic interaction between the CB_1 receptor and voltage-dependent Ca^{2+} channels (VDCCs). These data contrast with the results from studies in cellular expression systems (Mackie et al., 1995; McAllister et al., 1999) and in cultured neurons (Deadwyler et al., 1995; Mu et al., 1999), where postsynaptic CB_1 receptors appear to interact with various classes of somatodendritic K^+ channels. Whereas several studies have found that cannabinoids regulate voltage-dependent K^+ channels in hippocampal pyramidal neurons (Deadwyler et al., 1995; Mu et al., 1999; Schweitzer, 2000), our own studies (Hoffman and Lupica, 2000) and those of Nicoll and colleagues (Wilson and Nicoll, 2001; Wilson et al., 2001) strongly suggest that the inhibition of GABA release in the hippocampus is due entirely to the CB_1-mediated inhibition of VDCCs on interneuron terminals.

CANNABINOID REGULATION OF SYNAPTIC GLUTAMATE RELEASE

Cannabinoid agonists also inhibit glutamate release throughout the CNS. To date, this effect has been demonstrated in the hippocampus (Shen et al., 1996; Sullivan, 1999), the dorsolateral and ventral striatum (Gerdeman and Lovinger, 2001; Robbe et al., 2001), the cerebellum (Levenes et al., 1998; Takahashi and Linden, 2000), the substantia nigra (Szabo et al., 2000), the prefrontal cortex (Auclair et al., 2000), and the spinal cord (Morisset and Urban, 2001). However, these effects have been more variable than those reported for GABA release in several respects. Whereas in some studies the cannabinoid-mediated inhibition of glutamate release is linked to the inhibition of VDCCs (Shen and Thayer, 1998; Sullivan, 1999), others have found that the inhibitory effects of WIN55212-2 on glutamate release are mediated by activation of a 4-aminopyridine-sensitive K^+ channel (Robbe et al., 2001). Curiously, our own studies in the shell of the nucleus accumbens found no evidence for presynaptic inhibition of glutamate release by WIN55212-2, despite this agonist's strong effect on GABA release in the same area (Hoffman and Lupica, 2001), and the inhibition of glutamate release in the nucleus accumbens core (Robbe et al., 2001). Also, we have found that cannabinoids inhibit glutamatergic, excitatory postsynaptic currents (EPSCs) in pyramidal neurons in the hippocampus, whereas this effect is absent when recording EPSCs from

TABLE 9.1
Summary of Effects of Cannabinoids at Excitatory and Inhibitory Synapses

Brain Region	Synapse	Preparation	Method(s)	CB Effect/Postulated Mechanisms	References
Hippocampus	Inhibitory	Slice (rat)	Intracellular	Presynaptic inhibition, voltage-dependent Ca^{2+} channels	(Hoffman and Lupica, 2000)
	Excitatory	Culture and slice (rat)	Intracellular, extracellular	Presynaptic inhibition, *N*-type Ca^{2+} channels, inhibition of LTP	(Shen et al., 1996; Shen and Thayer, 1998; Misner and Sullivan, 1999; Sullivan, 1999; Hajos et al., 2001)
		Mouse slice	Intracellular	Inhibition, even in CB_1 (–/–)	
Striatum	Excitatory	Slice (rat)	Intracellular, extracellular recordings	Presynaptic inhibition, *N*-type Ca^{2+} channels	(Gerdeman and Lovinger, 2001; Huang et al., 2001)
	Inhibitory	Slice (rat)	Intracellular	Inhibition	(Szabo et al., 1998)
Nucleus accumbens	Inhibitory	Slice (rat/mouse)	Intracellular	Presynaptic inhibition	(Hoffman and Lupica, 2001; Manzoni and Bockaert, 2001)
	Excitatory	Slice (rat)	Intracellular	Shell: Minimal effect, not presynaptic	(Hoffman and Lupica, 2001)
				Core: Presynaptic inhibition, activation of K^+ channels	(Robbe et al., 2001)
		Slice (mouse)	Intracellular, extracellular		
Substantia nigra pars reticulata	Inhibitory	Slice (rat)	Intracellular	Presynaptic inhibition, possibly due to inhibition of voltage-dependent Ca^{2+} channels	(Chan et al., 1998; Chan and Yung, 1998; Wallmichrath and Szabo, 2002)
	Excitatory	Slice (rat)	Intracellular	Presynaptic inhibition	(Szabo et al., 2000)
Rostral ventromedial medulla (RVM)	Inhibitory	Slice (rat)	Intracellular	Presynaptic inhibition	(Vaughan et al., 1999)
Periacqueductal gray (PAG)	Inhibitory and excitatory	Slice (rat)	Intracellular	Presynaptic inhibition, independent of Ca^{2+} (K^+ channels?)	(Vaughan et al., 2000)
Cerebellum	Excitatory	Slice (rat)	Intracellular	Presynaptic inhibition, possibly due to Ca^{2+} channels	(Levenes et al., 1998; Takahashi and Linden, 2000)
	Inhibitory	Slice (rat)	Intracellular	Presynaptic inhibition	(Takahashi and Linden, 2000)
Prefrontal cortex	Excitatory	Slice (rat)	Intracellular	Presynaptic inhibition	(Auclair et al., 2000)

GABAergic interneurons (Hoffman, Riegel, et al., 2003). However, GABA release onto both interneurons and pyramidal neurons is inhibited by presynaptic CB_1 receptors (Hoffman, Riegel, et al., 2003). Interestingly, Freund and colleagues (Hajos et al., 2001) have reported that glutamatergic EPSCs are inhibited by cannabinoid agonists in the hippocampus of mice lacking the CB_1 receptor gene and that the antagonist SR141716A can reverse this effect. Because most CB_1 receptor antibodies appear to only weakly label glutamatergic profiles in the hippocampus, it is possible that a third cannabinoid receptor that retains affinity for many of the currently available CB_1 ligands (but not the antagonist AM251; see in Hajos and Freund, 2002) is responsible for the inhibition of glutamate release from Schaffer collateral axons onto CA1 pyramidal neurons (Hajos et al., 2001). These data are also consistent with the finding that hippocampal EPSCs are, on average, less sensitive than inhibitory postsynaptic currents (IPSCs) to cannabinoid-mediated inhibition (Ohno-Shosaku et al., 2002). Such findings suggest a high degree of segregation of cannabinoid modulation of glutamatergic terminals according to cellular subtypes in both the hippocampus and nucleus accumbens, whereas the modulation of GABAergic inhibition via the activation of CB_1 receptors appears to be more ubiquitous among a variety of neuronal phenotypes.

ENDOCANNABINOID INHIBITION OF EXCITATORY AND INHIBITORY SYNAPSES

The studies described in the previous section demonstrate convincingly that exogenously applied cannabinoid agonists presynaptically inhibit GABAergic and glutamatergic synaptic transmission in many areas of the brain. However, much like the discovery of opiate receptors before endogenous opioid peptides (Snyder and Pasternak, 2003), a full physiological role for these cannabinoid receptors could not be elucidated until endogenous ligands for cannabinoid receptors had been identified. In the following sections, we will discuss recently discovered roles for endogenous cannabinoids (endocannabinoids) at specific synapses in the CNS. An important consideration for the reader will be the use of the term "synaptic plasticity," because the effects of endocannabinoids that will be discussed here will be related to this phenomenon. In the first case, the role of endocannabinoids in mediating two "short-term" forms of synaptic plasticity, termed depolarization-induced suppression of inhibition (DSI) and depolarization-induced suppression of excitation (DSE), will be discussed. This will be followed in a subsequent section by a discussion of the role of endocannabinoids in mediating a form of "long-term" synaptic plasticity termed long-term depression (LTD). In an effort to limit the scope of this chapter, we will primarily discuss the role of these synaptic plasticity paradigms in establishing a link between endocannabinoid release and synaptic transmission. Although we shall discuss some of the possible functional implications of these studies, the reader is also referred to several recent reviews encompassing a broader overview of the field of synaptic plasticity (Tao and Poo, 2001; Kemp and Bashir, 2001; Martin and Morris, 2002; Sheng and Kim, 2002; Zucker and Regehr, 2002).

DSI and DSE: A Role for Endocannabinoids in Short-Term Synaptic Plasticity

One of the earliest reports postulating a physiological role for an endocannabinoid at a specific CNS synapse came from a study by Stella and colleagues (Stella et al., 1997). It was found that high-frequency stimulation of Schaffer collateral and commissural fibers in the CA1 region of the hippocampus, similar to that commonly given during LTP induction, resulted in a fourfold increase in the production of the cannabinoid agonist, 2-arachidonylglycerol (2-AG). The release of 2-AG was blocked by the Na^+-channel blocker tetrodotoxin, and by removal of extracellular Ca^{2+}, suggesting that the release of 2-AG required increased neuronal activity. Moreover, the long-term potentiation (LTP) of Schaffer collateral and commissural glutamate inputs to CA1 pyramidal neurons was blocked in brain slices exposed to 2-AG prior to high-frequency stimulation. From these data, the authors hypothesized that 2-AG, released from nerve terminals during high-frequency stimulation, played a critical role in modulating hippocampal LTP. However, these results did not explain why exogenous 2-AG application blocked LTP, whereas LTP was still observed under conditions of high-frequency stimulation that released endogenous 2-AG. Because the authors did not directly evaluate the effect of either a CB_1 antagonist or endocannabinoid metabolic inhibitors on LTP, the true role of endocannabinoids in modulating synaptic plasticity remained unresolved. Nevertheless, these data were important in establishing possible conditions under which the endogenous cannabinoid system might be activated within a well-understood neuronal circuit. However, despite these encouraging findings, additional studies failed to observe an "endogenous cannabinoid tone" at individual synapses using electrophysiological assays under baseline conditions. For example, CB_1 receptor antagonists applied alone to slices were not observed to produce significant changes in baseline synaptic responses, despite their clear ability to reverse the effects of directly applied synthetic and endogenous cannabinoid agonists (Hoffman and Lupica, 2000; Wilson and Nicoll, 2001). Thus, the identification of a neural system in which endogenous cannabinoids could be readily released, and their effects clearly established, remained elusive at that time.

DSI at Hippocampal Synapses

Studies by Alger and colleagues over the past decade investigating the regulation of GABA release from inhibitory axon terminals impinging on pyramidal neurons in the hippocampus had identified and characterized the DSI phenomenon (Pitler and Alger, 1994; Alger and Pitler, 1995; Alger et al., 1996). This phenomenon involves the depolarization (2-5 s) of hippocampal pyramidal neurons while continuously monitoring GABAergic IPSCs in these same cells. It was found that after postsynaptic membrane depolarization, GABA release onto pyramidal neurons was transiently inhibited, and this inhibition could be maintained for up to 30s following the depolarization. A number of subsequent experiments (Pitler and Alger, 1994; Alger et al., 1996) revealed that DSI, (1) was dependent on a *postsynaptic* Ca^{2+} increase in the pyramidal cell; (2) was expressed *presynaptically* as a decrease in the probability of GABA release; (3) could be blocked by pertussis toxin, implying a role for G-protein-coupled receptors; and (4) was confined to a subset of large amplitude, rapidly decaying $GABA_A$ receptor-mediated IPSCs on pyramidal cells (Martin et al., 2001). Thus, DSI was proposed to represent a novel form of short-term synaptic plasticity, involving the release of a retrograde signaling molecule from the postsynaptic pyramidal cell, that activated a mechanism that inhibited GABA release from interneuron axon terminals. While the unequivocal identification of the retrograde messenger mediating DSI remained elusive, the hypothesis that an endocannabinoid was involved in this process was based on the convergence of data suggesting that cannabinoids inhibited synaptic GABA release onto pyramidal neurons (Hoffman and Lupica, 2000), and that endocannabinoids could be released in hippocampal circuits (Stella et al., 1997). This hypothesis was subsequently confirmed and reported in two independent studies (Ohno-Shosaku et al., 2001; Wilson and Nicoll, 2001). Wilson and Nicoll demonstrated that DSI could be blocked by preincubation of hippocampal slices with cannabinoid CB_1 receptor antagonists (SR141716A and AM251), and was occluded by the nonselective cannabinoid receptor agonist WIN55212-2 (Wilson and Nicoll, 2001). In addition, they reported that whereas 2-AG had only a minimal effect on IPSCs, pretreatment of the slices with the endocannabinoid uptake inhibitor AM404 led to a CB_1 antagonist-sensitive occlusion of DSI. This latter finding suggested that, under normal conditions, endogenous cannabinoid "tone" was regulated by a balance of release and uptake processes, thus explaining the earlier observation that CB_1 antagonists have no visible effect on baseline GABA release (Hoffman and Lupica, 2000; Irving et al., 2000; Wilson and Nicoll, 2001). Thus, when endocannabinoid uptake was blocked by AM404, the result was an accumulation of endogenous cannabinoid-activated presynaptic CB_1 receptors, thereby inhibiting GABA release and occluding DSI. In an elegant experiment, the authors also demonstrated that, whereas inhibitors of classical vesicular release machinery were ineffective at blocking DSI, flash photolysis-induced "uncaging" of Ca^{2+} within the pyramidal cell body produced cannabinoid receptor antagonist-sensitive DSI identical to that observed following depolarization (Wilson and Nicoll, 2001). Thus, these experiments confirmed that Ca^{2+} was required for endocannabinoid release from postsynaptic neurons. A subsequent study by this group demonstrated that DSI was also absent in mice containing a targeted deletion of the CB_1 receptor gene ("knockout") (Wilson et al., 2001), and that the inhibition of IPSCs onto pyramidal neurons by endocannabinoids was limited to a specific population of inhibitory inputs. That is, the same large amplitude, rapidly decaying $GABA_A$ receptor–mediated responses that had been described in earlier work on DSI (Wilson et al., 2001; Martin et al., 2001).

Many of these findings, including the Ca^{2+}-dependence of the endocannabinoid-mediated DSI, were confirmed by Ohno-Shosaku and collaborators using cultured hippocampal neurons (Ohno-Shosaku et al., 2001). In addition, these studies demonstrated that trains of action potentials elicited in cultured pyramidal cells could produce DSI, suggesting for the first time that physiologically meaningful patterns of hippocampal neuronal activity could result in endocannabinoid release. However, more recent *in vivo* studies (Hampson et al., 2003) failed to confirm DSI in response to a variety of behaviorally relevant neuronal activation patterns. Thus, it cannot presently be assumed that endocannabinoids are released under the same conditions *in vivo* as *in vitro*, and that these molecules are released via a DSI-like process in the intact mammalian brain. Furthermore, recent studies in our laboratory have demonstrated that not all neuronal phenotypes in the hippocampus exhibit the capacity for endocannabinoid-mediated DSI (Hoffman, Riegel, et al., 2003) (Figure 9.1). In this work we have found that whereas endocannabinoid-dependent DSI could be observed following the depolarization of hippocampal CA1 pyramidal cells, DSI was not observed in CA1 GABAergic interneurons possessing somata located either in stratum oriens or stratum radiatum. However, despite the absence of endocannabinoid-mediated DSI in these interneurons, direct application of WIN55212-2 reduced synaptic GABA release onto these cells. These data suggest that whereas interneurons do not appear to be capable of generating endocannabinoids following depolarization, the inhibitory axon terminals targeting these cells do express functional CB_1 receptors. This further implies that the endocannabinoid released during depolarization is primarily derived from pyramidal neurons in the hippocampus, and that spillover of this endocannabinoid from the pyramidal neurons to axon terminals impinging upon interneurons may also occur.

Endocannabinoid-Dependent DSI and DSE at Cerebellar Synapses

At the same time that studies were elucidating a role for endocannabinoids at hippocampal synapses, other groups were investigating similar mechanisms in the cerebellum, another brain area known to be rich in CB_1 receptors (Herkenham et al., 1990). As mentioned previously (Table 9.1; Levenes et al., 1998; Takahashi and Linden, 2000), both excitatory and inhibitory cerebellar synapses were known to be sensitive to exogenously applied cannabinoids, thereby providing a physiological explanation for the disruption of fine motor skills produced by marijuana. Given the presynaptic effects of cannabinoids in the cerebellum, and the knowledge that DSI had already been long established in this brain region (Llano et al., 1991; Marty and Llano, 1995), it was not surprising that the hippocampal studies were soon extended to the cerebellum, with strikingly similar results (Kreitzer and Regehr, 2001a; Diana et al., 2002). A more surprising finding occurred at the same time that the hippocampal DSI/endocannabinoid link was being firmly established. Kreitzer and Regehr found that both climbing fiber and parallel fiber excitatory inputs to Purkinje neurons could also be transiently suppressed following Purkinje cell depolarization, the DSE phenomenon. Much like DSI, this short-term form of plasticity is postsynaptically initiated by a rise in intracellular Ca^{2+} (i.e., it is blocked by the intracellular application of Ca^{2+} chelators) but is presynaptically expressed as a decrease in the probability of glutamate release (Kreitzer and Regehr, 2001b; Kreitzer and Regehr, 2002). A number of different pharmacological agents were tested to determine the nature of the retrograde messenger involved in DSE, and once again, it was shown that DSE could be blocked by AM251 and occluded by WIN55212-2. Thus, the actions of endocannabinoids as

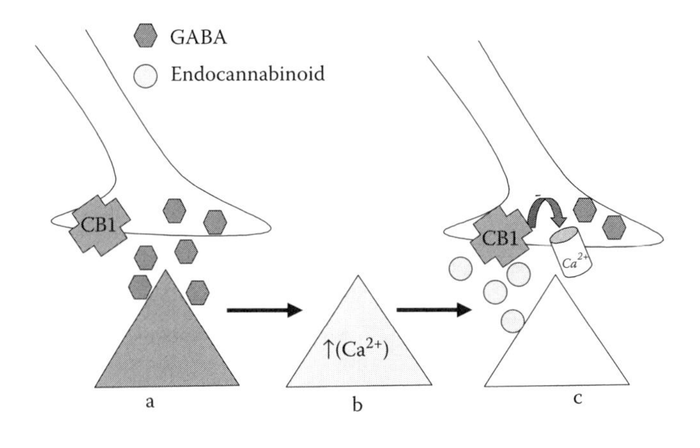

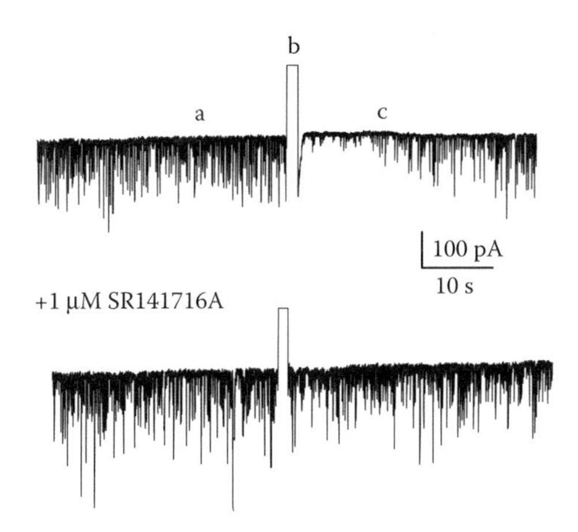

FIGURE 9.1 Summary of DSI at inhibitory synapses on CA1 hippocampal pyramidal cells. Traces in the lower portion of the figure represent spontaneous GABAergic IPSCs recorded in a whole-cell voltage clamp experiment in which a CA1 pyramidal neuron was held at 80 mV. (From Hoffman, A.F., Riegel, A.C., and Lupica, C.R. [2003] *Eur J Neurosci*, 18, 524–534. With permission.) Ongoing GABA release (a) onto the pyramidal cell is briefly suppressed by a short (2 s) depolarizing step from the holding potential to 0 mV. As discussed in the text, this process is dependent on a postsynaptic increase in intracellular Ca^{2+}. (b) In the same cell, in the presence of the CB_1 antagonist SR141716A, no inhibition is observed following the step.

retrograde, presynaptic inhibitors of GABAergic synapses in both the cerebellum and hippocampus now have been extended to excitatory cerebellar synapses as well (Kreitzer and Regehr, 2001b). Recently, Ohno-Shosaku and colleagues (Ohno-Shosaku et al., 2002) have attempted to determine if DSE can also be elicited at hippocampal synapses. Interestingly, it appears that hippocampal DSE is less robust and is only observed following much longer (>7 s) postsynaptic depolarizing steps. These authors suggest that this disparity may result from the lower sensitivity of excitatory synapses, relative to inhibitory synapses, to exogenous cannabinoid agonists (Ohno-Shosaku et al., 2002). Because the molecular identity of the receptor responsible for the effects of cannabinoids at excitatory terminals in the hippocampus has been the subject of recent debate (Hajos et al., 2001; Hajos and Freund, 2002), a more precise role of endocannabinoids and the unequivocal determination of their presynaptic targets, at excitatory hippocampal synapses, remains to be determined. Nevertheless, the combined studies of DSI and DSE in the hippocampus and cerebellum have firmly established a role for endogenous cannabinoids as a class of retrograde messengers capable of producing short-term changes in synaptic activity.

Modulation of the Release of Endocannabinoids by Other Neurotransmitters

The essential finding that postsynaptic increases in intracellular Ca^{2+} can induce endocannabinoid release has also led many investigators to an important corollary hypothesis: that activation of postsynaptic receptors known to be coupled to the mobilization of intracellular Ca^{2+} should also promote endocannabinoid release. At both GABAergic and glutamatergic synapses in the CNS, a number of G-protein-coupled receptor systems are known to be linked to Ca^{2+} mobilization, including metabotropic glutamate receptors (mGluRs; Anwyl, 1999; De Blasi et al., 2001) and muscarinic cholinergic receptors (Kerlavage et al., 1987). Thus, several recent studies have investigated whether activation of these receptor systems can also facilitate endocannabinoid release. In hippocampal slices, Varma et al. (2001) showed that activation of group I mGluRs (mGluR1 and/or mGluR5) reduced GABA release onto CA1 pyramidal cells and that this effect was blocked by a CB_1 receptor antagonist (Varma et al., 2001). The effects of the mGluR agonists on GABA release are absent in CB_1 receptor knockout mice, providing further evidence for a link between mGluR activation and endocannabinoid release. Finally, it was demonstrated that concentrations of the mGluR1/5 agonist, (S)-3,5-dihydroxyphenylglycine (DHPG) that were normally ineffective at reducing IPSCs could enhance DSI (Varma et al., 2001; Alger, 2002). A similar link between group I mGluR receptor activation and endocannabinoid release was described at climbing fiber-Purkinje cell excitatory synapses in the cerebellum (Maejima et al., 2001). Interestingly, intracellular chelation of Ca^{2+} by BAPTA did not block the effect of the mGluR agonist, leading the authors to suggest that endocannabinoid release in this instance results from a Ca^{2+}-independent mechanism (Maejima et al., 2001). It is possible that the activation of mGluRs on other nearby neurons could have led to Ca^{2+}-dependent endocannabinoid production in those cells (e.g., "spillover"; Alger, 2002), sufficient to overcome the effects of BAPTA. However, increases in Ca^{2+} fluorescence were not observed in neighboring dendrites during mGluR activation. In addition, the effects of the mGluR agonist were blocked by intracellular application of the nonhydrolyzable GTP analog, GTPγS, which should also be overcome if spillover of endocannabinoids had occurred.

Previous findings that muscarinic acetylcholine receptor activation enhanced DSI at hippocampal GABAergic synapses (Martin and Alger, 1999) led the same group to investigate whether endocannabinoid production was involved in this effect (Kim et al., 2002). Here, low concentrations ($< 1\ \mu M$) of carbachol were shown to enhance DSI without affecting baseline GABA release, an effect that was blocked by CB_1 antagonists and absent in CB_1 receptor knockout mice. Muscarinic receptor-mediated release of the endocannabinoid was also apparently dependent on a G-protein-coupled, but not a Ca^{2+}-dependent, signaling cascade, because the effects of carbachol were blocked by intracellular GTPγS but not by the Ca^{2+} chelator, BAPTA. As was the case in the cerebellum, the ability of intracellular GTPγS but not BAPTA to block the release of the endocannabinoid argues against the likelihood of spillover of endocannabinoid from neighboring pyramidal cells following carbachol application (Kim et al., 2002; Alger, 2002). Based on a recently published study by Ohno-Shosaku and colleagues (Ohno-Shosaku et al., 2003), it appears as though both postsynaptic M1 and M3 muscarinic receptors are required for the enhancement of DSI by acetylcholine.

In summary, it appears as though endocannabinoids are active at both excitatory and inhibitory synapses through either depolarization of the postsynaptic cell membrane or through activation of G-protein-coupled receptors on the postsynaptic cell. While increased postsynaptic Ca^{2+} levels appear necessary to observe endocannabinoid release in the former instance, it also appears that another form of Ca^{2+}-independent endocannabinoid release is present in the CNS. The G-protein-coupled release of endocannabinoid will no doubt remain of significant interest to future research because of the potentially pivotal role that a variety of neurotransmitter systems may play in the regulation of endocannabinoid signaling in the CNS.

Functional Implications of DSI/DSE

Both DSI and DSE studies have established a clear paradigm in which endocannabinoid release at both excitatory and inhibitory synapses in the CNS can be studied. The question that now must be addressed is what roles these forms of short-term synaptic plasticity play under normal physiological conditions. It is well established that activation of cannabinoid receptors by the primary psychoactive ingredient in marijuana, Δ^9-tetrahydrocannabinol (Δ^9-THC), can disrupt hippocampal and cerebellar function, as evidenced by marijuana's deleterious effects on memory, spatial performance, and motor coordination (Abel, 1970; Abood and Martin, 1992; Adams and Martin, 1996; Winsauer et al., 1999). However, the significance of endogenous cannabinoids in the moment-to-moment functioning of these synapses has only recently begun to be investigated. Despite the recent evidence that *in vitro* models of endocannabinoid release may not relate directly to *in vivo* function (Hampson et al., 2003), studies are ongoing to address the functional role of endocannabinoid release in regulating CNS activity. For example, a recent study by Alger's group (Carlson et al., 2002) determined that LTP at hippocampal excitatory synapses can be facilitated by endocannabinoid release. Thus, when weak stimulus trains that did not alone produce LTP were preceded by a DSI-inducing depolarizing step, LTP was induced. This form of DSI-paired LTP was blocked by the CB_1 receptor antagonist AM251, suggesting that the postsynaptic depolarization produced an endogenous cannabinoid-dependent inhibition of GABA release that was sufficient to facilitate glutamatergic transmission. Moreover, the ability to induce LTP was dependent on the interval between the depolarizing step and the stimulus train, as would be expected given the transient nature (<30 s) of the endogenous cannabinoid effect, and the phenomenon of DSI (Carlson et al., 2002). Thus, one possible role for endogenous cannabinoids may be to facilitate the induction of LTP at specific hippocampal synapses by transiently suppressing inhibitory inputs, thereby giving greater weight to excitatory inputs impinging upon specific pyramidal cell populations. The ability of endocannabinoids to facilitate LTP contrasts with earlier studies demonstrating blockade of hippocampal LTP following the application of cannabinoid agonists (Nowicky et al., 1987; Collins et al., 1994; Misner and Sullivan, 1999). However, because recordings from neuron pairs have demonstrated that DSI is spatially restricted to ≤20 μm in the hippocampus (Wilson and Nicoll, 2001), it can be assumed that endocannabinoids released by postsynaptic depolarization will only facilitate LTP within a circumscribed region surrounding the depolarized cell. In contrast, a reduction in LTP would tend to occur during periods of global CB_1 receptor activation (Nowicky et al., 1987; Collins et al., 1994).

Another possible consequence of endocannabinoid release has recently been described in the cerebellum (Kreitzer et al., 2002). Here, the authors found that, following depolarization of Purkinje cells, the firing of inhibitory interneurons was also suppressed, due to the activation of a barium-sensitive K^+ conductance. As a single interneuron can innervate Purkinje neurons over long distances (>100 μm), the significance of this finding is that the release of an endocannabinoid from a single Purkinje cell can, by reducing the firing rate of nearby interneurons, influence GABA release over a much longer range than would be predicted by the simple diffusion of the endocannabinoid. In other words, DSI in the cerebellum does not appear to be spatially restricted to the same extent as in the hippocampus, so that endocannabinoid released from a single cell may exert a widespread influence over cerebellar network output (Kreitzer et al., 2002). Clearly, despite the underlying fundamental role of endocannabinoids in DSI/DSE, a number of critical experiments remain in order to fully understand this unique form of synaptic plasticity, as well as its relevance to animal behavior.

LTD: A Role for Endocannabinoids in Long-Term Plasticity

Based on the presynaptic localization of CB_1 receptors and the effects of endocannabinoids on short-term synaptic plasticity, it was further hypothesized that these molecules might also be involved in long-term forms of synaptic plasticity. As described previously, initial studies demonstrated that directly applied endocannabinoids could block LTP in the hippocampus (Stella et al., 1997), or that endocannabinoids could facilitate LTP in this brain area when DSI was initiated (Carlson et al., 2002). However, unequivocal support for an essential role for endocannabinoid molecules in the establishment of long-term synaptic plasticity was not initially obtained in the hippocampus but rather in the dorsal striatum, where LTD was shown to be dependent upon endocannabinoid release (Gerdeman et al., 2002). Subsequently, an endocannabinoid was also shown to be critical to another form of LTD in the ventral striatum (i.e., the nucleus accumbens; see Robbe, Kopf, et al., 2002). There are compelling reasons to speculate that both the acute effects of exogenous cannabinoids on synaptic transmission in the nucleus accumbens (Hoffman and Lupica, 2001), as well as the effects of endocannabinoids on LTD in this region, are important in mediating the euphoric or addictive effects of marijuana. Specifically, it is believed that these endogenous and exogenous cannabinoids may, by altering ongoing synaptic activity within the dorsal and ventral striatum, influence the formation of habitual or stereotyped behavioral patterns that are associated with, and sustain, illicit drug use (Gerdeman et al., 2003). The implications for LTD in drug abuse and addiction will be discussed later in this section, and the reader is also referred to several recent reviews on this topic (Winder et al., 2002; Gerdeman et al., 2003; Thomas and Malenka, 2003) (Figure 9.2).

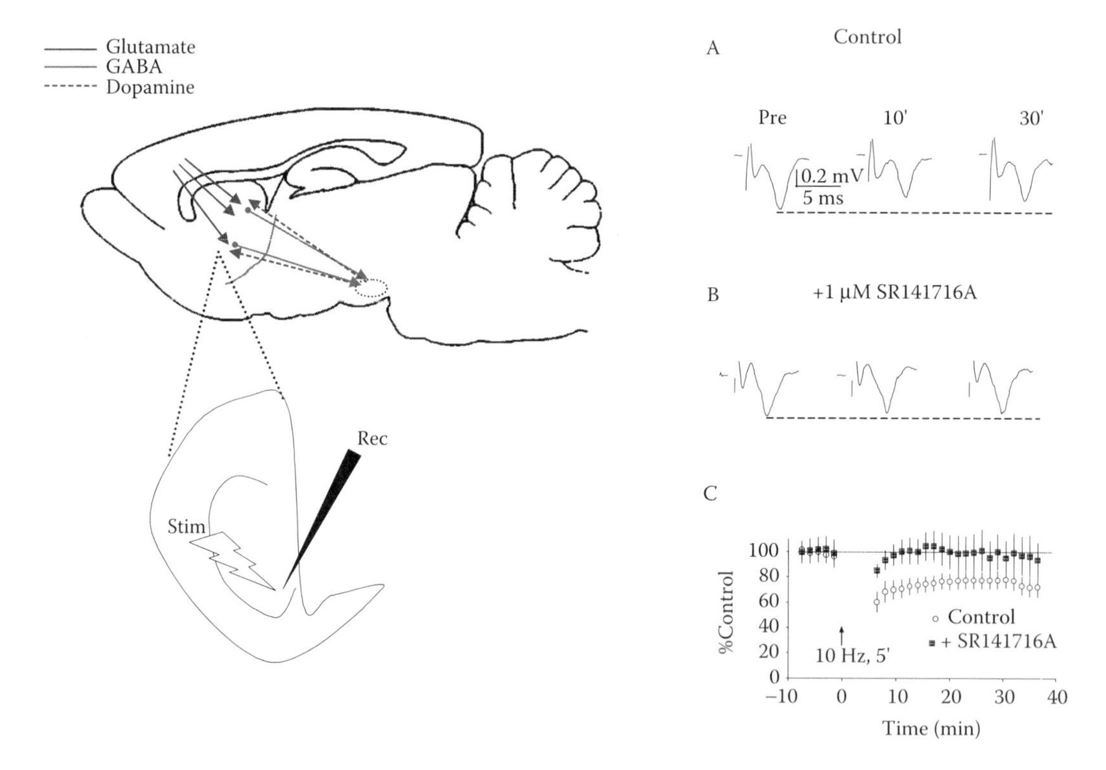

FIGURE 9.2 Long-term depression (LTD) in the nucleus accumbens is dependent on endogenous cannabinoids. Left panel illustrates the major components of the reward and habit-forming areas of the CNS in a sagittal section through the rat brain. Glutamatergic inputs from the cortex and dopaminergic inputs from the ventral tegemental area and substantia nigra (gray circle outline) project to GABAergic medium spiny neurons in the dorsal striatum and ventral striatum. Inset diagram illustrates a coronal section with an extracellular recording electrode (right) and a stimulating electrode (left) placed in the core subregion of the nucleus accumbens (NAc). Right panel, (A) In extracellular recordings from a brain slice containing the NAc (see Hoffman, Oz, et al., 2003 for experimental details), LTD of the glutamatergic population spike persists for over 30 min following delivery of repeated 0.1 ms stimuli (10 Hz, 5 min). (B) The inhibition is blocked by SR141716A. (C) Summary of LTD in the NAc and its block by SR141716A.

LTD at Glutamatergic Synapses in the Dorsal Striatum

The basal ganglia contain some of the highest densities of CB_1 receptors in the CNS (Herkenham et al., 1991; Herkenham, 1992). As was the case for other brain regions, pharmacological and electrophysiological techniques confirmed the presynaptic inhibitory effects of CB_1 receptors at both glutamatergic and GABAergic terminals within various basal ganglia nuclei (see Table 9.1). However, as was the case in the hippocampus and cerebellum, the fundamental link between the presence of these functional CB_1 receptors and their endogenous ligands remained to be established. Anandamide release in the dorsolateral striatum, and its enhancement by dopamine D2 receptor agonists, was first demonstrated using microdialysis techniques (Giuffrida et al., 1999). Such an approach, although well-suited to addressing global or widespread activation of endocannabinoid systems, cannot assess endocannabinoid release and activity at the synaptic level. Thus, whole-cell and extracellular recording techniques were needed in order to further address the conditions favoring endocannabinoid release. LTD of excitatory afferents to medium spiny neurons in the dorsal striatum has been demonstrated in several studies (Calabresi et al., 1992; Lovinger and Tyler, 1996; Choi and Lovinger, 1997b). In general, brief (1 s) trains of high-frequency stimuli (3–4 trains, 100 Hz, delivered once every 10 s) can be used to induce LTD, which persists for > 30 min in brain slices. Like DSI/DSE, a critical component of striatal LTD is its dependence on both postsynaptic induction via a Ca^{2+}-dependent mechanism and presynaptic expression as a decrease in the probability of glutamate release (Choi and Lovinger, 1997a; Choi and Lovinger, 1997b). Furthermore, because of this postsynaptic Ca^{2+} dependence, it was hypothesized that a retrograde messenger was also involved in the establishment and maintenance of striatal LTD. Because the activation of CB_1 receptors with exogenous agonists was known to presynaptically inhibit glutamate release in the striatum (Gerdeman and Lovinger, 2001), an enticing hypothesis emerged to suggest that LTD may involve the activation of presynaptic CB_1 receptors by an endocannabinoid released from medium spiny neurons during high-frequency stimulation. This hypothesis was subsequently confirmed using both pharmacological and genetic approaches (Gerdeman et al., 2002). Thus, striatal LTD was blocked in slices pretreated with SR141716A and was completely absent in mice lacking the CB_1 receptor gene. In addition, chelation of intracellular Ca^{2+} by EGTA blocked LTD, as would be expected if endocannabinoid release was dependent on a postsynaptic rise in intracellular Ca^{2+}. Interestingly, this effect could be overcome by the endocannabinoid uptake blocker, AM404, suggesting that by reducing the uptake of endocannabinoid released from neighboring (non-EGTA-filled) cells, LTD expression could occur even in those cells in which postsynaptic Ca^{2+} levels were limited. Finally, in an elegant study designed to determine which endocannabinoid was responsible for the observed LTD, the authors "loaded" cells with anandamide by including it in the intracellular pipette. The resulting decrease in EPSCs and the blockade of this effect by SR141716A confirmed that anandamide released from medium spiny neurons could indeed presynaptically inhibit glutamatergic synaptic transmission. Together, these results extended the critical role of endocannabinoids to yet another form of synaptic plasticity. In addition, they complemented, at the synaptic level, the earlier work demonstrating anandamide release in the striatum (Giuffrida et al., 1999). Because striatal LTD is also dependent on dopamine D2 receptor activation (Calabresi et al., 1997; Tang et al., 2001), these findings may also confirm the importance of the interplay between G-protein-coupled dopamine receptors and the anandamide releasing pathways (Giuffrida et al., 1999).

LTD of Glutamatergic Synapses in the Nucleus Accumbens

As in the dorsal striatum, both the acute presynaptic effects of cannabinoids (Hoffman and Lupica, 2001; Robbe et al., 2001) and LTD of excitatory afferents to the nucleus accumbens (NAc) (Kombian and Malenka, 1994; Robbe, 2002a) have been described. Therefore, it was of considerable interest to determine whether LTD in the NAc was similar to that described in the dorsal striatum, in terms of its dependence upon endocannabinoids. Similar to the dorsal striatum, LTD in the NAc was absent in CB_1 receptor knockout animals (Robbe, 2002b) and was blocked in wild-type animals by application of the CB_1 receptor antagonist SR141716A (Robbe, 2002b; Hoffman, Oz, et al., 2003). In addition, the studies in the NAc differed from the LTD studies in the dorsal striatum (Gerdeman et al., 2002) in several respects. First, the stimulus parameters used to produce endocannabinoid-dependent LTD were different, requiring stimulus intensities of long duration and much lower frequencies (10 min, 13 Hz; Robbe, 2002b), rather than the high-frequency activation used by Gerdeman et al. (2002). In fact, a similar paradigm using high-frequency stimulation in the NAc typically generates LTD that is dependent on presynaptic group II mGluRs (Robbe, 2002a). In our own laboratory, we have been unable to routinely produce LTD in the NAc using high-frequency stimulation parameters, but have recently found that a 10 Hz, 5 min stimulation can reliably produce endocannabinoid-dependent LTD (Hoffman, Oz, et al., 2003).

Another important distinction to be made between LTD studies in the dorsal striatum and the NAc is the possible involvement of postsynaptic group I mGluRs (mGluR5) in the release of endocannabinoids in the NAc (Robbe, 2002b). Because mGluR5 receptors are linked to postsynaptic increases in intracellular Ca^{2+} in medium spiny neurons (Shigemoto et al., 1993; Casabona et al., 1997; Testa et al., 1998), Robbe et al. (2002b) tested the hypothesis that activation of mGluR5 could stimulate endocannabinoid release in the NAc. Thus, whereas application of the mGluR5 agonist (S)-DHPG resulted in LTD that was blocked by SR141716A in brain slices from wild-type mice, this form of LTD was absent in CB_1 receptor knockout mice. In addition, endocannabinoid-mediated LTD was blocked by a selective mGluR5 antagonist in the NAc (Robbe, 2002b). In contrast, (S)-DHPG does not affect glutamatergic transmission in the striatum, although striatal LTD is blocked by an mGluR5 antagonist (Sung et al., 2001). The studies performed by Gerdeman et al. (Gerdeman et al., 2002) did not directly address the issue of whether endocannabinoid-dependent LTD relied upon mGluR5 receptor activation. Given the critical role of the NAc within the brain reward and motivational circuitry (Wise and Rompre, 1989; Wise, 1996), a link between endocannabinoid release and LTD in the NAc may be viewed as an important finding in terms of our understanding of the normal functioning of this nucleus and in terms of the addictive potential of marijuana.

LTD of GABAergic Synapses: Forgetting and the Endocannabinoids

Whereas the preceding sections highlight the role of endocannabinoid-mediated LTD at excitatory synapses within motor and reward circuitry, recent reports have begun to examine a novel form of LTD at GABAergic inhibitory synapses in the basolateral amygdala and the hippocampus. CB_1 receptors can inhibit GABA release in the basolateral amygdala (Katona et al., 2001), an area that plays an important role in the process of extinction of aversive memories (LeDoux, 2000). Marsicano and colleagues therefore performed a series of experiments designed to assess the role of the endocannabinoid system in the extinction process (Marsicano et al., 2002). Apart from the behavioral and biochemical findings that clearly delineated a role for endocannabinoids in aversive memory extinction (for details, see Marsicano et al., 2002), the authors also demonstrated that repetitive, low frequency stimulation in the basolateral amygdala resulted in LTD at GABAergic terminals. This form of LTD was apparently due to presynaptic mechanisms (e.g., a decrease in release probability) and was not present in slices from either mice lacking the CB_1 receptor gene or in slices from wild-type mice treated with SR141716A. Thus, the authors propose that LTD at GABAergic synapses in the basolateral amygdala is dependent on endocannabinoid release and that this may have a functional significance in terms of the ability of animals to extinguish aversive memories (Marsicano et al., 2002). In a more recently published report, LTD was also observed at GABAergic synapses onto hippocampal CA1 pyramidal neurons (Chevaleyre and Castillo, 2003). Here, the authors began with the observation that a high-frequency stimulation protocol that normally elicits LTP at excitatory synapses resulted in LTD of GABAergic synapses. Because LTD was only observed when glutamatergic fibers were activated during the stimulation and as ionotropic glutamate receptor subtypes were blocked in all experiments, the authors hypothesized that activation of an mGluR receptor was responsible for the LTD. In fact, blockade of group I mGluRs prevented the LTD, whereas activation of these receptors by the exogenous mGluR agonist (S)-DHPG occluded LTD. Moreover, LTD was blocked by AM251, and this same CB_1 receptor antagonist also blocked the "pharmacological" LTD produced by (S)-DHPG. This novel form of LTD can be blocked if AM251 is present between 1 and 5 min following stimulation (Chevaleyre and Castillo, 2003), consistent with the previous reports suggesting an enhanced, long-term elevation of 2-AG following high-frequency stimulation (Stella et al., 1997). The most striking finding of this work, however, was a direct comparison between endocannabinoid-mediated LTD and DSI. The authors reported that the same population of inhibitory inputs that expressed DSI also expressed LTD. However, whereas DSI could be generated by stimulation of either the pyramidal cell layer or the underlying stratum radiatum, LTD was only observed when the radiatum layer, and therefore, Schaffer collateral glutamatergic afferents, was stimulated. In addition, whereas DSI was confirmed to be dependent on an increase in postsynaptic intracellular Ca^{2+}, LTD was blocked by a diacylglycerol (DAG) lipase inhibitor that inhibits the conversion of DAG to 2-AG (Chevaleyre and Castillo, 2003). Thus, the activation of group I mGluR receptors by glutamate released following stimulation of Schaffer collateral afferent fibers and the resulting activation of DAG can generate an endocannabinoid (probably 2-AG) that promotes LTD at hippocampal GABAergic synapses. Whereas DSI can also result from mGluR activation (Varma et al., 2001), this is not a prerequisite for short-term depression, and DSI can clearly exist at the same synapses where LTD does not. The findings of Chevaleyre and Castillo (2003) are, therefore, highly significant, as they illustrate the capacity of endocannabinoids to function as mediators of both short- and long-term forms of synaptic plasticity at the same synapses (Figure 9.3).

Functional Significance of LTD

As the preceding studies indicate, it is clear that endocannabinoids are involved in the presynaptic expression of LTD in several brain areas. In the same way that endocannabinoids appear to function as mediators of short-term synaptic plasticity at both GABAergic and glutamatergic synapses in the hippocampus and cerebellum, so does LTD represent a means through which these molecules can affect long-term plasticity within neuronal circuits. Moreover, it is now evident that both forms of plasticity can be present within the same synaptic elements and that endocannabinoids can play a role in either or both, depending on the conditions. In the hippocampus, the significance of LTD may be to give greater weight to excitatory inputs, thereby helping to facilitate pyramidal cell excitability during periods of glutamatergic activation (Chevaleyre and Castillo, 2003). In this context, LTD might be viewed as an extended or more global form of DSI, which is already thought to facilitate LTP under some conditions (Carlson et al., 2002).

Within the basal ganglia and reward circuitry, the functional significance of LTD is still the subject of much speculation. However, several recent studies have highlighted a potential role for this form of long-term synaptic plasticity in drug abuse and addiction. For example, one current hypothesis is that addiction reflects a change in the normal cellular changes necessary for learning to occur (Gerdeman et al., 2003). Because LTD and other forms of synaptic plasticity are often used as models for learning processes, changes in synaptic efficacy following acute or chronic drug exposure may underlie part of the addictive process. Thus, it has been demonstrated that behavioral sensitization following repeated cocaine administration is accompanied by enhanced LTD in the nucleus accumbens (Thomas et al., 2001). In addition, the LTD of glutamatergic inputs to dopamine neurons in the ventral tegmental area is enhanced by a single exposure to cocaine (Ungless et al., 2001). Similarly, we have recently found that chronic exposure to Δ^9-THC or WIN55212-2 results in CB_1 receptor tolerance and the blockade of endocannabinoid-dependent LTD in the nucleus accumbens (Hoffman, Oz, et al., 2003), suggesting that repeated marijuana use may result in changes in the sensitivity of excitatory synapses to endocannabinoids and a reduction in long-term

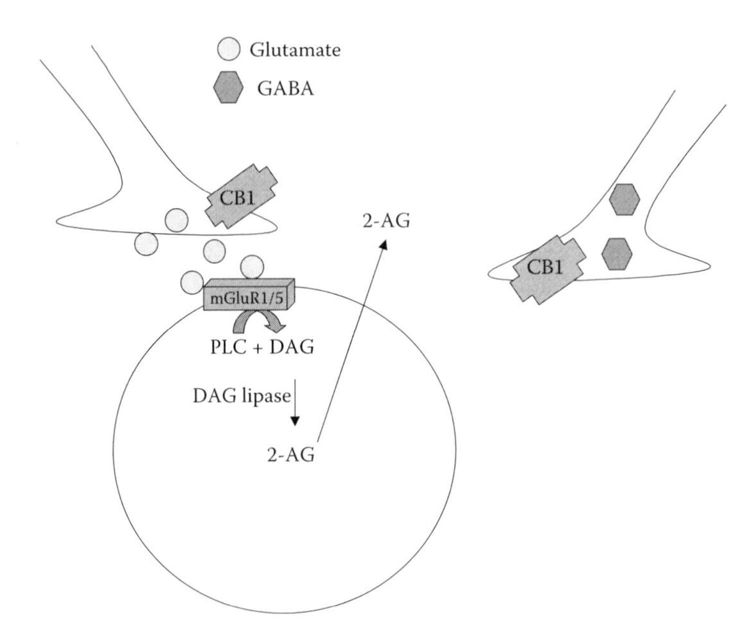

FIGURE 9.3 Hypothetical model of endocannabinoid release mediated through activation of metabotropic glutamate receptors (mGluR). Release of glutamate from excitatory terminals (data from Chevaleyre, V. and Castillo, P.E. [2003] *Neuron*, 38(3), 461–472) or direct application of an mGluR agonist activates group I mGluRs, mobilizing diacylglycerol (DAG) and phospholipase C (PLC). The release of 2-AG formed from DAG activates presynaptic CB_1 receptors, resulting in either transient inhibition (DSI; data from Varma, N. et al. [2001] *J. Neurosci.*, 21(24), RC188) or long-term depression (LTD; data from Chevaleyre, V. and Castillo, P.E. [2003] *Neuron*, 38(3), 461–472) at hippocampal inhibitory synapses. In the cerebellum, a similar mechanism has been proposed to underlie short-term depression at excitatory synapses. (DSE; data from Maejima, T. et al. [2001] *Neuron*, 31(3), 463–475)

plasticity at these synapses. Ongoing studies, in our laboratory and in many others, should help to elucidate both the role of LTD in drug abuse, as well as the importance of endocannabinoids in this form of synaptic plasticity.

CONCLUSIONS AND FUTURE DIRECTIONS

The development of specific pharmacological ligands for the cannabinoid system, coupled with sensitive electrophysiological assays of synaptic function, has permitted an extensive analysis of the role of endocannabinoids at fast amino-acid neurotransmitter (GABA and glutamate) synapses in the CNS. The data presented in recent literature suggest that endocannabinoids can be released from postsynaptic neurons through both Ca^{2+}-dependent, and Ca^{2+}-independent mechanisms, and that these molecules can diffuse in a retrograde manner to activate presynaptic CB_1 receptors that inhibit GABA or glutamate release from axon terminals. Various forms of neuronal stimulation can induce endocannabinoid release, ranging from physiological (neuronal firing or depolarization) to pharmacological (activation of postsynaptic G-protein-coupled receptors). Once generated, the spread of endocannabinoids appears to be limited by their diffusion or uptake away from sites of action. However, the ability of these lipid mediators to affect neuronal firing rates in some broader areas (Kreitzer et al., 2002) and to generate long-term changes in synaptic plasticity in others means that the actions of endocannabinoids are not necessarily restricted to single synapses, or to short-term actions.

At this point, it is premature to ascribe all of these synaptic effects to a single endocannabinoid molecule (2-AG, anandamide, noladin ether, etc.). Although evidence in the hippocampus seems to suggest that 2-AG is the dominant endocannabinoid species (Stella et al., 1997; Chevaleyre and Castillo, 2003), it remains a distinct possibility that other endocannabinoids may be active within different brain areas (Gerdeman et al., 2002). The continued development of pharmacological ligands with unique structure-activity relationships may be useful in shedding light on this issue. We also suggest, based on our work in the hippocampus, that not all neuronal subtypes release endocannabinoids under the same conditions (Hoffman, Riegel, et al., 2003). In addition, there may be varying degrees of sensitivity to endocannabinoids at different synapses, either due to the expression or function of cannabinoid receptors on presynaptic elements (Ohno-Shosaku et al., 2002; Hajos and Freund, 2002), or the ability of different populations of neurons to synthesize and release endocannabinoids (Hoffman, Riegel, et al., 2003). A combination of both functional and morphological mapping of receptors or enzymes involved in the synthesis and regulation of endocannabinoids will be needed in order to resolve this issue.

The involvement of endocannabinoids in mediating various forms of both short- and long-term plasticity appears to now be a well-established theme that extends across many brain regions. Based on these data alone, we now have a much better understanding of why cannabinoid receptors may have evolved and are expressed at such high levels in the mammalian CNS. If a major role of endocannabinoids is truly to "fine tune" synaptic transmission, then there appears to be a teleological basis for the presence of their receptors on discrete, fast, amino-acid-based synapses (glutamate and GABA), rather than on more diffuse elements such as catecholaminergic terminals. However, the goal of determining the relevance of such synaptic plasticity to the everyday behavior of both man and animal remains elusive, and as such, it appears as though the placement of endocannabinoids within the framework of global CNS function is just beginning.

REFERENCES

Abel, E.L. (1970) Marijuana and memory, *Nature,* 227(263), 1151–1152.

Abood, M.E. and Martin, B.R. (1992) Neurobiology of marijuana abuse, *Trends Pharmacol. Sci.,* 13(5), 201–206.

Adams, I.B. and Martin, B.R. (1996) Cannabis: pharmacology and toxicology in animals and humans, *Addiction,* 91(11), 1585–1614.

Alger, B.E. (2002) Retrograde signaling in the regulation of synaptic transmission: focus on endocannabinoids, *Prog. Neurobiol.,* 68(4), 247–286.

Alger, B.E. and Pitler, T.A. (1995) Retrograde signaling at GABAA-receptor synapses in the mammalian CNS, *Trends Neurosci.,* 18(8), 333–340.

Alger, B.E., Pitler, T.A., Wagner, J.J., Martin, L.A., Morishita, W., Kirov, S.A., and Lenz, R.A. (1996) Retrograde signalling in depolarization-induced suppression of inhibition in rat hippocampal CA1 cells, *J. Physiol,* 496 (Pt 1), 197–209.

Anwyl, R. (1999) Metabotropic glutamate receptors: electrophysiological properties and role in plasticity, *Brain Res. Brain Res. Rev.,* 29(1), 83–120.

Auclair, N., Otani, S., Soubrie, P., and Crepel, F. (2000) Cannabinoids modulate synaptic strength and plasticity at glutamatergic synapses of rat prefrontal cortex pyramidal neurons, *J. Neurophysiol.,* 83(6), 3287–3293.

Calabresi, P., Maj, R., Pisani, A., Mercuri, N.B., and Bernardi, G. (1992) Long-term synaptic depression in the striatum: physiological and pharmacological characterization, *J Neurosci.,* 12(11), 4224–4233.

Calabresi, P., Pisani, A., Centonze, D., and Bernardi, G. (1997) Synaptic plasticity and physiological interactions between dopamine and glutamate in the striatum, *Neurosci. Biobehav. Rev.,* 21, 519–523.

Carlson, G., Wang, Y., and Alger, B.E. (2002) Endocannabinoids facilitate the induction of LTP in the hippocampus, *Nat. Neurosci.,* 5(8), 723–724.

Casabona, G., Knopfel, T., Kuhn, R., Gasparini, F., Baumann, P., Sortino, M.A., Copani, A., and Nicoletti, F. (1997) Expression and coupling to polyphosphoinositide hydrolysis of group I metabotropic glutamate receptors in early postnatal and adult rat brain, *Eur. J. Neurosci.,* 9(1), 12–17.

Chan, P.K., Chan, S.C., and Yung, W.H. (1998) Presynaptic inhibition of GABAergic inputs to rat substantia nigra pars reticulata neurones by a cannabinoid agonist, *Neuroreport,* 9(4), 671–675.

Chan, P.K. and Yung, W.H. (1998) Occlusion of the presynaptic action of cannabinoids in rat substantia nigra pars reticulata by cadmium, *Neurosci. Lett.,* 249(1), 57–60.

Chevaleyre, V. and Castillo, P.E. (2003) Heterosynaptic LTD of hippocampal GABAergic Synapses. a novel role of endocannabinoids in regulating excitability, *Neuron,* 38(3), 461–472.

Choi, S. and Lovinger, D.M. (1997a) Decreased frequency but not amplitude of quantal synaptic responses associated with expression of corticostriatal long-term depression, *J. Neurosci.,* 17(21), 8613–8620.

Choi, S. and Lovinger, D.M. (1997b) Decreased probability of neurotransmitter release underlies striatal long-term depression and postnatal development of corticostriatal synapses, *Proc. Natl. Acad. Sci. U. S. A,* 94(6), 2665–2670.

Cohen, G.A., Doze, V.A., and Madison, D.V. (1992) Opioid inhibition of GABA release from presynaptic terminals of rat hippocampal interneurons, *Neuron,* 9(2), 325–335.

Collins, D.R., Pertwee, R.G., and Davies, S.N. (1994) The action of synthetic cannabinoids on the induction of long-term potentiation in the rat hippocampal slice, *Eur. J. Pharmacol.,* 259(3), R7–R8.

De Blasi, A., Conn, P.J., Pin, J., and Nicoletti, F. (2001) Molecular determinants of metabotropic glutamate receptor signaling, *Trends Pharmacol. Sci.,* 22(3), 114–120.

Deadwyler, S.A., Hampson, R.E., Mu, J., Whyte, A., and Childers, S. (1995) Cannabinoids modulate voltage sensitive potassium A-current in hippocampal neurons via a cAMP-dependent process, *J. Pharmacol. Exp. Ther.,* 273(2), 734–743.

Diana, M.A., Levenes, C., Mackie, K., and Marty, A. (2002) Short-term retrograde inhibition of GABAergic synaptic currents in rat Purkinje cells is mediated by endogenous cannabinoids, *J. Neurosci.,* 22(1), 200–208.

Doze, V.A., Cohen, G.A., and Madison, D.V. (1995) Calcium channel involvement in GABAB receptor-mediated inhibition of GABA release in area CA1 of the rat hippocampus, *J. Neurophysiol.,* 74(1), 43–53.

Gerdeman, G. and Lovinger, D.M. (2001) CB_1 cannabinoid receptor inhibits synaptic release of glutamate in rat dorsolateral striatum, *J. Neurophysiol.,* 85(1), 468–471.

Gerdeman, G.L., Partridge, J.G., Lupica, C.R., and Lovinger, D.M. (2003) It could be habit forming: drugs of abuse and striatal synaptic plasticity, *Trends Neurosci.,* 26(4), 184–192.

Gerdeman, G.L., Ronesi, J., and Lovinger, D.M. (2002) Postsynaptic endocannabinoid release is critical to long-term depression in the striatum, *Nat. Neurosci.,* 5(5), 446–451.

Giuffrida, A., Parsons, L.H., Kerr, T.M., Rodriguez, D.F., Navarro, M., and Piomelli, D. (1999) Dopamine activation of endogenous cannabinoid signaling in dorsal striatum, *Nat. Neurosci.,* 2(4), 358–363.

Hajos, N. and Freund, T.F. (2002) Pharmacological separation of cannabinoid sensitive receptors on hippocampal excitatory and inhibitory fibers, *Neuropharmacology,* 43(4), 503–510.

Hajos, N., Ledent, C., and Freund, T.F. (2001) Novel cannabinoid-sensitive receptor mediates inhibition of glutamatergic synaptic transmission in the hippocampus, *Neuroscience,* 106(1), 1–4.

Hampson, R.E., Zhuang, S.Y., Weiner, J.L., and Deadwyler, S.A. (2003) Functional significance of cannabinoid-mediated, depolarization induced suppression of inhibition (DSI) in the hippocampus, *J. Neurophysiol.,* 90, 55–64.

Herkenham, M. (1992) Cannabinoid receptor localization in brain: relationship to motor and reward systems, *Ann. N. Y. Acad. Sci.,* 654, 19–32.

Herkenham, M., Lynn, A.B., Johnson, M.R., Melvin, L.S., de Costa, B.R., and Rice, K.C. (1991) Characterization and localization of cannabinoid receptors in rat brain: a quantitative in vitro autoradiographic study, *J. Neurosci.,* 11(2), 563–583.

Herkenham, M., Lynn, A.B., Little, M.D., Johnson, M.R., Melvin, L.S., de Costa, B.R., and Rice, K.C. (1990) Cannabinoid receptor localization in brain, *Proc. Natl. Acad. Sci. U. S. A,* 87(5), 1932–1936.

Hoffman, A.F. and Lupica, C.R. (2000) Mechanisms of cannabinoid inhibition of GABA(A) synaptic transmission in the hippocampus, *J. Neurosci.,* 20(7), 2470–2479.

Hoffman, A.F. and Lupica, C.R. (2001) Direct actions of cannabinoids on synaptic transmission in the nucleus accumbens: a comparison with opioids, *J. Neurophysiol.,* 85(1), 72–83.

Hoffman, A.F., Oz, M., Caulder, T., and Lupica, C.R. (2003) Functional tolerance and blockade of long-term depression at synapses in the nucleus accumbens after chronic cannabinoid exposure, *J. Neurosci.,* 23(12), 4815.

Hoffman, A.F., Riegel, A.C., and Lupica, C.R. (2003) Functional localization of cannabinoid receptors and endogenous cannabinoid production in distinct neuron populations of the hippocampus, *Eur. J. Neurosci.,* 18, 524–534.

Huang, C.C., Lo, S.W., and Hsu, K.S. (2001) Presynaptic mechanisms underlying cannabinoid inhibition of excitatory synaptic transmission in rat striatal neurons, *J. Physiol.,* 532(Pt. 3), 731–748.

Irving, A.J., Coutts, A.A., Harvey, J., Rae, M.G., Mackie, K., Bewick, G.S., and Pertwee, R.G. (2000) Functional expression of cell surface cannabinoid CB(1) receptors on presynaptic inhibitory terminals in cultured rat hippocampal neurons, *Neuroscience,* 98(2), 253–262.

Jiang, Z.G. and North, R.A. (1992) Pre- and postsynaptic inhibition by opioids in rat striatum, *J. Neurosci.,* 12(1), 356–361.

Katona, I., Rancz, E.A., Acsady, L., Ledent, C., Mackie, K., Hajos, N., and Freund, T.F. (2001) Distribution of CB_1 cannabinoid receptors in the amygdala and their role in the control of GABAergic transmission, *J. Neurosci.,* 21(23), 9506–9518.

Kemp, N. and Bashir, Z.I. (2001) Long-term depression: a cascade of induction and expression mechanisms, *Prog. Neurobiol.,* 65(4), 339–365.

Kerlavage, A.R., Fraser, C.M., and Venter, J.C. (1987) Muscarinic cholinergic receptor structure: molecular biological support for subtypes, *Trends Pharmacol. Sci.,* 8(11), 426–431.

Kim, J., Isokawa, M., Ledent, C., and Alger, B.E. (2002) Activation of muscarinic acetylcholine receptors enhances the release of endogenous cannabinoids in the hippocampus, *J. Neurosci.,* 22(23), 10182–10191.

Kombian, S.B. and Malenka, R.C. (1994) Simultaneous LTP of non-NMDA- and LTD of NMDA-receptor-mediated responses in the nucleus accumbens, *Nature,* 368(6468), 242–246.

Kreitzer, A.C., Carter, A.G., and Regehr, W.G. (2002) Inhibition of interneuron firing extends the spread of endocannabinoid signaling in the cerebellum, *Neuron,* 34(5), 787–796.

Kreitzer, A.C. and Regehr, W.G. (2001a) Cerebellar depolarization-induced suppression of inhibition is mediated by endogenous cannabinoids, *J. Neurosci.,* 21(20), 174RC.

Kreitzer, A.C. and Regehr, W.G. (2001b) Retrograde inhibition of presynaptic calcium influx by endogenous cannabinoids at excitatory synapses onto Purkinje cells, *Neuron,* 29(3), 717–727.

Kreitzer, A.C. and Regehr, W.G. (2002) Retrograde signaling by endocannabinoids, *Curr. Opin. Neurobiol.,* 12(3), 324–330.

LeDoux, J.E. (2000) Emotion circuits in the brain, *Annu. Rev. Neurosci.,* 23, 155–184.

Levenes, C., Daniel, H., Soubrie, P., and Crepel, F. (1998) Cannabinoids decrease excitatory synaptic transmission and impair long-term depression in rat cerebellar Purkinje cells, *J. Physiol. (Lond.),* 510(Pt. 3), 867–879.

Llano, I., Leresche, N., and Marty, A. (1991) Calcium entry increases the sensitivity of cerebellar Purkinje cells to applied GABA and decreases inhibitory synaptic currents, *Neuron,* 6(4), 565–574.

Lovinger, D.M. and Tyler, E. (1996) Synaptic transmission and modulation in the neostriatum., *Int. Rev. Neurobiol.,* 39, 77–111.

Lupica, C.R. (1995) Delta and mu enkephalins inhibit spontaneous GABA-mediated IPSCs via a cyclic AMP-independent mechanism in the rat hippocampus, *J. Neurosci.,* 15(1 Pt. 2), 737–749.

Mackie, K., Lai, Y., Westenbroek, R., and Mitchell, R. (1995) Cannabinoids activate an inwardly rectifying potassium conductance and inhibit Q-type calcium currents in AtT20 cells transfected with rat brain cannabinoid receptor, *J. Neurosci.,* 15(10), 6552–6561.

Maejima, T., Hashimoto, K., Yoshida, T., Aiba, A., and Kano, M. (2001) Presynaptic inhibition caused by retrograde signal from metabotropic glutamate to cannabinoid receptors, *Neuron,* 31(3), 463–475.

Manzoni, O.J. and Bockaert, J. (2001) Cannabinoids inhibit GABAergic synaptic transmission in mice nucleus accumbens, *Eur. J. Pharmacol.,* 412(2), R3–R5.

Marsicano, G., Wotjak, C.T., Azad, S.C., Bisogno, T., Rammes, G., Cascio, M.G., Hermann, H., Tang, J., Hofmann, C., Zieglgansberger, W., Di, M.V., and Lutz, B. (2002) The endogenous cannabinoid system controls extinction of aversive memories, *Nature,* 418(6897), 530–534.

Martin, L.A. and Alger, B.E. (1999) Muscarinic facilitation of the occurrence of depolarization-induced suppression of inhibition in rat hippocampus, *Neuroscience,* 92(1), 61–71.

Martin, L.A., Wei, D.S., and Alger, B.E. (2001) Heterogeneous susceptibility of GABA(A) receptor-mediated IPSCs to depolarization-induced suppression of inhibition in rat hippocampus, *J. Physiol.,* 532(Pt. 3), 685–700.

Martin, S.J. and Morris, R.G. (2002) New life in an old idea: the synaptic plasticity and memory hypothesis revisited, *Hippocampus,* 12(5), 609–636.

Marty, A. and Llano, I. (1995) Modulation of inhibitory synapses in the mammalian brain, *Curr. Opin. Neurobiol.,* 5(3), 335–341.

McAllister, S.D., Griffin, G., Satin, L.S., and Abood, M.E. (1999) Cannabinoid receptors can activate and inhibit G protein-coupled inwardly rectifying potassium channels in a xenopus oocyte expression system, *J. Pharmacol. Exp. Ther.,* 291(2), 618–626.

Misner, D.L. and Sullivan, J.M. (1999) Mechanism of cannabinoid effects on long-term potentiation and depression in hippocampal CA1 neurons, *J. Neurosci.,* 19(16), 6795–6805.

Miyazaki, T. and Lacey, M.G. (1998) Presynaptic inhibition by dopamine of a discrete component of GABA release in rat substantia nigra pars reticulata, *J. Physiol. (Lond),* 513(Pt. 3), 805–817.

Morisset, V. and Urban, L. (2001) Cannabinoid-induced presynaptic inhibition of glutamatergic EPSCs in substantia gelatinosa neurons of the rat spinal cord, *J. Neurophysiol.,* 86(1), 40–48.

Mu, J., Zhuang, S.Y., Kirby, M.T., Hampson, R.E., and Deadwyler, S.A. (1999) Cannabinoid receptors differentially modulate potassium A and D currents in hippocampal neurons in culture, *J. Pharmacol. Exp. Ther.,* 291(2), 893–902.

Nicola, S.M. and Malenka, R.C. (1997) Dopamine depresses excitatory and inhibitory synaptic transmission by distinct mechanisms in the nucleus accumbens, *J. Neurosci.,* 17, 5697–5710.

Nowicky, A.V., Teyler, T.J., and Vardaris, R.M. (1987) The modulation of long-term potentiation by delta-9-tetrahydrocannabinol in the rat hippocampus, in vitro, *Brain Res. Bull.,* 19(6), 663–672.

Ohno-Shosaku, T., Maejima, T., and Kano, M. (2001) Endogenous cannabinoids mediate retrograde signals from depolarized postsynaptic neurons to presynaptic terminals, *Neuron,* 29(3), 729–738.

Ohno-Shosaku, T., Matsui, M., Fukudome, Y., Shosaku, J., Tsubokawa, H., Taketo, M.M., Manabe, T., and Kano, M. (2003) Postsynaptic M1 and M3 receptors are responsible for the muscarinic enhancement of retrograde endocannabinoid signalling in the hippocampus, *Eur. J. Neurosci.,* 18(1), 109–116.

Ohno-Shosaku, T., Tsubokawa, H., Mizushima, I., Yoneda, N., Zimmer, A., and Kano, M. (2002) Presynaptic cannabinoid sensitivity is a major determinant of depolarization-induced retrograde suppression at hippocampal synapses, *J. Neurosci.,* 22(10), 3864–3872.

Pitler, T.A. and Alger, B.E. (1994) Depolarization-induced suppression of GABAergic inhibition in rat hippocampal pyramidal cells: G protein involvement in a presynaptic mechanism, *Neuron,* 13(6), 1447–1455.

Rinaldi-Carmona, M., Barth, F., Heaulme, M., Shire, D., Calandra, B., Congy, C., Martinez, S., Maruani, J., Neliat, G., and Caput, D. (1994) SR141716A, a potent and selective antagonist of the brain cannabinoid receptor, *FEBS Lett.,* 350(2-3), 240–244.

Robbe, D., Alonso, G., Chaumont, S., Bockaert, J., and Manzoni, O.J. (2002a) Role of p/q-Ca2+ channels in metabotropic glutamate receptor 2/3-dependent presynaptic long-term depression at nucleus accumbens synapses, *J. Neurosci.,* 22(11), 4346–4356.

Robbe, D., Kopf, M., Remaury, A., Bockaert, J., and Manzoni, O.J. (2002b) Endogenous cannabinoids mediate long-term synaptic depression in the nucleus accumbens, *Proc. Natl. Acad. Sci. U. S. A.,* 99(12), 8384.

Robbe, D., Alonso, G., Duchamp, F., Bockaert, J., and Manzoni, O.J. (2001) Localization and mechanisms of action of cannabinoid receptors at the glutamatergic synapses of the mouse nucleus accumbens, *J. Neurosci.,* 21(1), 109–116.

Schweitzer, P. (2000) Cannabinoids decrease the K(+) M-current in hippocampal CA1 neurons, *J. Neurosci.,* 20(1), 51–58.

Shen, M., Piser, T.M., Seybold, V.S., and Thayer, S.A. (1996) Cannabinoid receptor agonists inhibit glutamatergic synaptic transmission in rat hippocampal cultures, *J. Neurosci.,* 16(14), 4322–4334.

Shen, M. and Thayer, S.A. (1998) The cannabinoid agonist Win55,212-2 inhibits calcium channels by receptor-mediated and direct pathways in cultured rat hippocampal neurons, *Brain Res.,* 783(1), 77–84.

Sheng, M. and Kim, M.J. (2002) Postsynaptic signaling and plasticity mechanisms, *Science,* 298(5594), 776.

Shigemoto, R., Nomura, S., Ohishi, H., Sugihara, H., Nakanishi, S., and Mizuno, N. (1993) Immunohistochemical localization of a metabotropic glutamate receptor, mGluR5, in the rat brain, *Neurosci. Lett.,* 163(1), 53–57.

Snyder, S.H. and Pasternak, G.W. (2003) Historical review: Opioid receptors, *Trends Pharmacol. Sci.,* 24(4), 198–205.

Stella, N., Schweitzer, P., and Piomelli, D. (1997) A second endogenous cannabinoid that modulates long-term potentiation, *Nature,* 388(6644), 773–778.

Sullivan, J.M. (1999) Mechanisms of cannabinoid-receptor-mediated inhibition of synaptic transmission in cultured hippocampal pyramidal neurons, *J. Neurophysiol.,* 82(3), 1286–1294.

Sung, K.W., Choi, S., and Lovinger, D.M. (2001) Activation of group I mGluRs is necessary for induction of long-term depression at striatal synapses, *J. Neurophysiol.,* 86(5), 2405–2412.

Szabo, B., Dörner, L., Pfreundtner, C., Nörenberg, W., and Starke, K. (1998) Inhibition of gabaergic inhibitory postsynaptic currents by cannabinoids in rat corpus striatum, *Neuroscience,* 85(2), 395–403.

Szabo, B., Wallmichrath, I., Mathonia, P., and Pfreundtner, C. (2000) Cannabinoids inhibit excitatory neurotransmission in the substantia nigra pars reticulata, *Neuroscience,* 97(1), 89–97.

Takahashi, K.A. and Linden, D.J. (2000) Cannabinoid receptor modulation of synapses received by cerebellar purkinje cells, *J. Neurophysiol.,* 83(3), 1167–1180.

Tang, K., Low, M.J., Grandy, D.K., and Lovinger, D.M. (2001) Dopamine-dependent synaptic plasticity in striatum during in vivo development, *Proc. Natl. Acad. Sci. U. S. A,* 98(3), 1255–1260.

Tao, H.W. and Poo, M. (2001) Retrograde signaling at central synapses, *Proc. Natl. Acad. Sci. U. S. A,* 98(20), 11009–11015.

Testa, C.M., Friberg, I.K., Weiss, S.W., and Standaert, D.G. (1998) Immunohistochemical localization of metabotropic glutamate receptors mGluR1a and mGluR2/3 in the rat basal ganglia, *J. Comp. Neurol.,* 390(1), 5–19.

Thomas, M.J., Beurrier, C., Bonci, A., and Malenka, R.C. (2001) Long-term depression in the nucleus accumbens: a neural correlate of behavioral sensitization to cocaine, *Nat. Neurosci.,* 4(12), 1217–1223.

Thomas, M.J. and Malenka, R.C. (2003) Synaptic plasticity in the mesolimbic dopamine system, *Philos. Trans. R. Soc. Lond B Biol. Sci.,* 358(1432), 815–819.

Uchimura, N. and North, R.A. (1991) Baclofen and adenosine inhibit synaptic potentials mediated by gamma-aminobutyric acid and glutamate release in rat nucleus accumbens, *J. Pharmacol. Exp. Ther.,* 258(2), 663–668.

Ungless, M.A., Whistler, J.L., Malenka, R.C., and Bonci, A. (2001) Single cocaine exposure in vivo induces long-term potentiation in dopamine neurons, *Nature,* 411(6837), 583–587.

Varma, N., Carlson, G.C., Ledent, C., and Alger, B.E. (2001) Metabotropic glutamate receptors drive the endocannabinoid system in hippocampus, *J. Neurosci.,* 21(24), RC188.

Vaughan, C.W., Connor, M., Bagley, E.E., and Christie, M.J. (2000) Actions of cannabinoids on membrane properties and synaptic transmission in rat periaqueductal gray neurons in vitro, *Mol. Pharmacol.,* 57(2), 288–295.

Vaughan, C.W., McGregor, I.S., and Christie, M.J. (1999) Cannabinoid receptor activation inhibits GABAergic neurotransmission in rostral ventromedial medulla neurons in vitro, *Br. J. Pharmacol.,* 127(4), 935–940.

Wallmichrath, I. and Szabo, B. (2002) Analysis of the effect of cannabinoids on GABAergic neurotransmission in the substantia nigra pars reticulata, *Naunyn Schmiedebergs Arch. Pharmacol,* 365(4), 326–334.

Wilson, R.I., Kunos, G., and Nicoll, R.A. (2001) Presynaptic specificity of endocannabinoid signaling in the hippocampus, *Neuron,* 31(3), 453–462.

Wilson, R.I. and Nicoll, R.A. (2001) Endogenous cannabinoids mediate retrograde signalling at hippocampal synapses, *Nature,* 410(6828), 588–592.

Winder, D.G., Egli, R.E., Schramm, N.L., and Matthews, R.T. (2002) Synaptic plasticity in drug reward circuitry, *Curr. Mol. Med.,* 2(7), 667–676.

Winsauer, P.J., Lambert, P., and Moerschbaecher, J.M. (1999) Cannabinoid ligands and their effects on learning and performance in rhesus monkeys, *Behav. Pharmacol.,* 10(5), 497–511.

Wise, R.A. (1996) Neurobiology of addiction, *Curr. Opin. Neurobiol.,* 6(2), 243–251.

Wise, R.A. and Rompre, P.P. (1989) Brain dopamine and reward., *Annu. Rev. Psychol.,* 40, 191–225.

Zucker, R.S. and Regehr, W.G. (2002) Short-term synaptic plasticity, *Annu. Rev. Physiol.,* 64, 355–405.

10 Cannabinoids and the Central Serotonergic System

Marisela Morales

CONTENTS

Serotonin (5-hydroxytryptamine, 5-HT) is synthesized in neurons and in several types of peripheral cells (i.e., platelets, mast cells, and enterochromaffin cells) and participates in the regulation of diverse systems such as the cardiovascular system, gastrointestinal system, and central and peripheral systems. It is well known that serotonin plays a principal role in the control of nociception, vigilance state, thermoregulation, food consumption, and sexual behaviors. Within the brain, most of the 5-HT-producing neurons are distributed along the midline of the brainstem; however, the serotonergic axons innervate almost every region of the central nervous system (Molliver, 1987; Jacobs and Azmitia, 1992). Several studies suggest that the central serotonergic system interacts with cannabinoids, as it has been shown that cannabinoids affect the metabolism of serotonin in the adult (Holtzman et al., 1969; Sofia, Dixit, et al., 1971; Molina-Holgado et al., 1993; Egashira et al., 2002) and developing brain (Molina-Holgado et al., 1993; Egashira et al., 2002). In addition, behavioral, pharmacological, and electrophysiological observations suggest functional interactions between cannabinoids and 5-HT receptors (Fan, 1995; Kimura et al., 1998; Cheer et al., 1999; Darmani, 2001; Devlin and Christopoulos, 2002; Barann et al., 2002; Oz et al., 2002).

EFFECTS OF CANNABINOIDS ON THE CENTRAL SEROTONERGIC SYSTEM

Serotonin content and metabolism in the brain are affected by acute peripheral administration of Δ^9-THC. For example, a single administration of Δ^9-THC increases the concentration of serotonin in brains of adult mice (Holtzman et al., 1969) and rats (Sofia et al., 1971; Molina-Holgado et al., 1993; Egashira et al., 2002). In mice, increases in the whole-brain levels of serotonin occur following injections with doses of Δ^9-THC higher than 5 mg/kg. At a dose of 10 mg/kg, the increase in the amount of serotonin peaks at 45 min after Δ^9-THC administration and subsequently returns to control values between 3 and 6 h thereafter (Holtzman et al., 1969). In rats, the Δ^9-THC-induced elevation of 5-HT has been shown to be restricted to specific brain regions (Molina-Holgado et al., 1993; Egashira et al., 2002). Following acute treatment of 5 mg/kg, the 5-HT content increases in dorsal hippocampus (35%), substantia nigra (61%), and neostriatum (62%) but remains unchanged in cingulate cortex, raphe nuclei, locus ceruleus, and anterior hypothalamus. In the same study, Molina-Holgado et al. (1993) demonstrated that 5-HT metabolism, as measured by 5-hydroxyindole-3-acetic acid (5-HIAA) content, remains unchanged in dorsal hippocampus, substantia nigra, and neostriatum but decreases in raphe nucleus and anterior hypothalamus. Moreover, evaluation of the ratios of [5-HIAA]/[5-HT] in several brain areas of rats treated with Δ^9-THC are lower than those of the vehicle-treated rats. In addition, the ratios of [5-HIAA]/[5-HT] in several brain areas (dorsal hippocampus, anterior hypothalamus, neostriatum, substantia nigra, and raphe nucleus) are lower in Δ^9-THC-treated rats when compared with vehicle-treated rats. These observations led to the conclusion that acute Δ^9-THC administration affects the central serotonergic system by decreasing 5-HT metabolism (Molina-Holgado et al., 1993).

The Δ^9-THC-induced elevation in 5-HT content does not seem to be a result of decreasing degradation of 5-HT, because the activity of the enzyme responsible for 5-HT degradation (monoamine oxidase) is not affected by Δ^9-THC (Sofia, Dixit, et al., 1971). Alternatively, the authors suggested that increases in the levels of 5-HT induced by cannabinoids were the result of alterations of the vesicular membrane surrounding stored 5-HT. In support of this hypothesis, it was demonstrated that Δ^1-THC delayed the rate of reserpine-induced depletion of brain 5-HT (Sofia, Dixit, et al., 1971). In subsequent studies, the uptake of [^{3}H] 5-HT by synaptosomal preparation was found to be inhibited by Δ^1-THC (Sofia, Ertel, et al., 1971; Banerjee et al., 1975) and Δ^9-THC (Banerjee et al., 1975). This type of inhibition has been shown to be noncompetitive (Banerjee et al., 1975). Further characterization of the effects of Δ^9-THC on 5-HT accumulation and release in rat forebrain synaptosomes demonstrated that Δ^9-THC facilitates 5-HT release from the synaptic vesicle and delays its accumulation at the neuronal membrane (Johnson et al. 1976). Together, these studies suggest presynaptic effects of cannabinoids on serotonergic release and uptake. However, because these studies used synaptosomal preparations, it is unclear whether similar effects actually take place in the intact brain.

Recent evidences suggest the involvement of the central receptor for cannabinoids (CB_1) in the release of serotonin, as synthetic cannabinoid receptor agonist WIN55212-2 inhibited both the electrically induced and Ca^{2+}-induced release of [^{3}H] 5-HT in mouse brain cortex slices (Nakazi et al., 2000). However, this inhibition was relatively small in comparison with the release of [^{3}H] choline seen under similar experimental conditions (Nakazi et al., 2000). Interestingly, the cannabinoid antagonist SR141716 does not affect electrically induced or Ca^{2+}-induced release of [^{3}H] 5-HT, suggesting that CB_1 receptors responsible for the release of [^{3}H] 5-HT are not activated by endogenously formed cannabinoids (Nakazi et al., 2000). Participation of cannabinoids in the release of serotonin has also been suggested by *in vivo* brain microdialysis studies that showed 5-HT release decreases in the ventral hippocampus following peripheral administration of 6 mg/kg Δ^9-THC (Egashira et al., 2002). Given that the decrease of 5-HT release in the ventral hippocampus was accompanied by a significant increase in the amount of 5-HT, these results further imply that elevated 5-HT content results from decrease in 5-HT release.

Administration of Δ^9-THC does not affect the density of central 5-HT transporters, as [3H] paroxetine-binding sites (5-HT uptake sites) are similar in brain preparations of Δ^9-THC- and vehicle-treated rats (Molina-Holgado et al., 1993). Likewise, lack of changes in 5-HT transporter density has been found in BeoWo cells exposed to the cannabinoid agonist WIN55212-2 (Kenney et al., 1999). Despite the absence of changes in 5-HT transporter in BeoWo cells, WIN55212-2 induces a decrease in the activity of the 5-HT transporter. WIN55212-2 does not bind to 5-HT transporter, and the observed decrease in the function of this transporter is independent of alterations in intracellular cyclic adenosine monophosphate levels. Therefore, WIN55212-2 is likely to regulate the 5-HT transporter activity in this cell line by acting through a second-messenger pathway (Kenney et al., 1999). The effects of Δ^9-THC and endogenous cannabinoids on the brain 5-HT transporter activity remain to be addressed.

As with acute administration, Anton et al. (1974) reported a significant elevation in mouse brain 5-HT levels following 13 weeks of orally administered THC (50 mg/Kg per day for 5 d each week).

Similarly, chronic administration of Δ^9-THC (20 mg/kg for 6 d) produces an increase in the levels of 5-HT in the whole rat brain and hypothalamus plus midbrain (Sofia et al., 1977). In common with other effects of cannabinoids, the cannabinoid-induced increases in 5-HT content develop tolerance over time (Sofia et al., 1977).

In summary, although acute Δ^9-THC administration results in changes in the levels of 5-HT, 5-HIAA, and the ratio of 5-HIAA/5-HT, these changes are not accompanied by alterations in 5-HT transporter concentrations or monoamine oxidase activity. Although it has been suggested that CB_1 receptors located presynaptically on serotonergic nerve endings mediate 5-HT release, conclusive experimental evidence supporting the presence of these receptors on serotonergic neurons or terminals is lacking. Because it is well documented that CB_1 receptors are present in GABAergic neurons (Katona et al., 1999, 2001; Marsicano and Lutz, 1999; Tsou et al., 1999; Hajos et al., 2000; McDonald and Mascagni, 2001), cannabinoid inhibition of 5-HT release may be mediated through GABAergic neurons rather than throughout the direct action at the serotonergic nerve endings.

MATERNAL EXPOSURE TO CANNABINOIDS AND ITS EFFECTS ON THE SEROTONERGIC SYSTEM

The cannabinoids may affect the fetus by crossing the placenta during gestation (Vardaris et al., 1976; Martin et al., 1977; Bailey et al., 1987) and interacting with cannabinoid receptors located on the placenta membranes (Buckley et al., 1998; Kenney et al., 1999; Park et al., 2003). Postnatal exposure to Δ^9-THC can also occur through the milk during lactation (Jakubovic et al., 1973). Several studies indicate interactions of cannabinoids with the serotonergic system during development. For example, it has been reported that maternal exposure to Δ^9-THC (from gestation day 13 to postnatal day 7) produced significant changes in 5-HT and 5-HIAA content of the offspring, and that these changes persist to adulthood (Molina-Holgado et al., 1993). Alterations were not detected in 5-HT innervations, indicating that exposure to Δ^9-THC is not likely to alter the pattern of serotonergic fiber innervations. In a follow-up study, a daily dose of Δ^9-THC to pregnant rats from gestational day five to postnatal day one was found to decrease diencephalic levels of 5-HT, with males being more susceptible than females (Molina-Holgado et al., 1996). These observations were confirmed and expanded in a subsequent study, in which maternal exposure to Δ^9-THC from gestation day five to postnatal day one produced a region- and sex-depended effect in endogenous levels of indoleamine, particularly in the midbrain raphe nuclei (Molina-Holgado et al., 1997). Although these studies clearly demonstrated alteration in the serotonergic system due to maternal exposure to Δ^9-THC, the effects of endogenous cannabinoids on the development of the serotonergic circuitry have not been addressed.

INTERACTIONS BETWEEN CANNABINOIDS AND 5-HT RECEPTORS

In addition to the effects of cannabinoids on the serotonergic metabolism, cannabinoids may affect the function of 5-HT receptors. The receptors for serotonin include seven distinct classes ($5\text{-}HT_1$ to $5\text{-}HT_7$), with multiple members in each class (Hoyer et al., 2002). Although most of these receptors are G protein–coupled, the $5\text{-}HT_3$ receptor is the only ion channel receptor for serotonin in mammals (Derkach et al., 1989). Among the seven 5-HT receptors, only $5\text{-}HT_2$ and $5\text{-}HT_3$ receptors have to date been found to interact with cannabinoids.

INTERACTIONS BETWEEN CANNABINOIDS AND $5\text{-}HT_2$ RECEPTORS

Studies with rodents suggest functional interactions between cannabinoids and some behaviors induced by the $5\text{-}HT_2$ receptors. For instance, Δ^9-THC and several synthetic cannabinoid agonists (HU210, CP55940, and WIN55212-2) have been shown to reduce the frequency in mice of the head-twitch response and ear-scratch induced by the selective 5-$HT_{2a/c}$ receptor agonist DOI ((+/–)-1-(2,5-dimethoxy-4-iodophenyl)-2-aminopropane hydrochloride) (Darmani, 2001). In rats, Cheer et al. (1999) demonstrated a functional interaction between CB_1 and 5-HT_2 receptors, as the cannabinoid receptor agonist HU210 potentiated back muscle contraction induced by treatment with DOI. Significantly, the CB_1 antagonist SR141716A prevents this potentiation.

Interaction between CB_1 and 5-HT_2 receptors has been further suggested by analyzing 5-HT_2-receptor radioligand-binding sites in rat cerebellar membranes and by showing that HU210 enhances the high affinity binding of 5-HT_2 sites for 5-HT (Cheer et al., 1999). In addition, *in vitro* studies also suggest crosstalk between CB_1 and 5-HT_2 receptors in membranes of the rat cerebellum (Devlin and Christopoulos, 2002). However, further support of this hypothesis will benefit from efforts to demonstrate coexistence of these receptors at the cellular level. To date, although anatomical studies indicate abundant expression of the CB_1 receptor in granule cells (Mailleux and Vanderhaegen, 1992), there are no evidences of the presence of any of the three subclasses of 5-HT_2 receptors in these neurons. The 5-HT_{2A} (Mengod et al., 1990) and 5-HT_{2C} (Clemett et al., 2000) receptors appeared to be absent from cerebellum, whereas the 5-HT_{2B} receptors are confined to Purkinje cells (Duxon et al., 1997). Therefore, the specific subclass of 5-HT_2 receptor participating in interactions with CB_1 in the cerebellum remains to be determined.

The possibility of interactions between different 5-HT receptors, including 5-$HT_{2A/2B}$ receptors, and anandamide had been explored in crude synaptic membranes from bovine brain (Kimura et al., 1998). Concentrations of anandamide in the order of 0.01 to 0.1 μM do not significantly affect the specific binding for [^{3}H] 5-HT. However, anandamide at concentrations of 1, 10, and 100 μM reduces [^{3}H] 5-HT binding to 46%, 41%, and 29 %, respectively. Higher concentrations of anandamide are necessary to inhibit specific binding for the 5-$HT_{2A/2B}$ receptor radioligand [^{3}H] ketanserin. Anandamide at concentrations of 0.01 to 10 μM does not change the specific [^{3}H] ketanserin binding when compared with that of the controls, whereas at a higher concentration (100 μM) anandamide decreases [^{3}H] ketanserin binding to 43% of the control (Kimura et al., 1998). These observations suggest that anandamide may interact with different 5-HT receptors in the brain.

Therefore, while behavioral and radioligand-binding studies suggest interactions between cannabinoid agonist and 5-HT_2 receptors (Kimura et al., 1998; Cheer et al., 1999; Darmani, 2001; Devlin and Christopoulos, 2002), the phenotype of the neuronal components participating in these interactions are as yet unknown.

Interactions between Cannabinoids and 5-HT_3 Receptors

Pharmacological observations have shown that 5-HT_3 receptor antagonists and cannabinoid agonists share common effects, as both have analgesic and antiemetic properties (Gralla et al., 1984; Dewey, 1986; Fozard, 1992; Greenshaw, 1993; Walker et al., 1999; Darmani 2002). Although these observations suggest possible functional interactions between cannabinoids and 5-HT_3 receptors, it is currently unknown whether these interactions are direct or indirect. However, electrophysiological studies indicate that chemically different CB_1 agonists inhibit 5-HT_3 receptor–mediated currents in several biological preparations, suggesting that cannabinoids operate through 5-HT_3 receptors (Fan, 1995; Oz et al., 2002; Barann et al., 2002). For example, cannabinoid receptor agonists inhibit the 5-HT_3-receptor-mediated inward currents induced by 5-HT_3 receptor agonists in the rat nodose ganglion neurons (Fan, 1995). The highest inhibition of 5-HT_3-induced current was obtained with the cannabinoid receptor CP55940 (IC_{50} = 94 nM) followed by anandamide (IC_{50} = 190 nM), and WIN55212-2 (IC_{50} = 310 nM). This inhibition is slow developing, noncompetitive, not dependent on membrane potential, and not affected by adenosine 3′, 5′-cyclic monophosphate (cAMP) analogs and guanosine-5′-O-(2-thiodiphosphate). Thus, these observations provide indirect evidence of the lack of participation of CB_1 receptors in this cannabinoid-mediated inhibition. Although these data strongly suggested that endogenous 5-HT_3 receptors were a site of action of cannabinoid agonists in the rat nodose ganglion neurons, it did not discard the possibility of the presence of cannabinoid receptors in these cells (Fan, 1995).

The inhibitory effect of cannabinoids on 5-HT_3-receptor-mediated currents has been confirmed through heterologous systems lacking cannabinoid receptors but expressing homomeric 5-HT_{3A} receptors (Barann et al., 2002; Oz et al., 2002). When *Xenopus* oocytes were injected with neuroblastoma cell line 5-HT_{3A} mRNA, the expressing 5-HT_3 receptors were capable to mediate a 5-HT-induced ion current that was inactivated by anandamide (Oz et al., 2002). This inhibition was not dependent on the membrane potential, and anandamide did not have an effect on the reversal potential of 5-HT-induced currents. In the presence of anandamide, the maximum 5-HT-induced response was also inhibited. In these experiments, the respective EC50 values were 3.4 M and 3.1 M in the absence and presence of anandamide, indicating that anandamide acts as a noncompetitive antagonist on 5-HT_3 receptors. The CB_1 receptor antagonist SR141716A did not cause a significant change on the inhibition of 5-HT responses by anandamide; the effect of anandamide was independent of the activation of cAMP pathway and not mediated by the activation of pertussis toxin–sensitive G proteins. In addition, the effect of anandamide was not changed by preincubating the oocytes with 8-Br-cAMP, a membrane-permeable analog of cAMP, or Sp-cAMPS, a membrane-permeable protein kinase. Based on these experimental findings, it was concluded that anandamide inhibition of 5-HT-induced currents was cannabinoid-receptor independent and noncompetitive.

A combination of electrophysiological and radioligand-binding studies further provide strong evidence showing that cannabinoids allosterically modulate 5-HT_{3A} receptors expressed in HEK 293 cells (Barann et al., 2002). In these cells, cannabinoid receptor agonists were found to inhibit 5-HT-induced currents with different potency. The highest inhibition was obtained with Δ^9-THC (IC_{50} = 38.4 n*M*), followed by WIN55212-2 (IC_{50} = 103.5 n*M*), anandamide (IC_{50} = 129.6 n*M*), and JWH015 (IC_{50} = 146.5 n*M*); the lowest effect was seen with CP55940 (IC_{50} = 647.6 n*M*). The specificity of the actions of cannabinoids on the 5- HT_{3A} receptor was inferred from experiments in which it was noted that the potency of the cannabinoid antagonist inhibition of 5-HT-induced currents did not correlate with the degree to which they activate the cannabinoid receptors. In addition, the CB_1 receptor antagonist SR141716A fails to affect WIN55212-2–induced inhibition, and CB_1 receptor–binding sites were not detected in HEK 293 cells. As shown in nodose ganglia and oocytes preparation, the cannabinoid inhibition of 5-HT-induced currents was also not competitive in 5-HT_{3A}-expressing HEK 293 cells. Cannabinoids do not inhibit binding of the 5-HT_3 receptor radioligand [^{3}H] GR65630 to membranes and fail to inhibit the 5-HT-induced current when administered during but not prior to stimulation by 5-HT (Barann et al., 2002). Thus, it has been concluded that cannabinoids do not act at the ligand-recognition site of the 5-HT_3 receptor or within its channel pore. Because cannabinoid antagonists had to be present in the superfusion fluid for at least 15 min to establish their inhibitory effects in nodose ganglia neurons (Fan, 1995) or for 3 min to obtain their equilibrium effect in HEK 293 cells (Barann et al., 2002), it has been suggested that the slow development of inhibition is indicative of a transmembrane or cytosolic domain of the receptor protein as a site of action for cannabinoids, i.e., an allosteric modulatory site of the 5-HT_3 receptor (Barann et al., 2002).

Lastly, a recent study suggests that cannabinoid receptor agonists modulate the activity of the rat peripheral 5-HT_3 receptors on the terminals of cardiopulmonary afferent C-fibers *in vivo* (Godlewski et al., 2003). In this study, urethane-anesthetized rats pretreated intravenously with SR141716A received a bolus injection of the serotonin-5-HT_3-receptor agonist phenylbiguanide or the vanilloid VR1 receptor agonist capsaicin. Both agonists caused an immediate decrease in heart rate and mean arterial blood pressure (the von Bezold–Jarisch reflex). The phenylbiguanide-induced bradycardia was dose-dependently attenuated by the cannabinoid receptor agonists CP55940 and WIN55212-2. The extent of inhibition by the highest doses of cannabinoid receptor agonists was 50%. In contrast, both cannabinoid receptor agonists did not affect the capsaicin-evoked bradycardia. Hence, it was concluded that cannabinoid receptor agonists modulate the von Bezold–Jarisch reflex by inhibiting peripheral serotonin 5-HT_3 receptors (Godlewski et al., 2003).

NEURONAL COEXPRESSION OF 5-HT_{3A} AND CB_1 TRANSCRIPTS

Two subunits of the 5-HT_3 receptor have been cloned: 5-HT_{3A} (Maricq et al., 1991; Belelli et al., 1995) and 5-HT_{3B} (Davies et al., 1999; Dubin et al., 1999). Although expression of recombinant 5-HT_{3A} subunit produces functional homomeric 5-HT_3 receptors, expression of 5-HT_{3B} subunit alone does not result in functional receptors. However, in heteromeric receptor complexes, the 5-HT_{3B} subunit confers unique pharmacological and biophysical properties to the assembled receptor (Davies et al., 1999; Dubin et al., 1999). Recent studies have demonstrated that sensory neurons express high levels of 5-HT_{3A} and 5-HT_{3B} subunits (Morales et al., 2001). In contrast, rat central neurons, although expressing the 5-HT_{3A} subunit, appear to lack detectable levels of 5-HT_{3B} subunit transcripts (Ferezou et al., 2002; Morales and Wang, 2002).

Characterization of telencephalon 5-HT_{3A}-immunoreactive neurons has demonstrated that these neurons express 5-HT_{3A} mRNA (Morales, Battenberg, de Lecea, Sanna, and Bloom, 1996) and are GABAergic (Morales, Battenberg, de Lecea, and Bloom, 1996; Morales, Battenberg, de Lecea, L., Sanna, et al., 1996; Morales and Bloom, 1997). In addition, they may contain the peptide cholecystokinin but lack the calcium-binding protein parvalbumin (Morales and Bloom, 1997). Interestingly, similar phenotypic characteristics are shared by hippocampal interneurons expressing the central CB_1 cannabinoid receptor (Katona et al., 1999; Marsicano and Lutz, 1999; Tsou et al., 1999), suggesting some degree of overlap between hippocampal interneurons containing the 5-HT_3 receptor and those having the CB_1 receptor. In agreement with this suggestion, it has been found that a high degree of coexistence of 5-HT_{3A} subunit and CB_1 transcripts exists in the rat and mouse hippocampus (Hermann et al., 2002; Morales and Backman, 2002), and rat dentate gyrus (Morales and Backman, 2002) (Table 10.1, Figure 10.1). Colocalization of 5-HT_{3A} subunit and CB_1 transcripts was found to be restricted to GABAergic neurons (Morales et al., 2005). In this regard, electrophysiological studies have demonstrated that stimulation of the 5-HT3 receptor facilitates GABAergic neurotransmission in hippocampal slices (Ropert and Guy, 1991; Corradetti et al., 1992; Kawa, 1994; Maeda et al., 1994) and cell cultures

TABLE 10.1
Percentage of Neurons Expressing CB_1 and 5-HT_{3A} mRNA in the Total Population of CB_1 Expressing Neurons within the Hippocampus and Dentate Gyrus

Region	Rostral Regions Mean ± SEM[a]	Caudal Regions Mean ± SEM
CA1		
Stratum oriens	73 ± 5.2	84 ± 5.2
Stratum pyramidale	78 ± 6.0	88 ± 4.5
Stratum radiatum	75 ± 5.5	72 ± 4.9
Stratum lacunosum moleculare	80 ± 2.0	81 ± 2.6
CA3		
Stratum oriens	83 ± 2.6	80 ± 5.4
Stratum pyramidale	83 ± 4.9	75 ± 4.9
Stratum radiatum	74 ± 3.0	79 ± 2.7
Stratum lucidum	78 ± 3.6	82 ± 3.2
DG		
Subgranular layer	79 ± 2.7	83 ± 2.0

[a] Standard error of the mean.

Source: Modified from Morales, M. and Backman, C. (2002) Coexistence of serotonin 3 (5-HT3) and CB1 cannabinoid receptors in interneurons of hippocampus and dentate gyrus. *Hippocampus*, 12: 756–64.

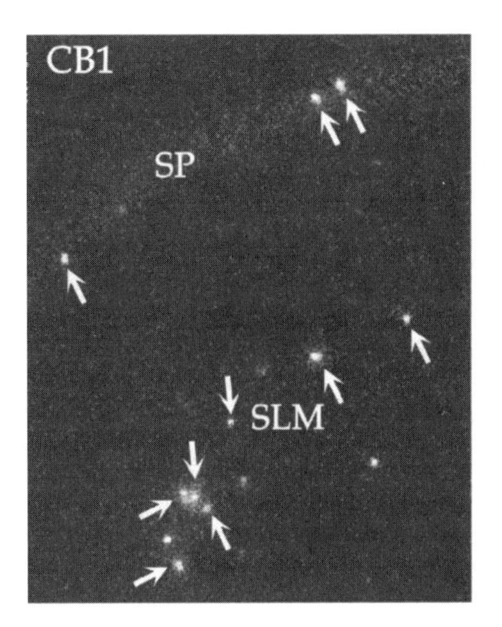

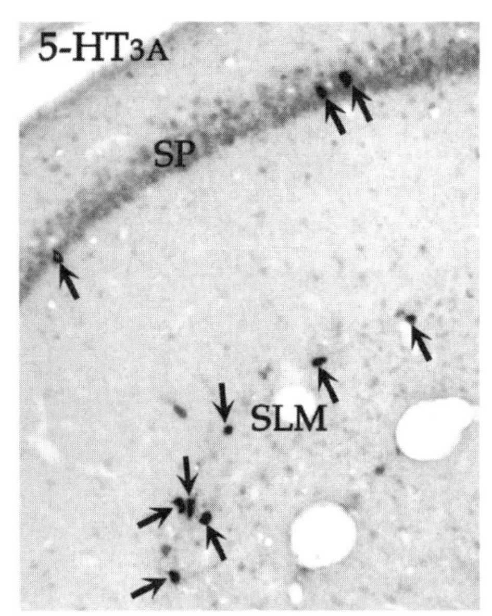

FIGURE 10.1 Coexpression of CB_1 and $5HT_{3A}$ transcripts in the hippocampus. Neurons coexpressing both transcripts are observed in the stratum pyramidale (SP) and stratum lacunosum moleculare (SLM).

(Yakel and Jackson, 1988). In contrast, endogenous cannabinoids, as well as synthetic CB_1 agonists, act on presynaptic CB_1 receptors to suppress GABA release in hippocampus (Katona et al., 1999, 2001; Hoffman and Lupica, 2000; Hajos et al., 2000; Ohno-Shosaku et al., 2001; Wilson and Nicoll, 2001), amygdala (Katona et al., 2001), substantia nigra pars reticulata (Chan et al., 1998; Wallmichrath and Szabo, 2002), striatum (Szabo et al., 1998), nucleus accumbens (Hoffman and Lupica, 2001), and Purkinje cells (Takahashi and Linden, 2000). Thus, coexpression of 5-HT_{3A} and CB_1 transcripts in interneurons of the telencephalon suggest that at least two mechanisms of cannabinoid-induced inhibition of GABA release may be functional in the telencephalon: direct activation of CB_1 receptors by cannabinoid agonists and cannabinoid-induced inhibition of 5-HT_3 receptors–mediated GABA release.

PARTICIPATION OF SEROTONIN ON CANNABINOID EFFECTS

Serotonin has been suggested to participate in several behaviors mediated by cannabinoids. The involvement of 5-HT in hypothermia induced by Δ^9-THC has been demonstrated in rodents. In rats, the 5-HT-selective reuptake inhibitors clomipramine and fluoxetine were found to modify hypothermia induced by Δ^9-THC (Fennessy and Taylor, 1978; Malone and Taylor, 1998). As pretreatment with fluoxetine 40 min before Δ^9-THC administration significantly reduces Δ^9-THC-induced hypothermia, it has been suggested that an increase in extracellular 5-HT can activate autoreceptors that may decrease serotonergic activity (Malone and Taylor, 1998). This decrease of serotonergic activity could account for the reduction of Δ^9-THC-induced hypothermia (Malone and Taylor, 1998). On the contrary, fluoxetine administered 40 min after Δ^9-THC significantly potentiated the Δ^9-THC-induced hypothermia; the reuptake block at this point in time may potentiate the already activated serotonergic system (Malone and Taylor, 1998). Further evidence of the participation of serotonin in cannabinoid-induced changes in temperature was provided by Davies and Graham (1980), who established that depletion in mice of 5-HT with p-chlorophenylalanine reduced the hypothermic response to Δ^9-THC, as did pretreatment with the 5-HT antagonist methysergide. Moreover, inhibition of the reuptake of serotonin with clomipramine was found to potentiate hypothermia following Δ^9-THC administration. Thus, the hypothermic effect of Δ^9-THC in the mice may be mediated to a large extent via serotonergic mechanisms.

In chronic studies, in which rats received intravenous injections of Δ^9-THC (2-6 mg/kg) twice daily over a 10-d period, the animals showed two main behavioral signs associated with physical dependence: writhing and backward kicking (Verberne et al., 1980). When rats that were chronically treated with Δ^9-THC received intraperitoneal injections with fluoxetine (5-HT reuptake inhibitor) on day 11 of the experiment, the frequency of writhes and backward kicks was increased. These findings suggest the involvement of 5-HT in the production of a withdrawal-like behavior after chronic Δ^9-THC treatment (Verberne et al., 1980).

Acute administration of Δ^9-THC (6 mg/kg) impaired spatial memory and significantly increased 5-HT content in the ventral hippocampus (Egashira et al., 2002). Furthermore, microdialysis experiments have shown that Δ^9-THC (6 mg/kg i.p.) decreased 5-HT release in the ventral hippocampus, suggesting that elevation of the amount of 5-HT results from inhibition of its release (Egashira et al., 2002). In addition, significant attenuation of Δ^9-THC-induced impairment of spatial memory has been found following treatment with several 5-HT-related drugs (5-HT precursor, 5-HT reuptake inhibitor, and 5-HT_2 receptor agonist). Collectively, these observations provide the basis for proposing that the 5-HT neuronal system may be involved in the Δ^9-THC-induced impairment of spatial memory (Egashira et al., 2002).

As detailed in the preceding text, results from synaptosomal preparations indicate that Δ^9-THC increases the levels of central 5-HT-without affecting 5-HT-transporter concentration or monoamine oxidase activity. Moreover, it has been suggested that CB_1 receptors located presynaptically on serotonergic nerve endings mediate 5-HT release. However, these receptors have not yet been morphologically identified in association with serotonergic neurons or terminals. Available experimental evidence also does not appear to support the participation of endogenous cannabinoids in the mechanisms of 5-HT release (Nakazi et al., 2000). Thus, efforts to establish the relationship between CB_1 receptor and the central serotonergic system would benefit from studies aimed at mapping the physical presence of this class of receptor within different cell phenotypes. Increasing evidence indicates that 5-HT receptors, in common with other receptors, are altered by cannabinoids. For example, synthetic cannabinoid agonists affect functions mediated by 5-HT_2 receptors. However, studies performed with the endogenous cannabinoid anandamide indicate that a high concentration of the endogenous cannabinoid is required to displace less than half of 5-HT_2-receptor-binding sites in crude synaptic membranes from bovine brain suggesting that anandamide may interact with different 5-HT receptors in the brain (Kimura et al., 1998). Further clarification of possible interactions between cannabinoids and 5-HT_2 receptors will thus profit from future electrophysiological and immunolocalization analysis. In regard to the 5-HT_3 receptor, it has been established that different CB_1 agonists, including anandamide, inhibit 5-HT_3 receptor-mediated currents in distinct cellular preparations. Interestingly, recent *in vivo* studies indicate that cannabinoid receptor agonists modulate the von Bezold–Jarisch reflex by inhibiting peripheral serotonin 5-HT_3 receptors (Godlewski et al., 2003). Thus, cannabinoid-mediated inhibition of 5-HT_3 receptor functions may also take place in the central nervous system. In addition, 5-HT_{3A} and CB_1 transcripts are present in a subset of GABAergic neurons, suggesting that two mechanisms of cannabinoid-induced inhibition of GABA release may operate within the same cell: direct activation of CB_1 receptors by cannabinoid agonists and cannabinoid-induced inhibition of 5-HT_3 receptors–mediated currents. Pharmacological approaches and the availability of different knockout mice will permit future exploration of *in vivo* contributions by 5-HT_3 and CB_1 receptors to the central and peripheral effects of cannabinoids.

REFERENCES

Anton, A.H., Serrano, A., Beyer, R.D., and Lavappa, K.S. (1974) Tetrahydrocannabinol, sesame oil and biogenic amines: a preliminary report. *Life Sci,* 14: 1741–6.

Bailey, J.R., Cunny, H.C., Paule, M.G., and Slikker, W., Jr. (1987) Fetal disposition of delta 9-tetrahydrocannabinol (THC) during late pregnancy in the rhesus monkey. *Toxicol Appl Pharmacol,* 90: 315–21.

Banerjee, S.P., Snyder, S.H., and Mechoulam, R. (1975) Cannabinoids: influence on neurotransmitter uptake in rat brain synaptosomes. *J Pharmacol Exp Ther,* 194: 74–81.

Barann, M., Molderings, G., Bruss, M., Bonisch, H., Urban, B.W., and Gothert, M. (2002) Direct inhibition by cannabinoids of human 5-HT3A receptors: probable involvement of an allosteric modulatory site. *Br J Pharmacol,* 137: 589–96.

Belelli, D., Balcarek, J.M., Hope, A.G., Peters, J.A., Lambert, J.J., and Blackburn, T.P. (1995) Cloning and functional expression of a human 5-hydroxytryptamine type 3AS receptor subunit. *Mol Pharmacol,* 48: 1054–62.

Buckley, N.E., Hansson, S., Harta, G., and Mezey, E. (1998) Expression of the CB1 and CB2 receptor messenger RNAs during embryonic development in the rat. *Neuroscience,* 82: 1131–49.

Chan, P.K., Chan, S.C., and Yung, W.H. (1998) Presynaptic inhibition of GABAergic inputs to rat substantia nigra pars reticulata neurones by a cannabinoid agonist. *Neuroreport,* 9: 671–5.

Cheer, J.F., Cadogan, A.K., Marsden, C.A., Fone, K.C., and Kendall, D.A. (1999) Modification of 5-HT2 receptor mediated behaviour in the rat by oleamide and the role of cannabinoid receptors. *Neuropharmacology,* 38: 533–41.

Clemett, D. A., Punhani, T., Duxon, M. S., Blackburn, T. P. and Fone, K. C. (2000) Immunohistochemical localisation of the 5-HT2C receptor protein in the rat CNS. *Neuropharmacology,* 39: 123–32.

Corradetti, R., Ballerini, L., Pugliese, A.M., and Pepeu, G. (1992) Serotonin blocks the long-term potentiation induced by primed burst stimulation in the CA1 region of rat hippocampal slices. *Neuroscience,* 46: 511–8.

Darmani, N.A. (2001) Cannabinoids of diverse structure inhibit two DOI-induced 5-HT(2A) receptor-mediated behaviors in mice. *Pharmacol Biochem Behav,* 68: 311–7.

Darmani, N.A. (2002) in *Biology of Marijuana: From Gene to Behavior,* Onaivi, E.S. Ed., Taylor and Francis, London, pp. 356–389.

Davies, J.A. and Graham, J.D. (1980) The mechanism of action of delta 9-tetrahydrocannabinol on body temperature in mice. *Psychopharmacology (Berl),* 69: 299–305.

Davies, P.A., Pistis, M., Hanna, M.C., Peters, J.A., Lambert, J.J., Hales, T.G. and Kirkness, E.F. (1999) The 5-HT3B subunit is a major determinant of serotonin-receptor function. *Nature,* 397: 359–63.

Derkach, V., Surprenant, A., and North, R. A. (1989) 5-HT3 receptors are membrane ion channels. *Nature,* 339: 706–9.

Devlin, M.G. and Christopoulos, A. (2002) Modulation of cannabinoid agonist binding by 5-HT in the rat cerebellum. *J Neurochem,* 80: 1095–102.

Dewey, W.L. (1986) Cannabinoid pharmacology. *Pharmacol Rev,* 38: 151–78.

Dubin, A.E., Huvar, R., D'Andrea, M.R., Pyati, J., Zhu, J.Y., Joy, K.C., Wilson, S.J., Galindo, J.E., Glass, C.A., Luo, L., Jackson, M.R., Lovenberg, T.W. and Erlander, M.G. (1999) The pharmacological and functional characteristics of the serotonin 5-HT(3A) receptor are specifically modified by a 5-HT(3B) receptor subunit. *J Biol Chem,* 274: 30799–810.

Duxon, M.S., Flanigan, T.P., Reavley, A.C., Baxter, G.S., Blackburn, T.P., and Fone, K.C. (1997) Evidence for expression of the 5-hydroxytryptamine-2B receptor protein in the rat central nervous system. *Neuroscience,* 76: 323–9.

Egashira, N., Mishima, K., Iwasaki, K., and Fujiwara, M. (2002) Intracerebral microinjections of delta 9-tetrahydrocannabinol: search for the impairment of spatial memory in the eight-arm radial maze in rats. *Brain Res,* 952: 239–45.

Fan, P. (1995) Cannabinoid agonists inhibit the activation of 5-HT3 receptors in rat nodose ganglion neurons. *J Neurophysiol,* 73: 907–10.

Fennessy, M.R. and Taylor, D.A. (1978) Antagonism of the effects on thermoregulation of delta9-tetrahydrocannabinol by clomipramine in the rat. *Br J Pharmacol,* 63: 267–73.

Ferezou, I., Cauli, B., Hill, E.L., Rossier, J., Hamel, E., and Lambolez, B. (2002) 5-HT3 receptors mediate serotonergic fast synaptic excitation of neocortical vasoactive intestinal peptide/cholecystokinin interneurons. *J Neurosci,* 22: 7389–97.

Fozard, J. (1992) Pharmacological relevance of 5-HT3 receptors. In: Langer SZ, Racagni G, Mendlewicz J, Eds. *Serotonin Receptor Subtypes: Pharmacological Significance and Clinical Implications.* Basel, Karger. p 44–45.

Godlewski, G., Gothert, M., and Malinowska, B. (2003) Cannabinoid receptor-independent inhibition by cannabinoid agonists of the peripheral 5-HT(3) receptor-mediated von Bezold-Jarisch reflex. *Br J Pharmacol,* 138: 767–74.

Gralla, R.J., Tyson, L.B., Bordin, L.A., Clark, R.A., Kelsen, D.P., Kris, M.G., Kalman, L.B. and Groshen, S. (1984) Antiemetic therapy: a review of recent studies and a report of a random assignment trial comparing metoclopramide with delta-9-tetrahydrocannabinol. *Cancer Treat Rep,* 68: 163–72.

Greenshaw, A.J. (1993) Behavioural pharmacology of 5-HT3 receptor antagonists: a critical update on therapeutic potential. *Trends Pharmacol Sci,* 14: 265–70.

Hajos, N., Katona, I., Naiem, S.S., MacKie, K., Ledent, C., Mody, I., and Freund, T.F. (2000) Cannabinoids inhibit hippocampal GABAergic transmission and network oscillations. *Eur J Neurosci,* 12: 3239–49.

Hermann, H., Marsicano, G., and Lutz, B. (2002) Coexpression of the cannabinoid receptor type 1 with dopamine and serotonin receptors in distinct neuronal subpopulations of the adult mouse forebrain. *Neuroscience,* 109: 451–60.

Hoffman, A.F. and Lupica, C.R. (2000) Mechanisms of cannabinoid inhibition of GABA(A) synaptic transmission in the hippocampus. *J Neurosci,* 20: 2470–9.

Hoffman, A.F. and Lupica, C.R. (2001) Direct actions of cannabinoids on synaptic transmission in the nucleus accumbens: a comparison with opioids. *J Neurophysiol,* 85: 72–83.

Holtzman, D., Lovell, R.A., Jaffe, J.H., and Freedman, D.X. (1969) 1-delta9-tetrahydrocannabinol: neurochemical and behavioral effects in the mouse. *Science,* 163: 1464–7.

Hoyer, D., Hannon, J.P., and Martin, G.R. (2002) Molecular, pharmacological and functional diversity of 5-HT receptors. *Pharmacol Biochem Behav,* 71: 533–54.

Jacobs, B.L. and Azmitia, E.C. (1992) Structure and function of the brain serotonin system. *Physiol Rev,* 72: 165–229.

Jakubovic, A., Hattori, T., and McGeer, P.L. (1973) Radioactivity in suckled rats after giving 14 C-tetrahydrocannabinol to the mother. *Eur J Pharmacol,* 22: 221–3.

Johnson, K.M., Ho, B.T., and Dewey, W.L. (1976) Effects of delta9-tetrahydrocannabinol on neurotransmitter accumulation and release mechanisms in rat forebrain synaptosomes. *Life Sci,* 19: 347–56.

Katona, I., Rancz, E.A., Acsady, L., Ledent, C., Mackie, K., Hajos, N., and Freund, T.F. (2001) Distribution of CB1 cannabinoid receptors in the amygdala and their role in the control of GABAergic transmission. *J Neurosci,* 21: 9506–18.

Katona, I., Sperlagh, B., Sik, A., Kafalvi, A., Vizi, E.S., Mackie, K., and Freund, T.F. (1999) Presynaptically located CB1 cannabinoid receptors regulate GABA release from axon terminals of specific hippocampal interneurons. *J Neurosci,* 19: 4544–58.

Kawa, K. (1994) Distribution and functional properties of 5-HT3 receptors in the rat hippocampal dentate gyrus: a patch-clamp study. *J Neurophysiol,* 71: 1935–47.

Kenney, S.P., Kekuda, R., Prasad, P.D., Leibach, F.H., Devoe, L.D., and Ganapathy, V. (1999) Cannabinoid receptors and their role in the regulation of the serotonin transporter in human placenta. *Am J Obstet Gynecol,* 181: 491–7.

Kimura, T., Ohta, T., Watanabe, K., Yoshimura, H., and Yamamoto, I. (1998) Anandamide, an endogenous cannabinoid receptor ligand, also interacts with 5-hydroxytryptamine (5-HT) receptor. *Biol Pharm Bull,* 21: 224–6.

Maeda, T., Kaneko, S., and Satoh, M. (1994) Inhibitory influence via 5-HT3 receptors on the induction of LTP in mossy fiber-CA3 system of guinea-pig hippocampal slices. *Neurosci Res,* 18: 277–82.

Mailleux, P. and Vanderhaeghen, J. J. (1992) Distribution of neuronal cannabinoid receptor in the adult rat brain: a comparative receptor binding radioautography and in situ hybridization histochemistry. *Neuroscience,* 48: 655–68.

Malone, D.T. and Taylor, D.A. (1998) Modulation of delta9-tetrahydrocannabinol-induced hypothermia by fluoxetine in the rat. *Br J Pharmacol,* 124: 1419–24.

Maricq, A.V., Peterson, A.S., Brake, A.J., Myers, R.M., and Julius, D. (1991) Primary structure and functional expression of the 5HT3 receptor, a serotonin-gated ion channel. *Science,* 254: 432–7.

Marsicano, G. and Lutz, B. (1999) Expression of the cannabinoid receptor CB1 in distinct neuronal subpopulations in the adult mouse forebrain. *Eur J Neurosci,* 11: 4213–25.

Martin, B.R., Dewey, W.L., Harris, L.S., and Beckner, J.S. (1977) 3H-delta9-tetrahydrocannabinol distribution in pregnant dogs and their fetuses. *Res Commun Chem Pathol Pharmacol,* 17: 457–70.

McDonald, A.J. and Mascagni, F. (2001) Localization of the CB1 type cannabinoid receptor in the rat basolateral amygdala: high concentrations in a subpopulation of cholecystokinin-containing interneurons. *Neuroscience,* 107: 641–52.

Mengod, G., Pompeiano, M., Martinez-Mir, M.I., and Palacios, J.M. (1990) Localization of the mRNA for the 5-HT2 receptor by in situ hybridization histochemistry. Correlation with the distribution of receptor sites. *Brain Res,* 524: 139–43.

Molina-Holgado, F., Molina-Holgado, E., Leret, M.L., Gonzalez, M.I., and Reader, T.A. (1993) Distribution of indoleamines and [3H]paroxetine binding in rat brain regions following acute or perinatal delta 9-tetrahydrocannabinol treatments. *Neurochem Res,* 18: 1183–91.

Molina-Holgado, F., Amaro, A., Gonzalez, M.I., Alvarez, F.J., and Leret, M.L. (1996) Effect of maternal delta 9-tetrahydrocannabinol on developing serotonergic system. *Eur J Pharmacol,* 316: 39–42.

Molina-Holgado, F., Alvarez, F.J., Gonzalez, I., Antonio, M.T., and Leret, M.L. (1997) Maternal exposure to delta 9-tetrahydrocannabinol (delta 9-THC) alters indolamine levels and turnover in adult male and female rat brain regions. *Brain Res Bull,* 43: 173–8.

Molliver, M.E. (1987) Serotonergic neuronal systems: what their anatomic organization tells us about function *J Clin Psychopharmacol,* 7: 3S–23S.

Morales, M., Battenberg, E., de Lecea, L., and Bloom, F.E. (1996) The type 3 serotonin receptor is expressed in a subpopulation of GABAergic neurons in the rat neocortex and hippocampus. *Brain Res,* 731: 199–202.

Morales, M., Battenberg, E., de Lecea, L., Sanna, P.P. and Bloom, F.E. (1996) Cellular and subcellular immunolocalization of the type 3 serotonin receptor in the rat central nervous system. *Brain Res Mol Brain Res,* 36: 251–60.

Morales, M. and Bloom, F.E. (1997) The 5-HT3 receptor is present in different subpopulations of GABAergic neurons in the rat telencephalon. *J Neurosci,* 17: 3157–67.

Morales, M., McCollum, N., and Kirkness, E.F. (2001) 5-HT(3)-receptor subunits A and B are co-expressed in neurons of the dorsal root ganglion. *J Comp Neurol,* 438: 163–72.

Morales, M. and Backman, C. (2002) Coexistence of serotonin 3 (5-HT3) and CB1 cannabinoid receptors in interneurons of hippocampus and dentate gyrus. *Hippocampus,* 12: 756–64.

Morales, M. and Wang, S. D. (2002) Differential composition of 5-hydroxytryptamine-3 receptors synthesized in the rat CNS and peripheral nervous system. *J Neurosci,* 22: 6732–41.

Morales, M., Shwun-De, W., Diaz-Ruiz, O., Hyun-Jin Jho, D., and Abara, C. (2004) Cannabinoid CB1 receptor and serotonin 3 receptor subunit A (5-HT3A) are co-expressed in GABA neurons in the telencephalon. *J Comp Neurol* 468: 205–16.

Nakazi, M., Bauer, U., Nickel, T., Kathmann, M., and Schlicker, E. (2000) Inhibition of serotonin release in the mouse brain via presynaptic cannabinoid CB1 receptors. *Naunyn Schmiedebergs Arch Pharmacol,* 361: 19–24.

Ohno-Shosaku, T., Maejima, T., and Kano, M. (2001) Endogenous cannabinoids mediate retrograde signals from depolarized postsynaptic neurons to presynaptic terminals. *Neuron,* 29: 729–38.

Oz, M., Zhang, L., and Morales, M. (2002) Endogenous cannabinoid, anandamide, acts as a noncompetitive inhibitor on 5-HT3 receptor-mediated responses in Xenopus oocytes. *Synapse,* 46: 150–156.

Park, B., Gibbons, H.M., Mitchell, M.D., and Glassa, M. (2003) Identification of the CB1 cannabinoid receptor and fatty acid amide hydrolase (FAAH) in the human placenta. *Placenta,* 24: 473–8.

Ropert, N. and Guy, N. (1991) Serotonin facilitates GABAergic transmission in the CA1 region of rat hippocampus in vitro. *J Physiol,* 441: 121–36.

Sofia, R.D., Dixit, B.N. and Barry, H., 3rd (1971) The effect of delta-1-tetrahydrocannabinol on serotonin metabolism in the rat brain. *Life Sci,* 10: 425–36.

Sofia, R.D., Ertel, R.J., Dixit, B.N., and Barry, H., 3rd (1971) The effect of 1-tetrahydrocannabinol on the uptake of serotonin by rat brain homogenates. *Eur J Pharmacol,* 16: 257–9.

Sofia, R.D., Dixit, B.N., and Barry, H., 3rd (1977) The effect of repeated administration of delta9-tetrahydrocannabinol on serotonin metabolism in the rat brain. *Arch Int Pharmacodyn Ther,* 229: 52–8.

Szabo, B., Dorner, L., Pfreundtner, C., Norenberg, W., and Starke, K. (1998) Inhibition of GABAergic inhibitory postsynaptic currents by cannabinoids in rat corpus striatum. *Neuroscience,* 85: 395–403.

Takahashi, K.A. and Linden, D.J. (2000) Cannabinoid receptor modulation of synapses received by cerebellar Purkinje cells. *J Neurophysiol,* 83: 1167–80.

Tsou, K., Mackie, K., Sanudo-Pena, M.C., and Walker, J.M. (1999) Cannabinoid CB1 receptors are localized primarily on cholecystokinin-containing GABAergic interneurons in the rat hippocampal formation. *Neuroscience,* 93: 969–75.

Vardaris, R.M., Weisz, D.J., Fazel, A., and Rawitch, A.B. (1976) Chronic administration of delta-9-tetrahydrocannabinol to pregnant rats: studies of pup behavior and placental transfer. *Pharmacol Biochem Behav,* 4: 249–54.

Verberne, A.J., Taylor, D.A., and Fennessy, M.R. (1980) Withdrawal-like behaviour induced by inhibitors of biogenic amine reuptake in rats treated chronically with delta 9-tetrahydrocannabinol. *Psychopharmacology (Berl),* 68: 261–7.

Walker, J.M., Hohmann, A.G., Martin, W.J., Strangman, N.M., Huang, S.M., and Tsou, K. (1999) The neurobiology of cannabinoid analgesia. *Life Sci,* 65: 665–73.

Wallmichrath, I. and Szabo, B. (2002) Analysis of the effect of cannabinoids on GABAergic neurotransmission in the substantia nigra pars reticulata. *Naunyn Schmiedebergs Arch Pharmacol,* 365: 326–34.

Wilson, R.I. and Nicoll, R.A. (2001) Endogenous cannabinoids mediate retrograde signalling at hippocampal synapses. *Nature,* 410: 588–92.

Yakel, J.L. and Jackson, M.B. (1988) 5-HT3 receptors mediate rapid responses in cultured hippocampus and a clonal cell line. *Neuron,* 1: 615–21.

11 Endocannabinoids and Dopamine-Related Functions in the CNS

Javier Fernández-Ruiz, Rosario de Miguel, Mariluz Hernández, Maribel Cebeira, and José A. Ramos

CONTENTS

INTRODUCTION: THE CONTROL OF NEUROTRANSMITTER ACTIVITY BY ENDOCANNABINOIDS

The endocannabinoid system plays a modulatory role in several physiological processes, mainly in the brain (Di Marzo et al., 1998; Fernández-Ruiz et al., 2000, 2002; Iversen, 2003; van der Stelt and Di Marzo, 2003), although also in peripheral functions such as immune regulation (Parolaro, 1999), the cardiovascular system (Hillard, 2000), reproductive endocrine processes (Wenger, Toth, et al., 1999), and the control of energetic metabolism (Guzmán and Sánchez, 1999). In the brain, endocannabinoids participate in processes such as the control of movement (Consroe, 1998; Sañudo-Peña et al., 1999; Romero et al., 2002; Fernández-Ruiz et al., 2002), learning and memory (Hampson and Deadwyler, 1999; Castellano et al., 2003), brain reward (Gardner and Vorel, 1998; Basavarajappa and Hungund, 2002; Parolaro and Rubino, 2002), nociception (Walker et al., 1999, 2002; Pertwee, 2001), control of appetite (Berry and Mechoulam, 2002), and emesis (Mechoulam and Hanus, 2001; Di Carlo and Izzo, 2003). They also play an important role in various events related to brain development (Fernández-Ruiz et al., 2000). This can be concluded from numerous data that showed (1) the relevant presence of cannabinoid CB_1 receptors and their endogenous ligands, mainly anandamide and 2-arachidonoylglycerol, in the brain structures involved in the above processes (Breivogel and Childers, 1998; Bisogno et al., 1999; Freund et al., 2003), (2) the pharmacological effects reported after the administration of synthetic, plant-derived, or endogenous cannabinoids in humans and laboratory animals (Pertwee, 1997; Chaperon and Thiebot, 1999; Iversen, 2003), and (3) the functional changes found in mice lacking CB_1 receptor or fatty acid amide hydrolase (FAAH) gene expression (Ledent et al., 1999; Zimmer et al., 1999; Cravatt et al., 2001; for a review, see Kunos and Batkai, 2001). In addition, the involvement of the endocannabinoid signaling system in relevant functions in the brain implies that it may be considered of therapeutic relevance for different pathologies related to these brain functions (Iversen, 2003), which explains the increasing development of the cannabinoid pharmacology in recent years (Pertwee, 2000; Piomelli et al., 2000; Giuffrida et al., 2001; Palmer et al., 2002).

On the other hand, the involvement of the endocannabinoid system in this large list of brain functions is likely the consequence of its capability to interact with specific neurotransmitters in several brain regions (Chaperon and Thiebot, 1999; Freund et al., 2003). Thus, the frequent, although not exclusive, presynaptic location of CB_1 receptors allows endocannabinoids to directly influence presynaptic events such as synthesis, release, or reuptake for specific transmitters, mainly glutamate, opioid peptides, and GABA (as CB_1 receptors are frequently located on neurons containing these neurotransmitters in the brain) and also for acetylcholine and serotonin (for a review, see Schlicker and Kathmann, 2001). Also concordant with this presynaptic location of cannabinoid receptors is the recent proposal that endocannabinoids may act as retrograde signal molecules at synapses (for reviews, see Kreitzer and Regehr, 2002; Iversen, 2003).

Dopamine is one of the neurotransmitters that is frequently affected by cannabinoids. Possibly, these effects are not direct because, in general, dopaminergic neurons in the CNS do not contain CB_1 receptors, at least in the adult brain. However, these receptors are abundantly expressed in brain regions that are innervated by dopamine-releasing neurons (Herkenham, Lynn, Little, et al., 1991; Mailleux and Vanderhaeghen, 1992; Tsou, Brown, et al., 1998); therefore, it can be assumed that, through indirect mechanisms, dopaminergic neurons, whose cell bodies are frequently located in the reticular formation of the midbrain, are susceptible to being influenced by the endocannabinoid signaling. These dopaminergic cell bodies project to different forebrain structures, namely, the caudate-putamen (nigrostriatal pathway), the nucleus accumbens, and the prefrontal cortex (mesocorticolimbic pathway), exerting a regulatory action on different effector neurons in these structures and influencing processes such as the control of movement (Pollack, 2001; Brooks, 2001) and various cognitive functions (Spanagel and Weiss, 1999; Nieoullon, 2002). Deficiency or overactivity of these neurons can result in disorders such as Parkinson's disease (Dawson and Dawson, 2003) and schizophrenia (Lewis and Lieberman, 2000). On the other hand, the modification of dopaminergic transmission, although exerted via indirect mechanisms, is a necessary condition associated with the motor and limbic effects of cannabinoids (for a review, see van der Stelt and Di Marzo, 2003). Another important dopaminergic pathway that is influenced by endocannabinoid signaling through indirect mechanisms is the group of intrahypothalamic dopaminergic neurons, the so-called tuberoinfundibular system (Ben-Jonathan and Hnasko, 2001). The modification of this system underlies the neuroendocrine effects produced by cannabinoids and results in changes in the synthesis and release of various anterior pituitary hormones, mainly prolactin (Ben-Jonathan and Hnasko, 2001). It is also well established that dopamine attenuation in this pathway is associated with hyperprolactinemia (Ben-Jonathan and Hnasko, 2001).

Therefore, assuming that these three major dopaminergic pathways do not contain CB_1 receptors, with some exceptions that will be described in the following text (Hernández et al., 2000; Wenger et al., 2003), how can the relationship between endocannabinoids and dopamine be explained? There are two explanations. First, as mentioned in the preceding text, CB_1 receptors are present in GABAergic, glutamatergic, or opioidergic projections located in the closest vicinity of dopaminergic neurons (Herkenham, Lynn, Little, et al., 1991; Mailleux and Vanderhaeghen, 1992; Tsou, Brown, et al., 1998; Breivogel and Childers, 1998). Second, there is an increasing role of vanilloid TRPV1 receptors for some effects of certain cannabinoids (for reviews, see Zygmunt, Julius, et al., 2000; Di Marzo et al., 2002). Vanilloid TRPV1 receptors have been detected in dopaminergic neurons (Mezey et al., 2000) in various endocannabinoid effects. Considering these two aspects, the present chapter aims to review the roles played by endocannabinoids on three adult brain functions, where dopamine is a key regulatory neurotransmitter: (1) the control of motor function at the basal ganglia level, (2) the expression of some cognitive functions, including emotionality, motivation, and brain reward, and (3) the hypothalamic regulation of anterior pituitary hormone secretion. In addition, we also review the role of endocannabinoids in the expression of key genes for dopaminergic neurotransmission during brain development, which represent an example of a direct interaction between dopamine and endocannabinoids. Other interactions between cannabinoids and dopamine have been claimed for memory consolidation (Costanzi et al., 2004) and retina function (Schlicker et al., 1996; Weber and Schlicker, 2001) in mammals. There is also evidence of their interactions in the brain of invertebrates (Stefano et al., 1997). However, they have been scarcely explored, so they will not be addressed in the present chapter.

INTERACTIONS BETWEEN CANNABINOIDS AND DOPAMINE AT THE BASAL GANGLIA

Dopamine is a key neurotransmitter in the basal ganglia circuitry. This can be asserted from both physiological (Pollack, 2001) and therapeutic (Hornykiewicz, 1998) points of view. So, the activation of dopamine transmission in this circuitry is generally associated with an increase of movement, whereas the inhibition is followed by hypokinesia (Brooks, 2001). In fact, the basal ganglia disorder with the highest prevalence in the human population, Parkinson's disease, is a consequence of the progressive degeneration of nigrostriatal dopaminergic neurons, resulting in slowing of movement (bradykinesia), rigidity, and tremor (Dawson and Dawson, 2003). Cannabinoids are hypokinetic substances, thus producing motor depression and even catalepsy (see Romero et al., 2002; Fernández-Ruiz et al., 2002), and it has been largely speculated that this hypokinetic effect of cannabinoids might be produced by reducing dopaminergic activity. This assumption is correct, but the role of CB_1 receptors, which are not located on dopaminergic neurons (Herkenham, Lynn, de Costa, et al., 1991), has not been completely elucidated in regard to this effect. Meanwhile, the recent relationship of TRPV1 receptors with endocannabinoids (Zygmunt et al., 1999; Smart et al., 2000), as well as the location of these receptors on dopaminergic neurons (Mezey et al., 2000), have opened interesting novel aspects to discuss about the role of the endocannabinoid signaling in the basal ganglia (de Lago, de Miguel, et al., 2004), from both basic and clinical perspectives, aspects that will be addressed later in this text.

FUNCTION OF THE ENDOCANNABINOID SYSTEM IN THE BASAL GANGLIA

The abundant presence of CB_1 receptors and their endogenous ligands in brain regions related to the control of movement, such as the caudate-putamen, the globus pallidus, the substantia nigra, and the cerebellum (Herkenham, Lynn, Little, et al., 1991; Mailleux and Vanderhaeghen, 1992;

Tsou, Brown, et al., 1998; Bisogno et al., 1999), suggests that the endocannabinoid system is strongly related to the control of movement (for reviews, see Consroe, 1998; Sañudo-Peña et al., 1999; Fernández-Ruiz et al., 2002; Romero et al., 2002). Plant-derived, synthetic, or endogenous cannabinoid agonists produce dose-dependent motor inhibition in both humans and laboratory animals (see Fernández-Ruiz et al., 2002, for a review). Thus, low doses reduce spontaneous activity, whereas high doses may even produce catalepsy (Consroe, 1998; Sañudo-Peña et al., 1999; Fernández-Ruiz et al., 2002; Romero et al., 2002). Similar results were obtained by administering inhibitors of endocannabinoid inactivation, the so-called indirect cannabinoid agonists (González, Romero, et al., 1999; Beltramo et al., 2000; Lastres-Becker, Hansen, et al., 2002; de Lago et al., 2002 and de Lago, Ligresti, et al., 2004). By contrast, the administration of SR141716, a selective antagonist of CB_1 receptors, reversed these hypokinetic effects and even produced by itself a certain degree of hyperlocomotion due to its function as inverse agonist (Compton et al., 1996). Another piece of evidence supporting the involvement of the CB_1 receptor in the control of movement derives from the observation of motor anomalies in mice lacking the CB_1 receptor gene (Ledent et al., 1999; Zimmer et al., 1999), despite certain conflicting observations between the two models of knockout mice developed.

Involvement of Dopamine in Motor Effects of Cannabinoids

The motor effects of cannabinoid-based compounds have an explanation in the capability of these substances to influence the activity of several neurotransmitters acting at the basal ganglia circuitry. Striatal projection GABAergic neurons and subthalamonigral glutamatergic neurons, both containing CB_1 receptors (Herkenham, Lynn, de Costa, et al., 1991; Mailleux and Vanderhaeghen, 1992; Tsou, Brown, et al., 1998), are the major neurochemical substrates for the action of cannabinoids. Cannabinoids block GABA reuptake, thus potentiating GABA action (Maneuf et al., 1996; Romero, de Miguel, et al., 1998), or inhibit glutamate release (Szabo et al., 2000). CB_1 receptors located in GABAergic and glutamatergic neurons in the cerebellum have also been involved in motor effects of cannabinoids, in particular, with their effects on posture and balance, but the neurochemical basis for these effects has been poorly explored (see Iversen, 2003, for a review).

Dopamine transmission is also affected by cannabinoids in the basal ganglia circuitry. This can be concluded *a priori* from pharmacological studies reporting that cannabinoids potentiated reserpine-induced hypokinesia (Moss et al., 1981), mimicking dopaminergic antagonist-induced catalepsy (Anderson et al., 1996), while reducing amphetamine-induced hyperactivity (Gorriti et al., 1999). Despite the lack of selectivity of these two drugs, it appears obvious that both reserpine and amphetamine behaved in the basal ganglia circuitry mainly as dopamine-acting drugs, cannabinoids being able to influence the magnitude of these effects. Neurochemical studies also support the notion that cannabinoids are able to reduce the activity of nigrostriatal dopaminergic neurons (Romero, Garcia, et al., 1995 and Romero, de Miguel, et al., 1995; Cadogan et al., 1997; see Romero et al., 2002; van der Stelt and Di Marzo, 2003, for reviews), although some studies have shown increases rather than decreases (Sakurai-Yamashita et al., 1989; see Romero et al., 2002; van der Stelt and Di Marzo, 2003, for reviews).

In our laboratory, we found that anandamide reduced the activity of tyrosine hydroxylase in the caudate-putamen and the substantia nigra (Romero, Garcia, et al., 1995 and Romero, de Miguel, et al., 1995), where this enzyme is only located on dopamine-containing neurons, an effect that is compatible with the hypokinesia caused by anandamide and by other cannabinoid agonists (Romero, de Miguel, et al., 1995). However, the effects of anandamide on tyrosine hydroxylase activity were small and transient (Romero, Garcia, et al., 1995 and Romero, de Miguel, et al., 1995), possibly because CB_1 receptors are not located on nigrostriatal dopaminergic neurons (Herkenham, Lynn, de Costa, et al., 1991). In concordance with this idea, cannabinoid agonists and antagonists failed to inhibit electrically evoked dopamine release in the striatum (Szabo et al., 1999), although the matter is not clarified because other studies proved opposite effects (increases vs. decreases) of cannabinoids on striatal dopamine release *in vitro* (see van der Stelt and Di Marzo, 2003, for details). The lack of CB_1 receptors in nigrostriatal dopaminergic neurons would support the observation that the changes in the activity of these neurons caused by cannabinoids *in vivo* were originated by previous changes in GABAergic influences reaching the substantia nigra (Maneuf et al., 1996; Romero et al., 1998). As mentioned in the preceding text, striatal GABAergic projection neurons do contain CB_1 receptors (Herkenham, Lynn, de Costa, et al., 1991) and their activation stimulates GABA transmission (Maneuf et al., 1996; Romero, de Miguel, et al., 1998), which would result in a greater inhibition of nigral dopamine neurons.

However, several recent discoveries have provided new elements to reevaluate the idea that endocannabinoid effects on dopamine transmission in the basal ganglia are indirect. It has been demonstrated that anandamide and some analogs but not classic cannabinoids are also able to behave as full agonists for the vanilloid TRPV1 receptors (Zygmunt et al., 1999; Smart et al., 2000). These receptors are molecular integrators of nociceptive stimuli, abundant on sensory neurons, but they have also been located in the basal ganglia circuitry, possibly on nigrostriatal dopaminergic neurons (Mezey et al., 2000), thus representing another target for the action of anandamide in the basal ganglia. We have recently reported that: (1) the activation of vanilloid-like receptors with their classic agonist, capsaicin, or with other potential ligands produced hypokinesia in rats (Di Marzo et al., 2001) and (2) the antihyperkinetic activity of several cannabinoid-based compounds, such as AM404, in rat models of hyperkinetic disorders, such as Huntington's disease, is dependent on their capability to activate vanilloid-like receptors rather than CB_1 receptors (Lastres-Becker, Hansen, et al., 2002 and Lastres-Becker, de Miguel, De Petrocellis, et al., 2003), despite earlier evidence that suggested a major role of CB_1 receptors in the AM404 effects (Beltramo et al., 2000). More recently, we have also observed that the hypokinetic and dopamine-lowering effects of anandamide were reversed by capsazapine, an antagonist of vanilloid-like receptors. This effect has also been confirmed *in vitro* using perifused striatal fragments (de Lago, de Miguel, et al., 2004). Classic cannabinoids, such as Δ^9-tetrahydrocannabinol (Δ^9-THC), which do not bind to vanilloid-like receptors, are not able to produce the same effect. This is in concordance with the observation that anandamide reduced dopamine release from striatal slices (Cadogan et al., 1997), although the authors also found a dopamine-lowering effect after application of a classic cannabinoid such as CP55940 (Cadogan et al., 1997).

Extending the above findings that mainly indicate the occurrence of a regulation by cannabinoids of dopaminergic activity in the basal ganglia, there are also proposals of a bidirectional regulation. Thus, in striatal GABAergic projection neurons, CB_1 receptors colocalize with D_1 (striatonigral pathway) or with D_2 (striatopallidal pathway) receptors (Hermann et al., 2002), which allow additional postsynaptic interactions between endocannabinoids and dopamine at the level of G-protein and adenylyl cyclase signal transduction (Giuffrida et al., 1999; Meschler, Conley, et al., 2000 and Meschler, Clarkson, et al., 2000; Meschler and Howlett, 2001). In addition, there is also evidence of a regulation by dopaminergic D_2 receptors of anandamide production in the striatum (Giuffrida et al., 1999), which was interpreted by the authors as indicative of a role for the endocannabinoid system as an inhibitory feedback mechanism counteracting dopamine-induced facilitation of psychomotor activity (Giuffrida et al., 1999). This observation agrees with the results reported by Mailleux and Vanderhaeghen (1993) that endocannabinoid signaling in the basal ganglia is under a negative control exerted by dopamine, which might be relevant for Parkinson's disease (see the following text). The authors found that chronic administration with dopaminergic antagonists or the lesion of nigrostriatal dopaminergic neurons with 6-hydroxydopamine upregulated CB_1 receptors (Mailleux and Vanderhaeghen, 1993). Similar observations were obtained in 6-hydroxydopamine-lesioned rats by our group (Romero et al., 2000) and others (Herkenham, Lynn, de Costa, et al., 1991; Gubellini et al., 2002). These observations were extended to other models of dopamine deficiency, such as MPTP-treated marmosets (Lastres-Becker, Cebeira, et al., 2001) and reserpine-treated rats (Di Marzo et al., 2000). Interestingly, dopaminergic replacement with L-dopa in these models reversed endocannabinoid overactivity (Lastres-Becker, Cebeira, et al., 2001; Macarrone et al., 2003).

Therapeutic Implications of Cannabinoid–Dopamine Interactions in the Basal Ganglia

Based on these observations that strongly support endocannabinoids modulating the activity of dopamine and other neurotransmitters at the basal ganglia by acting at CB_1 or TRPV1 receptors, it has been postulated that the pharmacological management of the endocannabinoid system, which might result in normalizing dopamine transmission, could be useful to alleviate motor symptoms in various basal ganglia disorders characterized by either dopaminergic degeneration or malfunctioning (for reviews, see Consroe, 1998; Sañudo-Peña et al., 1999; Fernández-Ruiz et al., 2002; Romero et al., 2002, van der Stelt and Di Marzo, 2003). Most studies are preclinical and have provided the first experimental evidences using animal models (for a review, see Fernández-Ruiz et al., 2002). However, in some cases, there are also some clinical trials presently in progress or already finalized (Sieradzan et al., 2001; Fox, Kellett, et al., 2002; Müller-Vahl et al., 2002, 2003; Zajicek et al., 2003).

Hyperkinetic Disorders: Huntington's Disease

Direct or indirect (inhibitors of endocannabinoid inactivation) agonists of CB_1 receptors have been proposed as having therapeutic value in hyperkinetic disorders, such as Huntington's chorea (Lastres-Becker, Hansen, et al., 2002 and Lastres-Becker, de Miguel, De Petrocellis, et al., 2003; for a review, see Lastres-Becker, De Miguel, Fernández-Ruiz, et al., 2003) or Gilles de la Tourette's syndrome (Müller-Vahl et al., 1998, 2002, 2003; for a review, see Müller-Vahl, 2003). In the latter, cannabinoid agonists have been demonstrated to reduce tics and improve behavioral problems in patients (for a review, see Müller-Vahl, 2003), but there are no data on the status of the endocannabinoid signaling in patients or in animal models of this disease. Additionally, there is no information on the neurochemical substrates underlying the beneficial effects of cannabinoids. By contrast, in Huntington's disease, analysis of postmortem brains revealed a marked loss of CB_1 receptor binding in the substantia nigra, in the lateral part of the globus pallidus and, to a lesser extent, in the putamen (Glass et al., 1993, 2000; Richfield and Herkenham, 1994). *A priori*, this loss might be a mere side-effect caused by the degeneration of striatal medium-spiny GABAergic neurons that contain CB_1 receptors. However, studies in patients affected by Huntington's disease at different stages of the disease have revealed that the loss of CB_1 receptors occurs earlier than the loss of other neurotransmitter receptors, even in presymptomatic phases of this disease, before the appearance of major Huntington's disease symptomatology and when the cell death is minimal (grades 0 and 1) (Glass et al., 2000). This has also been found in different models of transgenic mice that express mutated forms of huntingtin, as in the human pathology (Denovan-Wright and Robertson, 2000; Lastres-Becker, Berrendero, et al., 2002). It is remarkable that these changes always occurred in the absence of the cell death, paralleling the data observed in the early grades of the human disease (Glass et al., 2000) when cell loss is minimal. All these observations, collectively, have supported the opinion that CB_1 receptor losses might be involved in the pathogenesis of Huntington's disease.

Huntington's chorea may also be generated in rats after local or systemic injection of 3-nitropropionic acid, a toxin that selectively destroys striatal projection neurons. These animals also exhibited a marked loss of CB_1 receptors and their endogenous ligands in the basal ganglia (Page et al., 2000; Lastres-Becker, Fezza, et al., 2001 and Lastres-Becker, Hansen, et al., 2002), compatible with the notion that endocannabinoid transmission in the basal ganglia becomes hypofunctional in Huntington's disease, which contributes to some extent to the hyperkinesia typical of this disease. In this case, the changes might be mere side effects caused by the strong death of striatal medium-spiny GABAergic neurons originated by the neurotoxin, a situation comparable to the pattern of cell loss that occurs in symptomatic stages of the human disease (grades 2 to 4). This GABA reduction is accompanied by a profound dopamine deficit in the basal ganglia (Lastres-Becker, Hansen, et al., 2002).

Despite the evidence of a decreased endocannabinoid transmission in the basal ganglia in Huntington's disease and the suggestion that CB_1 receptor activation might play an instrumental role in this disease (Lastres-Becker, Bizat, et al., 2003), the administration of cannabinoid agonists in humans increased choreic movements in Huntington's disease (Müller-Vahl et al., 1998, 1999; Consroe, 1998). It is possible that this is related to the fact that only those cannabinoid-based compounds having an additional profile as TRPV1 receptor agonists were really effective in animal models of this disease (Lastres-Becker, de Miguel, De Petrocellis, et al., 2003), thus stressing a relevant role for TRPV1 rather than CB_1 receptors in this disease. This was the case of AM404, an inhibitor of the endocannabinoid transporter (Beltramo et al., 1997) that also exhibits affinity for the TRPV1 receptor (Zygmunt, Chuang, et al., 2000). This compound, in contrast with other transporter inhibitors devoid of TRPV1 receptor affinity, behaved as antihyperkinetic in Huntington's disease animals (Lastres-Becker, Hansen, et al., 2002 and Lastres-Becker, de Miguel, De Petrocellis, et al., 2003), and this effect was presumably produced by the activation of TRPV1 receptors that originates a normalization of neurochemical deficits, including a rise in GABA and dopamine activities in the basal ganglia (Lastres-Becker, Hansen, et al., 2002 and Lastres-Becker, de Miguel, De Petrocellis, et al., 2003). These effects did not occur in normal rats (Lastres-Becker, Hansen, et al., 2002 and Lastres-Becker, de Miguel, De Petrocellis, et al., 2003).

Hypokinetic Disorders: Parkinson's Disease

Cannabinoid-based compounds might also be useful in other basal ganglia disorders more directly related to a disturbance of dopaminergic transmission. This is the case of the dopamine deficiency typical of Parkinson's disease. In this disorder, CB_1 receptor agonists or antagonists have both been proposed to be of therapeutic value, alone or as coadjuvants, to alleviate different motor symptoms characteristic of this disease (Brotchie, 2000; Romero et al., 2000; Di Marzo et al., 2000; Lastres-Becker, Cebeira, et al., 2001; Fox, Henry, et al., 2002). In particular, CB_1 receptor agonists have been proposed for: (1) the reduction of tremor associated with an overactivity of the subthalamic nucleus (Sañudo-Peña et al., 1999), (2) the reduction or delay of the dyskinetic states associated with long-term therapy of dopaminergic replacement with L-dopa (Sierazdan et al., 2001), and (3) the protection against dopaminergic cell death in the case of those compounds having antioxidant properties (Lastres-Becker et al., 2005). By contrast, the blockade of CB_1 receptors with rimonabant (SR141716) or other selective antagonists has been proposed for its capability to reduce bradykinesia (Romero et al., 2000; Lastres-Becker, Cebeira, et al., 2001; Gubellini et al., 2002; Fernández-Espejo et al., 2004) and also L-dopa-induced dyskinesia (Brotchie, 2000, 2003). The evidence for the usefulness of CB_1 receptor antagonists in this disease is based on studies reporting that dysfunctions of nigrostriatal dopaminergic neurons, such as those caused by reserpine (Di Marzo et al., 2000) or by dopaminergic antagonists (Mailleux and Vanderhaeghen, 1993), or the degeneration of these neurons caused by the local application of 6-hydroxydopamine (Mailleux and Vanderhaeghen, 1993; Romero et al., 2000; Gubellini et al., 2002; Fernández-Espejo et al., 2004) or MPTP (Lastres-Becker, Cebeira, et al., 2001) are associated with an increased endocannabinoid transmission in the basal ganglia (see preceding text). This suggests the occurrence of an unbalance between dopamine and endocannabinoids at the basal ganglia that supports the usefulness of CB_1 receptor antagonists in this disease. In agreement with this, the classic dopaminergic replacement therapy with L-dopa reversed endocannabinoid anomalies (Lastres-Becker, Cebeira, et al., 2001; Maccarrone et al., 2003).

Despite this evidence, the first pharmacological studies that have examined the capability of rimonabant to reduce hypokinesia in animal models of Parkinson's disease have offered conflicting results (Di Marzo et al., 2000; Meschler et al., 2001). It is possible that the blockade of CB_1 receptors might be effective only at very advanced phases of the disease. In this sense, recent evidence provided by Fernández-Espejo and co-workers (2004) is in favor of this option, which presents an additional advantage as it would allow an antiparkinsonism compound in a stage of the disease when the classic dopaminergic therapy has generally failed. In addition, one must also consider the implication of TRPV1 receptors, in view of their recently demonstrated role in regulating dopamine release from nigral neurons (de Lago, de Miguel, et al., 2004), as well as the evidence of neuroprotective effects ascribed to cannabinoid agonists in this (Lastres-Becker et al., 2005) and in other neurodegenerative diseases (for a review, see Grundy et al., 2001; Mechoulam et al., 2002; van der Stelt et al., 2002), effects that would not be directly related to the influence of the endocannabinoid signaling system on dopamine transmission.

Other Motor-Related Disorders

There are other neurological diseases where the manipulation of endocannabinoid signaling has been proposed as therapeutically promising, and where these therapeutic effects might also be related to an influence on dopamine transmission (for a review, see Fernández-Ruiz et al., 2002). These disorders are not directly related to basal ganglia degeneration but are associated with dysfunction in several neurotransmitters in this area, thus resulting in motor deterioration.

Multiple Sclerosis

One relevant case is multiple sclerosis whose origin is immune but progresses with neurological deterioration mainly affecting the motor system (Pertwee, 2002; Baker and Pryce, 2003). Cannabinoid agonists have been proposed as clinically promising in this disease for the management of motor-related symptoms such as spasticity, tremor, ataxia, and others (for reviews, see Pertwee, 2002; Baker and Pryce, 2003). Thus, Baker et al. (2000) reported that plant-derived, synthetic, and endogenous cannabinoid agonists were able to reduce some signs (spasticity and tremor) in a mouse model of multiple sclerosis. They also demonstrated that these effects were mediated by cannabinoid CB_1 and, to a lesser extent, CB_2 receptors (Baker et al., 2000). Using this mouse model, they have also described antispastic effects of compounds that are able to inhibit the process of termination of the biological action of endocannabinoids (Baker et al., 2001; Brooks et al., 2002; de Lago, Ligresti, et al., 2004). These data were concordant with the observation of increased contents of endocannabinoids in the brain and spinal cord of these animals (Baker et al., 2001). Using a rat model of multiple sclerosis, other authors (Lyman et al., 1989; Wirguin et al., 1994) have also demonstrated that a chronic administration of plant-derived cannabinoids may reduce or delay the incidence and severity of clinical signs in these rats.

This pharmacological evidence obtained in animal models has provided solid experimental support to previous anecdotal, uncontrolled, or preclinical data that suggested a beneficial effect for marijuana when smoked by multiple sclerosis patients to alleviate such symptoms as spasticity, dystonia, tremor, ataxia, and pain (for a review, see Consroe, 1998). In this line, a clinical trial has recently finalized in the U.K. using oral administration of placebo, cannabis extract, or Δ^9-THC in a population of 667 patients with stable multiple sclerosis and muscle spasticity. The results of this trial have suggested that cannabinoids did not have a beneficial effect on spasticity in multiple sclerosis patients but increased the patient's perception of improvement of other signs including pain (Zajicek et al., 2003).

Despite the progress in the pharmacological evaluation of cannabinoid-based medicines in multiple sclerosis both in patients and animal models, there are no data on the possible changes in CB_1 and CB_2 receptors in the postmortem brain of patients with multiple sclerosis, whereas only a few studies have examined the status of the endocannabinoid transmission in animal models of this disease (Baker et al., 2001; Berrendero et al., 2001). Thus, as mentioned above, Baker and coworkers reported an increase of endocannabinoid levels in the brain and the spinal cord in the mouse model of multiple sclerosis (Baker et al., 2001) that was interpreted by these authors as indicative of an endocannabinoid influence in the control of some symptoms of multiple sclerosis in an environment of existing neurological damage (see Baker and Pryce, 2003, for a review). In our laboratory, using the rat model of this disease, we reported a decrease of CB_1 receptor binding and mRNA levels (Berrendero et al., 2001), although the decreases in CB_1 receptors were mainly circumscribed to the basal ganglia (lateral and medial caudate-putamen) and to a lesser extent to cortical regions, which was related to the fact that motor deterioration is one of the most prominent neurological signs in these rats (Berrendero et al., 2001) and also in the human disease (for a review, see Baker and Pryce, 2003). We have also demonstrated that this decrease was accompanied by a reduction of endocannabinoid levels that may be recorded also in other brain structures (Cabranes et al., 2005). Based on this fact, we hypothesized that the changes in CB_1 receptors and their ligands in the basal ganglia might be associated with disturbances in several neurotransmitters, in particular in dopamine transmission, acting at this circuitry. If this were the case, the well-known effects of cannabinoid agonists on these neurotransmitters might underlie the improving effects of these compounds in motor symptoms of multiple sclerosis (see Fernández-Ruiz et al., 2002, for a review). However, our hypothesis failed because we did not record any changes in dopamine, serotonin, GABA, or glutamate in the basal ganglia of EAE rats (Cabranes et al., 2005). So, there was no option for examining whether cannabinoid administration might influence these neurotransmitters in the rat model of multiple sclerosis. This also indicates that the endocannabinoid signaling in the basal ganglia has an important relevance in these rats, as the changes in receptors (Berrendero et al., 2001) and the changes in ligands (Cabranes et al., 2005) represent the only transmitter disturbance associated with motor deterioration in these animals. Assuming this particular relevance, it was expected that the pharmacological manipulation of this system might be beneficial in the rat model of multiple sclerosis, as previous studies revealed (Lyman et al., 1989; Wirguin et al., 1994). In this sense, we recently analyzed the effects of various inhibitors of endocannabinoid transport that were capable of elevating endocannabinoid levels and found that they were able to reduce the neurological decline typical of these rats, although we also documented a relevant involvement of TRPV1 receptors in these effects (Cabranes et al., 2005).

Alzheimer's Disease

A second disorder is Alzheimer's disease, which, as with multiple sclerosis, is not a disorder of the basal ganglia, but where extrapyramidal signs, possibly caused by the degeneration of glutamate cortical afferents to the caudate-putamen, are frequently observed (for a review, see Kurlan et al., 2000). No data exist on the status of endocannabinoid transmission in the basal ganglia or in other brain regions in animal models of this disease. However, studies with postmortem brain regions of patients affected by this disease have revealed a significant loss of CB_1 receptors that seem to be notably circumscribed to the basal ganglia, in particular to the caudate nucleus, medial globus pallidus, and substantia nigra (Westlake et al., 1994). Brain regions other than the basal ganglia were less affected or did not exhibit any changes for the CB_1 receptor, except the hippocampus, which also showed significant reductions (Westlake et al., 1994). However, it is important to remark that the authors considered their results related more to increasing age rather than to an effect selectively associated with the pathology characteristic of Alzheimer's disease (Westlake et al., 1994). Also using postmortem tissue from Alzheimer's patients, Benito et al. (2003) recently reported a significant induction of CB_2 receptors in activated microglia that surrounds the senile plaques, which would suggest both a role of this receptor subtype in a part of the pathogenesis in this disease and a potential therapeutic application of selective compounds targeting this receptor. The latter would be related to recent evidences concerning the possible utilization of cannabinoid-based compounds in Alzheimer's disease due to their potential as neuroprotectants (Milton, 2002; Iuvone et al., 2004), which adds to previous evidence of the symptomatic role of nabilone or other cannabinoid agonists in reducing anorexia and improving behavioral disturbances in Alzheimer's patients (Volicer et al., 1997).

INTERACTIONS BETWEEN CANNABINOIDS AND DOPAMINE AT THE CORTICOLIMBIC STRUCTURES

Mesocorticolimbic dopaminergic neurons subserve cognitive processes (Nieoullon, 2002), motivated behavior (McAllister, 2000), the central stress response (Koob and Le Moal, 1997), and the pleasure produced by natural (i.e., sex, food) or unnatural (i.e., drugs of abuse) reinforcers, or by compulsive activities such as gambling, overeating, and sex dependence (for reviews, see Comings and Blum, 2000; Gardner, 2002). In fact, dopamine is possibly the neurotransmitter more studied as a potential target for the pharmacological effects of habit-forming drugs (Self and Nestler, 1994; Wise, 1996; Spanagel and Weiss, 1999; Gardner, 2002), including cannabis derivatives (Tanda et al., 1997; Gardner and Vorel, 1998; Diana, Melis, Muntoni, et al., 1998; Gardner, 2002). This is possibly the consequence of the existence of interactions between endocannabinoids and dopamine in relation to both brain reward stimulation of mesocorticolimbic dopaminergic neurons (Tanda et al., 1997; Gardner and Vorel, 1998; Diana, Melis, Gessa, 1998; Gardner, 2002) and abstinence or dysphoric responses to inhibition of this circuit (Diana, Melis, Muntoni, et al., 1998; Gardner, 2002). However, the interactions of endocannabinoids and dopamine are probably largest at this level and involve different cortical and subcortical structures and their specific functions. Thus, cannabinoids, depending on doses and duration of treatment, produce euphoria (Gardner, 2002), stimulate brain reward (Gardner et al., 1988; Gardner and Vorel, 1998), are anxiolytic (Berrendero and Maldonado, 2002), and decrease motivation and arousal while increasing emotionality (Navarro et al., 1993). Although, in most of these effects, the researchers have recorded parallel alterations in dopamine transmission in the mesocorticolimbic pathway, it is generally accepted that these effects would be indirect (Cheer et al., 2003). They would involve glutamatergic or GABAergic inputs to the nucleus accumbens or prefrontal cortex and ventral-tegmental area (Pistis et al., 2001 and Pistis, Muntoni, et al., 2002; Cheer et al., 2003) that would contain CB_1 receptors regulating presynaptic events (for a review, see Schlicker and Kathmann, 2001). However, Wenger and co-workers (2003) have recently presented evidence of colocalization of CB_1 receptors and tyrosine hydroxylase in the ventral-tegmental area, the regions containing the cell bodies of mesocorticolimbic dopaminergic neurons, thus opening the possibility of a direct influence by cannabinoids on the major substrate for processes such as reward, motivation, emotionality, and other limbic functions.

Function of the Endocannabinoid System in the Corticolimbic Structures

As mentioned in the preceding text, presynaptic CB_1 receptors located in cortical and subcortical (limbic) structures might be involved in two well-known effects of exogenously added cannabinoids: (1) cognitive effects and (2) reinforcing properties, which necessarily represent two key functions for the endocannabinoid signaling system in corticolimbic structures. Concerning the first of these effects, it is well established that CB_1 receptors are moderately abundant in the cerebral cortex; in particular, they are located in superficial and deep layers, presumably on GABAergic interneurons (Herkenham, Lynn, Little, et al., 1991; Mailleux and Vanderhaeghen, 1992; Breivogel and Childers, 1998; Tsou, Brown, et al., 1998). Endocannabinoids are also detected in the cerebral cortex (Bisogno et al., 1999). These data suggest a role for the endocannabinoid signaling system in the control of the sleep-waking cycle, performance of complex cognitive tasks, working memory, temporal organization of behavior, adaptation of behavioral strategies, sensory perception, and other cognitive functions, whose control resides mainly in the cerebral cortex (for a review, see Iversen, 2003). In fact, these functions are related to many of the major subjective effects and cognitive impairments experienced by cannabis consumers. Thus, the acute administration of cannabinoids reversibly impairs cognitive functions both in humans and laboratory animals (for a review, see Iversen, 2003), whereas long-term marijuana abuse might be associated with severe irreversible deficits in cognitive function and precipitation of psychiatric symptoms, such as psychosis, anxiety, or depression, particularly when marijuana is consumed by young people (see a recent study in Patton et al., 2002), although the evidence is still confusing (see Iversen, 2003, for a review).

The endocannabinoid system seems to also play a role in brain reward circuitry that is activated by different types of reinforcers, among them the habit-forming drugs (Comings et al., 1997; Manzanares et al., 1999; Basavarajappa and Hungund, 2002; Parolaro and Rubino, 2002). This role would explain why cannabis consumption is associated with marked changes in endocannabinoid transmission, mainly at the level of CB_1 receptors, as numerous studies in laboratory animals have demonstrated (see Maldonado, 2002, for a recent review). It would also explain the changes reported for the endocannabinoid transmission in states of dependence or abstinence for other drugs of abuse (see Comings et al., 1997; Manzanares et al., 1999; Basavarajappa and Hungund, 2002; Parolaro and Rubino, 2002, and the following text). Different types of studies have provided experimental support for the role of the endocannabinoid system in brain reward. Thus, neuroanatomical studies have demonstrated that CB_1 receptors and their endogenous ligands are present in the different limbic structures that form the brain reward circuitry (Herkenham, Lynn, Little, et al., 1991; Mailleux and Vanderhaeghen, 1992; Tsou, Brown, et al., 1998 and Tsou, Nogueron, et al., 1998; Bisogno et al., 1999; Wenger et al., 2003). In the same line, biochemical studies have proved the occurrence of changes in the density of CB_1 receptors and in the levels of their endogenous ligands in these regions of animals rendered dependent to the most frequently-consumed habit-forming drugs, such as opioids (González, Fernández-Ruiz, et al., 2002 and Gonzalez et al., 2003; Vigano et al., 2003), cocaine (González, Fernández-Ruiz, et al., 2002 and González, Cascio, et al., 2002), nicotine (González, Cascio, et al., 2002), and alcohol (Basavarajappa and Hungund, 1999a, 1999b, 2002; González, Fernández-Ruiz, et al., 2002 and González, Cascio, et al., 2002). Finally, studies using strategies of either pharmacological management of the endocannabinoid system or abolition of the gene expression for key proteins of this system have demonstrated relevant changes in processes related to drug addiction, which are directly or indirectly related to dopamine transmission. Among them, individual vulnerability, reinforcing potential, degree of dependence, abstinence responses, or risk for relapse for different drugs of abuse such as opioids (Mas-Nieto et al., 2001; De Vries et al., 2003; Solinas et al., 2003; Fattore et al., 2003), alcohol (see Basavarajappa and Hungund, 2002, for a review), nicotine (Cohen et al., 2002; Castañé et al., 2002; Valjent et al., 2002), or cocaine (De Vries et al., 2001; see van den Brink and van Ree, 2003, for a review) can be observed.

Involvement of Dopamine in Corticolimbic Effects of Cannabinoids

As mentioned above, dopamine and also opioids have been involved as critical neurobiological substrates for brain reward (Comings and Blum, 2000). Like other addictive drugs, cannabinoids (mainly Δ^9-THC) have been reported to potentiate brain self-stimulation in rats (Gardner and Vorel, 1998; Gardner, 2002) by increasing mesolimbic dopaminergic activity (Gessa et al., 1998; Yamamoto and Takada, 2000). This is supported by the following observations obtained after the treatment of laboratory animals with different cannabinoid agonists (see Gardner, 2002, for a recent review): (1) elevation of D_1 receptor density and contents of dopamine and its metabolites in various limbic structures (Bowers and Hoffman, 1986; Navarro et al., 1993); (2) stimulation of dopamine release in the nucleus accumbens (Chen, Paredes, Li, et al., 1990; Tanda et al., 1997; Gardner and Vorel, 1998); and (3) elevation of the firing rate of mesolimbic dopaminergic neurons in the A10 region (French et al., 1997; Wu and French, 2000). Additional evidence comes from the study of the reinforcing properties of different drugs in knockout mice for the CB_1 receptor. These mice exhibited reduced voluntary alcohol consumption (Hungund et al., 2003), morphine self-administration (Mascia et al., 1999; Cossu et al., 2001), and absence of rewarding effects of nicotine evaluated in the conditioned place preference test (Castañé et al., 2002) but did not show equivalent changes for cocaine or nicotine reinforcement in the self-administration paradigm (Cossu et al., 2001). In the case of alcohol, morphine, and nicotine, the involvement of dopamine seems robust because the drug-induced dopamine release in the nucleus accumbens is completely absent in CB_1 receptor knockout mice (Mascia et al., 1999; Cohen et al., 2002; Hungund et al., 2003).

Cannabinoid agonists also augmented the activity of those dopaminergic neurons that, coming from the ventral-tegmental area, project specifically to the prefrontal cortex (Chen, Paredes, Lowinson, et al., 1990; Jentsch et al., 1997; Diana, Melis, Gessa, 1998). As mentioned above, this is a region involved in many cognitive functions, including working memory, temporal organization of behavior, and adaptation of behavioral strategies (Goldman-Rakic, 1995; Rolls, 2000) and contains CB_1 receptors in a moderate density (Herkenham, Lynn, Little, et al., 1991; Mailleux and Vanderhaeghen, 1992). The acute administration of cannabinoids increased dopamine release in rat prefrontal cortex, measured by *in vivo* microdialysis, accompanied by an increase of glutamate release and a decrease of GABA. All these effects were CB_1 receptor mediated because they were reversed by SR141716 (Pistis, Ferraro, et al., 2002). However, the repeated administration of cannabinoid agonists, rather than increasing, decreased dopamine turnover in the prefrontal cortex (Jentsch et al., 1998; Verrico et al., 2003), an effect that persisted even after a drug-free period of 2 weeks (Verrico et al., 2003). This did not occur in the nucleus accumbens or the striatum (Verrico et al., 2003).

In addition to the data obtained in laboratory animals, there is also evidence of changes in mesocorticolimbic dopaminergic activity following cannabinoid exposure in humans (Voruganti et al., 2001; Steffens et al., 2004). Thus, in a recent study using postmortem brain samples, Steffens and co-workers (2004) proved that CB_1 receptors might be located presynaptically on dopaminergic nerve terminals, which would agree with the data obtained in rats by Wenger and co-workers (2003). The authors concluded that these receptors would be tonically activated by endocannabinoids to modulate dopamine release (Steffens et al., 2004).

Despite the above evidence, the molecular mechanisms by which cannabinoids activate mesocorticolimbic dopaminergic activity are not well understood, and the scarce evidence has yielded conflicting results. Thus, in a recent report, Patel and Hillard (2003) demonstrated that cannabinoid administration was able to induce c-fos expression in A10 dopaminergic neurons and that this effect was enhanced by noradrenergic influences from the locus coeruleus. This observation contrasts with the data reported by Alonso et al. (1999), who observed an increased c-fos expression in mesocorticolimbic dopaminergic regions with SR141716, the selective CB_1 receptor antagonist, instead of agonists. On the other hand, there is no discrepancy concerning the observation that the activation of CB_1 receptors is involved in all the effects produced by cannabinoids on mesocorticolimbic activity. However, as mentioned in the preceding text, these effects have been considered, so far, as exerted indirectly (Cheer et al., 2003). Possibly, they would be caused through modifying GABAergic influences to the ventral-tegmental area and the nucleus accumbens. This hypothesis is based, as mentioned in the previous text, on the assumption that CB_1 receptors would be located in these GABA neurons rather than in mesocorticolimbic ones (Herkenham, Lynn, de Costa, et al., 1991; Szabo et al., 1999; see Breivogel and Childers, 1998, for a review), although they colocalize with D_1 and D_2 receptors in limbic regions such as the olfactory tubercle, but not in the neocortex where CB_1 receptors colocalize with serotonin receptor subtypes (Hermann et al., 2002). These GABA neurons tonically inhibit dopamine-containing neurons, and inhibition of GABA release by cannabinoid agonists via presynaptic CB_1 receptors would be expected to increase the activity of dopaminergic neurons (Cheer et al., 2000; Pistis et al., 2001 and Pistis, Muntoni, et al., 2002; Hoffman and Lupica, 2001; Manzoni and Bockaert, 2001; Szabo et al., 2002; for a review, see Schlicker and Kathman, 2001). An alternative is that CB_1 receptors are located in the excitatory glutamatergic inputs to the GABA-containing neurons that project from the nucleus accumbens to the ventral-tegmental area, so that activation of these receptors would result in a decrease of glutamate release followed by reduction in GABA activity and, again, in an increase in the activity of dopaminergic neurons (Robbe et al., 2001; Pistis, Muntoni, et al., 2002; for a review, see Schlicker and Kathman, 2001). Additional evidence on the presence of CB_1 receptors in glutamatergic neurons in the ventral-tegmental area has been recently presented by Melis and co-workers (2004). However, a recent study by Wenger and co-workers has demonstrated for the first time, using double immunohistochemistry, that CB_1 receptors colocalize with tyrosine hydroxylase in the nucleus accumbens (Wenger et al., 2003), which opens the possibility of a direct action of cannabinoids on the major neurochemical substrate of brain reward, dysphoria-mediated drug craving, and drug relapse.

Therapeutic Implications of Cannabinoid–Dopamine Interactions in Corticolimbic Structures

As in the case of the basal ganglia, the demonstrated capability of the endocannabinoid system to influence dopamine transmission in corticolimbic structures supports the observation that the pharmacological management of this system might have therapeutic value in those diseases involving anomalies of mesocorticolimbic dopamine transmission, including addictive states, schizophrenia, and other psychoses. The working hypothesis is that normalizing dopamine transmission with either cannabinoid agonists or antagonists, depending on the type of dysfunction, would result in reducing addictive processes or producing antipsychotic effects.

Drug Addiction

The endocannabinoid transmission has been related to signs of drug addiction such as individual vulnerability, craving, degree of dependence and intensity of abstinence, or risk to relapse for different types of drugs (Comings et al., 1997; Colombo et al., 1998; Diana, Melis, Muntoni, et al., 1998; Tanda et al., 1999; De Vries et al., 2001, 2003; Fattore et al., 2003), which has opened the possibility of using cannabinoid-related substances in the treatment of different addictive states (for reviews, see Chaperon and Thiebot, 1999; Basavarajappa and Hungund, 2002; van den Brink and van Ree, 2003). Because of the role played by dopamine in addictive states, normalization of dopamine transmission may underlie the potential beneficial effects of cannabinoid-related substances in drug addiction, particularly in those processes more directly related to dopamine transmission such as reinforcement or relapse (Tanda et al., 1997) or withdrawal and dysphoria-mediated drug craving (Diana, Melis, Gessa, 1998). A compound that deserves a special mention is SR141716 (rimonabant), a selective CB_1 receptor antagonist, which has been suggested to be of therapeutic value for reducing drug-seeking behaviors involving various types of abused substances. The selective blockade of CB_1 receptors by SR141716 impaired the perception of the reinforcing potential of different habit-forming drugs (Chaperon and Thiebot, 1999), indicating that positive incentive and motivational processes could be under the permissive control of CB_1 receptor-related mechanisms (Chaperon and Thiebot, 1999). Similar findings were derived from studies in CB_1 receptor knockout mice (Ledent et al., 1999) and using electrical brain stimulation in rats (Deroche-Gamonet et al., 2001). The authors suggested that endocannabinoid transmission would facilitate the effects of several reinforcers (Deroche-Gamonet et al., 2001), a fact also found for drug relapse (De Vries et al., 2001, 2003). However, other studies revealed only modest effects using electrical brain stimulation in rats (Arnold et al., 2001).

Rimonabant might be effective in the case of alcohol (Colombo et al., 1998; Freedland et al., 2001; Serra et al., 2002; Hungund et al., 2003), morphine (Mas-Nieto et al., 2001; Fattore et al., 2003; Solinas et al., 2003; De Vries et al., 2003), cocaine (De Vries et al., 2001), and nicotine (Cohen et al., 2002). A clinical trial is presently underway for the use of this compound in tobacco addiction, based on preclinical evidence indicating that: (1) SR141716 blocked the motivational and dopamine-releasing effect of nicotine in rats (Cohen et al., 2002), (2) chronic nicotine administration increased endocannabinoid contents and CB_1 receptor densities in various brain regions involved in classic effects of nicotine (González, Cascio, et al., 2002), (3) CB_1 receptor knockout mice exhibited enhanced motor and antinociceptive responses following acute nicotine administration, whereas the rewarding effects of this drug disappeared without affecting abstinence responses (Castañé et al., 2002), and (4) by contrast, Δ^9-THC-induced hypothermia, antinociception, and hypolocomotion, as well as Δ^9-THC tolerance and physical dependence, were facilitated by nicotine (Valjent et al., 2002). However, not all studies have focused on CB_1 receptor antagonists. Agonists have also been reported to be useful in the treatment of specific signs of addictive states (for reviews, see Manzanares et al., 1999; Cichewicz and Welch, 2003), particularly those involving drugs such as opioids, which are mainly depressors of the nervous system. The capability of cannabinoid agonists to be anxiolytic enables these compounds to be used to reduce morphine withdrawal signs (Vela et al., 1995). However, it is unlikely that this effect is related to influencing dopamine transmission.

One of the options where cannabinoid-based compounds might be useful is with reward deficiency syndrome, a disorder characterized by an increase of affected individuals in the risk to become addicted *per se* or through increasing the influence of other risk factors (for a review, see Comings and Blum, 2000). Reward deficiency syndrome is a poligenic disorder caused by the occurrence of genetic variants in proteins involved in dopamine function, such as tyrosine hydroxylase, monoamine oxidase, dopamine transporter, or D_1 and D_2 receptors, and also for other neurotransmitters connected with dopamine including endocannabinoids (for a review, see Comings and Blum, 2000). In fact, polymorphisms of the CB_1 receptor and FAAH have been recently described (Gadzicki et al., 1999; Sipe et al., 2002; for a review, see Onaivi et al., 2002), and their higher occurrence in several disorders, including addiction to specific drugs (alcohol, opioids) documented (Comings et al., 1997; Sipe et al., 2002; Schmidt et al., 2002; Ponce et al., 2003; Preuss et al., 2003).

Schizophrenia

Dopaminergic anomalies are classic in several psychotic disorders, mainly schizophrenia, a disease which has also been related to cannabinoids from genetic, pharmacological, and pathological points of view (Hollister, 1998). By one side, the brain structures involved in the pathogenesis of schizophrenia (limbic areas and prefrontal cortex) contain a moderate but significant density of CB_1 receptors (Herkenham, Lynn, Little, et al., 1991; Mailleux and Vanderhaeghen, 1992; Tsou, Brown, et al., 1998). They also contain relevant amounts of endocannabinoids (Bisogno et al., 1999). Second, cannabis consumption could induce or exacerbate psychosis, and it has been claimed as a factor of risk in the incidence of this disorder, particularly in the case of heavy abusers and predisposed individuals (Andreasson et al., 1987; Linszen et al., 1994; Hollister, 1998; Negrete and Gill, 1999; Johns, 2001). However, this has not been confirmed in other studies (Arseneault et al., 2004). Possibly, this is the result of an inadequate study of the association between cannabis and psychosis, as the consumption of cannabis might be considered both as a factor eliciting the disease (Andreasson et al., 1987; Linszen et al., 1994; Hollister, 1998; Negrete and Gill, 1999; Johns, 2001) or as a self-medicine used by psychotic patients to overcome the unpleasant feelings produced by antipsychotic therapy (Voruganti et al., 1997) as they usually do with tobacco (Vanable et al., 2003). In this respect, a recent *in vivo* SPECT study by Voruganti et al. (2001) has demonstrated that cannabis smoking in schizophrenic patients produced an immediate calming effect, although followed by a worsening of psychotic symptoms. The calming effect correlated with a reduction in striatal D_2 receptor binding that the authors interpreted as suggestive of increased dopamine activity (Voruganti et al., 2001). All this evidence is indicative of the occurrence of an interaction between cannabinoids and dopamine in schizophrenia at the level of regions involved in psychotic symptoms (Dean et al., 2003). In addition, several studies have demonstrated that cannabinoids interfere with classic antipsychotic therapy, thus reducing their capability to block D_2 receptors (Knudsen and Vilmar, 1984; Negrete and Gill, 1999). This is compatible with the idea that cannabinoids increase dopamine transmission at the nucleus accumbens (see preceding text) and that dopaminergic hyperactivity in the nucleus accumbens is a key event in schizophrenia (for a review, see Lewis and Lieberman, 2000).

As described in the preceding text for dopamine-cannabinoid interactions at the basal ganglia level, there is also a bidirectional interaction here. Thus, it has been also suggested that the pharmacological blockade of CB_1 receptor might provide some type of benefit in schizophrenia and related psychosis (D'Souza and Kosten, 2001), which is also supported by the fact that the levels of anandamide in the CSF (Leweke et al., 1999) or in the blood (De Marchi et al., 2003), as well as the density of CB_1 receptors in cortical and subcortical structures (Dean et al., 2001; Zavitsanou et al., 2004), were increased in schizophrenic patients. In addition, association between frequency of a polymorphism of the CB_1 receptor gene and increased susceptibility to schizophrenia has been also reported (Leroy et al., 2001; Ujike et al., 2002), thus suggesting that the endocannabinoid signaling system might play a role in the pathogenesis of schizophrenia, at least in a subgroup of schizophrenic individuals (Emrich et al., 1997). This is the so-called endocannabinoid hypothesis of schizophrenia. According to this hypothesis, the increase in endocannabinoid transmission in cortical and subcortical (limbic) regions might be associated with hyperactivity of dopaminergic neurons (positive symptoms) and hypoactivity of glutamate neurons (negative symptoms) (for a review, see van der Stelt and Di Marzo, 2003). In contrast with all the above data, mainly related to humans, recent evidence has proposed that CB_1 receptor knockout mice might theoretically be a useful model of schizophrenia, based on the observation that these mice exhibited behavioral alterations that parallel schizophrenic symptoms, as well as considering the existence of correlations between CB_1 and D_2 receptor genes (Fritzsche, 2001). However, there is no further experimental work addressed to evaluate this proposal.

INTERACTIONS BETWEEN CANNABINOIDS AND DOPAMINE AT THE HYPOTHALAMIC LEVEL

Dopamine is also an important regulatory neurotransmitter in the neuroendocrine hypothalamus, where it belongs to the group of different hypothalamic factors that, released in the median eminence to the portal-hypophysial blood supply, are able to reach the anterior pituitary to control the synthesis and release of anterior pituitary hormones, mainly prolactin (for a review, see Ben-Jonathan and Hnasko, 2001). Therefore, it represents a target neurotransmitter for those compounds, either endogenous or exogenous, that interfere with anterior pituitary hormones. This is the case with cannabinoids, which are able to decrease prolactin and gonadotrophin secretion while increasing corticotrophin release in laboratory animals (Rettori et al., 1988; Romero et al., 1994; Wenger et al., 1994; Weidenfeld et al., 1994; Fernández-Ruiz et al., 1997; de Miguel et al., 1998) and also in humans (for a review, see Brown and Dobs, 2002).

FUNCTION OF THE ENDOCANNABINOID SYSTEM IN THE NEUROENDOCRINE HYPOTHALAMUS

As mentioned in the preceding text, classic cannabinoids, such as Δ^9-THC, and also endogenous cannabinoids, such as anandamide, decreased the release of most of anterior pituitary hormones, including prolactin (Rettori et al., 1988; Romero et al., 1994; Wenger et al., 1994; Fernández-Ruiz et al., 1997; de Miguel et al., 1998; for reviews, see Murphy et al., 1998; Wenger and Moldrich, 2002), luteinizing hormone (Murphy et al., 1990, 1994; Fernández-Ruiz et al., 1997; de Miguel et al., 1998; for reviews, see Murphy et al., 1998; Wenger and Moldrich, 2002), growth hormone (Rettori et al., 1988; for reviews, see Murphy et al., 1998; Wenger and Moldrich, 2002), thyrotropin (Hillard et al., 1984; for a review, see Murphy et al., 1998), and, to a lesser extent, follicle-stimulating hormone (Fernández-Ruiz et al., 1997; de Miguel et al., 1998; for a review, see Murphy et al., 1998; Wenger and Moldrich, 2002), with only the exception of corticotrophin, which was stimulated by cannabinoids (Weidenfeld et al., 1994; for a review, see Murphy et al., 1998). This occurred mainly in male animals, the effects in females being slightly different and depending on the ovarian cycle (Bonnin et al., 1993; Rodríguez de Fonseca et al., 1994; Scorticati et al., 2003). Numerous data have demonstrated that these effects likely originate through the capability of cannabinoids to interfere with hypothalamic factors controlling anterior pituitary hormones, among them the tubero-infundibular dopaminergic system but also other neurotransmitters (Rodríguez de Fonseca et al., 1992; Bonnin et al., 1993; Murphy et al., 1994; de Miguel et al., 1998; for reviews, see Murphy et al., 1998; Wenger and Moldrich, 2002). These effects are produced through the activation of CB_1 receptors, as revealed by experiments with selective antagonists for this receptor subtype (Fernández-Ruiz et al., 1997; de Miguel et al., 1998). These receptors are present in several hypothalamic nuclei, in particular in the ventromedial hypothalamic nucleus (Herkenham, Lynn, Little, et al., 1991; Fernández-Ruiz et al., 1997; Romero, Wenger, et al., 1998). However, a direct action of cannabinoids on anterior pituitary cells seems to be also possible. This is based on the demonstration of endocannabinoid synthesis (González, Manzanares, et al., 1999) and the presence of CB_1 receptor binding (Lynn and Herkenham, 1994), gene expression (González, Manzanares, et al., 1999), and immunoreactivity (Wenger, Fernandez-Ruiz, et al., 1999; González et al., 2000) in the anterior pituitary gland. The immunoreactivity for the CB_1 receptor is located in prolactin- and luteinizing-hormone-containing cells (Wenger, Fernandez-Ruiz, et al., 1999; González et al., 2000). In addition, *in vitro* experiments have revealed that cannabinoid agonists were able to exert direct effects on the release of different anterior pituitary hormones from cultured cells or pituitary explants (Murphy et al., 1991; Rodríguez de Fonseca et al., 1999). In any case, independently of the mechanisms used by cannabinoids to alter anterior pituitary hormone secretion, direct effects, and modulation of hypothalamic influences to the anterior pituitary, it appears well demonstrated that these compounds are able to control the circulating levels of these hormones and then to affect the activity of different peripheral glands.

Involvement of Dopamine in Neuroendocrine Effects of Cannabinoids

In a similar way as with basal ganglia, the involvement of dopamine in the neuroendocrine effects of cannabinoid agonists is indirect, as it does not appear that CB_1 receptors identified in the medial basal hypothalamus are located in tuberoinfundibular dopaminergic neurons. In fact, the phenotype of neurons containing CB_1 receptors in this region has remained elusive. It is known that (1) these neurons are intrinsic to the hypothalamus because the deafferentation of this structure does not accompany any decrease in the density of CB_1 receptors (Romero, Wenger, et al., 1998); (2) CB_1 receptor density is scarce but detectable in the arcuate nucleus (Fernández-Ruiz et al., 1997) and it is also present in other nondopaminergic structures of the hypothalamus, such as the paraventricular and periventricular nuclei, and the medial preoptic area (Fernández-Ruiz et al., 1997); and (3) the cell bodies of these neurons seem to be exclusively located in the ventromedial hypothalamus nucleus, as this structure, though exhibiting scarce specific binding for CB_1 receptors (Herkenham, Lynn, Little, et al., 1991), excels by a significant amount of mRNA transcripts for this receptor that are not present in other hypothalamic structures (Mailleux and Vanderhaeghen, 1992). Tuberoinfundibular dopaminergic neurons would adjust the two first conditions but not the last one; therefore, it is little probable that they contain CB_1 receptors. These receptors might be better located in GABAergic neurons (de Miguel et al., 1998) or, alternatively or complementarily, in opioidergic neurons (Drolet et al., 2001), which are also intrinsic to the hypothalamus. Opioidergic neurons are located in the ventromedial hypothalamus nucleus (Drolet et al., 2001), correlating with the presence of CB_1 receptor–mRNA (Mailleux and Vanderhaeghen, 1992). In addition, cannabinoids are able to modify the gene expression for opioid peptides in the hypothalamus (Corchero et al., 1999), which would support the option of location of CB_1 receptors in opioidergic neurons. However, no double-labeling studies have been performed yet in order to clarify this question. In any case, it appears clear that the hypothalamic neurons containing CB_1 receptors would be intimately connected with dopaminergic ones, as the activation of these receptors have been reported to significantly activate hypothalamic dopamine inputs, as reflected by an increase in tyrosine hydroxylase activity (Romero et al., 1994), the contents of dopamine and metabolites (Rodríguez de Fonseca et al., 1992; Bonnin et al., 1993; Fernández-Ruiz et al., 1997; Arévalo et al., 2001), and dopamine turnover (Rodríguez de Fonseca et al., 1992; Scorticati et al., 2003). Therefore, it can be concluded that the effects of cannabinoids on prolactin secretion were mediated by an activation of intrahypothalamic dopaminergic neurons, although the effects were not directly exerted on these neurons (Fernández-Ruiz et al., 1997) and involved alterations in other hypothalamic neurotransmitters (de Miguel et al., 1998; Corchero et al., 1999; Scorticati et al., 2003).

Therapeutic Implications of Cannabinoid–Dopamine Interactions in the Hypothalamus

This is probably another interesting aspect of the cannabinoid-dopamine interaction, because dysfunctions in the activity of dopaminergic neurons in the medial basal hypothalamus, among other causes, have been involved in the development of several types of pituitary disorders, including the development of prolactin-secreting adenomas (for a review, see Colao et al., 2002). In this sense, Di Marzo and co-workers (Pagotto et al., 2001) and our group (González et al., 2000) have recently reported the occurrence of changes in hypothalamic–pituitary endocannabinoid activity associated with the development of pituitary tumors. Unfortunately, there are no relevant pharmacological data yet supporting the observation that the interactions between cannabinoids and dopamine at the hypothalamic-pituitary level might have therapeutic applications. However, based on the effects of cannabinoids on dopamine, it should be expected that cannabinoid agonists might be useful to attenuate the dopaminergic deficiency accompanying the hyperprolactinemic disorders (see Colao et al., 2002, for a review). This hypothesis will have to be explored in the future.

INTERACTIONS BETWEEN CANNABINOIDS AND DOPAMINE DURING BRAIN DEVELOPMENT

In contrast with the adult brain, where it is assumed that CB_1 receptors are not located on dopaminergic neurons (with the only exception being the dopaminergic endings reaching the nucleus accumbens) (Wenger et al., 2003), these neurons seem to contain CB_1 receptors during specific periods of brain development, especially around the last week of gestation in rats (Hernández et al., 2000). Interestingly, the activation of these receptors would be associated with an induction of the gene-encoding tyrosine hydroxylase, the rate-limiting enzyme in the synthesis of dopamine, and other catecholamines (Hernández et al., 1997, 2000; see Fernández-Ruiz et al., 2000, for a review). This action, in addition to the control of other key genes for neural development such as proenkephalin (Pérez-Rosado et al., 2000), the neural adhesion molecule L1 (Gómez et al., 2003), or the apoptosis regulators bcl-2 and bax (Fernández-Ruiz et al., 2004), might be one of the specific events of neural development where the endocannabinoid system would play a regulatory role (Berrendero et al., 1998; Hernández et al., 2000; Fernández-Ruiz et al., 2000). In addition, this might be the mechanism by which plant-derived cannabinoids are able to interfere with the maturation of dopaminergic neurons when consumed during the perinatal period (for a review, see Fernández-Ruiz et al., 1999).

Cannabinoids and Tyrosine Hydroxylase during Brain Development

Most of the neurotoxicological studies on the effects of cannabinoids when administered during the perinatal life have focused on dopamine transmission and, in particular, on their effects on tyrosine hydroxylase (for reviews, see Fernández-Ruiz et al., 2000). These studies have allowed us to suggest in the preceding text that the regulation of gene expression and function of tyrosine hydroxylase would be one of the roles played by the endocannabinoid signaling system during brain development (for a review, see Fernández-Ruiz et al., 2000). In favor of this option is the fact that tyrosine hydroxylase-containing cells, which mostly synthesize dopamine as mediator, emerge early during brain development (Spetch et al., 1981) and play a neurotrophic action (for a review, see Levitt et al., 1997). Tyrosine hydroxylase enzyme appears around the gestational day 12 in rats and is related to the differentiation of several groups of dopaminergic neurons (Specht et al., 1981). The enzyme is present in the growing axons before their contact with their target neurons and seems to play an important role, together with active receptors located onto the target neurons, in the formation of connectivity (axonal guidance, neuronal recognition, and synaptogenesis) (for a review, see Insel, 1995). There is evidence that some specific groups of tyrosine hydroxylase-containing neurons only exist during brain development (for reviews, see Emerit et al., 1992; Levitt et al., 1997), having a neurotrophic action in the maturation of dopaminergic neurons themselves and of other neurotransmitters (Berger et al., 1987; Emerit et al., 1992; Levitt et al., 1997; Viggiano et al., 2003). This makes tyrosine hydroxylase gene a potential target for the action of substances, such as cannabinoids, susceptible to influencing brain development when consumed during the perinatal life (for a review, see Levitt, 1998). If this hypothesis is correct, the exposure to plant-derived cannabinoids, by mimicking the effects of natural ligands of cannabinoid receptors, would interfere with the sequence of events in which the expression of tyrosine hydroxylase gene is involved during brain development, so that these potential imbalances caused by the exposure to plant-derived cannabinoids might contribute to the abnormal pre- and postnatal development of tyrosine hydroxylase–containing neurons themselves and other neurotransmitters (for a review, see Fernández-Ruiz et al., 1999).

We examined this hypothesis and found that cannabinoids were able to modify tyrosine hydroxylase gene expression showing effects that appeared early, even before the complete differentiation and maturation of dopaminergic projections into their target areas (Bonnin et al., 1996), in particular during the last third of gestation and first week after birth. Thus, we have seen that Δ^9-THC increased tyrosine hydroxylase gene expression in *in vivo* studies (Bonnin et al., 1996) and also *in vitro* by using cultured mesencephalic neurons obtained from fetal brains at gestational day 14 (Hernández et al., 2000). In these cells, cannabinoids increased the activity of tyrosine hydroxylase (Hernández et al., 1997) and elevated tyrosine hydroxylase-mRNA levels (Hernández et al., 2000), which coincide with the effects observed *in vivo* (Bonnin et al., 1996). During the postnatal and, in particular, the peripubertal period, there was a tendency towards a recovery of a normal expression and activity for this enzyme, although some effects of cannabinoids could still be evident, but they differed from those found at earlier ages and exhibited a notorious region-dependency (Bonnin et al., 1996; see Fernández-Ruiz et al., 1999, for a review). In any case, they were completely absent when animals reached adulthood (see Fernández-Ruiz et al., 1999, for a review), although we also found that animals that had been perinatally exposed to Δ^9-THC, mostly having similar basal indices for this enzyme, exhibited an abnormal ability to respond to drugs such as α-methyl-p-tyrosine, which when combined with reserpine, is able to demand the novo synthesis of tyrosine hydroxylase (for a review, see Fernández-Ruiz et al., 1999). The same differences were found for the pharmacological responses to other dopamine-acting agents (García-Gil et al., 1996), in particular for dopaminergic antagonists, which might be related to region-dependent changes in dopamine parameters other than tyrosine hydroxylase, such as D_1 and D_2 receptor density (Rodríguez de Fonseca et al., 1991), and dopamine and DOPAC contents (Rodríguez de Fonseca et al., 1991). This supports the existence of irreversible although silent changes in the adult functionality of dopamine-containing neurons due to early contact with Δ^9-THC, but that later did not depend on the presence of this cannabinoid.

On the other hand, these early effects of Δ^9-THC were probably mediated through the activation of CB_1 receptors, which emerge and are operative early in brain development (Berrendero et al., 1998). In addition, mesencephalic neurons contain CB_1 receptors which colocalized with tyrosine hydroxylase at gestational day 14 (Hernández et al., 2000). As these cells do not contain CB_1 receptors in the adult brain (Herkenham, Lynn, de Costa, et al., 1991), this indicates the transient nature of this atypical location and its potential relation with a specific role of these receptors in neural development.

CONCLUDING REMARKS

In this chapter, we have reviewed the recent advances on cannabinoid-dopamine interactions, emphasizing those processes in which dopamine has been proposed as a key neurotransmitter, such as basal ganglia functionality, corticolimbic processes, and neuroendocrine regulation. We have explored the mechanisms underlying these interactions, which represent ways for plant-derived cannabinoids to interfere with these processes. In most of the cases, we have concluded that dopaminergic neurons do not contain CB_1 receptors, with some exceptions, but these receptors are located on neurons present in regions innervated by dopaminergic neurons, which allows relevant bidirectional interactions. Lastly, we have reviewed those diseases characterized by either deficiency or overactivity of dopamine transmission and where cannabinoids might be of therapeutic potential possibly through actions that facilitate, among others, a normalization of dopamine transmission. These diseases included basal ganglia disorders (Huntington's and Parkinson's diseases), other motor-related disorders (multiple sclerosis and Alzheimer's disease), drug addiction, and schizophrenia and related psychoses.

REFERENCES

Alonso R, Voutsinos B, Fournier M, Labie C, Steinberg R, Souilhac J, Le Fur G, Soubrie P. Blockade of cannabinoid receptors by SR141716 selectively increases Fos expression in rat mesocorticolimbic areas via reduced dopamine D2 function. *Neuroscience* 91, 607–620 (1999).

Anderson JJ, Kask AM, Chase TN. Effects of cannabinoid receptor stimulation and blockade on catalepsy produced by dopamine receptor antagonists. *Eur. J. Pharmacol.* 295, 163–168 (1996).

Andreasson S, Allebeck P, Engstrom A, Rydberg U. Cannabis and schizophrenia. A longitudinal study of Swedish conscripts. *Lancet* 2, 1483–1486 (1987).

Arévalo C, de Miguel R, Hernandez-Tristan R. Cannabinoid effects on anxiety-related behaviours and hypothalamic neurotransmitters. *Pharmacol. Biochem. Behav.* 70, 123–131 (2001).

Arnold JC, Hunt GE, McGregor IS. Effects of the cannabinoid receptor agonist CP55,940 and the cannabinoid receptor antagonist SR 141716 on intracranial self-stimulation in Lewis rats. *Life Sci.* 70, 97–108 (2001).

Arseneault L, Cannon M, Witton J, Murray RM. Causal association between cannabis and psychosis: examination of the evidence. *Br. J. Psychiatry* 184, 110–117 (2004).

Baker D, Pryce G, Croxford JL, Brown P, Pertwee RG, Huffman JW, Layward L. Cannabinoids control spasticity and tremor in a multiple sclerosis model. *Nature* 404, 84–87 (2000).

Baker D, Pryce G, Croxford JL, Brown P, Pertwee RG, Makriyannis A, Khanolkar A, Layward L, Fezza F, Bisogno T, Di Marzo V. Endocannabinoids control spasticity in a multiple sclerosis model. *FASEB J.* 15, 300–302 (2001).

Baker D, Pryce G. The therapeutic potential of cannabis in multiple sclerosis. *Expert Opin. Invest. Drugs* 12, 561–567 (2003).

Basavarajappa BS, Hungund BL. Down-regulation of cannabinoid receptor agonist-stimulated [^{35}S]GTPγS binding in synaptic plasma membrane from chronic ethanol exposed mouse. *Brain Res.* 815, 89–97 (1999a).

Basavarajappa BS, Hungund BL. Chronic ethanol increases the cannabinoid receptor agonist anandamide and its precursor N-arachidonoylphosphatidylethanolamine in SK-N-SH cells. *J. Neurochem.* 72, 522–528 (1999b).

Basavarajappa BS, Hungund BL. Neuromodulatory role of the endocannabinoid signaling system in alcoholism: an overview. *Prostag. Leukot. Essent. Fatty Acid* 66, 287–299 (2002).

Beltramo M, Stella N, Calignano A, Lin SY, Makriyannis A, Piomelli D. Functional role of high-affinity anandamide transport, as revealed by selective inhibition. *Science* 277, 1094–1097 (1997).

Beltramo M, de Fonseca FR, Navarro M, Calignano A, Gorriti MA, Grammatikopoulos G, Sadile AG, Giuffrida A, Piomelli D. Reversal of dopamine D2 receptor responses by an anandamide transport inhibitor. *J. Neurosci.* 20, 3401–3407 (2000).

Benito C, Nunez E, Tolon RM, Carrier EJ, Rabano A, Hillard CJ, Romero J. Cannabinoid CB2 receptors and fatty acid amide hydrolase are selectively overexpressed in neuritic plaque-associated glia in Alzheimer's disease brains. *J. Neurosci.* 23, 11136–11141 (2003).

Ben-Jonathan N, Hnasko R. Dopamine as a prolactin (PRL) inhibitor. *Endocr. Rev.* 22, 724–763 (2001).

Berger B, Verney C, Gaspar P, Febvret A. Transient expression of tyrosine hydroxylase immunoreactivity in some neurones of the rat neocortex during postnatal development. *Dev. Brain Res.* 23, 141–147 (1987).

Berrendero F, Garcia-Gil L, Hernandez ML, Romero J, Cebeira M, de Miguel R, Ramos JA, Fernandez-Ruiz JJ. Localization of mRNA expression and activation of signal transduction mechanisms for cannabinoid receptor in rat brain during fetal development. *Development* 125, 3179–3188 (1998).

Berrendero F, Sánchez A, Cabranes A, Puerta C, Ramos JA, García-Merino A, Fernández-Ruiz JJ, Changes in cannabinoid CB1 receptors in striatal and cortical regions of rats with experimental allergic encephalomyelitis, an animal model of multiple sclerosis. *Synapse* 41, 195–202 (2001).

Berrendero F, Maldonado R. Involvement of the opioid system in the anxiolytic-like effects induced by Δ^9-tetrahydrocannabinol. *Psychopharmacology* 163, 111–117 (2002).

Berry EM, Mechoulam R. Tetrahydrocannabinol and endocannabinoids in feeding and appetite. *Pharmacol. Ther.* 95, 185–190 (2002).

Bisogno T, Berrendero F, Ambrosino G, Cebeira M, Ramos JA, Fernández-Ruiz JJ, Di Marzo V. Brain regional distribution of endocannabinoids: implications for their biosynthesis and biological function. *Biochem. Biophys. Res. Commn.* 256, 377–380 (1999).

Bonnin A, Ramos JA, Rodríguez de Fonseca F, Cebeira M, Fernandez-Ruiz JJ. Acute effects of Δ^9-tetrahydrocannabinol on tuberoinfundibular dopamine activity, anterior pituitary sensitivity to dopamine and prolactin release vary as a function of estrous cycle. *Neuroendocrinology* 58, 280–286 (1993).

Bonnin A, de Miguel R, Castro JG, Ramos JA, Fernandez-Ruiz JJ. Effects of perinatal exposure to Δ^9-tetrahydrocannabinol on the fetal and early postnatal development of tyrosine hydroxylase-containing neurons in rat brain. *J. Mol. Neurosci.* 7, 291–308 (1996).

Bowers MB Jr., Hoffman FJ Jr. Regional brain homovanillic acid following Δ^9-tetrahydrocannabinol and cocaine. *Brain Res.* 366, 405–407 (1986).

Breivogel CS, Childers SR. The functional neuroanatomy of brain cannabinoid receptors. *Neurobiol. Dis.* 5, 417–431 (1998).

Brooks DJ. Functional imaging studies on dopamine and motor control. *J. Neural Transm.* 108, 1283–1298 (2001).

Brooks JW, Pryce G, Bisogno T, Jaggar SI, Hankey DJ, Brown P, Bridges D, Ledent C, Bifulco M, Rice AS, Di Marzo V, Baker D. Arvanil-induced inhibition of spasticity and persistent pain: evidence for therapeutic sites of action different from the vanilloid TRPV1 receptor and cannabinoid CB1/CB2 receptors. *Eur. J. Pharmacol.* 439, 83–92 (2002).

Brotchie JM. The neural mechanisms underlying levodopa-induced dyskinesia in Parkinson's disease. *Ann. Neurol.* 47, S105–S112 (2000).

Brotchie JM. CB1 cannabinoid receptor signalling in Parkinson's disease. *Curr. Opin. Pharmacol.* 3, 54–61 (2003).

Brown TT, Dobs AS. Endocrine effects of marijuana. *J. Clin. Pharmacol.* 42, 90S–96S (2002).

Cabranes A, Venderova K, de Lago E, Fezza F, Sánchez A, Mestre, L, Valenti, M, García-Merino A, Ramos JA, Di Marzo V, Fernández-Ruiz JJ. Decreased endocannabinoid levels in the brain and beneficial effects of agents activating cannabinoid and/or vanilloid receptors in a rat model of multiple sclerosis. *J. Neuroimmunol.* In press (2005).

Cadogan AK, Alexander SP, Boyd EA, Kendall DA. Influence of cannabinoids on electrically evoked dopamine release and cyclic AMP generation in the rat striatum. *J. Neurochem.* 69, 1131–1137 (1997).

Castañé A, Valjent E, Ledent C, Parmentier M, Maldonado R, Valverde O. Lack of CB1 cannabinoid receptors modifies nicotine behavioural responses, but not nicotine abstinence. *Neuropharmacology* 43, 857–867 (2002).

Castellano C, Rossi-Arnaud C, Cestari V, Costanzi M. Cannabinoids and memory: animal studies. *Curr. Drug Target CNS Neurol. Disord.* 2, 389–402 (2003).

Chaperon F, Thiebot MH. Behavioral effects of cannabinoid agents in animals. *Crit. Rev. Neurobiol.* 13, 243–281 (1999).

Cheer JF, Marsden CA, Kendall DA, Mason R. Lack of response suppression follows repeated ventral tegmental cannabinoid administration: an in vitro electrophysiological study. *Neuroscience* 99, 661–667 (2000).

Cheer JF, Kendall DA, Mason R, Marsden CA. Differential cannabinoid-induced electrophysiological effects in rat ventral tegmentum. *Neuropharmacology* 44, 633–641 (2003).

Chen JP, Paredes W, Li J, Smith D, Lowinson J, Gardner EL. Δ^9-tetrahydrocannabinol produces naloxone-blockable enhancement of presynaptic basal dopamine efflux in nucleus accumbens of conscious, freely-moving rats as measured by intracerebral microdialysis. *Psychopharmacology* 102, 156–162 (1990).

Chen JP, Paredes W, Lowinson JH, Gardner EL. Δ^9-Tetrahydrocannabinol enhances presynaptic dopamine efflux in medial prefrontal cortex. *Eur. J. Pharmacol.* 190, 259–262 (1990).

Cichewicz DL, Welch SP. Modulation of oral morphine antinociceptive tolerance and naloxone-precipitated withdrawal signs by oral Δ^9-tetrahydrocannabinol. *J. Pharmacol. Exp. Ther.* 305, 812–817 (2003).

Cohen C, Perrault G, Voltz C, Steinberg R, Soubrie P. SR141716, a central cannabinoid (CB1) receptor antagonist, blocks the motivational and dopamine-releasing effects of nicotine in rats. *Behav. Pharmacol.* 13, 451–463 (2002).

Colao A, di Sarno A, Pivonello R, di Somma C, Lombardi G. Dopamine receptor agonists for treating prolactinomas. *Expert. Opin. Invest. Drugs* 11, 787–800 (2002).

Colombo G, Agabio R, Fa M, Guano L, Lobina C, Loche A, Reali R, Gessa GL. Reduction of voluntary ethanol intake in ethanol-preferring sP rats by the cannabinoid antagonist SR141716. *Alcohol Alcohol.* 33, 126–130 (1998).

Comings DE, Muhleman D, Gade R, Johnson P, Verde R, Saucier G, MacMurray J. Cannabinoid receptor gene (CNR1): association with i.v. drug use. *Mol. Psychiatry.* 2, 161–168 (1997).

Comings DE, Blum K. Reward deficiency syndrome: genetic aspects of behavioral disorders. *Prog Brain Res.* 126, 325–341 (2000).

Compton DR, Aceto MD, Lowe J, Martin BR. In vivo characterization of a specific cannabinoid receptor antagonist (SR141716A): inhibition of Δ^9-tetrahydrocannabinol-induced responses and apparent agonist activity. *J Pharmacol. Exp. Ther.* 277, 586–594 (1996).

Consroe P. Brain cannabinoid systems as targets for the therapy of neurological disorders. *Neurobiol. Dis.* 5, 534–551 (1998).

Corchero J, Manzanares J, Fuentes JA. Repeated administration of Δ^9-tetrahydrocannabinol produces a differential time related responsiveness on proenkephalin, proopiomelanocortin and corticotropin releasing factor gene expression in the hypothalamus and pituitary gland of the rat. *Neuropharmacology* 38, 433–439 (1999).

Costanzi M, Battaglia M, Rossi-Arnaud C, Cestari V, Castellano C. Effects of anandamide and morphine combinations on memory consolidation in cd1 mice: Involvement of dopaminergic mechanisms. *Neurobiol. Learn. Mem.* 81, 144–149 (2004).

Cossu G, Ledent C, Fattore L, Imperato A, Bohme GA, Parmentier M, Fratta W. Cannabinoid CB1 receptor knockout mice fail to self-administer morphine but not other drugs of abuse. *Behav. Brain Res.* 118, 61–65 (2001).

Cravatt BF, Demarest K, Patricelli MP, Bracey MH, Giang DK, Martin BR, Lichtman AH. Supersensitivity to anandamide and enhanced endogenous cannabinoid signaling in mice lacking fatty acid amide hydrolase. *Proc. Natl. Acad. Sci. U.S.A.* 98, 9371–9376 (2001).

Dawson TM, Dawson VL. Molecular pathways of neurodegeneration in Parkinson's disease. *Science* 302, 819–822 (2003).

de Lago E, Fernandez-Ruiz J, Ortega-Gutierrez S, Viso A, Lopez-Rodriguez ML, Ramos JA. UCM707, a potent and selective inhibitor of endocannabinoid uptake, potentiates hypokinetic and antinociceptive effects of anandamide. *Eur. J. Pharmacol.* 449, 99–103 (2002).

de Lago E, de Miguel, Lastres-Becker I, Ramos JA, Fernández-Ruiz JJ. Involvement of vanilloid-like receptors in the effects of anandamide on motor behavior and nigrostriatal dopaminergic activity: in vivo and in vitro evidence. *Brain Res.* 1007, 152–159 (2004).

de Lago E, Ligresti A, Ortar G, Morera E, Cabranes A, Pryce G, Bifulco M, Baker D, Fernandez-Ruiz J, Di Marzo V. In vivo pharmacological actions of two novel inhibitors of anandamide cellular uptake. *Eur. J. Pharmacol.* 484, 249–257 (2004).

De Marchi N, De Petrocellis L, Orlando P, Daniele F, Fezza F, Di Marzo V. Endocannabinoid signalling in the blood of patients with schizophrenia. *Lipids Health Dis.* 2, 5 (2003).

de Miguel R, Romero J, Munoz RM, Garcia-Gil L, Gonzalez S, Villanua MA, Makriyannis A, Ramos JA, Fernandez-Ruiz JJ. Effects of cannabinoids on prolactin and gonadotrophin secretion: involvement of changes in hypothalamic γ-aminobutyric acid (GABA) inputs. *Biochem. Pharmacol.* 56, 1331–1338 (1998).

De Vries TJ, Shaham Y, Homberg JR, Crombag H, Schuurman K, Dieben J, Vanderschuren LJMJ, Schoffelmeer ANM. A cannabinoid mechanism in relapse to cocaine seeking. *Nature Med.* 7, 1151–2254 (2001).

De Vries TJ, Homberg JR, Binnekade R, Raaso H, Schoffelmeer AN. Cannabinoid modulation of the reinforcing and motivational properties of heroin and heroin-associated cues in rats. *Psychopharmacology* 168, 164–169 (2003).

Dean B, Sundram S, Bradbury R, Scarr E, Copolov D. Studies on [^{3}H]CP-55940 binding in the human central nervous system: regional specific changes in density of cannabinoid-1 receptors associated with schizophrenia and cannabis use. *Neuroscience* 103, 9–15 (2001).

Dean B, Bradbury R, Copolov DL. Cannabis-sensitive dopaminergic markers in postmortem central nervous system: changes in schizophrenia. *Biol. Psychiatry* 53, 585–592 (2003).

Denovan-Wright EM, Robertson HA. Cannabinoid receptor messenger RNA levels decrease in a subset of neurons of the lateral striatum, cortex and hippocampus of transgenic Huntington's disease mice. *Neuroscience* 98, 705–713 (2000).

Deroche-Gamonet V, Le Moal M, Piazza PV, Soubrie P. SR141716, a CB1 receptor antagonist, decreases the sensitivity to the reinforcing effects of electrical brain stimulation in rats. *Psychopharmacology* 157, 254–259 (2001).

Di Carlo G, Izzo AA. Cannabinoids for gastrointestinal diseases: potential therapeutic applications. *Expert Opin. Invest. Drugs* 12, 39–49 (2003).

Di Marzo V, Melck D, Bisogno T, De Petrocellis L. Endocannabinoids: endogenous cannabinoid receptor ligands with neuromodulatory action. *Trends Neurosci.* 21, 521–528 (1998).

Di Marzo V, Hill MP, Bisogno T, Crossman AR, Brotchie JM. Enhanced levels of endogenous cannabinoids in the globus pallidus are associated with a reduction in movement in an animal model of Parkinson's disease. *FASEB J.* 14, 1432–1438 (2000).

Di Marzo V, Lastres-Becker I, Bisogno T, De Petrocellis L, Milone A, Davis JB, Fernandez-Ruiz JJ. Hypolocomotor effects in rats of capsaicin and two long chain capsaicin homologues. *Eur. J. Pharmacol.* 420, 123–131 (2001).

Di Marzo V, De Petrocellis L, Fezza F, Ligresti A, Bisogno T. Anandamide receptors. *Prostaglandins Leukot. Essent. Fatty Acid.* 66, 377–391 (2002).

Diana M, Melis M, Muntoni AL, Gessa GL, Mesolimbic dopaminergic decline after cannabinoid withdrawal. *Proc. Natl. Acad. Sci. U.S.A.* 95, 10269–10273 (1998).

Diana M, Melis M, Gessa GL. Increase in meso-prefrontal dopaminergic activity after stimulation of CB1 receptors by cannabinoids. *Eur. J. Neurosci.* 10, 2825–2830 (1998).

Drolet G, Dumont EC, Gosselin I, Kinkead R, Laforest S, Trottier JF. Role of endogenous opioid system in the regulation of the stress response. *Prog. Neuropsychopharmacol. Biol. Psychiatry* 25, 729–741 (2001).

D'Souza DC, Kosten TR. Cannabinoid antagonists: a treatment in search of an illness. *Arch. Gen. Psychiatry* 58, 330–331 (2001).

Emerit MB, Riad M, Hamon M. Trophic effects of neurotransmitters during brain maturation. *Biol. Neonate* 62, 193–201 (1992).

Emrich HM, Leweke FM, Schneider U. Towards a cannabinoid hypothesis of schizophrenia: cognitive impairments due to dysregulation of the endogenous cannabinoid system. *Pharmacol. Biochem. Behav.* 56, 803–807 (1997).

Fattore L, Spano MS, Cossu G, Deiana S, Fratta W. Cannabinoid mechanism in reinstatement of heroin-seeking after a long period of abstinence in rats. *Eur. J. Neurosci.* 17, 1723–1726 (2003).

Fernández-Espejo E, Caraballo I, Rodríguez de Fonseca F, El Banoua F, Ferrer B, Flores JA, Galán-Rodríguez B. Cannabinoid CB_1 antagonists possess antiparkinsonian efficacy only in rats with very severe nigral lesion in experimental parkinsonism. *Neurobiol. Dis.* 18, 591–601 (2005).

Fernández-Ruiz JJ, Muñoz RM, Romero J, Villanua MA, Makriyannis A, Ramos JA. Time course of the effects of different cannabimimetics on prolactin and gonadotrophin secretion: evidence for the presence of CB1 receptors in hypothalamic structures and their involvement in the effects of cannabimimetics. *Biochem. Pharmacol.* 53, 1919–1928 (1997).

Fernández-Ruiz JJ, Berrendero F, Hernandez ML, Romero J, Ramos JA. Role of endocannabinoids in brain development. *Life Sci.* 65, 725–736 (1999).

Fernández-Ruiz JJ, Berrendero F, Hernández ML, Ramos JA. The endogenous cannabinoid system and brain development. *Trends Neurosci.* 23, 14–20 (2000).

Fernández-Ruiz JJ, Lastres-Becker I, Cabranes A, González S, Ramos JA. Endocannabinoids and basal ganglia functionality. *Prostag. Leukot Essent. Fatty Acid.* 66, 263–273 (2002).

Fernández-Ruiz J, Gómez M. Hernández ML, de Miguel R, Ramos JA. Cannabinoids and gene expression during brain development. *Neurotox. Res.* 6, 389–401 (2004).

Fox SH, Kellett M, Moore AP, Crossman AR, Brotchie JM. Randomised, double-blind, placebo-controlled trial to assess the potential of cannabinoid receptor stimulation in the treatment of dystonia. *Mov. Disord.* 17, 145–149 (2002).

Fox SH, Henry B, Hill M, Crossman A, Brotchie J. Stimulation of cannabinoid receptors reduces levodopa-induced dyskinesia in the MPTP-lesioned nonhuman primate model of Parkinson's disease. *Mov. Disord.* 17, 1180–1187 (2002).

Freedland CS, Sharpe AL, Samson HH, Porrino LJ. Effects of SR141716A on ethanol and sucrose self-administration. *Alcohol Clin. Exp. Res.* 25, 277–282 (2001).

French ED, Dillon K, Wu X. Cannabinoids excite dopamine neurons in the ventral tegmentum and substantia nigra. *Neuroreport* 8, 649–652 (1997).

Freund TF, Katona I, Piomelli D. Role of endogenous cannabinoids in synaptic signaling. *Physiol. Rev.* 83, 1017–1066 (2003).

Fritzsche M. Are cannabinoid receptor knockout mice animal models for schizophrenia? *Med. Hypotheses* 56, 638–643 (2001).

Gadzicki D, Muller-Vahl K, Stuhrmann M. A frequent polymorphism in the coding exon of the human cannabinoid receptor (CNR1) gene. *Mol. Cell. Probes* 13, 321–323 (1999).

Garcia-Gil L, de Miguel R, Ramos JA, Fernandez-Ruiz JJ. Perinatal Δ^9-tetrahydrocannabinol exposure in rats modifies the responsiveness of midbrain dopaminergic neurons in adulthood to a variety of challenges with dopaminergic drugs. *Drug Alcohol Depend.* 42, 155–166 (1996).

Gardner EL, Paredes W, Smith D, Donner A, Milling C, Cohen D, Morrison D. Facilitation of brain stimulation reward by Δ^9-tetrahydrocannabinol. *Psychopharmacology* 96, 142–144 (1988).

Gardner EL, Vorel SR. Cannabinoid transmission and reward-related events. *Neurobiol. Dis.* 5, 502–533 (1998).

Gardner EL. Addictive potential of cannabinoids: the underlying neurobiology. *Chem. Phys. Lipids* 121, 267–290 (2002).

Gessa GL, Melis M, Muntoni AL, Diana M. Cannabinoids activate mesolimbic dopamine neurons by an action on cannabinoid CB1 receptors. *Eur. J. Pharmacol.* 341, 39–44 (1998).

Giuffrida A, Parsons LH, Kerr TM, Rodriguez de Fonseca F, Navarro M, Piomelli D. Dopamine activation of endogenous cannabinoid signaling in dorsal striatum. *Nat. Neurosci.* 2, 358–363 (1999).

Giuffrida A, Beltramo M, Piomelli D. Mechanisms of endocannabinoid inactivation: biochemistry and pharmacology. *J. Pharmacol. Exp. Ther.* 298, 7–14 (2001).

Glass M, Faull RL, Dragunow M. Loss of cannabinoid receptors in the substantia nigra in Huntington's disease. *Neuroscience* 56, 523–527 (1993).

Glass M, Dragunow M, Faull RL. The pattern of neurodegeneration in Huntington's disease: a comparative study of cannabinoid, dopamine, adenosine and GABA-A receptor alterations in the human basal ganglia in Huntington's disease. *Neuroscience* 97, 505–519 (2000).

Goldman-Rakic PS. Cellular basis of working memory. *Neuron* 14, 477–485 (1995).

Gómez M, Hernández ML, Johansson B, de Miguel R, Ramos JA, Fernández-Ruiz J. Prenatal cannabinoid exposure and gene expression for neural adhesion molecule L1 in the fetal rat brain. *Dev. Brain Res.* 147, 201–207 (2003).

González S, Romero J, de Miguel R, Lastres-Becker I, Villanua MA, Makriyannis A, Ramos JA, Fernandez-Ruiz JJ. Extrapyramidal and neuroendocrine effects of AM404, an inhibitor of the carrier-mediated transport of anandamide. *Life Sci.* 65, 327–336 (1999).

González S, Manzanares J, Berrendero F, Wenger T, Corchero J, Bisogno T, Romero J, Fuentes JA, Di Marzo V, Ramos JA, Fernández-Ruiz JJ. Identification of endocannabinoids and cannabinoid CB1 receptor mRNA in the pituitary gland. *Neuroendocrinology* 70, 137–144 (1999).

González S, Mauriello-Romanazzi G, Berrendero F, Ramos JA, Franzoni MF, Fernandez-Ruiz J. Decreased cannabinoid CB1 receptor mRNA levels and immunoreactivity in pituitary hyperplasia induced by prolonged exposure to estrogens. *Pituitary* 3, 221–226 (2000).

González S, Fernández-Ruiz JJ, Sparpaglione V, Parolaro D, Ramos JA. Chronic exposure to morphine, cocaine or ethanol in rats produced different effects in brain cannabinoid CB1 receptor binding and mRNA levels. *Drug Alcohol. Depend.* 66, 77–84 (2002).

González S, Cascio MG, Fernandez-Ruiz J, Fezza F, Di Marzo V, Ramos JA. Changes in endocannabinoid contents in the brain of rats chronically exposed to nicotine, ethanol or cocaine. *Brain Res.* 954, 73–81 (2002).

González S, Schmid PC, Fernández-Ruiz JJ, Krebsbach R, Schmid HHO, Ramos JA. Region-dependent changes in endocannabinoid transmission in the brain of morphine-dependent rats. *Addict Biol.* 8, 159–166 (2003).

Gorriti MA, Rodriguez de Fonseca F, Navarro M, Palomo T. Chronic (-)-Δ^9-tetrahydrocannabinol treatment induces sensitization to the psychomotor effects of amphetamine in rats. *Eur. J. Pharmacol.* 365, 133–142 (1999).

Grundy RI, Rabuffeti M, Beltramo M. Cannabinoids and neuroprotection. *Mol. Neurobiol.* 24, 29–52 (2001).

Gubellini P, Picconi B, Bari M, Battista N, Calabresi P, Centonze D, Bernardi G, Finazzi-Agro A, Maccarrone M. Experimental parkinsonism alters endocannabinoid degradation: implications for striatal glutamatergic transmission. *J. Neurosci.* 22, 6900–6907 (2002).

Guzmán M, Sanchez C. Effects of cannabinoids on energy metabolism. *Life Sci.* 65, 657–664 (1999).

Hampson RE, Deadwyler SA. Cannabinoids, hippocampal function and memory. *Life Sci.* 65, 715–723 (1999).

Herkenham M, Lynn AB, Little MD, Melvin LS, Johnson MR, de Costa DR, Rice KC. Characterization and localization of cannabinoid receptors in rat brain: a quantitative in vitro autoradiographic study. *J. Neurosci.* 11, 563–583 (1991).

Herkenham M, Lynn AB, de Costa BR, Richfield EK. Neuronal localization of cannabinoid receptors in the basal ganglia of the rat. *Brain Res.* 547, 267–264 (1991).

Hermann H, Marsicano G, Lutz B. Coexpression of the cannabinoid receptor type 1 with dopamine and serotonin receptors in distinct neuronal subpopulations of the adult mouse forebrain. *Neuroscience* 109, 451–460 (2002).

Hernández ML, Garcia-Gil L, Berrendero F, Ramos JA, Fernandez-Ruiz JJ. Δ^9-Tetrahydrocannabinol increases activity of tyrosine hydroxylase in cultured fetal mesencephalic neurons. *J. Mol. Neurosci.* 8, 83–91 (1997).

Hernández ML, Berrendero F, Suarez I, Garcia-Gil L, Cebeira M, Mackie K, Ramos JA, Fernández-Ruiz JJ. Cannabinoid CB1 receptors colocalize with tyrosine hydroxylase in cultured fetal mesencephalic neurons and their activation increases the levels of this enzyme. *Brain Res.* 857, 56–65 (2000).

Hillard CJ, Farber NE, Hagen TC, Bloom AS. The effects of Δ^9-tetrahydrocannabinol on serum thyrotropin levels in the rat. *Pharmacol. Biochem. Behav.* 20, 547–550 (1984).

Hillard CJ. Endocannabinoids and vascular function. *J. Pharmacol. Exp. Ther.* 294, 27–32 (2000).

Hoffman AF, Lupica CR. Direct actions of cannabinoids on synaptic transmission in the nucleus accumbens: a comparison with opioids. *J. Neurophysiol.* 85, 72–83 (2001).

Hollister LE. Health aspects of cannabis: revisited. *Int. J. Neuropsychopharmacol* 1, 71–80 (1998).

Hornykiewicz O. Biochemical aspects of Parkinson's disease. *Neurology* 51, S2–S9 (1998).

Hungund BL, Szakall I, Adam A, Basavarajappa BS, Vadasz C. Cannabinoid CB1 receptor knockout mice exhibit markedly reduced voluntary alcohol consumption and lack alcohol-induced dopamine release in the nucleus accumbens. *J. Neurochem.* 84, 698–704 (2003).

Insel TR. The development of brain and behavior. in *Psychopharmacology: The Fourth Generation of Progress.* Bloom FE and Kupfer DJ (Eds.). Raven Press, New York, pp. 683–694 (1995).

Iuvone T, Esposito G, Esposito R, Santamaria R, Di Rosa M, Izzo AA. Neuroprotective effect of cannabidiol, a non-psychoactive component from Cannabis sativa, on beta-amyloid-induced toxicity in PC12 cells. *J. Neurochem.* 89, 134–141 (2004).

Iversen L. Cannabis and the brain. *Brain* 126, 1252–1270 (2003).

Jentsch JD, Andrusiak E, Tran A, Bowers MB Jr, Roth RH. Δ^9-tetrahydrocannabinol increases prefrontal cortical catecholaminergic utilization and impairs spatial working memory in the rat: blockade of dopaminergic effects with HA966. *Neuropsychopharmacology* 16, 426–432 (1997).

Jentsch JD, Verrico CD, Le D, Roth RH. Repeated exposure to Δ^9-tetrahydrocannabinol reduces prefrontal cortical dopamine metabolism in the rat. *Neurosci Lett.* 246, 169–172 (1998).

Johns A. Psychiatric effects of cannabis. *Br. J. Psychiatry* 178, 116–122 (2001).

Knudsen P, Vilmar T. Cannabis and neuroleptic agents in schizophrenia. *Acta Psychiatr. Scand.* 69, 162–174 (1984).

Koob GF, Le Moal M. Drug abuse: hedonic homeostatic dysregulation. *Science* 278, 52–58 (1997).

Kreitzer AC, Regehr WG. Retrograde signaling by endocannabinoids. *Curr. Opin. Neurobiol.* 12, 324–330 (2002).

Kunos G, Batkai S. Novel physiologic functions of endocannabinoids as revealed through the use of mutant mice. *Neurochem. Res.* 26, 1015–1021 (2001).

Kurlan R, Richard IH, Papka M, Marshall F. Movement disorders in Alzheimer's disease: more rigidity of definitions is needed. *Mov. Disord.* 15, 24–29 (2000).

Lastres-Becker I, Cebeira M, de Ceballos M, Zeng B-Y, Jenner P, Ramos JA, Fernández-Ruiz JJ. Increased cannabinoid CB1 receptor binding and activation of GTP-binding proteins in the basal ganglia of patients with Parkinson's syndrome and of MPTP-treated marmosets. *Eur. J. Neurosci.* 14, 1827–1832 (2001).

Lastres-Becker I, Fezza F, Cebeira M, Bisogno T, Ramos JA, Milone A, Fernández-Ruiz JJ, Di Marzo V. Changes in endocannabinoid transmission in the basal ganglia in a rat model of Huntington's disease. *Neuroreport* 12, 2125–2129 (2001).

Lastres-Becker I, Hansen HH, Berrendero F, de Miguel R, Pérez-Rosado A, Manzanares J, Ramos JA, Fernández-Ruiz JJ. Alleviation of motor hyperactivity and neurochemical deficits by endocannabinoid uptake inhibition in a rat model of Huntington's disease. *Synapse* 44, 23–35 (2002).

Lastres-Becker I, Berrendero F, Lucas JJ, Martín-Aparicio E, Yamamoto A, Ramos JA, Fernández-Ruiz JJ. Loss of mRNA levels, binding and activation of GTP-binding proteins for cannabinoid CB_1 receptors in the basal ganglia of a transgenic model of Huntington's disease. *Brain Res.* 929, 236–242 (2002).

Lastres-Becker I, de Miguel R, De Petrocellis L, Makriyannis A, Di Marzo V, Fernández-Ruiz JJ. Compounds acting at the endocannabinoid and/or endovanilloid systems reduce hyperkinesia in a rat model of Huntington's disease. *J. Neurochem.* 84, 1097–1109 (2003).

Lastres-Becker I, De Miguel R, Fernández-Ruiz JJ. The endocannabinoid system and Huntington's disease. *Curr. Drug Target CNS Neurol. Disord.* 2, 335–347 (2003).
Lastres-Becker I, Bizat N, Boyer F, Hantraye P, Brouillet E, Fernández-Ruiz J. Effects of cannabinoids in the rat model of Huntington's disease generated by an intrastriatal injection of malonate. *Neuroreport* 14, 813–816 (2003).
Lastres-Becker I, Molina-Holgado F, Ramos JA, Mechoulam R, Fernández-Ruiz J. Cannabinoids provide neuroprotection against 6-hydroxydopamine toxicity in vivo and in vitro: Relevance to Parkinson's disease. *Neurobiol. Dis.* 19, 96–107 (2005).
Ledent C, Valverde O, Cossu G, Petitet F, Aubert JF, Beslot F, Böhme GA, Imperato A, Pedrazzini T, Roques BP, Vassart G, Fratta W, Parmentier M. Unresponsiveness to cannabinoids and reduced addictive effects of opiates in CB1 receptor knockout mice. *Science* 283, 401–404 (1999).
Leroy S, Griffon N, Bourdel MC, Olie JP, Poirier MF, Krebs MO. Schizophrenia and the cannabinoid receptor type 1 (CB1): association study using a single-base polymorphism in coding exon 1. *Am. J. Med. Genet.* 105, 749–752 (2001).
Levitt P, Harvey JA, Friedman E, Simansky K, Murphy EH. New evidence for neurotransmitter influences on brain development. *Trends Neurosci.* 20, 269–274 (1997).
Levitt P. Prenatal effects of drugs of abuse on brain development. *Drug Alcohol Depend.* 51, 109–125 (1998).
Leweke FM, Giuffrida A, Wurster U, Emrich HM, Piomelli D. Elevated endogenous cannabinoids in schizophrenia. *Neuroreport* 10, 1665–1669 (1999).
Lewis DA, Lieberman JA. Catching up on schizophrenia: natural history and neurobiology. *Neuron* 28, 325–334 (2000).
Linszen DH, Dingemans PM, Lenior ME. Cannabis abuse and the course of recent-onset schizophrenic disorders. *Arch. Gen. Psychiatry* 51, 273–279 (1994).
Lyman WD, Sonett JR, Brosnan CF, Elkin R, Bornstein MB. Δ^9-Tetrahydrocannabinol: a novel treatment for experimental autoimmune encephalomyelitis. *J. Neuroimmunol.* 23, 73–81 (1989).
Lynn AB, Herkenham M. Localization of cannabinoid receptors and nonsaturable high-density cannabinoid binding sites in peripheral tissues of the rat: implications for receptor-mediated immune modulation by cannabinoids. *J. Pharmacol. Exp. Ther.* 268, 1612–1623 (1994).
Maccarrone M, Gubellini P, Bari M, Picconi B, Battista N, Centonze D, Bernardi G, Finazzi-Agro A, Calabresi P. Levodopa treatment reverses endocannabinoid system abnormalities in experimental parkinsonism. *J. Neurochem.* 85, 1018–1025 (2003).
Mailleux P, Vanderhaeghen JJ. Distribution of neuronal cannabinoid receptor in the adult rat brain: a comparative receptor binding radioautography and in situ hybridization histochemistry. *Neuroscience* 48, 655–668 (1992).
Mailleux P, Vanderhaeghen JJ. Dopaminergic regulation of cannabinoid receptor mRNA levels in the rat caudate-putamen: an in situ hybridization study. *J. Neurochem.* 61, 1705–1712 (1993).
Maldonado R. Study of cannabinoid dependence in animals. *Pharmacol. Ther.* 95, 153–164 (2002).
Maneuf YP, Crossman AR, Brotchie JM. Modulation of GABAergic transmission in the globus pallidus by the synthetic cannabinoid WIN 55,212-2. *Synapse* 22, 382–385 (1996).
Manzanares J, Corchero J, Romero J, Fernández-Ruiz JJ, Ramos JA, Fuentes JA. Pharmacological and biochemical interactions between opioids and cannabinoids. *Trends Pharmacol. Sci.* 20, 287–294 (1999).
Manzoni OJ, Bockaert J. Cannabinoids inhibit GABAergic synaptic transmission in mice nucleus accumbens. *Eur. J. Pharmacol.* 412, R3–R5 (2001).
Mas-Nieto M, Pommier B, Tzavara ET, Caneparo A, Da Nascimento S, Le Fur G, Roques BP, Noble F. Reduction of opioid dependence by the CB1 antagonist SR141716A in mice: evaluation of the interest in pharmacotherapy of opioid addiction. *Br. J. Pharmacol.* 132, 1809–1816 (2001).
Mascia MS, Obinu MC, Ledent C, Parmentier M, Bohme GA, Imperato A, Fratta W. Lack of morphine-induced dopamine release in the nucleus accumbens of cannabinoid CB1 receptor knockout mice. *Eur. J. Pharmacol.* 383, R1–R2 (1999).
McAllister TW. Apathy. *Semin. Clin. Neuropsychiatry* 5, 275–282 (2000).
Mechoulam R, Hanus L. The cannabinoids: an overview. Therapeutic implications in vomiting and nausea after cancer chemotherapy, in appetite promotion, in multiple sclerosis and in neuroprotection. *Pain Res. Manag.* 6, 67–73 (2001).
Mechoulam R, Panikashvili A, Shohami E. Cannabinoids and brain injury: therapeutic implications. *Trends Mol. Med.* 8, 58–61 (2002).

Melis M, Pistis M, Perra S, Muntoni AL, Pillolla G, Gessa GL. Endocannabinoids mediate presynaptic inhibition of glutamatergic transmission in rat ventral tegmental area dopamine neurons through activation of CB1 receptors. *J. Neurosci.* 24, 53–62 (2004).

Meschler JP, Conley TJ, Howlett AC. Cannabinoid and dopamine interaction in rodent brain: effects on locomotor activity. *Pharmacol. Biochem. Behav.* 67, 567–573 (2000).

Meschler JP, Clarkson FA, Mathews PJ, Howlett AC, Madras BK. D2, but not D1 dopamine receptor agonists potentiate cannabinoid-induced sedation in nonhuman primates. *J. Pharmacol. Exp. Ther.* 292, 952–959 (2000).

Meschler JP, Howlett AC. Signal transduction interactions between CB1 cannabinoid and dopamine receptors in the rat and monkey striatum. *Neuropharmacology* 40, 918–926 (2001).

Meschler JP, Howlett AC, Madras BK. Cannabinoid receptor agonist and antagonist effects on motor function in normal and 1-methyl-4-phenyl-1,2,5,6-tetrahydropyridine (MPTP)-treated non-human primates. *Psychopharmacology* 156, 79–85 (2001).

Mezey E, Toth ZE, Cortright DN, Arzubi MK, Krause JE, Elde R, Guo A, Blumberg PM, Szallasi A. Distribution of mRNA for vanilloid receptor subtype 1 (TRPV1), and TRPV1-like immunoreactivity, in the central nervous system of the rat and human. *Proc. Natl. Acad. Sci. U.S.A.* 97, 3655–3660 (2000).

Milton NG. Anandamide and noladin ether prevent neurotoxicity of the human amyloid-beta peptide. *Neurosci Lett.* 332, 127–130 (2002).

Moss DE, McMaster SB, Rogers J. Tetrahydrocannabinol potentiates reserpine-induced hypokinesia. *Pharmacol. Biochem. Behav.* 15, 779–783 (1981).

Müller-Vahl KR, Kolbe H, Schneider U, Emrich HM. Cannabinoids: possible role in patho-physiology and therapy of Gilles de la Tourette syndrome. *Acta Psychiatr. Scand.* 98, 502–506 (1998).

Müller-Vahl KR, Schneider U, Emrich HM. Nabilone increases choreatic movements in Huntington's disease. *Mov. Disord.* 14, 1038–1040 (1999).

Müller-Vahl KR, Schneider U, Koblenz A, Jobges M, Kolbe H, Daldrup T, Emrich HM. Treatment of Tourette's syndrome with Δ^9-tetrahydrocannabinol (THC): a randomized crossover trial. *Pharmacopsychiatry* 35, 57–61 (2002).

Müller-Vahl KR. Cannabinoids reduce symptoms of Tourette's syndrome. *Expert Opin. Pharmacother.* 4, 1717–1725 (2003).

Müller-Vahl KR, Schneider U, Prevedel H, Theloe K, Kolbe H, Daldrup T, Emrich HM. Δ^9-Tetrahydrocannabinol (THC) is effective in the treatment of tics in Tourette syndrome: a 6-week randomized trial. *J. Clin. Psychiatry* 64, 459–65 (2003).

Murphy LL, Steger RW, Smith MS, Bartke A. Effects of Δ^9-tetrahydrocannabinol, cannabinol and cannabidiol, alone and in combinations, on luteinizing hormone and prolactin release and on hypothalamic neurotransmitters in the male rat. *Neuroendocrinology* 52, 316–321 (1990).

Murphy LL, Newton SC, Dhali J, Chavez D. Evidence for a direct anterior pituitary site of Δ^9-tetrahydrocannabinol action. *Pharmacol. Biochem. Behav.* 40, 603–607 (1991).

Murphy LL, Gher J, Steger RW, Bartke A. Effects of Δ^9-tetrahydrocannabinol on copulatory behavior and neuroendocrine responses of male rats to female conspecifics. *Pharmacol. Biochem. Behav.* 48, 1011–1017 (1994).

Murphy LL, Munoz RM, Adrian BA, Villanua MA. Function of cannabinoid receptors in the neuroendocrine regulation of hormone secretion. *Neurobiol. Dis.* 5, 432–446 (1998).

Navarro M, Fernandez-Ruiz JJ, de Miguel R, Hernandez ML, Cebeira M, Ramos JA. An acute dose of Δ^9-tetrahydrocannabinol affects behavioral and neurochemical indices of mesolimbic dopaminergic activity. *Behav. Brain Res.* 57, 37–46 (1993).

Negrete JC, Gill K. Cannabis and schizophrenia. in *Marihuana and Medicine* (Nahas GG, Sutin KM, Harvey DJ, Agurell S, Eds.), Humana Press, New Jersey, 1999.

Nieoullon A. Dopamine and the regulation of cognition and attention. *Prog. Neurobiol.* 67, 53–83 (2002).

Onaivi ES, Leonard CM, Ishiguro H, Zhang PW, Lin Z, Akinshola BE, Uhl GR. Endocannabinoids and cannabinoid receptor genetics. *Prog. Neurobiol.* 66, 307–344 (2002).

Page KJ, Besret L, Jain M, Monaghan EM, Dunnett SB, Everitt BJ. Effects of systemic 3-nitropropionic acid-induced lesions of the dorsal striatum on cannabinoid and mu-opioid receptor binding in the basal ganglia. *Exp. Brain Res.* 130, 142–150 (2000).

Pagotto U, Marsicano G, Fezza F, Theodoropoulou M, Grubler Y, Stalla J, Arzberger T, Milone A, Losa M, Di Marzo V, Lutz B, Stalla GK. Normal human pituitary gland and pituitary adenomas express cannabinoid receptor type 1 and synthesize endogenous cannabinoids: first evidence for a direct role of cannabinoids on hormone modulation at the human pituitary level. *J. Clin. Endocrinol. Metab.* 86, 2687–2696 (2001).

Palmer SL, Thakur GA, Makriyannis A. Cannabinergic ligands. *Chem. Phys. Lipids.* 121, 3–19 (2002).

Parolaro D. Presence and functional regulation of cannabinoid receptors in immune cells. *Life Sci.* 65, 637–644 (1999).

Parolaro D, Rubino T. Is cannabinoid transmission involved in rewarding properties of drugs of abuse? *Br. J. Pharmacol.* 136, 1083–1084 (2002).

Patel S, Hillard CJ. Cannabinoid-induced Fos expression within A10 dopaminergic neurons. *Brain Res.* 963, 15–25 (2003).

Patton GC, Coffey C, Carlin JB, Degenhardt L, Lynskey M, Hall W. Cannabis use and mental health in young people: cohort study. *Br. Med. J.* 325, 1195–1198 (2002).

Pérez-Rosado A, Manzanares J, Fernandez-Ruiz J, Ramos JA. Prenatal Δ^9-tetrahydrocannabinol exposure modifies proenkephalin gene expression in the fetal rat brain: sex-dependent differences. *Dev. Brain Res.* 120, 77–81 (2000).

Pertwee RG. Pharmacology of cannabinoid CB1 and CB2 receptors. *Pharmacol. Ther.* 74, 129–180 (1997).

Pertwee RG. Cannabinoid receptor ligands: clinical and neuropharmacological considerations, relevant to future drug discovery and development. *Expert. Opin. Invest. Drugs* 9, 1553–1571 (2000).

Pertwee RG. Cannabinoid receptors and pain. *Prog. Neurobiol.* 63, 569–611 (2001).

Pertwee RG. Cannabinoids and multiple sclerosis. *Pharmacol. Ther.* 95, 165–174 (2002).

Piomelli D, Giuffrida A, Calignano A, Rodriguez de Fonseca F. The endocannabinoid system as a target for therapeutic drugs. *Trends Pharmacol. Sci.* 21, 218–224 (2000).

Pistis M, Porcu G, Melis M, Diana M, Gessa GL. Effects of cannabinoids on prefrontal neuronal responses to ventral tegmental area stimulation. *Eur. J. Neurosci.* 14, 96–102 (2001).

Pistis M, Muntoni AL, Pillolla G, Gessa GL. Cannabinoids inhibit excitatory inputs to neurons in the shell of the nucleus accumbens: an in vivo electrophysiological study. *Eur. J. Neurosci.* 15, 1795–1802 (2002).

Pistis M, Ferraro L, Pira L, Flore G, Tanganelli S, Gessa GL, Devoto P. Δ^9-tetrahydrocannabinol decreases extracellular GABA and increases extracellular glutamate and dopamine levels in the rat prefrontal cortex: an in vivo microdialysis study. *Brain Res.* 948, 155–158 (2002).

Pollack AE. Anatomy, physiology, and pharmacology of the basal ganglia. *Neurol. Clin.* 19, 523–534 (2001).

Ponce G, Hoenicka J, Rubio G, Ampuero I, Jimenez-Arriero MA, Rodriguez-Jimenez R, Palomo T, Ramos JA. Association between cannabinoid receptor gene (CNR1) and childhood attention deficit/hyperactivity disorder in Spanish male alcoholic patients. *Mol. Psychiatry* 8, 466–467 (2003).

Preuss UW, Koller G, Zill P, Bondy B, Soyka M. Alcoholism-related phenotypes and genetic variants of the CB1 receptor. *Eur. Arch. Psychiatry Clin. Neurosci.* 253, 275–280 (2003).

Rettori V, Wenger T, Snyder G, Dalterio S, McCann SM. Hypothalamic action of Δ^9-tetrahydrocannabinol to inhibit the release of prolactin and growth hormone in the rat. *Neuroendocrinology* 47, 498–503 (1988).

Richfield EK, Herkenham M. Selective vulnerability in Huntington's disease: preferential loss of cannabinoid receptors in lateral globus pallidus. *Ann. Neurol.* 36, 577–584 (1994).

Robbe D, Alonso G, Duchamp F, Bockaert J, Manzoni OJ. Localization and mechanisms of action of cannabinoid receptors at the glutamatergic synapses of the mouse nucleus accumbens. *J. Neurosci.* 21, 109–116 (2001).

Rodriguez de Fonseca F, Cebeira M, Fernandez-Ruiz JJ, Navarro M, Ramos JA. Effects of pre- and perinatal exposure to hashish extracts on the ontogeny of brain dopaminergic neurons. *Neuroscience* 43, 713–723 (1991).

Rodriguez De Fonseca F, Fernandez-Ruiz JJ, Murphy LL, Cebeira M, Steger RW, Bartke A, Ramos JA. Acute effects of Δ^9-tetrahydrocannabinol on dopaminergic activity in several rat brain areas. *Pharmacol. Biochem. Behav.* 42, 269–275 (1992).

Rodriguez de Fonseca F, Cebeira M, Ramos JA, Martin M, Fernandez-Ruiz JJ. Cannabinoid receptors in rat brain areas: sexual differences, fluctuations during estrous cycle and changes after gonadectomy and sex steroid replacement. *Life Sci.* 54, 159–170 (1994).

Rodriguez de Fonseca F, Wenger T, Navarro M, Murphy LL. Effects of Δ^9-THC on VIP-induced prolactin secretion in anterior pituitary cultures: evidence for the presence of functional cannabinoid CB1 receptors in pituitary cells. *Brain Res.* 841, 114–122 (1999).

Rolls ET. The orbitofrontal cortex and reward. *Cereb. Cortex* 10, 284–294 (2000).

Romero J, García-Gil L, Ramos JA, Fernández-Ruiz JJ. The putative cannabinoid receptor ligand, anandamide, stimulates tyrosine hydroxylase and inhibits prolactin release. *Neuroendocrinol. Lett.* 16, 159–165 (1994).

Romero J, Garcia L, Cebeira M, Zadrozny D, Fernandez-Ruiz JJ, Ramos JA. The endogenous cannabinoid receptor ligand, anandamide, inhibits the motor behavior: role of nigrostriatal dopaminergic neurons. *Life Sci.* 56, 2033–2040 (1995).

Romero J, de Miguel R, García-Palomero E, Fernández-Ruiz JJ, Ramos JA. Time-course of the effects of anandamide, the putative endogenous cannabinoid receptor ligand, on extrapyramidal function. *Brain Res.* 694, 223–232 (1995).

Romero J, de Miguel R, Ramos JA, Fernández-Ruiz JJ. The activation of cannabinoid receptors in striatonigral GABAergic neurons inhibited GABA uptake. *Life Sci.* 62, 351–363 (1998).

Romero J, Wenger T, De Miguel R, Ramos JA, Fernández-Ruiz JJ. Cannabinoid receptor binding did not vary in several hypothalamic nuclei after hypothalamic deafferentation. *Life Sci.* 63, 351–356 (1998).

Romero J, Berrendero F, Pérez-Rosado A, Manzanares J, Rojo A, Fernández-Ruiz JJ, de Yébenes JG, Ramos JA. Unilateral 6-hydroxydopamine lesions of nigrostriatal dopaminergic neurons increased CB1 receptor mRNA levels in the caudate-putamen. *Life Sci.* 66, 485–494 (2000).

Romero J, Lastres-Becker I, de Miguel R, Berrendero F, Ramos JA, Fernández-Ruiz JJ. The endogenous cannabinoid system and the basal ganglia. biochemical, pharmacological, and therapeutic aspects. *Pharmacol. Ther.* 95, 137–152 (2002).

Sakurai-Yamashita Y, Kataoka Y, Fujiwara M, Mine K, Ueki S. Δ^9-Tetrahydrocannabinol facilitates striatal dopaminergic transmission. *Pharmacol. Biochem. Behav.* 33, 397–400 (1989).

Sañudo-Peña MC, Tsou K, Walker JM. Motor actions of cannabinoids in the basal ganglia output nuclei. *Life Sci.* 65, 703–713 (1999).

Schlicker E, Fink K, Zentner J, Gothert M. Presynaptic inhibitory serotonin autoreceptors in the human hippocampus. *Naunyn Schmiedebergs Arch. Pharmacol.* 354, 393–396 (1996).

Schlicker E, Kathmann M. Modulation of transmitter release via presynaptic cannabinoid receptors. *Trends Pharmacol. Sci.* 22, 565–572 (2001).

Schmidt LG, Samochowiec J, Finckh U, Fiszer-Piosik E, Horodnicki J, Wendel B, Rommelspacher H, Hoehe MR. Association of a CB1 cannabinoid receptor gene (CNR1) polymorphism with severe alcohol dependence. *Drug Alcohol Depend.* 65, 221–224 (2002).

Scorticati C, Mohn C, De Laurentiis A, Vissio P, Fernandez Solari J, Seilicovich A, McCann SM, Rettori V. The effect of anandamide on prolactin secretion is modulated by estrogen. *Proc. Natl. Acad. Sci. U.S.A.* 100, 2134–2139 (2003).

Self DW, Nestler EJ. Molecular mechanisms of drug reinforcement and addiction. *Annu. Rev. Neurosci.* 18, 463–495 (1994).

Serra S, Brunetti G, Pani M, Vacca G, Carai MA, Gessa GL, Colombo G. Blockade by the cannabinoid CB1 receptor antagonist, SR 141716, of alcohol deprivation effect in alcohol-preferring rats. *Eur. J. Pharmacol.* 443, 95–97 (2002).

Sieradzan KA, Fox SH, Hill M, Dick JP, Crossman AR, Brotchie JM. Cannabinoids reduce levodopa-induced dyskinesia in Parkinson's disease: a pilot study. *Neurology* 57, 2108–2111 (2001).

Sipe JC, Chiang K, Gerber AL, Beutler E, Cravatt BF. A missense mutation in human fatty acid amide hydrolase associated with problem drug use. *Proc. Natl. Acad. Sci. U.S.A.* 99, 8394–8399 (2002).

Smart D, Gunthorpe MJ, Jerman JC, Nasir S, Gray J, Muir AI, Chambers JK, Randall AD, Davis JB. The endogenous lipid anandamide is a full agonist at the human vanilloid receptor (hVR1). *Br. J. Pharmacol.* 129, 227–230 (2000).

Solinas M, Panlilio LV, Antoniou K, Pappas LA, Goldberg SR. The cannabinoid CB1 antagonist N-piperidinyl-5-(4-chlorophenyl)-1-(2,4-dichlorophenyl)-4-methylpyrazole-3-carboxamide (SR-141716A) differentially alters the reinforcing effects of heroin under continuous reinforcement, fixed ratio, and progressive ratio schedules of drug self-administration in rats. *J. Pharmacol. Exp. Ther.* 306, 93–102 (2003).

Spanagel R, Weiss F. The dopamine hypothesis of reward: past and current status. *Trends Neurosci.* 22, 521–527 (1999).

Specht LA, Pickel VM, Joh TH, Reis DJ. Light-microscopic immunocytochemical localization of tyrosine hydroxylase in prenatal rat brain. I. Early ontogeny. *J. Comp. Neurol.* 199, 233–253 (1981).

Stefano GB, Salzet B, Rialas CM, Pope M, Kustka A, Neenan K, Pryor S, Salzet M. Morphine- and anandamide-stimulated nitric oxide production inhibits presynaptic dopamine release. *Brain Res.* 763, 63–68 (1997).

Steffens M, Engler C, Zentner J, Feuerstein TJ. Cannabinoid CB1 receptor-mediated modulation of evoked dopamine release and of adenylyl cyclase activity in the human neocortex. *Br. J. Pharmacol.* 141, 1193–1203 (2004).

Szabo B, Muller T, Koch H. Effects of cannabinoids on dopamine release in the corpus striatum and the nucleus accumbens in vitro. *J. Neurochem.* 73, 1084–1089 (1999).

Szabo B, Wallmichrath I, Mathonia P, Pfreundtner C. Cannabinoids inhibit excitatory neurotransmission in the substantia nigra pars reticulata. *Neuroscience* 97, 89–97 (2000).

Szabo B, Siemes S, Wallmichrath I. Inhibition of GABAergic neurotransmission in the ventral tegmental area by cannabinoids. *Eur. J. Neurosci.* 15, 2057–2061 (2002).

Tanda G, Pontieri FE, Di Chiara G. Cannabinoid and heroin activation of mesolimbic dopamine transmission by a common mu1 opioid receptor mechanism. *Science* 276, 2048–2050 (1997).

Tanda G, Loddo P, Di Chiara G. Dependence of mesolimbic dopamine transmission on Δ^9-tetrahydrocannabinol. *Eur. J. Pharmacol.* 376, 23–26 (1999).

Tsou K, Brown S, Sañudo-Peña MC, Mackie K, Walker JM. Immunohistochemical distribution of cannabinoid CB1 receptors in the rat central nervous system. *Neuroscience* 83, 393–411 (1998).

Tsou K, Nogueron MI, Muthian S, Sanudo-Pena MC, Hillard CJ, Deutsch DG, Walker JM. Fatty acid amide hydrolase is located preferentially in large neurons in the rat central nervous system as revealed by immunohistochemistry. *Neurosci. Lett.* 254, 137–140 (1998).

Ujike H, Takaki M, Nakata K, Tanaka Y, Takeda T, Kodama M, Fujiwara Y, Sakai A, Kuroda S. CNR1, central cannabinoid receptor gene, associated with susceptibility to hebephrenic schizophrenia. *Mol. Psychiatry* 7, 515–518 (2002).

Valjent E, Mitchell JM, Besson MJ, Caboche J, Maldonado R. Behavioural and biochemical evidence for interactions between Δ^9-tetrahydrocannabinol and nicotine. *Br. J. Pharmacol.* 135, 564–578 (2002).

van den Brink W, van Ree JM. Pharmacological treatments for heroin and cocaine addiction. *Eur. Neuropsychopharmacol.* 13, 476–487 (2003).

van der Stelt M, Veldhuis WB, Maccarrone M, Bar PR, Nicolay K, Veldink GA, Di Marzo V, Vliegenthart JF. Acute neuronal injury, excitotoxicity, and the endocannabinoid system. *Mol. Neurobiol.* 26, 317–346 (2002).

van der Stelt M, Di Marzo V. The endocannabinoid system in the basal ganglia and in the mesolimbic reward system: implications for neurological and psychiatric disorders. *Eur. J. Pharmacol.* 480, 133–150 (2003).

Vanable PA, Carey MP, Carey KB, Maisto SA. Smoking among psychiatric outpatients: relationship to substance use, diagnosis, and illness severity. *Psychol. Addict. Behav.* 17, 259–265 (2003).

Vela G, Ruiz-Gayo M, Fuentes JA. Anandamide decreases naloxone-precipitated withdrawal signs in mice chronically treated with morphine. *Neuropharmacology* 34, 665–668 (1995).

Verrico CD, Jentsch JD, Roth RH. Persistent and anatomically selective reduction in prefrontal cortical dopamine metabolism after repeated, intermittent cannabinoid administration to rats. *Synapse* 49, 61–66 (2003).

Vigano D, Cascio MG, Rubino T, Fezza F, Vaccani A, Di Marzo V, Parolaro D. Chronic morphine modulates the contents of the endocannabinoid, 2-arachidonoyl glycerol, in rat brain. *Neuropsychopharmacology* 28, 1160–1167 (2003).

Viggiano D, Ruocco LA, Pignatelli M, Grammatikopoulos G, Sadile AG. Prenatal elevation of endocannabinoids corrects the unbalance between dopamine systems and reduces activity in the Naples High Excitability rats. *Neurosci. Biobehav. Rev.* 27, 129–139 (2003).

Volicer L, Stelly M, Morris J, McLaughlin J, Volicer BJ. Effects of dronabinol on anorexia and disturbed behavior in patients with Alzheimer's disease. *Int. J. Geriatr. Psychiatry* 12, 913–919 (1997).

Voruganti LN, Heslegrave RJ, Awad AG. Neuroleptic dysphoria may be the missing link between schizophrenia and substance abuse. *J. Nerv. Ment. Dis.* 185, 463–465 (1997).

Voruganti LN, Slomka P, Zabel P, Mattar A, Awad AG. Cannabis induced dopamine release: an in-vivo SPECT study. *Psychiatry Res.* 107, 173–177 (2001).

Walker JM, Hohmann AG, Martin WJ, Strangman NM, Huang SM, Tsou K. The neurobiology of cannabinoid analgesia. *Life Sci.* 65, 665–673 (1999).

Walker JM, Krey JF, Chu CJ, Huang SM. Endocannabinoids and related fatty acid derivatives in pain modulation. *Chem. Phys. Lipids* 121, 159–172 (2002).

Weber B, Schlicker E. Modulation of dopamine release in the guinea-pig retina by Gi- but not by Gs- or Gq-protein-coupled receptors. *Fundam. Clin. Pharmacol.* 15, 393–400 (2001).

Weidenfeld J, Feldman S, Mechoulam R. Effect of the brain constituent anandamide, a cannabinoid receptor agonist, on the hypothalamo-pituitary-adrenal axis in the rat. *Neuroendocrinology* 59, 110–113 (1994).

Wenger T, Fragakis G, Probonas K, Toth BE, Yiannakakis N. Anandamide (endogenous cannabinoid) affects anterior pituitary hormone secretion in adult male rats. *Neuroendocrinol. Lett.* 16, 295–303 (1994).

Wenger T, Toth BE, Juaneda C, Leonardelli J, Tramu G, The effects of cannabinoids on the regulation of reproduction. *Life Sci.* 65, 695–701 (1999).

Wenger T, Fernandez-Ruiz JJ, Ramos JA. Immunocytochemical demonstration of CB1 cannabinoid receptors in the anterior lobe of the pituitary gland. *J. Neuroendocrinol.* 11, 873–878 (1999).

Wenger T, Moldrich G. The role of endocannabinoids in the hypothalamic regulation of visceral function. *Prostaglandins Leukot. Essent. Fatty Acid* 66, 301–307 (2002).

Wenger T, Moldrich G, Furst S. Neuromorphological background of cannabis addiction. *Brain Res. Bull.* 61, 125–128 (2003).

Westlake TM, Howlett AC, Bonner TI, Matsuda LA, Herkenham M. Cannabinoid receptor binding and messenger RNA expression in human brain: an in vitro receptor autoradiography and in situ hybridization histochemistry study of normal aged and Alzheimer's brains. *Neuroscience* 63, 637–652 (1994).

Wirguin I, Mechoulam R, Breuer A, Schezen E, Weidenfeld J, Brenner T. Suppression of experimental autoimmune encephalomyelitis by cannabinoids. *Immunopharmacology* 28, 209–214 (1994).

Wise RA. Addictive drugs and brain stimulation reward. *Annu. Rev. Neurosci.* 19, 319–340 (1996).

Wu X, French ED. Effects of chronic Δ^9-tetrahydrocannabinol on rat midbrain dopamine neurons: an electrophysiological assessment. *Neuropharmacology* 39, 391–398 (2000).

Yamamoto T, Takada K. Role of cannabinoid receptor in the brain as it relates to drug reward. *Jpn. J. Pharmacol.* 84, 229–236 (2000).

Zajicek J, Fox P, Sanders H, Wright D, Vickery J, Nunn A, Thompson A. Cannabinoids for treatment of spasticity and other symptoms related to multiple sclerosis (CAMS study): multicentre randomised placebo-controlled trial. *Lancet* 362, 1517–1526 (2003).

Zavitsanou K, Garrick T, Huang XF. Selective antagonist [^{3}H]SR141716A binding to cannabinoid CB1 receptors is increased in the anterior cingulate cortex in schizophrenia. *Prog. Neuropsychopharmacol. Biol. Psychiatry* 28, 355–360 (2004).

Zimmer A, Zimmer AM, Hohmann AG, Herkenham M, Bonner TI. Increased mortality, hypoactivity, and hypoalgesia in cannabinoid CB1 receptor knockout mice. *Proc. Natl. Acad. Sci. U.S.A.* 96, 5780–5785 (1999).

Zygmunt PM, Petersson J, Andersson DA, Chuang H, Sorgard M, Di Marzo V, Julius D, Hogestatt ED. Vanilloid receptors on sensory nerves mediate the vasodilator action of anandamide. *Nature* 400, 452–457 (1999).

Zygmunt PM, Julius I, Di Marzo I, Hogestatt ED. Anandamide — the other side of the coin. *Trends Pharmacol. Sci.* 21, 43–44 (2000).

Zygmunt PM, Chuang H, Movahed P, Julius D, Hogestatt ED. The anandamide transport inhibitor AM404 activates vanilloid receptors. *Eur. J. Pharmacol.* 396, 39–42 (2000).

12 Pharmacology of the Oxidative Metabolites of Endocannabinoids

Ruth A. Ross

CONTENTS

INTRODUCTION

The endocannabinoid system comprises two known receptors (CB_1 and CB_2); a family of endogenous ligands; and specific molecular machinery for the synthesis, transport, and inactivation of CB ligands (Pertwee and Ross, 2002). Anandamide and 2-arachidonoylglycerol (2-AG) are rapidly hydrolyzed by the microsomal enzyme, fatty acid amide hydrolase (FAAH). In addition, anandamide and 2-AG can also be metabolized by a range of oxygenase enzymes that are already known to convert arachidonic acid to potent biologically active compounds (see Kozak and Marnett, 2002). These include cyclooxygenase (COX), lipoxygenase (LOX), and the cytochrome P450 enzymes. The presence of such pathways have important implications for the pharmacology of endocannabinoids because it may be directed by tissue-specific differences in the balance of these metabolic enzymes. Oxygenation of anandamide and 2-AG leads to the production of a range of novel lipid products whose physiological role has yet to be elucidated. This chapter will cover recent developments

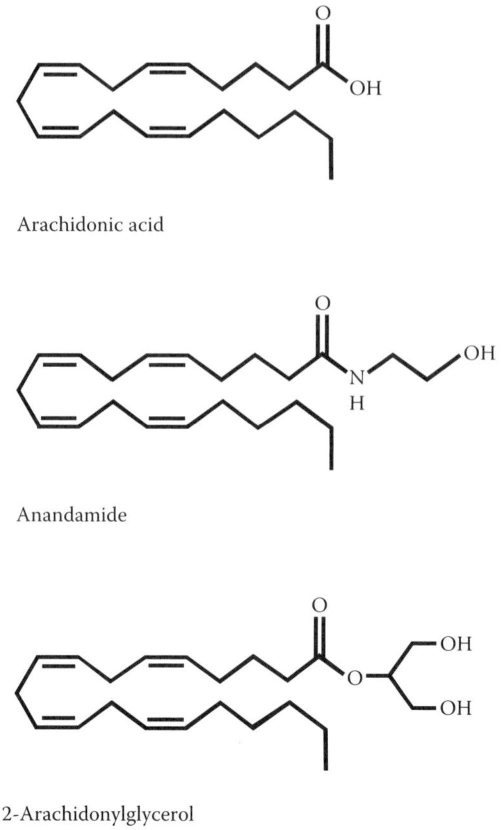

FIGURE 12.1 Structures of arachidonic acid, anandamide, and 2-arachidonylglycerol.

in the pharmacology and diverse functions of these oxygenated derivatives of endocannabinoids (Figure 12.1).

COX METABOLISM OF ENDOCANNABINOIDS

COX exists in two isoforms (COX-1 and COX-2) and catalyzes the conversion of arachidonic acid to prostaglandin endoperoxide (PGH_2), which serves as a precursor for prostaglandins and thromboxanes. Whereas COX-1 is constitutively active, COX-2 is inducible and controls prostaglandin biosynthesis during inflammation and in the brain and spinal cord.

ANANDAMIDE

Formation and Catabolism

Yu et al. (1997) demonstrated that anandamide is oxygenated by COX-2, whereas COX-1 does not display any detectable activity with anandamide as substrate. The major product of COX-2 metabolism of anandamide is PGE_2 ethanolamide. Furthermore, anandamide is oxygenated by COX-2 in the physiological environment: in HFF cells that express both COX isoforms, prostaglandin ethanolamides are products of anandamide metabolism; in THP cells that express COX-1 only, there is no detectable metabolism. In RAW 264.7 macrophages, PGE_2 ethanolamide is synthesized from anandamide, and pretreatment of the cells with lipopolysaccharide (LPS), which is an inducer of COX-2 expression, leads to a significant enhancement of the production of this metabolite (Burstein et al., 2000). A recent study further investigated the nature of the prostaglandin ethanolamides that can be formed as a consequence of COX-2 metabolism of anandamide (Kozak, Crews, et al., 2002). In HCA-7 cells that constitutively express COX-2, COX-2 oxygenation of anandamide leads to the production of the endoperoxide intermediate PGH_2 ethanolamide, the major metabolites of which are PGE_2 and PGD_2 ethanolamide. Bovine prostacyclin synthase catalyzes isomerization of PGH_2 ethanolamide to prostacyclin ethanolamide. Following COX-2 metabolism of anandamide, the generation of PGD_2, PGE_2, and PGI_2 ethanolamides by the action of the corresponding prostaglandin synthase occurs at rates comparable to those observed with arachidonic acid (Figure 12.2).

It is well established that prostaglandins are rapidly catabolized *in vivo*, which means that these compounds act as autocrine and paracrine messengers and not in an endocrine capacity. In contrast to the prostaglandins, the prostaglandin ethanolamides appear to be resistant to catabolism (Kozak, Crews, et al., 2002). Consequently, PGE_2 ethanolamide is not hydrolyzed in plasma; it is stable in cerebrospinal fluid and is not efficiently oxidized by 15-hydroxyprostaglandin dehydrogenase, which is responsible for the oxidation of the 15-hydroxyl group of prostaglandins to produce 15-keto derivatives that are not biologically active. Furthermore, PGE_2 ethanolamide has a long half-life *in vivo*, being detectable in plasma even 2 h after administration. The implication of these findings is that prostaglandin ethanolamides may have sufficient stability to act in a systemic endocrine manner.

Pharmacology

PGE_2 ethanolamide does not bind to CB_1 receptors (in rat brain membranes), although it has a low affinity for CB_2 receptors (in human tonsilar membranes) (Berglund et al., 1999). The compound does, however, activate G-proteins in a CB_1-receptor-independent manner and stimulates cyclic AMP

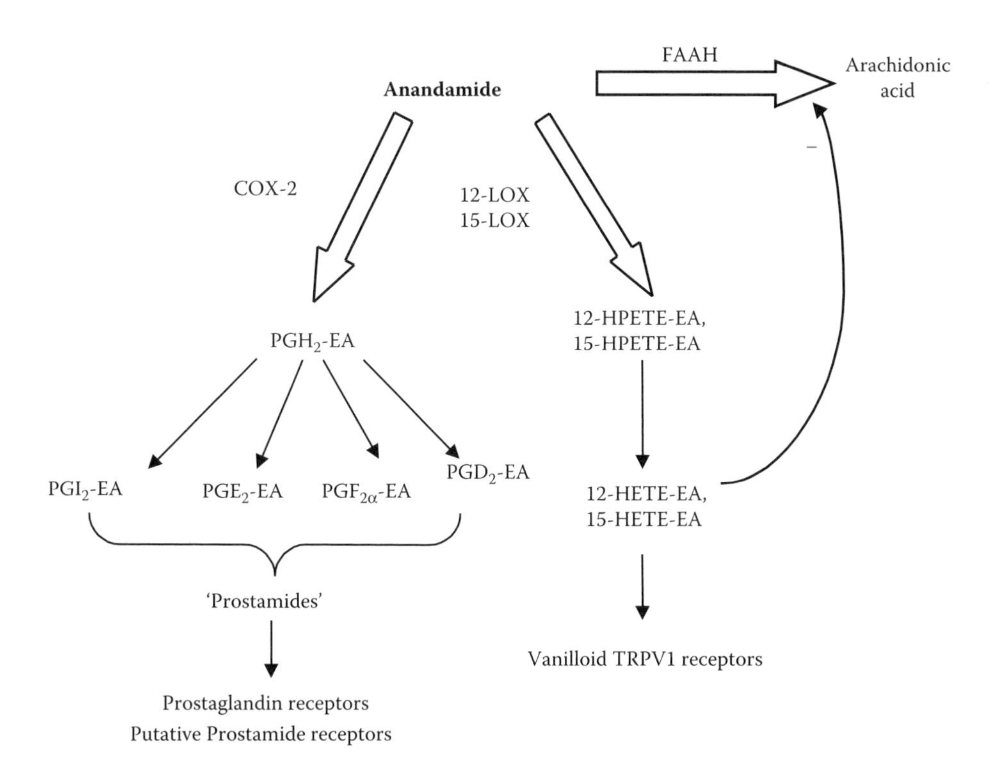

FIGURE 12.2 Oxidative metabolism of anandamide by cyclooxygenase (COX) and lipoxygenase (LOX) to yield ethanolamide (EA) derivatives.

production (Berglund et al., 1999). This is in line with the compound activating G-protein-coupled EP receptors that are sensitive to PGE_2. The activation of the four EP receptor subtypes by PGE_2 ethanolamide has been investigated in depth using *in vitro* pharmacology (Ross et al., 2002). These studies reveal that it has a low affinity for EP receptor subtypes: in radioligand-binding studies using recombinant cell lines expressing EP1, EP2, EP3, and EP4 receptors, the K_i values for PGE_2 ethanolamide are 500- to 600-fold higher than those obtained for PGE_2. Ross et al. (2002) performed a characterization of the pharmacology of PGE_2 ethanolamide using a range of *in vitro* preparations known to express the various EP receptor subtypes. In the rabbit jugular vein, which is thought to contain EP4 receptors mediating relaxation (Coleman et al., 1994), PGE_2 ethanolamide is around 200-fold less potent than PGE_2. In the guinea-pig vas deferens preparation, which is reported to contain EP3 receptors, PGE_2 ethanolamide is 45-fold less potent than PGE_2, the EC_{50} values for inhibition of electrically evoked contractions being 37 n*M* and 0.82 n*M* respectively. The inhibitory effect of PGE_2 ethanolamide is not blocked by the CB_1 receptor antagonist SR141716A or the TRPV1 receptor antagonist capsazepine. This is in line with the observations of Berglund et al. (1999), demonstrating that this compound does not have affinity for CB_1 receptors. In the presence of an FAAH inhibitor, the relative potency of these compounds is unaltered, indicating that PGE_2 ethanolamide is not inhibiting the electrically evoked contractions via conversion to PGE_2. The affinity of PGE_2 for hEP3 receptors (K_i = 0.48 n*M*) is similar to that in the guinea-pig vas deferens (EC_{50} = 0.82 n*M*). In contrast, the potency of PGE_2 ethanolamide is higher than that predicted by the low affinity of this compound for the hEP3 receptor (K_i = 250 n*M*). At present, EP3 receptor antagonists are not available. In the guinea-pig trachea, PGE_2 ethanolamide exhibited a bell-shaped concentration–response relationship, as has previously been demonstrated for PGE_2 (Dong et al., 1986). PGE_2 and PGE_2 ethanolamide produced a maximal contraction of the preparation at 100 n*M* and 1 μ*M* respectively, followed by a relaxation of the tissue at higher concentrations. The contractile action of both compounds is abolished by SC-51089, indicating that they induce contractions via an interaction with the EP1 receptor. When the tracheal preparations are precontracted with histamine in the presence of SC-51089, both PGE_2 and PGE_2 ethanolamide produce concentration-related relaxation of the tissue. In this case, PGE_2 ethanolamide is only 15-fold less potent than PGE_2, the EC_{50} values being 76.9 n*M* and 5.19 n*M*, respectively. Previous pharmacological analysis indicates that this relaxant action of PGE_2 is mediated by the EP2 receptor subtype (Coleman et al., 1990). The relatively high potency of PGE_2 ethanolamide at the EP2 receptor in the trachea is not in line with the low affinity of this compound for the EP2 receptor subtype. As with the EP3 receptor, selective EP2 receptor antagonists are not yet available. Thus, the potency of PGE_2 ethanolamide in both the vas deferens and the trachea (relaxation) is higher than predicted from binding assays using human receptors. This may be accounted for by species differences. It is possible that the higher relative potency of this compound in some tissues may be due to its resistance to metabolism by the enzymes that are responsible for the rapid inactivation of PGE_2. Alternatively, this compound may be interacting with other as yet uncharacterized receptors for prostaglandin ethanolamides.

Evidence for novel receptors for prostaglandin ethanolamides (also known as *prostamides*) is obtained from studies of the pharmacology of $PGF_{2\alpha}$ ethanolamide (prostamide $F_{2\alpha}$). $PGF_{2\alpha}$ and $PGF_{2\alpha}$ ethanolamide reduce intraocular pressure; although the $PGF_{2\alpha}$ effect is mediated by FP receptors, $PGF_{2\alpha}$ ethanolamide does not bind to recombinant FP receptors or exhibit functional responses in tissue preparations that contain FP receptors (Woodward et al., 2001). Consequently, the pharmacology of $PGF_{2\alpha}$ ethanolamide has prompted the suggestion of the existence of a novel prostamide receptor. This hypothesis is corroborated by the pharmacology of bimatoprost, a novel $PGF_{2\alpha}$ analog in which the carboxylic acid has been replaced with an ethylamide substituent (Woodward et al., 2003). Bimatoprost exhibits a novel pharmacological profile and is potent in reducing intraocular pressure in both ocular normotensive and hypertensive situations. Bimatoprost has low affinity for the FP receptor in cells expressing either human or feline recombinant FP receptors, the K_i values being 4.7 μ*M* and 8.9 μ*M* (Woodward et al., 2003), respectively, and 9.2 μ*M* using bovine corpus luteum (Sharif, Kelly, and Williams, 2003). This compound also has low potency in functional FP-receptor-mediated responses. In studies of mobilization of intracellular Ca^{2+} in cells expressing human recombinant FP receptors, bimatoprost is found have an EC_{50} value of 3 μM (Sharif, Kelly, and Williams, 2003). In natively expressing Swiss 3T3 cells, it is found to be inactive at concentrations less than 10 μM (Woodward et al., 2002) or to have an EC_{50} value of 3.1 μM (Kelly et al., 2003). In tissues sensitive to $PGF_{2\alpha}$, including rat colon, gerbil colon, rat fundus, and mouse ileum, bimatoprost has little agonist activity. In contrast to the low potency of bimatoprost as compared to $PGF_{2\alpha}$ in functional FP preparations, this compound is 25-fold more potent than $PGF_{2\alpha}$ in producing contractions of the feline lung parenchymal preparation, the EC_{50} values being 40 nM and 1 μM for bimatoprost and $PGF_{2\alpha}$, respectively (Woodward et al., 2002). Furthermore, $PGF_{2\alpha}$, prostamide $F_{2\alpha}$, and bimatoprost all contract the cat iris sphincter preparation (Woodward et al., 2001). There has been debate around the possibility that prostamide $F_{2\alpha}$ and bimatoprost act as prodrugs, being hydrolyzed to $PGF_{2\alpha}$ and 17-phenyl-trinor $PGF_{2\alpha}$, respectively, which act on the FP receptor to reduce intraocular pressure. This is of particular importance because 17-phenyl-trinor $PGF_{2\alpha}$ has a high affinity for, and potency at, recombinant FP receptors, reported K_i and EC_{50} values being 59 nM and 15 nM respectively (Sharif, Kelly, and Williams, 2003). Likewise, in Swiss 3T3 cells, 17-phenyl-trinor $PGF_{2\alpha}$, has an EC_{50} value of 49 nM for Ca^{2+} mobilization (Kelly et al., 2003). In isolated human trabecular meshwork cells, bimatoprost stimulates PI turnover with an EC_{50} of 6.9 μM compared to 120 nM for $PGF_{2\alpha}$. The free acid form of bimatoprost was found to be more potent with an EC_{50} of 112 nM (Sharif, Kelly, and Crider, 2003). These studies involve a 60-min incubation, and the low potency of bimatoprost compared to the free acid suggests that the compound is not hydrolyzed in these cells. Furthermore, studies of the pharmacokinetics and metabolism of bimatoprost reveal that it is not converted to a free acid metabolite in ocular tissue from monkeys or humans (Woodward et al., 2001; Woodward et al., 2003). Consequently, this supports the hypothesis that bimatoprost and related prostaglandin ethanolamides are acting at a novel non-FP receptor site. In line with the hypothesis that prostaglandin ethanolamides are not subject to hydrolysis, PGE_2 ethanolamide is not an inhibitor of FAAH metabolism of anandamide (Ross et al., 2002), and it is resistant to hydrolysis in plasma and cerebrospinal fluid (Kozak, Crews, et al., 2002). The novel receptor hypothesis is further corroborated by recent studies of the actions of various compounds on the expression of two genes Cyr61 and CTGF (Liang et al., 2003): whereas $PGF_{2\alpha}$ and bimatoprost upregulate Cyr61 mRNA expression in cat iris tissue and human ciliary muscle cells, only $PGF_{2\alpha}$ upregulates CTFG mRNA expression. Furthermore, in cells isolated from the cat iris sphincter, bimatoprost and $PGF_{2\alpha}$ elicit Ca^{2+} signaling in distinct populations of cells (Spada et al., 2002).

Prostanoids are known to both directly activate sensory neurons and to sensitize them to other potent nociceptive agents such as bradykinin. PGE_2 has been shown to activate a subpopulation of small-diameter capsaicin-sensitive DRG neurons and to potentiate the bradykinin-evoked increases in $[Ca^{2+}]$i. Both these actions of PGE_2 appear to involve PKA-dependent mechanisms (Smith et al., 2000). Smith et al. found that 1 μM PGE_2 evoked an increase in $[Ca^{2+}]_i$ in 16% of capsaicin-sensitive DRG neurons. PGI_2 and $PGF_{2\alpha}$ (1 μM) also evoke calcium transients in 26 and 29% of DRG neurons, respectively. Similarly, Ross et al. (2002) found that PGE_2 ethanolamide (3 μM) evokes an increase in $[Ca^{2+}]$i in 21% of small-diameter capsaicin-sensitive DRG neurons. The possibility that PGE_2 ethanolamide shares the ability of PGE_2 to sensitize DRG neurons to capsaicin (Lopshire and Nicol, 1997) and bradykinin (Smith et al., 2000) is the subject of ongoing investigations. The receptor mechanisms underlying these actions of the prostanoids have not yet been investigated, and it remains to be established whether the prostanoids are acting through the same or distinct sites to activate DRG neurons.

2-ARACHIDONYLGLYCEROL

FORMATION AND CATABOLISM

Kinetic analysis reveals that 2-AG is oxygenated by both human and mouse COX-2, with K_m values similar to those obtained with arachidonic acid (Kozak et al., 2000). For human COX-2 the K_m value is 60-fold higher than that reported for anandamide (Yu et al.,1997). *In vitro*, the main glycerol

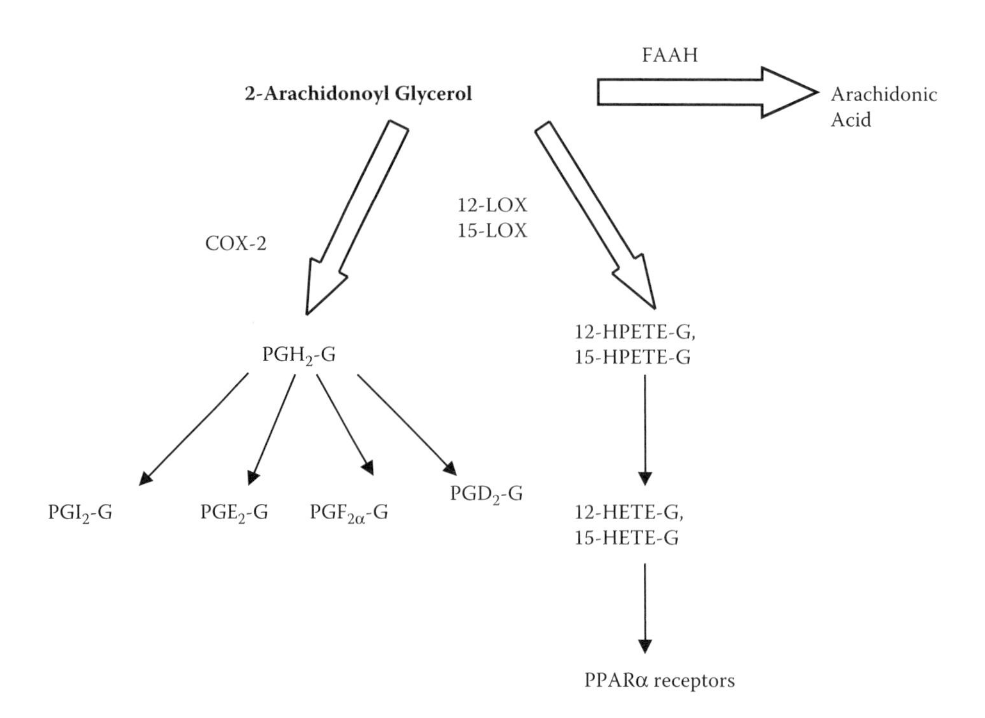

FIGURE 12.3 Oxidative metabolism of 2-arachidonoylglycerol by cyclooxygenase (COX) and lipoxygenase (LOX) to yield glycerol ester (G) derivatives.

ester prostaglandin products of the oxygenation of 2-AG appear to be PGH_2, PGE_2, and PGD_2 glycerol ester. In intact inactivated RAW264.7 macrophages that do not express COX-2, 2-AG is not oxygenated. In the same cells in which COX-2 expression is induced with LPS and IFL-γ, PGD_2 glycerol ester is synthesized and released. Furthermore, stimulation of release of 2-AG *in vitro* via ionomycin treatment leads to the synthesis of PGD_2 glycerol ester in activated RAW264.7 cells. The detection of glycerol prostaglandins in stimulated macrophages indicates that a biosynthetic pathway for 2-AG exists in these cells and, crucially, demonstrates that endogenous 2-AG is a substrate for COX-2 (Figure 12.3).

A recent study further investigates the nature of the prostaglandin glycerol esters that can be formed as a consequence of COX-2 metabolism of 2-AG (Kozak, Gupta, et al., 2002). The profile of metabolites is similar to that observed when anandamide is the substrate. In HCA-7 cells that constitutively express COX-2, COX-2 oxygenation of 2-AG leads to the production of the endoperoxide intermediate PGH_2 glycerol ester, the major metabolites of which are PGE_2 and PGD_2 glycerol ester. Bovine prostacyclin synthase catalyzed isomerization of PGH_2 glycerol ester to prostacyclin glycerol ester. Following COX-2 metabolism of 2-AG, the generation of PGD_2, PGE_2, and PGI_2 glycerol esters by the action of the corresponding prostaglandin synthase occurs at rates comparable to those observed with arachidonic acid. Furthermore, a combination of recombinant human COX-2 and thromboxane synthase yields thromboxane A2 glycerol esters, but the rate of production is dramatically lower than when the substrate is arachidonic acid. Thus, PGH_2 glycerol ester would appear to be a poor substrate for thromboxane synthase.

Similar to the prostaglandin ethanolamides, prostaglandin glycerol esters are resistant to catabolism. PGE_2 glycerol ester is stable in cerebrospinal fluid and is a poor substrate for 15-hydroxyprostaglandin dehydrogenase, which accounts for the initial inactivation of prostaglandins *in vivo*.

N-ARACHIDONOYLGLYCINE

Formation and Catabolism

One putative pathway for the synthesis of *N*-arachidonoylglycine is the oxidative metabolism of anandamide via alcohol dehydrogenase (Burstein et al., 2000). Though this pathway may exist, it is clear that *N*-arachidonoylglycine is also synthesized from the precursors arachidonic acid and glycine (Huang et al., 2001). The compound is degraded by FAAH and displays higher potency as an FAAH inhibitor than anandamide. Among the tissues analyzed, the highest levels were present in the brain, spinal cord, small intestine, kidney, and skin (Huang et al., 2001). In addition to *N*-arachidonoylglycine, other arachidoloyl amino acids have been found in the brain, including *N*-arachidonoyl-gamma-aminobutyric acid (GABA) and *N*-arachidonoylalanine (Huang et al., 2001).

Tissues that are known to express COX-2, including the brain and spinal cord, contain high levels of *N*-arachidonoylglycine. Although this compound is resistant to metabolism by COX-1, COX-2 metabolites of *N*-arachidonoylglycine have been isolated and represent a novel class of lipid-signaling molecules (Prusakiewicz et al., 2002). Additionally, *N*-arachidonoylglycine is more selective for COX-2 than either 2-AG or anandamide. Anandamide and 2-AG are neutral cannabinoids and, thus, *N*-arachidonoylglycine is the first charged COX-2 selective substrate. The products of *N*-arachidonoylglycine COX-2 metabolism are PGH_2 glycine and HETE glycine.

Pharmacology

Although structurally similar to anandamide, *N*-arachidonoylglycine does not have affinity for cannabinoid receptors (Burstein et al., 2000). Limited analysis of *N*-arachidonoylglycine suggests that this lipid exhibits a different pharmacological profile from anandamide (Burstein et al., 2000). In tests for catalepsy and antinociception tested using hot plate and immobility assays, *N*-arachidonoylglycine displays significantly lower cataleptic activity and greater antinociception than THC. In inflammatory models of pain, this compound produced a significant antinociceptive effect (Walker et al., 2002). Furthermore, *N*-arachidonoylglycine enhances the level of anandamide, possibly by inhibiting FAAH; in RAW264.7 macrophages, the compound elevates basal levels of anandamide, and oral administration of *N*-arachidonoylglycine significantly increases the levels of circulating anandamide in rats *in vivo* (Burstein et al., 2002).

LOX METABOLISM OF ENDOCANNABINOIDS

LOX enzymes catalyze the hydroperoxidation of polyunsaturated fatty acids, and six LOXs are known to exist in humans. Hydroperoxidation by LOX is stereospecific and regiospecific.

ANANDAMIDE

Formation and Catabolism

In vitro, 12- and 15-LOX convert anandamide to 12- and 15-hydroperoxyeicosatetraenoyl acid (HPETE) ethanolamide, respectively, the reaction rates being similar to those for arachidonic acid (see Kozak and Marnett, 2002). 12- and 15-HPETE-ethanolamide are produced when anandamide is incubated with porcine leukocyte 12-LOX and rabbit reticulocyte 15-LOX, respectively (Ueda et al., 1995). In line with this, 12-LOX isolated from the rat pineal gland effectively metabolizes anandamide (Edgemond et al., 1998). In these studies, the reaction rates for anandamide metabolism were found to be not significantly different from those obtained with arachidonic acid as substrate. However, anandamide does not appear to be oxygenated by human platelet 12-LOX or by porcine 5-LOX (Ueda et al., 1995). In platelets stimulated with A23187, 15-HETE(*S*)-ethanolamide and 12(*S*)-HETE-ethanolamide are formed as the principal oxygenated derivatives of anandamide (Edgemond et al., 1998).

Pharmacology

Radioligand-binding assays reveal that 12(*S*)-HETE-ethanolamide has K_i values at CB_1 and CB_2 receptors that are not dissimilar to those obtained for anandamide (Edgemond et al., 1997; van der Stelt et al., 2002). 12(*R*)-HETE-ethanolamide and 15-HETE(*S*)-ethanolamide have K_i values of 416 n*M* and ~700 n*M* for the CB_1 receptor, but >1 μ*M* for the CB_2 receptor (Edgemond et al., 1997; van der Stelt et al., 2002). 5-HETE(*S*) ethanolamide does not bind to CB_1 or CB_2 receptors, whereas 15-HETE(*R*)-ethanolamide has low affinity for both receptors (K_i ~700 n*M*).

Specific inhibitors of 5-LOX markedly enhance the level of hydrolysis of anandamide by FAAH in human mast cells (Maccarrone, Fiorucci, et al., 2000). Thus, Maccarrone et al. hypothesize that 5-LOX metabolites of anandamide may act as endogenous inhibitors of FAAH. To complement this hypothesis, they have demonstrated that various HPETE and HETE ethanolamides are potent endogenous inhibitors of FAAH (Maccarrone, Salvati, et al., 2000; van der Stelt et al., 2002). Although all the derivatives are inhibitors of FAAH, 12-HETE(*S*)-ethanolamide is the most potent, the K_i value being threefold higher than that obtained for anandamide. Such findings raise the possibility that LOX metabolites of anandamide may act as "entourage" compounds that enhance endocannabinoid signaling by inhibiting FAAH metabolism of anandamide.

The vanilloid VR1 or TRPV1 receptor is part of a family of transient receptor potential (TRP) channels (see Benham et al., 2002), whose expression is largely associated with small-diameter primary afferent fibers. This receptor is a nonselective cation channel that integrates multiple noxious stimuli and is associated with the pathophysiology of various major diseases (Szallasi, 2002). The search for endogenous TRPV1 receptor activators or "endovanilloids" is ongoing, and recent advances suggest that anandamide may be one such compound (Di Marzo et al., 2002). The LOX metabolites of arachidonic acid, particularly 12-(*S*)-hydroperoxyeicosatetraenoyl acid (HPETE), 5-(*S*)HETE (hydroxyeicoatetraenoic acid), and leukotriene B4 (LTB4), are agonists of the TRPV1 receptor (Hwang et al., 2000; Piomelli, 2001). Recent studies of the action of the potent inflammatory mediator bradykinin provide more compelling evidence for the role of LOX metabolites in the activation of TRPV1. Thus, bradykinin activation of TRPV1 receptors in both cultured DRG neurons and skin is significantly attenuated by LOX inhibitors (Shin et al., 2002). Furthermore, extracellular recording from C-fiber-receptive fields in guinea-pig isolated airways reveals that LOX inhibitors dramatically inhibit bradykinin-induced action potentials, which are TRPV1-receptor mediated (Carr et al., 2002).

As an alternative substrate for LOX enzymes, anandamide may attenuate the production of metabolites of arachidonic acid that are TRPV1 receptor agonists. Alternatively, anandamide may activate TRPV1 via the LOX metabolites of arachidonic acid formed subsequent to its metabolism by FAAH. A further possibility is that the LOX metabolites of anandamide may themselves activate TRPV1 receptors. In the bronchus the TRPV1-receptor-mediated contractile action of anandamide is little affected by FAAH inhibitors, but is significantly attenuated by LOX inhibitors (Craib et al., 2001). This suggests that, in this tissue, anandamide may be metabolized to hydroperoxyeicosatetraenoyl ethanolamides (HPETEEs) and LTB4 ethanolamides that, similar to the hydroperoxy derivatives of arachidonic acid (HPETEs) and LTB4, may be vanilloid receptor agonists. In hTRPV1-HEK cells, the LOX metabolites of anandamide, 11(*S*)-HPETEE and 5(*S*)-HPETEE, do not induce an increase in $[Ca^{2+}]i$, and 15(*S*)-HPETEE has only a modest effect at concentrations above 10 μ*M* (De Petrocellis et al., 2001). However, it is feasible that these compounds and related molecules are more potent as TRPV1 agonists if produced intracellularly via metabolism of arachidonic acid or anandamide close to the intracellular TRPV1 agonist-binding site.

In the bronchus, LOX inhibitors modestly attenuate the contractile action of capsaicin (Craib et al., 2001), raising the possibility that the increase in intracellular calcium elicited by TRPV1 receptor activation leads to the release of arachidonic acid or anandamide or both, whose hydroxylation by LOX may lead to the formation of compounds that are themselves vanilloid agonists.

Indeed, mass spectrometric analysis shows that capsaicin and depolarization (KCl) induce significant release of anandamide in DRG cultures (Ahluwalia et al., 2003). It is perhaps notable that the capsaicin-evoked release of anandamide is significantly attenuated when the FAAH inhibitor MAFP is excluded from the buffer, demonstrating that anandamide is rapidly metabolized in DRG neurons. Metabolic products of anandamide have also been implicated in anandamide-induced depolarization of the guinea-pig isolated vagus nerve, which is TRPV1-receptor mediated (Kagaya et al., 2002). In this preparation, depolarization by anandamide but not capsaicin is inhibited by LOX inhibitors, but only in the presence of calcium. It is not clear, however, whether these active metabolites are produced via direct LOX metabolism of anandamide or via metabolism of arachidonic acid, because these experiments did not include FAAH inhibitors.

Potent FAAH inhibitors have recently been synthesized that enhance the levels of anandamide significantly, and these compounds may have considerable therapeutic potential (Boger et al., 2000). In the event of inhibition of FAAH metabolism of anandamide, increased levels of endogenous anandamide may lead to the production of significant levels of the HPETE ethanolamides. In addition, LOX metabolism producing metabolites that are FAAH inhibitors (van der Stelt et al., 2002) of anandamide may enhance TRPV1 receptor activation by increasing the levels of available anandamide.

2-ARACHIDONOYLGLYCEROL

Formation and Catabolism

2-AG is oxygenated to 12-HPETE glycerol ester by porcine leukocyte 12-LOX and in COS-7 cells transfected with leukocyte 12-LOX, but not by platelet-type 12-LOX (Moody et al., 2001). It is not metabolized by potato or human 5-LOX (Kozak, Gupta, et al., 2002). 2-AG is metabolized by 15-LOX, the rates indicating that it is preferred to arachidonic acid as a substrate. 15(*S*)-HPETE glycerol ester is formed via 15-LOX metabolism of 2-AG in human epidermal keratinocytes (Kozak, Gupta, et al., 2002).

Pharmacology

Peroxisome proliferator-activated receptors (PPARs) are ligand-activated transcription factors that increase transcription of target genes by binding to a specific nucleotide sequence in the gene's promoter. Three different PPAR isotypes can be distinguished: α, β, and γ. PPARs have an important role in lipid metabolism and energy homeostasis (PPAR-α and PPAR-γ) and in epidermal maturation and skin-wound repair (PPAR-α and PPAR-β). The 15-LOX products of arachidonic acid and 2-AG display differing pharmacological profiles at these receptors: whereas 15(*S*)-HPETE is an agonist of the PPAR-γ receptor, 15(*S*)-HPETE glycerol ester is an agonist of the PPAR-α subtype but does not activate PPAR-γ (Kozak, Gupta, et al., 2002). PPAR-α is expressed mainly in skeletal muscle, heart, liver, and kidney and regulates many genes involved in the β-oxidation of fatty acids. 15(*S*)-HPETE glycerol ester therefore represents a novel potent PPAR-α agonists that may have implications for lipid metabolism.

CONCLUSIONS

The metabolism of anandamide is a variable process that differs in various cell types. It is metabolized by COX and LOX to form families of novel lipids that are pharmacologically active. Some enzymes display low activity toward anandamide; nevertheless, in the absence of other metabolic alternatives, such pathways may become significant. The physiological importance of these novel metabolic pathways remains to be established.

REFERENCES

Ahluwalia, J., Yaqoob, M., Urban, L., Bevan, S., and Nagy, I. (2003). Activation of capsaicin-sensitive primary sensory neurons induces anandamide production and release. *J. Neurochem.*, 84, 585–591.

Benham, C.D., Davis, J.B., and Randall, A.D. (2002). Vanilloid and TRP channels: a family of lipid-gated cation channels. *Neuropharmacology*, 42, 873–888.

Berglund, B.A., Boring, D.L., and Howlett, A.C. (1999). Investigation of structural analogs of prostaglandin amides for binding to and activation of CB1 and CB2 cannabinoid receptors in rat brain and human tonsils. *Adv. Exp. Med. Biol.*, 469, 527–533.

Boger, D.L., Sato, H., Lerner, A.E., Hedrick, M.P., Fecik, R.A., Miyauchi, H., Wilkie, G.D., Austin, B.J., Patricelli, M.P., and Cravatt, B.F. (2000). Exceptionally potent inhibitors of fatty acid amide hydrolase: the enzyme responsible for degradation of endogenous oleamide and anandamide. *Proc. Natl. Acad. Sci. USA*, 97, 5044–5049.

Burstein, S.H., Huang, S.M., Petros, T.J., Rossetti, R.G., Walker, J.M., and Zurier, R.B. (2002). Regulation of anandamide tissue levels by N-arachidonylglycine. *Biochem. Pharmacol.*, 64, 1147–1150.

Burstein, S.H., Rossetti, R.G., Yagen, B., and Zurier, R.B. (2000). Oxidative metabolism of anandamide. *Prostaglandins Other Lipid Mediat.*, 61, 29–41.

Coleman, R.A., Kennedy, I., Humphrey, P.P.A., Bunce, K., and Lumley, P. (1990). Prostanoids and their receptors. in *Comprehensive Medicinal Chemistry*, J.C. Emmett, Ed., Oxford, U.K. Pergamon Press, 3: 643–714.

Coleman, R.A., Smith, W.L., and Narumiya, S. (1994) International Union of Pharmacology classification of prostanoid receptors: properties, distribution and structure of the receptors and their subtypes. *Pharmacol. Rev.*, 46: 205–229.

Carr, M.J., Kollarik, M., Meeker, S.N., and Undem, B.J. (2002). A role for TRPV1 in bradykinin induced excitation of vagal airway afferent nerve terminals. *J. Pharmacol. Exp. Ther.*, 304, 1275–1279.

Craib, S.J., Ellington, H.C., Pertwee, R.G., and Ross, R.A. (2001). A possible role of lipoxygenase in the activation of vanilloid receptors by anandamide in the guinea-pig bronchus. *Br. J. Pharmacol.*, 134, 30–37.

De Petrocellis, L., Bisogno, T., Maccarrone, M., Davis, J.B., Finazziagro, A., and Di Marzo, V. (2001). The activity of anandamide at vanilloid VR1 receptors requires facilitated transport across the cell membrane and is limited by intracellular metabolism. *J. Biol. Chem.*, 276, 12856–12863.

Dong, Y.L., Jones, R.L., and Wilson, N.H. (1986). Prostaglandin E subtypes in smooth muscle: agonists activities of stable prostacyclin analogues. *Br. J. Pharmacol.*, 87: 97–107.

Di Marzo, V., Blumberg, P.M., and Szallasi, A. (2002). Endovanilloid signaling in pain. *Curr. Opin. Neurobiol.*, 12, 372–379.

Edgemond, W.S., Hillard, C.J., Falck, J.R., Kearn, C.S., and Campbell, W.B. (1998). Human platelets and polymorphonuclear leukocytes synthesize oxygenated derivatives of arachidonylethanolamide (anandamide): their affinities for cannabinoid receptors and pathways of inactivation. *Mol. Pharmacol.*, 54, 180–188.

Hampson, A.J., Hill, W., Zanphillips, M., Makriyannis, A., Leung, E., Eglen, R.M., and Bornheim, L.M. (1995). Anandamide hydroxylation by brain lipoxygenase: metabolite structures and potencies at the cannabinoid receptor. *Biochim. Biophys. Acta Lipids Lipid Metab.*, 1259, 173–179.

Huang, S.M., Bisogno, T., Petros, T.J., Chang, S.Y., Zavitsanos, P.A., Zipkin, R.E., Sivakumar, R., Coop, A., Maeda, D.Y., De Petrocellis, L., Burstein, S., Di Marzo, V., and Walker, J.M. (2001). Identification of a new class of molecules, the arachidonyl amino acids, and characterization of one member that inhibits pain. *J. Biol. Chem.*, 276, 42639–42644.

Hwang, S.W., Hawoon, C., Kwak, J., Lee, S.-Y., Kang, C.-J., Jung, J., Cho, S., Min, K.H., Suh, Y.-G., Kim, G., and Oh, U. (2000). Direct activation of capsaicin receptors by products of lipoxygenase: endogenous capsaicin-like substances. *Proc. Natl. Acad. Sci. USA*, 97, 6155–6160.

Kagaya, M., Lamb, J., Robbins, J., Page, C.P., and Spina, D. (2002). Characterization of the anandamide induced depolarization of guinea-pig isolated vagus nerve. *Br. J. Pharmacol.*, 137, 39–48.

Kelly, C.R., Williams, G.W., and Sharif, N.A., (2003). Real-time intracellular Ca2+ mobilization by travoprost acid, bimatoprost, unoprostone, and other analogs via endogenous mouse, rat, and cloned human FP prostaglandin receptors. *J. Pharmacol. Exp.Ther.*, 304, 238–245.

Kozak, K.R., Crews, B.C., Morrow, J.D., Wang, L.H., Ma, Y.H., Weinander, R., Jakobsson, P.J., and Marnett, L.J. (2002). Metabolism of the endocannabinoids, 2-arachidonylglycerol and anandamide, into prostaglandin, thromboxane, and prostacyclin glycerol esters and ethanolamides. *J. Biol. Chem.*, 277, 44877–44885.

Kozak, K.R., Gupta, R.A., Moody, J.S., Ji, C., Boeglin, W.E., Dubois, R.N., Brash, A.R., and Marnett, L.J. (2002). 15-lipoxygenase metabolism of 2-arachidonylglycerol — generation of a peroxisome proliferator-activated receptor alpha agonist. *J. Biol. Chem.,* 277, 23278–23286.

Kozak, K.R. and Marnett, L.J. (2002). Oxidative metabolism of endocannabinoids. *Prostaglandins Leukot. Essent. Fatty Acids,* 66, 211–220.

Kozak, K.R., Prusakiewicz, J.J., Rowlinson, S.W., Schneider, C., and Marnett, L.J. (2001). Amino acid determinants in cyclooxygenase-2 oxygenation of the endocannabinoid 2-arachidonylglycerol. *J. Biol. Chem.,* 276, 30072–30077.

Kozak, K.R., Rowlinson, S.W., and Marnett, L.J. (2000). Oxygenation of the endocannabinoid, 2-arachidonylglycerol, to glyceryl prostaglandins by cyclooxygenase-2. *J. Biol. Chem.,* 275, 33744–33749.

Liang, Y., Li, C., Guzman, V.M., Evinger, A.J., Protzman, C.E., Krauss, A.H-P., and Woodward, D.F. (2003). Comparison of PGF2a, bimatoprost (prostamide) and butaprost (EP2 agonist) on Cyr61 and CTFG gene expression. *J. Biol. Chem.,* 278, 27267–27277.

Lopshire, J.C. and Nicol, G.D. (1997). Activation and recovery of the PGE2-mediated sensitisation of the capsaicin response in rat sensory neurons. *J. Neurophysiol.,* 78: 3152–3164.

Maccarrone, M., Fiorucci, L., Erba, F., Bari, M., Finazziagro, A., and Ascoli, F. (2000). Human mast cells take up and hydrolyze anandamide under the control of 5-lipoxygenase and do not express cannabinoid receptors. *FEBS Lett.,* 468, 176–180.

Maccarrone, M., Salvati, S., Bari, M., and Finazzi-Agro, A. (2000). Anandamide and 2-arachidonoylglycerol inhibit fatty acid amide hydrolase by activating the lipoxygenase pathway of the arachidonate cascade. *Biochem. Biophys. Res. Commn.,* 278, 576–583.

Moody, J.S., Kozak, K.R., Ji, C., and Marnett, L.J., (2001). Selective oxygenation of the endocannabinoid 2AG by leukocyte-type 12-lipoxygenase. *Biochemistry* 40, 861–866.

Pertwee, R.G. and Ross, R.A. (2002). Cannabinoid receptors and their ligands. *Prostaglandins Leukot. Essent. Fatty Acids,* 66, 101–121.

Piomelli, D. (2001). The ligand that came from within. *Trends Pharmacol. Sci.,* 22, 17–19.

Ross, R.A., Craib, S.J., Stevenson, L.A., Pertwee, R.G., Henderson, A., Toole, J., and Ellington, H.C. (2002). Pharmacological characterization of the anandamide cyclooxygenase metabolite: prostaglandin E2 ethanolamide. *J. Pharmacol. Exp. Ther.,* 301, 900–907.

Sharif, N.A., Kelly, C.R., and Crider, J.Y., (2003). Human trabecular meshwork cell responses induced by Bimatoprost, Travoprost, Unoprostone and other FP prostaglandin receptor agonist analogues. *Invest. Ophthalmol. Vis. Sci.,* 44, 715–721.

Sharif, N.A., Kelly, C.R., and Williams, G.W. (2003). Bimatoprost (lumigan) is an agonist at the cloned human ocular FP prostaglandin receptor: real-time FLIPR-based intracellular Ca^{2+} mobilization studies. *Prostaglandins Leukot. Essent. Fatty Acids,* 68, 27–33.

Shin, J., Cho, H., Hwang, S.W., Jung, J., Shin, C.Y., Lee, S-Y., Kim, S.H., Lee, M.G., Choi, Y.H., Kim, J., Haber, N.A., Reichling, D.B., Khasar, S., Levine, J.D., and Oh, U. (2002). Bradykinin-12-lipoxygenase-VR1 signalling pathway for inflammatory hyperalgesia. *Proc. Natl. Acad. Sci. USA,* 99, 10150–10155.

Smith, J.A.M., Davis, C.L., and Burgess, G.M., (2000). Prostaglandin E2-induced sensitization of bradykinin-evoked responses in rat dorsal root ganglion neurons is mediated by cAMP-dependent protein kinase A. *Eur. J. Neurosci.,* 12, 3250–3258.

Spada, C.S., Woodward, D.F., Krauss, A.H-P., Neives, A.L., Wheeler, L.A., Scott, D., and Sachs, G. (2002). Bimatoprost and PGF2-alpha selectively stimulate calcium signals in different cat iris sphincter cells. *Exp. Eye Res.,* 72, Suppl. 2, 78.

Szallasi, A. (2002). Vanilloid (capsaicin) receptors in health and disease. *Am. J. Clin. Pathol.,* 118, 110–121.

Ueda, N., Kurahashi, Y., Yamamoto, K., Yamamoto, S., and Tokunaga, T. (1996). Enzymes for anandamide biosynthesis and metabolism. *J. Lipid Mediat. Cell Signal.,* 14, 57–61.

Ueda, N., Yamamoto, K., Yamamoto, S., Tokunaga, T., Shirakawa, E., Shinkai, H., Ogawa, M., Sato, T., Kudo, I., Inoue, K., Takizawa, H., Nagano, T., Hirobe, M., Matsuki, N., and Saito, H. (1995). Lipoxygenase-catalyzed oxygenation of arachidonylethanolamide, a cannabinoid receptor agonist. *Biochim. Biophys. Acta Lipids Lipid Metabol.,* 1254, 127–134.

Van der Stelt, M., Van Kuik, J.A., Bari, M., Van Zadelhoff, G., Leeflang, B.R., Veldink, G.A., Finazzi-Agro, A., Vliegenthart, J.F.G., and Maccarrone, M. (2002). Oxygenated metabolites of anandamide and 2-arachidonoylglycerol: conformational analysis and interaction with cannabinoid receptors, membrane transporter, and fatty acid amide hydrolase. *J. Med. Chem.,* 45, 3709–3720.

Walker, J.M. and Huang, S.M. (2002). Endocannabinoids in pain modulation. *Prostaglandins Leukot. Essent. Fatty Acids,* 66, 235–242.

Woodward, D.F., Krauss, A.H-P., Chen, J., Lai, R.K., Spada, C.S., Burk, R.M., Andrews, S.W., Shi, L., Liang, Y., Kedzie, K.M., Chen, R., Gil, D.W., Kharlamb, A., Archeampong, A., Ling, J., Madhu, C., Ni, J., Rix, P., Usansky, J., Usansky, H., Weber, A., Welty, D., Yang, W., Tang-Liu, D.D.S., Garst, M.E., Brar, B., Wheeler, L.A., and Kaplan, L.J. (2001). The pharmacology of bimatoprost (Lumigan (TM)). *Surv. Ophthalmol.,* 45, S337–S345.

Woodward, D.F., Krauss, A.H.P., Chen, J., Liang, Y., Li, C., Protzman, C.E., Borgardus, R., Chen, R., Kedzie, K.M., Gil, D.W., Kharlamb, A., Wheeler, L.A., Babusis, D., Welty, D., Tang-Liu, D., Cherukury, M., Andrews, M., Burk, R.M., and Garst, M.E. (2003). Pharmacological characterization of a novel antiglaucoma agent, Bimatoprost (AGN 192024). *J. Pharmacol. Exp. Ther.,* 305, 772–785.

Yu, M., Ives, D., and Ramesha, C.S. (1997). Synthesis of prostaglandin E2 ethanolamide from anandamide by cyclooxygenase-2. *J. Biol. Chem.,* 272: 21181–21186.

13 Behavioral Effects of Endocannabinoids

Marie-Hélène Thiébot, Frédérique Chaperon, Ester Fride, and Emmanuel S. Onaivi

CONTENTS

INTRODUCTION

Marijuana is a widely used drug that contains natural cannabinoids, of which Δ^9-THC is the active constituent. These cannabinoids, together with synthetic cannabinoids, are structurally diverse groups of compounds that mediate their behavioral and psychotomimetic effects by activating the cannabinoid and vanilloid receptors directly or indirectly through endogenous ligands. These endocannabinoids and endovanilloids are known to produce diverse physiological effects by activating not only the cannabinoid receptors but also by acting on other systems, including the vanilloid receptors. Accumulating evidence on the mechanism of cannabinoids' action on multiple biological systems is increasing our understanding of the myriad behavioral effects produced by smoking marijuana or induced in animals by the administration of endocannabinoids. Thus, endocannabinoids can mimic the action of exogenous cannabinoids to produce cannabinoid-induced behaviors. The dose dependency of the behavioral effects of endogenous and exogenous cannabinoids is receptor and nonreceptor mediated. This review presents additional knowledge of the behavioral effects of endocannabinoids induced by the activation of cannabinoid receptors and other systems. Because of the diversity of the behavioral effects of endocannabinoids, we limit our coverage to affective, aggressive, appetitive, motivational, and drug- and alcohol-addictive behaviors. We also explore the cellular and molecular mechanisms underlying pain modulation, motor activity, learning, and memory produced by endocannabinoids, which underscore the functional significance of this previously unknown but ubiquitous endocannabinoid physiological control system (EPCS). A number of studies have used *in vivo* and *in vitro* systems to investigate why smoking marijuana is associated with acute impairment of learning and memory. Behavioral data suggest that a disruption of normal hippocampal function contributes to learning and memory impairments produced by cannabinoids (Sullivan, 2000). *In vitro* experiments find that cannabinoids and endocannabinoids produce persistent changes in memory-related neuronal activity (Gerdeman and Lovinger, 2003). Smoking marijuana probably causes the released endocannabinoids and other endogenous molecules to stimulate appetite and intake of palatable foods, supporting a link between cannabinoid signaling and feeding behavior (Soderstrom et al., 2004). In addition, prenatal exposure to marijuana induces subtle but definite deficiencies in memory, motor and addictive behaviors, and in higher cognitive (executive) function in human offspring (Fride, 2004). As reviewed here, a large body of literature indicates that endocannabinoids and cannabinoids suppress behavioral responses to acute and persistent noxious stimulation. This is supported by behavioral, neurophysiological, and neuroanatomical evidence of cannabinoids' role in suppressing nociceptive transmission at central, spinal, and peripheral levels (Hohmann, 2002). It appears that patients find smoked marijuana more powerful in alleviating pain and inducing appetite than administration of cannabinoid agonists. Although systematic clinical reports supporting such anecdotal reports are scanty, it would be of interest to study the differential mechanism of action of cannabinoid agonists, endocannabinoids, and smoked marijuana. We, however, speculate that numerous cannabinoid constituents in marijuana have differing effects on biological systems, some of which are opposing. These diverse effects of different cannabinoids in marijuana may be important in the reported powerful effects of smoked marijuana in pain suppression and induction of appetite in comparison to the administration of cannabinoids like marinol or dronabinol. Another tantalizing alternative is the manipulation of enzymes responsible for the termination of endocannabinoid signaling, including inhibitors, to augment endocannabinoid activity, which might exhibit superior selectivity in their elicited behavioral effects compared to the direct cannabinoid agonists (Cravatt and Lichtman, 2003). The high density of cannabinoid receptors in the basal ganglia and cerebellum is consistent with the role of EPCS in the regulation of motor activity. This can be characterized by alterations in such motor behaviors as catalepsy and ataxia produced in animals after acute administration of cannabinoids and endocannabinoids such as anandamide (Fride and Mechoulam, 1993). The presynaptic localization of CB_1 cannabinoid receptors has been associated with the intricate and complex interaction of the EPCS in modulating the release of neurotransmitters from axon terminals by retrograde signaling. The functional implication and potential therapeutic target of the EPCS arise from this tight association with brain circuits involved in affective behaviors such as anxiety, aggressiveness, schizophrenia, depression, obsessive compulsive disorders, and neurological disturbances. For example, cannabinoid receptors are expressed in brain reward circuit areas and modulate DA release, with consequences in psychiatric disorders, including a role of the endocannabinoid system, which could be exploited in the treatment of addictions, obesity, and psychosis. Data indicating a high comorbidity of smoking marijuana with affective disorders (Soyka, 2003) underscore the role of EPCS in mental health and conjure the self-medication hypothesis of substance abuse. This chapter focuses on the myriad behavioral effects of endocannabinoids, using the behavioral effects of cannabinoid agents in animals (Chaperon and Thiébot, 1999) as a template.

THE ENDOCANNABINOID PHYSIOLOGICAL CONTROL SYSTEM

The advancements in marijuana and cannabinoid research indicate the existence of a previously unknown but elaborate and ubiquitous EPCS in the human body and brain, whose role is unfolding. This remarkable progress includes identification of genes encoding CB_1 (Matsuda et al., 1990; Chakrabarti et al., 1995) and CB_2 receptors (Munro et al., 1993), isolation of endocannabinoids (Devane et al., 1992; Hanus et al., 2001; Mechoulam et al., 1998; Porter et al., 2002; Sugiura et al., 1995) and entourage ligands (Ben-Shabat et al., 1998), and functional identification of transporters and enzymes for the biosynthesis and degradation of these endogenous substances, which represent this EPCS (for a review see Onaivi, 2002; Onaivi et al., 2002). Although the CB_1 turned out to be one of the most abundant neuromodulatory receptors in the brain, both CB_1 and CB_2 receptors are widely distributed in peripheral tissues with CB_2 particularly enriched in immune tissues (Wilson and Nicol, 2002; Sugiura and Waku, 2002). Despite this wealth of information and major advances, little information is available at the molecular level about cannabinoid receptor gene structure, regulation, function, and polymorphisms. Consequently, the contribution of CB receptor genes and the way they are regulated in the manifestation of the behavioral and physiological effects of cannabinoids are poorly understood.

AFFECTIVE BEHAVIOR

ANXIETY-RELATED BEHAVIOR

In humans, contradictory findings have been reported regarding involvement of the cannabinoid system in anxiety disorders. Thus, although experienced users report a mild relaxation after marijuana smoking, anxiety is the most common unpleasant side effect of occasional cannabis use and seems a reason for discontinuation of use (Hall and Solowij, 1998). A double-blind study gave evidence of a large increase in the level of anxiety in healthy volunteers after ingestion of Δ^9-THC (Zuardi et al., 1982). However, other double-blind placebo-controlled studies showed that chronic treatment with nabilone, a synthetic cannabinoid, decreased anxiety score (Hamilton Anxiety Scale) in anxious patients (Fabre and McLendon, 1981; Ilaria et al., 1981). Similar discrepancies have been described in animals. Indeed, cannabinoids are able to display both anxiogenic- and anxiolytic-like effects depending upon doses, animal models, specific test conditions, and strains (Onaivi et al., 1990; Onaivi et al., 1996; Rodríguez de Fonseca et al., 1996). Overall, low doses of cannabinoid agonists usually induce an anxiolytic-like effect, whereas higher doses cause the opposite response. Some results suggest that the endocannabinoid system is involved in the control of emotional behavior via CB_1 receptors. Neuroanatomical studies showed that this receptor is expressed at high levels in brain regions involved in the control of fear and anxiety, such as the basolateral amygdala, the anterior cingulated cortex, the prefrontal cortex, and the paraventicular nucleus (PVN) of the hypothalamus (Mailleux and Vanderhaeghen, 1992; Tsou et al., 1998). Furthermore, both exogenous and endogenous cannabinoids have been found to activate the hypothalamic–pituitary–adrenal (HPA) axis, the neuroendocrine system implicated in responses to emotional stress (Weidenfeld et al., 1994; Wenger et al., 1997). Acute intracerebroventricular (i.c.v.) administration of anandamide or Δ^9-THC stimulates the release of adreno-corticotropin hormone (ACTH) and corticosterone, probably via a central mechanism, which involves the secretion of corticotropin releasing factor (CRF). Consistent with the fact that the CRF antagonist D-Phe CRF_{12-41} prevented anxiogenic-like effects of HU210, a synthetic cannabinoid agonist (Rodríguez de Fonseca et al., 1996), these results suggest that HPA activation plays an important role in the mediation of cannabinoid–induced anxiogenic patterns. However, a recent study pointed out a mechanism of rapid glucocorticoid feedback inhibition of the HPA involving a release of endocannabinoid and the activation of CB_1 receptors in the PVN (Di et al., 2003). Moreover, mice lacking CB_1 receptors (KO) exhibited an increased basal level of anxiety (Haller et al., 2002; Maccarrone et al., 2002; Martin et al., 2002), and the highly selective CB_1 receptor antagonist rimonabant (SR141716) induced anxiety-like responses in rats subjected to the defensive withdrawal test (Navarro et al., 1997) and the elevated plus maze (Arévalo et al., 2001). These data suggest the existence of an intrinsic anxiolytic tone mediated by endogenous cannabinoids. In agreement with this hypothesis, the blockade of anandamide hydrolysis by URB597 and URB532, two fatty acid amide hydrolase (FAAH) inhibitors, induced anxiolytic-like effects in rats subjected to the elevated zero maze and in rat pups in isolation-induced ultrasonic vocalizations. These effects were prevented by rimonabant (Kathuria et al., 2003). FAAH inhibitors may modulate anxiety-related behaviors by enhancing the tonic action of anandamide on a subset of CB_1 receptors, which may normally be engaged in controlling emotions and stress.

However, some conflicting data have been reported, suggesting possible species difference in the role of endocannabinoid system and CB_1 receptors in mechanisms of anxiety. Although rimonabant exerted anxiogenic-like effects in rats, as mentioned in the previous paragraph (Navarro et al., 1997; Rodríguez de Fonseca et al., 1997), as well as in DBA/2 mice (Akinshola et al., 1999), surprisingly, it reduced anxiety-like behavior in ICR and $C_{57}Bl_6$ mice subjected to the light-dark box and the elevated plus maze (Akinshola et al., 1999; Haller et al., 2002). Interestingly, Rodgers et al. (2003) reported that rimonabant produced some anxiolytic-like effects in maze-experienced mice but not in naïve mice, emphasizing the crucial importance of the animals' experimental history in the consequences of a blockade of CB_1 receptors. These results suggest that initial exposure to the plus maze fails to recruit the endocannabinoid system. In contrast, the reexposure to the maze may activate the endocannabinoid substrates, which should induce a qualitatively different basal anxiety via an inhibition of GABA release (see Schlicker and Kathmann, 2001). Consistent with a role of endocannabinoids in memory and learning processes, Marsicano et al. (2002) demonstrated that reexposure of mice to learned fear cues selectively increased endocannabinoid levels in the basolateral amygdala. Thus, during the second plus-maze exposure, fear cues acquired during the first trial would activate the endocannabinoid system, producing an elevation of basal anxiety level. Under these conditions, a blockade of CB_1 receptors would result in an apparent anxiolytic profile.

Interestingly, it has been reported that four-month-old CB_1 KO mice display decreased anandamide levels in the hippocampus compared with young (one-month-old) KO or age-matched wild-type mice, an effect correlated with a mild reduction of anxiety-related behavior in the light-dark box (Maccarrone et al., 2002). This suggests that CB_1 receptors are involved in age-dependent adaptive changes of endocannabinoid metabolism, which appear to correlate with waning of anxiety-like behavior exhibited by young CB_1 KO mice. However, consistent with the weak strength of this effect, it appears that anxiety-related behaviors may be less dependent on an activation of CB_1 receptors than previously thought. Furthermore, a recent study demonstrated that rimonabant was able to reduce anxiety in the elevated plus maze in both wild-type and CB_1 KO animals (Haller et al., 2002). This suggests that a novel rimonabant-sensitive non-CB_1 neuronal cannabinoid receptor would be involved in anxiety mechanisms (Di Marzo et al., 2000; Breivogel et al., 2001). The existence of two different central cannabinoid receptors, having opposing influences on the expression of anxiety-like behavior, may explain the contradictory literature.

SCHIZOPHRENIA

Schizophrenia is a complex psychiatric disorder characterized by symptoms such as delusions, hallucinations, deterioration of social functioning, and cognitive deficits. Although dopaminergic and glutamatergic neurotransmissions are thought to be abnormal, the neurobiological and neurobiochemical bases of this disease are still poorly understood. Several clinical findings suggest that schizophrenia may be associated with functional anomalies in the endocannabinoid signaling system. First, the prevalence of regular or problematic cannabis use is higher in schizophrenic patients than in the general population (Bersani et al., 2002). Second, similarities between some effects of cannabinoid intoxication and some symptoms of schizophrenia, especially regarding cognitive disturbances, hallucinations, perceptual distortion, and paranoia, have been shown in many reports (Wylie et al., 1995; Hall and Solowij, 1998; Leweke, Schneider, et al., 1999; Skosnik et al., 2001; Fergusson et al., 2003). Third, heavy abuse of cannabis can be considered as a factor eliciting relapse in patients with schizophrenia and possibly a premorbid precipitant (Linszen et al., 1994; Linszen et al., 1997). Clinical data indicate that cannabis consumption might place vulnerable subjects at particular risk of developing psychotic symptoms and, perhaps, lasting psychotic disorders (Zammit et al., 2002; Degenhart, 2003). Otherwise, a recent genetic study on CB_1 receptor gene polymorphisms reported that certain alleles or genotypes of this gene might confer a susceptibility to the hebephrenic type of schizophrenia (Ujike et al., 2002). Finally, elevated levels of anandamide and palmitylethanolamide (PEA), another endogenous cannabinoid, have been measured in the cerebrospinal fluid (CSF) of ten schizophrenic patients compared with controls (Leweke, Giuffrida, et al., 1999).

Although an effect of antipsychotic medication cannot be excluded, it seems that such functional abnormalities in endogenous cannabinoid signaling may participate in the pathogenesis of schizophrenia. Leweke, Giuffrida, et al. (1999) considered several mechanisms that could account for this elevation of brain anandamide and PEA concentrations. An *in vivo* microdialysis study showed that activation of DA D2-like receptors by quinpirole increased anandamide release in rat dorsal striatum (Giuffrida et al., 1999). Thus, the high levels of anandamide found in schizophrenics' CSF might result from an overstimulation of D2-like receptors, due to the activation of DA neurotransmission in these patients. Alternatively, increased CSF anandamide levels may reflect a primary "hypercannabinergic" state, which may occur in schizophrenic patients. This dysregulation of endocannabinoid systems may then modify glutamate and DA neurotransmission. Indeed, evidence exists that CB_1 receptor activation could reduce glutamate release in rat striatum (Gerdeman and Lovinger, 2001; Huang et al., 2001). Other findings suggested that complex interactions exist between DA and cannabinoid systems. In rats, autoradiography and *in situ* hybridization studies showed that CB_1 receptors are highly expressed in basal ganglia, limbic structures (hippocampus, olfactory bulbs, and septum), and cerebellum (Herkenham et al., 1991; Mailleux and Vanderhaeghen, 1992).

In human brain, high densities of CB_1 receptors are also found in basal ganglia, limbic system, and the cerebral cortex (Glass et al., 1997), providing an opportune anatomical substrate for functional interactions between endocannabinoid and dopaminergic systems. In mice, a recent study demonstrated that systemic administration of the synthetic cannabinoid agonists CP55940 and WIN55212-2 induced Fos expression within A10 DA neurons. This effect, probably mediated by CB_1 receptors because it was prevented by rimonabant, also appeared to be dependent upon an activation of noradrenergic neurotransmission (Patel and Hillard, 2003). These results are consistent with earlier works showing that cannabinoids increased the firing rate of DA neurons in the ventral tegmental area (VTA) and the substantia nigra (SN) (French et al., 1997) and enhanced DA release in the medial prefrontal cortex (mPFC) and the nucleus accumbens (NAcc) (Chen, Paredes, Li, et al., 1990; Chen, Paredes, Lowinson, et al., 1990). Interestingly, an *in vivo* microdialysis study in rats indicated that rimonabant significantly increased the efflux of DA selectively in the mPFC but not in the NAcc (Tzavara et al., 2003). As a decrease in PFC function has been proposed to contribute to the physiopathology of schizophrenia (Grace, 1991; Knable and Weinberger, 1997),

CB_1 receptor antagonists, by enhancing mesocortical dopaminergic neurotransmission, might have some therapeutic potential in psychoses.

However, such a proposal is challenged by recent results obtained in rats subjected to the prepulse inhibition (PPI) procedure, claimed to model the sensorimotor gating deficit found in schizophrenics. Indeed, in this paradigm, the startle response to an acoustic stimulus is significantly blunted by preexposure to a priming stimulus presented a few milliseconds earlier, and this effect is disrupted by psychotomimetics such as amphetamine or apomorphine. Distinct to a variety of clinically effective antipsychotic drugs, rimonabant failed to reverse the PPI disruption induced by apomorphine, MK 801, and amphetamine (Martin et al., 2003). This result, together with the failure of CP55940 to disrupt PPI at doses producing no severe decrease of startle amplitude (Mansbach et al., 1996; Martin et al., 2003), suggests that CB_1 receptors exert only a modest control over DA and glutamate pathways involved in sensory motor gating mechanisms. On the other hand, Poncelet et al. (1999) demonstrated that rimonabant, like clozapine, antagonized the hyperactivity induced in habituated gerbils by drugs such as cocaine, amphetamine, and WIN55212-2, known to produce or exacerbate schizophrenic symptoms, and suggested that the endocannabinoid system might play a role in the psychostimulant effect of these drugs. Nevertheless, data from a phase IIa metatrial performed in schizophrenic patients indicated that, unlike haloperidol, rimonabant was not different from placebo on any efficacy endpoint (Arvanitis et al., 2001).

It clearly appears that if the endocannabinoid system really plays a role in the physiopathology of schizophrenia, the activation of CB_1 receptors does not seem to be a crucial point. Interestingly, CB_1 is not the sole CB receptor subtype expressed in the brain. Some studies using CB_1 KO mice supported the existence of G-protein-coupled non-CB_1 and non-CB_2 cannabinoid receptors, sensitive to endocannabinoids, synthetic agonists, and rimonabant (Di Marzo et al., 2000; Breivogel et al., 2001; Fride et al., 2003). In addition, anandamide has been found to activate both CB_1 and vanilloid VR_1 receptors (for review see Di Marzo et al., 2002). However, the localizations and functions of such types of receptors are unclear, and their putative role in schizophrenia remains to be demonstrated. Although cannabis use appears to be neither a sufficient nor a necessary cause for psychosis, the current status of research on cannabis-associated psychosis (Leweke et al., 2004) is inconclusive, and further research is needed to understand the mechanisms by which cannabis is associated with psychosis.

Depression

Whereas several preclinical findings support cannabinoids' role in anxiety-related behaviors (as mentioned earlier) the implication of EPCS in the regulation of affective disorders seems more difficult to establish. In humans, the comorbidity of cannabis abuse and depression is relatively common in clinical and community populations (Angst, 1996; Agosti et al., 2002). However, the existence of a causal association between cannabis use and depression remains controversial (Degenhart et al., 2001; Chen et al., 2002). Some longitudinal studies in adults reported that cannabis abuse may increase the risk of developing depressive symptoms (Weller and Halikas, 1985; Angst, 1996; Bovasso, 2001; Degenhart, 2003). In addition, it has been found that early marijuana use in childhood or adolescence was related to major depressive disorders in adulthood (late 20s) (Green and Ritter, 2000; Brook et al., 2002; Patton et al., 2002; Fergusson et al., 2003). These later findings suggest that depression may follow cannabis abuse rather than vice versa. In contrast, other findings suggest that preexisting depressive symptoms might raise the likelihood of cannabis use through a mechanism of self-medication. Thus, anecdotal reports indicated that cannabis would induce an antidepressant effect and would be also useful in the treatment of bipolar disorders (Gruber et al., 1996; Grinspoon and Bakalar, 1998). In line with these findings, in the oncology population, Δ^9-THC could have positive effects to enhance mood in addition to the recognized antinausea and analgesic action (Walsh et al., 2003). Finally, to the best of our knowledge, a possible role of the endocannabinoid system in the physiopathology of depression has never been suggested in humans. The preliminary findings from the laboratory of Onaivi et al., 2004 (unpublished reports) indicate that CB_2 receptors can be found in the brain of naïve mice and that the expression of the CB_2 receptors is enhanced in the brain of mice subjected to chronic mild stress.

Animal studies also provided contradictory results. In the mouse forced–swimming or tail-suspension tests, two procedures predictive of an antidepressant activity (Porsolt et al., 1977; Stéru et al., 1987), rimonabant and AM251, a CB_1 receptor antagonist/inverse agonist, induced an antidepressant-like reduction of the time spent immobile, at doses that did not affect locomotor activity (Shearman et al., 2003; Tzavara et al., 2003). Consistent with the CB_1 receptors' role in such effects, the AM251-induced reduction of immobility did not occur in CB_1 KO mice subjected to the forced-swimming test (Shearman et al., 2003). Moreover, *in vivo* microdialysis experiments also showed that rimonabant increased the efflux of norepinephrine (NE), DA, and 5-HT in the mPFC (Tzavara et al., 2003), an effect which has been proposed to play a role in the antidepressant-like effect of clinically effective drugs such as desipramine and fluoxetine (Bymaster et al., 2002). However, opposite results have been reported with CB_1 KO mice; compared to their wild-type counterparts, mutant mice exhibited higher depressive-like responses in the chronic unpredictable mild stress (CMS) procedure (Martin et al., 2002). This paradigm has been proposed as a valid animal model of depression, especially of anhedonia (Willner, 1997). Defined as a loss of interest or pleasure, anhedonia is one of the core symptoms of a major depressive episode (DMS-IV-TR, 2000). CMS has been shown to cause an antidepressant-reversible reduction of consumption of sucrose solutions, which is hypothesized to reflect a decrease in the perception of the rewarding value of sucrose. In this procedure, CB_1 KO mice showed an enhanced sensitivity in developing a depressive-like state. Together, these results suggest that an endogenous cannabinoid tone may contribute to the maintenance of mood, probably through a modulation of monoaminergic pathways. However, as shown previously (Chaperon and Thiébot, 1999), the perception of the motivational value of positive reinforcers seems to require the stimulation of CB_1 receptors. Although baseline levels of sucrose intake were similar in CB_1 KO and wild-type mice (Tzavara et al., 2003; Martin et al., 2002), it cannot be excluded that the larger reduction of sucrose consumption observed in stressed mutants reflected a lowered perception of the rewarding value of sucrose rather than an increased sensitivity to the chronic stress regimen *per se*. Therefore, further studies are necessary to elucidate the exact relationship between the endocannabinoid system and mood control.

Aggressive Behavior

Aggressive behavior is not a unitary phenomenon and accordingly is more complex and difficult to analyze and interpret than several other behaviors. Many diverse definitions of aggression have been proposed, particularly as the term is applied to human behavior. Only a relatively few of these definitions have been applied to animal research paradigms. Initial studies demonstrated that cannabinoids might alter aggressiveness in both humans and animals. However, the numerous work performed during the 1970s and 1980s have shown some inconsistent results. In normal animals, it seems that acute administration of cannabis or Δ^9-THC might reduce aggressive behavior, probably due to a suppressant effect on locomotor activity and a depression of general motivation (for a review, see Abel, 1975; Frischknecht, 1984). For example, in pigeons, the rate of pecking maintained by access to a stuffed target bird that could be attacked was reduced by Δ^9-THC (0.5 and 1 mg/kg), at doses which did not affect key pecking intermittently reinforced with food (Cherek et al., 1980). Moreover, Miczek (1978) showed that when a resident animal (mouse, rat, or squirrel monkey) is confronted by an intruder conspecific, Δ^9-THC (0.25 to 2 mg/kg) decreased species-specific attack behavior. More recently, a study by Sulcova et al. (1998) demonstrated that singly housed timid male mice exhibited aggressive reactions toward a nonaggressive intruder following a 7-d chronic treatment with anandamide at a very small dose (0.01 mg/kg.d). A larger dose (1 mg/kg.d) produced no noticeable effect, whereas in similar conditions, anandamide (10 mg/kg.d) reduced agonistic behavior and enhanced defensive conducts in otherwise spontaneously aggressive mice, an effect perhaps linked to motor deficits induced by such a high dose. This biphasic pattern of effect seems to be a feature of cannabinoids and especially of anandamide. However, it seems that under particularly stressful experimental conditions, Δ^9-THC can also facilitate aggressive behavior (Abel, 1975). For instance, when administered to rats already stressed by a 4-d REM sleep deprivation, cannabis extract and Δ^9-THC provoked irritability and aggressiveness and impaired defensive–submissive behavioral pattern (Carlini, 1977). In isolated rats, food deprived for 22 h and then fed *ad libitum* for a 3-h period, a single injection of Δ^9-THC (11 mg/kg) induced mouse-killing (muricidal) behavior with enhanced aggressiveness, as indicated by the dramatic increase in the number of attacks on the dead mouse until it was completely torn in pieces (Bac et al., 1998). These authors also showed that Δ^9-THC, at doses (2, 4, and 8 mg/kg) inactive to induce muricidal behavior in control rats, became efficient in rats suffering a magnesium (Mg^{2+}) deprivation for 6 weeks. A severe Mg^{2+} deficiency (50-ppm diet) induced killing behavior by itself, and Δ^9-THC exacerbated further attacks on the dead mouse. A moderate Mg^{2+}-deficient diet (150-ppm) alone did not produce muricidal behavior, but all the rats became mouse killers when given Δ^9-THC, whatever the dose. These results suggest a potentiation between both treatments to elicit aggressiveness. Δ^9-THC would act as a trigger to induce aggression in Mg^{2+}-deficient rats and reciprocally Mg^{2+} deficiency would reveal the potential neurotoxicity of a low dose of Δ^9-THC (Bac et al., 2002). Surprisingly, an increase in aggressive responses has been observed in CB_1 KO mice subjected to the resident-intruder test, but this effect occurred exclusively during the first interaction session (Martin et al., 2002). This suggests that an absence of CB_1 receptor–mediated endocannabinoid tone may lead to emergence of aggressive reactions. However, an enhanced basal level of anxiety (as measured in the light–dark test) in CB_1 KO mice can also facilitate the increase of aggressiveness during the first session of the test.

Clinical assays devoted to study the effects of oral Δ^9-THC or smoked marijuana on human aggression have also produced inconsistent results. Using a procedure in which the subjects received intense provocation (electric shock administered by an opponent), Myerscough and Taylor (1985) showed that a low dose (0.1 mg/kg) of oral Δ^9-THC tended to increase aggressive responses. In contrast, subjects receiving a larger dose (0.4 mg/kg) behaved in a relatively nonaggressive manner throughout the experimental session. On the other hand, using the point-subtraction aggression paradigm, Cherek and Dougherty (1995) found that smoked marijuana reduced the enhanced rate of aggressive responding induced by a shift from low to high level of provocation.

The mechanism involved in the alterations of aggressive behavior by CB receptor ligands, exclusively in certain stress-forming situations, is not yet precisely understood. A link between central 5-HT levels and aggressive behavior has been established in humans and animals (for review see Chiavegatto and Nelson, 2003). High aggression in humans is correlated with low CSF concentrations of the 5-HT metabolite 5-hydroxyindoleacetic acid (5-HIAA), suggesting a diminution of 5-HT turnover. In aggressive laboratory animals, brain 5-HT turnover is also reduced, and pharmacological manipulations of the serotoninergic system substantiate a negative correlation between 5-HT neurotransmission and aggressive behavior. Accordingly, mice lacking the 5-HT transporter (5-HTT KO) were found to be less aggressive than wild-type controls (Holmes et al., 2002).

Interestingly, the CB receptor agonist CP55940 inhibited the production of nitric oxide (NO), an effect probably mediated through CB_1 receptors, because it was reversed by rimonabant (Waksman et al., 1998). NO modulates many behavioral and neuroendocrine responses, and its synthetic enzyme NO synthase has been found in high densities within emotion-regulating brain regions (Nelson et al., 1997). Recently, the excessive aggressive and impulsive traits of neuronal NO synthase knockout mice ($nNOS^{-/-}$) have been attributed to reductions in 5-HT turnover (Chiavegatto and Nelson, 2003). This suggests that possible modifications of the endogenous cannabinoid system would contribute, via NO-related mechanisms, to local changes in 5-HT levels leading to aggressive behavior. Furthermore, REM sleep deprivation and diet restriction have been shown to reduce 5-HT levels in the medial medullary reticular formation and in the hypothalamus and hippocampus, respectively (Hao et al., 2000; Blanco-Centurion and Salin-Pascual, 2001). Anandamide also decreased the concentration of 5-HT in the hippocampus, and rimonabant was found to increase brain 5-HT levels and turnover (Hao et al., 2000; Darmani et al., 2003; Tzavara et al., 2003). Together these data suggest that some categories of stress (diet restriction, REM sleep deprivation) combined with an alteration of the endocannabinoid control system could lead to a reduction in 5-HT transmission, which may favor the development of aggressive behavior. Unfortunately, the possible involvement of endocannabinoids in aggression has not been investigated so far, although this may represent an exciting research field. In particular, it would be very interesting to study if, whether or not in humans, a latent dysfunction of the endocannabinoid system could explain the propensity of reputedly violent individuals to react aggressively in many stressful circumstances (aggression as trait characteristic of an individual).

APPETITE AND FEEDING BEHAVIOR

There is increasing evidence of endocannabinoids' role in the regulation of appetite. Exogenous cannabinoids (smoked marijuana, Δ^9-THC, dronabinol) have been shown to stimulate eating in humans (Abel, 1971; Foltin et al., 1988; Mattes et al., 1994), and this property may be used to enhance appetite in patients with cancer, AIDS, and Alzheimer's disease, and even in the elderly (Plasse et al., 1991; Volicer et al., 1997; Balog et al., 1998; Berry and Marcus, 2000). In rodents, exogenous and endogenous cannabinoids also promote overeating (see the following text), although some discrepancies exist in this respect (Crawley et al., 1993; Graceffo and Robinson, 1998; Giuliani et al., 2000).

The effects of cannabinoid agonists on feeding behavior are easier to observe when spontaneous eating is low, i.e., in animals fed *ad libitum* or when they are provided highly palatable food. For example, in free-feeding Lewis rats, given the choice between chow and highly palatable food, Δ^9-THC administered either systemically (0.5 to 1 mg/kg i.p.) or centrally (1 to 25 μg i.c.v.) induced a short-lasting (30 to 60 min), binge-like pattern of palatable food eating (Koch and Matthews, 2001). However, higher doses of Δ^9-THC (2 or 2.5 mg/kg i.p.) were inactive or even induced hypophagia. In presatiated rats, Δ^9-THC (0.5-2 mg/kg p.o.) induced an early large increase in usual food consumption, which was subsequently offset, so that a 24-h intake was similar to that of controls (Williams et al., 1998). Likewise, acute systemic injections of anandamide (up to 10 mg/kg s.c. or i.p.) or the synthetic CB receptor agonist, WIN55212-2 (0.4-2 mg/kg i.p.), enhanced food intake in presatiated rats (but not in 24-h food-deprived animals) and, curiously, were inactive on central administration (Williams and Kirkham, 1999; Gómez et al., 2002; Kirkham et al., 2002). The stimulation of feeding by Δ^9-THC, anandamide, and 2-AG was mediated by CB_1 receptors because it was selectively blocked by rimonabant but not SR144528, a selective CB_2 receptor antagonist (Williams and Kirkham, 1999; Kirkham et al., 2002; Williams and Kirkham, 2002). In mice subjected to a 40% diet restriction, a 1-week chronic administration of anandamide at a very low dose (0.001 mg/kg.d i.p.) enhanced food intake during daily 2.5-h feeding periods (Hao et al., 2000), but not at higher doses (0.7 and 4 mg/kg). The biphasic dose-response modulation of feeding, frequently observed in these studies, might be related to motor deficits reported at Δ^9-THC and anandamide doses above 2 to 3 mg/kg (Crawley et al., 1993; Romero et al., 1995). Such kind of effect might explain why acute and subacute administration of high doses of the potent CB receptor agonist HU210 (25 to 100 μg/kg i.p.) produced dose- and time-dependent decreases of body weight. However, the reduction of food intake induced by the highest dose of HU210 was still present 7 d after the last of 4 daily injections, suggesting that perhaps other factors unrelated to ingestive behavior are involved in this anorectic effect (Giuliani et al., 2000).

On the other hand, the selective blockade of the CB_1 receptors by acute or subacute administrations of rimonabant reduced feeding in nonobese animals, whether nonfasted or food restricted. In some of these studies, rimonabant selectively reduced the consumption of palatable food (or drink), while having little effects on regular food or water intake (Arnone et al., 1997; Colombo et al., 1998; Simiand et al., 1998; Gallate and McGregor, 1999), suggesting that CB_1 receptors have a role not only in the consummatory aspect of feeding but also in the hedonic evaluation of food.

However, other studies showed that food deprivation or high palatability are not necessary to observe suppressive effects of rimonabant on food intake (Freedland et al., 2000; Rowland et al., 2001; Gómez et al., 2002; McLaughlin et al., 2003; Verty et al., 2004). Another selective CB_1 receptor antagonist, AM281 (20 to 40 μg i.c.v.), also blocked chow intake in fasting rats (Werner and Koch, 2003). This indicates that an endogenous cannabinoid tone might normally stimulate feeding behavior (Berry and Mechoulam, 2002). However, the control of ingestive behavior by endocannabinoids might be age dependent, because the usual food intake following an 18-h deprivation period was reduced by rimonabant in young (6 to 10 weeks) but not old (26 to 48 weeks) mice (Wang et al., 2003).

The endocannabinoid system may also play a key role in overeating behavior and obesity. In mice with high fat diet-induced obesity, chronic administration of rimonabant induced a transient reduction of food intake and a sustained reduction of body weight gain and adiposity (Ravinet Trillou et al., 2003). In free-feeding Zucker rats, rimonabant caused dose-dependent reductions of eating and weight gain which were greater in genetically obese (fa/fa) animals than in their lean counterparts (Vickers et al., 2003). Interestingly, in both studies, hypophagia was limited to the first week of treatment, whereas the reduction of body weight gain in obese animals was maintained throughout the 4-week (rats) or 5-week (mice) chronic administration (Ravinet Trillou et al., 2003; Vickers et al., 2003). Rimonabant discontinuation resulted in a rebound hyperphagia, and weights rapidly returned to (and perhaps exceeded) vehicle levels (Vickers et al., 2003). On the other hand, unrestricted obese db/db mice given a 7-d rimonabant (3 mg/kg.d i.p.) treatment also exhibited acute reduction of food intake, resulting in weight loss, but tolerance developed rapidly, so that by day six, weights were similar to vehicle-injected db/db controls (Di Marzo et al., 2001).

Studies using CB_1 KO mice gave further evidence of the involvement of CB_1 receptors in the regulation of feeding, because these mice ate less than their wild-type littermates in response to an 18-h fasting period, or an i.c.v. infusion of the orexigenic neuropeptide Y (NPY) (Di Marzo et al., 2001; Poncelet et al., 2003). CB_1 KO mice also displayed less weight gain, essentially because of lower body fat mass, regardless of what they were fed *ad libitum*: with regular chow (Cota et al., 2003); but see Ledent et al., 1999; Zimmer et al., 1999), or with an obesity-promoting diet (Cota et al., 2003; Ravinet Trillou et al., 2003). Moreover, rimonabant reduced the hyperphagic response to fasting, NPY, and high fat diet in wild-type but not CB_1 KO mice (Di Marzo et al., 2001; Poncelet et al., 2003; Ravinet Trillou et al., 2003). Altogether, these results indicate that, by activating CB_1 receptors, endocannabinoids tonically stimulate eating and participate in the regulation of body weight.

Apparently, endocannabinoid levels in brain are sensitive to a variety of physiological and biochemical changes. In particular, acute food deprivation (20 to 24 h) induced moderate increases (from 50% to 150%) of anandamide and/or 2-AG levels in rat brain areas involved in feeding control (Kirkham et al., 2002), as well as of 2-AG in mouse brain (Hanus et al., 2003). In contrast, a 12-d diet restriction has been reported to lower 2-AG levels in mouse hippocampus and hypothalamus, an effect that might precipitate in some anorectic-like coping strategies when food is scarce (Hanus et al., 2003). In starving rats, a sevenfold increase of anandamide was also found in the small intestine, which was normalized by refeeding (Gómez et al., 2002). Interestingly, a capsaicin-induced sensory deafferentation prevented the stimulation of food intake by systemic administration of anandamide or WIN55212-2 and the anorectic effect of rimonabant (Gómez et al., 2002). According to Cota et al. (2003), the lower body fat mass in CB_1 KO mice might be determined by both hypothalamic alterations and impaired adipocytes functions. Together, these results suggest that, in addition to central receptors, peripheral CB_1 receptors participated in the regulation of ingestive behavior.

A possible reciprocal link may exist between endocannabinoid mechanisms and leptin, an appetite-suppressing peptide hormone. Leptin, released from adipocytes, is believed to signal the nutritional status to brain areas controlling appetite. It directly stimulates the action of anorexigenic agents (such as α-melanocyte-stimulating hormone) and suppresses the effects of orexigenic peptides (primarily, NPY), thereby decreasing appetite. Defective leptin signaling — as observed in genetically obese fa/fa Zucker rats (non functional leptin receptor), ob/ob mice (leptin deficient) or db/db mice (defective leptin receptor) — is associated with elevated 2-AG levels in the hypothalamus but not the cerebellum (an area not directly related to feeding), as compared to respective lean controls, and hypothalamic anandamide seems also marginally enhanced in db/db mice (Di Marzo et al., 2001). The administration of leptin, in the range of doses active to reduce food intake and body weight in nonobese animals fed *ad libitum* (Halaas et al., 1997), lowered the hypothalamic levels of 2-AG and anandamide in both normal rats and ob/ob mice (Di Marzo et al., 2001), suggesting that endocannabinoids may be under the negative control of this hormone. Consistent with 2-AG's role in feeding or nutritional status, in nonobese rats, hypothalamic levels of this endocannabinoid undergo feeding-related changes. They were enhanced in fasting rats, reduced in free-feeding animals engaged in eating a highly palatable diet, and identical to those of controls fed *ad libitum*, when rats were allowed to eat a palatable diet to satiety. In contrast, hypothalamic levels of anandamide did not change in any experimental condition (Kirkham et al., 2002). The blockade of CB_1 receptors by rimonabant reduced food intake in nonfasted ob/ob or db/db mice (3 mg/kg i.p.), and fa/fa Zucker rats (10 to 30 mg/kg p.o.), suggesting that the hypothalamic levels of endocannabinoids may contribute to overeating, which leads to obesity in these animals (Di Marzo et al., 2001; Vickers et al., 2003). Because a deficient leptin function has been reported in diet-induced obesity (Lin et al., 2000), it would be interesting to determine the levels of endocannabinoids in eating-related brain structures of animals subjected to such food regimen. Taken altogether, these data suggest that hypothalamic endocannabinoids may play an important role in mediating the appetite-suppressant action of leptin.

Evidence also exists of interactions between the cannabinoid and opioid systems in the regulation of ingestive behavior. The reduction by Δ^9-THC of electrical thresholds for hypothalamic stimulation-induced feeding in satiated rats was attenuated by naloxone (Trojniar and Wise, 1991). Overeating produced by Δ^9-THC or anandamide was reversed by naloxone (Williams and Kirkham, 1999, 2002), and the orexigenic effect of morphine was attenuated by rimonabant (Verty et al., 2003). In addition, subanorectic doses of rimonabant and naloxone acted synergistically to depress food intake (Kirkham and Williams, 2001; Rowland et al., 2001). Leptin is one of the postulated factors (together with CCK, MSH, etc.) that could be involved in the regulation of cannabinoid and opioid interactions, and it has been hypothesized that CB_1 receptors may control opioid-regulated feeding through an action on hypothalamic leptin levels (Verty et al., 2003).

The endocannabinoid system may be of critical importance in the initiation of suckling. Animal and human milk contains 2-AG and large amounts of 2-palmitoyl glycerol and 2-linoleoyl glycerol, two fatty acid glycerol esters, which do not bind to CB receptors but enhance the activity of 2-AG by a so-called "entourage effect" (Ben-Shabat et al., 1998). Recent studies by Fride and co-workers have shown that a high dose of rimonabant, administered once to one-d-old mice, or twice a day between postnatal days two and eight, completely impaired milk ingestion, thus causing death within few days. The first postnatal day seems the most critical period since almost 100% of the pups died after a single injection on day one, and 50% when the injection was done on day two. Such an effect was almost completely suppressed by the coadministration of Δ^9-THC, and 2-AG injected with its entourage molecules delayed the mortality rate. This indicates that endocannabinoids, through the activation of CB_1 receptors, are critical to the initiation of suckling (Fride et al., 2001). Interestingly, CB_1 KO mouse pups also failed to ingest milk during their first day of life, but this effect was transient and suckling was initiated by days two to three (Fride et al., 2003). Despite the absence of CB_1 receptors, rimonabant had a partial detrimental effect on milk intake and survival in these animals, suggesting the existence of a third CB receptor, which can be blocked by rimonabant. Overall, these data indicate that a compensatory mechanism enables the CB_1 KO newborn mice to start drinking (Fride et al., 2003).

MOTIVATIONAL PROCESSES

The activation of the mesolimbic DA neurons is a common feature of drugs of abuse (i.e., cocaine, opioids, amphetamine), and it appears that cannabinoids indirectly activate the same dopaminergic pathways.

ENDOCANNABINOIDS IN THE REWARD PATHWAY

The endocannabinoid system has closed connections with the dopaminergic systems, and this may account for several of its effects. The mesocorticolimbic DA system, and in particular the NAcc, is known to play an important role in the establishment of behaviors that are reinforced by natural- or drug-rewarding stimuli. Mainly located within the A10 cell group in the VTA, DA neurons project to the PFC and to limbic forebrain structures such as the NAcc and the amygdala. The NAcc also receives dense excitatory glutamatergic inputs from the basolateral amygdala, the hippocampus and the PFC. The A10 DA neurons are tonically inhibited by GABAergic interneurons and a long-loop feedback projection from GABAergic medium spiny neurons located in the NAcc. DA itself inhibits both GABAergic and glutamatergic inputs to these NAcc neurons.

Pharmacological studies showed that systemic administrations of cannabinomimetics increase DA neuron firing and DA release within the NAcc and the PFC, and that these effects are counteracted by rimonabant (Chen et al., 1990a; Chen et al., 1990b; French, 1997; Gessa et al., 1998; Wu and French, 2000). Interestingly, Δ^9-THC and WIN 55,212-2 have been reported to enhance extracellular levels of DA selectively in the shell, but not the core, of the NAcc, and this effect claimed to correlate with the abuse liability of drugs, was impaired by rimonabant (Tanda et al., 1997). However, there is no expression of CB_1 receptor mRNA or protein in A10 DA neurons, thus ruling out a direct control of endocannabinoids over DA release in these forebrain structures (Herkenham et al., 1991; Tsou et al., 1998; Patel and Hillard, 2003), or a trans-synaptic reduction of local GABA release, through presynaptic CB_1 receptors on GABA terminals in the VTA. Instead, the activation of presynaptic CB_1 receptors may modulate both glutamatergic and GABAergic neurotransmission in the NAcc. Indeed, CB_1 receptors are present on glutamatergic afferents to the NAcc (Robbe et al., 2001), and their activation, by reducing the excitatory transmission, and thus the firing rate of the GABAergic efferent pathway to the VTA, in turn results in a disinhibition of DA neuron activity (Pistis et al., 2002b). In fact, it seems that endocannabinoids could be released from medium spiny neurons in the NAcc and, by a retrograde activation of presynaptic CB_1 receptors, induce long-term depression of the glutamate input to this structure (Robbe et al., 2002). Some results also indicate that the stimulation of CB_1 receptors may indirectly facilitate mesolimbic DA transmission through an activation of the endogenous opioid. Indeed, in addition to numerous behavioral data in favor of functional interactions between opioid and cannabinoid systems (see below), it has been shown that opioid receptor antagonists can block the CB_1 receptor-mediated release of DA in the NAcc shell (Chen et al., 1990a; Tanda et al., 1997).

REINFORCING EFFECTS OF CANNABINOIDS

In animals, motivational properties of drugs can be approached by three main behavioral procedures. Responding for intravenous (iv) or intracerebroventricular (icv) *self-administration*, which gives direct evidence that drugs serve as reinforcing stimuli on their own, has been demonstrated as a reasonable predictor of abuse liability in humans (Goudie, 1991). The *intracranial self-stimulation (ICSS)* is a self-delivered electrical stimulation of CNS reward circuits. Drug-induced reduction in ICSS reward thresholds which occurs at a variety of brain reward loci, may provide a quantitative index of their reinforcing properties; on the contrary, elevations in ICSS reward thresholds represent an indirect index for the negative affective state during withdrawal (Stellar and Rice, 1989). *Place conditioning procedures* allow the assessment of the perception by the animals of the motivational value of a reinforcer; these procedures are based upon the principle that animals would learn to approach or avoid previously neutral environmental stimuli that have been repeatedly paired with rewarding or aversive events, respectively (Carr et al., 1989). Beside these procedures, *drug discrimination paradigms* are believed to evaluate the presence or absence of subjective and perceptive effects of drugs, though not necessarily reinforcing effects (see e.g. Goudie and Smith, 1999; Dekeyne and Millan, 2003). In such paradigms, animals are trained to detect whether they receive an active drug or saline in order to determine, based on the specific interoceptive stimuli induced by the drug, which of two operant responses will deliver food reward.

Drug discrimination studies indicate that Δ^9-THC and synthetic CB receptor agonists produce subjective effects in rats and monkeys. These effects were blocked by rimonabant, but not by SR 140098, a CB_1 receptor antagonist that does not cross the blood-brain barrier, suggesting that they are mediated by central CB_1 receptors (Wiley et al., 1995b; Pério et al., 1996; Wiley et al., 1997; Wiley et al., 1998; De Vry and Jentzsch, 2002, 2003). Substitution tests showed that, whereas a large majority of non-cannabinoid psychotropic drugs did not produce cannabinoids discriminative stimulus effects (Barrett et al., 1995), a selective cross-discrimination existed between natural and synthetic cannabinoids (Wiley et al., 1995b; Wiley et al., 1995c). In rats or monkeys, anandamide hardly substituted for Δ^9-THC or CP 55,940, probably due to its pharmacokinetic properties (Wiley et al., 1995a; Wiley et al., 1997; Järbe et al., 2001), but its metabolically stable analogs [*(R)*-methanandamide, methylfluoroanandamide] could produce Δ^9-THC discriminative stimulus effect, at least when animals were trained to low doses of Δ^9-THC (Burkey and Nation, 1997; Wiley et al., 1997; Järbe et al., 1998; Järbe et al., 2001). In addition, *(R)*-methanandamide has been used as a discriminative stimulus in rats, and Δ^9-THC produced a complete generalization (Järbe et al., 2001). Because both discriminative stimulus effects of various cannabinoids and marijuana-intoxication symptoms in humans were found to highly correlate with CB_1 receptor binding, it was suggested that the rat model of drug discrimination might be used to predict cannabinoid intoxication in humans (Balster and Prescott, 1992). Rimonabant has been established as a discriminative stimulus in rhesus monkeys given chronic administration of Δ^9-THC, an effect that might relate to a Δ^9-THC withdrawal-like state (McMahon and France, 2003). On the other hand, two studies mentioned unsuccessful attempts to establish rimonabant as a discriminative stimulus on its own, in rats and pigeons (Mansbach et al., 1996; Pério et al., 1996), suggesting that a blockade of the CB_1 receptor-mediated basal activity of the endocannabinoid system does not produce discernible subjective effects. These results are in keeping with the report that, in humans, rimonabant (90 mg orally, i.e., circa 1 to 1.5 mg/kg) has no discernible effect (Huestis et al., 2001). However, discriminative stimulus effects of rimonabant have been reported by using a discriminated taste aversion procedure, which allows measuring the capacity of a drug to serve as a discriminative stimulus predicting that drinking water will result in sickness (due to a post-session injection of lithium chloride). It was the case for rimonabant (though this occurred at a dose which also reduced unconditioned drinking) and in rats that discriminated rimonabant from vehicle, AM-251, another CB_1 receptor antagonist, but not SR 144528, a CB_2 receptor antagonist, substituted for rimonabant. As a corollary to these data, Δ^9-THC dose-dependently attenuated the stimulus effect of rimonabant, indicating that the blockade of CB_1 receptors has consequences that can be perceived by rats (Järbe et al., 2004). It cannot be excluded that peripheral effects such as those eliciting grooming and scratching (Järbe et al., 2002) play a role in rats' ability to discriminate rimonabant from water.

The reinforcing properties of cannabinoids have been rather difficult to establish in animals, and until the last few years, most of the studies failed to give evidence for cannabinoid rewarding effects.

In self-administration procedures, numerous attempts to demonstrate reliable and persistent reinforcing effects of Δ^9-THC and synthetic cannabinoids, in several animal species, were unsuccessful, even when animals had already learned to self-administer other drugs of abuse (for a review see Chaperon and Thiébot, 1999; Tanda and Goldberg, 2003). The first positive results were obtained in experimental conditions different from those used for other drugs of abuse. For instance, WIN 55,212-2 or HU-210, were iv self-administered in an experimental paradigm which used pairs of active and yoked-passive *drug-naïve* mice subjected to a *single* self-administration session (Martellotta et al., 1998; Navarro et al., 2001), and Δ^9-THC iv self-administration was obtained in monkeys with a cocaine self-administration history (Tanda et al., 2000). More recent studies have shown that drug-naive rats can acquire and maintain self-administration of WIN 55,212-2 (iv) (Fattore et al., 2001) and CP 55,940 or Δ^9-THC (icv) (Braida et al., 2001b; 2004), and squirrel monkeys with no prior exposure to other drugs can learn to respond for Δ^9-THC as well as anandamide and methanandamide iv infusions (Justinová et al., 2003; 2005). Interestingly, Δ^9-THC self-administration behavior in monkeys was obtained within a range of very low doses, similar to those obtained in humans after a single marijuana cigarette (Tanda et al., 2000; Justinová et al., 2003). As for most self-administered drugs, bell-shaped relationships existed between the unit dose and the response frequency (Martellotta et al., 1998; Braida et al., 2001b, 2004; Justinová et al., 2003, 2005), and substitution of saline for Δ^9-THC or anandamide resulted in a reduction of injections per session (Tanda et al., 2000; Justinová et al, 2003, 2005). CB_1 receptors seem to be involved in these effects since cannabinoid self-administration was not observed in CB_1 KO mice (Ledent et al., 1999; Sanchis-Segura et al., 2004) and was suppressed by rimonabant (Martellotta et al., 1998; Braida et al., 2001b, 2004; Fattore et al., 2001; Justinová et al., 2005). On its own, rimonabant is not self-administered by mice and monkeys (Martellotta et al., 1998; Navarro et al., 2001; Beardsley et al., 2002). However, non-cannabinoid mechanisms may also participate in cannabinoid reinforcing effects, in particular the opioid system since in rats, naloxone or naltrexone, non selective opiate receptor antagonists, attenuated Δ^9-THC or CP 55,940 self-administration (Braida et al., 2001b; 2004; Justinová et al., 2004).

Until now, only few studies investigated the effects of cannabinoids in ***intracranial self-stimulation procedures***. Overall, the results are controversial since some studies indicated a small facilitation of median forebrain bundle (MFB) self-stimulation (i.e., a reduction of reward thresholds) by Δ^9-THC (Gardner et al., 1988; Lepore et al., 1996), whereas others failed to demonstrate any facilitatory effect of CP 55,940 on MFB self-stimulation in the same Lewis rat strain (Arnold et al., 2001) and WIN 55,212-2 on lateral hypothalamus (LH) self-stimulation, in drug-naïve or WIN-experienced Sprague-Dawley rats (Vlachou et al., 2003). On the other hand, an "anhedonic-like" elevation of MFB self-stimulation thresholds has been reported, under drug-free condition, 24 hours after a *single* injection of Δ^9-THC (Gardner and Vorel, 1998; Gardner, 2002), but not CP 55,940 (Arnold et al., 2001). Since elevations in ICSS electrical thresholds, proposed as a model for withdrawal-induced dysphoria or depression, have been observed upon withdrawal from *chronic*, but not acute, administration of several drugs of abuse (cocaine, morphine, amphetamine, ethanol, nicotine), possible anhedonic-like consequences of cannabinoid discontinuation warrant further research. Especially, it would be interesting to investigate whether cannabinoids can elevate reward thresholds upon withdrawal from chronic administration. Recent studies have shown that moderate to high doses of rimonabant decreased rats' sensitivity to MFB-stimulation, suggesting that endocannabinoids might facilitate ICSS-related reward processes (Deroche-Gamonet et al., 2001; De Vry et al., 2004); (but see Arnold et al., 2001).

In ***place conditioning procedures*** also, studies with cannabinoids are not always consistent. Indeed, numerous reports indicate that Δ^9-THC or synthetic CB receptor agonists produced conditioned place *aversion* (Parker and Gillies, 1995; McGregor et al., 1996; Sañudo-Peña et al., 1997; Chaperon et al., 1998; Hutcheson et al., 1998; Mallet and Beninger, 1998; Cheer et al., 2000). Other studies, using particular experimental conditions (spaced schedule of injection, prior experience with the drug, biased procedure) revealed conditioned place preference and suggested that the reinforcing action of CB receptor agonists may follow biphasic dose-response curves, low doses being appetitive, an effect that disappeared or even turned to be aversive at higher doses (Lepore et al., 1995; Valjent and Maldonado, 2000; Braida et al., 2001a; Ghozland et al., 2002; Braida et al., 2004; Castañé et al., 2004; Lopez-Moreno et al., 2004). This might be due to predominant non-selective aversive effects (hypotension, nausea, anxiety, etc.) and/or to a post-drug dysphoric rebound, masking the expression of a rewarding component. However, such a bidirectional effect of CB receptor agonists did not always occur since WIN 55,212-2 failed to support conditioned place preference, even at very low doses, or in rats with short (4 days) or long (21 days) prior experience in their home cage, or given an additional injection before the test session in an attempt to counteract a putative withdrawal-like dysphoric effect (Chaperon et al., 1998; Chaperon and Thiébot, 1999). In addition, WIN 55,212-2 did not potentiate conditioned place preference supported by food or a low dose of cocaine, and even dose-dependently reduced the effect of the former, suggesting that, in this procedure, a rewarding potential is not disclosed in rats already given an appetitive reinforcer (Chaperon and Thiébot, 1999). However, it has been shown recently in mice, that non-rewarding doses of WIN 55,212-2 induced a sensitization to morphine conditioned place preference (Manzanedo et al., 2004). It is also interesting to note that anandamide did not induce place conditioning (preference or avoidance) over a large range of doses, even when its hydrolysis was reduced by prior injection of phenylmethyl-sulfonyl fluoride (PMSF) (Mallet and Beninger, 1998), and to the best of our knowledge, there is no published attempt to establish 2-AG as a stimulus able to support place conditioning.

Regarding the interactions between the cannabinoid and the opioid systems, place preference induced by a low dose of Δ^9-THC (1 mg/kg) in animals with prior experience with the drug, was not observed in mice lacking μ-opioid receptors, whereas place aversion supported by a higher dose (5 mg/kg), in non pre-exposed subjects, was abolished in κ-opioid receptor KO mice (Ghozland et al., 2002). Thus the opioid systems seem to play a role in controlling reward pathways, and the authors concluded that an opposing activity of μ- and κ-opioid receptors might form the basis for the dual dysphoric-euphoric effects of Δ^9-THC.

Role of Cannabinoid Receptors in Motivational Processes

In the absence of CB_1 receptor selective agonists, only selective antagonists allow to investigating the role of CB_1 receptors in motivational processes. However, such studies also provided controversial results. For instance, as a corollary to the aversive effect of cannabinomimetics frequently observed in animals (see above), rimonabant has been reported to support conditioned place preference in rats, and it has been proposed that an endogenous cannabinoid tone exists and serves normally to suppress reward or induce aversion (Sañudo-Peña et al., 1997; Cheer et al., 2000). In other rodent studies, however, rimonabant did not induce place conditioning at doses active to antagonize the effects of CB receptor agonists (Chaperon et al., 1998; Braida et al., 2001a; Mas-Nieto et al., 2001; Braida et al., 2004; Manzanedo et al., 2004; Singh et al., 2004), in line with its absence of both discriminative stimulus effect (Mansbach et al., 1996; Pério et al., 1996), and reinforcing effect in self-administration procedures (Martellotta et al., 1998; Navarro et al., 2001; Beardsley et al., 2002). Taken together, these results indicate that an endogenous cannabinoid tone, which would ensure a basal hedonic level to the organism, seems unlikely to exist. On the other hand, rimonabant prevented the acquisition of WIN 55,212-2-induced conditioned place aversion, providing strong support to the involvement of a CB_1-related mechanism in the aversive effect of this compound (Chaperon et al., 1998). Interestingly, impaired central cannabinoid transmission also interferes with appetitive behavior supported by non-cannabinoid reinforcers. For instance, CB_1 KO mice did not exhibit morphine-induced conditioned place preference (Martin et al., 2000; but see Rice et al., 2002), and seemed less prone to self-administer morphine than their wild-type counterparts (Ledent et al., 1999; Cossu et al., 2001). Accordingly, rimonabant prevented the establishment and the expression of conditioned place preference to morphine (Chaperon et al., 1998; Mas-Nieto et al., 2001; Navarro et al., 2001), blocked morphine and heroin self-administration (Krsiak and Sulcova, 1990; Fratta et al., 1998; Navarro et al., 2001; Caillé and Parsons, 2003; Solinas et al., 2003), and impaired the ability of a priming dose of heroin to reinstate heroin-seeking after long-term extinction of heroin self-administration (Fattore et al., 2003). Thus, these results point to an essential role of CB_1 receptors in the reinforcing effects of opiates, and further support the existence of specific functional interactions between cannabinoid and opioid systems. A role of endocannabinoids in the reinforcing effects of other drugs of abuse is more controversial. Indeed, cocaine, amphetamine and nicotine self-administration, as well as cocaine-induced conditioned place preference were not abolished in CB_1 KO mice (Martin et al., 2000; Cossu et al., 2001; Houchi et al., 2005), and the blockade of CB_1 receptors by rimonabant did not impair cocaine self-administration in rats and monkeys (Fattore et al., 1999; Tanda et al., 2000). By contrast, rimonabant reduced self-administration of methamphetamine (Vinklerová et al., 2002) and nicotine (Cohen et al., 2002), attenuated the reinstatement of extinguished responding for cocaine self-administration by a priming injection of cocaine or by cocaine-associated cues (De Vries et al., 2001), and prevented the establishment of conditioned place preference to cocaine (Chaperon et al., 1998), 3,4-methylenedioxymethamphetamine (MDMA, icv) (Braida et al., 2005) and nicotine (Forget et al., 2005). In CB_1 KO mice, consonant with lower ethanol preference (Poncelet et al., 2003; Wang et al., 2003), a reduction in ethanol-induced conditioned place preference has been reported (Houchi et al., 2005), as compared with their wild-type counterparts. Rimonabant decreased alcohol consumption by ethanol-preferring C57Bl6 mice, and Sardinian alcohol-preferring (sP) or Wistar rats (Arnone et al., 1997; Colombo et al., 1998; Gallate and McGregor, 1999; Rodríguez de Fonseca et al., 1999; Freedland et al., 2001; Poncelet et al., 2003). In sP rats, the blockade of CB_1 receptors also prevented the acquisition of alcohol drinking and abolished the rebound of alcohol intake after transient deprivation (Serra et al., 2001; 2002). Conversely, in Wistar rats, the CB receptor agonist WIN 55,212-2 induced a long-lasting increase of alcohol relapse during alcohol deprivation (Lopez-Moreno et al., 2004). Wang et al. (2003) showed that ethanol preference in young C57Bl6 mice was reduced by rimonabant to the level observed in their CB_1 KO littermates or in old wild-type mice, in both of which ethanol preference was unaltered by rimonabant, and they obtained a series of data suggesting that the age-dependent decline in the appetite for ethanol (as well as for food) might be linked to a parallel decline in CB_1 receptor coupling to G-proteins in the limbic forebrain areas. Therefore, these studies give evidence for an involvement of the endocannabinoid system in motivation for a variety of drugs of abuse.

Motivation for natural rewards seems also controlled by endocannabinoids. Indeed, rimonabant prevented the establishment of food-induced conditioned place preference in rats (Chaperon et al., 1998), and reduced the potentiation by a dopaminergic D3 receptor preferential agonist, quinelorane, of food-priming-induced reinstatement of extinguished responding for food, though it did not impair the reinstating effect of food-priming alone (Duarte et al., 2004). Some studies suggested that the blockade or the deletion of CB_1 receptors preferentially reduced the consumption of palatable ingesta (usually sweet food or solutions) by rodents and monkeys (Arnone et al., 1997; Simiand et al., 1998; Gallate et al., 1999; Sanchis-Segura et al., 2004). However, other reports indicated that high palatability is not necessary to observe suppressive effects of rimonabant on food intake (Freedland et al., 2000; Rowland et al., 2001; Gómez et al., 2002; McLaughlin et al., 2003; Verty et al., 2004). In rats implanted with a gastric chronic fistula (a model in which ingestion is exclusively motivated by food palatability since satiation processes are impaired by immediate stomach emptying), rimonabant did not decrease sucrose sham-feeding, a result which argues against a role for endocannabinoids in the *consummatory* aspects of food reward (see Kirkham and Williams, 2001a). The authors concluded that endocannabinoids are not involved in food reward during ingestion, and are not crucial to the pleasure derived from the orosensory characteristics of food. On the other hand, using a progressive ratio schedule of reinforcement, in which the number of responses required to obtain a reward (beer or sucrose solution) progressively increases during the experimental session, it has been shown that the breaking-point (i.e., the ratio at which rats stop to respond) was increased by CP 55,940 and lowered by rimonabant, as well as in CB_1 KO mice, suggesting a role for the endocannabinoid system in the processes which control the *motivation* to obtain the reinforcement (Gallate and McGregor, 1999; Gallate et al., 1999; Sanchis-Segura et al., 2004). In line with these results, it has been shown that (i) both Δ^9-THC and anandamide reduced eating latency and increased usual food consumed in pre-satiated or free-feeding rats (Kirkham and Williams, 2001a), (ii) exogenous or endogenous agonists and rimonabant induced opposite modifications of licking patterns in non-deprived rats ingesting a palatable sucrose solution (Higgs et al., 2003), (iii) rimonabant lengthened post-reinforcement pauses in operant responding for sucrose (Pério et al., 2001), and (iv) deletion of CB_1 receptors resulted in a reduction of the sensitivity to the rewarding properties of sucrose in non-restricted mice (Sanchis-Segura et al., 2004). Taken as a whole, these studies show that the perception of the *incentive value* of food depends, at least in part, on the stimulation of CB_1 receptors by endogenous or exogenous cannabinoids.

Interestingly, at odds with the effect of CB_1 receptor blockade on positive reinforcers, rimonabant did not prevent the establishment of conditioned place avoidance induced by FG 7142 (a β-carboline with anxiogenic effects), and tended to potentiate the aversive effects of lithium and naloxone (Chaperon and Thiébot, 1999), indicating that negative incentives do not depend on CB_1-related processes.)

Endocannabinoids in Alcohol-Related Behaviors

It appears that some of the pharmacological and behavioral effects of alcohol, including alcohol drinking and alcohol-preferring behavior, are mediated through the EPCS (Hungund and Basavarajappa, 2004). The association of the EPCS with the molecular events that are related to the reinforcing effects of alcohol are derived from the following observations:

1. The synthesis of endocannabinoids, 2-AG, and anandamide are increased by chronic ethanol administration that may result in the downregulation of the CB_1 receptor signaling system.
2. Chronic ethanol exposure leads to activation of Ca^{2+}-dependent and arachidonic acid — specific phospholipase A_2 (PLA_2), a key enzyme in the formation of endocannabinoids in neuronal cells and brain.
3. DBA/2 mice that are known to avoid alcohol intake have significantly reduced brain CB_1 receptor function.
4. Rimonabant, the cannabinoid antagonist, has been shown to inhibit alcohol drinking in rodents and recently in humans.
5. Chronic ethanol exposure decreased the B_{max} of CB_1 receptors and inhibited the ability of the CB_1 receptor agonist to stimulate GTPγS binding in mice.

These findings have led to the suggestion that the CB_1 receptor gene plays a role in excessive alcohol drinking behavior and development of alcoholism, and that agents that modulate the endocannabinoid signaling system may be implicated in the treatment of alcoholism (Hungund and Basavarajappa, 2004).

ENDOCANNABINOIDS IN THE MODULATION OF PAIN

Convergence between Opioid, Cannabinoid, and Vanilloid Systems in Pain Pathways

One major therapeutic indication of cannabinoids is analgesia, but it may be limited by their psychotropic effects. Over the years, strong evidence of the physiological and biochemical basis of the involvement of the EPCS in pain has emerged. The antinociceptive effects of CB receptor agonists have been demonstrated in animal models of acute, inflammatory, and nerve injury–induced pain, following peripheral, systemic, spinal, supraspinal, or intracerebral administrations (Goya et al., 2003). Thus, a number of preclinical studies have shown that cannabinoids reduce the hyperalgesia and allodynia associated with formalin, capsaicin, carrageenan, nerve injury, and visceral persistent pain (Karst et al., 2003). Evidence also exists that endocannabinoids suppress pain. Endocannabinoid mechanisms for pain suppression are known to exist at various levels of the pain pathway, in a system that is distinct from, but parallel to, that involving opiates (Iversen and Chapman, 2002). Parallel with the remarkable advances in opioid and cannabinoid research, groundbreaking work over the last decade has led to deeper insights about vanilloid–capsaicin receptors and mechanisms associated with the effects of capsaicin when consumed by humans and animals (Szallasi and Blumberg, 1999, Xue et al., 1999). Just as natural cannabinoids (e.g., Δ^9-THC in marijuana) activate CB receptors, capsaicin (the pungent chemical present in hot chili peppers) is a natural agonist of the cloned VR_1 vanilloid receptor, a heat-activated nonspecific ion channel protein, acting as a molecular integrator of nociceptive stimuli (Caterina et al., 1997). Although this chapter is mainly concerned with the behavioral effects of endocannabinoids, there is also an accumulating body of work indicating the existence of an endovanilloid system, whose interaction with the endocannabinoid system may have a significant implication in the modulation of pain pathways. The VR_1 receptors are known to be activated by capsaicin, endovanilloids, and cannabinoids, and there exists evidence of multiple vanilloid receptors, of which VR1 is the first member of the family of transient receptor potential (TRP) channels activated by capsaicin (Szallasi and Blumberg, 1999). There is also evidence that multiple splice variants of the VR1 receptor subtype exist (Xue et al., 1999). It appears the same might be true of the CB receptors (unfortunately, scanty information is available about CB receptor gene structure, isoforms, splice variants, and their regulation). Like endocannabinoids, endovanilloid ligands, e.g., *N*-arachidonoyldopamine (NADA) and *N*-oleoyldopamine (OLDA), have been identified and shown to activate both VR1 and CB_1 receptors differentially (Bisogno et al., 2000, Chu et al., 2003, Huang et al., 2002). Whereas capsaicin likely acts indirectly on CB receptors, natural cannabinoids, like Δ^9-THC and the nonpsychotropic compound, cannabidiol, weakly activate VR1 receptors (Bisogno et al., 2001). However, the synthetic cannabinoid agonists WIN55212-2, HU210, and CP55940 have no effect on rat and human VR1 receptors (Ralevic, 2003; Smart et al., 2001; Zygmunt et al., 1999). This might be of some advantage in analgesia studies, particularly if CB_1 and VR1 receptors have different sensitivities to endovanilloids and endocannabinoids, so that there may be preferential activation of one or the other of the receptors. Differential selective interaction of endovanilloids and endocannabinoids with the endogenous opioid system might implicate novel or multiple therapeutic targets for pain relief. This is, however, further complicated by the functional evidence of the existence of non-CB_1 and non-CB_2 receptors (Wiley and Martin, 2002) that might play a significant role in the classical opioid-induced analgesia.

Opioids and cannabinoids are distinct drug classes that have historically been used in combination to alleviate pain (Welch and Eads, 1999). Just like the cannabinoid and opioid receptors, vanilloid receptors are present in the brain and peripheral tissues (Huang et al., 2002, Mezey et al., 2000). There appear to be some differences and similarities between the actions of cannabinoid, opioid, and vanilloid agonists and antagonists. For example, smoking marijuana or administration of cannabinoids can activate CB_1 and VR_1 receptors through endocannabinoids in the central nervous system (CNS), whereas capsaicin, a habit-forming food substance present in hot peppers, can also activate CB_1 receptors indirectly and VR_1 receptors directly. Whereas the antinociception and antithermal effects of opioid and cannabinoid agonists are mediated by central and peripheral specific opioid and cannabinoid receptors, capsaicin mediates nociception and thermal effects by activating both central and peripheral VR_1 receptors. Complementary DNA for μ, CB_1 and VR_1 receptors have been cloned and sequenced and mutant mice have been generated with interesting phenotypes (Caterina et al., 2000; Ledent et al., 1999; Zimmer et al., 1999). The μ-receptor mutants neither exhibit the behavioral effects induced by opioids nor become physically dependent when given opioids (Cami and Farre, 2003). Cannabinoid knockout mice have confirmed major roles of brain cannabinoid systems controlling locomotion, pain, drug reward, and other features. CB_2 receptors' role in immune modulation has been confirmed, too (Buckley et al., 2000; Ledent et al., 1999; Zimmer et al., 1999). Impaired nociception and pain sensation has been reported in mice lacking VR_1 receptors (Caterina et al., 2000). In the aversive drinking test, it was demonstrated that VR_1 wild-type mice took a sip of the capsaicin-containing water and avoided further consumption, whereas VR_1 mutant mice showed no aversive response and continued to drink the capsaicin-adulterated water (Caterina et al., 2000), indicating that the VR_1 receptor is important in this action of capsaicin. Therefore, morphine in opium, cannabinoids in marijuana, and capsaicin in hot chili peppers are all natural compounds that may modulate pain through the opioid, cannabinoid, and vanilloid receptors. Whether a common link or downstream target exists for the modulation of pain by these natural products, for which endogenous counterparts have been identified in animals and humans, remains to be established.

Role of Endocannabinoids in the Modulation of Pain

Cannabinoid- and opioid-induced analgesia have been well characterized in animal models predictive of human analgesic action (Pertwee, 2001). The antinociceptive effects of VR_1 receptor blockers in two models of inflammatory hyperalgesia suggest that endovanilloids might be produced also by peripheral tissues and act in concert with locally enhanced temperature and acidity during inflammation (Huang et al., 2002). The discovery of endogenous molecules activating opioid, cannabinoid, and vanilloid receptors indicate the involvement of these receptors in pain modulation and may lead to the identification of novel neurobiological substrates in the regulation of central, spinal, and peripheral pain pathways. Most desirable analgesic agents will be those that do not cause addiction and other side effects but are potent in ameliorating intense and chronic pain. These targets will therefore include those mechanisms associated with pain relief involving peripheral, spinal, or central mechanisms or any combinations of the systems involved. This section, therefore, provides an overview of endocannabinoids in pain modulation. Analgesia and appetite stimulation for sweet foods are common features between cannabinoids and opiates. The emerging consensus is that endocannabinoids regulate pain (Walker and Huang, 2002) parallel with other systems, including opioids, at different levels. It can therefore be imagined that pain involves multiple convergent systems. Perhaps, it is of importance that endocannabinoids mediate retrograde signaling in neuronal tissues, with significant functional implications; however, it is of major importance that the activation of CB_2 receptors may hold promise for the treatment of pain and inflammation without CNS side effects (Malan et al., 2002). This is because the CB_2 receptor subtype is associated with immune-related effects of cannabinoids. Endocannabinoids activate cannabinoid receptors, and the generation of CB_1 and CB_2 receptor mutant mice implies the involvement of the EPCS, not only in stress-induced analgesia (SIA) but also in peripheral, spinal, and centrally mediated analgesia as discussed below.

Centrally Mediated Analgesia

Marijuana, marijuana extracts, and synthetic cannabinoids have been reported to be analgesic agents (Martin, 1991), and the identification of the EPCS has now provided additional evidence of the unique role of endocannabinoids in pain perception and pain suppression systems. The CNS areas associated with pain modulation and densely populated with CB_1 receptors are the rostral ventrolateral medulla, the amygdala, and the periacqueductal gray (PAG), which are involved in descending pain modulation. The induction of analgesia following injection of cannabinoid agonists into these areas supports the evidence that the major site of cannabinoid-induced analgesia is at the end of the supraspinal descending pathway and at the spinal level. To determine the involvement of endocannabinoids in the modulation of pain, one or more components of the EPCS have been examined (for a review see Walker and Huang, 2002). For example, *in vivo* electrical stimulation of rat PAG induced marked local release of anandamide and reduced the tail-flick response to thermal pain (Walker et al., 1999). Analgesia induced by PAG stimulation was inhibited by rimonabant, indicating that CB_1 receptors mediated this effect (Walker et al., 1999) and suggesting a role for endocannabinoids in pain suppression systems. The physiological relevance and contribution of the EPCS in pain management in humans remains attractive and plausible, because the EPCS plays a role in the control of somatosensory systems involved in pain perception (Valverde et al., 2000). Three of the endocannabinoids, anandamide, 2-AG, and noladin ether, have been shown to inhibit central pain perception (Fride, 2002). A fourth endocannabinoid, virodhamine, has been identified and shown to be a partial agonist with *in vivo* antagonistic activity at the CB_1 receptor and full agonist activity at the CB_2 receptor (Porter et al., 2002). Like most cannabinoids, virodhamine produced hypothermia in mice, yet its effect on pain perception has not been determined, although it is known that rimonabant produced hyperalgesia in rodents (Richardson et al., 1997; Strangman et al., 1998). A fifth molecule with arachidonic acid backbone, NADA, has been identified in rat and bovine brain and purportedly has antinociceptive properties (Walker et al., 2002). Therefore, there is an antinociceptive role of endocannabinoids in mechanical and sensory perception of pain, and different components of the EPCS present a therapeutic target for different types of pain management. This role of endocannabinoids in pain modulation is further supported by the presence of endocannabinoid markers (for instance, FAAH) in tissues that mediate pain response (Walker et al., 2002). Accordingly, the FAAH inhibitors, URB532 and URB597, which enhanced anandamide brain concentration, induced a rimonabant-sensitive mild analgesia in the mouse hot-plate test (Kathuria et al, 2003). Likewise, FAAH knockout mice exhibited a 15-fold increase in brain level of anandamide, a tonic elevation of pain thresholds and a profound enhancement of the antinociceptive potency of anandamide; the two latter effects were prevented by rimonabant (Cravatt et al., 2001). These results provide direct evidence of a CB_1-mediated role of endocannabinoids in pain reduction.

Spinal Analgesia

Cannabinoids produce antinociception through peripheral, spinal, and supraspinal cannabinoid receptors by suppressing nociceptive transmission and, perhaps, by noncannabinoid mechanisms. There is also some evidence of an endocannabinoid analgesic tone demonstrated by the antisense knockdown of spinal CB_1 receptors or by the ability of intrathecally administered rimonabant to produce hyperalgesia in the hot-plate test (Richardson, Aanonsen et al., 1998). Spinally administered CB receptor agonists are antinociceptive in a variety of models of acute and persistent pain (Kelly and Chapman, 2003). The antinociceptive effects of systemically and intrathecally administered cannabinoids are attenuated following spinal transaction, indicating a role for spinal mechanisms in cannabinoid analgesia (for a review see Hohmann, 2002). The use of exogenous cannabinoids in models of pain further emphasized the role of spinal cannabinoid mechanisms in modulating somatic inflammatory pain. For example, allodynia induced by adjuvant inflammation, carrageenan-evoked thermal hyperalgesia, and C-fiber-evoked responses in spinal dorsal horn neurons of inflamed rats are attenuated by intrathecally administered cannabinoids as reviewed by Hohmann (2002). In the spinal cord, CB_1 receptors are located on cell bodies of primary afferent neurons and in the laminae of the dorsal horn that are associated with nociceptive transmission. The expression of CB_1 receptor transcripts and the localization of this receptor subtype on cells in the dorsal root ganglion (DRG) and its differential distribution in rat lumbar and spinal cord further support a functional role of EPCS in spinal nociception and antinociceptive mechanisms. The availability of specific antibodies and ligands allowed immunocytochemical and *in situ* hybridization studies which have shown that CB_1 receptors are colocalized, in different cell populations or interneurons, with a number of molecular markers like GABA, μ-opioid receptors, substance P, calcitonin-gene-related peptide (CGRP), glutamate, tyrosine kinase A (trkA), NO synthase, and other neurotrophic factors that are involved in nociceptive processing. Therefore, further studies will undoubtedly identify the most relevant targets at the spinal level for cannabinoid suppression of nociceptive transmission. Although the hypothesis of an endogenous cannabinoid analgesic tone is attractive, it has not been firmly established (Rice et al., 2002).

Peripheral Analgesia

Evidence from animal models of inflammatory pain using pharmacological approaches indicate that intraplantar (not systemic) administration of cannabinoids and anandamide reduced carrageenan-induced thermal hyperalgesia, an effect reversed by rimonabant (Richardson, Kilo et al., 1998). Interestingly, selective activation of cannabinoid CB_2 receptors suppresses hyperalgesia evoked by intradermal capsaicin (Hohmann et al., 2002). Similarly, it was demonstrated that AM1241, a selective CB_2 receptor agonist, dose dependently reversed tactile and thermal hypersensitivity produced by ligation of the L5 and L6 spinal nerves in rats (Ibrahim et al., 2003). These effects were antagonized by CB_2, but not CB_1, receptor antagonists; AM1241 was also effective in CB_1 KO mice, suggesting the potential utility of selective CB_2 receptor agonists for neuropathy (Ibrahim et al., 2003). Pain inhibition by activation of CB_2 receptors that are not present in the brain is an attractive approach, as CB_2 receptor agonists are not likely to produce the euphoria and high associated with marijuana smoking or the use of synthetic cannabinoids that activate the CB_1 receptors. It has been suggested that whereas a proportion of the peripheral analgesic effect of endocannabinoids can be attributed to a neuronal mechanism acting through CB_1 receptors expressed by primary afferent neurons, the anti-inflammatory actions of endocannabinoids mediated through CB_2 receptors also appear to contribute to local analgesic action (Rice et al., 2002). The increasing interest in the potential role of CB_2 receptors is understandable, because neuropathic pain is one of the main challenges of pain therapy. Neuropathic pain is a condition currently without effective treatment (Ibrahim et al., 2003; Goya et al., 2003), and it is further complicated by lack of experimental animal models to mimic human neuropathy. The involvement of the endocannabinoid system in neuropathic pain deserves intensive study, as it provides novel targets for pain reduction. Another type of pain, for which current treatment is unsatisfactory and for which cannabis has anecdotal support, includes phantom limb pain following amputation and pain secondary to damaged nerves (Williamson and Evans, 2000). Other studies using endocannabinoids indicate that anandamide and palmitoylethanolamide (PEA) produce antinociception that was synergistic involving CB_1 and CB_2 receptors. PEA is not strictly an endocannabinoid, but it exhibits cannabinomimetic properties, including analgesic effects, which are blocked by SR144528, a selective CB_2 receptor antagonist. It has been hypothesized that PEA may be acting via a metabolite agonist at CB_2 receptors or through an entourage effect; thereby, it enhances the activity of other endocannabinoids at CB_2 receptors (Rice et al., 2002). For visceral inflammatory pain, anandamide and PEA attenuated the bladder hyperreflexia induced by intravesical administration of nerve growth factor (NGF), and the effects of PEA were sensitive to CB_2 receptor blockade. These data are consistent with the hypothesis that an activation of CB_2 receptors suppresses inflammatory hyperalgesia by attenuating NGF-induced mast cell degranulation and neutrophil accumulation (Rice et al., 2002). In a controlled clinical trial, the synthetic cannabinoid, CT-3 (1′, 1′dimethylheptyl-Δ^8-tetrahydrocannabinol-11-oic acid) has been reported to significantly alleviate chronic neuropathic pain in 21 patients without relevant side effects (Karst et al., 2003). The mechanism of action of CT-3 on neuropathic pain is unknown. However, CB_1 and CB_2 receptors, and perhaps other as yet unknown CB receptors, may be involved in mediating the analgesic and anti-inflammatory effects of this compound. It was also suggested that other actions of CT-3, e.g., inhibition of eicosanoid synthesis, downregulation of cyclooxygenase 2 and binding to the peroxisome proliferator-activated receptor γ (PPARγ) could be involved (Karst et al., 2003). A randomized double-blind crossover trial indicated that two cannabis extracts, Δ^9-THC alone and Δ^9-THC plus cannabidiol, reduced neuropathic pain due to damage of the brachial plexus and improved sleep quality in 48 patients (Berman et al., 2003). Although most of the available data will support the involvement of CB_1 and CB_2 receptors in pain of different origin, including neuropathy, it can be concluded that CB_2 receptor activation, which does not produce euphoria and high, may represent a novel therapeutic approach for pain management. However, the euphoria and high induced by smoking marijuana may be beneficial to the quality of life of some terminally ill patients, and putting such patients in jail for the "crime" of trying to relieve some of the misery caused by their illness is cruel. Like any medicine, the possible side effects of cannabinoids must be taken into account but should not preclude the legitimate medical use of cannabinoids.

Stress-Induced Analgesia

Various stressors produce a wide range of behavioral responses, such as increase in pain-threshold, catalepsy, and changes in locomotor activity and body temperature, which are sensitive to opioid receptor antagonists, such as naloxone (Yamada and Nabeshima, 1995). Opioids and cannabinoids induce similar behavioral responses, suggesting that endogenous opioid and cannabinoid control systems are involved in stress responses, including analgesia. Because SIA is reduced by naloxone, it can be speculated that it will also be sensitive to CB receptor antagonists. To the best of our knowledge, there is no pharmacological investigation on this issue as yet, but a reduction of SIA, but not of opiate-induced analgesia, has been demonstrated in mice lacking CB_1 receptors (Valverde et al., 2000), indicating the involvement of endocannabinoid mechanisms in SIA. In summary, the discovery of the involvement of EPCS in a number of biological processes, including pain, presents a unique opportunity for the development of compounds that can be targeted to any component of the EPCS for the treatment of a variety of pain disorders. However, anecdotal reports indicating that patients find smoked cannabis more powerful in alleviating pain than administration of CB receptor agonists have paved the way for research into the safe delivery of marijuana extracts or as patches.

COGNITION AND MEMORY

In humans, acute and chronic use of marijuana induces severe disruption of immediate recall, short-term memory and memory retrieval (see Davies et al., 2002). The cellular and molecular mechanisms underlying learning and memory deficits produced by cannabinoids, and the role of endocannabinoids in such mechanisms, have been investigated in a number of studies using *in vivo* and *in vitro* systems. Behavioral data support the hypothesis that the EPCS plays a role in memory processes and suggest that a disruption of normal hippocampal function contributes to learning and memory impairments produced by cannabinoids (Sullivan, 2000; Davies et al., 2002). *In vitro* experiments indicated that cannabinoids and endocannabinoids produce persistent changes in memory-related neuronal activity (Gerdeman and Lovinger, 2003). The dense localization of CB_1 receptors in the hippocampus and amygdala, which play an important role in learning and memory, may represent the anatomical substrate for cannabinoids to influence mnemonic processes. Δ^9-THC impairs memory in rodents and monkeys tested in a variety of experimental procedures (radial maze, instrumental discrimination tasks, Morris water maze). The effects exerted by CB receptor agonists, including Δ^9-THC, WIN-55,212-2, CP 55,940, and anandamide, are reversed by rimonabant, providing evidence for the involvement of CB_1-related mechanisms (Mallet and Beninger, 1998; Castellano et al., 2003). Although endocannabinoids mimic the pharmacological effects of cannabinoids, experiments carried out by the latter group have shown that anandamide impairs memory consolidation in random bred mice (CD1), and exerts genotype-dependent influences on memory in inbred strains of mice (C57BL/6 and DBA/2) (Castellano et al., 2003).

The Morris water maze task with hidden platform is widely accepted for examining spatial learning and memory, and a number of studies have shown that some phytocannabinoids and cannabinoids can impair memory (Castellano et al., 2003) or enhance memory functions in rodent models. Endocannabinoids mimic the pharmacological effects of cannabinoids. Furthermore, genetic and pharmacological manipulations have been used to evaluate the CB_1 KO mutant mice in Morris water maze and rimonabant in a delayed radial maze task (Castellano et al., 2003; Varvel and Lichtman, 2002). The hippocampus is involved in mnemonic processes learning and memory and the high concentration of cannabinoid CB_1 receptors in various regions makes this brain structure the hippocampus a major target for a major role of CB_1 cannabinoid receptors and endocannabinoids in learning and memory. The mechanism by which cannabinoids and endocannabinoids influence learning and memory may be by directly acting on the cannabinoid CB_1 receptors in the hippocampus or through the modulation of the release of other neurotransmitters, such as glutamate and acetylcholine. Indeed, cannabinoid agonists have been shown to disrupt long-term potentiation, a cellular substrate underlying memory processes (Castellano et al., 2003). Our new understanding of the role of endocannabinoids as intercellular signals by acting as retrograde messengers (Alger 2004) makes plausible the involvement of the components of the EPCS in the modification of cholinergic and glutamatergic mechanisms associated with the various types of learning and memory processes. Using CB_1 KO mutant mice, it has been shown that the EPCS plays a key role in the extinction of fear-related memories in which aversive stimuli are involved (Marsicano et al., 2002). Finally, it was speculated that a possible role for endocannabinoids and CB_1 cannabinoid receptors and endocannabinoids might be to regulate the storage (i.e., encoding) of information, as well as the means by which that information is retrieved (Hampson and Deadwyler, 1998). However, In this context, most while endocannabinoids, like exogenous cannabinoid agonists, impair learning and memory, probably through the cannabinoid CB_1 receptors in the hippocampus. Although the cognitive effects of the endocannabinoid, virodhamine on various memory tasks remains to be determined assessed, it can be hypothesized that the action of this endocannabinoid, endowed with a low agonist or antagonist activity, may be similar to that of exogenous cannabinoid CB_1 receptor antagonists, shown to improve consolidation processes, and thus may be useful in treating memory disorders (Wolff and Leander, 2003).

MOTOR ACTIVITY

Modulation of Motor Activity by Endocannabinoids

It is now well accepted that the control of movement is one of the more relevant physiological role of the endocannabinoids in the brain. Synthetic, plant-derived and endogenous cannabinoids have powerful actions on motor activity in animals (for a review, see Chaperon and Thiébot, 1999; Romero et al., 2002). In fact, these effects are bidirectional, depending on the dose. Large doses of cannabinoid reduced motor activity in a variety of behavioral tests and even produced strong catalepsy (for a review, see Chaperon and Thiébot, 1999; Fernández-Ruiz et al., 2002), whereas low doses stimulated motor activity as indicated by hyperlocomotion in intact animals (McGregor et al., 1996; Sañudo-Peña et al., 2000), and ipsilateral circling in rats with unilateral 6-hydroxydopamine (6-OH-DA) lesion of the substantia nigra (Sakurai et al., 1985). Likewise, low doses (0.01 mg/kg) of anandamide enhanced, and moderate or high doses (10-100 mg/kg) reduced motor activity in rodents (Mechoulam and Fride, 1995; Sulcova et al., 1998). However, although the overall pharmacological activity of endocannabinoids is similar to that of exogenous cannabinoids, there are also differences, and it is clear that anandamide has partial effects for some behavioral components (Mechoulam and Fride, 1995). Moreover, when different routes of anandamide administration were compared, a complex pattern of full and partial agonist activities was observed (Smith et al., 1994). Additional behavioral differences include the inhibition by very low doses (0.0001-0.01 mg/kg) of anandamide and docosahexaenylethanolamide, a synthetic endocannabinoid-like compound, but not Δ^9-THC, of the pharmacological effects of conventional doses of Δ^9-THC (Fride et al., 1995). This suggests that the physiological functions of endocannabinoids may be opposite to many of the effects observed in experimental conditions with high doses of cannabinoids. A speculative explanation was offered as a linkage of the CB_1 receptor to Gs proteins at different levels of CB_1 receptor activation. Thus, when agonist concentrations are high (which is usually the case in pharmacological experiments, or after intake of high amounts of cannabis), only Gi protein activation is observable, resulting overall, in a behavioral depression. In contrast, when agonist concentrations are low, activation of Gs proteins become apparent (Fride et al., 1995), in an analogous fashion to what has been found for opiate receptors (Cruciani et al., 1993). Direct evidence for such Gs protein linkage to the CB_1 receptor has been demonstrated, at least in neurons from corpus striatum and in CB_1 receptor transfected cells (Glass and Felder, 1997). The development and characterization of compounds behaving as inhibitors of endocannabinoid cellular transport, such as *N*-(4-hydroxyphenyl) arachidonylamide (AM404) or *N*-arachidonoyl-(2-methyl-4-hydroxyphenyl)amine (VDM-11) offer the advantage of potentiating the endogenous tone of endocannabinoids without direct activation of CB_1 receptors (Beltramo et al., 1997; De Petrocellis et al., 2000). In rats, AM404 has been reported to cause an elevation of plasmatic levels of anandamide, paralleled by a time-dependent reduction of locomotion, suggesting that AM404 may produce its behavioral effects by protecting endogenous anandamide from transport-mediated inactivation (Beltramo et al., 2000; Giuffrida et al., 2000). However, anandamide and AM404 are full agonists at vanilloid VR_1 receptors (De Petrocellis et al., 2000; Zygmunt et al., 2000), and recent observations suggest that AM404 might activate anandamide biosynthesis from VR_1-containing neurons (Giuffrida et al., 2000). Since anandamide seems to activate VR_1 receptors by acting at an intracellular site (De Petrocellis et al., 2001), it has been suggested that the effects of AM404 are mostly mediated by VR_1 receptors, whose activation can result in both enhanced anandamide production and hypolocomotion (Di Marzo et al., 2001b; Lastres-Becker et al., 2003). On the other hand, a novel inhibitor of anandamide cellular uptake, (S)-*N*-oleoyl-(1'-hydroxybenzyl)-2'-ethanolamide (OMDM-2), which does not activate VR_1 receptors (Ortar et al., 2003), was found to potentiate the hypokinetic effects of a subeffective dose of anandamide (de Lago et al., 2004), indicating that these motor effects may result from only an increase of brain extracellular concentrations of anandamide.

Rimonabant prevented the motor effects of CB receptor agonists (for a review see Chaperon and Thiébot, 1999; Järbe et al., 2002), and reversed the AM404-induced hypokinesia (Giuffrida et al., 2000), giving evidence for these effects to result from the activation of CB_1 receptors. Numerous studies indicated that this compound is devoid of intrinsic stimulant properties (Lichtman et al., 1998; Masserano et al., 1999; Poncelet et al., 1999; Tzavara et al., 2003; but see Compton et al., 1996). Therefore, a permanent endogenous cannabinoid tone, which would control processes involved in motor activity, seems unlikely to exist. Surprisingly, a modest stimulation of locomotion was observed in CB_1 KO mice placed in an open field (Ledent et al., 1999). This effect occurred on the first, but not subsequent, open field session, suggesting that such variations more likely reflected an adaptation to new surroundings than true motor effects. In addition, when spontaneous locomotion was recorded in small activity boxes, these mutant mice were similar to their wild-type littermates (Valverde et al., 2000). Using the same line of CB_1 KO mice generated on a CD1 genetic background, Naassila et al. (2004) found that male mutants only, not females, exhibited an increased basal locomotor activity compared to wild-type. By contrast, CB_1 KO mice obtained from a different mutant construct and bred on a C57BL/6J background were severely hypoactive, displaying low levels of activity and almost no rearings in an open field, though showing intact motor coordination in the rotarod test (Zimmer et al., 1999). Therefore, the discrepancies observed are likely accounted for by the different genetic background of the two CB_1 KO lines.

Interactions between Endocannabinoid System and GABA, Dopamine, and Glutamate Neurotransmissions in the Basal Ganglia

The basal ganglia are an important brain area for motor-related processes, and a modulation of GABA, DA, and glutamate neurotransmissions at this level would account for the cannabinoid-induced motor deficits (Fernández-Ruiz et al., 2002). There is a particularly high density of CB_1 receptors within the basal ganglia, where they occur both pre- and postsynaptically. Within the striatum, they are located both on GABAergic interneurons, and on striato-nigral and striato-pallidal GABAergic neuron terminals (Mailleux and Vanderhaeghen, 1992; Tsou et al., 1998), thus providing the possibility for a direct modulation of GABA inhibitory effect. Anandamide and 2-AG are present in the striatum, the globus pallidus and the substantia nigra in concentrations that are in general higher than those measured in the whole brain (Bisogno et al., 1999; Di Marzo et al., 2000a,b).

Several findings indicate that the stimulation of CB_1 receptors modulate GABA neurotransmission within the basal ganglia. Thus, *in vivo* electrophysiological studies suggested that cannabinoids may reduce the inhibitory control exerted by GABA on substantia nigra neurons, though these effects appear modest (Miller and Walker, 1995). On the other hand, some neurochemical data demonstrated that cannabinoids altered neither GABA synthesis nor release in these brain structures (Maneuf et al., 1996; Romero et al., 1998). However, in a rat model of Huntington's disease caused by bilateral intrastriatal injections of 3-nitropropionic acid (3-NP, a mitochondrial toxin that selectively destroys striatal GABAergic efferent neurons) the blockade of endocannabinoid membrane transport by AM404 tended to induce recovery from the GABA deficits in basal ganglia (Lastres-Becker et al., 2002b). Furthermore, Δ^9-THC and HU-210 were found to reduce GABA re-uptake into striatal efferent neurons to the globus pallidus and substantia nigra (Maneuf et al., 1996; Romero et al., 1998). Therefore, by increasing extracellular levels of GABA in basal ganglia, the stimulation of CB_1 receptors would result in an enhanced inhibitory influence of GABA (probably through the activation of GABAB receptors) upon neuronal activity, in particular on nigro-striatal DA neurons (Romero et al., 1996). On the other hand, the acute blockade of CB_1 receptors by rimonabant has been shown to enhance A9, but not A10, population response of DA neurons without affecting either their spontaneous firing rate or their inhibition by apomorphine. However, rimonabant prevented the amphetamine-induced inhibition of A9, but not A10, cell firing (Gueudet et al., 1995). Taken together, these results suggest that, under physiological conditions, a cannabinoid-like endogenous tone may facilitate GABA-mediated effects in the striato-nigral pathway.

The endocannabinoid system may also affect motor activity through the modulation of DA transmission in the basal ganglia. CB_1 receptor mRNAs are co-expressed with dopamine D1 and D2 receptor mRNAs in the striatum (Hermann et al., 2002), but CB_1 and DA receptors are not located on cell bodies or terminals of DA nigro-striatal neurons (Herkenham et al., 1991; Tsou et al., 1998), indicating that a direct control of endocannabinoids on DA neurons is unlikely. Anandamide reduced the electrically evoked DA release from striatum slices (Cadogan et al., 1997), and on acute administration, it also induced a modest reduction of striatal DA and DOPAC contents (Romero et al., 1995). Interestingly, in freely moving rats, anandamide was released by local depolarization in the striatum, and this effect was clearly potentiated by the stimulation of D2-like (but not D1-like) DA receptors (Giuffrida et al., 1999). Furthermore, AM404 reduced the stimulation of motor activity elicited by the D2 receptor agonist, quinpirole (Beltramo et al., 2000). Together with the fact that the blockade of CB_1 receptors by rimonabant enhanced the quinpirole-induced hyperactivity (Giuffrida et al., 1999), these findings suggest that an endocannabinoid system may act as a feedback mechanism inhibiting the DA-mediated facilitation of motor activity. Although, the mechanisms involved in this action remain unclear, it appears that the endocannabinoid system may offer a therapeutic target in the treatment of psychomotor disorders involving dysfunction of DA transmission.

Finally, cannabinoids might modulate glutamatergic transmission in the basal ganglia. Electrophysiological studies in the rat suggested that cannabinoids modify the activity of pallidal and nigral neurons through an inhibition of glutamate release from the subthalamo-nigral terminals (Szabo et al., 2000). In addition, Δ^9-THC and WIN 55,212-2 were reported to reduce glutamate release and $[^3H]$-glutamate uptake in striatal slices (Gerdeman and Lovinger, 2001). Both effects were prevented by rimonabant suggesting a role for CB_1 receptors in glutamatergic transmission. However, a recent study failed to support such an hypothesis, since the inhibition by WIN 55,212-2 of glutamate release from rat hippocampal synaptosomes was reversed by neither the CB_1 receptor antagonists, rimonabant and AM251, nor the VR_1 receptor antagonist, capsazepine, and was still observed in hippocampal synaptosomes from CB_1 KO mice (Köfalvi et al., 2003). In contrast, *in vivo* microdialysis studies showed that Δ^9-THC and WIN 55,212-2 induced a rimonabant-sensitive increase of extracellular levels of glutamate in rat prefrontal cortex (Pistis et al., 2002a; Antonelli et al., 2004). In addition, a recent *in vitro* study on brain slices demonstrated that anandamide facilitated glutamatergic input onto DA neurons in the substantia nigra pars compacta, probably via an activation of VR_1 receptors, because this effect occurred in the presence of a CB_1 receptor antagonist, AM281, and was inhibited by a VR_1 receptor antagonist, iodoresiniferatoxin. Accordingly, the blockade of VR_1 receptors *per se* reduced the frequency of spontaneous excitatory postsynaptic currents on DA neurons (Marinelli et al., 2003). Therefore, a tonic activation of VR_1

receptors, and thereby a facilitation of glutamate release, is likely to play an important role in the regulation of DA neuron activity. Collectively, these results suggest that the control of motor behavior through cannabinoid/glutamate interactions within the basal ganglia, more likely involves VR_1 than CB_1 receptors. However, further experiments are needed to understand the exact relationships between glutamate and endocannabinoid systems in the modulation of motor activity.

Role of Endocannabinoids in Motor Disorders

The consequences of the close connections between the endocannabinoid system and GABAergic, glutamatergic and dopaminergic neurotransmissions become evident in pathological conditions, particularly in neurological disorders, such as Huntington's and Parkinson's diseases, affecting the basal ganglia structures.

Huntington's Disease

Huntington's disease (HD) is an autosomal-dominant neurodegenerative disorder characterized primarily by the development of progressive involuntary choreiform movements, and secondarily by personality changes and cognitive decline. Motor disorders are due to the selective loss of GABAergic neurons projecting from the striatum to the globus pallidus and the substantia nigra (for a review see van der Stelt and Di Marzo, 2003). In post-mortem brain of HD patients a significant loss of CB_1 and DA receptors was observed in these two projection areas (Richfield and Herkenham, 1994; Glass et al., 2000). Consistent with these findings, a reduction of CB_1 receptor mRNA has been observed in the striatum of different transgenic mouse models of HD, which express exon 1 of the human HD gene, and which develop many features of HD such as striatum atrophy and progressive dystonia (Denovan-Wright and Robertson, 2000; Lastres-Becker et al., 2002a). A possible role of the endocannabinoid system in the progression of neurodegenerative processes is suggested by the dramatic reduction of CB_1 receptors observed in all regions of the human basal ganglia at pre-symptomatic and early degenerative grades of HD, occurring prior to the loss of co-localized receptors such as D1 and D2 receptors (Glass et al., 2000). Consonant with such an assumption, HD mice housed in an enriched laboratory environment displayed a delayed onset of motor disorders and a reduction in disease progression, which were correlated with delayed loss of cannabinoid CB_1 receptors (van Dellen et al., 2000). In the rat model of HD generated by intrastriatal infusions of 3-NP, injected animals exhibited biphasic motor disturbances, with an early (1-2 weeks) hyperactivity phase followed by a late (3-4 weeks) motor depression (Lastres-Becker et al., 2002b). At the hyperactivity stage, there was a marked reduction of CB_1 receptor mRNA levels in the caudate-putamen, along with a decrease of CB_1 receptor binding in the caudate-putamen and the globus pallidus. These data are compatible with a reduction of endocannabinoid transmission, as also suggested by decreased levels of anandamide and 2-AG in the striatum of 3-NP-lesioned rats (Lastres-Becker et al., 2001b). However a small population of CB_1 receptors seems to survive this degeneration, even in advanced stages of the disease in patients with HD or in lesioned rats, thus providing a possible therapeutic target for the treatment of hyperkinesia. Recently, Lastres-Becker and co-workers (2003) found that AM404 produced–antihyperkinetic effects in 3-NP-lesioned rats, an action blocked by a VR_1 (capsazepine), but not a CB_1 (rimonabant), receptor antagonist. Interestingly, in such animals, the VR_1 receptor agonist, capsaicin, exhibited a strong antihyperkinetic activity and attenuated the 3-NP-induced GABA and DA deficits in the basal ganglia (Lastres-Becker et al., 2003). Although a partial involvement of CB_1 receptors cannot be excluded, these data suggest that the antihyperkinetic response to endocannabinoids is produced via the stimulation of VR_1 receptors. Whatever the exact mechanism, it appears that an enhancement of endocannabinoid function could be a new therapeutic approach to treat HD.

Parkinson's Disease

Parkinson's disease (PD) is a chronic, progressive disorder of late life, characterized by rigidity, tremor and bradykinesia, caused by progressive degeneration of dopaminergic neurons in the nigrostriatal pathway. This initiates a cascade of neurochemical changes within the striatum and in other nuclei of the basal ganglia, downstream the striatum (see Brotchie, 2003). In particular, an abnormal hyperactivation of cortico-striatal glutamatergic neurons, and changes of GABA inhibitory control on the external segment of the globus pallidus and the subthalamic nucleus, are supposed to underlie motor symptoms of PD and L-DOPA-induced dyskinesia. Endocannabinoids might exert a modulatory role in these processes.

One study reports an increase in CB_1 receptor binding in basal ganglia from patients with PD, along with a marked increased efficacy, as indicated by WIN 55,212-2-induced G-protein activation (Lastres-Becker et al., 2001a). In 1-methyl-4-phenyl-1,2,3,6-tetrahydropyridine (MPTP)-lesioned marmosets, a non-human primate model of PD, similar changes also exist, which are attenuated by chronic administration of L-DOPA (Lastres-Becker et al., 2001a). Consistent with these results, studies using rat models of PD indicated that endocannabinoid transmission becomes overactive in basal ganglia. The complete suppression of locomotion induced by reserpine is accompanied by a marked increase of 2-AG in the globus pallidus (though not in other basal ganglia structures) (Di Marzo et al., 2000c). Anandamide level is enhanced and FAAH activity is reduced in the striatum of rats with unilateral nigral 6-OH-DA lesion, and both effects are completely reversed by chronic administration of L-DOPA (Maccarrone et al., 2003). It has been hypothesized that the augmentation of anandamide levels, likely due to a down-regulation of its degradation, may represent a compensatory mechanism trying to counteract the increased cortico-striatal glutamatergic transmission caused by DA denervation (Maccarrone et al., 2003). These authors proposed that the inhibition of anandamide hydrolysis might represent a novel pharmacological approach in the therapy of PD. Consistent with this hypothesis, a recent study showed that AM404 improved akinesia and sensorimotor orientation in 6-OH-DA-lesioned rats (Fernández-Espejo et al., 2004).

On the contrary, it has been also proposed that an increase in endocannabinoid levels may contribute, at least in part, to the pathophysiology of parkinsonian symptom. Indeed, in the reserpine model of PD, the increased 2-AG levels in the globus pallidus, as reported by Di Marzo et al. (2000c), may cause an enhancement of GABA transmission by reducing its reuptake in this structure (for a review see Brotchie, 2003). Accordingly, a blockade of CB_1 receptors would be useful for therapeutic treatment. However, rimonabant failed to alleviate the motor deficits in MPTP-lesioned monkeys (Meschler et al., 2001). Finally, a recent study in 6-OH-DA-lesioned rats indicated that a deficiency in endocannabinoid transmission in basal ganglia may contribute to L-DOPA–induced dyskinesias, and that these complications could be alleviated by the activation of CB_1 receptors (Ferrer et al., 2003). In agreement with this finding, a treatment with the cannabinoid agonist nabilone reduced the score of L-DOPA–induced dyskinesia in MTPT-lesioned monkeys, as well as in patient with PD (Sieradzan et al., 2001; Fox et al., 2002).

Multiple Sclerosis

Multiple sclerosis (MS), the most common chronic inflammatory, demyelinating disease of the CNS in humans, is characterized by neurological deficits including strong motor symptoms such as spasticity, tremor, ataxia (Compston and Coles, 2002). There is anecdotal and clinical evidence for cannabinoids to have beneficial effects on tremor and particularly spasticity (for a review see Pertwee, 2002). However, a recent multicentric randomized placebo-controlled study reported that treatment with Δ^9-THC or cannabis extract did not objectively improve spasticity, but did result in some benefit in secondary outcome measures, assessing mobility and patients' perception of the drug effects on spasticity (Zajicek et al., 2003). Results obtained in animal models of MS indicated that exogenous and endogenous cannabinoids were able to modulate spasticity and tremor. Thus, it has been demonstrated that Δ^9-THC inhibited the development of chronic relapsing experimental allergic encephalomyelitis (CREAE) in rodents (Lyman et al., 1989), and that, on acute treatment, exogenous and endogenous cannabinoids induced a transient reduction of both tremor and spasticity in mice with CREAE (Baker et al., 2000; Baker et al., 2001). On the contrary, in mildly spastic mice, these symptoms were exacerbated by rimonabant, and to a lesser extend by SR144528 a selective CB_2 receptor antagonist, and mice given both compounds became further spastic. These results support the idea that the endogenous cannabinoid system may be tonically active in the control of tremor and spasticity, and that both CB_1 and CB_2 receptors are involved in this effect (Baker et al., 2000). In agreement with this hypothesis, anandamide and 2-AG levels were increased in the brain and spinal cord of spastic mice but not of nonspastic CREAE remission animals, and spasticity was improved by systemic administration of the anandamide membrane transport inhibitors, VDM11 and AM404, and also by AM374, a selective inhibitor of anandamide hydrolysis (Baker et al., 2001).

Consistent results have been found in another model of MS, the Theiler's murine encephalomyelitis virus–induced demyelinating disease (TMEV-IDD). In this model, TMEV-treated mice develop progressive impaired motor coordination as evaluated by reduced rotarod performance. A 10-day treatment with either WIN 55,212-2, a non selective CB_1/CB_2 receptor agonist, ACEA (arachidonyl-2-chloroethylamide), a selective CB_1 agonist, or JWH-015, a preferring CB_2 agonist, improved motor coordination, and this effect was still present 25 days after treatment discontinuation. Interestingly, this functional recovery of TMEV-infected mice paralleled a reduction in CNS inflammation and extensive remyelination in the spinal cord (Arévalo-Martin et al., 2003).

Together, these findings suggest that the endocannabinoid system may be tonically involved in the control of both the spastic events and the development of neurodegeneration. In animal models of MS, cannabinoids seem to promote long-lasting functional recovery by interfering with the inflammatory demyelinating processes and by favoring myelin repair. Thus, an action at this level could provide a new therapeutic strategy in MS both for the management of motor symptoms and to slow the neuronal loss.

Gilles de la Tourette's Syndrome

Gilles de la Tourette's syndrome (GTS) is a complex neuropsychiatric disorder characterized by multiple motor and one or more vocal tics, associated with a variety of behavioral disorders such as obsessive compulsive symptoms (OCS), lack of impulse control, attention deficit hyperactivity disorder (ADHD), anxiety, depression and self-injurious behavior. Although the etiology of this disease still remains unknown, data support an involvement of the fronto-subcortical pathways. Abnormalities in dopaminergic neurotransmission have also been found in patient with GTS (for a review see Singer, 2000). There is some evidence that cannabinoids might have beneficial effects in this disease, particularly to reduce tics. One placebo-controlled trial in 24 patients with GTS showed that chronic Δ^9-THC induced significant improvement of motor and vocal tics and occasionally ameliorated premonitory urges, OCS and ADHD (Müller-Vahl et al., 2003). However, no direct link between endocannabinoid system and this disease has been established.

FUTURE DIRECTIONS

The ability of the endocannabinoid system to interact with multiple systems provides limitless signaling capabilities of cross talk within and possibly between receptor families. This limitless signaling capability of cannabinoids to interact with multiple systems to induce myriad behavioral changes requires continuous intensive investigation. In this era of the genome and of new technologies, the creation and availability of transgenic, knockout, and knockin mice will allow the determination of the mechanisms associated with specific behaviors induced by cannabinoids and endocannabinoids. As the identity of previously unknown components of the EPCS become available, the contribution of cannabinoid genetics in mediating the behavioral effects of cannabinoids and endocannabinoids will be valuable. Although several decades of irrational prejudice may have hampered basic and clinical research on the therapeutic potential of cannabinoids and endocannabinoids, new knowledge of the evolutionary conservation of the components of the EPCS and the high copy number of cannabinoid receptors in the brain and periphery underscores the numerous behaviors induced by cannabinoids and endocannabinoids. Undoubtedly, future research will uncover the role of the EPCS in diverse pathways associated with apoptosis, neurogenesis, epigenesis, neuroinflammation, and neuroprotection and how these processes affect behavior. Thus, cannabinoid research has experienced major breakthroughs over the last 25 years, since the discovery of the CB_1 and CB_2 receptors for the plant-derived Δ^9-THC and their endogenous ligands. No specific behavioral methods have been devised to capture cannabinoid-induced behaviors, except for the tetrad, which describes a relatively selective "cannabinoid profile." This is not surprising however, in view of the ubiquitous presence of CB_1 receptors and their ligands in the CNS and the rich interactions between the EPCS and other neurotransmitter/modulator systems, such as opiates, vanilloids and DA. We therefore envision progress in behavioral cannabinoid research as going in tandem with progress made in neurobehavioral techniques designed to measure the endocannabinoid as interacting with a variety of other neurochemical systems. When the smoke clears, the influence of the EPCS in feeding and other addictive behaviors may be exploited in the treatment of obesity and other disturbances regulated by endocannabinoid signaling.

REFERENCES

Abel, E. L. (1971) Effects of marihuana on the solution of anagrams, memory and appetite, *Nature,* 231:260–1.

Abel., E.L. (1975) Cannabis and aggression in animals, *Behav Biol,* 14:1–20.

Agosti, V., Nunes, E. and Levin, F. (2002) Rates of psychiatric comorbidity among U.S. residents with lifetime cannabis dependence, *Am J Drug Alcohol Abuse,* 28:643–52.

Akinshola, B.E., Chakrabarti, A. and Onaivi, E.S. (1999) In-vitro and in-vivo action of cannabinoids, *Neurochem Res,* 24:1233–40.

Alger, B.E. (2004) Endocannabinoids: Getting the message across, *Proc Natl Acad Sci USA,* 101:8512–8513.

Angst, J. (1996) Comorbidity of mood disorders: a longitudinal prospective study, *Br J Psychiatry Suppl,* 30:31–7.

Antonelli, T., Tanganelli, S., Tomasini, M.C., Finetti, S., Trabace, L., Steardo, L., Sabino, V., Carratu, M.R., Cuomo, V. and Ferraro, L. (2004) Long-term effects on cortical glutamate release induced by prenatal exposure to the canabinoid receptor agonist (R)-(+)-[2,3-dihydro-5-methyl-3-(4-morpholinyl-methyl)-pyrrolo[1,2,3-de]-1,4-benzoxazin-6-yl]-1-naphthalenylmethanone:an in vivo microdialysis study in the awake rat, *Neuroscience*, 124:367–375.

Arévalo, C., de Miguel, R. and Hernàndez-Tristàn, R. (2001) Cannabinoid effects on anxiety-related behaviors and hypothalamic neurotransmitters, *Pharmacol Biochem Behav,* 70:123–31.

Arévalo-Martin, A., Vela, J.M., Molina-Holgado, E., Borell, J. and Guaza, C. (2003) Therapeutic action of cannabinoids in a murine model of multiple sclerosis, *J Neurosci*, 23:2511–2516.

Arnold, J.C., Hunt, G.E. and McGregor, I.S. (2001) Effects of the cannabinoid receptor agonist CP 55,940 and the cannabinoid receptor antagonist SR 141716 on intracranial self-stimulation in Lewis rats, *Life Sci*, 70:97–108.

Arnone, M., Maruani, J., Chaperon, F., Thiébot, M.H., Poncelet, M., Soubrié, P. and Le Fur, G. (1997) Selective inhibition of sucrose and ethanol intake by SR 141716, an antagonist of central cannabinoid (CB_1) receptors, *Psychopharmacology (Berl),* 132:104–6.

Arvanitis, L., Bauer, D. and Rein, W. (2001) Efficacy and tolerability of four novel compounds in schizophrenia: results of the metatrial project. In *40th Annual Meeting of American College of Neuropsychopharmacology,* Abstract book, pp. 178.

Bac, P., Pages, N., Herrenknecht, C. and Paris, M. (1998) Measurement of the three phrases of muricidal behavior induced by D9-tetrahydrocannabinol in isolated, fasting rats, *Physiol Behav,* 63:815–20.

Bac, P., Pages, N., Herrenknecht, C., Dupont, C., Maurois, P., Vamecq, J. and Durlach, J. (2002) THC aggravates rat muricide behavior induced by two levels of magnesium deficiency, *Physiol Behav,* 77:189–95.

Baker, D., Pryce, G., Croxford, J.L., Brown, P., Pertwee, R.G., Makriyannis, A., Khanolkar, A., Layward, L., Fezza, F., Bisogno, T. and Di Marzo, V. (2001) Endocannabinoids control spasticity in a multiple sclerosis model, *FASEB J*, 15:300–302.

Balog, D.L., Epstein, M.E. and Amodio-Groton, M.I. (1998) HIV wasting syndrome: treatment update, *Ann Pharmacother,* 32:446–58.

Balster, R.L. and Prescott, W.R. (1992) Delta 9-tetrahydrocannabinol discrimination in rats as a model for cannabis intoxication, *Neurosci Biobehav Rev*, 16:55–62.

Barrett, R.L., Wiley, J.L., Balster, R.L. and Martin, B.R. (1995) Pharmacological specificity of delta 9-tetrahydrocannabinol discrimination in rats, *Psychopharmacology (Berl),* 118:419–424.

Beardsley, P.M., Dance, M.E., Balster, R.L. and Munzar, P. (2002) Evaluation of the reinforcing effects of the cannabinoid CB_1 receptor antagonist, SR141716, in rhesus monkeys, *Eur J Pharmacol,* 435:209–216.

Beltramo, M., Stella, N., Calignano, A., Lin, S.Y., Makriyannis, A. and Piomelli, D. (1997) Functional role of high-affinity anandamide transport, as revealed by selective inhibition, *Science*, 277:1094–1097.

Beltramo, M., Fonseca, F.R. d., Navarro, M., Galignano, A., Gorriti, M.A., Grammatikopoulos, G., Sadile, A. G., Giuffrida, A. and Piomelli, D. (2000) Reversal of dopamine D2 receptor responses by anandamide transport inhibitor, *J Neurosci*, 20:3401–3407.

Ben-Shabat, S., Fride, E., Sheskin, T., Tamiri, T., Rhee, M.H., Vogel, Z., Bisogno, T., De Petrocellis, L., Di Marzo, V. and Mechoulam, R. (1998) An entourage effect: inactive endogenous fatty acid glycerol esters enhance 2-arachidonoyl-glycerol cannabinoid activity, *Eur J Pharmacol,* 353:23–31.

Berman, J., Lee, J., Cooper, M., Cannon, A., Sach, J., McKerral, S., Taggart, M., Symonds, C., Fishel, K. and Birch, R. (2003) Efficacy of two cannabis-based extracts for relief of central neuropathic pain from brachial plexus avulsion: results of a randomized controlled trial, *Anaesthesia,* 58:938.

Berry, E.M. and Marcus, E.L. (2000) Disorders of eating in the elderly, *J Adult Dev,* 7:87–99.

Berry, E.M. and Mechoulam, R. (2002) Tetrahydrocannabinol and endocannabinoids in feeding and appetite, *Pharmacol Ther,* 95:185–90.

Bersani, G., Orlandi, V., Kotzalidis, G.D. and Pancheri, P. (2002) Cannabis and schizophrenia: impact on onset, course, psychopathology and outcomes, *Eur Arch Psychiatry Clin Neurosci,* 252:86–92.

Bisogno, T., Berrendero, F., Ambrosino, G., Cebeira, M., Ramos, J.A., Fernandez-Ruiz, J. and Di Marzo, V. (1999) Brain regional distribution of endocannabinoids: implications for their biosynthesis and biological function, *Biochem Biophys Res Commun*, 256:377–380.

Bisogno, T., Melck, D., Bobrov, M.Y., Gretskaya, N.M., Bezuglov, V.V., De Petrocellis, L. and Di Marzo, V. (2000) N-acyl-dopamines: novel synthetic CB_1 cannabinoid-receptor ligands and inhibitors of anandamide inactivation with cannabimimetic activity in vitro and in vivo, *Biochem J,* 351:817–824.

Bisogno, T., Hanus, L., De Petrocellis, L., Tchilibon, S., Ponde, D.E., Brandi, I., Moriello, A.S., Davis, J.B., Mechoulam, R. and Di Marzo, V. (2001) Molecular targets for cannabinoid and its synthetic analogues: effect on vanilloid VR_1 receptors and on the cellular uptake and enzymatic hydrolysis of anandamide, *Br J Pharmacol,* 134:845–852.

Blanco-Centurion, C. A. and Salin-Pascual, R. J. (2001) Extracellular serotonin levels in the medullary reticular formation during normal sleep and after REM sleep deprivation, *Brain Res,* 923:128–36.

Bovasso, G.B. (2001) Cannabis abuse as a risk factor for depressive symptoms, *Am J Psychiatry,* 158:2033–7.

Braida, D., Pozzi, M., Cavallini, R. and Sala, M. (2001a) Conditioned place preference induced by the cannabinoid agonist CP 55,940: interaction with the opioid system, *Neuroscience*, 104:923–926.

Braida, D., Pozzi, M., Parolaro, D. and Sala, M. (2001b) Intracerebral self-administration of the cannabinoid receptor agonist CP 55,940 in the rat: interaction with the opioid system, *Eur J Pharmacol*, 413:227–234.

Braida, D., Iosuè, S., Pegorini, S. and Sala, M. (2004) Δ9-Tetrahydrocannabinol-induced conditioned place preference and intracerebroventricular self-administration in rats, *Eur J Pharmacol*, 506:63–69.

Braida, D., Iosuè, S., Pegorini, S. and Sala, M. (2005) 3,4-Methylenedioxymethamphetamine-induced conditioned place preference (CPP) is mediated by endocannabinoid system, *Pharmacol Res*, 51:177–182.

Breivogel, C.S., Griffin, G., Di Marzo, V. and Martin, B.R. (2001) Evidence for a new G protein-coupled cannabinoid receptor in mouse brain, *Mol Pharmacol,* 60:155–63.

Brook, D.W., Brook, J.S., Zhang, C., Cohen, P. and Whiteman, M. (2002) Drug use and the risk of major depressive disorder, alcohol dependence, and substance use disorders, *Arch Gen Psychiatry,* 59:1039–44.

Brotchie, J.M. (2003) CB_1 cannabinoid receptor signalling in Parkinson's disease, *Curr Opin Pharmacol,* 3:54–61.

Buckley, N.E., McCoy, K.L., Mezey, E., Bonner, T., Zimmer, A., Felder, C.C., Glass, M. and Zimmer A. (2000) Immunomodulation by cannabinoids is absent in mice deficient for the cannabinoid CB2 receptor, *Eur J Pharmacol,* 396:141–149.

Burkey, R.T. and Nation, J.R. (1997) (R)-methanandamide, but not anandamide, substitutes for delta 9-THC in a drug-discrimination procedure, *Exp Clin Psychopharmacol,* 5:195–202.

Bymaster, F.P., Zhang, W., Carter, P., Shaw, J., Chernet, E., Phebus, L., Wong, D. and Perry, K.W. (2002) Fluoxetine, but not other selective serotonin uptake inhibitors, increases norepinephrine and dopamine extracellular levels in prefrontal cortex, *Psychopharmacology (Berl),* 160:353–61.

Cadogan, A.K., Alexander, S.P.H., Boyd, E.A. and Kendall, D.A. (1997) Influence of cannabinoids on electrically evoked dopamine release and cyclic AMP generation in the rat striatum, *J Neurochem,* 69:1131–1137.

Caillé, S. and Parsons, L.H. (2003) SR141716A reduces the reinforcing properties of heroin but not heroin-induced increases in nucleus accumbens dopamine in rats, *Eur J Neurosci,* 18:3145–3149.

Cami, J. and Farre, M. (2003) Drug addiction, *N Engl J Med,* 349:975–986.

Carlini, E.A. (1977) Further studies of the aggressive behavior induced by D9-tetrahydrocannabinol in REM sleep-deprived rats, *Psychopharmacology (Berl),* 53:135–45.

Carr, G.D., Fibiger, H.C. and Phillips, A.G. (1989) Conditioned place preference as a measure of drug reward. In *The neuropharmacological basis of reward* (Eds, Liebman, J.M. and Cooper, S.J.) Clarendon Press, Oxford, pp. 264–319.

Castañé, A., Maldonado, R. and Valverde, O. (2004) Role of different brain structures in the behavioural expression of WIN 55,212-2 withdrawal in mice, *Br J Pharmacol,* 142:1309–1317.

Castellano, C., Rossi-Arnaud, C., Cestari, V. and Costanzi, M. (2003) Cannabinoids and memory: animal studies, *Curr Drug Targets CNS Neurol Disord,* 2:389–402.

Caterina, M.J., Leffler, A., Malmberg, A.B., Martin, W.J., Trafton, J., Petersen-Zeitz, K.R., Koltzenburg, M., Basbaum, A.I. and Julius, D. (2000) Impaired nociception and pain sensation in mice lacking the capsaicin receptor, *Science,* 288:306–313.

Caterina, M.J., Schumacher, M.A., Tominaga, M., Rosen, T.A., Levine, J.D. and Julius, D. (1997) The capsaicin receptor: a heat-activated ion channel in the pain pathway, *Nature,* 398:816–824.

Chakrabarti, A., Onaivi, E.S. and Chaudhuri, G. (1995) Cloning and sequencing of a cDNA encoding the mouse brain-type cannabinoid receptor protein, *DNA Sequence,* 5:385–388.

Chaperon, F., Soubrié, P., Puech, A.J. and Thiébot, M.H. (1998) Involvement of central cannabinoid (CB_1) receptors in the establishment of place conditioning in rats, *Psychopharmacology (Berl),* 135: 324–332.

Chaperon, F and Thiébot, M-H (1999) Behavioral effects of cannabinoid agents in animals, *Crit Rev Neurobiol,* 13:243–281.

Cheer, J.F., Kendall, D.A. and Marsden, C.A. (2000) Cannabinoid receptor and reward in the rat: a conditioned place preference study, *Psychopharmacology (Berl),* 151:25–30.

Chen, C.Y., Wagner, F.A. and Anthony, J.C. (2002) Marijuana use and the risk of major depressive episode. Epidemiological evidence from the United States National Comorbidity Survey, *Soc Psychiatry Psychiatr Epidemiol,* 37:199–206.

Chen, J.P., Paredes, W., Li, J., Smith, D., Lowinson, J. and Gardner, E.L. (1990a) Δ^9-Tetrahydrocannabinol produces naloxone-blockable enhancement of presynaptic basal dopamine efflux in nucleus accumbens of conscious, freely-moving rats as measured by intracerebral microdialysis, *Psychopharmacology (Berl),* 102:156–62.

Chen, J.P., Paredes, W., Lowinson, J. and Gardner, E.L. (1990b) Δ^9-Tetrahydrocannabinol enhances presynaptic dopamine efflux in medial prefrontal cortex, *Eur J Pharmacol,* 190:259–62.

Cherek, D.R. and Dougherty, D.M. (1995) Provocation frequency and its role in determining the effects of smoked Marijuana on human aggressive responding, *Behav Pharmacol,* 6:405–12.

Cherek, D.R., Thompson, T. and Kelly, T. (1980) Chronic delta-9-tetrahydrocannabinol administration and schedule-induced aggression, *Pharmacol Biochem Behav,* 12:305–9.

Chiavegatto, S. and Nelson, R.J. (2003) Interaction of nitric oxide and serotonin in aggressive behavior, *Horm Behav,* 44:233–41.

Chu, C.J., Huang, S.M., De Petrocellis, L., Bisogno, T., Ewing, S.A., Miller, J.D., Zipkin, R.E., Daddario, N., Appendino, G., Di Marzo, V. and Walker, J.M. (2003) N-Oleoyldopamine, a novel endogenous capsaicin-like lipid that produces hyperalgesia, *J Biol Chem,* 278:13633–13639.

Cohen, C., Perrault, G., Voltz, C., Steinberg, R. and Soubrie, P. (2002) SR141716, a central cannabinoid (CB_1) receptor antagonist, blocks the motivational and dopamine-releasing effects of nicotine in rats, *Behav Pharmacol*, 13:451–463.

Colombo, G., Agabio, R., Fà, M., Guano, L., Lobina, C., Loche, A., Reali, R. and Gessa, G.L. (1998) Reducing voluntary ethanol intake in ethanol-preferring sP rats by the cannabinoid antagonist SR 141716, *Alcohol Alcohol,* 33:126–30.

Compston, A. and Coles, A. (2002) Multiple sclerosis, *Lancet*, 359:1221–1231.

Compton, D.R., Aceto, M.D., Lowe, J. and Martin, B.R. (1996) In vivo characterization of a specific cannabinoid receptor antagonist (SR 141716A): inhibition of Δ^9-tetrahydrocannabinol-induced responses and apparent agonist activity, *J Pharmacol Exp Ther*, 277:586–594.

Cossu, G., Ledent, C., Fattore, L., Imperato, A., Böhme, G.A., Parmentier, M. and Fratta, W. (2001) Cannabinoid CB_1 receptor knockout mice fail to self-administer morphine but not other drugs of abuse, *Behav Brain Res,* 118:61–65.

Cota., D, Marsicano, G., Tschop, M., Grubler, Y., Flachskamm, C., Schubert, M., Auer, D., Yassouridis, A., Thone-Reineke, C., Ortmann, S., et al. (2003) The endogenous cannabinoid system affects energy balance via central orexigenic drive and peripheral lipogenesis, *J Clin Invest,* 112:423–31.

Cravatt, B.F and Lichtman, A.H. (2003) Fatty acid amide hydrolase: an emerging therapeutic target in the endocannabinoid system, *Curr Opin Chem Biol*, 7:469–475.

Cravatt, B.F., Demarest, K., Patricelli, M.P., Bracey, M.H, Giang, D.K., Martin, B.R and Lichtman, A.H (2001) Supersensitivity to anandamide and enhanced endogenous cannabinoid signaling in mice lacking fatty acid amide hydrolase, *Proc Natl Acad Sci USA,* 98:9371–6.

Crawley, J.N., Corwin, R.L., Robinson, J.K., Felder, C.C., Devane, W.A. and Alxelrod, J. (1993) Anandamide, an endogenous ligand of the cannabinoid receptor, induces hypomotility and hypothermia in vivo in rodents, *Pharmacol Biochem Behav,* 46:967–72.

Cruciani, R.A., Dvorkin, B., Morris, S.A., Crain, S.M. and Makman, M.H. (1993) Direct coupling of opioid receptors to both stimulatory and inhibitory guanine nucleotide binding proteins in F-11 neuroblastoma-sensory neuron hybrid cells, *Proc Natl Acad Sci USA,* 90:3019–3023.

Darmani, N.A., Janoyan, J.J., Kumar, N. and Crim, J.L. (2003) Behaviorally active doses of the CB_1 receptor antagonist SR141716A increase brain serotonin and dopamine levels and turnover, *Pharmacol Biochem Behav,* 75:777–87.

Davies, S.N., Pertwee, R.G. and Riedel, G. (2002) Functions of cannabinoid receptors in the hippocampus, *Neuropharmacology*, 42:993–1007.

Degenhart, L. (2003) The link between cannabis use and psychosis: furthering the debate, *Psychol Med,* 33:3–6.

Degenhart, L., Hall, W. and Lynskey, M. (2001) The relationship between cannabis use, depression and anxiety among Australian adults: findings from the National Survey of Mental Health and Well-Being, *Soc Psychiatry Psychiatr Epidemiol,* 36:219–27.

Dekeyne, A. and Millan, M.J. (2003) Discriminative stimulus properties of antidepressant agents: a review, *Behav Pharmacol*, 14:391–407.

de Lago, E., Ligresti, A., Ortar, G., Morera, E., Cabranes, A., Pryce, G., Bifulco, M., Baker, D., Fernández-Ruiz, J. and Di Marzo, V. (2004) In vivo pharmacological actions of two novel inhibitors of anandamide cellular uptake, *Eur J Pharmacol*, 484:249–257.

Denovan-Wright, E.M. and Robertson, H.A. (2000) Cannabinoid receptor messenger RNA levels decrease in a subset of neurons of the lateral striatum, cortex and hippocampus of transgenic Huntington's disease mice, *Neuroscience*, 98:705–713.

De Petrocellis, L., Bisogno, T., Davis, J.B., Pertwee, R.G. and Di Marzo, V. (2000) Overlap between the ligand recognition properties of the anandamide transporter and the VR_1 vanilloid: inhibitors of anandamide uptake with negligible capsaicin-like activity, *FEBS Lett*, 483:52–56.

De Petrocellis, L., Bisogno, T., Maccarrone, M., Davis, J.B., Finazzi-Agro, A. and Di Marzo, V. (2001) The activity of anandamide at vanilloid VR_1 receptors requires facilitated transport across the cell membrane and is limited by intracellular metabolism, *J Biol Chem*, 276:12856–12863.

Deroche-Gamonet, V., Le Moal, M., Piazza, P.V. and Soubrie, P. (2001) SR141716, a CB_1 receptor antagonist, decreases the sensitivity to the reinforcing effects of electrical brain stimulation in rats, *Psychopharmacology (Berl)*, 157:254–259.

Devane, W.A., Hanus, L., Breuer, A, Pertwee, R.G., Stevenson, L.A., Griffin, G., Gibson, D., Mandel-baum, A., Etinger, A. and Mechoulam, R. (1992) Isolation and structure of a brain constituent that binds to the cannabinoid receptor, *Science*, 258:1946–1949.

De Vries, T.J., Shaham, Y., Homberg, J.R., Crombag, H., Schuurman, K., Dieben, J., Vanderschuren, L.J. and Schoffelmeer, A.N. (2001) A cannabinoid mechanism in relapse to cocaine seeking, *Nat Med*, 7:1151–1154.

De Vry, J. and Jentzsch, K.R. (2002) Discriminative stimulus effects of BAY 38 7271, a novel cannabinoid receptor agonist, *Eur J Pharmacol*, 457:147–152.

De Vry, J. and Jentzsch, K.R. (2003) Intrinsic activity estimation of cannabinoid CB_1 receptor ligands in a drug discrimination paradigm, *Behav Pharmacol*, 14:471–476.

De Vry, J., Schreiber, R., Eckel, G. and Jentzsch, K.R. (2004) Behavioral mechanisms underlying inhibition of food-maintained responding by the cannabinoid receptor antagonist/inverse agonist SR141716A, *Eur J Pharmacol*, 483:55–63.

Di Marzo, V., Melck, D., Bisogno, T., and De Petrocellis, L. (1998) Endocannabinoids: endogenous cannabinoid receptor ligands with neuromodulatory action, *Trends Neurosci*, 21:521–528.

Di Marzo, V., Berrendero, F., Bisogno, T., González, S., Cavaliere, P., Romero, J., Cebeira, M., Ramos, J.A. and Fernández-Ruiz, J. (2000a) Enhancement of anandamide formation in the limbic forebrain and reduction of endocannabinoid contents in the striatum of Δ^9-tetrahydrocannabinol-tolerant rats, *J Neurochem*, 74:1627–1635.

Di Marzo, V., Breivogel, C.S., Tao, Q., Bridgen, D.T., Razdan, R.K., Zimmer, A.M., Zimmer, A. and Martin, B.R. (2000b) Levels, metabolism, and pharmacolgical activity of anandamide in CB_1 cannabinoid receptor knockout mice: evidence for non CB_1, non CB_2 receptor-mediated actions of anandamide in mouse brain, *J Neurochem*, 75:2434–2444.

Di Marzo, V., Hill, M.P., Bisogno, T., Crossman, A.R. and Brotchie, J.M. (2000c) Enhanced levels of endogenous cannabinoids in the globus pallidus are associated with a reduction in movement in an animal model of Parkinson's disease, *FASEB J*, 14:1432–1438.

Di Marzo, V., Goparaju, S.K., Wang, L., Liu, J., Bátkai, S., Járai, Z., Fezza, F., Miura, G.I., Palmiter, R.D., Sugiura, T., et al. (2001a) Leptin-regulated endocannabinoids are involved in maintaining food intake, *Nature*, 410:822–5.

Di Marzo, V., Lastres-Becker, I., Bisogno, T., De Petrocellis, L., Milone, A., Davis, J.B. and Fernández-Ruiz, J. (2001b) Hypolocomotor effects in rats of capsaicin and two long chain capsaicin homologues, *Eur J Pharmacol*, 420:123–131.

Di Marzo, V., De Petrocellis, L., Fezza, F., Ligresti, A. and Bisogno, T. (2002) Anandamide receptors, *Prostaglandins Leukot Essent Fatty Acids*, 66:377–91.

Di, S., Malcher-Lopes, R., Halmos, K. and Tasker, J. (2003) Nongenomic glucocorticoid inhibition via endocannabinoid release in the hypothalamus: a fast feedback mechanism, *J Neurosci*, 23:4850–7.

Duarte, C., Alonso, R., Bichet, N., Cohen, C., Soubrié, P. and Thiébot, M.H. (2004) Blockade by the cannabinoid CB_1 receptor antagonist, rimonabant (SR141716), of the potentiation by quinelorane of food-primed reinstatement of food-seeking behavior, *Neuropsychopharmacology*, 29:911–920.

Fabre, L.F. and McLendon, D. (1981) The efficacy and safety of nabilone (a synthetic cannabinoid) in the treatment of anxiety, *J Clin Pharmacol*, 21 (8-9 Suppl):377S-82S.

Fattore, L., Martellotta, M.C., Cossu, G., Mascia, M.S. and Fratta, W. (1999) CB_1 cannabinoid receptor agonist WIN 55,212–2 decreases intravenous cocaine self-administration in rats, *Behav Brain Res*, 104:141–146.

Fattore, L., Cossu, G., Martellotta, C.M. and Fratta, W. (2001) Intravenous self-administration of the cannabinoid CB_1 receptor agonist WIN 55,212-2 in rats, *Psychopharmacology (Berl)*, 156:410-416.

Fattore, L., Spano, M.S., Cossu, G., Deiana, S. and Fratta, W. (2003) Cannabinoid mechanism in reinstatement of heroin-seeking after a long period of abstinence in rats, *Eur J Neurosci*, 17:1723-1726.

Fergusson, D.M., Horwood, L.J. and Swain-Campbell, N.R. (2003) Cannabis dependence and psychotic symptoms in young people, *Psychol Med*, 33:15–21.

Fernández-Espejo, E., Caraballo, I., Rodrígez de Fonseca, F., Ferrer, B., El Banoua, F., Flores, J.A. and Galan-Rodrígez, B. (2004) Experimental parkinsonism alters anandamide precursor synthesis, and functional deficits are improved by AM404: a modulator of endocannabinoid function, *Neurospychopharmacology*, 29:1134–1142.

Fernández-Ruiz, J., Lastres-Becker, I., Cabranes, A., González, S. and Ramos, J.A. (2002) Endocannabinoids and basal ganglia functionality, *Prostaglandins Leukot Essent Fatty Acids*, 66:257–267.

Ferrer, B., Asbrock, N., Kathuria, S., Piomelli, D. and Giuffrida, A. (2003) Effects of levodopa on endocannabinoid levels in rat basal ganglia: implications for the treatment of levodopa-induced dyskinesias, *Eur J Neurosci*, 18:1607–1614.

Foltin, R.W., Fischman, M.W. and Byrne, M.F. (1988) Effects of smoked marijuana on food intake and body weight of humans living in a residential laboratory, *Appetite,* 11:1–14.

Forget, B., Hamon, M. and ThiÈbot, M. H. (2005) Cannabinoid CB_1 receptors are involved in motivational effects of nicotine in rats, *Psychopharmacology (Berl),* 29: in press, DOI: 10.1007/s00213-00005-00015-00216.

Fratta, W., Martellotta, M.C., Cossu, G. and Fattore, L. (1998) Self-administration of the cannabinoid agonist WIN 55212-2 and morphine in mice: evidence for a common neurobiological mechanism, *Eur Neuropsychopharmacol,* 8 (suppl. 1):S41–S42.

Freedland, C.S., Poston, J.S. and Porrino, L.J. (2000) Effects of SR141716A, a central cannabinoid receptor antagonist, on food-maintained responding, *Pharmacol Biochem Behav,* 67:265–70.

Freedland, C.S., Sharpe, A.L., Samson, H.H. and Porrino, L.J. (2001) Effects of SR141716A on ethanol and sucrose self-administration, *Alcohol Clin Exp Res,* 25:277–282.

French, E.D. (1997) Delta9-Tetrahydrocannabinol excites rat VTA dopamine neurons through activation of cannabinoid CB1 but not opioid receptors, *Neurosci Lett,* 226:159–162.

French, E.D., Dillon, K. and Wu, X. (1997) Cannabinoids excite dopamine neurons in the ventral tegmentum and substantia nigra, *Neuroreport,* 8:649–52.

Fox, S.H., Henry, B., Hill, M., Crossman, A.R. and Brotchie, J.M. (2002) Stimulation of cannabinoid receptors reduces levodopa-induced dyskinesia in the MPTP-lesioned nonhuman primate model of Parkinson's disease, *Mov Disord,* 17:1180–1187.

Fride, E. (2004) The endocannabinoid-CB receptor system: Importance for development and in pediatric disease, *Neuro Endocrinol Lett,* 25:24–30.

Fride, E. (2002) Endocannabinoids in the central nervous system – an overview, *Prostaglandins Leukot Essent Fatty Acids,* 66:221–233.

Fride, E. and Mechoulam, R. (1993) Pharmacological activity of the cannabinoid receptor agonist, anandamide, a brain constituent, *Eur J Pharmacol,* 231:313–314.

Fride, E., Barg, J., Levy, R., Saya, D., Heldman, E., Mechoulam, R. and Vogel, Z. (1995) Low doses of anandamides inhibit pharmacological effects of delta-9-tetrahydrocannabinol, *J Pharmacol Exp Ther,* 272:699–707.

Fride, E., Foox, A., Rosenberg, E., Faigenboim, M., Cohen, V., Barda, L., Blau, H. and Mechoulam, R. (2003) Milk intake and survival in newborn cannabinoid CB_1 receptor knockout mice: evidence for a "CB_3" receptor, *Eur J Pharmacol,* 461:27–34.

Fride, E., Foox, A., Rosenberg, E., Faigenboim, M., Cohen, V., Barda, L., Blau, H. and Mechoulam, R. (2003) Milk intake and survival in newborn cannabinoid CB_1 receptor knockout mice: evidence for a "CB_3" receptor, *Eur J Pharmacol,* 461:27–34.

Fride, E., Ginzburg, Y., Breuer, A., Bisogno, T., Di Marzo, V. and Mechoulam, R. (2001) Critical role of the endogenous cannabinoid system in mouse pup suckling and growth, *Eur J Pharmacol,* 419:207–14.

Frischknecht, H.-R. (1984) Effects of cannabis drugs on social behavior of laboratory rodents, *Prog Neurobiol,* 22:39–58.

Gallate, J.E. and McGregor, I.S. (1999) The motivation for beer in rats: effects of ritanserin, naloxone and SR 141716, *Psychopharmacology (Berl),* 142:302–8.

Gallate, J.E., Saharov, T., Mallet, P.E. and McGregor, I.S. (1999) Increased motivation for beer in rats following administration of a cannabinoid CB_1 receptor agonist, *Eur J Pharmacol,* 370:233–240.

Gardner, E.L. (2002) Marijuana addiction and CNS reward-related events, in: *The Biology of Marijuana,* ed E. Onaivi (Harwood Academic Publishers, Reading) p. 75–109.

Gardner, E.L. and Vorel, S.R. (1998) Cannabinoid transmission and reward-related events, *Neurobiol Dis,* 5:502–533.

Gardner, E.L. and Lowinson, J.H. (1991) Marijuana's interaction with brain reward systems: update 1991, *Pharmacol Biochem Behav,* 40:571–580.

Gardner, E.L., Paredes, W., Smith, D., Donner, A., Milling, C., Cohen, D. and Morisson, D. (1988) Facilitation of brain stimulation reward by delta-9 tetrahydrocannabinol, *Psychopharmacology (Berl),* 96:142–144.

Gerdeman, G. and Lovinger, D.M. (2001) CB1 cannabinoid receptor inhibits synaptic release of glutamate in rats dorsolateral striatum, *J Neurophysiol,* 85:468–71.

Gerdeman, G.L., and Lovinger, D.M. (2003) Emerging roles for endocannabinoids in long-term synaptic plasticity, *Br J Pharmacol,* 140:781–789.

Gessa, G.L., Melis, M., Muntoni, A.L. and Diana, M. (1998) Cannabinoids activate mesolimbic dopamine neurons by an action on cannabinoid CB_1 receptors, *EurJ Pharmacol*, 341:39–44.

Ghozland, S., Matthes, H.W., Simonin, F., Filliol, D., Kieffer, B.L. and Maldonado, R. (2002) Motivational effects of cannabinoids are mediated by mu-opioid and kappa-opioid receptors, *J Neurosci*, 22:1146–1154.

Giuffrida, A., Parsons, L.H., Kerr, T.M., Rodríguez de Fonseca, F., Navarro, M. and Piomelli, D. (1999) Dopamine activation of endogenous cannabinoid signaling in dorsal striatum, *Nat Neurosci*, 2:358–63.

Giuffrida, A., Rodrígez de Fonseca, F., Nava, F., Loubet-Lescoulié, P. and Piomelli, D. (2000) Elevated circulating levels of anandamide after administration of the transport inhibitor, AM404, *Eur J Pharmacol*, 408:161–168.

Giuliani, D., Ferrari, F. and Ottani, A. (2000) The cannabinoid agonist HU 210 modifies rat behavioral responses to novelty and stress, *Pharmacol Res*, 41:45–51.

Glass, M. and Felder, C.C. (1997) Concurrent stimulation of cannabinoid CB_1 and dopamine D2 receptors augments camp accumulation in striatal neurons: evidence for a Gs linkage to the CB_1 receptor, *J Neurosci*, 17:5327–5333.

Glass, M., Dragunow, M. and Faull, R.L.M. (1997) Cannabinoid receptors in the human brain: a detailed anatomical and quantitative autoradiographic study in the fetal, neonatal and adult human brain, *Neuroscience*, 77:299–318.

Glass, M., Dragunow, M. and Faull, R.L.M. (2000) The pattern of neurodegeneration in Huntington's disease: a comparative study of cannabinoid, dopamine, adenosine and $GABA_A$ receptor alterations in the human basal ganglia in Huntington's disease, *Neuroscience*, 97:505–519.

Goudie, A.J. (1991) Animal models of drug abuse and dependence. In *Behavioural models in psychopharmacology: theoretical, industrial and clinical perspectives* (Ed, Willner, P.) Cambridge University Press, Cambridge, pp. 453–484.

Goudie, A.J. and Smith, J.A. (1999) Discriminative stimulus properties of antipsychotics, *Pharmacol Biochem Behav*, 64:193–201.

Gómez, R., Navarro, M., Ferrer, B., Trigo, J.M., Bilbao, A., Del Arco, I., Cippitelli, A., Nava, F., Piomelli, D. and Rodriguez de Fonseca, F. (2002) A peripheral mechanism for CB_1 cannabinoid receptor-dependent modulation of feeding, *J Neurosci*, 22:9612–7.

Goya, P., Jagerovic, N, Hernandez-Folgado, L. and Martin, M.I. (2003) Cannabinoids and neuropathic pain, *Mini Reviews Med Chem*, 3:765–772.

Grace, A.A. (1991) Phasic versus tonic dopamine release and the modulation of dopamine system responsivity: a hypothesis for the etiology of schizophrenia, *Neuroscience*, 41:1–24.

Graceffo, T.J. and Robinson, J.K. (1998) Delta-9-tetrahydrocannabinol (THC) fails to stimulate consumption of a highly palatable food in the rat, *Life Sci*, 62:PL85-PL8.

Green, B.E. and Ritter, C. (2000) Marijuana use and depression, *J Health Soc Behav*, 41:40–9.

Grinspoon, L. and Bakalar, J.B. (1998) The use of cannabis as a mood stabilizer in bipolar disorder: anectodal evidence and the need for clinical research, *J Psychoactive Drugs*, 30:171–7.

Gruber, A.J., Pope, H.G.J. and Brown, M.E. (1996) Do patients use marijuana as an antidepressant? *Depression*, 4:77–80.

Gueudet, C., Santucci, V., Rinaldi-Carmona, M., Soubrié, P. and Le Fur, G. (1995) The CB_1 cannabinoid receptor antagonist SR 141716A affects A9 dopamine neuronal activity in the rat, *Neuroreport*, 6:1421–1425.

Halaas, J.L., Boozer, C., Blair-West, J., Fidahusein, N., Denton, D.A. and Friedman, J.M. (1997) Physiological response to long-term peripheral and central leptin infusion in lean and obese mice, *Proc Natl Acad Sci USA*, 94:8878–83.

Hall, W. and Solowij, N. (1998) Adverse effects of cannabis, *Lancet*, 352:1611–6.

Haller, J., Bakos, N., Szirmay, M., Ledent, C. and Freund, T.F. (2002) The effect of genetic and pharmacological blockade of CB_1 cannabinoid receptor on anxiety, *Eur J Neurosci*, 16:1395–8.

Hampson, R.E. and Deadwyler, S.A. (1998) Role of cannabinoid receptors in memory storage, *Neurobiol Dis*, 5:474–482.

Hanus, L, Abu-Lafi, S, Fride, E, Breuer, A., Vogel, Z., Shalev, D.E., Kustanovich, I. and Mechoulam, R. (2001) 2-Arachidonyl glyceryl ether, an endogenous agonist of the cannabinoid CB1 receptor, *Proc Natl Acad Sci USA*, 98, 3662–3665.

Hanus, L., Avraham, Y., Ben-Shushan, D., Zolotarev, O., Berry, E.M. and Mechoulam, R. (2003) Short-term fasting and prolonged semistarvation have opposite effects on 2-AG levels in mouse brain, *Brain Res,* 983:144–51.

Hao, S., Avraham, Y., Mechoulam, R. and Berry, E.M. (2000) Low dose anandamide affects food intake, cognitive function, neurotransmitter and corticosterone levels in diet-restricted mice, *Eur J Pharmacol,* 392:147–56.

Herkenham, M., Lynn, A.B., Johnson, M.R., Melvin, L.S., de Costa, B.R. and Rice, K.C. (1991) Characterization and localization of cannabinoid receptors in rat brain: a quantitative in vitro autoradiographic study, *J Neurosci,* 11:563–83.

Hermann, H., Marsicano, G. and Lutz, B. (2002) Coexpression of the cannabinoid receptor type 1 with dopamine and serotonin receptors in distinct neuronal subpopulations of the adult mouse forebrain, *Neuroscience*, 109:451–460.

Higgs, S., Williams, C.M. and Kirkham, T.C. (2003) Cannabinoid influences on palatability: microstructural analysis of sucrose drinking after delta9-tetrahydrocannabinol, anandamide, 2-arachidonoyl glycerol and SR141716, *Psychopharmacology (Berl)*, 165:370–377.

Hohmann, A.G. (2002). Spinal and peripheral mechanisms of cannabinoid antinociception: behavioral, neurophysiological and neuroanatomical perspectives, *Chem Phys Lipids*, 121:173–190.

Hohmann, A.G., Farthing, J.N., Zvonok, A.M. and Makriyannis, A. (2004) Selective activation of cannabinoid CB2 receptors suppresses hyperalgesia evoked by intradermal capsaicin, *J Pharmacol Exp Ther*, 308:446–453.

Holmes, A., Murphy, D.L. and Crawley, J.N. (2002) Reduced aggression in mice lacking the serotonin transporter, *Psychopharmacology (Berl),* 161:160–7.

Houchi, H., Babovic, D., Pierrefiche, O., Ledent, C., Daoust, M. and Naassila, M. (2005) CB_1 receptor knockout mice display reduced ethanol-induced conditioned place preference and increased striatal dopamine D2 receptors, *Neuropsychopharmacology*, 30:339–349.

Huang, C.C., Lo, S.W. and Hsu, K.S. (2001) Presynaptic mechanisms underlying cannabinoid inhibition of excitatory synaptic transmission in rat striatal neurons, *J Physiol,* 532:731–48.

Huang, S.M., Bisogno, T., Trevisani, M., Al-Hayani, A., De Petrocellis, L., Fezza, F., Tognetto, M., Petros, T.J., Krey, J.F., Chu, C.J., Miller, J.D., Davis, S.N., Geppetti, P., Walker, J.M. and Di Marzo, V. (2002) An endogenous capsaicin-like substance with high potency at recombinant and native vanilloid VR_1 receptors, *Proc Natl Acad Sci USA,* 99:8400–8405.

Huestis, M.A., Gorelick, D.A., Heishman, S.J., Preston, K.L., Nelson, R.A., Moolchan, E.T. and Franck, R.A. (2001) Blockade of the effects of smoked marijuana by CB_1-selective cannabinoid receptor antagonist SR141716, *Arch Gen Psychiatry*, 58:330–331.

Hungund, B.L. and Basavarajappa, B.S. (2004) Role of endocannabinoids and cannabinoid CB_1 receptors in alcohol-related behaviors, *Ann NY Acad Sci*, 1025:515–527.

Hutcheson, D.M., Tzavara, E.T., Smadja, C., Valjent, E., Roques, B.P., Hanoune, J. and Maldonado, R. (1998) Behavioural and biochemical evidence for signs of abstinence in mice chronically treated with Δ^9-tetrahydrocannabinol, *Br J Pharmacol*, 125:1567–1577.

Ibrahim, M.M., Deng, H., Zvonok, A., Cockayne, D.A., Kwan, J., Mata, H.P., Vanderah, T.W., Lai, J., Porreca, F., Makriyannis, A. and Malan, T.P. (2003). Activation of CB_2 cannabinoid receptors by AM1241 inhibits experimental neuropathic pain: Pain inhibition by receptors not present in the CNS, *Proc Natl Acad Sci USA,* 100:10529–10533.

Ilaria, R.L., Thornby, J.L. and Fann, W.E. (1981) Nabilone, a cannabinoid derivative, in the treatment of anxiety neuroses, *Curr Ther Res,* 29:943–9.

Iversen, L. and Chapman, V. (2002). Cannabinoids: a real prospect for pain relief? *Curr Opin Pharmacol,* 2:50–55.

Järbe, T.U.C., Lamb, R.J., Makriyannis, A., Lin, S. and Goutopoulos, A. (1998) Delta 9-THC training dose as a determinant for (R)-methanandamide generalization in rats, *Psychopharmacology (Berl),* 140:519–522.

Järbe, T.U.C., Lamb, R.J., Lin, S. and Makriyannis, A. (2001) (R)-methanandamide and Delta 9-THC as discriminative stimuli in rats: tests with the cannabinoid antagonist SR-141716 and the endogenous ligand anandamide, *Psychopharmacology (Berl),* 156:369–380.

Järbe, T.U.C., Andrzejewski, M.E. and DiPatrizio, N.V. (2002) Interactions between the CB_1 receptor agonist Δ^9-THC and the CB_1 receptor antagonist SR 141716 in rats: open-field revisited, *Pharmacol Biochem Behav*, 73:911–919.

Järbe, T.U.C., Harris, M.Y., Li, C., Liu, Q. and Makriyannis, A. (2004) Discriminative stimulus effects in rats of SR-141716 (rimonabant), a cannabinoid CB1 receptor antagonist, *Psychopharmacology (Berl),* 177:35–45.

Justinová, Z., Tanda, G., Redhi, G.H. and Goldberg, S.R. (2003) Self-administration of delta9-tetrahydrocannabinol (THC) by drug naive squirrel monkeys, *Psychopharmacology (Berl)*, 169:135–140.

Justinová, Z., Tanda, G., Munzar, P. and Goldberg, S.R. (2004) The opioid antagonist naltrexone reduces the reinforcing effects of Δ^9-tetrahydrocannabinol (THC) in squirrel monkeys, *Psychopharmacology (Berl)*, 173:186–194.

Justinová, Z., Solinas, M., Tanda, G., Redhi, G.H. and Goldberg, S.R. (2005) The endogenous cannabinoid anandamide and its synthetic analog R(+)-methanandamide are intravenously self-administered by squirrel monkeys, *J Neurosci,* 25:5645–5650.

Karst, M., Salem, K., Burstein, S., Conrad, I., Hoy, L. and Schneider, U. (2003) Analgesic effect of the synthetic cannabinoid CT-3 on chronic neuropathic pain. A randomized controlled trial, *JAMA*, 290:1757–1762.

Kathuria, S., Gaetani, S., Fegley, D., Valino, F., Duranti, A., Tontini, A., Mor, M., Tarzia, G., La Rana, G., Calignano, A., et al. (2003) Modulation of anxiety through blockade of anandamide hydrolysis, *Nat Med,* 9:76–81.

Kelly, S. and Chapman, V. (2003) Cannabinoid CB_1 receptor inhibition of mechanically evoked responses of spinal neurons in control rats, but not in rats with hindpaw inflammation, *Eur J Pharmacol,* 474:209–216.

Kirkham, T.C. and Williams, C.M. (2001) Synergistic effects of opioid and cannabinoid antagonists on food intake, *Psychopharmacology (Berl),* 153:267–70.

Kirkham, T.C. and Williams, C.M. (2001) Endogenous cannabinoids and appetite, *Nutr Res Rev*, 14:65–86.

Kirkham, T.C., Williams, C.M., Fezza, F. and Di Marzo, V. (2002) Endocannabinoid levels in rat limbic forebrain and hypothalamus in relation to fasting, feeding and satiation: stimulation of eating by 2-arachidonoyl glycerol, *Br J Pharmacol,* 136:550–7.

Knable, M.B. and Weinberger, D.R. (1997) Dopamine, the prefrontal cortex and schizophrenia, *J Psychopharmacol,* 11:123–31.

Koch, J.E. and Matthews, S.M. (2001) Δ^9-Tetrahydrocannabinol stimulates palatable food intake in Lewis rats: effects of peripheral and central administration, *Nutr Neurosci,* 4:179–87.

Köfalvi, A., Vizi, E.S., Ledent, C. and Sperlàgh, B. (2003) Cannabinoids inhibit the release of [^{3}H] glutamate from rodent hippocampal synaptosomes via a novel CB1 receptor-independent action, *Eur J Neurosci*, 18:1973–1978.

Krsiak, M. and Sulcova, A. (1990) Differential effects of six structurally related benzodiazepines on some ethological measures of timidity, aggression and locomotion in mice, *Psychopharmacology (Berl)*, 101:396–402.

Lastres-Becker, I., Cebeira, M., de Ceballos, M.L., Zeng, B.-Y., Jenner, P., Ramos, J.A. and Fernández-Ruiz, J. (2001a) Increased cannabinoid CB_1 receptor binding and activation of GTP-binding proteins in the basal ganglia of patients with parkinson's syndrome and of MPTP-treated marmosets, *Eur J Neurosci*, 14:1827–1832.

Lastres-Becker, I., Fezza, F., Cebeira, M., Bisogno, T., Ramos, J.A., Milone, A., Fernández-Ruiz, J. and Di Marzo, V. (2001b) Changes in endocannabinoid transmission in the basal ganglia in a rat model of Huntington's disease, *Neuroreport*, 12:2125–2129.

Lastres-Becker, I., Berrendero, F., Lucas, J.J., Martin-Aparicio, E., Yamamoto, A., Ramos, J.A. and Fernández-Ruiz, J. (2002a) Loss of mRNA levels, binding and activation of GTP-binding proteins for cannabinoid CB_1 receptors in the basal ganglia of a transgenic model of Huntington's disease, *Brain Res*, 929:236–242.

Lastres-Becker, I., Hansen, H.H., Berrendero, F., de Miguel, R., Pérez-Rosado, A., Manzanares, J., Ramos, J.A. and Fernández-Ruiz, J. (2002b) Alleviation of motor hyperactivity and neurochemical deficits by endocannabinoid uptake inhibition in a rat model of Huntington's disease, *Synapse*, 44:23–35.

Lastres-Becker, I., de Miguel, R., De Petrocellis, L., Makriyannis, A., Di Marzo, V. and Fernández-Ruiz, J. (2003) Compounds acting at the endocannabinoid and/or endovanilloid systems reduce hyperkinesia in a rat model of Huntington's disease, *J Neurochem*, 84:1097–1109.

Ledent, C., Valverde, O., Cossu, G., Petiter, F., Aubert, J.F., Beslot, F., Bohme, G.A., Imperato, A., Pedrazzini, T., Roqes, B.P., Vassart, G., Fratta, W. and Parmentier, M. (1999) Unresponsiveness to cannabinoids and reduced addictive effects of opiates in CB_1 receptor knockout mice. ,*Science*, 283:401–404.

Lepore, M., Vorel, S.R., Lowinson, J. and Gardner, E.L. (1995) Conditioned place preference induced by Δ^9-tetrahydrocannabinol: comparison with cocaine, morphine, and food reward, *Life Sci*, 56: 2073–2080.

Lepore, M., Liu, X., Savage, V., Matalon, D. and Gardner, E.L. (1996) Genetic differences in delta 9-tetrahydrocannabinol-induced facilitation of brain stimulation reward as measured by a rate-frequency curve-shift electrical brain stimulation paradigm in three different rat strains, *Life Sci*, 58:PL365-PL372.

Leweke, F.M., Gerth, C.W. and Klosterkotter, J (2004) Cannabis-associated psychosis: current status of research, *CNS Drugs*, 18:895–910.

Leweke, F.M., Giuffrida, A., Wurster, U., Emrich, H.M. and Piomelli, D. (1999a) Elevated endogenous cannabinoids in schizophrenia, *Neuroreport*, 10:1665–9.

Leweke, F.M., Schneider, U., Thies, M., MÅnte, T.F. and Emrich, H.M. (1999b) Effects of synthetic Δ^9-tetrahydrocannabinol on binocular depth inversion of natural and artificial objects in man, *Psychopharmacology* (Berl), 142:230-5.

Lichtman, A.H., Wiley, J.L., LaVecchia, K.L., Neviaser, S.T., Arthur, D.B., Wilson, D.M. and Martin, B.R. (1998) Effects of SR 141716A after acute and chronic cannabinoid administration in dogs, *Eur J Pharmacol*, 357:139–148.

Lin, S., Thomas, T.C., Storlien, L.H. and Huang, X.F. (2000) Development of high fat diet-induced obesity and leptin resistance in C57Bl/6J mice, *Int J Obes Relat Metab Disord*, 24:639–46.

Linszen, D.H., Dingemans, P.M. and Lenior, M.A. (1994) Cannabis abuse and the course of recent-onset schizophrenic disorders, *Arch Gen Psychiatry*, 51:273–9.

Linszen, D.H., Dingemans, P.M., Nugter, M.A., Van der Does, A.J., Scholte, W.F. and Lenior, M.A. (1997) Patient attributes and expressed emotion as risk factor for psychotic relapse, *Schizophr Bull*, 23:119–30.

Lopez-Moreno, J.A., Gonzalez-Cuevas, G., Rodriguez de Fonseca, F. and Navarro, M. (2004) Long-lasting increase of alcohol relapse by the cannabinoid receptor agonist WIN 55,212-2 during alcohol deprivation, *J Neurosci*, 24:8245–8252.

Lyman, W.D., Sonett, J.R., Brosman, C.F., Elkin, R. and Bornstein, M.B. (1989) Delta 9-tetrahydrocannabinol: a novel treatment for experimental autoimmune encephalomyelitis, *J Neuroimmunol*, 23:73–81.

Maccarrone, M., Valverde, O., Barbaccia, M.L., Castañe, A., Maldonado, R., Ledent, C., Parmentier, M. and Finazzi-Agro, A. (2002) Age-related changes of anandamide metabolism in CB_1 cannabinoid receptor knockout mice: correlation with behavior, *Eur J Neurosci*, 15:1178–86.

Maccarrone, M., Gubellini, P., Bari, M., Picconi, B., Battista, N., Centonze, D., Bernardi, G., Finazzi-Agró, A. and Calabresi, P. (2003) Levodopa treatment reverses endocannabinoid system abnormalities in experimental parkinsonism, *J Neurochem*, 85:1018–1025.

Mailleux, P. and Vanderhaeghen, J.J. (1992) Distribution of neural cannabinoid receptor in the adult rat brain: a comparative receptor binding radioautography and in situ hybridization histochemistry, *Neuroscience*, 48:655–68.

Malan, T.P., Ibrahim, M.M., Vanderah, T.W., Makriyannis, A. and Porreca, F. (2002). Inhibition of pain responses by activation of CB(2) cannabinoid receptors, *Chem Phys Lipids*, 121:191–200.

Mallet, P.E and Beninger, R.J. (1998) The cannabinoid CB_1 receptor antagonist SR141716A attenuates the memory impairment produced by delta 9 tetrahydrocannabinol or anandamide, *Psychopharmacology (Berl)*, 140:11–19.

Mallet, P.E. and Beninger, R.J. (1998) Δ^9-Tetrahydrocannabinol, but not the endogenous cannabinoid receptor ligand anandamide, produces conditioned place avoidance, *Life Sci*, 62:2431–2439.

Maneuf, Y.P., Nash, J.E., Crossman, A.R. and Brotchie, J.M. (1996) Activation of the cannabinoid receptor by Δ^9-tetrahydrocannabinol reduces g aminobutyric acid uptake in the globus pallidus, *Eur J Pharmacol*, 308:161–164.

Mansbach, R.S., Rovetti, C.C., Winston, E.N. and Lowe III, J.A. (1996) Effects of the cannabinoid CB_1 receptor antagonist SR-141716A on the behavior of pigeons and rats, *Psychopharmacology (Berl)*, 124:315–22.

Manzanedo, C., Aguilar, M.A., Rodriguez-Arias, M., Navarro, M. and Minarro, J. (2004) Cannabinoid agonist-induced sensitisation to morphine place preference in mice, *Neuroreport*, 15:1373–1377.

Marinelli, S., Di Marzo, V., Berretta, N., Matias, I., Maccarrone, M., Bernardi, G. and Mercuri, N. (2003) Presynaptic facilitation of glutamatergic synapses to dopaminergic neurons of the rat substantia nigra by endogenous stimulation of vanilloid receptors, *J Neurosci*, 23:3136–3244.

Marsicano, G., Wotjak, C.T., Azad, S.C., Bisogno, T., Rammes, G., Cascio, M.G., Hermann, H., Tang, J., Hofmann, C., Zieglgansberger, W., et al. (2002) The endogenous cannabinoid system controls extinction of aversive memories, *Nature,* 418:530–4.

Martellotta, M.C., Cossu, G., Fattore, L., Gessa, G.L. and Fratta, W. (1998) Self-administration of the cannabinoid receptor agonist WIN 55,212-2 in drug naive mice, *Neuroscience*, 85:327–330.

Martin, B.R., Compton, D.R., Thomas, B.F., Prescott, W.R., Little, P.J., Razdan, R.K., Johnson, M.R., Melvin, L.S., Mechoulam, R. and Ward, S.J. (1991) Behavioral, biochemical, and molecular modeling evaluations of cannabinoid analogs, *Pharmacol Biochem Behav*, 40:471–478.

Martin, M., Ledent, C., Parmentier, M., Maldonado, R. and Valverde, O. (2000) Cocaine, but not morphine, induces conditioned place preference and sensitization to locomotor responses in CB_1 knockout mice, *Eur J Neurosci*, 12:4038–4046.

Martin, M., Ledent, C., Parmentier, M., Maldonado, R. and Valverde, O. (2002) Involvement of CB_1 cannabinoid receptors in emotional behavior, *Psychopharmacology (Berl),* 159:379–87.

Martin, R.S., Secchi, R.L., Sung, E., Lemaire, M., Bonhaus, D.W., Hedley, L. R. and Lowe, D.A. (2003) Effects of cannabinoid receptor ligands on psychosis-relevant behavior models in the rat, *Psychopharmacology (Berl),* 165:128–35.

Mas-Nieto, M., Pommier, B., Tzavara, E.T., Caneparo, A., Da Nascimento, S., Le Fur, G., Roques, B.P. and Noble, F. (2001) Reduction of opioid dependence by the CB_1 antagonist SR141716A in mice: evaluation of the interest in pharmacoptherapy of opioid addiction, *Br J Pharmacol*, 132:1809–1816.

Masserano, J.M., Karoum, F. and Wyatt, R.J. (1999) SR141716, a CB_1 cannabinoid receptor antagonist, potentiates the locomotor stimulant effects of amphetamine and apomorphine, *Behav Pharmacol*, 10:429–432.

Matsuda, L.A., Lolait, T.I., Brownstein, M.J., Young, A.C., and Bonner, T.I. (1990) Structure of a cannabinoid receptor and functional expression of the cloned cDNA, *Nature*, 346:561–564.

Mattes, R.D, Engelman, K., Shaw, L.M. and Elsohly, M.A. (1994) Cannabinoids and appetite stimulation, *Pharmacol Biochem Behav,* 49:187–95.

McGregor, I.S., Issakidis, C.N. and Prior, G. (1996) Aversive effects of the synthetic cannabinoid CP 55,940 in rats, *Pharmacol Biochem Behav*, 53:657–664.

McLaughlin, P.J., Winston, K., Swezey, L., Wisniecki, A., Aberman, J., Tardif, D.J., Betz, A.J., Ishiwari, K., Makriyannis, A. and Salamone, J.D. (2003) The cannabinoid CB_1 antagonists SR 141716A and AM 251 suppress food intake and food-reinforced behavior in a variety of tasks in rats, *Behav Pharmacol,* 14:583–8.

McMahon, L.R. and France, C.P. (2003) Discriminative stimulus effects of the cannabinoid antagonist, SR 141716A, in delta9-tetrahydrocannabinol-treated rhesus monkeys, *Exp Clin Psychopharmacol*, 11:286–293.

Mechoulam, R. and Fride, E. (1995) The unpaved road to the endogenous brain cannabinoid ligands, the anandamides, in: *Cannabinoid receptors*, ed. R.G. Pertwee (Academic Press, London) p. 233–238.

Mechoulam, R., Ben-Shabat S., Hanus, L., Ligumsky, M., Kaminski, N.E., Schatz, N.E., Gopher, A., Almog, S., Martin, B.R., Compton, D.R., Pertwee, R.G., Griffin, G., Bayewitch, M., Barge, J. and Vogel, Z. (1995) Identification of an endogenous 2-monoglyceride, present in canine gut, that binds to cannabinoid receptors, *Biochem Pharmacol*, 50:83–90.

Mechoulam, R., Fride, E. and Di Marzo, V. (1998) Endocannabinoids, *Eur J Pharmacol*, 359:1–18.

Meschler, J.P., Howlett, A.C. and Madras, B.K. (2001) Cannabinoid receptor agonist and antagonist effects on motor function in normal and 1-methyl-4-phenyl-1,2,5,6-tetrahydropyridine (MPTP)-treated non-human primates, *Psychopharmacology (Berl)*, 156:79–85.

Mezey, E., Toth, Z.E., Cortright, D.N., Arzubi, M.K., Krause, J.E., Elde, R., Guo, A., Blumberg, P.M. and Szallasi, A. (2000) Distribution of mRNA for vanilloid receptor subtype (VR_1), and VR_1-like immunoreactivity, in the central nervous system of the rat and human, *Proc Natl Acad Sci USA,* 97:3655–3660.

Miczek, K.A. (1978) Delta-9-tetrahydrocannabinol: antiaggressive effects in mice, rats, and squirrel monkeys, *Science,* 199:1459–61.

Miller, A.S. and Walker, J.M. (1995) Effects of a cannabinoid on spontaneous and evoked neuronal activity in the substantia nigra, *Eur J Pharmacol*, 279:179–185.

Müller-Vahl, K.R., Schneider, U., Prevedel, H., Theloe, K., Kolbe, H., Daldrup, T. and Emrich, H.M. (2003) Delta 9-tetrahydrocannabinol (THC) is effective in the treatment of tics in Tourette syndrome: a 6-week randomized trial, *J Clin Psychiatry*, 64:459–465.

Munro, S., Thomas, K.L. and Abu-Shaar, M. (1993) Molecular characterization of a peripheral cannabinoid receptor, *Nature*, 365:61–65.

Myerscough, R. and Taylor, S. (1985) The effects of marijuana on human physical aggression, *J Pers Soc Psychol,* 49:1541–6.

Naassila, M., Pierrefiche, O., Ledent, C. and Daoust, M. (2004) Decreased alcohol self-administration and increased alcohol sensitivity and withdrawal in CB_1 receptor knockout mice, *Neuropharmacology,* 46:243–253.

Navarro, M., Hernández, E., Muñoz, R.M., del Arco, I., Villanúa, M.A., Carrera, M.R.A. and Rodríguez de Fonseca, F. (1997) Acute administration of the CB_1 cannabinoid receptor antagonist SR-141716A induces anxiety-like responses in the rat, *Neuroreport,* 8:491–6.

Navarro, M., Carrera, M.R., Fratta, W., Valverde, O., Cossu, G., Fattore, L., Chowen, J.A., Gomez, R., del Arco, I., Villanua, M.A., Maldonado, R., Koob, G.F. and Rodríguez de Fonseca, F. (2001) Functional interaction between opioid and cannabinoid receptors in drug self-administration, *J Neurosci,* 21:5344–5350.

Nelson, R.J., Kriegsfeld, L.J., Dawson, V.L. and Dawson, T.M. (1997) Effects of nitric oxide on neuroendocrine function and behavior, *Front Neuroendocrinol,* 18:463–91.

Onaivi, E.S. (ed) (2002) *Biology of Marijuana: From gene to behavior,* London and New York: Taylor and Francis.

Onaivi, E.S., Chakrabarti, A., Gwebu, E.T. and Chaudhuri, G. (1996) Neurobehavioral effects of Δ^9-THC and cannabinoid (CB_1) receptor gene expression in mice, *Behav Brain Res,* 72:115–25.

Onaivi, E.S., Green, M.R. and Martin, B.R. (1990) Pharmacological characterization of cannabinoids in the elevated plus maze, *J Pharmacol Exp Ther,* 253:1002–9.

Onaivi, E.S., Leonard, C.M., Ishiguro, H., Zhang, P-W., Lin, Z., Akinshola, B.E. and Uhl, G. (2002) Endocannabinoids and cannabinoid receptor genetics, *Prog. Neurobiol*, 66:307–344.

Ortar, G., Ligresti, A., De Petrocellis, L., Morera, E. and Di Marzo, V. (2003) Novel selective and metabolically stable inhibitors of anandamide cellular uptake, *Biochem Pharmacol*, 65:1473–1481.

Parker, L.A. and Gillies, T. (1995) THC-induced place and taste aversion in Lewis and Sprague-Dawley rats, *Behav Neurosci*, 109:71–78.

Patel, S. and Hillard, C.J. (2003) Cannabinoid-induced Fos expression within A10 dopaminergic neurons, *Brain Res,* 963:15–25.

Patton, G.C., Coffey, C., Carlin, J.B., Degenhart, L., Lynskey, M. and Hall, W. (2002) Cannabis use and mental health in young people: cohort study, *BMJ,* 325:1195–8.

Pério, A., Rinaldi-Carmona, M., Maruani, J., Barth, F., Le Fur, G. and Soubrié, P. (1996) Central mediation of the cannabinoid cue: activity of a selective CB_1 antagonist, SR 141716A, *Behav Pharmacol,* 7:65–71.

Pério, A., Barnouin, M.C., Poncelet, M. and Soubrié, P. (2001) Activity of SR141716 on post-reinforcement pauses in operant responding for sucrose reward in rats, *Behav Pharmacol*, 12:641–645.

Pertwee, R.G. (1972) The ring test: a quantitative method for assessing the 'cataleptic' effect of cannabis in mice, *Br J Pharmacol*, 46:753–763.

Pertwee, R.G. (2001) Cannabinoid receptors and pain, *Prog Neurobiol*, 63:569–611.

Pertwee, R.G. (2002) Cannabinoids and multiple sclerosis, *Pharmacol Ther*, 95:165–174.

Pistis, M., Ferraro, L., Pira, L., Flore, G., Tanganelli, S., Gessa, G.L. and Devoto, P. (2002) Δ^9-Tetrahydrocannabinol decreases extracellular GABA and increases extracellular glutamate and dopamine levels in the rat prefrontal cortex: an in vivo microdialysis study, *Brain Res*, 948:155–158.

Pistis, M., Muntoni, A.L., Pillolla, G. and Gessa, G.L. (2002) Cannabinoids inhibit excitatory inputs to neurons in the shell of the nucleus accumbens: an in vivo electrophysiological study, *Eur J Neurosci,* 15:1795–1802.

Plasse, T.F., Gorter, R.W., Krasnow, S.H., Lane, M., Shepard, K.V. and Wadleigh, R.G. (1991) Recent clinical experience with dronabinol, *Pharmacol Biochem Behav,* 40:695-700.

Poncelet, M., Barnouin, M.C., Breliere, J.C., Le Fur, G. and Soubrie, P. (1999) Blockade of cannabinoid (CB_1) receptors by SR141716 selectively antagonizes drug-induced reinstatement of exploratory behavior in gerbils, *Psychopharmacology (Berl),* 144:144–50.

Poncelet, M., Maruani, J., Calassi, R. and Soubrié, P. (2003) Overeating, alcohol and sucrose consumption decrease in CB_1 receptor deleted mice, *Neurosci Lett,* 343:216–8.

Porsolt, R.D., Le Pichon, M. and Jalfre, M. (1977) Depression: a new animal model sensitive to antidepressant treatments, *Nature,* 266:730–2.

Porter, A.C., Sauer, J.M., Knierman, M.D., Becker, G.W., Berna, M.J., Bao, J., Nomikos, G.G., Carter, P., Bymaster, F.P., Leese, A.B. and Felder, C.C. (2002) Characterization of a novel endocannabinoid, virodhamine, with antagonist activity at the CB1 receptor, *J Pharmacol Exp Ther,* 301:1020–1024.

Ralevic, V. (2003) Cannabinoid modulation of peripheral autonomic and sensory neurotransmission, *Eur J Pharmacol*, 472:1–21.

Ravinet Trillou, C., Arnone, M., Delgorge, C., Gonalons, N., Keane, P., Maffrand, J.P. and Soubrie, P. (2003) Anti-obesity effect of SR141716, a CB_1 receptor antagonist, in diet-induced obese mice, *Am J Physiol Regul Integr Comp Physiol,* 284:R345–53.

Rice, A.S.C., Farquhar-Smith, W.P. and Nagy, I. (2002). Endocannabinoids and pain: spinal and peripheral analgesia in inflammation and neuropathy, *Prostaglandins Leukot Essent Fatty Acids,* 66:243–256.

Rice, O.V., Gordon, N. and Gifford, A.N. (2002) Conditioned place preference to morphine in cannabinoid CB_1 receptor knockout mice, *Brain Res*, 945:135–138.

Richardson, J.D., Aanonsen, L. and Hargreaves, K.M. (1997) SR 141716A, a cannabinoid receptor antagonist, produces hyperalgesia in untreated mice, *Eur J Pharmacol*, 319:R3–4.

Richardson, J.D., Aanonsen, L., and Hargreaves, K.M. (1998a). Hypoactivity of the spinal cannabinoid system results in NMDA-dependent hyperalgesia, *J Neurosci*, 18:451–457.

Richardson, J.D., Kilo, S., and Hargreaves, K.M. (1998b). Cannabinoids reduce hyperalgesia and inflammation via interaction with peripheral CB_1 receptors. *Pain* 75:111–119.

Richfield, E.K. and Herkenham, M. (1994) Selective vulnerability in Huntington's disease: preferential loss of cannabinoid receptors in lateral globus-pallidus, *Ann Neurol*, 36:577–584.

Robbe, D., Alonso, G., Duchamp, F., Bockaert, J. and Manzoni, O.J. (2001) Localization and mechanisms of action of cannabinoid receptors at the glutamatergic synapses of the mouse nucleus accumbens, *J Neurosci*, 21:109–116.

Robbe, D., Kopf, M., Remaury, A., Bockaert, J. and Manzoni, O.J. (2002) Endogenous cannabinoids mediate long-term synaptic depression in the nucleus accumbens, *Proc Natl Acad Sci USA*, 99:8384–8388.

Rodgers, R.J., Haller, J., Halasz, J. and Mikics, E. (2003) "One-trial sensitization" to the anxiolytic-like effects of cannabinoid receptor antagonist SR141716A in the mouse elevated plus-maze, *Eur J Neurosci,* 17:1279–86.

Rodríguez de Fonseca, F., Carrera, M.R.A., Navarro, M., Koob, G.F. and Weiss, F. (1997) Activation of corticotropin-releasing factor in the limbic system during cannabinoid withdrawal, *Science,* 276: 2050–4.

Rodríguez de Fonseca, F., Rubio, P., Menzaghi, F., Merlo-Pich, E., Rivier, J., Koob, G.F. and Navarro, M. (1996) Corticotropin-releasing factor (CRF) antagonist [D-Phe^{12}, $Nle^{21,38}$, $C^{a}MeLeu^{37}$]CRF attenuates the acute actions of the highly potent cannabinoid receptor agonist HU-210 on defensive-withdrawal behavior in rats, *J Pharmacol Exp Ther,* 276:56–64.

Rodríguez de Fonseca, F., Roberts, A.J., Bilbao, A., Koob, G.F. and Navarro, M. (1999) Cannabinoid receptor antagonist SR141716A decreases operant ethanol self administration in rats exposed to ethanol-vapor chambers, *Acta Pharmacol Sin*, 20:1109–1114.

Romero, J., García, L., Cebeira, M., Zadrozny, D., Fernández-Ruiz, J.J. and Ramos, J.A. (1995) The endogenous cannabinoid receptor ligand, anandamide, inhibits the motor behavior: role of nigrostriatal dopaminergic neurons, *Life Sci,* 56:2033–40.

Romero, J., García-Palomero, E., Fernández-Ruiz, J.J. and Ramos, J.A. (1996) Involvement of $GABA_B$ receptors in the motor inhibition produced by agonists of brain cannabinoid receptors, *Behav Pharmacol*, 7:299–302.

Romero, J., de Miguel, R., Ramos, J.A. and Fernández-Ruiz, J.J. (1998) The activation of cannabinoid receptors in striatonigral GABAergic neurons inhibited GABA uptake, *Life Sci*, 62:351–363.

Romero, J., Lastres-Becker, I., de Miguel, R., Berrendero, F., Ramos, J.A. and Fernández-Ruiz, J. (2002) The endogenous cannabinoid system and the basal ganglia: biochemical, pharmacological and therapeutic aspects, *Pharmacol Ther*, 95:137–152.

Rowland, N.E., Mukherjee, M. and Robertson, K. (2001) Effects of the cannabinoid receptor antagonist SR 141716, alone and in combination with dexfenfluramine or naloxone, on food intake in rats, *Psychopharmacology (Berl),* 159:111–6.

Sakurai, Y., Ohta, H., Shimazoe, T., Kataoka, Y., Fujiwara, M. and Ueki, S. (1985) Δ^9-Tetrahydrocannabinol elicited ipsilateral circling behavior in rats with unilateral nigral lesion, *Life Sci*, 37:2181–2185.

Sanchis-Segura, C., Cline, B.H., Marsicano, G., Lutz, B. and Spanagel, R. (2004) Reduced sensitivity to reward in CB_1 knockout mice, *Psychopharmacology (Berl)*, 176:223–232.
Sañudo-Peña, M.C., Tsou, K., Delay, E.R., Hohman, A.G., Force, M. and Walker, J.M. (1997) Endogenous cannabinoids as an aversive or counter-rewarding system in the rat, *Neurosci Lett*, 223:125–128.
Sañudo-Peña, M.C., Romero, J., Seale, G.E., Fernández-Ruiz, J. and Walker, J.M. (2000) Activational role of cannabinoids on movement, *Eur J Pharmacol*, 391:269–274.
Schlicker, E. and Kathmann, M. (2001) Modulation of transmitter release via presynaptic cannabinoid receptors, *Trends Pharmacol Sci*, 22:565–71.
Serra, S., Carai, M.A., Brunetti, G., Gomez, R., Melis, S., Vacca, G., Colombo, G. and Gessa, G.L. (2001) The cannabinoid receptor antagonist SR 141716 prevents acquisition of drinking behavior in alcohol-preferring rats, *Eur J Pharmacol*, 430:369–371.
Serra, S., Brunetti, G., Pani, M., Vacca, G., Carai, M.A., Gessa, G.L. and Colombo, G. (2002) Blockade by the cannabinoid CB_1 receptor antagonist, SR 141716, of alcohol deprivation effect in alcohol-preferring rats, *Eur J Pharmacol*, 443:95–97.
Shearman, L.P., Rosko, K.M., Fleischer, R., Wang, J., Xu, S., Tong, X.S. and Rocha, B.A. (2003) Antidepressant-like and anorectic effects of the cannabinoid CB_1 receptor inverse agonist AM251 in mice, *Behav Pharmacol*, 14:573–82.
Sieradzan, K.A., Fox, S.H., Hill, M., Dick, J.P.R., Crossman, A.R. and Brotchie, J.M. (2001) Cannabinoids reduce levodopa-induced dyskinesia in Parkinson's disease: a pilot study, *Neurology*, 57:2108–2111.
Simiand, J., Keane, M., Keane, P.E. and Soubrié, P. (1998) SR-141716, a CB_1 cannabinoid receptor antagonist, selectively reduces sweet food intake in marmoset, *Behav Pharmacol*, 9:179–81.
Singer, H.S. (2000) Current issues in Tourette Syndrome, *Mov Disord*, 15:1051–1063.
Singh, M.E., Verty, A.N., McGregor, I.S. and Mallet, P.E. (2004) A cannabinoid receptor antagonist attenuates conditioned place preference but not behavioural sensitization to morphine, *Brain Res*, 1026:244–253.
Skosnik, P.D., Spatz-Glenn, L. and Park, S. (2001) Cannabis use is associated with schizotypy and attentional disinhibition, *Schizophr Res*, 48:83–92.
Smart, D., Jerman, J.C., Gunthorpe, M.J., Brough, S.J., Ranson, J., Cairns, W., Hayes, P.D., Randall, A.D. and Davis, J.B. (2001) Characterization using FLIPR of human vanilloid VR1 receptor pharmacology, *Eur J Pharmacol*, 417:51–58.
Smith, P.B., Compton, D.R., Welch, S.P., Razdan, R.K., Mechoulam, R. and Martin, B.R. (1994) The pharmacological activity of anandamide, a putative endogenous cannabinoid in mice, *J Pharmacol Exp Ther*, 270:219–227.
Soderstrom, K., Tian, Q., Valenti, M and Di Marzo, V. (2004) Endocannabinoids link feeding state and auditory perception-related gene expression, *J Neuroscience*, 24:10013–10021.
Solinas, M., Panlilio, L.V., Antoniou, K., Pappas, L.A. and Goldberg, S.R. (2003) The cannabinoid CB_1 antagonist N-piperidinyl-5-(4-chlorophenyl)-1-(2,4-dichlorophenyl) -4-methylpyrazole-3-carboxamide (SR-141716A) differentially alters the reinforcing effects of heroin under continuous reinforcement, fixed ratio, and progressive ratio schedules of drug self-administration in rats, *J Pharmacol Exp Ther*, 306:93–102.
Soyka, M. (2003) Cannabinoids and mental health, *Schmerz*, 17:268–273.
Stellar, J.S. and Rice, M.B. (1989) Pharmacological basis of intracranial self-stimulation reward. In *The neuropharmacological basis of reward* (Eds, Liebman, J.M. and Cooper, S.J.) Clarendon Press, Oxford, pp. 14–65.
Stéru, L., Chermat, R., Thierry, B., Mico, J.A., Lenegre, A., Steru, M., Simon, P. and Porsolt, R.D. (1987) The automated Tail Suspension Test: a computerized device which differentiates psychotropic drugs, *Prog Neuropsychopharmacol Biol Psychiatry*, 11:659–71.
Strangman, N.M., Patrick, S.L., Hohmann, A.G., Tsou, K. and Waker, J.M. (1998) Evidence for a role of endogenous cannabinoids in the modulation of acute and tonic pain sensitivity, *Brain Res*, 813: 323–328.
Sugiura, T. and Waku, K. (2002) Cannabinoid receptors and their endogenous ligands, *J Biochem*, 132:7–12.
Sugiura, T., Kondo, S., Sukagawa, A., Nakane, A., Shinoda, A., Itoh, K., Yamashita, A. and Waku, K. (1995) 2-Arachidonoylglycerol: A possible endogenous cannabinoid ligand in brain, *Biochem Biophys Res Commun*, 215:89–97.
Sulcova, A., Mechoulam, R. and Fride, E. (1998) Biphasic effects of anandamide, *Pharmacol Biochem Behav*, 59:347–52.

Sullivan, J.M. (2000) Cellular and molecular mechanisms underlying learning and memory impairments produced by cannabinoids, *Learn Mem*, 7:132–139.

Szabo, B., Wallmichrath, I., Mathonia, P. and Pfreundtner, C. (2000) Cannabinoids inhibit excitatory neurotransmission in the substantia nigra pars reticulata, *Neuroscience*, 97:89–97.

Szallasi, A. and Blumberg, P.M. (1999) Vanilloid (capsaicin) receptors and mechanisms, *Pharmacological Rev*, 51:159–211.

Tanda, G., Pontieri, F.E. and Di Chiara, G. (1997) Cannabinoid and heroin activation of mesolimbic dopamine transmission by a common μ1 opioid receptor mechanism, *Science,* 276:2048–2050.

Tanda, G., Munzar, P. and Goldberg, S.R. (2000) Self-administration behavior is maintained by the psychoactive ingredient of marijuana in squirrel monkeys, *Nat Neurosci*, 3:1073–1074.

Tanda, G. and Goldberg, S.R. (2003) Cannabinoids: reward, dependence, and underlying neurochemical mechanisms — a review of recent preclinical data, *Psychopharmacology (Berl)*, 169:115–134.

Trojniar, W. and Wise, R.A. (1991) Facilitory effect of Δ^9-tetrahydrocannabinol on hypothalamically induced feeding, *Psychopharmacology (Berl),* 103:172–6.

Tsou, K., Brown, S., Sanudo-Pena, M.C., Mackie, K. and Walker, J.M. (1998) Immunohistochemical distribution of cannabinoid CB_1 receptors in the rat central nervous system, *Neuroscience,* 83:393–411.

Tzavara, E.T., Davis, R.J., Perry, K.W., Li, X., Salhoff, C., Bymaster, F.P., Witkin, J.M. and Nomikos, G.G. (2003) The CB_1 receptor antagonist SR141716A selectively increases monoaminergic neurotransmission in the medial prefrontal cortex: implications for therapeutic actions, *Br J Pharmacol,* 138:544–53.

Ujike, H., Takaki, M., Nakata, K., Tanaka, Y., Takeda, T., Komada, M., Fujiwara, Y., Sakai, A. and Kuroda, S. (2002) CNR1, central cannabinoid receptor gene, associated with susceptibility to hebephrenic schizophrenia, *Mol Psychiatry,* 7:515–8.

Valjent, E. and Maldonado, R. (2000) A behavioural model to reveal place preference to Δ^9-tetrahydrocannabinol in mice, *Psychopharmacology (Berl)*, 147:436–438.

Valverde, O., Ledent, C., Beslot, F., Parmentier, M. and Roques, B.P. (2000) Reduction of stress-induced analgesia but not of exogenous opioid effects in mice lacking CB_1 receptors, *Eur J Neurosci*, 12:533–539.

van Dellen, A., Blackemore, C., Deacon, R., York, D. and Hannan, A.J. (2000) Delaying the onset of Huntington's disease in mice, *Nature*, 404:721–722.

van der Stelt, M. and Di Marzo, V. (2003) The endocannabinoid system in the basal ganglia and in the mesolimbic reward system: implications for neurological and psychiatric disorders, *Eur J Pharmacol*, 480:133–150.

Varvel, S.A. and Lichtman, A.H. (2002) Evaluation of CB_1 receptor knockout mice in the Morris water maze, *J. Pharmacol Exp Ther,* 301:915–924.

Verty, A.N., Singh, M.E., McGregor, I.S. and Mallet, P.E. (2003) The cannabinoid receptor antagonist SR 141716 attenuates overfeeding induced by systemic or intracranial morphine, *Psychopharmacology (Berl),* 168:314–23.

Verty, A.N., McGregor, I.S. and Mallet, P.E. (2004) Consumption of high carbohydrate, high fat, and normal chow is equally suppressed by a cannabinoid receptor antagonist in non-deprived rats, *Neurosci Lett,* 354:217–20.

Vickers, S.P., Webster, L.J., Wyatt, A., Dourish, C.T. and Kennett, G.A. (2003) Preferential effects of the cannabinoid CB_1 receptor antagonist, SR 141716, on food intake and body weight gain of obese (*fa/fa*) compared to lean Zucker rats, *Psychopharmacology (Berl),* 167:103–11.

Vinklerová, J., Nováková, J. and Sulcová, A. (2002) Inhibition of methamphetamine self-administration in rats by cannabinoid receptor antagonist AM 251, *J Psychopharmacol*, 16:139–143.

Vlachou, S., Nomikos, G.G. and Panagis, G. (2003) WIN 55,212-2 decreases the reinforcing actions of cocaine through CB_1 cannabinoid receptor stimulation, *Behav Brain Res*, 141:215–222.

Volicer, L., Stelly, M., Morris, J., McLaughlin, J. and Volicer, B.J. (1997) Effects of dronabinol on anorexia and disturbed behavior in patients with Alzheimer's disease, *Int J Geriatr Psychiatry,* 12:913–9.

Waksman, Y., Olson, J.M., Carlisle, S.J. and cabral, G.A. (1998) The central cannabinoid receptor (CB) mediates inhibition of nitric oxide production by rat microglial cells, *J Pharmacol Exp Ther,* 288:1357–66.

Walker, J.M and Huang S.M. (2002) Endocannabinoids in pain modulation, *Prostaglandins Leukot Essent Fatty Acids*, 66:235–242.

Walker, J.M., Huang, S.M., Strangman, N.M., Tsou, K. and Sanudo-Pena, M.C. (1999) Pain modulation by release of the endogenous cannabinoid anandamide, *Proc. Natl. Acad Sci USA*, 96:12198–12203

Walker, J.M., Krey, J.F., Chu, C.J. and Huang, S.M. (2002) Endocannabinoids and related fatty acid derivatives in pain modulation, *Chem Phys Lipids*, 121:159–172.

Walsh, D., Nelson, K.A. and Mahmoud, F.A. (2003) Established and potential therapeutic applications of cannabinoids in oncology, *Support Care Cancer,* 11:137–43.

Wang, L., Liu, J., Harvey-White, J., Zimmer, A. and Kunos, G. (2003) Endocannabinoid signaling via cannabinoid receptor 1 is involved in ethanol preference and its age-dependent decline in mice, *Proc Natl Acad Sci USA,* 100:1393–8.

Weidenfeld, J., Feldman, S. and Mechoulam, R. (1994) Effect of the brain constituent anandamide, a cannabinoid receptor agonist, on the hypothalamo-pituitary-adrenal axis in the rat, *Neuroendocrinology,* 59:110–2.

Welch, S.P and Eads, M. (1999) Synergistic interactions of endogenous opioids and cannabinoid systems, *Brain Res,* 848:183–190.

Weller, R.A. and Halikas, J.A. (1985) Marijuana use and psychiatric illness: a follow-up study, *Am J Psychiatry,* 142:848–50.

Wenger, T., Jamali, K.A., Juaneda, C., Leonardelli, J. and Tramu, G. (1997) Arachidonyl ethanolamide (anandamide) activates the parvocellular part of hypothalamic paraventricular nucleus, *Biochem Biophys Res Commun,* 237:724–8.

Werner, N.A. and Koch, J.E. (2003) Effects of the cannabinoid antagonists AM281 and AM630 on deprivation-induced intake in Lewis rats, *Brain Res,* 967:290–2.

Wiley, J.L. and Martin, B.R. (2002) Cannabinoid pharmacology: implications for additional cannabinoid receptor subtypes, *Chem Phys Lipids* 121:57–63.

Wiley, J.L., Balster, R.L. and Martin, B.R. (1995a) Discriminative stimulus effects of anandamide in rats, *Eur J Pharmacol,* 276:49–54.

Wiley, J.L., Barrett, R.L., Lowe, J., Balster, R.L. and Martin, B.R. (1995b) Discriminative stimulus effects of CP 55,940 and structurally dissimilar cannabinoids in rats, *Neuropharmacology,* 34:669–676.

Wiley, J.L., Huffman, J.W., Balster, R.L. and Martin, B.R. (1995c) Pharmacological specificity of the discriminative stimulus effects of delta 9-tetrahydrocannabinol in rhesus monkeys, *Drug Alcohol Depend,* 40:81–86.

Wiley, J.L., Golden, K.M., Ryan, W.J., Balster, R.L., Razdan, R.K., Martin, B.R. (1997) Evaluation of cannabimimetic discriminative stimulus effects of anandamide and methylated fluoroanandamide in rhesus monkeys, *Pharmacol Biochem Behav,* 58:1139–1143.

Wiley, J.L., Ryan, W.J., Razdan, R.J. and Martin, B.R. (1998) Evaluation of cannabimimetic effects of structural analogs of anandamide, *Eur J Pharmacol,* 355:113–118.

Williams, C.M. and Kirkham, T.C. (1999) Anandamide induces overeating: mediation by central cannabinoid (CB_1) receptor, *Psychopharmacology (Berl),* 143:315–7.

Williams, C.M. and Kirkham, T.C. (2002) Reversal of Δ^9-THC hyperphagia by SR141716 and naloxone but not dexfenfluramine, *Pharmacol Biochem Behav,* 71:315–7.

Williams, C.M., Rodgers, P.J. and Kirkham, T.C. (1998) Hyperphagia in pre-fed rats following oral Δ^9-THC, *Physiol Behav,* 65:343–6.

Williamson, E.M. and Evans, F.J. (2000). Cannabinoids in Clinical Practice, *Drugs*, 60:1303–1314.

Willner, P. (1997) Validity, reliability and utility of the chronic mild stress model of depression: a 10-year review and evaluation, *Psychopharmacology (Berl),* 134:319–29.

Wilson, R.I. and Nicol, R.A. (2002) Endocannabinoid Signaling in the Brain, *Science,* 296:678–682.

Wolff, M.C. and Leander J.D. (2003) SR141716A, a cannabinoid CB_1 receptor antagonist, improves memory in a delayed radial maze task, *Eur J Pharmaco*, 477:213–217.

Wu, X. and French, E.D. (2000) Effects of chronic Δ9-tetrahydrocannabinol on rat midbrain dopamine neurons: an electrophysiological assessment, *Neuropharmacology,* 39:391–398.

Wylie, A.S., Scott, R.T. and Burnett, S.J. (1995) Psychosis due to "skunk", *BMJ,* 311:125.

Xue, Q., Yu, Y., Trilk, S.L., Jong, B.E. and Schumacher, M.A. (1999) The genomic organization of the gene encoding the vanilloid receptor: Evidence for multiple splice variants, *Genomics,* 76:14–20.

Yamada, K. and Nabeshima, T. (1995) Stress-induced behavioral responses and multiple opioid systems in the brain, *Behav Brain Res* 67:133–145.

Zajicek, J., Fox, P., Sanders, H., Wright, D., Vickery, J., Nunn, A. and Thompson, A. (2003) Cannabinoids for treatment of spasticity and other symptoms related to multiple sclerosis (CAMS study): multicentre randomised placebo-controlled trial, *Lancet*, 362:1517–1526.

Zammit, S., Allebeck, P., Andreasson, S., Lundberg, I. and Lewis, G. (2002) Self reported cannabis use as a risk factor for schizophrenia in Swedish conscripts of 1969: historical cohort study, *BMJ,* 325:1–5.

Zimmer, A., Zimmer, A.M., Hohmann, A.G., Herkenham, M. and Bonner, T.I. (1999) Increased mortality, hypoactivity, and hypoalgesia in cannabinoid CB_1 receptor knockout mice, *Proc Natl Acad Sci USA,* 96:5780–5.

Zuardi, A.W., Shirakawa, I., Finkelfarb, E. and Karniol, I.G. (1982) Action of cannabidiol on the anxiety and other effects produced by Δ^9-THC in normal subjects, *Psychopharmacology (Berl),* 76:245–50.

Zygmunt, P.M., Petersson, J., Andersson, D.A., Chuang, H., Sorgard, M., Di Marzo, V., Julius, D. and Hogestatt, E.D. (1999) Vanilloid receptors on sensory nerves mediate the vasodilator action of anandamide, *Nature,* 400:452–457.

Zygmunt, P.M., Chuang, H., Movahed, P., Julius, D. and Hogestatt, E.D. (2000) The anandamide transport inhibitor AM404 activates vanilloid receptors, *Eur J Pharmacol*, 396:39–42.

Part V

Endocannabinoids in CNS Pathology

14 Neuroprotection: Glutamate and Endocannabinoids

Harald S. Hansen, Gitte Petersen, and Henrik H. Hansen

CONTENTS

INTRODUCTION

Brain injury triggered by trauma or ischemia is a major cause of death and morbidity in Western societies. There is an ongoing effort to find efficient medical treatment. However, many compounds that were neuroprotective in animal models did not succeed in clinical trials (1,2). With the discovery of cannabinoid receptors (3), and their endogenous ligands (4–6), and the finding that activation of cannabinoid receptors inhibits glutamate release from neurons in culture (7), the cytoprotective impact of cannabinoids/endocannabinoids has acquired increasing research interest. A number of reviews on this concept have been published (8–18). This chapter discusses endogenous cannabinoid receptor agonists as well as a number of nonendocannabinoid NAEs that may constitute a putative cytoprotective system (19, 20).

GLUTAMATE AND EXCITOTOXICITY

Glutamate is the principal excitatory neurotransmitter in the brain, which is stored in presynaptic vesicles. Upon stimulation, glutamate is released into the synaptic cleft, where it activates postsynaptic glutamate receptors. The glutamatergic receptors encompass the ionotropic receptors (NMDA, AMPA, and kainate receptors) and the metabotropic receptors (mGluR1–mGluR8). Following its action, glutamate is taken up by neurons and astrocytes via high-efficacy glutamate transporters (EAAT1–EAAT5) (21). Excessive vesicular release of glutamate or reversal of glutamate transporters can result in hyperstimulation of neuronal ionotropic glutamate receptors, resulting in excessive Ca^{2+} influx. The increased intracellular Ca^{2+} concentration may lead to oxidative stress and mitochondrial injury, which eventually results in neuronal death (2,22,23). This mode of neuronal demise is also associated with activation of microglial cells and an inflammation-like response in the brain (24,25). Generally, this phenomenon is referred to as excitotoxicity. Thus, NMDA receptors and Ca^{2+}-permeable AMPA receptors can cause cell death in many neuropathological conditions when intensively or chronically activated, and increased extracellular glutamate levels are the primary cause of neuronal death following acute injury such as ischemic stroke, mechanical trauma, or seizure (26–28). The excitotoxicity hypothesis may also explain some aspects of neuronal death observed in many chronic neurodegenerative diseases (22,29,30). On the other hand, NMDA receptors also have neuroprotective functions (2,31). It has been suggested that it is primarily the extrasynaptic NMDA receptors, as opposed to the synaptic NMDA receptors, that are involved in glutamate-stimulated cell death, and synaptic-NMDA-receptor-dependent Ca^{2+} influx mediates the expression of prosurvival genes (2). Several different rodent models have been employed to study excitotoxicity, and developmental hypersensitivity to glutamate receptor activation in these rodent models is well established (32–37).

ENDOCANNABINOIDS AND THEIR FORMATION IN THE BRAIN

All the endocannabinoids are of lipid nature, and they are derived from membrane phospholipids. Figure 14.1 shows the four compounds that are suggested to be endogenous agonists for the cannabinoid receptors. Of these four compounds, the biological significance of noladin ether (2-arachidonyl glyceryl ether) and virodhamine (*O*-arachidonoylethanolamine) is at present questionable. Noladin ether has been reported to be present in pig and rat brain (38,39), although this has not been found by others (40,41). Virodhamine is reported to be normally present in rat and human

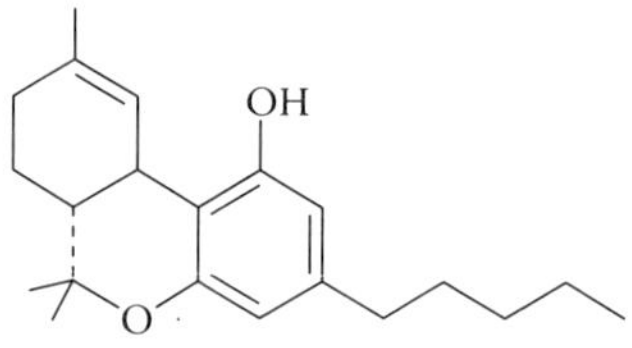

Δ^9-THC

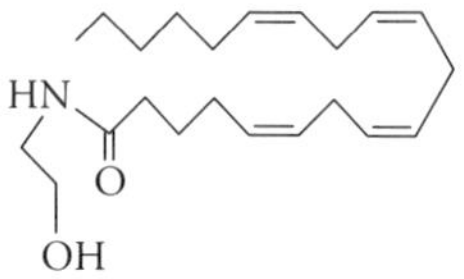

Anandamide

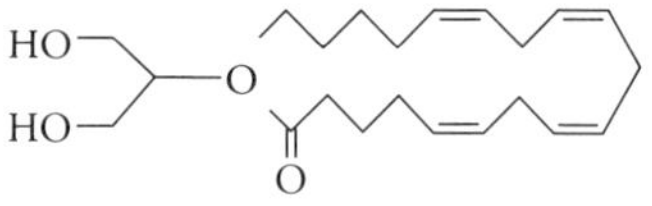

2-AG

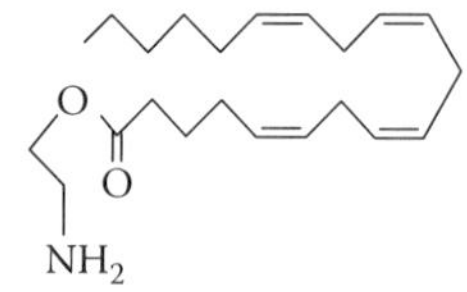

Virodhamine

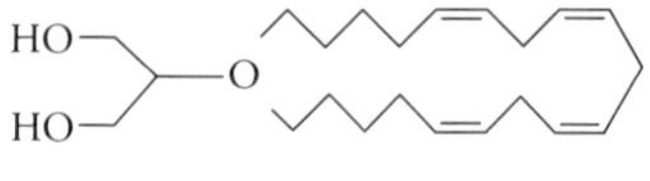

Noladin ether

FIGURE 14.1 Four endogenous substances suggested as being endocannabinoids: anandamide (*N*-arachidonoylethanolamine), 2-AG (2-arachidonoylglycerol), noladin ether (2-arachidonylglyceryl ether), and virodhamine (*O*-arachidonoylethanolamine). The structure of the principal cannabinoid of plant origin, Δ^9-THC (Δ^9-tetrahydrocannabinol), is inserted.

brain (42), although under certain conditions it can also be formed artificially from anandamide (43). The enzymes that catalyze the formation of noladin ether and virodhamine are presently unknown.

Anandamide was the first endogenous activator of the cannabinoid type-I receptor (CB_1 receptor) to be discovered (4). Along with the isolation of anandamide, two other anandamide-like compounds were also found in brain tissue, *N*-dihomo-γ-linolenoylethanolamine and *N*-docosatetraenoylethanolamine (44), indicating concurrent formation of several structurally related NAEs. These two polyunsaturated NAEs showed the same biological activity as anandamide on smooth muscle, although with slightly lower potency (45). Anandamide is formed from a precursor phospholipid *N*-arachidonoylethanolamine phospholipid (NAPE), which is hydrolyzed by the enzyme *N*-acylphosphatidylethanolamine-hydrolyzing phospholipase D (NAPE-PLD). NAPE-PLD does not discriminate between different *N*-acyl groups (46, 47) and appears to be catalytically different from other known mammalian phospholipase D enzymes in that it cannot catalyze the transphosphatidylation reaction using ethanol instead of water in the hydrolytic reaction (48). The *N*-acyl group of NAPEs originates from the *sn*-1 group of phosphatidylcholine. This intermolecular transfer reaction is catalyzed by the enzyme *N*-acyltransferase, which does also not discriminate between acyl groups (47,49,50). As predicted by the *sn*-1 acyl composition of phosphatidylcholine, the palmitoyl, stearoyl, and oleoyl moieties make up the bulk of the *N*-acyl groups in both NAPEs and NAEs (see Figure 14.2) in the mammalian brain. Anandamide usually accounts for 1–5% of the NAE species (51–54).

Remodeling of the fatty acid composition of the *sn*-1 position of phosphatidylcholine or *N*-acyltransferase substrate phospholipids will result in a changed composition of the NAEs. This is seen in severe neuronal injury, when the *N*-stearoylethanolamine accumulates to a greater extent than the other NAEs (55,56).

Anandamide and other NAEs are degraded by a fatty acid amide hydrolase (FAAH) (18,57,58–60). Inhibition or knockout of FAAH results in increased tissue levels of anandamide

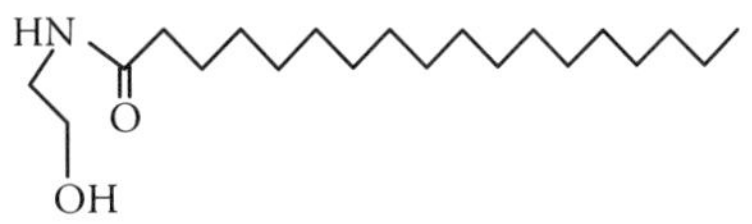

HN O OH

N-stearoylethanolamine

HN O OH

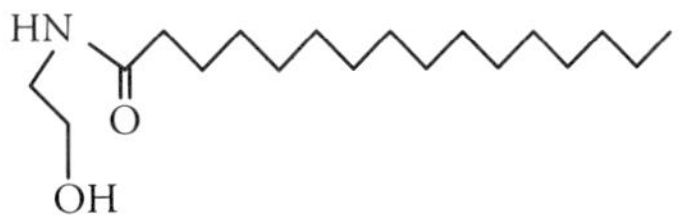

N-palmitoylethanolamine

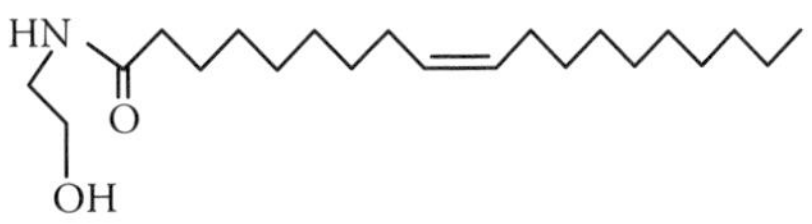

HN O OH

N-oleoylethanolamine

FIGURE 14.2 Molecular structure of nonendocannabinoid NAEs.

and the other NAEs (61, 62). Due to the lipophilicity of anandamide, it is largely bound to proteins, e.g., albumin, when transported in the aqueous phase (63–65), and it does not seem to use a specific membrane transporter for its uptake into cells (66).

2-AG was reported to be a ligand for the cannabinoid receptors shortly after the discovery of anandamide (5,6). In comparison with anandamide, 2-AG is slightly more water soluble due to the extra hydroxy group. Furthermore, it behaves as a full agonist for the cannabinoid receptors, whereas anandamide is a partial agonist (67,68). 2-AG can be formed from several sources of arachidonic-acid-rich membrane phospholipids. In general, arachidonic acid is esterified to the *sn*-2 position of phospholipids and is found in high amounts in phosphatidylinositides and phosphatidylethanolamine (69). 2-AG can be formed through several pathways (70–72), although the formation of 2-AG from 1-stearoyl-2-arachidonoyl-glycerol derived from inositol lipid turnover (73) seems to be the most plausible pathway under physiological conditions. Therefore, the principal enzyme for 2-AG synthesis is diacylglycerol lipase, liberating the fatty acid in the *sn*-1 position of 1-stearoyl-2-arachidonoyl-glycerol, resulting in generation of 2-AG (74–76). Thus, 2-AG as well as anandamide are not stored in vesicles but are formed in response to a stimulus, and these endocannabinoids are therefore believed to be released from their precursors on demand. Tissue levels of 2-AG are usually found in nanomole per gram of tissue, whereas anandamide levels are present in picomole per gram of tissue (54). 2-AG is inactivated by hydrolysis (77,78), which can be catalyzed by FAAH *in vitro* (79,80). However, the use of FAAH knockout mice suggests that FAAH is not the principal enzyme degrading 2-AG *in vivo* (81). The enzymes hydrolyzing 2-AG *in vivo* have not been identified. Both anandamide and 2-AG are also substrates for cyclooxygenase, lipoxygenases, and cytochrome P450 hydroxylases *in vitro* (82), but the biological significance of these alternative endocannabinoid metabolic pathways is unknown.

2-AG accumulates during tissue sampling procedures (83), probably due to initial diacylglycerol accumulation (84), as well as due to a glycerol:arachidonoyl-CoA acyltransferase activity (85) triggered by exceedingly high tissue levels of glycerol and arachidonoyl-CoA during brain injury (86,87). Some researches have found that 2-AG production increases during brain injury *in vivo* (88), although others have not seen increased 2-AG levels (56,89,90). This difference may be explained by methodological differences, e.g., the use of different animal models of brain damage, variations in sampling mode, and the timing of tissue sampling for 2-AG analysis.

The CB_1 receptor is primarily expressed in neural tissues (91) where the receptor is located in high densities at presynaptic nerve terminals, in particular, on GABAergic basal ganglia projections and hippocampal interneurons (92,93) and, to a lesser extent, on glutamatergic neurons (94,95).

Cannabinoid receptor type-II (CB_2 receptor) is mainly present on immune cells (91). Furthermore, microglial cells, as well as macrophages, express CB_2 receptors upon stimulation (96), especially at the leading edge of migrating microglial cells (97), which also express the so-called abnormal cannabidiol-sensitive receptor (CB_{abn} receptor) (97). The CB_{abn} receptor was first demonstrated on endothelial cells, where it could be activated by a cannabidiol derivative (abnormal cannabidiol) as well as anandamide (98). In these cells, 2-AG behaves as a full agonist and anandamide as a partial agonist on CB_{abn} receptors (97). It is, therefore, accepted that anandamide is a partial agonist on all three types of cannabinoid receptors, whereas 2-AG provides a full agonist response on the receptors. By contrast, anandamide activates the vanilloid VR_1 receptor as a full agonist (99–101), whereas 2-AG has no stimulatory activity (99). The VR_1 receptor is a nonselective cation channel primarily expressed on peripheral sensory C and Aδ fibers (102,103) and also, although to a lesser extent, in the brain (104,105). Anandamide also affects L-type calcium channels (106), gap junctions (107), TASK-1 K^+ channels (108), and NMDA receptors (109) *in vitro*, but the physiological importance of these effects is not known. There is some preliminary evidence for the existence of other anandamide-activated receptors in the brain (110–115); however, the nature and identity of these putative receptor sites are far from clear.

ENDOCANNABINOIDS AND NEUROPROTECTION

Exogenous cannabinoids inhibit glutamate release upon binding to presynaptic CB_1 receptors *in vitro* (116,117). Endocannabinoids are increasingly believed to be inhibitors of neurotransmitter release via binding to presynaptic CB_1 receptors following their retrograde transport from a postsynaptic locus of synthesis. This retrograde endocannabinoid function has been demonstrated to modulate GABA release (118–120) and glutamate release (121–124) in different brain regions of the rat. Activation of CB_1 receptors may thus inhibit excitation. Depending on the anatomical site of action, the CB_1 receptors may also cause disinhibition of excitatory synapses via modulation of GABA release, which results in increased glutamate release. In accordance with this notion, the principal plant-derived cannabinoid, Δ^9-tetrahydrocannabinol (Δ^9-THC, see Figure 14.1), decreases extracellular GABA levels and elevates extracellular glutamate levels in the prefrontal cortex of the rat (125). This intriguing mechanism that triggers depolarization-induced postsynaptic synthesis and release of endocannabinoids may likely be activation of postsynaptic metabotropic glutamate receptors by glutamate itself (121,126), which provides a loop in which the extent of glutamate release *per se* controls the suppressive action of endocannabinoids on glutamatergic and GABAergic synapses.

At glutamatergic synapses, postsynaptic metabotropic glutamate receptors are likely to be involved in inositol lipid turnover (126), resulting in formation of 1-stearoyl-2-arachidonoyl-glycerol. The retrograde messenger effect of endocannabinoids has so far only been assessed by indirect measures (using cannabinoid receptor agonists and FAAH inhibitors). However, we hypothesize that 2-AG may prove to be the most important inhibitory retrograde mediator of neurotransmitter release; 2-AG is present in the brain at levels that are 200-fold higher than those of anandamide, and 2-AG behaves in some *in vitro* assays as a full agonist at CB_1 receptors, whereas anandamide is a partial agonist (see section titled "Endocannabinoids and Their Formation in the Brain").

During our study of NAE and NAPE formation in cultured cortical neurons, we confirmed that these compounds accumulate in response to cell injury (51,127,128), whether mediated by activation of the NMDA receptor or by inhibition of the respiratory function of mitochondria (127,128). In these experiments, anandamide was also detected (9), presumably triggered by increased intracellular concentration of Ca^{2+} (8,57). Elevated intracellular Ca^{2+} concentrations activate *N*-acyltransferase, the enzyme that catalyzes the formation of NAPEs (129). Ca^{2+} ions have also been reported to stimulate NAPE-PLD *in vitro* (130), but it is uncertain whether this also occurs *in vivo*. In order to investigate whether the excessive formation of NAEs and NAPEs also occurs *in vivo*, we determined the response to brain damage in neonatal rats because they have high *N*-acyltransferase activity (131), and at this specific developmental stage the rat brain is exceedingly sensitive to NMDA receptor overactivation (32). We found a massive accumulation of NAPEs (55) and NAEs, including anandamide (56), in response to NMDA-induced excitotoxicity and, although to a lesser degree, following traumatic brain injury. Figure 14.3 shows the increase in NAEs 4 h and 24 h after intrastriatal NMDA exposure.

By contrast, we found no changes in 2-AG levels in neonatal rat brains (56). Using different rodent models of brain injury, others have also measured endocannabinoid levels in the brain. Van der Stelt et al. (90) found no changes in anandamide and 2-AG levels following ouabain-induced excitotoxic damage inflicted on the neonatal rat brain, whereas Panikashvili et al. (88) reported elevated 2-AG levels, but no change in anandamide levels, in adult murine brains exposed to traumatic brain injury. In these two studies, other NAEs were not determined. By contrast, we did not find increased levels of the NAPEs in mice during pentylenetetrazol-induced seizures in which some neuronal damage probably occurs. This apparent discrepancy may be explained by the developmental profile of *N*-acyltransferase (low in adult rodents) and NAPE-PLD (high in adult rodents) activity, which favors accumulation of the NAEs instead of their precursors (132). In the first human case study to be reported on the impact of brain damage on endocannabinoid homeostasis, levels of anandamide and other NAEs were found increased (as determined by microdialysis) in a patient with hemispheric stroke (133).

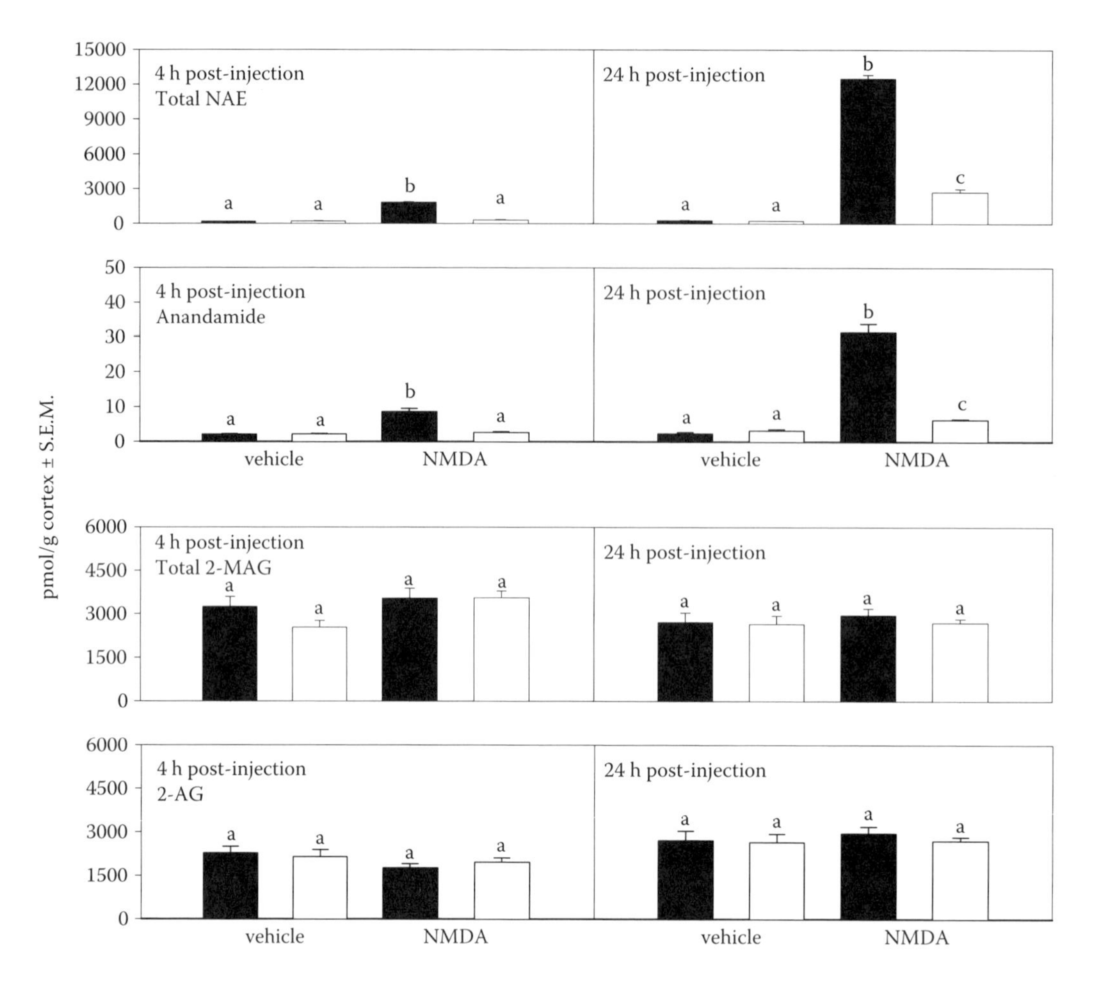

FIGURE 14.3 Developmental NMDA-induced excitotoxicity *in vivo* results in progressively high accumulation of *N*-acylethanolamines (NAEs), including anandamide. 2-MAG, 2-monoacyl glycerol. Values are means (pmol/g cortex) ± SEM, and data with a letter in common are not statistically significant. (From Hansen, H. H. et al. [2001] *J. Neurochem.* **78,** 1415–1427. With permission.)

NAEs, including anandamide, accumulate in the mammalian brain *post mortem* (52,53), which indicates that prolonged tissue sampling time can possibly mask any response differences in experimental groups. It is conceivable that many *N*-acylethanolamines accumulate in the rodent brain in response to injury depending on species, degree and site of injury, timing schedule, and brain developmental stage. Furthermore, as acyl group remodeling occurs during tissue injury, which favors accumulation of the saturated NAEs, anandamide levels will decrease when expressed as a percentage of the total NAEs (55,56,134). Thus, because anandamide is just a minor constituent of the NAEs, measuring anandamide levels alone may not reveal a general increase in the bulk volume of NAEs. Diacylglycerol levels, partially generated from inositol lipids, increase in response to different types of brain injury (84,89,135–138), and cerebral 2-AG levels rapidly increase following decapitation (83). Thus, whether 2-AG levels increase during brain injury may depend on the same experimental conditions as suggested earlier for anandamide.

Because activation of CB_1 receptors inhibits vesicular glutamate release, there is great interest in finding CB_1 receptor ligands that exhibit neuroprotective potential, not only by exogenous administration, but also by stimulation of intrinsic pathways leading to elevated synaptic levels of endocannabinoids. The *in vivo* experiments published so far (Table 14.1) have not given a clear answer.

TABLE 14.1
Summary of the *In Vivo* Effects of (Endo)cannabinoid Receptor Ligands in Animal Models of Brain Damage

In vivo Animal Model	Species	Effect of Exogenous CB_1 Receptor Ligands	Effect of CB_1 Receptor Antagonist Alone	Knockout Models	Reference
Transient global ischemia	Rats	WIN55212-2 protective	SR141716A without effect	—	(143)
Permanent focal ischemia	Rats	WIN55212-2 protective	SR141716A without effect	—	(143)
Transient global ischemia	Gerbils	CP55940 protective	SR1417161A neurotoxic	—	(194)
Transient global ischemia	Rats	Chronic Δ^9-THC protective	Not determined	—	(195)
Transient focal ischemia	Rats	WIN55212-2 without effect	SR1417161A and LY320135 protective	—	(139)
Ouabain-induced excitotoxicity	Neonatal rats	Anandamide protective	SR1417161A without effect	—	(90)
Ouabain-induced excitotoxicity	Neonatal rats	Δ^9-THC protective	SR141716A without effect	—	(142)
Ouabain-induced excitotoxicity	Neonatal rats	Anandamide, and mixed CB1/VR_1 agonist protective	SR141716A, VR_1 antagonist without effect	—	(196)
Permanent focal ischemia	Rats	Bay 38-7271 protective	Not determined	—	(197)
Closed head injury	Mice	2-AG protective	SR141716A without effect	—	(88)
NMDA-induced excitotoxicity	Neonatal rats	WIN55212-2 without effect	SR141716A protective	—	(140)
Permanent focal ischemia	Mice	—	—	CB_1 knockouts more sensitive	(141)
NMDA-induced excitotoxicity	Mice	—	—	CB_1 knockouts more sensitive	(141)
Kainate-induced seizures	Mice	Anandamide neurotoxic and proconvulsive	SR141716A proconvulsive[a]	FAAH knockouts more sensitive	(144)
Bicuculline-induced seizures	Mice	Anandamide neurotoxic and proconvulsive	SR141716A proconvulsive[a]	FAAH knockouts more sensitive	(144)

SR141716A is proconvulsive in FAAH knockout mice, but this effect is not mediated via CB_1 receptors (144).

Exogenous cannabinoids are neuroprotective in many of the studies published to date, although some have indicated no effect (139,140). If endogenous endocannabinoids are neuroprotective via activation of CB_1 receptors, it should be expected that: (1) CB_1 receptor knockout mice would display increased sensitivity to a neurodegenerative insult as compared to wild-type mice, (2) blockade of intrinsic CB_1 receptor activity with a CB_1 receptor antagonist would be protoxic in events of brain damage, and (3) FAAH knockout mice would display higher resistance to neurodegenerative stimuli as compared to wild-type mice. In this context, CB_1 receptor knockout mice have been found to have increased severity of stroke (141); however exposure to the CB_1 receptor antagonist SR141716A did not exacerbate the neurotoxic insult in four different animal models of neurodegeneration (88,142,143). Moreover, SR141716A partially prevented brain damage induced by ischemia or glutamatergic excitoxicity (139, 140). In FAAH knockout mice in which anandamide levels are increased, exogenously applied anandamide is proconvulsive and induces neuronal death in the hippocampus (144). The explanation for these discrepancies remains to be discovered, and the cerebral actions of exo- and endocannabinoids may well prove to be very complex. Thus, the glutamate-release-inhibition model of CB_1 receptor-induced neuroprotection may well be too simplistic. Endocannabinoids also modulate synaptic GABA release (as mentioned earlier in this section). Depending on the arrangement of GABAergic and glutamatergic interconnections, blockade of GABA release could result in disinhibition of glutamatergic efferents, thereby increasing neurotoxic glutamatergic tonus in the brain regions at risk. Furthermore, CB_1 and CB_2 (and CB_{abn}) receptors are found on activated microglial cells (96,97), where they are involved in several aspects of microglial cell function that may be either proinflammatory, e.g., recruiting microglial cells toward neuroinflammatory lesion sites, or anti-inflammatory, e.g., inhibiting nitric oxide formation (97,145,146). Thus, at present, it is unclear whether endocannabinoid action on microglial cells has a neuroprotective element. Vascular function is also of great importance in aspects of brain injury, and CB_1 receptors are found on the vascular endothelium, where they are suggested to be involved in vasopressor inhibition (147). A vasodilatory effect of CB_1 receptor activation may be of importance in neuroprotection by maintaining some oxygen supply to tissues surrounding ischemic brain areas (141,148).

PUTATIVE NEUROPROTECTIVE EFFECTS OF OTHER NAES

Anandamide is always formed in parallel with other NAEs and is only a minor component (54,149). In addition to the agonist activity of anandamide on cannabinoid and vanilloid receptors, saturated and monounsaturated NAEs (see Figure 14.2) exhibit biological activity that may be both dependent and independent of cannabinoid receptor function. In this connection, Mechoulam and coworkers have introduced the term *entourage effect* to explain the phenomenon by which receptor-inactive NAE congeners enhance receptor-mediated effects of 2-AG and anandamide (150). Thus, several saturated and monounsaturated NAEs enhance the effect of anandamide via interference with vanilloid receptor or anandamide degradation or both (65,151–153). Thus, in events in which anandamide induces neuroprotection, this effect may be enhanced by other NAEs present in the vicinity of the anandamide action via FAAH substrate competition. In theory, nonendocannabinoid NAEs may exert neuroprotective actions by several other mechanisms: e.g., (1) by inhibiting the necrosis of injured cells, (2) by stimulating the injured cells or neighboring cells or both to activate apoptotic mechanisms in order to stop the spread of harmful necrosis, and (3) by inhibiting release of mediators that promote necrosis and inflammation. Furthermore, whenever NAEs are formed, a stoichiometric amount of phosphatidic acid is also formed in the enzymatic process (54). Phosphatidic acid is a signaling molecule within the cell (154–157), and it is a precursor for lysophosphatidic acid. This lysophospholipid is an agonist for several G-protein-coupled receptors (158) of importance in the developing nervous system and promotes the survival of myelinated Schwann cells (159).

N-Stearoylethanolamine is the major *N*-acylethanolamine species generated in injured rat brain (56). Recently, it has been shown that *N*-stearoylethanolamine has specific binding sites in the mouse brain, which appear to be different from CB_1 and vanilloid receptors (147). In C6 glioma cells, *N*-stearoylethanolamine can promote apoptosis *in vitro* (160). Induction of apoptosis will be neuroprotective if the alternative is necrosis.

N-Palmitoylethanolamine is also a major NAE species in injured rat brain (56). This NAE has been demonstrated to inhibit proinflammatory properties of mast cells (161–165), and it reduces seizure severity in a number of animal models of epilepsy (166). Some of the biological effects of *N*-palmitoylethanolamine can be antagonized by the CB_2 receptor antagonist SR144528 (148,167–169), although *N*-palmitoylethanolamine has no affinity for this receptor (67,170). It is not known whether these effects of *N*-palmitoylethanolamine can be explained by the existence of an unknown cannabinoid-like receptor or an entourage effect on endocannabinoids. At the level of the individual injured cell, a variety of saturated and monounsaturated NAEs may stabilize injured mitochondria, thereby preventing Ca^{2+} leakage (171,172), which is interesting because mitochondrial dysfunction is an important aspect of excitotoxicity (22). Furthermore, *N*-palmitoylethanolamine has been reported to be neuroprotective against excitotoxic death in brain cell cultures (173,174), which presumably involves some of these mechanisms.

N-Oleoylethanolamine is also a predominant NAE species in the injured rat brain (56), and it has also been found to be the major NAE species in a human brain that has suffered a hemispheric stroke (133). As early as 1975 (175), *N*-oleoylethanolamine was synthesized as an inhibitor of ceramidase, the enzyme that degrades ceramide. Ceramide is involved in the regulation of apoptosis and cell proliferation (176,177). Cannabinoid-induced apoptosis in glioma cells is mediated via formation of ceramide (178–181). On a tentative basis, it can be suggested that anandamide-induced apoptosis may be aggravated by the presence of *N*-oleoylethanolamine because this leads to increased formation of ceramide. There are numerous studies in which *N*-oleoylethanolamine has been shown to facilitate the apoptosis-inducing effect of different compounds (182–188) mediated via increased ceramide levels. However, it has also been reported that *N*-oleoylethanolamine decreases ceramide levels in JB6 P^+ cells by an unknown mechanism (189). Recently, *N*-oleoylethanolamine has been shown to have a CB_1-receptor-independent anorexic effect by inhibiting some intestinal neuronal functions in the rat (190,191), and it also causes vasodilation in rat mesenteric arterial segments by an unknown receptor mechanism (192). Whether these recently discovered biological effects of *N*-oleoylethanolamine are of significance for neuroprotection is not known, but it indicates that nonendocannabinoid NAEs are also bioactive molecules with potential for cerebral actions.

FINAL REMARKS

2-AG is a full agonist for the CB_1, CB_2, and CB_{abn} receptors, and it may fulfill the role of an endocannabinoid involved in neurotransmitter modulation. Anandamide is a partial agonist for the same receptors as well as for the vanilloid receptor, but the enzymatic synthesis of anandamide is always accompanied by the cosynthesis of a vast amount of saturated and monounsaturated NAEs as well as phosphatidic acid. Although there are indications that formation of NAEs, including anandamide, may occur in response to receptor activation *in vitro* (193), it is, however, evident that neuronal injury is a powerful triggering mechanism for NAE production and extracellular release. Several of the different biochemical responses seen during neuronal injury *in vivo* have been suggested to represent a neuronal defense system (20), and the massive formation of NAEs, including anandamide, during neuronal injury has been suggested to serve a similar defensive function (8,54).

However, recent studies with FAAH and CB_1 receptor knockout mice indicate both the neuroprotective and neurotoxic effects of endocannabinoids. The clarification of the exact role of the saturated and monounsaturated NAEs must await further studies.

ACKNOWLEDGMENTS

The studies conducted in the authors' laboratory have been supported by the the Carlsberg Foundation, the Novo Nordisk Foundation, the Augustinus Foundation, the Lundbeck Foundation, the Alfred Benzon Foundation, Direktør Ib Henriksen Foundation, the Danish Medical Research Council, and Frænkel's Memorial Foundation.

REFERENCES

1. De Kreyser, J., Sulter, G., and Luiten, P. G. (1999). Clinical trials with neuroprotective drugs in acute ischaemic stroke: are we doing the right thing? *Trends Neurosci* **22,** 535–540.
2. Hardingham, G. E. and Bading, H. (2003). The yin and yang of NMDA receptor signalling. *Trends Neurosci* **26,** 81–89.
3. Devane, W. A., Dysarz, F. A., Johnson, M. R., Melvin, L. S., and Howlett, A. C. (1988). Determination and characterization of a cannabinoid receptor in rat brain. *Mol. Pharmacol.* **34,** 605–613.
4. Devane, W. A., Hanus, L., Breuer, A., Pertwee, R. G., Stevenson, L. A., Griffin, G., Gibson, D., Mandelbaum, A., Etinger, A., and Mechoulam, R. (1992). Isolation and structure of a brain constituent that binds to the cannabinoid receptor. *Science* **258,** 1946–1949.
5. Mechoulam, R., Ben-Shabat, S., Hanus, L., Ligumsky, M., Kaminski, N. E., Schatz, A. R., Gopher, A., Almog, S., Martin, B. R., Compton, D. R. et al. (1995). Identification of an endogenous 2-monoglyceride, present in canine gut, that binds to cannabinoid receptors. *Biochem. Pharmacol.* **50,** 83–90.
6. Sugiura, T., Kondo, S., Sukagawa, A., Nakane, S., Shinoda, A., Itoh, K., Yamashita, A., and Waku, K. (1995). 2-Arachidonoylglycerol: a possible endogenous cannabinoid receptor ligand in brain. *Biochem. Biophys. Res. Commn.* **215,** 89–97.
7. Shen, M. X. and Thayer, S. A. (1998). Cannabinoid receptor agonists protect cultured rat hippocampal neurons from excitotoxicity. *Mol. Pharmacol.* **54,** 459–462.
8. Hansen, H. S., Lauritzen, L., Moesgaard, B., Strand, A. M., and Hansen, H. H. (1998). Formation of *N*-acyl-phosphatidylethanolamines and *N*-acylethanolamines — proposed role in neurotoxicity. *Biochem. Pharmacol.* **55,** 719–725.
9. Hansen, H. S., Moesgaard, B., Hansen, H. H., Schousboe, A., and Petersen, G. (1999). Formation of *N*-acyl-phosphatidylethanolamine and *N*-acylethanolamine (including anandamide) during glutamate-induced neurotoxicity. *Lipids* **34** Suppl., 327–330.
10. Farooqui, A. A., Horrocks, L. A., and Farooqui, T. (2000). Glycerophospholipids in brain: their metabolism, incorporation into membranes, functions, and involvement in neurological disorders. *Chem. Phys. Lipids* **106,** 1–29.
11. Grundy, R. I., Rabuffetti, M., and Beltramo, M. (2001). Cannabinoids and neuroprotection. *Mol. Neurobiol.* **24,** 29–51.
12. Porter, A. C. and Felder, C. C. (2001). The endocannabinoid nervous system: unique opportunities for therapeutic intervention. *Pharmacol. Ther.* **90,** 45–60.
13. Fride, E. and Shohami, E. (2002). The endocannabinoid system: function in survival of the embryo, the newborn and the neuron. *Neuroreport* **13,** 1833–1841.
14. Hansen, H. S., Moesgaard, B., Petersen, G., and Hansen, H. H. (2002). Putative neuroprotective actions of *N*-acyl-ethanolamines. *Pharmacol. Ther.* **95,** 119–126.
15. Mechoulam, R., Panikashvili, D., and Shohami, E. (2002). Cannabinoids and brain injury: therapeutic implications. *Trends Mol Med* **8,** 58–61.
16. Romero, J., Lastres-Becker, I., De Miguel, R., Berrendero, F., Ramos, J. A., and Fernández-Ruiz, J. (2002). The endogenous cannabinoid system and the basal ganglia: biochemical, pharmacological, and therapeutic aspects. *Pharmacol. Ther.* **95,** 137–152.
17. Van der Stelt, M., Veldhuis, W. B., Maccarrone, M., Bär, P. R., Nicolay, K., Veldink, G. A., Di Marzo, V., and Vliegenthart, J. F. G. (2002). Acute neuronal injury, excitotoxicity, and the endocannabinoid system. *Mol. Neurobiol.* **26,** 317–346.
18. Fowler, C. J. (2003). Plant-derived, synthetic and endogenous cannabinoids as neuroprotective agents — non-psychoactive cannabinoids, "entourage" compounds and inhibitors of *N*-acyl ethanolarnine breakdown as therapeutic strategies to avoid psychotropic effects. *Brain Res. Rev.* **41,** 26–43.

19. Dirnagl, U., Simon, R. P., and Hallenbeck, J. M. (2003). Ischemic tolerance and endogenous neuroprotection. *Trends Neurosci* **26,** 248–254.
20. Sapolsky, R. M. (2001). Cellular defenses against excitotoxic insults. *J. Neurochem.* **76,** 1601–1611.
21. Gegelashvili, G. and Schousboe, A. (1998). Cellular distribution and kinetic properties of high-affinity glutamate transporters. *Brain Res. Bull.* **45,** 233–238.
22. Atlante, A., Calissano, P., Bobba, A., Giannattasio, S., Marra, E., and Passarella, S. (2001). Glutamate neurotoxicity, oxidative stress and mitochondria. *FEBS Lett.* **497,** 1–5.
23. Lee, J. M., Grabb, M. C., Zipfel, G. J., and Choi, D. W. (2000). Brain tissue responses to ischemia. *J. Clin. Invest.* **106,** 723–731.
24. Hanisch, U. K. (2002). Microglia as a source and target of cytokines. *Glia* **40,** 140–155.
25. Aschner, M., Allen, J. W., Kimelberg, H. K., LoPachin, R. M., and Streit, W. J. (1999). Glial cells in neurotoxicity development. *Annu. Rev. Pharmacol. Toxicol.* **39,** 151–173.
26. Lee, J. M., Zipfel, G. J., and Choi, D. W. (1999). The changing landscape of ischaemic brain injury mechanisms. *Nature* **399** Suppl., A7–A14.
27. Dirnagl, U., Iadecola, C., and Moskowitz, M. A. (1999). Pathobiology of ischaemic stroke: an integrated view. *Trends Neurosci* **22,** 391–397.
28. Olney, J. W., Collins, R. C., and Sloviter, R. S. (1986). Excitotoxic mechanism of epileptic brain damage. *Adv. Neurol.* **44,** 857–877.
29. Nicotera, P. and Lipton, S. A. (1999). Excitotoxins in neuronal apoptosis and necrosis. *J. Cereb. Blood Flow Metab.* **19,** 583–591.
30. Kaul, M., Garden, G. A., and Lipton, S. A. (2001). Pathways to neuronal injury and apoptosis in HIV-associated dementia. *Nature* **410,** 988–994.
31. Ikonomidou, C., Bosch, F., Miksa, M., Bittigau, P., Vöckler, J., Dikranian, K., Tenkova, T. I., Stefovska, V., Turski, L., and Olney, J. W. (1999). Blockade of NMDA receptors and apoptotic neurodegeneration in the developing brain. *Science* **283,** 70–74.
32. Ikonomidou, C., Mosinger, J. L., Salles, K. S., Labruyere, J., and Olney, J. W. (1989). Sensitivity of the developing rat brain to hypobaric/ischemic damage parallels sensitivity to *N*-methyl-aspartate neurotoxicity. *J. Neurosci.* **9,** 2809–2818.
33. Isagai, T., Fujimura, N., Tanaka, E., Yamamoto, S., and Higashi, H. (1999). Membrane dysfunction induced by in vitro ischemia in immature rat hippocampal CA1 neurons. *J. Neurophysiol.* **81,** 1866–1871.
34. Ishimaru, M. J., Ikonomidou, C., Tenkova, T. I., Der, T. C., Dikranian, K., Sesma, M. A., and Olney, J. W. (1999). Distinguishing excitotoxic from apoptotic neurodegeneration in the developing rat brain. *J. Comp. Neurol.* **408,** 461–476.
35. Maruoka, N., Murata, T., Omata, N., Fujibayashi, Y., Waki, A., Yoshimoto, M., Yano, R., Yonekura, Y., and Wada, Y. (2001). Greater resistance and lower contribution of free radicals to hypoxic neurotoxicity in immature rat brain compared to adult brain as revealed by dynamic changes in glucose metabolism. *Dev. Neurosci.* **23,** 412–419.
36. Gurd, J. W., Bissoon, N., Beesley, P. W., Nakazawa, T., Yamamoto, T., and Vannucci, S. J. (2002). Differential effects of hypoxia-ischemia on subunit expression and tyrosine phosphorylation of the NMDA receptor in 7- and 21-day-old rats. *J. Neurochem.* **82,** 848–856.
37. Mikati, M. A., Abi-Habib, R. J., El Sabban, M. E., Dbaibo, G. S., Kurdi, R. M., Kobeissi, M., Farhat, F., and Asaad, W. (2003). Hippocampal programmed cell death after status epilepticus: evidence for NMDA-receptor and ceramide-mediated mechanisms. *Epilepsia* **44,** 282–291.
38. Hanus, L., Abu-Lafi, S., Fride, E., Breuer, A., Vogel, Z., Shalev, D. E., Kustanovich, I., and Mechoulam, R. (2001). 2-Arachidonyl glyceryl ether, an endogenous agonist of the cannabinoid CB_1 receptor. *Proc. Natl. Acad. Sci. USA* **98,** 3662–3665.
39. Fezza, F., Bisogno, T., Minassi, A., Appendino, G., Mechoulam, R., and Di Marzo, V. (2002). Noladin ether, a putative novel endocannabinoid: inactivation mechanisms and a sensitive method for its quantification in rat tissues. *FEBS Lett.* **513,** 294–298.
40. Sugiura, T. and Waku, K. (2002). Cannabinoid receptors and their endogenous ligands. *J. Biochem.* **132,** 7–12.
41. Oka, S., Tsuchie, A., Tokumura, A., Muramatsu, M., Suhara, Y., Takayama, H., Waku, K., and Sugiura, T. (2003). Ether-linked analogue of 2-arachidonoylglycerol (noladin ether) was not detected in the brains of various mammalian species. *J. Neurochem.* **85,** 1374–1381.

42. Porter, A. C., Sauer, J. M., Knierman, M. D., Becker, G. W., Berna, M. J., Bao, J. Q., Nomikos, G. G., Carter, P., Bymaster, F. P., Leese, A. B. et al. (2002). Characterization of a novel endocannabinoid, virodhamine, with antagonist activity at the CB1 receptor. *J. Pharmacol. Exp. Ther.* **301,** 1020–1024.
43. Markey, S. P., Dudding, T., and Wang, T. C. L. (2000). Base- and acid-catalyzed interconversions of O-acyl- and N-acylethanolamines: a cautionary note for lipid analyses. *J. Lipid Res.* **41,** 657–662.
44. Hanus, L., Gopher, A., Almog, S., and Mechoulam, R. (1993). Two new unsaturated fatty acid ethanolamides in brain that bind to the cannabinoid receptor. *J. Med. Chem.* **36,** 3032–3034.
45. Pertwee, R., Griffin, G., Hanus, L., and Mechoulam, R. (1994). Effects of two endogenous fatty acid ethanolamides on mouse vasa deferentia. *Eur. J. Pharmacol.* **259,** 115–120.
46. Schmid, P. C., Reddy, P. V., Natarajan, V., and Schmid, H. H. O. (1983). Metabolism of *N*-acylethanolaminc phospholipids by a mammalian phosphodiesterase of the phospholipase D type. *J. Biol. Chem.* **258,** 9302–9306.
47. Sugiura, T., Kondo, S., Sukagawa, A., Tonegawa, T., Nakane, S., Yamashita, A., Ishima, Y., and Waku, K. (1996). Transacylase-mediated and phosphodiesterase-mediated synthesis of *N*-arachidonoylethanolamine, an endogenous cannabinoid-receptor ligand, in rat brain microsomes—comparison with synthesis from free arachidonic acid and ethanolamine. *Eur. J. Biochem.* **240,** 53–62.
48. Petersen, G. and Hansen, H. S. (1999). *N*-Acylphosphatidylethanolamine-hydrolysing phospholipase D lacks the ability to transphosphatidylate. *FEBS Lett.* **455,** 41–44.
49. Natarajan, V., Schmid, P. C., Reddy, P. V., Zuarte-Augustin, M. L., and Schmid, H. H. O. (1983). Biosynthesis of *N*-acylethanolamine phospholipids by dog brain preparations. *J. Neurochem.* **41,** 1303–1312.
50. Sugiura, T., Kondo, S., Sukagawa, A., Tonegawa, T., Nakane, S., Yamashita, A., and Waku, K. (1996). Enzymatic synthesis of anandamide, an endogenous cannabinoid receptor ligand, through N-acylphosphatidylethanolamine pathway in testis: involvement of Ca^{2+}-dependent transacylase and phosphodiesterase activities. *Biochem. Biophys. Res. Commn.* **218,** 113–117.
51. Schmid, H. H. O., Schmid, P. C., and Natarajan, V. (1990). *N*-Acylated glycerophospholipids and their derivatives. *Prog. Lipid Res.* **29,** 1–43.
52. Schmid, P. C., Krebsbach, R. J., Perry, S. R., Dettmer, T. M., Maasson, J. L., and Schmid, H. H. O. (1995). Occurrence and postmortem generation of anandamide and other long-chain *N*-acylethanolamines in mammalian brain. *FEBS Lett.* **375,** 117–120.
53. Schmid, P. C., Krebsbach, R. J., Perry, S. R., Dettmer, T. M., Maasson, J. L., and Schmid, H. H. O. (1996). Corrigendum to: Occurrence and postmortem generation of anandamide and other long-chain *N*-acylethanolamines in mammalian brain. *FEBS Lett.* **385,** 125–126.
54. Hansen, H. S., Moesgaard, B., Hansen, H. H., and Petersen, G. (2000). *N*-Acylethanolamines and precursor phospholipids — relation to cell injury. *Chem. Phys. Lipids* **108,** 135–150.
55. Hansen, H. H., Ikonomidou, C., Bittigau, P., Hansen, S. H., and Hansen, H. S. (2001). Accumulation of the anandamide precursor and other *N*-acylethanolamine phospholipids in infant rat models of *in vivo* necrotic and apoptotic neuronal death. *J. Neurochem.* **76,** 39–46.
56. Hansen, H. H., Schmid, P. C., Bittigau, P., Lastres-Becker, I., Berrendero, F., Manzanares, J., Ikonomidou, C., Schmid, H. H. O., Fernández-Ruiz, J. J., and Hansen, H. S. (2001). Anandamide, but not 2-arachidonoylglycerol, accumulates during *in vivo* neurodegeneration. *J. Neurochem.* **78,** 1415–1427.
57. Schmid, P. C., Zuarte-Augustin, M. L., and Schmid, H. H. O. (1985). Properties of rat liver *N*-acylethanolamine amidohydrolase. *J. Biol. Chem.* **260,** 14145–14149.
58. Deutsch, D. G. and Chin, S. A. (1993). Enzymatic synthesis and degradation of anandamide, a cannabinoid receptor agonist. *Biochem. Pharmacol.* **46,** 791–796.
59. Cravatt, B. F. and Lichtman, A. H. (2002). The enzymatic inactivation of the fatty acid amide class of signaling lipids. *Chem. Phys. Lipids* **121,** 135–148.
60. Deutsch, D. G., Ueda, N., and Yamamoto, S. (2002). The fatty acid amide hydrolase (FAAH). *Prostag Leukot. Essent. Fatty Acids* **66,** 201–210.
61. Cravatt, B. F., Demarest, K., Patricelli, M. P., Bracey, M. H., Giang, D. K., Martin, B. R., and Lichtman, A. H. (2001). Supersensitivity to anandamide and enhanced endogenous cannabinoid signaling in mice lacking fatty acid amide hydrolase. *Proc. Natl. Acad. Sci. USA* **98,** 9371–9376.
62. Kathuria, S., Gaetani, S., Fegley, D., Valiño, F., Duranti, A., Tontini, A., Mor, M., Tarzia, G., La Rana, G., Calignano, A. et al. (2003). Modulation of anxiety through blockade of anandamide hydrolysis. *Nature Med.* **9,** 76–81.

63. Giuffrida, A., Rodriguez de Fonseca, F., Nava, F., Loubet-Lescoulié, P., and Piomelli, D. (2000). Elevated circulating levels of anandamide after administration of the transport inhibitor, AM404. *Eur. J. Pharmacol.* **408,** 161–168.
64. Bojesen, I. N. and Hansen, H. S. (2003). Binding of anandamide to bovine serum albumin. *J. Lipid Res.* **44,** 1790–1794.
65. De Petrocellis, L., Davis, J. B., and Di Marzo, V. (2001). Palmitoylethanolamide enhances anandamide stimulation of human vanilloid VR1 receptors. *FEBS Lett.* **506,** 253–256.
66. Glaser, S. T., Abumrad, N. A., Fatade, F., Kaczocha, M., Studholme, K. M., and Deutsch, D. G. (2003). Evidence against the presence of an anandamide transporter. *Proc. Natl. Acad. Sci. USA* **100,** 4269–4274.
67. Sugiura, T., Kondo, S., Kishimoto, S., Miyashita, T., Nakane, S., Kodaka, T., Suhara, Y., Takayama, H., and Waku, K. (2000). Evidence that 2-arachidonoylglycerol but not *N*-palmitoylethanolamine or anandamide is the physiological ligand for the cannabinoid CB_2 receptor — comparison of the agonistic activities of various cannabinoid receptor ligands in HL-60 cells. *J. Biol. Chem.* **275,** 605–612.
68. Sugiura, T., Kodaka, T., Nakane, S., Miyashita, T., Kondo, S., Suhara, Y., Takayama, H., Waku, K., Seki, C., Baba, N. et al. (1999). Evidence that the cannabinoid CB_1 receptor is a 2-arachidonoylglycerol receptor — Structure–activity relationship of 2-arachidonoylglycerol, ether-linked analogues, and related compounds. *J. Biol. Chem.* **274,** 2794–2801.
69. Farooqui, A. A., Horrocks, L. A., and Farooqui, T. (2000). Deacylation and reacylation of neural membrane glycerophospholipids — A matter of life and death. *J. Mol. Neurosci.* **14,** 123–135.
70. Sugiura, T., Kobayashi, Y., Oka, S., and Waku, K. (2002). Biosynthesis and degradation of anandamide and 2-arachidonoylglycerol and their possible physiological significance. *Prostag Leukot. Essent. Fatty Acids* **66,** 173–192.
71. Giuffrida, A., Beltramo, M., and Piomelli, D. (2001). Mechanisms of endocannabinoid inactivation: biochemistry and pharmacology. *J. Pharmacol. Exp. Ther.* **298,** 7–14.
72. Di Marzo, V. (1998). 'Endocannabinoids' and other fatty acid derivatives with cannabimimetic properties: biochemistry and possible physiopathological relevance. *Biochim. Biophys. Acta.* **1392,** 153–175.
73. Rhee, S. G. (2001). Regulation of phosphoinositide-specific phospholipase C. *Annu. Rev. Biochem.* **70,** 281–312.
74. Allen, A. C., Gammon, C. M., Ousley, A. H., McCarthy, K. D., and Morell, P. (1992). Bradykinin stimulates arachidonic acid release through the sequential actions of an *sn*-1 diacylglycerol lipase and a monoacylglycerol lipase. *J. Neurochem.* **58,** 1130–1139.
75. Stella, N., Schweitzer, P., and Piomelli, D. (1997). A second endogenous cannabinoid that modulates long-term potentiation. *Nature* **388,** 773–778.
76. Hasegawa-Sasaki, H. (1985). Early changes in inositol lipids and their metabolites induced by platelet-derived growth factor in quiescent Swiss mouse 3T3 cells. *Biochem. J.* **232,** 99–109.
77. Di Marzo, V., Bisogno, T., Sugiura, T., Melck, D., and De Petrocellis, L. (1998). The novel endogenous cannabinoid 2-arachidonoylglycerol is inactivated by neuronal- and basophil-like cells: connections with anandamide. *Biochem. J.* **331,** 15–19.
78. Di Marzo, V., Bisogno, T., De Petrocellis, L., Melck, D., Orlando, P., Wagner, J. A., and Kunos, G. (1999). Biosynthesis and inactivation of the endocannabinoid 2-arachidonoylglycerol in circulating and tumoral macrophages. *Eur. J. Biochem.* **264,** 258–267.
79. Goparaju, S. K., Ueda, N., Yamaguchi, H., and Yamamoto, S. (1998). Anandamide amidohydrolase reacting with 2-arachidonoylglycerol, another cannabinoid receptor ligand. *FEBS Lett.* **422,** 69–73.
80. Lang, W. S., Qin, C., Lin, S. Y., Khanolkar, A. D., Goutopoulos, A., Fan, P. S., Abouzid, K., Meng, Z. X., Biegel, D., and Makriyannis, A. (1999). Substrate specificity and stereoselectivity of rat brain microsomal anandamide amidohydrolase. *J. Med. Chem.* **42,** 896–902.
81. Lichtman, A. H., Hawkins, E. G., Griffin, G., and Cravatt, B. F. (2002). Pharmacological activity of fatty acid amides is regulated, but not mediated, by fatty acid amide hydrolase in vivo. *J. Pharmacol. Exp. Ther.* **302,** 73–79.
82. Kozak, K. R. and Marnett, L. J. (2002). Oxidative metabolism of endocannabinoids. *Prostag Leukot. Essent. Fatty Acids* **66,** 211–220.
83. Sugiura, T., Yoshinaga, N., and Waku, K. (2001). Rapid generation of 2-arachidonoylglycerol, an endogenous cannabinoid receptor ligand, in rat brain after decapitation. *Neurosci. Lett.* **297,** 175–178.

84. Reddy, T. S. and Bazan, N. G. (1987). Arachidonic acid, stearic acid, and diacylglycerol accumulation correlates with the loss of phosphatidylinositol 4,5-bisphosphate in cerebrum 2 seconds after electroconvulsive shock: complete reversion of changes 5 minutes after stimulation. *J. Neurosci. Res.* **18,** 449–455.
85. Lee, D. P., Deonarine, A. S., Kienetz, M., Zhu, Q. S., Skrzypczak, M., Chan, M., and Choy, P. C. (2001). A novel pathway for lipid biosynthesis: the direct acylation of glycerol. *J. Lipid Res.* **42,** 1979–1986.
86. Frykholm, P., Hillered, L., Långström, B., Persson, L., Valtysson, J., Watanabe, Y., and Enblad, P. (2001). Increase of interstitial glycerol reflects the degree of ischaemic brain damage: a PET and microdialysis study in a middle cerebral artery occlusion–reperfusion primate model. *J. Neurol. Neurosurg. Psychiatry* **71,** 455–461.
87. Deutsch, J., Rapoport, S. I., and Purdon, A. D. (1997). Relation between free fatty acid and acyl-CoA concentrations in rat brain following decapitation. *Neurochem. Res.* **22,** 759–765.
88. Panikashvili, D., Simeonidou, C., Ben-Shabat, S., Hanus, L., Breuer, A., Mechoulam, R., and Shohami, E. (2001). An endogenous cannabinoid (2-AG) is neuroprotective after brain injury. *Nature* **413,** 527–531.
89. Kunievsky, B., Bazan, N. G., and Yavin, E. (1992). Generation of arachidonic acid and diacylglycerol second messengers from polyphosphoinositides in ischemic fetal brain. *J. Neurochem.* **59,** 1812–1819.
90. Van der Stelt, M., Veldhuis, W. B., Van Haaften, G. W., Fezza, F., Bisogno, T., Bär, P. R., Veldink, G. A., Vliegenthart, J. F. G., Di Marzo, V., and Nicolay, K. (2001). Exogenous anandamide protects rat brain against acute neuronal injury *in vivo*. *J. Neurosci.* **21,** 8765–8771.
91. Howlett, A. C., Barth, F., Bonner, T. I., Cabral, G., Casellas, P., Devane, W. A., Felder, C. C., Herkenham, M., Mackie, K., Martin, B. R. et al. (2002). International Union of Pharmacology. XXVII. Classification of cannabinoid receptors. *Pharmacol. Rev.* **54,** 161–202.
92. Tsou, K., Mackie, K., Sañudo-Peña, M. C., and Walker, J. M. (1999). Cannabinoid CB1 receptors are localized primarily on cholecystokinin-containing gabaergic interneurons in the rat hippocampal formation. *Neuroscience* **93,** 969–975.
93. Irving, A. J., Coutts, A. A., Harvey, J., Rae, M. G., Mackie, K., Bewick, G. S., and Pertwee, R. G. (2000). Functional expression of cell surface cannabinoid CB_1 receptors on presynaptic inhibitory terminals in cultures of hippocampal neurons. *Neuroscience* **98,** 253–262.
94. Tsou, K., Brown, S., Sañudo-Peña, M. C., Mackie, K., and Walker, J. M. (1998). Immunohistochemical distribution of cannabinoid CB1 receptors in the rat central nervous system. *Neuroscience* **83,** 393–411.
95. Rodriguez de Fonseca, F., Gorriti, M. A., Bilbao, A., Escuredo, L., García-Segura, L. M., Piomelli, D., and Navarro, M. (2001). Role of the endogenous cannabinoid system as a modulator of dopamine transmission: implications for Parkinson's disease and schizophrenia. *Neurotox. Res.* **3,** 25–35.
96. Carlisle, S. J., Marciano-Cabral, F., Staab, A., Ludwick, C., and Cabral, G. A. (2002). Differential expression of the CB_2 cannabinoid receptor by rodent macrophages and macrophage-like cells in relation to cell activation. *Immunopharmacology* **2,** 69–82.
97. Walter, L., Franklin, A., Witting, A., Wade, C., Xie, Y. H., Kunos, G., Mackie, K., and Stella, N. (2003). Nonpsychotropic cannabinoid receptors regulate microglial cell migration. *J. Neurosci.* **23,** 1398–1405.
98. Járai, Z., Wagner, J. A., Varga, K., Lake, K. D., Compton, D. R., Martin, B. R., Zimmer, A. M., Bonner, T. I., Buckley, N. E., Mezey, É. et al. (1999). Cannabinoid-induced mesenteric vasodilation through an endothelial site distinct from CB_1 or CB_2 receptors. *Proc. Natl. Acad. Sci. USA* **96,** 14136–14141.
99. Zygmunt, P. M., Petersson, J., Andersson, D. A., Chuang, H. H., Sorgård, M., Di Marzo, V., Julius, D., and Högestätt, E. D. (1999). Vanilloid receptors on sensory nerves mediate the vasodilator action of anandamide. *Nature* **400,** 452–457.
100. Premkumar, L. S. and Ahern, G. P. (2000). Induction of vanilloid receptor channel activity by protein kinase c. *Nature* **408,** 985–990.
101. Smart, D., Gunthorpe, M. J., Jerman, J. C., Nasir, S., Gray, J., Muir, A. I., Chambers, J. K., Randall, A. D., and Davis, J. B. (2000). The endogenous lipid anandamide is a full agonist at the human vanilloid receptor (hVR1). *Br. J. Pharmacol.* **129,** 227–230.
102. Di Marzo, V., Blumberg, P. M., and Szallasi, A. (2002). Endovanilloid signaling in pain. *Curr. Opin. Neurobiol.* **12,** 372–379.
103. Szallasi, A. (2002). Vanilloid (capsaicin) receptors in health and disease. *Am. J. Clin. Pathol.* **118,** 110–121.
104. Mezey, É., Tóth, Z. E., Cortright, D. N., Arzubi, M. K., Krause, J. E., Elde, R., Guo, A., Blumberg, P. M., and Szallasi, A. (2000). Distribution of mRNA for vanilloid receptor subtype 1 (VR1), and VR1-like immunoreactivity, in the central nervous system of the rat and human. *Proc. Natl. Acad. Sci. USA* **97,** 3655–3660.

105. Al-Hayani, A., Wease, K. N., Ross, R. A., Pertwee, R. G., and Davies, S. N. (2001). The endogenous cannabinoid anandamide activates vanilloid receptors in the rat hippocampal slice. *Neuropharmacology* **41,** 1000–1005.
106. Johnson, D. E., Heald, S. L., Dally, R. D., and Janis, R. A. (1993). Isolation, identification and synthesis of an endogenous arachidonic amide that inhibits calcium channel antagonist 1,4-dihydropyridine binding. *Prostag Leukot. Essent. Fatty Acids* **48,** 429–437.
107. Boger, D. L., Patterson, J. E., Guan, X. J., Cravatt, B. F., Lerner, R. A., and Gilula, N. B. (1998). Chemical requirements for inhibition of gap junction communication by the biologically active lipid oleamide. *Proc. Natl. Acad. Sci. USA* **95,** 4810–4815.
108. Maingret, F., Patel, A. J., Lazdunski, M., and Honoré, E. (2001). The endocannabinoid anandamide is a direct and selective blocker of the background K^+ channel TASK-1. *EMBO J.* **20,** 47–54.
109. Hampson, A. J., Bornheim, L. M., Scanziani, M., Yost, C. S., Gray, A. T., Hansen, B. M., Leonoudakis, D. J., and Bickler, P. E. (1998). Dual effects of anandamide on NMDA receptor-mediated responses and neurotransmission. *J. Neurochem.* **70,** 671–676.
110. Facchinetti, F., Del Giudice, E., Furegato, S., Passarotto, M., and Leon, A. (2003). Cannabinoids abate release of TNFα in rat microglial cells stimulated with lypopolysaccharide. *Glia* **41,** 161–168.
111. Di Marzo, V., De Petrocellis, L., Fezza, F., Ligresti, A., and Bisogno, T. (2002). Anandamide receptors. *Prostag Leukot. Essent. Fatty Acids* **66,** 377–391.
112. Hájos, N. and Freund, T. F. (2002). Pharmacological separation of cannabinoid sensitive receptors on hippocampal excitatory and inhibitory fibers. *Neuropharmacol.* **43,** 503–510.
113. Monory, K., Tzavara, E. T., Lexime, J., Ledent, C., Parmentier, M., Borsodi, A., and Hanoune, J. (2002). Novel, not adenylyl cyclase-coupled cannabinoid binding site in cerebellum of mice. *Biochem. Biophys Res. Commn.* **292,** 231–235.
114. Onaivi, E. S., Leonard, C. M., Ishiguro, H., Zhang, P. W., Lin, Z. C., Akinshola, B. E., and Uhl, G. R. (2002). Endocannabinoids and cannabinoid receptor genetics. *Prog. Neurobiol.* **66,** 307–344.
115. Zygmunt, P. M., Andersson, D. A., and Högestätt, E. D. (2002). Delta^9-tetrahydrocannabinol and cannabinol activate capsaicin-sensitive sensory nerves via a CB_1 and CB_2 cannabinoid receptor-independent mechanism. *J. Neurosci.* **22,** 4720–4727.
116. Shen, M. X., Piser, T. M., Seybold, V. S., and Thayer, S. A. (1996). Cannabinoid receptor agonists inhibit glutamatergic synaptic transmission in rat hippocampal cultures. *J. Neurosci.* **16,** 4322–4334.
117. Shen, M. X. and Thayer, S. A. (1999). Delta^9-Tetrahydrocannabinol acts as a partial agonist to modulate glutamatergic synaptic transmission between rat hippocampal neurons in culture. *Mol. Pharmacol.* **55,** 8–13.
118. Wilson, R. I. and Nicoll, R. A. (2001). Endogenous cannabinoids mediate retrograde signalling at hippocampal synapses. *Nature* **410,** 588–592.
119. Ohno-Shosaku, T., Maejima, T., and Kano, M. (2001). Endogenous cannabinoids mediate retrograde signals from depolarized postsynaptic neurons to presynaptic terminals. *Neuron* **29,** 729–738.
120. Wilson, R. I., Kunos, G., and Nicoll, R. A. (2001). Presynaptic specificity of endocannabinoid signaling in the hippocampus. *Neuron* **31,** 453–462.
121. Varma, N., Carlson, G. C., Ledent, C., and Alger, B. E. (2001). Metabotropic glutamate receptors drive the endocannabinoid system in hippocampus. *J. Neurosci.* **21,** NIL14–NIL18.
122. Kreitzer, A. C. and Regehr, W. G. (2001). Retrograde inhibition of presynaptic calcium influx by endogenous cannabinoids at excitatory synapses onto Purkinje cells. *Neuron* **29,** 717–727.
123. Kreitzer, A. C., Carter, A. G., and Regehr, W. G. (2002). Inhibition of interneuron firing extends the spread of endocannabinoid signaling in the cerebellum. *Neuron* **34,** 787–796.
124. Gerdeman, G. L., Ronesi, J., and Lovinger, D. M. (2002). Postsynaptic endocannabinoid release is critical to long-term depression in the striatum. *Nat. Neurosci.* **5,** 446–451.
125. Pistis, M., Ferraro, L., Pira, L., Flore, G., Tanganelli, S., Gessa, G. L., and Devoto, P. (2002). Delta^9-tetrahydrocannabinol decreases extracellular GABA and increases extracellular glutamate and dopamine levels in the rat prefrontal cortex: an in vivo microdialysis study. *Brain Res.* **948,** 155–158.
126. Maejima, T., Hashimoto, K., Yoshida, T., Aiba, A., and Kano, M. (2001). Presynaptic inhibition caused by retrograde signal from metabotropic glutamate to cannabinoid receptors. *Neuron* **31,** 463–475.
127. Hansen, H. S., Lauritzen, L., Strand, A. M., Moesgaard, B., and Frandsen, A. (1995). Glutamate stimulates the formation of *N*-acylphosphatidylethanolamine and *N*-acylethanolamine in cortical neurons in culture. *Biochim. Biophys. Acta* **1258,** 303–308.

128. Hansen, H. S., Lauritzen, L., Strand, A. M., Vinggaard, A. M., Frandsen, A., and Schousboe, A. (1997). Characterization of glutamate-induced formation of *N*-acylphosphatidylethanolamine and *N*-acylethanolamine in cultured neocortical neurons. *J. Neurochem.* **69,** 753–761.
129. Natarajan, V., Reddy, P. V., Schmid, P. C., and Schmid, H. H. O. (1982). *N*-Acylation of ethanolamine phospholipids in canine myocardium. *Biochim. Biophys. Acta* **712,** 342–355.
130. Ueda, N., Liu, Q., and Yamanaka, K. (2001). Marked activation of the *N*-acylphosphatidylethanolamine-hydrolyzing phosphodiesterase by divalent cations. *Biochim. Biophys. Acta* **1532,** 121–127.
131. Moesgaard, B., Petersen, G., Jaroszewski, J. W., and Hansen, H. S. (2000). Age dependent accumulation of *N*-acyl-ethanolamine phospholipids in ischemic rat brain: a ^{31}P NMR and enzyme activity study. *J. Lipid Res.* **41,** 985–990.
132. Moesgaard, B., Hansen, H. H., Hansen, S. L., Hansen, S. H., Petersen, G., and Hansen, H. S. (2003). Brain levels of *N*-acylethanolamine phospholipids in mice during pentylenetetrazol-induced seizure. *Lipids* **38,** 387–390.
133. Schäbitz, W. R., Giuffrida, A., Berger, C., Aschoff, A., Schwaninger, M., Schwab, S., and Piomelli, D. (2002). Release of fatty acid amides in a patient with hemispheric stroke — a microdialysis study. *Stroke* **33,** 2112–2114.
134. Kondo, S., Sugiura, T., Kodaka, T., Kudo, N., Waku, K., and Tokumura, A. (1998). Accumulation of various *N*-acylethanolamines including *N*-arachidonoylethanolamine (anandamide) in cadmium chloride-administered rat testis. *Arch. Biochem. Biophys.* **354,** 303–310.
135. Homayoun, P., De Turco, E. B. R., Parkins, N. E., Lane, D. C., Soblosky, J., Carey, M. E., and Bazan, N. G. (1997). Delayed phospholipid degradation in rat brain after traumatic brain injury. *J. Neurochem.* **69,** 199–205.
136. Homayoun, P., Parkins, N. E., Soblosky, J., Carey, M. E., De Turco, E. B. R., and Bazan, N. G. (2000). Cortical impact injury in rats promotes a rapid and sustained increase in polyunsaturated free fatty acids and diacylglycerols. *Neurochem. Res.* **25,** 269–276.
137. Katsura, K., De Turco, E. B. R., Kristian, T., Folbergrová, J., Bazan, N. G., and Siesjö, B. K. (2000). Alterations in lipid and calcium metabolism associated with seizure activity in the postischemic brain. *J. Neurochem.* **75,** 2521–2527.
138. De Turco, E. B. R., Tang, W., Topham, M. K., Sakane, F., Marcheselli, V. L., Chen, C., Taketomi, A., Prescott, S. M., and Bazan, N. G. (2001). Diacylglycerol kinase ε regulates seizure susceptibility and long-term potentiation through arachidonoylinositol lipid signaling. *Proc. Natl. Acad. Sci. USA* **98,** 4740–4745.
139. Muthian, S. and Hillard, C. J. (2000). CB_1 receptor antagonists are neuroprotective in focal cerebral ischemia-reperfusion injury. *Proceedings of the ICRS Symposium on the Cannabinoids.* p. 107. International Cannabinoid Research Society. Burlington.
140. Hansen, H. H., Azcoitia, I., Pons, S., Romero, J., García-Segura, L. M., Ramos, J. A., Hansen, H. S., and Fernández-Ruiz, J. (2002). Blockade of cannabinoid CB_1 receptor function protects against *in vivo* disseminating brain damage following NMDA-induced excitotoxicity. *J. Neurochem.* **82,** 154–158.
141. Parmentier-Batteur, S., Jin, K. L., Mao, X. O., Xie, L., and Greenberg, D. A. (2002). Increased severity of stroke in CB1 cannabinoid receptor knock-out mice. *J.Neurosci.* **22,** 9771–9775.
142. Van der Stelt, M., Veldhuis, W. B., Bär, P. R., Veldink, G. A., Vliegenthart, J. F. G., and Nicolay, K. (2001). Neuroprotection by Delta^9-tetrahydrocannabinol, the main active compound in marijuana, against ouabain-induced *in vivo* excitotoxicity. *J. Neurosci.* **21,** 6475–6479.
143. Nagayama, T., Sinor, A. D., Simon, R. P., Chen, J., Graham, S. H., Jin, K. L., and Greenberg, D. A. (1999). Cannabinoids and neuroprotection in global and focal cerebral ischemia and in neuronal cultures. *J. Neurosci.* **19,** 2987–2995.
144. Clement, A. B., Hawkins, E. G., Lichtman, A. H., and Cravatt, B. F. (2003). Increased seizure susceptibility and proconvulsant activity of anandamide in mice lacking fatty acid amide hydrolase. *J. Neurosci.* **23,** 3916–3923.
145. Stefano, G. B., Liu, Y., and Goligorsky, M. S. (1996). Cannabinoid receptors are coupled to nitric oxide release in invertebrate immunocytes, microglia, and human monocytes. *J. Biol. Chem.* **271,** 19238–19242.
146. Waksman, Y., Olson, J. M., Carlisle, S. J., and Cabral, G. A. (1999). The central cannabinoid receptor (CB1) mediates inhibition of nitric oxide production by rat microglial cells. *J. Pharmacol. Exp. Ther.* **288,** 1357–1366.

147. Liu, J., Gao, B., Mirshahi, F., Sanyal, A. J., Khanolkar, A. D., Makriyannis, A., and Kunos, G. (2000). Functional CB1 cannabinoid receptors in human vascular endothelial cells. *Biochem. J.* **346,** 835–840.
148. Bouchard, J. F., Lépicier, P., and Lamontagne, D. (2003). Contribution of endocannabinoids in the endothelial protection afforded by ischemic preconditioning in the isolated rat heart. *Life Sci.* **72,** 1859–1870.
149. Schmid, H. H. O., Schmid, P. C., and Berdyshev, E. V. (2002). Cell signaling by endocannabinoids and their congeners: questions of selectivity and other challenges. *Chem. Phys. Lipids* **121,** 111–134.
150. Ben-Shabat, S., Fride, E., Sheskin, T., Tamiri, T., Rhee, M. H., Vogel, Z., Bisogno, T., De Petrocellis, L., Di Marzo, V. and Mechoulam, R. (1998). An entourage effect: inactive endogenous fatty acid glycerol esters enhance 2-arachidonoyl-glycerol cannabinoid activity. *Eur. J. Pharmacol.* **353,** 23–31.
151. Maccarrone, M., Cartoni, A., Parolaro, D., Margonelli, A., Massi, P., Bari, M., Battista, N., and Finazzi-Agrò, A. (2002). Cannabimimetic activity, binding, and degradation of stearoylethanolamide within the mouse central nervous system. *Mol. Cell. Neurosci.* **21,** 126–140.
152. Smart, D., Jonsson, K. O., Vandevoorde, S., Lambert, D. M., and Fowler, C. J. (2002). "Entourage" effects of *N*-acyl ethanolamines at human vanilloid receptors. Comparison of effects upon anandamide-induced vanilloid receptor activation and upon anandamide metabolism. *Br. J. Pharmacol.* **136,** 452–458.
153. Jonsson, K. O., Vandevoorde, S., Lambert, D. M., Tiger, G., and Fowler, C. J. (2001). Effects of homologues and analogues of palmitoylethanolamide upon the inactivation of the endocannabinoid anandamide. *Br. J. Pharmacol.* **133,** 1263–1275.
154. Baillie, G. S., Huston, E., Scotland, G., Hodgkin, M., Gall, I., Peden, A. H., MacKenzie, C., Houslay, E. S., Currie, R., Pettitt, T. R. et al. (2002). TAPAS-1, a novel microdomain within the unique N-terminal region of the PDE4A1 cAMP-specific phosphodiesterase that allows rapid, Ca^{2+}-triggered membrane association with selectivity for interaction with phosphatidic acid. *J. Biol. Chem.* **277,** 28298–28309.
155. Jones, J. A. and Hannun, Y. A. (2002). Tight binding inhibition of protein phosphatase-1 by phosphatidic acid — specificity of inhibition by the phospholipid. *J. Biol. Chem.* **277,** 15530–15538.
156. Grange, M., Sette, C., Cuomo, M., Conti, M., Lagarde, M., Prigent, A. F., and Némoz, G. (2000). The cAMP-specific phosphodiesterase PDE4D3 is regulated by phosphatidic acid binding—consequences for cAMP signaling pathway and characterization of a phosphatidic acid binding site. *J. Biol. Chem.* **275,** 33379–33387.
157. El Bawab, S., Macovschi, O., Sette, C., Conti, M., Lagarde, M., Nemoz, G., and Prigent, A. F. (1997). Selective stimulation of a cAMP-specific phosphodiesterase (PDE4A5) isoform by phosphatidic acid molecular species endogenously formed in rat thymocytes. *Eur. J. Biochem.* **247,** 1151–1157.
158. Chun, J., Goetzl, E. J., Hla, T., Igarashi, Y., Lynch, K. R., Moolenaar, W., Pyne, S., and Tigyi, G. (2002). International Union of Pharmacology. XXXIV. Lysophospholipid receptor nomenclature. *Pharmacol. Rev.* **54,** 265–269.
159. Hla, T., Lee, M. J., Ancellin, N., Paik, J. H., and Kluk, M. J. (2001). Lysophospholipids — receptor revelations. *Science* **294,** 1875–1878.
160. Maccarrone, M., Pauselli, R., Di Rienzo, M., and Finazzi-Agrò, A. (2002). Binding, degradation and apoptotic activity of stearoylethanolamide in rat C6 glioma cells. *Biochem. J.* **366,** 137–144.
161. Facci, L., Dal Toso, R., Romanello, S., Buriani, A., Skaper, S. D., and Leon, A. (1995). Mast cells express a peripheral cannabinoid receptor with differential sensitivity to anandamide and palmitoylethanolamide. *Proc. Natl. Acad. Sci. USA* **92,** 3376–3380.
162. Mazzari, S., Canella, R., Petrelli, L., Marcolongo, G., and Leon, A. (1996). *N*-(2-Hydroxyethyl)hexadecanamide is orally active in reducing edema formation and inflammatory hyperalgesia by down-modulating mast cell activation. *Eur. J. Pharmacol.* **300,** 227–236.
163. Ross, R. A., Brockie, H. C., and Pertwee, R. G. (2000). Inhibition of nitric oxide production in RAW264.7 macrophages by cannabinoids and palmitoylethanolamide. *Eur. J. Pharmacol.* **401,** 121–130.
164. Granberg, M., Fowler, C. J., and Jacobsson, S. O. P. (2001). Effects of the cannabimimetic fatty acid derivatives 2-arachidonoylglycerol, anandamide, palmitoylethanolamide and methanandamide upon IgE-dependent antigen-induced β-hexosaminidase, serotonin and TNF α release from rat RBL-2H3 basophilic leukaemia cells. *Naunyn Schmiedebergs Arch. Pharmacol.* **364,** 66–73.
165. Costa, B., Conti, S., Giagnoni, G., and Colleoni, M. (2002). Therapeutic effect of the endogenous fatty acid amide, palmitoylethanolamide, in rat acute inflammation: inhibition of nitric oxide and cyclo-oxygenase systems. *Br. J. Pharmacol.* **137,** 413–420.

166. Lambert, D. M., Vandevoorde, S., Diependaele, G., Govaerts, S. J., and Robert, A. R. (2001). Anticonvulsant activity of *N*-palmitoylethanolamide, a putative endocannabinoid, in mice. *Epilepsia* **42,** 321–327.
167. Calignano, A., La Rana, G., and Piomelli, D. (2001). Antinociceptive activity of the endogenous fatty acid amide, palmitylethanolamide. *Eur. J. Pharmacol.* **419,** 191–198.
168. Farquhar-Smith, W. P., and Rice, A. S. C. (2001). Administration of endocannabinoids prevents a referred hyperalgesia associated with inflammation of the urinary bladder. *Anesthesiology* **94,** 507–513.
169. Conti, S., Costa, B., Colleoni, M., Parolaro, D., and Giagnoni, G. (2002). Antiinflammatory action of endocannabinoid palmitoylethanolamide and the synthetic cannabinoid nabilone in a model of acute inflammation in the rat. *Br. J. Pharmacol.* **135,** 181–187.
170. Lambert, D. M., DiPaolo, F. G., Sonveaux, P., Kanyonyo, M., Govaerts, S. J., Hermans, E., Bueb, J. L., Delzenne, N. M., and Tschirhart, E. J. (1999). Analogues and homologues of *N*-palmitoylethanolamide, a putative endogenous CB_2 cannabinoid, as potential ligands for the cannabinoid receptors. *Biochim. Biophys. Acta* **1440,** 266–274.
171. Epps, D. E., Palmer, J. W., Schmid, H. H. O., and Pfeiffer, D. R. (1982). Inhibition of permeability-dependent Ca^{2+} release from mitochondria by *N*-acylethanolamines, a class of lipids synthesized in ischemic heart tissue. *J. Biol. Chem.* **257,** 1383–1391.
172. Szewczyk, A. and Wojtczak, L. (2002). Mitochondria as a pharmacological target. *Pharmacol. Rev.* **54,** 101–127.
173. Skaper, S. D., Facci, L., Romanello, S., and Leon, A. (1996). Mast cell activation causes delayed neurodegeneration in mixed hippocampal cultures via the nitric oxide pathway. *J. Neurochem.* **66,** 1157–1166.
174. Skaper, S. D., Buriani, A., Dal Toso, R., Petrelli, L., Romanello, S., Facci, L., and Leon, A. (1996). The ALIAmide palmitoylethanolamide and cannabinoids, but not anandamide, are protective in a delayed postglutamate paradigm of excitotoxic death in cerebellar granule neurons. *Proc. Natl. Acad. Sci. USA* **93,** 3984–3989.
175. Sugita, M., Williams, M., Dulaney, J., and Moser, H. (1975). Ceramidase and ceramide synthesis in human kidney and cerebellum. Description of a new alkaline ceramidase. *Biochim. Biophys. Acta* **398,** 125–133.
176. Van Blitterswijk, W. J., Van der Luit, A. H., Veldman, R. J., Verheij, M., and Borst, J. (2003). Ceramide: second messenger or modulator of membrane structure and dynamics? *Biochem. J.* **369,** 199–211.
177. Hannun, Y. A. and Obeid, L. M. (2002). The ceramide-centric universe of lipid-mediated cell regulation: Stress encounters of the lipid kind. *J. Biol. Chem.* **277,** 25847–25850.
178. Del Pulgar, T. G., Velasco, G., Sánchez, C., Haro, A., and Guzmán, M. (2002). *De novo*-synthesized ceramide is involved in cannabinoid-induced apoptosis. *Biochem. J.* **363,** 183–188.
179. Guzmán, M., Galve-Roperh, I., and Sánchez, C. (2001). Ceramide: a new second messenger of cannabinoid action. *Trends Pharmacol. Sci.* **22,** 19–22.
180. Sánchez, C., Rueda, D., Ségui, B., Galve-Roperh, I., Levade, T., and Guzmán, M. (2001). The CB_1 cannabinoid receptor of astrocytes is coupled to sphingomyelin hydrolysis through the adaptor protein fan. *Mol. Pharmacol.* **59,** 955–959.
181. Galve-Roperh, I., Sánchez, C., Cortés, M. L., Del Pulgar, T. G., Izquierdo, M., and Guzmán, M. (2000). Anti-tumoral action of cannabinoids: involvement of sustained ceramide accumulation and extracellular signal-regulated kinase activation. *Nature Med.* **6,** 313–319.
182. Huwiler, A., Dorsch, S., Briner, V. A., Van den Bosch, H., and Pfeilschifter, J. (1999). Nitric oxide stimulates chronic ceramide formation in glomerular endothelial cells. *Biochem. Biophys. Res. Commn.* **258,** 60–65.
183. Spinedi, A., Di Bartolomeo, S., and Piacentini, M. (1999). *N*-Oleoylethanolamine inhibits glucosylation of natural ceramides in CHP-100 neuroepithelioma cells: Possible implications for apoptosis. *Biochem. Biophys. Res. Commn.* **255,** 456–459.
184. Chmura, S. J., Nodzenski, E., Kharbanda, S., Pandey, P., Quintáns, J., Kufe, D. W., and Weichselbaum, R. R. (2000). Down-regulation of ceramide production abrogates ionizing radiation-induced cytochrome c release and apoptosis. *Mol. Pharmacol.* **57,** 792–796.
185. Burek, C., Roth, J., Koch, H. G., Harzer, K., Los, M., and Schulze-Osthoff, K. (2001). The role of ceramide in receptor- and stress-induced apoptosis studied in acidic ceramidase-deficient Farber disease cells. *Oncogene* **45,** 6493–6502.
186. Wiesner, D. A., Kilkus, J. P., Gottschalk, A. R., Quintáns, J., and Dawson, G. (1997). Anti-immunoglobulin-induced apoptosis in WEHI 231 cells involves the slow formation of ceramide from sphingomyelin and is blocked by bcl-x_L. *J. Biol. Chem.* **272,** 9868–9876.

187. Wiesner, D. A. and Dawson, G. (1996). Programmed cell death in neurotumour cells involves the generation of ceramide. *Glycoconjugate J.* **13,** 327–333.
188. Wiesner, D. A. and Dawson, G. (1996). Staurosporine induces programmed cell death in embryonic neurons and activation of the ceramide pathway. *J. Neurochem.* **66,** 1418–1425.
189. Berdyshev, E. V., Schmid, P. C., Dong, Z. G., and Schmid, H. H. O. (2000). Stress-induced generation of *N*-acylethanolamines in mouse epidermal JB6 P^+ cells. *Biochem. J.* **346,** 369–374.
190. Gaetani, S., Oveisi, F., and Piomelli, D. (2003). Modulation of meal pattern in the rat by anorexic lipid mediator oleoylethanolamide. *Neuropsychopharmacology* 1–6.
191. Rodriguez de Fonseca, F., Navarro, M., Gómez, R., Escuredo, L., Nava, F., Fu, J., Murillo-Rodríguez, E., Giuffrida, A., LoVerme, J., Gaetani, S. et al. (2001). An anorexic lipid mediator regulated by feeding. *Nature* **414,** 209–212.
192. Kunos, G., Bátkai, S., Offertáler, L., Mo, F., Liu, J., Karcher, J., and Harvey-White, J. (2002). The quest for a vascular endothelial cannabinoid receptor. *Chem. Phys. Lipids* **121,** 45–56.
193. Stella, N. and Piomelli, D. (2001). Receptor-dependent formation of endogenous cannabinoids in cortical neurons. *Eur. J. Pharmacol.* **425,** 189–196.
194. Braida, D., Pozzi, M., and Sala, M. (2000). CP 55,940 protects against ischemia-induced electroencephalographic flattening and hyperlocomotion in mongolian gerbils. *Neurosci. Lett.* **296,** 69–72.
195. Louw, D. F., Yang, F. W., and Sutherland, G. R. (2000). The effect of δ-9-tetrahydrocannabinol on forebrain ischemia in rat. *Brain Res.* **857,** 183–187.
196. Veldhuis, W. B., Van der Stelt, M., Wadman, M. W., Van Zadelhoff, G., Maccarrone, M., Fezza, F., Veldink, G. A., Vliegenthart, J. F. G., Bar, P. R., Nicolay, K. et al. (2003). Neuroprotection by endogenous cannabinoid anandamide and arvanil against in vivo excitotoxicity in the rat: role of vanilloid receptors and lipoxygenases. *J. Neurosci.* **23,** 4127–4133.
197. Mauler, F., Mittendorf, J., Horváth, E., and De Vry, J. (2002). Characterization of the diarylether sulfonylester (-)-(*R*)-3-(2-hydroxymethylindanyl-4-oxy)phenyl-4,4,4-trifluoro-1-sulfonate (BAY 38-7271) as a potent cannabinoid receptor agonist with neuroprotective properties. *J. Pharmacol. Exp. Ther.* **302,** 359–368.

15 Neuropsychiatry: Schizophrenia, Depression, and Anxiety

Ester Fride and Ethan Russo

CONTENTS

INTRODUCTION

An association between cannabis use and neuropsychiatric conditions is of major importance because of the potential for the use of cannabis as a recreational drug or medicine and also for understanding the biological mechanisms by which cannabinoids act on brain structure and function. In this chapter we review evidence for a number of hypotheses which have been proposed to describe the putative role of cannabis use in the precipitation of psychosis. The majority of studies support the *vulnerability hypothesis*, especially with respect to schizophrenia. The hypothesis that the consumption of cannabis is a form of *self-medication* for the symptoms of schizophrenia and depression has also gained support.

Evidence for and against long-term neuropsychiatric impairment due to cannabis consumption has to be considered, keeping in mind both the time interval between cannabis use and the neuropsychological assessments and the age of onset of cannabis use.

Finally, recent progress in the investigation of the role of the endocannabinoid–CB_1 receptor system in mental disease in animal models as well as in humans, especially with respect to its interaction with the dopamine neurotransmitter system, opens up new avenues for the understanding and treatment of the major mental disorders.

The history and science of cannabis and cannabinoids and their relation to mental health and disease are fraught with controversy, ambivalence, and contradictory claims. Perhaps this is inevitable when one considers that endocannabinoids serve a modulatory function in many neurochemical and psychopharmacological processes, and deficiencies or excesses in any of these may produce manifestations of psychopathology. The different aspects of this controversy regarding the role of cannabis in psychiatric disorders have been reviewed previously in detail

(Grinspoon, 1977; Grinspoon and Bakalar, 1997; Grinspoon, Bakalar, and Russo, 2005; Russo, 2001; Russo, 2004; Russo et al., 2001).

HISTORICAL HIGHPOINTS

The first well-documented description of the effects of cannabis on mood may derive from Mesopotamia (see Figure 15.1). Clay tablets discovered in the Assyrian library of Ashurbanipal seem to derive from Sumerian and Akkadian documents of the 22nd century B.C.E., wherein cannabis is described as being ingested as a remedy for grief or "depression of spirits" (Thompson, 1949).

```
   Chinese Emperor Shên-Nung prescribes cannabis for senility
            Sumerian/Akkadian, cannabis use fornissati,grief
                      Atharva Veda,bhang for grief
                              Qunnapu,Babylonian incense
                                       Herodotus, cannabis as Scythian funerary
                                         Tombs of Pazyryk, Scythian burned cannabis
                                         Diodorus Siculus, use in Egypt
                                       Galen, cannabis as inebriant
                                            Tantric use of cannabis in India
                                                    Jabir ibn Hayyan, psychoactive
                                               Hildegard von Bingen
                                                         Avicenna, Persia
ANCIENT/MIDDLE AGES                                        Maimonides, Egypt
_________________________________________________________________________________________
3000 BCE       2000          1000          0           1000       1500 CE


             Garcia da Orta, India, psychoactive
                           Robert Burton,Anatomy of Melancholy,ecstatic
                                         Rumphius, Indonesia, inebriation
                                               Kaempfer, psychotropic in Persia and India
RENAISSANCE
_________________________________________________________________________________________
1500                    1600                    1700                    1800 CE


O'Shaughnessy, Indian hemp
  Clendinning,England,anxiety,depression,morphine withdrawal
  Lallemand,Le Hachcych,utopian visionary
    Moreau, Du Haschich et de l'Alientation Mentale
             Fitz Hugh Ludlow,The Hasheesh Eater
              McMeens Report, Ohio, bipolar disease
                     Tyrell, delirium tremens, opiate addiction
                      Reynolds, depression, senile restlessness
                        Polli, Italy, melancholia and anxiety
                                  Strange, melancholia, depression, insomnia
                                 Aulde, delirium tremens
                                      Mattison,addiction cocaine,opiates
                                          India Hemp Drugs Commission
19thCENTURY                                    Dixon, smoked for pain/work/appetite
_________________________________________________________________________________________
1840   1850    1860    1870    1880    1890     1900 CE


    Panama Canal Zone Commissions
                LaGuardia Commission (USA)
                               Mechoulam/Edry isolation and synthesis of THC
                                        Mikuriya, cannabis in alcoholism
                                        Controlled Substances Act (USA)
                                            Regelson, mood elevator, tranquilizer in cancer
                                              Chronic use studies, Costa Rica, Jamaica
                                                       CB1receptor defined: Devane/Howlett
                                                 Anandamide: Devane/Mechoulam
                                                         Grinspoon: psychiatric uses
                                                 2-AG/noladine discovered
                                                     Volicer, Alzheimer
20th CENTURY                                                   Muller-Vahl,TS,OCD
                                                               Entourage effect-Israel
_________________________________________________________________________________________
1930     1940     1950     1960     1970     1980     1990     2000 CE
```

FIGURE 15.1 Cannabis Psychiatric Time Line by Ethan Russo, M.D. (Adapted and expanded from Russo, 2004.)

In the Indian *Atharva Veda* (passage 11,6,15) circa 1600 B.C.E., cannabis, or *bhanga*, is one of five herbs employed to "release us from anxiety." (Commission, 1894, Appendix 3, p. 286).

Perhaps older yet is the *Pên Tsao Ching,* written in the first or second century but attributed to Emperor Shên Nung in the third millennium B.C.E., which noted that excessive ingestion of cannabis flowers produced hallucinations (literally, "seeing devils") (Li, 1974, p. 446).

The Greek historian Herodotus noted the use of burned cannabis flowers in the 5th century B.C.E. by the Scythian peoples of central Asia as a funerary rite causing them to "shriek with delight at the fumes" (Herodotus, 1998, p. 259, Book 4, Passage 75).

Hyperbole has often surrounded any discussion of cannabis, and this is true even in the early accounts of its dangers: Al-Ghazzi, quoting Az-Zarkashi in the 14th century, indicated that cannabis "causes sudden death or madness" (Hamarneh, 1978, p. 288), although he acknowledged its therapeutic role in medicine.

The benefits of Indian hemp in depression were noted in Europe by the 17th century (Burton, 1907) but were not studied more formally until the 19th. Jacques-Joseph Moreau investigated cannabis both as a model psychosis-inducing and psychotherapeutic agent (Moreau, 1845). In North America in 1859, McMeens (1860) described a case study of a man with "hysterical insanity" who had cycles of manic energy during which he thought himself a great inventor. These were interspersed with bouts of melancholy and inertia — symptoms of what we would likely recognize today as bipolar disease. A tincture of *Cannabis indica* seemed to even his mood, presaging similar anecdotal evidence of such benefit noted by Grinspoon in modern-day bipolar patients (Grinspoon and Bakalar, 1998).

Controversy continued, however, and led to the first in-depth governmental investigation by the British that resulted in the Indian Hemp Drugs Commission Report (Indian Hemp Drugs Commission, 1894), which exceeded 3000 pages in length. The commission scoured the asylums of the subcontinent to assess the widely assumed role of cannabis as a precipitant of mental illness. It was observed that documentation was poor, and attribution to *ganja* was often assigned arbitrarily in cases where no other explanation was forthcoming (Kaplan, 1969). No actual association of cannabis and mental illness was apparent, and only 61 cases in the entire region were identified in which the herb could be etiologically implicated. Equivalence of symptomatology was observed in cases with or without such exposure, and affliction associated with cannabis usage appeared to be self-limiting. In contrast in Egypt, one author felt that hashish was a frequent cause of insanity (Warnock, 1903), but this was not based on any epidemiological investigation.

Another extensive evaluation was undertaken in India (Chopra and Chopra, 1939) in which the authors noted the prevalent usage of cannabis as a stress reliever and accessory to hard physical labor. They even acknowledged a benefit in individuals suffering from hypochondriasis or neurosis. They summarized the issue of etiological causation of mental illness as follows (p. 103): "It does not necessarily produce insanity except in perhaps those who have predisposition to it."

After extensive study, the LaGuardia Commission noted (Wallace and Cunningham, 1944, p. 218): "Furthermore, those who have been smoking marihuana for a period of years showed no mental or physical deterioration which may be attributed to the drug."

In Morocco, a "cannabis psychosis" was noted among *kif* smokers (Benabud, 1957), but (see Grinspoon, Bakalar, and Russo, 2005) its purported incidence (5/1000) is well below the baseline rate of schizophrenia in this and other populations worldwide.

THE LAST 40 YEARS

The widespread use of cannabis in western industrial nations began in the 1960s and led the National Institute on Drug Abuse (NIDA) to fund an extensive series of studies on the chronic use of cannabis in nations with such experience. In Jamaica (Rubin and Comitas, 1975), chronic ganja use was studied in 30 users and matched controls. One nonuser showed signs of depression on neuropsychological testing and another was assessed to be a borderline case of depression. No signs of active psychosis were observed. In all, no significant differences were observed between the groups with respect to mood, thought, or behavior.

In Greece, hashish smokers were studied (Stefanis et al., 1977), and though a greater incidence of psychopathology was observed in users as compared to controls, most were accounted for by "personality disorders." Interestingly, more psychiatric abnormalities were observed in moderate as opposed to heavy users. In Costa Rica (Carter, 1980), cannabis smokers believed it to be of benefit for depression and malaise. No significant indicators of adverse sequelae were observed in the personality or performance of cannabis users.

More recently, studies have attempted to test several distinct hypotheses about the relationship between cannabis use and psychiatric disease. By far, the majority have dealt with schizophrenia and only a minority with depression and anxiety. In general, the following hypotheses were investigated:

1. Cannabis use causes psychosis.
2. Cannabis use precipitates a psychotic attack in vulnerable individuals.
3. Cannabis use worsens (positive) schizophrenic symptoms.
4. Cannabis use is comorbid with schizophrenia.
5. Cannabis is used as self-medication for the negative symptoms of schizophrenia.
6. Cannabis is used as self-medication for anxiety and depression.

Before discussing the evidence for and against these hypotheses, a separate but relevant issue which has remained unresolved for over a 100 yr (Iversen, 2000) must be discussed: Does cannabis induce (symptoms of) schizophrenia, or does the drug induce a separate entity (marijuana psychosis, [MP]) of psychotic symptoms that persist well after its consumption?

Nunez and Gurpegui (2002) compared spontaneous attacks of schizophrenia (acute schizophrenia) to chronic, heavy cannabis-use–induced schizophrenia (MP). Although there was a partial overlap of the symptoms, the symptoms of MP and acute schizophrenia could be distinguished from each other. Moreover, all MP patients completely recovered with neuroleptic treatment. Therefore, the findings of this study argue for the existence of MP as a separate disorder.

In two studies on healthy subjects (Dumas et al., 2002; Skosnik et al., 2001), schizotypic traits were compared among three groups: nonusers, regular current users, and an intermediate group consisting of past or occasional users. In both studies, regular use was associated with schizotypal personality traits. In the study by Dumas and colleagues, occasional and past users were similar to regular users, whereas, in contrast, in the study sample of Skosnik et al., past cannabis users were similar to nonusers. In the Skosnik study, the intermediate group included only past users (i.e., with no cannabis use for at least 45 d prior to assessment), whereas in the Dumas study, the intermediate group included occasional as well as past users; when data from both studies are combined, the findings argue against a residual effect of cannabis and are consistent with the position that MP is simply an expression of increased vulnerability to cannabis-induced intoxication (Nunez and Gurpegui, 2002). It should be noted, however, that "past users" in the Skosnik study may have included light or one-time users, who are not at risk of developing MP.

Whether or not MP is a separate psychotic entity, the effect of cannabis use on the onset and development of schizophrenia has been widely investigated but still remains controversial.

A variety of research designs have been used to investigate the cannabis–psychoses association, including retrospective *vs.* prospective designs, clinical *vs.* nonclinical, and with small *vs.* large samples. In these studies, timing is critical in two respects: (1) Were the symptoms measured during, immediately after, or long after cessation of cannabis use? and (2) What was the age of onset of cannabis use? Despite the widely different approaches, the findings of the majority of the studies support at least some association between cannabis smoking and psychosis.

Grinspoon studied 41 patients in Massachusetts with first-break acute schizophrenia (Grinspoon, 1977); six (15%) had a history of cannabis use. Upon close examination, he found that in four patients, the development of psychosis was quite remote from the exposure. The role, if any, of cannabis in the remaining two patients was quite unclear.

In a recent study from Sweden on schizophrenia (Zammit *et al.*, 2002), the authors reexamined Swedish conscripts from 1969, the same cohort studied earlier by Andreasson et al., (1987). In the current study, the group defined as the "frequent users" (> 50 times) was found to have an incidence of schizophrenia of 5.7% as opposed to 0.6% in nonusers of cannabis. These cannabis smokers reportedly had not used any other drug.

Another recent study examined a cohort of young New Zealanders for cannabis use and the development of adult psychosis (Arsenault et al., 2002). In this article, the controls were defined as those who had used cannabis 0 to 2 times, whereas "cannabis users" were those who had taken the drug 3 times or more by age 15 and continued at some unspecified rate of intake through age 18. Smoking cannabis was found to increase the incidence of psychosis in adults, and importantly, psychosis was more likely the earlier the age of onset of cannabis use. This age of onset was also determined to be relevant in a study by Nunez and Gurpegui (2002), who found that in heavy cannabis users who developed MP, a large proportion (64%) had started to smoke cannabis between the ages of 13 and 15.

In another study and its follow-up, performed on 232 schizophrenia patients, 13% of which had a history of cannabis use, the relationship between the age of onset of cannabis smoking and the onset of schizophrenia was analyzed (Hambrecht and Hafner, 2000; Buhler et al., 2002). Although the distinction between cannabis use and drug (other than alcohol) use was not clearly defined, the authors estimated that the great majority (88%) of the drug users predominantly but not solely used cannabis. It appeared that whereas overall drug or cannabis use was twice as high in the study sample as in the general population of that area (Germany), the cannabis users with schizophrenia in the study group could be divided into equal groups of: (1) those with a history of having used cannabis for years before the onset of psychosis; 2) those in whom cannabis use and schizophrenia started at the same time, and 3) those who had commenced cannabis use only after the onset of schizophrenia. Therefore, these observations support the causation, the vulnerability, as well as the self-medication hypotheses (see following text). A recent evaluation of the data (Degenhardt and Hall, 2002) indicates support for the concept that cannabis may serve as a precipitant in vulnerable patients and increase relapse rates.

On the other hand, evidence for a causal relationship between cannabis use and psychoses in otherwise low-risk persons is much more scarce (Johns, 2001; Hall and Degenhardt, 2000; Mass et al., 2001). Convincingly arguing against the causation hypothesis is the finding that in Australia no increased rate of incidence of schizophrenia has been reported over a period of several decades despite a dramatic increase in cannabis use during that period (Degenhardt, Hall and Lynskey, 2003).

In agreement with this, Joy, Watson, and Benson (1999) of the Institute of Medicine observed (p. 106) that "people with schizophrenia or with a family history of schizophrenia are likely to be at greater risk for adverse psychiatric effects from the use of cannabinoids," and "there is little evidence that cannabis alone produces a psychosis that persists after the period of intoxication."

Interestingly, in a cohort at very high risk for schizophrenia (defined by the presence of subthreshold psychotic symptoms and a family history of schizophrenia), no relationship was found between cannabis use and the incidence of schizophrenia (Phillips et al., 2002). Is it possible that the likelihood of developing schizophrenia was already so high that cannabis use could not add to the final outcome?

In support of the self-medication hypothesis, it is known that cannabis may ameliorate certain symptoms of psychosis (Warner et al., 1994), including activation symptoms and the subjective complaints of depression, anxiety, insomnia, and pain. In a sample of patients with chronic schizophrenia, Bersani and colleagues (2002) found evidence for a precipitating influence of cannabis in the development of schizophrenia, and in the same sample, a subgroup was identified that used cannabis to self-medicate for amelioration of negative symptoms. This finding is supported by observations in a nonclinical sample of a significant association between cannabis use and the negative symptoms of schizophrenia (Verdoux et al., 2003).

The proposition that cannabidiol (CBD), a nonpsychoactive component of the cannabis plant, possesses considerable antipsychotic activity (Zuardi and Guimaraes, 1997) is supportive of the self-medication role of cannabis in the disorder, but this action is not via the cannabinoid CB_1 receptor.

Taken together, most studies confirm the vulnerability hypothesis for cannabis use and schizophrenia. Thus, schizophrenia patients should probably not use cannabis because a psychotic episode can be induced in someone with a preexisting disorder and, indeed, increased hospitalization rates and symptom exacerbation have been demonstrated (Caspari, 1999). Increased rates of psychosis are also observed in those meeting the criteria for cannabis dependence (Fergusson, Horwood, and Swain-Campbell, 2003). Hollister, (1986) summarized in his review (p. 6–7): "It would seem reasonable to assume that cannabis might unmask latent psychiatric disorders."

MOOD DISORDERS

An acute depressive reaction, occasionally observed in cannabis users, has most often been encountered in those with underlying depression (Grinspoon and Bakalar, 1997). The same authors, however, have published several case reports of improvement in bipolar symptoms with cannabis usage (Grinspoon and Bakalar, 1998). These clinical observations are partly supported by epidemiologic investigations which have approached the question (analogous to the questions on etiology in schizophrenia; see section titled "The Last 40 Years") of whether cannabis use increases the risk for depression or, conversely, whether depression results in more frequent use of cannabis in an attempt to self-medicate. Cross-sectional studies tend to suggest that depression induces cannabis use, and longitudinal (prospective) studies suggest the opposite, i.e., that cannabis consumption precipitates depression in later life (up to a fourfold increase in risk) (Bovasso, 2001; Patton et al., 2002). However, in nonclinical samples, no association or very weak associations were found between various measures of depression and cannabis smoking (Chen et al., 2002; Green and Ritter, 2000; Tournier et al., 2003). Suicide attempts were significantly higher in cannabis smokers but this association lost statistical significance after sociodemographic factors, history of psychiatric symptoms in childhood, and concurrent psychiatric symptoms were controlled (Beautrais et al., 1999).

Thus, taken together, the epidemiological evidence does not support a causative or precipitating role for cannabis in chronic anxiety or depression. With respect to self-medication, among four patients with serious chronic diseases (in the Compassionate Use Investigational New Drug Program) who utilized high amounts of cannabis daily for many years for symptom control (Russo et al., 2002), scores on the Beck Depression Inventory were very low. Similarly, of the 2480 patients surveyed who had used cannabis for medical indications in one large practice in California over a number of years, 660 (26.6%) did so primarily for treatment of mood disorders; this included 162 (6.5%) for depression, 73 (3%) for anxiety, 34 (1.4%) for bipolar disorder, and even 26 (1%) for schizophrenia (Gieringer, 2001).

Although there are no reports of controlled clinical trials on the use of cannabis for mood disorders, benefits have been noted in depression measures in cancer patients treated with THC (Regelson et al., 1976). In addition, CBD has demonstrated benefit in the treatment of anxiety (Zuardi and Guimaraes, 1997), allaying anxiety in experimental subjects with no significant side effects (Zuardi et al., 1993). Controlled clinical trials seem indicated with this agent. Interestingly, a recent study on rats, in which extinction of cocaine- or amphetamine-induced place preference, a form of motivational learning, was enhanced, suggests that CBD may be involved in emotion-relevant processes (Parker et al., 2003).

EFFECTS OF LONG-TERM CANNABIS USE ON COGNITION AND BEHAVIOR

A major concern with long-term use of cannabis, whether consumed for recreational or medicinal purposes, is the possibility of irreversible damage to brain structure or function. There is agreement among researchers, however, that long-term use of cannabis does not result in structural brain damage or gross cognitive deficits (Solowij, 2002). However, the possibility that cannabis use results in subtle and/or specific cognitive impairments has been the subject of considerable controversy.

Hall and Solowij (1998) have reported detrimental neurocognitive effects in long-term heavy cannabis users. However, these subjects were investigated while they were current users and therefore the deficits could have been a direct effect of intoxication. Until recently, no studies were available which observed a period of abstinence of more than a few days (Solowij, 2002). In an additional study performed by Solowij and colleagues (2002), very long-term cannabis users (median 24 yr of use) showed deficits in tests for memory and attention. However, users defined as short-term users, but whose mean duration of use was nevertheless 10.2 yr, did not differ, in general, from controls. Moreover, abstinence intervals were minimal (median 17 h). As recovery of function is likely to involve changes in plasticity, such as CB_1 receptor densities, such a short interval is unlikely to allow for significant reversals.

More recently, investigations have been performed which allowed for considerably longer abstinence periods (1 to 3 months). These investigations in general do not support the existence of a residual effect of cannabis that persists well after cessation of consumption. Thus, in a long-term prospective investigation of a sample of children of mothers who smoked cannabis during pregnancy (Fried et al., 2002), these children themselves became heavy users as young adults, displaying a 4-point decrease in IQ (as compared to an earlier assessment made at the age of 9–12 yr). In contrast, heavy users who had abstained from cannabis use for at least 3 months gained IQ points, which was similar to the findings in light current users or nonusers: these groups gained 3.5, 5.8, and 2.6 points respectively. In a study by Pope and colleagues (2001), a 28-d-long wash-out period was sufficient to eliminate the deficits in neuropsychological performance observed in current heavy users. It is possible, however, that a more detailed analysis of cognitive function would reveal residual effects of cannabis smoking. Perhaps future studies will more definitively answer this question.

In accordance with the latter two investigations, a recent meta-analytic study of residual neurocognitive effects of cannabis use did not reveal detrimental effects, except for a modest impairment in memory tests. The authors conclude that when medical use of cannabinoids is being considered, the health benefits should be carefully weighed against the very modest potential for cognitive decline (Grant et al., 2003).

A meta-analytic study of 3206 subjects on the behavioral traits associated with cannabis use and abuse (Gorman and Derzon, 2002) demonstrated merely an association with "unconventionality" and not more serious disorders, mirroring much older observations (Goode, 1970). Cannabis use and good mental health are not mutually exclusive. A recent British study has shown that drug experimentation is associated with high self-esteem (Regis, 2001). Another British study showed no causal relationship between cannabis use and delinquency in young people (Hammersley et al., 2003).

In short, chronic cannabis use does not appear, after cessation of drug use, to result in a significant cognitive decline or behavioral symptomatology. However, as discussed for cannabis use as a risk factor for schizophrenia (see section titled "The Last 40 Years"), age of onset of cannabis use may be critical in determining the outcome of cannabis consumption later in life. Thus, in a recent study by Pope and colleagues (2003), heavy cannabis users who had started before the age of 17 differed from controls in neuropsychological tests, mainly on verbal IQ.

BIOLOGICAL EVIDENCE FOR A ROLE OF THE ENDOCANNABINOID–CB_1 RECEPTOR SYSTEM IN NEUROPSYCHIATRIC DISORDERS

The prefrontal cortex (PFC) is thought to integrate cognitive and emotional functions and, as the target area for the mesocortical dopamine system, may be the primary dysfunctional area in schizophrenia and the site of action for antischizophrenic drugs (Thierry et al., 1978). This area is also exquisitely responsive to stress (Thierry et al., 1976), and it is well established that schizophrenia is frequently triggered by stressful life events (Carlson, 2001). The density of CB_1 receptors in the PFC is high, at least relative to other G-protein-coupled receptors (Herkenham et al., 1990). Close interactions between the dopamine system and exo- and also endocannabinoids have been shown repeatedly. For example, THC was shown to increase presynaptic dopamine efflux and utilization in the PFC (Chen et al., 1990; Jentsch et al., 1997). In the dorsal striatum also, an interaction has been observed but in the opposite direction: activation of D_2 dopamine receptors caused increased outflow of anandamide (Giuffrida et al., 1999).

A link between stress and cannabinoids has also been shown repeatedly. Thus, administration of cannabinoids or anandamide produced anxiety-like responses in rats (Rodriguez de Fonseca et al., 1997; Navarro et al., 1997) and mice (Chakrabarti et al., 1998) and activation of the hypothalamic–pituitary–adrenal "stress hormone" axis in rats (Weidenfeld et al., 1994; Rodriguez de Fonseca et al., 1996) and calves (Zenor et al., 1999). We have shown that acute noise stress induces a fourfold increase in anandamide levels in the PFC but not in the hippocampus of adult mice (Fride and Sanudo-Pena, 2002). Taken together, these studies suggest that activation of the endocannabinoid–CB_1 receptor system in the PFC, either by stress or by cannabis use, may trigger symptoms of schizophrenia.

Thus far, only a few studies have been performed on the putative connection between the endocannabinoid–CB_1 receptor system and schizophrenia in humans. Higher concentrations of CB_1 receptors have been found in the dorsolateral PFC of deceased schizophrenic patients but not in other regions such as the hippocampus and the caudate-putamen (Dean et al., 2001). Levels of anandamide were higher in the cerebrospinal fluid of schizophrenic patients, but only in 3 out of 10 patients (Leweke et al., 1999). In the same study, an elevation of the non-CB_1 receptor-binding palmitoyl ethanol amide was also found in a subset of 4 of the 10 patients. It seems that only an analysis of both these findings combined produced statistical significance (Leweke et al., 1999). Further, healthy volunteers intoxicated with cannabis resin displayed perceptual abnormalities similar to that of schizophrenic patients who did not receive cannabis (Emrich et al., 1997). This observation was interpreted as support for the hypothesis that schizophrenia is characterized by a disturbance of the endocannabinoid–CB_1 receptor system. Taken together these preliminary findings support the possibility that schizophrenia may be characterized by an overactive endocannabinoid–CB_1 receptor system in the PFC. Fritsche (2000) has hypothesized that CB_1 receptor-deficient (knockout) mice may display similarities to patients with schizophrenia and may be a model for the disease, implying that schizophrenia is characterized by a lack of CB_1 receptors. Although much more work needs to be done, the data emanating from various approaches largely point to a major role for the endocannabinoid–CB_1 receptor system in schizophrenia, in close interaction with the dopamine neurotransmitter system.

CONCLUSIONS

Cannabis has been used in medicine and psychiatry for many centuries and in a variety of cultures. A role for cannabis in the etiology and precipitation of mental disease has been frequently suggested and has obtained wide support. However, the nature of this influence has remained controversial. Most research has focused on schizophrenia. The vulnerability hypothesis, according to which, in individuals at risk for psychosis, cannabis use may trigger the disease, has obtained extensive support. A simple causative effect by which cannabis use may induce schizophrenia in otherwise normal individuals is hardly tenable. The theory that consuming cannabis is an effort to self-medicate, mainly for the negative symptoms of schizophrenia, has also gained some support and remains an attractive hypothesis, especially in view of the biological evidence for an intimate connection between the endocannabinoid–CB_1 receptor system, the dopamine neurotransmitter system, stress, and schizophrenia.

Although most of the evidence does not support the existence of adverse effects persisting beyond the cessation of long-term cannabis use, it seems that when cannabis consumption is started at a young (teen or preteen) age, it may be prudent to exercise more caution, although current data are far from clear. Both a greater vulnerability to schizophrenia as well as claims of possible permanent cognitive damage have been reported in such young "starters." Finally, the dramatic progress which has been made over the last decade in unraveling the mechanisms by which the cannabinoids affect brain function, both as a healing agent and as a detrimental factor, will open new avenues for their medical application and for understanding the impact of cannabis use in neuropsychiatry.

REFERENCES

Andreasson, S., Allebeck, P., Engstrom, A., and Rydberg, U. (1987): Cannabis and schizophrenia. A longitudinal study of Swedish conscripts *Lancet* 2(8574): 1483–1486.

Arsenault, L., Cannon, M., Poulton, R., Murray, R., Caspi, A., and Moffitt, T.E. (2002): Cannabis use in adolescence and risk for adult psychosis: longitudinal prospective study. *British Medical Journal* 325: 1212–1213.

Beautrais, A.L., Joyce, P.R., and Mulder, R.T. (1999): Cannabis abuse and serious suicide attempts. *Addiction* 94(8): 1155–1164.

Benabud, A. (1957): Psychopathological aspects of the cannabis situation in Morocco: statistical data for 1956. *Bulletin of Narcotics* 9: 2.

Bersani, G., Orlandi, V., Kotzalidis, G.D., and Pancheri, P. (2002): Cannabis and schizophrenia: impact on onset, course, psychopathology and outcomes. *European Archives of Psychiatry and Clinical Neuroscience* 252: 86–92.

Bovasso, G.B. (2001): Cannabis abuse as a risk factor for depressive symptoms. *American Journal of Psychiatry* 158: 2033–7.

Buhler, B., Hambrecht, M., Loffler, W., an der Heiden, W., and Hafner, H. (2002): Precipitation and determination of the onset and course of schizophrenia by substance abuse — a retrospective and prospective study of 232 population-based first illness episodes. *Schizophrenia Research* 54: 243–251.

Burton, R. (1907): *The Anatomy of Melancholy.* Chatto and Windus, London.

Carlson, N.R. (2001): *Physiology of Behavior,* 7th ed. Allyn and Bacon, Boston.

Carter, W.E. (1980): *Cannabis in Costa Rica: A Study of Chronic Marihuana Use.* Institute for the Study of Human Issues, Philadelphia.

Caspari, D. (1999): Cannabis and schizophrenia: results of a follow-up study. *Eur Archives of Psychiatry and Clinical Neuroscience* 249: 45–49.

Chakrabarti, A., Ekuta, J.E., and Onaivi, E.S. (1998): Neurobehavioral effects of anandamide and cannabinoid receptor gene expression in mice. *Brain Research Bulletin* 45: 67–74.

Chen, J., Paredes, W., Lowinson, J.H., and Gardner, E.L. (1990): Delta 9-tetrahydrocannabinol enhances presynaptic dopamine efflux in medial prefrontal cortex. *European Journal of Pharmacology* 190: 259–262.

Chen, C.Y., Wagner, F.A., and Anthony, J.C. (2002): Marijuana use and the risk of Major Depressive Episode. Epidemiological evidence from the United States National Comorbidity Survey. *Society for Psychiatry and Psychiatric Epidemiology* 37: 199–206.

Chopra, I.C. and Chopra, R.N. (1939): The present position of hemp drug addiction in India. *Indian Medical Research Memoirs* 31: 1–119.

Commission, I.H.D. (1894): *Report of the Indian Hemp Drugs Commission, 1893–94.* Govt. Central Print. Office, Simla.

Dean, B., Sundram, S., Bradbury, R., Scarr, E., and Copolov, D. (2001): Studies on [3H]CP-55940 binding in the human central nervous system: regional specific changes in density of cannabinoid-1 receptors associated with schizophrenia and cannabis use *Neuroscience* 103(1): 9–15.

Degenhardt, L. and Hall, W. (2002): Cannabis and psychosis. *Current Psychiatry Reports* 4(3): 191–6.

Degenhardt, L., Hall, W., and Lynskey, M. (2003): Testing hypotheses about the relationship between cannabis use and psychosis. *Drug Alcohol Depend* 71(1): 37–48.

Dumas, P., Saoud, M., Bouafia, S., Gutknecht, C., Ecochard, R., Dalery, J., Rochet, T., and d'Amato, T. (2002): Cannabis use correlates with schizotypal personality traits in healthy students. *Psychiatry Research* 109: 27–35.

Emrich, H.M., Leweke, F.M., and Schneider, U. (1997): Towards a cannabinoid hypothesis of schizophrenia: cognitive impairments due to dysregulation of the endogenous cannabinoid system. *Pharmacology Biochemistry and Behavior* 56: 803–7.

Fergusson, D.M., Horwood, L.J., and Swain-Campbell, N.R. (2003): Cannabis dependence and psychotic symptoms in young people. *Psychological Medicine* 33: 15–21.

Fride, E. and Sanudo-Pena, C. (2002): Cannabinoids and endocannabinoids: behavioral and developmental aspects. in *The Biology of Marijuana.* Ed. E. Onaivi, Harwood Academic, Reading, MA.

Fried, P., Watkinson, B., James, D., and Gray, R. (2002): Current and former marijuana use: preliminary findings of a longitudinal study of effects on IQ in young adults. *Canadian Medical Association Journal* 166: 887–891.

Fritzsche, M. (2000): Are cannabinoid receptor knockout mice animal models for schizophrenia? *Medical Hypotheses.* 56(6): 638–643.

Gieringer, D. (2001): Medical use of cannabis: experience in California. in *Cannabis and Cannabinoids: Pharmacology, Toxicology, and Therapeutic Potential.* Eds. F. Grotenhermen and E. Russo. pp. 153–170, Haworth Press, Binghamton, NY.

Goode, E. (1970): *The Marijuana Smokers.* Basic Books, New York.

Gorman, D.M. and Derzon, J.H. (2002): Behavioral traits and marijuana use and abuse: a meta-analysis of longitudinal studies *Addictive Behaviors* 27: 193–206. Giuffrida, A., Parsons, L.H., Kerr, T.M., Rodriguez de Fonseca, F., Navarro, M., and Piomelli, D. (1999): Dopamine activation of endogenous cannabinoid signaling in dorsal striatum *Nature Neuroscience* 2: 358–363.

Grant, I., Gonzalez, R., Carey, C.L., Natarajan, L., and Wolfson, T. (2003): Non-acute (residual) neurocognitive effects of cannabis use: a meta-analytic study. *Journal of the International Neuropsychological Society* 9: 679–689.

Green, B.E. and Ritter, C. (2000): Marijuana use and depression. *Journal of Health and Social Behavior* 41: 40–49.

Grinspoon, L. (1977): *Marihuana Reconsidered,* 2nd edition. Harvard University Press, Cambridge, MA.

Grinspoon, L. and Bakalar, J.B. (1997): *Marihuana, the Forbidden Medicine* (rev. and exp. ed.). Yale University Press, New Haven.

Grinspoon, L., and Bakalar, J.B. (1998): The use of cannabis as a mood stabilizer in bipolar disorder: anecdotal evidence and the need for clinical research. *Journal of Psychoactive Drugs* 30: 171–177.

Grinspoon, L., Bakalar, J.B., and Russo, E. (2005): Marihuana. in *Substance Abuse: A Comprehensive Textbook.* J.H. Lowinson, L. Ruiz, R.B. Millman, and J.G. Langrod, Eds., Lippincott Williams and Wilkins, NY.

Hall, W. and Degenhardt, L. (2000): Cannabis use and psychosis: a review of clinical and epidemiological evidence. *Australian and New Zealand Journal of Psychiatry* 34(1): 26–34.

Hall, W. and Solowij, N. (1998): Adverse effects of cannabis. *Lancet* 352(9140): 1611–1616.

Hamarneh, S. (1978): Medicinal plants, therapy, and ecology in Al-Ghazzi's book on agriculture. *Studies in the History of Medicine* 2: 223–263.

Hambrecht, M. and Hafner, H. (2000): Cannabis, vulnerability, and the onset of schizophrenia: an epidemiological perspective. *Australian and New Zealand Journal of Psychiatry* 34(3): 468–475.

Hammersley, R., Marsland, L., and Reed, M. (2003): *Substance Use by Young Offenders: the Impact of the Normalisation of Drug Use in the Early Years of the 21st Century.* edited., p. 90, Home Office Research, Development and Statistics Directorate, London.

Herkenham, M., Lynn, A.B., Little, M.D., Johnson, M.R., Melvin, L.S., de Costa, B.R., and Rice, K.C. (1990): Cannabinoid receptor localization in brain. *Proceedings of the National Academy of Sciences U S A* 87: 1932–1936.

Herodotus (1998): *The Histories.* Oxford University Press, Oxford [England]; New York.

Hollister, L.E. (1986): Health aspects of cannabis *Pharmacological Reviews* 38(1): 1–20.

Iversen, L. (2000): *The science of marijuana.* Oxford Academic Press, Oxford.

Jentsch, J.D., Andrusiak, E., Tran, A., Bowers, M.B., Jr., and Roth, R.H. (1997): Delta 9-tetrahydrocannabinol increases prefrontal cortical catecholaminergic utilization and impairs spatial working memory in the rat: blockade of dopaminergic effects with HA966 *Neuropsychopharmacology* 16: 426–432.

Johns, A. (2001): Psychiatric effects of cannabis *British Journal of Psychiatry* 178: 116–122.

Joy, J.E., Watson, S.J., and Benson, J.A., Jr. (1999): *Marijuana and Medicine: Assessing the Science Base,* Institute of Medicine, Washington, D.C.

Kaplan, J. (1969): *Marijuana. Report of the Indian Hemp Drugs Commission, 1893–1894.* Thomas Jefferson Publishing Co., Silver Spring, MD.

Leweke, F.M., Giuffrida, A., Wurster, U., Emrich, H.M., and Piomelli, D. (1999): Elevated endogenous cannabinoids in schizophrenia. *Neuroreport* 10(8): 1665–1669.

Li, H.-L. (1974): An archaeological and historical account of cannabis in China. *Economic Botany* 28: 437–448.

Mass, R., Bardong, C., Kindl, K., and Dahme, B. (2001): Relationship between cannabis use, schizotypal traits, and cognitive function in healthy subjects. *Psychopathology* 34(4): 209–214.

McMeens, R.R. (1860): Report of the Ohio State Medical Committee on *Cannabis indica,* pp. 75–100, Ohio State Medical Society, White Sulphur Springs, OH.

Moreau, J.-J. (1845): *Du hachisch et de L'aliénation Mentale: Études Psychologiques*. Fortin Masson, Paris.

Navarro, M., Hernandez, E., Munoz, R.M., del Arco, I., Villanua, M.A., Carrera, M.R., and Rodriguez de Fonseca, F. (1997): Acute administration of the CB1 cannabinoid receptor antagonist SR 141716A induces anxiety-like responses in the rat. *Neuroreport* 8(2): 491–496.

Nunez, L.A. and Gurpegui, M. (2002): Cannabis-induced psychosis: a cross-sectional comparison with acute schizophrenia. *Acta Psychiatrica Scandinavica* 105(3): 173–178.

Parker, L., Mechoulam, R., Burton, P., and Yakiwchuck, C. (2003): Δ^9-Tetrahydrocannabinol and cannabidiol potentiate extinction of cocaine- and amphetamine-induced place preference learning. 2003 Symposium on the Cannabinoids, Burlington, Vermont, International Cannabinoid Research Society, p. 66.

Patton, G.C., Coffey, C., Carlin, J.B., Degenhardt, L., Lynskey, M., and Hall, W. (2002): Cannabis use and mental health in young people: cohort study. *British Medical Journal* 325: 1195–1198.

Phillips, L.J., Curry, C., Yung, A.R., Yuen, H.P., Adlard, S., and McGorry, P.D. (2002): Cannabis use is not associated with the development of psychosis in an 'ultra' high-risk group. *Australian and New Zealand Journal of Psychiatry* 36(6): 800–806.

Pope, H.G., Gruber, A.J., Hudson, J.I., Cohane, G., Huestis, M.A., and Yurgelun-Todd, D. (2003): Early-onset cannabis use and cognitive deficits: what is the nature of the association? *Drug and Alcohol Dependence* 69(3): 303–310.

Pope, H.G., Jr., Gruber, A.J., Hudson, J.I., Huestis, M.A., and Yurgelun-Todd, D. (2001): Neuropsychological performance in long-term cannabis users. *Archives of General Psychiatry* 58(10): 909–915.

Regelson, W., Butler, J.R., Schulz, J., Kirk, T., Peek, L., Green, M.L., and Zalis, M.O. (1976): Delta 9-Tetrahydrocannabinol as an effective antidepressant and appetite-stimulating agent in advanced cancer patients. in Braude M.C., Szara S., Eds. *Pharmacology of Marihuana.* Vol. 2. New York, Raven Press, pp. 763–776.

Regis, E. (2001): Evaluating the threat. *Scientific American*, 285(6): 21–23.

Rodriguez de Fonseca, F., Carrera, M.R., Navarro, M., Koob, G.F., and Weiss, F. (1997): Activation of corticotropin-releasing factor in the limbic system during cannabinoid withdrawal *Science* 276(5321): 2050–2054.

Rodriguez de Fonseca, F., Rubio, P., Menzaghi, F., Merlo-Pich, E., Rivier, J., Koob, G.F., and Navarro, M. (1996): Corticotropin-releasing factor (CRF) antagonist [D-Phe12,Nle21,38,C alpha MeLeu37]CRF attenuates the acute actions of the highly potent cannabinoid receptor agonist HU-210 on defensive-withdrawal behavior in rats *Journal of Pharmacologist and Experimental Therapeutics* 276: 56–64.

Rubin, V.D. and Comitas, L. (1975): *Ganja in Jamaica: A Medical Anthropological Study of Chronic Marihuana Use*. Mouton, The Hague.

Russo, E.B. (2001): *Handbook of Psychotropic Herbs: A Scientific Analysis of Herbal Remedies for Psychiatric Conditions*. Haworth Press, Binghamton, NY.

Russo, E.B. (2004): The history of cannabis as medicine, in *Medicinal Uses of Cannabis and Cannabinoids.* B.A. Whittle, G.W. Guy, and P. Robson, Eds. Pharmaceutical Press, London.

Russo, E.B., Mathre, M.L., Byrne, A., Velin, R., Bach, P.J., Sanchez-Ramos, J., and Kirlin, K.A. (2002): Chronic cannabis use in the compassionate investigational new drug program: an examination of benefits and adverse effects of legal clinical cannabis. *Journal of Cannabis Therapeutics* 2: 3–57.

Skosnik, P.D., Spatz-Glenn, L., and Park, S. (2001): Cannabis use is associated with schizotypy and attentional disinhibition *Schizophrenia Research* 48(1): 83–92.

Solowij, N. (2002) Marijuana and cognitive function. in *The Biology of Marijuana,* E. Onaivi, Ed., Harwood Academic, Reading, MA. pp.

Solowij, N., Stephens, R.S., Roffman, R.A., Babor, T., Kadden, R., Miller, M., Christiansen, K., McRee, B., and Vendetti, J. (2002): Cognitive functioning of long-term heavy cannabis users seeking treatment *Journal of the American Medical Association* 287: 1123–12.

Stefanis, C.N., Dornbush, R.L., and Fink, M. (1977): *Hashish: Studies of Long-Term Use*. Raven Press, New York.

Thierry, A.M., Tassin, J.P., Blanc, G., and Glowinski, J. (1976): Selective activation of mesocortical DA system by stress. *Nature* 263(5574): 242–244.

Thierry, A.M., Tassin, J.P., Blanc, G., and Glowinski, J. (1978): Studies on mesocortical dopamine systems. *Advances in Biochemical Psychopharmacology* 19: 205–216.

Thompson, R.C. (1949): *A Dictionary of Assyrian Botany.* British Academy, London.

Tournier, M., Sorbara, F., Gindre, C., Swendsen, J.D., and Verdoux, H. (2003): Cannabis use and anxiety in daily life: a naturalistic investigation in a non-clinical population. *Psychiatry Research* 118: 1–8.

Verdoux, H., Sorbara, F., Gindre, C., Swendsen, J.D., and van Os, J. (2003): Cannabis use and dimensions of psychosis in a nonclinical population of female subjects. *Schizophrenia Research* 59: 77–84.

Wallace, G.B. and Cunningham, E.V. (1944): *The Marihuana Problem in the City of New York; Sociological, Medical, Psychological and Pharmacological Studies.* New York (NY). Mayor's Committee on Marihuana. The Jaques Cattell Press, Lancaster, PA.

Warner, R., Taylor, D., Wright, J., Sloat, A., Springett, G., Arnold, S., and Weinberg, H. (1994): Substance use among the mentally ill: prevalence, reasons for use, and effects on illness. *American Journal of Orthopsychiatry* 64: 30–39.

Warnock, J. (1903): Insanity from hasheesh. *Journal of Mental Science* 49: 96–100.

Weidenfeld, J., Feldman, S., and Mechoulam, R. (1994): Effect of the brain constituent anandamide, a cannabinoid receptor agonist, on the hypothalamo-pituitary-adrenal axis in the rat. *Neuroendocrinology* 59(2): 110–112.

Zammit, S., Allebeck, P., Andreasson, S., Lundberg, I., and Lewis, G. (2002): Self reported cannabis use as a risk factor for schizophrenia in Swedish conscripts of 1969: historical cohort study. *British Medical Journal* 325: 1199–1203.

Zenor, B.N., Weesner, G.D., and Malven, P.V. (1999): Endocrine and other responses to acute administration of cannabinoid compounds to non-stressed male calves. *Life Sciences* 65: 125–133.

Zuardi, A.W., Cosme, R.A., Graeff, F.G., and Guimaraes, F.S. (1993): Effects of ipsapirone and cannabidiol on human experimental anxiety. *Journal of Psychopharmacology* 7: 82–88.

Zuardi, A.W. and Guimaraes, F.S. (1997): Cannabidiol as an anxiolytic and antipsychotic, in *Cannabis in Medical Practice: A Legal, Historical and Pharmacological Overview of the Therapeutic Use of Marijuana.* M.L. Mathre, Ed. pp. 133–141, McFarland, Jefferson, NC.

16 The Role of Endocannabinoids in the Development, Progression, and Treatment of Neurodegenerative Diseases

Michelle Glass

CONTENTS

INTRODUCTION

Long before the cloning of the first cannabinoid receptor (CB_1) over a decade ago (Matsuda et al., 1990), there were reported uses of cannabis as a therapy for a range of neurological disorders. The discovery of an endogenous cannabinoid system consisting of at least two receptors (Matsuda et al., 1990; Munro et al., 1993) and an ever increasing array of putative endogenous cannabinoids (Devane et al., 1992; Di Marzo et al., 1994; Hanus et al., 2001; Porter et al., 2002), has provided a scientific basis for these therapies, as well as an increasing range of therapeutic targets. It is likely that the major advances in the next 10 years will be improved delivery of highly receptor-selective cannabinoid compounds and selective inhibitors of endocannabinoid synthesis, transport, and metabolism that will enable precise modulation of the endocannabinoid system.

This chapter will focus on the existing evidence that cannabinoids may be useful therapies in neurological disease and will examine the suggestions that modulation of the endocannabinoid system may be contributing to disease processes and symptomology. Neuropsychiatric disorders, such as schizophrenia and anxiety, will be covered elsewhere in this volume; therefore, this chapter will focus on the neurodegenerative disorders of Huntington's disease (HD), Parkinson's disease (PD), and multiple selerosis (MS). In addition to these disorders, many review articles have proposed that cannabinoids may have a potential role in Alzheimer's disease due to their potent effects in inhibiting memory formation. However, to date, little evidence exists to validate this hypothesis; therefore, this disorder will not be further discussed.

MULTIPLE SCLEROSIS

Among all the neurodegenerative diseases discussed within this chapter, perhaps the most compelling evidence for a primary role for cannabinoid-based treatments exists for MS. MS is characterized by a loss of myelin, which normally forms an outer cover or insulation layer around nerve fibers, and axonal loss. This results in an inability of the neurons to conduct nerve impulses efficiently. Demyelination is considered to occur as a consequence of a chronic inflammation (Noseworthy et al., 2000) that may be triggered by an initial viral exposure (Johnson, 1994). The symptoms presenting depend on the location of the myelin and axonal loss, but include painful muscle spasms, tremor, ataxia, weakness or paralysis, difficulty speaking, constipation, and loss of bladder control. The clinical symptoms often fluctuate unpredictably and worsen with age. There is an accumulating body of evidence to suggest that cannabis may be effective in suppressing the symptoms of pain and spasticity associated with this disorder.

For decades, there have been anecdotal reports of MS patients benefiting from cannabis use, more recently supported by a survey of patients believed to be self-medicating with cannabis (Consroe et al., 1997). The survey found a consistently reported improvement in most types of pain and spasticity following cannabis ingestion. These reports are supported by a small number of clinical trials, generally on small numbers of patients. Clinical trials were carried out with orally administered THC (Petro and Ellenberger, 1981; Clifford, 1983; Ungerleider et al., 1987; Killestein et al., 2002), orally administered nabilone (a synthetic cannabinoid agonist) (Martyn et al., 1995; Hamann and di Vadi, 1999), or inhaled cannabis (Meinck et al., 1989; Schon et al., 1999). Generally, these trials have found that patients reported improvements in pain reduction as well as objectively measurable improvements in spasticity, rigidity, and tremor. Several phase 2 and phase 3 clinical trials assessing the safety, tolerability, and effectiveness of inhaled THC and THC:CBD (1:1), have recently been completed in larger numbers of patients (http://www.gwpharm.com/corp_glan_phas_2.html). It is likely that publication of these studies in peer-reviewed journals will shed more light on the effectiveness of cannabinoid therapies in these disorders. The sublingual spray developed by GW Pharmaceuticals provides a significant step towards well-designed clinical trials, as orally administered cannabinoids have been hampered by highly variable bioavailability in patients, with slow unreliable absorption followed by hepatic first pass metabolism to active and inactive metabolites.

As cannabinoids appear be effective in the treatment of symptoms of the disease, significant work has been carried out utilizing animal models to establish whether the endocannabinoid system may be compromised by or contributing to the disease process. Murine models of MS are based on the inflammatory or immune nature of the disorder and include virally induced CNS demyelination in animals (e.g., Theilers murine encephalomyelitis virus (TMEV) (Dal Canto and Lipton, 1977) or inflammatory responses induced by sensitization to myelin antigens, leading to either acute (EAE) or chronic relapsing models (CREAE) (Baker et al., 1990). THC has been demonstrated to delay the onset and significantly reduce clinical signs of EAE in rodents (Lyman et al., 1989; Wirguin et al., 1994), probably by generating immunosuppression, which prevents the conditions that lead to the development of paralysis, rather than having a direct effect on the paralysis itself. A significant development in the field occurred when Baker et al. (2000) developed a CREAE model in Biozzi ABH mice, who in addition to developing demyelination and axonal loss, also developed spasticity and tremor, enabling the investigation of the effects of cannabinoids on these behavioral symptoms. The authors reported that limb spasticity and tremor could be readily suppressed by the cannabinoid receptor agonist WIN55,212 and THC. Suppression of spasticity and tremor was apparently achieved by a combined CB_1/CB_2-receptor-mediated mechanism because pretreatment with either receptor antagonist led to attenuation of the agonist effect. Although CB_2 receptors are clearly involved in immune function, they are not believed to be localized in the CNS, so the mechanism by which they influence CREAE is unclear. A "CB_2-like" receptor has been suggested by several studies (Calignano et al., 1998; Calignano et al., 2001). This receptor appears to respond to palmitoylethanolamide and the "CB_2" antagonist SR144528. Therefore, the receptor may be responsible for the effects observed by Baker et al. (2000) as they found palmitoylethanolamide to be effective in reducing spasticity, despite this ligand not activating either CB_1 or CB_2 receptors (Showalter et al., 1996; Sugiura et al., 2000). This area requires further investigation.

An intriguing aspect of the Baker study was the finding that the cannabinoid antagonists alone could increase spasticity in mice. Both of these compounds have previously been demonstrated to be inverse agonists at their respective receptors (Rinaldi-Carmona et al., 1998; Bouaboula et al., 1999); (Coutts et al., 2000). Thus, it was proposed that this effect might be mediated by blocking constitutively active receptors or, alternatively, by antagonizing the effect of the endogenous cannabinoids. Consistent with the latter hypothesis, spastic CREAE mice have elevated concentrations of the endocannabinoids anandamide and 2-arachidonyl glycerol in their brains and spinal cords, and increased palmitoylethanolamide in their spinal cords (Baker et al., 2001). Furthermore, inhibitors of endocannabinoid membrane transport and hydrolysis have been demonstrated to ameliorate spasticity in these animals (Baker et al., 2001). In contrast to the apparent hyperactive system suggested by these studies, CB_1 receptors have been suggested to be decreased in motor pathways of CREAE animals (Berrendero et al., 2001).

Two recent studies have examined the impact of cannabinoids on Theilers virus infection induced demyelination (Arevalo-Martin et al., 2003; Croxford and Miller, 2003). These studies reported a functional recovery even following disease establishment. Cannabinoids were reported to reduce microglial activation and diminish inflammation in a manner that paralleled remyelination (Arevalo-Martin et al., 2003). The CB_2 selective agonist JWH015 also produced similar effects (Arevalo-Martin et al., 2003), suggesting that such processes may have been regulated by these receptors. However, this compound does exhibit low affinity for CB_1 and, therefore, future studies are required to understand the mechanism of this effect. Treatment of primary astrocytes from 1-d-old postnatal mouse cerebral cortex, infected with TMEV, with anandamide has previously been demonstrated to induce production of interleukin-6 by a CB_1-mediated process (Molina-Holgado et al., 1998). Interleukin-6 has been suggested to have neuroprotective properties (see Molina-Holgado et al., 1998) and to reduce inflammation and demyelination in TMEV-infected animals (Rodriguez et al., 1994), potentially providing a mechanism for cannabinoid-mediated protection. A final complicating factor in the investigation of the mechanism of cannabinoid-induced modulation of spasticity comes from a recent study that demonstrated that Arvanil, a structural hybrid between capsaicin and anandamide (Brooks et al., 2002), produced potent inhibition of spasticity in CREAE animals in a noncannabinoid, nonvanilloid receptor-mediated mechanism, suggesting that an alternative target may exist for these compounds.

Thus, the evidence for cannabinoid involvement in both the progression and treatment of MS is compelling if not conclusive. Further studies are clearly warranted, both to investigate suitable therapeutic strategies (synthetic receptor selective agonists and alternative forms of delivery of THC) and to understand the mechanisms by which cannabinoids produce their apparent protection and improvement in this condition.

CANNABINOIDS AND MOVEMENT DISORDERS

Cannabinoid receptors in the brain appear to be intimately involved in the control of movement. For example, the administration of plant-derived, synthetic, or endogenous cannabinoids produces a complex pattern of movement disorders with low-dose cannabinoids producing hyperactivity, whereas high doses produce a marked catalepsy. This effect is paralleled by changes in the levels of a range of neurotransmitters in basal ganglia, including GABA, dopamine, and glutamate (Romero et al., 1995; Maneuf et al., 1996a,b; Romero et al., 1998; Szabo et al., 2000), all of which are important in the control of movement. These actions of cannabinoids are mediated by cannabinoid CB_1 receptors, which are densely localized in the basal ganglia (Herkenham et al., 1991; Glass et al., 1997). Specifically, CB_1 receptors are located presynaptically on striatal GABAergic projections to the substantia nigra, globus pallidus, and entopenduncular nucleus (Herkenham et al., 1991). In addition, CB_1 receptors are localized to the terminals of the subthalamo-nigral glutamatergic pathway (Mailleux and Vanderhaeghen, 1992). Therefore, the CB_1 receptors are colocalized with both D1 and D2 dopamine receptors and show complex patterns of interactions with dopamine, a critical neuromodulator in the control of movement (Glass and Felder, 1997; Giuffrida et al., 1999; Beltramo et al., 2000; Meschler and Howlett, 2001).

Systemic cannabinoid exposure produces an overall inhibition of movement. This may, however, be misleading in the determination of the role of endogenous cannabinoids in the control of movement, because studies have clearly shown that cannabinoids can produce markedly different effects depending on which part of the basal ganglia they are activating (Sanudo-Pena et al., 1996; Miller and Walker, 1998; Sanudo-Pena et al., 1998a; Sanudo-Pena and Walker, 1998a). Cannabinoids activate movement when they are microinjected into the striatum (Sanudo-Pena et al., 1998a) or substantia nigra reticulata (Sanudo-Pena et al., 1996) and inhibit movement when microinjected into the globus pallidus (Sanudo-Pena and Walker, 1998a) or into the subthalamic nucleus (Miller and Walker, 1998).

Cannabinoids and Parkinson's Disease

Parkinson's disease (PD) is caused by degeneration of the dopaminergic projection from the substantia nigra to the striatum. Thus, dopamine replacement therapy, predominantly in the form of levodopa (L-DOPA), has been the mainstay of PD treatment for decades. PD is a progressive movement disorder characterized by muscle rigidity, tremor, and impaired ability to initiate movement. Given both the interactions of cannabinoids with dopamine and their ability to modulate movement, extensive research has been carried out investigating the putative role of cannabinoids in the disease symptomology and potential therapy. Several studies have examined cannabinoid receptor distribution and activity in animal models of PD and in humans affected by PD. These studies have suggested that the endocannabinoid system may be modified by loss of dopaminergic input to the striatum. Increased CB_1 receptor-specific binding and mRNA expression have been demonstrated in the striatum of PD patients, MPTP-treated marmosets (Lastres-Becker et al., 2001), and rats with unilateral lesions of the nigral dopaminergic neurons caused by 6-hydroxydopamine (6OHDA) (Romero et al., 2000). However, other studies in 6OHDA depleted rats have found no alterations in CB_1 receptors (Zeng et al., 1999; Gubellini et al., 2002). Intriguingly, the therapy of dopaminergic replacement with L-DOPA has been demonstrated to diminish the increase in CB_1 receptors observed in MPTP-treated primates (Lastres-Becker et al., 2001).

Although increased receptor expression is often a compensatory response to decreased levels of endogenous ligand, recent studies have demonstrated that the levels of the endocannabinoids are increased in the basal ganglia of 6OHDA-lesioned (Gubellini et al., 2002; Maccarrone et al., 2003) and reserpinized rats (Di Marzo et al., 2000) with a corresponding decrease in FAAH activity (Gubellini et al., 2002; Maccarrone et al., 2003). Chronic treatment of 6OHDA-treated rats with L-DOPA completely reversed the anomalies in the endocannabinoid components. Furthermore, enhancement of glutamatergic transmission in these animals was restored by pharmacological inhibition of FAAH in 6OHDA-treated rats (Maccarrone et al., 2003). Taken together with the changes in CB_1 receptors being returned to normal in L-DOPA-treated animals (Lastres-Becker et al., 2001), these studies suggest that the disruption of the endocannabinoid system may be contributing to the symptomology observed in these animals and that protection or restoration of the cannabinoid system may be beneficial in treating this disorder.

Whether the altered receptor and endocannabinoid expression in the striatum reflects compensation for the loss of dopaminergic input or itself contributes to the symptomology is not clear. The complex behavioral responses to cannabinoids make this hard to predict; an increase in cannabinoids in the striatum may be expected to increase movement (Sanudo-Pena et al., 1998a), implying a compensatory mechanism, but studies to date have not investigated whether this alteration is matched by further changes in the receptor or endogenous ligand level in the remainder of the circuitry of the basal ganglia. This may prove to be a critical point, as the influence of cannabinoids in the different nuclei has been demonstrated to be differentially altered in 6OHDA-lesioned animals (Sanudo-Pena et al., 1998b; Sanudo-Pena and Walker, 1998b), suggesting that the relative contribution of the individual nuclei following systemic administration may be altered by the disease state.

Given that the overall effect of systemic cannabinoids is to reduce movement, many researchers have proposed that the cannabinoid compounds of therapeutic value in PD would be receptor antagonists rather than agonists (Brotchie, 1998; Consroe, 1998). In agreement with this, CB_1 antagonism was capable of restoring locomotion in reserpinized rats when injected in combination with D2 receptor agonists (Di Marzo et al., 2000). Denervated striata express a dramatic overactivity of glutamatergic transmission (Calabresi et al., 1993; Tang et al., 2001), reflecting a loss of D2-receptor-mediated control of corticostriatal transmission (Cepeda et al., 2001). A recent study (Gubellini et al., 2002) has demonstrated that pharmacological inhibition of FAAH, leading to increased endocannabinoids, produces a much stronger depression of striatal glutamatergic activity, restoring normal corticostriatal function in lesioned compared to naïve rats. This study, therefore, implies that compounds which increase the brain concentration of endocannabinoids may be beneficial in PD. This apparent paradox may arise due to the distribution of FAAH in the basal ganglia. Although an in-depth analysis of FAAH localization in the basal ganglia has not yet been published, one study has suggested moderate expression in the rat striatum with fairly low expression in other basal ganglia nuclei (Tsou et al., 1998). Thus, drugs that target FAAH may preferentially increase endocannabinoids in the striatum over other nuclei.

A major adverse effect of long-term L-DOPA therapy is the development of dyskinesia. Cannabinoid agonists have been proposed as an adjunct to L-DOPA therapy capable of reducing or delaying the incidence of dyskinesia. This has been tested recently in reserpine-treated rats (Segovia et al., 2003), with complex results as both agonists and antagonists modulated behaviors induced by long-term L-DOPA treatment. However, encouragingly, a recent pilot study in humans has suggested that the cannabinoid agonist nabilone significantly improved dyskinesia in humans (Sieradzan et al., 2001), and likewise nabilone treatment in conjunction with L-DOPA in an MPTP nonhuman primate model was associated with significantly less total dyskinesia than L-DOPA treatment alone (Fox et al., 2002). Clearly, further trials in humans may help to establish the usefulness of cannabinoids as treatment strategies in humans.

HUNTINGTON'S DISEASE

HD is an inherited, autosomal dominant disorder in which there is progressive neurodegeneration, affecting particularly the basal ganglia and cerebral cortex of the brain that invariably leads to death. The disease mutation consists of an unstable expanded CAG trinucleotide repeat of the 5′ coding region of the huntingtin gene that encodes a stretch of polyglutamines (The Huntington's Disease Collaborative Research Group, 1993). How this aberrant protein then leads to the characteristic HD pattern of cell death, predominantly of the striatal medium spiny neurons, is the subject of intense investigation (Reddy et al., 1999). The disease is characterized by motor disturbances such as chorea, dystonia, and cognitive decline.

In humans suffering from HD, one of the earliest neurochemical alterations observed is the loss of cannabinoid CB_1 receptors from the basal ganglia nuclei (Glass et al., 1993; Richfield and Herkenham, 1994; Glass et al., 2000). Although this receptor loss is consistent with the localization of the CB_1 receptor on striatal projection neurons that degenerate in the disease progress, loss of CB_1 receptors appears to occur prior to terminal degeneration, or loss of colocalized dopamine receptors or neuropeptides (Glass et al., 2000).

Numerous animal models exist for HD. A pattern of cell death similar to that observed in the disease progress can be induced by chemical lesion either with quinolinic acid (Beal et al., 1986) or 3-nitropropioinic acid (3-NP) (Beal et al., 1993). The discovery of the HD gene has further led to the development of transgenic mice containing a portion of an abnormal human HD chromosome (Mangiarini et al., 1996; Yamamoto et al., 2000). Several of these mouse lines develop a progressive abnormal neurological phenotype, characterized by increasingly poor motor coordination, loss of body weight, and neuronal inclusions (Mangiarini et al., 1996; Davies et al., 1997; Sathasivam et al., 1999) consistent with human HD. Consistent with the early loss of CB_1 receptor in humans, loss of CB_1 receptor mRNA has been observed in the striatum of transgenic HD mice prior to the onset of motor-related HD-like symptoms in these mice and preceding neural degeneration (Denovan-Wright and Robertson, 2000; Lastres-Becker et al., 2002a). Similar results have been observed following local (Lastres-Becker et al., 2002b) or systemic 3-NP (Page et al., 2000) injection.

Exposure of transgenic mice to a stimulating, enriched environment from an early age helps to prevent the loss of cerebral volume and delays the onset of motor disorders by an as yet undefined mechanism in R6/1 (van Dellen et al., 2000) and R6/2 (Hockly et al., 2002) lines of HD mice. Other changes in housing conditions and handling can also affect survival of the R6/2 mice (Carter et al., 2000). Recently, it was demonstrated that HD mice exposed to a stimulating environment show equivalent loss of D1 and D2 receptors as their "nonenriched" counterparts; in contrast, these animals show significantly less depletion of CB_1 receptors (Glass et al., in press). Thus, there may be a correlation between CB_1 receptor loss and cell death that requires future investigation.

CONCLUSION

A wealth of evidence now exists for putative roles of cannabinoids in neurodegenerative disease. Too often, though, these are small, isolated, and often contradictory studies. The need exists to focus on larger, fully controlled clinical trials and in-depth mechanistic studies into the role of cannabinoids in neurological disease. Nevertheless, the existing studies paint an interesting picture of a neuromodulator that is intricately involved in the control of mood and movement and a potentially exciting target for future therapeutic exploitation.

REFERENCES

Arevalo-Martin, A., J.M. Vela, E. Molina-Holgado, J. Borrell, and C. Guaza (2003). Therapeutic action of cannabinoids in a murine model of multiple sclerosis. *J Neurosci* **23**(7): 2511–2516.

Baker, D., J.K. O'Neill, S.E. Gschmeissner, C.E. Wilcox, C. Butter, and J.L. Turk (1990). Induction of chronic relapsing experimental allergic encephalomyelitis in Biozzi mice. *J Neuroimmunol* **28**(3): 261–270.

Baker, D., G. Pryce, J.L. Croxford, P. Brown, R.G. Pertwee, J.W. Huffman, and L. Layward (2000). Cannabinoids control spasticity and tremor in a multiple sclerosis model. *Nature* **404**(6773): 84–87.

Baker, D., G. Pryce, J.L. Croxford, P. Brown, R.G. Pertwee, A. Makriyannis, A. Khanolkar, L. Layward, F. Fezza, T. Bisogno, and V. Di Marzo (2001). Endocannabinoids control spasticity in a multiple sclerosis model. *FASEB J* **15**(2): 300–302.

Beal, M.F., E. Brouillet, B.G. Jenkins, R.J. Ferrante, N.W. Kowall, J.M. Miller, E. Storey, R. Srivastava, B.R. Rosen, and B.T. Hyman (1993). Neurochemical and histologic characterization of striatal excitotoxic lesions produced by the mitochondrial toxin 3-nitropropionic acid. *J Neurosci* **13**(10): 4181–4192.

Beal, M.F., N.W. Kowall, D.W. Ellison, M.F. Mazurek, K.J. Swartz, and J.B. Martin (1986). Replication of the neurochemical characteristics of Huntington's disease by quinolinic acid. *Nature* **321**(6066): 168–171.

Beltramo, M., F.R. de Fonseca, M. Navarro, A. Calignano, M.A. Gorriti, G. Grammatikopoulos, A.G. Sadile, A. Giuffrida, and D. Piomelli (2000). Reversal of dopamine D(2) receptor responses by an anandamide transport inhibitor. *J Neurosci* **20**(9): 3401–3407.

Berrendero, F., A. Sanchez, A. Cabranes, C. Puerta, J. A. Ramos, A. Garcia-Merino, and J. Fernandez-Ruiz (2001). Changes in cannabinoid CB(1) receptors in striatal and cortical regions of rats with experimental allergic encephalomyelitis, an animal model of multiple sclerosis. *Synapse* **41**(3): 195–202.

Bouaboula, M., N. Desnoyer, P. Carayon, T. Combes, and P. Casellas (1999). Gi protein modulation induced by a selective inverse agonist for the peripheral cannabinoid receptor CB2: implication for intracellular signalization cross-regulation. *Mol Pharmacol* **55**(3): 473–480.

Brooks, J.W., G. Pryce, T. Bisogno, S.I. Jaggar, D.J. Hankey, P. Brown, D. Bridges, C. Ledent, M. Bifulco, A.S. Rice, V. Di Marzo, and D. Baker (2002). Arvanil-induced inhibition of spasticity and persistent pain: evidence for therapeutic sites of action different from the vanilloid VR1 receptor and cannabinoid CB(1)/CB(2) receptors. *Eur J Pharmacol* **439**(1–3): 83–92.

Brotchie, J.M. (1998). Adjuncts to dopamine replacement: a pragmatic approach to reducing the problem of dyskinesia in Parkinson's disease. *Mov Disord* **13**(6): 871–876.

Calabresi, P., N.B. Mercuri, G. Sancesario, and G. Bernardi (1993). Electrophysiology of dopamine-denervated striatal neurons. Implications for Parkinson's disease. *Brain* **116**(Pt. 2): 433–452.

Calignano, A., G. La Rana, A. Giuffrida, and D. Piomelli (1998). Control of pain initiation by endogenous cannabinoids. *Nature* **394**(6690): 277–281.

Calignano, A., G. La Rana, and D. Piomelli (2001). Antinociceptive activity of the endogenous fatty acid amide, palmitylethanolamide. *Eur J Pharmacol* **419**(2-3): 191–198.

Carter, R.J., M.J. Hunt, and A.J. Morton (2000). Environmental stimulation increases survival in mice transgenic for exon 1 of the Huntington's disease gene. *Mov Disord* **15**(5): 925–937.

Cepeda, C., R.S. Hurst, K.L. Altemus, J. Flores-Hernandez, C.R. Calvert, E.S. Jokel, D.K. Grandy, M.J. Low, M. Rubinstein, M.A. Ariano, and M.S. Levine (2001). Facilitated glutamatergic transmission in the striatum of D2 dopamine receptor-deficient mice. *J Neurophysiol* **85**(2): 659–670.

Clifford, D.B. (1983). Tetrahydrocannabinol for tremor in multiple sclerosis. *Ann Neurol* **13**(6): 669–671.

Consroe, P. (1998). Brain cannabinoid systems as targets for the therapy of neurological disorders. *Neurobiol Dis* **5**(6 Pt. B): 534–551.

Consroe, P., R. Musty, J. Rein, W. Tillery, and R. Pertwee (1997). The perceived effects of smoked cannabis on patients with multiple sclerosis. *Eur Neurol* **38**(1): 44–48.

Coutts, A.A., N. Brewster, T. Ingram, R.K. Razdan, and R.G. Pertwee (2000). Comparison of novel cannabinoid partial agonists and SR141716A in the guinea-pig small intestine. *Br J Pharmacol* **129**(4): 645–652.

Croxford, J.L. and S.D. Miller (2003). Immunoregulation of a viral model of multiple sclerosis using the synthetic cannabinoid R+WIN55,212. *J Clin Invest* **111**(8): 1231–1240.

Dal Canto, M.C. and H.L. Lipton (1977). Multiple sclerosis. Animal model: Theiler's virus infection in mice. *Am J Pathol* **88**(2): 497–500.

Davies, S.W., M. Turmaine, B.A. Cozens, M. DiFiglia, A.H. Sharp, C.A. Ross, E. Scherzinger, E.E. Wanker, L. Mangiarini, and G.P. Bates (1997). Formation of neuronal intranuclear inclusions underlies the neurological dysfunction in mice transgenic for the HD mutation. *Cell* **90**(3): 537–548.

Denovan-Wright, E.M. and H.A. Robertson (2000). Cannabinoid receptor messenger RNA levels decrease in a subset of neurons of the lateral striatum, cortex and hippocampus of transgenic Huntington's disease mice. *Neuroscience* **98**(4): 705–713.

Devane, W. A., L. Hanus, A. Breuer, R.G. Pertwee, L.A. Stevenson, G. Griffin, D. Gibson, A. Mandelbaum, A. Etinger, and R. Mechoulam (1992). Isolation and structure of a brain constituent that binds to the cannabinoid receptor. *Science* **258**(5090): 1946–1949.

Di Marzo, V., A. Fontana, H. Cadas, S. Schinelli, G. Cimino, J.C. Schwartz, and D. Piomelli (1994). Formation and inactivation of endogenous cannabinoid anandamide in central neurons. *Nature* **372**(6507): 686–691.

Di Marzo, V., M.P. Hill, T. Bisogno, A.R. Crossman, and J.M. Brotchie (2000). Enhanced levels of endogenous cannabinoids in the globus pallidus are associated with a reduction in movement in an animal model of Parkinson's disease. *FASEB J* **14**(10): 1432–1438.

Fox, S.H., M. Kellett, A.P. Moore, A.R. Crossman, and J.M. Brotchie (2002). Randomised, double-blind, placebo-controlled trial to assess the potential of cannabinoid receptor stimulation in the treatment of dystonia. *Mov Disord* **17**(1): 145–149.

Giuffrida, A., L.H. Parsons, T.M. Kerr, F. Rodriguez de Fonseca, M. Navarro, and D. Piomelli (1999). Dopamine activation of endogenous cannabinoid signaling in dorsal striatum. *Nat Neurosci* **2**(4): 358–363.

Glass, M., M. Dragunow, and R.L. Faull (1997). Cannabinoid receptors in the human brain: a detailed anatomical and quantitative autoradiographic study in the fetal, neonatal and adult human brain. *Neuroscience* **77**(2): 299–318.

Glass, M., M. Dragunow, and R.L. Faull (2000). The pattern of neurodegeneration in Huntington's disease: a comparative study of cannabinoid, dopamine, adenosine and GABA(A) receptor alterations in the human basal ganglia in Huntington's disease. *Neuroscience* **97**(3): 505–519.

Glass, M., R.L. Faull, and M. Dragunow (1993). Loss of cannabinoid receptors in the substantia nigra in Huntington's disease. *Neuroscience* **56**(3): 523–527.

Glass, M. and C.C. Felder (1997). Concurrent stimulation of cannabinoid CB1 and dopamine D2 receptors augments cAMP accumulation in striatal neurons: evidence for a Gs linkage to the CB1 receptor. *J Neurosci* **17**(14): 5327–5333.

Glass, M., A. van Dellen, C. Blakemore, A.J. Hannan, and R.L.M. Faull (2004). Delayed onset of Huntington's disease in mice in an enriched environment correlates with delayed loss of Cannabinoid CB1 receptors. *Neuroscience* **123**: 207–212.

Gubellini, P., B. Picconi, M. Bari, N. Battista, P. Calabresi, D. Centonze, G. Bernardi, A. Finazzi-Agro, and M. Maccarrone (2002). Experimental parkinsonism alters endocannabinoid degradation: implications for striatal glutamatergic transmission. *J Neurosci* **22**(16): 6900–6907.

Hamann, W. and P.P. di Vadi (1999). Analgesic effect of the cannabinoid analogue nabilone is not mediated by opioid receptors. *Lancet* **353**(9152): 560.

Hanus, L., S. Abu-Lafi, E. Fride, A. Breuer, Z. Vogel, D.E. Shalev, I. Kustanovich, and R. Mechoulam (2001). 2-arachidonyl glyceryl ether, an endogenous agonist of the cannabinoid CB1 receptor. *Proc Natl Acad Sci U S A* **98**(7): 3662–3665.

Herkenham, M., A.B. Lynn, B.R. de Costa, and E.K. Richfield (1991). Neuronal localization of cannabinoid receptors in the basal ganglia of the rat. *Brain Res* **547**(2): 267–274.

Hockly, E., P.M. Cordery, B. Woodman, A. Mahal, A. van Dellen, C. Blakemore, C.M. Lewis, A.J. Hannan, and G.P. Bates (2002). Environmental enrichment slows disease progression in R6/2 Huntington's disease mice. *Ann Neurol* **51**(2): 235–242.

Johnson, R.T. (1994). The virology of demyelinating diseases. *Ann Neurol* **36** Suppl: S54–60.

Killestein, J., E.L. Hoogervorst, M. Reif, N.F. Kalkers, A.C. Van Loenen, P.G. Staats, R.W. Gorter, B.M. Uitdehaag, and C.H. Polman (2002). Safety, tolerability, and efficacy of orally administered cannabinoids in MS. *Neurology* **58**(9): 1404–1407.

Lastres-Becker, I., F. Berrendero, J.J. Lucas, E. Martin-Aparicio, A. Yamamoto, J.A. Ramos, and J.J. Fernandez-Ruiz (2002a). Loss of mRNA levels, binding and activation of GTP-binding proteins for cannabinoid CB(1) receptors in the basal ganglia of a transgenic model of Huntington's disease. *Brain Res* **929**(2): 236–242.

Lastres-Becker, I., H.H. Hansen, F. Berrendero, R. De Miguel, A. Perez-Rosado, J. Manzanares, J.A. Ramos, and J. Fernandez-Ruiz (2002b). Alleviation of motor hyperactivity and neurochemical deficits by endocannabinoid uptake inhibition in a rat model of Huntington's disease. *Synapse* **44**(1): 23–35.

Lastres-Becker, I., M. Cebeira, M.L. de Ceballos, B.Y. Zeng, P. Jenner, J.A. Ramos, and J.J. Fernandez-Ruiz (2001). Increased cannabinoid CB1 receptor binding and activation of GTP-binding proteins in the basal ganglia of patients with Parkinson's syndrome and of MPTP-treated marmosets. *Eur J Neurosci* **14**(11): 1827–1832.

Lyman, W.D., J.R. Sonett, C.F. Brosnan, R. Elkin, and M.B. Bornstein (1989). Delta 9-tetrahydrocannabinol: a novel treatment for experimental autoimmune encephalomyelitis. *J Neuroimmunol* **23**(1): 73–81.

Maccarrone, M., P. Gubellini, M. Bari, B. Picconi, N. Battista, D. Centonze, G. Bernardi, A. Finazzi-Agro, and P. Calabresi (2003). Levodopa treatment reverses endocannabinoid system abnormalities in experimental parkinsonism. *J Neurochem* **85**(4): 1018–1025.

Mailleux, P. and J.J. Vanderhaeghen (1992). Distribution of neuronal cannabinoid receptor in the adult rat brain: a comparative receptor binding radioautography and in situ hybridization histochemistry. *Neuroscience* **48**(3): 655–668.

Maneuf, Y.P., A.R. Crossman, and J.M. Brotchie (1996a). Modulation of GABAergic transmission in the globus pallidus by the synthetic cannabinoid WIN 55,212-2. *Synapse* **22**(4): 382–385.

Maneuf, Y.P., J.E. Nash, A.R. Crossman, and J.M. Brotchie (1996b). Activation of the cannabinoid receptor by delta 9-tetrahydrocannabinol reduces gamma-aminobutyric acid uptake in the globus pallidus. *Eur J Pharmacol* **308**(2): 161–164.

Mangiarini, L., K. Sathasivam, M. Seller, B. Cozens, A. Harper, C. Hetherington, M. Lawton, Y. Trottier, H. Lehrach, S.W. Davies, and G.P. Bates (1996). Exon 1 of the HD gene with an expanded CAG repeat is sufficient to cause a progressive neurological phenotype in transgenic mice. *Cell* **87**(3): 493–506.

Martyn, C.N., L.S. Illis, and J. Thom (1995). Nabilone in the treatment of multiple sclerosis. *Lancet* **345**(8949): 579.

Matsuda, L.A., S.J. Lolait, M.J. Brownstein, A.C. Young, and T.I. Bonner (1990). Structure of a cannabinoid receptor and functional expression of the cloned cDNA. *Nature* **346**(6284): 561–564.

Meinck, H.M., P.W. Schonle, and B. Conrad (1989). Effect of cannabinoids on spasticity and ataxia in multiple sclerosis. *J Neurol* **236**(2): 120–122.

Meschler, J.P. and A.C. Howlett (2001). Signal transduction interactions between CB1 cannabinoid and dopamine receptors in the rat and monkey striatum. *Neuropharmacology* **40**(7): 918–926.

Miller, A.S. and J.M. Walker (1998). Local effects of cannabinoids on spontaneous activity and evoked inhibition in the globus pallidus. *Eur J Pharmacol* **352**(2-3): 199–205.

Molina-Holgado, F., E. Molina-Holgado, and C. Guaza (1998). The endogenous cannabinoid anandamide potentiates interleukin-6 production by astrocytes infected with Theiler's murine encephalomyelitis virus by a receptor-mediated pathway. *FEBS Lett* **433**(1-2): 139–142.

Munro, S., K.L. Thomas, and M. Abu-Shaar (1993). Molecular characterization of a peripheral receptor for cannabinoids. *Nature* **365**(6441): 61–65.

Noseworthy, J.H., C. Lucchinetti, M. Rodriguez, and B.G. Weinshenker (2000). Multiple sclerosis. *N Engl J Med* **343**(13): 938–952.

Page, K.J., L. Besret, M. Jain, E.M. Monaghan, S.B. Dunnett, and B.J. Everitt (2000). Effects of systemic 3-nitropropionic acid-induced lesions of the dorsal striatum on cannabinoid and mu-opioid receptor binding in the basal ganglia. *Exp Brain Res* **130**(2): 142–150.

Petro, D.J. and C. Ellenberger, Jr. (1981). Treatment of human spasticity with delta 9-tetrahydrocannabinol. *J Clin Pharmacol* **21**(8-9 Suppl): 413S–416S.

Porter, A.C., J.M. Sauer, M.D. Knierman, G.W. Becker, M.J. Berna, J. Bao, G.G. Nomikos, P. Carter, F.P. Bymaster, A.B. Leese, and C.C. Felder (2002). Characterization of a novel endocannabinoid, virodhamine, with antagonist activity at the CB1 receptor. *J Pharmacol Exp Ther* **301**(3): 1020–1024.

Reddy, P.H., M. Williams, and D.A. Tagle (1999). Recent advances in understanding the pathogenesis of Huntington's disease. *Trends Neurosci* **22**(6): 248–255.

Richfield, E.K. and M. Herkenham (1994). Selective vulnerability in Huntington's disease: preferential loss of cannabinoid receptors in lateral globus pallidus. *Ann Neurol* **36**(4): 577–584.

Rinaldi-Carmona, M., A. Le Duigou, D. Oustric, F. Barth, M. Bouaboula, P. Carayon, P. Casellas and G. Le Fur (1998). Modulation of CB1 cannabinoid receptor functions after a long-term exposure to agonist or inverse agonist in the Chinese hamster ovary cell expression system. *J Pharmacol Exp Ther* **287**(3): 1038–1047.

Rodriguez, M., K.D. Pavelko, C.W. McKinney, and J.L. Leibowitz (1994). Recombinant human IL-6 suppresses demyelination in a viral model of multiple sclerosis. *J Immunol* **153**(8): 3811–3821.

Romero, J., F. Berrendero, A. Perez-Rosado, J. Manzanares, A. Rojo, J.J. Fernandez-Ruiz, J.G. de Yebenes, and J.A. Ramos (2000). Unilateral 6-hydroxydopamine lesions of nigrostriatal dopaminergic neurons increased CB1 receptor mRNA levels in the caudate-putamen. *Life Sci* **66**(6): 485–494.

Romero, J., R. de Miguel, J.A. Ramos, and J.J. Fernandez-Ruiz (1998). The activation of cannabinoid receptors in striatonigral GABAergic neurons inhibited GABA uptake. *Life Sci* **62**(4): 351–363.

Romero, J., L. Garcia, M. Cebeira, D. Zadrozny, J.J. Fernandez-Ruiz, and J.A. Ramos (1995). The endogenous cannabinoid receptor ligand, anandamide, inhibits the motor behavior: role of nigrostriatal dopaminergic neurons. *Life Sci* **56**(23-24): 2033–2040.

Sanudo-Pena, M.C., M. Force, K. Tsou, A.S. Miller, and J.M. Walker (1998a). Effects of intrastriatal cannabinoids on rotational behavior in rats: interactions with the dopaminergic system. *Synapse* **30**(2): 221–226.

Sanudo-Pena, M.C., S.L. Patrick, S. Khen, R.L. Patrick, K. Tsou, and J.M. Walker (1998b). Cannabinoid effects in basal ganglia in a rat model of Parkinson's disease. *Neurosci Lett* **248**(3): 171–174.

Sanudo-Pena, M.C., S.L. Patrick, R.L. Patrick, and J.M. Walker (1996). Effects of intranigral cannabinoids on rotational behavior in rats: interactions with the dopaminergic system. *Neurosci Lett* **206**(1): 21–24.

Sanudo-Pena, M.C. and J.M. Walker (1998a). Effects of intrapallidal cannabinoids on rotational behavior in rats: interactions with the dopaminergic system. *Synapse* **28**(1): 27–32.

Sanudo-Pena, M.C. and J.M. Walker (1998b). A novel neurotransmitter system involved in the control of motor behavior by the basal ganglia. *Ann N Y Acad Sci* **860**: 475–479.

Sathasivam, K., C. Hobbs, L. Mangiarini, A. Mahal, M. Turmaine, P. Doherty, S.W. Davies, and G.P. Bates (1999). Transgenic models of Huntington's disease. *Philos Trans R Soc Lond B Biol Sci* **354**(1386): 963–969.

Schon, F., P.E. Hart, T.L. Hodgson, A.L. Pambakian, M. Ruprah, E.M. Williamson, and C. Kennard (1999). Suppression of pendular nystagmus by smoking cannabis in a patient with multiple sclerosis. *Neurology* **53**(9): 2209–2210.

Segovia, G., F. Mora, A.R. Crossman, and J.M. Brotchie (2003). Effects of CB1 cannabinoid receptor modulating compounds on the hyperkinesia induced by high-dose levodopa in the reserpine-treated rat model of Parkinson's disease. *Mov Disord* **18**(2): 138–149.

Showalter, V.M., D.R. Compton, B.R. Martin, and M.E. Abood (1996). Evaluation of binding in a transfected cell line expressing a peripheral cannabinoid receptor (CB2): identification of cannabinoid receptor subtype selective ligands. *J Pharmacol Exp Ther* **278**(3): 989–999.

Sieradzan, K.A., S.H. Fox, M. Hill, J.P. Dick, A.R. Crossman, and J.M. Brotchie (2001). Cannabinoids reduce levodopa-induced dyskinesia in Parkinson's disease: a pilot study. *Neurology* **57**(11): 2108–2111.

Sugiura, T., S. Kondo, S. Kishimoto, T. Miyashita, S. Nakane, T. Kodaka, Y. Suhara, H. Takayama, and K. Waku (2000). Evidence that 2-arachidonoylglycerol but not N-palmitoylethanolamine or anandamide is the physiological ligand for the cannabinoid CB2 receptor. Comparison of the agonistic activities of various cannabinoid receptor ligands in HL-60 cells. *J Biol Chem* **275**(1): 605–612.

Szabo, B., I. Wallmichrath, P. Mathonia, and C. Pfreundtner (2000). Cannabinoids inhibit excitatory neurotransmission in the substantia nigra pars reticulata. *Neuroscience* **97**(1): 89–97.

Tang, K., M.J. Low, D.K. Grandy, and D.M. Lovinger (2001). Dopamine-dependent synaptic plasticity in striatum during in vivo development. *Proc Natl Acad Sci U S A* **98**(3): 1255–1260.

The Huntington's Disease Collaborative Research Group (1993). A novel gene containing a trinucleotide repeat that is expanded and unstable on Huntington's disease chromosomes. *Cell* **72**(6): 971–983.

Tsou, K., M. I. Nogueron, S. Muthian, M.C. Sanudo-Pena, C.J. Hillard, D.G. Deutsch, and J.M. Walker (1998). Fatty acid amide hydrolase is located preferentially in large neurons in the rat central nervous system as revealed by immunohistochemistry. *Neurosci Lett* **254**(3): 137–140.

Ungerleider, J.T., T. Andyrsiak, L. Fairbanks, G.W. Ellison, and L.W. Myers (1987). Delta-9-THC in the treatment of spasticity associated with multiple sclerosis. *Adv Alcohol Subst Abuse* **7**(1): 39–50.

van Dellen, A., C. Blakemore, R. Deacon, D. York, and A.J. Hannan (2000). Delaying the onset of Huntington's in mice. *Nature* **404**(6779): 721–722.

Wirguin, I., R. Mechoulam, A. Breuer, E. Schezen, J. Weidenfeld, and T. Brenner (1994). Suppression of experimental autoimmune encephalomyelitis by cannabinoids. *Immunopharmacology* **28**(3): 209–214.

Yamamoto, A., J.J. Lucas, and R. Hen (2000). Reversal of neuropathology and motor dysfunction in a conditional model of Huntington's disease. *Cell* **101**(1): 57–66.

Zeng, B.Y., B. Dass, A. Owen, S. Rose, C. Cannizzaro, B.C. Tel, and P. Jenner (1999). Chronic L-DOPA treatment increases striatal cannabinoid CB1 receptor mRNA expression in 6-hydroxydopamine-lesioned rats. *Neurosci Lett* **276**(2): 71–74.

Part VI

Endocannabinoids in Peripheral Organ Systems

17 Endocannabinoids and Gastrointestinal Function

Nissar A. Darmani

CONTENTS

INTRODUCTION

Cannabis refers to products of the plant *Cannabis sativa*, commonly known as marijuana. Cannabis has been used medically for thousands of years in Asian and Middle Eastern countries and as an intoxicant for many hundreds of years in India and the Middle East. Well-documented uses of cannabis range from treatment of cramps, migraine, convulsions, and neuralgia to attenuation of nausea and vomiting, decreased intestinal motility during diarrhea, and appetite stimulation (Mechoulam, 1986). During 1842 to 1854, cannabis preparations were introduced into Europe and the United States. However, its medicinal use declined in the last century because of lack of any standardized preparation of the plant product, erratic and inadequate oral absorption, and development of more potent drugs used in treatment of conditions for which cannabis was then being used. Basic research on cannabis continued through 1940s (Adams, 1942), ultimately leading to the isolation and chemical characterization of the major psychoactive constituent of marijuana plant

Δ^9-THC in 1964 (Mechoulam et al., 1998). Cannabis synthesizes at least 400 chemicals, of which more than 60 are structurally related to Δ^9-THC. Today, several hundred cannabinoid agonists, including active and inactive metabolites and related structures are available. Renewed interest in the therapeutic potential of cannabis products is clearly apparent as European (GW Pharmaceuticals) and American (Solvay Pharmaceuticals) pharmaceutical industries and universities are in the process of developing better delivery techniques (sublingual spray, sublingual tablet, or inhaler) to administer Δ^9-THC and its related products. In addition to being licensed as an antiemetic for cancer patients receiving chemotherapy and as an appetite stimulant in AIDS patients, claims are made for Δ^9-THC efficacy in treating pain, multiple sclerosis, epilepsy, glaucoma, Tourette's syndrome, Crohn's disease, and ulcerative colitis. Moreover, the French pharmaceutical company Sanofi-Synthelabo will be introducing Rimonabant (SR141716A, a cannabinoid CB_1 receptor antagonist) as a new class of appetite-suppressive weight loss agents.

The development of novel and potent analogs of Δ^9-THC has played a major role in the characterization and cloning of cannabinoid CB_1 and CB_2 receptors in the early 1990s (Pertwee, 1997). The discovery of cannabinoid receptors was followed in 1992 and 1995 by the demonstration of the existence of endogenous cannabinoid receptor agonists such as arachidonoyl ethanolamide (anandamide) and 2-AG. Other, less well-studied putative endocannabinoids are 2-arachidonoylglycerol ether (noladin ether) (Hanus et al., 2001) and 0-arachidonoyl ethanolamine (virodhamine) (Porter et al., 2002). Today, cannabinoid agonists can be classified according to their chemical structure into four main groups. The first of these is the "classical cannabinoid" group, which is made of dibenzopyran derivatives and includes Δ^9-THC. The second "nonclassical group" consists of bicyclic and tricyclic analogs of Δ^9-THC that lack a pyran ring, such as CP55940. The third group comprises aminoalkylindoles, and the prototype of this group is WIN55212-2. The fourth group is the "eicosanoids," which contains arachidonic acid (AA) derivatives such as anandamide and 2-AG. Δ^9-THC and CP55940 exhibit little difference in their affinities for CB_1 and CB_2 receptors, whereas anandamide exhibits marginal selectivity for CB_1 receptors, and WIN55212-2 shows modest selectivity for CB_2 receptors (Pertwee, 1997). Selective CB_1 (e.g., methanandamide)- and CB_2 (e.g., JWH133)-receptor agonists are also available. In addition, development of selective CB_1 (e.g., SR141716A and LY320135) and CB_2 (e.g., SR144528) receptor antagonists has revolutionized the field of cannabinoid research. This chapter summarizes the possible mechanisms and the role of endogenous, plant–derived, and synthetic cannabinoids in GI motility, secretion, appetite, and emesis.

ESTABLISHED NEUROTRANSMITTERS OF THE GUT

The gastrointestinal tract (GIT) is in a state of perpetual contractile and secretory activity. The patterns of motility of the GIT include mixing and propulsive movements that are confined to regions and organized patterns of movement such as swallowing and esophageal peristalsis, migrating complexes, vomiting, and defecation, which involve large sections of the digestive tract (Olsson and Holmgren, 2001; Thomson et al., 2001). These activities are controlled in part by the muscle itself, the local enteric nervous system (ENS), and by the central nervous system (CNS) acting via both autonomic (sympathetic and parasympathetic) and somatic innervation as well as humoral pathways. However, basically, most functions of the intestine are autonomous and are largely controlled by the ENS (Kunze and Furness, 1999). The ENS lies entirely in the wall of the gut, beginning in the esophagus and extending all the way to the anus. The number of neurons in the enteric system is approximately 100 million, which almost equals the number of neurons in the spinal cord. The ENS is organized into two plexi: the myenteric (or Auerbach's) plexus, which controls motor activity and is found between the circular and longitudinal muscle layers of the GIT, and the submucosal (or Meisner's) plexus, situated below the epithelium, which regulates secretion. The myenteric plexus also contains interstitial cells of Cagal (ICC), which play a significant role in mediating neurogenic influence on the GI smooth muscle. The parasympathetic innervation is supplied primarily by the vagus and pelvic nerves, whereas sympathetic innervation involves nerves that run between the spinal cord and the prevertebral ganglia and between these ganglia and the organs of the gut. Intestinal motility is regulated by ascending excitatory and descending inhibitory reflexes that modulate the activity of intestinal circular muscle. The postganglionic neurons of the parasympathetic system are located in the myenteric plexus, and stimulation of the parasympathetic nerves causes a general increase in the activity of the entire enteric system. This, in turn, enhances the activity of most but not all neurons because some of the enteric neurons are inhibitory. Stimulation of the sympathetic nervous system inhibits activity in the GIT, causing many effects opposite to those of the parasympathetic system. The principle neurotransmitters involved in these motility reflexes include acetylcholine, tachykinins (TKs), 5-hydroxytryptamine (5-HT), nitric oxide (NO), and vasoactive intestinal peptide (VIP).

The primary neurotransmitter of excitatory motor neurons is acetylcholine, which acts upon muscarinic M_2 and M_3 receptors to increase GIT motility via an increase in smooth muscle intracellular Ca^{2+} levels. Acetylcholine plays an important role in modulating active ion transport and motility in mammalian intestine. The TKs, substance P, and neurokinin A are excitatory cotransmitters of enteric cholinergic neurons that are coreleased with acetylcholine following depolarization of enteric neurons. TK release from extrinsic afferents can specifically be elicited by the excitotoxin capsaicin via vanilloid VR1 receptors that are expressed on spinal trigeminal and vagal afferent neurons (Holzer, 2000). TKs act via three receptors (NK_1, NK_2, and NK_3) to both stimulate and inhibit GI motility, and the net response depends upon the type and site of TK receptors that are activated. Nerve-independent facilitation of GI motor activity is mediated by NK_1 receptors on ICC and NK_2 as well as NK_1 receptors on muscle cells. NK_3 receptors are largely confined to enteric neurons and mediate cholinergic contraction of the GIT. However, some NK_1 and NK_3 receptors are also present on inhibitory motor pathways of the ENS and depress motor activity via release of the inhibitory transmitters NO and VIP.

5-HT is another excitatory neurotransmitter within the myenteric plexus; its effects are complex and involve a variety of serotonergic receptor subtypes. These actions are mediated via 5-HT_3 and 5-HT_4 receptors that synapse with cholinergic primary motor neurons in the myenteric plexus. Antagonism of 5-HT_3 receptors on the nonadrenergic, noncholinergic inhibitory neurons leads to enhancement of GIT motility, an effect that can be also achieved by agonists of 5-HT_4 receptors on excitatory cholinergic interneurons. Although 5-HT receptors (e.g., 5-HT_{2A}) are also present on the smooth muscle cells of the GIT, their effects may not be seen unless high concentrations of 5-HT are present in the vicinity of these cells.

NO is not stored in vesicles but is produced when the nerve cell is stimulated and then immediately diffuses into the target cell. Subsequent activation of guanylate cyclase and increased levels of c-GMP lead to lowered Ca^{2+}-levels in the cytoplasm, and K^+ efflux from the cell results in the inhibition of GI motility. NO may also mediate the inhibitory effects of other neurotransmitters such as GABA and ATP (Olsson and Holmgren, 2001). Significant amounts of dopamine are present in the GIT whose function appears to be suppression of GI motility, including reduction of lower esophageal sphincter and intragastric pressure. These effects stem from the activation of the D_2 subtype of dopamine receptor, which ultimately leads to the suppression of acetylcholine release from the myenteric motor neurons. Thus, by antagonizing the effect of dopamine on myenteric motor neurons, several dopamine D_2 receptor antagonists, such as metoclopramide and domperidone, were developed as prokinetic agents (Tonini, 1996).

ESTABLISHED EMETIC NEUROTRANSMITTERS AND MORE NOVEL MEDIATORS OF VOMITING

Vomiting (emesis) is a reflex, developed to different degrees in diverse species, which allows an animal to rid itself of ingested toxins. Emesis is a complex process and requires coordination by the vomiting center (VC) (Veyrat-Follet, 1997). The VC is a collection of recipient and effector nuclei in the brainstem that includes part of the nucleus tractus solitarius (NTS), the dorsal motor nucleus of the vagus nerve, and the area postrema. The VC receives input from the chemoreceptor trigger zone (CTZ), from vestibular apparatus, from higher brain structures, and from visceral afferents arising from the GIT.

Several well-established emetic neurotransmitters (such as acetylcholine, dopamine, histamine, and serotonin) act via specific receptors to induce emesis. Indeed, selective activation of serotonin 5-HT_3-, dopamine D_2-, or muscarinic M_1-receptors in the CTZ and histamine H_1 receptor in the NTS can induce vomiting. Recently, more novel receptors such as neurokinin NK_1 receptors are shown to be involved in emesis (Diemunsch and Grelot, 2000). In addition to their central antiemetic effect, 5-HT_3- and NK_1-receptor antagonists block their corresponding receptors located on vagal terminals in the GIT which is a major peripheral site of action of these agents (Veyrat-Follet, 1997; Diemunsch and Grelot, 2000).

In addition to the discussed established neurotransmitter systems in emetic circuits, a number of other less well-recognized mediators appear to play an important role in the production of emesis. We have recently shown that AA, the common precursor of prostaglandins, thromboxanes, hydroxyeicosatetraenoic acids (HETEs), hydroperoxyeicosatetraenoic acids (HpETEs), and prostacyclins, is a potent emetic agent in the least shrew (Darmani, 2002b). Moreover, both prostaglandins PGE_2 and PGF_2 induce vomiting in humans (Karim and Filshe, 1970; Wislicki, 1982), pigs (Wechsung, 1996; De Saedeleer et al., 1992), and the least shrew (Darmani, unpublished findings). Furthermore, prostaglandin synthesis inhibitors such as dexamethasone are used clinically as additive antiemetics for the prevention of vomiting produced by both chemo- and radiotherapy (Ioanidis, 2000; Maranzano, 2001). Stimulation of vanilloid VR1 receptors by agonists such as capsaicin and resiniferatoxin induce vomiting in shrews, whereas rapid desensitization of these receptors following the initial exposure of animals to such VR1 agonists produce antiemetic effects (Andrews et al., 1993, 2000; Rudd and Wai, 2001).

PRESENCE OF ENDOCANNABINOIDS AND CANNABINOID CB_1 RECEPTORS IN GUT–BRAIN CIRCUITS AFFECTING GIT MOTILITY AND EMESIS

Though most motility functions of the GIT are mediated by the local ENS, the brainstem plays an important role in both the initiation and coordination of GIT motility (Krowicki and Hornby, 1995). As will be discussed later in the text, both exogenous cannabinoids and endocannabinoids affect GIT motility functions and emesis via cannabinoid CB_1 receptors. The first endocannabinoid to be identified in a peripheral tissue was 2-AG. Although Mechoulam et al. (1995) detected 2-AG but not anandamide in the canine small intestine, more recent studies show that relatively large amounts (44.1 ± 4 nmol/g tissue and 36.4 ± 6.1 pmol/g tissue, respectively) of both endocannabinoids are present in the small intestine of mice (Izzo, Fezza, et al., 2001). The mouse small intestine is highly enriched with 2-AG, as its concentration exceeds the levels found in other peripheral tissues (liver, spleen, lung, and kidney) by 37 to 55 times and in various brain regions by 3- to 20-fold (Bisogno et al., 1999; Izzo, Fezza, et al., 2001; Sugiura et al., 2002). On the other hand, the levels of anandamide both in the CNS and peripheral tissues of several species can be similar, lower, or greater than that present in the mouse small intestine. Both 2-AG and anandamide are differentially distributed in several regions of the rat brain, where the highest concentrations occur in the brainstem (87 ± 45 pmol/g tissue and 14 ± 7.1 nmol/g tissue, respectively), and lowest amounts are found in the diencephalon (10.2 ± 2.3 pmol/g tissue and 2 nmol/g tissue, respectively). Anandamide may be responsible for specific functions in particular areas of the GIT because its colonic tissue concentration is 10-fold higher than that found in the small intestine, whereas 2-AG concentration appears to be approximately 2 times lower in the mouse colon relative to basal levels in small intestine (Pinto, Capasso, et al., 2002). Anandamide is probably more rapidly metabolized in the mouse colon as the main enzyme responsible for its degradation is also more concentrated in this tissue. 2-AG levels both in the brain of several species and in the mouse small intestine are more than 100- to 1000-fold higher than those of anandamide. The amounts of these endocannabinoids in both the CNS and small intestine are likely to yield tissue concentrations similar to or higher than the K_i values for the activation of CB_1 receptors (Izzo, Fezza, et al., 2001). Lynn and Herkenham (1994) utilized ^{3}H-CP55940 to visualize cannabinoid CB_1 binding sites in Peyer's patches located in rat jejunum, ileum, and rectum by autoradiography. However, CB_1 sites were not detected in the rat stomach, duodenum, cecum, or colon. These initial findings suggested that cannabinoid CB_1 receptors are highly localized in discrete regions such as nerve terminals in the myenteric and submucosal plexuses. On the other hand, more sensitive techniques such as immunohistochemistry and RT polymerase chain reaction have shown the presence of the CB_1 receptor or its markers on neurons throughout the GIT of several animal species including humans (Buckley et al., 1998; Casu et al., 2003; Kulkarni-Narla and Brown, 2000; Pinto, Izzo, et al., 2002; Ross et al., 1998; Storr et al., 2002; Shire et al., 1995). However, CB_1 receptors seem to be differentially distributed along the length of the GIT, with the stomach and colon being highly enriched with these sites (Casu et al., 2003). Although the brainstem appears to be relatively CB_1 receptor sparse, the NTS, the area postrema, and the dorsal motor nucleus of the vagus in the brainstem contain significantly moderate amounts of cannabinoid CB_1 receptors in nonemetic (Herkenham et al., 1991) and several emetic species including the ferret (Van Sickle et al., 2001), least shrew (*Cryptotis parva*) (Darmani et al., 2003) and humans (Glass et al., 1997). The NTS and area postrema are among important loci within the brainstem that control emesis. The brainstem via the vagus also controls gastric motor and secretory activities.

ENDOCANNABINOIDS AND GI MOTILITY

In Vitro Studies on Intestinal, Longitudinal, and Circular Muscle Preparations

Over the past three decades, Pertwee and his coinvestigators have been the leading contributors to the understanding of the *in vitro* effects of Δ^9-THC, its synthetic analogs, and anandamide on intestinal motility functions (Pertwee, 2001). Their studies evolved from the observation that Δ^9-THC inhibits the contractile response produced by electrically evoked contractions of isolated whole ileum or isolated myenteric plexus-longitudinal muscle (MPLM) preparation of the guinea-pig ileum. In these preparations, the contractile response is thought to be due to the electrically induced release of acetylcholine from enteric cholinergic nerves. Cannabinoids of diverse structure and activity attenuate the induced contractions with a potency order (CP55940 > WIN55212-2 > nabilone > Δ^9-THC > cannabinol > anandamide) that is similar to their affinity rank order for the cannabinoid CB_1 receptor (Breivogel and Childers, 2000; Breivogel et al., 2001). Moreover, the inhibitory effect of cannabinoids can be competitively and specifically antagonized by the selective CB_1 receptor antagonist/inverse agonist SR141716A (Pertwee et al., 1996; Coutts and Pertwee, 1997; Croci et al., 1998; Pertwee, 2001; Pinto, Izzo, et al., 2002). Anandamide also inhibits electrically evoked acetylcholine release (Mang et al., 2001) and electrically induced contractions of both MPLM (Pertwee et al., 1995; Mang et al., 2001) and circular muscle preparations of the guinea-pig ileum (Izzo et al., 1998; Pertwee et al., 1995). However, the role of CB_1 receptors on the inhibitory effect of anandamide on MPLM is questioned because a large concentration of SR141716A (1 μM) was required to reverse the induced inhibition (Mang et al., 2001) though the CB_1 antagonist was more effective in the circular muscle preparation (Izzo et al., 1998). These, as well as similar findings in other test systems (see Mang et al., 2001) suggest: (1) the existence of subtypes of CB_1 receptors to which SR141716A binds with a lower affinity and (2) that anandamide acts on both CB_1 and non-CB_1 receptors to inhibit the discussed electrically evoked effects on GI motility. In addition, anandamide is also an endovanilloid and stimulates both basal acetylcholine release and basal MPLM tone in the guinea-pig ileum via the activation of vanilloid VR1 receptor (Mang et al., 2001). Moreover, other VR1 agonists such as capsaicin and piperine cause an initial contraction of guinea-pig isolated ileum followed by blockade of contraction of circular and longitudinal muscles of this tissue evoked either by electrical field stimulation (ESF) or mesenteric nerve stimulation (Takai et al., 1990; Jin et al., 1990). Pertwee (2001) has summarized published evidence supporting the notion that cannabinoids act selectively and prejunctionally via CB_1 receptors to inhibit electrically evoked contractions of guinea-pig MPLM. The following are the characteristics of cannabinoid actions in this tissue:

1. Inhibition of evoked fast excitatory synaptic potentials of myenteric neurons by CB_1 agonists.
2. Presence of CB_1 receptor markers in the myenteric and submucosal plexuses.
3. Cannabinoid-induced CB_1-receptor-mediated inhibition of acetylcholine release.
4. Blockade of the evoked contractions by tetrodotoxin.
5. Only CB_1 receptor antagonists counter the inhibitory effects of cannabinoids.
6. Cannabinoid agonists do not reduce the contractile response of such preparations to exogenously added acetylcholine, carbachol, substance P, or histamine, each of which directly acts on the smooth muscle. However, cannabinoids prevent the ability of 5-HT or GABA to produce contractions of these muscle preparations as the latter agents act presynaptically to increase acetylcholine release.
7. The more stable analog of anandamide, methanandamide, does not affect indomethacin-induced phasic contractions of guinea-pig ileum that are generated at the muscle level as they are left unchanged by tetrodotoxin (Heinemann et al., 1999).
8. CB_1 receptor stimulation does not directly suppress smooth muscle activity (Izzo et al., 1998; Coutts and Pertwee, 1997).

Although 2-AG was the first endocannabinoid to be discovered in the intestine, virtually nothing has been published on its effects in the GI system. A recent study shows that 2-AG in the presence of a nonselective COX inhibitor (indomethacin) and an NO synthase inhibitor (N^G-nitro-L-arginine) causes concentration-dependent and tetrodotoxin-sensitive contractions of longitudinal muscle strips of isolated distal colon of the guinea pig (Kojima et al., 2002). Thus, the observed contractions appear to be due to the stimulation of enteric nerves rather than the activation of release of prostanoids or NO. The induced contractions were markedly attenuated by atropine and partially blocked by either a ganglionic blocker (hexamethonium) or a lipoxygenase inhibitor (nordihydroguaiaretic acid) pretreatment but were not affected by cannabinoid CB_1 (SR141716A)-, CB_2 (SR144528)- or vanilloid VR1 (capsazepine)-receptor antagonists. In this study, anandamide also produced similar but less robust neurogenic contractions, whereas the synthetic cannabinoid WIN55212-2 had no effect. In addition, both 2-AG and anandamide failed to contract circular smooth muscle preparations of the guinea-pig distal colon. These results suggest that 2-AG and anandamide induce contractions of the longitudinal muscle preparations of the guinea-pig colon mainly via stimulation of myenteric cholinergic neurons and that neither cannabinoid CB_1/CB_2 or vanilloid VR1 receptors contribute to the contractile response, but the effect may be mediated by one or more lipoxygenase metabolites of 2-AG. For example, leukotrienes LTC_4, LTD_4, and LTE_4 (but not LTB_4) cause concentration-dependent contractions of guinea-pig ileum (Gardiner et al., 1990). Downstream COX metabolites of endocannabinoids such as prostaglandins (PGs) PGE_1, PGE_2, 8-iso-PGE_2, and $PGE_{2\alpha}$ also cause contractions of isolated guinea-pig ileum (Grobovic and Radmanovic, 1987; Ishizawa and Miyaraki, 1975; Sametz et al., 2000).

In Vitro Peristalsis Studies in Isolated Intestine

From the earlier discussion, it is apparent that GI motility, in general, and peristalsis, in particular, are mainly controlled by the ENS. Peristalsis is a coordinated motor activity that allows the intestine to propel its content anal-ward. Peristalsis consists of a preparatory phase in which the longitudinal muscle contracts in response to effluent infusion and a subsequent emptying phase in which coordinated contractions of circular muscle expel the effluent from the mouth to the anal end.

Recent studies by the leading investigator in this field (Izzo, Mascolo, et al., 2000) have demonstrated that synthetic cannabinoids (CP55940 and WIN55212-2) inhibit peristalsis induced in segments of guinea pig isolated ileum by continuous luminal fluid infusion. Indeed, during the preparatory phase of peristalsis, these CB_1/CB_2 cannabinoid agonists decreased both the longitudinal smooth muscle reflex contraction and resistance of the intestinal wall to the infused liquid (compliance) but increased threshold pressure and the volume required to elicit peristalsis in a selective and SR141716A-sensitive manner. In addition, the cited agents decreased maximal ejection pressure during the emptying phase of peristalsis, which was also countered by SR141716A pretreatment. The inhibitory effect of cannabinoids seems to be extended to inhibition of electrically evoked peristaltic activity in isolated distal colon of mice, because WIN55212-2 reduced intraluminal pressure, longitudinal displacement of intestinal segment (shortening), and ejected fluid volume during peristaltic contractions in a dose-dependent and SR141716A-sensitive manner (Mancinelli et al., 2001). Heinemann and co-workers (1999) have recently shown that methanandamide also inhibits distension-induced propulsive motility of luminally perfused guinea-pig isolated ileum by increasing peristaltic pressure threshold in a SR141716A-sensitive and a dose-dependent fashion. To explain this effect, the authors have proposed that methanandamide stimulates CB_1 receptors to: (1) activate inhibitory enteric motor neurons that oppose distension-induced peristalsis by releasing both NO- and apamine-sensitive neurotransmitters, as L-NAME (an NO synthase inhibitor) and apamine (an inhibitor of fast inhibitory junction potentials mediated by transmitters of inhibitory motor neurons in the guinea-pig small intestine) attenuate the induced peristaltic activity and (2) inhibit excitatory enteric motor neurons that mediate ascending enteric reflex (AER) contraction of guinea-pig intestinal circular muscle in response to distension. Heinemann et al. (1999) further suggest that activation of CB_1 receptors in the guinea-pig ileum suppresses propulsive peristalsis and AER contractions by inhibiting both cholinergic and noncholinergic transmission, because these inhibitory effects of methanandamide were not only preserved but enhanced under conditions in which peristalsis is maintained via endogenous TKs (i.e., the blockade of cholinergic transmission with atropine or hexamethonium and the production of peristalsis by naloxone via release of endogenous TKs (Holzer et al., 1998). Supporting evidence for the latter notion comes from the ability of synthetic cannabinoid WIN55212-2 to inhibit cholinergic and nonadrenergic noncholinergic contractions evoked by electrical stimulation of circular muscle of the guinea-pig ileum in a SR141716A-sensitive manner (Izzo et al., 1998). Following an initial contraction, the vanilloid VR1 agonist capsaicin can depress intestinal peristalsis via the release of NO (Bartho and Holzer et al., 1995), an effect is also shared by anandamide (Stefano et al., 1997). Although the effects of anandamide on *in vitro* intestinal peristalsis remain unknown, it does stimulate basal release of acetylcholine and longitudinal muscle tone in the guinea-pig ileum preparation via VR1 receptor stimulation (see earlier). Capsaicin-induced neurogenic contraction of the smooth muscle of guinea-pig intestine also involves release of acetylcholine and TKs from enteric neurons (Bartho and Holzer, 1995). In addition, although both anandamide and piperine (a vanilloid VR1 agonist) inhibit GI transit *in vivo* (Izzo, Capasso, et al., 2001), the VR1 receptor is not involved in anandamide-induced inhibition, as the effect was sensitive to SR141716A but not to pretreatment with either capsazepine (a VR1 receptor antagonist) or capsaicin (another VR1 agonist).

The effect of 2-AG on peristaltic activity has not yet been reported. However, downstream metabolites of both endocannabinoids can differentially affect propulsive peristalsis in fluid-perfused segments of the guinea-pig small intestine (Shahbazian et al., 2002). These findings indicate that PGs PGE_1 and PGE_2 decrease peristaltic performance, whereas leukotriene LTD_4 and PG PGD_2 increase peristalsis pressure threshold.

In Vivo Intestinal Motility Studies

The basic propulsive movement of the GIT is peristalsis in which a contractile ring appears around the gut and then moves forward. The usual stimulus for peristalsis is distension of the gut wall, which stimulates the ENS to initiate peristaltic movement thus allowing the food to be propelled analward.

The discussed *in vitro* studies have well established that CB_1/CB_2 agonists inhibit electrically evoked contractions of ileum, guinea-pig MPLM contractions as well as peristalsis in isolated intestine through the activation of prejunctional cannabinoid CB_1 receptors via inhibition of release of acetylcholine and other excitatory neurotransmitters from the enteric, cholinergic, nonadrenergic, and noncholinergic nerves, respectively. *In vivo* studies investigating the effects of drugs on GIT motility employ techniques which measure the passage of a nonadsorbable transit marker or glass beads in the GIT of several species, and these agents are administered either orally, intraduodenally, or in distal colon.

In line with the *in vitro* findings, *in vivo* studies show that structurally diverse exogenous cannabinoids [Δ^8-THC, Δ^9-THC, nabilone, cannabinol, WIN55212-2, CP55940, HU210, and arachidonoyl-2′-chloroethylamide (ACEA, a selective CB_1 agonist)] attenuate gut motility, intestinal transit, and intestinal secretions in several species (Casu et al., 2003; Pertwee, 2001; Pinto, Capasso, et al., 2002; Pinto, Izzo, et al., 2002; Izzo, Capasso, et al., 2001; Landi et al., 2002). The inhibitory *in vivo* GIT effects of these cannabinoids were also SR141716A sensitive, implying an important role for the cannabinoid CB_1 receptor. The endocannabinoid anandamide causes similar inhibitory actions on intestinal motility, as it was shown to dose-dependently and selectively reduce the passage of both charcoal marker in the upper GIT (Calignano et al., 1997; Izzo, Capasso, et al., 2001) and glass beads in distal colon (Pinto, Izzo, et al., 2002) of mice in an SR141716A-sensitive manner. Moreover, the selective CB_1 receptor agonist and an analog of anandamide, ACEA, more potently inhibited the colonic propulsion. As expected, a selective inhibitor of anandamide cellular reuptake, VDM11, with little affinity for either CB_1 or CB_2 receptors, was shown to significantly inhibit colonic propulsion in a manner sensitive to SR141716A. However, another inhibitor of anandamide reuptake, AM404, which is structurally similar and equipotent to VDM11 on the anandamide transporter, was found to be inactive on small intestine motility in mice (Calignano et al., 1997). This lack of effect on intestinal motility is puzzling and is probably either due to a tissue specificity of the tonic action of endocannabinoids or because AM404 also stimulates vanilloid VR1 receptors, activation of which might have masked the inhibitory action of AM404 on intestinal motility (Pinto, Izzo et al., 2002). The latter notion appears less likely because the same group of investigators, in a previous study, has shown that exogenous administration of the endocannabinoid/endovanilloid agonist anandamide reduces intestinal motility in mice via cannabinoid CB_1 sites and not vanilloid VR1 sites (Izzo, Capasso, et al., 2001). However, the *in vivo* effects of VR1 receptor agonists seem to be complex as piperine inhibited but capsaicin failed to alter GIT motility. The effect of piperine is thought to involve capsaicin-sensitive neurons but not vanilloid VR1 receptors, because the inhibitory effect of piperine (Izzo, Capasso, et al., 2001) was strongly inhibited in capsaicin-pretreated mice. In addition, the effects of capsaicin differ in different sections of GIT (Shibata et al., 2002).

The basal anandamide level in the mouse colon is about 10 times greater than its concentration in the small intestine (Pinto, Izzo, et al., 2002), which implies that the colon requires more anandamide to control its motility and other functions. In line with this notion, the antitransit activity of Δ^9-THC appears to be somewhat selective for the small intestine relative to the colon (Shook and Burks, 1989). Noladin ether (2-arachidonoylglycerol ether) is a putative endocannabinoid which also possesses antitransit properties because it reduces defecation in mice (Hanus et al., 2001). Although 2-AG is present both in small intestine and colon (Pinto, Izzo, et al., 2002), the *in vivo* effects of 2-AG on intestinal motility have not yet been investigated. However, downstream metabolites of endocannabinoids (e.g., PGs) have region- and muscle-layer-specific *in vivo* effects in different areas of the GIT. For example, PGE_2 in the proximal colon inhibits myoelectric and mechanical activity at low doses but at higher doses causes marked excitation, whereas in the distal colon only excitation is observed (Burakoff and Percy, 1992).

In Vitro Isolated Gastric Tissue and *In Vivo* Gastric Emptying Studies

Emptying of the stomach is controlled to a moderate degree by stomach factors (such as degree of filling and excitatory effect of gastrin) but mainly by feedback signals from the duodenum including enterogastric nervous system feedback reflexes and hormonal feedback. Stomach emptying is promoted by the intense peristaltic contractions of the stomach antrum. In general, vagal stimulation increases the force and frequency of contractions, whereas sympathetic stimulation decreases both of these parameters. The endocannabinoid anandamide has recently been shown to attenuate cholinergically mediated contractions of isolated rat gastric smooth muscle preparation produced by EFS (Storr et al., 2002). Anandamide also attenuated nonadrenergic, noncholinergic-mediated relaxant neural responses (in the presence of atropine and guanethidine) produced by EFS in this preparation. Because the depressant effects of anandamide were countered by the CB_1 antagonist AM630, it is suggested that CB_1 receptors play an important role in the attenuation of excitatory cholinergic and inhibitory NANC neurotransmission in the rat isolated gastric fundus preparation. The aminoalkylindole cannabinoid WIN55212-2 also produces similar depressant effects in this fundus preparation. However, AM630 failed to counter WIN55212-2's inhibitory effect, suggesting further cannabinoid receptors may be involved. The effect of cannabinoid agonists seems to be prejunctional and not postjunctional because: (1) they did not alter NO- and VIP-induced relaxation of the stomach preparation and (2) CB_1 receptor markers are found in prejunctional nerve terminals rather than the stomach tissue (see preceding text). To date, the effect of endocannabinoids on *in vivo* models of gastric emptying has not been reported. However, in line with the above *in vitro* findings, several *in vivo* studies (Izzo, Mascolo, Pinto, et al., 1999; Izzo, Mascolo, Capasso, et al., 1999; Landi et al., 2002; Shook and Burks, 1989) have shown that structurally diverse exogenous cannabinoids can delay gastric emptying in several species with an ED_{50} order of potency (CP55940 > WIN55212-2 >)^9-THC > cannabinol) that is similar to: (1) their potency rank order in attenuating electrically evoked contractions of isolated ileum preparations and (2) their affinity rank order for cannabinoid CB_1 receptor (see earlier). In addition, the cited studies show that the CB_1 receptor antagonist SR141716A but not the CB_2 antagonist SR144528, in a dose-dependent manner, counters the inhibitory effects of these CB_1/CB_2 receptor agonists on gastric emptying in several species. In addition to delaying gastric emptying, such cannabinoids (e.g., Δ^9-THC) reduce intragastric pressure, pyloric contractility, and greater curvature contractile activity of the stomach in an SR141716A-sensitive manner (Krowicki et al., 1999). Besides preventing such parameters of gastric contractility, some cannabinoids (e.g., WIN55212-2) have been reported to inhibit transient lower esophageal sphincter relaxation (TLESR) and gastroesophageal reflux (major aberrant factors in gastroesophageal disease) in a dose-dependent and SR141716A-sensitive manner (Lehmann et al., 2002).

POSSIBLE EFFECTS OF ENDOCANNABINOIDS ON GI SECRETIONS

To date, there is no direct evidence on the effects of endocannabinoids on GIT secretions. However, a number of studies provide indirect evidence for possible antisecretory effects of endocannabinoids. For example, the selective cannabinoid CB_1 (SR141716A)- but not the CB_2 (SR144528)-receptor antagonist is able to increase intestinal fluid volume (Izzo, Mascolo, Borrelli, et al., 1999) and fecal water content (Izzo, Mascolo, Pinto, et al., 1999) in rats. One possible explanation offered for these effects is that SR141716A antagonizes the actions of endocannabinoids. As SR141716A is also able to increase acetylcholine release from enteric nerves (Coutts and Pertwee, 1997), the prosecretory actions of the CB_1 antagonist appears to be cholinergically mediated. Indeed, acetylcholine plays an important role in modulating ion transport in the mammalian intestine (Brown and Miller, 1991), and atropine (a cholinergic antagonist) abolishes the secretory effects of SR141716A (Izzo, Mascolo, Borrelli, et al., 1999). Supporting these findings, it has been shown that WIN55212-2 inhibits both neurally evoked ileal secretion in mucosal preparation of isolated sheets of rat small intestine (Tyler et al., 2000) and *in vivo* fluid accumulation in the rat intestine (Izzo, Mascolo, Borrelli, et al., 1999). In addition, noneffective doses of SR141716A have been shown to counter the antisecretory effects of WIN55212-2, thus supporting a role for the cannabinoid CB_1 receptor in controlling intestinal secretion.

Evidence is also accumulating that structurally diverse cannabinoids (Δ^9-THC, WIN55212-2, HU210) inhibit stimulated but not basal gastric acid secretion (Rivas-V and Garcia, 1980; Coruzzi et al., 1999; Adami et al., 2000, 2002) as well as reducing the volume of gastric juice (Sofia et al., 1978). In the latter study, acid concentration was not changed, however, gastric acid output was reduced in people who have chronically smoked cannabis (Nalin et al., 1978). The gastric antisecretory effect of these cannabinoids appears to be CB_1-receptor mediated, because the selective CB_1 agonist HU210 dose-dependently decreased acid secretion produced by both pentagastrin and 2-deoxy-D-glucose (Adami et al., 2000, 2002; Coruzzi et al., 1999). In addition, the CB_1/CB_2 agonist WIN55212-2 also caused similar antisecretory effects; however, both its inactive enantiomer (WIN55212-3) and the selective CB_2 agonist JWH015 were ineffective. Moreover, selective CB_1 receptor antagonists such as SR141716A and LY320135 were shown to counter WIN55212-2-induced antisecretory effects, whereas the CB_2 antagonist SR144528 was ineffective. By themselves, neither SR141716A, LY320135, nor SR144528 affected either basal or pentagastrin-stimulated gastric acid secretion. The reported reduction of the antisecretory effect of HU210 via bilateral cervical vagotomy or ganglionic blockade suggests that the mechanism underlying inhibition of acid secretion by CB_1 agonists is mediated via the suppression of vagal drive to the stomach through activation of CB_1 receptors, located on pre- and postganglionic cholinergic pathways. However, because atropine failed to block the antisecretory effects of HU210, the current results suggest that the release of noncholinergic excitatory neurotransmitters is probably regulated by presynaptic CB_1 receptors.

Although it has been previously shown that a high dose of Δ^9-THC (20 μM) prevents histamine-induced but not basal gastric acid release in the isolated stomach preparation of the rat (Rivas-V and Garcia, 1980), recent *in vivo* studies in anesthetized rats with lumen-perfused stomach show that both HU210 (1.5 μmol/kg) and WIN55212-2 (4 μmol/kg) failed to reduce gastric acid release induced by continuous infusion of histamine (Adami et al., 2002). These findings suggest that although cannabinoids at usual doses may not block gastric acid secretion caused by direct acting histamine agonists, they do prevent the action of prosecretory agents such as pentagastrin, which in part act indirectly to release histamine (Watanabi et al., 1996).

Because cannabinoid CB_1 receptor agonists reduce stimulated gastric acid secretion, they are expected to possess gastric antiulcer effects. Indeed, a high oral dose of Δ^9-THC was shown to prevent gastric ulcer formation in rats after ligation of the pylorus (Sofia et al., 1978). A more recent study shows that low intraperitoneal doses (e.g., 0.1 mg/kg) of WIN55212-2, but not its inactive enantiomer (WIN55212-3), reduce gastric ulcers produced by cold or restraint stress in a SR141716A-sensitive manner (Germano et al., 2001).

ENDOCANNABINOIDS IN RELATION TO FEEDING AND APPETITE

Several nuclei in the hypothalamus and brainstem act as input stations for hormonal and GI information and control several aspects of feeding (Halford and Blundell, 2000; Chiesi et al., 2001). During and after a meal, several peripheral satiety factors (such as cholesystokinin [CCK], bombesin, gastrin-releasing peptide, and glucagons) are released from GI secretory cells in response to the physical and chemical presence of food in the GIT. Of these factors, CCK is thought to activate CCK_1 receptors on vagal afferents that transmit signals to the hindbrain, particularly to the NTS. The NTS communicates with several hypothalamic nuclei that play critical roles in appetite regulation (Halford and Blundell, 2000; Broberger and Hokfelt, 2001; Chiesi et al., 2001). In addition, the hormone leptin, which is secreted from adipose tissue, enters the CNS and stimulates the arcuate nucleus within the hypothalamus. Leptin is the primary signal through which the hypothalamus senses nutritional state and modulates food intake. Leptin directly affects neurons in which either the anorexigenic (appetite-reducing) peptides pro-opiomelanocortin (POMC) and cocaine- and amphetamine-regulated transcript (CART) or the orexigenic (appetite-stimulating) peptides neuropeptide Y (NPY) and agouti-related protein (AGRP) are colocalized. These two neuronal populations project in parallel into other brain centers where they exert antagonistic actions on food intake. Thus, the NPY/AGRP expressing neurons increase feeding, whereas those that express POMC/CART inhibit feeding. The stimulatory effects of cannabis products on appetite has been anecdotally known for centuries. Δ^9-THC as well as other exogenous cannabinoids also produce hyperphagic effects under laboratory conditions in most human and animal models (Kirkham and Williams, 2001; Croxford, 2003; Berry and Mechoulam, 2002, Mattes et al., 1994); however, no effect (Graceffo and Robinson, 1998) or a reduction in food intake (Miczek and Dixit, 1980) has been reported. The hyperphagia induced by Δ^9-THC is cannabinoid CB_1-receptor mediated as it is reversed by the selective CB_1 antagonist SR141716A and not by SR144528, an antagonist for the CB_2 receptor (Williams and Kirkham, 2002). In addition, following temporary food restriction, CB_1 knockout mice eat less than their wild-type littermates (Di Marzo et al., 2001). In line with the latter findings, most but not all studies indicate that SR141716A suppresses appetite and food intake in rodents (Arnone et al., 1997; Simiand et al., 1998; Colombo, Agabio, Diaz, et al., 1998; Williams and Kirkham, 1999; Koch and Werner, 2000, Jamshidi and Taylor, 2001). These findings suggest an important role for endocannabinoids in feeding behaviors. Indeed, both anandamide and 2-AG induce overeating in laboratory animals in an SR141716A-sensitive manner (Williams and Kirkham, 1999; 2001; Hao et al., 2000; Jamshidi and Taylor, 2001). Leptin administration reduces food intake, in part, via the inhibition of the actions of the cited orexigenic neurotransmitters (Ammar et al., 2000). In regard to endocannabinoids, elevated hypothalamic levels of anandamide and 2-AG have been found in hyperphagic, genetically obese rodents with defective leptin signaling (Di Marzo et al., 2001). Moreover, leptin treatment was shown to reduce the hypothalamic anandamide content. Paradigms with palatable rich diets used to induce obesity in rats have been shown to reduce cannabinoid CB_1 receptors in several brain areas, including those loci involved in the hedonic aspects of eating (e.g., nucleus accumbens and hippocampus) (Harrold et al., 2002). However, the CB_1 receptor density was not altered in the hypothalamus of these obese rats. Taken together, the discussed data strongly suggest that the endocannabinoids' contribution to the stimulation of feeding and enhancement of appetite occurs via the activation of cannabinoid CB_1 receptors. The rationale for the potential use of CB_1 receptor ligands in the clinic is clearly based on observations that Δ^9-THC (Dronabinol or Marinol) improves appetite and food intake in acquired immunodeficiency syndrome (AIDS) and cancer patients suffering from anorexia/cachexia (Beal et al., 1995; 1997; Mattes et al., 1994; Nelson et al., 1994; Struwe et al., 1993), whereas SR141716A (Rimonabant) is undergoing clinical trials as an appetite suppressant.

ENDOCANNABINOIDS AND VOMITING

Emesis is the forceful expulsion of intestinal and gastric contents through the mouth. In humans emesis is often associated with the feeling of nausea. Emesis is also preceded in humans by retching, which is a pattern of motor activity that overrides antireflux mechanisms in the GIT. Initially, a wave of reverse peristalsis begins in the distal small intestine moving intestinal content orad. The antiperistalsis may begin as far down in the GIT as the ileum, which pushes intestinal contents to the duodenum. Distension of the upper GIT, especially duodenum, becomes an excitatory factor in initiating the process of vomiting. Eventually, stronger retches develop, and a sudden strong contraction of abdominal muscles raises the diaphragm high into the thorax, and the increased intrathoracic pressure forces the GIT contents out of the mouth.

Aside from vomiting being initiated by irritative stimuli in the GIT, emesis can also be caused by different emetic neurotransmitters stimulating various areas of the brain outside the VC such as the NTS, CTZ, and vestibular apparatus (see earlier). Because of the presence of both endocannabinoids and CB_1 receptors in structures that control GIT motility and emesis, it is tempting to suggest that endocannabinoids may modulate the process of vomiting. Indeed, anandamide and its more stable analog methanandamide can protect shrews and ferrets from vomiting caused by 2-AG (Darmani, 2002b) or morphine (Van Sickle et al., 2001). The antiemetic activity of anandamide and methanandamide is in line with their discussed ability to attenuate: (1) *in vitro* electrically evoked contractions of both isolated MPLM preparation of guinea-pig ileum as well as circular muscle preparation of guinea-pig ileum, (2) *in vitro* distension-induced propulsive motility of luminally perfused guinea pig isolated ileum, (3) *in vitro* electrically evoked contractions of gastric smooth muscle preparation, and (4) *in vivo* passage of charcoal marker in the upper GIT and glass beads in distal colon of mice. In addition, as noted earlier, a large amount of anandamide is found in the brainstem, an area which is generally sparse with cannabinoid CB_1 receptors. However, specific emetic loci of brainstem such as the NTS and CTZ contain significant amounts of CB_1 sites as well as fatty acid amide hydrolase (FAAH) that metabolizes anandamide (Darmani, Sim-Selly, et al., 2003; Van Sickle et al., 2001; Glass et al., 1997). Spurred on by our initial publication that high doses of the cannabinoid CB_1 receptor antagonist SR141716A produces vomiting in the least shrew (Darmani, 2001a), we and other investigators have shown that in several animal models of vomiting, structurally diverse cannabinoids (HU210, CP55, 950, WIN55212-2, Δ^9-THC and anandamide) prevent vomiting produced by various emetic stimuli such as: (1) SR141716A (Darmani, 2001a), (2) 2-AG (Darmani 2002b), (3) cisplatin (Darmani, 2001b, 2001c; Darmani, 2002a; Darmani, Sim-Selly, et al., 2003), (4) AA (Darmani, 2002b), 5-hydroxytryptophan (5-HTP) (Darmani, 2002c), (5) apomorphine (London et al., 1979), (6) morphine (Simoneau et al., 2001; Van Sickle et al., 2001), and 7) anticipatory nausea and vomiting (Parker and Kemp, 2001). From this diverse array of antiemetic activity, one may propose that Δ^9-THC and synthetic cannabinoids act as broad-spectrum antiemetic agonists. In addition, several of the cited studies have shown that the antiemetic activity of Δ^9-THC and its synthetic analogs can be dose-dependently countered by nonemetic doses of SR141716A (and not SR144528), which implies an important role for the cannabinoid CB_1 receptor in the antiemetic activity of cannabinoids. Moreover, at higher doses (>10 mg/kg, i.p. or >40 mg/kg, s.c.), SR141716A by itself induces emesis in the least shrew in a dose-dependent fashion. In addition, the emetic activity of SR141716A and several other stimuli were potently blocked by structurally diverse cannabinoids with an antiemetic ID_{50} potency order (CP55940 > WIN55212-2 > Δ^9-THC) that is similar and is highly correlated (Darmani, Sim-Selly, et al., 2003) to their: (1) rank affinity order for cannabinoid CB_1 receptors in shrew brain homogenates and (2) EC_{50} potency order for GTPγS stimulation. In addition, the antiemetic potency of each tested cannabinoid appears to be dependent upon the stimulus-producing emesis. For example, although both WIN55212-2 and Δ^9-THC are equipotent against cisplatin-induced emesis, WIN55212-2 is 4 to 9 times more potent than Δ^9-THC in preventing SR141716A- and 2-AG-induced vomiting (Darmani, 2001a, 2001b, 2001c; Darmani, Sim-Selly, et al., 2003). CP55940 appears to be 4 to 20 times more potent than WIN55212-2 against the cited emetic stimuli (Darmani, 2001a, 2001b, 2001c; Darmani, Sim-Selly, et al., 2003). However, both Δ^9-THC and WIN55212-2 provide significant emesis protection at nonsedative doses, whereas in most studies CP55940 prevented vomiting produced by the cited diverse emetic stimuli at motor suppressive doses (Darmani, 2001a, 2001b, 2001c; Darmani, Sim-Selly, et al., 2003). Overall, these findings suggest that structurally diverse classes of xenobiotic cannabinoids (and possibly the endocannabinoid anandamide) possess significant broad-spectrum antiemetic activity via stimulation of cannabinoid CB_1 receptors.

Another mechanism via which anandamide may prevent emesis can be through the stimulation of vanilloid VR1 receptor, as this endocannabinoid is also a full agonist at the latter site (Szallasi and Di Marzo, 2000). VR1 agonists such as capsaicin and resiniferatoxin initially induce emesis in shrews in a potent manner that can be blocked by VR1 antagonists (Andrews et al., 2000; Rudd and Wai, 2001). However, rapid desensitization occurs following the initial exposure to both emetic and other effects of resiniferatoxin as well as to the emetic action of other stimuli. Although the possible VR1 component of anandamide's emetic/antiemetic actions have not yet been tested directly, it is clear that anandamide produces emesis at a high dose whereas its lower doses may provide protection against some emetic stimuli (Darmani, 2002b; Van Sickle et al., 2001). Thus, anandamide shares with other vanilloid agonists emetic and antiemetic properties. In addition, some downstream metabolites of both 2-AG and anandamide [such as 12 (S) HETE and 15 (S) HETE] have similar or higher affinity for cannabinoid CB_1 receptors, whereas others [e.g., 12 (S) HpETE, 15 (S) HpETE, and leukotriene B_4] are more potent than anandamide at VR1 receptors (Pertwee and Ross, 2002). These metabolites may also exert emetic and antiemetic effects by themselves.

Unlike anandamide and exogenous cannabinoid agonists, the less well investigated endocannabinoid, 2-AG, is a potent emetogenic agent (ED_{50} 0.48-1.13 mg/kg, i.p.) in the least shrew (Darmani, 2002b). It is unlikely that the emetic activity of 2-AG is mediated via cannabinoid CB_1 receptors because indomethacin (a nonselective cyclooxygenase inhibitor) pretreatment prevented the emetogenic effect of 2-AG. This finding implies that one or more downstream metabolites of 2-AG causes vomiting. Indeed, the major metabolite of 2-AG, AA, was found to be a highly potent emetogen ($ED_{50} = 0.58 \pm 2$ mg/kg, i.p.) (Darmani, 2002b). The ability of AA to induce emesis was also blocked by indomethacin, suggesting that further downstream metabolites of 2-AG induce vomiting. As detailed elsewhere in this book, free AA is metabolized to oxygenated products by distinct enzyme systems including cyclooxygenases, one of several lipoxygenases, or cytochrome 450s (Morrow and Roberts, 2001). Cyclooxygenase is a bifunctional enzyme exhibiting both cyclooxygenase and peroxidase activities. The cyclooxygenase component converts AA to PGG_2, and the peroxidase component reduces PGG_2 to PGH_2, which is the precursor of PGs, thromboxanes, and prostacyclins. The 5-, 12- or 15-lipoxygenases are enzymes that catalyze conversion of AA to leukotrienes, HETEs, and HpETEs. Cytochrome P450 enzymes convert AA to a number of epoxy-eicosatrienoic products (EpETrEs). Overall, free AA can be converted to over 100 products, and the pharmacology of many of these metabolites has not yet been investigated. Our ongoing preliminary screening assays on some of these products in the least shrew show that though a number of AA metabolites are emetogenic (PGE_2; 20 (OH) PGE_2; PGG_2; PGH_2; $PGF_{2\alpha}$; PG 20-(OH) $F_{2\alpha}$; 20-HETE; (±) 5 (6)-EpETrE; (±) 11 (12)-EpETrE), other tested metabolites do not produce emesis at 2 mg/kg (PGD_2, PGI_2; tetramer PGFM; 5 (S) HpETE; 13, 14-dihydro-15-keto $PGF_{2\alpha}$; 19 (R)-OH $PGF_{2\alpha}$; 15 keto $PGF_{2\alpha}$). There is substantial published support for our cited preliminary findings. For example: (1) *Staphylococcus-aureus* enterotoxin B produces vomiting in monkeys which is associated in a time-dependent manner with increases in plasma concentrations of several AA metabolites (PGF_2, leukotriene B_4, and 5-HETE) (Jett et al., 1990), (2) significant correlation exists between the rise in maternal PGE_2 serum levels and the symptoms of nausea and vomiting in early pregnancy (Gadsby et al., 2000), (3) administration of PGE_2 or $PGF_{2\alpha}$ produces vomiting and diarrhea in both piglets (Wechsung, 1996) and patients (Wislicki, 1982), and (4) PG synthesis inhibitors such as dexamethasone are used as additive antiemetics for the prevention of vomiting produced by both chemo- and radiotherapy (Ioanidis et al., 2000).

An important question remains to be resolved: Why is 2-AG a potent emetogen whereas anandamide mainly possess antiemetic activity? Although the exact mechanism is unclear, several factors may contribute to the phenomenon of differential effects of endocannabinoids on emesis:

1. Though both endocannabinoids are metabolized to a common potent emetic metabolite (AA), 2-AG is metabolized by FAAH several-fold faster than anandamide (Pertwee and Ross, 2002).
2. Besides FAAH, other enzymes such as monoglycerol lipase also convert 2-AG to AA (Sugiura et al., 2002).
3. 2-AG acts as a natural AA precursor whose metabolism by COX and generation of PGE_2 (an emetic agent) participates in positive modulation of inflammatory mediators such as NO production and COX-2 induction, which may induce certain types of vomiting (Chang et al., 2001; Jeon et al., 1996).
4. In the latter studies, anandamide was shown to inhibit production of proinflammatory agents such as NO, PGE_2, and IL-6, which may help to avert production of emesis under certain conditions.

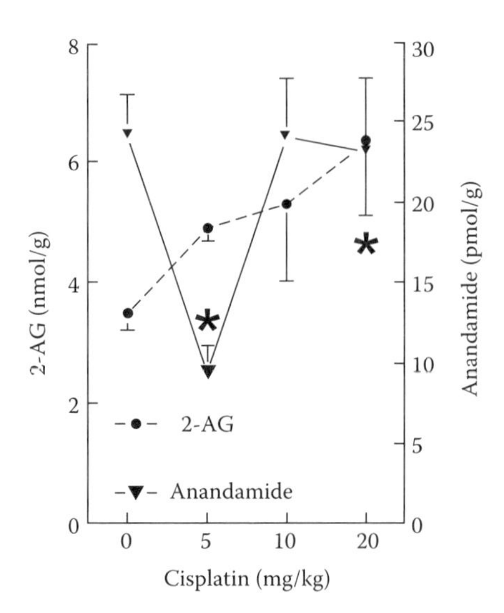

FIGURE 17.1 Effects of cisplatin exposure on shrew brain contents of 2-AG and anandamide. Different groups of shrews (*Cryptotis parva*) were injected with varying doses of cisplatin (0, 5, 10, and 20 mg/kg, i.p., n = 4 per group) and 30 min later, the animals were sacrificed by rapid decapitation, and whole brains were frozen at −80°C until endocannabinoid analysis. *Significantly different from corresponding vehicle-treated control group.

5. The more stable analogs of both 2-AG (noladin ether, also a putative endocannabinoid) and anandamide (methanandamide) do not induce vomiting (Darmani, 2002b and Darmani, unpublished findings).
6. As discussed earlier, anandamide but not 2-AG can bind vanilloid VR1 receptors whose activation can lead to emetic/antiemetic effects, a property which anandamide shares.
7. Though the tissue levels of both anandamide and 2-AG can be simultaneously altered by drug administration or in pathological states, there is also strong evidence that the release of 2-AG can be independently modified, which can be region specific (Pertwee and Ross, 2002).

In the light of these findings, preliminary results of our recent collaboration with the Italian endocannabinoid research group suggest that a specific increase in brain 2-AG (but not anandamide) levels may contribute to cisplatin-induced vomiting (N.A. Darmani, G. Ambrosino, M.G. Cascio, V. Di Marzo). Indeed, a 30-min prior treatment with the chemotherapeutic agent cisplatin (0, 5, 10, and 20 mg/kg), specifically and dose-dependently, increased ($p < .05$) 2-AG but not anandamide levels in shrew whole brain (Figure 17.1). In fact, anandamide tissue level was decreased in the 5 mg/kg cisplatin dose group. This preliminary finding has been confirmed by our more recent extensive studies (Darmani et al., 2005). These findings, in conjunction with our previously published report on the potent emetic activity of 2-AG and its downstream metabolites (Darmani, 2002b), strongly suggest an important role for 2-AG and its metabolites in mediation of chemotherapy-induced vomiting.

IN VITRO AND *IN VIVO* INVERSE AGONIST/ANTAGONIST EFFECTS OF SR141716A (RIMONABANT) ON GI MOTILITY, SECRETION, FEEDING, AND VOMITING

Numerous lines of diverse evidence suggest that a number of G-protein-coupled receptors (such as the cannabinoid CB_1 receptor) may exist in a state that allows a tonic level of stimulation *in vivo* in the absence of an activating ligand, i.e., such receptors are constitutively active and therefore would signal without agonist stimulation (Leurs et al., 1998). Cannabinoid CB_1 receptor antagonists such as SR141716A and LY320135 not only reverse CB_1-mediated effects in the GIT and other tissues both *in vitro* and *in vivo* (Pertwee, 1999), but they also elicit responses by themselves in some but not all CB_1-receptor-mediated assays that are opposite in direction (i.e., inverse agonism) from those produced by CB_1 agonists. Thus, both SR141716A and LY320135 behave as silent (or pure) antagonists in some assays, whereas in other CB_1 test systems, they act as inverse agonist or antagonist compounds. The direct antagonist component of actions of these CB_1 ligands is attributable to a competitive blockade of responses elicited at CB_1 receptors by endogenously released endocannabinoids, whereas their inverse agonist activity would arise from their ability to directly reduce the constitutive activity of CB_1 receptors.

Regarding GIT motility, several *in vitro* studies have shown that when administered by itself at low n*M* concentrations, SR141716A potentiates electrically evoked contractions of longitudinal and circular smooth muscles of the guinea-pig small intestine (Pertwee, 2001; Pinto, Capasso, et al., 2002). The excitatory effects of SR141716A are probably due to postjunctional release of acetylcholine from the myenteric plexus nerve endings because the CB_1 antagonist has been found to augment electrically evoked release of this cholinergic excitatory neurotransmitter but not the direct effects of exogenous acetylcholine (Coutts and Pertwee, 1997). Likewise, in intestinal preparations, SR141716A increases: (1) maximal ejection pressure during the empty phase of peristalsis in the guinea-pig small intestine (Izzo, Mascolo, et al., 2000), (2) distension-induced AER contractions of guinea-pig ileal circular muscle (Heinemann et al., 1999), and (3) tonic and phasic peristaltic activities in longitudinal smooth muscle of mouse colon (Mancinelli et al., 2001). In line with these *in vitro* findings, most but not all *in vivo* studies support an excitatory role for SR141716A on intestinal motility. Indeed, SR141716A has been shown to increase the upper GIT transit (as indicated by propulsion of nonabsorbable markers), defecation, and fecal water content in mice (Casu et al., 2003; Colombo, Agabio, Lobina, et al., 1998; Costa et al., 1999; Izzo, Mascolo, Borrelli, et al., 1999; Izzo, Pinto, et al., 2000, Izzo, Fezza, et al., 2001; Mascolo et al., 2002). Likewise, SR141716A enhances defecation, upper GIT motility, and small intestinal fluid accumulation in rats (Izzo, Mascolo, Pinto, et al., 1999; Izzo, Mascolo, Borrelli, 1999; Costa and Colleoni, 1999; Costa et al., 1999). Although SR141716A increases intestinal fluid accumulation and the fecal content of water (Izzo, Mascolo, Pinto, et al., 1999; Izzo, Mascolo, Borrelli, 1999), it has no effect on stimulated gastric acid release (Adami et al., 2002; Coruzzi et al., 1999). Although most studies have shown that SR141716A suppresses food intake in rodents (Arnone et al., 1997; Simiand et al., 1998; Colombo, Agabio, Diaz, et al., 1998; Kirkham and Williams, 2001), others have failed to show such suppression of appetite (Jamshidi and Taylor, 2001; Koch and Werner, 2000). Differences in methodology may account for the observed diverse effects of SR141716A on appetite.

Both 2-AG and anandamide, as well as their major metabolic enzyme (FAAH) (an uptake mechanism for endocannabinoids), and cannabinoid CB_1 receptors are all found in the GIT and brain circuits that control the above effects (Pertwee, 2001, also see earlier), it is likely that the observed *in vitro* and *in vivo* actions of SR141716A are a consequence of reducing the background inhibitory endocannabinoid tone that probably exists in the GI tissue. Indeed, when the basal intestinal tissue level of anandamide was increased by acetic acid in a mouse model of paralytic ileus, SR141716A was less effective in stimulating intestinal motility relative to vehicle-exposed control group (Mascolo et al., 2002). Although the ability of nonselective (AM404) and selective (VDM11) inhibitors of endocannabinoid uptake to potentiate, respectively, the inhibitory *in vitro* and *in vivo* effects of anandamide on GIT motility further support the notion of an inhibitory cannabinoid tone (Pertwee, 2001; Pinto, Izzo, et al., 2002), other studies (Pertwee, 2001) do not support this view because: (1) not all CB_1 antagonists promote GI motility in every published study, (2) the FAAH inhibitor, phenylenethylsuphonyl fluoride, by itself does not mimic the inhibitory effect of anandamide on electrically evoked contractions of guinea-pig ileum, and (3) SR141716A does not increase several other parameters of contractility in the guinea-pig ileum. Thus, the inverse agonist rather than the pure silent antagonist nature of SR141716A may also be a contributing factor to the observed direct effects of the CB_1 antagonist. The results of a study on rat hippocampal synaptosomes clearly support the latter possibility (Gifford et al., 2000). In synaptosomes in which endocannabinoids are efficiently removed by a superfusion stream resulting in no accumulation of 2-AG or anandamide in the biophase of receptors, SR141716A nonetheless facilitates acetylcholine release. Thus, the SR141716A effect cannot be explained by the interruption of a tonic inhibition of acetylcholine release by endocannabinoids.

It is of interest to note that, generally, SR141716A stimulates GI motility by itself at higher doses than those sufficient to antagonize the inhibitory effects of CB_1 agonists (Pertwee, 2001). In line with this notion, Sim-Selley et al. (2001) have suggested that SR141716A behaves as a competitive CB_1 antagonist at n*M* concentrations, and at μ*M* concentrations it probably acts as an inverse agonist. Likewise, we have observed that lower doses (<10 mg/kg, i.p.) of SR141716A counters the antiemetic effects of structurally diverse cannabinoids (see preceding text), whereas at larger doses the CB_1 antagonist (>10 mg/kg, i.p.) by itself induces vomiting (Darmani, 2001a). However, the ability of SR141716A to induce vomiting and other behaviors (such as serotonergically mediated head-twitch and ear-scratching behaviors) could be also due to enhancements in release and turnover of serotonin and dopamine in the CNS (Darmani, Janoyan, et al., 2003; Darmani and Pandya, 2000).

CENTRAL VS. PERIPHERAL COMPONENTS OF ACTIONS OF CANNABINOID CB_1 LIGANDS ON GI MOTILITY, FEEDING, AND EMESIS

As mentioned earlier in this chapter, the CNS has a significant role in the initiation and coordination of feeding, GI motility, and vomiting. The CB_1 receptor and its endogenous ligands are found within the brainstem and the peripheral ENS, both of which control these processes. Because initial studies had shown that intravenously (IV) active doses of Δ^9-THC were required to inhibit rat intestinal transit when it was administered via the intracerebroventricular (ICV) route, Shook and Burks (1989) concluded that cannabinoids act via a peripheral site to produce their antimotility effects on the GIT. However, more recent studies show that structurally diverse cannabinoid agonists (WIN55212-2, cannabinol, ACEA) and the CB_1 antagonist SR141716A are all significantly more potent on GIT function when administered ICV than when given intraperitoneally (IP) (Izzo, Pinto, et al., 2000; Pinto, Izzo, et al., 2002). Further support for a central site of cannabinoid action comes from the ability of: (1) IP-administered hexamethonium (a ganglionic blocker) to counter both the prokinetic GI effects of ICV-administered SR141716A (Izzo, Pinto, et al., 2000) and the antipropulsive effects of ICV injected ACEA (Pinto, Izzo, et al., 2002), (2) orally or ICV-administered SR141716A to prevent the delay in gastric emptying and intestinal transit caused by ICV-administered WIN55212-2 (Landi et al., 2002), (3) ganglionic blockade or vagotomy to abolish the gastric motor effects of peripherally administered Δ^9-THC (Krowicki et al., 1999), and (4) ICV-administered pertussis toxin to reduce inhibition of gastric emptying and the antitransit intestinal effects of centrally administered WIN55212-2 (Landi et al., 2002). However, other evidence suggests that central cannabinoid CB_1 receptors are not the major contributor for the effects of peripherally administered cannabinoid CB_1 ligands because: (1) the antipropulsive effects of peripherally administered ACEA were resistant to hexamethonium (IP) in mice (Pinto, Izzo, et al., 2002), (2) both ICV-administered SR141716A and pertussis toxin failed to counter the antipropulsive activity of IP-administered WIN55212-2 on gastric emptying and GIT transit in rats (Landi et al., 2002), and (3) peripheral administration of hexamethonium failed to modify the promotility GIT effects of IP-administered SR141716A in mice (Izzo, Mascolo, Borrelli, et al., 1999). Overall, the cited published studies suggest that peripheral cannabinoid CB_1 receptors play a primary (but not exclusive) role in the GI transit effects of peripherally administered cannabinoid agonists and antagonists.

Likewise, systemically administered cannabinoid CB_1 ligands (both agonists and antagonists) appear to affect food intake and appetite predominantly by engaging peripheral CB_1 receptors because: (1) central administration of SR141716A does not affect food intake in presatiated or food-deprived animals (Jamshidi and Taylor, 2001; Gomez et al., 2002), (2) capsaicin-induced sensory deafferentation prevents the hyperphagic effects of peripherally administered anandamide and WIN55212-2 as well as the hypophagic effects of SR141716A (Gomez et al., 2002), and (3) food deprivation was shown to produce a sevenfold increase in anandamide content in the small intestine but not in the brain or stomach (Gomez et al., 2002). Moreover, refeeding normalized intestinal anandamide levels.

It is generally well-accepted that the antiemetic effects of 5-HT_3- and neurokinin NK_1-receptor antagonists can be mediated via both central and peripheral mechanisms (Veyrat-Follet et al., 1997; Diemunsch and Grelot, 2000; Darmani, 1998). Although, in elegant studies, Van Sickle et al. (2001) have shown the presence of both CB_1 receptors and FAAH in the dorsal vagal complex (which contains several emetic nuclei), their conclusion that cannabinoids inhibit emesis solely via brainstem CB_1 receptors may be premature. In a recent study, we have observed that Δ^9-THC more potently suppresses the ability of the serotonin precursor 5-HTP to induce vomiting in shrews in the presence of carbidopa (a peripheral decarboxylase inhibitor) relative to its absence (Darmani, 2002c). This finding implies that relatively less Δ^9-THC is required to prevent vomiting produced via central 5-HT_3 receptors (i.e., in the presence of carbidopa) compared to emesis produced by 5-HTP via both central and peripheral mechanisms (i.e., in absence of carbidopa). Thus, cannabinoids probably prevent vomiting via the activation of both peripheral and central CB_1 receptors.

POSSIBLE ROLE OF ENDOCANNABINOIDS IN GI PATHOLOGICAL STATES

Whenever the GIT is challenged by infection, allergy, inflammation, or other injury, the GI immune system is called into operation, which releases a host of proinflammatory mediators including cytokines, PGs, leukotrienes, NO, etc. (Holzer, 2001). Paralytic ileus results from activation of a network of macrophages that reside in intestinal muscularis (Kalff et al., 2003). Release of some of the cited proinflammatory agents from the macrophage network in turn causes the sluggish electrical and contractile responses of the muscularis that are the hallmark of paralytic ileus. Recently, the role of endocannabinoids has been suggested in the pathogenesis of paralytic ileus induced by intraperitoneal administration of acetic acid in mice (Mascolo et al., 2002). Indeed, relative to the vehicle-treated control group, the pathological state was accompanied by increased: (1) tissue levels of intestinal anandamide, and (2) number of cannabinoid CB_1 receptors in myenteric plexes and nerve bundles of fibers of the external muscles of the jejunum. The authors suggest that increased levels of anandamide coupled with the overexpression of CB_1 receptors result in overactivation of the endocannabinoid system, which inhibits the release of the excitatory cholinergic neurotransmitter acetylcholine leading to the development of paralytic ileus. The CB_1 antagonist SR141716A alleviated the induced hypomotility as it restored normal intestinal motility. Furthermore, the selective inhibitor of anandamide membrane transporter, VDM11, significantly delayed GI transit in acetic acid-treated mice, which further supports the role of anandamide in this pathological state. The effect of acetic acid on the tissue levels of 2-AG in this model of paralytic ileus remains unknown. In regard to feeding, the levels of both 2-AG and anandamide have been shown to be elevated in the hypothalamus of genetically obese rats with hyperphagic syndrome (Di Marzo et al., 2001), which indicates aberrant activity of the endocannabinoid system in appetitive disorders.

In another animal model of intestinal inflammation (caused by oral administration of croton oil in mice), an increased expression in the density of intestinal cannabinoid CB_1 receptors has also been observed (Izzo, Fezza, et al., 2001). In line with increased CB_1 expression, this and other studies (Izzo, Pinto, et al., 2000) show that cannabinoid CB_1 agonists (CP55940, WIN55212-2, and cannabinol) more potently delay intestinal motility in croton-oil-treated mice than in control mice.

SR141716A pretreatment countered these inhibitory effects, but when administered alone it increased intestinal motility to the same extent in both control and croton-oil-exposed mice. Moreover, anandamide and 2-AG level assay showed no significant difference between inflamed and noninflamed intestinal tissue. The lack of increase in endocannabinoid levels in the inflamed intestine could be due to a more rapid turnover of both 2-AG and anandamide, which could be due to upregulation of anandamide amidohydrolase as the activity of this enzyme was significantly increased in the inflamed intestine.

As discussed earlier, studies in this laboratory have shown that 2-AG is a potent emetogenic agent, whereas anandamide possesses antiemetic and emetic effects (Darmani, 2002b). Xenobiotic cannabinoids prevent 2-AG-induced emesis as well as vomiting produced by other emetic stimuli. The emetic activity of 2-AG was also blocked by the nonselective cyclooxygenase inhibitor indomethacin, suggesting downstream metabolites of endocannabinoids are responsible for the induced emesis. The chemotherapeutic agent cisplatin not only produces vomiting but also causes intestinal mucosal damage and inflammation, diarrhea, and ileus, as well as release of emetic agents such as serotonin, dopamine, and downstream emetic metabolites of endocannabinoids such as PGE_2 and $PGF_{2\alpha}$ (Darmani, 2001b, 2001c; Goto et al., 1998; Ito et al., 1994; Endo et al., 1990; Bostrom, 1988; Dewit et al., 1987). In collaboration with Di Marzo's laboratories, we have shown that IP administration of cisplatin, specifically and dose dependently, increases 2-AG but not anandamide levels in shrew brains (see Figure 17.1). These results further support a role for endocannabinoids in GI-related pathological states.

SUMMARY AND CONCLUSIONS

The neural control of GI functions such as motility, secretion, appetite, and emesis involve to varying degrees both central and peripheral mechanisms. In these control circuits, the presence of 2-AG and anandamide, cannabinoid CB_1 receptors, endocannabinoid metabolic enzymes, and the endocannabinoid reuptake system have largely been confirmed. Although relatively extensive published findings already exist on the GIT effects of exogenous cannabinoids, currently a limited amount of information is also available on the effects of anandamide. Though 2-AG was the first endocannabinoid to be discovered in the intestine, thus far, a search for its role on GI functions has virtually been ignored. Thus, anandamide (or its more stable analog, methanandamide) and structurally diverse exogenous cannabinoids (Δ^9-THC, WIN55212-2, and CP55940) inhibit evoked GI contractility in several *in vitro* and *in vivo* models and delay the process of gastric emptying in different species. The antimotility effects of cannabinoids are thought to be due to inhibition of evoked acetylcholine release (and also release of other excitatory neurotransmitters such as substance P) from enteric cholinergic neurons (and nonadrenergic noncholinergic neurons) via the activation of cannabinoid CB_1 receptors. Though the effects of endocannabinoids on GI secretions remain unknown, exogenous cannabinoids inhibit both basal and evoked intestinal fluid accumulation and evoked but not basal gastric acid secretion via CB_1 receptors. These receptors are also involved in orexigenic effects of both endogenous and exogenous cannabinoids and in the anorexigenic effect of the selective CB_1 receptor antagonist and inverse agonist SR141716A. 2-AG is a potent emetogen that produces vomiting following rapid conversion to its downstream emetic metabolites, whereas anandamide and exogenous cannabinoids exert antiemetic activity via CB_1 receptors. SR141716A by itself enhances GI motility as well as producing anorexigenic activity and vomiting. Some direct effects of SR141716A are partly explained in terms of either inverse agonist activity or antagonism of an endocannabinoid tone on the GIT. Endocannabinoids may have important functions in the GIT because several models of GI pathological states exhibit an overexpression of enteric intestinal cannabinoid CB_1 receptors accompanied with or without increases in tissue endocannabinoid levels and turnover. The current literature suggests that both cannabinoid CB_1 agonists and antagonists produce these diverse GI effects mainly via peripheral enteric CB_1 receptors, though central CB_1 sites in the brainstem and the hypothalamus may also contribute, in varying degrees, to some of these effects. In addition to CB_1 receptors, anandamide can induce stimulatory GI effects via vanilloid VR1 receptors, whereas downstream metabolites of both endocannabinoids (e.g., PGs) produce intestinal contractility as well as emesis.

ACKNOWLEDGMENTS

This work was supported in part by grants from the National Institute on Drug Abuse (DA 12605) and Solvay Pharmaceuticals, Inc. The author would like to thank Professor R. Theobald for proofing and helpful suggestions and R. Chronister for typing the manuscript.

REFERENCES

Adami, M., Bertini, S., Frati, P., Soldani, G., and Coruzzi, G. (2000) Cannabinoid CB_1 receptors are involved in the regulation of net gastric acid secretion, *Pharmacol. Commn.* **6**: 273–275.

Adami, M. Frati, P., Bertini, S., Kulkarni-Narla, A., Brown, D.R., de Caro, G., Coruzzi, G., and Soldani, G. (2002) Gastric antisecretory role and immunohistochemical localization of cannabinoid receptors in the rat stomach, *Br. J. Pharmacol.* **135**: 1598–1606.

Adams, R. (1942) Maharihuana, *Harvey Lecture* **37**: 168–197.

Ammar, A.A., Sederholm, F., Saito, T.R., Scheurink, A.J.W., Johnson, A-E., and Sodersten, P. (2000) NPY-leptin: opposing effects on appetitive and consumatory ingestive behavior and sexual behavior, *Am. J. Physiol.* **278**: R1627–R1633.

Andrews, P.L.R. and Bhandari, P. (1993) Resiniferatoxin, an ultra potent capsaicin analogue, has antiemetic properties in the ferret, *Neuropharmacology* **32**: 799–806.

Andrews, P.L.R., Okada, F., Woods, A.J., Hagiwara, H., Kakaimoto, S., Toyoda, M., and Matsuki, N. (2000) The emetic and antiemetic affects of the capsaicin analog resiniferatoxin in *Suncus murinus*, the house musk shrew, *Br. J. Pharmacol.* **130**: 1247–1254.

Arnone, M., Maruani, J., Chaperon, F., Thiebot, M.H., Poncelet, M., Soubrie, P., and Le Fur, G. (1997) Selective inhibition of sucrose and ethanol intake by SR 141716A, an antagonist of central cannabinoid (CB1) receptors, *Psychopharmacology* **132**: 104–106.

Bartho, L. and Holzer, P. (1995) The inhibitory modulation of guinea-pig intestinal peristalsis caused by capsaicin involves calcitonin gene related peptide and nitric oxide, *Naunyn Schmied Arch Pharmacol.* **353**: 102–109.

Beal, J.E., Olson, R., Laubenstein, L., Morales, J.O., Bellman, P., Yangco, B., Lefkowitz, L., Plasse T.F., and Shepard, K.V. (1995) Dronabinol as a treatment for anorexia associated with weight loss in patients with AIDS, *J. Pain Symptom Manage.* **10**: 89–97.

Beal, J.E., Olson, R. Lefkowitz, L., Laubenstein, L., Bellman, P., Yangco, B., Morales, J.O., Murphy, R., Powderly, W., Plasse, T.F., Mosdell, K.W., and Shepard, K. (1997) Long-term efficacy and safety of dronabinol for acquired immunodeficiency syndrome-associated anorexia, *J. Pain Symptom. Manage.* **14**: 7–14.

Berry, E.M. and Mechoulam, R. (2002) Tetrahydrocannabinol and endocannabinoids in feeding and appetite, *Pharmacol. Ther.* **95**: 185–190.

Bisogno, T., Berrendero, F., Ambrosino, G., Cebeira, M., Ramos, J.A., Fernandez-Ruiz, J.J., and Di Marzo, V. (1999) Brain regional distribution of endocannabinoids: implications for their biosynthesis and biological function, *Biochem. Biophys. Res. Comm.* **256**: 377–380.

Bostrom, B. (1988) Severe ileus from cisplatin and vinblastine infusion in neuroblastoma, *J. Clin. Oncol.* **6**: 1356.

Breivogel, C.S. and Childers, S.R. (2000) Cannabinoid signal transduction in rat brain: comparison of cannabinoid agonists in receptor binding, G-protein activation and adenylate cyclase inhibition, *J. Pharmacol. Exp. Ther.* **295**: 328–336.

Breivogel, C.S., Griffin, G., Di Marzo, V., and Martin, B.R. (2001) Evidence for a new G protein-coupled cannabinoid receptor in the mouse brain, *Mol. Pharmacol.* **60**: 155–163.

Broberger, C. and Hökfelt, T. (2001) Hypothalamic and vagal neuropeptide circuitries regulating food intake, *Physiol. Behav.* **74**: 669–682.

Brown, D.R. and Miller, R.J. (1991) Neurohormonal control of fluid and electrolyte transport in intestinal mucosa, in *Handout of Physiology, The Gastrointestinal System IV, Gastrointestinal Physiology: Absorptive and Secretory Process of the Intestine*, M. Field and R.A. Frizzel, Eds. pp. 527–589. Bethesda, MD: American Physiological Society.

Buckley, N.E., Hansson, S., Harta, G., and Mezey, E. (1998) Expression of the CB_1 and CB_2 receptor messenger RHAs during embryonic development in the rat, *Neuroscience* **82**: 1131–1149.

Burakoff, R. and Percy, W.H. (1992) Studies in vivo and in vitro on effects of PGE2 on colonic motility in rabbits, *Am. J. Physiol.* **262** (1 pt. 1): G23–G229.

Calignano, A., La Rana, G., Makriyannis, A., Lin, S.Y., Beltramo, M., and Piomelli, D. (1997) Inhibition of intestinal motility by anandamide, an endogenous cannabinoid, *Eur. J Pharmacol.* **340**: R7–R8.

Casu, M.A., Porcella, A., Ruiu, S., Saba, P., Marchese, G., Carai, M.A.M., Reali, R., Gessa, G.L., and Pani, L. (2003) Differential distribution of functional cannabinoid CB_1 receptors in the mouse gastroenteric tract, *Eur. J. Pharmacol.* **459**: 97–105.

Chang, Y.-H., Lee, S.T., and Lin, W.-W. (2001) Effects of cannabinoids on LPS-stimulated inflammatory mediator release from macrophages: involvement of eicosanoids, *J. Cell Biochem.* **81**: 715–723.

Chiesi, M., Huppertz, C., and Hofbauer, K.G. (2001) Pharmacotherapy of obesity: targets and prospectives, *Trends Pharmacol. Sci.* **22**: 247–254.

Colombo, G., Agabio, R., Diaz, G., Lobina, C., Reali, R., and Gessa, G.L. (1998) Appetite suppression and weight loss after the cannabinoid antagonist SR 141716A, *Life Sci.* **63**: 113–117.

Colombo, G., Agabio, R., Lobina, C., Reali, R., and Gessa, G.L. (1998) Cannabinoid modulation of intestinal propulsion in mice, *Eur. J. Pharmacol.* **344**: 67–69.

Coruzzi, G., Adami, M., Coppelli, G., Frati, P., and Soldani, G. (1999) Inhibitory effect of the cannabinoid receptor agonist WIN 55, 212-2 on pentagastrin-induced acid secretion in the anesthetized rat, *Naunyn-Schmied. Arch. Pharmacol.* **360**: 715–718.

Costa, B. and Colleoni, M. (1999) SR 141716A induces in rats a behavioral pattern opposite to that of CB_1 receptor agonists, *Zhongguo Yao Li Xue Bao* **20**: 1103–1108.

Costa, B., Vailati, S., and Colleoni, M. (1999) SR 141716A, a cannabinoid receptor antagonist, reverses the behavioral effects of anandamide-treated rats, *Behav. Pharmacol.* **10**: 327–331.

Coutts, A.A. and Pertwee, R.G. (1997) Inhibition by cannabinoid receptor agonists of acetylcholine release from the guinea-pig myenteric plexus, *Br. J. Pharmacol.* **121**: 1557–1566.

Croci, T., Manara, L., Aureggi, G., Guagnini, F., Rinaldi-Carmona, M., Maffrand, J-P., Le Fur, G., Mukenge, S., and Ferla, G. (1998) In vitro functional evidence of neuronal cannabinoid CB_1 receptors in human ileum, *Br. J. Pharmacol.* **125**: 1393–1395.

Croxford, J.L. (2003) Therapeutic potential of cannabinoids in CNS disease, *CNS Drugs* **17**: 179–202.

Darmani, N.A. (2002a) Antiemetic action of Δ^9-tetrahydrocannabinold and synthetic cannabinoids in chemotherapy-induced nausea and vomiting, in *Biology of Marijuana: From Gene to Behavior*, E.S. Onaivi, Ed., pp. 356–389. London, U.K.: Taylor and Francis.

Darmani, N.A. (2002b) The potent emetogenic effects of the endocannabinoid, 2-AG (2-arachidonoylglycerol) are blocked by Δ^9-tetrahydrocannabinol and other cannabinoids, *J. Pharmacol. Exp. Ther.* **300**: 34–42.

Darmani, N.A. (2002c) Δ^9-THC pretreatment concomitantly prevents both 5-HTP-induced $5\text{-}HT_3$-receptor mediated vomiting and $5\text{-}HT_{2A}$ receptor-mediated head-twitching in the least shrew, Program No. 738.8, Abstract Viewer/Itinerary Planner, Washington, D.C., Society for Neuroscience.

Darmani, N.A. (2001a) Δ^9-Tetrahydrocannabinol and synthetic cannabinoids prevent emesis produced by the cannabinoid CB_1 receptor antagonist/inverse agonist SR 141716A, *Neuropsychopharmacology* **24**: 198–203.

Darmani, N.A. (2001b) Delta-9-tetrahydrocannabinol differentially suppresses cisplatin-induced emesis and indices of motor function via cannabinoid CB_1 receptors in the least shrew, *Pharmacol. Biochem. Behav.* **69**: 239–249.

Darmani, N.A. (2001c) The cannabinoid antagonist/inverse agonist SR 141716A reverses the antiemetic and motor depressant action of WIN 55, 212-2 in the least shrew, *Eur. J. Pharmacol.* **430**: 49–58.

Darmani, N.A. (1998) Serotonin $5\text{-}HT_3$ receptor antagonists prevent cisplatin-induced emesis in *Cryptotis parva*: a new experimental model of emesis, *J. Neural Trans.* **105**: 1143–1154.

Darmani, N.A. and Pandya, D.K. (2000) Involvement of other neurotransmitters in behaviors induced by the selective cannabinoid CB_1 receptor antagonist/inverse agonist in naive mice, *J. Neural Trans.* **107**: 931–945.

Darmani, N.A., McClanahan, B.A., Trinh, C., Petrosino, S., Valenti, M., and Di Marzo, V. (2005) Cisplatin increases brain 2-arachidonoylglycerol (2-AG) and concomitantly reduces intestinal 2-AG and anandamide levels in the least shrew. *Neuropharmacology* (in press).

Darmani, N.A., Sim-Selly, L.J., Martin, B.R., Janoyan, J.J., Crim, J.L., and Breivogel, C.S. (2003) Antiemetic and motor depressive actions of CP55, 940: Cannabinoid CB_1 receptor characterization, distribution and G-protein activation, *Eur. J. Pharmacol.* **459**: 83–95.

Darmani, N.A., Janoyan, J.J., Kumar, N., and Crim, J.L. (2003) Behaviorally active doses of the CB_1 receptor antagonist SR 141716A increase brain serotonin and dopamine levels and turnover, *Pharmacol. Biochem. Behav.* 75: 777–787.

De Saedeleer, V., Wechsung, E., and Houvenaghel, A. (1992) Effect of infusion of a diarrhoegenic dose of PGE2 on gastrointestinal electrical activity in the conscious piglet, *Prostag Leukot. Essent. Fatty Acids* **45**: 67–70.

Dewit, L., Oussoren, Y., and Bartelink, H. (1987) Early and late damage in the mouse rectum after irradiation and cis-diamine dichloroplatinum (II), *Radiother. Oncol.* **8**: 57–69.

Diemunsch, P. and Grelot, L. (2000) Potential of substance P antagonists as antiemetics, *Drugs* **60**: 533–546.

Di Marzo, V., Goparaju, S.K., Wang, L., Liu, J., Batkai, S., Jarai, Z., Fezza, F., Miura, G.I., Palmiter, R.D., Sugiura, T., and Kunos, G. (2001) Leptin-regulated endocannabinoids are involved in maintaining food intake, *Nature* **410**: 822–825.

Endo, T., Minami, M., Mona, Y., Saito, H., and Takeuchi, M. (1990) Emesis-related biochemical and histopathological changes induced by cisplatin in the ferret, *J. Toxicol.* **15**: 235–244.

Gadsby, R., Barnie-Ashead, A., Grammatoppoulos, D., and Gadsby, P. (2000) Nausea and vomiting in pregnancy: An association between symptoms and maternal prostaglandin E_2, *Gynecol. Obstet. Invest.* **50**: 149–152.

Gardiner, P.J., Abram, T.S., and Cuthbert, N.J. (1990) Evidence for two leukotriene receptor types in the guinea-pig isolated ileum, *Eur. J. Pharmacol.* **182**: 291–299.

Germano, M.P., D'Angelo, V., Mondello, M.R., Pergolizzi, S., Capasso, F., Capasso, R., Izzo, A.A., Mascolo, N., and De Pasquale, R. (2001) Cannabinoid CB_1-mediated inhibition of stress-induced gastric ulcers in rats, *Naunyn-Schmied. Arch. Pharmacol.* **363**: 241–244.

Gifford, A.N., Bruneus, M., Gatley, S.J., and Volkow, N.D. (2000) Cannabinoid receptor-mediated inhibition of acetylcholine release from hippocampal and cortical synaptosomes, *Br. J. Pharmacol.* **131**: 645–650.

Glass, M., Dragunow, M., and Faull, R.L.M. (1997) Cannabinoid receptors in the human brain: A detailed anatomical and quantitative autoradiographic study in the fetal, neonatal and adult human brain, *Neuroscience* **77**: 299–318.

Gomez, R., Navarro, M., Ferrer, B., Trigo, J.M., Bilbao, A., Del Arco, I., Cippitelli, A., Nava, F., Piomelli, D., Rodriguez de Fonseca, F. (2002) A peripheral mechanism for CB1 cannabinoid receptor-dependent modulation of feeding, *J. Neurosci.* **22**: 9612–9617.

Goto, H., Tachi, K., Arisawa, T., Niwa, Y., Hayakawa, T., and Sugiyama, S. (1998) Effects of gamma-glutamylcysteine ethyl ester in cisplatin-induced changes in prostanoid concentrations in rat gastric and colonic mucosa, *Cancer Detect. Prev.* **22**: 153–160.

Graceffo, T.J. and Robinson, J.K. (1998) Delta-9-tetrahydrocannabinol (THC) fails to stimulate consumption of a highly palatable food in the rat, *Life Sci.* **62**: PL85–PL88.

Grobovic, L. and Radmanovic, B.Z. (1987) The action of prostaglandin E_2 on the rhythmic activity of circular muscle of the isolated guinea-pig ileum, *Arch. Int. Pharmacodyn. Ther.* **285**: 166–176.

Halford, J.C.G. and Blundell, J.E. (2000) Separate systems for serotonin and leptin in appetite control, *Ann. Med.* **32**: 222–232.

Hanus, L., Alu-Lafi, S., Fride, E., Breuer, A., Vogel, Z., Shalev, D.E., Kustanovich, I., and Mechoulam, R. (2001) 2-Arachidonyl glycerol ether, an endogenous agonist of the cannabinoid CB_1 receptor, *Proc. Natl. Acad. Sci. U S A* **98**: 3662–3665.

Hao S., Avraham, Y., Mechoulam, R., and Berry, E.M. (2000) Low dose anandamide affects food intake, cognitive function, neurotransmitter and corticosterone levels in diet-restricted mice, *Eur. J. Pharmacol.* **392**: 147–156.

Harrold, J.A., Elliot, J.C., King, P.J., Widdowson, P.S., and Williams, G. (2002) Down-regulation of cannabinoid-I (CB-I) receptor in specific extrahypothalamic regions of rats with dietary obesity: a role for endogenous cannabinoids in driving appetite for palatable food, *Brain Res.* **952**: 232–238.

Heinemann, A., Shabazian, A. and Holzer, P. (1999) Cannabinoid inhibition of guinea-pig intestinal peristalsis via inhibition of excitatory and activation of inhibitory neural pathways, *Neuropharmacology* **38**: 1289–1297.

Herkenham, M., Lynn, A.B., Johnson, M.R., Melvin, L.S., de Costa, B.R., and Rice K.C. (1991) Characterization and localization of cannabinoid receptors in rat brain: a quantitative in vitro autoradiographic study, *J. Neurosci.* **11**: 563–583.

Holzer, P. (2000) Tachykinins: from basic concepts to therapeutic implications, in *Drug Development Molecular Targets for GI Diseases*, Eds. S. Gaginella and A. Guglieta, pp. 113–146. New Jersey: Hamana Press.

Holzer, P. (2001) Gastrointestinal afferents as targets of novel drugs for the treatment of functional bowel disorders and visceral pain, *Eur. J. Pharmacol.* **429**: 177–193.

Holzer, P., Lippe, I.T., Heinemann, A., and Barthó, L. (1998) Tachykinin NK1 and NK2 receptor-mediated control of peristaltic propulsion in the guinea-pig small intestine in vitro, *Neuropharmacology* **37**: 131–138.

Ioanidis, J.P.A., Hesketh, P.J., and Lau, J. (2000) Contributions of dexamethasone to control chemotherapy-induced nausea and vomiting: a meta-analysis of randomized evidence, *J. Clin. Oncol.* **18**: 3409–3422.

Ishizawa, M. and Miyaraki, E. (1975) Effect of prostaglandin $F_{2\alpha}$ on propulsive activity of the isolated segmental colon of the guinea-pig, *Prostaglandins* **10**: 759–768.

Ito, H., Komaki, R., and Milas, L. (1994) Protection of WR-2721 against radiation plus cis-diamine dichloroplatinum II caused injury to colonic epithelium in mice, *Int. J. Radiat.* **28**: 899–903.

Izzo, A.A., Capasso, R., Pinto, L., Di Carlo, G., Mascolo, N., and Capasso, F. (2001) Effect of vanilloid drugs on gastrointestinal transit in mice, *Br. J. Pharmacol.* **132**: 1411–1416.

Izzo, A.A., Fezza, F., Capasso, R., Bisogno, T., Pinto, L., Iuvone, T., Eposito, G., Mascolo, N. De Marzo, V., and Capasso, F. (2001) Cannabinoid CB_1 receptor mediated regulation of gastrointestinal motility in mice in a model of intestinal inflammation, *Br. J. Pharmacol.* **134**: 563–570.

Izzo, A.A., Mascolo, N., Borelli, F., and Capasso, F. (1998) Excitatory transmission to the circular muscle of the guinea-pig ileum: evidence for the involvement of cannabinoid CB_1 receptors, *Br. J. Pharmacol.* **124**: 1363–1368.

Izzo, A.A., Mascolo, N., Pinto, L., Capasso, R., and Capasso, F. (1999) The role of cannabinoid receptor in intestinal motility, defaecation and diarrhea in rats, *Eur. J. Pharmacol.* **384**: 37–42.

Izzo, A.A., Mascolo, N., Capasso, R., Germano, M.P., De Pasquale, R., and Capasso, F. (1999) Inhibitory effect of cannabinoid agonists on gastric emptying in the rat, *Naunyn-Schmied. Arch. Pharmacol.* **360**: 221–223.

Izzo, A.A., Mascolo, N., Borrelli, F., and Capasso, F. (1999) Defaecation, intestinal fluid accumulation and motility in rodents: implications of cannabinoid CB_1 receptors, *Naunyn-Schmied. Arch. Pharmacol.* **359**: 65–70.

Izzo, A.A., Mascolo, N., Tonini, M., and Capasso, F. (2000) Modulation of peristalsis by cannabinoid CB_1 ligands in the isolated guinea-pig ileum, *Br. J. Pharmacol.* **129**: 984–990.

Izzo, A.A., Pinto, L., Borrelli, F., Capasso, R., Mascolo, N., and Capasso, F. (2000) Central and peripheral cannabinoid modulation of gastrointestinal transit in physiological states and during diarrhea induced by croton oil, *Br. J. Pharmacol.* **129**: 1027–1032.

Jamshidi, N. and Taylor, D.A. (2001) Anandamide administration into the ventromedial hypothalamus stimulates appetite in rats, *Br. J. Pharmacol.* **134**: 1151–1154.

Jeon, Y.J., Yang, K.H. and Pulsaki, J.T. (1996) Attenuation of inducible nitric oxide synthase gene-expression by delta-9-tetrahydrocannabinol is mediated through the inhibition of nuclear factor-kappa-/rel activation, *Mol. Pharmacol.* **50**: 334–341.

Jett, M., Brinkley, W., Neill, R., Gemski, P., and Hunt, R. (1990) Staphylococcus aureus enterotoxin B challenge of monkeys: correlation of plasma levels of arachidonic acid cascade products with occurrence of illness, *Infect. Immun.* **58**: 3494–3499.

Jin, J.G., Takai, M., and Nakayama, S. (1990) The inhibitory effect of capsaicin on the ascending pathway of the guinea-pig ileum and antagonism of this effect by ruthenium red, *Eur. J. Pharmacol.* **180**: 13–90.

Kalff, J.C., Turler, A., Schwarz, N.T., Schraut, W.H., Lee, K.K.W., Tweardy, D.J., Billiar, T.R., Simmons, R.L., and Bauer, A.J. (2003) Intra-abdominal activation of a local inflammatory response within the human muscularis externa during laparotomy, *Ann. Surg.* **237**: 301–315.

Karim, S.M. and Filshe, G.M. (1970) Use of prostaglandin E2 for therapeutic abortion, *Med. J.* **111**: 198–200.

Kirkham, T.C. and Williams, C.M. (2001) Endogenous cannabinoids and appetite, *Nutr. Res. Rev.* **14**: 65–86.

Koch, J.E. and Werner, N.A. (2000) Effects of the cannabinoid antagonists AM630 and AM281 on deprivation-induced food intake in Lewis rats, *Soc. Neurosci. Abstr.* **26**: 1528.

Kojima, S.-I., Sugiura, T., Waku, D., and Kamikawa, Y. (2002) Contractile response to a cannabimimetic eicosanoid, 2-arachidonoylglycerol, of longitudinal smooth muscle from the guinea-pig distal colon in vitro, *Eur. J. Pharmacol.* **444**: 203–207.

Krowicki, Z.K. and Hornby, P.J. (1995) Hindbrain neuroactive substance controlling gastrointestinal functions, in *Regulatory Mechanisms in Gastrointestinal Function*, Ed. T.S. Gaginella, pp. 277–319. New York: CRC Press.

Krowicki, Z.K., Moerchbaecher, J.M., Winsauer, P.J., Digavalli, S.V. and Hornby, P.J. (1999) Δ^9-Tetrahydrocannabinol inhibits gastric motility in the rat through cannabinoid CB_1 receptors, *Eur. J. Pharmacol.* **371**: 187–196.

Kulkarni-Narla, A. and Brown, D.R. (2000) Localization of CB_1-cannabinoid receptor immunoreactivity in the porcine enteric nervous system, *Cell Tissue Res.* **302**: 73–80.

Kunze, W.A.A. and Furness, J.B. (1999) The enteric nervous system and regulation of intestinal motility, *Ann. Rev. Physiol.* **61**: 117–142.

Landi, M., Croci, T., Rinaldi-Carmona, M., Maffrand, J.-P., Le Fur, G., and Manara, L. (2002) Modulation of gastric emptying and gastrointestinal transit in rats through intestinal cannabinoid CB_1 receptors, *Eur. J. Pharmacol.* **450**: 77–83.

Lehmann, A., Blackshaw, L.A., Brändén, L., Carlsson, A., Jensen, J., Nygren, E., and Smid, S.D. (2002) Cannabinoid receptor agonism inhibits transient lower esophageal sphincter relaxations and reflux in dogs, *Gastroenterology* **123**: 1129–1134.

Leurs, R., Smit, M.J., Alewijnse, A.E., and Timmerman, H. (1998) Agonist-independent regulation of constitutively active receptors, *TIBS* **23**: 418–422.

London, S.W., McCarthy, L.E., and Borison, H.L. (1979) Suppression of cancer chemotherapy-induced vomiting in the cat by nabilone, a synthetic cannabinoid, *Proc. Soc. Exp. Biol. Med.* **160**: 437–440.

Lynn, A.B. and Herkenham, M. (1994) Localization of cannabinoid receptors and nonsaturable high-density cannabinoid binding sites in peripheral tissues of the rat: implications for receptor-mediated immune modulation by cannabinoids, *J. Pharmacol. Exp. Ther.* **268**: 1612–1623.

Mancinelli, R., Fabrizi, A., Del Monaco, S., Azzena, G.B., Vargiu, R., Colombo, G.C., and Gessa, G.L. (2001) Inhibition of peristaltic activity of cannabinoids in the isolated distal colon of mouse, *Life Sci.* **69**: 101–111.

Mang, C.F., Erbelding, D., and Kilbinger, H. (2001) Different effects of anandamide on acetylcholine release in the guinea-pig ileum mediated via vanilloid non-CB_1 cannabinoid receptors, *Br. J. Pharmacol.* **134**: 161–167.

Maranzano, E. (2001) Radiation-induced emesis: a problem with many open questions, *Tumori* **87**: 213–218.

Mascolo, N., Izzo, A.A., Ligresti, A., Costagliola, A., Pinto, L., Cascio, M.G., Maffia, P., Cecio, A., Capasso, F., and Di Marzo, V. (2002) The endocannabinoid system and the molecular basis of paralytic ileus in mice, *FASEB J.* **16**: 1973–1975.

Mattes, R.D., Engelman, K., Shaw, L.M., and Elsohly, M.A. (1994) Cannabinoids and appetite stimulation, *Pharmacol. Biochem. Behav.* **49**: 187–195.

Mechoulam, R. (1986) The pharmaco-history of Cannabis Sativa, in *Cannabinoids as Therapeutic Agents*, Ed. R. Mechoulam, pp. 1–20. Boca Raton: CRC Press.

Mechoulam, R., Ben-Shabat, S., Hanus, L., Ligumsky, M., Kaminski, N.E., Schatz, A.R. Gopher, A., Almog, S., Martin, B.R. Compton, D.R., Pertwee, R.G., Griffen, G., Bayewitch, M., Barg, J., and Vogel, Z. (1995) Identification of an endogenous 2-monoglyceride, present in canine gut, that binds to cannabinoid receptor, *Biochem. Pharmacol.* **50**: 83–90.

Mechoulam, R., Hanus, L., and Fride, E. (1998) Towards cannabinoid drugs-revisited, *Progress Med. Chem.* **35**: 199–243.

Miczek, K.A. and Dixit, B.N. (1980) Behavioral and biochemical effects of chronic delta-9-tetrahydrocannabinol in rats, *Psychopharmacology* **67**: 195–202.

Morrow, J.D. and Roberts, L.J. (2001) Lipid derived autocoids. Eicosanoids and platelet activating factor, in *Goodman and Gillman's, The Pharmacological Basis of Therapeutics*, Eds. J.G. Hardman and L.E. Limbird, pp. 669–685. New York: McGraw-Hill.

Nalin, D.R., Levine, M.M., Rhead, J., Berguist, E., Rennels, M., Hughes, T., O'Donnel, S., and Hornick, R.B. (1978) Cannabis, hypochlorhydia and cholera, *Lancet* **2**: 859–862.

Nelson, K., Walsh, D., Deeter, P., and Sheehan, F. (1994) A phase II study of delta-9-tetrahydrocannabinol for appetite stimulation in cancer-associated anorexia, *J. Palliat. Care* **10**: 14–18.

Olsson, C. and Holmgren, S. (2001) The control of gut motility, *Comparative Biochem. Physiol.* **128**: 481–503.

Parker, L.A. and Kemp, S.W. (2001) Tetrahydrocannabinol (THC) interferes with conditioned retching in *Suncus murinus*: an animal model of anticipatory nausea and vomiting (ANV), *Neuroreport* **12**: 749–751.

Pertwee, R.G. (2001) Cannabinoids and the gastrointestinal tract, *Gut* **48**: 859–867.

Pertwee, R.G. (1999) Pharmacology of cannabinoid receptor ligands, *Curr. Med. Chem.* **6**: 635–664.

Pertwee, R.G. (1997) Pharmacology of cannabinoid CB_1 and CB_2 receptors, *Pharmacol. Ther.* **74**: 129–180.

Pertwee, R.G., Gernando, S.R., Griffin, G., Abadji, V., and Makriyannis, A. (1995) Effect of phenylmethylsulphonyl fluoride on the potency of anandamide as an inhibitor of electrically evoked contractions in two isolated tissue preparations, *Eur. J. Pharmacol.* **272**: 73–78.

Pertwee, R.G., Fernando, S.R., Nash, J.E., and Coutts, A.A. (1996) Further evidence for the presence of cannabinoid CB_1 receptors in guinea-pig small intestine, *Br. J. Pharmacol.* **118**: 2199–2005.

Pertwee, R.G. and Ross, R.A. (2002) Cannabinoid receptors and their ligands, *Prostag Leukot. Essen. Fatty Acids* **66**: 101–121.

Pinto, L., Capasso, R., Di Carlo, G., and Izzo, A.A. (2002) Endocannabinoids and the gut, *Prostag Leukot. Essent. Fatty Acids* **66**: 333–341.

Pinto, L., Izzo, A.A., Cascio, M.G., Bisogno, T., Hospodar-Scott, K., Brown, D.R., Mascolo, N., Di Marzo, V., and Capasso, F. (2002) Endocannabinoids as physiological regulators of colonic propulsion in mice, *Gastroenterology* **123**: 227–234.

Porter, A.C., Saur, J-M., Knierman, M.D., Becker, G.W., Berna, M.J., Bao, J., Nomikos, G.G.,, Carter, P., Bymaster, F.P., Leese, A.B., and Felder, C.C. (2002) Characterization of a novel endocannabinoid, virodhamine, with antagonist activity at the CB_1 receptor, *J. Pharmacol. Exp. Ther.* **301**: 1020–1024.

Rivas, J.F. and Garcia, R. (1980) Inhibition of histamine-stimulated gastric acid secretion by Δ^9-tetrahydrocannabinol in rat isolated stomach, *Eur. J. Pharmacol.* **65**: 317–318.

Ross, R.A., Brockie, H.C., and Fernando, S.R. (1998) Comparison of cannabinoid binding sites in guinea-pig forebrain and small intestine, *Br. J. Pharmacol.* **125**: 1345–1351.

Rudd, J.A. and Wai, M.K. (2001) Genital grooming and emesis induced by vanilloids in *Suncus murinus*, the house musk shrew, *Eur. J. Pharmacol.* **422**: 185–195.

Sametz, W., Hennerbichler, S., Glaser, S., Wintersteiger, R., and Juan, H. (2000) Characterization of prostanoid receptors mediating actions of the isoprostanes, 8-iso-PGE_2 and 8-iso-$PGE_{2\alpha}$, in some isolated smooth muscle preparations, *Br. J. Pharmacol.* **130**: 1903–1910.

Shahbazian, A., Heinemann A., Peskar, B.A., and Holzer, P. (2002) Differential peristaltic motor effects of prostanoid (DP, EP, IP, TP) and leukotriene receptor agonists in guinea-pig isolated small intestine, *Br. J. Pharmacol.* **137**: 1047–1054.

Shibata, C., Jin, X.L., Naito, H., Matsuno, S., and Sasaki, I. (2002) Intraileal capsaicin inhibits gastrointestinal contractions via a neural reflex in conscious dogs, *Gastroenterology* **123**: 1904–1911.

Shire, D., Carillon, C., Kaghad, M., Calandra, B., Rinaldi-Carmona, M., Le Fur, G., Caput, D., and Ferrara, P. (1995) An amino terminal variant of the central cannabinoid receptor resulting from alternative splicing, *J. Biol. Chem.* **270**(8): 3726–3731.

Shook, J.E. and Burks, T.F. (1989) Psychoactive cannabinoids reduce gastrointestinal propulsion and motility in rodents, *J. Pharmacol. Exp. Ther.* **249**: 444–449.

Simiand, J., Keane, M., Keane, P.E., and Soubrie, P. (1998) SR 141716A, a CB1 cannabinoid receptor antagonist, selectively reduces sweet food intake in marmoset, *Behav. Pharmacol.* **9**: 179–181.

Sim-Selley, L.J., Brunk, L.K., and Selley, D.E. (2001) Inhibitory effects of SR 141716A on G-protein activation in rat brain, *Eur. J. Pharmacol.* **414**: 135–143.

Simoneau, I.I., Hamza, M.S., Mata, H.P., Siegel, E.M., Vanderah, T.W., Porreca, F., Makriyannis, A., and Malan, T.P. (2001) The cannabinoid agonist WIN 55, 212-2 suppresses opioid-induced emesis in ferrets, *Anesthesiology* **94**: 882–887.

Sofia, R.D., Diamantis, W., Harrison, J.E., and Melton, J. (1978) Evaluation of antiulcer activity of Δ^9-tetrahydrocannabinol in the Shay Rat Test, *Pharmacology* **17**: 173–177.

Stefano, G.B, Salzet, B., Rialas, C.M., Pope, M., Kustka, A., Neenan, D., Pryor, S., and Salzet, M. (1997) Morphine and anandamide-stimulated nitric oxide production inhibits presynaptic dopamine release, *Brain Res.* **763**: 36–38.

Storr, M., Gaffal, E., Saur, D., Schusdziarra, V., and Allescher, H.D. (2002) Effect of cannabinoids on neural transmission in rat gastric fundus, *Can. J. Physiol.* **80**: 67–76.

Struwe, M., Kaempfer, S.H., Geiger, C., Pavia, A.T., Plasse, T.F., Shepard, K.V., Ries, K., and Evans, T.G. (1993) Effect of dronabinol on nutritional status in HIV infection, *Ann. Pharmacother.* **27**: 827–831.

Sugiura, T., Kobayashi, Y., Oka, S., and Waku, K. (2002) Biosynthesis and degradation of anandamide and 2-arachidonoylglycerol and their possible physiological significance, *Prostaglandins Leukot. Essent. Fatty Acids* **66**: 173–192.

Szallasi, A. and Di Marzo, U. (2000) New perspectives on enigmatic vanilloid receptors, *Trends Neurosci.* **23**: 491–497.

Takai, M., Jin, J.-G., Lu, Y.-F., and Nakayama, S. (1990) Effects of piperine on the motility of the isolated guinea-pig ileum: comparison with capsaicin, *Eur. J. Pharmacol.* **186**: 71–77.

Thomson, A.B.R., Keelan, M., Thiesen, A., Clandinin, M.T., Propeleski, M., and Wild, G.E. (2001) Small bowel review normal physiology Part 2, *Digestive Dis. Sci.* **46**: 2588–2607.

Tonini, M. (1996) Recent advances in the pharmacology of gastrointestinal prokinetics, *Pharmacol. Res.* **33**: 217–226.

Tyler, L., Hilland, C.J., and Greenwood-Van Meerveld, B. (2000) Inhibition of small intestinal secretion by cannabinoids is CB_1 receptor-mediated in rats, *Eur. J. Pharmacol.* **409**: 207–211.

Van Sickle, M.D., Oland, L.D., Ho, W., Hillard, C.J., Mackie, K., Davison, J.S., and Sharkey, K.W. (2001) Cannabinoids inhibit emesis through CB1 receptors in the brainstem of the ferret, *Gastroenterology* **121**: 767–774.

Veyrat-Follet, C., Farinoti, R., and Palmer, J.L. (1997) Physiology of chemotherapy-induced emesis and antiemetic therapy. Predictive models for evaluation of new compounds, *Drugs* **53**: 206–234.

Watanabi, H., Mochizaki, T., and Maeyama, K. (1996) Transient increase of blood histamine level induced by pentagastrin. Continuous monitoring by in vivo microdialysis, *Scand. J. Gastroenterol.* **31**: 1144–1150.

Wechsung, E. (1996) The involvement of prostaglandins in the inhibiting effect of endotoxin on the myoelectric activity of the gastrointestinal system in pigs, *Verh K. Acad. Geneeskd Belg* **58**: 711–738.

Williams, C.M. and Kirkham, T.C. (1999) Anandamide induces overeating: mediation by central cannabinoid (CB1) receptors, *Psychopharmacology* **143**: 315–317.

Williams, C.M. and Kirkham, T.C. (2001) Hyperphagia induced by intra-accumbens injection of 2-arachidonoylglycerol. Symposium on the Cannabinoids, Burlington, Vermont, *Int. Cannabinoid Res. Soc.* P103.

Williams, C.M. and Kirkham, T.C. (2002) Observational analysis of feeding induced by Δ^9-THC and anandamide, *Physiol. Behav.* **76**: 241–250.

Wislicki, L. (1982) Systemic adverse reactions to prostaglandin F2 (PGF2 alpha, dinoprostone, prostin F2 alpha, prostalmon F), *Int. J. Biol. Res. Pregnancy* **3**: 158–160.

18 Endocannabinoids and the Cardiovascular System

Michael D. Randall

CONTENTS

CARDIOVASCULAR EFFECTS OF CANNABINOIDS

From the discovery of anandamide in the central nervous system and the identification of "classical" cannabinoid CB_1 and CB_2 receptors, both centrally and peripherally, the cardiovascular actions have received interest. To date there have been a number of studies which point to endocannabinoids having vasodilator actions, whereas the *in vivo* effects are less clear. One key point to emerge is that endocannabinoids may act via a range of mechanisms, of which action at vanilloid receptors is now well established.

ACTIONS OF CANNABINOIDS *IN VITRO*

Vasorelaxation

The overwhelming theme from *in vitro* studies is that endogenous cannabinoids cause vasorelaxation. However, as will be seen later, this does not simply translate into the *in vivo* situation. The first *in vitro* report that anandamide was a vasodilator came from Ellis et al. (1995), who demonstrated that anandamide caused cerebrovascular vasodilatation. Subsequent studies also demonstrated that anandamide was a vasorelaxant in the rat isolated mesenteric and coronary vasculatures (Randall et al., 1996; Randall and Kendall, 1997).

The Role of Prostanoids

In the Ellis et al. study, anandamide was shown to act via the release of vasodilator prostanoids, and Δ^9-tetrahydrocannabinol (Δ^9-THC) also acted in this way. Furthermore, Fleming et al. (1999) found that the cyclooxygenase inhibitor, diclofenac, abolished vasorelaxation to anandamide in rat mesenteric arterial vessels. In addition, Grainger and Boachie-Ansah (2001) reported that, in the sheep coronary artery, anandamide caused relaxations and this involved cyclooxygenase-dependent metabolism to vasodilator prostanoids. Despite these observations, most other studies have ruled out a major role for prostanoids in anandamide-induced relaxation (Randall et al., 1996; Randall et al., 1997; Plane et al., 1997; White and Hiley, 1997).

The Role of the Endothelium

The role of the endothelium in vasorelaxation to endogenous cannabinoids varies between vascular beds and tissues. Most studies have shown that the vasorelaxant responses to anandamide are endothelium-independent (Randall et al., 1996; White and Hiley, 1997; White et al., 2001) or only partly endothelium-dependent (Chaytor et al., 1999). However, in the bovine coronary artery, anandamide induces relaxations that are strictly endothelium-dependent (Pratt et al., 1998). This was explained by the endothelial cells metabolizing exogenous anandamide, via a cytochrome P450-dependent mono-oxygenase, to vasoactive metabolites. Similarly, Grainger and Boachie-Ansah (2001) reported endothelium-dependent metabolism of anandamide via the cyclooxygenase pathway underpinning vasorelaxation.

In 1999, Wagner and colleagues proposed that anandamide acted, in part, via an endothelial anandamide receptor in rat mesenteric arterial vessels. This was based on the observation that relaxation to anandamide was partly sensitive to both removal of the endothelium and the CB_1 receptor antagonist SR141716A, but when the endothelium was removed, the sensitivity to the antagonist was lost. This led to the proposal that anandamide acted at a cannabinoid receptor that was sensitive to SR141716A but as it was not the CB_1 receptor, it was termed the "anandamide receptor." An additional observation was that the exogenous cannabinoid Δ^9-THC did not cause vasorelaxation. Subsequent work by that group demonstrated that the endothelial cannabinoid receptor was also activated by the neurobehaviorally inactive "abnormal cannabidiol" (abn-cbd), which caused vasorelaxation (Jarai et al., 1999). One possibility to arise from the identification of the SR141716A-sensitive, endothelium-dependent component is that anandamide acts in part via EDHF and that SR141716A is acting via inhibition of EDHF activity (e.g., through blockade of myoendothelial gap junctions, Chaytor et al., 1999).

In 2003, Offertáler et al. provided further evidence for the "endothelial anandamide" receptor. Specifically, they reported that a novel cannabidiol analog, O-1918, opposed the relaxant effects of anandamide and abn-cbd, the *in vivo* hypotensive effects of abn-cbd, and the phosphorylation of p42/44 MAP kinase induced by abn-cbd in endothelial cells. These actions of O-1918 were independent of classical cannabinoid and vanilloid receptors, and this led the authors to conclude that O-1918 was a selective antagonist of the "endothelial anandamide" receptor. It was suggested that the endothelium-dependent relaxation to abn-cbd and anandamide is G protein coupled to MAP kinase activation and charybdotoxin-sensitive potassium channels but not to nitric oxide. Taken together, the authors proposed that the novel receptor may be coupled to the release of the EDHF.

The Role of the EDHF

The identification of EDHF is controversial, with several agents being proposed as representing EDHF. These include cytochrome P450-derived metabolites of arachidonic acid, potassium ions, and mediation via myoendothelial gap junctions (Busse et al., 2002). In relation to myoendothelial gap junctions, Chaytor et al. (1999) demonstrated that the actions of anandamide in rabbit mesenteric vessels were partly endothelium-dependent and sensitive to gap junctional inhibitors, and that SR141716A is a gap junctional inhibitor. These findings certainly point to anandamide acting in part via EDHF-type relaxation, which involves myoendothelial gap junctions. Furthermore, in the rat mesenteric arterial bed, Harris et al. (2002) reported that some but not all gap junction inhibitors opposed responses to anandamide, although an alternative possibility was that these inhibitors were inhibiting the sodium pump.

It was originally proposed that anandamide itself was an EDHF (Randall et al., 1996). This was partly based on the observation that SR141716A opposed EDHF-type relaxations. This inhibitory action was explained by the work of Chaytor et al. (1999), which demonstrated that the cannabinoid CB_1 receptor antagonist SR141716A was also a gap junctional inhibitor and was thus blocking EDHF activity at this level, and that EDHF-type relaxations are involved in the responses to anandamide. Similarly, sensitivity of relaxant responses to potassium channel blockers, including cytochrome P450 inhibitors (Randall et al., 1997) charybdotoxin and iberiotoxin (Plane et al., 1997), and the combination of charybdotoxin and apamin (Randall and Kendall, 1998) can be explained by EDHF's mediating part of the relaxation to anandamide. However, in the guinea pig carotid artery, the anandamide-induced hyperpolarization, which was insensitive to charybdotoxin plus apamin, was blocked by the ATP-sensitive potassium channel inhibitor, glibenclamide (Chataigneau et al., 1998), which may implicate these channels in the responses to anandamide, an action independent of EDHF.

The Role of Nitric Oxide

Anandamide has been shown to act via the release of endothelium-derived nitric oxide in the rat kidney (Deutsch et al., 1997). A range of human blood vessels and the right atrium have also been shown to release nitric oxide in response to anandamide (Bilfinger et al., 1998). However, in many instances (see Randall et al., 1996; White and Hiley, 1997; Jarai et al., 1999), vasorelaxant responses to anandamide are insensitive to inhibition of nitric oxide synthase. In HUVECs, Maccarrone et al. (2000) reported that anandamide and the CB agonist, HU210, both cause an upregulation of the expression and activity of the inducible nitric oxide synthase. Mukhopadhyay et al. (2002) demonstrated that the endothelium-dependent component was G protein coupled and mediated via nitric oxide, whereas the endothelium-independent component was due to activation of vanilloid receptors, at least in rabbit aortic rings.

The Role of Calcium Channels

Endocannabinoids have been shown to inhibit vascular, smooth-muscle calcium channels (Gebremedhin et al., 1999). Specifically, in feline cerebral vessels, it was shown that endocannabinoids and synthetic cannabinoid agonists act via G-protein-coupled CB_1 receptors to cause inhibition of voltage-sensitive calcium channels, leading to vasodilatation. This action was proposed to contribute towards vasodilatation in cerebral hypoxia, which was associated with the release of endocannabinoids.

Action at Vanilloid Receptors

One of the most attractive and novel proposals to account for the vasodilatation in response to anandamide has been that it acts as a vanilloid agonist, because anandamide shares structural similarities with the vanilloid agonist olvanil. This led Zygmunt et al. (1999) to investigate the role of vanilloid receptors in the vascular actions of anandamide. In this respect, they reported that relaxation to anandamide (but not 2-AG, palmitoylethanolamide, or synthetic cannabinoid receptor agonists) was essentially abolished by depletion of the sensory nerves of calcitonin gene-related peptide (CGRP) by capsaicin in guinea pig basilar, rat hepatic, and rat mesenteric arteries. Furthermore, relaxation to anandamide was sensitive to the vanilloid receptor antagonist capsazepine and also to CGRP receptor antagonism with CGRP (8-37). Clearly, anandamide can evoke the release of neurotransmitters from sensory nerves leading to vasorelaxation. This conclusion was supported by the demonstration that anandamide is an agonist at the cloned rat vanilloid receptor (rVR1) (Zygmunt et al., 1999). This observation was later confirmed at the human vanilloid receptor (hVR1) (Smart et al., 2000).

Similar observations have been made with the analog of anandamide, methanandamide, which was also shown to cause capsaicin- and capsazepine-sensitive vasorelaxation in the rat mesenteric arterial bed and isolated mesenteric arteries (Ralevic et al., 2000). However, in the same vascular bed, Harris et al. (2002) reported that vasorelaxation to anandamide was only partly sensitive to capsaicin pretreatment. Moreover, in the presence of NO synthase blockade, vasorelaxation due to anandamide was insensitive to capsaicin pretreatment and thus does not occur exclusively via sensory nerves. In rat isolated coronary arteries and in the intact vasculature, White et al. (2001)

and Ford et al. (2002) have reported that vasorelaxation to anandamide is independent of vanilloid VR1 receptors and appears to be mediated via a novel cannabinoid receptor. Accordingly, the activation of sensory nerves by anandamide may only explain part of the actions of anandamide and only under some circumstances.

More recent work by Zygmunt et al. (2002) has shown in rat mesenteric vessels that Δ^9-THC and cannabinol cause relaxation via action on sensory nerves but that this involves a site of action which is not VR_1. From this, they propose that there is a novel cannabinoid receptor or target which may control sensory nerve transmission.

The fact that the hypotensive action of anandamide is absent in knockout mice lacking CB_1 receptors (Ledent et al., 1999) suggests that any action via vanilloid receptors on sensory nerves is only of minor importance in the hemodynamic profile of systemically administered cannabinoids. In urethane-anesthetized rats, there is a triphasic cardiovascular response to systemic intravenous administration of anandamide, but it appears that only the initial rapid hypotensive and bradycardic phase involves VR_1 receptors (Malinowska et al., 2001, Smith and McQueen, 2001), and this phase is not associated with vasodilatation (Gardiner et al., 2002a).

A Unifying Mechanism of Vasorelaxation?

The preceding sections have dealt with a range of possible mechanisms of vasorelaxation to endogenous cannabinoids. There is clearly a diverse range of putative mechanisms, and this may well reflect tissue differences. Because it is known that EDHF activity tends to be greatest in smaller resistance vessels, it might be speculated that EDHF might play a significant role in these vessels, and the work of Offertáler et al. (2003) supports the notion that the release of EDHF is coupled to a novel endothelial cannabinoid receptor. Undoubtedly, VR_1 receptors on sensory nerves also play a substantial role in mediating responses to endogenous cannabinoids; once again, contribution may differ between different arteries and, indeed, prevailing conditions. In relation to tissue selectivity, Andersson and colleagues (2002) have addressed this issue and, in this regard, suggest that differences in the abundance of the cannabinoid transporter, and the relative content of cannabinoid and vanilloid receptors, coupled with agonist efficacy, may determine which mechanisms prevail.

Vasoconstriction

Contractions mediated by anandamide and synthetic cannabinoids have been reported in rat isolated small mesenteric arteries (White and Hiley, 1998). The responses were small, and it was suggested that only low levels of calcium were released from intracellular stores, which were insufficient to stimulate extracellular calcium entry. In that rat aorta, Δ^9-THC has been shown to cause contractile responses that are mediated by cyclooxygenase metabolites (O'Sullivan et al., 2005). In the anesthetized rat, anandamide, despite causing widespread vasodilatation, has recently been shown to cause SR141716A-insensitive vasoconstriction in the spleen (Wagner et al., 2001b). It is possible that synergistic interactions of cannabinoids with circulating and locally released contractile mediators account for the pronounced, albeit transient, pressor response observed upon systemic administration of cannabinoids in anesthetized animals.

Vascular Cannabinoid Receptors

The ability of cannabinoids to cause vascular effects implies that the vasculature contains a molecular target. Evidence to date suggests that there may be vascular cannabinoid receptors, which may either fall into the classical CB_1/CB_2 classification or represent a new subtype. As stated above, the biphasic hypotension in response to anandamide is absent in CB_1 receptor knockout mice (Ledent et al., 1999). This clearly points to the involvement of the CB_1 receptor. However, it should be noted that this does not conclusively identify the cannabinoid receptors as being on the vascular smooth muscle or associated with neuronal tissue. The sensitivity of vasorelaxant responses to CB receptor antagonists has been controversial, with some studies indicating that the responses are opposed by SR141716A (Randall et al., 1996; White and Hiley, 1997) and others demonstrating that they are insensitive to this antagonist (Plane et al., 1997). The insensitivity to SR141716A might reflect noncannabinoid receptor actions (Pratt et al., 1998; Chaytor et al., 1999), whereas Jarai et al. (1999) have suggested the presence of a novel vascular CB receptor.

Using reverse transcriptase–polymerase chain reaction, the gene product encoding for CB_1 receptors has been located in renal endothelial cells, mesenteric resistance arterioles, and cerebral micro vessels, which is consistent with the expression of CB_1 receptors in the vasculature (Deutsch et al., 1997; Darker et al., 1998; Randall et al., 1999). Others have also identified mRNA in human endothelial cells (Sugiura et al., 1998; Liu et al., 2000) and Liu et al. (2000) have identified CB_1 receptor binding sites by radioligand studies. Similarly, immunoreactivity to the CB_1 receptor has been identified on human saphenous vein endothelial cells (Bilfinger et al., 1998).

Cannabinoid CB_1 receptors have also been localized to cat cerebral arterial smooth muscle (Gebremedhin et al., 1999). In this study, it was demonstrated that feline vascular smooth muscle contained CB_1 receptors together with cDNA, showing very close homology to that associated with neuronal CB_1 receptors.

Cardiac Actions

Not only do endocannabinoids affect vascular function, but they may also have direct cardiac actions. Ford et al. (2002) reported in the rat isolated heart that anandamide had a negative inotropic effect and reduced left ventricular pressure. This action appeared to be due to action at a novel cannabinoid receptor. In human atrial muscle, anandamide has also been shown to exert negative inotropic effects but, in this case, via the activation of CB_1 receptors (Bonz et al., 2003).

Actions *In Vivo*

In vivo, the cardiovascular effects of exogenous cannabinoids are variable, with both vasodilator and vasoconstrictor actions being reported (Stark and Dews, 1980). In humans, acute administration of cannabinoids is associated with tachycardia and a small pressor effect; whereas long-term use is associated with hypotension and bradycardia (Benowitz and Jones, 1975; Benowitz et al., 1979). In 1996, Vidrio and colleagues reported that the cannabinoid agonist HU210, following intraperitoneal administration, caused prolonged bradycardia and hypotension, in both conscious and anesthetized rats. The hypotension in response to HU210 in anesthetized rats was subsequently reported by Wagner et al. (2001b) as being due to a reduction in cardiac output without effects on vascular resistance. Work in pithed rabbits by Niederhoffer and Szabo (1999), in which sympathetic tone was evoked by continuous electrical stimulation, has demonstrated that intravenous injection of CB_1 receptor agonists (CP55940 and WIN55212-2) causes prejunctional inhibition of sympathetic activity leading to hypotension. By contrast, Gardiner et al. (2002b) demonstrated in conscious rats that WIN55212-2 and HU210 caused pressor and regional vasoconstrictor effects. These effects were sensitive to the cannabinoid CB_1 receptor antagonist AM251 and appeared to be mediated via increased sympathetic activity. In addition, the cannabinoid agonists also caused hindquarters vasodilatation via the activation of β_2-adrenoceptors. These clear differences with previous findings were ascribed to the confounding effects of generic anesthetic agents used in most *in vivo* studies.

The cardiovascular effects of endogenous cannabinoids are similarly complex; with respect to anandamide in anesthetized rats, it has been shown to cause bradycardia (with brief secondary hypotension), then a transient pressor effect, which is followed by a delayed but long-lasting depressor action (Varga et al., 1995; Lake et al., 1997). The initial bradycardia and associated hypotension are believed to be vagally mediated, as it is abolished by atropine treatment or cervical vagotomy (Varga et al., 1995).

In the anesthetized rat, the second depressor effect, which follows the transient pressor phase, is believed to be mediated by CB_1 receptor prejunctional inhibition of sympathetic outflow in the periphery as the effect is attenuated by cervical spinal transection, α-adrenoceptor, and cannabinoid receptor antagonists (Varga et al., 1995; Lake et al., 1997).

In conscious rats, the cardiovascular effects of anandamide are markedly different from those reported in studies carried out in anesthetized animals. In conscious rats, Stein et al. (1996) reported that anandamide caused bradycardia, with a transient hypotensive effect, followed by a longer pressor phase, and only at the higher doses was there delayed hypotension. It seems likely that the greater pressor effect obscures the hypotension. Furthermore, Gardiner and colleagues (2002a) reported that intravenously administered anandamide led to transient pressor effects associated with mesenteric, renal, and hindquarters vasoconstriction. High doses of anandamide were associated with initial bradycardia, and the hindquarters vasoconstriction was followed by vasodilatation. These complex cardiovascular actions were insensitive to the CB_1 receptor antagonist, AM251 and thus, not mediated via cannabinoid CB_1 receptors. The bradycardia was atropine-sensitive, and the apparent hindquarters vasodilatation appeared to be mediated via β_2-adrenoceptors. It was speculated that this may be due to the release of adrenaline via adrenal vanilloid receptors.

In mice, both anandamide and synthetic cannabinoid receptor agonists cause biphasic hypotension (a depressor response, followed by a more sustained hypotensive phase but without a pressor component), which is thought to be entirely CB_1 receptor mediated, as the responses are absent in CB_1 receptor knockout mice (Ledent et al., 1999).

Endocannabinoids and Pathophysiology

Wagner et al. (1997) demonstrated in a rat model of hemorrhagic shock that activated macrophages release anandamide. In endotoxic shock, the synthesis of 2-AG in platelets is increased and anandamide is only detectable in macrophages after exposure to lipopolysaccharide (Varga et al., 1998). *In vitro,* mouse J774 macrophages also release both 2-AG and anandamide, and participate in their degradation (Di Marzo et al., 1999). In patients with endotoxic shock, increases in plasma anandamide and 2-AG have now been reported (Wang et al., 2001). These findings certainly point to the genesis of endocannabinoids in blood cells, which is enhanced in shock and contributes towards the cardiovascular sequelae.

In the context of septic shock, the induction of the inducible nitric oxide synthase and excessive production of nitric oxide are widely implicated. Interestingly, Ross et al. (2000) demonstrated that the cannabinoid agonist WIN55212, acting via CB_2 receptors, actually inhibited lipopolysaccharide-induced nitric oxide release from macrophages.

In terms of hemorrhagic shock, Wagner et al. (1997) demonstrated that the accompanying hypotension, in part due to macrophage-derived endocannabinoids, was reversed by the cannabinoid receptor antagonist SR141716A. Similarly in endotoxic shock, the synthesis of 2-AG in platelets and anandamide in macrophages is increased (Varga et al., 1998). It is possible that the activated blood cells could also stimulate the release of endocannabinoids from the endothelium or other vascular sites, contributing further towards the hypotension. The release of anandamide by central neurones under hypoxic conditions, leading to improved blood flow and protection against ischemia, has also been advanced as a pathophysiological role for anandamide (Gebremedhin et al., 1999).

There is also the possibility that pathophysiological conditions may alter the normal responses to endocannabinoids. In this regard, Mendizabal and colleagues (2001) reported that the vasorelaxant responses to anandamide were enhanced in mesenteric vessels from rats rendered hypertensive through chronic treatment with a nitric oxide synthase inhibitor but not in rats rendered hypertensive through aortic coarctation. Enhanced responses following chronic nitric oxide synthase inhibition were confirmed by Tep-areenan et al. (2002). Others have also reported that depressor responses to anandamide are enhanced in anesthetized, spontaneously hypertensive rats (SHR) (Li et al., 2003). It is thus possible that the endocannabinoid system may become altered to compensate for this loss of nitric oxide.

Endotoxemia is another circumstance in which responses to anandamide may be altered, as Orliac et al. (2003) reported that vasorelaxation to anandamide was enhanced in mesenteric arterial beds from endotoxemic rats. They concluded that this upregulation might involve a metabolite of anandamide and also activation of vanilloid receptors.

Endocannabinoids and Cardiac Ischemia

As commented upon in the preceding text, endocannabinoids may exert cardiac effects, and there has been interest in the influence of cannabinoids on cardiac ischemia. Lagneux and Lamontagne (2001) reported that cardioprotection of the rat heart against ischemia by pretreatment with lipopolysaccharide involved endocannabinoids. In this respect, pretreatment with lipopolysaccharide was found to protect hearts against ischemia, but this was sensitive to blockade via the cannabinoid CB_2 receptor antagonist SR144528 but not the CB_1 receptor antagonist SR141716A. The implication from this work was that lipopolysaccharide-induced cardioprotection involved the release of endocannabinoids, which caused protection via cannabinoid CB_2 receptors. Subsequent work by that group also reported that palmitoylethanolamide and 2-arachidonoyl glycerol both caused cardioprotection via CB_2 receptor activation and that this involved p38, ERK1/2, and protein kinase C activation (Lepicier et al., 2003).

Related to cardioprotection, Wagner and colleagues (2001a) have reported that in a rat model of myocardial infarction due to coronary ligation *in vivo*, there is the release of anandamide and 2-arachidonoyl glycerol from monocytes and platelets. This release of endocannabinoids was associated with systemic hypotension but decreased mortality, as administration of SR141716A reduced the hypotension but increased mortality. This might suggest that the release of endocannabinoids may be protective. In this model, the release of endocannabinoids did not reduce the size of myocardial infarction. Subsequent studies have examined the effects of treatment of the CB_1 receptor antagonist AM251 or the CB agonist HU210 for 12 weeks after myocardial infarction (Wagner et al., 2003). The key findings from this study were that cannabinoid receptor antagonism promotes remodeling and that cannabinoid agonists may prevent endothelial dysfunction and hypotension. Therefore, it is possible that the release of endocannabinoid in postmyocardial infarction might play a role in opposing the deleterious process of remodeling, which leads to long-term complication such as the development of heart failure. It was also observed that the administration of the cannabinoid agonist led to preservation of endothelial function.

CONCLUDING REMARKS

Endocannabinoids exert potent and complex cardiovascular effects. The findings in isolated arterial vessels overwhelmingly support the view that endocannabinoids are vasorelaxants and act via several different mechanisms. These include release of vasorelaxant neurotransmitters from sensory nerves via vanilloid receptor activation, the release of EDHF coupled to the activation of a novel endothelial cannabinoid receptor, and actions at "classical" cannabinoid receptors. The cardiovascular effects of endocannabinoids *in vivo* are, however, complex, as the *in vitro* vasorelaxant actions do not translate into simple responses. Indeed, the responses observed appear dependent on the prevailing conditions, e.g., the absence or presence of anesthetic. Roles are emerging for endocannabinoids in pathophysiological conditions, where the release may be involved in some of the circulatory changes and adaptive responses; there is the possibility that endocannabinoids may lead to cardioprotection in ischemia.

ACKNOWLEDGMENTS

The author thanks the British Heart Foundation for financial support.

REFERENCES

Andersson, D.A., Adner, M.A., Hogestatt, E.D., and Zygmunt, P.M. (2002). Mechanisms underlying tissue selectivity of anandamide and other vanilloid receptor agonists. *Molecular Pharmacology,* **62:** 705–713.

Benowitz, N.L. and Jones, R.T. (1975). Cardiovascular effects of prolonged delta-9-tetrahydrocannabinol ingestion. *Clinical Pharmacology and Therapeutics,* 18: 287–297.

Benowitz, N.L., Rosenberg, J., Rogers, W., Bachman, J., and Jones, R.T. (1979). Cardiovascular effects of intravenous delta-9- tetrahydrocannabinol: autonomic nervous mechanisms. *Clinical Pharmacology and Therapeutics,* 25: 440–446.

Bilfinger, T.V., Salzet, M., Fimiani, C., Deutsch, A., Tramu, G., and Stefano, G.B. (1998). Pharmacological evidence for anandamide amidase in human cardiac vascular tissues. *International Journal of Cardiology,* **64** (Suppl. 1): S15–S22.

Bonz, A., Laser, M., Kullmer, S., Kniesch, S., Babin-Ebell, J., Popp, V., Ertl, G., and Wagner, J.A. (2003). Cannabinoids acting on CB_1 receptors decrease contractile performance in human atrial muscle. *Journal of Cardiovascular Pharmacology,* **41:** 657–664.

Busse R., Edwards G., Feletou M., Fleming I., Vanhoutte P.M., and Weston A.H. (2002). EDHF: bringing the concepts together. *Trends In Pharmacological Sciences,* **23:** 374–380.

Chataigneau, T., Feletou, M., Thollon, C., Villeneuve, N., Vilaine, J.P., Duhault, J. et al. (1998). Cannabinoid CB1 receptor and endothelium-dependent hyperpolarization in guinea-pig carotid, rat mesenteric and porcine coronary arteries. *British Journal of Pharmacology,* **123:** 968–974.

Chaytor, A.T., Martin, P.E.M., Evans, W.H., Randall, M.D., and Griffith, T.M. (1999). The endothelial component of cannabinoid-induced relaxation in rabbit mesenteric artery depends on gap junctional communication. *Journal of Physiology,* **520:** 539–550.

Darker, I.T., Millns, P.J., Selbie, L., Randall, M.D., S-Baxter, G., and Kendall, D.A. (1998). Cannabinoid (CB_1) receptor expression is associated with mesenteric resistance vessels but not thoracic aorta in the rat. *British Journal of Pharmacology,* **125:** 95 p.

Deutsch D.G., Goligorsky M.S., Schmid P.G., Krebsbach R.J., Schmid H.H.O., Das S.K. et al. (1997). Production and physiological actions of anandamide in the vasculature of the rat kidney. *Journal of Clinical Investigation,* **100:** 1538–1546.

Di Marzo, V., Bisogno, T., DePetrocellis, L., Melck, D., Orlando, P., Wagner, J.A. et al. (1999). Biosynthesis and inactivation of the endocannabinoid 2-arachidonoylglycerol in circulating and tumoral macrophages. *European Journal of Biochemistry,* **264:** 258–267.

Ellis, E.F., Moore, S.F., and Willoughby, K.A. (1995). Anandamide and Δ^9-THC dilation of cerebral arterioles is blocked by indomethacin. *American Journal of Physiology,* **269:** H1859–H1864.

Fleming, I., Schermer, B., Popp, R., and Busse, R. (1999). Inhibition of the production of endothelium-derived hyperpolarizing factor by cannabinoid receptor agonists. *American Journal of Physiology,* **126:** 949–960.

Ford, W.R., Honan, S.A., White, R., and Hiley, C.R. (2002). Evidence of a novel site mediating anandamide-induced negative inotropic and coronary vasodilator responses in rat isolated hearts. *British Journal of Pharmacology,* **135**: 1191–1198.

Gardiner, S.M., March, J.E., Kemp, P.A., and Bennett, T. (2002a). Complex regional haemodynamic effects of anandamide in conscious rats. *British Journal of Pharmacology,* **135:** 1889–1896.

Gardiner, S.M., March, J.E., Kemp, P.A., and Bennett, T. (2002b). Influence of the CB1 receptor antagonist, AM 251, on the regional haemodynamic effects of WIN 55212-2 or HU 210 in conscious rats. *British Journal of Pharmacology,* **136:** 581–587.

Gebremedhin, D., Lange, A.R., Campbell, W.B., Hillard, C.J., and Harder, D.R. (1999). Cannabinoid CB_1 receptor of cat cerebral arterial muscle functions to inhibit L-type Ca^{2+} channel current. *American Journal of Physiology,* **45:** H2085–H2093.

Grainger, J. and Boachie-Ansah, G. (2001). Anandamide-induced relaxation of sheep coronary arteries: the role of the vascular endothelium, arachidonic acid metabolites and potassium channels. *British Journal of Pharmacology,* **134:** 1003–1012.

Harris, D., McCulloch, A.I., Kendall, D.A., and Randall, M.D. (2002). Characterization of vasorelaxant responses to anandamide in the rat mesenteric arterial bed. *Journal of Physiology,* **539,** 893–902.

Jarai, Z., Wagner, J.A., Varga, K., Lake, K., Compton, D.R., Martin, B.R. et al. (1999). Cannabinoid-induced mesenteric vasodilation through an endothelial site distinct from CB_1 or CB_2 receptors. *Proceedings of the National Academy of Sciences,* **96,** 14136–14141.

Lagneux, C. and Lamontagne, D. (2001). Involvement of cannabinoids in the cardioprotection induced by lipopolysaccharide. *British Journal of Pharmacology* **132:** 793–796.

Lake, K., Martin, B.R., Kunos, G., and Varga, K. (1997). Cardiovascular effects of anandamide in anesthetized and conscious normotensive rats. *Hypertension,* **29:** 1204–1210.

Ledent, C., Valverde, O., Cossu, C. Petitet, F., Aubert, L.F., Beslot, F. et al. (1999). Unresponsiveness to cannabinoids and reduced additive effects of opiates in CB_1 receptor knockout mice. *Science,* **283:** 401–404.

Lepicier, P., Bouchard, J-F., Lagneux, C., and Lamontagne, D. (2003). Endocannabinoids protect the rat isolated heart against ischemia. *British Journal of Pharmacology,* **139:** 805–815.

Li, J. Kaminski, N.E. and Wang, D.H. (2003). Anandamide-induced depressor effect in spontaneously hypertensive rats. *Hypertension,* **41:** 757–762.

Liu, J., Gao, B., Mirshahi, F., Sanyal, A.J., Khanolkar, A.D., Makriyannis, A., and Kunos, G. (2000). Functional CB_1 cannabinoid receptors in human vascular endothelial cells. *Biochemical Journal,* **346:** 835–840.

Maccarrone, M., Bari, M., Lorenzon, T., Bisogno, T., Di Marzo, V., and Finazzi-Agro, A. (2000). Anandamide uptake by human endothelial cells and its regulation by nitric oxide. *Journal of Biological Chemistry,* **275:** 13484–13492.

Malinowska, B., Kwolek, G., and Gothert, M. (2001). Anandamide and methanandamide induce both vanilloid VR_1- and cannabinoid CB_1 receptor-mediated changes in heart rat and blood pressure in anesthetized rats. *Naunyn-Schmiedeberg's Archives of Pharmacology,* **364:** 562–569.

Mendizabal, V.E., Orliac, M.L., Adler-Graschinsky, E. (2001). Long-term inhibition of nitric oxide synthase potentiates effects of anandamide in the rat mesenteric bed. *European Journal of Pharmacology,* **427,** 251–262.

Mombouli, J.V., Schaeffer, G., Holzmann, S., Kostner, G.M., and Graier, W.F. (1999). Anandamide-induced mobilization of cytosolic Ca^{2+} in endothelial cells. *British Journal of Pharmacology,* **126:** 1593–1600.

Mukhopadhyay, S., Chapnick, B.M., and Howlett, A.C. (2002). Anandamide-induced vasorelaxation in rabbit aortic rings has two components: G protein dependent and independent. *American Journal of Physiology,* **282:** H2046–H2054.

Niederhoffer, N. and Szabo, B. (1999). Effect of the cannabinoid receptor agonist WIN55212-2 on sympathetic cardiovascular regulation. *British Journal of Pharmacology,* **126:** 457–466.

Offertáler, L., Mo, F-M, Bátkai, S., Liu, J., Begg, M., Razdan, R.K., Martin, B.R., Bukoski, R.B., and Kunos, G. (2003). Selective ligands and cellular effectors of a G protein–coupled endothelial cannabinoid receptor. *Molecular Pharmacology,* **63:** 699–705.

Orliac, M.L., Peroni, R., Celuch, S.M., and Adler-Graschinsky, E. (2003). Potentiation of anandamide effects in mesenteric beds isolated from endotoxaemic rats. *Journal of Pharmacology and Experimental Therapeutics,* **304:** 179–184.

O'Sullivan, S.E., Kendall, D.A., and Randall, M.D. (2005). The effects of Δ^9-tetrahydrocannabinol (THC), anandamide and N-arachidonoyldopamine (NADA) in the rat isolated aorta. *European Journal of Pharmacology,* **145:** 514–526.

Plane, F., Holland, M., Waldron, G.J., Garland, C.J., and Boyle, J.P. (1997). Evidence that anandamide and EDHF act via different mechanisms in rat isolated mesenteric arteries. *British Journal of Pharmacology,* **121:** 1509–1512.

Pratt, P.F., Hillard, C.J., Edgemond, W.S., and Campbell, W.B. (1998). *N*-arachidonylethanolamide relaxation of bovine coronary artery is not mediated by CB_1 cannabinoid receptor. *American Journal of Physiology,* **274:** H375–H381.

Ralevic, V., Kendall, D.A., Randall, M.D., Zygmunt, P.M., Movahed, P., and Högestatt, E.D. (2000). Vanilloid receptors on capsaicin-sensitive nerves mediate relaxation to methanandamide in the rat isolated mesenteric bed. *British Journal of Pharmacology,* **130:** 1483–1488.

Randall M.D., Alexander, S.P.H., Bennett, T., Boyd, E.A., Fry, J.R., Gardiner, S.M. et al. (1996) An endogenous cannabinoid as an endothelium-derived vasorelaxant. *Biochemical Biophysical Research Communications,* **229:** 114–120.

Randall, M.D., Harris, D., Darker, I.T., Millns, P.J., and Kendall, D.A (1999). Endocannabinoids : endothelium-derived vasodilators. in *Endothelium-dependent Hyperpolarizations,* Ed. P.M. Vanhoutte pp. 149–155. Amsterdam: Harcourt.

Randall, M.D. and Kendall, D.A. (1997). The involvement of an endogenous cannabinoid in EDHF-mediated vasorelaxation in the rat coronary vasculature. *European Journal of Pharmacology,* **335:** 205–209.

Randall, M.D. and Kendall, D.A. (1998). Anandamide and endothelium-derived hyperpolarizing factor act via a common vasorelaxant mechanism in rat mesentery. *European Journal of Pharmacology,* **346:** 51–53.

Randall, M.D., McCulloch, A.I., and Kendall, D.A. (1997). Comparative pharmacology of endothelium-derived hyperpolarizing factor and anandamide in rat isolated mesentery. *European Journal of Pharmacology,* **333:** 191–197.

Ross, R.A., Brockie, H.C., and Pertwee, R.G. (2000). Inhibition of nitric oxide production in RAW264.7 macrophages by cannabinoids and palmitoylethanolamine. *European Journal of Pharmacology,* **401:** 121–130.

Smart, D., Gunthorpe, M.J., Jerman, J.C., Nasir, S., Gray, J., Muir, A.I., Chambers, J.K., Randall, A.D., and Davis, J.B. (2000). The endogenous lipid anandamide is a full agonist at the human vanilloid receptor (hVR1). *British Journal of Pharmacology,* **129:** 227–230.

Smith P.J.W. and McQueen D.S. (2001). Anandamide induces cardiovascular and respiratory reflexes via vasosensory nerves in the anaesthetized rat. *British. Journal. of Pharmacology,* **134:** 655–663.

Stark, P. and Dews, P.B. (1980). Cannabinoids. II. Cardiovascular effects. *Journal of Pharmacology and Experimental Therapeutics,* **214:** 131–138.

Stein, E.A., Fuller, S.A., Edgemond, W.S., and Campbell, W.B. (1996). Physiological and behavioural effects of the endogenous cannabinoid, arachidonylethanolamine (anandamide), in the rat. *British Journal of Pharmacology,* **119:** 107–114.

Sugiura, T., Kodaka, T., Nakane, S., Kishimoto, S., Kondo, S., and Waku, K. (1998). Detection of an endogenous cannabimimetic molecule, 2-arachidonoylglycerol, and cannabinoid CB_1 receptor mRNA in human vascular cells: is 2-arachidonoylglycerol a possible vasomodulator? *Biochemical Biophysical Research Communications,* **243:** 838–843.

Tep-areenan, P., March, J.E., Kemp, P.A., Randall, M.D., Kendall, D.A., Bennett, T., and Gardiner, S.M. (2002). Effects of chronic *in vivo* treatment with a nitric oxide synthase inhibitor on vasorelaxant responses to anandamide in rat isolated arteries. *British Journal of Pharmacology,* **137:** 55 p.

Varga, K., Lake, K., Martin, B.R., and Kunos, G. (1995). Novel antagonist implicates CB_1 cannabinoid receptor in the hypotensive action of anandamide. *European Journal of Pharmacology,* **278:** 279–283.

Varga, K., Wagner, J.A., Bridgen, D.T., and Kunos, G. (1998). Platelet- and macrophage-derived endogenous cannabinoids are involved in endotoxin-induced hypotension. *FASEB Journal,* **12:** 1035–1044.

Vidrio, H., Sanchez-Salvatori, M.A., and Medina, M. (1996). Cardiovascular effects of (-)-11-*OH*-Δ^8-tetrahydrocannabinol-dimethylheptyl in rats. *Journal of Cardiovascular Pharmacology,* **28:** 332–336.

Wagner, J.A., Hu, K., Bauersachs, J., Karcher, J., Wiesler, M., Goparaju, S.K, Kunos, G., and Ertl, G. (2001a). Endogenous cannabinoids mediate hypotension after experimental myocardial infarction. *Journal of the American College of Cardiology,* **38:** 2048–2054.

Wagner, J.A., Jaria, Z., Batkari, S., and Kunos, G. (2001b). Haemodynamic effects of cannabinoids: coronary and cerebral vasodilation mediated by cannabinoid CB_1 receptors. *European Journal of Pharmacology,* **423:** 203–210.

Wagner, J.A., Hu, K., Karcher, J., Bauersachs, J., Schafer, A., Laser, M., Han, H., and Ertl, G. (2003). CB_1 cannabinoid receptor antagonism promotes remodeling and cannabinoid treatment prevents endothelial dysfunction and hypotension in rats with myocardial infarction. *British Journal of Pharmacology,* **138:** 1251–1258.

Wagner, J.A., Varga, K., Ellis, E.F., Rzigalinski, B.A., Martin, B.R., and Kunos, G. (1997). Activation of peripheral CB_1 cannabinoid receptors in haemorrhagic shock. *Nature,* **390:** 518–521.

Wagner, J.A., Varga, K., Jarai, Z., and Kunos, G. (1999). Mesenteric vasodilation mediated by endothelial anandamide receptors. *Hypertension,* **33:** 429–434.

Wang, Y., Liu, Y., Hashiguchi, T., Kitajima, I., Yamakuchi, M., Shimizu, H., Matsuo, S., Imaizuma, H., and Maruyama, I. (2001). Simultaneous measurement of anandamide and 2-arachidonoylglycerol by polymyxin B-selective adsorption and subsequent high-performance liquid chromatography analysis: increase in endogenous cannabinoids in the sera of patients with endotoxic shock. *Analytical Biochemistry,* **294:** 73–82.

White, R. and Hiley, C.R. (1997). A comparison EDHF-mediated and anandamide-induced relaxations in the rat isolated mesenteric artery. *British Journal of Pharmacology,* **122:** 1573–1584.

White, R. and Hiley, C.R. (1998). The actions of some cannabinoid receptor ligands in the rat isolated mesenteric artery. *British Journal of Pharmacology,* **125:** 533–541.

White, R., Ho W.S.V., Bottrill, F.E., Ford, W.R., and Hiley, C.R. (2001). Mechanisms of anandamide-induced vasorelaxation in rat isolated coronary arteries. *British Journal of Pharmacology,* **134:** 921–929.

Zygmunt, P.M., Anderssson, D.A., and Hogestatt, E.D. (2002). Δ^9-tetrahydrocannabinol and cannabinol activate capsaicin-sensitive sensory nerves via a CB_1 and CB_2 cannabinoid receptor-independent mechanism. *Journal of Neuroscience,* **22:** 4720–4727.

Zygmunt, P.M., Petersson, J., Anderssson, D.A., Chuang, H-h., Sorgard, M., Di Marzo, V. et al. (1999). Vanilloid receptors on sensory nerves mediate the vasodilator action of anandamide. *Nature,* **400:** 452–457.

19 Endocannabinoids in Inflammation and Immune Response

Evgeny V. Berdyshev

CONTENTS

INTRODUCTION

Endocannabinoids represent a heterogenic group of lipid messenger molecules that bind to and activate cannabinoid receptors. The two best-known endogenous ligands for cannabinoid receptors, anandamide (*N*-arachidonoylethanolamine, 20:4 NAE) and 2-AG, belong to two distinct groups of neutral lipids, NAEs and monoacylglycerols (MGs), respectively. However, the similarity of action of anandamide and 2-AG in the initiation of cannabinoid-receptor-mediated signaling has resulted in their collective grouping under the term *endocannabinoids*. In addition, their receptor-inactive saturated and monounsaturated congeners are also included in the same group of compounds.

NAEs attracted significant attention in the 1980s when the stress-related activation of their biosynthesis was discovered (Schmid et al., 1990). Despite extensive efforts and thorough elucidation of NAE metabolism, the physiological roles of NAEs remain unclear. During this period, anandamide was not being extensively studied due to difficulties in its detection and its as yet unknown physiological significance. Interest in NAEs dramatically increased after the identification of anandamide as the endogenous ligand for CB_1 cannabinoid receptors (Devane et al., 1992). Three years later, 2-AG was also reported to be a ligand for cannabinoid receptors (Lee et al., 1995; Mechoulam et al., 1995; Sugiura et al., 1995). These findings followed the discovery of central CB_1 (Devane et al., 1988) and peripheral CB_2 (Munro et al., 1993) cannabinoid receptors and raised questions about the role of the endocannabinoid system in the regulation of cell function and physiological homeostasis.

Characterization of expression of CB_1 cannabinoid receptors in the central nervous system (CNS) and CB_2 cannabinoid receptors within cells of the immune system (Berdyshev, 2000) defined the major directions of cannabinoid research. Presently, considerable information exists regarding the neurophysiological and behavioral effects of cannabinoids including the role of CB_1 cannabinoid receptors in regulation of neurotransmission. Immunomodulation by cannabinoids has to date received much less attention.

In recent years, however, significant progress has been made in defining the role of CB_2-cannabinoid-receptor-mediated signaling in the regulation of immune cell function, but this research is largely limited and has produced some contradictory findings. The purpose of this chapter is to give an overview of the role of endocannabinoids as immunomodulatory molecules, with special attention being given to recently published information.

ENDOCANNABINOIDS AND INFLAMMATION

It was about 50 yr ago that the NAEs first became of interest to immunologists. Coburn et al. (1954) first reported the antianaphylactic property of an alcohol-soluble fraction derived from egg yolk. This finding led to the identification of the active component in 1957 as 16:0 NAE (Kuehl et al., 1957). Additional studies found 16:0 NAE to possess the most antianaphylactic activity when compared to several synthetic structural analogs (Kuehl et al., 1957). Antibacterial and antiviral properties of 16:0 NAE have also been reported (Perlík et al., 1971a, 1971b, and 1973; Rašková and Mašek, 1967; Rašková et al., 1972). These properties formed the basis for the development of the drug called *impulsin* from 16:0 NAE (Masek et al., 1974; Kahlich et al., 1979). Unfortunately, the antibacterial and antiviral properties of 16:0 NAE did not receive further attention and, therefore, its mechanism of action remained unclear. Only after the identification of 20:4 NAE as an endogenous ligand for cannabinoid receptors (Devane et al, 1992) did the mechanistic principle of endocannabinoid action become more defined. Subsequent studies, however, examining the role of endocannabinoids and their receptors in inflammation in the periphery produced clarifications as well as confusion concerning the exact mechanisms of the *in vivo* action of endocannabinoids.

There is an increasing body of evidence showing that, *in vivo*, anandamide (20:4 NAE) and palmitoylethanolamide (16:0 NAE) can diminish the development and intensity of the inflammatory response. In 1993 Aloe et al. demonstrated that long-chain and short-chain NAEs (16:0 NAE and 4:0 NAE) could decrease substance-P-induced mast cell degranulation in rats when applied subcutaneously. The authors suggested that 16:0 NAE may act as a local autocoid, negatively controlling mast cell activation. In subsequent studies, orally administered 16:0 NAE (1 mg/kg) was found to significantly diminish substance-P-induced mast cell degranulation and plasma extravasation in a rat model (Mazzari et al., 1996). It was also reported that both 16:0 NAE and 20:4 NAE could attenuate formalin-induced cutaneous inflammatory pain in rats in a dose-dependent manner, and the effect of 16:0 NAE could be inhibited by the CB_2 receptor antagonist SR144528 (Calignano et al., 1998; Jaggar et al., 1998). Conti et al. (2002) also showed the anti-inflammatory properties of 16:0 NAE when administered orally in a model of an acute, carrageenan-induced rat hind paw inflammation. In this study, the CB_2 receptor antagonist SR144528 was once again shown to prevent 16:0 NAE-induced anti-inflammatory action. However, in another study, 16:0 NAE administered orally to rats (1–10 mg/kg), decreased carrageenan-induced edema in a dose-dependent manner, but this effect could not be reversed by SR144528 (Costa et al., 2002). When 16:0 NAE and anandamide were compared in a model of formalin-induced inflammatory pain, the *in vivo* effects of 16:0 NAE and anandamide were found to be different (Calignano et al., 1998). Thus, 16:0 NAE could decrease both early- and late-phase responses to formalin, whereas the effect of anandamide was detected only during the early-phase response to inflammatory stimulation. Once again, the effects of 16:0 NAE could be inhibited by the CB_2 receptor antagonist SR144528 but not the CB_1 receptor antagonist SR141716. The effect of anandamide was inhibited only by the CB_1 receptor antagonist SR141716. In support of this observation, Richardson et al. (1998) found that anandamide exhibits CB_1-receptor-mediated antihyperalgesic action when injected locally into the inflamed hind paw of the rat. Thus, the *in vivo* data suggest a role for both CB_1 and CB_2 cannabinoid receptors in mediating the effects of anandamide and 16:0 NAE on dermal inflammatory processes. These observations seem to be in agreement with the data obtained from *in vivo* experiments with synthetic cannabinoids. Clayton et al. (2002) demonstrated that systemic administration of the CB_2-receptor-selective agonist GW 405833 partially prevented carrageenan-induced inflammation-mediated hyperalgesia. The other CB_2-receptor-selective agonists HU308 (Hanus et al., 1999) and AM1241 (Quartilho et al., 2003) were also shown to inhibit arachidonic-acid-induced ear edema and carrageenan-induced rat hind paw edema and inflammatory hyperalgesia, respectively.

Surprisingly, straightforward and mostly coherent *in vivo* experiments showing the anti-inflammatory properties of NAEs cannot yet be explained mechanistically from the point of view of the exact cellular targets and the mechanisms implicated in the effects of endocannabinoids in these models.

Mast cells play a significant role in the initiation and development of the inflammatory response in skin, but the experiments with isolated primary mast cells or basophilic cell lines in *in vitro* experiments do not confirm their role in mediating the *in vivo* effects of endocannabinoids. In contrast to the original *in vivo* observations by Aloe et al. (1993) showing the potency of 16:0 NAE to diminish substance-P-induced mast cell degranulation, 16:0 NAE and 20:4 NAE as well as synthetic cannabinoids, failed to inhibit anti-IgE-induced (Lau and Chow, 2003) or c48/80-induced (Bueb et al., 2001) histamine release by rat peritoneal mast cells *in vitro*. Moreover, anandamide at 10^{-5} *M* was capable of inducing histamine release from nonactivated mast cells, and WIN5512-2 and HU210 at 10^{-5} *M* increased anti-IgE-induced cell stimulation (Lau and Chow, 2003). In experiments with RBL-2H3 cells, 16:0 NAE slightly decreased antigen-induced serotonin release only at high (>100 μM) concentrations, whereas anandamide had no effect (Granberg et al., 2001). In addition, 2-AG and methanandamide even increased the antigen-provoked release of serotonin and β-hexosaminidase. In other experiments with the human HMC-1 mast cell line, neither 16:0 NAE nor 20:4 NAE affected tryptase release elicited by the calcium ionophore A23187 (Maccarone et al., 2000). Thus, it seems unlikely that the anti-inflammatory properties of endocannabinoids observed in *in vivo* models of skin inflammation and pain are mediated by their effect on mast cells. It should also be noted that cultured RBL-243 and HMC-1 cells lack the ability to respond to substance P (Fewtrell et al., 1982) and c48/80 (Wierecky et al., 2000) and are thus inappropriate models. In addition to models of skin inflammation, endocannabinoids have also been tested in *in vivo* models of bronchopulmonary inflammation elicited with aerosolized LPS (Berdyshev et al., 1998). Once again, 16:0 NAE and 20:4 NAE showed quantitative and qualitative differences in their ability to affect bronchopulmonary inflammation. Intranasally administered anandamide decreased TNF-α levels in bronchoalveolar lavage fluid and diminished neutrophil recruitment in the lung of mice at low (0.075 μmol/kg) dose but not at higher doses, whereas 16:0 NAE decreased TNF-α production only at the highest dose tested (0.75 μmol/kg). Endocannabinoids may also play a regulatory role in intestinal inflammation. The presence of both 20:4 NAE and 2-AG have been reported in the intestinal tissues of croton oil-irritated mouse small intestine (Izzo et al, 2001). Although their levels did not differ from that in noninflamed intestine, this finding may indicate elevated turnover of both endocannabinoids, as inflamed small intestine has significantly elevated activity of FAAH as well as expression of CB_1 cannabinoid receptors (Izzo et al, 2001).

It should be noted that when evaluating *in vivo* experiments with orally administered NAEs, NAE bioavailability and pharmacokinetics are rarely taken into consideration. It is known that intestinal mucosa, liver (Schmid et al., 1985), and small intestine (Katayama et al., 1997) have a high level of FAAH activity. Thus, it is likely that dietary NAEs are readily degraded in the intestinal tract. In this context, it is interesting to note that the original paper about the anti-inflammatory properties of NAEs attributed them to the ethanolamine moiety of the molecule (Kuehl et al., 1957). Mazzari et al. (1996), however, did not find any anti-inflammatory effect of orally administered ethanolamine and palmitic acid in a model of substance-P-induced extravasation. Thus, there are multiple contradictions in the available information on the systemic anti-inflammatory properties of NAE that require further investigation.

ENDOCANNABINOID EFFECT ON IMMUNE CELL FUNCTIONS

It would be correct to say that recent progress in the investigation of the endocannabinoid system, including signaling effects provided through the activation of CB_1 and CB_2 cannabinoid receptors, now modifies the general understanding of how this system functions. A few years ago, the statement that the activation of cannabinoid receptors results in immunosuppressive signals would not have been disputed; to date most studies have been performed with natural and synthetic cannabinoids, and the effects of endocannabinoids have only been sporadically tested. The increasing volume of detailed studies on signaling mechanisms and gene expression following cannabinoid receptor activation and the growing attention being given to endocannabinoids and CB_1/CB_2 receptor-inactive endocannabinoids such as 16:0 NAE allow us to talk about the endocannabinoid system as an entity that regulates rather than suppresses the immune response. There is no doubt that the abuse of marijuana results in general immunosuppression and decreased resistance to bacterial and viral infections. Extensive reviews on this subject were recently provided by Roth et al., 2002 and Klein et al., 2003. At the same time, understanding the mechanisms of endocannabinoid-related immunosuppression opened up the possibility of selectively targeting CB_2 cannabinoid receptors to treat pathologies related to the hyperactivation of the Th1-type response, as cannabinoids are known to switch the balance toward a Th2-type response (Roth et al., 2002; Yuan et al., 2002). For the purpose of the present chapter, it would be confusing to describe in detail all the available evidence on the effects of cannabinoids on the immune response. Instead, a brief overview of the immunomodulatory properties of cannabinoids is presented in Table 19.1. The following subsections describe the properties of endocannabinoids as immunomodulators.

ENDOCANNABINOIDS AND HEMATOPOIETIC CELL GROWTH AND MIGRATION

One of the earliest reports regarding the positive effects of cannabinoid receptor ligands on the functioning of hematopoietic cells were the findings of Luo et al. (1992) and Derocq et al. (1995), who demonstrated the stimulation of mouse splenocyte (Luo et al., 1992) and human B-cell (Derocq et al., 1995) proliferation by low nanomolar concentrations of cannabinoids. These findings suggest that endogenous ligands for cannabinoid receptors may be important growth-factor-like molecules or cofactors controlling cell proliferation and differentiation. In fact, Valk et al. (1997) showed that nanomolar and low micromolar concentrations of anandamide markedly potentiated interleukin-3 (IL-3)-induced mouse bone marrow cell colony growth. Similarly, the growth of multiple murine hematopoietic growth factors (HGFs)-dependent cell lines was also dramatically enhanced by nanomolar concentrations of anandamide in the absence of serum (Valk et al., 1997). The difference in the effect of anandamide on CB_2 cannabinoid receptor- or empty vector-transfected 32D/G-CSF-R cells indicated the necessity of CB_2 cannabinoid receptors for the transmission of anandamide-elicited signals. Interestingly, synthetic and natural cannabinoids fail to provoke similar responses. Moreover, they even inhibited cell proliferation when serum was present in the culture medium. The authors suggested that anandamide is a novel and synergistic growth stimulant for hematopoietic cells. Unfortunately, this work does not bring additional insights into the role of the CB_2 cannabinoid receptor in mediating the observed effects of anandamide, nor does it compare the effect of anandamide with that of the cannabinoid-receptor-inactive congeners. In fact, later work by Derocq et al. (1998) suggests that the growth-inducing effect of anandamide is independent of the cannabinoid receptors, as arachidonic acid, by itself, mimicked the effect of anandamide, with pertussis toxin being unable to block its effect.

Additional controversy has also resulted from the work by Lee et al. (1995), who compared 2-AG and anandamide in ability to affect mouse splenocyte proliferation and immunogenic

TABLE 19.1
Immunomodulatory Effects of Natural and Synthetic Cannabinoids

Compound	Tested Concentration	Model	Parameter	Effect	Reference
Δ^9-THC	3.2–22 μ*M*	EL4.IL-2 cells	IL-2 synthesis, cAMP accumulation, AP-1 activation	Inhibition	Condie R. et al., *J. Biol. Chem.*; 1996, 271, 13175.
Δ^9-THC, CP55940	0.0001–1 μ*M*	Mϕ-antigen processing and T-cell activation	IL-2 secretion by T-cells	Inhibition	McCoy K.L. et al., *J. Pharmacol. Exp. Ther.*, 1999, 289, 1620.
Cannabinol, Cannabidiol CP55940 Δ^9-THC	0.01–20 μ*M*	EL4 cells	Suboptimal stimulation with PMA of IL-2 mRNA expression and IL-2 secretion	Stimulation	Jan T-R. et al., *Mol. Pharmacol.*, 2002, 61, 446.
Δ^9-THC	5–10 μg/ml	NKB61A2 NK cell line	IL-2 Receptor subunit expression	Increase in IL2Rβ mRNA, Decrease in IL2Rγ mRNA	Zhu et al., *J. Pharmacol. Exp. Ther.*, 1995, 274, 1001.
Δ^9-THC	3.2–22 μ*M*	EL4.IL-2 cells	IL-2 synthesis, cAMP accumulation, AP-1 activation	Inhibition	Condie R. et al., *J. Biol. Chem.*, 1996, 271, 13175.
Cannabinol, WIN55212-2	1–20 μ*M*	Mouse splenocytes	PMA/Io-stimulated ERK activation	Inhibition	Kaplan B.L.F. and Kaminski N. *Int. Immunopharmacol.*, 2003, 3, 1503.
Δ^9-THC	3.2–22 μ*M*	Mouse splenocytes	cAMP-dependent signaling	Inhibition	Koh W.S. et al., *Biochem. Pharmacol.* 1997, 53, 1477.
Δ^9-THC, Cannabidiol	2.5–10 μg/ml	CD8+ NK cells, HUT-78 cells, HTLV-1-positive B-cells	Constitutive and PMA-stimulated cytokine production	Inhibition and stimulation	Srivastava M.D. et al., *Immunopharmacology*, 1998, 40, 179.
Δ^9-THC	0–5 μg/ml	Human T-cells in the presence of allogenic dendritic cells	T-cell proliferation IFN-γ production TH1/TH2 ratio	Inhibition, decrease in the ratio	Yuan M et al., *J. Neuroimmunol.*, 2002, 133, 124.
Cannabinol	5–15 μ*M*	Mouse thymocytes	CREB/ATF phosphorylation, CRE and κB DNA binding activity	Inhibition	Herring A.C. et al., *Cell. Signaling*, 2001, 13, 241.

(continued)

TABLE 19.1 (Continued)
Immunomodulatory Effects of Natural and Synthetic Cannabinoids

Compound	Tested Concentration	Model	Parameter	Effect	Reference
WIN55212-2 CP55940, HU210	1–10 μ*M*	Rat cortical microglial cells	LPS-induced TNF-α release	Inhibition	Facchinetti F. et al., *Glia*, 2003, 41, 161.
Δ⁹-THC, CP55940	0.0001–1 μ*M*	Mϕ-antigen processing and T-cell activation	IL-2 secretion by T-cells	Inhibition	McCoy K.L. et al., *J. Pharmacol. Exp. Ther.*, 1999, 289, 1620.
Δ⁹-THC	1–10 μ*M*	J774 macrophages	NO, PGE2, IL-6	Decrease	Chang Y. et al., *J. Cell Biochem.*, 2001, 81, 715.
CP55940	0.001–1 μ*M*	Rat peritoneal macrophages	fMLP-induced chemotaxis	Inhibition	Sacerdote P. et al. *J. Neuroimmunol.*, 2000, 109, 155.
CP55940 WIN55212-2	0.1–30 μ*M*	RAW246.7 macrophages	LPS-stimulated NO production	Inhibition	Ross R.A. et al., *Eur. J. Pharmacol.*, 2000, 401, 121.
Δ⁹-THC	3–10 μg/ml	Mouse peritoneal macrophages	LPS-stimulated IL-1 synthesis and bioactivity	Stimulation	Zhu W. et al., *J. Pharmacol. Exp. Ther.* 1994, 270,1334.
HU210	50 n*M*	U373 MG human astrocytoma cells	ERK activation	Stimulation	Galve-Roperh I. et al., *Mol. Pharmacol.*, 2002, 62, 1385.
JWH015 Δ⁹-THC, Δ⁸-THC	1–10 μ*M*	THP-1 cells	LPS-induced TNF-α, IL-1β secretion	Inhibition	Klegeris A. et al., *Br. J. Pharmacol.*, 2003, 139, 775.
Δ⁹-THC	0.5–5 μg/ml	Mouse peritoneal macrophages	LPS-IFN-γ-induced NO production	Inhibition	Coffey R.G. et al., *Biochem. Pharmacol.*, 1996, 52, 743.
Δ⁹-THC	0.003–3 μ*M*	Human PBMC	LPS- and PHA-induced cytokine release	Variable	Berdyshev E.V. et al., *Eur. J. Pharmacol.*, 1997, 330, 231.
Ajulemic acid	1–30 μ*M*	Human monocytes	LPS-induced IL-1β and TNF-α production	IL-1β — inhibition TNF-α — no effect	Zurier R.B. et al., *Biochem. Pharmacol.*, 2003, 65, 649.
WIN55212-2	0.01–1 μ*M*	Human CD4+ lymphocytes, microglial cells	HIV-1 expression	Inhibition	Peterson P.K. et al., *J. Neuroimmunol.*, 2004, 147, 123.
Δ⁹-THC	1–100 μg/ml	Herpes simplex virus *in vitro* replication	Replication	Inhibition	Lancz G. et al., *Proc. Soc. Exp. Biol., Med.* 1991, 196, 401.

CP55940 JWH015 WIN55212-2 Δ^9-THC	0.1 nM–100 μ*M*	HT29 human colonic epithelial cells	TNF-α-induced IL-8 synthesis	Inhibition	Ihenetu K. etl al., *Eur. J. Pharmacol.*, 2003, 458, 207.
WIN55212-2	5–20 mg/kg	Mouse, TMEV-induced demyelinating disease	T-cell proliferation, antiviral TH1- and TH2-type cytokines	Inhibition	Croxford J.L. and Miller S.D., *J. Clin. Invest.*, 2003, 111, 1231.
Δ^9-THC, WIN55212-2, HU210	3.125–50 mg/kg	Mouse, control or primed with *Corynebacterium parvum*	Circulating levels of TNF-α, IL-12, IL-10	Decrease in TNF-α, IL-12 Increase in IL-10	Smith S.R. et al., *J. Pharmacol. Exp. Ther.*, 2000, 293, 136.
Δ^9-THC	8 mg/kg	Mouse, *Legionella pneumophila* injection	Mortality, Circulating levels of IFN-γ, IL-12, IL-4; TH1/TH2 ratio	Increased mortality, decrease in TH1/TH2 ratio	Klein T.W. et al., *J. Immunol.*, 2000, 164, 6461.
Cannabis smoking	Moderate, heavy smoking	Human	Lymphocyte profiling, IL-2, IL-10, TGFβ1	Decrease in NK cells, IL-2, increase in IL-10 and TGFβ1	Pacific R. et al., *JAMA*, 2003, 289, 15, 1929.

responses. The compounds were tested in micromolar concentrations (1–36 μ*M*) in the presence of serum. 2-AG was found to produce a dose-related inhibition of anti-CD3 mAB-induced T-cell proliferation and LPS-induced B-cell proliferation. Anandamide had no effect. Interestingly, the effects of 2-AG were, in part, related to cell density, as at a high density 2-AG enhanced lymphoproliferation, whereas at a low cell density proliferation was inhibited. It is possible that these observations were affected by the constant presence of 2-AG and anandamide during cell culture, as both endocannabinoids have been shown to be present at high levels in FBS (Berdyshev et al., 2000). Also, metabolic oxidation of 2-AG and formation of bioactive prostaglandin-like molecules cannot be ruled out as a factor affecting proliferation.

Although the results observed for the role of endocannabinoids and peripheral cannabinoid receptors in hematopoietic cell proliferation are not absolutely conclusive, hematopoietic cell migration appears to be clearly related to CB_2-cannabinoid-receptor-specific activation. Of the several ligands tested, 2-AG is the most efficient at inducing cell migration. Jorda et al. (2002) investigated the induction of chemotaxis and chemokinesis by 2-AG in 32D/G-CSF-R cells transfected with exon-1B/exon-2 CB_2 complementary DNA and in the myeloid cell line NCF78. The effect of 2-AG was compared with that of anandamide, 16:0 NAE, 18:1 NAE, and natural and synthetic cannabinoids. The authors found that only 2-AG, and to lesser extent anandamide, induced cell migration in the nanomolar to low-micromolar concentration ranges. None of the other NAEs tested possessed chemotactic or chemokinetic properties. Also, in these studies the presence of CB_2 cannabinoid receptors was shown to be critical for mediating the effect of 2-AG. Interestingly, such cannabinoids as Δ^9-THC, WIN55212-2, and CP55940 not only failed to induce cell migration but also antagonized the effects of 2-AG. Surprisingly, overexpression of the CB_2 cannabinoid receptors in 32D/G-CSF-R cells also resulted in cell neutrophilic maturation arrest that occurred even without the presence of exogenously added 2-AG (Jorda et al., 2003). It is possible that endogenously synthesized and secreted 2-AG was responsible for the observed effect of CB_2 cannabinoid receptor transfection as the CB_2 receptor antagonist SR144528 and not the CB_1 receptor antagonist SR141716A reestablished G-CSF-induced cell neutrophilic differentiation. Investigation examining signaling events implicated in the mediation of the effect of CB_2 receptor overexpression on cell neutrophilic maturation revealed participation of MEK/ERK and PI3-kinase pathways (Jorda et al., 2003).

Similar induction of migration by 2-AG was recently demonstrated for differentiated HL60 cells, U937 cells, differentiated THP-1 cells, and human monocytes (Kishimoto et al., 2003). Once again, 2-AG was a more potent inducer of cell migration than the synthetic cannabinoids CP55940 or WIN55212-2, and 2-arachidonoyl glycerol ether (2-AG ether). Of the other MGs tested, only 2-eicosa-5,8,11-trienoylglycerol and 2-eicosa-5,8,11,14,17-pentaenoylglycerol induced HL60 cell migration. Anandamide had no effect. The induction of cell migration was CB_2 cannabinoid receptor specific as shown by the ability of pertussis toxin or SR144528, but not SR141716A, to block the effect of 2-AG on this migration of differentiated HL60 cells and human monocytes. Rho kinase, MEK, and p38 MAP kinases were shown to mediate 2-AG-elicited migration.

Microglial cells were also found to migrate upon stimulation with 2-AG (Walter et al., 2003). Migration of these cells was also induced, although to a lesser extent, by anandamide but not by 16:0 NAE, and the effect of 2-AG was inhibited by cannabinol and cannabidiol. The study suggested that 2-AG stimulated cell migration through CB_2- and abnormal cannabidiol-sensitive receptors with subsequent activation of the ERK1/2 signaling pathway. Franklin et al. (2003) found that 16:0 NAE potentiates anandamide-induced migration of mouse microglial BV-1 cells. The involvement of a new $G_{i/o}$-protein-coupled receptor was suggested. This was based on several observations: first, these receptors had to be different from the classical CB_1/CB_2 cannabinoid receptors or abnormal-CBD receptors as pertussis toxin could block the inhibitory effect of 16:0 NAE on forskolin-induced cAMP accumulation but neither SR141716A nor SR144528 could influence the observed effect of 16:0 NAE. Second, 16:0 NAE is known to be unable to bind to both cannabinoid receptors. Third, cannabidiol did not mimic the effect of 16:0 NAE, and the 16:0-NAE-induced response was not affected by compound O-1918 — the antagonist for abnormal-CBD receptors. Thus, these data reveal a selective specific functional role for 16:0 NAE in microglial cell physiology.

Endocannabinoid Effect on Leukocyte and Lymphocyte Functional Responses

Immune cells undergo dramatic metabolic and functional changes upon stimulation. The outcome of such stimulation depends on the type of stimuli and is specific for individual subsets of immune cells. Cannabinoids have been shown to have a marked deleterious effect on immune cell function (see reviews by Berdyshev, 2000; Roth et al., 2002; and Klein et al., 2003). In general, endocannabinoids follow the natural and synthetic cannabinoids in their ability to decrease immunogenic responses, but much more work is needed to clarify the regulatory mechanisms that link endocannabinoid metabolism with the functional outcomes of up- or downregulation of endocannabinoid production.

Macrophages represent the first line of defense against bacterial and viral infections; they process and present antigen to $CD4^+$ T-lymphocytes. High doses of anandamide (20–80 mg/kg, i.p.) were shown to suppress TNF-α-dependent killing of L929 fibroblasts by macrophages isolated from mice that had received *Propionebacterium acnes* for cell activation (Cabral et al., 1995). Another endocannabinoid, 2-AG, was also shown to decrease LPA-upregulated circulating TNF-α levels in mice when injected intraperitoneally (10 mg/kg) along with LPS (Gallily et al., 2000). Also, coinjection of 2-AG with the cannabinoid-receptor-inactive MGs, 2-palmitoylglycerol (5 mg/kg) and 2-linoleoylglycerol (10 mg/kg) resulted in almost complete inhibition of TNF-α production. *In vitro*, 2-AG demonstrated dose-dependent (0.05–50 μg/ml) inhibition of TNF-α-mediated killing of BALB/c CL.7 cells by mouse peritoneal macrophages when it was present during cell coculture for 24 h. It also suppressed zymozan-induced ROS generation in macrophages but not LPS/IFN-γ-elicited NO production (Gallily et al., 2000). Both anandamide and 2-AG, at relatively high (10 μ*M*) concentration, were found to attenuate LPS-stimulated TNF-α synthesis by rat microglial cells (Facchinetti et al., 2003). Interestingly, in microglia-like THP-1 cells, anandamide (5–10 μ*M*) did not inhibit LPS- and IFN-γ-induced TNF-α secretion but significantly attenuated IL-1β secretion (Klegeris et al., 2003), thus showing that the outcome of cell activation with anandamide may differ and be selective depending on the particular cell type or model employed. The data described by Molina-Holgado et al. (1998) additionally demonstrated that the functional outcome of anandamide (10–25 μ*M*) treatment may even be positive as it was found in the model of Theiler's murine encephalomyelitis virus-induced IL-6 synthesis by mouse astrocytes in culture. In this model, anandamide was found to stimulate virus-induced IL-6 synthesis. In LPS-stimulated J744 macrophages, IL-6 and NO production were suppressed by high (3–30 μ*M*) concentrations of anandamide, whereas 2-AG (3–30 μ*M*) inhibited IL-6 production but slightly increased NOS-dependent NO production (Chang et al., 2001). This discrepancy between the effects of the two endocannabinoids was suggested to be determined by the generation of bioactive metabolites of 2-AG, i.e., arachidonic acid and PGE_2, both of which activated iNOS and COX-2. Similarly, 2-AG was found to stimulate NO production in human monocytes when they were subjected to 2-AG (0.1–10 μ*M*) treatment without co-stimulation (Stefano et al., 2000). In this case, however, the effect of 2-AG seems to be purely CB_1 cannabinoid receptor dependent as SR141716A could block its effect.

Ross et al. (2000) addressed the question of the possible involvement of cannabinoid receptors in mediating the inhibitory effect of CB_1/CB_2 cannabinoid-receptor-inactive 16:0 NAE on LPS-induced NO production in RAW264.7 macrophages. In these cells, 10–30 μ*M* of 16:0 NAE was shown to diminish NO production significantly. However, this action of 16:0 NAE was not mediated by any of the known cannabinoid receptors as it did not affect forskolin-induced cAMP production in cells and its action was pertussis toxin- and CB_1/CB_2 receptor antagonist-insensitive.

The functional response of human peripheral blood mononuclear cells (PBMC), which predominantly comprise monocytes and lymphocytes, was also affected (although with some discrepancies) by anandamide and 16:0 NAE (Berdyshev et al., 1997). We found that both compounds have approximately the same ability to downregulate LPS-induced synthesis of TNF-α, IL-6, IL-8, and PHA-induced IL-4 synthesis at nanomolar (3–300 n*M*) concentrations, but anandamide was a more potent inhibitor of TNF-α synthesis at higher (3 μ*M*) concentrations. Also, anandamide and 16:0 NAE happened to differ in their ability to affect PHA-induced IFN-γ production. Palmitoylethanolamide could not modulate PHA-stimulated IFN-γ synthesis by PBMC, although anandamide inhibited its production. Thus, this study also demonstrated the ability of 16:0 NAE to affect immune cell functional responses in a way different from that of anandamide.

Lee et al. (1995) found that 2-AG (5–25 μ*M*), but not anandamide, could inhibit mixed lymphocyte response, anti-CD3 mAB-induced T-cell proliferation, and LPS-induced B-cell proliferation at low cell density (1×10^6 cells per ml). On the contrary, at high cell density 2-AG enhanced cell proliferation, possibly due to increased 2-AG metabolism and the formation of PGE_2 and other arachidonic acid metabolites. Although this report did not address the mechanistic questions that could explain the observed difference in the properties of the two major endocannabinoids, it clearly demonstrated the discrepancy in their functions. Meanwhile, it has been proposed that 2-AG rather than anandamide is the endogenous ligand for both CB_1 and CB_2 cannabinoid receptors (Sugiura et al., 1999, 2000; Gonsiorek et al., 2000). When comparing the ability of 2-AG, anandamide, and 16:0 NAE to affect $[Ca^{2+}]_i$ in HL60 cells, only 2-AG behaved as a full agonist at CB_2 cannabinoid receptors expressed in HL60 cells; anandamide was found to be a weak partial agonist, and 16:0 NAE was devoid of any activity (Sugiura et al., 2000). Unfortunately, until now there have been only a few studies that have thoroughly investigated the mechanistic differences in the immunomodulatory properties of both of the major groups of endocannabinoids. 2-AG has been shown to downregulate PMA/ionomycin-induced IL-2 synthesis by mouse splenocytes and EL4.IL-2 cells expressing CB_2 cannabinoid receptors (Ouyang et al., 1998). Although the question of mediation of the 2-AG effect by CB_1 or CB_2 cannabinoid receptors was not addressed in this work, it provides in-depth knowledge concerning 2-AG's effect on NF-AT, NF-κB, AP-1, octamer, and cAMP-response element binding protein-dependent regulation of IL-2 synthesis in these cells. It was found that only NF-AT binding and promoter activity and, to a lesser degree, NF-κB/Rel binding and promoter activity were dose-dependently (5–50 μ*M*) inhibited by 2-AG. Signaling events leading to the inhibition of NF-AT and NF-κB DNA binding and promoter activity by 2-AG are yet to be defined.

A recent study by Sancho et al. (2003) is one of the most intriguing in terms of confirmation of cannabinoid-receptor-independent inhibition of TNF-α-induced NF-κB activation by anandamide in Jurkat-cell-line-derived 5.1 clone. In this report anandamide (5–25 μ*M*) was demonstrated to inhibit IκB degradation and IKK activation without any effect on ERK or p38 phosphorylation. This effect of anandamide was suggested to be cannabinoid-receptor- and anandamide-hydrolysis-independent as in the 5.1 clone, which only expresses CB_2 cannabinoid receptors; the inhibition of FAAH did not affect anandamide-induced response; and in KBF-Luc-transfected A549 cells, another cell line used in this study that expresses only CB_1 cannabinoid receptors, FAAH inhibitor and CB_1 receptor antagonist SR141716A did not modify anandamide-induced inhibition of NF-κB. It was also suggested, based on indirect evidence, that vanilloid VR1 receptors are not involved in mediating the effect of anandamide, which is known to bind to and activate VR1 receptors (Zygmunt et al., 1999). In this context it is worth noting that anandamide, 16:0 NAE, 18:1 NAE, and 18:2 NAE were previously shown to slightly activate ERK and AP-1-dependent transcriptional activity in epithelial JB6 cell line through a cannabinoid-receptor-independent mechanism (Berdyshev, Schmid, Krebsbach, Hillard, et al., 2001). It is possible that this particular cannabinoid-receptor-independent NF-κB targeting by anandamide, and possibly by other NAEs, plays a role in the previously observed induction of apoptosis by anandamide in T- and B-lymphocytes (Schwarz et al., 1994), in murine lymphoma EL4 cell line, and in human T-lymphoblastic leukemia Molt-4 cell line (McKallip et al., 2002). Clearly, the exact mechanism of the cannabinoid- and vanilloid-receptor-independent action of NAEs deserves further and thorough investigation.

REGULATION OF ENDOCANNABINOID METABOLISM IN IMMUNE CELLS

It is obvious that endocannabinoid levels in tissues, cells, and biological fluids are controlled through the regulation of their metabolism. Of the two groups of endocannabinoids, NAEs and 2-MGs, only NAE metabolism is fully elucidated; the exact routes of 2-AG biosynthesis and catabolism and anabolism, in relation to immunogenic or excitatory stimulation, are still obscure. In this chapter we will not discuss general questions of endocannabinoid metabolism but rather refer readers to several recent reviews in this area (see reviews by Schmid, 2000; Schmid et al., 2002; Schmid and Berdyshev, 2002; Sugiura et al., 2002), briefly overview general metabolic pathways, and address the question of regulation of endocannabinoid levels in the cells of the immune system.

NAE biosynthesis (Figure 19.1) is a two-step process that was recently referred to as the *Schmid pathway* (Farooqui et al., 2000; Sugiura et al., 2002) and is catalyzed by two enzymes: a Ca^{2+}-dependent energy-independent transacylase that generates *N*-acyl-phosphatidylethanolamine (*N*-acyl PE) (Natarajan et al., 1982; Reddy, Natarajan, et al., 1983; Reddy, Schmid, et al., 1983; Reddy et al., 1984), and a phosphohydrolase of the phospholipase-D type that forms NAE (including anandamide) and phosphatidic acid (Schmid et al., 1983). This biosynthetic pathway of NAE generation was described by Schmid's group in the early 1980s and was shown later to be the same one that results in the generation of anandamide (Bisogno et al., 1997; Cadas et al., 1996, 1997; Di Marzo et al., 1994, Di Marzo, De Petrocellis, Sepe, et al., 1996; Sugiura et al., 1996). Because the initial step of NAE biosynthesis includes acyl transfer from the *sn*-1 position of the donor phospholipid to an ethanolamine moiety of PE, it determines the low percentage of anandamide formed in cells due to a predominant phospholipid esterification with arachidonate at the *sn*-2 position. The phosphodiesterase that cleaves *N*-acyl PE to PA and NAE is different from other phospholipase-D-type enzymes because it does not catalyze the transphosphatidylation reaction (Peterson and Hansen, 1999). Finally, NAE is hydrolyzed by amidohydrolase, which was first characterized in rat liver (Schmid et al., 1985) and later referred to as fatty acid amidohydrolase (FAAH) due to its ability to hydrolyze NAEs as well as primary fatty acid amides (Maurelli et al., 1995; Patricelli and Cravatt, 2001).

It is difficult to say which step of NAE metabolism is rate limiting and controls the up- and downregulation of NAE levels in cells. It is well documented that NAE levels are dramatically increased during stress and *post mortem* — events related to the loss of Ca^{2+} homeostasis (reviewed

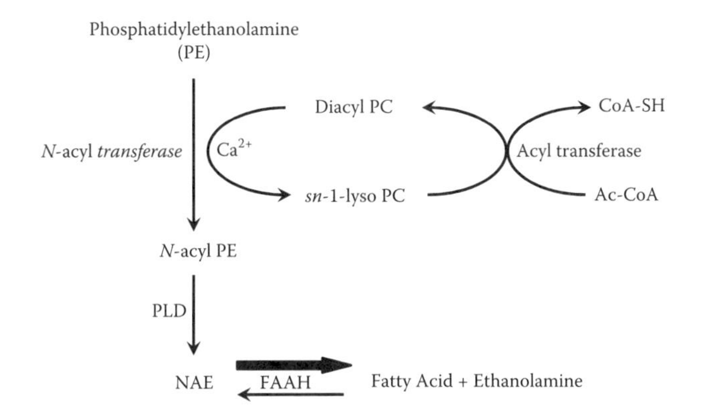

FIGURE 19.1 General scheme of NAE biosynthesis and catabolism.

by Schmid, 2000; Schmid et al., 2002; Schmid and Berdyshev, 2002). Ionomycin and Ca^{2+} have been shown to enhance anandamide and *N*-acyl PE levels in cells of the immune system (Bisogno et al., 1997; Di Marzo, De Petrocellis, Sepe, et al., 1996; Pestonjamasp and Burstein, 1998), but this information has limited relevance to specific immunogenic-receptor-mediated stimulation of cells. In this respect a recent study by Liu et al. (2003) provides a detailed look at the dynamics of anandamide generation by mouse RAW264.7 macrophages stimulated with LPS. In these cells, the CD14 protein/toll-like receptor-4-mediated activation resulted in a sustained increase (> tenfold) in the anandamide level, peaking 1.5 to 2 h after stimulation. The increase in anandamide levels was receptor specific and ERK MAP kinase- and NF-κB-dependent as it was blocked by cell preincubation with CD14 receptor antibodies and ERK- or NF-κB inhibitors. The authors also found that LPS stimulation resulted in increase in the activity of both *N*-acyltransferase and PLD-type phosphodiesterase. However, FAAH activity was also upregulated 90 min after stimulation, but FAAH expression was significantly increased only 4 h after LPS treatment. Thus, the actual increase in detected levels of anandamide and *N*-acyl PE in RAW264.7 macrophages reflects the change in the balance among the activities of each enzyme of NAE biosynthesis and catabolism. In this respect a crucial observation was made by Kuwae et al. (1999), who evaluated the rate of *N*-acyl PE and NAE turnover in mouse peritoneal macrophages with the use of $H_2{}^{18}O$. The rationale for this approach was based on the fact that, metabolically, *N*-acyl PE and NAE can get the ^{18}O label from $H_2{}^{18}O$ only through the step of reacylation of the donor lipid used to provide fatty acid for *N*-acyl transacylation (see Figure 19.1). It was found that even a 30-min incubation of mouse peritoneal macrophages in media containing 40% $H_2{}^{18}O$ is enough to find a substantial (up to 20%) amount of ^{18}O atom excess in their amide-linked fatty acids (at least in some molecular species). The most plausible explanation for such a high level of ^{18}O in *N*-acyl PE and NAE might be the existence of a small pool of acyl donor phospholipids exhibiting a rapid rate of turnover, thus providing ^{18}O-labelled 1-O-acyl groups for transacylation to the amino groups of PE (Kuwae et al., 1999). It is conceivable that this high rate of *N*-acyl PE–NAE turnover might even be increased upon specific cell stimulation. Altogether, these observations indicate that *N*-acyl PE–NAE levels in cells should be evaluated not from the point of view of actual levels or concentrations but rather as a momentum "slice" from the very fast process of *N*-acyl PE–NAE synthesis and degradation with the possibility that this process can be dramatically up- and downregulated. In long-lasting pathological conditions, such as endotoxic shock, this regulation of the entire process of NAE metabolism may be differentially controlled at the transcriptional level as is indicated by the data provided by Liu et al. (2003). A recent study by Walter et al. (2002) demonstrates that anandamide levels in cells actually can be upregulated selectively (compared to other NAEs). When mouse astrocytes were stimulated with endothelin-1 for 2.5 min, anandamide levels doubled, whereas the levels of the other NAEs tested (16:0 NAE, 18:1 NAE, 20:3 NAE, and 22:6 NAE) were not affected. The same treatment with endothelin-1 also resulted in the upregulation of 2-AG production (Walter and Stella, 2003).

Anandamide may escape metabolic degradation by FAAH through its enzymatic oxygenation. It was shown that anandamide may be metabolized by lipoxygenases (Ueda et al., 1995; Hampson et al., 1995; Edgemond et al., 1998), COX-2 (Yu et al., 1997), or P450 hydroxylates (Bornheim et al., 1993,1995). The functional significance of such metabolic conversion is yet to be identified.

The question of metabolic regulation of 2-AG production in immune cells is much more complicated because of the existence of multiple pathways that result in the generation of 2-AG in cells (Figure 19.2; reviewed by Sugiura et al., 2002). One of the major routes of 2-AG production may be the hydrolysis of arachidonic-acid-rich phospholipids, such as phosphatidylinositol, polyphosphoinositides, and phosphatidylcholine by phospholipase C and then by diacylglycerol lipase (Sugiura et al., 1995, Di Marzo, De Petrocellis, Sugiura, et al., 1996). Direct hydrolysis of arachidonic-acid-containing lysophosphatidic acid by phosphatases is also possible (Sugiura et al., 1995). There is no doubt that biosynthetic pathways for 2-AG differ depending on the type of tissue and cell as well as the applied stimuli. In platelets, 2-AG synthesis was induced by LPS (Varga et al., 1998)

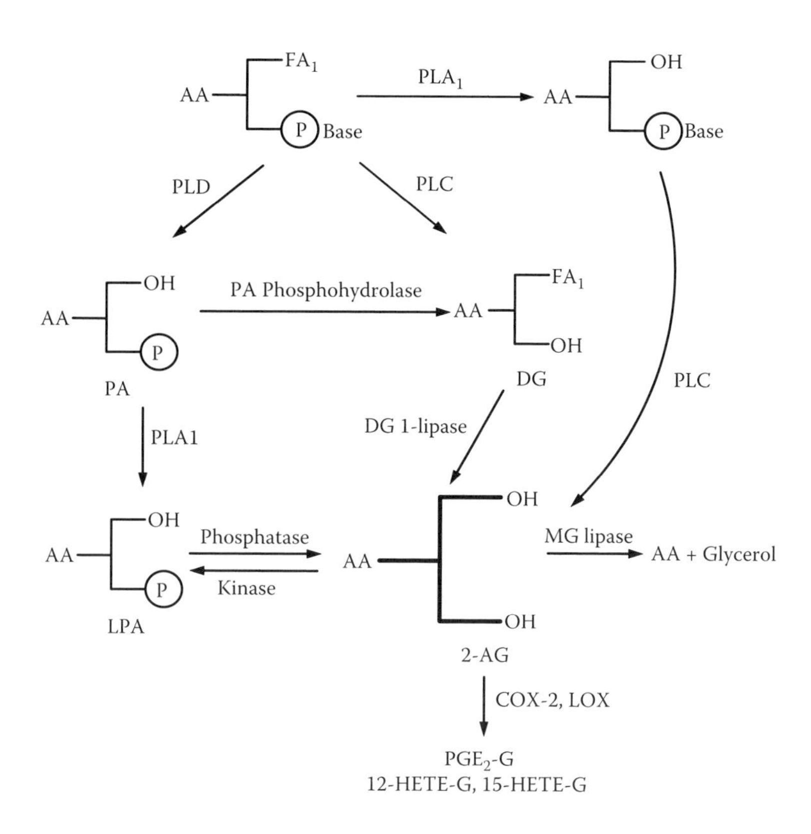

FIGURE 19.2 Pathways of 2-AG metabolism.

and platelet-activating factor PAF (Berdyshev, Schmid, Krebsbach, Schmid, 2001). Most probably, the original observation of AG formation in thrombin-induced platelets by Prescott and Majerus (1983) is relevant to 2-AG production. LPS also stimulated 2-AG production in J774 macrophage-like cells (2.5-fold) and in rat-circulating macrophages (~twofold) when measured 90 min after stimulation (Di Marzo et al., 1999). LPS was also used to induce human dendritic cell maturation, and the amount of 2-AG, but not anandamide or other NAEs, was increased 2.8-fold 24 h after stimulation (Matias et al., 2002). We found that PAF (30 n*M*) can provoke a rapid (within seconds) and profound (10- to 20-fold) increase in 2-AG, but not anandamide or other NAE, levels in P388D1 macrophages. 2-AG was quickly released into the medium and then probably reuptaken and metabolized. Importantly, 2-AG production was specific, as PAF did not affect the level of other cannabinoid-receptor-inactive saturated and unsaturated MGs and only slightly increased the level of 1-AG. In contrast, platelet stimulation with PAF was nonselective not only stereochemically but also in terms of the elevation of the level of MGs other than 2-AG (Berdyshev, Schmid, Krebsbach, Schmid, 2001). The use of pharmacological inhibitors indicated the involvement of PI(P)- and PC-specific phospholipase C in the generation of 2-AG in P388D1 macrophages in response to PAF, but further investigation is needed to confirm this hypothesis.

Liu et al. (2003) explored the possibility that LPS could induce 2-AG formation in RAW264.7 macrophages. They did not find any changes in 2-AG levels after cell stimulation with LPS (in contrast, anandamide levels were dramatically increased as discussed earlier in this chapter). However, RAW264.7 macrophages did respond to stimulation with PAF (30 n*M*) by a modest, but statistically significant, increase in the level of 2-AG. Surprisingly, a CB_1 receptor antagonist SR141716A could block PAF-induced hypotensive response in anesthetized rats, thus confirming a physiological link between PAF receptor activation *in vivo*, 2-AG production, and the cardiovascular effects of both classes of lipid-signaling molecules. LPS is known to induce PAF production in macrophages (Braquet and Rola-Pleszczynski, 1987). PAF also participates in the autocrine upregulation of LPS-induced signaling responses (Szabo et al., 1993; Bulger et al., 2002). Thus, PAF is the most probable link between LPS stimulation and 2-AG production in macrophages but has no relation to LPS-induced sustained anandamide synthesis in these cells.

2-AG formed in response to cell stimulation is rapidly metabolized through catabolic degradation by lipase or anabolic reintegration into lipid-signaling pools, most probably through regeneration of LPA, PA, and other phospholipids (reviewed by Sugiura et al., 2002). Another route of 2-AG metabolism is its enzymatic oxygenation. 2-AG is a good substrate for COX-2 (Kozak et al., 2000), and the formation of COX-2-mediated prostaglandin H_2 glycerol ether (PGH_2-G) and prostaglandin D_2 glycerol ether (PGD_2-G) was demonstrated *in vitro* by cultured RAW264.7 macrophages (Kozak et al., 2000). 15-Lipoxygenase metabolism of 2-AG has also been described (Kozak et al., 2002). Thus, oxidative metabolism of 2-AG may generate additional signaling molecules acting through their own targets to provide a concerted signaling response to the primary stimulus (see also reviews by Kozak and Marnett, 2002; Kozak et al., 2004).

CONCLUSION AND OUTLOOK

Research on the metabolic regulation of endocannabinoid production, interaction of 2-AG/NAE with cannabinoid receptors, and the resulting modulation of the immune response has greatly advanced our knowledge of these unique lipid-signaling molecules. Present findings clearly demonstrate the functional heterogeneity within endocannabinoids and also the complexity of the signaling responses elicited by cannabinoid-receptor-active and inactive NAEs. At least part of the NAE-elicited physiological response is cannabinoid-receptor-independent and mediated by their direct intracellular targets (Figure 19.3). Endocannabinoids have a complex mode of regulation of the cells of the immune system. On the one hand, they promote immune cell proliferation and induce chemotaxis and migration, and on the other hand, they downregulate proinflammatory responses by affecting

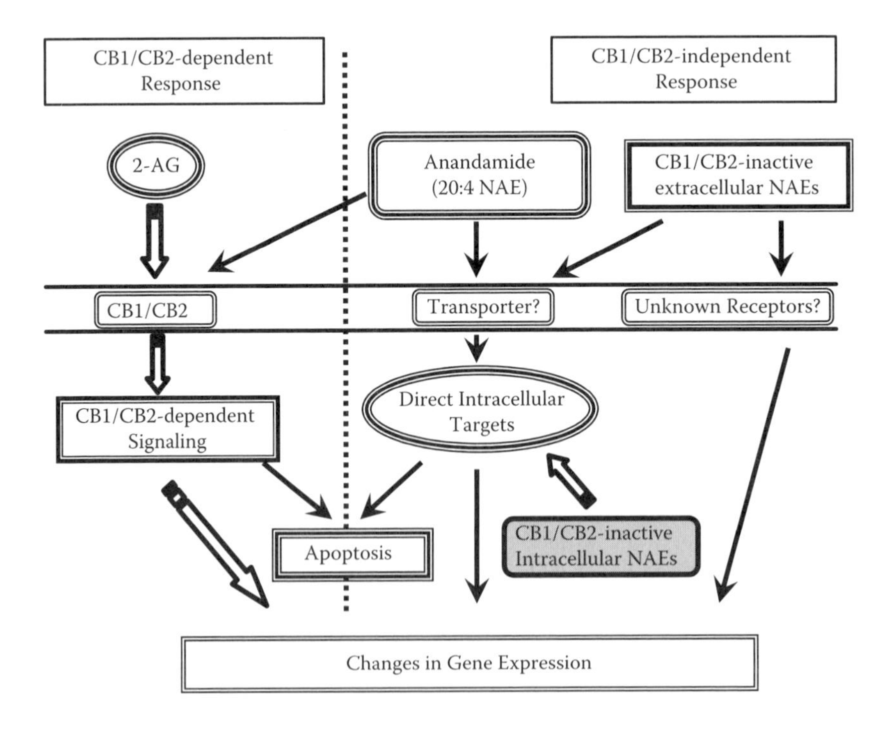

FIGURE 19.3 Schematic representation of endocannabinoid regulation of cell functions.

transcription, translation, and secretion of pro- and anti-inflammatory cytokines. Endocannabinoids may be considered as autocoids that affect "on demand" cell–cell interaction and inflammatory processes locally as well as systemically. It becomes clear that 2-AG and anandamide are differentially linked to diverse proinflammatory stimuli, thus providing multiple signaling responses. The emerging role of cannabinoid-receptor-inactive NAEs as independent signaling molecules rather than as "entourage" components of anandamide-mediated signaling provides additional support for the significance of a complex view of the reasons and consequences of NAE generation in cells. More studies will be needed to define the exact role of endocannabinoids in maintaining immunological homeostasis as well as in the development of immune system disorders.

ACKNOWLEDGMENTS

The author would like to thank Dr. Doug Bibus, Dr. Irina Gorshkova, and Dr. Craig Travis for helpful discussions and aid in preparing this manuscript. The author's work was supported in part by the NIH grant HL71152.

ACRONYMS

NAE – *N*-Acylethanolamine
N-acyl PE – *N*-Acyl phosphatidylethanolamine (may consist of diacyl as well as alkylacyl- and alk-1-enylacyl subclasses)
CB_1 – Neuronal cannabinoid receptor
CB_2 – Peripheral cannabinoid receptor
FAAH – Fatty acid amidohydrolase
SR141716A – *N*-Piperidino-5-(5-chlorophenyl)-1-(2,4-dichlorophenyl)-4-methyl-3-pyrazole-carboxamide
SR144528 – *N*-[1(*S*)-Endo-1,3,3-trimethyl bicyclo [2.2.1] heptan-2-yl]-5-(4-chloro-3-methyl-phenyl)-1-(4-methylbenzyl)-pyrazole-3-carboxamide
FBS – Fetal bovine serum
ERK – Extracellular-signal-regulated protein kinase
AA – Arachidonic acid
COX-2 – Cyclooxygenase-2
LOX – Lipoxygenase
PG-G – Glyceryl prostaglandins
HETE – Hydroxyeicosapentaenoic acid
NF-κB – Nuclear factor for immunoglobulin κ chain in B cells
PLD – Phospholipase D
PLC – Phospholipase C
PLA_1 – Phospholipase A_1

REFERENCES

Aloe, L., Leon, A., Levi-Montalcini, R., 1993. A proposed autocoid mechanism controlling mastocyte behavior. *Agents Actions* 39, C145–C147.

Berdyshev E.V., 2000. Cannabinoid receptors and the regulation of immune response. *Chem. Phys. Lipids* 108, 169–190.

Berdyshev, E.V., Boichot, E., Germain, N, Allain, N., Anger, J.P., Lagente, V., 1997. Influence of fatty acid ethanolamides and Δ^9-tetrahydrocannabinol on cytokine and arachidonate release by mononuclear cells. *Eur. J. Pharmacol.* 330, 231–240.

Berdyshev, E., Boichot, E., Corbel, M., Germain, N., Lagente, V., 1998. Effects of cannabinoid receptor ligands on LPS-induced pulmonary inflammation in mice. *Life Sci.* 63, PL125–PL129.

Berdyshev, E.V., Schmid, P.C., Dong, Z., Schmid, H.H.O., 2000. Stress-induced generation of *N*-acylethanolamines in mouse epidermal JB6 P^+ cells. *Biochem. J.* 346, 369–374.

Berdyshev, E.V., Schmid, P.C., Krebsbach, R.J., Hillard, C.J., Huang, C., Chen, N., Dong, Z., Schmid, H.H.O., 2001. Cannabinoid-receptor-independent cell signalling by *N*-acylethanolamines. *Biochem. J.* 360, 67–75.

Berdyshev, E.V., Schmid, P.C., Krebsbach, R.J., Schmid, H.H.O., 2001. Activation of PAF receptors results in enhanced synthesis of 2-arachidonoylglycerol (2-AG) in immune cells. *FASEB J.* 15, 2171–2178.

Bisogno, T., Maurelli, S., Melck, D., De Petrocellis, L., Di Marzo, V., 1997. Biosynthesis, uptake and degradation of anandamide and palmitoylethanolamide in leukocytes. *J. Biol. Chem.* 272, 3315–3323.

Bornheim, L.M., Kim, K.Y., Chen, B., Correia, M.A., 1993. The effect of cannabidiol on mouse hepatic microsomal cytochrome P450-dependent anandamide metabolism. *Biochem. Biophys. Res. Commn.* 197, 740–746.

Bornheim, L.M., Kim, K.Y., Chen, B., Correia, M.A., 1995. Microsomal cytochrome P450-mediated liver and brain anandamide metabolism. *Biochem. Pharmacol.* 50, 677–686.

Braquet, P., Rola-Pleszczynski, M., 1987. The role of PAF in immunological responses: a review. *Prostaglandins* 34, 143–148.

Bueb, J.L., Lambert, D.M., Tschirhart, E.J., 2001. Receptor-independent effects of natural cannabinoids in rat peritoneal mast cells in vitro. *Biochim. Biophys. Acta* 1538, 252–259.

Bulger, E.M., Arbabi, S., Garcia, I., Maier, R.V., 2002. The macrophage response to endotoxin requires platelet activating factor. *Shock* 17, 173–179.

Cabral, G.A., Toney, D.M., Fischer-Stenger, K., Harrison, M.P., Marciano-Cabral, F., 1995. Anandamide inhibits macrophage-mediated killing of tumor necrosis factor-sensitive cells. *Life Sci.* 56, 2065–2072.

Cadas, H., Gaillet, S., Beltramo, M., Venance, L., Piomelli, D., 1996. Biosynthesis of an endogenous cannabinoid precursor in neurons and its control by calcium and cAMP. *J. Neurosci.* 16, 3934–3942.

Cadas, H., di Tomaso, E., Piomelli, D., 1997. Occurrence and biosynthesis of endogenous cannabinoid precursor, *N*-arachidonoyl phosphatidyl-ethanolamine, in rat brain. *J. Neurosci.* 17, 1226–1242.

Calignano, A., La Rana, G., Giuffrida, A., Piomelli, D., 1998. Control of pain initiation by endogenous cannabinoids. *Nature* 394, 277–281.

Chang, Y.H., Lee, S.T., Lin, W.W., 2001. Effects of cannabinoids on LPS-stimulated inflammatory mediator release from macrophages: involvement of eicosanoids. *J. Cell Biochem.* 81, 715–723.

Clayton, N., Marshall, F.H., Bountra, C., O'Shaughnessy, C.T., 2002. CB_1 and CB_2 cannabinoid receptors are implicated in inflammatory pain. *Pain* 96, 253–260.

Coburn, A.F., Graham, C.E., Hahinger J., 1954. Effect of egg yolk in diets on anaphylactic arthritis (passive Arthus phenomenon) in the guinea pig. *J. Exp. Med.* 100, 425–435.

Conti, S., Costa, B., Colleoni, M., Parolaro, D., Giagnoni, G., 2002. Antiinflammatory action of endocannabinoid palmitoylethanolamide and the synthetic cannabinoid nabilone in a model of acute inflammation in the rat. *Br. J. Pharmacol.* 135, 181–187.

Costa, B., Conti, S., Giagnoni, G., Colleoni, M., 2002. Therapeutic effect of the endogenous fatty acid amide, palmitoylethanolamide, in rat acute inflammation: inhibition of nitric oxide and cyclo-oxygenase systems. *Br. J. Pharmacol.* 137, 413–420.

Derocq, J.M., Segui, M., Marchand, J., Le Fur, G., Casellas, P., 1995. Cannabinoids enhance human B-cell growth at low nanomolar concentrations. *FEBS Lett.* 369, 177–182.

Derocq, J.M., Bouaboula, M., Marchand, J., Rinaldi-Carmona, M., Segui, M., Casellas, P., 1998. The endogenous cannabinoid anandamide is a lipid messenger activating cell growth via a cannabinoid receptor-independent pathway in hematopoietic cell lines. *FEBS Lett.* 425, 419–425.

Devane, W.A., Dysarz, F.A., Johnson, M.R., Melvin, L.S., Howlett, A.C., 1988. Determination and characterization of a cannabinoid receptor in rat brain. *Mol. Pharmacol.* 34, 605–613.

Devane, W.A., Hanus, L., Breuer, A., Pertwee, R.G., Stevenson, L.A., Griffin, G., Gibson, D., Mandelbaum, A., Etinger, A., Mechoulam R., 1992. Isolation and structure of a brain constituent that binds to the cannabinoid receptor. *Science* 258, 1946–1949.

Di Marzo, V., Fontana, A., Cadas, H., Schinelli, S., Cimino, G., Schwartz, J.-C., Piomelli, D., 1994. Formation and inactivation of endogenous cannabinoid anandamide in central neurons. *Nature* 372, 686–691.

Di Marzo, V., De Petrocellis, L., Sepe, N., Buono, A., 1996. Biosynthesis of anandamide and related acylethanolamides in mouse J774 macrophages and N18 neuroblastoma cells. *Biochem. J.* 316, 977–984.

Di Marzo, V., De Petrocellis, L., Sugiura, T., Waku, K., 1996. Potential biosynthetic connections between the two cannabimimetic eicosanoids, anandamide and 2-arachidonoyl-glycerol, in mouse neuroblastoma cells. *Biochem. Biophys. Res. Commn.* 227, 281–288.

Di Marzo, V., Bisogno, T., De Petrocellis, L., Melck, D., Orlando, P., Wagner, J.A., Kunos, G., 1999. Biosynthesis and inactivation of the endocannabinoid 2-arachidonoylglycerol in circulating and tumoral macrophages. *Eur. J. Biochem.* 264, 258–267.

Edgemond, W.S., Hillard, C.J., Falck, J.R., Kearn, C.S., Campbell, W.B., 1998. Human platelets and polymorphonuclear leukocytes synthesize oxygenated derivatives of arachidonylethanolamide (anandamide): their affinities for cannabinoid receptors and pathways of inactivation. *Mol. Pharmacol.* 54, 180–188.

Facchinetti, F., Del Giudice, E., Furegato, S., Passarotto, M., Leon, A., 2003. Cannabinoids ablate release of TNFα in rat microglial cells stimulated with lypopolysaccharide. *Glia* 41, 161–168.

Farooqui, A.A., Horrocks, L.A., Farooqui, T., 2000. Glycerophospholipids in brain: their metabolism, incorporation into membranes, functions, and involvement in neurological disorders. *Chem. Phys. Lipids* 106, 1–29.

Fewtrell, C.M., Foreman, J.C., Jordan, C.C., Oehme, P., Renner, H., Stewart, J.M., 1982. The effects of substance P on histamine and 5-hydroxytryptamine release in the rat. *J. Physiol.* 330, 393–411.

Franklin, A., Parmentier-Batteur, S., Walter, L., Greenberg, D.A., Stella, N., 2003. Palmitoylethanolamide increases after focal cerebral ischemia and potentiates microglial cell motility. *J. Neurosci.* 23, 7767–7775.

Gallily, R., Breuer, A., Mechoulam, R., 2000. 2-Arachidonylglycerol, an endogenous cannabinoid, inhibits tumor necrosis factor-α production in murine macrophages, and in mice. *Eur. J. Pharmacol.* 406, R5–R7.

Gonsiorek, W., Lunn, C., Fan, X., Narula, S., Lundell, D., Hipkin, R.W., 2000. Endocannabinoid 2-arachidonyl glycerol is a full agonist through human type 2 cannabinoid receptor: antagonism by anandamide. *Mol. Pharmacol.* 57, 1045–1050.

Granberg, M., Fowler, C.J., Jacobsson, S.O., 2001. Effects of the cannabimimetic fatty acid derivatives 2-arachidonoylglycerol, anandamide, palmitoylethanolamide and methanandamide upon IgE-dependent antigen-induced beta-hexosaminidase, serotonin and TNF-α release from rat RBL-2H3 basophilic leukaemic cells. *Naunyn Schmiedebergs Arch. Pharmacol.* 364, 66–73.

Hampson, A.J., Hill, W.A., Zan-Phillips, M., Makriyannis, A., Leung, E., Eglen, R.M., Bornheim, L.M., 1995. Anandamide hydroxylation by brain lipoxygenase:metabolite structures and potencies at the cannabinoid receptor. *Biochim. Biophys. Acta* 1259, 173–179.

Hanus, L., Breuer, A., Tchilibon, S., Shiloah, S., Goldenberg, D., Horowitz, M., Pertwee, R.G., Ross, R.A., Mechoulam, R., Fride, E., 1999. HU-308: a specific agonist for CB(2), a peripheral cannabinoid receptor. *Proc. Natl. Acad. Sci. USA* 96, 14228–14233.

Izzo, A.A., Fezza, F., Capasso, R., Bisogno, T., Pinto, L., Iuvone, T., Esposito, G., Mascolo, N., Di Marzo, V., Capasso, F., 2001. Cannabinoid CB_1-receptor mediated regulation of gastrointestinal motility in mice in a model of intestinal inflammation. *Br. J. Pharmacol.* 134, 563–570.

Jaggar, S.I., Hasnie, F.S., Sellaturay, S., Rice, A.S., 1998. The anti-hyperalgesic actions of the cannabinoid anandamide and the putative CB_2 receptor agonist palmitoylethanolamide in visceral and somatic inflammatory pain. *Pain* 76, 189–199.

Jorda, M.A., Verbakel, S.E., Valk, P.J., Vankan-Berkhoudt, Y.V., Maccarrone, M., Finazzi-Agro, A., Lowenberg, B., Delwel, R., 2002. Hematopoietic cells expressing the peripheral cannabinoid receptor migrate in response to the endocannabinoid 2-arachidonoylglycerol. *Blood* 99, 2786–2793.

Jorda, M.A., Lowenberg, B., Delwel, R., 2003. The peripheral cannabinoid receptor CB_2, a novel oncoprotein, induces a reversible block in neutrophilic differentiation. *Blood* 101, 1336–1343.

Kahlich R., Klima J., Cihla F., Frankova V., Masek K., Rosicky M., Matousek F., Bruthans J., 1979. Studies on prophylactic efficacy of *N*-2-hydroxyethyl palmitamide (Impulsin) in acute respiratory infections. Serologically controlled field trials. *J. Hyg. Epidemiol. Microbiol. Immunol.* 23, 11–24.

Katayama, K., Ueda, N., Kurahashi, Y., Suzuki, H., Yamamoto, S., Kato, I., 1997. Distribution of anandamide amidohydrolase in rat tissues with special reference to small intestine. *Biochim. Biophys. Acta* 1347, 212–218.

Kishimoto, S., Gokoh, M., Oka, S., Muramatsu, M., Kajiwara, T., Waku, K., Sugiura, T., 2003. 2-arachidonoylglycerol induces the migration of HL-60 cells differentiated into macrophage-like cells and human peripheral blood monocytes through the cannabinoid CB2 receptor-dependent mechanism. *J. Biol. Chem.* 278, 24469–24475.

Klegeris, A., Bissonnette, C.J., McGeer, P.L., 2003. Reduction of human monocytic cell neurotoxicity and cytokine secretion by ligands of the cannabinoid-type CB2 receptor. *Br. J. Pharmacol.* 139, 775–786.

Klein, T.W., Newton, C., Larsen, K., Lu, L., Perkins, I., Nong, L., Friedman, H., 2003. The cannabinoid system and immune modulation. *J. Leukoc. Biol.* 74, 486–496.

Kozak, K.R., Rowlinson, S.W., Marnett, L.J., 2000. Oxygenation of the endocannabinoid, 2-arachidonylglycerol, to glyceryl prostaglandins by cyclooxygenase-2. *J. Biol. Chem.* 275, 33744–33749.

Kozak, K.R., Marnett, L.J., 2002. Oxidative metabolism of endocannabinoids. *Prostag Leukot. Essent. Fatty Acids* 66, 211–220.

Kozak, K.R., Gupta, R.A., Moody, J.S., Ji, C., Boeglin, W.E., DuBois, R.N., Brash, A.R., Marnett, L.J., 2002. 15-Lipoxygenase metabolism of 2-arachidonylglycerol. Generation of a peroxisome proliferator-activated receptor α agonist. *J. Biol. Chem.* 277, 23278–23286.

Kozak, K.R., Prusakiewicz, J.J., Marnett, L.J., 2004. Oxidative metabolism of endocannabinoids by COX-2. *Curr. Pharm. Des.* 10, 659–667.

Kuehl, Jr., F.A., Jacob, T.A., Ganley O.H., Ormond, R.E., Meisinger, M.A.P., 1957. The identification of *N*-(2-hydroxyethyl)-palmitamide as a naturally occurring anti-inflammatory agent. *J. Am. Chem. Soc.* 79, 5577–5578.

Kuwae, T., Shiota, Y., Schmid, P.C., Krebsbach, R., Schmid, H.H.O., 1999. Biosynthesis and turnover of anandamide and other *N*-acylethanolamines in peritoneal macrophages. *FEBS Lett.* 459, 123–127.

Lau, A.H. and Chow, S.S., 2003. Effects of cannabinoid receptor agonists on immunologically induced histamine release from rat peritoneal mast cells. *Eur. J. Pharmacol.* 464, 229–235.

Lee, M., Yang, K.H., Kaminski, N.E., 1995. Effects of putative cannabinoid receptor ligands, anandamide and 2-arachidonyl-glycerol, on immune function in B6C3F1 mouse splenocytes. *J. Pharmacol. Exp. Ther.* 275, 529–536.

Liu, J., Batkai, S., Pacher, P., Harvey-White, J., Wagner, J.A., Cravatt, B.F., Gao, B., Kunos, G., 2003. Lipopolysaccharide induces anandamide synthesis in macrophages via CD14/MAPK/phosphoinositide 3-kinase/NF-κB independently of platelet-activating factor. *J. Biol. Chem.* 278, 45034–45039.

Luo, Y.D., Patel, M.K, Wiederhold, M.D., Ou, D.W., 1992. Effects of cannabinoids and cocaine on the mitogen-induced transformations of lymphocytes of human and mouse origins. *Int. J. Immunopharmacol.* 14, 49–56.

Maccarrone, M., Fiorucci, L., Erba, F., Bari, M., Finazzi-Agro, A., Ascoli, F., 2000. Human mast cells take up and hydrolyze anandamide under the control of 5-lipoxygenase and do not express cannabinoid receptors. *FEBS Lett.* 468, 176–180.

Masek K., Perlik F., Klima J., Kahlich R., 1974. Prophylactic efficacy of *N*-2-hydroxyethyl palmitamide (impulsin) in acute respiratory tract infections. *Eur. J. Clin. Pharmacol.* 7, 415–419.

Matias, I., Pochard, P., Orlando, P., Salzet, M., Pestel, J., Di Marzo, V., 2002. Presence and regulation of the endocannabinoid system in human dendritic cells. *Eur. J. Biochem.* 269, 3771–3778.

Maurelli, S., Bisogno, T., De Petrocellis, L., Di Luccia, A., Marino, G., Di Marzo, V., 1995. Two novel classes of neuroactive fatty acid amides are substrates for mouse neuroblastoma "anandamide amidohydrolase." *FEBS Lett.* 377, 82–86.

Mazzari, S., Canella, R., Petrelli, L., Marcolongo, G., Leon, A., 1996. *N*-(2-hydroxyethyl)hexadecanamide is orally active in reducing edema formation and inflammatory hyperalgesia by down-modulating mast cell activation. *Eur. J. Pharmacol.* 300, 227–236.

McKallip, R.J., Lombard, C., Fisher, M., Martin, B.R., Ryu, S., Grant, S., Nagarkatti, P.S., Nagarkatti, M., 2002. Targeting CB2 cannabinoid receptors as a novel therapy to treat malignant lymphoblastic disease. *Blood* 100, 627–634.

Mechoulam, R., Ben-Shabat, S., Hanus, L., Ligumsky, M., Kaminski, N.E., Schatz, A.R., Gopher, A., Almog, S., Martin, B.R., Compton, D.R., Pertwee, R.G., Griffin, G., Bayewitch, M., Barg, J., Vogel, Z., 1995. Identification of an endogenous 2-monoglyceride, present in canine gut, that binds to cannabinoid receptors. *Biochem. Pharmacol.* 50, 83–90.

Molina-Holgado, F., Molina-Holgado, E., Guaza, C., 1998. The endogenous cannabinoid anandamide potentiates interleukin-6 production by astrocytes infected with Theiler's murine encephalomyelitis virus by a receptor-mediated pathway. *FEBS Lett.* 433, 139–142.

Munro, S., Thomas, K.L., Abu-Shaar, M., 1993. Molecular characterization of a peripheral receptor for cannabinoids. *Nature*, 365, 61–65.

Natarajan, V., Reddy, P.V., Schmid, P.C., Schmid, H.H.O., 1982. *N*-Acylation of ethanolamine phospholipids in canine myocardium. *Biochim. Biophys. Acta* 712, 342–355.

Ouyang, Y., Hwang, S.G., Han, S.H., Kaminski, N.E., 1998. Suppression of interleukin-2 by the putative endogenous cannabinoid 2-arachidonyl-glycerol is mediated through down-regulation of the nuclear factor of activated T cells. *Mol. Pharmacol.* 53, 676–683.

Patricelli, M.P., Cravatt, B.F., 2001. Proteins regulating the biosynthesis and inactivation of neuromodulatory fatty acid amides. *Vitam. Horm.* 62, 95–131.

Pestonjamasp, V.K., Burstein, S.H., 1998. Anandamide synthesis is induced by arachidonate mobilizing agonists in cells of the immune system. *Biochim. Biophys. Acta* 1394, 249–260.

Perlík, F., Elis, F., Rašková, H., 1971a. Anti-inflammatory activity of *N*-(2-hydroxyethyl)-palmitamide. *Acta Physiol. Hung.* 39, 123–124.

Perlík, F., Elis, F., Rašková, H., 1971b. Anti-inflammatory properties of *N*-(2-hydroxyethyl)-palmitamide. *Acta Physiol. Hung.* 39, 395–400.

Perlík, F., Krejci, J, Elis, F., Pekárek J., Svejcar, J., 1973. The effect of *N*-(2-hydroxyethyl)-palmitamide on delayed hypersensitivity in guinea-pig. *Experientia* 29, 67–68.

Peterson G., Hansen H.S., 1999. *N*-Acylphosphatidylethanolamine-hydrolysing phospholipase D lacks the ability to transphosphatidylate. *FEBS Lett.* 455, 41–44.

Prescott, S.M., Majerus, P.W., 1983. Characterization of 1,2-diacylglycerol hydrolysis in human platelets. Demonstration of an arachidonoyl-monoacylglycerol intermediate. *J. Biol. Chem.* 258, 764–769.

Quartilho, A., Mata, H.P., Ibrahim, M.M., Vanderah, T.W., Porreca, F., Makriyannis, A., Malan, T.P. Jr., 2003. Inhibition of inflammatory hyperalgesia by activation of peripheral CB2 cannabinoid receptors. *Anesthesiology* 99, 955–960.

Rašková, H., Mašek, K., 1967. Nouvelles possibilités d'augmentation de la résistance non-spécifique. *Thérapie* 22, 1241–1246.

Rašková, H., Mašek, K., Linet, O., 1972. Nonspecific resistance induced by palmitoylethanolamide. *Toxicon* 10, 485–490.

Reddy, P.V., Natarajan, V., Schmid, P.C., Schmid, H.H.O., 1983. *N*-Acylation of dog heart ethanolamine phospholipids by transacylase activity. *Biochim. Biophys. Acta* 750, 472–480.

Reddy, P.V., Schmid, P.C., Natarajan, V., Schmid, H.H.O., 1983. The role of cardiolipin as an acyl donor in *N*-acylethanolamine phospholipid biosynthesis. *Biochim. Biophys. Acta* 751, 241–246.

Reddy, P.V., Schmid, P.C., Natarajan, V., Muramatsu, T., Schmid, H.H.O., 1984. Properties of canine myocardial phosphatidylethanolamine *N*-acyltransferase. *Biochim. Biophys. Acta* 795, 130–136.

Richardson, J.D., Kilo, S., Hargreaves, K.M., 1998. Cannabinoids reduce hyperalgesia and inflammation via interaction with peripheral CB1 receptors. *Pain* 75, 111–119.

Ross, R.A., Brockie, H.C., Pertwee, R.G., 2000. Inhibition of nitric oxide production in RAW264.7 macrophages by cannabinoids and palmitoylethanolamide. *Eur. J. Pharmacol.* 401, 121–130.

Roth, M.D., Baldwin, G.C., Tashkin, D.P., 2002. Effects of delta-9-tetrahydrocannabinol on human immune function and host defense. *Chem. Phys. Lipids* 121, 229–239.

Sancho, R., Calzado, M.A., Di Marzo, V., Appendino, G., Munoz, E., 2003. Anandamide inhibits nuclear factor-κB activation through a cannabinoid receptor-independent pathway. *Mol. Pharmacol.* 63, 429–438.

Schmid, H.H.O., 2000. Pathways and mechanisms of *N*-acylethanolamine biosynthesis: can anandamide be generated selectively? *Chem. Phys. Lipids* 08, 71–87.

Schmid, P.C., Reddy, P.V., Natarajan, V., Schmid, H.H.O., 1983. Metabolism of *N*-acylethanolamine phospholipids by a mammalian phosphodiesterase of the phospholipase D type. *J. Biol. Chem.* 258, 9302–9306.

Schmid, P.C., Zuzarte-Augustin, M.L., Schmid, H.H.O., 1985. Properties of rat liver *N*-acylethanolamine amidohydrolase. *J. Biol. Chem.* 260, 14145–14149.

Schmid, H.H.O., Schmid, P.C. and Natarajan, V., 1990. *N*-acylated glycerophospholipids and their derivatives. *Prog. Lipid Res.* 29, 1–43.

Schmid, H.H.O., Berdyshev, E.V., 2002. Cannabinoid receptor-inactive *N*-acylethanolamines and other fatty acid amides. Metabolism and function. *Prostaglandins Leukot. Essent. Fatty Acids* 66, 363–376.

Schmid, H.H.O., Schmid, P.C., Berdyshev, E.V., 2002. Cell signaling by endocannabinoids and their congeners: questions of selectivity and other challenges. *Chem. Phys. Lipids* 121, 111–134.

Schwarz, H., Blanco, F.J., Lotz, M., 1994. Anandamide, an endogenous cannabinoid receptor agonist inhibits lymphocyte proliferation and induces apoptosis. *J. Neuroimmunol.* 55, 107–115.

Stefano, G.B., Bilfinger, T.V., Rialas, C.M., Deutsch, D.G., 2000. 2-arachidonyl-glycerol stimulates nitric oxide release from human immune and vascular tissues and invertebrate immunocytes by cannabinoid receptor 1. *Pharmacol. Res.* 42, 317–322.

Sugiura, T., Kondo, S., Sukagawa, A., Nakane, S., Shinoda, A., Itoh, K., Yamashita, A., Waku, K., 1995. 2-Arachidonoylglycerol: a possible endogenous cannabinoid receptor ligand in brain. *Biochem. Biophys. Res. Commn.* 215, 89–97.

Sugiura, T., Kondo, S., Sukagawa, A., Tonegawa, T., Nakane, S., Yamashita, A., Ishima, Y., Waku, K., 1996. Transacylase-mediated and phosphodiesterase-mediated synthesis of *N*-arachidonoylethanolamide, an endogenous cannabinoid receptor ligand, in rat brain microsomes. Comparison with synthesis from free arachidonic acid and ethanolamine. *Eur. J. Biochem.* 240, 53–62.

Sugiura, T., Kodaka, T., Nakane, S., Miyashita, T., Kondo, S., Suhara, Y., Takayama, H., Waku, K., Seki, C., Baba, N., Ishima, Y., 1999. Evidence that the cannabinoid CB1 receptor is a 2-arachidonoylglycerol receptor. Structure-activity relationship of 2-arachidonoylglycerol, ether-linked analogues, and related compounds. *J. Biol. Chem.* 274, 2794–2801.

Sugiura, T., Kondo, S., Kishimoto, S., Miyashita, T., Nakane, S., Kodaka, T., Suhara, Y., Takayama, H., Waku, K., 2000. Evidence that 2-arachidonoylglycerol but not *N*-palmitoylethanolamine or anandamide is the physiological ligand for the cannabinoid CB2 receptor. Comparison of the agonistic activities of various cannabinoid receptor ligands in HL-60 cells. *J. Biol. Chem.* 275, 605–612.

Sugiura, T., Kobayashi, S., Oka, S., Waku, K., 2002. Biosynthesis and degradation of anandamide and 2-arachidonoylglycerol and their possible physiological significance. *Prostag Leukot. Essent. Fatty Acids* 66, 173–192.

Szabo, C., Wu, C.C., Mitchell, J.A., Gross, S.S., Thiemermann, C., Vane, J.R., 1993. Platelet-activating factor contributes to the induction of nitric oxide synthase by bacterial lipopolysaccharide. *Circ. Res.* 73, 991–999.

Ueda, N., Yamamoto, K., Yamamoto, S., Tokunaga, T., Shirakawa, E., Shinkai, H., Ogawa, M., Sato, T., Kudo, I., Inoue, K., Takizawa, H., Nagano, H., Hirobe M., Matsuki, N., Saito, H., 1995. Lipoxygenase-catalyzed oxygenation of arachidonylethanolamide, a cannabinoid receptor agonist. *Biochim. Biophys. Acta* 1254, 127–134.

Valk, P., Verbakel, S., Vankan, Y., Hol, S., Mancham, S., Ploemacher, R., Mayen, A., Lowenberg, B., Delwel, R., 1997. Anandamide, a natural ligand for the peripheral cannabinoid receptor is a novel synergistic growth factor for hematopoietic cells. *Blood* 90, 1448–1457.

Varga, K., Wagner, J.A., Bridgen, D.T., Kunos, G., 1998. Platelet- and macrophage-derived endogenous cannabinoids are involved in endotoxin-induced hypotension. *FASEB J.* 12, 1035–1044.

Walter, L., Stella, N., 2003. Endothelin-1 increases 2-arachidonoyl glycerol (2-AG) production in astrocytes. *Glia* 44, 85–90.

Walter, L., Franklin, A., Witting, A., Moller, T., Stella, N., 2002. Astrocytes in culture produce anandamide and other acylethanolamides. *J. Biol. Chem.* 277, 20869–20876.

Walter, L., Franklin, A., Witting, A., Wade, C., Xie, Y., Kunos, G., Mackie, K., Stella, N., 2003. Nonpsychotropic cannabinoid receptors regulate microglial cell migration. *J. Neurosci.* 23, 1398–1405.

Wierecky, J., Grabbe, J., Wolff, H.H., Gibbs, B.F., 2000. Cytokine release from a human mast cell line (HMC-1) in response to stimulation with anti-IgE and other secretagogues. *Inflamm. Res.* 49, S7–S8.

Yu, M., Ives, D., Ramesha, C.S., 1997. Synthesis of prostaglandin E2 ethanolamide from anandamide by cyclooxygenase-2. *J. Biol. Chem.* 272, 21181–21186.

Yuan, M., Kiertscher, SM., Cheng, Q., Zoumalan, R., Tashkin, D.P., Roth, M.D., 2002. Delta 9-tetrahydrocannabinol regulates Th1/Th2 cytokine balance in activated human T cells. *J. Neuroimmunol.* 133, 124–131.

Zygmunt, P.M., Petersson, J., Andersson, D.A., Chuang, H., Sorgard, M., Di Marzo, V., Julius, D., Hogestatt, E.D., 1999. Vanilloid receptors on sensory nerves mediate the vasodilator action of anandamide. *Nature* 400, 452–457.

20 Involvement of the Endocannabinoid System in Cancer

Mauro Maccarrone

CONTENTS

INTRODUCTION

Two main molecular targets of Δ^9-tetrahydrocannabinol (Δ^9-THC; Figure 20.1), the psychoactive principle of *Cannabis sativa*, are type-1 (CB_1) and type-2 (CB_2) cannabinoid receptors (Pertwee, 2001; McAllister and Glass, 2002). In the past few years many endogenous agonists of CB receptors have been characterized and are collectively called the *endocannabinoids* (Di Marzo, 1998). They are amides, esters, and ethers of long-chain polyunsaturated fatty acids isolated from brain and peripheral tissues (Mechoulam et al., 2002). Although structurally different from plant cannabinoids, endocannabinoids share critical pharmacophores with Δ^9-THC (Pertwee, 2001; Van der Stelt et al., 2002). Two arachidonate derivatives, *N*-arachidonoylethanolamine (anandamide, AEA; Figure 20.1) and 2-AG, were shown to mimic Δ^9-THC by functionally activating CB receptors, and these are the endocannabinoids whose biological activity has been best characterized to date (Pertwee and Ross, 2002; Sugiura et al., 2002). Also, an ether-type endocannabinoid, i.e., 2-arachidonoyl glyceryl ether (noladin ether), has also been discovered (Hanus et al., 2001). Because ethers are generally stable *in vivo*, whereas AEA (an amide) and 2-AG (an ester) are rapidly hydrolyzed, noladin ether might lead to drug development. More recently, two novel endocannabinoids have been discovered, *O*-arachidonoylethanolamine (virodhamine) (Porter et al., 2002), and *N*-arachidonoyl-dopamine (Huang et al., 2002). In contrast, *N*-palmitoylethanolamine (PEA), *N*-oleoylethanolamine (OEA), and *N*-stearoylethanolamine (SEA; Figure 20.1) are "endocannabinoid-like" compounds, which are present in human, mouse, and rat brain in considerable amounts (Maccarrone and Finazzi-Agrò, 2002). Besides being involved in some biological activities, either cannabimimetic or noncannabimimetic, and having a potential structural role in lipid bilayers, these endocannabinoid-like molecules can have an additional *entourage effect*, i.e., they may potentiate the activity of AEA or 2-AG by inhibiting their degradation (Di Marzo, 1998; Ben-Shabat et al., 1998). In just one decade,

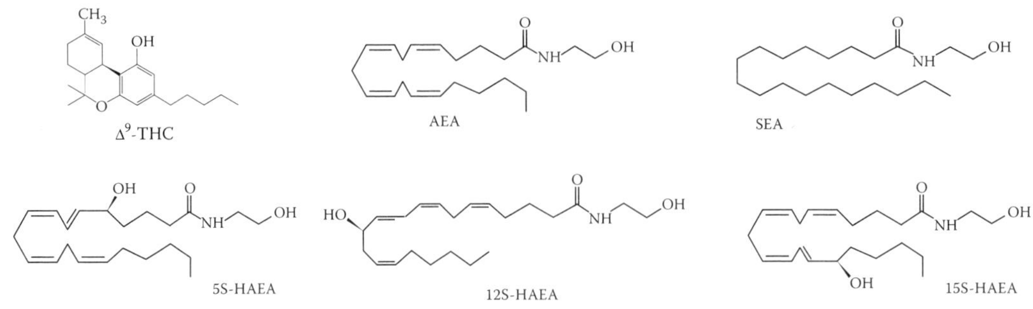

FIGURE 20.1 Chemical structures of exogenous (Δ^9-tetrahydrocannabinol, Δ^9-THC) and endogenous (anandamide, AEA, and *N*-stearoylethanolamine, SEA) cannabinoids, and of some oxygenated derivatives of AEA.

the endocannabinoids have been shown to play manifold roles, both in the central nervous system and in the periphery, and these have been recently reviewed by Maccarrone and Finazzi-Agrò (2002). In particular, it is noteworthy that they are the only neurotransmitters as yet known to act as retrograde synaptic messengers (MacDonald and Vaughan, 2001; Wilson and Nicoll, 2002), although the meaning of this retrograde signaling in neuronal networks needs further investigation.

THE ENDOCANNABINOID SYSTEM

Unlike classical neurotransmitters and neuropeptides, AEA and 2-AG are not stored in intracellular compartments but are produced on demand by receptor-stimulated cleavage of lipid precursors. The AEA precursor is an *N*-arachidonoyl-phosphatidylethanolamine (NArPE), which is cleaved by an as yet uncharacterized *N*-acylphosphatidylethanolamine (NAPE)-specific phospholipase D (E.C. 3.1.4.4; PLD), releasing AEA and phosphatidic acid (Di Marzo, 1998; Hansen et al., 2000). A similar route may be operational also for the synthesis of the other cannabimimetic *N*-acylethanolamines, because their precursors, *N*-acylethanolamine phospholipids, are ubiquitous constituents of animal and human cells, tissues, and body fluids (Di Marzo, 1998; Hansen et al., 2000).

The biological activity of AEA at CB receptors is terminated by its removal from the extracellular space, which occurs through a two-step process: (1) cellular uptake by a high-affinity transporter, followed by (2) intracellular degradation by an FAAH. Several properties of a selective AEA membrane transporter (AMT) have been characterized, although its molecular structure remains unknown (Hillard and Jarrahian, 2000). An interesting feature of AMT is its functional coupling to CB_1R, mediated by nitric oxide (NO): Activation of CB_1R by AEA releases NO, which in turn activates AMT (perhaps by nitrosylating a cysteine residue at the binding site), thus stimulating the removal of AEA from the extracellular space (Maccarrone and Finazzi-Agrò, 2002). This regulatory loop represents a "timer," by which activation of CB_1R by AEA triggers the termination of the activity of AEA at the receptor itself. Another interesting feature of AMT that deserves further investigation is its ability to work in reverse, i.e., to extrude AEA outside the cell. This activity, recently demonstrated in human endothelial cells (Maccarrone, Bari, et al., 2002), might be critical in regulating AEA-mediated retrograde signaling (MacDonald and Vaughan, 2001; Wilson and Nicoll, 2002). Once taken up by cells, AEA is a substrate for the enzyme FAAH (*N*-arachidonoylethanolamine amidohydrolase, EC 3.5.1.4), which breaks the amide bond and releases arachidonic acid (AA) and ethanolamine (Fowler et al., 2001). FAAH is a membrane-bound enzyme, which has been recently crystallized and analyzed at a 2.8 Å resolution (Bracey et al., 2002). It has been proposed that FAAH controls the cellular uptake of AEA by creating (or maintaining) an inward concentration gradient that drives the facilitated diffusion of AEA via AMT (Deutsch et al., 2001). Although FAAH may not be the only factor controlling AEA transport, its pivotal role in AEA degradation may explain why FAAH is modulated in several pathophysiological conditions

(Maccarrone and Finazzi-Agrò, 2002). In this context, it should be pointed out that the relationship between AMT and FAAH is still under debate, because FAAH might not need a transporter to contact AEA (Bracey et al., 2002). In addition, a recent report has questioned even the existence of AMT, suggesting that AEA uptake is a FAAH-driven process of simple diffusion (Glaser et al., 2003). However, it has been demonstrated that FAAH but not AMT is activated by progesterone, and that AMT, but not FAAH, is activated by NO and peroxynitrite (reviewed in Maccarrone and Finazzi-Agrò, 2002). Moreover, estrogen activates AMT but inhibits FAAH (Maccarrone, Bari, et al., 2002), whereas ethanol inhibits AMT without affecting FAAH (Basavarajappa et al., 2003). These data appear to favor the existence of an AEA transporter distinct from AEA hydrolase and the development of new drugs able to selectively inhibit AMT corroborates this hypothesis (Lopez-Rodriguez et al., 2003; Ortar et al., 2003). Similarly, it is noteworthy that the hydro(pero)xides generated from AEA by lipoxygenase (LOX) activity inhibit FAAH but not AMT with apparent inhibition constants in the low micromolar range (Maccarrone, Van der Stelt, et al., 2002). In fact, these hydro(pero)xy-anandamides (see the 5-, 12-, and 15-hydroxy-AEAs in Figure 20.1) are the most powerful natural inhibitors of FAAH yet discovered and may control the tone of AEA *in vivo* (Maccarrone, Van der Stelt, et al., 2002).

The molecular targets of AEA and 2-AG are: (1) the type-1 cannabinoid receptors (CB_1Rs), present mainly on central and peripheral neurons, (2) the type-2 cannabinoid receptors (CB_2Rs), expressed predominantly by immune cells, (3) the non-CB_1/non-CB_2 receptors, (4) the noncannabinoid receptors, and (5) the vanilloid receptors (Breivogel et al., 2001; Pertwee and Ross, 2002; McAllister and Glass, 2002). CB_1R and CB_2R belong to the family of the "seven *trans*-membrane-spanning receptors," and are coupled to G proteins, particularly those of the $G_{i/o}$ family (Pertwee, 2001; McAllister and Glass, 2002). Signal transduction pathways regulated by CBR-coupled G proteins include the inhibition of adenylyl cyclase (AC), the regulation of ionic currents (inhibition of voltage-gated L, N, and P/Q-type Ca^{2+} channels, and activation of K^+ channels), and the activation of focal adhesion kinase, mitogen-activated protein kinase, cytosolic phospholipase A_2, and NO synthase (Howlett and Mukhopadhyay, 2000). Evidence has emerged that in addition to CB_1R and CB_2R there are other molecular targets through which the endocannabinoids might induce a biological activity. In particular, a new target of AEA that is attracting great interest is the type-1 vanilloid receptor (VR1), a "six *trans*-membrane-spanning protein" with intracellular N- and C-terminals and a pore loop between the fifth and sixth transmembrane helices (Jordt and Julius, 2002). The intracellular binding site of VR1 is activated by vanilloid ligands such as capsaicin and also by noxious stimuli such as heat and acids, and thus, it can be viewed as a molecular integrator of noxious stimuli in the peripheral terminals of primary sensory neurons (Di Marzo, Bisogno, et al., 2001). In the last 3 yr, numerous studies have pointed toward a physiological role for AEA as a VR1 agonist, leading to the concept that AEA, besides being an endocannabinoid, is also a true "endovanilloid" (De Petrocellis et al., 2001; Jordt and Julius, 2002). Along the same lines, *N*-arachidonoyl-dopamine has been recently found to act as an endogenous capsaicin-like substance in bovine brain, where it activates VR1 receptors with a potency similar to that of capsaicin itself (Huang et al., 2002). Together with AEA and congeners, PLD, AMT, FAAH, CB, and non-CB receptors form the endocannabinoid system, which is schematically depicted in Figure 20.2.

ENDOCANNABINOIDS AND CANCER

The antiproliferative properties of the active principle of marijuana, Δ^9-THC, have been known for decades (Parolaro et al., 2002). Among cancer cells, leukemia cells were the most sensitive to Δ^9-THC, which reduced their growth *in vitro* through inhibition of DNA synthesis and arrest of cell differentiation (Melck et al., 2002). A number of studies showed that the antiproliferative potency of Δ^9-THC analogs did not correlate with their affinity for CBR, suggesting that these receptors were not necessarily activated upon cannabinoid-induced growth retardation (De Petrocellis et al., 2000). In addition to exogenous cannabinoids such as Δ^9-THC, recent attention has been focused

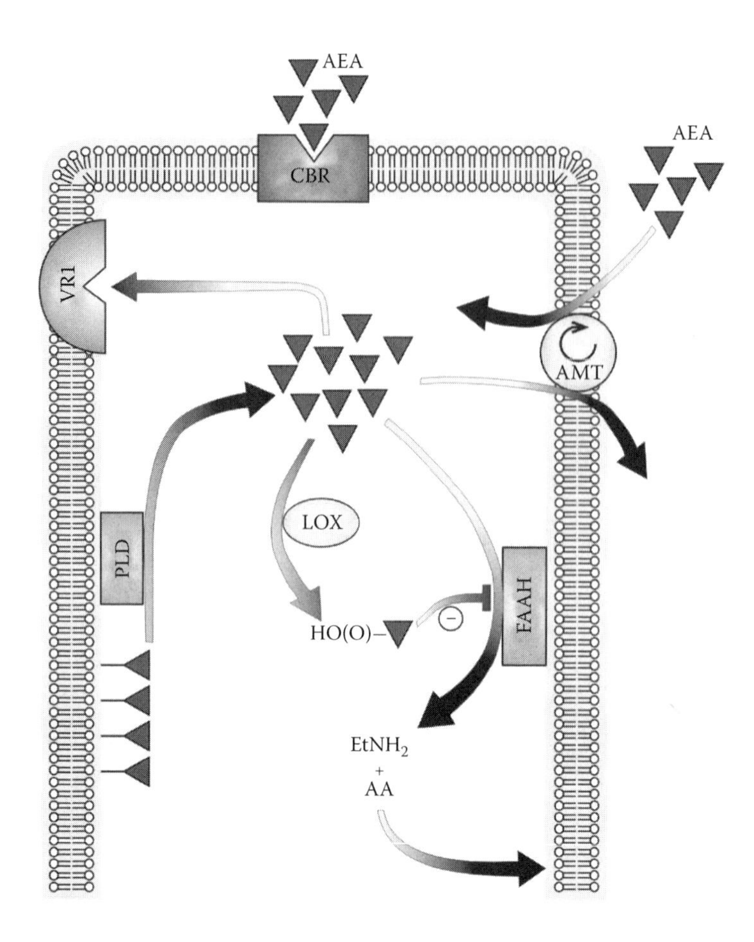

FIGURE 20.2 The endocannabinoid system. The synthesis of anandamide (AEA, triangles) from membrane *N*-arachidonoylphosphatidylethanolamines is catalyzed by a yet uncharacterized phospholipase D (PLD), which releases AEA and phosphatidic acid. AEA is transported in both directions through the cell membrane by an AEA membrane transporter (AMT) and, once taken up, is hydrolyzed by fatty acid amide hydrolase (FAAH) to ethanolamine ($EtNH_2$) and arachidonic acid (AA). The latter is immediately reincorporated into membrane lipids. Alternatively, intracellular AEA can be oxidized by lipoxygenase (LOX), which produces hydro(pero)xy-anandamides able to back-inhibit FAAH. The main targets of AEA are type-1 or type-2 cannabinoid receptors (CBRs), showing an extracellular binding site, and type-1 vanilloid receptors (VR1s), showing an intracellular binding site.

on the possibility that AEA and other endocannabinoids may also regulate cell growth and differentiation, thus accounting for some of the pathophysiological effects of these lipids. Modulation of the endocannabinoid system may interfere with cancer cell proliferation either by inhibiting mitogenic autocrine or paracrine loops or by directly inducing programmed cell death (apoptosis).

AEA has been shown to inhibit *in vitro* the growth of human breast and prostate cancer cells, in a concentration- and time-dependent manner (De Petrocellis et al., 2000; Melck et al., 2002). This antiproliferative action of AEA was not due to loss in cell viability but rather to growth arrest at the G_1/S transition of the cell cycle and was mostly attributable to activation of CB_1-like receptors (De Petrocellis et al., 1998). In turn, these receptors triggered simultaneous inhibition of AC and activation of the Raf1/extracellular signal-regulated kinase (ERK) transduction pathway (Melck et al., 1999). The signaling pathways linked to these two enzymes were shown to disrupt the mitogenic properties of at least two growth factors critical for cancer cell growth, i.e., prolactin (PRL) and nerve growth factor (NGF), leading to a lower expression of their respective receptors PRL_R

(high-molecular weight form; De Petrocellis et al., 1998) and TrkA (Melck et al., 2000). Overall, the data on human breast and prostate cancer cells suggest that AEA is devoid of direct cytotoxicity in these experimental models and exerts its antiproliferative effects by interfering with autocrine and paracrine mitogenic stimuli through a CB_1-like receptor-mediated mechanism. In this context, it seems noteworthy that Δ^9-THC, which also activates CB_1R, failed to affect human breast cancer cell growth (Ruh et al., 1997), suggesting that other non-CB_1 receptors may be involved. In addition, a recent report has demonstrated that AEA can inhibit PRL release from the pituitary gland of male rats in an estrogen-dependent manner (Scorticati et al., 2003), increasing the number of players involved in the endocannabinoid control of autocrine and paracrine signals. Recently, the antitumor effect of AEA has been also demonstrated *in vivo*, where it appears to inhibit the activity of the K-*ras* oncogene product $p21^{ras}$, thereby leading to the inhibition of the *ras* cascade-dependent tumor growth (Bifulco et al., 2001). These events are depicted in Figure 20.3.

On the other hand, growth arrest by AEA may be due to induction of apoptosis (Guzman et al., 2001). Preliminary evidence that the immunosuppressive effects of AEA may be associated with inhibition of lymphocyte proliferation and induction of programmed cell death was reported shortly after the discovery of this endocannabinoid (Schwarz et al., 1994), and since then growing evidence has accumulated demonstrating that AEA does indeed have proapoptotic activity *in vitro* (Sarker et al., 2000). These observations extend to endocannabinoids the earlier findings on Δ^9-THC, which has been shown to induce apoptosis in glioma cells (Sànchez et al., 1998), primary neurons (Chan et al., 1998), hippocampal slices (Chan et al., 1998), prostate cells (Ruiz et al., 1999), and glioma tumors (Galve-Roperh et al., 2000). However, the mechanisms of AEA-induced apoptosis remain to be elucidated.

The proapoptotic activity of AEA in different cellular models has been shown to occur through the activation of different receptors, as summarized in Table 20.1. On the one hand, programmed death of glioma cells *in vitro* has been shown to involve activation of CB_1R followed by ceramide accumulation and Raf1/ERK activation (Galve-Roperh et al., 2000) and, on the other hand, the activation of CB_2R appears to be the critical event leading to inhibition of glioma growth *in vivo* (Sànchez, de Ceballos, et al., 2001). In rat cortical astrocytes and human astrocytoma cells, AEA activates CB_1 receptors leading to sphingomyelin breakdown through the adaptor protein FAN, suggesting a CB_1-receptor-mediated proapoptotic signaling independent of $G_{i/o}$ proteins (Sànchez, Rueda, et al., 2001). In the same cells, CB_1 receptor activation also leads to long-term activation of c-Jun N-terminal kinase (JNK) and p38 mitogen-activated protein kinase (MAPK), suggesting that a threshold might exist above which endocannabinoid-induced JNK and p38 MAPK activation would lead to cell death (Rueda et al., 2000). More recently, the JNK/p38 MAPK signaling cascade triggered by AEA has been shown to depend on the apoptosis signal-regulating kinase 1 (ASK1), a key player in the control of cell survival and death (Sarker et al., 2003). In general, it may be speculated that AEA binding to CB_1 receptors modulates the balance among ERK, JNK, and p38 MAPK, thus regulating the cell's choice between proliferation and death. In addition to the modulation of ceramide degradation through neutral sphingomyelinase (Galve-Roperh et al., 2000) and of ceramide synthesis through serine palmitoyltransferase (Gómez del Pulgar, Velasco, et al., 2002), it has been shown that cannabinoids are able to modulate, again through CB_1 receptors, the phosphatidylinositol 3-kinase/protein kinase B (PI3K/PKB) pathway, which serves as a pivotal antiapoptotic signal (Gómez del Pulgar et al., 2000, Gómez del Pulgar, De Ceballos, et al., 2002). This finding is of particular interest, because it points toward a protective role for cannabinoid receptors against programmed cell death, a concept that has found fresh support in studies involving human astrocytoma cells (Galve-Roperh et al., 2002) and human umbilical vein endothelial cells (Yamaji et al., 2003). The first demonstration that activation of cannabinoid receptors by AEA has a protective role was reported in human neuroblastoma and lymphoma cells, where AEA was shown to induce apoptosis through vanilloid receptors (Maccarrone, Lorenzon, et al., 2000). This effect of AEA occurs through a series of events including increased intracellular calcium concentration, activation of the arachidonate cascade along the cyclooxygenase and the LOX pathways, uncoupling

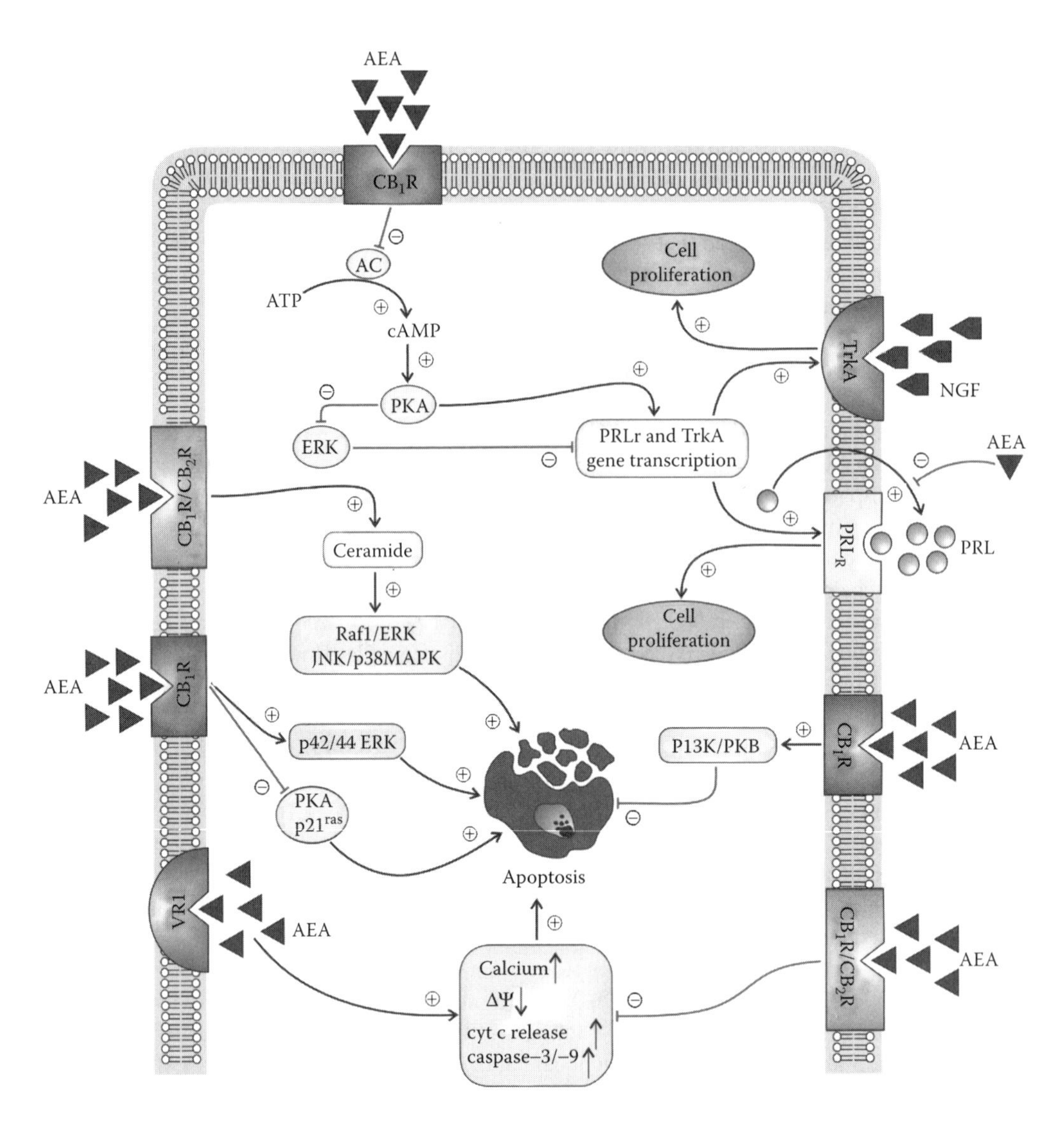

FIGURE 20.3 Signaling pathways underlying the antitumor effects of endocannabinoids. Activation of type-1 cannabinoid receptors (CB_1R) by extracellular anandamide (AEA, triangles) inhibits paracrine or autocrine stimulation of cell growth, by downregulating the expression of nerve growth factor (NGF) receptor TrkA and of prolactin receptor (PRL_R), respectively. These effects are due to inhibition of protein kinase A (PKA) signaling (via inhibition of adenylyl cyclase (AC) and of cAMP synthesis) and to either relief of inhibition or direct stimulation of extracellular signal-regulated kinase (ERK). In addition, AEA inhibits the release of PRL, further reducing its autocrine action. On the other hand, binding of AEA to CB_1R or to type-2 cannabinoid receptors (CB_2R) triggers different signal transduction pathways leading to apoptosis, depending on the cell type. Activation of CB_1R or CB_2R increases ceramide, which activates Raf1/ERK cascade and engages c-Jun N-terminal kinase (JNK)/p38 mitogen-activated protein kinase (MAPK) along the pathway leading to apoptosis. In addition, binding of AEA to CB_1R can trigger activation of p42/p44 ERK and inhibition of PKA, and of the K-*ras* oncogene product $p21^{ras}$, all leading to apoptosis. In astrocytes, binding of AEA to CB_1R can activate the phosphatidylinositol 3-kinase/protein kinase B (PI3K/PKB) pathway, resulting in protection against apoptosis. Alternatively, AEA can activate type-1 vanilloid receptors (VR1) by binding to an intracellular site, thus triggering a proapoptotic series of mitochondia-related events, including elevation of intracellular calcium, drop in mitochondrial potential ($\Delta\Psi$), increased release of cytochrome c (cyt c), and activation of caspase-3 and caspase-9. These effects of AEA at VR1 are prevented by simultaneous activation of CB_1R (in neuronal or endothelial cells) or CB_2R (in immune cells).

TABLE 20.1
Cells Forced into Apoptosis by AEA and Congeners, and the Receptors Involved

Cellular Model	Receptor Involved (Role of Activation)	Reference (First Report)
Human lymphocytes	Not determined	Schwarz et al., 1994
Rat hippocampal neurons	CB_1 (proapoptotic)	Chan et al., 1998
Human prostate PC-3 cells	No receptor involved	Ruiz et al., 1999
Rat glioma C6 cells *in vitro*	CB_1 (proapoptotic)	Galve-Roperh et al., 2000
	VR1 (proapoptotic), CB_1 (antiapoptotic)	Maccarrone, Lorenzon, et al., 2000
	VR1 (proapoptotic), CB_1 (weak proapoptotic)	Jacobsson et al., 2001
	SBS[a] (proapoptotic), CB_1 (proapoptotic)	Maccarrone, Pauselli, et al., 2002
Rat glioma C6 cells *in vivo*	CB_2 (proapoptotic)	Sànchez, de Ceballos, et al., 2001
Human astrocytoma U373MG cells	CB_1 (proapoptotic)	Rueda et al., 2000
	CB_1 (antiapoptotic)	Gómez del Pulgar et al., 2000
Rat pheochromocytoma PC12 cells	CB_1 (proapoptotic)	Sarker et al., 2000
Human neuroblastoma CHP100 cells	VR1 (proapoptotic)	Maccarrone, Lorenzon, et al., 2000
Human lymphoma U937 cells	VR1 (proapoptotic)	Maccarrone, Lorenzon, et al., 2000
Human leukemia DAUDI cells	VR1 (proapoptotic), CB_2 (antiapoptotic)	Maccarrone, Lorenzon, et al., 2000
Rat astrocytes	CB_1 (antiapoptotic)	Sànchez, Rueda, et al., 2001
Mouse Ki Mol cells	CB_1 (proapoptotic)	Bifulco et al., 2001
Human Jurkat, Mol-4 and Sup-T1 cells	CB_2 (proapoptotic)	McKallip et al., 2002
Mouse and human skin tumors	CB_1 (proapoptotic), CB_2 (proapoptotic)	Casanova et al., 2003
Mouse Sertoli cells	CB_2 (antiapoptotic)	Maccarrone et al., 2003
Human umbilical vein endothelial cells	VR1 (proapoptotic), CB_1 (antiapoptotic)	Yamaji et al., 2003

[a] SBS, SEA-binding site.

of mitochondria and release of cytochrome c, and activation of caspases 3 and 9 (Maccarrone, Lorenzon, et al., 2000). It is remarkable that the same series of events, and the same roles for CB_1R and VR1, have been recently reported in human umbilical vein endothelial cells (Yamaji et al., 2003), suggesting that, among the different apoptotic pathways potentially triggered by AEA, human neuronal, immune, and endothelial cells choose one, depending on mitochondrial signals. In addition, AEA-induced death in rat pheochromocytoma PC12 cells also involves mitochondrial disruption and mitochondria-associated events, which are prevented by the powerful antiapoptotic protein Bcl-2 (Sarker et al., 2003). It should be noted that some of these events are typical of different, unrelated proapoptotic stimuli (Wahl and Carr, 2001), which indicates that AEA shares with other inducers common signaling pathways. On the other hand, it seems noteworthy that AEA exerts opposite effects, i.e., a proapoptotic activity at vanilloid receptors and an antiapoptotic action at cannabinoid receptors, but it remains to be clarified if the different localization of the binding sites of these receptors (intracellular for VR1 and extracellular for CBR; see Figure 20.2) plays a role in deciding the effect. It is tempting to speculate that modulation of intracellular and extracellular levels of AEA through fine-tuning of the activity of FAAH (and possibly of AMT) is a "checkpoint," as suggested by several observations on the pivotal role of this enzyme in controlling AEA metabolism (Maccarrone and Finazzi-Agrò, 2002; Bracey et al., 2002). At any rate, of interest is the finding that activation of CB_1R (in human neuronal and endothelial cells) or CB_2R (in human immune cells) prevents AEA-induced apoptosis (Maccarrone, Lorenzon, et al., 2000; Yamaji et al., 2003), a concept that has been extended also to mouse Sertoli cells, where activation of CB_2R prevents the apoptosis induced by AEA (Maccarrone et al., 2003). Similarly, an interesting report has recently shown that activation of CB_1 receptors protects rat glioma cells against HIV-1 Tat-induced cytotoxicity (Esposito et al., 2002). An overall picture of the intracellular events triggered by AEA along different apoptotic pathways is depicted in Figure 20.3. It has to be mentioned that the relative involvement of the cannabinoid and vanilloid receptors in the induction of apoptosis by AEA has also been investigated recently in rat glioma C6 cells, where the proapoptotic effect of AEA through activation of VR1 has been confirmed and has been shown to indicate induction of oxidative stress and activation of calpain (Jacobsson et al., 2001). The latter finding is in keeping with the notion that AEA-induced apoptosis in PC12 cells requires CB_1R-mediated production of superoxide anions (Sarker et al., 2000). In this context, it should be recalled that LOX, a redox stress sensor implicated in various death programs (Maccarrone et al., 2001), generates AEA hydro(pero)xides that inhibit the degradation of AEA itself (Maccarrone, Van der Stelt, et al., 2002). Therefore, it can be proposed that these compounds may play a role in the redox balance controlling the apoptotic activity of AEA. Furthermore, a recent report has shown that activation of CB receptors prevents the growth of skin tumors of mice and humans by inducing apoptosis and blocking angiogenesis (Casanova et al., 2003). Keeping in mind that the incidence of both benign and malignant skin neoplasms has been rising at an alarming rate over the past years and that nonmelanoma skin cancer is one of the most common malignancies in humans, it can be proposed that local administration of (endo)cannabinoids may constitute a novel therapy for skin tumors.

Unlike AEA, 2-AG, OEA, or PEA are not able to modulate cell survival and death, either in human neuronal CHP100 cells or in human lymphoma U937 cells (Maccarrone, Lorenzon, et al., 2000). Rat glioma C6 cells seem to be an exception, because the antiproliferative potency of 2-AG in these cells has been shown to be similar to that of AEA, although PEA was ineffective even at tenfold higher concentrations (Jacobsson et al., 2001). Instead, SEA was almost as effective as AEA in inducing apoptosis, acting at a specific SEA-binding site different from CBR or VR1 (Maccarrone, Pauselli, et al., 2002). It seems noteworthy that human brain tumors such as meningioma and glioblastoma have an approximately fivefold lower content of AEA and an approximately tenfold lower content of SEA than healthy controls; the levels of 2-AG are, however, equal (Table 20.2).

TABLE 20.2
Endocannabinoid Levels in Healthy Human Brain and Brain Tumors

Endocannabinoids (nmol/mg protein)	Healthy Human Brain	Human Meningioma	Human Glioblastoma
AEA	0.16 ± 0.05 (100%)	0.03 ± 0.01[a] (19%)	0.04 ± 0.01[a] (25%)
2-AG	0.10 ± 0.05 (100%)	0.10 ± 0.05[b] (100%)	0.11 ± 0.05[b] (110%)
SEA	0.29 ± 0.08 (100%)	0.03 ± 0.01[a] (10%)	0.04 ± 0.01[a] (14%)
PEA	1.04 ± 0.37 (100%)	0.12 ± 0.03[a] (11%)	0.10 ± 0.03[a] (10%)

[a] Denotes a two-tailed p value $< .05$ vs. healthy brain, considered significant.
[b] denotes a p value $> .05$ vs. healthy brain, considered not significant.

Therefore, it is tempting to suggest that downregulation of endocannabinoids might normally take place in brain tumors (Maccarrone and Finazzi-Agrò, 2002). A lower AEA content and an enhanced expression of cannabinoid receptors, found in malignant cells such as transformed thyroid cells (Bifulco et al., 2001) and gliomas (Sànchez, de Ceballos, et al., 2001), are indicative of a role for the endocannabinoid system in the tonic suppression of cancer growth. A reduced level of AEA may favor cell proliferation, because exogenous (endo)cannabinoids have an *ex vivo* antitumoral action in rat and mice models (Galve-Roperh et al., 2000; Bifulco et al., 2001; Casanova et al., 2003). However, a final answer regarding the relative contribution of AEA synthesis, degradation, and binding to different receptors to the cell's choice between survival and death awaits further investigation. In addition, it should be pointed out that endocannabinoid-like compounds such as PEA have been shown to potentiate the antiproliferative activity of AEA by inhibiting its degradation by FAAH (Di Marzo, Melck, et al., 2001) and by enhancing its VR1-mediated effects (De Petrocellis et al., 2002). Because PEA content is approximately tenfold lower in brain tumors than in healthy brain (Table 20.2), these findings also demonstrate that the entourage effect of endocannabinoid (-like) compounds might be pivotal in the control of tumor cell proliferation by AEA and active congeners.

THERAPEUTIC POTENTIAL OF OXIDATIVE METABOLITES OF AEA

Oxidative metabolism of AEA can be performed by cytochrome P_{450}, yielding at least 20 different derivatives (Bornheim et al., 1995), and by cyclooxygenase-2, which converts AEA into prostaglandin E_2 (PGE_2)-ethanolamide (Ross et al., 2002); cyclooxygenase-1, however, is inactive (Yu et al., 1997). Interestingly, the chemically synthesized putative cyclooxygenase products of AEA, i.e., ethanolamides of PGE_2, PGA_2, PGB_1, and PGB_2, fail to bind to the CB_1 receptor (Pinto et al., 1994). More recently, it has also been shown that 2-AG is converted by cyclooxygenase-2, but not by cyclooxygenase-1, into PGH_2-glycerol ester *in vitro* and in cultured macrophages (Kozak et al., 2000, 2002).

AEA is also a substrate for porcine leukocyte, rabbit reticulocyte, and rat pineal gland LOXs, which are nonheme iron-containing dioxygenases (Hampson et al., 1995; Ueda et al., 1995). LOXs catalyze the conversion of substrates containing one or more (1*Z*,4*Z*)-pentadiene systems into 1-hydroperoxy-2*Z*,4*E*-pentadiene derivatives, which are readily reduced to their hydroxy derivatives. Human platelets convert AEA into 12-hydroxyanandamide (12-HAEA), and human polymorphonuclear leukocytes oxygenate AEA to 12-HAEA and 15-hydroxyanandamide (15-HAEA; Figure 20.1) (Edgemond et al., 1998). Also, 2-AG can be converted by porcine leukocyte 12-LOX *in vitro* and in transfected COS-7 cells to generate 12*S*-hydroxyeicosa-5,8,10,14-tetraenoyl glycerol ester (Moody et al., 2001). An interesting point is that the affinities of 12*S*-HAEA for rat CB_1R and human CB_2R were similar to those of AEA, whereas 15*S*-HAEA was approximately sixfold less potent than AEA at rat CB_1 receptors (Table 20.3). We have recently extended these observations to other LOX metabolites, showing that introduction of a hydroxyl function at various positions of the conjugated diene system of AEA by different LOXs reduces the capacity of AEA derivatives to compete for rat CB_1R binding (Van der Stelt et al., 2002). In particular, 5*S*-HAEA fails to bind to CB_1R or CB_2R with high affinity, showing an inhibition constant (K_i) greater than 1 μM (Table 20.3).

An interesting feature of the oxygenated metabolites of AEA is that they are good inhibitors of FAAH in human lymphoma U937 cells, showing K_i values in the low micromolar range, but they do not inhibit the transport of AEA ($K_i > 10$ μM) into the same cells through AMT (Table 20.3). Introduction of the 1-hydroxy-2*E*,4*Z*-pentadiene system increased the ability of all AEA derivatives to competitively inhibit the hydrolysis of AEA by U937 cells, e.g., 5-*S*HAEA and 12*S*-HAEA have an approximately threefold higher K_i than AEA, whereas 15*S*-HAEA is approximately fourfold more potent (Table 20.3). If LOXs could oxygenate AEA *in vivo*, they would generate selectivity in the endocannabinoid system. It has been shown that 5-LOX inhibitors "disclose" a cryptic FAAH activity in human mast cells, which is undetectable in their absence (Maccarrone, Fiorucci, et al., 2000).

TABLE 20.3
Inhibition Constants (K_is) of AEA and Its Hydroxides for CB_1R/CB_2R Binding, FAAH Activity, and AMT Activity

Compound	CB_1R (n*M*)[a]	CB_2R (n*M*)[a]	FAAH (μM)[b]	AMT (μM)[c]
AEA	90 ± 20	360 ± 50	> 10	> 10
5*S*-HAEA	> 1000	> 1000	2.52 ± 0.13	> 10
12*S*-HAEA	150 ± 30	500 ± 60	2.90 ± 0.15	> 10
15*S*-HAEA	600 ± 120	> 1000	0.63 ± 0.03	> 10

[a] CB_1R and CB_2R were from rat brain and rat spleen, respectively.
[b] All compounds were competitive inhibitors of FAAH activity in human lymphoma U937 cells.
[c] AMT activity was assayed in human lymphoma U937 cells.

Furthermore, the increased capability of HAEAs to inhibit FAAH may be a way to enhance endocannabinoid signaling, which is in keeping with previous findings (Ben-Shabat et al., 1998). Along the same lines, it is noteworthy that recently we have demonstrated that rat brain and blood cells do convert AEA into 12-HAEA and 15-HAEA (Veldhuis et al., 2003), suggesting that endogenous LOX metabolites of AEA can be present in the cell. In support of this hypothesis, recent data raise the possibility that the contractile action of AEA in guinea pig bronchus may be due, at least in part, to its oxygenated metabolites (Craib et al., 2001). On the other hand, the products of AEA generated by cyclooxygenase-2 do not appear to affect FAAH activity (Maccarrone, Van der Stelt, et al., 2002).

The molecular recognition of the oxidative derivatives of AEA by the proteins of the endocannabinoid system is an important issue from the perspective of exploiting this system for the development of therapeutics for various disorders, including cancer. This may indeed facilitate the design of selective and potent drugs directed against a selective target. Crystal or nuclear magnetic resonance structures of the proteins of the endocannabinoid system are not yet available; therefore, a structural comparison of active and inactive ligands has been used to derive pharmacophore models (Thomas et al., 1996; Tong et al., 1998; Craib et al., 2001). The characterization of the active conformations of a ligand allows the construction of a good pharmacophore model, in particular, in the case of AEA, which is very flexible and can assume many different conformations (Fichera et al., 2000; Reggio and Traore, 2000). The introduction of the 1-hydroperoxy-2*Z*,4*E*-pentadiene system in AEA at various positions restricts the flexibility of the acyl chain and leads to different CB_1R binding profiles (Hampson et al., 1995; Edgemond et al., 1998). Therefore, we have recently used molecular dynamic (MD) simulations of AEA and its oxygenated derivatives to obtain insights into the structural requirements for the selective interaction of the acyl chain with the CB_1 receptor (Van der Stelt et al., 2002). Our MD simulations in a box of explicit water molecules indicate that the ability of AEA to adopt a tightly folded conformation is not the only essential feature of the acyl chain enabling CB_1R binding (Reggio and Traore, 2000). In fact, AEA derivatives, both active and inactive at CB_1R and CB_2R, can adopt a folded conformation in which the pharmacophores of AEA match those of CP55940 (Thomas et al., 1996). The critical differences among the various derivatives are located in the conformational details of the loop between C_2 and C_{15}, e.g., the position and orientation of the hydroxyl group (Van der Stelt et al., 2002). In particular, the pentyl tail orientations of the high-affinity ligands AEA and 12*S*-HAEA may be different from those of the low-affinity and inactive ligands (Van der Stelt et al., 2002).

Mounting evidence suggests that degradation of endocannabinoids is a critical step in regulating their activity at receptor and nonreceptor targets. FAAH seems to be a checkpoint in the degradation process, as indicated also by FAAH knockout mice, which have approximately 15-fold augmented levels of AEA (Cravatt et al., 2001). These FAAH knockout animals are less sensitive to pain than controls, strongly corroborating the key role of the enzyme in pain perception (Cravatt et al., 2001) and leading to the suggestion that FAAH inhibitors could be developed as new painkilling drugs (Gurwitz and Weizman, 2001; Walker and Huang, 2002). These findings extend previous reports showing that a defective synthesis of FAAH in human lymphocytes is associated with increased plasma levels of AEA, and notably, that it is predictive of spontaneous abortion in women (Maccarrone and Finazzi-Agrò, 2002). In addition, human lymphocytes challenged with lipopolysaccharide respond by increasing FAAH activity and expression, whereas the AMT and CB receptors remain unchanged (Maccarrone and Finazzi-Agrò, 2002). Taken together, FAAH seems indeed to be the target of choice for new therapeutic agents, and we suggest that hydroxy derivatives of AEA might be lead compounds to develop drugs directed against FAAH. Because the oxidative metabolites of AEA generated by LOX are the only powerful inhibitors of FAAH of natural origin as yet reported, these hydro(pero)xy-anandamides might find broad therapeutic applications in cancer (Bifulco and Di Marzo, 2002; Walsh et al., 2003) as well as in other pathologic conditions involving dysregulation of the endocannabinoid system, such as inflammation (Richardson et al., 1998; Melck et al., 2002), pain (Gurwitz and Weizman, 2001; Walker and Huang, 2002), excitotoxic insult (Van der Stelt et al., 2001), and brain trauma (Mechoulam et al., 2002). The future holds the promise of new agents that might act as "magic bullets" for endocannabinoid-oriented anticancer, anti-inflammatory, antipain, and antitrauma drugs.

CONCLUSIONS

A role for the endogenous cannabinoid system in several aspects of human (patho)physiology has been proposed, through the activation of cannabinoid and vanilloid receptors and/or via nonreceptor-mediated actions. In the case of AEA, the control of cell fate seems to be the core of its biological activity. Here, we have briefly described the endocannabinoid system in order to put in perspective its involvement in cancer cell proliferation and death. We have also reviewed the role of AEA and congeners in apoptosis and have presented the different molecular targets and signal transduction pathways that are so far known to be involved in this activity. Overall, the available evidence suggests that AEA and some of its congeners, such as SEA, may play a role as modulators of cell survival and death. These findings, although not yet generalizable, seem to be relevant also from the perspective of targeting the endocannabinoid system in cancer therapy (Bifulco and Di Marzo, 2002; Walsh et al., 2003). Indeed, recent evidence suggests that targeting CB_2Rs can be a novel form of therapy in malignant lymphoblastic diseases (McKallip et al., 2002). In particular, we suggest that natural inhibitors of FAAH, such as the oxidative metabolites of AEA generated by LOX, may be lead compounds to develop new anticancer drugs, in view of the critical role of this hydrolase in regulating the tone and biological activity of AEA. It can be also anticipated that investigations aimed at elucidating how the endocannabinoid system is integrated with the hormone and cytokine networks in the central nervous system and at the periphery will clarify the relevance of these lipid mediators for human (patho)physiology.

ACKNOWLEDGMENTS

We wish to thank Prof. Alessandro Finazzi-Agrò for helpful discussions and continuous support, Drs. Monica Bari, Natalia Battista, and Valeria Gasperi for their expert assistance with the experimental work, and Mr. Graziano Bonelli for excellent production of the artwork. This investigation was supported by Ministero dell'Istruzione, dell'Università e della Ricerca (Cofin 2003), Rome.

ACRONYMS

AA – arachidonic acid
AC – adenylyl cyclase
AEA – *N*-arachidonoylethanolamine (anandamide)
2-AG – 2-arachidonoylglycerol
AMT – Anandamide membrane transporter
ASK1 – Apoptosis signal-regulating kinase 1
$CB_{1/2}R$ – type 1/2 cannabinoid receptor
Δ^9-THC – Δ^9-tetrahydrocannabinol
ERK – Extracellular signal-regulated kinase
FAAH – Fatty acid amide hydrolase
HAEA – hydroxyanandamide
JNK – c-Jun N-terminal kinase
LOX – lipoxygenase
MAPK – Mitogen-activated protein kinase
NAPE – *N*-acylphosphatidylethanolamine
NArPE – *N*-arachidonoylphosphatidyl-ethanolamine
NGF – Nerve growth factor
NO – Nitric oxide
OEA – *N*-oleoylethanolamine
PEA – *N*-palmitoylethanolamine
PG – Prostaglandin
PKA/B – Protein kinase A/B
PI3K – Phosphatidylinositol 3-kinase
PLD – Phospholipase D
$PRL_{(R)}$ – prolactin (receptor)
SEA – *N*-stearoylethanolamine
VR – Vanilloid receptor

REFERENCES

Basavarajappa, B.S., Saito, M., Cooper, T.B., and Hungund, B.L. (2003) Chronic ethanol inhibits the anandamide transport and increases extracellular anandamide levels in cerebellar granule neurons, *European Journal of Pharmacology* **466**: 73–83.

Ben-Shabat, S., Fride, E., Sheskin, T., Tamiri, T., Rhee, M. H., Vogel, Z. et al. (1998) An entourage effect: inactive endogenous fatty acid glycerol esters enhance 2-arachidonoyl-glycerol cannabinoid activity, *European Journal of Pharmacology* **353**: 23–31.

Bifulco, M. and Di Marzo, V. (2002) Targeting the endocannabinoid system in cancer therapy: a call for further research, *Nature Medicine* **8**: 547–550.

Bifulco, M., Laezza, C., Portella, G., Vitale, M., Orlando, P., De Petrocellis, L. et al. (2001) Control by the endogenous cannabinoid system of ras oncogene-dependent tumor growth, *The FASEB Journal* **15**: 2745–2747.

Bornheim, L.M., Kim, K.Y., Chen, B., and Correia, M.A. (1995) Microsomal cytochrome P450-mediated liver and brain anandamide metabolism, *Biochemical Pharmacology* **50**: 677–686.

Bracey, M.H., Hanson, M.A., Masuda, K.R., Stevens, R.C., and Cravatt, B.F. (2002) Structural adaptations in a membrane enzyme that terminates endocannabinoid signaling, *Science* **298**: 1793–1796.

Breivogel, C.S., Griffin, G., Di Marzo, V., and Martin, B.R. (2001) Evidence for a new G protein-coupled cannabinoid receptor in mouse brain, *Molecular Pharmacology* **60**: 155–163.

Casanova, M.L., Blazquez, C., Martinez-Palacio, J., Villanueva, C., Fernandez-Acenero, M.J., Huffman, J.W. et al. (2003) Inhibition of skin tumor growth and angiogenesis in vivo by activation of cannabinoid receptors, *The Journal of Clinical Investigation* **111**: 43–50.

Chan, G.C.-K., Hinds, T.R., Impey, S., and Storm, D.R. (1998) Hippocampal neurotoxicity of Δ^9-tetrahydrocannabinol, *Journal of Neuroscience* **18**: 5322–5332.

Craib, S.J., Ellington, H.C., Pertwee, R.G., and Ross, R.A. (2001) A possible role of lipoxygenase in the activation of vanilloid receptors by anandamide in the guinea-pig bronchus, *British Journal of Pharmacology* **134**: 30–37.

Cravatt, B.F., Demarest, K., Patricelli, M.P., Bracey, M.H., Giang, D.K., Martin, B.R. et al. (2001) Supersensitivity to anandamide and enhanced endogenous cannabinoid signaling in mice lacking fatty acid amide hydrolase, *Proceedings of the National Academy of Sciences of the USA* **98**: 9371–9376.

De Petrocellis, L., Bisogno, T., Ligresti, A., Bifulco, M., Melck, D., and Di Marzo, V. (2002) Effect on cancer cell proliferation of palmitoylethanolamide, a fatty acid amide interacting with both the cannabinoid and vanilloid signalling systems, *Fundamental Clinical Pharmacology* **16**: 297–302.

De Petrocellis, L., Bisogno, T., Maccarrone, M., Davis, J.B., Finazzi-Agrò, A., and Di Marzo, V. (2001) The activity of anandamide at vanilloid VR1 receptors requires facilitated transport across the cell membrane and is limited by intracellular metabolism, *Journal of Biological Chemistry* **276**: 12856–12863.

De Petrocellis, L., Melck, D., Bisogno, T., and Di Marzo, V. (2000) Endocannabinoids and fatty acid amides in cancer, inflammation and related disorders, *Chemistry and Physics of Lipids* **108**: 191–209.

De Petrocellis, L., Melck, D., Palmisano, A., Bisogno, T., Laezza, C., Bifulco, M. et al. (1998) The endogenous cannabinoid anandamide inhibits human breast cancer cell proliferation, *Proceedings of the National Academy of Scieces of the USA* **95**: 8375–8380.

Deutsch, D.G., Glaser, S.T., Howell, J.M., Kunz, J.S., Puffenbarger, R.A., Hillard, C.J. et al. (2001) The cellular uptake of anandamide is coupled to its breakdown by fatty acid amide hydrolase (FAAH), *Journal of Biological Chemistry* **276**: 6967–6973.

Di Marzo, V. (1998) Endocannabinoids and other fatty acid derivatives with cannabimimetic properties: biochemistry and possible physiopathological relevance, *Biochimica et Biophysica Acta* **1392**: 153–175.

Di Marzo, V., Bisogno, T., and De Petrocellis, L. (2001) Anandamide: some like it hot, *Trends in Pharmacological Sciences* **22**: 346–349.

Di Marzo, V., Melck, D., Orlando, P., Bisogno, T., Zagoory, O., Bifulco, M. et al. (2001) Palmitoylethanolamide inhibits the expression of fatty acid amide hydrolase and enhances the anti-proliferative effect of anandamide in human breast cancer cells, *Biochemical Journal* **358**: 249–255.

Edgemond, W.S., Hillard, C.J., Falck, J.R., Kearn, C.S., and Campbell, W.B. (1998) Human platelets and polymorphonuclear leukocytes synthesize oxygenated derivatives of arachidonylethanolamide (anandamide): their affinities for cannabinoid receptors and pathways of inactivation, *Molecular Pharmacology* **54**: 180–188.

Esposito, G., Ligresti, A., Izzo, A.A., Bisogno, T., Ruvo, M., Di Rosa, M. et al. (2002) The endocannabinoid system protects rat glioma cells against HIV-1 Tat protein-induced cytotoxicity. Mechanism and regulation, *Journal of Biological Chemistry* **277**: 50348–50354.

Fichera, M., Cruciani, G., Bianchi, A., and Musumarra, G. (2000) A 3D-QSAR study on the structural requirements for binding to CB(1) and CB(2) cannabinoid receptors, *Journal of Medicinal Chemistry* **43**: 2300–2309.

Fowler, C.J., Jonsson, K.-O., and Tiger, G. (2001) Fatty acid amide hydrolase: biochemistry, pharmacology, and therapeutic possibilities for an enzyme hydrolyzing anandamide, 2-arachidonoylglycerol, palmitoyethanolamide, and oleamide, *Biochemical Pharmacology* **62**: 517–526.

Galve-Roperh, I., Rueda, D., Gómez del Pulgar, T., Velasco, G., and Guzman, M. (2002) Mechanism of extracellular signal-regulated kinase activation by the CB_1 cannabinoid receptor, *Molecular Pharmacology* **62**: 1385–1392.

Galve-Roperh, I., Sànchez, C., Cortes, M.L., Gómez del Pulgar, T., Izquierdo, M., and Guzman, M. (2000) Anti-tumoral action of cannabinoids: involvement of sustained ceramide accumulation and extracellular signal-regulated kinase activation, *Nature Medicine* **6**: 313–316.

Glaser, S.T., Abumrad, N.A., Fatade, F., Kaczocha, M., Studholme, K.M., and Deutsch, D.G. (2003) Evidence against the presence of an anandamide transporter, *Proceedings of the National Academy of Sciences of the USA* **100**: 4269–4274.

Gómez del Pulgar, T., De Ceballos, M.L., Guzman, M., and Velasco, G. (2002) Cannabinoids protect astrocytes from ceramide-induced apoptosis through the phosphatidylinositol 3-kinase/protein kinase B pathway, *Journal of Biological Chemistry* **277**: 36527–36533.

Gómez del Pulgar, T., Velasco, G., and Guzman, M. (2000) The CB_1 cannabinoid receptor is coupled to the activation of protein kinase B/Akt, *Biochemical Journal* **347**: 369–373.

Gómez del Pulgar, T., Velasco, G., Sànchez, C., Haro, A., and Guzman, M. (2002) *De novo*-synthesized ceramide is involved in cannabinoid-induced apoptosis, *Biochemical Journal* **363**: 183–188.

Gurwitz, D. and Weizman, A. (2001) Fatty acid amide hydrolase inhibitors and the marijuana debate, *Lancet* **358**: 1548.

Guzman, M., Sànchez, C., and Galve-Roperh, I. (2001) Control of the cell survival/death decision by cannabinoids, *Journal of Molecular Medicine* **78**: 613–625.

Hampson, A.J., Hill, W.A., Zan-Phillips, M., Makriyannis, A., Leung, E., Eglen, R.M. et al. (1995) Anandamide hydroxylation by brain lipoxygenase:metabolite structures and potencies at the cannabinoid receptor, *Biochimica et Biophysica Acta* **1259**: 173–179.

Hansen, H.H., Hansen, S.H., Schousboe, A., and Hansen, H.S. (2000) Determination of the phospholipid precursor of anandamide and other *N*-acylethanolamine phospholipids before and after sodium azide-induced toxicity in cultured neocortical neurons, *Journal of Neurochemistry* **75**: 861–871.

Hanus, L., Abu-Lafi, S., Fride, E., Breuer, A., Vogel, Z., Shalev, D.E. et al. (2001) 2-Arachidonyl glyceryl ether, an endogenous agonist of the cannabinoid CB_1 receptor, *Proceedings of the National Academy of Sciences of the USA* **98**: 3662–3665.

Hillard, C.J. and Jarrahian, A. (2000) The movement of *N*-arachidonoylethanolamine (anandamide) across cellular membranes, *Chemistry and Physics of Lipids* **108**: 123–134.

Howlett, A.C. and Mukhopadhyay, S. (2000) Cellular signal transduction by anandamide and 2-arachidonoylglycerol, *Chemistry and Physics of Lipids* **108**: 53–70.

Huang, S.M., Bisogno, T., Trevisani, M., Al-Hayani, A., De Petrocellis, L., Fezza, F. et al. (2002) An endogenous capsaicin-like substance with high potency at recombinant and native vanilloid VR1 receptors, *Proceedings of the National Academy of Sciences of the USA* **99**: 8400–8405.

Jacobsson, S.O., Wallin, T., and Fowler, C.J. (2001) Inhibition of rat C6 glioma cell proliferation by endogenous and synthetic cannabinoids. Relative involvement of cannabinoid and vanilloid receptors, *The Journal of Pharmacology and Experimental Therapeutics* **299**: 951–959.

Jordt, S.E. and Julius, D. (2002) Molecular basis for species-specific sensitivity to "hot" chili peppers, *Cell* **108**: 421–430.

Kozak, K.R., Crews, B.C., Morrow, J.D., Wang, L.H., Ma, Y.H., Weinander, R. et al. (2002) Metabolism of the endocannabinoids, 2-arachidonyl-glycerol and anandamide, into prostaglandin, thromboxane, and prostacyclin glycerol esters and ethanolamides, *Journal of Biological Chemistry* **277**: 44877–44885.

Kozak, K.R., Rowlinson, S.W., and Marnett, L.J. (2000) Oxygenation of the endocannabinoid, 2-arachidonylglycerol, to glyceryl prostaglandins by cyclooxygenase-2, *Journal of Biological Chemistry* **275**: 3744–3749.

Lopez-Rodriguez, M.L., Viso, A., Ortega-Gutierrez, S., Fowler, C.J., Tiger, G., de Lago, E. et al. (2003) Design, synthesis, and biological evaluation of new inhibitors of the endocannabinoid uptake: comparison with effects on fatty acid amidohydrolase, *Journal of Medicinal Chemistry* **46**: 1512–1522.

Maccarrone, M. and Finazzi-Agrò, A. (2002) Endocannabinoids and their actions, *Vitamins and Hormones* **65**: 225–255.

Maccarrone, M., Bari, M., Battista, N., and Finazzi-Agrò, A. (2002) Estrogen stimulates arachidonoylethanolamide release from human endothelial cells and platelet activation, *Blood* **100**: 4040–4048.

Maccarrone, M., Cecconi, S., Rossi, G., Battista, N., Pauselli, R., and Finazzi-Agrò, A. (2003) Anandamide activity and degradation are regulated by early postnatal ageing and follicle-stimulating hormone in mouse Sertoli cells, *Endocrinology* **144**: 20–28.

Maccarrone, M., Fiorucci, L., Erba, F., Bari, M., Finazzi-Agrò, A., and Ascoli, F. (2000) Human mast cells take up and hydrolyze anandamide under the control of 5-lipoxygenase, and do not express cannabinoid receptors, *FEBS Letters* **468**: 176–180.

Maccarrone, M., Lorenzon, T., Bari, M., Melino, G., and Finazzi-Agrò, A. (2000) Anandamide induces apoptosis in human cells via vanilloid receptors. Evidence for a protective role of cannabinoid receptors, *Journal of Biological Chemistry* **275**: 31938–31945.

Maccarrone, M., Melino, G., and Finazzi-Agrò, A. (2001) Lipoxygenases and their involvement in programmed cell death, *Cell Death and Differentiation* **8**: 776–784.

Maccarrone, M., Pauselli, R., Di Rienzo, M., and Finazzi-Agrò, A. (2002) Binding, degradation and apoptotic activity of stearoylethanolamide in rat C6 glioma cells, *Biochemical Journal* **366**: 137–144.

Maccarrone, M., Van der Stelt, M., Veldink, G.A., and Finazzi-Agrò, A. (2002) Inhibitors of endocannabinoid degradation as potential therapeutic agents, *Current Medicinal Chemistry — Anti-Inflammatory and Anti-Allergy Agents* 1: 103–113.

MacDonald, J.C. and Vaughan, C.W. (2001) Cannabinoids act backwards, *Nature* **410**: 527–530.

McAllister, S.D. and Glass, M. (2002) CB_1 and CB_2 receptor-mediated signalling: a focus on endocannabinoids, *Prostaglandins Leukotrienes and Essential Fatty Acids* **66**: 161–171.

McKallip, R.J., Lombard, C., Fisher, M., Martin, B.R., Ryu, S., Grant, S. et al. (2002) Targeting CB_2 cannabinoid receptors as a novel therapy to treat malignant lymphoblastic disease, *Blood* 100: 627–634.

Mechoulam, R., Panikashvili, D., and Shohami, E. (2002) Cannabinoids and brain injury: therapeutic implications, *Trends in Molecular Medicine* **8**: 58–61.

Melck, D., Bisogno, T., De Petrocellis, L., Beaulieu, P., Rice, A.S., and Di Marzo, V. (2002) Cannabimimetic eicosanoids in cancer and inflammation: an update, *Advances in Experimental Medicine and Biology* **507**: 381–386.

Melck, D., De Petrocellis, L., Orlando, P., Bisogno, T., Laezza, C., Bifulco, M. et al. (2000) Suppression of nerve growth factor Trk receptors and prolactin receptors by endocannabinoids leads to inhibition of human breast and prostate cancer cell proliferation, *Endocrinology* **141**: 118–126.

Melck, D., Rueda, D., Galve-Roperh, I., De Petrocellis, L., Guzman, M., and Di Marzo, V. (1999) Involvement of the cAMP/protein kinase A pathway and of mitogen-activated protein kinase in the anti-proliferative effects of anandamide in human breast cancer cells, *FEBS Letters* **463**: 235–240.

Moody, J. S., Kozak, K. R., Ji, C., and Marnett, L.J. (2001) Selective oxygenation of the endocannabinoid 2-arachidonylglycerol by leukocyte-type 12-lipoxygenase, *Biochemistry* **40**: 861–866.

Ortar, G., Ligresti, A., De Petrocellis, L., Morera, E., and Di Marzo, V. (2003) Novel selective and metabolically stable inhibitors of anandamide cellular uptake, *Biochemical Pharmacology* **65**: 1473–1481.

Parolaro, D., Massi, P., Rubino, T., and Monti, E. (2002) Endocannabinoids in the immune system and cancer, *Prostaglandins Leukotrienes and Essential Fatty Acids* **66**: 319–332.

Pertwee, R.G. (2001) Cannabinoid receptors and pain, *Progress in Neurobiol*ogy **63**: 569–611.

Pertwee, R.G. and Ross, R.A. (2002) Cannabinoid receptors and their ligands, *Prostaglandins Leukotrienes and Essential Fatty Acids* **66**: 101–121.

Pinto, J.C., Potie, F., Rice, K.C., Boring, D., Johnson, M.R., Evans, D.M. et al. (1994) Cannabinoid receptor binding and agonist activity of amides and esters of arachidonic acid, *Molecular Pharmacology* **46**: 516–522.

Porter, A.C., Sauer, J.M., Knierman, M.D., Becker, G.W., Berna, M.J., Bao, J. et al. (2002) Characterization of a novel endocannabinoid, virodhamine, with antagonist activity at the CB1 receptor, *The Journal of Pharmacology and Experimental Therapeutics* **301**: 1020–1024.

Reggio, P.H. and Traore, H. (2000) Conformational requirements for endocannabinoid interaction with the cannabinoid receptors, the anandamide transporter and fatty acid amidohydrolase, *Chemistry and Physics of Lipids* **108**: 15–35.

Richardson, J.D., Kilo, S., and Hargreaves, K.M. (1998) Cannabinoids reduce hyperalgesia and inflammation via interaction with peripheral CB1 receptors, *Pain* **75**: 111–119.

Ross, R.A., Craib, S.J., Stevenson, L.A., Pertwee, R.G., Henderson, A., Toole, J. et al. (2002) Pharmacological characterization of the anandamide cyclooxygenase metabolite: prostaglandin E2 ethanolamide, *The Journal of Pharmacology and Experimental Therapeutics* **301**: 900–907.

Rueda, D., Galve-Roperh, I., Haro, A., and Guzman, M. (2000) The CB_1 cannabinoid receptor is coupled to the activation of c-Jun N-terminal kinase, *Molecular Pharmacology* **58**: 814–820.

Ruh, M.F., Taylor, J.A., Howlett, A.C., and Welshons, W.V. (1997) Failure of cannabinoid compounds to stimulate estrogen receptors, *Biochemical Pharmacology* **53**: 35–41.

Ruiz, L., Miguel, A., and Diaz-Laviada, I. (1999) Delta9-tetrahydrocannabinol induces apoptosis in human prostate PC-3 cells via a receptor-independent mechanism, *FEBS Letters* **458**: 400–404.

Sànchez, C., de Ceballos, M.L., Gómez del Pulgar, T., Rueda, D., Corbacho, C., Velasco, G. et al. (2001) Inhibition of glioma growth in vivo by selective activation of the CB_2 cannabinoid receptor, *Cancer Research* **61**: 5784–5789.

Sànchez, C., Galve-Roperh, I., Canova, C., Brachet, P., and Guzman, M. (1998) $Delta^9$-tetrahydrocannabinol induces apoptosis in C6 glioma cells, *FEBS Letters* **436**: 6–10.

Sànchez, C., Rueda, D., Segui, B., Galve-Roperh, I., Levade, T., and Guzman, M. (2001) The CB_1 cannabinoid receptor of astrocytes is coupled to sphingomyelin hydrolysis through the adaptor protein FAN, *Molecular Pharmacology* **59**: 955–959.

Sarker, K.P., Biswas, K.K., Yamakuchi, M., Lee, K.Y., Hahiguchi, T., Kracht, M. et al. (2003) ASK1-p38 MAPK/JNK signaling cascade mediates anandamide-induced PC12 cell death, *Journal of Neurochemistry* **85**: 50–61.

Sarker, K.P., Obara, S., Nakata, M., Kitajima, I., and Maruyama, I. (2000) Anandamide induces apoptosis of PC-12 cells: involvement of superoxide and caspase-3, *FEBS Letters* **472**: 39–44.

Schwarz, H., Blanco, F.J., and Lotz, M. (1994) Anandamide, an endogenous cannabinoid receptor agonist, inhibits lymphocyte proliferation and induces apoptosis, *Journal of Neuroimmunology* **55**: 107–115.

Scorticati, C., Mohn, C., De Laurentiis, A., Vissio, P., Fernandez Solari, J., Seilicovich, A. et al. (2003) The effect of anandamide on prolactin secretion is modulated by estrogen, *Proceedings of the National Academy of Sciences of the USA* **100**: 2134–2139.

Sugiura, T., Kobayashi, Y., Oka, S., and Waku, K. (2002) Biosynthesis and degradation of anandamide and 2-arachidonoylglycerol and their possible physiological significance, *Prostaglandins Leukotrienes and Essential Fatty Acids* **66**: 173–192.

Thomas, B.F., Adams, I.B., Mascarella, S.W., Martin, B.R., and Razdan, R.K. (1996) Structure-activity analysis of anandamide analogs: relationship to a cannabinoid pharmacophore, *Journal of Medicinal Chemistry* **39**: 471–479.

Tong, W., Collantes, E.R., Welsh, W.J., Berglund, B.A., and Howlett, A.C. (1998) Derivation of a pharmacophore model for anandamide using constrained conformational searching and comparative molecular field analysis, *Journal of Medicinal Chemistry* **41**: 4207–4215.

Ueda, N., Yamamoto, K., Yamamoto, S., Tokunaga, T., Shirakawa, E., Shinkai, H. et al. (1995) Lipoxygenase-catalyzed oxygenation of arachidonylethanolamide, a cannabinoid receptor agonist, *Biochimica et Biophysica Acta* **1254**: 127–134.

Van der Stelt, M., Veldhuis, W.B., Bar, P.R., Veldink, G.A., Vliegenthart, J.F.G., and Nicolay, K. (2001) Neuroprotection by Delta9-tetrahydrocannabinol, the main active compound in marijuana, against ouabain-induced in vivo excitotoxicity, *Journal of Neuroscience* **21**: 6475–6479.

Van der Stelt, M., Van Kuik, J.A., Bari, M., van Zadelhoff, G., Leeflang, B.R., Veldink, G.A. et al. (2002) Oxygenated metabolites of anandamide and 2-arachidonoyl-glycerol: conformational analysis and interaction with cannabinoid receptors, membrane transporter and fatty acid amide hydrolase, *Journal of Medicinal Chemistry* **45**: 3709–3720.

Veldhuis, W.B., Van der Stelt, M., Wadman, M.W., Van Zadelhoff, G., Maccarrone, M., Fezza, F. et al. (2003) Neuroprotection by the endogenous cannabinoid anandamide and arvanil against *in vivo* excitotoxicity in the rat: role of vanilloid receptors and lipoxygenases, *Journal of Neuroscience* **23:** 4127–4133.

Wahl, G.M. and Carr, A.M. (2001) The evolution of diverse biological responses to DNA damage: insights from yeast and p53, *Nature Cell Biology* **3**: 277–286.

Walker, J.M. and Huang, S.M. (2002) Endocannabinoids in pain modulation, *Prostaglandins Leukotrienes and Essential Fatty Acids* **66**: 235–242.

Walsh, D., Nelson, K.A., and Mahmoud, F.A. (2003) Established and potential therapeutic applications of cannabinoids in oncology, *Support Care Cancer* **11**: 137–143.

Wilson, R.I. and Nicoll, R.A. (2002) Endocannabinoid signaling in the brain, *Science* **296**: 678–682.

Yamaji, K., Sarker, K.P., Kawahara, K., Iino, S., Yamakuchi, M., Abeyama, K. et al. (2003) Anandamide induces apoptosis in human endothelial cells: its regulation system and clinical implications, *Thrombosis and Haemostasis* **89**: 875–884.

Yu, M., Ives, D., and Ramesha, C.S. (1997) Synthesis of prostaglandin E_2 ethanolamide from anandamide by cyclooxygenase-2, *Journal of Biological Chemistry* **272**: 21181–21186.

21 Endocannabinoids and Endocrine Function

Laura L. Murphy

CONTENTS

INTRODUCTION

Marijuana and its cannabinoids affect most aspects of endocrine function. Since the early 1970s, when it was reported that chronic marijuana smoking in men led to reduced sperm count and gynecomastia (Kolodny et al., 1974, 1976), laboratory and clinical studies have focused on identifying the extent of marijuana's effects on various endocrine systems. Laboratory studies initially utilized delta-9-tetrahydrocannabinol (THC), the principal psychoactive cannabinoid component in marijuana, and found that endocrine function is dramatically altered, most notably the neuroendocrine, pituitary, gonadal, and adrenal hormone systems, with consequent effects on the physiological systems they regulate, including reproduction, lactation, metabolism, and stress homeostasis (for reviews see Murphy et al., 1998 and Brown and Dobs, 2002).

There has been considerable focus on elucidating the sites at which cannabinoids act to alter endocrine function; the CNS and hypothalamus, the pituitary gland, and the target endocrine tissues are all candidate sites. There is increasing evidence that cannabinoids elicit their effects on the neuroendocrine axis and endocrine system via interaction with the CB_1 cannabinoid receptor, a G-protein-linked membrane receptor discussed in detail in this book. The endocannabinoids, such as anandamide and 2-arachidonylglycerol, modulate neurotransmission in the brain, and the endocannabinoid system is intricately linked to a number of neurotransmitter and neuropeptide systems involved in neuroendocrine regulation of pituitary hormone secretion, appetite, and several aspects of reproduction and early development. This chapter will address how cannabinoids affect endocrine function and the role the endocannabinoid system may play in normal and abnormal endocrine function.

NEUROENDOCRINE SYSTEM

The brain is the most likely site at which cannabinoids act to alter hormone secretion (Murphy et al., 1998). The administration of THC or anandamide directly into the brain produces significant decreases in LH, prolactin, and GH secretion (Wenger et al., 1987; Rettori et al., 1988;

Scorticati et al., 2003, 2004), and increases in adreno-corticotropin hormone (ACTH) and corticosterone release in rats (Weidenfeld et al., 1994; Manzanares et al., 1999). Cannabinoids such as THC and anandamide bind to dense populations of cannabinoid CB_1 receptors in the brain and, notably, the hypothalamus (Herkenham et al., 1991). Autoradiography studies have demonstrated the presence of CB_1 receptors in several hypothalamic nuclei involved in neuroendocrine modulation, including the medial preoptic area and the paraventricular nuclei (PVN), predominant sites of GnRH and CRH neuronal cell bodies, respectively (Fernández-Ruiz et al., 1997).

The ability of specific cannabinoid receptor ligands, i.e., CP55940, WIN55212, HU210, and anandamide, to inhibit LH and prolactin and stimulate ACTH release in laboratory animals indicates that cannabinoid receptors most likely mediate the neuroendocrine actions of THC/cannabinoids (Murphy et al., 1998; Scorticati et al., 2003, 2004). Treatment with HU210 produced a dose-related decrease in plasma GH levels, indicating that cannabinoid receptors most likely mediate effects on GH release as well (Martín-Calderón et al., 1998). There is good evidence that the neurotransmitter/neuropeptide systems involved in regulation of GnRH, CRH, and somatostatin release, i.e., norepinephrine, dopamine, serotonin, glutamate, GABA, and opioids, are also significantly altered following acute cannabinoid exposure and that cannabinoid receptors mediate these effects as well (Breivogel and Childers, 1998; Martín-Calderón et al., 1998; de Miguel et al., 1998; Murphy et al., 1998; Arevalo et al., 2001). Pretreatment with the CB_1 receptor antagonist SR141716A reverses the ability of CP55940 to increase serotonin and dopamine levels in the hypothalamus (Arevalo et al., 2001). Treating rats with either anandamide, CP55940, or THC produced a rapid increase in expression of the immediate-early gene c-*fos* (Wenger et al., 1997), CRH, and POMC (Corchero et al., 2001) in stress-responsive nuclei of the rat brain, i.e., the paraventricular hypothalamus and central nucleus of the amygdala. These studies suggest that cannabinoids, either directly or indirectly via neuromodulators, cause neuronal activation within brain nuclei leading to neurohormone activation and release.

Activation of the tuberoinfundibular dopaminergic (TIDA) and GABAergic neurons, the primary hypothalamic inputs regulating pituitary prolactin secretion, are considered to be largely responsible for the decrease in plasma prolactin levels observed in male rats and cycling or lactating females after acute cannabinoid exposure (Ayalon et al., 1977; Asch et al., 1979; Hughes et al., 1981; Tyrey and Hughes, 1984; Mendelson et al., 1985; de Miguel et al., 1998). Autoradiographic analyses have demonstrated the presence of CB_1 receptors in the medial basal hypothalamus and arcuate nucleus, hypothalamic regions where TIDA and GABAergic neurons are located (Fernández-Ruiz et al., 1997; de Miguel et al., 1998). Acute cannabinoid exposure significantly increases the activity of the TIDA and GABAergic neuronal systems, increases dopamine release and hypothalamic GABA content, and decreases prolactin secretion from the pituitary (Rodríguez de Fonseca, Fernández-Ruiz, et al., 1992; Martín-Calderón et al., 1998).

Oxytocin is a peptide hormone secreted by the posterior pituitary gland that stimulates milk letdown and uterine contractions in females, modulates food intake, and regulates neurotransmitter release. Suckling-induced discharges of prolactin and oxytocin play key roles in maintaining normal lactation. When repeatedly administered to rodents during pregnancy or the first few days postpartum, THC interferes with normal lactation, as evidenced by increased pup mortality due to starvation (Borgen et al., 1971; Hatoum et al., 1981). Acute THC administration inhibits milk ejection, indicative of a suppression in oxytocin release (Tyrey and Murphy, 1988), and blocks prolactin release in lactating, suckled rats (Tyrey and Hughes, 1984). Oxytocin also acts centrally to inhibit food and water intake (Arletti et al., 1990) and facilitate penile erection (Andersson, 2001). Oxytocin is synthesized in the magnocellular neurons of the supraoptic nuclei (SON) and PVN of the hypothalamus, where it is either locally released to modulate neurotransmission or the neurons project to the posterior pituitary, where oxytocin is released into circulation. These magnocellular neurons contain presynaptic CB_1 receptors (Herkenham et al., 1991; Hirasawa et al., 2004), and exogenous cannabinoid administration may alter neurotransmitter activity that regulates oxytocin synthesis/release (Hirasawa et al., 2004). Interactions between cannabinoid and oxytocin signaling may be involved in the consummatory activities (Verty et al., 2004) and penile erection (Melis et al., 2004) induced by oxytocin, as well.

The secretion of GH by the anterior pituitary gland is regulated by two hormones from the hypothalamus, GH-releasing hormone (GHRH), which stimulates GH secretion, and somatostatin, which inhibits the release of GH. Systemic or intracerebroventricular administration of THC or HU210 decreases plasma GH levels in experimental animals (Dalterio et al., 1981; Rettori et al., 1988; Martín-Calderón et al., 1998). *In vitro* studies have demonstrated that THC causes significant stimulation of somatostatin release from hypothalamic fragments (Rettori et al., 1988), suggesting that cannabinoid-induced stimulation of somatostatin is at least partly responsible for the inhibition of GH by cannabinoids.

PITUITARY GLAND

Cannabinoids may also regulate pituitary hormone secretion by a direct action on the pituitary gland. A diffuse population of CB_1 receptors has been identified in the anterior pituitary(González et al., 1999, 2000; Pagotto et al., 2001), and the endogenous cannabinoids anandamide and 2-arachidonoylglycerol have been detected in rat and human anterior pituitary tissues (González et al., 1999; Pagotto et al., 2001). Pituitary levels of cannabinoid receptor and endocannabinoid can be regulated by the sex steroid hormones estrogen and testosterone (González et al., 2000), and it has been shown that human pituitary adenomas have higher levels of CB_1 receptors and endocannabinoids than normal pituitary tissue (Pagotto et al., 2001). Cannabinoid receptors have consistently been identified on prolactin-secreting pituitary cells (lactotrophs), as well as lactotroph-derived GH_4C_1 cells, and cannabinoids can directly modulate basal and stimulated prolactin secretion from prolactin-secreting cells *in vitro* (Murphy et al., 1991; Wenger et al., 1999; Ho et al., 2000). CB_1 receptors have also been detected on gonadotropin- (gonadotrophs), TSH- (thyrotrophs), ACTH- (corticotrophs), and GH-secreting (somatotrophs) pituitary cells (Wenger et al., 1999; Pagotto et al., 2001; Cesa et al., 2002), although the number of cell types expressing CB_1 receptors depends on the species studied. Aside from prolactin, there is little evidence of direct cannabinoid effects on the release of other pituitary hormones from normal rodent tissue (Murphy et al., 1998; Barna et al., 2004). However, in cultured pituitary adenomas obtained from acromegalic, Cushing's, and hyperprolactinemic patients, basal or stimulated GH, ACTH, and prolactin release, respectively, were altered by WIN55212 treatment, suggesting a direct effect of cannabinoids in the regulation of human pituitary hormone secretion (Pagotto et al., 2001).

PERIPHERAL ENDOCRINE TISSUES

GONADS

Cannabinoids may have a direct gonadal action that, particularly with long-term cannabinoid exposure, may play a significant role in the occurrence of lowered sex steroid hormone levels in males and females and also abnormal sperm number and morphology in the male. This is supported by findings that high-dose marijuana smoking in humans has been reported to decrease sperm number in the absence of a corresponding decrease in LH, FSH, or testosterone levels (Hembree et al., 1976). That cannabinoids directly affect gonadal function can be shown in *in vitro* studies in which treatment with THC directly inhibited the Sertoli cell response to FSH in male rats (Newton et al., 1993), inhibited testosterone production from isolated mouse Leydig cells (Burstein et al., 1978; Dalterio et al., 1985), and directly suppressed female rat granulosa cell function (Adashi et al., 1983). Cannabinoid receptor mRNA is found within human testes (Gerard et al., 1991) and ovaries (Galiègue et al., 1995), and CB_1 receptors have been localized in Leydig and Sertoli cells (Wenger et al., 2001; Maccarrone et al., 2003) and in mature sperm (Schuel et al., 1994). Sperm cells obtained from the sea urchin or humans exhibited a dose-related inhibition of the acrosome reaction and in motility when exposed to cannabinoids that most likely involves a cannabinoid-receptor-mediated mechanism (Schuel et al., 1994, 2002).

Thyroid

CB_1 receptor mRNA has been detected in rat thyroid gland (Buckley et al., 1998; Porcella et al., 2002). Moreover, immunohistochemical localization demonstrated positive staining of CB_1 receptor in both follicular and parafollicular cells of the thyroid. Treatment of rats with WIN55212 significantly increases serum levels of the thyroid hormones 3,5,3-tri-iodothyonine (T3) and thyroxine (T4), in the absence of change in pituitary TSH release (Porcella et al., 2002). Moreover, CP55940 and anandamide were shown to decrease activity of thyroid parafollicular cells and calcitonin release (Winnicka et al., 2003). These studies support a direct effect of cannabinoid on rat thyroid hormonal activity.

Adrenal

CB_1 receptor mRNA has been detected in the adrenal gland of the embryonic and adult rat (Galiègue et al., 1995; Buckley et al., 1998). Acute cannabinoid treatment increased plasma corticosterone levels in rats and cortisol in humans (Cone et al., 1986; Rodríguez de Fonseca et al., 1992; Weidenfeld et al., 1994; Jackson and Murphy, 1997) and decreased adrenal medulla contents of both norepinephrine and epinephrine in rats (Rodríguez de Fonseca et al., 1991) and plasma epinephrine levels in rabbits (Niederhoffer et al., 2001). In isolated rabbit adrenal glands, electrically evoked epinephrine release was inhibited by WIN55212 in a manner reversed by the CB_1 receptor antagonist SR141716A (Niederhoffer et al., 2001). These results suggest that adrenal medullary function is directly affected by cannabinoids; however, there is no evidence of direct cannabinoid action on adrenal cortical synthesis or release of glucocorticoids.

ROLE OF ENDOCANNABINOIDS IN ENDOCRINE FUNCTION

As described above, CB_1 receptors are found in most hormone-secreting tissues, and acute treatment with the endocannabinoid anandamide affects hormone secretion, notably LH, prolactin, ACTH, and corticosterone. To determine if the endocannabinoid system helps regulate tonic hormone secretion, treatment with cannabinoid receptor antagonists SR141716A or AM251 has been utilized in experimental animals. Acute treatment with SR141716 increases basal and stress-induced plasma levels of ACTH and corticosterone in rodents (González et al., 2004; Patel et al., 2004). Moreover, in the CB_1 receptor knockout mouse ($CB_1^{-/-}$), CRH expression in the PVN of the hypothalamus is significantly upregulated (Cota et al., 2003), and basal levels of ACTH and corticosterone are increased (Barna et al., 2004; Haller et al., 2004). These findings are consistent with a role of the endocannabinoid system in maintaining tonic inhibition of the HPA axis at the hypothalamic level.

Hyperactivity of the HPA axis can have secondary consequences on reproduction and behavioral indices. Prolonged elevations in CRH can lead to suppression of GnRH release and decreased LH levels (Jackson and Murphy, 1997). Male $CB_1^{-/-}$ mice exhibit decreased basal LH and testosterone levels (Wenger et al., 2001). Acute treatment with SR141716A or AM251 either decreased (Scorticati et al., 2004) or had no effect (de Miguel et al., 1998) on LH levels, depending on route of drug administration and gonadal hormone status. $CB_1^{-/-}$ mice also exhibit increased anxiety in conjunction with altered HPA activity (Barna et al., 2004; Haller et al., 2004; Uriguen et al., 2004) and increased stress hormone responses to novel stressors (Barna et al., 2004; Uriguen et al., 2004). Lastly, as CRH exerts anorexigenic effects, elevated CRH activity in $CB_1^{-/-}$ mice may be responsible for the decreased food intake and reduced body weight exhibited by these mutant mice (Cota et al., 2003).

Although the endocannabinoid system appears to play a role in the maintenance of endocrine function, circulating hormones may, in turn, feed back to affect the activity of the endocannabinoid system. The steroid hormones corticosterone, estrogen, and progesterone modulate the expression of CB_1 receptors in the hypothalamus and CNS (Rodríguez de Fonseca et al., 1994; González et al., 2000; Paria et al., 2001) and endocannabinoid levels in the CNS, pituitary, and reproductive tract (González et al., 2000; Paria et al., 2001; Di et al., 2003). Changes in gonadal steroid levels that occur naturally in the cycling female coincide with changes in neuroendocrine responsivity to endogenous and exogenous cannabinoids (Mendelson et al., 1986; Corchero et al., 2001, 2002). Moreover, adrenalectomy and acute or chronic stress significantly alter CB_1 receptor and endocannabinoid levels and neuroendocrine responsivity to exogenous cannabinoids (Jackson and Murphy, 1997; Hill et al., 2003; Patel et al., 2004). Thus, it can be hypothesized that a novel neuroendocrine feedback system may exist whereby steroids modulate endocannabinoid levels and CB_1 receptor expression, which, in turn, help to maintain the "tone" of the neuroendocrine axis.

REFERENCES

Adashi, E.Y., Jones, P.B.C., and Hsueh, A.J.W. (1983) Direct antigonadal activity of cannabinoids: suppression of rat granulose cell functions. *Am J Physiol* 244: E177–185.

Andersson, K.E. (2001) Pharmacology of penile erection. *Pharmacol Rev* 53: 417–450.

Arevalo, C., de Miguel, R., and Hernandez-Tristan, R. (2001) Cannabinoid effects on anxiety-related behaviours and hypothalamic neurotransmitters. *Pharmacol Biochem Behav* 70: 123–131.

Arletti, R., Benelli, A., and Bertolini, A. (1990) Oxytocin inhibits food and fluid intake in rats. *Physiol Behav* 48: 825–830.

Asch, R.H., Smith, C.G., Siler-Khodr, T.M., and Pauerstein, C. (1979) Acute decreases in serum prolactin concentrations caused by delta-9-tetrahydrocannabinol in nonhuman primates. *Fertil Steril* 32: 571–574.

Ayalon, D., Nir, I., Cordova, T., Bauminger, S., Puder, M., and Naor, Z. (1977) Acute effects of delta-1-tetrahydrocannabinol on the hypothalamo-pituitary-ovarian axis in the rat. *Neuroendocrinology* 23: 31–42.

Barna, I., Zelena, D., Arszovszki, A.C., and Ledent, C. (2004) The role of endogenous cannabinoids in the hypothalamo-pituitary-adrenal axis regulation: in vivo and in vitro studies in CB_1 receptor knockout mice. *Life Sci* 75: 2959–2970.

Borgen, L.A., Davis W.M., and Pace H.B. (1971) Effects of synthetic delta-9-tetrahydrocannabinol on pregnancy and offspring in the rat. *Toxicol Appl Pharmacol* 20: 480–486.

Breivogel, C.S. and Childers, S.R. (1998) The functional neuroanatomy of brain cannabinoid receptors. *Neurobiol Dis* 5: 417–431.

Brown, T.T. and Dobs, A.S. (2002) Endocrine effects of marijuana. *J Clin Pharmacol* 42: 90S–96S.

Buckley, N.E., Hansson, S., Harta, G., and Mezey, E. (1998) Expression of CB_1 and CB_2 receptor messenger RNAs during embryonic development in the rat. *Neuroscience* 82: 1131–1149.

Burstein, S., Hunter, S.A., Shoupe, T.S., and Taylor, P. (1978) Cannabinoid inhibition of testosterone synthesis by mouse Leydig cells. *Res Commn Chem Pathol Pharmacol* 19: 557–560.

Cesa, R., Guastalla, A., Cottone, E., Mackie, K., Beltramo, M., and Franzoni, M.F. (2002) Relationships between CB_1 cannabinoid receptors and pituitary endocrine cells in *Xenopus laevis:* an immunohistochemical study. *Gen Comp Endocrinol* 125: 17–24.

Cone, E.J., Johnson, R.E., Moore, J.D., and Roache, J.D. (1986) Acute effects of smoking marijuana on hormones, subjective effects and performance in male human subjects. *Pharmacol Biochem Behav* 24: 1749–1754.

Corchero, J., Manzanares, J., and Fuentes, J.A. (2001) Role of gonadal steroids in the corticotropin-releasing hormone andproopiomelanocortin gene expression response to delta(9)-tetrahydrocannabinol in the hypothalamus of the rat. *Neuroendocrinology* 74: 185–192.

Corchero, J., Fuentes, J.A., and Manzanares, J. (2002) Gender differences in proenkephalin gene expression response to delta-9-tetrahydrocannabinol in the hypothalamus of the rat. *J Psychopharmacol* 16: 283–289.

Cota, D., Marsicano, G., Tschop, M., Grubler, Y., Flachskamm, C., Schubert, M., Auer, D., Yassouridis, A., Thone-Reineke, C., Ortmann, S., Tomassoni, F., Cervino, C., Nisoli, E., Linthorst, A.C.E., Pasquali, R., Lutz, B., Stalla, G.K., and Pagotto, U. (2003) The endogenous cannabinoid system affects energy balance via central orexigenic drive and peripheral lipogenesis. *J Clin Invest* 112: 423–431.

Dalterio, S.L., Michael, S.D., Macmillan, B.T., and Bartke, A. (1981) Differential effects of cannabinoid exposure and stress on plasma prolactin, growth hormone and corticosterone levels in male mice. *Life Sci* 28: 761–766.

Dalterio, S., Bartke, A., and Mayfield, D. (1985) Effects of delta-9-tetrahydrocannabinol on testosterone production in vitro: influence of Ca^{++}, Mg^{++}, or glucose. *Life Sci* 37: 605–612.

de Miguel, R., Romero, J., Muñoz, R.M., Garcia-Gil, L., Gonzalez, S., and Villanúa, M.A. (1998) Effects of cannabinoids on prolactin and gonadotropin secretion: involvement of changes in hypothalamic γ-aminobutyric acid (GABA) inputs. *Biochem Pharmacol* 56: 1331–1338.

Di, S., Malcher-Lopes, R., Halmos, K.C., and Tasker, J.G. (2003) Nongenomic glucocorticoid inhibition via endocannabinoid release in the hypothalamus: a fast feedback mechanism. *J Neurosci* 23: 4850–4857.

Fernández-Ruiz, J.J., Muñoz, R.M., Romero, J., Villanúa, M.A., Makriyannis, A., and Ramos, J.A. (1997) Time course of the effects of different cannabimimetics on prolactin and gonadotrophin secretion: evidence for the presence of CB_1 receptors in hypothalamic structures and their involvement in the effects of cannabimimetics. *Biochem Pharmacol* 53: 1919–1927.

Galiègue, S., Mary, S., Marchand, J., Dussossoy, D., Carrière, D., and Carayon, P. (1995) Expression of central and peripheral cannabinoid receptors in human immune tissues and leukocyte subpopulations. *Eur J Biochem* 232: 54–61.

Gerard, C.M., Mollereau, C., Vassart, G., and Parmentier, M. (1991) Molecular cloning of a human cannabinoid receptor which is also expressed in testis. *Biochem J* 279: 129–134.

González, S., Manzanares, J., Berrendero, F., Wenger, T., Corchero, J., and Bisogno, T. (1999) Identification of endocannabinoids and cannabinoid CB_1 receptor mRNA in the pituitary gland. *Neuroendocrinology* 70: 137–145.

González, S., Bisogno, T., Wenger, T., Manzanares, J., Milone, A., and Berrendero, F. (2000) Sex steroid influence on cannabinoid CB_1 receptor mRNA and endocannabinoid levels in the anterior pituitary gland. *Biochem Biophys Res Commn* 270: 260–266.

González, S., Fernandez-Ruiz, J., Di Marzo, V., Hernandez, M., Arevalo, C., Nicanor, C., Cascio, M.G., Ambrosio, E., and Ramos, J.A. (2004) Behavioral and molecular changes elicited by acute administration of SR141716 to delta-9-tetrahydrocannabinol-tolerant rats: an experimental model of cannabinoid abstinence. *Drug Alcohol Depend* 10: 159–170.

Haller, J., Varga, B., Ledent, C., Barna, I., and Freund, T.F. (2004) Context-dependent effects of CB_1 cannabinoid gene disruption on anxiety-like and social behaviour in mice. *Eur J Neurosci* 19: 1906–1912.

Hatoum, N.S., Davis, W.M., Elsohly, M.A., and Turner, C.E. (1981) Perinatal exposure to cannabichromene and delta 9-tetrahydrocannabinol: separate and combined effects on viability of pups and on male reproductive system at maturity. *Toxicol Lett* 8: 141–146.

Hembree, W.C., Zeidenberg, P., and Nahas, G.G. (1976) Marihuana's effects on human gonadal function, in *Marihuana, Chemistry, Biochemistry and Cellular Effects,* G.G. Nahas, Ed., pp. 521–532. New York: Springer-Verlag.

Herkenham, M., Lynn, A.B., Johnson, M.R., Melvin, L.S., deCosta, B.R., and Rice, K.C. (1991) Characterization and localization of cannabinoid receptors in rat brain: a quantitative in vitro autoradiographic study. *J Neurosci* 11: 563–583.

Hill, M.N., Carrier, E., Patel, S., Ormerod, B.K., Hollard, C.J., and Gorzalka, B.B. (2003) Chronic stress elicits differential regulation of cannabinoid type 1 (CB_1) receptor binding and expression in the hippocampus and limbic forebrain. Abstract-Soc *Neurosci* 829.18.

Hirasawa, M., Schwab, Y., Natah, S., Hillard, C.J., Mackie, K., Sharkey, K.A., and Pittman, Q.J. (2004) Dendritically released transmitters cooperate via autocrine and retrograde actions to inhibit afferent excitation in rat brain. *J Physiol* 559: 611–624.

Ho, B.Y., Stadnicka, A., Prather, P.L., Buckley, A.R., Current, L.L., Bosnjak, Z.J., and Kwok, W-M. (2000) Cannabinoid CB_1 receptor-mediated inhibition of prolactin release and signaling mechanisms in GH_4C_1 cells. *Endocrinology* 141: 1675–1685.

Hughes, C.L., Everett, J.W., and Tyrey, L. (1981) Delta-9-tetrahydrocannabinol suppression of prolactin secretion in the rat: lack of direct pituitary effect. *Endocrinology* 109: 876–880.

Jackson, A.L. and Murphy, L.L. (1997) Role of the hypothalamic-pituitary-adrenal axis in the suppression of luteinizing hormone release by delta-9-tetrahydrocannabinol. *Neuroendocrinology* 65: 446–452.

Kolodny, R.C., Masters, W.H., Kolodny, R.M., and Toro, G. (1974) Depression of plasma testosterone levels after chronic intensive marihuana use. *N Engl J Med* 290: 872–874.

Kolodny, R.C., Lessin, P., Toro, G., Masters, W.H., and Cohen, S. (1976) Depression of plasma testosterone with acute marihuana administration, in *The Pharmacology of Marijuana,* M.C. Braude and S. Szara, Eds., pp. 217–225, New York: Raven Press.

Maccarrone, M., Cecconi, S., Rossi, G., Battista, N., Pauselli, R., and Finazzi-Agro, A. (2003) Anandamide activity and degradation are regulated by early postnatal aging and follicle-stimulating hormone in mouse Sertoli cells. *Endocrinology* 144: 20–28.

Manzanares, J., Corchero, J., and Fuentes, J.A. (1999) Opioid and cannabinoid receptor-mediated regulation of the increase in adrenocorticotropin hormone and corticosterone plasma concentrations induced by central administration of delta(9)-tetrahydrocannabinol in rats. *Brain Res* 839: 173–179.

Martín-Calderón, J.L., Muñoz, R.M., Villanúa, M.A., del Arco, I., Moreno, J.L., Rodríguez de Fonseca, F., and Navarro, M. (1998) Characterization of the acute endocrine actions of (-)-11-hydroxy-delta-8-tetrahydrocannabinol-dimethylheptyl (HU-210), a potent synthetic cannabinoid in rats. *Eur J Pharmacol* 344: 77–86.

Melis, M.R., Succu, S., Mascia, M.S., and Argiolas, A. (2004) Antagonism of cannabinoid CB_1 receptors in the paraventricular nucleus of male rats induces penile erection. *Neurosci Lett* 359: 17–20.

Mendelson, J.H., Mello, N.K., and Ellingboe, J. (1985) Acute effects of marihuana smoking on prolactin levels in human females. *J Pharmacol Exp Ther* 232: 220–222.

Mendelson, J.H., Mello, N.K., Ellingboe, J., Skupny, A.S.T., Lex, B.W., and Griffin, M. (1986) Marihuana smoking suppresses luteinizing hormone in women. *J Pharmacol Exp Ther* 237: 862–866.

Murphy, L.L., Newton, S.C., Dhali, J., and Chavez, D. (1991) Evidence for a direct anterior pituitary site of delta-9-tetrahydrocannabinol action. *Pharmacol Biochem Behav* 40: 603–608.

Murphy, L.L., Muñoz, R.M., Adrian, B.A., and Villanúa, M.A. (1998) Function of cannabinoid receptors in the neuroendocrine regulation of hormone secretion. *Neurobiol Dis* 5: 432–446.

Niederhoffer, N., Hansen, H.H., Fernandez-Ruiz, J.J., and Szabo, B. (2001) Effects of cannabinoids on adrenaline release from adrenal medullary cells. *Br J Pharmacol* 134: 1319–1327.

Newton, S.C., Murphy, L.L., and Bartke, A. (1993) In vitro effects of psychoactive and non-psychoactive cannabinoids on immature rat Sertoli cell function. *Life Sci* 53: 1429–1437.

Pagotto, U., Marsicana, G., Fezza, F., Theodoropoulou, M., Grubler, Y., Stalla, J., Arzberger, T., Milone, A., Losa, M., Di Marzo, V., Lutz, B., and Stalla, G.K. (2001) Normal human pituitary gland and pituitary adenomas express cannabinoid receptor type I and synthesize endogenous cannabinoids: first evidence for a direct role of cannabinoids on hormone modulation at the human level. *J Clin Endocrinol Metab* 86: 2687–2696.

Paria, B.C., Song, H., Wang, X., Schmid, P.C., Krebsbach, R.J., Schmid, H.H., Bonner, T.I., Zimmer, A., and Dey, S.K. (2001) Dysregulated cannabinoid signaling disrupts uterine receptivity for embryo implantation. *J Biol Chem* 276: 20523–20528.

Patel, S., Roelke, C.T., Rademacher, D.J., Cullinan, W.E., and Hillard, C.J. (2004) Endocannabinoid signaling negatively modulates stress-induced activation of the hypothalamic-pituitary-adrenal axis. *Endocrinology* **145:** 5431–5438.

Porcella, A., Marchese, G., Casu, M.A., Rocchitta, A., Lai, M.L., Gessa, G.L., and Pani, L. (2002) Evidence for functional CB_1 cannabinoid receptor expressed in the rat thyroid. *Eur J Endocrinol* 147: 255–261.

Rettori, V., Wenger, T., Snyder, G., Dalterio, S., and McCann, S.M. (1988) Hypothalamic action of delta-9-tetrahydrocannabinol to inhibit the release of prolactin and growth hormone in rat. *Neuroendocrinology* 47: 498–503.

Rodríguez de Fonseca, F., Fernández-Ruiz, J.J., Murphy, L.L., Eldridge, J.C., Steger, R.W., and Bartke, A. (1991) Effects of delta-9-tetrahydrocannabinol exposure on adrenal medullary function: evidence of an acute effect and development of tolerance in chronic treatments. *Pharmacol Biochem Behav* 40: 593–598.

Rodríguez de Fonseca, F., Fernández-Ruiz, J.J., Murphy, L.L., Cebeira, M., Steger, R.W., and Bartke, A. (1992) Acute effects of delta-9-tetrahydrocannabinol on dopaminergic activity in several rat brain areas. *Pharmacol Biochem Behav* 42: 269–275.

Rodríguez de Fonseca, F., Murphy, L.L., Bonnin, A., Eldridge, J.C., Bartke, A., and Fernández-Ruiz, J.J. (1992) Delta-9-tetrahydrocannabinol administration affects anterior pituitary, corticoadrenal and adrenomedullary functions in male rats. *Neuroendocrinology (Life Sci Advances)* 11: 147–156.

Rodríguez de Fonseca, F., Cebeira, M., Ramos, J.A., Martin, M., and Fernández-Ruiz, J.J. (1994) Cannabinoid receptors in rat brain areas: sexual differences, fluctuations during estrous cycle and changes after gonadectomy and sex steroid replacement. *Life Sci* 54: 159–170.

Schuel, H., Goldstein, E., Mechoulam, R., Zimmerman, A.M., and Zimmerman, S. (1994) Anandamide (arachidonylethanolamide), a brain cannabinoid receptor agonist, reduces fertilizing capacity in sea urchins by inhibiting the acrosome reaction. *Proc Natl Acad Sci USA* 91: 7678–7682.

Schuel, H., Burkman, L.J., Lippes, J., Crickard, K., Mahony, M.C., Giuffrida, A., Picone, R.P., and Makriyannis, A. (2002) Evidence that anandamide-signaling regulates human sperm functions required for fertilization. *Mol Reprod Dev* 63: 376–387.

Scorticati, C., Mohn, C., De Laurentiis, A., Vissio, P., Fernandez Solai, J., Seilicovich, A., McCann, S.M., and Rettori, V. (2003) The effect of anandamide on prolactin secretion is modulated by estrogen. *Proc Natl Acad Sci USA* 100: 2134–2139.

Scorticati, C., Fernandez-Solari, J., De Laurentiis, A., Mohn, C., Prestifilippo, J.P., Lasaga, M., Seilicovich, A., Billi, S., Franchi, A., McCann, S.M., and Rettori, V. (2004) The inhibitory effect of anandamide on luteinizing hormone-releasing hormone secretion is reversed by estrogen. *Proc Natl Acad Sci USA* 101: 11891–11896.

Tyrey, L. and Hughes, C.L. (1984) Inhibition of suckling-induced prolactin secretion by delta-9-tetrahydrocannabinol, in *The Cannabinoids: Chemical, Pharmacologic and Therapeutic Aspects,* S. Agurell, W.L. Dewey, and R.E. Wilette, Eds., pp. 487–495, San Diego, CA: Academic Press.

Tyrey, L. and Murphy, L.L. (1988) Inhibition of suckling-induced milk ejections in the lactating rat by delta 9-tetrahydrocannabinol. *Endocrinology* 123: 469–472.

Uriguen, L., Perez-Rial, S., Ledent, C., Palomo, T., and Manzanares, J. (2004) Impaired action of anxiolytic drugs in mice deficient in cannabinoid CB_1 receptors. *Neuropharmacology* 46: 966–973.

Verty, A.N.A., McFarlane, J.R., McGregor, I.S., and Mallet, P.E. (2004) Evidence for an interaction between CB_1 cannabinoid and oxytocin receptors in food and water intake. *Neuropharmacology* 47: 593–603.

Weidenfeld, J., Feldman, S., and Mechoulam, R. (1994) Effect of the brain constituent anandamide, a cannabinoid receptor agonist, on the hypothalamo-pituitary-adrenal axis in the rat. *Neuroendocrinology* 59: 110–112.

Wenger, T., Rettori, V., Snyder, G.D., Dalterio, S., and McCann, S.M. (1987) Effects of delta-9-tetrahydrocannabinol on the hypothalamic-pituitary control of luteinizing hormone and follicle-stimulating hormone secretion in adult male rats. *Neuroendocrinology* 46: 488–493.

Wenger, T., Jamali, K.A., Juaneda, C., Leonardelli, J., and Tramu, G. (1997) Arachidonylethanolamide (anandamide) activates the parvocellular part of hypothalamic paraventricular nucleus. *Biochem Biophys Res Commn* 237: 724–728.

Wenger, T., Fernández-Ruiz, J.J., and Ramos, J.A. (1999) Immunocytochemical demonstration of CB_1 cannabinoid receptors in the anterior lobe of the pituitary gland. *J Neuroendocrinol* 11: 873–878.

Wenger, T., Ledent, C., Csernus, V., and Gerendai, I. (2001) The central cannabinoid receptor inactivation suppresses endocrine reproductive functions. *Biochem Biophys Res Commn* 284: 363–368.

Winnicka, M.M., Zbucki, R.R., Sawicki, B., Hryniewicz, A., Kosiorek, P., Bialuk, I., and Puchalski, Z. (2003) An immunohistochemical study of the thyroid parafollicular (C) cells in rats treated with cannabinoids — preliminary investigations. *Folia Morphol (Warsz)* 62: 419–421.

22 Endocannabinoids in Fertilization, Pregnancy, and Development

Herbert Schuel and Lani J. Burkman

CONTENTS

DEDICATION

Dedicated to the memory of the late Regina Schuel and Elizabeth Johnson.

INTRODUCTION

Historically, the pharmacological effects elicited by plant products provided the first evidence for the existence of endogenous signaling systems. For example, the demonstration of opioid receptors in brain (Pert and Snyder, 1973) led to the identification of their endogenous ligands, peptide neurotransmitters known as endorphins (Hughes et al., 1975). A similar situation applies to cannabinoids, the pharmacologically active constituents of marijuana (Di Marzo and De Petrocellis, 1997). Cannabis has been used for millennia for its medicinal and psychoactive properties (Abel, 1980; Mechoulam, 1986; Ahokas, 2002). The primary psychoactive cannabinoid in marijuana is (−)delta-9-tetrahydrocannabinol ((−)Δ^9-THC) (Mechoulam et al., 1991). Cannabinoids are extremely hydrophobic compounds and readily partition into biological membranes

(Makriyannis and Rapaka, 1990). The demonstration of cannabinoid receptors (CBRs) in brain provided the basis for the subsequent identification of AEA as the first endogenous agonist (endocannabinoid) (Devane et al., 1988; Devane et al., 1992). AEA mimics many of the pharmacological effects of (–)Δ^9-THC (Mechoulam and Hanus, 2000; Mechoulam, 2002). Because AEA is the ethanolamide derivative of arachidonic acid, it is considered to be a novel member of the eicosanoid and the *N*-acylethanolamide (NAE) families of lipid signal molecules (Burstein et al., 1995; Schmid, 2000). Independently, other investigators isolated AEA as an endogenous modulator for Ca^{2+} channels in brain (Johnson et al., 1993).

CBRs are members of the G-protein receptor family (Matsuda, 1997; McAllister and Glass, 2002). Two subtypes of CBRs have been cloned and characterized. CB_1 was originally cloned from rat and human brain (Matsuda et al., 1990; Gerard et al., 1991) and is widely distributed in neural and nonneural cells in peripheral organs (Schuel et al., 1999; Schuel et al., 2002a,b; Pertwee and Ross, 2002). CB_2 was originally cloned from human promyelocytic leukemia HL60 cells and is expressed in macrophages in the marginal zones of rat spleen (Munro et al., 1993). It has important roles in modulating immune responses (Berdyschev, 2000). Evidence in the literature suggests that additional CBR subtypes may exist (Kunos and Batkai, 2001; Kunos et al., 2002; Pertwee and Ross, 2002; Di Marzo et al., 2002; Wiley and Martin, 2002; Fride, Foox, et al., 2003; Piomelli, 2003), such as ionotropic CBRs in sensory neurons that are activated by (–)Δ^9-THC, CBN, mustard oil, and capsaicin but not by AEA or 2-AG (Jordt et al., 2004). The ligand binding sites for CBRs are located within the lipid bilayer of biological membranes (Makriyannis and Rapaka, 1990). Hydrophobic ligands such as (–)Δ^9-THC and AEA readily partition into membrane phospholipids and move within the lipid bilayer to react with the receptor's binding site (Makriyannis and Rapaka, 1990; Song and Bonner, 1996; Lynch and Reggio, 2005). Hence, local agonist concentrations in close proximity to the receptor's binding site within the membrane may be significantly higher than in the aqueous environment outside the cell (Schuel, et al., 2002a). CBRs regulate signal transduction mechanisms in neurons and other somatic cells (Matsuda, 1997; Berdyshev, 2000; McAllister and Glass, 2002; Schmid et al., 2002), inhibit neurotransmitter release in the central and peripheral nervous systems (Elphick and Egertová, 2001; Alger, 2002; Pertwee and Ross, 2002; Piomelli, 2003; Wang et al., 2004), inhibit acrosomal exocytosis in sea urchin sperm (Chang and Schuel, 1991; Chang et al., 1993; Schuel, et al., 1991a,b, 1993, 1994), and modulate capacitation and fertilizing potential of human sperm *in vitro* (Schuel, et al., 2002a; Rossato et al., 1995).

AEA, PEA, and OEA are enzymatically released together by actions of Ca^{2+}-transacylase and phospholipase D (PLD) from membrane phospholipid precursors (*N*-acyl-phosphatidylethanolamines [NAPE]) when cells are stimulated by neurotransmitters, hormones, and depolarizing agents (Di Marzo et al., 1994; Bisogno et al., 1998; Piomelli et al., 1998; Berdyshev, 2000; De Petrocellis et al., 2000; Schmid, 2000; Stella and Piomelli, 2001; Piomelli 2003; Okamoto et al., 2004). Released NAEs are quickly taken up by cells and hydrolyzed by FAAH, indicative of possible roles in cell signaling (Deutch and Chin, 1993; Bisogno, et al., 1997a; Piomelli et al., 1998; Ueda et al., 2000; Giuffrida at al., 2000; Cravatt and Lichtman, 2002; Kathuria et al., 2002; Bracey et al., 2003; Glasser et al., 2003; Piomelli 2003; Maccarrone and Finazzi-Agro, 2004). PEA is a potent anti-inflammatory and neuroprotective agent (Kuehl et al., 1957; Schmid et al., 1990; Facci et al., 1995; Mazzari et al., 1996; Skaper et al., 1996; Conti et al., 2002). PEA may elicit these effects via uncharacterized CB_2-like receptors (Facci et al., 1995; Skaper et al., 1996; Piomelli et al., 1998; Conti et al., 2002), but this hypothesis remains controversial (reviewed by Schuel, et al., 2002b). Related NAEs, such as dihomo-γ-linolenoylethanolamide, 7,10,13,16-docosatetraenoylethanolamide, mead ethanolamide, etc., have been identified as endogenous agonists for CBRs (Hanus et al., 1993; Priller et al., 1995; Mechoulam and Hanus, 2000; Schmid, 2000; Piomelli et al., 1998; Salzet et al., 2000). Monoacylglycerols such as 2-AG, and 2-AG ether (noladin ether) are enzymatically released from phospholipids when cells are stimulated and are rapidly degraded by FAAH and monoacylglycerol lipases; these are endogenous ligands for CBRs (Mechoulam et al., 1995; Stella et al., 1997; Mechoulam and Hanus, 2000;

Sugiura et al., 1998; Schmid, 2000; Hanus et al., 2001; Sugiura and Waku, 2000, 2002; Pertwee and Ross, 2002; Sugiura et al., 2002). NAEs (PEA, OEA, etc.), oleamide, and monoacylglycerols (2-palmitoyl glycerol [2-PG] and 2-linoleoyl glycerol) that do not bind with high affinity to CBRs can produce cannabimimetic effects as entourage compounds by protecting endocannabinoids from enzymatic hydrolysis via FAAH and monoacylglycerol lipases (Ben-Shabat et al., 1998; Mechoulam et al., 1998; Lambert and Di Marzo, 1999; Lambert et al., 1999; Jonsson et al., 2001; Cravatt and Lichtman 2002). *O*-Arachidonoyl ethanolamide (virodhamine) is a novel endocannabinoid in brain and peripheral organs and is the first endogenous antagonist for AEA at CB_1 receptors to be discovered (Porter et al., 2002). Virodhamine also is a partial agonist at CB_1, a full agonist at CB_2, and an inhibitor of [^{3}H]AEA uptake by RBL-2H3 cells. As a potential endogenous antagonist for CB_1 receptors, virodhamine provides the first glimpse of a new form of regulation within the endocannabinoid system (Porter et al., 2002). It is possible that additional endocannabinoids remain to be discovered.

AEA and 2-AG have higher affinity for CB_1 than for CB_2 receptors (Khanolkar et al., 1996; Shohami et al., 1996; Palmer et al., 2002) but differ in their efficacy to elicit signals via these receptors (Breivogel and Childers, 2000; Gonsiorke et al., 2000; Hillard, 2000; Howlett and Mukhopadhyay, 2000; Mukhopadhyay et al., 2002). AEA and (−)Δ^9-THC are partial agonists at CB_1 receptors (Mackie and Hille, 1992; Mackie et al., 1993; Sugiura et al.,1999), whereas AEA is a partial agonist at CB_2 (Sugiura et al., 2000). In contrast, 2-AG is a potent full agonist for both CB_1 and CB_2 (Sugiura et al., 1999; 2000). Because tissue levels of 2-AG are much higher than AEA in the brain and somatic and reproductive organs (Sugiura and Waku, 2000; Sugiura et al., 2002; Burkman, Sugiura, et al., 2003), 2-AG may be the most physiologically significant endocannabinoid identified to date. AEA also is an agonist at ion-type vanilloid receptors (Zygmunt et al., 1999; Van der Stelt and Di Marzo, 2004). It also directly (1) interacts with L-type Ca^{2+} and K^+ channels (Di Marzo et al.,2002); (2) modulates gap junctions in astrocytes (Venance et al., 1995); (3) activates glutamate ionotropic (NMDA) receptors (Di Marzo et al., 2002); (4) directly inhibits voltage-gated Ca^{2+} channels in T-tubule membranes of rabbit skeletal muscle (Oz et al., 2000). Finally, both 2-AG and the metabolically stable AEA analog R-methanandamide (AM356) directly block α_7-nicotinic acetylcholine-receptor-mediated responses in *Xenopus* oocytes (Oz et al., 2004).

While the attention of most investigators remains focused upon the psychoactive properties of cannabinoids and their effects on brain functions (Felder and Glass, 1998), it is now clear that CBRs also are located in nonneural cells outside of the nervous system, including (1) phagocytic cells and lymphocytes in the immune system (Zimmerman et al., 1991; Bouaboula et al., 1993; Galiegue et al., 1995; Stefano et al., 1996; Bisogno, Maurelli, et al., 1997; Cabral, 1999; Berdyschev, 2000); (2) type-II alveolar cells in the lung (Rice et al., 1997); (3) smooth muscle and endothelial cells in the vascular system (Varga et al., 1995; Randall et al., 1996; Sugiura et al., 1998; Kunos et al., 2002; Hiley and Ford, 2004); (4) human breast cancer and prostatic carcinoma cells (De Petrocellis et al., 2000; Di Marzo et al., 2000; Nithipatikom et al., 2004; Sarfaraz et al., 2005); as well as (5) renal endothelial cells and mesangial cells in the kidney (Deutsch et al., 1997). With respect to reproductive organs, endocannabinoids have been detected in (1) the pituitary gland (Pagotto et al., 2001); (2) testis (Gerard et al., 1991; Galiegue et al., 1995; Sugiura et al., 1996; Brown et al., 2002; Burkman, Sugiura, et al., 2003); (3) Leydig cells (Wenger et al., 2001); (4) Sertoli cells (Maccarrone et al., 2003); (5) epididymis (Burkman, Sugiura, et al., 2003); (6) prostate (Burkman, Sugiura, et al., 2003; Ruiz-Llorente et al., 2003, 1994; Sanchez et al., 2003); (7) sea urchin and human sperm (Chang et al., 1993; Schuel, Burkman, Lippes, Crickard, Mahony, et al., 2002; Rossato et al., 2005); (8) ovary, uterus, oviduct, and preimplantation embryo (Galiegue et al., 1995; Paria and Dey, 2000; Burkman, et al., 2003a; Dennedy et al., 2004; Wang et al., 2004); and (9) placenta and fetus (Buckley et al., 1998; Kenney et al., 1998). High levels of NAPE-PLD, the enzyme that releases AEA and its congeners from membrane phospholipids, are found in rat testis (Okamoto et al., 2004) along with high levels of NAEs in testis and uterus (Hansen et al., 2000; Schmid et al., 1997) and FAAH (Hansen et al., 2000).

The distribution of CBRs and endocannabinoids in reproductive organs is directly related to the biological effects produced by marijuana smoke and cannabinoids (Schuel et al., 2002b; Park et al., 2004). Exogenous cannabinoids affect all reproductive functions studied thus far in humans and laboratory animals by (1) inhibiting secretion of gonadotrophic hormones by the pituitary gland and affecting secretion of steroids by the gonads (Kolodny et al., 1974; Bloch et al., 1978; Burstein et al., 1978; Dalterio et al., 1981, 1983; Harclerode, 1984; Smith et al., 1984; Tyre, 1984; Smith and Asch, 1984, 1987; Maykut, 1985; Murphy et al., 1991; Brown and Dobs, 2002; Habayeb et al., 2002; Park et al., 2004); (2) inhibiting ovulation followed by the development of tolerance (Powell and Fuller, 1983; Field and Tyre, 1986; Mueller et al., 1990); (3) inhibiting sperm production and increasing the incidence of sperm with abnormal nuclei and acrosomes in humans and experimental animals (Kolodny et al., 1974; Zimmerman et al., 1978, 1979, 1986; Huang et al., 1978; Issidorides, 1978; Hembree et al., 1979; Dalterio et al., 1982; Morrill et al., 1983; Tilak and Zimmerman, 1984; De Celis et al., 1996; Silva, 1990; Nahas et al., 2002); (4) reducing the size of male accessory sex glands and the volume of seminal plasma ejaculated (Morrill et al., 1983; Harclerode, 1984; Burkman et al., 2003b); (5) reducing copulatory behavior in male rats (Merari et al., 1973; Murphy et al., 1994) and producing impotence in men after chronic intensive marijuana use (Kolodny et al., 1974); (6) increasing sexual receptivity in female rats (Mani et al., 2001); (7) inhibiting fertilization in sea urchins (Schuel et al., 1987; Schuel et al., 1991a,b; Schuel et al., 1994; Chang et al., 1993); (8) affecting capacitation and fertilizing potential of human sperm (Schuel, et al., 2002a); (9) affecting early embryonic development, hatching of the blastocyst from the ZP, and implantation of the blastocyst into the uterine endometrium (Das et al., 1995; Paria et al., 1995, 1996, 1998, 2001, 2002; Schmid et al., 1997; Wang et al., 1999; Paria and Dey, 2000; Habayeb et al., 2002; Maccarrone, et al., 2002a; Park et al., 2004; Wang et al., 2004); (10) reducing intrauterine weight gain by the fetus, promoting congenital anomalies, and altering the delivery process (Tennes, 1984); (11) reducing the number of pregnancies carried to term (Rosenkrantz, 1978; Powell and Fuller, 1983; Abel, 1984; Maccarrone et al., 2000, 2002b); (12) inhibiting milk ejection in lactating rats (Tyre and Murphy, 1988); and (13) regulating suckling in newborn mice (Fride et al., 2003a,b; Fride, 2004a,b). This review will concentrate on regulation of fertilization, pregnancy, and development by endocannabinoid signaling.

FERTILIZATION

Exchange of chemical signals is the universal language for communication between cells from bacteria and protozoa to both neural and nonneural cells in mammals (Csaba 1985; Sastry and Sadavongvivad, 1979; Schulz, 1992; Schuel et al., 1999; Miller et al., 2001; Meizel, 2004). Neurons activate target cells by exocytotic release of neurotransmitters at synapses. Sperm are "lean, mean swimming machines" designed to find and inject their genetic information into unfertilized eggs (Darszon et al., 1999, 2001; Meizel, 2004). They typically contain a single secretory granule (acrosome) in the anterior region of the sperm head, a nucleus with a haploid complement of condensed chromosomes, two centrioles, mitochondria, and a flagellum. Acrosomal exocytosis or acrosome reaction (AR) at the egg's surface is required for gamete fusion and egg activation during normal fertilization. Unfertilized eggs are surrounded by extracellular investments which must be penetrated by the fertilizing sperm before it can activate the egg: vitelline layer and jelly coat in sea urchins, which are analogous to the ZP and cumulus oophorus in mammals (Wassarman, 1987; Garbers, 1989; Yanagimachi, 1994). The species-specific ligand that stimulates AR in sea urchins is a homopolymer of sulfated polysaccharide in the egg's jelly (EJ) coat (Alves et al., 1997), and in mammals is an *O*-linked carbohydrate on glycoprotein ZP3 in the egg's ZP (Wassarman, 1987; Van Duin et al., 1994; Yanagimachi, 1994; Primakoff and Myles, 2002). Binding of these agonists with their receptors on sperm triggers signal cascades involving ion fluxes and synthesis of cAMP and other second messengers similar to those associated with exocytosis in neurons (Garbers, 1989; Darszon et al., 1999, 2001; Kopf et al., 1999; Roldan, 1999; Meizel, 2004).

Invertebrate and mammalian sperm express receptors for many neurotransmitters, e.g., acetylcholine (nicotinic and muscarinic types) (Sastry and Sadavongvivad, 1979; Nelson et al., 1980; Meizel, 2004); adenosine and ATP (Meizel, 2004); serotonin, and catecholamines (Nelson and Cariello, 1989; Bandivdekar et al., 1991; Meizel, 2004); prostaglandins and leukotrienes (Basuray et al., 1990; Schaefer et al., 1998); progesterone (Meizel, 1997, 2004) and estrogen (Aquila et al., 2004); amino acids (Meizel, 1984); peptides (Schulz, 1992; Darszon et al., 1999, 2001; Anderson et al., 1995; Carrell et al., 1995); and odorants (Meizel, 2004). Sperm also express receptors for psychoactive drugs, e.g., nicotine (Sastry and Sadavongvivad, 1979; Nelson, 1978, 1985; Nelson et al., 1980; Meizel, 2004), cocaine (Yazigi et al., 1991; Yelian et al., 1994), opioids (Cariello et al., 1986), and cannabinoids (Chang et al., 1993; Schuel, et al., 2002a; Rossato et al., 2005). Receptors for these neuroactive agents modulate normal sperm functions essential for fertilization including respiration, motility, chemotaxis, capacitation, and AR. These signal processes are critically important aspects of sperm physiology and have been highly conserved for over 600 million years of evolutionary history.

In studies on human populations of marijuana users or in experiments on laboratory mammals, it is difficult to discriminate between reduced fertility resulting from a direct effect on fertilization and indirect effects on other reproductive functions or behaviors. Although large numbers of human sperm can be obtained from volunteers, eggs are not readily available because human females normally produce only one egg each month, and a surgical procedure is required to collect it. Furthermore, significant ethical concerns preclude the use of live human eggs for research purposes.

Endocannabinoid Signaling Modulating Sea Urchin Sperm Fertility

Sea urchin gametes have been extensively used for over a century as an *in vitro* model system to study fertilization and early embryonic development because (1) these processes normally take place externally in sea water, (2) large quantities of mature gametes can be easily collected from adult animals, and (3) eggs undergo synchronous fertilization and development following insemination *in vitro* (Lillie, 1919; Harvey, 1956; Rothschild, 1956; Hagstrom and Lonning, 1973; Schuel, 1984, 1985). These features make it possible to directly treat eggs and sperm with a drug to assess its effects on fertilization processes under carefully controlled laboratory conditions.

Inhibition of Fertilization

During normal fertilization in sea urchins and in mammals, the egg is penetrated by a single sperm (Schuel, 1984; Hoodbhoy and Talbot, 1994). Penetration of the egg by more than one sperm during fertilization results in abnormal development and eventual death of the embryo. Previous studies with sea urchin gametes showed that psychoactive drugs such as morphine, cocaine, nicotine, the volatile anesthetic halothane, propranolol (a β-adrenergic receptor blocker), etc., promoted polyspermic fertilization in sea urchins (Hertwig and Hertwig 1887; Clark, 1936; Harvey, 1956; Cardasis and Schuel, 1976; Schuel, 1984; Hinkley and Wright, 1986; Nicotra and Schatten, 1996). Based upon these observations, we postulated that cannabinoids would likewise promote polyspermy. To test this hypothesis, *Strongylocentrotus purpuratus* eggs were pretreated with (–)Δ^9-THC (5 to 400 μ*M*) and then inseminated with excess sperm (5 to 10 × 10^7/ml) (Schuel et al., 1987). Unexpectedly, this treatment failed to promote polyspermy. The incidence of polyspermy in (–)Δ^9-THC-treated eggs was less than that of controls. These data suggested that the drug was inhibiting fertilization. This possibility was examined by pretreating eggs or sperm with (–)Δ^9-THC (0.1 to 10 μ*M*) for 5 min prior to insemination. A minimal sperm density just sufficient to fertilize ~90% of eggs cultured in sea water was used in these experiments. Eggs pretreated with (–)Δ^9-THC did not show a reduction in receptivity to untreated sperm. However, pretreatment of sperm with (–)Δ^9-THC resulted in a concentration-dependent reduction in fertilizing potential (IC_{50} 1.1 ± 1.1 μ*M*). At 10 μ*M* (–)Δ^9-THC, sperm fertility was reduced by 99.8 ± 0.4%. The fertilizing potential of sperm treated with (–)Δ^9-THC under these conditions depends upon sperm density and duration

TABLE 22.1
Inhibitory Effects of Acyl Ethanolamides and Cannabinoids on Sperm Fertilizing Capacity and Egg-Jelly-Stimulated Acrosome Reactions in Sea Urchins

Ligands	*S. intermedius* Fertilizing Capacity[a] IC_{50} [μM]	*S. purpuratus* Fertilizing Capacity[b] IC_{50} [μM]	*S. purpuratus* Acrosome Reaction IC_{50} [μM]
Anandamide	0.56 ± 0.12	1.23 ± 0.60	8.66 ± 2.12[b]
Linoleoylethanolamide (18:2)	0.99 ± 0.25	ND	ND
Oleoylethanolamide (18:1)	0.90 ± 0.10	ND	ND
(−)Δ^9-THC	0.22 ± 0.13	0.41 ± 0.14	4.62 ± 0.19[b]
(+)Δ^9-THC[e]	ND	ND	9.1 ± 0.55[ce]
CP55940	ND	ND	1.4 ± 0.86[c]
WIN55212-2	2.44 ± 0.69	ND	7.00 ± 3.31[d]
SR141716A	0.53 ± 0.1	ND	ND

Note: ND = not determined.

[a] Berdyshev EV. 1999. *Comp Biochem Physiol* Part C 122: 327–330.
[b] Schuel H et al. 1994. *Proc Natl Acad Sci USA* 91: 7678–7682.
[c] Chang MC et al. 1993. *Mol Reprod Dev* 36: 507–516.
[d] Schuel & Goldstein, unpublished data.
[e] $p < .025$ vs. (−)Δ^9-THC.

of pretreatment. The adverse effects elicited by (−)Δ^9-THC on sperm fertility were completely reversible. To determine whether (−)Δ^9-THC (10 μM) could affect cleavage, eggs were inseminated with higher concentrations of sperm, sufficient to fertilize all the eggs. Under these conditions, first division was not delayed in zygotes that had been fertilized with sperm pretreated with 10 μM (−)Δ^9-THC. Two other nonpsychoactive cannabinoids commonly found in marijuana, CBN, and cannabidiol (CBD), produced similar effects on sperm fertility (Schuel et al., 1987). Subsequent studies using *S. purpuratus* (Schuel et al., 1991a; Schuel et al., 1994; Schuel et al., 1999; Chang et al., 1993) and *S. intermedius* (Berdyshev, 1999) gametes showed that other cannabinergic ligands including WIN55212-2, SR141716A, AEA, OEA, and linoleoylethanolamide (LEA) likewise inhibit fertilization by affecting the sperm (Table 22.1). However, AEA was the only unsaturated NAE detected in sea urchin eggs (Bisogno et al., 1997b). Hence, the physiological significance of OEA and LEA in sea urchin fertilization remains to be determined (Berdyshev, 1999). During 90-min observation, sperm cultured with (−)Δ^9-THC and AEA swam longer and more vigorously than control sperm in sea water (Schuel et al., 1999). Because sea urchin eggs are fertilized within seconds after insemination (Schuel and Schuel, 1981; Schatten and Hulser, 1983), these data suggested that AEA and cannabinoids might inhibit fertilization in sea urchins by blocking the AR.

The *acrosome reaction* in sea urchins is stimulated by a sulfated polysaccharide in the EJ coat (Garbers, 1989; Alves et al., 1997). AEA, (−)Δ^9-THC, and other cannabinoids inhibit EJ-stimulated AR in a dose-dependent manner (Table 22.1). Furthermore, (−)Δ^9-THC also inhibits spontaneous ARs in sea urchin (Schuel et al., 1991a) and human sperm (Rosatto et al., 2005). These responses in sperm mimic those observed in neurons, in which CBR agonists inhibit evoked and spontaneous secretion of neurotransmitters (Pertwee and Ross, 2002). Sperm regain their ability to undergo AR and to fertilize eggs upon removal of cannabinergic ligands (Schuel et al., 1991a, 1994). These findings show that AEA and cannabinoids reduce sperm fertilizing capacity by means of blocking ligand-stimulated AR.

CBR signaling modulates ion channels in neurons and other somatic cells to inhibit exocytotic release of chemical messengers (Elphick and Egertová, 2001; Alger, 2002; Pertwee and Ross, 2002;

Wang et al., 2004). EJ-stimulated AR in sea urchin sperm is associated with opening of ligand- and voltage-gated ion channels resulting in net influx of Ca^{2+} and Na^+ and net efflux of H^+ and K^+, which changes membrane potential and elevates pH_i (Darszon et al., 1999, 2001). AR also can be induced artificially, using ionophores that bypass the ligand–receptor reaction with EJ, to trigger the AR by transporting specific cations across the sperm's plasma membrane and by NH_4OH that increases pH_i (Darszon et al., 1999, 2001). We observed that (–)Δ^9-THC and AEA do not block AR in sea urchin sperm, artificially induced by A23187 and ionomycin that promote Ca^{2+} influx, by monensin that promotes Na^+ influx, by nigericin that promotes K^+ efflux, and by NH_4OH that increases pH_i (Schuel et al., 1991a, 1994). AEA does not block ionomycin-induced ARs in human sperm (Rosatto et al., 2005). These results suggest that CBR signaling affects stimulation–secretion–coupling events in sea urchin sperm prior to opening of ion channels.

Ultrastructural observations on sea urchin sperm showed that the membrane fusion step in exocytosis of the acrosomal granule is blocked by cannabinoids (Chang and Schuel, 1991). This finding is consistent with observations that cannabinoid agonists and anandamides inhibit secretion of neurotransmitters by neurons in the central and peripheral nervous system (Alger, 2002; Pertwee and Ross, 2002), of hormones by the pituitary gland (Murphy et al., 1991), and of inflammatory mediators such as serotonin by mast cells (Facci et al., 1995). Taken together these observations suggest a general role for endocannabinoid signaling in modulating stimulation–secretion–coupling mechanisms by cells (Schuel et al., 1991a, 1994, 1999). The AR in sea urchin sperm is an ideal model system to study this process.

Surprisingly, our ultrastructural observations also revealed that cannabinoids cause the formation of lipid deposits in the subacrosomal fossa and in the centriolar fossa of treated sperm. The nuclear envelope is fragmented in close proximity to these lipid deposits (Chang and Schuel, 1991). The lipid deposits disappear from treated sperm after the cannabinoids are removed by washing. The sperm regain their capacity to undergo EJ-stimulated AR and to fertilize eggs. These unexpected findings suggested that the lipid deposits in sea urchin sperm may represent hydrolysis products derived from the fragmented nuclear envelope, possibly reflecting cannabinoid-induced activation of phospholipases within the sperm. Cannabinoids were known to activate phospholipase A_2 (PLA_2) in mammalian somatic cells (Burstein and Hunter, 1981). Our group showed that cannabinoids activate Ca^{2+}-dependent PLA_2 activity in homogenates of sea urchin sperm (Chang et al., 1991a). Arachidonic acid that is released from membrane phospholipids by the actions of PLA_2 or its metabolites are known to be potent lipid-signal molecules in somatic cells (Axelrod et al., 1988; Burstein, 1992). Similar processes regulate motility and AR in mammalian sperm (Bartoszewicz et al., 1975; Meizel and Turner, 1984; Bennet et al., 1987; Joyce et al., 1987; Basuray et al., 1990; Roldan, 1999). Furthermore, (–)Δ^9-THC is significantly more potent than (+)Δ^9-THC in activating PLA_2 in sea urchin sperm homogenates (Chang et al., 1991b). Stereoselectivity is a characteristic feature of receptor-mediated reactions (Matsuda, 1997). Cannabinoids are known to activate PLA_2 in mammalian somatic cells by receptor-mediated reactions (Burstein, 1992; Burstein et al., 1994). Collectively, these findings suggested that sea urchin sperm may contain functional CBRs.

Radioligand Binding

The potent bicyclic synthetic cannabinoid [^{3}H]CP55,940 has been used to demonstrate the presence of CBRs in brain and peripheral organs (Devane et al., 1988; Herkenham et al., 1990; Gerard et al., 1991; Matsuda, 1997). We showed that specific binding of [^{3}H]CP55,940 to live sea urchin sperm is saturable: K_D 5.16 ± 1.02 n*M*; Hill coefficient 0.98 ± 0.004; B_{max} 2.22 ± 0.42 pmoles/mg protein (Chang et al., 1993). These data suggest that sea urchin sperm contain a single class of cannabinoid binding sites and that significant cooperative interactions are absent. Sea urchin sperm contain 712 ± 122 receptors/cell. The rank order of potency to inhibit specific binding of [^{3}H]CP55,940 to live sea urchin sperm and to inhibit the DJ-stimulated AR is CP55,940 > (–)Δ^9-THC > (+)Δ^9-THC (Table 22.2). Again, stereoselectivity is a characteristic of receptor-mediated reactions. The observed differences in the potency of (–) and (+) enantiomers of Δ^9-THC in displacing [^{3}H]CP55,940 from

TABLE 22.2
Comparison of the Potency of Cannabinergic Ligands in Displacing [^{3}H]CP55940 from Sperm Receptors with Their Potency in Blocking the Egg-Jelly-Stimulated Acrosome Reaction in the Sea Urchin *Strongylocentrotus Purpuratus*

Ligand	Displacement K_i [n*M*][a]	Acrosome Reaction IC_{50} [μ*M*]
CP55940	2.18 ± 0.5	1.4 ± 0.86[a]
(–)Δ^9-THC	830 ± 180	4.62 ± 0.19[a]
(+)Δ^9-THC	3700 ± 520[d]	9.1 ± 0.55[ae]
Anandamide	ND	8.66 ± 2.12[b]
WIN55212-2	ND	7.00 ± 3.31[c]

Note: ND = not determined.

[a] Chang MC et al. 1993. *Mol Reprod Dev* 36: 507–516.
[b] Schuel H et al. 1994. *Proc Natl Acad Sci USA* 91: 7678–7682.
[c] Schuel & Goldstein, unpublished data.
[d] $p < .025$ vs. (–)Δ^9-THC.
[e] $p < .02$ vs. (–)Δ^9-THC.

its binding site on sperm, and in blocking EJ-stimulated ARs are statistically significant ($p < .02$ and $p < .025$, respectively). Together, these observations indicate that cannabinergic ligands inhibit the AR in sea urchin sperm by a receptor-mediated process and suggest that endocannabinoid signaling may modulate sperm fertility (Chang et al., 1993; Schuel et al., 1994 and 1999).

Receptor Characterization

An orthologue of vertebrate CBR has recently been cloned in the urochordate *Ciona intestinalis* (Elphick et al., 2003). The relationship between the CBR in sea urchin sperm and mammalian CB_1 and CB_2 remains to be determined. SR141716 is a potent selective inverse agonist/antagonist for CB_1 receptors and prevents CB_1 receptor–mediated effects both *in vivo* and *in vitro* (Rinaldi-Carmona et al., 1995; Palmer et al., 2002). Berdyshev (1999) used SR141716 to characterize the CBR in sea urchin sperm. Surprisingly, SR141716 mimics the effects of AEA, (–)Δ^9-THC, and WIN55212-2 in producing dose-dependent inhibitory effects on fertilizing capacity of *S. intermedius* sperm (Figure 22.1A). Furthermore, it did not block the inhibitory effect of (–)Δ^9-THC on sperm fertility. SR141716 actually potentiated the reduction in sperm fertilizing potential elicited by (–)Δ^9-THC (Figure 22.1B). These findings suggest that (–)Δ^9-THC and SR141716 may act at the same molecular target in sea urchin sperm, possibly a CB_1-like receptor. Intriguingly, SR141716 mimics inhibitory effects of AEA and 2-AG on opening voltage-gated Ca^{2+} channels in differentiated NG108-15 cells (Sugiura at al., 1997). Additional work is required to determine if SR141716A and AM251 are protean agonists that elicit inverse responses in some systems and positive responses in others (Kenakin, 2001) and whether they also produce biological effects via non-CB_1-mediated processes. Additional work is required to resolve this issue.

Physiological Role for NAE Signaling during Fertilization

The prevention of polyspermy is a critical event in normal fertilization (Schuel, 1984). Many sperm are likely to be in the vicinity of an unfertilized egg during the process of fertilization. As soon as the first sperm activates the egg, other sperm in the vicinity of the egg represent a potential hazard

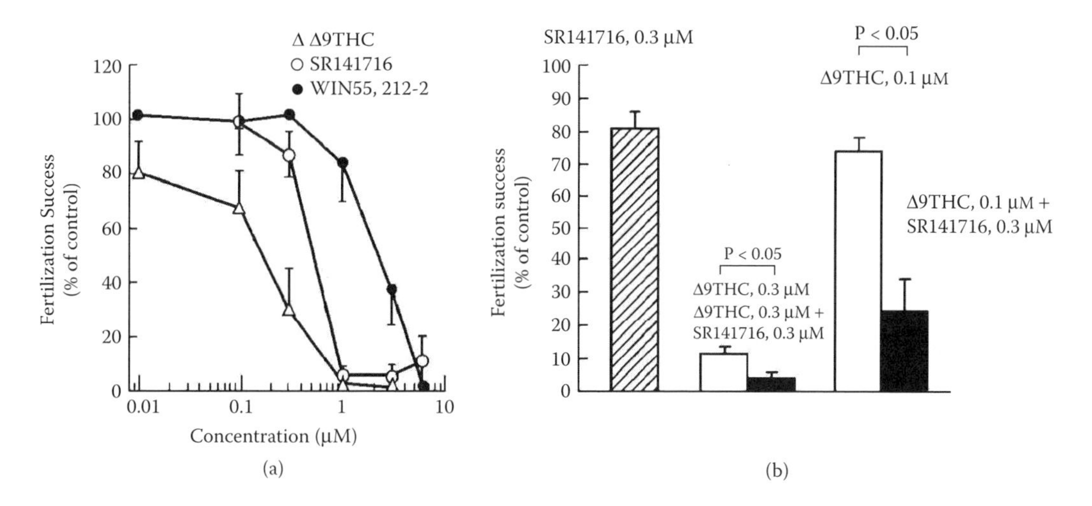

FIGURE 22.1 Effects of cannabinergic ligands on fertilization in *S. intermedius*. (A) Concentration-dependent reduction of sperm fertilizing potential by (−)Δ⁹-THC, SR141716, and WIN55212-2 (IC_{50} 0.22 ± 0.13, 0.53 ± 0.1, and 2.44 ± 0.69, respectively). (B) SR141716 does not block the inhibitory effects of (−)Δ⁹-THC on sperm fertilizing potential. The inhibitory effects of SR141716 were augmented by (−)Δ⁹-THC. Spermatozoa (2.5×10^6 cells/ml) were preincubated with indicated concentrations of cannabinergic ligands for 20 min. Then eggs (1000 cells/ml sperm suspension) were added. Fertilization was stopped 5 min later by glutaraldehyde fixation. Fertilization success, defined as the presence of fertilization envelope, was assessed by light microscopy. Data expressed as the percentage of control fertilization (0.1% ethanol) ± SEM. (Adapted from, Berdyshev EV. 1999. *Comp Biochem Physiol* Part C 122: 327–330. With permission.)

to normal development. Secretion of the egg's cortical granules is one of the early responses of the egg to stimulation by the fertilizing spermatozoon. This process (cortical reaction) begins at the site of binding of the fertilizing spermatozoon to the egg's vitelline coat and spreads in a wave-like manner around the surface of the egg. This reaction promotes detachment of the vitelline coat to form the fertilization envelope, which acts as a mechanical barrier to further sperm penetration (Schuel, 1978, 1984, 1985). In sea urchins such as *S. purpuratus*, the cortical reaction is completed within about 60 sec following insemination. Because most sea urchin eggs are activated within ~1 sec after insemination, they are potentially vulnerable to penetration by additional sperm until elevation of the fertilization envelope is completed (Longo et al., 1974; Schatten and Hulser, 1983). Other processes operate to limit sperm penetration during this period, including a rapid Na^+-dependent depolarization of the egg's plasma membrane (Jaffe 1976; Schuel and Schuel, 1981); release of H_2O_2 by the egg (Boldt et al., 1981, 1984; Coburn et al., 1981; Schuel and Schuel 1986, 1987, 1988); as well as processes involving prostaglandins and leukotrienes (Schuel et al., 1984, 1985).

AEA signaling may be another polyspermy-preventing process in sea urchins (Chang et al., 1993; Schuel et al., 1994). This hypothesis is supported by the Di Marzo group's work showing that (1) ovarian sea urchin eggs, but not sperm, can produce AEA; (2) lipid extracts of ovarian eggs contain AEA, PEA, and stearoyl-ethanolamide (SEA), as well as NAPEs, which are the phospholipid precursors of NAEs in mammalian tissues; (3) homogenates of ovarian eggs contain a PLD-like activity capable of releasing [^{3}H]AEA from synthetic [^{3}H]*N*-arachidonoyl-phosphatidyl-ethanolamine; (4) homogenates of ovarian eggs also contain a FAAH activity that catalyses the hydrolysis of AEA, PEA, and SEA (Bisogno et al., 1997a).

A proposed model for retrograde AEA signaling in modulating sperm fertility and prevention of polyspermy in sea urchins is depicted in Figure 22.2. Binding of the jelly-coat ligand to its receptor on the sperm surface stimulates opening of Ca^{2+} channels, resulting in Ca^{2+} influx that

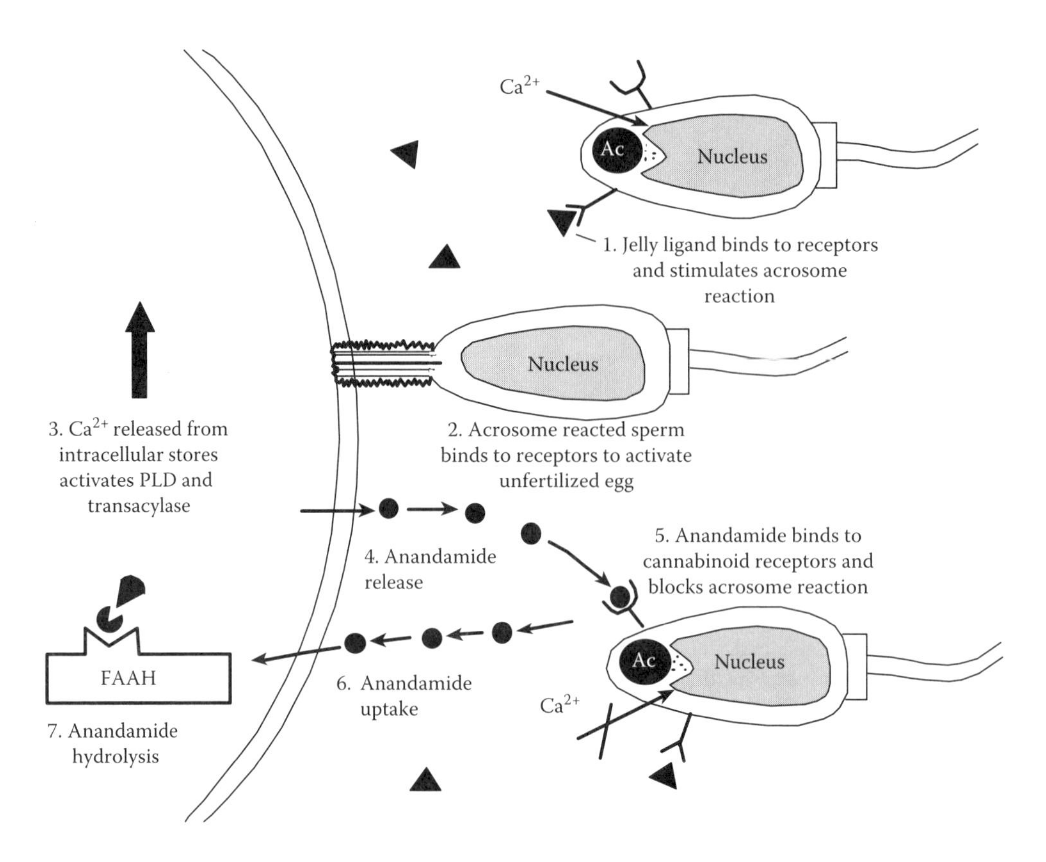

FIGURE 22.2 Schematic diagram depicting the postulated role for retrograde anandamide-signaling in down-regulating sperm fertility. Anandamide released from the egg following activation by the fertilizing spermatozoan binds to its receptors on nearby sperm to modulate (inhibit) the acrosome reaction, thereby preventing that spermatozoan from refertilizing the egg prior to completion of the cortical reaction.

promotes acrosomal exocytosis and polymerization of monomeric acting in the subacrosomal fossa to form the acrosomal filament (1). Binding of sperm to its receptor on the egg's surface activates the egg (2). Stimulation of the egg by fertilizing the spermatozoon promotes release of Ca^{2+} from intracellular stores (Whitaker and Steinhardt, 1982), resulting in activation of transacylase and PLD in the egg (3). AEA is released from the egg's membrane lipids (4). AEA binds to its receptor on the surface of other nearby sperm, inhibiting opening of Ca^{2+} channels, thereby blocking the AR and preventing that spermatozoon from refertilizing the egg (5). Reuptake of AEA by eggs is likely to be facilitated by a transporter system, analogous to those described in mammalian somatic tissues (6) and is followed by AEA hydrolysis mediated by FAAH within the egg (7). Additional research is required to establish the validity of this model.

These processes in sea urchin gametes closely resemble those operating in the mammalian brain, where retrograde AEA signals from depolarized postsynaptic neurons inhibit neurotransmitter release at excitatory synapses (Elphick and Egertová, 2001; Alger, 2002; Pertwee and Ross, 2002; Piomelli, 2003). A model for retrograde endocannabinoid signaling at synapses is depicted in Figure 22.3. Neurotransmitter secreted by the presynaptic cell binds to its receptor on the surface of postsynaptic neuron. In response to this stimulus, there is an increase in Ca^{2+}_i within the postsynaptic neuron which activates Ca^{2+}-dependent transacylase and PLD to release AEA into the synaptic cleft. (2). AEA then binds to CB_1 on the presynaptic neuron, leading a to G-protein-mediated signal cascade that reduces cAMP production, elevates K^+_i and reduces Ca^{2+}_i (4), which in turn inhibits secretion of neurotransmitter by the presynaptic neuron (5). A similar role has been proposed for 2-AG in

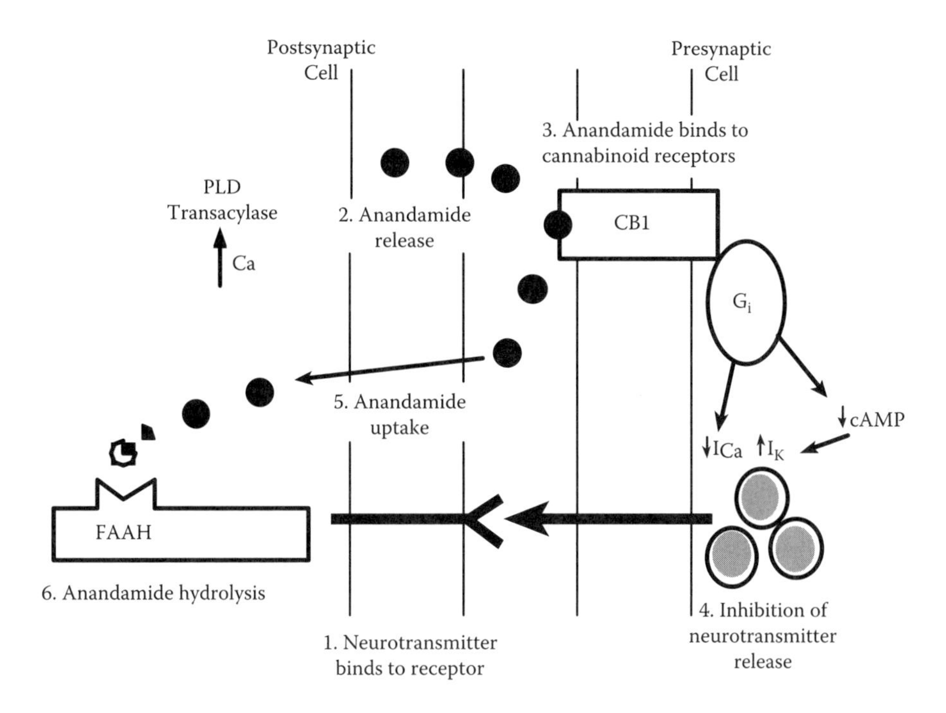

FIGURE 22.3 Schematic diagram depicting a model for synaptic endocannabinoid signaling, in which anandamide functions as a retrograde-signaling molecule that modulates (inhibits) the release of classical anterograde neurotransmitters by presynaptic terminals. (Adapted from Elphick MR, Egertová M. 2001. *Philos Trans R Soc London* B 356: 381–408. With permission.)

modulating neurotransmission (Sugiura and Waku, 2000). Within this context, neurons can be viewed as sperm cells without tails.

Endocannabinoid Signaling in the Modulation of Human Sperm

The gene for the human CB_1 also is expressed in the human testis (Gerard et al., 1991). This finding, taken together with our observations on sea urchin sperm, led us to postulate that human sperm would contain similar CBRs (Chang et al., 1993; Schuel et al., 1994). Subsequent studies showed that AEA is synthesized in rodent testis, uterus, and oviduct (Paria et al., 1995; Sugiura et al., 1996; Schmid et al., 1997; Paria and Dey, 2000). The mouse uterus was reported to contain higher levels of AEA than any mammalian organ including the brain (Schmid et al., 1997). FAAH, the enzyme that hydrolyzes AEA, also is present in the mouse uterus (Paria et al., 1996). Collectively, these observations suggested that sperm may normally be exposed to AEA as they transit human reproductive tracts and that AEA signaling may regulate sperm functions *in vivo*.

Unlike sea urchin sperm, which are able to fertilize eggs immediately after dispersal into sea water, freshly ejaculated sperm from humans or other mammals, bathed in male secretions forming the seminal plasma, are not yet capable of fertilizing an egg (Yanagimachi, 1994). Sperm acquire the capacity to fertilize eggs following removal from seminal plasma by exposure for several hours to female reproductive tract fluids *in vivo* or by incubation in appropriate culture medium *in vitro* (Meizel, 1985, 1997; Burkman, 1990, 1995; Yanagimachi, 1994; Kopf et al., 1999). "Capacitated" sperm exhibit vigorous hyperactivated motility (HA), required for fertilization, and can undergo physiological ARs at the egg's surface (Wassarman 1987; Burkman, 1990, 1995; Yanagimachi, 1994). Fertilization *in vivo* requires that the proper number of sperm arrive at the oviductal ampulla in the appropriate physiological state, coincident with a viable unfertilized egg. Human sperm penetrate the cervical mucus and enter the upper female reproductive tract only at midcycle and may be stored in reservoirs located in the cervix and oviductal isthmus (Zinaman et al., 1989; Overstreet and Katz, 1990; Bielfeld et al., 1991; Mortimer, 1995; Suarez, 1998). Could AEA signaling help regulate sperm capacitation, timely escape from sperm reservoirs in the cervix or caudal portion of the oviduct, sperm selection, and fertilizing potential *in vivo*? Evidence supporting such a mechanism has been obtained by radioligand binding studies on human sperm, analysis of reproductive tract fluids for the presence of AEA and related NEAs, and *in vitro* studies on sperm capacitation and fertilizing potential (Schuel et al., 2002a,b; Burkman et al., 2003a; Rossato et al., 2005).

Radioligand Binding

[^{3}H]CP55,940 binds to human sperm in a saturable manner, which suggests that they contain CBRs (Schuel, Burkman, Lippes, Crickard, Mahony, et al., 2002). Furthermore, the binding properties of [^{3}H]CP55,940 to putative CBRs in human sperm measured thus far are remarkably similar to those previously obtained with sea urchin sperm and mammalian somatic tissues (Table 22.3). [^{3}H]CP55,940 binds with high affinity to both CB_1 (K_i 0.6 n*M*) and CB_2 (K_i 0.7 n*M*) receptors (Palmer et al., 2002). Transcripts (mRNA) for human CB_1 are expressed in human and dog testis (Gerard et al., 1991; Galiegue et al., 1995), and mRNA for CB_2 is expressed in rat testis (Brown et al., 2002). CB_1 (but not CB_2) has been detected in human sperm extracts by Western blots (Rossato et al., 2005). Antibodies raised against the C-terminal peptide of CB_1 show that the receptor is present in the sperm head and midpiece. The distribution of CB_1 in human sperm reflects its functions in regulating acrosomal status and motility (Schuel, Burkman, Lippes, Crickard, Mahony, et al., 2002; Rossato et al., 2005). The detection of CBRs in human sperm implied the presence of endocannabinoids in human reproductive tract fluids.

Endocannabinoids in Reproductive Tract Fluids and Organs

Samples of human seminal plasma, midcycle oviductal fluid, and follicular fluid were analyzed by high-performance liquid chromatography/mass spectrometry (Schuel, Burkman, Lippes, Crickard, Mahony, et al., 2002; Schuel, Burkman, Lippes, Crickard, Forester, et al., 2002) and quantitated by isotope dilution (Giuffrida et al., 2000). AEA, OEA, and PEA are present in human seminal plasma, midcycle oviductal fluid, and follicular fluid (Table 22.4). The levels of PEA and OEA are significantly higher than AEA in these reproductive tract fluids ($p < 0.005$ to < 0.001), and OEA was the most abundant NAE in follicular fluid ($p < 0.005$). We also detected these NAEs in oviductal fluids collected from rabbits and goats at the time of ovulation. These findings are very different from those obtained with whole mouse uterus, in which AEA represents up to 95% of total NAEs (Schmid et al., 1997). Progesterone and prostaglandin E in follicular fluid are known to be produced by granulosa cells in ovarian follicles, as well as by granulosa cells in the cumulus matrix surrounding ovulated eggs (Meizel et al., 1990; Schaefer et al., 1998). It is tempting to speculate that NAEs may also be synthesized by granulosa cells at these sites. Consistent with previous observations on brain and other somatic organs (Hansen et al., 2000; Sugiura et al., 2002), we recently found that 2-AG is present in much higher concentrations than AEA in whole rabbit testis, epididymis, vas deferens, prostate, seminal vesicles, uterus, ovary, and oviduct (Burkman et al., 2003b). Because CBRs are present in the human ovary (Galiegue et al., 1995) and $(-)\Delta^9$-THC inhibits cAMP accumulation by cultured rat granulosa cells (Treinen et al., 1993), endocannabinoid signaling may help regulate follicular maturation and development. These factors may account, in part, for the adverse effects of marijuana and cannabinoids on ovulation (Powell and Fuller, 1983; Park et al., 2004).

Sperm are sequentially exposed to seminal plasma, oviductal fluid, follicular fluid, and secretions of granulosa cell eggs as they move from the vagina to the site of fertilization in the oviductal

TABLE 22.3
Specific Binding of [^{3}H]CP55940 to Cannabinoid Receptors in Sperm and Mammalian Somatic Tissues

Binding Properties	Human Sperm[a]	Sea Urchin Sperm[b]	Rat Brain (Minced) CB1[c]	Cloned Human Brain CB1[d]	Cloned Human CB2[e]	Human U937 Monocytes CB1[f]	Mouse Spleen Cells[g]
K_D (n*M*)	9.71 ± 1.04	5.16 ± 1.02	15.0 ± 3.0	3.3 ± 0.7	1.6 ± 0.5	0.1 ± 0.1	0.91
B_{max} (p moles/mg protein)	1.193 ± 0.15	2.44 ± 0.42	0.9	7.0 ± 0.5	NR	0.525	NR
Hill coefficient	1.3 ± 0.102	0.98 ± 0.004	NR	NR	NR	NR	~1
Receptors/cell	NR	712.0 ± 122	NR	NR	NR	NR	~1000

Note: NR = not reported.

[a] Schuel H et al. 2002. *Mol Reprod Dev* 63: 376–387.
[b] Chang MC et al. 1993. *Mol Reprod Dev* 36: 507–516.
[c] Herkenham et al., 1990.
[d] Felder et al., 1992.
[e] Munro S, et al., 1993. *Nature* 365: 61–65.
[f] Bouaboula M et al. 1993. *Eur J Biochem* 214: 173–180.
[g] Kaminski NE et al. 1992. *Mol Pharmacol* 42: 736–742.

Source: Adapted from Schuel H et al. 2002. *Mol Reprod Dev* 63: 376–387. With permission.

TABLE 22.4
***N*-Acylethanolamines in Human Reproductive Tract Fluids [n*M*]**

Source	AEA	PEA	OEA
Seminal plasma (N=15)	12.1 ± 2.1	31.5 ± 7.3 [a]($p < .0001$)	32.9 ± 4.7 [b]($p <.0001$) [c]($p > .3$)
Midcycle oviductal fluid (N=15)	10.7 ± 2.5	30.4 ± 6.9 [a]($p < .005$)	36.9 ± 7.5 [b]($p < .001$) [c]($p > .05$)
Follicular fluid (N = 9)	2.9 ± 0.9	11.3 ± 1.3 [a]($p = .0001$)	19.3 ± 2.9 [b]($p < .0001$) [c]($p < .005$)

[a] AEA vs. PEA.
[b] AEA vs. OEA.
[c] PEA vs. OEA.

Source: From Schuel H et al. 2002. *Chem Phys Lipids* 121: 211–227. With permission.

ampulla (Burkman, 1990; Yanagimachi, 1994; Mortimer, 1995). The presence of NAEs in these locations and the detection of CBRs in human sperm (Schuel et al., 2002a,b) implied that endocannabinoid signaling may regulate sperm capacitation and fertilizing potential within human reproductive tracts. Evidence supporting this hypothesis has been obtained by examining the effects of cannabinergic ligands on these processes *in vitro* (Schuel et al., 2002a; Burkman et al., 2003a; Rossato et al., 2005). AEA is rapidly degraded by cellular amidases (Deutsch and Chin, 1993). However, R-methanandamide AM356, a compound with close structural similarity to AEA, is a metabolically stable substitute for AEA (Abadji et al., 1994; Khanolkar et al., 1996; Palmer et al., 2002). Both ligands have higher affinities for CB_1 than for CB_2. Studies on *in vitro* effects of cannabinoid agonists on sperm function were performed using AM356 as a model for AEA in human reproductive fluids, and (–)Δ^9-THC was used to evaluate putative roles of sperm in these processes and as a model for possible effects associated with abuse of marijuana.

Motility

Before fertilization can occur, vigorous HA enables sperm to arrive at the egg surface and assists penetration of the ZP (Burkman 1990; Yanagimachi, 1994). Effects of AM 356 on HA in swim-up sperm were quantitatively evaluated, using a Hamilton-Thorne IVOS computerized semen analyzer, to identify motile sperm that simultaneously exhibited curvilinear velocity ≥ 100 μm/sec; head amplitude ≥ 7.5 μm; linearity < 65 (Schuel et al., 2002a). These characteristics define a sperm with high velocity and nonlinear swimming, which correlates with fertilizing potential (Burkman, 1990, 1995; Sukcharoen et al., 1995) and pregnancy (Burkman et al., unpublished data).

The effects of AM356 on HA during capacitation are depicted in Figure 22.4. Consistent with previous findings (Burkman, 1984), the incidence of HA in semen was 4.8 ± 0.7%. Control sperm rapidly reached their peak incidence of HA (32.2 ± 5%) after 1 h (Figure 22.4A). As expected (Burkman, 1990), HA then gradually declined to 20.7 ± 5% by 6 h. AM356 produced biphasic effects on HA (Figure 22.4B). An inhibitory effect of AM356 on HA was observed at 2.5 n*M*, whereas a stimulatory effect of AM356 on HA was seen at 0.25 n*M*, particularly at 4 to 5 h ($p < .05$, ANOVA). Subsequent studies showed that 2-AG also produced a potent biphasic response over several hours (Burkman et al., 2003a). In contrast, AEA elicited a weak and brief biphasic response

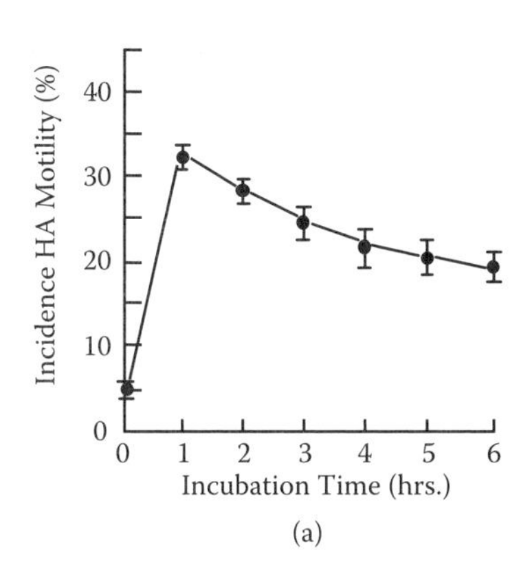

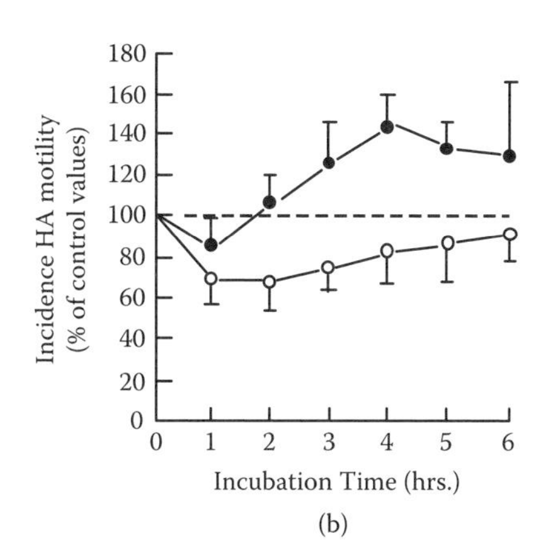

FIGURE 22.4 Effects of AM356 on hyperactivated motility sperm (HA) during incubation for 6 h. (a) Incidence of HA in sperm cultured in control medium. (b) Incidence of HA in sperm treated with AM356 (0.25 n*M*) (closed circles) and (2.5 n*M*) (open circles). Data expressed as percent of control values. The effects of AM356 on HA at 2.5 n*M* were significantly different from those produced by AM356 at (0.25 n*M*) ($p < .05$, ANOVA). N = 6 experiments using sperm from 3 different donors. (From Schuel H et al. 2002. *Mol Reprod Dev* 63: 376–387. With permission.)

on HA. These findings differ from those reported by Rosatto's group (2004) that AEA significantly inhibits motility of human sperm (expressed as percentage of motile sperm observed by light microscopy) in a dose-dependent manner (IC_{50} about 0.1 μ*M*) and was toxic between 1 to 10 μ*M*. AEA also depolarized the sperm's plasma membrane. These AEA-induced effects were blocked by SR141716, which suggests CB_1-mediated responses.

The biphasic responses for HA in human sperm that are produced by AM356 and AEA are similar to those previously reported by other investigators on Ca^{2+} channels in neurons (Mackie et al., 1993), testosterone secretion *in vitro* and *in vivo* (Burstein et al., 1978: Dalterio et al., 1981, 1983), trophoblast outgrowth in culture (Wang et al., 1999), phagocytosis in neutrophils, and behavior in mice (Sulcova et al., 1998). Biphasic responses are a general feature of cannabinoid as well as other G-protein-coupled receptors (Hoyer and Boddeke, 1993; Glass and Felder, 1997; Sulcova et al., 1998). These phenomena may be explained by agonist-induced allosteric modulation receptor conformation, by a large receptor reserve, or by CBR coupling to both G_i and G_s proteins, etc. Furthermore, Ca^{2+} channels in neurons and other somatic cells are modulated by (−)Δ^9-THC, AEA, and 2-AG (Mackie and Hille, 1992; Johnson et al., 1993; Mackie et al., 1993; Sugiura et al., 1999, 2000). Studies on demembranated mammalian sperm suggest that a rise in free intracellular Ca^{2+} alters flagellar beat patterns characteristic of HA (Lindemann and Goltz, 1988). It is possible that the observed effects of AEA, AM356, and 2-AG on HA may be mediated via regulation of Ca^{2+} channels. Furthermore, the biphasic effects of AM356 and AEA on HA are consistent with previous results reported by Burkman and colleagues on modulation of rabbit sperm HA in the oviduct (Johnson et al., 1981; Burkman et al., 1984). These data pointed to an unknown inhibitory substance in the isthmus, which was negated upon tenfold dilution. Factors in the rabbit isthmic environment provide regulatory control, initially inducing sperm quiescence and, at the appropriate time, stimulating vigorous HA motility, thus enabling sperm to escape and swim to the ampulla. There are regional differences in composition of fluids within the lumen of bovine and porcine oviducts (Anderson and Killian, 1994; Hunter, 1990). In addition, bovine isthmic and ampullary fluids produce different effects on motility, AR, zona binding, and fertility of bull sperm (Grippo et al., 1995; Topper et al., 1999). In this connection, it is interesting to note that localized differences in AEA levels in the uterus are associated with the potential implantation sites (Schmid et al., 1997).

AM404 is a potent AEA reuptake inhibitor in cultured cortical neurons (EC_{50} = 1 μ*M*) and astrocytes (EC_{50} = 5 μ*M*), which potentiates AEA-mediated responses both *in vitro* and *in vivo* (Piomelli et al., 1999; Beltramo et al., 1997; Palmer et al., 2002). However, recent kinetic studies on neuoblastoma and astrocytoma cells question the existence of an AM404-sensitive transporter and indicate that AM404 prevents clearance of AEA from the extracellular space by inhibiting the hydrolysis of AEA by FAAH (Glasser et al., 2003). AM404 stimulated HA in human sperm in the absence of added AEA (Burkman et al., 2003a). These results suggest that human sperm may produce and degrade their own AEA to modulate their own swimming behavior via their own CBRs. If correct, this would be analogous to the cholinergic autoregulatory system, operating via nicotinic receptors, which is known to control motility of mammalian and sea urchin sperm (Nelson, 1978, 1985; Stewart and Forrester, 1978; Sastry and Sadavongvivad, 1979; Sastry et al., 1981; Young and Laing, 1991; Meizel, 2004).

Sperm in ejaculates from men who are heavy marijuana smokers show abnormally high incidence of HA, which persist after wash and swim-up (Burkman et al., 2003b). These sperm may burn out quickly, which reduces their fertility (DeLamirande and Gagnon, 1993). For women marijuana smokers, exogenous cannabinoids might affect sperm functions in the vagina, cervix, uterus, and oviduct. Additional research is required to examine these possibilities.

Acrosomal Status

Acquisition of acrosomal competence during capacitation represents an intermediate state preparing sperm to respond to physiological stimuli at the egg's surface (Lee, Kopf, et al., 1987; Kligman et al., 1991; Kim, Cha, et al., 2001; Kim, Foster, et al., 2001). This process appears to be correlated with alteration of acrosomal caps of human, bovine, and rodent sperm *in vitro*, and *in vivo* within the isthmus, ampulla, and cumulus (Hunter et al., 1991; Kligman et al., 1991; Kim, Cha, et al., 2001; Kim, Foster, et al., 2001; Lee, Trucco, et al., 1987; Lee, Kopf, et al., 1987; Nolan et al., 1992; Yanagimachi and Philips, 1984). Here, membrane destabilization and localized transient fusion events expose acrosomal matrix proteins required for sperm–zona binding and acrosomal exocytosis (Kim, Cha, et al., 2001; Kim, Foster, et al., 2001; Kim and Gerton, 2003). Using the triple stain procedure to evaluate acrosomal status and sperm viability (Talbot and Chacon, 1981), we observed a time-dependent increase in the percentage of live human sperm with altered acrosomal caps (Schuel et al., 2002a). Punctated white spots were observed within altered acrosomal caps of some sperm. Initially ~20%, then ~50% (at 2 h), and finally ~70% (at 6 h) of viable sperm displayed altered caps. AM356 (1.0 and 2.5 n*M*) or (–)Δ^9-THC (150 and 1500 n*M*) did not affect alteration of acrosomal caps between 0 and 2 h. In contrast, (–)Δ^9-THC and AM356 at these concentrations completely blocked acrosomal cap alterations in another subpopulation of sperm that could respond between 2 and 6 h. Both (–)Δ^9-THC and AM356 potently inhibited acrosomal modifications in a concentration-dependent manner during the later period of capacitation: IC_{50} 5.9 ± 0.6 p*M* for AM356 and 3.5 ± 1.5 n*M* for (–)Δ^9-THC (Figure 22.5). These results suggest that signaling via CBRs can regulate acrosomal cap modifications in human sperm during capacitation.

Sperm Fertilizing Potential

Previous studies showed that AEA and cannabinoids reduce the fertilizing capacity of sea urchin sperm (Schuel et al., 1987, 1991, 1994; Chang et al., 1993; Berdyshev,1999). Ethical concerns preclude similar experiments on live human eggs. Nevertheless, fertilizing potential of human sperm can be determined on the basis of tight binding of capacitated sperm to the ZP in the Hemizona assay (Burkman et al., 1988; Burkman, 1995; Franken et al., 1989). This assay is highly predictive of sperm fertility during human *in vitro* fertilization. The Hemizona assay is a reflection of zona-stimulated AR, because tightly bound sperm >80% are acrosome-reacted (Franken et al., 1991; Fulgham, et al., 1992). In this assay, one half of a bisected zona was inseminated with sperm that had been capacitated in culture medium containing AM356, and the matching half was inseminated with sperm incubated in medium containing vehicle to serve as an internal control (Schuel et al., 2002a). Tight binding of sperm was reduced 50% by (1 n*M*) AM356 under these conditions ($p < 0.001$) (Table 22.5).

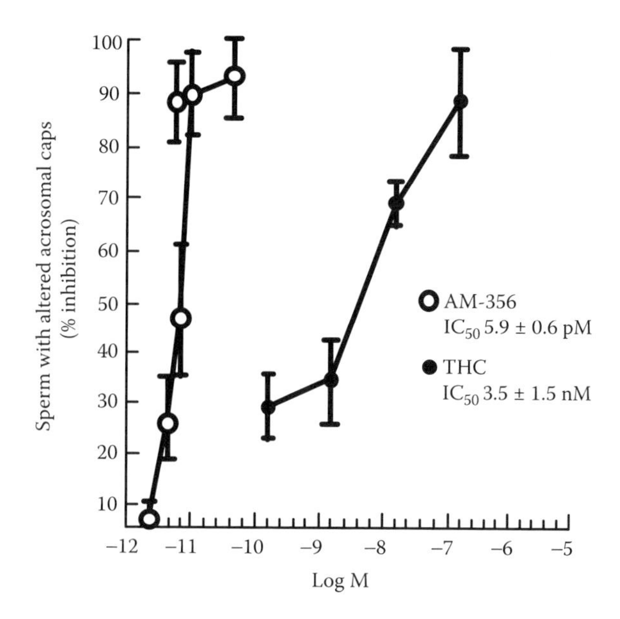

FIGURE 22.5 Dose-response curves for inhibition of acrosomal cap alterations in human sperm between 2 and 6 h of culture in AM356 and (−)Δ^9-THC. N = 4 experiments using sperm from four different donors. (From Schuel H. et al. 2002. *Mol Reprod Dev* 63: 376–387. With permission.)

These results, together with the effects of AM356 and (−)Δ^9-THC on modification of acrosomal caps described above, provide the first evidence that endocannabinoid signaling directly modulates development of acrosomal competence during capacitation in human sperm and thereby regulates sperm fertilizing potential. These CBR-mediated processes are potential targets for exogenous cannabinoids delivered by marijuana smoke.

NAEs in reproductive tract fluids also may produce physiological responses that do not involve sperm CBRs. For example, prostaglandins (derived from the seminal vesicles) regulate functions of epithelia and contractility of smooth muscle in the female reproductive tract (Luke and Coffey, 1994). Further, semen is a potential vehicle for the introduction of infectious organisms into the female. As foreign cells, sperm are vulnerable to immunological attack in the female reproductive tract. Sperm are

TABLE 22.5
AM356 [1.0 n*M*] Inhibits Tight Binding of Sperm to Human Zona Pellucida in the Hemizona Assay

Control Sperm (Bound/Hemizona)	AM356 Treated Sperm (Bound/Hemizona)	Hemizona Index	Percent Inhibition
63.0 ± 10.1	32.5 ± 7.1[a]	50.2 ± 5.5	49.8 ± 5.1

Note: Sperm were continuously exposed to either AM356 or to vehicle from swim-up to the end of coincubation with hemizonae.

N = 12 experiments with sperm from 5 different donors. A total of 1146 tightly bound sperm were counted.

[a] $p < .001$ vs. control sperm (Student's paired "T" test).

Source: From Schuel H. et al. 2002. *Mol Reprod Dev* 63: 376–387. With permission.

also very sensitive to damage by reactive oxygen radicals and lipid peroxidation (De Lamirande and Gagnon, 1999). PEA and OEA are potent anti-inflammatory, antioxidant, and antimicrobial agents (Schmid et al., 1990; Berdyshev et al., 1996; Mazzari et al., 1996; Berdyshev, 2000; De Petrocellis et al., 2000; Conti et al., 2002). Interestingly, these properties are shared with certain cannabinoids (Zimmerman et al., 1991; Hampson et al., 1998; Burstein, 1992; Cabral, 1999; Conti et al., 2002). Collectively, these observations suggest that NAEs may perform multiple roles in reproductive tract fluids by modulating sperm capacitation and fertilizing potential, regulating reproductive tract function, protecting against infection, and maintaining sperm viability.

PREGNANCY

Previous studies showed that marijuana smoke and (–)Δ^9-THC adversely affect fetal growth and development and the number of pregnancies carried to term (Harbison et al., 1972; Maykut, 1985; Asch and Smith, 1986; Smith and Asch, 1987), which implied that endocannabinoid signaling may regulate these processes. We now know that (1) CB_1 receptors are expressed in the mouse uterus (Das et al., 1995; Paria and Dey, 2000), whereas transcripts for both CB_1 and CB_2 are expressed in human uterus (Galiegue et al., 1995; Dennedy et al., 2004; (2) NAPE-PLD is expressed in mouse uterus (Maccarrone et al., 2004) and oviduct (Wang et al., 2004); (3) AEA and 2-AG are present in uterus and oviduct of mice (Das et al., 1995; Schmid et al., 1997; Maccarrone et al., 2004) and rabbits (Burkman et al., 2003a); (4) both CB_1 and CB_2 are expressed in preimplantation mouse embryos (Paria et al., 1995); (5) FAAH is expressed in mouse uterus and preimplantation embryos (Paria et al., 1996, 1999); (6) [^{3}H]-AEA binds in a specific manner to uterine epithelium and stromal cells (Das et al., 1995), as well as to embryos from the one cell stage to mature blastocysts (Paria et al., 1995); and (7) forskolin-stimulated cAMP production is inhibited by AEA and (–)Δ^9-THC in the uterus and the blastocyst, and this process is prevented by pertussis toxin (Das et al., 1995; Paria et al., 1995). These findings suggest multiple roles for endocannabinoid-signaling in pregnancy.

Oviductal transport of mouse embryos is regulated by interactions between AEA and β-adrenergic signaling: (1) CB_1 receptors colocalize with β2-adrenergic receptors in mouse oviductal muscularis; (2) large numbers of embryos are retained in oviducts of CB_1-deficient mice and in wild-type mice treated with SR141716 (Wang et al., 2004); and (3) adverse effects of SR141716 on oviductal embryo transport in wild-type mice are reversed by pretreatment with AM356 or by isoproternol a β-adrenergic agonist. These findings suggest that a basal endocannabinoid tone collaborates with β-adrenergic signaling to regulate oviductal motility required for effective embryo transport. Aberrant functioning of this system might lead to ectopic pregnancy in the fallopian tube. Significantly, women who smoke cigarettes show an increased incidence of ectopic pregnancy (Stillman et al., 1986), which correlates with direct effects of tobacco smoke and nicotine on oviductal transport of embryos in laboratory animals (DiCarlantonio and Talbot, 1999). It remains to be determined whether women who smoke both marijuana and tobacco are at higher risk for ectopic pregnancy. Although both 2-AG and AM356 directly block cholinergic signaling via cloned α_7-nicotinic receptor expressed in *Xenopus* oocytes (Oz et al., 2004), evidence for similar interactions between endocannabinoid- and nicotinic signaling in reproductive physiology has yet to be reported.

AEA, 2-AG, CP55940, WIN55212-2, and (–)Δ^9-THC (but not (+)Δ^9-THC) inhibit cleavage, hatching of the blastocyst from the ZP, and implantation of the blastocyst into the uterine mucosa (Paria et al., 1995, 1998; Yang et al., 1996). Prevention of implantation *in vivo* and detection of (–)Δ^9-THC in the uterus occurred only if the drug was infused along with inhibitors of cytochrome P450 (Paria et al., 1998, 2001). These effects were blocked by the CB_1 antagonists SR141716 and AM251, but not by the CB_2 antagonist SR144528. The CB_2 selective agonist AM663 failed to affect embryo development. Regional differences in the levels of AEA in the uterus determine potential implantation sites; downregulation of AEA is associated with uterine receptivity, while upregulation is correlated with uterine refractoriness to embryo implantation (Schmid et al., 1997). Consistent with these findings, low levels of AEA promote trophoblast outgrowth in culture, whereas higher concentrations of AEA are inhibitory (Wang et al., 1999). Preimplantation embryo development is asynchronous in mutant mice lacking CB_1 or CB_2, which suggests that ligand-receptor signaling via endocannabinoids coordinates embryo development with uterine receptivity for implantation (Paria et al., 2001).

Uterine physiology and pregnancy appear to be regulated by cross-talk between signaling by hormones and endocannabinoids. FAAH levels in the mouse uterus are modulated by progesterone and estrogen (Maccarrone et al., 2000a, 2002a). Consistent with these findings, levels of AEA, 2-AG, and *N*-arachidonoylglycine in the rat uterus change as a function of hormonal status during the estrous cycle (Bradshaw and Walker, 2003). In women, low FAAH levels and high AEA levels are associated with failure to achieve an ongoing pregnancy after IVF and embryo transfer and increased incidence of miscarriage (Maccarrone et al., 2000a; 2002b). Leptin knockout (ob/ob) mice are obese and infertile (Cunningham et al., 1999; Ahima and Flier, 2000). These mice show elevated levels of uterine AEA and 2-AG, the former resulting lower FAAH activity and the latter resulting from elevated synthesis via higher diacylglycerol lipase activity and lower removal via monoacyl glycerol lipase activity (Maccarrone et al., 2004). Injection of exogenous leptin into leptin knockout mice restores these constituents of the endocannabinoid system to normal levels (Maccarrone et al., 2004) along with fertility (Chehab et al., 1996).

In situ hybridization studies show that mRNAs for CB_1 and CB_2 are expressed in the placental cone and smooth muscle of the uterus in rats, rodent embryo and fetus, and term human placenta, as well as in human BeWo choricocarcinoma cells (Buckley et al., 1998; Kenney et al., 1999). CB_1 receptor and FAAH immuno-reactivity are present in human placenta and fetal membranes (Park et al., 2003). Human amniotic fluid contains AEA, OEA, and PEA (Schuel et al., 2002b). The biological functions of these NAEs in amniotic fluids remain to be determined. However, (–)Δ^9-THC crosses the placenta and accumulates in chorionic and amniotic fluids and in all fetal organs (Harbison et al., 1972; Martin et al., 1977; Asch and Smith, 1986; Bailey et al., 1987). Maternal plasma contains significant amounts of 11-nor-9-carboxy THC, a major metabolite of (–)Δ^9-THC, but this metabolite is not detected in amniotic fluid or the fetus (Bailey et al., 1987). These data indicate that 11-nor-9-carboxy THC does not cross the placenta and that the fetus cannot metabolize (–)Δ^9-THC. Hence, the fetus is extremely vulnerable to (–)Δ^9-THC's actions. The placenta in rhesus monkeys chronically exposed to (–)Δ^9-THC during pregnancy shows gross morphological abnormalities and vascular infarctions (Sassenrath et al., 1977). Cannabinoid agonists such as (–)Δ^9-THC and WIN55212-2 inhibit uptake of amino acids by slices of term human placenta and inhibit the serotonin transporter in human BeWo choricocarcinoma cells (Fisher et al., 1987). These observations indicate that the human placenta is a direct target for cannabinoids during pregnancy. Furthermore, the embryo and fetus contain CBRs. Indeed, CB_1 mRNA is expressed in the neural tube, as well as in the fetal central nervous system, retina, autonomic ganglia, enteric ganglia, thyroid, and adrenal glands (Buckley et al., 1998). Brain CB_1 receptors have been detected at week 14 of gestation in humans (Biegon and Kerman, 2001). The developing brain in rat fetus and in newborns contains CB_1 receptors and endogenous ligands (AEA and 2-AG), indicative of possible roles for endocannabinoid signaling in brain development (Berrendero et al., 1999). CB_2 mRNA is expressed exclusively in the liver as early as embryonic stage-E13. The region-specific expression of CB_1 and CB_2 mRNAs suggest functional roles for these receptors during embryogenesis.

Cannabis has been used medicinally since ancient times to assist women during labor (Mechoulam, 1986). AEA signaling has been implicated in the onset and modulation of uterine contractions during labor. Human myometrial biopsy specimens, obtained from women undergoing cesarean delivery (38 to 40 weeks gestation), contain mRNA transcripts for CB_1 and CB_2 (Dennedy et al., 2004). Plasma levels of AEA are highest during the first trimester of pregnancy, decline in the second and third trimesters, and rise at term before the onset of labor (Habayeb et al., 2004). In laboring women, additional increases in plasma AEA levels are related to the duration of contractions and cervical dilation. Both (–)Δ^9-THC and AEA inhibit oxytocin-stimulated contraction in human myometrial strips (Dennedy et al., 2004). The myometrial relaxant effects of these ligands were blocked by the CB_1 antagonist SR141716 but not by the CB_2 antagonist SR144528.

Taken together, these findings indicate that endocannabinoid signaling may directly mediate diverse functions from early embryonic development to birth. All these processes are potential targets for exogenous cannabinoids, which may account for many adverse effects of marijuana and cannabinoids on pregnancy outcome.

POSTNATAL DEVELOPMENT

Nursing infants receive complete nutrition and protection from environmental pathogens from their mother's milk (Newman, 1995; Kelly and Coutts, 2000; Korhonen et al., 2000). Human, bovine, and goat milk contain 2-AG, 2-PG, 2-linoleoyl glycerol (2-LG), oleamide, AEA, PEA, and OEA (Fride et al., 2001; Schuel et al., 2002b). 2-AG is present in milk at much higher concentrations than AEA, whereas 2-PG and 2-LG are present at the highest levels (Fride et al., 2001). OEA is the most abundant NEA in human milk, whereas AEA is present at lower levels than OEA and PEA (Schuel et al., 2002b). OEA reduces feeding in rats (Rodriguez de Fonseca et al., 2001). PEA also produces anorexic effects but is less potent than OEA. The inhibitory effect of OEA on feeding is not blocked by CB_1- and CB_2-specific antagonists (SR141716 and SR144528, respectively), suggesting that OEA does not act via currently known CBRs. PEA and OEA are potent anti-inflammatory, antioxidant, and antimicrobial agents (Schmid et al., 1990; Berdyshev et al., 1996; Mazzari et al., 1996; Conti et al., 2002). These properties are shared with certain cannabinoids (Zimmerman et al., 1991; Hampson et al., 1998; Burstein, 1999; Cabral, 1999; Conti et al., 2002). It is tempting to speculate that OEA and PEA in milk may help protect the infant from infectious pathogens. Oleamide is a potent sleep inducing agent (Cravatt et al., 1995), exhibits some cannabimimetic activity, and binds with low affinity to CB_1 and CB_2 (Lambert and Di Marzo, 1999; Leggett et al., 2003). SR141716 blocks the soporific action of oleamide (Mendelson and Basile, 2001) as well as its inhibitory effects on locomotor activity in rats (Leggett et al., 2003), indicating that oleamide may act via CB_1 receptors. Cannabis has been used since ancient times to stimulate appetite and induce sleep (Abel, 1980; Mechoulam, 1986; Mechoulam and Hanus, 2001; Ahokas, 2002), properties that are shared by (−)Δ^9-THC and AEA (Haney et al., 1999; Williams and Kirkham, 1999; Hao et al., 2000; Rowland et al., 2001). These results indicate that milk-derived endocannabinoids may influence infant physiology and behavior, a notion supported by common parental experience that nursing babies get sleepy as they become satiated.

Endocannabinoid signaling via CB_1 receptors has a critical role in nursing (Fride et al., 2001). Injection of the CB_1 antagonist SR141716 into newly born mouse pups produces devastating effects on suckling and milk ingestion, followed by death within 4 to 8 d. CB_1 receptor blockade must occur within 24 h after birth to produce these effects (Fride, Foox, et al., 2003). Coadministration of (−)Δ^9-THC protects pups from the adverse effects of SR141716 on suckling and survival, suggesting involvement of CB_1 receptors. This conclusion is supported by similar but more moderate deficiencies observed in (untreated) $CB_1^{-/-}$ knockout pups (Fride, Foox, et al., 2003). Significantly, smaller litters and high rates of cannibalism (infanticide) were observed in knockout compared to wild-type mice, which indicate important roles for CB_1 receptors in fertility and normal maternal behavior. CB_1 knockout pups did not ingest milk on the first day of life, began to ingest on the second day, and were less sensitive than wild-type pups to SR141716. Antagonist-induced CB_1 receptor blockage in wild-type pups suppressed ultrasonic vocalizations used to attract maternal attention and impaired the pups' ability to attach to the nipple and suck (Fride, Ezra, et al., 2003). Reduction in body temperature is a characteristic response elicited by CB_1 agonists (Mechoulam et al., 1995). Because SR141716 is an inverse agonist/antagonist for CB_1 receptors (Palmer et al., 2002), it might impair nursing by causing hyperthermia (fever-induced "sickness behavior"). However, SR141716 actually produced persistent hypothermia in the pups (Fride, Ezra, et al., 2003). Thus, failure of CB_1 antagonist-treated pups to nurse is not the result of hyperthermia. Administration of 2-AG to pups did not overcome the adverse effects of SR141716 on suckling and survival (Fride et al., 2001). However, administration of 2-AG plus 2-PG and 2-LG reduced mortality of SR141716-treated pups. Although 2-PG and 2-LG do not bind to CB_1, they can function as entourage compounds and enhance the actions of 2-AG by inhibiting its hydrolysis (Ben-Shabat et al., 1998). Oral administration of AEA and 2-AG to adult mice can produce psychotropic effects, indicating that these endocannabinoids reach the brain (Di Marzo et al., 1998). However, administration of 2-AG, WIN55212-2, and CP55940 to $CB_1^{-/-}$ knockout pups failed to promote survival or weight gain (Fride, Foox, et al., 2003). Normally, brain concentrations of 2-AG peak on the first day of life (Berrendero et al., 1999), and may be required to initiate suckling and milk ingestion (Fride, Foox, et al., 2003). From day two, when 2-AG levels in brain begin to decline (Berrendero et al., 1999), endocannabinoids (2-AG and AEA) obtained from maternal milk (Fride et al., 2001; Schuel et al., 2002b) may maintain suckling (Fride, Foox, et al., 2003). These observations indicate a critical role for endocannabinoid signaling in the initiation of suckling, support the existence of an unknown CBR with partial control over milk ingestion in newborns, imply that $CB_1^{-/-}$ neonates may possess a compensatory mechanism which helps them overcome the lack of CB_1 receptors (possibly a CB_3 receptor?), and also suggest the involvement of CB_1 receptors in modulating fertility and maternal behavior.

Can marijuana smoking by nursing mothers affect their babies? Infants exposed to secondary smoke are obviously vulnerable. Furthermore, studies on humans, monkeys, and rodents showed that (–)Δ^9-THC administered to nursing mothers accumulates in milk and is transferred to their offspring (Chao et al., 1976; Dalterio and Bartke, 1979; Frischknecht et al., 1980; Perez-Reyes and Wall, 1982; Asch and Smith, 1986; Hutchings et al., 1989; Astley and Little, 1990). Dairy animals grazing in regions where Cannabis is part of the natural vegetation produce milk contaminated with cannabinoids, which can be transferred to human infants (Ahmad and Ahmad, 1990). Postnatal growth, motor development, and behavior are adversely affected in babies fed (–)Δ^9-THC-contaminated milk (Dalterio et al., 1984b; Frischknecht et al., 1980; Astley and Little, 1990). Furthermore, (–)Δ^9-THC inhibits suckling-induced milk ejection in lactating rats (Tyre and Murphy, 1988). As a whole, these findings suggest that (–)Δ^9-THC in milk can directly affect newborns and that it also may produce secondary effects in infants by reducing maternal milk availability.

Evidence in the literature suggests that *in utero* exposure to marijuana smoke or (–)Δ^9-THC adversely affects postnatal growth, development, and behavior. These observations imply possible roles for endocannabinoid signaling in these processes. For example, the chronic administration of (–)Δ^9-THC to pregnant rodents resulted in placental transfer to fetuses (Vardaris et al., 1976; Dalterio and Bartke, 1979). Rat pups from drug-treated dams showed deficits in acquisition of passive avoidance response behavior (Vardaris et al., 1976). Other studies showed that oral administration of (–)Δ^9-THC to female mice in late pregnancy and during early lactation alters body weight regulation and pituitary gonadal function and suppresses copulatory activity in male offspring (Dalterio and Bartke, 1979; Dalterio et al., 1984a). Prenatal exposure to marijuana smoke in humans negatively impacts development of executive functions (attention, behavior, visual analysis, hypothesis testing, higher order cognitive processes, and impulse control) in children and adolescents (Goldschmidt et al., 2000; Fried and Smith, 2001; Fried, 2002). Although these findings have been attributed to long-term consequences of in utero exposure to cannabinoids, the contributions of continued exposure to cannabinoids delivered via mother's milk and from secondary smoke also need to be considered, especially for children raised in households where marijuana smoking is prevalent.

CONCLUSIONS

"Marijuana and Sex: Strange Bed Partners" was the title of a major review article written by Powell and Fuller (1983). The evidence summarized in this review shows that an intimate association exists at the molecular level between endocannabinoid signaling and a broad spectrum of reproductive processes in humans and other organisms. Elucidation of these phenomena requires additional experimentation and will enhance our understanding of normal reproductive biology, clarify certain currently unexplained types of human infertility, and eventually may provide the basis for the development of novel drugs for use in reproductive medicine. Does marijuana smoking reduce human fertility significantly, and are adverse effects on reproductive functions reversible? The available literature on the reproductive effects of cannabinoids in humans and laboratory animals raises serious concerns. Limited epidemiological evidence suggests that chronic heavy marijuana smokers may experience fertility problems, but this view remains controversial (Chopra and Jandu, 1976; Smith and Asch, 1984; Mueller et al., 1990; Grotenhermen, 1999). New large-scale epidemiological studies are required to provide more definitive answers.

How could exogenous cannabinoids derived from marijuana smoke impact endocannabinoid signaling in reproductive organs? Normal operation of an endogenous signaling system requires regulated rapid release and removal of endogenous agonists. A classic example of this phenomenon is the hydrolysis of acetylcholine by acetylcholine esterase immediately after its release at synaptic endings. A similar situation applies to signals mediated by AEA, other NAEs, 2-AG, and oleamide, which are hydrolyzed by FAAH (Piomelli et al., 1998; Giuffrida et al., 2001; Hillard, 2000). 2-AG also can be removed by mono acyl glycerol lipase (Schmid, 2000; Sugiura and Waku, 2000). In contrast, drugs such as (−)Δ^9-THC flood endocannabinoid-signal systems because they are slowly metabolized, accumulate in fat stores, and produce persistent effects with potentially damaging consequences, especially in chronic marijuana smokers (Nahas et al., 2002). For example, mammalian testis contains NAPE-PLD (enzyme that releases NAEs from membrane phospholipids) (Okamato et al., 2004), endocannabinoids (AEA and 2-AG) (Sugiura et al., 1996; Sugiura and Waku, 2000; Burkman et al., 2003a), CB_1 and CB_2 receptors (Gerard et al., 1991; Galiegue et al., 1995; Wenger et al., 2001; Brown et al., 2002; Maccarrone et al., 2003), and FAAH (Ueda et al., 2000). These findings suggest possible roles for AEA signaling via CBRs in modulating sperm production and may account, at least in part, for the adverse effects of marijuana smoke and cannabinoids on sperm production in humans and laboratory animals (Schuel et al., 1999; Schuel et al., 2002a).

Prior to the discovery of CBRs and their endogenous ligands, it was generally believed that the peripheral pharmacological effects of cannabinoids resulted from their actions within the central nervous system. Direct drug-induced effects on nonneural cells outside the central nervous system were considered to be unlikely. These opinions reflected the primary focus of investigators on the psychoactive properties of marijuana and its classification as a dangerous illicit substance. However, recent studies show that CBRs are expressed by epithelial, smooth muscle, and immune cells in peripheral organs, as well as by sperm and preimplantation embryos (Chang et al., 1993; Galiegue et al., 1995; Bisogno, Ventriglia, et al., 1997; Deutsch et al., 1997; Rice et al., 1997; Schuel et al., 1999; Schuel, Burkman, Lippes, Crickard, Mahony, et al., 2002; Berdyshev, 2000; De Petrocellis et al., 2000; Paria et al., 2001; Rossato et al., 2005; Wang et al., 2004). Endocannabinoids are likewise produced and degraded locally by nonneural somatic and reproductive cells. With the wisdom of hindsight, it is now clear that marijuana's peripheral effects were red flags announcing the presence of endocannabinoid-signaling systems outside the nervous system.

Why do we have CBRs? The binding of lipid-signal molecules such as AEA and 2-AG to their receptors in all of the body's organ systems regulates how particular cells respond to stimulation. Evidence for endocannabinoid signaling have been obtained in mammals (Elphick and Egertova, 2001); birds (Sonderstrom and Johnson, 2001); amphibians (Soderstrom et al., 2000); fish (Elphick, 2002); the urochordate *Ciona intestinalis* (Elphick et al., 2003); various invertebrates (Salzet and Stefano, 2002), including sea urchins (Chang et al., 1993; Schuel et al., 1994; Bisogno, Ventriglia, et al., 1997), mollusks (Stefano et al., 1996; Sepe et al., 1998), hydra (De Petrocellis et al., 1999), and the leech (Salzet et al., 1997); microorganisms; and higher plants (Schmid et al., 1990; Di Tormaso et al., 1998; Chapman, 2000). Furthermore, cannabinoids affect cyclic nucleotide metabolism in Tetrahymena (Zimmerman et al., 1981), which has recently been shown to express FAAH (Zafiriou et al., 2004). This protozoan contains receptors for a variety of signal molecules that we call neurotransmitters and hormones (Csaba, 1985). Together these findings emphasize that endocannabinoids and their endogenous ligands have an ancient origin in evolutionary history, which may predate the origin of multicellular animals and plants (Chang et al., 1993; Schuel et al., 1994, 1999, 2002). Within this context, endocannabinoid signaling has far greater biological significance than the psychotropic properties of marijuana.

ACKNOWLEDGMENTS

Supported in part by a Multi-Disciplinary Research Pilot Project grant from the State University of New York, Buffalo (LJB and HS), a Moir P. Tanner Foundation grant (LJB), and additional funds provided by the Division of Anatomy and Cell Biology, Department of Gynecology and Obstetrics, and the School of Medicine and Biomedical Sciences, State University of New York, Buffalo.

NOTE ADDED IN PROOF

Recent radioligand binding and immunochemical studies show that boar sperm express receptors that are activated by AEA, including CB_1 as well as type 1 vanilloid receptors which are now called transient receptor potential channel vanilloid receptor subunit 1 [TRPV1] (Maccarrone et al., 2005). A weak immunochemical signal for CB_2 also was detected in porcine sperm. Consistent with these observations, approximately 85% of specific binding of [^{3}H]CP55,940 to boar sperm was blocked by the CB_1 selective antagonist SR141716A, while ~15% was blocked by the CB_2 selective antagonist SR144528. Porcine sperm also have biochemical machinery necessary to produce and degrade their own AEA, including AEA-synthesizing NAPE-PLD, AEA transporter (AMT) to facilitate reuptake, and FAAH to hydrolyze AEA. CB_1, TRPV1, and FAAH are co-localized to the head, middle piece, and tail of boar sperm by immunocytochemistry, which supports their roles in modulating acrosomal status and motility (Schuel et al., 2002a; Rossato et al., 2005). Significantly, levels of CB_1 binding, TRPV1 binding, NAPE-PLD activity, AEA reuptake activity, and FAAH activity in boar sperm decline during capacitation. Consistent with previous studies on human sperm (Schuel et al., 2002a), R-methanandamide (AM356) inhibits capacitation of boar sperm evaluated by chlortetracycline fluorescence staining, inhibits adenylate cyclase activity in sperm, and AR-stimulated by solubilized zona pellucida (Maccarrone et al., 2005). These effects of AM356 on boar sperm are blocked by the CB_1 selective antagonist SR141716A. Consistent with previous observations on sea urchin and human sperm (Schuel et al., 1991a; Rossato et al., 2005), AM-356 also inhibits spontaneous ARs in boar sperm during capacitation (Maccarrone et al., 2005). This effect appears to be mediated via activation of TRPV1 receptors. These findings, together with data showing that the AMT inhibitor AM404 stimulates HA in the absence of added AEA (Burkman et al., 2003b), indicate that mammalian sperm produce and degrade their own AEA to modulate their own functions via cannabinoid and TRPV1 receptors. This situation appears to be analogous to the cholinergic autoregulatory system that modulates motility of sea urchin and mammalian sperm (Meizel, 2004), and opens new vistas to understand the roles of endocannabinoid-signaling in reproductive physiology.

REFERENCES

Abadji V, Lin S, Taha G, Griffin G, Stevenson LA, Pertwee RG, Makriyannis A. 1994. (R)-methanandamide: a chiral novel anandamide possessing high potency and metabolic stability. *J Med Chem* 37: 1889–1893.

Abel, EL. 1980. *Marihuana, the First Twelve Thousand Years*. Plenum Press, NY.

Abel EL. 1984. Effects of Δ^9THC on pregnancy and offspring in rats. *Neurobehav Toxicol Teratol* 6: 29–32.

Ahima RS, Flier JS. 2000. Leptin. *Annu Rev Physiol* 62: 413–437.

Ahokas, H. 2002. Cultivation of *Brassica* species and *Cannabis* by ancient Finnic peoples, traced by linguistic, historical and ethnological data; revision of *Brassica napus* as *B. radice-rapi*. *Acta Bot Fennica* 172: 1–32.

Ahmad GR, Ahmad N. 1990. Passive consumption of marijuana through milk: a low level chronic exposure to delta-9-tetrahydrocannabinol. *J Toxicol Clin Toxicol* 28: 255–260.

Alger BE. 2002. Retrograde signaling in the regulation of synaptic transmission: focus on endocannabinoids. *Prog Neurobiol* 68: 241–286.

Alves A-P, Mulloy B, Diniz JA, Maurao PAS. 1997. Sulfated polysaccharides from the egg jelly layer are species specific inducers of acrosomal reactions in sperms of sea urchins. *J Biol Chem* 272: 6965–6971.

Anderson SH, Killian GJ. 1994. Effect of macromolecules from oviductal conditioned medium on bovine sperm motion and capacitation. *Biol Reprod* 51: 759–799.

Anderson RA, Feathergill KA, Rawlins RG, Mack SR, Zaneveld LJD. 1995. Atrial natriuretic peptide: a chemoattractant of human spermatozoa by a guanylate cyclase-dependent pathway. *Mol Reprod Dev* 40: 371–378.

Aquila S, Sisci D, Gentile M, Middea E, Catalano S, Carpino A, Rago V, Ando S. 2004. Estrogen receptor (ER) alpha and ER beta are both expressed in human ejaculated spermatozoa: evidence of their direct interaction with phosphatidylinositol-3-OH kinase/Akt pathway. *J Clin Endocrinol Metab* 89: 1443–1451.

Asch RH, Smith CG. 1986. The effects of delta-9-THC, the principal psychoactive component of marijuana, during pregnancy in the rhesus monkey. *J Reprod Med* 31: 1071–1081.

Astley SJ, Little RE. 1990. Maternal marijuana use during lactation and infant development at one year. *Neurotoxicol Teratol* 12: 161–168.

Axelrod J, Burch RM, Jelsema CL. 1988. Receptor-mediated activation of phospholipase A_2 via GTP-binding proteins: arachidonic acid and its metabolites as second messengers. *Trends Neurosci* 11: 117–123.

Bailey JL, Storey BT. 1994. Calcium influx into mouse spermatozoa activated by solubilized mouse zona pellucida, monitored with the calcium fluorescent indicator, Fluo-3. Inhibition of the influx by three inhibitors of the zona pellucida induced acrosome reaction: tyrphostin A28, pertussis toxin, and 3-quinclidinyl benzilate. *Mol Reprod Dev* 39: 297–308.

Bailey JR, Cunny HC, Paule MG, Slikker W. 1987. Fetal disposition of delta-9-tetrahydrocannabinol (THC) during late pregnancy in the rhesus monkey. *Toxicol Appl. Pharmacol* 90: 315–321.

Bandivdekar AH, Segal SJ, Koide SS. 1991. Binding of 5-hydroxytryptamine analogs by isolated *Spisula* sperm membrane. *Invert Reprod Dev* 21: 43–46.

Barann M, Molderings G, Bruss M, Bonisch H, Urban BW, Gothert M. 2002. Direct inhibition by cannabinoids of human 5-HT_{3A} receptors: probable involvement of an allosteric modulatory site. *Br J Pharmacol* 137: 589–596.

Bartoszewicz W, Dandekar WP, Glass RH, Gordon M. 1975. Localization of prostaglandin on the plasmalemma of rabbit sperm. *J Exp Zool* 191: 151–160.

Basuray R, De Jong C, Zaneveld LJD. 1990. Evidence for a role for cysteinyl leukotrienes in mouse and human sperm function. *J Androl* 11: 47–51.

Beltramo M, Stella N, Calignano A, Lin SY, Makriyannis A. 1997. Functional role of high-affinity anandamide transport, as revealed by selective inhibition. *Science* 277: 1094–1097.

Beltran C, Zapata O, Darszon A. 1996. Membrane potential regulates sea urchin sperm adenylylcyclase. *Biochemistry* 35: 7591–7598.

Bennet PJ, Moatti JP, Mansat A, Ribbes H, Cayrac JC, Pontonnier F, Chap H, Douste-Blazy L. 1987. Evidence for the activation of phospholipases during acrosome reaction of human sperm elicited by calcium ionophore A23187. *Biochim Biophys Acta* 919: 255–265.

Ben-Shabat S, Fride E, Sheskin T, Tamiri T, Rhee MH, Vogel Z, Bisogno T, De Petrocellis L, Di Marzo V, Mechoulam R. 1998. An entourage effect; inactive endogenous fatty acid glycerol esters enhance 2-arachidonoyl-glycerol cannabinoid activity. *Eur J Pharmacol* 353: 23–31.

Berdyshev EV. 1999. Inhibition of sea urchin fertilization by fatty acid ethanolamides and cannabinoids. *Comp Biochem Physiol* Part C 122: 327–330.

Berdyshev EV. 2000. Cannabinoid receptors and regulation of immune response. *Chem Phys Lipids* 108: 169–190.

Berdyshev EV, Boichot E. Lagente V. 1996. Anandamide — a new look on fatty acid ethanolamides. *J Lipid Mediat Cell Signal* 15: 49–67.

Berrendero F, Sepe S, Ramos JA, Di Marzo V, Fernandez-Ruiz, J. 1999. Analysis of cannabinoid receptor binding and mRNA expression and endogenous cannabinoid contents in the developing rat brain during late gestation and early postnatal period. *Synapse* 33: 181–191.

Biegon A, Kerman IA. 2001. Autoradiographic study of pre- and post-natal distribution of cannabinoid receptors in human brain. *Neuroimage* 14: 1463–1468.

Bielfeld P, Jeyendran RS, Zaneveld LJD. 1991. Human sperm do not undergo the acrosome reaction during storage in the cervix. *Int J Fertil* 36: 302–306.

Bisogno T, Maurelli S, Melch D, De Petrocellis L, Di Marzo V. 1997a. Biosynthesis, uptake, and degradation of anandamide and palmitoylethanolamide in leukocytes. *J Biol Chem* 272: 3315–3323.

Bisogno T, Ventriglia M, Milone A, Mosca M, Cimino G, Di Marzo V. 1997b. Occurrence and metabolism of anandamide and related acyl-ethanolamides in ovaries of the sea urchin *Paracentrotus lividus*. *Biochim. Biophys. Acta* 1345: 338–348.

Bisogno, T., Katayama, K., Melck, D., Ueda, N., De Petrocellis, L., Di Marzo, V., 1998. Biosynthesis and degradation of bioactive fatty acid amides in human breast cancer and rat pheochromocytoma cells — implications for cell proliferation and differentiation. *Eur J Biochem* 254: 634–642.

Bleil JD, Wassarman PM. 1983. Sperm-egg interactions in the mouse: sequence of events and induction of the acrosome reaction by a zona pellucida glycoprotein. *Dev Biol* 95: 317–324.

Bloch E, Thysen B, Morrill GA, Gardner E, Fujimoto G. 1978. Effects of cannabinoids on reproduction and development. *Vitam Horm* 36: 203–258.

Boldt J, Schuel H, Schuel R, Dandekar P, Troll W. 1981. Reaction of sperm with egg-derived hydrogen peroxide helps prevent polyspermy during fertilization in the sea urchin. *Gamete Res* 4: 365–377.

Boldt J, Alliegro MC, Schuel H. 1984. A separate catalase and peroxidase in sea urchin sperm. *Gamete Res* 10: 267–281.

Bouaboula M, Rinaldi M, Carayon P, Carillon C, Delpech B, Shire D, LeFur G, Casellas P. 1993. Cannabinoid-receptor expression in human leukocytes. *Eur J Biochem* 214: 173–180.

Bracey MH, Hanson MA, Masuda KR, Stevens RC, Cravatt BF. 2002. Structural adaptations in a membrane enzyme that terminates endocannabinoid signaling. *Science* 298: 1793–1796.

Bradshaw HB, Walker M. 2003. Levels of endocannabinoids and related lipid mediators in the female rat reproductive tract change as function of hormonal status. 2003 Symposium on the Cannabinoids. Burlington, VT: International Cannabinoid Research Society, p. 35.

Bray C, Son JH, Meizel S. 2002. A nicotinic acetylcholine receptor is involved in the acrosome reaction of human sperm initiated by recombinant human ZP3. *Biol Reprod* 67: 782–788.

Breivogel CS, Childers SR. 2000. Cannabinoid agonist signal transduction in brain: comparison of cannabinoid agonists in receptor binding, G-protein activation, and adenylyl cyclase inhibition. *J Pharmacol Exp Ther* 295: 328–336.

Brown TT, Dobs AS. 2002. Endocrine effects of marijuana. *J Clin Pharmacol* 42: 90S–96S.

Brown SM, Wager-Miller J, Mackie K. 2002. Cloning and molecular characterization of the rat CB_2 cannabinoid receptor. *Biochim Biophys Acta* 1576: 255–264.

Buckley NE, Hansson S, Harta G, Mezne E. 1998. Expression of the CB_1 and CB_2 receptor messenger RNAs during embryonic development of the rat. *Neuroscience* 8: 1131–1149.

Burkman LJ. 1984. Characterization of hyperactivated motility by human spermatozoa during capacitation: comparison of fertile and oligospermic populations. *Arch Androl* 13: 153–165.

Burkman LJ. 1990. Hyperactivated motility of human spermatozoa during in vitro capacitation, in Gagnon E, Ed. *Controls of Sperm Motility*. Boca Raton, FL: CRC Press, pp. 304–329.

Burkman LJ. 1995. New assays for evaluating sperm function, in Centola GM and Ginsburg KA, Eds. *Evaluation and Treatment of the Infertile Male*. Cambridge, U.K.: Cambridge University Press. pp. 108–129.

Burkman LJ, JW Overstreet, DF Katz. 1984. A possible role for potassium and pyruvate in the modulation of sperm motility in the rabbit oviductal isthmus. *J Reprod* 71: 367–372.

Burkman LJ, Coddington C, Franken C, Kruger T, Rosenwaks Z, Hodgen G. 1988. The hemizona assay (HZA): development of a diagnostic test for the binding of human spermatozoa to human hemizona pellucida to predict fertilization potential. *Fertil Steril* 49: 688–697.

Burkman LJ, Coddington CC, Franken DR, Oehninger SC, Hodgen GD. 1990. The hemizona assay (HZA): assessment of fertilizing potential by means of human sperm binding to the human zona pellucida. in Keel BA and Webster B, Eds. *Laboratory Diagnosis and Treatment of Infertility*. Boca Raton, FL: CRC Press. pp. 213–228.

Burkman LJ, Sugiura T, Makriyannis A, Schuel H. 2003a. Anandamide and 2-arachidonoylglyerol are present in rabbit reproductive organs, and can modulate motility of human sperm. 2003 Symposium on the Cannabinoids. Burlington, VT: International Cannabinoid Research Society, p. 36.

Burkman LJ, Bodziak ML, Schuel H, Palaszewski D, Gurunatha R. 2003b. Marijuana (MJ) impacts sperm function both *in vivo* and *in vitro*: semen analyses from men smoking marijuana. *Fertil Steril* 80(3S): S231.

Burstein S. 1992. Eicosanoids as mediators of cannabinoid action, in Murphy L and Bartke A, Eds. *Marijuana/Cannabinoids: Neurobiology and Neurophysiology.* Boca Raton, FL: CRC Press, pp. 73–91.

Burstein SH. 1999. The cannabinoid acids: nonpsychoactive derivatives with therapeutic potential. *Pharmacol Ther* 82: 87–96.

Burstein S, Hunter SA. 1981. Prostaglandins and cannabis-VII. Elevation of phospholipase A_2 activity in whole cells and subcellular preparations. *J Clin Pharmacol* 21: 240S–248S.

Burstein S, Hunter SA, Shoupe TS, Taylor P. 1978. Cannabinoid inhibition of testosterone synthesis by mouse Leydig cells. *Res Commn Chem Pathol Pharmacol* 19: 557–560.

Burstein S, Budrow J, Debatis M, Hunter SA, Subramanian A. 1994. Phospholipase participation in cannabinoid-induced release of free arachidonic acid. *Biochem Pharmacol* 48: 1253–1264.

Burstein S, Young JK, Wright GE. 1995. Relationships between eicosanoids and cannabinoids. Are eicosanoids cannabimimetic agents? *Biochem Pharmacol* 50: 1735–1742.

Cabral G. 1999. Marihuana and the immune system, in Nahas G, Sutin KM, Agurell S, Eds. *Marihuana and Medicine,* Totowa, NJ: Humana Press, pp. 317–325.

Calignano A, La Rana G, Giuffrida A, Piomelli D. 1998. Control of pain initiation by endogenous cannabinoids. *Nature* 394: 277–281.

Cardasis C, Schuel H. 1976. The sea urchin egg as a model system to study the effects of narcotics on secretion, in Ford DH, Clouet D, Eds. *Tissue Responses to Addictive Drugs,* New York: Spectrum Press, pp. 631–640.

Carrell DT, Peterson CM, Urry RL. 1995.The binding of recombinant human relaxin to human spermatozoa. *Endocrine Res* 21: 697–707.

Cariello L, Zanetti L, Spagnuolo A, Nelson L. 1986. Effects of opioids and antagonists on the rate of sea urchin sperm progressive motility. *Biol Bull* 171: 208–216.

Chang MC, Schuel H. 1991. Reduction of the fertilizing capacity of sea urchin sperm by cannabinoids derived from marihuana. II. Ultrastructural changes associated with inhibition of the acrosome reaction. *Mol Reprod Dev* 29: 60–71.

Chang MC, Berkery D, Laychock SG, Schuel H. 1991a. Reduction of the fertilizing capacity of sea urchin sperm by cannabinoids derived from marihuana. III. Activation of phospholipase A_2 in sperm homogenate by Δ^9-tetrahydrocannabinol. *Biochem Pharmacol* 42: 899–904.

Chang MC, Laychock SG, Berkery D, Schuel R, Zimmerman AM, Zimmerman S. 1991b. Effects of (–) and (+) enantiomers of delta-9-tetrahydrocannabinol on sea urchin sperm function. *J Cell Biol* 115: 320A.

Chang MC, Berkery D, Schuel R, Laychock SG, Zimmerman AM, Zimmerman S, Schuel H. 1993. Evidence for a cannabinoid receptor in sea urchin sperm and its role in blockade of the acrosome reaction. *Mol Reprod Dev* 36: 507–516.

Chao FC, Green DE, Forrest IS, Kaplan JN, Winship-Ball A, Braude M. 1976. The passage of ^{14}C-Δ^9-tetrahydrocannabinol into milk of lactating monkeys. *Res Commn Chem Pathol Pharmacol* 15: 303–317.

Chapman KD. 2000. Emerging physiological roles for *N*-acylphosphatidylethanolamine metabolism in plants: signal transduction and membrane protection. *Chem Phys Lipids* 108: 221–230.

Chehab FF, Lim ME, Lu R. 1996. Correction of the sterility defect in homozygous obese female mice by treatment with the human recombinant leptin. *Nat Genet* 12: 318–320.

Chopra GS, Jandu BS. 1976. Psychoclinical effects of long-term marijuana use in 275 Indian chronic users. A comparative assessment of effects in Indian and USA users. *Ann NY Acad Sci* 282: 95–108.

Clark JM. 1936. An experimental study of polyspermy. *Biol Bull* 70: 361–384.

Coburn M, Schuel H, Troll W. 1981. A hydrogen peroxide block to polyspermy in the sea urchin *Arbacia punctulata*. *Dev Biol* 84: 235–238.

Conti S, Costa B, Colleoni M, Parolaro D, Giagnoni G. 2002. Antiinflammatory action of endocannabinoid palmitoylethanolamide and the synthetic cannabinoid nabilone in a model of acute inflammation in the rat. *Br J Pharmacol* 135: 181–187.

Cravatt BF, Lichtman AH. 2002. The enzymatic inactivation of the fatty acid amide class of signaling lipids. *Chem Phys Lipids* 121: 135–148.

Cravatt BF, Prospero-Garci, O, Siuz G., Gillula NB, Henriksen SJ, Boger DL, Lerner RA. 1995. Chemical characterization of a family of brain lipids that induce sleep. *Science* 268: 1506–1509.

Csaba G. 1985. The unicellular *Tetrahymena* as a model cell for receptor research. *Int Rev Cytol* 95: 327–377.

Cunningham MJ, Clifton DK, Steiner RA. 1999. Leptin's actions on the reproductive axis: perspectives and mechanisms. *Biol Reprod* 60: 216–222.

Dalterio S, Bartke A. 1979. Perinatal exposure to cannabinoids alter male reproductive function in mice. *Science* 205: 1420–1422.

Dalterio S, Bartke A, Mayfield D. 1981. Δ^9-tetrahydrocannabinol increases plasma testosterone concentrations in mice. *Science* 213: 581–583.

Dalterio S, Badr F, Bartke A, Mayfield D. 1982. Cannabinoids in male mice: effects on fertility and spermatogenesis. *Science* 216: 315–316.

Dalterio S, Bartke A, Mayfield D. 1983. Cannabinoids stimulate and inhibit testosterone production *in vitro* and *in vivo*. *Life Sci* 32: 605–612.

Dalterio S, Steger R, Mayfield D, Bartke A. 1984a. Early cannabinoid exposure influences neuroendocrine and reproductive functions in male mice: I. Prenatal exposure. *Pharmacol Biochem Behav* 20: 107–113.

Dalterio S, Steger R, Mayfield D, Bartke A. 1984b. Early cannabinoid exposure influences neuroendocrine and reproductive functions in male mice: II. Postnatal exposure. *Pharmacol Biochem Behav* 20: 115–123.

Darszon A, Labarca P, Nishigaki T, Espinosa F. 1999. Ion channels in sperm physiology. *Physiol Rev* 79: 481–510.

Darszon A, Beltran C, Felix R, Nishigaki T, Trevino CL. 2001. Ion transport in sperm signaling. *Dev Biol* 240: 1–14.

Das SK, Paria BC, Chakraborty I, Dey SK. 1995. Cannabinoid ligand-receptor signaling in the mouse uterus. *Proc Natl Acad Sci USA* 92: 4332–4336.

De Celis R, Pedron-Nuevo N, Feria-Velasco A. 1996. Toxicology of male reproduction in animals and humans. *Arch Androl* 37: 201–218.

De Lamirande E, Gagnon C. 1993. Human sperm hyperactivation in whole semen and its association with low superoxide scavenging capacity in seminal plasma. *Fertil Steril* 59: 1291–1295.

De Lamirande E, Gagnon C. 1999. The dark and bright sides of reactive oxygen species on sperm function. in Gagnon C, Ed. *The Male Gamete: From Basic Science to Clinical Applications*. Vienna, IL: Cache River Press, pp. 455–467.

De Petrocellis L, Melck D, Bisogno T, Milone A, Di Marzo V. 1999. Finding of the endocannabinoid signaling system in Hydra, a very primitive organism: possible role in the feeding response. *Neuroscience* 92: 377–387.

De Petrocellis L, Melck D, Bisogno T, Di Marzo V. 2000. Endocannabinoids and fatty acid amides in cancer, inflammation and related disorders. *Chem Phys Lipids* 108: 191–209.

Dennedy MC, Friel AM, Houlihan DD, Broderick VM, Smith T, Morrison JJ. 2004. Cannabinoids and the human uterus during pregnancy. *Am J Obstet Gynecol* 190: 2–9.

Deutsch D, Chin SA. 1993. Enzymatic synthesis and degradation of anandamide, a cannabinoid receptor agonist. *Biochem Pharmacol* 46: 791–796.

Deutsch DG, Goligorsky MS, Schmid PC, Krebsbach RJ, Das SK, Dey SK, Arreaza G, Thorup C, Stefano G, Moore LC. 1997. Production and physiological actions of anandamide in the vasculature of the rat kidney. *J Clin Invest* 100: 1538–1546.

Devane WA, Dysarz FA, Johnson MR, Melvin LS, Howlett AC. 1988. Determination and characterization of a cannabinoid receptor in rat brain. *Mol Pharmacol* 34: 605–613.

Devane WA, Hanus L, Breuer A, Partwee, RG, Stevenson LA, Griffin G, Gibson D, Mandelbaum A, Etinger A, Mechoulam R. 1992. Isolation and structure of a brain constituent that binds to the cannabinoid receptor. *Science* 258: 1946–1949.

Di Carlantonio G, Talbot P. 1999. Inhalation of mainstream and sidestream cigarette smoke retards embryo transport and slows muscle contraction in oviducts of hamsters (*Mesocricetus auratus*). *Biol Reprod* 61: 651–656.

Di Marzo V, De Petrocellis L. 1997. The endogenous cannabinoid signaling system: chemistry, biochemistry and physiology. www.Netsci-journal.com/v197007/index.htm.

Di Marzo V, Fontana A, Cadas H, Schinelli S, Cimino G, Schwartz JC, Piomelli D. 1994. Formation and inactivation of endogenous cannabinoid anandamide in central neurons. *Nature* 372: 686–691.

Di Marzo V, Sepe N, De Petrocellis L, Berger A, Crozier G, Fride E, Mechoulam R. 1998. Trick or treat from food endocannabinoids? *Nature* 396: 636–637.

Di Marzo V, Melck D, De Petrocellis L, Bisogno T. 2000. Cannabimimetic fatty acid derivatives in cancer and inflammation. *Prostag Other Lipid Mediat* 61: 43–61.

Di Marzo V, De Petrocellis L, Fezza F, Ligresti A, Bisogno T. 2002. Anandamide receptors. *Prostag Leukot Essent Fatty Acids* 66: 377–391.

Di Tormaso E, Beltramo M, Piomelli D. 1996. Brain cannabinoids in chocolate. *Nature* 382: 677–678.

Elphick, MR. 2002. Evolution of cannabinoid receptors in vertebrates: identification of a CB_2 gene in the puffer fish *Fugu rubripes*. *Biol Bull* 202: 104–107.

Elphick MR, Egertová M. 2001. The neurobiology and evolution of cannabinoid signaling. *Philos Trans R Soc London* B 356: 381–408.

Elphick MR, Egertová M. 2005. The phylogenetic distribution and evolutionary origins of endocannabinoid signalling, in Pertwee R. (ed.), *Cannabinoids — Handbook of Experimental Pharmacology*. Heidelberg, Germany: Springer, 168: 276–290.

Elphick MR, Satou Y, Satoh N. 2003. The invertebrate ancestry of endocannabinoid signaling: an orthologue of vertebrate cannabinoid receptors in the urochordate *Ciona intestinalis*. *Gene* 302: 95–101.

Facci L, Toso RD, Romanello S, Buriani A, Skaper SD, Leon A. 1995. Mast cells express a peripheral cannabinoid receptor with differential sensitivity to anandamide and palmitoylethanolamide. *Proc. Natl Acad Sci USA* 92: 3376–3380.

Felder CC, Glass M. 1998. Cannabinoid receptors and their endogenous agonists. *Annu Rev Pharmacol Toxicol* 38: 179–200.

Field E, Tyre L. 1986. Blockade of first ovulation in pubertal rats by delta-9-tetrahydrocannabinol: requirement for advanced treatment due to early initiation of the critical period. *Biol Reprod* 34: 512–516.

Fisher SE, Atkinson M, Chang B. 1987. Effect of delta-9-tetrahydrocannabinol on the *in vitro* uptake of alpha-amino isobutyric acid by term human placental slices. *Pediatr Res* 21: 104–107.

Franken DR, Burkman LJ, Oehninger SC, Coddington CC, Veeck LL, Kruger TF, Rosenwaks Z, Hodgen GD. 1989. Hemizona assay using salt-stored human oocytes: evaluation of zona pellucida capacity for binding human spermatozoa. *Gamete Res* 22: 15–26.

Franken DR, Oosthuizen WT, Cooper S, Kruger TF, Burkman LJ, Coodington CC, Hodgen CD. 1991. Electron microscopic evidence on the acrosomal status of bound sperm and their penetration into human hemizonae pellucida after storage in a buffered salt solution. *Andrologia* 23: 205–208.

Fride E. 2004a. The endocannabinoid-CB receptor system: importance for development and in pediatric disease. *Neuroendocrinol Lett* 25: 24–30.

Fride E. 2004b. The endocannabinoid-CB_1 receptor system in pre- and postnatal life. *Eur J Pharmacol* doi:10.1016/j.ejphar2004.2004.07.033.

Fride E, Shohami E. 2002. The endocannabinoid system: function in survival of the embryo, the newborn and the neuron. *Neuroreport* 13: 1833–1841.

Fride E, Ginsburg Y, Breuer A, Bisogno T, Di Marzo, Mechoulam R. 2001. Critical role of the endogenous cannabinoid system in mouse pup suckling and growth. *Eur J Pharmacol* 419: 207–214.

Fride E, Foox A, Rosenberg E, Faigenboim M, Cohen V, Barda L, Blau H, Mechoulam R. 2003. Milk intake and survival in newborn cannabinoid CB_1 receptor knockout mice: evidence for a "CB3" receptor. *Eur J Pharmacol* 461: 27–34.

Fride E, Ezra D, Suris R, Weisblum R, Blau H, Feigin C. 2003. Role of CB_1 receptors in newborn feeding and survival: maintenance of ultrasonic distress calls and body temperature. 2003 Symposium on the Cannabinoids. Burlington, VT: International Cannabinoid Research Society, p. 37.

Fried PA. 2002. Adolescents prenatally exposed to marijuana: examination of facets of complex behaviors and comparisons with the influence of in utero cigarettes. *J Clin Pharmacol* 42: 97S–102S.

Fried PA, Smith AM. 2001. A literature review of the consequences of prenatal marihuana exposure. An emerging theme of a deficiency in aspects of executive function. *Neurotoxicol Teratol*. 23: 1–11.

Frischknecht HR, Sieber B, Waser PG. 1980. Behavioral effects of hashish in mice. II. Nursing behavior and development of the sucklings. *Psychopharmacology* 70: 155–161.

Fulgham DL, Coddington CC, Johnson D, Kerr J, Alexander N, Hodgen GD. 1992. Human sperm acrosome reaction rate on zona pellucida: a time course study. *ARTA* 3: 25–34.

Galiegue S, Mary S, Marchand J, Dussossoy D, Carriere D, Carayon P, Bouaboula M, Shire D, Le Fur G, Casellas P. 1995. Expression of central and peripheral cannabinoid receptors in human immune tissues and leukocyte populations. *Eur J Biochem* 232: 54–61.

Garbers DL. 1981. The elevation of cyclic AMP concentrations in flagella-less sea urchin sperm heads. *J Biol Chem* 256: 620–624.

Garbers DL. 1989. Molecular basis of fertilization. *Annu Rev Biochem* 58: 719–742.

Garcia-Sato J, Gonzalez-Martinez M, De La Torre L, Darszon A. 1987. Internal pH can regulate Ca^{2+} uptake and the acrosome reaction in sea urchin sperm. *Dev Biol* 120: 112–120.

Gerard CM, Mollereau C, Vassart G, Parmentier M. 1991. Molecular cloning of a human cannabinoid receptor which is also expressed in testis. *Biochem J* 279: 129–134.

Gifford AN, Bruneus M, Gatley SJ, Lan R, Makriyannis A. 1999. Large receptor reserve for cannabinoid actions in the central nervous system. *J Pharmacol Exp Ther* 288: 478–483.

Giuffrida A, De Fonseca FR, Piomelli D. 2000. Quantification of bioactive acylethanolamides in rat plasma by electrospray mass spectrometry. *Anal Biochem* 280: 87–93.

Giuffrida A, Beltramo M, Piomelli D. 2001. Mechanisms of endocannabinoid inactivation: biochemistry and pharmacology. *J Pharmacol Exp Ther* 298: 7–14.

Glaser ST, Abumrad NA, Fatade F, Kaczocha M, Studholme KM, Deutsch DG. 2003. Evidence against the presence of an anandamide transporter. *Proc Natl Acad Sci USA* 100: 4269–4274.

Glass M, Felder CC. 1997. Concurrent stimulation of cannabinoid CB_1 and dopamine D_2 receptors augments cAMP accumulation in striatal neurons: evidence for a G_s linkage to the CB_1 receptor. *J Neurosci* 17: 5327–5333.

Godlewski G, Gothert M, Malinowska B. 2003. Cannabinoid receptor-independent inhibition by cannabinoid agonists of the peripheral 5-HT_3 receptor-mediated von Bezold-Jarisch reflex. *Br J Pharmacol* 138: 767–774.

Goldschmidt L, Day NL, Richardson GA. 2000. Effects of prenatal marijuana exposure on child behavior problems at age 10. *Neurotoxicol Teratol* 22: 325–326.

Gonsiorek W, Lunn C, Fan X, Narula S, Lundell D, Hipkin RW. 2000. Endocannabinoid 2-arachidonyl glycerol is a full agonist through type 2 cannabinoid receptor: antagonism by anandamide. *Mol Pharmacol* 57: 1045–1050.

Grotenhermen F. 1999. The effects of cannabis and THC. *Forsch Komplementarmedizin* 6 Suppl: 7–11.

Guerrero A, Darszon A. 1989. Evidence for the activation of two different Ca^{2+} channels during the egg jelly-induced acrosome reaction of sea urchin sperm. *J Biol Chem* 264: 19593–19599.

Grippo AA, Way AL, Killian GJ. 1995. Effect of bovine ampullary and isthmic fluid on motility, acrosome reaction, and fertility of bull spermatozoa. *J Reprod Fert* 105: 57–64.

Habayeb OMH, Bell SC, Konje JC. 2002. Endogenous cannabinoids: metabolism and their role in reproduction. *Life Sci* 70: 1963–1977.

Habayeb OMH, Taylor AH, Evans MD, Cooke MS, Taylor DJ, Bell SC, Konje JC. 2004. Plasma levels of the endocannabinoid anandamide in women — a potential role in pregnancy maintenance and labor? *J Clin Endocrinol Metab* 89: 5482–5487.

Hagstrom BE, Lonning S. 1973. The sea urchin egg as a testing object in toxicology. Acta Pharmacol Toxicol 32 (Suppl. 1): 1–49.

Hampson AJ, Crimald, M, Axelrod J, Wink D. 1998. Cannabidiol and $(-)\Delta^9$ tetrahydrocannabinol are neuroprotective antioxidants. *Proc Natl Acad Sci USA* 95: 8268–8273.

Haney M, Ward AS, Comer SD, Foltin RW, Fischman MW. 1999. Abstinence symptoms following oral THC administration to humans. *Psychopharmacology* 141: 385–394.

Hansbrough JR, Garbers DL. 1981. Sodium dependent activation of sea urchin spermatozoa by speract and monensin. *J Biol Chem* 256: 2235–2241.

Hansen HS, Mosegaard B, Hansen HH, Petersen G. 2000. *N*-acylethanolamines and precursor phospholipids — relation to cell injury. *Chem Phys Lipids* 108: 135–150

Hanus L, Gopher A, Almog S, Mechoulam R. 1993. Two new unsaturated fatty acid ethanolamides in brain that bind to the cannabinoid receptor. *J Med Chem* 36: 3032–3034.

Hanus L, Abu-Lafi S, Fride E, Breuer A, Vogel Z, Shalev DE, Kustanovich I, Mechoulam R. 2001. 2-Arachidonyl glycerol ether, an endogenous agonist of the cannabinoid receptor CB_1 receptor. *Proc Natl Acad Sci USA* 98: 3662–3665.

Hao S, Avraham Y, Mechoulam R, Berry EM. 2000. Low dose anandamide affects food intake, cognitive function, neurotransmitter and corticosterone levels in diet-restricted mice. *Eur J Pharmacol.* 392: 147–156.

Harbison RD, Mantilla-Plata B, Lubin DJ. 1972. Prenatal toxicity, maternal distribution, and placental transfer of tetrahydrocannabinol. *J Pharmacol Exp Ther* 180: 446–453.

Harclerode J. 1984. Endocrine effects of marijuana in the male: preclinical studies. *NIDA Res Monogr* 44: 46–64.

Harvey EB. 1956. *The American Arbacia and Other Sea Urchins*. Princeton NJ: Princeton University Press.

Hembree WC, Nahas GG, Zeidenberg P, Huang HFS. 1979. Changes in human spermatozoa associated with high dose marihuana smoking, in Nahas GG, Paton WDM, Eds. *Marihuana: Biological Effects*. Oxford: Pergamon Press, pp. 429–439.

Herkenham M, Lynn AB, Little MD, Johnson MR, Melvin, LS, De Costa BR, Rice KC, 1990. Cannabinoid receptor localization in brain. *Proc Natl Acad Sci USA* 87: 1932–1936.

Hertwig O, Hertwig R. 1887. Uber den befruchtungs-und teilungsvorgang des eies unter dem einfluss ausserer agentien. *Jena Zeitsch NF* 13: 120–241 and 477–510.

Hiley CR, Ford WR. 2004. Cannabinoid pharmacology in the cardiovascular system: potential protective mechanisms through lipid signaling. *Biol Rev* 79: 187–205.

Hillard CJ. 2000. Biochemistry and pharmacology of the endocannabinoids arachidonylethanolamide and 2-arachidonylglycerol. *Prostagland Other Lipid Mediat* 61: 3–18.

Hinkley RE, Wright BD. 1986. Effects of the volatile anesthetic halothane on fertilization and early development in the sea urchin *Lytechinus variegatus*: evidence that abnormal development is due to polyspermy. *Teratology* 34: 291–301.

Hoodbhoy T, Talbot P. 1994. Mammalian cortical granules: contents, fate, and function. *Mol Reprod Dev* 39: 439–448.

Howlett AC, Mukhopadhyay S. 2000. Cellular signal transduction by anandamide and 2-arachidonoylglycerol. *Chem Phys Lipids* 108: 53–70.

Hoyer D, Boddeke HWGM. 1993. Partial agonists, full agonists, antagonists: dilemmas of definition. *Trends Pharmacol Sci* 14: 270–275.

Huang HFS, Nahas GG, Hembree WC. 1978. Effects of marihuana inhalation on spermatogenesis of the rat, in Nahas GG, Paton WDM, Eds. *Marihuana: Biological Effects*. Oxford, U.K.: Pergamon Press, pp. 419–427.

Hughes J, Smith TW, Kosterlitz HW, Fothergill LA, Morgan BA, Morris HR. 1975. Identification of two related pentapeptides from the brain with potent opiate agonist activity. *Nature* 258: 577–579.

Hunter RHF. 1990. Physiology of the fallopian tubes, with special reference to gametes, embryos and microenvironments, in Evers JHL, Heineman MJ, Eds. *From Ovulation to Implantation*. Amsterdam: Elsevier Publishers (Biomedical Division), pp. 101–119.

Hunter RHF, Flechon B, Flechon JE. 1991. Distribution, morphology and epithelial interactions of bovine spermatozoa in the oviduct before and after ovulation: a scanning electron microscope study. *Tissue Cell* 23: 641–656.

Hutchings DE, Martin BR, Gamagaris N, Miller N, Fico T. 1989. Plasma concentrations of delta-9-tetrahydrocannabinol in dams and fetuses following acute or multiple prenatal dosing in rats. *Life Sci* 44: 697–701.

Ishac EJN, Jiang L, Lake KD, Varga K, Abood ME Kunos G. 1996. Inhibition of exocytotic noradrenaline release by presynaptic cannabinoid receptors on peripheral sympathetic nerves. *Br J Pharmacol* 118: 2023–2028.

Issidorides MR. 1978. Observations in chronic hashish users: nuclear aberrations in blood and sperm and abnormal acrosomes in spermatozoa, in Nahas GG, Paton WDM, Eds. *Marihuana: Biological Effects*. Oxford: Pergamon Press, pp. 377–388.

Jaffe L. 1976. Fast block to polyspermy in sea urchins is electrically mediated. *Nature* 261: 68–71.

Johnson LL, Katz DF, Overstreet JW. 1981. The movement characteristics of rabbit spermatozoa before and after activation. *Gamete Res* 4: 275–282.

Johnson DE, Heald, SL, Dally RD, Janis RA (1993). Isolation, identification and synthesis of an endogenous arachidonic amide that inhibits calcium channel antagonist 1, 4-dihydropyridine binding. *Prostag Leukot Essent Fatty Acids* 48: 429–437.

Jonsson KO, Vandevoodrde S, Lambert DM, Tiger G, Fowler, CJ. 2001. The effects of homologs and analogues of palmitoylethanolamide upon the inactivation of the endocannabinoid anandamide. *Br J Pharmacol* 133: 1263–1275.

Jordt S-E, Bautista DM, Chuang H, McKemy DD, Zygmunt PM, Hogestatt ED, Meng ID, Julius D. 2004. Mustard oils and cannabinoids excite sensory nerve fibers through the TRP channel ANKTM1. *Nature* 427: 260–265.

Joyce CL, Nuzzo NA, Wilson L, Zaneveld LJD. 1987. Evidence for role of cyclooxygenase (prostaglandin synthetase) and prostaglandins in the sperm acrosome reaction and fertilization. *J Androl* 8: 74–82.

Kaminski NE, Abood ME, Kessler FK, Martin BR, Schatz AR. 1992. Identification of a functionally relevant cannabinoid receptor in mouse spleen cells that is involved in cannabinoid-mediated immune modulation. *Mol Pharmacol* 42: 736–742.

Kathuria S, Gaetani S, Fegley D, Valino F, Duranti A, Tontini A, Mor M, Tarzia G, La Rana G, Calignano A, Giustino A, Tattoli M, Palmery M, Cumo V, Piomelli D. 2002. Modulation of anxiety through blockade of anandamide hydrolysis. *Nat Med* 9: 76–81.

Kelly D, Coutts AG. 2000. Early nutrition and the development of immune function in the neonate. *Proc Nutr. Soc* 59: 177–185.

Kenakin T. 2001. Inverse, protean and ligand-selective agonism: matters of receptor conformation. *FASEB J* 15: 598–611.

Kenney SP, Kekuda R, Prasad PD, Leibach FH, Devoe LD, Ganapath, V. 1999. Cannabinoid receptors and their role in the regulation of the serotonin transporter in human placenta. *Am J Obstet Gynecol* 181: 491–497.

Khanolkar AD, Abadji V, Lin S, Hill WAG, Taha G, Abouzid K, Meng Z, Fan P, Makriyannis A. 1996. Head group analogs of arachidonylethanolamide, the endogenous cannabinoid ligand. *J Med Chem* 39: 4515–4519.

Kligman I, Glassner M, Storey BT, Kopf GS. 1991. Zona-pellucida-mediated acrosomal exocytosis in mouse spermatozoa: characterization of an intermediate stage prior to the completion of the acrosome reaction. *Dev Biol* 145: 344–355.

Kim KS, Gerton GL 2003. Differential release of soluble and matrix components: evidence for intermediate states of secretion during spontaneous acrosomal exocytosis in mouse sperm. *Dev Biol* 264: 141–152.

Kim KS, Cha MC, Gerton GL. 2001a. Mouse sperm protein sp56 is a component of the acrosomal matrix. *Biol Reprod* 64: 36–43.

Kim KS, Foster JA, Gerton GL. 2001b. Differential release of guinea pig sperm acrosomal components during exocytosis. *Biol Reprod* 64: 148–156.

Koga D, Santa T., Fukushima T, Homma H, Iami K, 1997. Liquid chromatographic-atmospheric pressure chemical ionization mass spectrometric determination of anandamide and its analogs in rat brain and peripheral tissues. *J Chromatogr B Biomed Sci Appl* 690: 7–13.

Kolodny RC, Masters WH, Kolodner RM, Toro G. 1974. Depression of plasma testosterone levels after chronic intensive marijuana use. *New Engl J Med* 29: 872–874.

Kopf GS, Ning XP, Visconti PE, Purdon M, Galantino-Homer H, Fornes M. 1999. Signaling mechanisms controlling mammalian sperm fertilization competence and activation, in Gagnon C, Ed. *The Male Gamete: From Basic Science to Clinical Applications*. Vienna, IL: Cache River Press, pp. 105–118.

Kolodny RC, Masters WH, Kolodner RM, Toro G. 1974. Depression of plasma testosterone levels after chronic intensive marihuana use. *New Engl J Med* 290: 872–874.

Korhonen H, Marnila P, Gill HS. 2000. Bovine milk antibodies for health. *Br J Nutr* 84: S135–146.

Kuehl FA, Jacob TA, Ganley OH, Ormond RE, Meisinger MAP. 1957. The identification of N-(2-hydroxyethyl)-palmitamide as a naturally occurring anti-inflammatory agent. *J Am Chem Soc* 79: 5577–5578.

Kunos G, Batkai S. 2001. Novel physiologic functions of endocannabinoids as revealed through the use of mutant mice. *Neurochem Res* 26: 1015–1021.

Kunos G, Batkai S, Offertaler L, Mo F, Liu J, Karcher J, Harvey-White J. 2002. The quest for a vascular endothelial cannabinoid receptor. *Chem Phys Lipids* 121: 45–56.

Lambert DM, Di Marzo V. 1999. The palmitoylethanolamide and oleamide enigmas: are these two fatty acid amides cannabimimetic? *Curr Med Chem* 6: 757–773.

Lambert DM, Di Paolo FG, Sonveaux P, Kanyono M, Govaerts SJ, Hermans E, Bueb JL, Delzenne NM, Tschirhart EJ, 1999. Analogs and homologues of *N*-palmitoylethanolamide, a putative endogenous CB_2 cannabinoid, as potential ligands for the cannabinoid receptors. *Biochim Biophys Acta* 1440: 266–274.

Lee MA, Trucco GS, Bechtol KB, Wummer N, Kopf GS, Blasco L, Storey BT. 1987a. Capacitation and acrosome reactions in human spermatozoa monitored by a chlortetracycline fluorescence assay. *Fertil Steril* 48: 649–658.

Lee MA, Kopf GS, Storey BT. 1987b. Effects of phorbol esters and a diacylglycerol on the mouse sperm acrosome reaction induced by the zona pellucida. *Biol Reprod* 36: 617–627.

Leggett J, Beckett S, Kendall D. 2003. Oleamide binds to, activates and produces behavioral effects via the cannabinoid CB_1 receptor. 2003 Symposium on the Cannabinoids. Burlington, VT: International Cannabinoid Research Society, p. 36.

Lillie FR. 1919. *Problems of Fertilization*. Chicago, IL: University of Chicago Press.

Lindemann CB, Goltz JS. 1988. Calcium regulation of flagellar curvature and swimming pattern in Triton X-100-extracted sperm. *Cell Motil Cytoskel* 10: 420–431.

Longo FJ, Schuel H, Wilson WL 1974. Mechanism of soybean trypsin inhibitor induced polyspermy as determined by an analysis of re-fertilized sea urchin (*Arbacia punctulata*) eggs. *Dev Biol* 41: 193–201.

Luke MC, Coffey DS, 1994. The male sex accessory tissues. Structure, androgen action, and physiology, in Knobil E and Neil JD, Eds. *The Physiology of Reproduction,* 2nd ed. New York: Raven Press, Vol. I, pp. 1435–1487.

Lynch DL, Reggio PH. 2005. Molecular dynamics simulations of the endocannabinoid N-arachidonoylethanolamine (anandamide) in a phospholipid bilayer: probing structure and dynamics. *J Med Chem* 48: 4824–4833.

Maccarrone M, Finazzi-Agro A. 2004. Anandamide hydrolase: a guardian angel of human reproduction. *Trends Pharmacol Sci* 25: 353–357.

Maccarrone M, Valensise H, Bari M, Lazzarin N, Romanini C, Finazzi-Agro A. 2000a. Relation between decreased anandamide hydrolase concentrations in human lymphocytes and miscarriage. *Lancet* 355: 1326–1329.

Maccarrone M, De Felici M, Bari M, Klinger F, Siracusa G, Finazzi-Argo A. 2000b. Down–regulation of anandamide hydrolase in mouse uterus by sex hormones. *Eur J Biochem* 267: 2991–2997.

Maccarrone M, Valensise H, Bari M, Lazzarin N, Romanini C, Finazzi-Agro A. 2001. Progesterone up-regulates anandamide hydrolase in human lymphocytes: role of cytokines and implications for fertility. *J Immunol* 166: 7183–7189.

Maccarrone M, Falciglia K, Di Rienzo M, Finazzi-Argo A. 2002a. Endocannabinoids, hormone-cytokine networks and human fertility. *Prostaglandins Leukot Essential Fatty Acids* 66: 309–317.

Maccarrone M, Bisogno T, Valensise H, Lazzarin N, Fezza F, Manna C, Di Marzo V, Finazzi-Agro A. 2002b. Low fatty acid amide hydrolase and high anandamide levels are associated with failure to achieve an ongoing pregnancy after IVF and embryo transfer. *Mol Hum Reprod* 8: 188–195.

Maccarrone M, Cecconi S, Rossi G, Battista N, Pauselli R, Finazzi-Agro A. 2003. Anandamide activity and degradation are regulated by early postnatal aging and follicle-stimulating hormone in mouse Sertoli cells. *Endocrinology* 144: 20–28.

Maccarrone M, Fride E, Bisogno T, Bari M, Cascio MG, Battista N, Finazzi-Argo A, Suris R, Mechoulam R, Di Marzo V. 2004. Up-regulation of the endocannabinoid system in the uterus of leptin knockout (*ob/ob*) mice and implications for fertility. *Mol Hum Reprod* 11: 21–28.

Maccarrone, M, Barboni, B, Paradisi, A, Bernabo, N, Gasperi, V, Pistilli, MG, Fezza, F, Lucidi, P, Mattioli, M. 2005. Characterization of the endocannabinoid system in boar spermatozoa and implications for sperm capacitation and acrosome reaction. *J Cell Sci* in press.

Mackie K, Hille B. 1992. Cannabinoids inhibit N-type calcium channels in neuroblastoma-glioma cells. *Proc Natl Acad Sci USA* 89: 3825–3829.

Mackie K, Devane WA, Hille B. 1993. Anandamide, an endogenous cannabinoid, inhibits calcium currents as a partial agonist in N18 neuroblastoma cells. *Mol Pharmacol* 44: 498–503.

Makriyannis A, Rapaka RS. 1990. The molecular basis of cannabinoid activity. *Life Sci* 47: 2173–2184.

Mani SK, Mitchell A, O'Malley BW. 2001. Progesterone receptor and dopamine receptors are required in Δ^9-tetrahydrocannabinol modulation of sexual receptivity in female rats. *Proc Natl Acad Sci USA* 98: 1249–1254.

Martin BR, Dewey WL, Harris LS, Beckner JS. 1977. ^{3}H-Δ^9-tetrahydrocannabinol distribution in pregnant dogs and their fetuses. *Res Commn Chem Pathol Pharmacol* 17: 457–470.

Matsuda LA. 1997. Molecular aspects of cannabinoid receptors. *Crit Rev Neurobiol* 11: 143–166.

Matsuda LA, Lolait SJ, Brownstein MJ, Young AC, Bonner TI. 1990. Structure of a cannabinoid receptor and functional expression of the cloned cDNA. *Nature* 346: 561–564.

Matias I, McPartland JM, Di Marzo V. 2005. Occurrence and possible biological role of the endocannabinoid system in the sea squirt *Crona intestinalis*. *J Neurochem* 93: 1141–1156.

Maykut MO. 1985. Health consequences of acute and chronic marihuana use. *Prog Neuro-Pyschopharmacol Biol Psychiatry* 9: 209–238.

Mazzari S, Canella R, Petrelli L, Marcolongo G, Leon A. 1996. N-(2hydroxyethyl) hexadecanamide is orally active in reducing edema formation and inflammatory hyperalgesis by down-modulating mast cell activation. *Eur J Pharmacol* 300: 227–236.

McAllister SD, Glass M. 2002. CB_1 and CB_2 receptor-mediated signaling: a focus on endocannabinoids. *Prostaglandins Leukot Essent Fatty Acids* 66: 161–171.

Mechoulam R. 1986. The pharmacohistory of *Cannabis sativa*, in Mechoulam R, Ed. *Cannabinoids as Therapeutic Agents*. Boca Raton: FL, CRC Press, pp. 1–19.

Mechoulam R. 2002. Discovery of endocannabinoids and some random thoughts on their possible roles in neuroprotection and aggression. *Prostaglandins Leukot Essent Fatty Acids* 66: 93–99.

Mechoulam R, Hanus L. 2000. A historical overview of chemical research on cannabinoids. *Chem Phys Lipids* 108: 1–13.

Mechoulam R, Hanus L. 2001. The cannabinoids: an overview. Therapeutic implications in vomiting and nausea after cancer chemotherapy, in appetite promotion, in multiple sclerosis and in neuroprotection. *Pain Res Manage* 6: 67–73.

Mechoulam R, Devane A, Breuer A, Zahalka J. 1991. A random walk through a cannabis field. *Pharmacol Biochem Behav* 40: 461–464.

Mechoulam R, Ben-Shabat S, Hanus L, Ligumsky, M, Kaminski NE, Schatz AR, Gopher A, Almog S, Martin BR, Compton DR, Pertwee RG, Griffin G, Bayewitch M, Barg J, Vogel Z. 1995. Identification of an endogenous 2-monoglyceride, present in canine gut, that binds to cannabinoid receptors. *Biochem Pharmacol* 50: 83–90.

Mechoulam R, Fride E, Di Marzo V. 1998. Endocannabinoids. *Eur J Pharmacol* 359: 1–18.

Meizel S. 1985. Molecules that initiate or help stimulate the acrosome reaction by their reaction with the mammalian sperm surface. *Am J Anat* 174: 285–302.

Meizel S. 1997. Amino acid neurotransmitter receptor/chloride channels of mammalian sperm and the acrosome reaction. *Biol Reprod* 56: 569–574.

Meizel S. 2004. The sperm, a neuron with a tail: "neuronal" receptors in mammalian sperm. *Biol Rev* 79: 713–732.

Meizel S, Turner KO. 1984. The effects of products and inhibitors of arachidonic acid metabolism on the hamster sperm acrosome reaction. *J Exp Zool* 231: 283–288.

Meizel S, Pillai MC, Perez ED, Thomas P. 1990. Initiation of the human sperm acrosome reaction by components of human follicular fluid and cumulus secretions including steroids, in Bavister BD, Cummins J, Roldan ERS, Eds. *Fertilization in Mammals*. Norwell, MA: Serono Symposia, USA. pp. 205–222.

Mendelson JH, Mello NK. 1999. Marijuana effects on pituitary and gonadal hormones in women, in Nahas G, Sutin KM, Agurell S. Eds. *Marihuana and Medicine*. Totowa, NJ: Humana Press. pp. 385–392.

Mendelson WB, Basile AS. 2001. The hypnotic actions of the fatty acid amide, oleamide. *Neuropsychopharmacology* 25: S36–S39.

Merari A, Barak A, Plaves M. 1973. Effects $\Delta^{1(2)}$tetrahydrocannabinol on copulation in the male rat. *Psychopharmacologia* 28: 243–246.

Miller MB, Bassler BL. 2001. Quorum sensing in bacteria. *Annu Rev Microbiol* 55: 165–99.

Morrill GA, Kostellow AB, Ziegler DH, Fujimoto GI. 1983. Effects of cannabinoids on function of testis and secondary sex organs in the Fischer rat. *Pharmacology* 26: 20–28.

Mortimer D. 1995. Sperm transport in the female genital tract, in Grudzinskas JG, Yovich JL, Eds. *Gametes-the Spermatozoan*. Cambridge: Cambridge University Press. pp. 157–174.

Mueller BA, Daling JR, Weiss NS, Moore DE. 1990. Recreational drug use and the risk of primary infertility. *Epidemiology* 1: 195–200.

Mukhopadhyay S, Shim JY, Assi AA, Norford D, Howlett AC. 2002. CB_1 cannabinoid receptor-G protein association: a possible mechanism for differential signaling. *Chem Phys Lipids* 121: 91–109.

Munro S, Thomas KL, Abu-Shaar M. 1993. Molecular characterization of a peripheral receptor for cannabinoids. *Nature* 365: 61–65.

Murphy LL, Newton SC, Dhali J, Chavez D. 1991. Evidence for a direct anterior pituitary site of delta-9-tetrahydrocannabinol action. *Pharmacol Biochem Behav* 40: 603–607.
Murphy LL, Cher J, Steger RW, Bartke A. 1994. Effects of Δ^9-tetrahydrocannabinol on copulatory behavior and neuroendocrine responses of male rats to female conspecifics. *Pharmacol Biochem Behav* 48: 1011–1017.
Nahas GG, Frick HC, Lattimer JK, Latour C, Harvey D. 2002. Pharmacokinetics of THC in brain and testis, male gametotoxicity and premature apoptosis of spermatozoa. *Hum Psychopharmaol Clin Exp* 17: 103–113.
Nelson L. 1978. Chemistry and neurochemistry of sperm motility control. *Fed Proc* 37: 2543–2547.
Nelson L. 1985. Sperm cell enzymes, in Metz CB, Monroy A, Eds. *Biology of Fertilization*. Orlando, FL: Academic Press, Vol. 2, pp. 215–234.
Nelson L, Cariello L. 1989. Adrenergic stimulation of sea urchin sperm cells. *Gamete Res* 24: 291–302.
Nelson L, Young MJ, Gardner ME. 1980. Sperm motility and calcium transport: a neurochemically controlled process. *Life Sci* 26: 1739–1749.
Newman J. 1995. How breast milk protects newborns. *Sci Am* 273: 76–79.
Nolan JP, Graham JK, Hammerstedt RH. 1992. Artificial induction of exocytosis in bull sperm. *Arch Biochem Biophys* 292: 311–322.
Nicotra A, Schatten G. 1996. Propranolol induces polyspermy during sea urchin fertilization. *Mol Reprod Dev* 43: 387–391.
Nithipatikom K, Endsley MP, Isbell MA, Falck JR, Iwamoto Y, Hillard CJ, Campbell WB. 2004. 2-arachidonoylglycerol: a novel inhibitor of androgen-independent prostate cancer cell invasion. *Cancer Res* 64: 8826–8830.
Okamoto Y, Morishita J, Tsuboi K, Tonai T, Ueda N. 2004. Molecular characterization of a phospholipase D generating anandamide and its congeners. *J Biol Chem* 279: 5298–5305.
Overstreet JW, Katz D. 1990. Interaction between the female reproductive tract and spermatozoa, in Gagnon C, Ed. *Controls of Human Sperm Motility: Biological and Clinical Aspects*. Boca Raton, FL: CRC Press, pp. 64–75.
Oz M, Tchugunova YB, Dunn SMJ. 2000. Endogenous cannabinoid anandamide directly inhibits voltage-dependent Ca2+ fluxes in rabbit T-tubule membranes. *Eur J Pharmacol* 404: 13–20.
Oz M, Zhang L, Ravindran A, Morales M, Lupica CR. 2004. Differential effects of endogenous and synthetic cannabinoids on α_7-nictonic acetylcholine receptor-mediated responses in Xenopus oocytes. *J Pharmacol Exp Ther* 310: 1152–1160.
Pagotto U, Marsican G., Fezza F. Theodoropoulou M, Grubler Y, Stalla J, Arzberger T, Milone A, Losa M, Di Marzo V, Lutz B, Stalla GK. 2001. Normal human pituitary gland pituitary adenomas express cannabinoid receptor type 1 and synthesize endogenous cannabinoids: first evidence for a direct role of cannabinoids on hormone modulation at the human pituitary level. *J Clin Endocrinol Metabol* 86: 2687–2696.
Palmer SL, Thakur GA, Makriyannis A. 2002. Cannabinergic ligands. *Chem Phys Lipids* 121: 3–19.
Paria BC, Dey SH. 2000. Ligand-receptor signaling with endocannabinoids in preimplantation embryo development and implantation. *Chem Phys Lipids* 108: 211–220.
Paria BC, Das KS, Dey KS. 1995. The preimplantation mouse embryo is a target for cannabinoid ligand-receptor signaling. *Proc Natl Acad Sci USA* 92: 9460–9464.
Paria BC, Deutsch DG, Dey SK. 1996. The uterus is a potential site for anandamide synthesis and hydrolysis: differential profiles of anandamide synthase and hydrolase activities in the mouse uterus during the periimplantation period. *Mol Reprod Dev* 45: 183–192.
Paria BC, Ma W, Andrenyak DM, Schmid PC, Schmid HHO, Moody DE, Deng H, Makriyannis A, Dey SK. 1998. Effects of cannabinoids on preimplantation mouse embryo development and implantation are mediated by brain-type cannabinoid receptors. *Biol Reprod* 58: 1490–1495.
Paria BC, Zhao X, Wang J, Das SK, Dey SK. 1999. Fatty-acid amide hydrolase is expressed in the mouse uterus and embryo during the periimplantation period. *Biol Reprod* 60: 1151–1157.
Paria BC, Song H, Wang X, Schmid PC, Krebsbach RJ, Schmid HHO, Bonner TL, Zimmer A, Dey SK. 2001. Dysregulated cannabinoid signaling disrupts uterine receptivity for embryo implantation. *J Biol Chem* 276: 20523–20528.
Paria BC, Wang H, Dey SK. 2002. Endocannabinoid signaling in synchronizing embryo development and uterine receptivity for implantation. *Chem Phys Lipids* 121: 201–210.
Park B, McPartland JM, Glass M. 2004. Cannabis, cannabinoids and reproduction. *Prostaglandins Leukot Essent Fatty Acids* 70: 189–197.

Perez-Reyes M, Wall ME. 1982. Presence of Δ^9-tetrahydrocannabinol in human milk. *New Engl J Med* 307: 819–820.

Pert CB, Snyder SH. 1973. Opiate receptor: demonstration in nervous tissue. *Science* 179: 1011–1014.

Pertwee RG. 1999. Cannabinoid receptors and their ligands in brain and other tissues, in Nahas GG, Sutin KM, Harvey DJ, Agurell S, Eds. *Marihuana and Medicine*. Totowa, NJ: Humana Press. pp. 177–185.

Pertwee RG, Ross A. 2002. Cannabinoid receptors and their ligands. *Prostaglandins Leukot Essent Fatty Acids* 66: 101–121.

Pertwee R, Griffin G, Hanus L, Mechoulam R. 1994. Effects of two endogenous fatty acid ethanolamides on mouse vasa deferentia. *Eur J Pharmacol* 259: 115–120.

Piomelli D. 2003. The molecular logic of endocannabinoid signaling. *Nature Med* 4: 873–884.

Piomelli D, Beltramo M, Giuffrida A. 1998. Endogenous cannabinoid signaling. *Neurobiol Dis* 5: 462–473.

Piomelli D, Beltramo M, Glasnapp S, Lin SY, Goutopoulos A, Xie XQ, Makriyannis A. 1999. Structural determinants for recognition and translocation by the anandamide transporter. *Proc Natl Acad Sci USA* 96: 5802–5807.

Porter AC, Sauer JM, Knierman MD, Becker GW, Berna MJ, Bao J, Nomikos GG, Carter P, Bymaster FP, Leese AB, Felder CC. 2002. Characterization of a novel endocannabinoid, virodhamine, with antagonist activity at the CB_1 receptor. *J Pharmacol Exp Ther* 301: 1020–1024.

Powell DJ, Fuller RW. 1983. Marijuana and sex: strange bed partners. *J Psychoactive Drugs* 15: 269–280.

Priller J, Briley EM, Mansouri J, Devane WA, Mackie K, Felder CC. 1995. Mead ethanolamide, a novel eicosanoid, is an agonist for central (CB_1) and peripheral (CB_2) cannabinoid receptors. *Mol Pharmacol* 48: 288–292.

Primakoff P, Myles DG. 2002. Penetration, adhesion and fusion in mammalian sperm-egg interaction. *Science* 296: 2183–2185.

Randall MD, Alexander SPH, Bennett T, Boyd EA, Fry JR, Gardiner SM, Kemp PA, McCulloch AI, Kendall DA. 1996. An endogenous cannabinoid as an endothelium-derived vasorelaxant. *Biochem Biophys Res Commn* 229: 114–120.

Rice W, Shannon JM, Burton R, Fieldeldey D. 1997. Expression of brain type cannabinoid receptor (CB_1) in alveolar type-II cells in the lung — regulation by hydrocortisone. *Eur J Pharmacol* 327: 227–232.

Rinaldi-Carmona M, Barth F, Heaulume M, Alonso R, Shire D, Congy C, Soubrie P, Breliere JC, Le Fur G. 1995. Biochemical and pharmacological characterization of SR141716A, the first potent and selective brain cannabinoid receptor antagonist. *Life Sci* 56: 1941–1947.

Rinaldi-Carmona M, Barth F, Millan J, Derocq JM, Casellas P, Congy C, Oustric D, Sarran M, Bouaboula M, Calandra B, Portier M, Shire D, Breliere JC, Le Fur GL. 1998. SR144528, the first potent and selective antagonist of the CB_2 cannabinoid receptor. *J Pharmacol Exp Ther* 284: 644–650.

Rodriguez de Fonseca F, Navarro M, Gomez R, Escuredo L., Nava F, Fu J, Murillo-Rodriguez E, Giuffrida A, LoVerme J, Gaetani S, Kathuria S, Gall C, Piomelli D. 2001. An anorexic lipid mediator regulated by feeding. *Nature* 414: 209–212.

Roldan ERS. 1999. Signaling for exocytosis: lipid second messengers, phosphorylation cascades and cross-talks, in Gagnon C. Ed. *The Male Gamete: From Basic Science to Clinical Applications*. Vienna IL: Cache River Press, pp. 127–138.

Rosenkrantz H. 1978. Effects of cannabinoids on fetal development in rodents, in Nahas G and Paton WDM (Eds.) *Marihuana: Biological Effect*. Oxford: Pergamon Press, pp. 479–500.

Rossato M, Popa IF, Ferigo M, Clari G, Foresta C. 2005. Human sperm express cannabinoid receptor CB_1 which activation inhibits motility, acrosome reaction and mitochondrial function. *J Clin Endocrinol Metabol* 90: 984–991.

Roth SH. 1978. Stereospecific presynaptic inhibitory effect of Δ^9-tetrahydrocannabinol on cholinergic transmission in the myenteric plexus of the guinea pig. *Can J Physiol Pharmacol* 56: 968–975.

Rothschild L. 1956. *Fertilization*. London: Methuen.

Rowland NE, Mukherjee M, Robertson K. 2001. Effects of the cannabinoid receptor antagonist SR-141716, alone and in combination with dexfenfluramine or naloxone, on food intake in rats. *Psychopharmacology* 159: 111–116.

Ruiz-Llorente L, Sanchez MG, Carmena MJ, Prieto JC, Sanchez-Chapado M, Izquierdo A, Diaz-Laviada I. 2003. Expression of functionally active cannabinoid receptor CB_1, in the human prostate gland. *The Prostate* 54: 95–102.

Salzet M, Stefano GB. 2002. The endocannabinoid system in invertebrates. *Prostag Leukot Essent Fatty Acids* 66: 353–361.

Salzet M, Breton C, Bisogno T, Di Marzo V. 2000. Comparative biology of the endocannabinoid system. Possible role in immune response. *Eur J Biochem* 267: 4917–4927.
Sanchez MG, Sanchez AM, Ruiz-Lorente L, Diaz-Lavida I. 2003. Enhancement of androgen receptor expression induced by (R)-methanandamide in prostate LNCaP cells. *FEBs Lett* 555: 561–566.
Sarfaraz, S, Afaq F, Adhami VM, Mukhtar H. 2005. Cannabinoid receptor as a novel target for the treatment of prostate cancer. *Cancer Res* 65: 1635–1641.
Sassenrath EN, Chapman LF, Goo GP. 1979. Reproduction in rhesus monkeys chronically exposed to delta-9-tetrahydrocannabinol, in Nahas GG, Paton WDM, Eds. *Marihuana: Biological Effects*. Oxford, U.K.: Pergamon Press, pp. 501–512.
Sastry BVR, Sadavongvivad C. 1979. Cholinergic systems in non-nervous systems. *Pharmacol Rev* 30: 65–132.
Sastry BVR, Janson VE, Chaturvedi AK. 1981. Inhibition of human sperm motility by inhibitors of choline acetyltransferase. *J Pharmacol Exp Ther* 216: 378–384.
Schackman RW, Christen R, Shapiro BM. 1984. Measurement of plasma membrane and mitochondrial potentials in sea urchin sperm. Changes upon activation and induction of the acrosome reaction. *J Biol Chem* 259: 13914–13922.
Schaefer M, Hofmann T, Schultz G, Gudermann T. 1998. A new prostaglandin E receptor mediates calcium influx and acrosome reaction in human spermatozoa. *Proc Natl Acad Sci USA* 95: 3008–3013.
Schatten G, Hulser D. 1983. Timing the early events during sea urchin fertilization. *Dev Biol* 100: 244–248.
Schmid, HHO. 2000. Pathways and mechanisms of N-acylethanolamine biosynthesis: can anandamide be generated selectively? *Chem Phys Lipids* 108: 71–87.
Schmid HHO, Schmid PC, Natarajan V. 1990. *N*-Acylated glycerophospholipids and their derivatives. *Prog Lipid Res* 29: 1–43.
Schmid PC, Paria BC, Krebsbach RJ, Schmid HHO, Dey SK. 1997. Changes in anandamide levels in mouse uterus are associated with uterine receptivity for embryo implantation. *Proc Natl Acad Sci USA* 94: 4188–4192.
Schmid, HO, Schmid PC, Berdyshev EV. 2002. Cell signaling by endocannabinoid and their congeners: questions of selectivity and other challenges. *Chem Phys Lipids* 121: 114–134.
Schuel H. 1978. Secretory functions of egg cortical granules in fertilization and development: a critical review. *Gamete Res* 1: 299–382.
Schuel H. 1984. The prevention of polyspermic fertilization in sea urchins. *Biol Bull* 167: 271–309.
Schuel H. 1985. Function of egg cortical granules, in Metz CB, Monroy A, Eds. *Biology of Fertilization*. New York: Academic Press, Vol. 3, pp. 1–43.
Schuel H, Schuel R. 1981. A rapid sodium dependent block to polyspermy in sea urchin eggs. *Dev Biol* 87: 249–258.
Schuel H, Schuel R. 1986. Sea urchin sperm peroxidase is competitively inhibited by benzohydroxamic acid and phenylhydrazine. *Biochem Cell Biol* 64: 1333–1338.
Schuel, H, Schuel, R. 1987. Benzohydroxamic acid induces polyspermic fertilization in the sea urchin *Arbacia punctulata*. *Cell Biol Int Rep* 11: 189–196.
Schuel, H, Schuel, R. 1988. Peroxide block to polyspermy in sea urchins: H_2O_2 inhibits sperm acrosome reaction. *J Cell Biol* 107: 166a.
Schuel H, Traeger E, Schuel R, Boldt J. 1984. Anti-inflammatory drugs promote polyspermic fertilization in sea urchins. *Gamete Res* 10: 9–19.
Schuel H, Moss R, Schuel R. 1985. Induction of polyspermic fertilization in sea urchins by the leukotriene antagonist FPL-55712 and the 5-lipoxygenase inhibitor BW755C. *Gamete Res* 11: 41–50, 1985.
Schuel H, Schuel R, Zimmerman AM, Zimmerman S. 1987. Cannabinoids reduce fertility of sea urchin sperm. *Biochem Cell Biol* 65: 130–136.
Schuel H, Berkery D, Schuel R, Chang MC, Zimmerman AM, Zimmerman S. 1991a. Reduction of the fertilizing capacity of sea urchin sperm by cannabinoids derived from marihuana. I. Inhibition of the acrosome reaction induced by egg jelly. *Mol Reprod Dev* 29: 51–59.
Schuel H, Chang MC, Berkery D, Schuel R, Zimmerman AM, Zimmerman S. 1991b. Cannabinoids inhibit fertilization in sea urchins by reducing the fertilizing capacity of sperm. *Pharmacol Biochem Behav* 40: 609–615.
Schuel H, Goldstein E, Mechoulam R, Zimmerman AM, & Zimmerman S. 1994. Anandamide (arachidonylethanolamide), a brain cannabinoid receptor agonist, reduces fertilizing capacity in sea urchins by inhibiting the acrosome reaction. *Proc Natl Acad Sci USA* 91: 7678–7682.
Schuel H, Chang MC, Burkman LJ, Picone RP, Makriyannis A, Zimmerman AM, Zimmerman S. 1999. Cannabinoid receptors in sperm, in Nahas G, Sutin KM, Agurell S. Eds. *Marihuana and Medicine*. Totowa, NJ: Humana Press pp. 335–345.

Schuel H, Burkman, LJ, Lippes J, Crickard K, Mahony MC, Giuffrida A, Picone R, Makriyannis A. 2002a. Evidence that anandamide-signaling regulates human sperm functions required for fertilization. *Mol Reprod Dev* 63: 376–387.

Schuel H, Burkman LJ, Lippes J, Crickard K, Forester E, Piomelli D, Giuffrida A. 2002b. *N*-acylethanolamines in human reproductive fluids. *Chem Phys Lipids* 121: 211–227.

Schulz S. 1992. Guanylyl cyclase a cell-surface receptor throughout the animal kingdom. *Biol Bull* 183: 155–158.

Sepe N, De Petrocellis L, Montanaro F, Cimino G, Di Marzo V. 1998. Bioactive long chain N-acylethanolamines in five species of edible bivalve molluscs. Possible implications for mollusc physiology and sea food industry. *Biochim Biophys Acta* 1389: 101–111.

Shohami E, Weidenfeld J, Ovadia H, Vogel Z, Hanus, L, Fride E, Bruer A, Ben-Shabat S, Sheskin T, Mechoulam, R. 1996. Endogenous and synthetic cannabinoids: recent developments. *CNS Drug Rev* 2: 429–451.

Silva OL. 1990. Metabolic effects of tobacco and cannabis, in Becker KL, Ed. *Principles and Practice of Endocrinology and Metabolism*. Philadelphia, PA: Lippincott, pp. 1684–1686.

Skaper SD., Buriani A, Toso RD, Petrelli L, Romanello S, Facci L, Leon A. 1996. The ALIAmide palmitoylethanolamide and cannabinoids, but not anandamide, are protective in a delayed postglutamate paradigm of excitotoxic death in cerebellar granule neurons. *Proc Natl Acad Sci USA* 93: 3984–3989.

Smith CG, Asch RH. 1984. Acute, short-term, and chronic effects of marijuana on the female primate reproductive function. NIDA *Res Monogr* 44: 82–96.

Smith CG, Asch RH. 1987. Drug abuse and reproduction. *Fertil Steril* 48: 355–373.

Smith CG, Almirez RG, Scher PM, Asch RH. 1984. Tolerance to the reproductive effects of Δ^9-tetrahydrocannabinol. Comparison of the acute, short-term, and drug chronic effects on menstrual cycle hormones, in Agurell S, Dewey W, Willette R., Eds. *The Cannabinoids: Chemical, Pharmacologic and Therapeutic Aspects*. New York: Academic Press, pp. 471–485.

Soderstrom K, Johnson F. 2001. Zebra finch CB_1 cannabinoid receptor: pharmacology and *in vivo* and *in vitro* effects of activation. *J Pharmacol Exp Ther* 297: 189–197.

Soderstrom K, Leid M, Moore FL, Murray TF. 2000. Behavioral, pharmacological and molecular characterization of an amphibian cannabinoid receptor. *J Neurochem* 75: 413–423.

Song ZH, Bonner TI. 1996. A lysine residue of the cannabinoid receptor is critical for receptor recognition by several agonists, but not by WIN55,212-2. *Mol Pharmacol* 49: 891–896.

Stefano GB, Liu Y, Goligorsky MS. 1996. Cannabinoid receptors are coupled to nitric oxide release in invertebrate immunocytes, microglia, and human monocytes. *J Biol Chem* 271: 19238–19242.

Stella N, Piomelli D. 2001. Receptor-dependent formation of endogenous cannabinoids in cortical neurons. *Eur J Pharmacol* 425: 189–196.

Stella N, Schweltzer P, Piomelli D. 1997. A second endogenous cannabinoid modulates long-term potentiation. *Nature* 388: 773–778.

Stewart TA, Forrester IT. 1978. Acetylcholinesterase and choline acetyltransferase in ram spermatozoa. *Biol Reprod* 19: 271–279.

Stillman RJ, Rosenberg MJ, Sachs BP. 1986. Smoking and reproduction. *Fertil Steril* 46: 545–566.

Suarez SS. 1998. The oviductal sperm reservoir in mammals: mechanisms of formation. *Biol Reprod* 58: 1105–1107.

Suarez SS. 1999. Regulation of sperm transport in the mammalian oviduct, in Gagnon C, Ed. *The Male Gamete: From Basic Science to Clinical Applications*, Vienna, IL: Cache River Press. pp. 71–80.

Sugiura, T., Waku, K. 2000. 2-Arachidonoylglycerol and the cannabinoid receptors. *Chem Phys Lipids* 108: 89–106.

Sugiura T, Waku K. 2002. Cannabinoid receptors and their endogenous ligands. *J Biochem* 132: 7–12.

Sugiura T, Kondo S, Sukagawa A, Tonegawa T, Nakane S, Yamashita A, Waku K. 1996. Enzymatic synthesis of anandamide, an endogenous cannabinoid receptor ligand, through N-acylphosphatidylethanolamine pathway in testis: involvement of Ca^{2+}-dependent transacylase and phosphodiesterase activities. *Biochem Biophys Res Commn* 218: 113–117.

Sugiura T, Kodaka T, Nakane S, Kishimoto S, Kondo S, Waku K. 1997. Detection of an endogenous cannabimimetic molecule, 2-arachidonoylglycerol, and cannabinoid CB_1 receptor mRNA in human vascular cells: is 2-arachidonoylglycerol a possible vasomodulator? *Biochem Biophys Res Commn* 243: 838–843.

Sugiura T, Kodaka T, Nakane S, Miyashita T, Kondo S, Suhara Y, Takayama H, Waku K, Seki C, Baba N, Ishima Y. 1999. Evidence that the cannabinoid CB_1 receptor is a 2-arachidonolylglycerol receptor.

Structure-activity relationship of 2-arachidonoylglycerol, ether-linked analogues, and related compounds. *J Biol Chem* 274: 2794–2801.

Sugiura T, Kondo S, Kishimoto S, Miyashita T, Nakane S, Kodaka T, Suhara Y, Takayama H, Waku K. 2000. Evidence that 2-arachidonoylgylcerol but not *N*-palmitoylethanolamide or anandamide is the physiological ligand for the cannabinoid CB_2 receptor. Comparison of the agonistic activities of various cannabinoid receptor ligands in HL-60 cells. *J Biol Chem* 275: 605–612.

Sugiura T, Kobayashi Y, Oka S, Waku K. 2002. Biosynthesis and degradation of anandamide and 2-arachidonoylglycerol and their possible physiological significance. *Prostag Leukot Essent Fatty Acids* 66: 173–192.

Sukcharoen N, Keith J, Irvine DS, Atken RJ. 1995. Definition of the optimal criteria for identifying hyperactivated spermatozoa at 25 Hz using in vitro fertilization as a functional end-point. *Hum Reprod* 10: 2928–2937.

Sulcova E, Mechoulam R, Fride E. 1998. Biphasic effects of anandamide. *Pharmacol Biochem Behav* 59: 347–352.

Talbot P, Chacon RS. 1981. A triple-stain technique for evaluating normal acrosome reactions of human sperm. *J Exp Zool* 215: 201–208.

Tennes K. 1984. Effects of marijuana on pregnancy and fetal development in the human. *NIDA Res Monogr* 44: 115–123.

Tesarik J, Carreras A, Mendoza C. 1993. Differential sensitivity of progesterone- and zona pellucida-induced acrosome reactions to pertussis toxin. *Mol Reprod Dev* 34: 183–189.

Tilak Sk, Zimmerman AM. 1984. Effects of cannabinoids on macromolecular synthesis in isolated spermatogenic cells. *Pharmacol* 29: 343–350.

Topper EK, Killian GJ, Way A, Engel B, Woelders H. 1999. Influence of capacitation and fluids from the male and female genital tract on the zona binding ability of bull spermatozoa. *J Reprod Fertil* 115: 175–183.

Treinen KA, Sneeden JL, Heindel JJ. 1993. Specific inhibition of cAMP accumulation by Δ^9-tetrahydrocannabinol in cultured rat granulosa cells. *Toxicol Appl Pharmacol* 118: 53–57.

Tyre L. 1984. Endocrine aspects of cannabinoid action in female subprimates. *NIDA Res Monogr* 44: 65–81.

Tyre L, Murphy L. 1988. Inhibition of suckling-induced milk ejections in the lactating rat by Δ^9-tetrahydrocannabinol. *Endocrinology* 123: 469–472.

Ueda N., Puffenbarger, RA, Yamamato S, Deutsch DG. 2000. The fatty acid amide hydrolase. *Chem Phys Lipids* 108: 107–121.

Van der Stelt, Di Marzo, V. 2004. Endovanilloids putative endogenous ligands of transient receptor potential vanilloid 1 channels. *Eur J Biochem* 271: 1827–1834.

Van Duin VM, Polman JEM, De Breet ITM, Van Ginneken K, Bunschoten H, Grootenhuis A, Brindle J, Aitken RJ. 1994. Recombinant human zona pellucida protein ZP3 produced by chinese hamster ovary cells induces the human sperm acrosome reaction and promotes sperm-egg fusion. *Biol Reprod* 51: 607–617.

Vardaris RM, Weisz DJ, Fazel A, Rawitch AB. 1976. Chronic administration of delta-9-tetrahydrocannabinol to pregnant rats: studies of pup behavior and placental transfer. *Pharmacol. Biochem. Behav.* 4: 249–254.

Venance L, Piomelli D, Glowinski J, Giaume C. 1995. Inhibition by anandamide of gap junctions and intracellular calcium signaling in striatal astrocytes. *Nature* 376: 590–594.

Wang J, Paria BC, Dey SK, Armant DR. 1999. Stage-specific excitation of cannabinoid receptor exhibits differential effects on mouse embryonic development. *Biol Reprod* 60: 839–844.

Wang H, Guo Y, Wang D, Kingsley PJ, Marnett LJ, Das SK, DuBois RN, Dey SK. 2004. Aberrant cannabinoid signaling impairs oviductal transport of embryos. *Nature Med.* 10: 1074–1080.

Ward CR, Kopf GS. 1993. Molecular events mediating sperm activation. *Dev Biol* 158: 9–34.

Wassarman PM. 1987. The biology and chemistry of fertilization. *Science* 235: 553–560.

Wenger T, Ledent C, Csernus V, Gerendai I. 2001. The central cannabinoid receptor inactivation suppresses endocrine reproductive functions. *Biochem Biophys Res Commn* 284: 363–368.

Whitaker MJ, Steinhardt RA. 1982. Ionic regulation of egg activation. *Quart Rev Biophys* 15: 593–666.

Wiley JL, Martin BR. 2002. Cannabinoid pharmacology: implications for additional cannabinoid receptor subtypes. *Chem Phys Lipids* 121: 57–63.

Williams CM, Kirkham TC. 1999. Anandamide induces overeating; mediation by central cannabinoid receptors. *Psychopharmacology* 143: 315–317.

Yanagimachi R. 1994. Mammalian fertilization, in Knobil E, Neil JD, Eds. *The Physiology of Reproduction,* 2nd ed. New York: Raven Press, Vol 1, pp. 189–317.

Yanagimachi R, Phillips, DM. 1984. The status of acrosomal caps of hamster spermatozoa immediately before fertilization *in vivo*. *Gamete Res* 9: 1–19.

Yang ZM, Paria BC, Dey SK, 1996. Activation of brain-type cannabinoid receptors interferes with preimplantation mouse embryo development. *Biol Reprod* 55: 756–761.

Yazigi RA, Odem RR, Polakoski KL. 1991. Demonstration of specific binding of cocaine to human spermatozoa. *JAMA* 266: 1956–1959.

Yelian FD, Sacco AG, Ginsburg KA, Doerr PA, Armant DR. 1994. The effects of in vitro cocaine exposure on human sperm motility, intracellular calcium, and oocyte penetration. *Fert Steril* 61: 915–921.

Young RJ, Laing JC. 1991. The binding characteristics of cholinergic sites in rabbit spermatozoa. *Mol Reprod Dev* 28: 55–61.

Zafiriou PM, Karava V, Boutou E, Vorigas CE, Maccarrone M, Siafaka-Kapadai A. Partial purification and characterization of a fatty acid amidohydrolase (FAAH) from Tetrahymena pyriformis. 2004 Annual Symposium on the Cannabinoids, International Cannabinoid Research Society, Burlington VT, p. 146.

Zinaman M, Drobnis EZ, Morales P, Brazil C, Kiel M, Cross NL, Hanson FW, Overstreet JW. 1989. The physiology of sperm recovered from the human cervix: acrosomal status and response to inducers of the acrosome reaction. *Biol Reprod* 41: 790–797.

Zimmerman AM, Zimmerman S, Raj AY. 1978. Effects of cannabinoids on spermatogenesis in mice, in Nahas GG, Paton WDM, Eds. *Marihuana: Biological Effects*. Oxford, U.K.: Pergamon Press, pp. 407–418.

Zimmerman AM, Bruce WR, Zimmerman S. 1979. Effects of cannabinoids on sperm morphology. *Pharmacology* 18: 143–147.

Zimmerman S, Zimmerman AM, Laurence H. 1981. Effect of Δ^9-tetrahydrocannabinol on cyclic nucleotides in synchronously dividing *Tetrahymena*. *Can J Biochem* 59: 489–493.

Zimmerman AM, Murer-Orlando ML, Richer CL. 1986. Effects of cannabinoids on spermatogenesis *in vivo:* a cytological study. *Cytobios* 45: 7–15.

Zimmerman AM, Titishov N, Mechoulam R, Zimmerman S. 1991. Effects of stereospecific cannabinoids on the immune system. *Adv Exp Med Biol* 288: 71–80.

Zygmunt PM, Petersson J, Andersson DA, Chuang HH, Sorgard M, Di Marzo V, Julius D, Hogestatt ED. 1999. Vanilloid receptors on sensory nerves mediate the vasodilator action of anandamide. *Nature* 400: 452–457.

Part VII

Endocannabinoid Phylogenetics

23 Distribution of Endocannabinoids and Their Receptors and Enzymes on the Tree of Life

John M. McPartland

CONTENTS

OVERVIEW

The endocannabinoid system's phylogenetic distribution is not well known, and coevolution among its components has not been analyzed. The ligands cannot be directly investigated by molecular methods because they are not polypeptides. Ligand phylogenetics can be inferred, nonetheless, by the extraction of ligands from a range of extant organisms. A chemotaxonomic meta-analysis of ligand extraction studies was concatenated with molecular studies of cannabinoid receptors (CBRs), vanilloid receptors, and ligand-catabolizing enzymes such as fatty acid amid hydrolase (FAAH), monoglyceride lipase (MAGL), *N*-acyl-phosphatidylethanolamine phospholipase D (NAPE-PLD), and two diacylglycerol lipases (DAGLα and DAGLβ). Thanks to the advent of high-throughput DNA sequencing, the number of fully sequenced organisms has become sufficient to perform phylogenetic analyses of their coding sequences (*phylogenomics*). The ages of receptors and enzymes have been estimated based on congruent chemotaxonomic and phylogenomic data. In summary, evidence suggests that endocannabinoid ligands evolved before CBRs, and that the ligands evolved independently multiple times, representing homoplasy rather than homology.

INTRODUCTION

Di Marzo and Fontana (1995) characterized endocannabinoids as endogenous ligands capable of binding to and functionally activating CBRs. Endocannabinoids are analogous to endorphins, the endogenous agonists of opioid receptors. The endocannabinoid system consists of ligands, CBRs (currently defined as CB_1 and CB_2), and ligand-metabolizing enzymes (primarily amidases and lipases). This discrete circumscription, however, belies a rather fuzzy reality. Interacting around the endocannabinoid system are ligands that do not affect CB_1 or CB_2 yet are checked by CB_1 and CB_2 antagonists, non-CB_1 and non-CB_2 receptors that express affinity for endocannabinoids, and enzymes whose perceived physiological substrates lie elsewhere. These entourage ligands, analogous receptors, and other enzymes will be discussed in the following sections. Di Marzo and Fontana (1995) differentiated endocannabinoids from cannabinoids, the latter defined by Mechoulam and Gaoni (1967) as C_{21} terpenophenolic compounds uniquely produced by cannabis. The best-known cannabinoid is Δ^9-tetrahydrocannabinol (Δ^9-THC). Δ^9-THC and endocannabinoids display a similar (but not identical) profile of biological activities, such as inhibition of adenylate cyclase and calcium channels, hypothermia, analgesia, hypomobility, and catalepsy (reviewed by Felder and Glass, 1998).

The purpose of this paper is to explore the phylogenetic distribution of the endocannabinoid system in six categories of organisms: (a) mammals, (b) other vertebrates, (c) invertebrates that do not molt, (d) invertebrates that molt, (e) plants, and (f) other organisms (sponges, fungi, protozoans, archaens, and bacteria). The names of organisms discussed in this review and their places on the phylogenetic "Tree of Life" (http://tolweb.org) are given in Figure 23.1. Presence of the endocannabinoid system in these organisms, in part or whole, will be weighed by two types of evidence: pharmacological and molecular. Pharmacological evidence includes *in vitro* ligand-extraction chemistry, ligand-binding work, receptor transduction studies, and enzymatic assays, or *in vivo* drug-response studies. Molecular evidence depends on the identification of homologous genes in other organisms, using *in vitro* cDNA recombinant PCR techniques or *in silico* whole-genome methods.

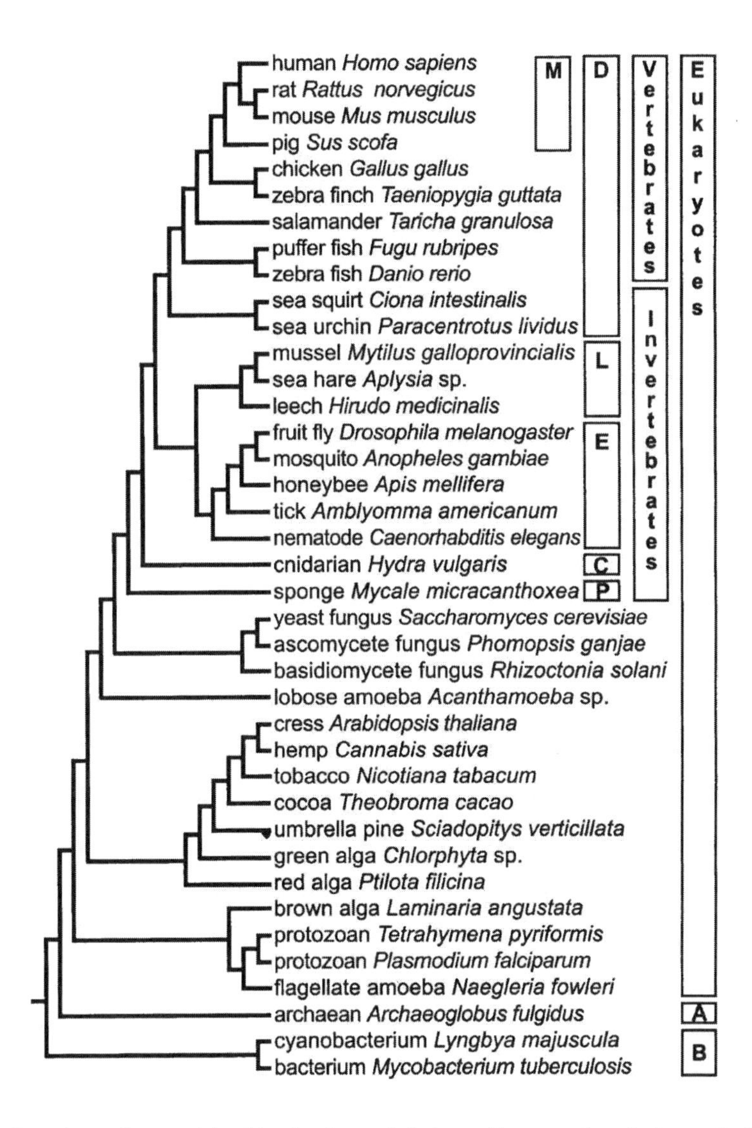

FIGURE 23.1 Organisms discussed in this chapter and their positions on the phylogenetic "Tree of Life." Mammals are marked by a bar labeled "M." Animals are divided into "Vertebrates" and "Invertebrates," and also divided into five physiological groups, D = Deuterostomes, L = Lophotrochozoans, E = Ecdysozoans, C = Cnidarians, and P = Poriferans. The three domains (supraphyla) are Eukaryotes, Archaens (A), and Bacteria (B).

MAMMALS

ENDOCANNABINOID LIGANDS

The mammalian endocannabinoid family includes four lipid *N*-fatty acyl-ethanolamines (NAEs), two *N*-acyl-dopamines, a lipid ester, and a lipid ether. The NAEs are *N*-arachidonoyl ethanolamine (anandamide, AEA) (Devane et al., 1992), docosatetraenoyl-ethanolamide and di-homo-γ-linolenoyl-ethanolamide (Hanus et al., 1993), and *O*-arachidonoyl ethanolamide (virodhamine) (Porter et al., 2002). The *N*-acyl-dopamines are *N*-arachidonoyl-dopamine (NADA) and *N*-oleoyl-dopamine (Huang et al., 2002). The lipid ester is sn-2 arachidonoyl glycerol (2-AG) (Mechoulam et al., 1995; Sugiura et al., 1995). The lipid ether is sn-2 arachidonoyl glyceryl ether (noladin ether) (Hanus et al., 2001).

Endocannabinoids act as agonists at CB_1 and CB_2 (reviewed by Pertwee and Ross, 2002), although virodhamine was originally portrayed as a CB_1 antagonist (Porter et al., 2002). This portrayal is somewhat semantic, because virodhamine behaved as a partial agonist at CB_1 in Porter's GTPγS binding assay, with an efficacy of 61% compared to AEA. This nearly equals that of Δ^9-THC, which demonstrated 57% efficacy compared to AEA in a similar CB_1/Gi/Sf9 assay (Glass and Northup, 1999), and Δ^9-THC is not normally considered a CBR antagonist.

Endocannabinoids are derived from membrane phospholipids that contain an arachidonic acid (AA) moiety, such as phosphatidylethanolamine (PE). According to the current model, PE undergoes a transacylase reaction to *N*-arachidonoyl-phosphatidylethanolamine (NAPE). AA esterified at the sn-1 position of NAPE is transferred to the N-position (Di Marzo et al., 1994), and then cleaved by a NAPE-specific phospholipase D-type enzyme (Okamoto et al., 2003) to yield AEA. 2-AG is cleaved from diacylglycerol by two specific sn-1 diacylglycerol lipases, DAGLα and DAGLβ (Bisogno et al., 2003). The abundance of 2-AG depends on the presence of AA at the 2-position of membrane phosphoglycerides (reviewed by Schmid et al., 2002).

AEA has been extracted from at least six mammals: pig, human, rat, mouse, sheep, and dog (Devane et al., 1992; Di Marzo et al., 1994; Di Marzo et al., 1996; and Felder et al., 1996). 2-AG has been isolated from dog gut (Mechoulam et al., 1995), rat brain (Sugiura et al., 1995), mouse neuroblastoma cells (Bisogno et al., 1997a), and human, bovine, and goat milk (Di Marzo et al., 1998). The concentration of 2-AG is much higher than that of AEA, which reflects the relative prevalences of their AA precursors; AA is commonly esterified at the 2-position in phospholipids, but rarely at the sn-1 position. Docosatetraenoyl-ethanolamide, di-homo-γ-linolenoyl-ethanolamide, and noladin ether were extracted from pig brain (Hanus et al., 1993; Hanus et al., 2001). NADA was extracted from bovine and rat brain (Huang et al., 2002).

Endocannabinoid Receptors

CBR homologs are divided into two types: *Paralogs* are homologs found in a given organism, derived by a gene-duplication event (such as the genes for CB_1 and CB_2). *Orthologs* are homologs found in different organisms, derived by descent from a common ancestor. CB_1 orthologs have been cloned and sequenced from 62 species of placental mammals, sampled across all extant orders within that phylogenetic clade (Murphy et al., 2001). Orthologs of human CB_2 have only been investigated in the human, mouse, and rat (Shire et al., 1996; Griffin et al., 2000; Brown et al., 2002). Sequence divergences among CB_2 orthologs are greater than those among CB_1 orthologs. For example, human CB_2 and rat CB_2 share 81% identity, whereas human CB_1 and rat CB_1 share 97% sequence identity; rat and mouse CB_2 share 93% identity, whereas rat and mouse CB_1 share 97% identity. This suggests that CB_2 is under less evolutionary pressure to conserve its sequence than CB_1.

Endocannabinoid-Metabolizing Enzymes

AEA is biosynthesized by NAPE-PLD (Okamoto et al., 2003). NAPE-PLD orthologs have been cloned from rat and mouse (Okamoto et al., 2003), NAPE-PLD sequences have been identified in genomes of human, rat, and mouse (McPartland, Pruitt et al., 2005), and NAPE-PLD activity has been reported in dog, cat, rabbit, and pig (Moesgaard et al., 2002). 2-AG is biosynthesized by DAGLα and DAGLβ; both enzymes were reported from human and mouse (Bisogno et al., 2003).

Fatty acid amide hydrolase (FAAH), formerly known as anandamide amidohydrolase (Duetsch and Chin, 1993), catabolizes AEA and, under some circumstances, 2-AG. Using PCR-based molecular methods, orthologs of human FAAH have been cloned from rat, mouse, and pig (reviewed by Ueda et al., 2000). Lipases may also be responsible for the hydrolysis of 2-AG (Goparaju et al., 1999). Dinh et al. (2002) identified monoglyceride lipase (MAGL) as the primary lipase acting on 2-AG. Orthologs of human MAGL were purified from bovine brain (Goparaju et al., 1999) and cloned from rodents (Karlsson et al., 1997; Dinh et al., 2002).

OTHER VERTEBRATES

ENDOCANNABINOID LIGANDS

No reports of AEA or 2-AG in nonmammalian vertebrates could be located in the literature, although Natarajan et al. (1985) extracted the precursor to AEA, *N*-arachidonoyl phosphatidylethanolamine, from the pike (*Esox lucius*) and the carp (*Cyprinus carpio*). This work was done before NAEs were recognized as ligands for CBRs.

ENDOCANNABINOID RECEPTORS

CB_1 orthologs have been cloned and sequenced from several lower vertebrates, including the zebra finch (*Taeniopygia guttata*) (Soderstrom and Johnson, 2000), the newt salamander (*Taricha granulose*) (Soderstrom et al., 2000), and the Japanese puffer fish (*Fugu rubripes*) (Yamaguchi et al., 1996). The puffer fish expressed a pair of paralogs, FCB_1A and FCB_1B. Howlett et al. (1990) suggested the absence of CBRs in a lower vertebrate; they reported no specific binding of the synthetic CBR ligand [^{3}H]CP55,940 in a lamprey (reported as *Ichthyomyzon intercostus* but probably *Ichthyomyzon unicuspis*).

A CB_2 ortholog in the puffer fish was identified by homology screening (Elphick, 2002). This challenged the results of Yamaguchi et al. (1996), who reported CB_2 absent in puffer fish. McPartland and Glass (2003) subjected the putative puffer fish CB_2 sequence to functional mapping and found that it exhibited crippling amino acid substitutions at six of eight critical residues known to confer CB_2 specificity. They concluded that the sequence coded for a CBR but possibly not for CB_2.

ENDOCANNABINOID-METABOLIZING ENZYMES

Moesgaard et al. (2002) reported *in vitro* NAPE-PLD activity in frog microsomes. McPartland, Pruitt et al. (2005) identified an NAPE-PLD ortholog in the puffer fish genome. Bisogno et al. (2003) reported orthologs of DAGLα and DAGLβ in the genomes of chicken and zebra fish. McPartland, Pruitt et al. (2005) also identified DAGLα and DAGLβ orthologs in the puffer fish genome.

In the absence of any *in vitro* FAAH research in lower vertebrates, McPartland (2004) searched for molecular orthologs using the whole-genome approach and identified FAAH orthologs in chicken and puffer fish. Both sequences expressed a single substitution in a motif known to impart FAAH specificity, the PPLP motif. Both sequences substituted as PPIP. However, the L $\rightarrow$ I substitution is conservative (BLOSUM62 substitution matrix, Henikoff and Henikoff, 1992), suggesting that proteins expressing this substitution may function as FAAH enzymes. Similarly, McPartland (2004) searched for MAGL orthologs in the genomes of nonmammalian vertebrates. A very high-identity ortholog was expressed by the zebra fish and a moderate-identity ortholog occurred in the puffer fish. Functional mapping showed that both sequences expressed a triad of critical amino acid residues known to impart specificity to MAGL: the GXSXG motif, D239, and H269 (Karlsson et al., 1997).

INVERTEBRATES THAT DO NOT MOLT

Molecular evidence has recently changed the face of invertebrate taxonomy. Aguinaldo et al. (1997) split metazoan invertebrates into four clades, in descending phylogenetic order: the Deuterostoma (e.g., sea squirts and sea urchins, animals developing from radial indeterminate cleavage and a pouched enterocoelom, in common with vertebrates), the Lophotrochozoa (e.g., leeches and mussels, lophophore-bearing animals with spiral determinate cleavage and a split schizocoelom), the Ecdysozoa (e.g., insects and nematodes, invertebrates that molt, also with spiral cleavage and a shizocoelom), and the Cnidaria (e.g., hydra, back-to-radial cleavage, but retaining this pattern into mature stages).

Endocannabinoid Ligands

AEA has been isolated from the sea squirt *Ciona intestinalis* (Matias et al., 2005), sea urchin *Paracentrotus lividus* (Bisogno et al., 1997b), the leeches *Hirudo medicinalis* and *Theromyzon tessulatum* (Matias et al., 2001), and many mollusks, including an oyster *Crassosterea* sp., the mussel *Mytilus galloprovincialis*, the clam *Tapes dicussatus* (Sepe et al., 1998), and the sea hare *Aplysia* sp. (Di Marzo et al., 1999). AEA has even been detected in a very primitive animal with a nerve network, the cnidarian *Hydra vulgaris* (De Petrocellis et al., 1999). 2-AG has been isolated from *C. intestinalis* (Matias et al., 2005), *P. lividus* (Bisogno et al., 1997b), *H. medicinalis* (Matias et al., 2001), Aplysia sp. (Di Marzo et al., 1999), and hydra (De Petrocellis et al., 1999).

Endocannabinoid Receptors

Elphick et al. (2003) identified a CB_1 ortholog in the genome of (*C. intestinalis*). This sequence shared 29% identity with human CB_1 and 24% identity with human CB_2. These divergences are greater than those between human CB_1 and CB_2 (47% identity), leading Elphick and colleagues to suggest that the ancestor of the sea squirt ortholog evolved prior to the CB_1–CB_2 duplication event. McPartland and Glass (2003) subjected the sea squirt sequence to functional mapping and identified substitutions at F3.36 and W5.43, two amino acid motifs required for mammalian CBRs to fully function (McAllister et al., 2003). Despite these substitutions, sea squirt neural tissues demonstrated high-affinity binding with [^{3}H]CP55,940 and [^{3}H]SR141716A (Matias et al., 2005; McPartland, Agraval et al., 2005). Overall, the sea squirt sequences shared more similarity with human CB_1 than CB_2, suggesting that the ancestral CBR may have functioned more like the present-day CB_1 than CB_2 (McPartland and Glass, 2003). Another deuterostome, the sea urchin *Strongylocentrotus purpuratus*, demonstrated specific binding with [^{3}H]CP55,940 (Chang et al., 1993).

Among Lophotrochozoa, evidence for CBRs exists in leeches, mollusks, and earthworms. A controversial sequence said to be a partial CB_1 gene was cloned from the leech *Hirudo medicinalis* (Stefano et al., 1997), although other authors have suggested that the sequence may be a product of PCR contamination (Elphick, 1998; Elphick and Egertová, 2001). Leech homogenates nevertheless demonstrated specific binding with [^{3}H]AEA (Stefano et al., 1997). Salzet and Stefano (2002) also proposed evidence for a CB_2-like receptor in *H. medicinalis* but did not present their data. The mollusk *Mytilus edulis* reportedly bound [^{3}H]AEA (Stefano et al., 1996), yet Howlett et al. (1990) reported no specific binding of [^{3}H]CP55,940 in a related mollusk, *Aplysia californica*. Neural tissues from the earthworm *Lumbricus terrestris* demonstrated high-affinity binding with [^{3}H]CP55,940 displaced by the CB_1-selective antagonist SR141716A (McPartland, Agraval et al., 2005). De Petrocellis et al. (1999) reported specific binding of [^{3}H]SR141716A in the basal metazoan *Hydra vulgaris* (Cnidaria, Class Hydrozoa). McPartland, Agraval et al. (2005) found no specific binding of [^{3}H]CP55,940 in the related sea anemone *Actinothoe albocincta* (Cnidaria, Class Anthozoa).

Endocannabinoid-Metabolizing Enzymes

Using phylogenomic analysis, Matias et al. (2005) identified orthologs of FAAH, MAGL, DAGLα, and DAGLβ in the sea squirt genome (but no NAPE-PLD ortholog). The FAAH ortholog showed sensitivity to specific FAAH inhibitors and exhibited subcellular/tissue distributions and pH-dependency similar to mammalian FAAHs. Its affinity for AEA was low, with an apparent $K_m = 450 \pm 35\ \mu M$, based on the enzymatic formation of [^{14}C]ethanolamide from [^{14}C]AEA. Another deuterostome, the sea urchin *P. lividus*, demonstrated enzymatic activity suggestive of FAAH (Bisogno et al., 1997b).

Matias et al. (2001) detected immunostaining in neural tissue of the leeches *H. medicinalis* and *T. tessulatum* using antibodies directed against the FAAH catalytic core (GGSSGGEGALI). The leech enzyme metabolized AEA, but it differed from mammalian FAAH in pH-dependency (optimal activity at pH 7 instead of pH 9 to 10), subcellular distributions (primarily cytosolic fractions instead of membrane fractions), produced primarily a 46-kDa band on Western blot (instead of 64 kDa), and showed no sensitivity to specific FAAH inhibitors. Taken together, these data suggested that leeches expressed a non-FAAH amidase (Matias et al., 2001). At least 80 non-FAAH amidases are known in the large amidase signature family (Patricelli and Cravatt, 2000), expressed in vertebrates, invertebrates, plants, fungi, and bacteria. Other pharmacological studies reported FAAH activity in invertebrates, but these may have inadvertently detected non-FAAH amidases in *M. galloprovincialis* (Sepe et al., 1998), *M. edulis* (Stefano et al., 1998), Aplysia sp. (Di Marzo et al., 1999), and hydra (De Petrocellis et al., 1999).

INVERTEBRATES THAT MOLT (ECDYSOZOA)

ENDOCANNABINOID LIGANDS

McPartland et al. (2001) did not detect AEA in brains dissected from the honeybee (*Apis mellifera*) or in heads of the fruit fly *Drosophila melanogaster*. Perhaps the lack of AEA in insects evolved from a relative paucity of AA. Insects produce very little AA (Stanley–Samuelson and Pedibhotla, 1996), in contrast to invertebrates that do not molt (such as sea urchins, mollusks, and hydra). Salivary glands of the lone star tick (*Amblyomma americanum*) contained no measurable amounts of AEA unless they were stimulated with AA (Fezza et al., 2003).

2-AG has been extracted from the beetle *Agabus affinis* (Schaaf and Dettner, 2000), *A. mellifera* and *D. melanogaster* (McPartland et al., 2001), and *A. americanum* (Fezza et al., 2003). Apparently, the modest amounts of AA present in insects are found on the 2-position of phospholipids, a situation consistent with other animals. No one has assayed *Caenorhabditis elegans* or other nematode worms for endocannabinoids, but these invertebrates deserve scrutiny. The cell membranes of *C. elegans* contain the precursors to AEA, such as PE and AA (Tanaka et al., 1996).

ENDOCANNABINOID RECEPTORS

McPartland et al. (2000a) and Elphick and Egertová (2001) screened the *D. melanogaster* genome and found no CB_1 or CB_2 orthologs. The *Anopheles gambiae* (mosquito) genome also lacked genes with significant identity to CBRs (McPartland, 2004). Similarly, McPartland and Glass (2001) found no orthologs in the genome of the nematode *C. elegans*. Taken together, these findings led McPartland et al. (2001) to propose that CBRs evolved prior to the hydra-bilaterian divergence, but that they were secondarily lost in the Ecdysozoa. Indeed, G-protein-coupled receptors (GPCRs) gated by other lipids such as prostaglandins and lysophosphatidic acids are not well represented in the Ecdysozoa (unpublished results using HomoloGene, www.ncbi.nlm.nih.gov/entrez/query.fcgi). It is difficult to prove a negative, but the null molecular studies were initially supported by pharmacological data. McPartland et al. (2001) found no specific binding of [^{3}H]CP55,940 or [^{3}H]SR141716A in a panel of insects including *D. melanogaster, A. mellifera, Gerris marginatus, Spodoptera frugiperda,* and *Zophobas atratus*. These negative studies were corroborated by the inability of insect brains to activate GTPγS by Δ^9-THC or HU210 (McPartland et al., 2001). Lack of CBRs in ecdysozoans would imply that they are not a prerequisite for consciousness. After all, if CBRs are absent in honeybees, then they were not required for the evolution of spatial memory, goal-directed desires, elements of deception, and symbolic communication through dance.

Conflicting research has negated the Ecdysozoa hypothesis; Egertová et al. (1998) reported 5% specific binding of [^{3}H]CP55,940 in muscles of the locust *Schistocerca gregaria*. Neural tissues from the velvet worm *Peripatoides novae-zealandiae* (Ecdysozoa-Onychophora-Peripatidae), rock lobster *Jasus edwardi* (Ecdysozoa-Arthropoda-Crustacea), and beer mat nematode *Panagrellus redivivus* (Ecdysozoa-Nematoda-Rhabditida) demonstrated high-affinity binding with [^{3}H]CP55,940 displaced by the CB_1-selective antagonist SR141716A (McPartland, Agraval et al., 2005).

Endocannabinoid-Metabolizing Enzymes

No orthologs of NAPE-PLD or DAGLβ were reported in the genomes of *D. melanogaster* and *C. elegans*, although DAGLα was identified in both the species (McPartland, Pruitt, et al., 2005). Initial molecular research did not report FAAH orthologs in the genomes of *D. melanogaster* and *C. elegans* (McPartland, 2004). However, FAAH orthologs were reported from these species in a subsequent study that utilized protein predictor programs (Pfam, Prosite, PSORT, and TMHMM) coupled with functional mapping and neighbor-joining phylogenetic trees constructed with ClustalX and TreeView (McPartland, Pruitt et al., 2005). Egertová et al. (1998) described "FAAH-like activity" in the locust *Schistocerca gregaria*, using [^{3}H]oleamide as a substrate. The initial report of MAGL lacking in the genomes of *D. melanogaster* and *C. elegans* (McPartland, 2004) was confirmed in a rigorous subsequent study (McPartland, Pruitt et al., 2005).

PLANTS

Endocannabinoid Ligands

Higher plants (angiosperms and gymnosperms) produce polyunsaturated fatty acids (PUFAs) with acyl tails limited to 18 carbons in length, with up to four double bonds (18:4) (Zank et al., 2000). Hence, reports of PUFAs in plants with longer acyl tails, such as AA and endocannabinoids (all 20:4), are problematic. DiTomaso et al. (1996) detected AEA in chocolate and cocoa powder (from seeds of *Theobroma cacao*). Their results were challenged by Di Marzo et al. (1998), who did not detect AEA or 2-AG in cocoa. Nakane et al. (2000) reportedly extracted cis-5,11,14-eicosatrienoic acid (sciadonic acid, 20:3) from seeds of a conifer, the umbrella pine *Sciadopitys verticillata*. This analog of 2-AG exhibited cannabimimetic activity at concentrations as low as 10 n*M* in NG108-15 neuroblastoglioma cells expressing CB_1 (Nakane et al., 2000). Unlike higher plants, lower nonvascular plants such as club mosses, mosses, and algae express Δ^6-elongase enzymes, and they are capable of producing AA (Zank et al., 2000). Soderstrom et al. (1998) extracted over a dozen cannabimimetic lipid extracts from unidentified green algae (Chlorophyta).

Endocannabinoid Receptors

McEno et al. (1991) proposed that humans acquired CBR genes from plants via horizontal gene transfer (HGT); McEno and colleagues conjectured that Δ^9-THC originally served as a ligand for CBRs in cannabis (their hypothesis predated the discovery of AEA). HGT represents the nonsexual transmission of DNA between genomes of unrelated, reproductively isolated organisms. Over 200 human genes may have been obtained via HGT (International Human Genome Sequencing Consortium, 2001). HGT can be vectored by parasites capable of bridging both hosts. The bacterium *Agrobacterium tumefaciens* is a potential vector; it normally acts as a plant pathogen and readily infects cannabis (McPartland et al., 2000). *A. tumefaciens* also infects humans (Hulse et al., 1993) and is capable of vectoring DNA into mammalian nuclei (Ziemienowicz et al., 1999). The plant-to-human HGT hypothesis was falsified when McPartland and Pruitt (2002) found no orthologs of human CBRs in *Cannabis sativa*, the entire genome of *Arabidopsis thaliana*, or any other plant cDNA sequences deposited at GenBank. These negative molecular results have been corroborated by Tripathy et al. (2003), who found no specific binding of [^{3}H]AEA in *A. thaliana*, tobacco (*Nicotiana tabacum*), or alfalfa (*Medicago truncatula*).

Endocannabinoid-Metabolizing Enzymes

Gerwick et al. (1997) described an enzyme from the red alga *Ptilota filicina*, which they named *polyenoic fatty acid isomerase*. The enzyme was capable of converting AEA into novel substances such as conjugated triene anandamide. Chapman (2000) described FAAH-like amidohydrolase activity in tobacco plants. The *A. thaliana* genome did not express orthologs of FAAH, although related amidases were found (McPartland, 2004; McPartland, Pruitt et al., 2005). No orthologs of MAGL, DAGLα, DAGLβ, and NAPE-PLD were found in the genome of *A. thaliana* (McPartland, Pruitt et al., 2005).

OTHER ORGANISMS

For sake of expediency, a polyphyletic group will be lumped for discussion here, including the sponges (Poriferans), fungi, protozoans, archeans, and bacteria.

ENDOCANNABINOID LIGANDS

Soderstrom et al. (1998) extracted five nitrogen-containing lipids from the sponge *Mycale micracanthoxea*. The lipids showed affinity for CB_1, with IC_{50} values as low as 12.6 μM. These researchers also extracted cannabimimetic compounds from the brown alga *Laminaria angustata* (an organism more closely related to protozoans than to green algae). Protozoans and fungi have not been tested for the presence of endocannabinoids, but they should be. The protozoan *Naegleria fowleri* synthesized AA and expressed a phospholipase A(2) enzyme that was inhibited by methylarachidonyl fluorophosphonate, an analog of AEA (Barbour and Marciano-Cabral, 2001). *Saccharomyces cerevisiae* and other fungi express Δ^6-elongase enzymes enabling them to produce AA (Zank et al., 2000). Soderstrom et al. (1998) extracted but did not isolate a cannabimimetic compound from the marine cyanobacterium (blue-green alga) *Lyngbya majuscula*. At the same time, Sitachitta and Gerwick (1998) extracted grenadamide, a cyclopropyl-containing fatty acid, from *L. majuscula*. Grenadamine exhibited modest affinity for CB_1 ($K_i = 4.7$ μM).

ENDOCANNABINOID RECEPTORS

McPartland, Agraval et al. (2005) reported no specific binding of [^{3}H]CP55,940 in the golf ball sponge, *Tethya aurantium*. McPartland and Glass (unpublished data) detected no specific binding in two fungi, *Phomopsis ganjae* (Ascomycota) and *Rhizoctonia solani* (Basidiomycota), using [^{3}H]CP55940 and [^{3}H]SR141716A. The genome of *S. cerevisiae* did not express orthologs of CB_1 or CB_2 (McPartland, 2004; McPartland, Pruitt et al., 2005). These results suggested a lack of CBRs in fungi, despite the fact that Δ^9-THC exerted antifungal effects (McPartland, 1984).

Onaivi et al. (2002) reported CBR activity in the cyanobacterium *Lyngbya majuscula*, but this may have been a misreading of the original literature (Sitachitta and Gerwick, 1998). McPartland (2004) and McPartland, Pruitt et al. (2005) found no sequences resembling CB1 in the genomes of protozoans (*Plasmodium falciparum*, *Tetrahymena thermophila*), archaens (*Archaeoglobus fulgidus*), or bacteria (*Mycobacterium tuberculosis*). Congruous pharmacological results are reported by F. Marciano-Cabral (personal communication, 2002), who found no specific binding of tritiated cannabinoid ligands in protozoan *Acanthamoeba* species.

ENDOCANNABINOID-METABOLIZING ENZYMES

In a whole-genome study, no orthologs of MAGL, DAGLα, DAGLβ, and NAPE–PLD were found in the genomes of *P. falciparum*, *T. thermophila*, *A. fulgidus*, and *M. tuberculosis*, although an FAAH ortholog was found in the *T. thermophila* genome (McPartland, Pruitt et al., 2005). An *in vitro* kinetics study demonstrated that *T. pyriformis* rapidly metabolized [^{3}H]AEA into [^{3}H]AA and ethanolamide, and the enzyme followed kinetics similar to those reported for human FAAH, expressed optimal activity at pH 9 to 10, was sensitive to FAAH inhibitors, and its activity was Ca^{2+} and Mg^{2+} independent (Karava et al., 2001).

ENTOURAGE LIGANDS, ANALOGOUS RECEPTORS, AND OTHER ENZYMES

The endocannabinoid system is impugned upon by nonendocannabinoid ligands, non-CBR receptors, and other enzymes. These compounds and structures may provide us with phylogenetic information regarding the endocannabinoid system itself.

ENTOURAGE LIGANDS

N-palmitylethanolamide (PEA, 16:0), a short-chain saturated NAE congener of AEA, has been a source of some confusion. PEA lacks affinity at CBRs, so it is not a true endocannabinoid, yet it reduces hyperalgesia in mice, and this action is blocked by the CBR antagonist SR141716A (Calignano et al., 2001). Calignano and colleagues proposed that PEA activated an unidentified CB_2-like receptor. Alternatively, PEA may activate CBRs indirectly by inhibiting the hydrolysis of endocannabinoids by FAAH via substrate competition. This entourage effect would allow nonmetabolized endocannabinoids to accumulate at CBRs (Mechoulam et al., 1998).

At least two other NAEs are metabolized by FAAH and may also act as entourage compounds: cis-9-octadecenoamide (oleamide, 18:1), a sleep-inducing substance (Cravatt et al., 1995), and *N*-stearoylethanolamide (SEA, 18:0), an NAE congener of AEA (Maccarrone et al., 2002). However, the inhibition of FAAH may not be physiologically relevant; *N*-lauroylethanolamide (LEA, 12:0) caused very little inhibition of FAAH, yet LEA generated the greatest potentiation of AEA at the receptor (Smart et al., 2002). Perhaps LEA inhibits other enzymes that weakly metabolize AEA, such as N-PEA hydrolase (Ueda et al., 2001).

The phylogenetic distribution of the entourage compounds is not well known, but they seem to be everywhere anyone has looked. PEA was originally extracted from chicken (Kuehl et al., 1957) and rodents (Bachur et al., 1965); its anti-inflammatory properties were recognized before the discovery of the endocannabinoid system. PEA has been extracted from invertebrates that synthesize AEA, such as *P. lividus* (Bisogno et al., 1997b), *H. medicinalis* (Matias et al., 2001), and the bivalves *M. galloprovincialis*, *T. dicussatus*, and *Crassosterea* sp. (Sepe et al., 1998), as well as from invertebrates that do not synthesize AEA, such as *A. mellifera* and *D. melanogaster* (McPartland et al., 2001). PEA even occurs in seeds of corn (*Zea mays*), cotton (*Gossypium hirsutum*), and tobacco (*N. tabacum*) (Chapman, 2000).

SEA has been extracted from human and rodent brains (Maccarrone et al., 2002), and from the tick *A. americanum* (Fezza et al., 2003). Oleamide is present in the cerebrospinal fluid of humans, rats, and cats (Cravatt et al., 1995), and in the seeds of *T. cacao* (Di Marzo et al., 1998), *Z. mays*, and *G. hirsutum* (Chapman, 2000). Short-chain NAEs such as *N*-lauroylethanolamine (LEA, 12:0) and *N*-myristoylethanolamine (MEA, 14:0) have been extracted from seeds of *N. tabacum* and *M. truncatula* (Chapman, 2000).

ANALOGOUS RECEPTORS

Short-chain NEAs such as PEA and MEA signal a transduction system in plants that is parallel to the endocannabinoid system in animals. Tripathy et al. (2003) detected saturable, high-affinity specific binding of [^{3}H]MEA in *A. thaliana, N. tabacum*, and *M. truncatula*, and this binding was displaced by the CB_1-specific antagonist AM281 and CB_2-specific antagonist SR144528. Previously, Chapman (2000) demonstrated that MEA suppressed the alkalinization response in tobacco, which is an immune response based on rapid Ca^{2+} influx and K^+ efflux. This response is homologous to CBR-mediated inhibition of Ca^{2+} and K^+ channels in human immune cells (Mackie et al., 1995). Suppression of the alkalinization response by MEA was reversed by AM281 and SR144528. This led Tripathy et al. (2003) to propose the existence of a CBR-like membrane protein that mediated signaling of short-chain NAEs in plants. Due to the paucity of GPCRs in plants, they proposed that the plant receptor might resemble the vanilloid receptor (VR1, aka TRPV1).

VR1 is a ligand-gated ion channel (LGIC), otherwise known as an *ionotropic cation channel* (Caterina et al., 1997). Although the topology of VR1 differs significantly from that of CBRs (which are metabotropic GPCRs), Zygmunt et al. (1999) found that VR1 had affinity for AEA, with a K_i value nearly equal to AEA's affinity for CB_1. NADA exhibited five times greater potency than AEA at VR1 (Huang et al., 2002). Speculation by Tripathy et al. (2003) that plants may express VR1 orthologs was addressed by McPartland (2004), who found no sequences with significant resemblance to VR1 in the genome of the plant *A. thaliana*. Genome studies suggest that VR1 orthologs are limited to mammals (McPartland, Pruitt et al., 2005).

OTHER ENZYMES

Because endocannabinoids are eicosanoids, they can be metabolized by enzymes that act on AA, such as cyclooxygenase (COX) and lipooxygenase (LOX). Human COX-2 (but not COX-1) metabolized AEA into an array of oxygenated metabolites including prostaglandin E_2 ethanolamide (Yu et al., 1997). Subsequent work demonstrated that COX-2 selectively oxygenated 2-AG as a preferred substrate, leading Kozak and Marnett (2002) to suggest that the striking catalytic efficiency of the 2-AG/COX-2 reaction provided a *raison d'etre* for the evolution of two distinct COX isoforms. A genome study identified COX-2 orthologs in human, mouse, puffer fish, and sea squirt genomes (McPartland, Pruitt et al., 2005a). Previously, Knight et al. (1999) predicted COX-2 in the sea squirt; they inhibited *Ciona* prostaglandin synthesis with COX-2-selective drugs (e.g., etodolac), with less inhibition by nonselective indomethacin and little inhibition by COX-1-selective drugs (e.g., resveratrol).

Ueda et al. (1995) described two LOX isoforms in pig and rabbit, 12-LOX and 15-LOX, that oxygenated AEA. Similar results have been reported with rat 12-LOX and human 15-LOX. In contrast, pig 5-LOX displayed no activity. These results are consistent with experiments using 2-AG (reviewed by Kozak and Marnett, 2002). Surprisingly, 5-LOX in plant species such as barley (*Hordeum vulgare*) and tomato (*Lycopersicon esculentum*) oxidized AEA into 11S-hydroperoxy-5,8,12,14-eicosatetraenoylethanolamide (11S-HPANA); thus, with AEA as a substrate, plant 5-LOX behaved similar to 11-LOX (van Zadelhoff et al., 1998). These workers also found that soybean 15-LOX could metabolize AEA into 15S-HPANA. Soybean (*Glycine max*) 15-LOX also metabolizes 2-AG (Kozak and Marnett, 2002).

ESTIMATING THE AGE OF CBRs

We can estimate the evolutionary age of CBRs with molecular data from extant organisms. CBR ortholog sequences vary from species to species because of accumulated mutations. McPartland and Pruitt (2002) utilized similarity among CBR ortholog sequences (percentage identity with human CB_1 and CB_2) to construct a CBR gene tree. The tree topography (branch placements and branch lengths) was nearly identical to a CBR gene tree constructed with a measure of dissimilarity (i.e., sequence divergence) (McPartland, Pruitt et al., 2005). An updated version is presented in Figure 23.2. Note that the topography of the CBR gene tree nearly mirrors the Tree of Life species tree (Figure 23.1), except for the placement of the leech sequence reported by Stefano et al. (1997), which incongruently clades with the vertebrates.

The CBR gene tree can provide a "molecular clock" for the timing of evolutionary events when it is calibrated with lineage estimates (e.g., Feng et al., 1997; Suga et al., 1999; Benton and Ayala, 2003), measured in terms of millions of years ago (MYA). Thus calibrated, the CBR gene tree is rooted by a primordial CBR gene that evolved prior to the divergence between sea squirts and vertebrates (590 MYA). If the tritiated binding study by De Petrocellis et al. (1999) indicates that a CBR is expressed by *Hydra vulgaris*, this would push back CBR evolution to the metazoan-bilaterian divergence (i.e., between extant hydra and leech) about 790 MYA. This dates to the evolution of early multicellular animals, in concert with their new needs for cell-to-cell communications.

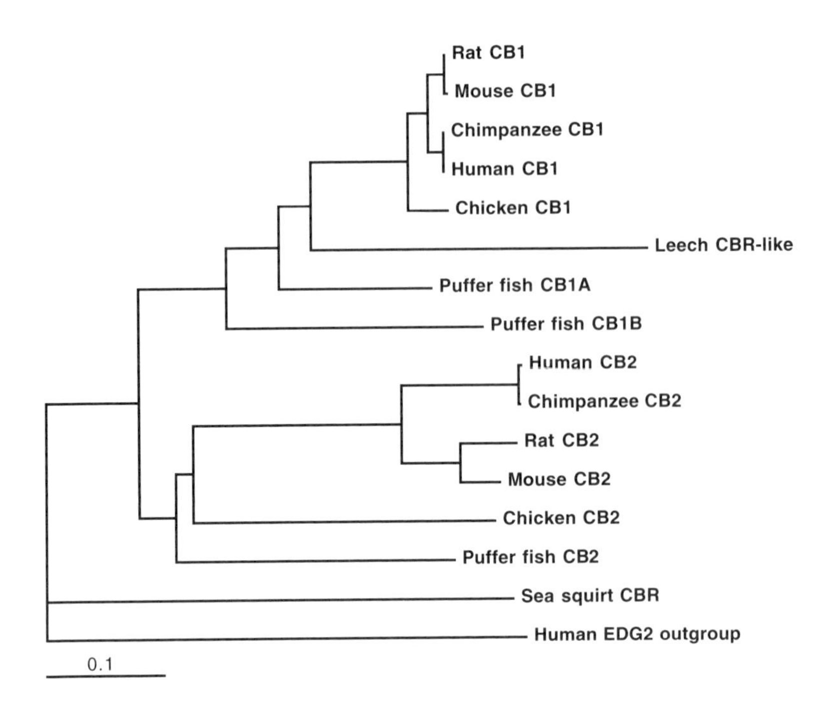

FIGURE 23.2 Phylogram (gene tree) of CB_1 and CB_2 orthologs and paralogs, based on sequence divergence calculated by a distance (neighbor-joining) algorithm, using ClustalX and TreeView.

The primordial CBR diverged from a closely related GPCR, perhaps EDG-1 (McPartland and Pruitt, 2002). The EDG-1 sequence shares closer identity with CB_1 than any other GPCR (P. Reggio, personal communication, 1998).

ESTIMATING THE AGE OF ENDOCANNABINOID LIGANDS

A molecular clock cannot be created for endocannabinoid ligands because they are not polypeptides; they cannot be sequenced. The antiquity of the ligands can be inferred, nonetheless, by their extraction from organisms with ancient lineages. AEA and 2-AG have been extracted from *Hydra vulgaris* (De Petrocellis et al., 1999). AEA-related fatty acids, however, have been extracted from sponges, brown algae (Soderstrom et al., 1998), and cyanobacteria (Sitachitta and Gerwick, 1998), whose ancestors diverged at least 930 MYA, 1500 MYA, and 2000 MYA, respectively. All these lineages predate the evolution of known CBRs. Bacteria are the oldest organisms known to express the precursors to endocannabinoids, PEs, in their cytoplasmic membranes (Schroder-Borm et al., 2003).

The relative ages of the ligands may also be inferred by the presence of endocannabinoid-metabolizing enzymes in the genomes of extant organisms. *Tetrahymena thermophila* expresses a FAAH ortholog (McPartland, Pruitt et al., 2005), which agrees with kinetic studies of the related species *T. pyriformis* (Karava et al., 2001). FAAH probably diverged from another member of the amidase signature family, conserved in plants, fungi, and even bacteria (Patricelli and Cravatt, 2000); these antediluvian lineages diverged 3000 MYA.

ADDRESSING CHICKEN-OR-EGG QUESTIONS

The question of whether CBRs or their ligands evolved first was addressed by McPartland and Pruitt (2002), who brought VR1 into the picture. They estimated that AEA evolved prior to CBRs. McPartland and Guy (2004) speculated on another chicken-or-egg question by suggesting that the primordial

CBR was originally gated by 2-AG and not AEA. They based their inference on affinity and efficacy studies of 2-AG and AEA (Pertwee and Ross, 2002). If our molecular clock is correct and the primordial CBR evolved at least 590 MYA, then CBRs evolved before cannabis, which is certainly not more than 34 million years old (McPartland and Nicholson, 2003). This chicken-or-egg question has fascinated researchers for years, even before the discovery of CBRs (e.g., McEno et al., 1991). It is intriguing to consider that dinosaurs probably expressed CBRs, but they became extinct before cannabis evolved. This is a good thing; imagine *Tyrannosaurus rex* with the munchies.

CONCLUSIONS

Evidence presented in this chapter suggests that the ability to synthesize endocannabinoids evolved prior to the development of CBRs. Indeed, insects can synthesize 2-AG despite the loss of CBR genes; there is greater evolutionary pressure to conserve ligands than to conserve receptors (Hoyle, 1999). In the absence of CBRs, endocannabinoids could serve other roles. AEA, for example, modulates 5-HT_{3A} receptors (Barann et al., 2002), inhibits Shaker-related K^+ channels (Poling et al., 1996), inhibits TASK-1 K^+ channels (Maingret et al., 2001), inhibits T-type Ca^{2+} channels (Chemin et al., 2001), and stimulates ERK phosphorylation and AP-1 transcription activity (Berdyshev et al., 2001). McPartland et al. (2001) suggested that endocannabinoids synthesized by insects served as feeding deterrents against organisms that do express CBRs. Fezza et al. (2003) demonstrated that ectoparasitic ticks expressed endocannabinoids in their salivary glands to help them elude detection by mammalian hosts. Hence, the evolution of endocannabinoids in organisms without CBRs may represent homoplasy rather than homology, known as *convergent evolution*. A case of parallel evolution can be seen in higher plants — they do not have CBRs, nor do they produce endocannabinoids, yet plants produce endocannabinoid-like entourage compounds, and these compounds activate CB-like receptors.

Further genomic and chemotaxonomic work must be done to ascertain some branches of the Tree of Life currently missing from our analysis. Whole-genome studies will benefit from some of the organisms currently in the sequencing pipeline, such as the honeybee and *Oxytricha trifallax* (a ciliated protozoan). Lastly, ligand extraction studies should be completed on bacteria, protozoans, fungi, and nematodes. The entire endocannabinoid system presents a fascinating weave of ligands, receptors, and enzymes, the "evolutionary arabesque" described by McPartland and Guy (2004).

ACKNOWLEDGMENTS

This work was supported by an unrestricted grant from GW Pharmaceuticals, Salisbury, U.K. The author thanks P.L. Pruitt for a careful review of the manuscript.

REFERENCES

Aguinaldo AM, Turbeville JM, Linford LS, Rivera MC, Garey JR, Raff RA, Lake JA. (1997) Evidence for a clade of nematodes, arthropods and other moulting animals. *Nature* 387: 489–493.

Bachur NR, Maser K, Melmon KL, Udenfriend S. (1965) Fatty acid amides of ethanolamine in mammalian tissues. *Journal Biological Chemistry* 240: 1019–1024.

Barann M, Molderings G, Bruss M, Bonisch H, Urban BW, Gothert M. (2002) Direct inhibition by cannabinoids of human 5-HT_{3A} receptors: probable involvement of an allosteric modulatory site. *British Journal Pharmacology* 137: 589–596.

Barbour SE, Marciano-Cabral F. (2001) Naegleria fowleri amoebae express a membrane-associated calcium-independent phospholipase A(2). *Biochemica Biophysica Acta* 1530: 123–133.

Benton MJ, Ayala FJ. (2003) Dating the tree of life. *Science* 300: 1698–1700.

Berdyshev EV, Schmid PC, Krebsbach RJ, Hillard CJ, Huang C, Chen N, Dong Z, Schmid HHO. (2001) Cannabinoid-receptor-independent cell signalling by *N*-acylethanolamines. *Biochemical Journal* 360: 67–75.

Bisogno T, Sepe N, Melck D, Maurelli S, De Petrocellis L, Di Marzo V. (1997a) Biosythesis, release and degradation of the novel indogenous cannabimimetic metabolite 2-arachidonoylglycerol in mouse neuroblastoma cells. *Biochemical Journal* 322: 671–677.

Bisogno T, Ventriglia M, Milone A, Mosca M, Cimino G, Di Marzo V. (1997b) Occurrence and metabolism of anandamide and related acyl-ethanolamides in ovaries of the sea urchin *Paracentrotus lividus*. *Biochemica Biophysica Acta* 1345: 338–348.

Bisogno T, Howell F, Williams G, Minassi A, Cascio MG, Ligresti A, Matias I, Schiano-Moriello A, Paul P, Williams EJ, et al. (2003) Cloning of the first sn1-DAG lipases points to the spatial and temporal regulation of endocannabinoid signaling in the brain. *Journal of Cell Biology* 163: 463–468.

Brown SM, Wager-Miller J, Mackie K. (2002) Cloning and molecular characterization of the rat CB2 cannabinoid receptor. *Biochemica Biophysica Acta* 1576: 255–264.

Calignano A, La Rana G, Piomelli D. (2001) Antinociceptive activity of the endogenous fatty acid amide, palmitylethanolamide. *European Journal Pharmacology* 419: 191–198.

Caterina MJ, Schumacher MA, Tominaga M, Rosen TA, Levine JD, Julius J. (1997) The capsaicin receptor: a heat-activated ion channel in the pain pathway. *Nature* 389: 816–824.

Chapman KD. (2000) Emerging physiological roles for *N*-acylphosphatidylethanolamine metabolism in plants: signal transduction and membrane protection. *Chemistry and Physics of Lipids* 108: 221–230.

Chang MC. Berkery D, Schuel R, Laychock SG, Zimmerman AM, Zimmerman S, Schuel H. (1993) Evidence for a cannabinoid receptor in sea urchin sperm and its role in blockade of the acrosome reaction. *Mol Reprod Dev* 36: 507–516.

Chemin J, Monteil A, Perez-Reyes E, Nargeot J, Lory P (2001) Direct inhibition of T-type calcium channels by the endogenous cannabinoid anandamide. *EMBO Journal* 17: 7033–7040.

Cravatt BF, Prospero-Garcia O, Siuzdak G, Gilula NB, Henriksen SJ, Boger DL, Lerner RA. (1995) Chemical characterization of a family of brain lipids that induce sleep. *Science* 268: 1506–1509.

De Petrocellis L, Melck D, Bisogno T, Milone A, Di Marzo V. (1999) Finding of the endocannabinoid signalling system in Hydra, a very primitive organism: possible role in the feeding response. *Neuroscience* 92: 377–387.

Devane WA, Hanus L, Breuer A, Pertwee RG, Stevenson LA, Griffin G, Gibson D, Mandelbaum A, Etinger A, Mechoulam R. (1992) Isolation and structure of a brain constituent that binds to the cannabinoid receptor. *Science* 258: 1946–1949.

Di Marzo V, Fontana A. (1995) Anandamide, an endogeonous cannabinomimetic eicosanoid: 'killing two birds with one stone.' *Prostaglandins Leukotrienes and Essential Fatty Acids* 53: 1–11.

Di Marzo V, Fontana A, Cadas H, Schinelli S, Cimino G, Schwartz JC, Piomelli D. (1994) Formation and inactivation of endogenous cannabinoid anandamide in central neurons. *Nature* 372: 686–691.

Di Marzo V, De Petrocellis L, Sepe N, Buono A. (1996) Biosynthesis of anandamide and related acylethanolamides in mouse J774 macrophages and N_{18} neuroblastoma cells. *Biochemical Journal* 316: 977–984.

Di Marzo V, Sepe N, De Petrocellis L, Berger A, Crozier G, Fride E, Mechoulam R. (1998) Trick or treat from food endocannabinoids? *Nature* 396: 636–637.

Di Marzo V, De Petrocellis L, Bisogno T, Melck D. (1999) Metabolism of anandamide and 2-arachidonoylglycerol: an historical overview and some recent developments. *Lipids* 34: S319–25.

Dinh TP, Freund TF, Piomelli D. (2002) A role for monoglyceride lipase in 2-arachidonoylglycerol inactivation. *Chemistry and Physics of Lipids* 121: 149–158.

Di Tomaso E, Beltramo M, Piomelli D. (1996) Brain cannabinoids in chocolate. *Nature* 382: 677–678.

Deutsch DG, Chin S. (1993) Enzymatic synthesis and degradation of anandamide, a cannabinoid receptor agonist. *Biochemistry and Pharmacology* 46: 791–796.

Egertová M, Cravatt BF, Elphick MR. (1998) Phylogenetic analysis of cannabinoid signalling. 1998 Symposium on the Cannabinoids, Burlington, VT: International Cannabinoid Research Society 1998: 101.

Elphick MR. (1998) An invertebrate G-protein coupled receptor is a chimeric cannabinoid/melanocortin receptor. *Brain Research* 780: 170–173.

Elphick MR. (2002) Evolution of cannabinoid receptors in vertebrates: identification of a CB(2) gene in the puffer fish Fugu rubripes. *Biological Bulletin* 202: 104–107.

Elphick MR, Egertová M. (2001) The neurobiology and evolution of cannabinoid signalling. *Philosophical Transactions of the Royal Society of London Biological Sciences* 356: 381–408.

Elphick MR, Satou Y, Satoh N. (2003) The invertebrate ancestry of endocannabinoid signalling: an orthologue of vertebrate cannabinoid receptors in the urochordate Ciona intestinalis. *Gene* 302: 95–101.

Felder CC, Nielsen A, Briley EM, Palkovits M, Priller J, Axelrod J, Nguyen DN, Richardson JM, Riggin RM, Koppel GA, Paul SM, Becker GW. (1996) Isolation and measurement of the endogenous cannabinoid receptor agonist, anandamide, in the brain and peripheral tissues of human and rat. *FEBS Letters* 393: 231–235.

Felder CC, Glass M. (1998) Cannabinoid receptors and their endogenous agonists. *Annual Review of Pharmacology and Toxicology* 38: 179–200.

Feng DF, Cho G, Doolittle RF. (1997) Determining divergence times with a protein clock: update and reevaluation. *Proceedings of the National Academy Science USA* 94: 13028–13033.

Fezza F, Dillwith JW, Bisogno T, Tucker JS, Di Marzo V, Sauer JR. (2003) Endocannabinoids and related fatty acid amides, and their regulation, in the salivary glands of the lone star tick. *Biochemica Biophysica Acta,* in press.

Fride E. (2002) Endocannabinoids in the central nervous system — an overview. *Prostaglandins, Leukotrienes and Essential Fatty Acids* 66: 221–233.

Gerwick WH, Wise ML, Sonderstrom K, Murray TF. (1997) Biosynthesis and cannabinoid receptor affinity of the novel eicosanoid, conjuncated triene anandamide. *Advances in Experimental Medicine and Biology* 407: 329–334.

Goparaju SK, Ueda N, Taniguchi K, Yamamoto S. (1999) Enzymes of porcine brain hydrolyzing 2-arachidonoylglycerol, an endogenous ligand of cannabinoid receptors. *Biochemistry and Pharmacology* 57: 417–423.

Glass M, Northup JK. (1999) Agonist selective regulation of G proteins by cannabinoid CB(1) and CB(2) receptors. *Molecular Pharmacology* 56: 1362–1369.

Griffin G, Tao Q, Abood ME. (2000) Cloning and pharmacological characterization of the rat CB(2) cannabinoid receptor. *Journal Pharmacology Experimental Therapeutics* 292: 886–894.

Hanus L, Gopher A, Almog S, Mechoulam R. (1993) Two new unsaturated fatty acid ethanolamides in brain that bind to the cannabinoid receptor. *Journal Medicinal Chemistry* 36: 3032–3034.

Hanus L, Abu-Lafi S, Fride E, Breuer A, Vogel, Z, Shalev, DE, Kustanovich, I, and Mechoulam R. (2001) 2-arachidonyl glyceryl ether, an endogenous agonist of the cannabinoid CB1 receptor. *Proceedings of the National Academy of Science USA* 98: 3662–3665.

Henikoff, S., Henikoff, J.G., 1992. Amino acid substitution matrices from protein blocks. *Proceedings of the National Academy Science USA* 89: 10915–10919.

Howlett AC, Bidaut-Russell, M, Devane, WA, Melvin, LS, Johnson, MR, and Herkenham M. (1990) The cannabinoid receptor: biochemical, anatomical and behavioral characterization. *Trends in Neuroscience* 13: 420–423.

Howlett AC, Mukhopadhyay, S, Wilken, GH, and Neckamyer WS. (2000) A cannabinoid receptor in Drosophila is pharmacologically unique. *Society for Neuroscience Abstracts* 26: 2165.

Hoyle CH. (1999) Neuropeptide families and the receptors: evolutionary perspectives. *Brain Research* 848: 1–25.

Huang SM, Bisogno T, Trevisani M, Al-Hayani A, De Petrocellis L, Fezza F, Tognetto M, Petros TJ, Krey JF, Chu CJ, Miller JD, Davies SN, Geppetti P, Walker JM, Di Marzo V. (2002) An endogenous capsaicin-like substance with high potency at recombinant and native vanilloid VR1 receptors. *Proceedings of the National Academy Science USA* 99: 8400–8405.

Hulse M, Johnson S, and Ferrieri P. (1993) Agrobacterium infections in humans: experience at one hospital and review. *Clinical Infectious Diseases* 16: 112–117.

International Human Genome Sequencing Consortium. (2001) Initial sequencing and analysis of the human genome. *Nature* 409: 860–920.

Karava V, Fasia L, Siafaka-Kapadai A. (2001) Anandamide amidohydrolase activity, released in the medium by Tetrahymena priformis. Identification and partial characterization. *FEBS Letters* 508: 327–331.

Karlsson M, Contreras JA, Hellman U, Tornqvist H, Holm C. (1997) cDNA cloning, tissue distribution, and identification of the catalytic triad of monoglyceride lipase. Evolutionary relationship to esterases, lysophospholipases, and haloperoxidases. *Journal Biological Chemistry* 272(43): 27218–27223.

Knight J, Taylor GW, Wright P, Clare AS, and Rowley AF. (1999) Eicosanoid biosynthesis in an advanced deuterostomate invertebrate, the sea squirt (*Ciona intestinalis*). *Biochim Biophys Acta* 1436: 467–478

Kozak KR, Marnett LJ. (2002) Oxidative metabolism of endocannabinoids. *Prostaglandins, Leukotrienes and Essential Fatty Acids* 66: 211–220.

Kuehl FA, Jacob TA, Ganley OH, Osmond RE, Meisinger MAP. (1957) The identification of N-2(hydroxy-ethyl)-palmitamide as a naturally occurring anti-inflammatory agent. *Journal American Chemistry Society* 79: 5577–5578.

Maccarrone M, Cartoni A, Parolaro D, Margonelli A, Massi P, Bari M, Battista N, Finazzi-Agro A. (2002) Cannabimimetic activity, binding, and degradation of stearoylethanolamide within the mouse central nervous system. *Molecular Cell Neuroscience* 21: 126.

Mackie K, Lai Y, Westenbroek R, Mitchell R. (1995) Cannabinoids activate an inwardly rectifying potassium conductance and inhibit Q-type calcium currents in AtT20 cells transfected with rat brain cannabinoid receptor. *Journal of Neuroscience* 15: 6552–6561.

Maingret F, Patel AJ, Lazdunski M, Honore E (2001) The endocannabinoid anandamide is a direct and selective blocker of the background K^+ channel TASK-1. *EMBO Journal* 20: 47–54.

Matias I, Bisogno T, Melck D, Vandenbulcke F, Verger-Bocquet M, De Petrocellis L, Sergheraert C, Breton C, Di Marzo V, and Salzet M. (2001) Evidence for an endocannabinoid system in the central nervous system of the leech *Hirudo medicinalis. Brain Research. Molecular Brain Research* 87: 145–159.

Matias I, DiMarzo V, McPartland JM (2005) Occurrence and possible biological role of the endocannabinoid system in the sea squirt *Ciona intestinalis. Journal of Neurochemistry* 93:1141–1156.

McAllister SD, Rizvi G, Anavi-Goffer S, Hurst DP, Barnett-Norris J, Lynch DL, Reggio PH, and Abood ME (2003) An aromatic microdomain at the cannabinoid CB(1) receptor constitutes an agonist/inverse agonist binding region, *Journal of Medical Chemistry* 46: 5139–5152.

McEno J, McPartland JM, Doctor B, and Canapasemi G. (1991) *Cannabis Ecology.* AMRITA Press, Middlebury, VT.

McPartland JM. (1984) Pathogenicity of Phomopsis ganjae on Cannabis sativa and the fungistatic effect of cannabinoids produced by the host. *Mycopathologia* 87: 149–153.

McPartland, JM (2004) Phylogenomic and chemotaxonomic analysis of the endocannabinoid system. *Brain Research Review* 45:18–29.

McPartland JM, Glass M. (2001) The nematocidal effects of Cannabis may not be mediated by cannabinoid receptors. *New Zealand Journal Crop and Horticultural Science* 29: 301–307.

McPartland JM, Glass M. (2003) Functional mapping of cannabinoid receptor orthologs and paralogs. *Gene*, accepted for publication.

McPartland JM, Guy G. (2004) The evolution of Cannabis and coevolution with the cannabinoid receptor — a hypothesis, in Guy, G.; Robson, R.; Strong, K.; Whittle, B. Eds. *The Medicinal Use of Cannabis*, pp. 71–102. London: Royal Society of Pharmacists.

McPartland JM, Nicholson J. (2003) Using parasite databases to identify potential nontarget hosts of biological control organisms. *New Zealand J Botany* 41(4): 699—706.

McPartland JM, Pruitt PL. (2002) Sourcing the code: searching for the evolutionary origins of cannabinoid receptors, vanilloid receptors, and anandamide. *Journal Cannabis Therapeutics* 2: 73–103.

McPartland JM, Glass M, and Mercer A. (2000a) Cannabis as a botanical pesticide: are cannabinoid receptors involved? ESA Program Abstracts, Entomological Society of America, Lanham, MD, p. 34.

McPartland JM, Pruitt PL, Matias I, DiMarzo V, Glass G. (2005) Phylogenomic analysis of nine endocannabinoid genes. Brain Research manuscript under review.

McPartland JM, Agraval J, Gleeson D, Heasman K, Glass M. (2005) Evidence for cannabinoid receptors in invertebrates. Journal of Evolutionary Biology manuscript under review.

McPartland JM, Clarke RC, and Watson DP. (2000) *Hemp Diseases and Pests: Management and Biological Control*. Wallingford, UK: CABI Publishing.

McPartland JM, Di Marzo, V, De Petrocellis, L, Mercer A, and Glass M. (2001) Cannabinoid receptors are absent in insects. *Journal of Comparative Neurology* 436: 423–429.

Mechoulam R, Gaoni Y. (1967) Recent advances in the chemistry of hashish. *Fortschritte der Chemie Organischer Naturstoffe* 25: 175–213.

Mechoulam R, Ben-Shabat, S, Hanus L, Ligumsky M, Kaminski, NF, Schatz AR, et al. (1995) Identification of an endogenous 2-monoglyceride, present in canine gut, that binds to cannabinoid receptors. *Biochemical Pharmacology* 50: 83–90.

Mechoulam R, Fride E, Di Marzo V. (1998) Endocannabinoids. *European Journal of Pharmacology* 359: 1–18.

Moesgaard B, Petersen G, Mortensen SA, and Hansen HS. (2002) Substantial species differences in relation to formation and degradation of *N*-acyl-ethanolamine phospholipids in heart tissue: an enzyme activity study. *Comp Biochem Physiol B Biochem Mol Biol.* 131: 475–82.

Murphy WJ, Eizirik E, Johnson WE, Zhang YP, Ryder OA, and O'Brien SJ. (2001) Molecular phylogenetics and the origins of placental mammals. *Nature* 409: 614–618.

Nakane S, Tanaka T, Satouchi K, Kobayashi Y, Waku K, and Sugiura T. (2000) Occurrence of a novel cannabimimetic molecule 2-sciadonoylglycerol (2-eicosa-5′,11′,14′-trienoylglycerol) in the umbrella pine Sciadopityl verticillata seeds. *Biological and Pharmaceutical Bulletin* 23: 758–761.

Natarajan V, Schmid PC, Reddy PV, Zuzarte-Augustin ML, Schmid HH. (1985) Occurrence of *N*-acylethanolamine phospholipids in fish brain and spinal cord. *Biochemica Biophysica Acta* 835: 426–433.

Okamoto Y, Morishita J, Tsuboi K, Tonai T, Ueda N. (2003) Molecular characterization of a phospholipase D generating anandamide and its congeners. *Journal of Biological Chemistry* 279: 5298–5305.

Onaivi ES, Leonard CM, Ishiguro H, Zhang PW, Lin Z, Akinshola BE, Uhl GR. (2002) Endocannabinoids and cannabinoid receptor genetics. *Progress in Neurobiology* 66: 307–344.

Patricelli MP, Cravatt BF. (2000) Clarifying the catalytic roles of conserved residues in the amidase signature family. *Journal Biological Chemistry* 275: 19177–19184.

Pertwee RG, Ross RA. (2002) Cannabinoid receptors and their ligands. *Prostaglandins, Leukotrienes and Essential Fatty Acids* 66: 101–121.

Poling JS, Rogawski MA, Salem N Jr, Vicini S (1996) Anandamide, an endogenous cannabinoid, inhibits Shaker-related voltage-gated K^+ channels. *Neuropharmacology* 35: 983–991.

Porter AC, Sauer JM, Knierman MD, Becker GW, Berna MJ, Bao J, Nomikos GG, Carter P, Bymaster FP, Leese AB, Felder CC. (2002) Characterization of a novel endocannabinoid, virodhamine, with antagonist activity at the CB1 receptor. *Journal Pharmacology Experimental Therapeutics* 301: 1020–1024.

Salzet M, Stefano GB. (2002) The endocannabinoid system in invertebrates. *Prostaglandins, Leukotrienes and Essential Fatty Acids* 66: 353–361.

Schaaf O, Dettner K. (2000) Polyunsaturated monoglycerides and a pregnadiene in defensive glands of the water beetle *Agabus affinis*. *Lipids* 35: 543–550.

Schmid HHO, Schmid PC, Berdyshev EV. (2002) Cell signaling by endocannabinoids and their congeners: questions of selectivity and other challenges. *Chemistry and Physics of Lipids* 121: 111–134.

Schroder-Borm H, Willumeit R, Brandenburg K, Andra J. (2003) Molecular basis for membrane selectivity of NK-2, a potent peptide antibiotic derived from NK-lysin. *Biochimica Biophysica Acta* 1612: 164–171.

Sepe N, De Petrocellis L, Montanaro F, Cimino G, Di Marzo V. (1998) Bioactive long chain *N*-acylethanolamines in five species of edible bivalve molluscs. Possible implications for mollusc physiology and sea food industry. *Biochemica Biophysica Acta* 1389(2): 101–111.

Shire D, Calandra B, Rinaldi-Carmona M, Oustric D, Pessegue B, Bonnin-Cabanne O, Le Fur G, Caput D, Ferrara P (1996) Molecular cloning, expression and function of the murine CB2 peripheral cannabinoid receptor. *Biochemica Biophysica Acta*, 132–136.

Sitachitta N, Gerwick WH. (1998) Grenadadiene and grenadamide, cyclopropyl-containing fatty acid metabolites from the marine cyanobacterium Lyngbya majuscula. *Journal Natural Products* 61: 681–684.

Smart D, Jonsson KO, Vandevoorde S, Lambert DM, Fowler CJ. (2002) 'Entourage' effects of *N*-acyl ethanolamines at human vanilloid receptors. Comparison of effects upon anandamide-induced vanilloid receptor activation and upon anandamide metabolism. *British Journal of Pharmacology* 136: 452–458.

Soderstrom K, Johnson F. (2000) CB1 cannabinoid receptor expression in brain regions associated with zebra finch song control. *Brain Research* 857: 151–157.

Soderstrom K, Leid M, Moore FL, Murray TF. (2000) Behavioral, pharmacological, and molecular characterization of an amphibian cannabinoid receptor. *Journal Neurochemistry* 75: 413–423.

Soderstrom K, Murray TF, Yoo HD, Ketchum S, Milligan, K, Gerwick, W, Ortega, MJ, Salva J. (1998) Discovery of novel cannabinoid receptor ligands from diverse marine organisms. *Advances in Experimental Medicine and Biology* 433: 73–77.

Stanley-Samuelson DW, Pedibhotla VK. (1996) What can we learn from prostaglandins and related eicosanoids in insects? *Insect Biochemistry and Molecular Biology* 26: 223–234.

Stefano GB, Liu Y, Goligorsky MS. (1996) Cannabinoid receptors are coupled to nitric oxide release in invertebrate immunocytes, microglia, and human monocytes. *Journal Biological Chemistry* 271: 19238–19242.

Stefano GB, Salzet B, Salzet M. (1997) Identification and characterization of the leech CNS cannabinoid receptor: coupling to nitric oxide release. *Brain Research* 753: 219–724.

Stefano GB, Rialas CM, Deutsch DG, Salzet M. (1998) Anandamide amidase inhibition enhances anandamide-stimulated nitric oxide release in invertebrate neural tissues. *Brain Research* 793: 341–345.

Suga H, Koyanagi M, Hoshiyama D, Ono K, Iwabe N, Kuma KI, Miyata T. (1999) Extensive gene duplication in the early evolution of animals before the parazoan-eumetazoan split demonstrated by G proteins and protein tyrosine kinase from sponge and hydra. *Journal of Molecular Evolution* 48: 646–653.

Sugiura T, Kondo S, Sukagawa A, Nakane S, Shinoda A, Itoh K, Yamashita A, Waku K. (1995) 2-Arachidonylglycerol: a possible endogenous cannabinoid receptor ligand in brain. *Biochemical and Biophysical Research Communication* 215: 89–97.

Tanaka T, Ikita K, Ashida T, Motoyama Y, Yamaguchi Y, Satouchi K. (1996) Effects of growth temperature on the fatty acid composition of the free living nematode Caenorhabditis elegans. *Lipids* 31: 1173–1178.

Tripathy S, Kleppinger-Sparace K, Dixon RA, Chapman KD. (2003) *N*-acylethanolamine signaling in tobacco is mediated by a membrane-associated, high-affinity binding protein. *Plant Physiology* 131: 1781–1791.

Ueda N, Yamamoto K, Kurahashi Y, Yamamoto S, Ogawa M, Matsuki N, Kudo I, Shinkai H, Shirakawa E, Tokunaga T. (1995) Oxygenation of arachidonylethanolamide (anandamide) by lipozygenases. *Advances in Prostaglandin Thromboxane and Leukotriene Research* 23: 163–165.

Ueda N, Puffenbarger RA, Yamamoto S, Deutsch DG. (2000) The fatty acid amide hydrolase (FAAH). *Chemistry and Physics of Lipids* 108: 107–121.

Ueda N, Yamanaka K, Yamamoto S. (2001) Purification and characterization of an acid amidase selective for *N*-palmitoylethanolamine, a putative endogenous anti-inflammatory substance. *Journal Biological Chemistry* 276: 35552–35557.

van Zadelhoff, G, Veldink, GA, and Vliegenhart FG. (1998) With anandamide as substrate plant 5-lipoxygenases behave like 11-lipoxygenases. *Biochemical and Biophysical Research Communications* 248: 33–38.

Yamaguchi F, Macrae AD, and Brenner S. (1996) Molecular cloning of two cannabinoid type 1-like receptor genes from the puffer fish Fugu rubripes. *Genomics* 35: 603–605.

Yu M, Ives D, Ramesha CS. (1997) Synthesis of prostaglandin E_2 ethanolamide from anandamide by cyclooxygenase-2. *Journal Biological Chemistry* 273: 21181–21186.

Zank TK, Zähringer U, Lerchl J, Heinz E. (2000) Cloning and functional expression of the first plant fatty acid elongase specific for Δ^6-polyunsaturated fatty acids. *Biochemical Society Transactions* 28: 654–658.

Ziemienowicz A, Gorlich D, Lanka E, Hohn B, and Rossi L. (1999) Import of DNA into mammalian nuclei by proteins originating from a plant pathogenic bacterium. *Proceedings National Academy Sciences USA* 96(7): 3729–3733.

Zygmunt PM, Petersson J, Andersson DA, Chuang, H, Sorgard M, Di Marzo V, Julius D, Hogestatt ED. (1999) Vanilloid receptors on sensory nerves mediate the vasodilator action of anandamide. *Nature* 400: 452–457.

Part VIII

Perspective

24 Endocannabinoid-Based Molecules as Potential Therapeutic Drugs

Alessia Ligresti and Vincenzo Di Marzo

CONTENTS

INTRODUCTION

The discovery of the metabolic pathways underlying endocannabinoid formation and inactivation does not simply represent an important step toward the understanding of the cannabinoid-receptor-mediated endogenous signaling system. In fact, the knowledge of the mechanisms through which the endogenous levels of the endocannabinoids, and hence the state of activation of cannabinoid receptors, are regulated may also have an enormous impact on the development of new therapeutic drugs. This concept is based on the original assumption, now being supported by an ever-increasing number of experimental reports, that the symptoms of several central and peripheral disorders are due to or cause changes in endocannabinoid biosynthetic and degradative pathways, and, subsequently, pathologically altered activation of either CB_1 or CB_2 cannabinoid receptors (see Chapters 7 and 14 through 22). If it is found that the symptoms of a certain disorder are caused, for example, by an increase in endocannabinoid levels, then substances that selectively inhibit endocannabinoid biosynthesis may be used, along with cannabinoid receptor antagonists, to improve these symptoms. Conversely, if the symptoms are due to defective endocannabinoid levels, then substances that selectively inhibit endocannabinoid inactivation may produce beneficial effects. The potential consequences of this line of reasoning may be even more important from the clinical point of view if one imagines that the onset or progress of some disorders such as multiple sclerosis (Pryce et al., 2003; Arevalo-Martin et al., 2003), and not merely their symptoms, may be inhibited by an enhanced endocannabinoid tone.

ENDOCANNABINOID BIOSYNTHESIS AND ITS PHARMACOLOGICAL INHIBITION

To date, the biosynthetic pathways of only the two best-studied endocannabinoids, anandamide (AEA) and 2-arachidonoyl-glycerol (2-AG), have been almost completely elucidated (Figure 24.1). It is well accepted that these pathways provide for the "on demand" formation of endocannabinoids with neuromodulatory action in the central and peripheral nervous systems. However, only seldom have the proteins involved in these processes been cloned or even characterized from a molecular point of view, which makes it difficult to develop selective inhibitors of endocannabinoid formation. The Ca^{2+}-dependent acyltransferase responsible for the biosynthesis of AEA precursor *N*-arachidonoylphosphatidylethanolamine (NarPE) (Figure 24.1) may be related to non-acylCoA-dependent acyltransferase enzymes, although in this case, the enzyme catalyzes the transfer of arachidonic acid (and other fatty acids) from the *sn*-1 position of phospholipids to the nitrogen atom in phosphatidylethanolamine (and hence not to an oxygen atom as with other acyltransferases). So far, this protein has been characterized only enzymatically (Sugiura et al., 1996; Cadas et al., 1997), and nonselective protease inhibitors have been found to block its activity with varying potency. Also, the enzyme catalyzing hydrolysis of NarPE to AEA has been characterized (Sugiura et al., 1996), and its recent cloning (Okamoto et al., 2004) established that this phosphodiesterase is distinct from phospholipases D (PLDs) that recognize as substrates the conventional phosphoglyceride classes (Schmid et al., 1996). No inhibitor has yet been found for this protein. In fact, this enzyme is a member of the zinc metallohydrolase family of the beta-lactamase fold, and bears little homology with the cloned PLD enzymes (Park et al., 1977). Finally, an enzyme that possibly catalyzes the formation of NarPE precursors, i.e., phospholipids with arachidonic acid esterified on the *sn*-1 position, has been recently characterized (Sugimoto and Yamashita, 1999).

Regarding 2-AG biosynthesis (Figure 24.1), two key isozymes catalyzing the formation of this compound from the selective hydrolysis of fatty acid esterified to the the *sn*-1 position of the corresponding diacylglycerol (DAG), also known as *sn*-1-selective DAG lipase (DAGL) (Bisogno, Sepe, et al., 1997; Stella et al., 1997; Bisogno et al., 1999), have been recently cloned and characterized (Bisogno et al., 2003). These two enzymes recognize DAGs with a wide range of different fatty acids in either the *sn*-1 or 2- position, but they preferentially cleave the fatty acid from the *sn*-1 position of the DAG substrate, thus selectively yielding the corresponding 2-acylglycerol. It was shown that inhibition of the two enzymes with somewhat selective DAGL inhibitors leads to blockade of 2-AG biosynthesis and release from intact cells, and prevents a typically endocannabinoid-mediated response in cerebellar granule cells (Bisogno et al., 2003). These substances might pave the way for the development of more selective and potent inhibitors of 2-AG biosynthesis. Because 2-AG is more abundant than AEA in most tissues, blocking its biosynthesis would strongly influence the output of the endocannabinoid system during physiological or pathological conditions.

Very little is known of the biosynthetic pathways of the other putative endocannabinoids, i.e., noladin ether, virodhamine, and *N*-arachidonoyl-dopamine (NADA). Although the actual occurrence of noladin ether in mammalian tissues is still controversial (Hanus et al., 2001; Fezza et al., 2002; Oka et al., 2003), also because of the lack of a working hypothesis regarding a plausible biosynthetic route for a 2-glyceryl ether, the formation of virodhamine has been related to that of AEA, from which the compound can be derived in a nonenzymatic reaction (Markey et al., 2000). In the case of NADA, preliminary data from our laboratory seem to rule out one of the two possible original routes proposed for this compound, i.e., that it might be derived from the action on *N*-arachidonoyl-tyrosine

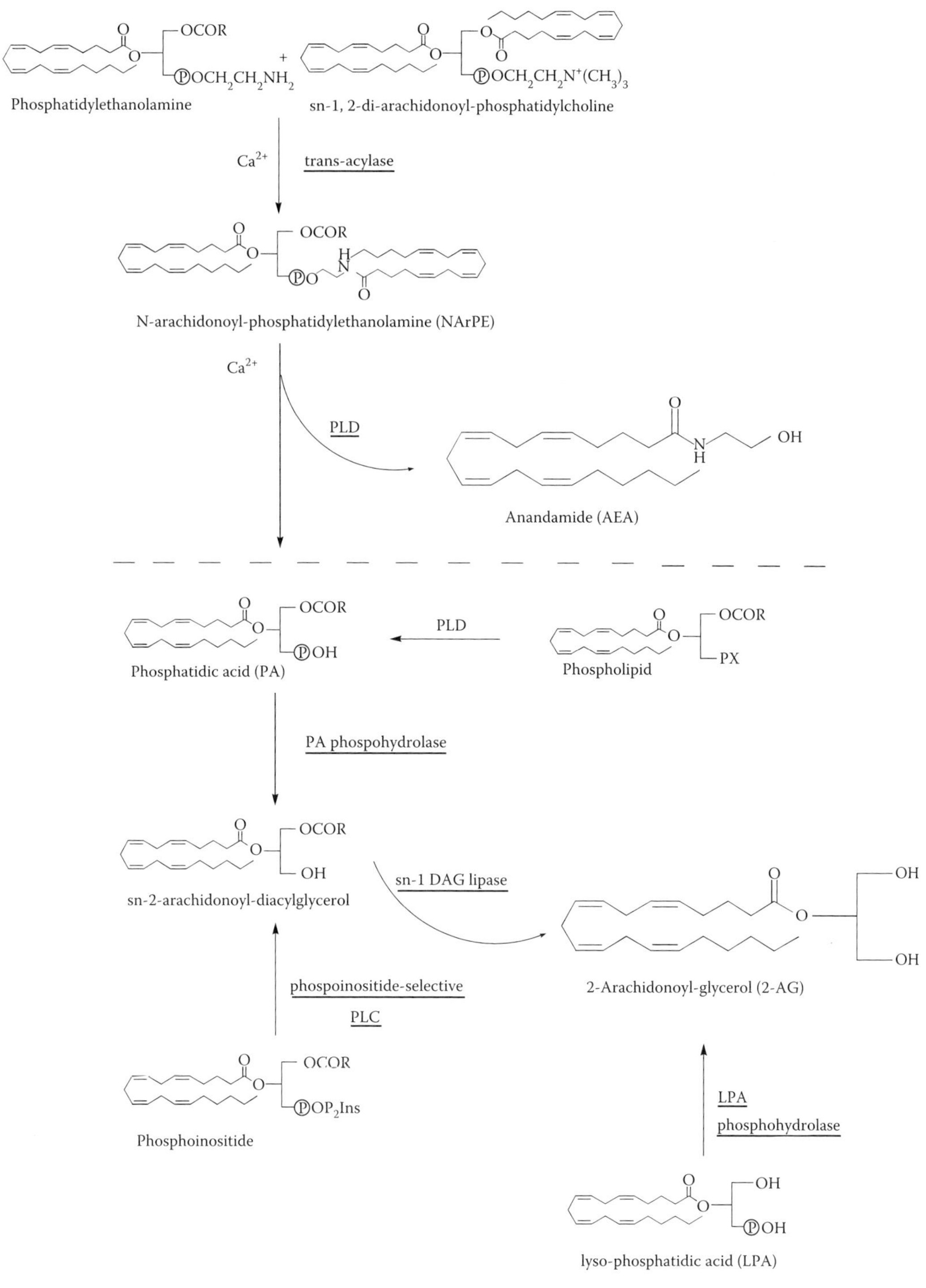

FIGURE 24.1 Biosynthetic pathways of anandamide (AEA) and 2-arachidonoyl-glycerol (2-AG) in intact cells.

of dopamine-synthesizing enzymes (Huang et al., 2002). Instead, the compound appears to be produced from the reaction between arachidonic acid and dopamine, and possibly in the same way as *N*-arachidonoyl-glycine is produced from arachidonic acid and glycine (Huang et al., 2001). The hypothesis of the existence of one or more "synthase" enzymes for amides between long-chain fatty acids and amino acids or bioactive amines and of their sensitivity to typical acylCoA-dependent synthase inhibitors, still needs to be tested.

ENDOCANNABINOID INACTIVATION AND ITS PHARMACOLOGICAL INHIBITION

The inactivation pathways of AEA and 2-AG, as well as of other endocannabinoids, seem to have one or more steps in common (Figure 24.2). A membrane transporter (see Hillard and Jarrahian, 2000, for a review) has been suggested to facilitate the cellular uptake of all endocannabinoids, depending on the gradient of concentration across the cell membrane. This would be in all cases the first step of inactivation because all the endocannabinoid-degrading enzymes found so far are intracellular. The suggestion of a common mechanism for the cellular uptake of endocannabinoids was based on several observations including: (1) AEA, 2-AG, noladin, and NADA are rapidly taken up by both neuronal and nonneuronal cells in a saturable, temperature-dependent (and in the case of AEA, energy-independent) manner (Di Marzo et al., 1994; Hillard et al., 1997; Beltramo and Piomelli, 2000; Bisogno et al., 2001; Fezza et al., 2002; Huang et al., 2002), (2) 2-AG, noladin, NADA, and virodhamine compete with AEA cellular uptake, and AEA competes with 2-AG, noladin, and NADA uptake (Beltramo and Piomelli, 2000; Bisogno et al., 2000, 2001; Fezza et al., 2002), and (3) AEA and 2-AG uptake is subject to selective inhibition by some AEA analogs and to stimulation by nitric oxide (Beltramo et al., 1997; Bisogno et al., 2001; Maccarrone et al., 1998, 2000).

The second and final step of inactivation depends on the chemical nature of the endocannabinoid. AEA and 2-AG are degraded via enzymatic hydrolysis of their respective amide and ester bonds. The enzymes mostly responsible for these reactions have been cloned, and they are: (1) the fatty acid amide hydrolase (FAAH) (Cravatt et al., 1996; see Ueda et al., 2000, for review), which catalyzes the hydrolysis of both AEA and 2-AG, and (2) the monoacylglycerol lipase (MAGL) (Karlsson et al., 1997; Dinh et al., 2002), which instead is selective for 2-AG and other monoacylglycerols. Although it is likely that virodhamine also is degraded via enzymatic hydrolysis, in the case of noladin, whose ether bond cannot be hydrolyzed by an enzyme, intracellular metabolism occurs through its incorporation into phosphoglycerides (Fezza et al., 2002). Finally, NADA is also refractory to enzymatic hydrolysis but appears to be easily methylated by the dopamine-inactivating enzyme catechol-*O*-methyl-transferase (Huang et al., 2002), and it is likely to be oxidized both enzymatically and nonenzymatically much in the same way dopamine is. The polyunsaturated fatty acid moiety of the endocannabinoids, particularly AEA and 2-AG, can also be oxidized by enzymes of the arachidonate cascade, such as lipoxygenases and cyclooxygenase-2 (see Kozak and Marnett, 2002, for review). Whether this reaction occurs *in vivo*, and with the only purpose of inactivating the endocannabinoid signal, is still under investigation.

The putative endocannabinoid transporter has not been cloned yet. This, and the fact that FAAH and MAGL play an important role in determining the rate of AEA and 2-AG cellular uptake (Bisogno et al., 2001; Deutsch et al., 2001), may suggest that endocannabinoid transport across plasma membranes is not carrier mediated and that it occurs uniquely via passive diffusion dependent on the rate of intracellular metabolism (Glaser et al., 2003). Accordingly, some inhibitors of AEA cellular uptake are also FAAH inhibitors (Jarrahian et al., 2000) and *vice versa* (Deutsch et al., 2001; Glaser et al., 2003). However, several indirect observations support the existence of an endocannabinoid transporter, and in particular (1) some substances selectively inhibit AEA cellular uptake without inhibiting FAAH (Di Marzo et al., 2002; Ortar et al., 2003; Lopez-Rodriguez et al., 2003), (2) FAAH inhibitors enhance and AEA uptake inhibitors inhibit AEA accumulation in cells (Kathuria et al., 2003), (3) cells that do not express FAAH rapidly take up AEA (Di Marzo et al., 1999; Deutsch et al., 2001), (4) inhibitors of AEA cellular uptake enhance the effects of AEA that are mediated by CB receptors but inhibit those that are exerted on the cytosolic side of membrane proteins, such as the stimulation of vanilloid VR1 receptors (De Petrocellis et al., 2001) or the inhibition of T-type

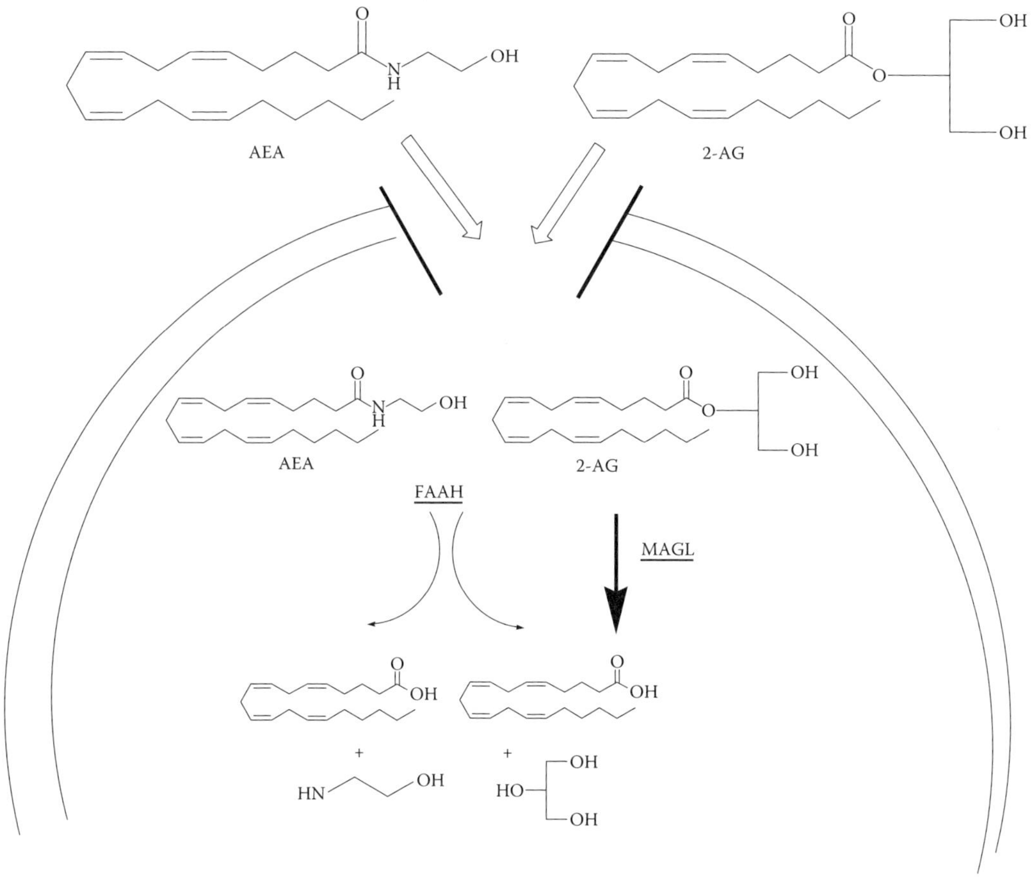

FIGURE 24.2 Cellular uptake and enzymatic hydrolysis: the major inactivation pathway of anandamide (AEA) and 2-arachidonoyl-glycerol (2-AG).

Ca^{2+} channels (Chemin et al., 2001), (5) NADA and noladin are rapidly taken up by cells, yet they are either very stable or refractory to enzymatic hydrolysis, respectively (Fezza et al., 2002; Huang et al., 2002), and (6) lipopolysaccharide inhibits FAAH expression without affecting AEA cellular uptake (Maccarrone et al., 2001); conversely, nitric oxide, peroxynitrite, and superoxide anions stimulate AEA cellular reuptake (Maccarrone et al., 2000), whereas acute or chronic ethanol treatment inhibits this process (Basavarajappa et al., 2003) without affecting FAAH activity.

As mentioned earlier, somewhat specific inhibitors of endocannabinoid cellular uptake (Beltramo et al., 1997; Melck et al., 1999; Jarrahian et al., 2000; De Petrocellis et al., 2000; Lopez-Rodriguez et al., 2001) and of FAAH (De Petrocellis et al., 1997; Deutsch et al., 1997; Bisogno et al., 1998; Boger et al., 2000; Martin et al., 2000; Kathuria et al., 2003; Segall et al., 2003) have been developed (Table 24.1 and Table 24.2). Although uptake inhibitors have often been found to be unselective over vanilloid VR1 receptors or FAAH (De Petrocellis et al., 2000), many FAAH inhibitors also inhibit phospholipase A_2 (PLA_2) enzymes or bind to cannabinoid CB_1 receptors. Nevertheless, in many instances these compounds were shown to enhance AEA pharmacological actions *in vitro* or *in vivo*. AM374, arachidonoyl-serotonin, O-1624, and URB-597 appear to be particularly efficacious *in vivo*, although only in a few cases have these compounds been shown to exert, as expected, their pharmacological actions via the enhancement of AEA levels (see the following sections). No selective inhibitor of MAGL has been developed yet, and the only example

TABLE 24.1
Inhibitors of Anandamide (AEA) Cellular Uptake Most Often Used *In Vivo*

Chemical Structure	IC_{50}	Cell Type Tested	References	Selectivity
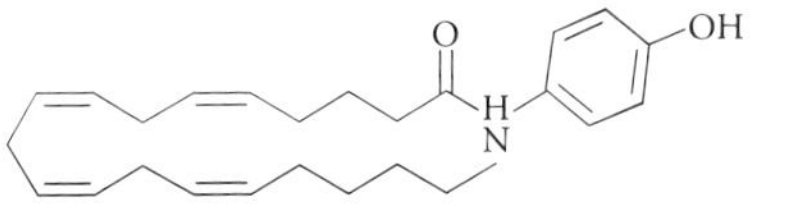 AM-404	10.2 μ*M* 8.1 μ*M* 1.0 μ*M*	C_6 glioma cells RBL-2H3 cells Rat brain neurons and astrocytes	De Petrocellis et al., 2000 Beltramo et al., 1997	Binds to vanilloid receptors and inhibits FAAH
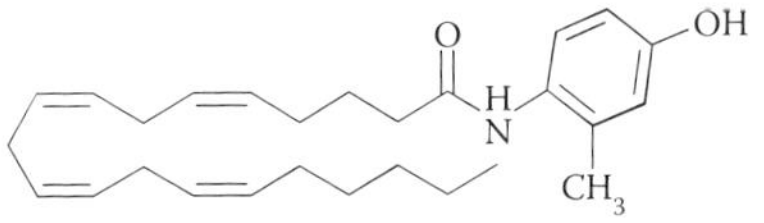 VDM-11	10.2 μ*M* 11.2 μ*M*	C_6 glioma cells RBL-2H3 cells	De Petrocellis et al., 2000	Selective over VR1 receptors and FAAH
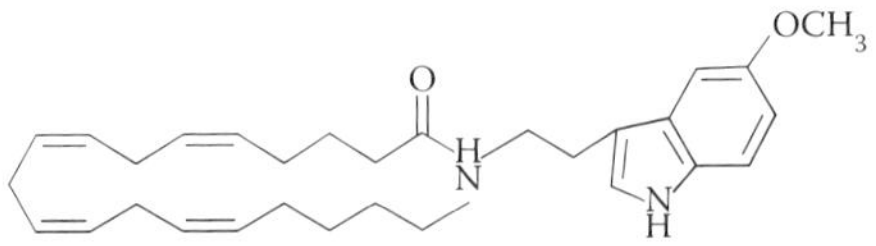 VDM-13	12.0 μ*M* 12.0 μ*M*	C_6 glioma cells RBL-2H3 cells	De Petrocellis et al., 2000	Selective over VR1 receptors and FAAH
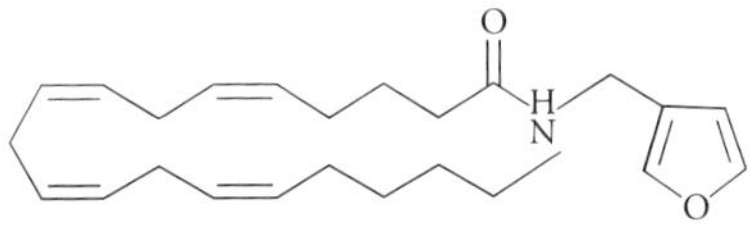 UCM-707	0.8 μ*M*	U937 lymphoma cells	Lopez-Rodriguez et al., 2001	Selective over VR1 receptors and FAAH
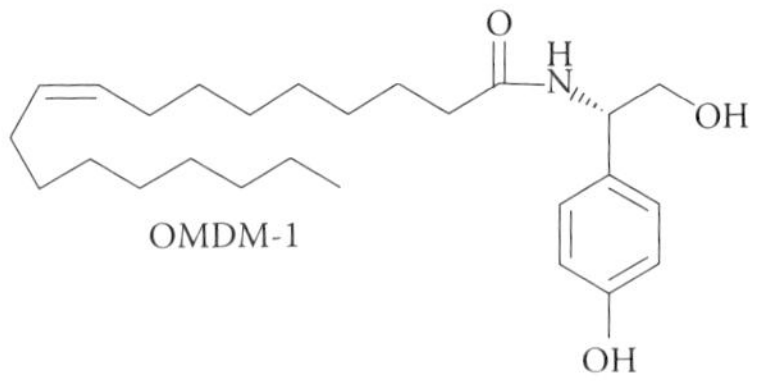 OMDM-1	2.4 μ*M*[a]	RBL-2H3 cells	Ortar et al., 2003	Selective over VR1 receptors and FAAH
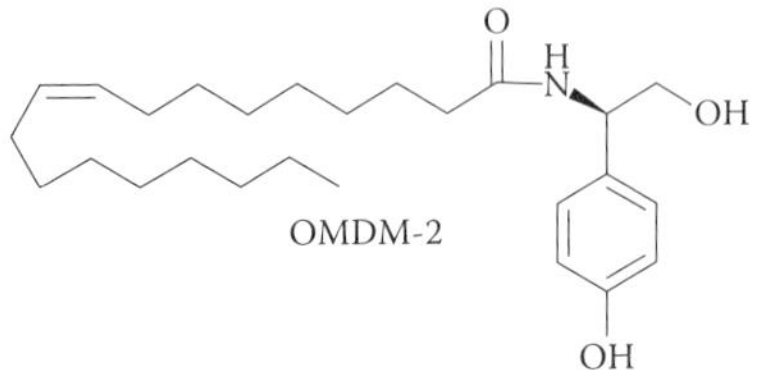 OMDM-2	3.0 μ*M*[a]	RBL-2H3 cells	Ortar et al., 2003	Selective over VR1 receptors and FAAH

[a] These data are reported as K_i values.

of potentiation of 2-AG effects exerted in part by inhibition of its inactivation is represented by the "entourage" effect, in which cannabinoid-receptor-inactive monoacylglycerols, which are normally produced by cells together with 2-AG, enhance its effects when administered together with the endocannabinoid (Ben-Shabat et al., 1998; Panikashvili et al., 2001). In the case of the endocannabinoid transporter, the most widely used compounds to date are AM404 (Beltramo et al., 1997), VDM-11, and VDM-13 (De Petrocellis et al., 2000) (Table 24.1). Olvanil, arvanil, and linvanil (Melck et al.,1999), which are relatively potent but unselective (similar to AM404, and unlike VDM-11 or VDM-13, they all activate vanilloid VR1 receptors), have been used sporadically *in vitro*, whereas others, such as UCM-707, OMDM-1, and OMDM-2, have just been developed

TABLE 24.2
Inhibitors of Fatty Acid Amide Hydrolase (FAAH) Most Often Used *In Vivo*

Chemical Structure	IC_{50}	Cell Type Tested	References	Selectivity
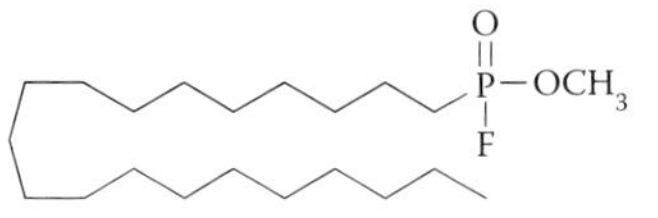 O1624	137 n*M*	Brain homogenate	Martin et al., 2000	Selective over CB_1; not tested on PLA_2
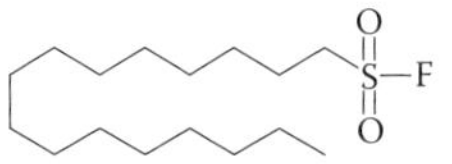 AM374	13 n*M*	Brain homogenate	Deutsch et al., 1997	Selective over CB_1; not tested on PLA_2
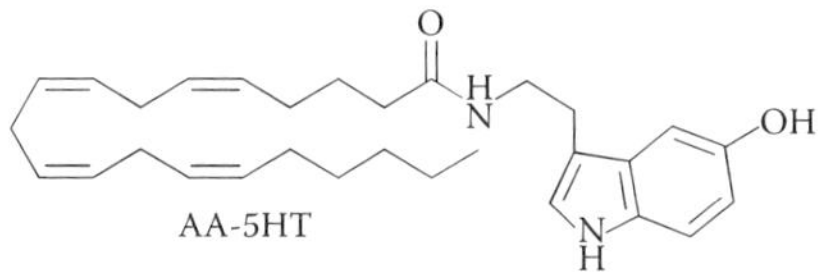 AA-5HT	5.6 μ*M* 1.0 μ*M*	RBL-2H3 cells Brain homogenate	Bisogno et al., 1998 Fowler et al., 2003	Selective over CB_1 and PLA_2
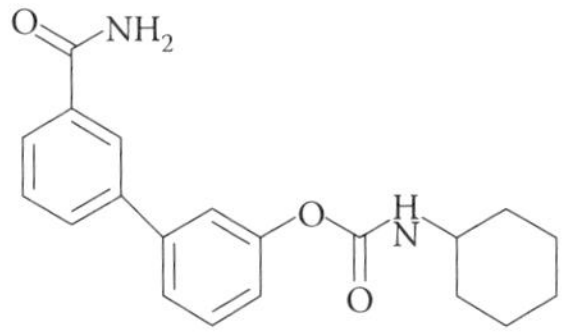 URB-597	4.6 n*M*	Brain homogenate	Kathuria et al., 2003	Selective over CB_1 and several hydrolases

Note: PLA_2 - phospholipase A_2.

and appear to be as selective as VDM-11, although they are more potent. Preliminary data exist so far for the possible use of the latter compounds *in vivo* (de Lago et al., 2002, 2004). Finally, although the AEA congener *N*-palmitoyl-ethanolamide does not share with AEA the capability of being a substrate for the putative endocannabinoid transporter (Bisogno, Maurelli, et al., 1997), analogs of these compounds, and particularly palmitoylisopropylamide, were found to inhibit the cellular uptake of AEA quite selectively (Jonsson et al., 2001).

The *in vivo* pharmacological actions of the endocannabinoid AEA that have been shown to be enhanced by inhibitors of endocannabinoid inactivation concern most often the typical analgesic and hypolocomotor actions of AEA. Thus, for example, O-1624 and URB-597 enhance the antinociceptive effects of AEA in tests of acute pain (Martin et al., 2000; Kathuria et al., 2003), and so do AM404, VDM-11, UCM-707, and OMDM-1 (Beltramo et al., 1997; de Lago et al., 2002, 2004). UCM-707 and OMDM-2 enhance the motor-inhibitory effect of the endocannabinoid (de Lago et al., 2002; de Lago et al., 2004). Interestingly, many FAAH inhibitors are capable of exerting analgesic effects *per se*, which is in agreement with the lower threshold of pain perception observed in transgenic mice with an inactive FAAH gene (Cravatt et al., 2001), in which endogenous AEA levels are enhanced exactly as they are after administration of URB-597 to rats (Kathuria et al., 2003). Indeed, some substances that possess cannabimimetic effects *in vivo*, such as the endogenous sleep-inducing factors *cis*-9-octadecenoaminde (oleamide) (Cravatt et al., 1995) and 2-octyl-γ-bromoacetoacetate (Boger et al., 1998), the former of which has been shown to mimic AEA in tests of antinociception, hypolocomotion, catalepsy, and hypothermia (Mechoulam et al., 1997), and the analgesic lipid

N-arachidonoyl-glycine (Huang et al., 2001) have been found to be substrates for and inhibitors of FAAH. Therefore, these compounds were suggested to act by inhibition of the inactivation of endogenous AEA, with subsequent enhancement of the levels of this compound and induction of its sedative or analgesic actions (Mechoulam et al., 1997; Burstein et al., 2002). Although this hypothesis has been ruled out, at least in part, for oleamide, by showing that this compound still produces some cannabimimetic effect in FAAH knockout mice (Lichtman et al., 2002), a recent observation has confirmed that inhibition of FAAH and subsequent enhancement of endocannabinoid levels can produce strong endocannabinoid-like pharmacological actions different from antinociception (Patel et al., 2003). In fact, it was found that the general anesthetic propofol (1) produces an enhancement of endocannabinoid levels within a time frame compatible with the peak of its sedative action in mice, (2) inhibits AEA hydrolysis by FAAH but does not inhibit AEA cellular uptake, and (3) induces the loss of righting reflex (an index of the sedative action of general anesthetics in mice) in a way that is antagonized by a selective dose of the CB_1 receptor antagonist SR141716A; the antagonist, instead, did not inhibit the sedative effect of another general anesthetic thiopental, which, accordingly, was unable to enhance mouse brain endocannabinoid levels. These data, together with the discovery of a clear relationship between the efficacy of the sedative effects of a series of propofol analogs and their potency as FAAH inhibitors, strongly suggested that the sedative effects of this compound, as well as some of its other cannabimimetic actions, are indeed due to FAAH inhibition and the subsequent enhancement of endocannabinoid levels and activation of cannabinoid CB_1 receptors. Furthermore, these observations have confirmed that some beneficial effects of THC and other "direct" CB_1 receptor agonists may be also obtained "indirectly," by using substances that inhibit the inactivation processes of the endocannabinoids.

USE OF ENDOCANNABINOID-BASED SUBSTANCES IN ANIMAL MODELS OF DISORDERS

The potential therapeutic value of *Cannabis sativa* and of its major components, THC and cannabidiol, is currently being reevaluated (Hall and Degenhardt, 2003). In the case of THC and its synthetic analogs capable of activating cannabinoid CB_1 receptors in the brain, the possible therapeutic applications are limited by their psychotropic side effects. For this reason, the possibility that substances capable of activating cannabinoid CB_1 receptors indirectly, by enhancing endocannabinoid levels, may represent templates for the development of new therapeutic drugs is being actively investigated. These inhibitors are, in fact, more likely to increase the concentrations of endocannabinoids only at sites where their biosynthesis or inactivation have been altered with pathological consequences, without greatly altering the activity of cannabinoid receptors at other sites. Based on this line of reasoning, several selective inhibitors of FAAH and the putative endocannabinoid transporter have been tested in animal models of both central and peripheral disorders, and the results obtained are briefly reviewed in this section.

MULTIPLE SCLEROSIS

The first example of a possibly successful use of such inhibitors in a disease was reported by Baker and co-workers (2001) in the case of chronic relapsing experimental allergic encephalomyelitis (CREAE) in mice. This is a disorder that very much resembles multiple sclerosis in its signs and development. Mice with CREAE undergo relapses of spasticity of the limbs and tremors, alternating with phases of absence of these signs. Spasticity and tremors in these mice can be alleviated by cannabinoid receptor agonists and worsened by antagonists (Baker et al., 2000), which suggests the existence of a tone of endocannabinoids that alleviate these signs. Indeed, the brains and, particularly, spinal cords of CREAE mice contain elevated levels of endocannabinoids only during the spasticity phase of the disorder, and these compounds, as with THC and other agonists, inhibit spasticity (Baker et al., 2001). The existence of an antispasticity tone of the endocannabinoids and the possibility of its exploitation for therapeutic purposes was strongly supported by the finding that a FAAH inhibitor, AM374, and two inhibitors of endocannabinoid cellular uptake, VDM-11 and AM404, also inhibited the spasticity in CREAE mice, their beneficial effect being blocked by cannabinoid receptor antagonists (Baker et al., 2001). Subsequently, OMDM-1 and OMDM-2 were also found to inhibit spasticity, the effect of the latter compound being slightly longer lasting than that of other inhibitors (de Lago et al., 2004). Preliminary evidence suggests that these compounds, as with cannabinoid receptor agonists (Arevalo-Martin et al., 2003), may also prove useful in slowing down the development of experimental allergic encephalomyelitis in mice (Mestre et al., 2005).

Protection against Neuronal Hyperactivity

Although the mechanism of their action was not elucidated, it is possible that the beneficial effect of endocannabinoid inactivation inhibitors against spasticity in CREAE was because the enhancement of the endocannabinoid signal in the spinal cord might eventually lead to counteraction of exaggerated activity of neuromuscular synapses. Indeed, compensatory activities of the endocannabinoids against neuronal hyperactivity have been explored. In a model of Parkinson's disease, the destruction in the rat brain of the nigrostriatal fibers with 6-hydroxy-dopamine (6-OHDA) was found to result in elevated striatal endocannabinoid levels (Gubellini et al., 2002). In view of the well-established inhibitory action of presynaptic CB_1 receptors on glutamate release in many neurons (see Davies et al., 2002, for a review), it was suggested that this was an adaptive response of the striatum to compensate for the hyperactivity of corticostriatal glutamatergic terminals. This hypothesis was recently supported by a study showing that (1) inhibitors of either FAAH (methyl-arachidonoyl-fluoro-phosphonate) or the endocannabinoid transporter (VDM-11 and AM404) reduced the hyperactivity of these neurons and (2) treatment of 6-OHDA-treated rats not only reversed the signs of Parkinson's disease but also reestablished normal levels of endocannabinoid in the striatum (Maccarrone et al., 2003). These data may suggest that endocannabinoid inactivation inhibitors might be developed into drugs that counter hypolocomotion in Parkinson's disease. However, it must be remembered that enhancement of endocannabinoid levels in another region of the basal ganglia, the globus pallidus, was suggested to be responsible for the hypolocomotion in reserpine-treated rats (Di Marzo et al., 2000). Therefore, the increase in endocannabinoid levels in the basal ganglia of animal models of Parkinson's disease may need to be region specific in order to bring about a beneficial effect on the symptoms of this disorder.

A typical example of neuronal hyperactivity is glutamate-induced excitotoxicity, which is one of the causes of seizures during epilepsy. This pathological state can be induced in mice with the administration of kainate, with typical consequences consisting of multiple seizures and, eventually, death. Using this model, it has been shown that AEA levels are enhanced in the hippo-campus with a peak 20 min after the injection of kainate, which decline 1 h after the administration. This observation, together with the finding that cannabinoid CB_1 knockout mice are more susceptible to kainate-induced seizures than wild-type mice and that they exhibit stronger neuronal hyperactivity in the hippocampus following kainate injections, suggested that endocannabinoids, produced on demand in the hippocampus, afford neuroprotection against the hyperactivity subsequent to the injection of kainate (Marsicano et al., 2003). This hypothesis was supported by the finding that the inhibitor of AEA reuptake UCM-707 ameliorated the neurological scores of wild-type mice treated with kainate, which also suggests that endocannabinoid inactivation inhibitors might be developed for neuroprotection against neuronal excitotoxicity (Marsicano et al., 2003).

Pain

The antinociceptive activity of the endocannabinoid transporter and, particularly, FAAH inhibitors has been mentioned earlier. This effect, however, has been mostly studied using the "hot plate" paradigm of acute pain, which is not very relevant to the most frequent types of chronic, inflammatory, and neuropathic pain occurring in humans. Several animal models of these types of pain have been developed, and in many but not all instances cannabinoid receptor (both CB_1 and CB_2) agonists and antagonists have been found to exert antihyperalgesic or hyperalgesic activity, respectively (for a review, see Walker and Huang, 2002). However, in no case were inhibitors of endocannabinoid inactivation tested, which would have provided further evidence for the postulated endocannabinoid tone controlling chronic and inflammatory pain. Instead, an important study recently revealed how the endocannabinoid system plays a major role in the phenomenon known as stress-induced analgesia. According to this study, animals submitted to a short period of stress, such as a non-noxious electrical shock, become hypoalgesic to stronger nociceptive stimuli, partly because of the activation of the endogenous opioid system. Using such a model, it was shown that endocannabinoids and CB_1 receptors mediate nonopioid stress-induced analgesia (Hohmann et al., 2002), and that the inhibitors of the endocannabinoid transporter, VDM-11 and AM404, and the FAAH blocker arachidonoyl-serotonin enhance stress-induced analgesia in a way inhibited by the CB_1 antagonist SR141716A (Suplita et al., 2003). Importantly, administration of arachidonoyl-serotonin intratechally or directly to the dorsal periaqueductal gray also produced a strong enhancement of stress-induced analgesia.

Anxiety and Aversive Memories (Posttraumatic Stress Disorders)

A series of inhibitors of FAAH were recently developed to obtain a new generation of anxiolytic drugs (Kathuria et al., 2003; Tarzia et al., 2003). The most potent compounds, URB-597 and URB-532, were shown to (1) selectively inhibit FAAH over the endocannabinoid transporter, (2) be inactive on cannabinoid receptors as well as on a series of esterases, including MAGL, (3) enhance AEA levels in isolated cells, (4) enhance AEA but not 2-AG levels after i.p. administration to rats, and (5) induce antinociceptive effects *per se* or potentiate the hypothermia induced by AEA. Most importantly, URB-597 and URB-532 were found to exert anxiolytic effects in two paradigms of anxiety in rats in a way that was opposed by the CB_1 receptor antagonist SR141716A. These findings not only suggest that endocannabinoids tonically control anxiety but also open the way for the development of new anxiolytic substances from URB-597 and its analogs.

Another possible therapeutic use of inhibitors of endocannabinoid inactivation may be against posttraumatic stress disorders, particularly if the recent finding of the involvement of the endocannabinoid system in the extinction of aversive memories in mice can be extrapolated to these disorders or to other types of phobias in humans. Marsicano and co-workers (2002) showed that CB_1 knockout mice or wild-type mice treated acutely with SR141716A were unable to extinguish the memory of conditioned fear. This memory was generated by the association of an electrical shock with an acoustic tune, and was probed every 24 h by administering the animals only with the tune (extinction trial) and measuring the duration of their subsequent fearful behavior (freezing response). It was also shown that, one day after the fear conditioning protocol and immediately before the following day's extinction trial, the amygdala of the mice produced a peak of endocannabinoids, which are likely to activate the CB_1-mediated extinction of the freezing response (Marsicano et al., 2002). These findings suggest that inhibitors of endocannabinoid degradation, by enhancing the levels of AEA and 2-AG in the amygdala (i.e., the brain region mainly responsible for the formation of aversive memories), might be used to treat those subjects with psychiatric disorders derived from the inability to "forget" very unpleasant life events and the associated feelings.

Diarrhea

Cannabis has been used for millennia against diarrhea and constipation, and the inhibitory effects of THC on gastric and intestinal hypersecretion and motility have been widely investigated (for reviews, see Izzo et al., 2001; Pertwee, 2001). Furthermore, the endocannabinoid system appears to control tonically both motility and acetylcholine signaling in the mouse colon (Pinto et al., 2002; Storr et al., 2003). A recent study showed that when mice are treated with cholera toxin to induce hypersecretion in the small intestine, and hence diarrhea, the levels of both CB_1 receptors and the endocannabinoid AEA are upregulated (Izzo et al., 2003). As CB_1-selective agonists and antagonists ameliorate or worsen, respectively, this lethal effect of the toxin, it was suggested that endocannabinoids and CB_1 receptors were overproduced following treatment of mice with cholera toxin in order to counteract its stimulatory action on intestinal hypersecretion and the subsequent diarrhea. Accordingly, VDM-11 was found to also inhibit the effect of cholera toxin in a way that was antagonized by SR141716A (Izzo et al., 2003). Thus, the future use of substances that inhibit AEA inactivation as therapeutic drugs against cholera-toxin-induced diarrhea is suggested.

Cancer

THC has long been used to stimulate body weight gain in AIDS patients and to induce appetite and inhibit emesis in cancer patients undergoing chemotherapy. Yet, recent findings (reviewed by Bifulco and Di Marzo, 2002) have suggested that stimulation of cannabinoid receptors via several molecular mechanisms might also lead to the death, or the inhibition of proliferation of cancer rather than normal cells, or blockade of cancer neoangiogenesis and metastasis (Casanova et al., 2003; Portella et al., 2003). Because the levels of the endocannabinoids and their receptors seem to be altered in malignant cells as compared with normal cells, it was suggested that the endocannabinoid system might represent an endogenous signaling mechanism controlling cancer cell proliferation and spreading (for a review, see Bifulco and Di Marzo 2002). Support for this hypothesis was recently (Ligresti et al., 2003) provided by the finding that (1) the levels of endocannabinoids (AEA and 2-AG) are increased in colorectal adenomatous polyps and carcinomas as compared with normal colorectal tissue, (2) the levels of endocannabinoids and CB_1 receptors and the antiproliferative effects of both synthetic and endogenous cannabinoids are higher in differentiating colorectal cancer Caco-2 cells, the higher the degree of malignancy of these cells (i.e., when they are not differentiated), (3) inhibitors of endocannabinoid inactivation (VDM-11, VDM-13, and arachidonoyl-serotonin) block the proliferation of undifferentiated Caco-2 cells with a potency identical to that on cellular uptake or FAAH; these effects are reversed by SR141716A and are accompanied by an increase in both AEA and 2-AG cellular levels. These findings suggest that endocannabinoids are indeed endogenous suppressors of colorectal carcinoma growth and that inhibitors of endocannabinoid inactivation might be used as nonpsychotropic anticancer drugs (Ligresti et al., 2003).

To investigate the possible use of endocannabinoid-based drugs against cancer cell proliferation, very recently experiments were carried out in our laboratory in an *in vivo* model of tumor growth. The experiments consisted of injecting rat thyroid cells transformed into malignant cells by the H-*ras* oncogene into athymic mice and measuring the size of the thyroid carcinoma subsequently formed under the mouse skin during the following three weeks. The test substances or the corresponding vehicles were injected twice a week for three weeks, and changes in tumor size caused by direct or indirect agonists of cannabinoid receptors were quantified (Bifulco et al., 2001). Using this protocol, we found that a metabolically stable AEA analog and a rather selective CB_1 receptor agonist, met-fluoro-anandamide, when administered intratumor at a relatively low dose (0.5 mg/kg), exerts a potent as well as CB_1-mediated tumor-growth inhibitory effect (Bifulco et al., 2001). Most importantly, we now have data showing that intratumor administration of a FAAH inhibitor (arachidonoyl-serotonin) or a blocker of the putative endocannabinoid membrane transporter (VDM-11) also inhibits thyroid carcinoma growth while causing a significant enhancement of tumor endocannabinoid levels (Bifulco et al., 2004).

Other Possible Targets: Huntington's Chorea, Anorexia, Emesis, and Inflammatory and Autoimmune Disorders

In principle, any disorder whose symptoms or progress are tonically counteracted by endocannabinoids or are caused by defective endocannabinoid biosynthesis or excessive degradation should be counteracted by agents that prolong endocannabinoid half-lives *in vivo*. Evidence has been accumulating, for example, for a role of endocannabinoids in the stimulation of food intake after food deprivation (Di Marzo et al., 2001; Kirkham et al., 2002), and CB_1 knockout mice have recently been shown to exhibit a lean phenotype and to consume less food than their wild-type littermates (Cota et al., 2003). Hence, it can be foreseen that endocannabinoid inactivation inhibitors might be used for the same purpose and much in the same way THC is used in the clinic to induce appetite in AIDS and cancer patients or also to alleviate anorexia nervosa. The same may apply to the control of emesis, which is potently inhibited by both synthetic and plant cannabinoids (Darmani, 2001; Van Sickle et al., 2001). Another possible example is Huntington's chorea, in which the first hyperkinetic phase of this neuromotor disorder in an animal model was suggested to be due to defective endocannabinoid signaling (Lastres-Becker et al., 2002). In this case, however, beneficial (antihyperkinetic) effect of the endocannabinoid uptake inhibitor AM404 was found to be due to stimulation of vanilloid receptors rather than inhibition of endocannabinoid cellular uptake as VDM-11 and AM374 were inactive in the same model (Lastres-Becker et al., 2003).

Finally, it has been suggested that some of the anti-inflammatory effects of endocannabinoids (see De Petrocellis et al., 2000, for review) may be due to their immunosuppressive effects. Therefore, it is possible that some inflammatory or autoimmune disorders may be due to a defective endocannabinoid system and that inhibitors of endocannabinoid inactivation might turn out to be useful anti-inflammatory drugs.

FUTURE DEVELOPMENTS AND CHALLENGES

As discussed in this chapter, there are now enough experimental data to suggest that inhibitors of endocannabinoid inactivation via either cellular reuptake or intracellular hydrolysis may be used as templates for the development of new therapeutic drugs. On the one hand, the beneficial actions found so far for these compounds still need to be fully investigated, in particular with the aim of demonstrating that they are indeed due to elevation of endocannabinoid levels in tissues, and on the other hand, there are several challenges for the years to come. First, selective inhibitors of 2-AG hydrolysis noladin and NADA and virodhamine degradation still need to be developed. In particular, MAGL blockers will have to be discovered because the efficacy of FAAH inhibitors may be limited by the fact that 2-AG hydrolysis can be catalyzed by MAGL also. In view of the selective changes that have been found in many instances for 2-AG and AEA levels during pathological conditions (such as during neuronal excitotoxicity or cholera-toxin-induced diarrhea), it will be important to develop substances that target selectively the two major endocannabinoids. On the other hand, for those conditions leading to changes in all endocannabinoids, inhibitors of the cellular uptake might provide a complete upstream block of their inactivation and a subsequent general upregulation of the "cannabinergic" signal. Although little is known of noladin, NADA, and, particularly, virodhamine catabolism, and even less work has been carried out to analyze the changes in the levels of these compounds during pathological conditions, it will be important to identify inhibitors of their intracellular degradation.

The cloning of enzymes that catalyze AEA and 2-AG biosynthesis will eventually result in the pharmacological manipulation of endocannabinoid levels at the level of their formation. Indeed, inhibitors of endocannabinoid formation might turn out to be useful therapeutic drugs in pathological conditions whose symptoms or onset are caused, in part, by excessive AEA or 2-AG levels in specific tissues. Possible examples of such conditions discovered so far in animal models are Parkinson's disease (high 2-AG levels in the globus pallidus of reserpinized rats [Di Marzo et al., 2000]), Alzheimer's disease (high 2-AG levels in the hippocampus of rats treated with a β-amyloid fragment [M. van der Stelt, G. Esposito, T. Iuvone, and V. Di Marzo, unpublished observations]), hyperphagia and obesity (high 2-AG levels in the hypothalamus of ob/ob and db/db mice and Zucker rats [Di Marzo et al., 2001]), and premature abortion (high AEA levels in the blood of women [Maccarrone et al., 2002] and in the uterus of mice [Schmid et al., 1997]). These and other disorders may, therefore, be targeted in the future by yet-to-be-developed inhibitors of NarPE-PLD and DAGL.

In conclusion, although the usefulness of cannabis and THC in medicine is still being debated (Hall and Degenhardt, 2003), and the possible pharmaceutical usefulness of cannabis-based medicines with an exact titer of THC and cannabidiol is being probed in several clinical trials (Wade et al., 2003; Berman et al., 2003), the future use of synthetic endocannabinoid-based substances in the clinic will require further preclinical research. Once this is successfully completed, the final challenge will be to render soluble in water, or at least more easily administrable to humans, those inhibitors of endocannabinoid inactivation or formation with the most promising and favorable pharmacological and toxicological profiles.

REFERENCES

Arevalo-Martin, A., Vela, J.M., Molina-Holgado, E., Borrell, J., and Guaza, C. (2003) Therapeutic action of cannabinoids in a murine model of multiple sclerosis. *J Neurosci.* 23: 2511–2516.

Baker, D., Pryce, G., Croxford, J.L., Brown, P., Pertwee, R.G., Makriyannis, A., Khanolkar, A., Layward, L., Fezza, F., Bisogno, T., and Di Marzo, V. (2001) Endocannabinoids control spasticity in a multiple sclerosis model. *FASEB J.* 15: 300–302.

Baker, D., Pryce, G., Croxford, J.L., Brown, P., Pertwee, R.G., Huffman, J.W., and Layward, L. (2000) Cannabinoids control spasticity and tremor in a multiple sclerosis model. *Nature*. 404: 84–87.

Basavarajappa, B.S., Saito, M., Cooper, T.B., and Hungund, B.L. (2003) Chronic ethanol inhibits the anandamide transport and increases extracellular anandamide levels in cerebellar granule neurons. *Eur. J. Pharmacol.* 466:73–83.

Beltramo, M. and Piomelli, D. (2000) Carrier-mediated transport and enzymatic hydrolysis of the endogenous cannabinoid 2-arachidonylglycerol. *Neuroreport* 11: 1231–1235.

Beltramo, M., Stella, N., Calignano, A., Lin, S.Y., Makriyannis, A., and Pomelli, D. (1997) Functional role of high-affinity anandamide transport, as revealed by selective inhibition. *Science,* 277: 1094–1097.

Ben-Shabat, S., Fride, E., Sheskin, T., Tamiri, T., Rhee, M.H., Vogel, Z., Bisogno, T., De Petrocellis, L., Di Marzo, V., and Mechoulam, R. (1998) An entourage effect: inactive endogenous fatty acid glycerol esters enhance 2-arachidonoyl-glycerol cannabinoid activity. *Eur. J. Pharmacol.* 353: 23–31.

Berman, J., Lee, J., Cooper, M., Cannon, A., Sach, J., McKerral, S., Taggart, M., Symonds, C., Fishe, K., and Birch, R. (2003) Efficacy of two cannabis-based medicinal extracts for relief of central neuropathic pain from brachial plexus avulsion: results of a randomised controlled trial. *Anaesthesia*, 58: 938.

Bifulco, M. and Di Marzo, V. (2002) Targeting the endocannabinoid system in cancer therapy: a call for further research. *Nat Med.* 8: 547–550.

Bifulco, M., Laezza, C., Portella, G., Vitale, M., Orlando, P., De Petrocellis, L., and Di Marzo, V. (2001) Control by the endogenous cannabinoid system of ras oncogene-dependent tumor growth. *FASEB J.* 15: 2745–2747.

Bifulco, M., Laezza, C., Valenti, M., Ligresti, A., Portella, G., and Di Marzo V. (2004) A new strategy to block tumor growth by inhibiting endocannabinoid inactivation. *FASEB J.* 18: 1606–1608.

Bisogno, T., Howell, F., Williams, G., Minassi, A., Cascio, M.G., Ligresti, A., Matias, I., Schiano Moriello, A., Paul, P., Williams, E.J., Gangadharan, U., Hobbs, C., Di Marzo, V., and Doherty, P. (2003) Cloning of the first sn1-DAG lipases points to the spatial and temporal regulation of endocannabinoid signalling in the brain. *J. Cell Biol.* 163: 463–468.

Bisogno, T., Maccarrone, M., De Petrocellis, L., Jarrahian, A., Finazzi-Agrò, A., Hillard, C., and Di Marzo, V. (2001) The uptake by cells of 2-arachidonoylglycerol, an endogenous agonist of cannabinoid receptors. *Eur. J. Biochem.* 268: 1982–1989.

Bisogno, T., Melck, D., Bobrov, M.Yu., Gretskaya, N.M., Bezuglov, V.V., De Petrocellis, L., and Di Marzo, V. (2000) N-acyl-dopamines: novel synthetic CB(1) cannabinoid-receptor ligands and inhibitors of anandamide inactivation with cannabimimetic activity in vitro and in vivo. *Biochem. J.* 351: 817–824.

Bisogno, T., Melck, D., De Petrocellis, L., and Di Marzo, V. (1999) Phosphatidic acid as the biosynthetic precursor of the endocannabinoid 2-arachidonoylglycerol in intact mouse neuroblastoma cells stimulated with ionomycin. *J. Neurochem.* 72: 2113–2119.

Bisogno, T., Melck, D., De Petrocellis, L., Bobrov, M.Yu., Gretskaya, N.M., Bezuglov, V.V., Sitachitta, N., Gerwick, W.H., and Di Marzo, V. (1998) Arachidonoylserotonin and other novel inhibitors of fatty acid amide hydrolase. *Biochem. Biophys. Res. Commn.* 248: 515–522.

Bisogno, T., Sepe, N., De Petrocellis, L., and Di Marzo, V. (1997a) Biosynthesis of 2-arachidonoyl-glycerol, a novel cannabimimetic eicosanoid, in mouse neuroblastoma cells. *Adv Exp. Med. Biol.* 433: 201–204.

Bisogno, T., Maurelli, S., Melck, D., De Petrocellis, L., and Di Marzo, V. (1997b) Biosynthesis, uptake, and degradation of anandamide and palmitoylethanolamide in leukocytes. *J Biol Chem.* 272: 3315–23.

Boger, D.L., Sato, H., Lerner, A.E., Hedrick, M.P., Fecik, R.A., Miyauchi, H., Wilkie, G.D., Austin, B.J., Patricelli, M.P., and Cravatt, B.F. (2000) Exceptionally potent inhibitors of fatty acid amide hydrolase: the enzyme responsible for degradation of endogenous oleamide and anandamide. *Proc Natl Acad Sci U S A.* 97: 5044–5049.

Boger, D.L., Henriksen, S.J., and Cravatt, B.F. (1998) Oleamide: an endogenous sleep-inducing lipid and prototypical member of a new class of biological signaling molecules. *Curr Pharm Des.* 4: 303–314.

Burstein, S.H., Huang, S.M., Petros, T.J., Rossetti, R.G., Walker, J.M., and Zurier, R.B. (2002) Regulation of anandamide tissue levels by N-arachidonylglycine. *Biochem Pharmacol.* 64: 1147–1150.

Cadas, H., di Tomaso, E., and Piomelli, D. (1997) Occurrence and biosynthesis of endogenous cannabinoid precursor, N-arachidonoyl phosphatidylethanolamine, in rat brain. *J Neurosci.* 17: 1226–1242.

Casanova, M.L., Blazquez, C., Martinez-Palacio, J., Villanueva, C., Fernandez-Acenero, M.J., Huffman, J.W., Jorcano, J.L., and Guzman, M. (2003) Inhibition of skin tumor growth and angiogenesis in vivo by activation of cannabinoid receptors. *J Clin Invest.* 111: 43–50.

Chemin, J., Monteil, A., Perez-Reyes, E., Nargeot, J., and Lory, P. (2001) Direct inhibition of T-type calcium channels by the endogenous cannabinoid anandamide. *EMBO J.* 20: 7033–40.

Cota, D., Marsicano, G., Tschöp, M., Grübler, Y., Flachskamm, C., Schubert, M., Auer, D., Yassouridis, A., Thöne-Reineke, C., Ortmann, S., Tomassoni, F., Cervino, C., Nisoli, E., Linthorst, A.C.E., Pasquali, R., Lutz, B., Stalla, G.K., and Pagotto, U. (2003) The endogenous cannabinoid system affects energy balance via central orexigenic drive and peripheral lipogenesis. *J. Clin. Invest.* 112: 423–431.

Cravatt, B.F., Demarest, K., Patricelli, M.P., Bracey, M.H., Giang, D.K., Martin, B.R., and Lichtman, A.H. (2001) Supersensitivity to anandamide and enhanced endogenous cannabinoid signaling in mice lacking fatty acid amide hydrolase. *Proc Natl Acad Sci U S A.* 98: 9371–9376.

Cravatt, B.F., Giang, D.K., Mayfield, S.P., Boger, D.L., Lerner, R.A., and Gilula, N.B. (1996) Molecular characterization of an enzyme that degrades neuromodulatory fatty-acid amides *Nature.* 384: 83–87.

Cravatt, B.F., Prospero-Garcia, O., Siuzdak, G., Gilula, N.B., Henriksen, S.J., Boger, D.L., and Lerner, R.A. (1995) Chemical characterization of a family of brain lipids that induce sleep. *Science.* 268: 1506–1509.

Darmani, N.A. (2001) Delta(9)-tetrahydrocannabinol and synthetic cannabinoids prevent emesis produced by the cannabinoid CB(1) receptor antagonist/inverse agonist SR 141716A. *Neuropsychopharmacology.* 24: 198–203.

Davies, S.N., Pertwee, R.G., and Riedel, G. (2002) Functions of cannabinoid receptors in the hippocampus. *Neuropharmacology.* 42: 993–1007.

de Lago, E., Ligresti, A., Ortar, G., Morera, E., Cabranes, A., Pryce, G., Bifulco, M., Baker, D., Fernandez-Ruiz, J., and Di Marzo, V. (2004) In vivo pharmacological actions of two novel inhibitors of anandamide cellular uptake. *Eur. J. Pharmacol.* 484: 249–257.

de Lago, E., Fernandez-Ruiz, J., Ortega-Gutierrez, S., Viso, A., Lopez-Rodriguez, M.L., Ramos, J.A. (2002) UCM707, a potent and selective inhibitor of endocannabinoid uptake, potentiates hypokinetic and antinociceptive effects of anandamide. *Eur J Pharmacol.* 449: 99–103.

De Petrocellis, L., Bisogno, T., Maccarrone, M., Davis, J.B., Finazzi-Agrò, A., and Di Marzo, V. (2001) The activity of anandamide at vanilloid VR1 receptors requires facilitated transport across the cell membrane and is limited by intracellular metabolism. *J Biol Chem.* 276: 12856–12863.

De Petrocellis, L., Bisogno, T., Davis, J.B., Pertwee, R.G., and Di Marzo, V. (2000) Overlap between the ligand recognition properties of the anandamide transporter and the VR1 vanilloid receptor: inhibitors of anandamide uptake with negligible capsaicin-like activity. *FEBS Lett.* 483: 52–56.

De Petrocellis, L., Melck, D., Bisogno, T., and Di Marzo, V. (2000) Endocannabinoids and fatty acid amides in cancer, inflammation and related disorders. *Chem Phys Lipids.* 108: 191–209.

De Petrocellis, L., Melck, D., Ueda, N., Maurelli, S., Kurahashi, Y., Yamamoto, S., Marino, G., and Di Marzo, V. (1997) Novel inhibitors of brain, neuronal, and basophilic anandamide amidohydrolase. *Biochem Biophys Res Commn.* 231: 82–88.

Deutsch, D.G., Glaser, S.T., Howell, J.M., Kunz, J.S., Puffenbarger, R.A., Hillard, C.J., and Abumrad, N. (2001) The cellular uptake of anandamide is coupled to its breakdown by fatty-acid amide hydrolase. *J Biol Chem.* 276: 6967–6973.

Deutsch, D.G., Lin, S., Hill, W.A., Morse, K.L., Salehani, D., Arreaza, G., Omeir, R.L., and Makriyannis, A. (1997) Fatty acid sulfonyl fluorides inhibit anandamide metabolism and bind to the cannabinoid receptor. *Biochem Biophys Res Commn.* 231: 217–221.

Di Marzo, V., Griffin, G., De Petrocellis, L., Brandi, I., Bisogno, T., Williams, W., Grier, M.C., Kulasegram, S., Mahadevan, A., Razdan, R.K., and Martin BR. (2002) A structure/activity relationship study on arvanil, an endocannabinoid and vanilloid hybrid. *J Pharmacol Exp Ther.* 300: 984–991.

Di Marzo, V., Goparaju, S.K., Wang, L., Liu, J., Batkai, S., Jarai, Z., Fezza, F., Miura, G.I., Palmiter, R.D., Sugiura, T., and Kunos, G. (2001) Leptin-regulated endocannabinoids are involved in maintaining food intake. *Nature.* 410: 822–825.

Di Marzo, V., Hill, M.P., Bisogno, T., Crossman, A.R., and Brotchie, J.M. (2000) Enhanced levels of endogenous cannabinoids in the globus pallidus are associated with a reduction in movement in an animal model of Parkinson's disease. *FASEB J.*, 14: 1432–1438.

Di Marzo, V., Bisogno, T., De Petrocellis, L., Melck, D., Orlando, P., Wagner, J.A., and Kunos, G. (1999) Biosynthesis and inactivation of the endocannabinoid 2-arachidonoylglycerol in circulating and tumoral macrophages. *Eur J Biochem.* 264: 258–267.

Di Marzo, V., Fontana, A., Cadas, H., Schinelli, S., Cimino, G., Schwartz, J.C., and Piomelli, D. (1994) Formation and inactivation of endogenous cannabinoid anandamide in central neurons. *Nature.* 372: 686–691.

Dinh, T.P., Carpenter, D., Leslie, F.M., Freund, T.F., Katona, I., Sensi, S.L., Kathuria, S., and Piomelli, D. (2002) Brain monoglyceride lipase participating in endocannabinoid inactivation. *Proc Natl Acad Sci U S A.* 99: 10819–10824.

Fezza, F., Bisogno, T., Minassi, A., Appendino, G., Mechoulam, R., and Di Marzo, V. (2002) Noladin ether, a putative novel endocannabinoid: inactivation mechanisms and a sensitive method for its quantification in rat tissues. *FEBS Lett.* 513: 294–298.

Fowler, C.J., Tiger, G., Lopez-Rodriguez, M.L., Viso, A., Ortega-Gutierrez, S., and Ramos, J.A., (2003) Inhibition of fatty acid amidohydrolase, the enzyme responsible for the metabolism of the endocannabinoid anandamide, by analogues of arachidonoyl-serotonin. *J Enzyme Inhib Med Chem.* 18: 225–231.

Glaser, S.T., Abumrad, N.A., Fatade, F., Kaczocha, M., Studholme, K.M., and Deutsch, D.G. (2003) Evidence against the presence of an anandamide transporter. *Proc Natl Acad Sci U S A.* 100: 4269–4274.

Gubellini, P., Picconi, B., Bari, M., Battista, N., Calabresi, P., Centone, D., Bernardi, G., Finazzi-Agrò, A., and Maccarrone, M. (2002) Experimental parkinsonism alters endocannabinoid degradation: implications for striatal glutamatergic transmission. *J Neurosci.* 22: 6900–6907.

Hall, W. and Degenhardt, L. (2003) Medical marijuana initiatives: are they justified? How successful are they likely to be? *CNS Drugs.*17: 689–697.

Hanus, L., Abu-Lafi, S., Fride, E., Breuer, A., Vogel, Z., Shalev, D.E., Kustanovich, I., and Mechoulam, R. (2001) 2-arachidonyl glyceryl ether, an endogenous agonist of the cannabinoid CB1 receptor. *Proc Natl Acad Sci U S A.* 98: 3662–3665.

Hillard, C.J. and Jarrahian, A. (2000) The movement of N-arachidonoylethanolamine (anandamide) across cellular membranes. *Chem Phys Lipids.* 108: 123–134.

Hillard, C.J., Edgemond, W.S., Jarrahian, A., and Campbell, W.B. (1997) Accumulation of N-arachidonoylethanolamine (anandamide) into cerebellar granule cells occurs via facilitated diffusion. *J Neurochem.* 69: 631–638.

Hohmann, A.G. Neely, M., Suplita, R., and Nackley, A.G. (2002) Endocannabinoid mechanism of stress-induced analgesia. *2002 Symposium on the Cannabinoids.* Burlington, Vermont, International Cannabinoid Research Society, p. 30.

Huang, S.M., Bisogno, T., Petros, T.J., Chang, S.Y., Zavitsanos, P.A., Zipkin, R.E., Sivakumar, R., Coop, A., Maeda, D.Y., De Petrocellis, L., Burstein, S., Di Marzo, V., and Walker J.M. (2001) Identification of a new class of molecules, the arachidonoyl amino acids, and characterization of one member that inhibits pain. *J Biol Chem.* 276: 42639–42644.

Huang, S.M., Bisogno, T., Trevisani, M., Al-Hayani, A., De Petrocellis, L., Fezza, F., Tognetto, M., Petros, T.J., Krey, J.F., Chu, C.J., Miller, J.D., Davies, S.N., Geppetti, P., Walker, J.M., and Di Marzo, V. (2002) An endogenous capsaicin-like substance with high potency at recombinant and native vanilloid VR1 receptors. *Proc Natl Acad Sci U S A.* 99: 8400–8405.

Izzo, A.A., Capasso, F., Costagliola, A., Bisogno, T., Marsicano, G., Ligresti, A., Matias, I., Capasso, R., Pinto, L., Borrelli, F., Cecio, A., Lutz, B., Mascolo, N., and Di Marzo, V. (2003) An endogenous cannabinoid tone attenuates colera toxin-induced fluid accumulation in mice. *Gastroenterology,* 125: 765–774.

Izzo, A.A., Mascolo, N., and Capasso, F. (2001) The gastrointestinal pharmacology of cannabinoids. *Curr Opin Pharmacol.* 1: 597–603.

Jarrahian, A., Manna, S., Edgemond, W.S., Campbell, W.B., and Hillard, C.J. (2000) Structure-activity relationships among N-arachidonylethanolamine (Anandamide) head group analogues for the anandamide transporter. *J Neurochem.* 74: 2597–2606.

Jonsson, K.O., Vandevoorde, S., Lambert, D.M., Tiger, G., and Fowler, C.J. (2001) Effects of homologues and analogues of palmitoylethanolamide upon the inactivation of the endocannabinoid anandamide. *Br J Pharmacol.* 133: 1263–1275.

Karlsson, M., Contreras, J.A., Hellman, U., Tornqvist, H., and Holm, C. (1997) cDNA cloning, tissue distribution, and identification of the catalytic triad of monoglyceride lipase. Evolutionary relationship to esterases, lysophospholipases, and haloperoxidases. *J Biol Chem.* 272: 27218–27223.

Kathuria, S., Gaetani, S., Fegley, D., Valino, F., Duranti, A., Tontini, A., Mor, M., Tarzia, G., La Rana, G., Calignano, A., Giustino, A., Tattoli, M., Palmery, M., Cuomo, V., and Piomelli D. (2003) Modulation of anxiety through blockade of anandamide hydrolysis. *Nat Med.* 9: 76–81.

Kirkham, T.C., Williams, C.M., Fezza, F., and Di Marzo, V. (2002) Endocannabinoid levels in rat limbic forebrain and hypothalamus in relation to fasting, feeding and satiation: stimulation of eating by 2-arachidonoyl glycerol. *Br J Pharmacol.* 136: 550–557.

Kozak, K.R. and Marnett, L.J. (2002) Oxidative metabolism of endocannabinoids. *Prostag Leukot Essent Fatty Acids.* 66: 211–220.

Lastres-Becker, I., de Miguel, R., De Petrocellis, L., Makriyannis, A., Di Marzo, V., and Fernandez-Ruiz, J. (2003) Compounds acting at the endocannabinoid and/or endovanilloid systems reduce hyperkinesia in a rat model of Huntington's disease. *J Neurochem.* 84: 1097–1109.

Lastres-Becker, I., Hansen, H.H., Berrendero, F., De Miguel, R., Perez-Rosado, A., Manzanares, J., Ramos, J.A., and Fernandez-Ruiz, J. (2002) Alleviation of motor hyperactivity and neurochemical deficits by endocannabinoid uptake inhibition in a rat model of Huntington's disease. *Synapse.* 44: 23–35.

Lichtman, A.H., Hawkins, E.G., Griffin, G., and Cravatt, B.F. (2002) Pharmacological activity of fatty acid amides is regulated, but not mediated, by fatty acid amide hydrolase in vivo. *J Pharmacol Exp Ther.* 302: 73–79.

Ligresti, A., Bisogno, T., Matias, I., De Petrocellis, L., Cascio, M.G., Cosenza, V., D'Argenio, G., Scaglione, G., Bifulco, M., Sorrentini, I., and Di Marzo, V. (2003) Possibile endocannabinoid control of colorectal cancer growth. *Gastroenterology,* 125: 677–687.

Lopez-Rodriguez, M.L., Viso, A., Ortega-Gutierrez, S., Fowler, C.J., Tiger, G., de Lago, E., Fernandez-Ruiz, J., and Ramos, J.A. (2003) Design, synthesis, and biological evaluation of new inhibitors of the endocannabinoid uptake: comparison with effects on fatty acid amidohydrolase. *J Med Chem.* 46: 1512–1522.

Lopez-Rodriguez, M.L., Viso, A., Ortega-Gutierrez, S., Lastres-Becker, I., Gonzalez, S., Fernandez-Ruiz, J., and Ramos, J.A. (2001) Design, synthesis and biological evaluation of novel arachidonic acid derivatives as highly potent and selective endocannabinoid transporter inhibitors. *J Med Chem.* 44: 4505–4508.

Maccarrone, M., Gubellini, P., Bari, M., Picconi, B., Battista, N., Centonze, D., Bernardi, G., Finazzi-Agrò, A., and Calabresi, P. (2003) Levodopa treatment reverses endocannabinoid system abnormalities in experimental parkinsonism. *J Neurochem.* 85: 1018–1025.

Maccarrone, M., Bisogno, T., Valensise, H., Lazzarin, N., Fezza, F., Manna, C., Di Marzo, V., and Finazzi-Agrò, A. (2002) Low fatty acid amide hydrolase and high anandamide levels are associated with failure to achieve an ongoing pregnancy after IVF and embryo transfer. *Mol Hum Reprod.* 8: 188–195.

Maccarrone, M., De Petrocellis, L., Bari, M., Fezza, F., Salvati, S., Di Marzo, V., and Finazzi-Agrò, A. (2001) Lipopolysaccharide downregulates fatty acid amide hydrolase expression and increases anandamide levels in human peripheral lymphocytes. *Arch Biochem Biophys.* 393: 321–328.

Maccarrone, M., Bari, M., Lorenzon, T., Bisogno, T., Di Marzo, V., and Finazzi-Agrò, A. (2000) Anandamide uptake by human endothelial cells and its regulation by nitric oxide. *J Biol Chem.* 275: 13484–13492.

Maccarrone, M., van der Stelt, M., Rossi, A., Veldink, G.A., Vliegenthart, J.F., and Finazzi-Agrò, A. (1998) Anandamide hydrolysis by human cells in culture and brain. *J Biol Chem.* 273: 32332–32339.

Markey, S.P., Dudding, T., and Wang, T.C. (2000) Base- and acid-catalyzed interconversions of O-acyl- and N-acyl-ethanolamines: a cautionary note for lipid analyses. *J Lipid Res.* 41: 657–662.

Marsicano, G. Goodenough, S., Monory, K., Hermann, H., Eder, M., Cannich, A., Azad, S.C., Cascio, M.G., Ortega Gutiérrez, S., van der Stelt, M., Lopez-Rodriguez, M.L., Casanova, E., Schütz, G., Zieglgänsberger, W., Di Marzo, V., Behl, C., and Lutz, B. (2003) CB1 cannabinoid receptors and on-demand defense against excitotoxicity. *Science,* 302: 84–88.

Marsicano, G., Wotjak, C.T., Azad, S.C., Bisogno, T., Rammes, G., Cascio, M.G., Hermann, H., Tang, J., Hofmann, C., Zieglgansberger, W., Di Marzo, V., and Lutz, B. (2002) The endogenous cannabinoid system controls extinction of aversive memories. *Nature.* 418: 530–534.

Martin, B.R., Beletskaya, I., Patrick, G., Jefferson, R., Winckler, R., Deutsch, D.G., Di Marzo, V., Dasse, O., Mahadevan, A., and Razdan, R.K. (2000) Cannabinoid properties of methylfluorophosphonate analogs. *J Pharmacol Exp Ther.* 294: 1209–1218.

Mechoulam, R., Fride, E., Hanus, L., Sheskin, T., Bisogno, T., Di Marzo, V., Bayewitch, M., and Vogel, Z. (1997) Anandamide may mediate sleep induction. *Nature.* 389: 25–26.

Melck, D., Bisogno, T., De Petrocellis, L., Chuang, H., Julius, D., Bifulco, M., and Di Marzo, V. (1999) Unsaturated long-chain N-acyl-vanillyl-amides (N-AVAMs): vanilloid receptor ligands that inhibit anandamide-facilitated transport and bind to CB1 cannabinoid receptors. *Biochem Biophys Res Commn.* 262: 275–284.

Mestre, L., Correa, F., Arevalo-Martin, A., Molina-Holgado, E., Valenti, M., Ortar, G., Di Marzo, V., and Guaza, C. (2005) Pharmacological modulation of the endocannabinoid system in a viral model of multiple sclerosis. *J. Neurochem.* 92: 1327–1339.

Oka, S., Tsuchie, A., Tokumura, A., Muramatsu, M., Suhara, Y., Takayama, H., Waku, K., and Sugiura, T. (2003) Ether-linked analogue of 2-arachidonoylglycerol (noladin ether) was not detected in the brains of various mammalian species. *J Neurochem.* 85: 1374–1381.

Ortar, G., Ligresti, A., De Petrocellis, L., Morera, E., and Di Marzo, V. (2003) Novel selective and metabolically stable inhibitors of anandamide cellular uptake. *Biochem Pharmacol.* 65: 1473–1481.

Panikashvili, D., Simeonidou, C., Ben-Shabat, S., Hanus, L., Breuer, A., Mechoulam, R., and Shohami, E. (2001) An endogenous cannabinoid (2-AG) is neuroprotective after brain injury. *Nature.* 413: 527–531.

Park, S.K., Provost, J.J., Bae, C.D., Ho, W.T., and Exton JH. (1997) Cloning and characterization of phospholipase D from rat brain. *J Biol Chem.* 272: 29263–29271.

Patel, S., Wohlfeil, E.R., Rademacher, D.J., Carrier, E.J., Perry, L.J., Kundu, A., Falck, J.R., Nithipatikom, K., Campbell, W.B., and Hillard, C.J. (2003) The general anesthetic propofol increases brain N-arachidonylethanolamine (anandamide) content and inhibits fatty acid amide hydrolase. *Br J Pharmacol.* 139: 1005–1013.

Pertwee, R.G. (2001) Cannabinoids and the gastrointestinal tract. *Gut.* 48: 859–867.

Pinto, L., Izzo, A.A., Cascio, M.G., Bisogno, T., Hospodar-Scott, K., Brown, D.R., Mascolo, N., Di Marzo, V., and Capasso, F.(2002) Endocannabinoids as physiological regulators of colonic propulsion in mice. *Gastroenterology.* 123: 227–234.

Portella, G., Laezza, C., Laccetti, P., De Petrocellis, L., Di Marzo, V., and Bifulco, M. (2003) Inhibitory effects of cannabinoid CB1 receptor stimulation on tumor growth and metastatic spreading: actions on signals involved in angiogenesis and metastasis. *FASEB J.* 17: 1771–1773.

Pryce, G., Ahmed, Z., Hankey, D.J., Jackson, S.J., Croxford, J.L., Pocock, J.M., Ledent, C., Petzold, A., Thompson, A.J., Giovannoni, G., Cuzner, M.L., and Baker D. (2003) Cannabinoids inhibit neurodegeneration in models of multiple sclerosis. *Brain.* 126: 2191–2202.

Schmid, H.H., Schmid, P.C., and Natarajan, V. (1996) The N-acylation-phosphodiesterase pathway and cell signalling. *Chem Phys Lipids* 80: 133–142.

Schmid, P.C., Paria, B.C., Krebsbach, R.J., Schmid, H.H., and Dey, S.K. (1997) Changes in anandamide levels in mouse uterus are associated with uterine receptivity for embryo implantation. *Proc Natl Acad Sci U S A*. 94: 4188–4192.

Segall, Y., Quistad, G.B., Nomura, D.K., and Casida, J.E. (2003) Arachidonylsulfonyl derivatives as cannabinoid CB1 receptor and fatty acid amide hydrolase inhibitors. *Bioorg Med Chem Lett*. 13: 3301–3303.

Stella, N., Schweitzer, P., and Piomelli, D. (1997) A second endogenous cannabinoid that modulates long-term potentiation. *Nature*. 388: 773–778.

Storr, M., Sibaev, A., Marsicano, G., Lutz, B., Schusdziarra, V., Timmermans, J.P., and Allescher, H.D. (2003) Cannabinoid receptor type 1 modulates excitatory and inhibitory neurotransmission in mouse colon. *Am J Physiol Gastrointest Liver Physiol*. 286: 110–117.

Sugimoto, H. and Yamashita, S. (1999) Characterization of the transacylase activity of rat liver 60-kDa lysophospholipase-transacylase. Acyl transfer from the sn-2 to the sn-1 position. *Biochim Biophys Acta*. 1438: 264–272.

Sugiura, T., Kondo, S., Sukagawa, A., Tonegawa, T., Nakane, S., Yamashita, A., and Waku K. (1996) Enzymatic synthesis of anandamide, an endogenous cannabinoid receptor ligand, through N-acylphosphatidylethanolamine pathway in testis: involvement of Ca(2+)-dependent transacylase and phosphodiesterase activities. *Biochem Biophys Res Commun*. 218: 113–117.

Suplita, R., Gutierrez, T., Farthing, J., Neely, M., and Hohmann, A.G. (2003) Manipulation of endocannabinoids alters non-opioid stress analgesia *2003 Symposium on the Cannabinoids*. Burlington, Vermont, International Cannabinoid Research Society, p. 127.

Tarzia, G., Duranti, A., Tontini, A., Piersanti, G., Mor, M., Rivara, S., Plazzi, P.V., Park, C., Kathuria, S., and Pomelli, D. (2003) Design, synthesis, and structure-activity relationships of alkylcarbamic acid aryl esters, a new class of fatty acid amide hydrolase inhibitors *J Med Chem*. 46: 2352–2360.

Ueda, N., Puffenbarger, R.A., Yamamoto, S., and Deutsch, D.G. (2000) The fatty acid amide hydrolase (FAAH). *Chem Phys Lipids*. 108: 107–121.

Van Sickle, M.D., Oland, L.D., Ho, W., Hillard, C.J., Mackie, K., Davison, J.S., and Sharkey, K.A. (2001) Cannabinoids inhibit emesis through CB1 receptors in the brainstem of the ferret. *Gastroenterology*. 121: 767–774.

Wade, D.T., Robson, P., House, H., Makela, P., and Aram, J. (2003) A preliminary controlled study to determine whether whole-plant cannabis extracts can improve intractable neurogenic symptoms. *Clin Rehab*. 17: 21–29.

Walker, J.M. and Huang, S.M. (2002) Endocannabinoids in pain modulation. *Prostag Leukot Essent Fatty Acids*. 66: 235–242.

Index

B

C

D

E

F

G

H

N

O

P

R

S

T

U

V

X